Praise for

ROBERT KEN
AND HIS TIMES

Winner of the National Book Award for Biography

"Arthur Schlesinger is the nation's premier builder of literary shrines to heroic presidents . . . In this, his finest work, he deals with the education (in the broadest sense) of one who might, but for an assassin's bullet, have become the first truly radical American president. Conjecture, yes. But the conclusion follows from this loving, witty, and richly documented book."

— *Foreign Affairs*

"Arthur Schlesinger has a kind of proprietary right to this subject. He was the craftsman of the framework within which Kennedys have been most often studied—the claim that Kennedys mature late; but that the maturity, when it comes, is spectacular . . . One comes to respect Mr. Schlesinger's confidence that Robert Kennedy does not need special pleading. All Kennedy critics are quoted frequently and fairly. The challenging, somewhat prickly charm of Mr. Kennedy is by this means more forcibly conveyed than in more protective works."

— *New York Times*

"This book [Schlesinger] calls a biography, and he strives to keep a historian's distance from his subject. But Robert Kennedy was a beloved friend, and there are sections here in which Schlesinger the biographer merges with Schlesinger the participant, the advocate . . . To read the story of Robert Kennedy and his times, as told by a distinguished historian, is to see from a new and revealing angle all the public issues we read about, argued about, guessed at, and grieved over in the '50s and '60s. The book is powerful—and sad."

— *American Heritage*

ROBERT KENNEDY
AND HIS TIMES

Books by Arthur M. Schlesinger, Jr.

ROBERT KENNEDY AND HIS TIMES

Arthur M. Schlesinger, Jr.

Mariner Books

HOUGHTON MIFFLIN COMPANY

BOSTON NEW YORK

Fortieth Anniversary Edition 2018

First Mariner Books edition 2002

Copyright © 1978, 2002 by Arthur M. Schlesinger, Jr.

Introduction to the Fortieth Anniversary Edition copyright © 2018 by Michael Beschloss

For information about permission to reproduce selections
from this book, write to trade.permissions@hmhco.com or to Permissions,
Houghton Mifflin Harcourt Publishing Company,
3 Park Avenue, 19th Floor, New York, New York 10016.

hmhco.com

Library of Congress Cataloging-in-Publication Data is available.
ISBN 978-1-328-56756-7 (paperback)

Printed in the United States of America
DOC 10 9 8 7 6 5 4 3 2 1

For Ethel, Jean and Steve

Great men, great nations, have not been boasters
and buffoons, but perceivers of the terror of life,
and have manned themselves to face it.

— EMERSON

Contents

Acknowledgments

I AM DEEPLY GRATEFUL to Jean Kennedy Smith and John Douglas for their careful reading of the manuscript. Though I did not adopt all their suggestions, their contribution to a clearer and more accurate text has been indispensable. I equally thank and absolve other friends who took time from overcrowded lives to read, correct and improve portions of the book—George W. Ball, Richard Boone, William Bundy, William B. Cannon, Ramsey Clark, Archibald C. Cox, Frederick W. Flott, Michael V. Forrestal, Richard Goodwin, David Hackett, Barbara Wendell Kerr, Mieczyslaw Maneli, Burke Marshall, Clark Mollenhoff, Lloyd Ohlin, Joseph L. Rauh, Jr., Pierre Salinger, Stephen C. Schlesinger, Frederick A. O. Schwarz, Jr., John Seigenthaler, and Stephen E. Smith.

Obviously the book could not have been written had it not been for the great generosity of Ethel Kennedy in permitting me unrestricted access to the papers of Robert F. Kennedy. I am also greatly indebted to the Kennedy family for letting me see the collection of family papers herein designated as the Hyannis Port Papers as well as the papers of Joseph P. Kennedy and Stephen Smith in New York. All these collections will go in due course to the Kennedy Library in Boston. Like all students of the recent political history of the United States, I have benefited immeasurably from the ready and expert cooperation of the directors and staffs of the presidential libraries—especially of Dan H. Fenn, Jr., John F. Stewart, William M. Moss, Joan-Ellen Marci and so many others at the Kennedy Library, which houses the papers of John F. Kennedy, Frank Mankiewicz, Burke Marshall, Theodore C. Sorensen and William vanden Heuvel, as well as the transcripts produced in the John F. Kennedy and Robert F. Kennedy Oral History Programs;

and also of archivists at the Lyndon B. Johnson Library in Austin, Texas, and the Herbert Hoover Library in West Branch, Iowa. Selections from Robert Kennedy's FBI files, made available to me under the Freedom of Information Act, are designated in the notes as 'RFK/FBI/FOIA release.' I thank Jules Feiffer, Mary Bailey Gimbel, Richard Goodwin, David Hackett, Thomas Johnston, Patricia Kennedy Lawford, Allard Lowenstein, William Manchester, Mieczyslaw Maneli, John Bartlow Martin, Barrett Prettyman, Jr., Abba Schwartz, James Stevenson, William C. Sullivan, Felicia Warburg, James Wechsler and Theodore H. White for their kindness in making personal papers available to me; A. J. P. Taylor and the Beaverbrook Foundation for sending me copies of the correspondence between Joseph P. Kennedy and Lord Beaverbrook from the Beaverbrook Papers; the late Herman Kahn and the Yale University Library for facilitating my consultation of the papers of Walter Lippmann and Chester Bowles; and John C. Broderick and the Manuscript Division of the Library of Congress for the papers of James M. Landis and Hugo Black.

I stand in particular debt to the oral history interviewers who have done so much to enrich and amplify the record of the time: Anthony Lewis, William Manchester, John Bartlow Martin and John Stewart for their interviews with Robert Kennedy; the host of volunteers who conducted interviews for the Kennedy Library after the death of John F. Kennedy; and the expert corps of Kennedy Library professionals who have interviewed close associates of both John and Robert Kennedy—notably Roberta Greene and L. J. Hackman, who between them conducted more than sixty interviews, and to many others. Jean Stein generously allowed me to see the oral history interviews she undertook for her invaluable book, edited in collaboration with George Plimpton, *American Journey* (New York, 1970). I thank especially the innumerable interviewees who kindly permitted me to quote from their transcripts as well as other persons, cited in the notes, who allowed me to interview them directly.

At Houghton Mifflin, Richard McAdoo watched the book stretch out in time and length with exemplary patience, and Helena Bentz Dorrance prepared the manuscript for the printer with exemplary thoroughness. I must also thank Luise Erdmann for reading the proofs and Julia Stair for an excellent index.

Once again I rejoice to express my unlimited gratitude to Gretchen Stewart and to Mary Chiffriller for the devoted and meticulous care they expended on typing several drafts of the manuscript, collating texts, checking references, getting the manuscript to the publisher and meanwhile keeping a busy office in a semblance of order. President Harold Proshansky and the Graduate School of the City University of New York, especially the efficient librarians, were helpful at all times. Above all, I thank Alexandra Emmet Schlesinger, who not only read the manuscript with a fastidious and unerring eye but suffered and sustained the author during the throes of composition; and I thank our children still at home, Robert Emmet Kennedy Schlesinger and Peter Cushing Allan, for putting up with it all.

ARTHUR M. SCHLESINGER, JR.

Introduction to the Fortieth
Anniversary Edition

MANY BOOKS have been written in the past four decades about Robert Kennedy, and to one extent or another, they have all been a response to this one. At the time it was published in 1978, on the tenth anniversary of RFK's death, Americans still largely viewed him through the prism of recent memory, press articles, and television documentaries, as well as his presence in books about John Kennedy's administration and the 1968 presidential campaign. In literary terms, *Robert Kennedy and His Times* took RFK from the realm of recent politics into history. Benefiting from full access to his private papers, interviews with his widow, family, and close friends, and the understanding he had gained from his own close relationship with the book's protagonist, Schlesinger brought the reader into Kennedy's life with the kind of detail and intimacy one usually finds only in a biography written much later, after the death of its subject.

When this book first appeared, its author was probably the best-known historian in the United States. Born in 1917, son of a revered Harvard history scholar, he was an undergraduate prodigy and published his Harvard senior thesis (on the New England transcendentalist Orestes Brownson) at the age of twenty-two. During World War II, he served as an intelligence analyst in the Office of Strategic Services. In 1944, he published *The Age of Jackson*, which viewed the seventh president as a forerunner of Franklin Roosevelt's New Deal liberalism and won a Pulitzer Prize for history. Unlike his austere father, the young Arthur, by then a fellow member of the Harvard history department, plunged into politics, which he once called, in his private journal, "the greatest spectator sport." In the 1950s, Schlesinger was chair of the powerful progressive group Americans for Democratic Action. He joined Adlai Stevenson's close circle during the campaigns of 1952 and 1956 as advisor and speechwriter. Late in that decade, he published the first three volumes (of a projected six) of *The Age of Roosevelt*, which attracted literary prizes and a wide national audience.

What changed his life was his relationship with the Kennedys. Schlesinger had been a 1938 Harvard classmate of Joseph P. Kennedy, Jr., and was an acquaintance of his brother Jack, of the class of 1940. As the historian gained more national prominence among liberal Democrats, JFK, as a senator and aspiring presidential candidate, paid him more and more serious attention. During the second Stevenson campaign, as he writes in this book, Schlesinger was chagrined to find himself in close proximity to the senator's brother Robert, whom Stevenson (who privately came to call RFK "the Black Prince") had taken on as a favor to Jack, and who was really there to study how a presidential campaign should be run, in order to help JFK in the future. As Schlesinger reveals in this volume, Robert found Stevenson, at close range, so incompetent that he quietly cast his vote that year for Eisenhower. Having previously viewed him mainly as a rough-edged former aide to Joseph McCarthy, the historian was pleasantly surprised to find unexpected chambers of Robert's character. No doubt he was also eager to make friends with the brother and most trusted confidant of a possible future Democratic president.

In January 1961, when JFK offered him the chance to enter the White House as advisor without portfolio, Schlesinger snapped it up. A public fixture of the New Frontier (he describes in this book the mini-furor over his jump, fully clothed, into Robert Kennedy's swimming pool during a party), he wrote some of John Kennedy's most trenchant speeches (for instance, at Amherst College, on the role of the artist in American society, a month before JFK's death), served as a conduit to liberals and academics, and had revealing private talks with the president, which he recorded in his private journal. Kennedy's assassination plunged him into unaccustomed despair and made him vow, as he told his diary, to "do everything I can to help" Robert Kennedy become president. In 1965, he published his widely read memoir of JFK in the White House, *A Thousand Days*, which landed him on the cover of *Time* magazine.

Declining to return to Harvard, he moved to Manhattan, where he took a chair at the City University of New York and became closer to Robert, by then senator from the Empire State, than he had ever been to Jack. As related in this book, he was deeply involved in RFK's struggle over whether to run against Lyndon Johnson in the Democratic primaries of 1968. Four days after Robert Kennedy was assassinated, Schlesinger wrote in his private journal:

> It is beyond belief, but it has happened—it has happened again . . . RFK began as a true believer; he acquired his sense of the

complexity of things from hard experience. He remained a true believer to the end but at a far deeper level; he had long since shucked away the external criteria and the received simplifications and got down as far as one can in politics to the human meaning of things . . . RFK had an astonishing capacity to identify himself with the casualties and victims of our society. When he went among them, these were *his* children, *his* scraps of food, *his* hovels . . . He was supposed to be hard, ruthless, unfeeling, unyielding, a grudge-bearer, a hater. In fact, he was an exceptionally gentle and considerate man, the most bluntly honest man I have ever encountered in politics, a profoundly idealistic man and an extremely funny man . . . There was for me such a poignancy about RFK — all the greater now that they killed him even before he had a chance to place his great gifts at the service of the nation in the presidency; Jack at least had two and a half years . . . What kind of a President would he have made? I think very likely a greater one than JFK.

After Schlesinger had spent the Fourth of July weekend of 1968 at Hyannis Port, while driving to the airport, Ethel Kennedy asked him, as he later told his diary, "whether I would do Bobby's biography." He went on, "Up to this point, I had rather dismissed this as a possibility and hoped that it would be worked out in some other way. But, when she spoke about it, I suddenly perceived that of course I should do it . . . It is owed to Bobby, whom I loved so much, and it is owed to the country, which ought to learn so much from his life and death." Four years later, Schlesinger named the son of his second marriage after Robert Kennedy.

Schlesinger grew so deeply involved in this project that its research and composition took twice as long as he had expected (in part, because he took time off to write his highly influential *The Imperial Presidency,* published in 1973, during Richard Nixon's Watergate scandal), yielding a manuscript that was much longer than the published version. Published when he was sixty, *Robert Kennedy and His Times,* a national bestseller and winner of the National Book Award, proved to be the last major book of Schlesinger's career. During the decades before his death at the age of eighty-nine, he published a column for the *Wall Street Journal* and a critique of American multiculturalism called *The Disuniting of America* (1991), which angered some of his old progressive friends, but, despite sporadic efforts, never resumed *The Age of Roosevelt.* In 1987, he explained in his journal:

> I spend my days answering letters, reading colloquium papers, writing recommendations, talking to people about the past (having bothered plenty of busy people in my own time for historical purposes, I feel I

must make myself equally available to historians and journalists today),
seeing foreign visitors whom I do not want to see and generally wasting
the most precious of commodities—time, which is inevitably running
out for me.

Although he was far more important as a historian than as a political figure,
Schlesinger cast a wistful eye backward, during his later years, on the 1960s,
the vanished time of his most intense political activism.

He was disappointed that *Robert Kennedy and His Times* did not receive the
hosannas that had greeted *A Thousand Days. Kirkus Reviews* called his book
"sentimental, rhetorical, partisan—and indispensable." In *Foreign Affairs,* the
historian Gaddis Smith said it was "loving, witty, and richly documented."
In the *New York Times Book Review,* Garry Wills called it a "learned and thorough, balanced yet affectionate book." Other reviewers, however, criticized
the author for serving as an attorney for Kennedy's defense—especially in
his coverage of the attorney general's role in CIA plotting against Fidel Castro—and being too incurious about his private life. In the *New York Review of
Books,* the journalist Marshall Frady called it "a kind of sedulously unflagging
apologia" and a "promotional pamphlet of exculpation and eulogy," while
conceding that the volume might show that "the tragedy of the loss of Jack
Kennedy in the nation's life may have only been exceeded by the loss of his
brother."

During his book tour, Schlesinger lamented into his journal, "All along the
way there were the cheap and obvious cracks about canonization, hagiography, *Love Story,* etc. . . . The great RFK themes—the plight of the underclass,
and our capacity and obligation to help them help themselves—are forgotten today. No one seems interested in the fact that an American politician
ten years ago cared deeply about the poor and the powerless." He found it
a difficult time for a biography written by someone who knew and loved his
subject, telling his diary, "In the aftermath of Woodward and Bernstein, the
hot thing to do is to expose scandals and ferret out secrets in the lives of public men . . . I believe in the exposure of public vice but the obsessions with
personal lives has degenerated to the point of prurience. All of this gives me
moments of depression about the reception of the book."

Although stung, Schlesinger was not entirely surprised. As he had written
in *The Age of Roosevelt,* he believed that the reputation of a historical figure
is often at its nadir about ten years after his or her death. To his mind, the
problem was augmented by the ebbing of 1960s liberalism. He believed in
his father's theory that American history unfolded in alternating historical
cycles, which ordained the start of a conservative wave in 1978. It was the
year of Proposition 13 in California, growing frustration with a Democratic

president, Jimmy Carter, a Republican surge in Congress, and the eve of Ronald Reagan's successful campaign for the White House — not a propitious time for a heartfelt biography of Robert Kennedy.

Without saying so outright, this volume makes the argument that Robert Kennedy should be admired as a hero in history. With a half century's distance and hindsight, that argument is more compelling now than it may have seemed when it first appeared. Fifty years after his death, RFK's leadership qualities have never been more needed, starting with his compassion and idealism; his concern for civil rights and the dispossessed; his hatred of braggadocio, lying, and corruption; his demands for a carefully calibrated foreign policy and sane management of world crises; his senses of humor and irony; and his ability to grow. And no biography does a better or more elegant job of showing us those qualities, written by a friend who saw them at close range, than the book you are about to read.

MICHAEL BESCHLOSS
2018

Foreword to the 2002 Edition

W HY ON EARTH should the death in the mid-twentieth century of a forty-two-year-old politician, whose career consisted of a mere four years in a president's cabinet and another four years in the Senate, strike such a nerve in the twenty-first century? Why does his face on the television screen still provoke such a keen sense of loss? Why do books about Robert Kennedy continue to appear? Why does his life have poignant meaning not just for those of us old enough to remember him in action but for so many young Americans who were not even born in his lifetime?

For young Americans, Robert Kennedy's older brother, John Fitzgerald Kennedy, is now a historical figure, a leader who belongs to the ages. The causes with which he is identified — the Cold War, the missile crisis, Vietnam — seem to the young almost as remote as the Spanish-American War was to my generation. But Robert Francis Kennedy has a contemporary feel about him, a sense of enduring identification with the woes and injustices of today's world. His causes — the growing disparities of income and opportunity in the United States, racial justice, the redemption of the dispossessed and humiliated — are with us every hour. Robert Kennedy represents certain unfulfilled possibilities — possibilities that we know in our hearts must be fulfilled if we are ever as a nation to redeem the promise of American life.

The two brothers could hardly have been closer. Yet they had very different personalities, and this too may in part explain the divergence in the way people think about them. John Kennedy was a man of reason; Robert, a man of passion. John was objective, analytical, invulnerable (except to the assassin's bullet). Robert was subjective, emotional and acutely vulnerable. John enjoyed his friends. Robert needed his friends. John was buoyant, Robert melancholy; John urbane, Robert brusque. As I have suggested in this book, John Kennedy was a realist brilliantly disguised as a romantic, Robert a romantic stubbornly disguised as a realist.

A striking thing about Robert Kennedy, and one reason for his continuing relevance, was his capacity for growth. His life may be divided into three acts. The first act could be entitled "His Father's Son." He was the smallest of the four boys, the gentlest and shyest, the least articulate and the least coordinated, the most dutiful. His father was absent during much of his childhood and when home was sometimes impatient with him. "I was the seventh of nine children," Bobby once said, "and when you come from that far down you have to struggle to survive."

Determined to win his father's respect and love, he began to harden his personality. The inner sensitivity remained, but a protective covering now formed over it. He became the Robert Kennedy who burst onto the public scene in the 1950s: an aggressive young fellow, opinionated, censorious, prickly, rigid, moralistic, inclined to tell people off and to get into fights. It was in this mood that he went to work for the infamous Joe McCarthy's Senate committee on investigations — a brief undertaking, soon terminated, with RFK returning to the committee as counsel for the Democratic minority and author of the minority report condemning McCarthy's investigation of the army.

The second act was soon to come — the act that could be entitled "His Brother's Brother." RFK made his political debut as manager of JFK's victorious campaign for the Senate from Massachusetts in 1952. The two brothers, separated by nearly nine years, had hardly had time to get to know each other. For all their personality differences, they now formed an intimate working partnership.

Under his older brother's influence, Robert began to lose his intolerance and rigidity. He grew more relaxed and rueful, acquired more ironic views of life and himself, developed a wry, self-mocking humor and in time displayed a personal charm against which editors warned their reporters. Drafted to run his brother's campaign for the presidency in 1960, Robert was tireless, intimidating and effective. Once elected, John persuaded a reluctant Robert to become attorney general.

Robert surprised the critics by assembling a staff of notable lawyers. He strove especially to bring the autocratic J. Edgar Hoover and his long-untouchable Federal Bureau of Investigation under control. Hoover was then a national idol, and his obsession was the pursuit of Communists. RFK, whose 1960 book *The Enemy Within* dealt not with Communism but with organized crime, thought this nonsense. The American Communist Party, he told a newspaperman, "couldn't be more feeble and less of a threat, and besides its membership consists largely of FBI agents." The fact that his brother was president gave him more leverage than most attorneys general, and he forced Hoover against his will to divert budget and agents into two new fields — organized crime and racial justice.

The spread across the South of violent resistance to court orders

made civil rights legislation by 1963 both a political possibility and, in
the view of the Kennedys, a moral necessity. Opponents charged that
Martin Luther King, Jr., the black civil rights leader, was controlled
by Communist agents. The attorney general acceded to Hoover's request
that King's telephone be wiretapped, confident that this would disprove
the allegations. Meanwhile he called for legislation, introduced in 1963 and
passed in 1964, outlawing discrimination, whether based on race, color, religion, national origin or (in the case of employment) sex, in public accommodations, employment, voting and education. It was the most far-reaching
civil rights law since Reconstruction.

His brother assigned Robert tasks considerably beyond the Department
of Justice. He especially valued Robert's ability to ask the tough questions
and get to the heart of difficult problems. After the Bay of Pigs fiasco, he
brought the attorney general increasingly into foreign affairs. Cuba became
a particular preoccupation. In 1961–62 Robert spurred on the Central Intelligence Agency to undertake Operation Mongoose, a foolish and ineffectual
pinprick program of covert action—infiltration, arms drops, sabotage—
against the Castro regime; it was not RFK's finest hour. There is no hard evidence, however, that he was aware, except as past history, of the CIA efforts,
originating in the Eisenhower administration, to assassinate Fidel Castro.

If he was, as some claim, obsessed with the goal of overthrowing the Castro regime, Castro himself provided the perfect pretext when he accepted
Soviet nuclear missiles in the summer of 1962. But during the missile crisis in
October, RFK led the opposition to a military solution. "All our heritage and
ideals," he said, "would be repugnant to such a sneak military attack," calling it a "Pearl Harbor in reverse." He negotiated a deal by which, if the Russians removed their missiles from Cuba, the Americans would in due course
remove their missiles from Turkey. The Soviet leader Nikita S. Khrushchev
noted in his memoirs that the Americans were "open and candid with us,
especially Robert Kennedy."

His brother's assassination in November 1963 devastated him. For weeks,
for months, he wandered in grief. Yet in a paradoxical sense it liberated him
too. He had repressed his inner self since childhood, first to prove himself
to his father, then to help his brother. In 1961 his father was disabled by a
stroke; now his brother was dead. The third act was beginning: "On His
Own." At last Robert Kennedy was free to become a voice and leader in his
own right.

After the new president rejected him as a running mate in 1964, he won
election to the Senate from New York. As senator, he welcomed the reforms
of Lyndon Johnson's Great Society but found himself in growing disagreement over foreign policy, and finally, as the Vietnam War swallowed the
Great Society, over domestic policy as well. The press often refused to accept

such disagreement on its merits, presenting it instead as a political maneuver by which ruthless Bobby Kennedy was out to reclaim the White House. This response both irritated and inhibited him.

He identified himself increasingly with the excluded and desolate in America—Indians on reservations, Latinos picking grapes in California, hungry blacks along the Mississippi Delta, migrant workers in squalid camps in upstate New York, despairing families in rat-infested tenements in New York City. His oldest daughter, Kathleen Kennedy Townsend, recalls his returning from the South as the Kennedy family was sitting down for dinner at Hickory Hill. "He entered our dining room ashen-faced and said, 'A whole family lives in a shack the size of this room. The children are covered with sores and their tummies stick out because they have no food. Do you know how lucky you are? Do you know how lucky you are? Do something for your country.'"

"Today in America," he said, "we are two worlds." His aim was to make the two worlds one. Always challenging, probing, testing, he sought new ways to empower people, to foster community self-development and self-reliance, to work out new structures by which people could sustain dignity and hope and devise workable means of helping themselves.

His sharp break with the administration came over foreign policy. He criticized United States intervention in the Dominican Republic in 1965 and concluded that Johnson had abandoned the reform aims of JFK's Alliance for Progress. "If we allow communism to carry the banner of reform," he warned after a tour of Latin America in 1966, "then the ignored and the dispossessed, the insulted and injured, will turn to it as the only way out of their misery."

As American involvement in Vietnam deepened, Kennedy called for bombing pauses and negotiation. When escalation continued, he evoked the "horror" of the war in urgent speeches. "Can we," he cried in 1968, "ordain to ourselves the awful majesty of God—to decide what cities and villages are to be destroyed, who will live and who will die, and who will join the refugees wandering in a desert of our creation?"

His critique of the Vietnam War pointed toward a challenge to Johnson's renomination in 1968. But Kennedy hung back, still fearing misconstruction of his motives, and antiwar Democrats rallied around Senator Eugene McCarthy of Minnesota. After McCarthy's success in the New Hampshire primary, Kennedy belatedly entered the contest. McCarthy fans denounced him as an opportunist. Johnson withdrew soon afterward, and Vice President Hubert Humphrey became the administration candidate.

Kennedy's was an uproarious campaign, filled with enthusiasm and fun. RFK had a wry wit and a satiric sense of the absurdities of politics and life. One newspaperman described it as a "huge, joyous adventure." At the same time, Kennedy was embarked on a desperately serious mission. Existing conditions in the United States, he said in speech after speech, were "not acceptable," and we diminished ourselves as a moral community when we accepted them.

His campaign generated wild enthusiasm as well as deep anger. His message of change brought hope to some, fear to others. Many saw him as a divisive figure, but he saw himself, especially after the assassination in April of Martin Luther King, Jr., as engaged in a mission of reconciliation, seeking to bridge the great schisms in American society—between white and nonwhite, between rich and poor, between age and youth, between order and dissent, between the past and the future. Recapitulating his message after he won the California primary, he took the shortcut through the kitchen of the Hotel Ambassador in Los Angeles.

What would have happened had he not been killed? He would certainly have had a rocky road to the nomination. The power of the Johnson administration and much of the party establishment was behind Humphrey. Still, the dynamism was behind Kennedy, and he might well have swept the convention. If nominated, he would most probably have beaten the Republican candidate, Richard M. Nixon.

Individuals do make a difference to history. A Robert Kennedy presidency would have brought a quick end to American involvement in the Vietnam War. Those thousands of Americans—and many thousands more Vietnamese and Cambodians—who were killed from 1969 to 1973 would have been at home with their families.

A Robert Kennedy presidency would have consolidated and extended the achievements of John Kennedy's New Frontier and Lyndon Johnson's Great Society. The liberal tide of the 1960s was still running strong enough in 1969 to affect Nixon's domestic policies. The Environmental Protection Act, the Occupational Safety and Health Act, the Comprehensive Employment and Training Act with its CETA employment program were all enacted under Nixon. If that still fast-flowing tide so influenced a conservative administration, what signal opportunities it would have given a reform president!

The confidence that both black and white working-class Americans had in Robert Kennedy would have created the possibility of progress toward racial reconciliation. His appeal to the young might have mitigated some of the under-thirty excesses of the time. And of course the election of Robert Kennedy would have delivered the republic from Watergate, with its attendant subversion of the Constitution and destruction of faith in government.

RFK joined idealism in his purposes with realism in his methods. He was a man of passionate conviction. He was at the same time a tough and experienced politician who understood the uses of power and government. And he was a compelling speaker with unusual capacity to inform, move and inspire the electorate and to rally popular support for his programs.

His blunt challenge to the complacencies of American society made many uncomfortable; but his insistence that any individual who "stands up for an ideal, or acts to improve the lot of others, or strikes out against injustice" can make a difference plucked a moral nerve, especially among the young. "Each of us can work to change a small portion of events, and in the total of all those acts will be written the history of this generation."

RFK, as David M. Shribman pointed out, was one of only five men in American political history to be known by his initials. The others —Theodore Roosevelt, Franklin Delano Roosevelt, John Fitzgerald Kennedy, Lyndon Baines Johnson—were all presidents.

Foreword

R OBERT KENNEDY died on June 6, 1968, at the age of forty-two—a
decade ago as I write today. He lived through a time of unusual turbulence in American history; and he responded to that turbulence more
directly and sensitively than any other political leader of the era. He was
equipped with certitudes of family and faith—certitudes that sustained him
till his death. But they were the premises, not the conclusions, of his life. For
he possessed to an exceptional degree what T. S. Eliot called an "experiencing nature." History changed him, and, had time permitted, he might have
changed history. His relationship to his age makes him, I believe, a "representative man" in Emerson's phrase—one who embodies the consciousness
of an epoch, who perceives things in fresh lights and new connections, who
exhibits unsuspected possibilities of purpose and action to his contemporaries.

He never had the chance to fulfill his own possibilites, which is why his
memory haunts so many of us now. Because he wanted to get things done,
because he was so often impatient and combative, because he felt simply
and cared deeply, he made his share of mistakes, and enemies. He was a
romantic and an idealist, and he was also prudent, expedient, demanding
and ambitious. Yet the insights he brought to politics—insights earned in a
labor of self-education that only death could stop—led him to see power not
as an end in itself but as the means of redeeming the powerless.

Any historian who has written about the 1930s as well as the 1960s must
recognize a terrible monotony in our national problems. The conditions of
misery and inequity that troubled Franklin Roosevelt and the New Dealers
troubled the Kennedys twenty years later—after an interval during which,
while misery and inequity persisted, people in power were untroubled. We
appear to be in another such interval today. Robert Kennedy's message of

commitment to the desolate and the disinherited is rarely sounded in the 1970s. But the problems about which he cared so much remain. The time will surely come again when the richest nation on earth will overcome its indifference to the degradation of its citizens. When that time returns, Robert Kennedy's journey will, I believe, be seen to have exemplary value.

Any book about Robert Kennedy is inextricably bound up with John Kennedy and his Presidency. Here I have tried to avoid retelling familiar stories. This work treats the Kennedy administration primarily as Robert Kennedy took part in events or as his papers cast particular light on them. On occasion I have added information that I did not know when I wrote *A Thousand Days* or that seemed inappropriate for publication in 1965. I have also taken the opportunity for a new look at some of the more controversial issues of the Kennedy years. The passage of time, the knowledge of consequences, the illumination of hindsight, the rise of new preoccupations—all give problems of the past new form and perplexity. While there would be little point, for example, in going through once again the day-to-day development of the Cuban missile crisis, there would appear considerable point in assessing latter-day interpretations of that crisis—or of the Indochina War, the civil rights struggle, the Central Intelligence Agency, the Federal Bureau of Investigation. In this sense, *Robert Kennedy and His Times* is a sort of sequel to *A Thousand Days*—though the earlier work was a personal memoir and this one is a biography.

It draws heavily on the personal papers of Robert Kennedy, made available to me through the generosity of Ethel Skakel Kennedy and of the Kennedy family. The Robert Kennedy Papers include diaries, kept for brief intervals at various times in his life, letters, *aides-mémoires* and oral history interviews. Robert Kennedy was a direct and candid man, often brusque in manner and opinion. He hated circumlocution. He distrusted what he considered high-flown or self-serving expressions. His instinct was to cloak idealism in throwaway phrases and laconic jokes. His deflationary, self-mocking, sometimes grim wit was employed often as a means of relieving tension in the midst of crisis. Some of the remarks in the Robert Kennedy Papers, wrenched from context, will no doubt be used against him. This is a risk that must be run. The serious reader will be able to set Robert Kennedy's words in the totality of his life. I am sure that he would have been entirely content to stand on the record. I must add that the Kennedy family made no attempt to influence my use of the papers and that the interpretations and judgments are exclusively my own.

Some readers may wonder about the extensive use made in these pages of oral history—i.e., of transcribed interviews conducted with participants after the fact. We all know that interviews can be no better than a person's memory and that little is more treacherous than that. Yet historians have rarely hesitated to draw on written reminiscences, which are no less self-promoting; nor have they hesitated, in order to impart immediacy to narrative, to quote conversations as recalled in diaries, letters and memoirs, when the content of the conversation is plausibly supported by context or other evidence. I have extended this tolerance to oral history and employed this literary convention with the same critical caution, I hope, illustrious predecessors have applied to written documents. It remains a convention. The recollected material cannot pretend to the exactitude of, say, the White House tapes of the Nixon years.

It is necessary to declare an interest. I was a great admirer and devoted friend of Robert Kennedy's. Association with him over the last decade of his life was one of the joys of my own life. It was not only that he deepened one's understanding of America, of its problems and possibilities, but that he was the occasion of such good and happy times. No one could be more fun than Robert Kennedy; no one more appealing, with those impulses of irony, bravado, gentleness and vulnerability so curiously intermingled in his vivid personality; no one in our politics, as I deeply believed, held out more promise for the future. I thought him the most creative man in American public life when he was killed. I am well aware that not everyone did. Some saw him as cold, vengeful, fanatical, 'ruthless.' But to adapt a phrase A. J. P. Taylor used in his life of Lord Beaverbrook, if it is necessary for a biographer of Robert Kennedy to regard him as evil, then I am not qualified to be his biographer.

Still sympathy may illumine as well as distort. In any case, the story of his life tells a great deal about the American ordeal in the middle decades of the twentieth century. I have tried to tell that story as accurately as I can, strengthened by the conviction that the last thing Robert Kennedy would have wished was hagiography. He was not a hard man, but he was a tough man. He valued bluntness, precision and truth. I have not hesitated to portray him in his frailty as well as in his valor. I can imagine no better way to serve his memory. What he had in the way of weakness was an essential part of his strength. For he was truly a perceiver of the terror of life, and he manned himself nobly to face it.

ROBERT KENNEDY
AND HIS TIMES

Prologue: 1968

THE TRAIN LEFT New York City at one o'clock on a sweltering June afternoon. In the last of twenty-two cars lay a coffin, covered by an American flag. A diversity of people filled the cars ahead — politicians and writers, government officials and entertainers, diplomats and trade unionists, astronauts and agitators, social workers and athletes, society figures and stenographers. They talked in somber tones, they rejoiced in the happier past, they despaired of the future, they stared out of the window.

Along the tracks people waited in the blazing sun. "For eight agonizing hours, as the train passed," said a senator, "I looked into their faces. I saw sorrow . . . bewilderment. I saw fury and I saw fright."[1] Some stood at rigid attention, hand over heart. Some waved. Some buried their faces in their hands. Some knelt. Some held up hand-printed signs — REST IN PEACE, ROBERT, or, ominously, WHO WILL BE THE NEXT ONE? or, desperately, WE HAVE LOST OUR LAST HOPE, or, starkly, PRAY FOR US, BOBBY. Many cried. Some threw roses at the last car. Some as if in a daze followed the train down the roadbed. As the train crossed rivers on its journey south, boats, clustered by the bridges, sounded their whistles. On a river before Newark, New Jersey, three men on a fireboat saluted; the boat bore the name *John F. Kennedy*.

The train slowed to a crawl with thousands — perhaps a million in all — crowding the way.[2] At Elizabeth, New Jersey, a northbound express, rounding a curve at high speed, plowed into mourners overflowing onto the tracks, killing a man and a woman. The bell of the funeral train rang incessantly. The whistle blew in long, piercing, melancholy blasts. As the train moved south through the Pennsylvania countryside, one occasionally saw a man or woman, far from any visible town, standing gravely and alone, enveloped in private grief. In Baltimore a great silent mass of people, mostly black, had waited for hours. When the train pulled in, they joined hands and sang "The Battle Hymn of the Republic." So, 103 years before, a crowd had sung in Baltimore as another funeral train had passed. The sun set, a baleful orange ball in the west. Shortly after nine o'clock the train reached Washington in the glimmering dark.

The Family

ROBERT FRANCIS KENNEDY was born in Brookline, Massachusetts, on November 20, 1925, the seventh child of Rose Fitzgerald and Joseph Patrick Kennedy. His great-grandparents had come from Ireland some three quarters of a century before. The legacy they brought to Boston was but mildly diluted in the next two generations. The nineteenth-century Irish saw themselves as the victims of history. Memories of dispossession and defeat filled their souls. They had lost their national independence, their personal dignity, their land, even their language, to intruders from across the sea. Joyce defined the Irish view: "History, Stephen said, is a nightmare from which I am trying to awake."[1]

I

They had kept trying to awake. Pessimism mingled with pride and romantic defiance. They countered fate by talk, incessant and extravagant; by irony and self-mockery; by fantasy and by drink. They evaded fate, as oppressed people do, by donning masks to deceive the oppressors. Some came to believe the blarney act they put on to bemuse the English. Others used it as a cloak for resistance. They fought fate, according to temperament, by terror or by politics. The struggle for Catholic emancipation had initiated the Irish in the arts of political organization well before they began to leave for the United States. Irish politics incorporated the code of the family. It was a tribal politics, dominated by expectations of unswerving loyalty and vindictive against what was often too eagerly seen as defection or betrayal.

Most important in securing life, along with the family, was the church. Life in Ireland was hard, and so was religion. Jansenism pervaded the Irish church, encouraging clerical tendencies toward censoriousness and bigotry. But the priests remained with the people, and faith was the ultimate consolation in a hopeless world. Catholicism absorbed and vindicated the Irish experience — the suffering, the bloodshed, the fatalism. It explained that man was weak and life more than a little absurd, unless redeemed by the grace of God mediated through the church.

The culture Robert Kennedy's great-grandparents transported to Boston was filled with this conviction of the bloodiness of life. The sense of disorder, tragedy and evil was not unlike that of the Puritans, three centuries earlier. But what the Puritans had placed on the isolated soul in the quest for salvation the Irish assigned to the family and the church. By the time the Fitzgeralds and Kennedys arrived, the descendants of the Puritans were turning away from Calvin and Jonathan Edwards to embrace the inevitability of progress and the perfectibility of man. In short, they were being Americanized. The Irish newcomers were set apart by their awareness of historic failure and human sinfulness and by their faith in the redemptive magic of family and church. They were set apart by their addiction to the scathing comedy that relieved the gloom of life. They were set apart by their lack of power, money, education and self-esteem — and by the complacent contempt in which the Yankees held them.

Separation preserved their legacy. But the Irish were being Americanized too. Robert Kennedy's grandfathers — Patrick Joseph Kennedy, born in Boston in 1858; John Francis Fitzgerald in 1863 — prospered in the new land. Kennedy was a quiet man, a saloon-keeper and ward boss, wielding influence softly, indifferent to publicity and to noise. Fitzgerald was an Irish talker, interested in everything and forever putting on a show ("Honey Fitz can talk you blind / On any subject you can find").[2] The two men were together in the state senate in the early 1890s and did not like each other then or later.

Rose, the oldest of Honey Fitz's six children, was born on July 22, 1890. She grew up in a great Dorchester house, with mansard roof and stained-glass windows. Her father took her on vacations to Old Orchard Beach in Maine and to Palm Beach in Florida and in 1911 gave her a spectacular coming-out party. By this time rich Irish Catholics were creating a social world of their own, paralleling the rich Yankee Protestant society that excluded them. Rose Fitzgerald, a bright and pretty girl, found life in this world enjoyable and fulfilling. For a moment, it is true, she had wanted to go to Wellesley College; but, as she said in later life, "in those days you didn't argue with your father."[3] Instead she was packed off to convents in Boston, New York and the Netherlands. She learned French and German in Europe and spoke them fluently for the rest of her life.

P. J. Kennedy's only son, Joseph Patrick, was born on September 6, 1888. Where the Fitzgeralds believed in duplicating the institutions of Yankee society, the Kennedys believed in infiltrating them. Young Kennedy went with the Yankee boys to the Boston Latin School and then on to Harvard College. Here he made a moderate mark — a rather considerable mark for an Irish Catholic — and graduated in 1912. In the meantime, Rose Fitzgerald and Joseph Kennedy had fallen in love. John F. Fitzgerald thought it a most unsuitable match and tried through domestic admonition and foreign travel to steer his daughter in other directions. The courtship was pursued clandestinely. In the end, Honey Fitz surrendered. Cardinal O'Connell married the young couple in October 1914.

A fortnight later Joseph Kennedy moved his bride into a house in Yankee Brookline. He had already begun his assault on Anglo-Saxon society. Instead of becoming a contractor in the Irish style, he chose banking, a Protestant preserve. Soon he went into utilities, forcing his way onto the board of the Massachusetts Electric Company; he explained to a friend, "Do you know a better way to meet people like the Saltonstalls?"[4] During the First World War he was assistant general manager of the Bethlehem Shipyards at Fore River, south of Boston. Here he met Franklin D. Roosevelt, then assistant secretary of the Navy. "Roosevelt was the hardest trader I'd ever run up against," Kennedy told Ernest K. Lindley twenty years later.[5] They ended on terms of mutual esteem. After the war, Kennedy joined the Yankee investment banking house of Hayden, Stone & Company and began his bold career in the stock market.

II

Kennedy was a loner who operated with a small staff and an astute instinct. He used to say, "Only a fool holds out for the top dollar" — a piece of wisdom that enabled him to escape the Wall Street crash relatively unscathed.[6] He played the game brilliantly. He also played it cynically. The game did not consume his life or even command his respect. It was a means to an end. By 1926, he was a millionaire and far more successful than his Yankee contemporaries. Still they denied him the acceptance he sought for himself and his family.

He resented exclusion. "They wouldn't have asked my daughters to join their debutante clubs," he told Joe McCarthy (the writer, not the senator) thirty years later.[7] In 1926 Joseph Kennedy abandoned Boston, taking his family by private railroad car to a house in Riverdale, an affluent New York suburb. Returning a decade later for the St. Patrick's Day dinner of the Clover Club, Kennedy observed that the Boston Irish "suffered under the handicap of not possessing family tradition adequate to win the respect and confidence of the Puritan neighbors. This Yankee pride of ancestry developed a boastfulness and a snobbishness which, though difficult to understand, explains many of the strange idiosyncrasies of Boston." He did not think that the dual society could last. "The influence of the Irish culture in this country," he told the Clover Club, "must be recognized as on the wane. Nor is it likely that anything or any person can change this process of cultural absorption."* If cultural absorption was inevitable, he was determined that his children — certainly his sons — should improve his own record.

When Robert Kennedy discussed apartheid in South Africa in 1966, he said that his father had left Boston because of the signs NO IRISH NEED APPLY. "Maybe," a skeptic named Michael Mooney wrote him. "But I can't help thinking of . . . the Irishman who left Boston (and all those signs) in his own Railway car." Mooney added, "It has always been a distinct right of all the Irish to invent for themselves mythological backgrounds — little lies that amuse their audience and serve their tellers well in diplomacy and love and politics."[8] Robert Kennedy replied, "Yes but — It was symbolic. The business establishment, the clubs, the golf course — at least that was what I was told at a very young age. . . . Undoubtedly lack of opportunity which had nothing to do with racial or religious background but were just Boston had something to do with it — But both my parents felt very strongly about the discrimination so you'll have to take the matter up with them. Excuse the scrawl — I'm in a plane. . . . [P.S.] Actually I don't write any better anywhere."[9]

No doubt there was an element of genuine hurt in Kennedy's attitude toward Boston. But he would very likely have left in 1926 even if he had been elected president of the Somerset Club. His Harvard classmate Ralph Lowell, a Boston banker, told R. J. Whalen,

---

* J. P. Kennedy, Clover Club speech (typescript), March 13, 1937, 2, 3, JPK Papers. His papers contain a garbled quotation from the autobiography of Charles Francis Adams, in which the tough old Yankee offered a view of Boston that the tough young Irish American obviously found sympathetic. "I can say little that is pleasant," Adams wrote, ". . . I have tried Boston socially on all sides: I have summered it and wintered it, tried it drunk and tried it sober; and, drunk or sober, there's nothing in it — save Boston!" The sheet with the garbled quotation is dated November 16, 1944. The exact quotation, used here, can be found in C. F. Adams, *An Autobiography* (Boston, 1916), 39.

"This city was a small, clear puddle. New York was a big, muddy one, and that's what Joe wanted." "If you want to make money," Joe Kennedy himself liked to say, "go where the money is."[10]

III

Joseph Kennedy was not quite thirty-eight when he left Boston, a tall, slender, reddish-haired man with heavy horn-rimmed glasses over penetrating blue eyes. The intensity of his personality made an immediate impression. He could be endlessly charming when he wished but could also be — and often was — blunt and brusque to the point of rudeness. In fact, he did not give a damn about most people. Perhaps, with his commitment to the Americanization of the Irish, he reacted against the stage-Irishman stereotype, all gregariousness and blarney, as embodied, for example, in his irrepressible father-in-law; at any rate, he had no trouble being as frosty as the frostiest Yankee. Other businessmen thought him rough and remorseless. "There are no big shots" was a Kennedy aphorism — a proposition deeply offensive to all who saw themselves as big shots. He had little regard for businessmen per se and few close friends in the business world, beyond a circle of cronies who stayed with him through the years.

When he gave affection, he was bountiful. But he expected fidelity in return. "His friendship," wrote Arthur Krock, the political pundit of the *New York Times* and for years a Kennedy intimate, ". . . is a mixture — of lavish generosity, fierce loyalty, excessive gratitude, harshly-articulated candor and sudden termination — that is without parallel in my experience."[11] He had a hot temper, but some considered it more admonitory than uncontrollable. "When he got mad," said his friend Morton Downey, the Irish American tenor, "I always suspected it was an act."[12]

He was a faithful Catholic, but described himself to his children as anticlerical. His daughter Eunice said, "He always trusted experience as the greatest creator of character. (My mother believed religion was.)"* He was personally abstemious — perhaps another reaction against the stereotype — smoked and drank little and did his best to induce his children to follow his example. His hobby was classical music — a taste formed at Harvard. One summer in Washington, while his family was in Hyannis Port, Kennedy played symphonies so incessantly that the cronies with whom he was sharing his house

---

* Edward M. Kennedy wrote in *The Fruitful Bough*, 207, "Although he said he was anti-clerical, two of his closest friends were churchmen." Eunice Kennedy Shriver's observation is on 218.

complained. "You dumb bastards don't appreciate culture," Kennedy said, putting on another symphony.[13] His children did not inherit his love of serious music any more than they inherited their mother's aptitude for foreign languages.

Making money was never enough. He had, said his perceptive friend Morton Downey, "a strange, almost turbulent interest in more seemingly unrelated topics and things than any man I ever knew."[14] Turbulent — a significant choice of word, suggesting the restlessness of personality that drove him from job to job and challenge to challenge. The ranging concern extended inescapably, for P. J. Kennedy's son, to politics. This was, in any case — far more than money — the Irish American road to power.

IV

In 1928, Kennedy had been too deeply engaged in business to do more than vote for the first Irish Catholic to run for the Presidency as the candidate of a major party. A year later the Great Depression began. "I am not ashamed to record," Kennedy recalled in 1936, "that in those days I felt and said I would be willing to part with half of what I had if I could be sure of keeping, under law and order, the other half. Then it seemed that I should be able to hold nothing for the protection of my family."[15]

An outsider in Wall Street as well as in the Back Bay, he had no deep allegiance to the old order. By this time his acquaintance from shipbuilding days in the First World War was the progressive governor of New York. Henry Morgenthau, Franklin D. Roosevelt's conservation commissioner, brought the two men together again.[16] Kennedy decided to back Roosevelt for the Democratic nomination. Why Roosevelt rather than Smith? Kennedy hardly knew Smith personally, but may have felt he knew Smith's Irish American generation too well. Very likely he was put off by the vociferous support Smith had in 1932 from unreconstructed Boston Irishmen like his father-in-law. "I knew that big drastic changes had to be made in our economic system," Kennedy told Joe McCarthy a quarter century later, "and I felt that Roosevelt was the one who could make these changes. I wanted him in the White House for my own security, and for the security of our kids, and I was ready to do anything to help elect him."[17]

Kennedy was an early and generous contributor to Roosevelt's campaign and served as a Roosevelt emissary to William Randolph Hearst, the newspaper publisher. Victory brought disappointment. Kennedy hoped to be Secretary of the Treasury. For whatever reason — partly because Louis Howe, Roosevelt's possessive and populist-minded adviser, disliked Wall Street speculators in general and Kennedy in particular — the post went to William H. Woodin, and, when Woodin retired in a few months because of ill health, to Morgenthau. Roosevelt remained amiable but evasive. Kennedy was privately upset. "If I were to send you a short summary of the whole situation in the country," he wrote Felix Frankfurter at Oxford in December 1933, "I would just say 'confusion.'"[18] Raymond Moley, a close Roosevelt adviser, noted after a holiday with Kennedy in Florida, "There must have been hundreds of dollars in telephone calls to provide an exchange of abuse of Roosevelt between Kennedy and W. R. Hearst."[19]

It must be supposed that Kennedy expected Frankfurter and Moley to report his discontent to the White House. But Roosevelt had an exact sense both of Kennedy's ability and of his intractability. Kennedy in the Treasury, for example, would have been a constant problem for a President with a penchant for experiment. Roosevelt proposed that Kennedy head a commission to negotiate reciprocal trade agreements in South America or that he become minister to Ireland. "I told him," Kennedy wrote his oldest son, "that I did not desire a position with the Government unless it really meant some prestige to my family."[20]

Later in 1934 Congress created the Securities and Exchange Commission (SEC) to oversee the stock market. Roosevelt decided to make Kennedy chairman. The appointment would inspire confidence in the financial community. At the same time, Kennedy knew all the tricks of the trade and was evidently ambitious to establish himself as a public servant. But word that a notorious speculator was scheduled to police Wall Street provoked a minor New Deal revolt. Roy Howard ran an editorial in the Scripps-Howard papers calling the proposal "a slap in the face" to Roosevelt's most loyal supporters.[21]

Unperturbed, FDR offered Kennedy the job. Kennedy said, "I don't think you ought to do this. I think it will bring down injurious criticism." Moley, who was present, asked whether there was anything in Kennedy's business career that could hurt the President. "With a burst of profanity," Moley recalled, "[Kennedy] defied anyone to question his devotion to the public interest. . . . What was more, he would give his critics — and here again the profanity flowed freely — an administration of the SEC that would be a credit to the country, the President, himself, and his family — clear down to the ninth child."[22]

The invocation of the family made an impression. It should have;

it was Kennedy's profoundest commitment. Harold Ickes, the Secretary of the Interior, sourly recounted the meeting in which Roosevelt told the cabinet. "Apparently he is going on the assumption that Kennedy ["a former stockmarket plunger"] would now like to make a name for himself for the sake of his family, but I have never known many of these cases to work out as expected."[23] This case did work out. The SEC was a complex and demanding assignment, requiring enormous knowledge, toughness and skill. Kennedy was an indefatigable worker. Over the next fifteen months he did a first-class job in establishing a vigilant and effective regulatory agency. In September 1935 Kennedy, having restored the capital market and secured the adoption of new trading rules, resigned. Later that month the Kennedys embarked for Europe.

v

In England Winston Churchill invited them to luncheon at Chartwell, his country house in Kent. Rose Kennedy thought that Churchill, "with his puckish face . . . looked more like a country squire than an English statesman." Unlike Lord Beaverbrook, who had asked countless questions two nights earlier, volunteering little on his own, Churchill "talked expansively, narrating, explaining and trying to convince us of the wisdom of his points." England and the United States, he said, should build a navy strong enough to police the world; this was the way to counter Nazism. But he acknowledged that this would be a hard idea to sell in America; there were too many isolationists, "too many Irish haters of England, too many people that would prefer to remain outside England's sphere." Mrs. Churchill asked whether Eleanor Roosevelt was an exhibitionist using her husband's high office to court publicity for herself. "I tried to convince her that I thought Mrs. Roosevelt was sincere," noted Rose Kennedy, and that "gradually people would become accustomed to her unconventional approaches."[24]

Later Churchill wrote Kennedy, "I have invited a distinguished company to meet you on the 28th and have already a great many acceptances." But Kennedy's second son, John, who had just entered the London School of Economics, was seriously ill with hepatitis. "I am deeply grieved at your anxiety about your son," Churchill continued, "and earnestly trust it will soon be relieved. Of course you must not think of being at all hampered by this engagement. On the whole however I think it would be well to carry on."[25] Kennedy nevertheless replied, "After a week of great concern and anxiety, I have sent my boy back to America this morning. I propose following him on Wednesday."[26]

Kennedy gave Roosevelt a gloomy report on conditions in Europe and then busied himself profitably as a consultant on corporate problems. As the 1936 election approached, he prepared to help in the campaign. In July, a newspaperman remarked at a presidential press conference, "Joe Kennedy is writing a book on what is going to happen to children under Roosevelt." FDR replied, "Very good; he has nine children and should qualify."[27] The newspaperman was only slightly off the mark. The first chapter of *I'm for Roosevelt*, published in August, was titled: "In Which the Father of Nine Children Explains What He Thinks This Election Means to Them."

The book, written with the help of Arthur Krock,[28] opened with a somewhat disingenuous disclaimer: "I have no political ambitions for myself or my children."* Kennedy's argument was that Roosevelt had saved capitalism from the capitalists. Disclosures of business malpractice, he wrote, had "completely shattered . . . the belief that those in control of the corporate life of America were motivated by honesty and the ideals of honorable conduct." Yet the New Deal was founded on "a basic belief in the efficacy of the capitalistic system." Roosevelt had been "at pains to protect the invested wealth of the nation." Business hatred of Roosevelt was therefore "the strangest hatred in history, because the most articulate and venomous among the opposition are those very persons whose fortunes have been repaired through Presidential activities since March, 1933." Kennedy found the business attack on government planning especially mindless. "An organized functioning society requires a planned economy. The more complex the society the greater the demand for planning." Mass unemployment was "the Achilles heel of freedom . . . the root of all the ills and ailments of subjugated people in Europe. . . . Planned action is imperative, or else capitalism and the American scheme of life will be in serious jeopardy."[29]

Businessmen were bitter in 1936. Marquis Childs of the *St. Louis Post-Dispatch* described the mood in his book *They Hate Roosevelt*. Kennedy was a rare and valued exception. To the idea, expounded by his old friend Hearst, that the New Deal was a communist plot, Kennedy replied contemptuously, "There has been scarcely a liberal piece of legislation during the last sixty years that has not been opposed as Communistic."[30] In a radio speech in October he praised the New Deal for having done away with the popular myth that business success was a guarantee of civic virtue. This was why, he

---

* A quarter century later, William Manchester asked him what he meant. Joseph Kennedy replied: "I wasn't thinking of the Presidency then. I just wanted them to be useful civic servants. If they'd become sheriffs or selectmen, that would've been OK with me." Manchester's comment is apt: "Sequiturwise, this was non." William Manchester, *Portrait of a President* (Boston, 1962), 173.

surmised, "men of wealth" hated Roosevelt. The businessman's "material position" had not been harmed,

> but his *moral* prestige is gone. . . . He is being judged by new standards
> which are quite unfamiliar to him. He feels exposed. He has been
> shaken in his own faith in himself. He is made to doubt secretly that he
> represents the American system in its most perfect expression. All this
> is unconscious and all the more frightful. Being unable — and certainly
> unwilling — to analyze with coolness the cause of his anxiety, he seeks
> relief in vituperation and hatred.[31]

VI

Kennedy had his forty-eighth birthday two months before the 1936
election. This proud, profane, imperious, frank, indiscreet, scheming, relentless, sensitive, vital man had had spectacular success in
business and government. He owned houses in Bronxville, Hyannis
Port and, after 1934, in Palm Beach. He was known in Europe as
well as in the United States. He had left far in his wake the Yankees
who had patronized him at Harvard and on State Street. Still he had
not quite achieved the goal he had long since set himself. When a
Boston newspaper described him as an Irishman, he exclaimed, "I
was born here. My parents were born here. What the hell do I have
to do to be called an American?"[32]

His complaint did not imply repudiation of his origins. He did not
resent being Irish but the condescending Yankee theory of what
being Irish involved. We must, he wrote Joseph I. Breen of the
Association of Motion Picture Producers, "see if we can't add a little
more lustre to the Irish reputation."[33] Most of his close friends were
of Irish descent. His character was formed by his Irish cultural
inheritance — the pessimism, the romantic defiance, the political
instinct, the irony, the sexual chauvinism, the Catholicism, the sense
that the world was a mess and not likely to improve very much —
though he fused these traits with a strictly American passion for
competition, achievement and victory. Still, powerful as this passion
was, it was conditioned by the Irish tradition; above all by his supreme
belief in the family — in *his* family.

"The measure of a man's success in life," Kennedy often said, "is
not the money he's made. It's the kind of family he has raised." One
day on the golf course at Palm Beach, Bernard Gimbel asked, "What
are you doing these days, Joe?" Kennedy replied, "My work is my
boys."[34] In the 1920s, when he had been away for long periods in
New York and later in Hollywood ("I did not think it was vital," his
wife wrote in her memoirs, "for my husband to be on hand for the
birth of the babies"),[35] the care of the young fell to Mrs. Kennedy.

As the children grew older, their father spent more time with them. LeMoyne Billings, John Kennedy's Choate roommate and an intimate of the household after 1934, could not remember visiting the Kennedys when Mr. Kennedy was not there.[36] The balance of authority shifted. "When he was there," recalled Eunice, "it was his personality that dominated. Our mother ... let him sort of take over."[37]

Joe Jr. was the child in whom the parents invested their deepest hope. He was twenty-one years old in 1936, a sophomore at Harvard after spending 1933–34 with Harold Laski at the London School of Economics. He had charm, confidence and gusto. His brother John recalled his "amazing intensity.... Even when still, there was always a sense of motion forcibly restrained."[38] Laski remembered abler students but not many with "either his eager zest for life or his gift of winning one's affection.... He has often sat in my study and submitted with that smile that was pure magic to relentless teasing about his determination to be nothing less than President of the United States."[39]

John, nineteen, was preparing to enter Harvard in the fall. He had spent much of his childhood sick in bed, surrounded by books. "Jack is far from being a well boy," his father had written as recently as the preceding November.[40] A family joke, evidently made up by his younger brother Robert: "If a mosquito bit Jack Kennedy, the mosquito would die."[41] There were occasional troubles between the older brothers. Joe was purposeful and methodical. Jack was casual, untidy and absent-minded. Joe was robust and, for all his gentleness with the younger children, hot-tempered. Jack was frail but quick-witted and provocative. Their mother described their childhood relationship as one of "friendly enmity."[42] Jack won the arguments, Joe the fights. When James MacGregor Burns asked Jack whether anything had really bothered him as a child, he could think only of his big brother. "He had a pugnacious personality," he replied laconically. "Later on it smoothed out but it was a problem in my boyhood." Burns added, "Bobby Kennedy to this day remembers cowering with his sisters upstairs while his older brothers fought furiously on the first floor."[43]

After the two brothers came four girls: Rosemary, 18 in 1936, who her parents now realized was mentally retarded; Kathleen, 16, gay and pretty; Eunice, 15, and Patricia, 12, still tomboys. There followed Robert, 11, Jean, 8, and Edward, the baby, who was 4. The family was the center of Joseph Kennedy's existence. Everything else took second place. His houses were designed not to impress his guests but to cherish his children. In Hyannis Port the Kennedys had an open, rambling, white Cape Cod house separated by eel grass from a fringe of beach along Nantucket Sound. The Palm Beach house was a white stucco 1920s villa overlooking the ocean. The Bronxville house, to which they had moved from Riverdale in 1929, was a large brick Georgian mansion on six acres of ground. All were notable for their large, comfortable rooms, their space and sunlight and flowers, their ample lawns easily converted into touch football fields and staging areas for pets.

Joseph Kennedy expended all the power, conscious and unconscious, of a demanding personality on pressing his children, especially the boys, to develop their potentialities to the full, on driving them beyond themselves. "He wanted a son to be something," said Kenneth O'Donnell, the long-time friend of John and Robert. ". . . Maybe it was to prove something for himself." [44] His insistence on competition and success became in time legendary — perhaps because it represented so quintessential a version of a basic American legend.

"We don't want any losers around here," he told the children. "In this family we want winners. . . . Don't come in second or third — that doesn't count — but win." [45] He subjected defeat to rigorous analysis and, if the postmortem revealed that the loser had not tried hard enough, registered severe paternal disapproval. "He tolerated a mistake once," his youngest son later testified, "but never a second time." [46] Losers had to keep their sorrows to themselves. "I don't want any sourpusses around here," he would say. ". . . We don't want any crying in this house" — a proposition his children came to render, somewhat sardonically, as "Kennedys never cry." [47]

The win-win-win compulsion dominated the household. Still it was not so bad as it sounded. The point was less to win than to do one's best to win. "We were to try harder than anyone else," Robert Kennedy wrote. "We might not be the best, and none of us were, but we were to make the effort to be the best. 'After you have done the best you can,' he used to say, 'the hell with it.'" [48] Failing to do your best was the mortal sin, not losing.

VII

The athletic code was accompanied by equally incessant pressure for intellectual development. Everyone knows about mealtime at the Kennedys, where Joseph Kennedy liked to preside over a family seminar. Among the children the two older boys did most of the talking. Their father sometimes took the other side in order to sharpen the discussion. "There was no frivolity," Lem Billings said of the family meals, "but they were always exhilarating and fun." [49]

The forbidden topic was business. "I have never discussed money with my wife and family," Kennedy once said, "and I never will."[50] The subject ultimately bored him. He did not wish to inflict it on his children. "He had made enough money not to need to make any more," Lem Billings commented; "he saw no reason for them to do what he had done."[51] The boys grew up cheerfully indifferent to business. "Listening to them talk about money," said Charles Spalding, another close friend of the brothers, "was like listening to nuns talk about sex."[52]

In any event, Joseph Kennedy thought, the reign of money per se was finished. "The boys might as well work for the government," he told his wife, "because politics will control the business of the country in the future."[53] Public affairs ruled the dinner table. "I can hardly remember a mealtime," Robert Kennedy later wrote, "when the conversation was not dominated by what Franklin D. Roosevelt was doing or what was happening in the world."[54] "Any of the children," recalled one of John's Choate friends, "could ask him questions and [he replied] in the greatest detail, no matter how simple or how stupid the questions might seem. . . . However, if an outsider, such as myself would ask the Ambassador a question . . . he would just answer it curtly as though he didn't want to be bothered with your questions. He was only concerned about educating his own children."[55]

His concern pursued the children everywhere. Excerpts from letters to John convey the flavor:

I strongly urge you to pay a little more attention to penmanship. Mine has always been pretty bad, so I am not a very good authority to speak about it, but yours is disgraceful and should get some attention.

I don't want to give the impression that I am a nagger, for goodness knows I think that is the worst thing any parent can be, and I also feel that you know if I didn't really feel you had the goods I would be most charitable in my attitude toward your failings. . . . Aren't you foolish not to get all there is out of what God has given you and what you can do with it yourself. . . . It is very difficult to make up fundamentals that you have neglected when you were very young and that is why I am always urging you to do the best you can.

Don't let me lose confidence in you again, because it will be pretty nearly an impossible task to restore it — I am sure it will be a loss to you and a distinct loss to me. The mere trying to do a good job is not enough — real honest-to-goodness effort is what I expect. Get yourself into shape so that a man can't write a halfway report on you. . . . You have the goods; why not try to show it?[56]

"I grew up in a very strict house," John Kennedy said in 1960, "and one where . . . there were no free riders, and everyone was expected to do, give their best to what they did. . . . There was a

constant drive for us for self-improvement."[57] Discipline was enforced, on the father's part, by moral rather than physical means.
"I'm sure Mr. Kennedy never slapped a child in his life," said Lem
Billings. "But his eyes would just take you right out of the
window."[58] That cold, appraising, steely glance over the top of his
glasses was known in the family as "Daddy's look."[59] The children's
greatest anguish came when Daddy was disappointed in them; their
greatest happiness lay in his approval. His commanding presence
awed their young friends, most of whom their father ignored but a
few of whom, like Billings and Spalding, were received into the
bosom of the family and became auxiliary brothers. Even they
thought Mr. Kennedy remote, a little humorless and infinitely intimidating. Billings found him "a most emotional man — but also a man
of complete self-control."[60] He was, said Spalding, "the first unbri-
dled force I ever encountered."[61]

VIII

Rose Kennedy was her husband's staunch ally. She was a remarkably
handsome, controlled, organized, and impervious woman. She was
also a cultivated woman, better read and in some ways more worldly
than her husband. Her father had instilled in her an enthusiasm for
history; she kept up her languages; all her life she remained a student, dedicated to self-improvement. She adored balls and trips to
Paris, while Joseph Kennedy, as she later said, considered parties a
waste of time, where he would either be solicited for tips on the stock
market or be required to argue about the New Deal. Pressed to go
out, he would say, "We eat better at home."[62] Nor did he much like
entertaining non-Kennedys even at home, except for the children's
friends.

Her husband's conventional religious faith paled next to her serene
and inexhaustible piety. This led to one of their few express disagreements. "I wanted all of the children to have at least a few years
in good Catholic schools," Rose Kennedy later wrote, "where . . . they
would receive thorough instruction in the doctrines of their religion
and intelligent answers to any doubts."[63] Joseph Kennedy told her
that children ought to learn their religion at home and at church; his
deeper view was that they should not be deflected from the mainstream of American life.

The result was predictable in a family culturally bound to regard
girls as less important than boys. The boys went in the main to
secular institutions; the girls first to convents and then to Catholic
colleges. Joe Jr. was a loyal Catholic in the style of his father. John

took religion, like everything else, with detachment. One day, passing a church, he said to his friend Charles Spalding, "How do you come out on all this?" Spalding, who had been born a Catholic, replied that he had never had time to work it all out; if he did, he supposed he would end by saying, "I don't know." John Kennedy observed that was about where he stood too.[64] Yet he said his prayers and remained generally faithful in religious observance. Robert, always the overachiever, was the most religious among the sons. As a small child in Bronxville, he aspired to be an altar boy and could be heard practicing his Latin in his room.[65] His devotion never, however, contrary to later suggestions, carried him to the point of contemplating the priesthood.[66]

No child could hope to escape father by appealing to mother. She was as articulate as he on the virtues of winning, as given to didactic questions at the dinner table, though hers went more to history and theology than to current affairs, as stern in her discipline, as unrelenting in her determination to make the children meet high standards. Letters suggest her maternal style:

> I do hope you are working harder, dearest Teddy. As I said before, the Kennedy boys worked hard and were usually at the top of their classes in Boston, and I am sure the first question most people ask about you is, "Are you smart like your father?" — not to mention that I graduated from good old Dorchester High School with honors myself — so please get on your horse.[67]

(To a daughter-in-law)

> I think in the future it might be well to review history and current events with him [a grandson] so that he may remember facts in that way. You might ask him questions like who the President of France is; what holidays are national holidays or church holidays, or both; the name of the Queen of England, etc.

> The reason I am sometimes reluctant about extra goodies for the children is that I read once that youngsters should get used to a few disappointments or jolts when they are young. They thus build up an immunity against them, so that when they encounter life's vicissitudes when they are older, they are not too shocked and they can take them with a smile.[68]

Her daughter Eunice later said, "It was this terrific drive, she wanted everyone to do their best. There was quite a little pressure around."[69]

IX

Everyone who has written about the Kennedys has discussed this pressure, the parental insistence, unceasing and unsparing, on competition, achievement, victory. The psychic burden must have been considerable. The desire to please parents forces children to some degree into roles set for them from above. Such a process can create doubts about worth and identity and lead to overcompensation or to despair. American life is littered with the wrecks of men and women who cared too much about pleasing their parents.

The mystery is why the parental demands did not crush the Kennedy children.* It would be hard to demonstrate that their upbringing disabled their capacity for affection, self-reliance and happiness. Rather the contrary: Charles Spalding, visiting Hyannis Port for the first time in 1940, registered a general impression when he described a family "moving in every direction and vitally involved and interested in what was going on. I had never seen . . . any family like that. . . . I thought to myself, 'Well, this is really the best possible way to approach life.'"[70] Winston Churchill and Franklin D. Roosevelt were incomparably greater public men than Joseph P. Kennedy. Yet Randolph Churchill and Franklin D. Roosevelt, Jr., youths of ability and ambition, never fulfilled their potentialities. The Kennedy sons, for whatever reason, pursued their capacities to the uttermost limits. Had the parents no part in this result?

I put this question to John Kennedy one evening at the White House. He said, "It was due to my father. He wasn't around as much as some fathers; but . . . he made his children feel that they were the most important things in the world to him."[71] When someone asked Franklin D. Roosevelt, Jr., what Joseph Kennedy was like as a father, he replied, no doubt out of some indecipherable mixture of memories, "From everything I saw he was a great father. He didn't leave a stone unturned to encourage his boys. . . . He loved doing things with his boys."[72]

In the spring of 1961, John Kennedy stopped off in Hyannis Port before leaving for Europe to see Khrushchev. A heavy fog was sweeping across the compound. He was walking to his father's house for dinner when Caroline, his daughter, came off the porch crying to him. As he started to comfort her, the kitchen door opened, and someone said, "Mr. President, they want you on the White House phone — they said it's important." Kennedy said, "Caroline, I'll be back in just a moment. Let me go take this phone call." When dinner began, there was an edgy silence. Finally Mr. Kennedy said, "Jack,

---

* Some writers think they were crushed. See, for example, Nancy G. Clinch, *The Kennedy Neurosis* (New York, 1973), especially 8–9, 15, 60–61, 108. This book is a splendid specimen of the heads-I-win-tails-you-lose school of psychohistory. Whatever the Kennedys did — one thing or its opposite — was grist to the analytical mill. If a Kennedy went up in a balloon, he would be an exhibitionist; if he refused to go up, he would be a coward.

I saw what happened outside. Caroline was in tears and came out. You had a call from the White House. I know there are a lot of things on your mind about your meeting with Khrushchev.... But let me tell you something: Nothing that will happen during your Presidency will be as important as how Caroline turns out. And don't forget it."[73] Joseph Kennedy had never forgotten it. But then he had never been President.

If pressure created problems for the children, love gave the children strength to meet those problems. "Mr. Kennedy was somehow able," LeMoyne Billings later said, "not to spoil his kids and yet not frighten them with his aggressive desires.... He did it with love, and yet with strictness."[74] It was a love that put "Daddy's look" and his eternal emphasis on competition in context. Still love by itself was not the answer to the mystery. Parental love is often suffocating. Joseph Kennedy's evidently was not.

His whole effort, in one aspect, was to make his children independent of him. "At an early age," Robert recalled, "we were sent on trips.... We were given tasks of responsibility. We embarked on study programs under teachers or political leaders who held views quite different from his."[75] A highly opinionated man, Joseph Kennedy did not require his children to adopt his opinions. What mattered was that they should be able to defend their own. "He always encouraged his children to argue with him," said Lem Billings, veteran of so many Kennedy dinner tables. "... He *never* wanted them to agree with him."[76]

The Laski connection was characteristic. Kennedy regarded Laski as "a nut and a crank. I disagreed with everything he wrote. We were black and white. But I never taught the boys to disapprove of some one just because I didn't accept his ideas. They heard enough from me, and I decided they should be exposed to someone of intelligence and vitality on the other side."[77] When Laski proposed taking young Joe to the Soviet Union, his father was delighted. "I think it would be wise for you to go with Laski," he wrote, "particularly because it would give you a very favorable viewpoint in consideration of things here."[78] Young Joe came back, "exultant," his mother later said, "in his enthusiasm" and engaged in vigorous debate with his father over capitalism and communism.[79] John Kennedy observed from the sidelines, "Joe seems to understand the situation a little better than Dad."[80] (Young Joe's enthusiasm infected Rose, and in 1936 she took Kathleen on a visit to the Soviet Union, concluding tolerantly that "many of its aims, if not many of the methods, were worthy of respect and discussion and study.")[81] Mr. Kennedy remained unperturbed. "If I were their age," he told his wife, "I probably would believe what they believe, but I am of a different background. . . . I don't care what the boys think about my ideas. I can always look out for myself. The important thing is that they should stick together."[82]

He was willing to take his chances. In the 1920s he had consulted his friend Hearst about setting up a trust fund for each child. Hearst warned him not to make them independent; when they got older they would leave him and go off with their money. Kennedy's comment was, "If that's the only way to hold them, I've been a lousy father."[83] He set up the trusts, fixing it, as he said, "so that any of my children, financially speaking, could look me in the eye and tell me to go to hell."[84] "If they want to be slobs," he told Frank Waldrop of the *Washington Times-Herald,* "they can, and I can't do anything about it."[85] It was family, not money, that would keep them together. "My father used his money," said Robert Kennedy, "to free us — not to hold us."[86]

"The generation that follows me," Joseph Kennedy once said, "may have to stand for everything that I stood against — and I realize that includes even my own sons. I made my choice among the philosophies offered when I was young. Each of them will have to make his or her choice."[87] He remained an Irish pessimist; but when his sons "were more optimistic about the future of life on this planet," Robert Kennedy remembered, "then in a more mellow mood he would say, 'If I were your age I would hold those views also.'"[88] Ideas were not the ultimate reality. The ultimate reality was family.

Joseph Kennedy's "fierce sense of family" (Morton Downey's phrase) was perhaps the key to the mystery.[89] It gave his children the ability to withstand his own pressure. And the family provided a protection against inordinate parental demands in part, it may be surmised, because it was an Irish Catholic family. In a Protestant family, where salvation was an individual responsibility, competition threw each child back on himself and parental demands could become an individual burden. In an Irish Catholic family, where salvation was culturally mediated through institutions, the destructive effect of competition, the suffocating effect of love, could be tempered by corporate solidity and mutual reinforcement. "The Kennedys," Joseph Alsop used to say, "are all in love with each other."[90] The Irish legacy had not faded away in the third American generation.

Of course, lest one appear to endorse theories of parental determinism, it is necessary to add that the character of the children themselves — their spirit, vitality, toughness, irreverence — had something to do with their capacity to survive. There was something very Irish about those traits too.

(2)

The Father

BY THE TIME of the seventh child, Rose Kennedy said, "Even the most enthusiastic parent may feel less excitement about 'another baby.'"[1] Robert Kennedy must have perceived that his older brothers and sisters enjoyed more parental attention. Apprehensions of inferiority were no doubt intensified when it became apparent that he would be a good deal shorter than his tall older brothers; even, in time, than his petted younger brother. His father, in addition, was away much of his early childhood. When at home, Joseph Kennedy was on occasion impatient and rough with him. His mother, compensating, protected and consoled him. Problems with his father increased a sense of inadequacy. They also inspired him to strive to the utmost to win his father's approval and love.

I

From all accounts, he was an extremely sweet child. "I loved Bobby from the day I set eyes on him," said Lem Billings, who saw him first when he was eight years old. "He was the nicest little boy I ever met." Once Billings commented to Joseph Kennedy on Bobby's exceptional generosity with possessions. Bobby's father grunted, "I don't know where he got that."[2] Luella Hennessey, who joined the household as the children's nurse when Bobby was twelve, called him "the most thoughtful and considerate of all the Kennedy children."[3]

A couple of letters written when he was nine are revealing:

Dear Daddy,
    I went to the circus Saturday. Dave, Frank, and Tommie went with me and we liked it very much. I wish you were with me. [Clyde] Beatty

the lion trainer wasn't there this year. I liked the horses jumping over the fire the best.

I will probably see you this week. I hope so.

Love,
Bobby[4]

Dear Mr. President,

I liked the stamps you sent me very much and the little book is very useful. I am just starting my collection and it would be great fun to see yours which mother says you have had for a long time.

I am going to frame your letter and I am going to keep it always in my room.

Daddy, mother, and all my brothers and sisters want to be remembered to you.

Bobby Kennedy[5]

The school administrator at the Bronxville public school, where he attended his third, fourth and fifth grades, remembered him thirty years later as a "nice little freckle-faced kid. . . . He needed no special handling."[6] And a report from Camp Winona in Maine where he went when he was eleven:

Unit Director's Comment: His progress has been fair. He shines in swimming and sailing. He is a boy who needs plenty of encouragement which he is now getting. . . .

Tent Counselor's Comment: Bob is erratic. At times he is as good as any boy in the tent group, but at other times he is quite diffident. Always obedient, he has a tendency to do only what he thinks is expected of him. . . .[7]

He was the gentlest and shyest in the roaring family, the least articulate orally and the least coordinated physically, the most conscientious and the most dutiful. "While they all knew what was expected of them," his mother recalled, "Bobby really worked at it."[8] Punctuality was much emphasized by the elder Kennedys. Once when Bobby was four, rushing to be on time for dinner, he careened into a heavy glass door and gashed his head. Something like this was always happening. Joseph Kennedy to Joe Jr. in 1937:

A boy threw some hot tar at Bobby and struck him in the eye. Through the grace of God, he has not lost the sight of his eye and will probably come out of it all right but it was a very narrow escape.[9]

"What I remember most vividly about growing up," Robert Kennedy told Jack Newfield in later years, "was going to a lot of different schools, always having to make new friends, and that I was very awkward. I dropped things and fell down all the time. I had to go to the hospital a few times for stitches in my head and my leg. And I was pretty quiet most of the time. And I didn't mind being alone."[10]

He kept rushing. He always had to prove himself. Nothing was easy for him. Slow in swimming, he threw himself off a yawl into Nantucket Sound as if determined to learn all at once or to drown. His brother Joe pulled him out. "It showed," John Kennedy observed, "either a lot of guts or no sense at all, depending on how you looked at it."[11] "I was," Robert said, "the seventh of nine children, and when you come from that far down you have to struggle to survive."[12] Struggle was the answer to the fear of inadequacy — struggle to fight his way, through sheer drive and nerve, into his father's attention and respect.

II

Early in 1937, Joseph Kennedy returned to Washington as chairman of the Maritime Commission. Roosevelt had other reasons, besides his desire for a strong executive in a tough job, to want Kennedy in the administration. The New Deal desperately needed friends in the business community. Kennedy was ready to tell businessmen at the New York Economic Club to "stop bellyaching." He faithfully backed Roosevelt's plan to enlarge the Supreme Court.[13]

Knowing how cordially most Harvard graduates detested Franklin D. Roosevelt, '04, the President must particularly have appreciated Kennedy's loyalty in telling Harvard '12, at the twenty-fifth class reunion in 1937, that big government "is not *foisted* on us by politicians. It has *grown* upon us because certain things have to be done which no one but government can do." If businessmen were disturbed by sit-down strikes and "bloody warfare at the factory gates," Kennedy said they should realize they were "reaping the whirlwind of a quarter century mishandling of labor relations. We complain about the lack of responsibility of labor and for years we did our best to render labor organizations impotent." Kennedy concluded by telling the anti–New Deal conclave that he could frankly see "nothing which justifies the hate and the despair which are all around."[14] Kennedy's classmates booed and jeered him.[15] He never went to a class reunion again. Still he remained, in an administration under mounting business attack, a precious symbol.

As Kennedy was a Roosevelt ambassador to the business community, so he was a Roosevelt ambassador to the Catholic Church. In the early thirties he had tried to keep Father Coughlin, the radio priest from Detroit, on the New Deal reservation. When Cardinal Pacelli (later Pope Pius XII) visited the United States in 1936, Kennedy arranged a luncheon for him with Roosevelt at Hyde Park and later had him to a family tea at Bronxville.[16] All this was quite nontheological. Kennedy did mention Leo XIII's encyclical "The Condition of the Working Classes" in *I'm for Roosevelt*. But he had no particular contact with those Catholic New Dealers, like Father John A. Ryan, who labored in the spirit of *De Rerum Novarum*.[17] Nor was a man who cheerfully sent his wife and a couple of children to godless Russia hewing to the Vatican's political line.

Roosevelt liked Kennedy, liked Rose Kennedy even more and was touched by Kennedy's preoccupation with his family. James A. Farley, it is true, claimed that Roosevelt "never liked" Kennedy.[18] But Farley, who felt snubbed by the Roosevelts, may have concluded that Roosevelt was socially opposed to all Irish Catholics. In fact, the Kennedys, unlike the Farleys, came often to the White House, and, what was more unusual, Roosevelt went on occasion to Kennedy's house along the Potomac, which he would never have done had he not counted on a good time. Morgenthau and Harold Ickes also noted with satisfaction Roosevelt's complaints that he always had to hold Joe Kennedy's hand.[19] But Roosevelt spent a great deal of time holding Morgenthau's and Ickes's hands too. His remark about Kennedy was among other things a warning, evidently too subtle, to them.

"I always think of the Roosevelts with great affection, as they all were very gracious to us as a family," Rose Kennedy wrote in a personal memorandum circa 1940. "The President had more charm than any man I ever met." A family that had no tolerance for self-pity could not but admire FDR's refusal to admit pain.

> There must have been times when the braces seemed insupportable, and his limbs ached, but still he would keep smiling. The only indication of suffering was in the stronger determination of that jaw. I recall the Review of the Fleet. We went to the house in New York to meet the President, and went to the pier where we were to board the President's yacht. . . . Amidst horns blowing and special police, we arrived at the pier which he slowly traversed amidst shouting applauding audiences still smiling and beaming. He again crossed a ramp onto the yacht and went above deck. Then, for a fleeting second, I caught a glimpse of him sinking exhausted into a chair saying, "For God's sake let's take off these darn braces."[20]

III

The Maritime Commission was a demanding job. But Kennedy found time to keep his eye on other matters. At the Securities and Exchange Commission he had struck up unexpected friendships with two young New Deal lawyers — James M. Landis of the Harvard Law School, who was a fellow commissioner, and William O. Douglas of the Yale Law School, whom he had brought down to head a study of corporate reorganization. After getting Roosevelt to appoint Landis his successor as chairman in 1935, Kennedy had got him to appoint Douglas to the SEC in 1936. Now his New Deal protégés had fallen out. Landis, scheduled to return to Cambridge in September 1937 as dean of the Harvard Law School, resolved to stay in the job long enough to keep the more radical Douglas, who had been named dean of the Yale Law School, from becoming the next chairman. Since Yale opened earlier than Harvard, Douglas thought he had lost. Two hours before the last train to New Haven, Kennedy told Douglas to expect a call from the White House; he had just left FDR, Kennedy said, and everything was back on the track. In a moment the President was on the phone to Douglas: "Unpack your bag — you are the new chairman." "If I had gone back to Yale," Douglas later wrote, "I . . . would not have been appointed to the [Supreme] Court."[21]

The episode illustrates something of Kennedy's complexity. He deeply liked both Landis and Douglas. They were outsiders too — one a missionary's son, born in Japan; the other, a kid from the wrong side of the tracks in Yakima, Washington. Both were men of impressive ability and fidelity. On the issues that now divided them, Douglas's position was probably, on the merits, less acceptable to Kennedy. Yet, as Kennedy had been loyal to Landis in 1935, so he was determined to be loyal to Douglas in 1937. He remained close to the able and taciturn Landis, but something especially delighted him about Douglas, with whom he shared a bluntness of speech, a skepticism about motives and an expansive scorn for the eastern establishment. Douglas had been five years old when his father died; and Kennedy, though only ten years his senior, filled a vacancy in his life. "Our friendship grew," Douglas later wrote, "until . . . we were almost father and son."[22]

As his work on the Maritime Commission drew to an end, Kennedy could well expect a higher and better assignment. He still wanted the Treasury. This was impossible. Morgenthau had tried to carry out New Deal policies in every respect, Roosevelt told Farley in 1938, while "Joe would want to run the Treasury in his own way, contrary to my plans and views." Declaring himself "thoroughly dissatisfied" with Daniel C. Roper, the Secretary of Commerce, Roosevelt said that this was the spot for Kennedy.[23] But Robert W. Bingham, the ambassador to England, was in poor health, and the Kennedys, Rose too, set their hearts on London. This ambition, Arthur Krock said, "was made known emphatically to the President."[24] FDR admired Kennedy's administrative ability, knew he could afford the post and was doubtless diverted by the idea of sending a Massachusetts Irishman to the Court of St. James's.

"The most thrilling, exciting, and interesting years of my life,"

Rose Kennedy wrote later, "were the ones I spent in London when Joe was Ambassador."[25] The mother and five children embarked on the *Manhattan* in March 1938. "The most tender parting," the Associated Press reported, "was that of Bobby and his Bronxville pal, Francis Deignan. They discussed schemes for raising $300 to get Francis to London and told each other solemnly, 'I'll write to you.'" As the ship prepared to sail, Robert launched a handmade parachute through the porthole in his stateroom.[26]

The family was soon established in the ambassadorial residence in Prince's Gate near Kensington Gardens. Felix Frankfurter wrote the new ambassador that his Massachusetts predecessors at the Court of St. James's, John Adams and Charles Francis Adams, had used their sons as personal assistants.[27] Encouraged by such weighty Yankee precedent, Joseph Kennedy put Joe Jr. on the embassy staff; John too, when he was in London on holiday. As always, the father was pushing his sons to move out on their own. Joe went to Czechoslovakia during the Munich crisis and to Spain in the last months of the Civil War, John to Czechoslovakia at the time of the German invasion in March 1939. "We were furious," George Kennan recalls, when the Prague embassy received word of John Kennedy's arrival on a fact-finding mission. "Joe Kennedy was not exactly known as a friend of the career service. . . . His son had no official status and was, in our eyes, obviously an upstart and an ignoramus. . . . That busy people should have their time taken up arranging his tour struck us as outrageous."[28]

Kathleen helped her mother deal with London society. Her younger sisters were distributed among local convents. The smaller boys went to Gibbs, a day school where Robert, now thirteen, did well in mathematics, poorly in Latin, which he finally dropped, and the best he could in cricket and rugby.[29] He also discharged occasional diplomatic tasks, such as laying a foundation stone at a newsboys' club. Here he delivered "a piece of oratory" that the clergyman who ran the club described improbably as "magnificent in its diction and delivery."[30]

Everyone adored the Kennedys in 1938, from the ambassador down. "His language," wrote the columnists Joseph Alsop and Robert Kintner that year, "is a cheerful torrent of profanity and picturesque expression. No one is better, or more ebullient company. Because he is rich, because he plays at being a rough diamond, because he is clever and because his brood of nine handsome children fascinate and astonish the British . . . he has been a vast success."[31] Harold Caccia of the Foreign Office noted that Kennedy had proved "he possesses every quality necessary for the success of his mission."[32]

Over Christmas, with her husband back in America, Rose took the children to St. Moritz. "Everybody had a wonderful time," Eddie Moore wrote back to the United States. The Kennedys had a special notion of a wonderful time. ". . . Fortunately [young] Joe's accident didn't amount to much, but it was necessary to have six stitches taken in his elbow, but while it kept him off the skis for three days, he skated in the meantime. Bobby didn't have any accident, but Teddy wrenched his knee."[33]

"Rose, this is a helluva long way from East Boston," the ambassador had said in the spring when they were spending a weekend with the royal family at Windsor Castle.[34] Indeed it was. "The children and I saw the Old Year out here at St. Moritz," Rose Kennedy wrote Franklin Roosevelt, "and we thought of all the interesting and stimulating experiences which we had enjoyed during 1938 . . . made possible through the honor which you had conferred upon Joe. And so we want to thank you again for 1938, and we want you to know that we are hoping and praying that you may enjoy every happiness in the New Year of 1939."[35]

IV

Nineteen thirty-nine did not turn out to be a year of much happiness for anyone. When he appointed Kennedy, Roosevelt seems not to have talked to him at length about Hitler. The President doubtless supposed that Kennedy would give the same generally loyal support in foreign policy he had given the New Deal at home. He was sure that Kennedy, as a good Irishman, would be ethnically immune to the blandishments of the Chamberlain government. And it was probably the case that Kennedy himself had no clear ideas about foreign affairs when he set forth for London.

The new ambassador had, however, one profound emotion — a dread of war. As a businessman, he believed that prosperity required a peaceful world. As a capitalist, he feared that war would lead to socialism; "a return even to our modified capitalism of today will hardly be feasible."[36] Above all, as a father, he could not bear the thought of his own sons going to war. When Hitler intensified his pressure on Czechoslovakia in the summer of 1938, Kennedy proposed saying in a speech, "I should like to ask you all if you know of any dispute or controversy existing in the world which is worth the life of your son, or of anyone else's son." Secretary of State Cordell Hull vetoed the thought.[37]

Kennedy found himself in chronic contention with the State Department. And, contrary to Roosevelt's expectations, Kennedy and

Neville Chamberlain hit it off famously. "I can talk Chamberlain's language," Kennedy bragged to Hull.[38] (Rose Kennedy liked the Prime Minister too; he charmed her by remarking that he had married his wife "because she had made good treacle.")[39] Soon Walter Lippmann was warning Kennedy that he was showing "some of the same symptoms which have appeared in every American ambassador [to London] I have known since Walter Hines Page, namely the tendency ... to be taken into camp by the government and the governing class."[40] "Who would have thought that the English could take into camp a red-headed Irishman?" Roosevelt said to Morgenthau. "Joe has been taken in by the British government people and the royal family," he told Farley.[41] This reading did Kennedy an injustice. His dread of war was leading him to a general view of America's role in the world — a view he advocated honestly and stubbornly at the cost of his political career and held for the rest of his life.

On Trafalgar Day in October 1938, a month after Chamberlain returned from Munich, Kennedy spoke again, this time eluding Hull's blue pencil by presenting his view as only "a pet theory of my own," not a policy of his government. "It has long been a theory of mine," he said,

> that it is unproductive for both the democratic and dictator countries to widen the division now existing between them by emphasizing their differences, which are now self-apparent. ... There is simply no sense, common or otherwise, in letting these differences grow into unrelenting antagonisms. After all, we have to live together in the same world, whether we like it or not.[42]

In the nuclear age these propositions would be considered defensible if not inevitable. In the age of Hitler they aroused alarm and outrage. "It must be," wrote the columnist Heywood Broun, "that the British have some secret drug which they slip into the tea of visiting celebrities. And in the case of our Ambassadors they give them double dosage." After all, Broun continued, Kennedy had come "from a ward where the kids are taught to twist the lion's tail even before they learn to roll their hoops." But he "has drunk the draught which makes the eagle cease to flap and induces it to lie down with the lion. Even his extensive and charming family has proved insufficient bodyguard."[43] Roosevelt tacitly disclaimed Kennedy's Trafalgar Day speech by saying a week later in a speech of his own that "peace by fear" had "no higher or more enduring quality than peace by the sword. ... There can be no peace if national policy adopts as a deliberate instrument the threat of war."[44]

Kennedy's pessimism grew in 1939. If war came, he was sure that

England and France would lose. At times, he was also sure that America would ultimately be dragged in.[45] At other times he thought it "nonsense to say that an Axis victory spells ruin for us"; America depended very little on foreign markets and could survive as "an intelligent self-contained national economy."[46] At still other times, he feared that an Axis victory would leave America "alone in a jealous and hostile world" where the "cost of maintaining 'splendid isolation' would be such as to bring about the destruction of all those values which the isolation policy had been designed to preserve."[47] His vacillations disabled him from developing a clear line of his own and prevented his differences with Washington from becoming impassable.

New Deal interventionists, however, became increasingly unhappy. Harold Ickes fed Roosevelt gossip about Kennedy's association with the Cliveden set, his disloyalty to the President, his own presidential ambitions. But Roosevelt, who was receiving ardently pro-allied messages from Bullitt in Paris, evidently saw value in hearing something different from London. In late August 1939, Ickes noted disappointedly in his diary, "The President said that Kennedy was as good as anyone in reporting carefully what was transpiring in England and in diplomatic circles and he gave no indication that he had an intention of removing him."[48]

The Soviet-Nazi Pact was announced on August 23. On September 1 Hitler invaded Poland. Telephoning the President to describe Chamberlain's speech declaring war on Germany, a distraught Kennedy cried, "It's the end of the world, the end of everything."[49]

V

"My family gets smaller and smaller," Kennedy wrote his wife in September 1939. Joe Jr. had sailed on the *Mauretania*; John was about to fly on the Clipper. Their father's letter showed his confusion of emotion:

> I think just as strong as ever that if England has a chance to make an honorable peace she had better do it. . . .
>
> If by any chance USA doesn't change present neutrality law then the position of the French and of British I think will be nothing short of disastrous. . . .
>
> I talk to Bullitt occasionally. He is more rattle brained than ever. His judgment is pathetic and I am afraid of his influence on F.D.R. because they think alike on many things.

He concluded that, with all the family safe in America, "I have no worries. I will miss you terribly but . . . with the state of the world I doubt if I could be happy in business. . . . This position at the minute is probably the most interesting and exciting in the world."[50]

Back in the United States, the children dispersed to boarding schools and convents. It appeared time for Robert, now almost fourteen, to go away. Exercising his paternal preference for secular education, Joseph Kennedy had enrolled him at St. Paul's in Concord, New Hampshire. But St. Paul's turned out not to be all that secular; and its unremitting Episcopalianism was too much for Rose. "The Protestant Bible was read at different times in the school," she later explained. "It was for this reason that I withdrew Bobby."[51] The ambassador cabled from London, BY ALL MEANS PUT BOBBY IN RIVERDALE IF YOU ARE MORE COMFORTABLE.[52] However, Rose Kennedy took advantage of her husband's absence to send Robert to Portsmouth Priory, a sound Benedictine school near Newport, Rhode Island, with morning and evening prayers, mass three times a week and high mass on Sundays.

The Priory more than made up for St. Paul's. "I just got out of about 3 hours of praying in Chapel," Robert wrote his parents at one point, "and so feel like a saint. Our retreat starts Thursday so I'm certainly going to be doing a lot of praying this week."[53] He became acquainted too with the liabilities of the faith. "I got another one of those post cards," he wrote his mother, "telling me like the last one how awful we Catholics are."[54] There were extracurricular excitements. He had hardly arrived at the school when a Boston newspaper reported: "Robert Kennedy, 13 year old son of Ambassador Joseph P. Kennedy, shared hero honors with a girl telephone operator, as fire destroyed the gymnasium at Portsmouth Priory."[55]

He did not distinguish himself in the classroom. "I guess it was because I wasn't studying as hard as I should have that I got such bad marks,"[56] he wrote his parents manfully. By February 1941 mathematics was his best subject. His worst grades were in history (60) and Christian doctrine (64).[57] Where his father had bombarded Joe and John with counsel and reprimand, he was now across the Atlantic and, perhaps, less interested. The letters of admonition came from Rose Kennedy, who was particularly concerned that Robert read more books. His replies provided exculpatory lists:

> I've read the following books
>   Finished "The Crisis"
>   Read "The Hurricane"
>   Reading "Men Against the Sea.". . .
> I'm reading "Berlin Diary" now . . . really a neat book.[58]

In the spring of 1940, the associate headmaster, thanking the ambassador for his contribution to the fund for a new gymnasium, added that Bobby was getting along better. "He is doing fairly well in his studies, and this Term has been playing on the Junior Basketball Squad. He seems to have settled down and appears happy."[59] A year later: "Bob's habit of not being neat is receiving considerable attention from all of his masters."[60]

Sports were his consuming interest. He tried his awkward hardest in school competitions:

I haven't got a chance for the [football] team but I'll try and stay on the squad.[61]

I don't play hockey very well but I've gotten much better since I started this year. There was one cut in the squad but I was lucky enough to stay on.[62]

He finally did succeed in becoming manager of the hockey team "which ought to be a lot of fun though I have to type all the letters I write to different schools because my writing is so bad."[63] His triumph was in sailing where he became vice commodore of the yacht club, in charge of racing.

His mother hoped for social graces and wanted to send him to dancing school in Providence. "Mother," Robert wrote back, "I don't think it would be worth going way up to Providence for dancing as it is quite far away and I wouldn't be able to play hockey and on most of the days I would probably miss study."[64] He won that one. In May 1941, Teddy was deposited at the Priory, though he was much younger than most of the students. He remembers complaining to his older brother that the bigger boys were picking on him. "You'll just have to look out for yourself," Bobby said crossly.[65] So three years passed without Robert Kennedy's making much impression on the school or the school on him.

VI

The first year of the European war was a lonely time for Joseph Kennedy — lonely and bitter. Returning to America for a visit in December 1939, he told a group in Boston, "This is not our fight."[66] The only victor, he predicted to a meeting of Army and Navy officers in Washington, would be communism.[67] He was appalled by the war, tired of being ambassador, homesick for his family. He spilled over to Robert Murphy, the counselor of the Paris embassy, about "his bad health, his discontent over returning to London, his belief that everything he could do there could just as well be done by a $50-a-month clerk, that he wanted to quit but didn't see how he gracefully could before the elections. He said the United States would be crazy to go into the war."[68]

As Kennedy talked in his indiscreet way, knowledge of his views spread around London. His popularity crumbled away. Harold Nicolson, in a blast relayed by Edward R. Murrow to the United States over CBS, wrote that Kennedy's return from his American visit was welcomed mainly by "the friends of Herr von Ribbentrop, the Nürembergers, the Munichois and the *disjecta membra* of former pro-Nazi organizations."[69] Most American newspapermen in London were pro-British and now had little use for their ambassador. His own military attaché, General Raymond E. Lee, wrote in his journal about his boss's "suave, toothy and inane grin" and found him "crude, blatant, and ignorant in everything he did or said."[70]

Sir Robert Vansittart, permanent head of the Foreign Office, had come to detest him. "Mr. Kennedy is a very foul specimen of double crosser and defeatist," he told the Foreign Secretary. "He thinks of nothing but his own pocket. I hope that this war will at least see the elimination of this type."[71] The head of the American desk pronounced him "malevolent and pigeon-livered."[72] The Foreign Office described Kennedy's "defeatist attitude" to Lord Lothian, the British ambassador in Washington, "in case it should later become necessary to ask you to drop a hint in the proper quarter." The British may even have considered declaring Kennedy persona non grata. In any event the Foreign Office resolved to minimize contact with him. When Churchill took over as Prime Minister in May, both governments felt that the American embassy in London had become an "undesirable channel."[73]

"You would be surprised how much anti-American [the English] have become," Kennedy had written his wife in March. ". . . I feel it strongly against me. . . . If the war gets worse which I am still convinced it will . . . I am sure they will all hate us more."[74] "Joe dear," Rose Kennedy wrote back with sublime tact, "I have a definite idea that it would be a wonderful feat if you could put over the idea that altho you are against America's entering the war — still you are encouraging help to England in some way. [Most Americans would be sympathetic,] and it would endear you to the hearts of the British."[75] "It is easy enough to say we should do something," Kennedy replied, "but the real difficulty is — what? My sympathies are completely with the Allies, but they must not run away with my judgment of what can happen to the United States. We may have to fight Hitler at some later date over South America, but we had better do it in our own backyard where we will be effective and not weaken ourselves by trying to carry on a fight over here." And a scrawled postscript: "Its Hell to be here without all of you. . . . I get news that you are more beautiful than ever. Maybe you do better away from me. All my love."[76]

He used the better-fight-in-our-own-backyard argument in cables to Washington. "The President and I were asking each other the same question," Cordell Hull later observed, "but we reached a different conclusion. . . . It seemed to us we should do better to keep the fighting away from our own back yard." [77] Kennedy's compulsive gloom was isolating him from Washington almost as much as it had already isolated him in London. "When real world forces come into conflict," Roosevelt wrote him in early May, trying to cheer him up, "the final result is never as dark as we mortals guess it in very difficult days." [78] It was to no avail.

VII

In May 1940 the Nazi armies swept into the Low Countries and France. On May 20 Kennedy sent his wife an urgent hand-written letter:

> Rose darling,
>   . . . The situation is terrible . . . I think the jig is up. The situation is more than critical. It means a terrible finish for the Allies. . . . The English will fight to the end but I just don't think they can stand up to the bombing indefinitely. What will happen then is probably a dictated peace with Hitler probably getting the British Navy, and we will find ourselves in a terrible mess.
>   My God how right I've been in my predictions. I wish I'd been wrong. Well darling its certainly been a great adventure. . . . I'll probably be home sooner than we thought. [79]

The only hope, he believed, was a negotiated peace. The English, he told Hull, "have not demonstrated one thing that would justify us in assuming any kind of partnership with them." [80]

Rose tried to divert him with news of the children. John Kennedy's book *Why England Slept* was out:

> Jack has certainly had the most astounding success. . . . Bobby is a different mold. He does not seem to be interested particularly in reading or sailing or his stamps. He does a little work in all three, but no special enthusiasm. He does follow the war news with interest and is reading "The Failure of a Mission." [Summer 1940]

> I am trying to get Bob to do some reading. He doesn't seem to care for sailing as much as the other boys. Of course he doesn't want to go to any of the dances. [July 11]

> Bobby has had a few boils and so it has held up his swimming a little. It is funny that he should have them this time of year, the same as Joe. We once thought it was from eating corn, but these days we do not have any corn. [August 9] [81]

She made a family recording for her husband's fifty-second birthday. "I have played it at least twenty times already," Kennedy wrote. "I can't say that any of my children's voices have improved in tone quality . . . but there is still plenty of pep in the sound."[82]

Not even the children distracted him for long. "I wonder," he said in the same letter, "what satisfaction any living person can be getting out of believing that this war will accomplish any of the things they thought it would accomplish. The so-called fight for Democracy and for liberty will enslave every nation in the world, and leave them nothing of either Democracy or Liberty. . . . I get sick to my stomach when I listen to all the drivel that is being poured out to the American public."[83]

The sourness of these days roused Kennedy's concern about Jewish pressure to keep the war going. It was a concern he shared with Neville Chamberlain. After the British guarantee to Poland, so Kennedy told Herbert Hoover in later years, Chamberlain had said to him that "he hoped the Americans and the Jews would now be satisfied but that he (Chamberlain) felt that he had signed the doom of civilization."[84] Kennedy later wrote in a draft autobiography that "a number of Jewish publishers and writers" assailed him after the Trafalgar Day speech. They wanted war, he thought. "They should not be condemned for such an objective. After all, the lives and fortunes of their compatriots were being destroyed by Hitler. Compromise could hardly cure that situation; only the destruction of Nazism could do so. . . . [But] they did not hesitate to resort to slander and falsehood to achieve their aims. . . . I was hardly prepared . . . for the full viciousness of this onslaught."[85]

On the other hand, he attacked religious persecution in the fascist states. He called the anti-Semitic outrages of November 1938 "the most terrible things I have ever heard of." He busied himself with schemes for the rescue of Jewish refugees to an extent that led the Arab National League of Boston to call him a "Zionist Charlie McCarthy."[86] Nor was he at all anti-Jewish in individual relations. In Palm Beach he played golf at a predominantly Jewish country club. "One of his most constant riding companions and closest collaborators in the Embassy," noted D. W. Brogan, who knew more about America than any other Briton, "was a Jew whose career in the Department was not helped by this part of his record."[87] Still, the ambassador's fear of war now moved him for a moment to speak of Jews collectively as a menacing influence.

By the autumn of 1940, Kennedy wanted to come home. Angry at being left out when Roosevelt and Churchill negotiated the exchange of overage American destroyers for British bases in the Western Hemisphere, Kennedy wired Hull that he did not enjoy being a dummy and threatened resignation. Roosevelt, handholding again, replied, "You are not only not a dummy but are essential to all of us both in the Government and in the Nation."[88]

The question arises why Roosevelt wished to keep so disaffected an ambassador in London. The answer lay in domestic politics. Nineteen forty was an election year. Knowing Kennedy's influence among businessmen and Catholics and the weight his London experience would give his criticisms of the administration's foreign policy, Roosevelt reckoned he would cause less trouble in England than in America.* Unless, that is, he came back as part of the campaign; but when FDR, at the suggestion of James F. Byrnes, offered Kennedy the national chairmanship after the resignation of James A. Farley (for whom Joseph P. Kennedy, Jr., had cast his delegate's vote at the Democratic convention), Kennedy declined.[89]

Kennedy — "primed," said General Lee, "for a vindictive and vigorous assault upon Roosevelt" — was determined to get back before the election.[90] The State Department put him off. In October, Kennedy (or so he later told his friend Krock) phoned Sumner Welles, the under secretary, to complain that he had been ignored for three months. He had sent a full account of his treatment to his office in New York, he continued, with instructions to release the story if he were not back in the United States by a specified date. The administration capitulated but took protective action. Roosevelt invited the Kennedys to spend the night at the White House, and Rose Kennedy met her husband at the airport to escort him to Washington. "The President sent you, a Roman Catholic, as Ambassador to London, which probably no other President would have done," she told him. ". . . You would write yourself down an ingrate in the view of many people if you resign now."[91]

When they arrived at the White House, they found the President at his desk, "shaking a cocktail shaker," Rose Kennedy recorded, "and reaching over for a few lumps of ice with his powerful hands." Senator Byrnes and his wife were there along with Marguerite LeHand, the President's secretary. After dinner, "Joe did most of the talking. The President looked rather pale, rather ashen, and I always noticed the nervous habit he had of nervously snapping his eyes."[92] FDR might well have been nervous. Joe Kennedy, as Byrnes noted, was "a forceful talker, and in his vocabulary are many words not found in dictionaries." He used them that night to denounce the humiliations inflicted on him by the State Department.

---

* See Alsop and Kintner in the *Boston Globe*, October 7, 1940: "The President regards Kennedy as likely to do less harm in London than in New York. . . . The President has repeatedly urged him to remain in London in order to keep him quiet."

Roosevelt was sympathetic, obliged by denouncing the Department himself and by the end of the evening had talked Kennedy around to making a radio speech advocating his reelection.[93] For all Kennedy's inner misgivings about its theme, his speech was accounted one of the most effective in the campaign. The day after Roosevelt won his third term Kennedy called at the White House and submitted his resignation.

VIII

A few days later in Boston, in an impassioned rush of words before a couple of newspapermen, Kennedy said, "I'm willing to spend all I've got to keep us out of the war." Democracy was "finished in England. It may be here." For the moment, "aid England as far as we can; that's our game. As long as she can hold out, give her what it takes . . . and don't expect anything back." But stay out of the war: "I know more about the European situation than anybody else, and it's up to me to see that the country gets it."[94] He carried his campaign to the west coast; returning to New York later in the month, he called on Herbert Hoover, whom he had opposed so energetically in 1932. Now a common fear of war brought them together. He told the former President, as Hoover later recorded it, "that if we went into the war, we would have a National Socialist state — he could see no return to democratic forms."* Kennedy also saw Charles A. Lindbergh, whose terrifying estimates of German air power had impressed him during the Munich crisis, and told him that the "best possible thing" for the British "would be a negotiated peace."[95]

Lindbergh was emerging as a leading spokesman for the passionately isolationist America First Committee; and attempts were made to enlist Kennedy. Kennedy kept away. Lindbergh's wartime journals report no further meeting. Kennedy was unwilling to take an irrevocable anti-Roosevelt, anti-Britain line. "I hope to God England wins," he summed up his position to a friend, "but I'm not willing to sacrifice everything we have left here to see it done, and that is what

---

* As for England, Kennedy said he had warned Roosevelt that aid to Britain was "a bet on a losing horse." The previous September, Kennedy told Hoover, Hitler had offered England peace through Sweden, guaranteeing the empire and the fleet. "Why didn't the British accept?" asked Hoover. "Nothing but Churchill's bullheadedness," said Kennedy; Churchill was "an entirely bellicose character" (Herbert Hoover, memorandum, November 22, 1940, Hoover Papers). Kennedy was probably referring to an approach through the president of the High Court of Sweden to the British minister in Stockholm at the request of Ludwig Weissauer, a Berlin lawyer who claimed to be an emissary of Hitler's.

war would do to us." [96] He therefore continued both to support military aid and to oppose military intervention.

This complicated his life when the lend-lease bill came up early in 1941. In a radio speech he simultaneously disclaimed the bill and advocated further aid to England. Robert Kennedy telegraphed his father from Portsmouth Priory, "Your speech was absolutely marvelous." [97] He wrote his mother, "I think it was the best speech that he ever made. I thought he really cleared himself from what people have been calling him and what he said about the President's bill will certainly change a lot of people's minds." [98] The columnist Dorothy Thompson, less sympathetic, said that Joseph Kennedy had out-Hamleted Hamlet: Instead of "to be or not to be," it was "to be *and* not to be." [99]

There was more of Hamlet when he testified on lend-lease before the House Foreign Affairs Committee. He criticized the bill "in its present form" on constitutional grounds because it involved the surrender of too much power by Congress to the Presidency, but went on to insist that he trusted this President, hated Nazism, supported aid to Great Britain short of war, was not an isolationist and even would favor the bill if Congress passed it. [100] Years later Robert claimed that his father "tipped the balance in favor of Lend-Lease" within the committee. "The America Firsters thereafter called him worse names than any the liberals have ever had for him." [101] This was an overstatement, but Kennedy's testimony did the bill no harm and honestly represented his own mixture of feelings.

What was consistent was his pessimism. "The death and destruction of everything we stand for is inevitable," he told the editors of *Collier's* in a characteristic outburst. "We'll get hell for saying this but that's what is going to happen to us. I have seen many a fellow in the stock market lose his fortune because he was right too soon, but it is true. . . . I think our way of life is finished — it is merely a question of what we can save for ourselves." [102] Disaster was coming, it appeared, whether England won or lost the war, whether America went in or stayed out; but it would be most certain and lasting if America went in.

By the spring of 1941 he had decided to give up the Bronxville house and settle in Palm Beach, maintaining his summer (and voting) residence in Hyannis Port. "I've been in New York for about eleven years," he explained, "and I haven't averaged two months a year there. In addition to that I think the day is gone when a man can maintain three houses. . . . With the economic and financial outlook as I visualize it a man will have all he can do to have one house." [103] He even had doubts about Palm Beach. Frank Waldrop, editor of the isolationist *Washington Times-Herald,* recalls Kennedy pointing at breaks in the sea wall and saying he wasn't going to fix the goddamn thing. Let it go. The whole world was falling apart, so why should he bother with a goddamn house in Palm Beach.[104] "I'm so terribly pessimistic about the outlook for us," he wrote a Boston friend in April 1941, "and yet realize that there's no constructive suggestion that I could make that would have the slightest chance of being accepted in Washington, that I just think I'd better keep my mouth shut."[105]

He did open his mouth the next month in a commencement address at Oglethorpe University. Here he reaffirmed his support of the policy "the overwhelming majority of our people have approved . . . to rearm as swiftly as possible, to give every possible aid to Great Britain, but to stay out of war." He condemned Nazism as the "new paganism" but scoffed at the idea that America was in danger of attack or that an Axis victory "spells ruin for us." Nothing could spell more ruin than war. He demanded on the part of the American government "the freest debate" and "the fullest disclosure." To all this he added a new theme — an attack on the idea of a military crusade for democracy.

> Democracy cannot be imposed by force or otherwise. It would not last even if we were able to present [other nations] with the most up-to-date constitutional system. In our very attempt at this colossal crusade we would end in failure and disgrace abroad, in disillusionment and bankruptcy at home.

The United States must not overestimate its power to decide the destiny of other nations.

> We cannot, my fellow Americans, divert the tides of the mighty revolution now sweeping Asia and Europe. They were not of our making and they will not be subject to our control, no matter how courageously or exhaustingly we strive to subject them.

Let us forswear ringing phrases, Kennedy said, and concentrate on the struggle for justice in our own society. "When the history of these times has been written, I hope and pray it will show that the American lads of 1941 did not pave the way for a radical change by fostering class antagonism . . . and that they realized and took vigorous steps to spread the gospel of equality."[106]

Interventionists bitterly criticized the speech. Three days later Roosevelt proclaimed a state of national emergency. In another few days at Notre Dame, Kennedy, observing darkly that "there is much in our Western civilization that does not deserve to survive" (he specified monopolistic capitalism and racketeering unionism), went on to say that the presidential proclamation deserved the "unlimited loyalty" of all Americans.[107] After the Nazi invasion of Russia on June 22, Hoover asked Kennedy to join with him, Alfred M. Landon, Robert M. Hutchins and others in a new manifesto against intervention. The western decision to aid Russia, the proposed statement began, made it "abundantly clear that this war is not a world conflict between tyranny and freedom." American entry into the war was "far more likely to destroy democracy in this country" than to establish democracy in Europe and Asia. Kennedy replied evasively. Hoover kept pressing him. Kennedy finally wired bluntly: I MUCH PREFER TO GO MY WAY ALONE. I CAN THEN TAKE MY POSITION ON ANY SUBJECT AND AT ANY TIME AS THE OCCASION DEMANDS WITHOUT CONSULTATION WITH ANYONE.[108]

IX

His older sons, each in his own style, adopted their father's position on the war. Joe Jr. now at Harvard Law School, was the activist. "As a member of the Harvard Committee against Marital [sic] Intervention," he wrote in January 1941 to John T. Flynn, a leading America Firster, "I want to extend an invitation to you. . . ."[109] A few weeks later he took the platform himself as a member of College Men for Defense First in a meeting in Washington; Senator Robert A. Taft was the main speaker.[110]

John Kennedy as usual remained detached. He intended to go to law school, his father wrote James Landis from London, "although he thought he would go to Yale, principally because he feels it would be better not to be constantly in direct competition with his brother, and I rather sympathize with him in that point of view. However, Jack has been having trouble for a number of years with his stomach and I now hear that the doctors advise him to take a year off."[111] *Why England Slept* was a singularly dispassionate statement to be flung into America's most passionate foreign policy debate of the century — so dispassionate, indeed, that it was impossible to conclude from the text whether the author was an interventionist or an isolationist. "I of course don't want to take sides too much," he had told his father.[112] "If Britain is defeated," he wrote in the book, ". . . we shall be at a definite disadvantage" in a world of dictatorships; but "we may be able to survive." If Britain wins, "we must be prepared to take our part in setting up a world order that will prevent the rise of a militaristic dictatorship."[113] The second brother was already finding a distinctive voice. The bit about setting up a world order was hardly in the paternal line. But his father cared less what his son had written in his book than that he had written a book at all, and a successful one at that.* On balance John Kennedy leaned toward the isolationist side in the great debate.

Still, whatever their hope that America could avoid involvement, the Kennedys were strong advocates of preparedness. Early in 1941 Joe Jr. decided to enlist. "I think," he wrote his father, "in that Jack is not doing anything, and with your stand on the war, that people will wonder what the devil I am doing back at school with everyone else working for national defense." Jack was not to be left behind. His father sought the assistance of Captain Alan Kirk, who had been his naval attaché in London and was now director of the Office of Naval Intelligence; and Kirk somehow contrived to spirit John Kennedy past the Navy's stringent physical standards.[114] When the Japanese struck on December 7, the oldest Kennedy sons were both in naval training.

---

* After reading the honors essay submitted to the Harvard government department, Joseph Kennedy relayed to his son the report that "one or two of those who have read it complain that you have gone too far in absolving the leaders of the National Government from responsibility for the state in which England found herself at Munich" (JPK to JFK, May 20, 1940, JFK Papers). John Kennedy adopted a number of his father's suggestions when he turned the paper into a book. Joseph Kennedy subsequently wrote Landis: "He has done a fine job on the book, I understand. I saw the original manuscript, but I know he has made a great many changes since then. I have always told him that to write a book at the age of 23 that commanded respect would be of great service to him in time to come" (JPK to Landis, August 6, 1940, Landis Papers).

The War

PEARL HARBOR found Robert Kennedy, just turned sixteen, in his third year at Portsmouth Priory. The school, only a few miles from the naval base at Newport, responded quickly to the tremors of war. The headmaster decreed a blackout. "We all had to go sit in the basement for 20 minutes."[1] When the aircraft carrier *Ranger* steamed into Narragansett Bay, the boys watched planes practicing dive-bombing in the hazy distance.[2] A fortnight after the Japanese attack, Robert went to Palm Beach for Christmas. He was "just the same," Rose Kennedy reported. "He plays a very good golf game and shot a 95 the other day with me. He also astonished me by his prowess at bridge the other night. . . . But he is very unsociable. . . . He absolutely refused to go to the Bath and Tennis [Club] and when he has gone out he doesn't seem to like any of the boys here."[3] After the holiday, Robert wrote from the Priory, "I've read about 8 books since I've been back at the school, and I'm now reading 'Botany Bay' which is very good."[4] His mother passed on the word that Robert "seems to have improved very much in his desire for reading," though in her letter she cut his estimate of books read to "five or six." She added, perhaps in gentle admonition to the others, "He also sent us two or three recommendations about serving Mass."[5]

I

Robert's marks arrived. "I was so disappointed in your report," his mother wrote him with appropriate sternness.

. . . The mark in Christian Doctrine was very low. . . . Certainly in a Catholic school you should be taught that above everything else and you should apply yourself to that subject. . . . Please get on your toes. . . . You have a good head and certainly you ought to use it. You owe it to yourself as well as to me to make the most of your advantages. Remem-

ber, too, that it is a reflection on my brains as the boys in the family are
supposed to get their intellect from their mother, and certainly I do not
expect my own little pet to let me down.[6]

The Priory was not accounted a great success. Joseph Kennedy,
worried that his third son was not learning enough to get into Harvard, decided to shift Robert in the autumn of 1942 to Milton Academy in Massachusetts. A new boy entering an upper form of a
boarding school does not have an easy time at best. For a shy Irish
Catholic boy, who had already attended six schools in ten years and
was now plunged into what a Milton friend described with precision
as "a consummate WASP school,"[7] it could have been an ordeal.

Robert lived in Forbes House, never complained and worked
doggedly on his studies. His grades were undistinguished. His father
reminded him that his mother had received honors in school. "I
don't know where I got my brains," Bobby wrote back ruefully, "but
its [sic] quite evident that I received them from neither my father
nor my mother."[8] Mathematics was now a particular trouble. "On
our last day of school," he told his father, "the math teacher made a
small speech to the class in which he said that two great things had
happened to him; one that Rommel was surrounded in Egypt and
2nd that Kennedy had passed a math test."[9]

When it came to public affairs, the Kennedy mealtime seminars
stood him in good stead. One classmate, Samuel Adams, said later
he had never met anyone his own age so well informed on international politics.[10] "We had Time's Current Events test here," Robert
wrote his parents, ". . . and though I wasn't too sharp I didn't do
badly, except for the questions on business and finance which I got
almost entirely wrong."[11] In spite, or perhaps in consequence, of his
deficiency, he insisted on finding a job in the summer of 1943. Let
Clem Norton, a famous Boston character — Charlie Hennessey in
Edwin O'Connor's The Last Hurrah — tell the story:

> At first, he wanted to go out and work on a fishing vessel. To really
> rough it. Asked me to have him signed on so that nobody would know
> who he was and would have to work just as hard as any fisherman. His
> ma objected and had her way saying, "Two of our boys are at the front
> now. This boy will have to go soon, so I want him as long as I can have
> him now." So young Bobby is working as a clerk at the East Boston
> bank where his dad started. . . . He is very punctual and very neat in
> the way he keeps his clothes about the room. He doesn't object to riding
> on the Elevated railway with common people, rather likes it.[12]

The bank job was as close to business as he ever got.

His letters home were sometimes assertions in a small way of
his own identity. After the Cocoanut Grove fire killed nearly five

hundred people in Boston on a Saturday night in November 1942, Robert noted that if Boston College had beaten Holy Cross in football in the afternoon, the B.C. team would have celebrated at the night-club on the fatal evening. "Losing that game," he wrote his father, "really saved their lives."[13] An unconscious protest against the eternal emphasis on winning? But he still wanted to win. Football was his chosen medium. "I am alternating first string now, but I hit my shoulder during blocking practice so I'm not sure whether I'll be able to scrimmage."[14] Football presented itself as some sort of moral test. "Bobby ran every practice play," Sam Adams said, "and tackled and blocked dummies as if he were in a hard fought game. To another sixteen year-old kid this Gung-ho attitude seemed a little weird. But, gradually, I became aware of the characteristic of Bobby's which I admired most — his ability to ridicule himself and his efforts and still go all out."[15] He watched his own performance with dispassion. "I played in the football game against St. Marks," he wrote his father, "but I was quite nervous and so did not do very well."[16]

His contemporaries remembered particularly his shyness, his determination and his self-mocking humor. Milton had a girls' school; and, while there was no coeducation, there was social contact. "He used to walk with his head way down, buried in his neck, like a bird in a storm," recalled Mary (Piedy) Bailey, a contemporary at the girls' school. "Everything was buried, including the bones of his face. He was moon-faced then; there was no hint of the angular frame that eventually came out. . . . He mostly stood on one foot with the other toe resting on top, his hands way down in his pockets." In his second year he used to take Piedy Bailey home after chapel on Sunday night, "walking four feet behind, his head down, looking out from underneath his forelock." This was a gain. The year before he had had a consuming crush on Ann Appleton but had never dared speak to her at all. Bobby, said Piedy Bailey, "was appealing, funny, larky, separate and a little bit solitary."[17]

Sam Adams, noting the disadvantage of coming in for two years when most of the class had been there at least four, said:

> Bobby certainly tried hard. He showed absolute determination; he decided to do something, he just gave it everything he had. He wasn't staging it; he was just doing it his way. He didn't care whether people noticed he was doing it, or kidded him because he was doing it. He wasn't trying for any particular reaction from people. He just did it because that's the way he was. He really did go absolutely all-out. Not exactly so at studies. The things he was interested in, he did go all-out.

He joined in bull sessions, but "he didn't participate too much when the conversation involved dirty jokes and that sort of thing. He was really bored to tears by that ... also angered, although he usually didn't say much; he just sort of crawled into his shell and ignored it." He disliked the adolescent sport of ganging up on one particular boy. "He wasn't one of those guys that stood there and said in some dramatic way: 'Don't lay a hand on this kid,' but he would make his feelings known and would walk away from it and say 'that was a horrible way to behave.'"[18]

He used to invite Adams to go to mass with him, and Adams concluded that his "absolute faith in God ... gave him faith in himself and appeared to make him oblivious of his lack of popularity among many of his classmates."[19] Bobby evidently hoped they might come to see the light. "Am now leading an underground movement to convert the school," he wrote his mother, "and am taking a lot of the boys to church on Sunday. Most of them like it very much, but I'm sorry to say I think the Protestant Ministers, who visit here are better speakers than the Priests at church."[20] This improved. "A Jesuit, Father Ford, is coming to speak to the school," he wrote,

> ... about the beliefs of the Catholic Church which should be very good. Eunice suggested that I have Father Keller come and speak ... so I am working on that.*

There were secular speakers too. Wendell Willkie addressed the school, and Robert persuaded his father to follow. "Dad made a great speech," Robert told his mother. "... Everybody was crazy about him."[21] His father wrote Kathleen that Bobby seemed "perfectly satisfied; in fact he said he thought it was better than Willkie's, believing that he was paying me a great compliment. I told him that if it weren't better than Willkie's I would have expected him to hiss me off the platform."[22]

II

He was not the most popular boy in the school. But in these years he formed his first strong friendship outside the family. David Hackett, a class behind him, had been at Milton since he was a small boy and had long since become, in Piedy Bailey's words, "the hero of our school and the star of everything." He was, said Sam Adams, "one of the greatest athletes that Milton ever had in that decade or other decades." "Hackett had such glamour [Piedy Bailey again] that the littlest kids from the lower school would just trail around as though

---

* RFK to his mother, n.d. [January 1944], Hyannis Port Papers. The Reverend George B. Ford was a notable figure at Columbia University. The Reverend James Keller was founder of the Christophers.

they were tin cans tied to him every place he went. . . . He'd never push you around or anything or be arrogant. He just had this incredible way about him so that people would stop and their jaws would hang open."[23] A portrait of Hackett is preserved in John Knowles's memorable novel about Exeter, where Hackett, an indifferent student, had gone one year to summer school. He was Phineas at Devon Academy in *A Separate Peace,* "the best athlete in the school," who "possessed an extra vigor, a heightened confidence in himself, a serene capacity for affection which saved him. Nothing as he was growing up at home, nothing at Devon, nothing even about the war had broken his harmonious and natural unity."[24]

Like Phineas, David Hackett was a boy with an uncommon gift for sympathy. "Maybe my first impression of him," Hackett said later of Robert Kennedy, "was that we were both, in a way, misfits."[25] It is not evident why Hackett considered himself a misfit — perhaps because he was a day boy from Dedham, more modestly circumstanced than the boarders at Milton, perhaps because, lacking the grade-getting knack, he underestimated his own perceptive intelligence. But it was evident why Robert Kennedy seemed a misfit. "He was," Hackett said later, "neither a natural athlete nor a natural student nor a natural success with girls and had no natural gift for popularity. Nothing came easily for him. What he had was a set of handicaps and a fantastic determination to overcome them. The handicaps made him redouble his effort." Because he had problems of his own, because he was "an underdog in sports, with studies and girls and as a Catholic," he was sensitive to the problems of others and went directly to their assistance however personally inconvenient it might be to himself or those with him. If a Sunday mass was short an altar boy, for example, he volunteered at once, though older and larger than the rest. Hackett admired Kennedy's readiness to do the things one knew one should do oneself but flinched from doing because they violated the iron conventions of adolescence. "From the moment I met him," Hackett said, "I knew that he would embarrass his friends" — and he loved him for it.[26]

They met on the football field, took to each other at once and soon grew inseparable. "They spent a lot of time wrestling," said Sam Adams, "and going in each other's room and throwing things out the window and wearing each other's clothes and generally horseplaying around."[27] Robert Kennedy was identified in the school as Hackett's friend, his best friend. This gave him status among his classmates. It also no doubt gave him new and needed reassurance about himself.

So Robert Kennedy's years at Milton passed tranquilly, at least as tranquilly as could be with the violent sounds of war in the background, V-mail letters from brothers overseas, a burning desire to get into uniform oneself. His housemaster was Albert Norris, who, as an earlier Miltonian, Cleveland Amory, put it, taught "Latin and Public Speaking and Tennis and Sarcasm" and could bring the class to order "with one fiercesome glower." [28] On the eve of Robert's own departure for the armed services, the fiercesome Norris summed up his Milton career to his mother:

> You have great reason to be proud of Bobby's record here. His studies have been an uphill battle all the way but he now (despite the D which I gave him myself in November!) has his scholastic feet well under him. If he were to be here until June, I believe that he would be on the Honor List. . . .
> Bobby also deserves a good pat on the back for the way he has adapted himself generally. Entering a group which has been together for several years is not easy, but Bobby has done it successfully, participating generously in school activities and winning himself warm friends. [29]

Looking back thirty years later, Norris recalled Robert Kennedy's particular qualities as "seriousness of purpose, hard work, and a refusal to be discouraged." He also recalled his sense of humor, "delightful" if sometimes "pretty caustic," his concern for "those who were less fortunate than he was" and his piety. [30] "I think he has been happy here," Arthur B. Perry, the principal, wrote Joseph Kennedy, "and I am very glad you decided to send him to Milton." [31] Joseph Kennedy was evidently glad too, for, when the time came, he sent his youngest son to Milton.

III

Joseph Kennedy had few enough things to be glad about in these years. The war he dreaded now enveloped his own country. When he offered the President his services immediately after Pearl Harbor (IN THIS GREAT CRISIS ALL AMERICANS ARE WITH YOU. NAME THE BATTLE POST. I'M YOURS TO COMMAND), Roosevelt did not reply. In March 1942, after trying in vain to reach the President by phone, Kennedy sent a stiff letter — "I feel that our relationship in the past certainly justifies my writing to you" — enclosing a copy of his Pearl Harbor telegram and concluding that, while he did not want to appear "in the role of a man looking for a job for the sake of getting an appointment, . . . Joe and Jack are in the service and I feel that my experience in these critical times might be worth something in some position." [32]

He met with Roosevelt. It is not altogether clear why the President, who brought so many inveterate opponents into the defense establishment, did not use a man of Kennedy's tested administrative ability. "When I saw Mr. Roosevelt," Kennedy wrote his faithful English friend Max Beaverbrook, "I was of the opinion that he intended to use me in the shipping situation, but the radicals and certain elements in the New Deal hollared [sic] so loud that I was not even considered." [33] This was only part of the story. Very likely, as James Roosevelt later surmised, Roosevelt himself had lost confidence in Kennedy's judgment during the London experience. [34] "I don't suppose it ever enters the head of a Joe Kennedy," said his one-time friend Felix Frankfurter, "that one who was so hostile to the whole war effort as he was all over the lot, and so outspoken in his foul-mouthed hostility to the President himself, barred his own way to a responsible share in the conduct of the war." [35]

When the matter came up at a meeting with congressional leaders toward the end of 1942, FDR, as Vice President Henry Wallace noted in his diary, "said something about how he always liked Joe Kennedy but how he especially liked Mrs. Kennedy." The President added that unfortunately Joe Kennedy told everything he knew to Arthur Krock and Frank Kent, both columnists notoriously unsympathetic to the administration. Wallace observed dryly that other high officials leaked to Krock and Kent, and Senator Lister Hill of Alabama suggested that Kennedy be put in charge of the problems of small business. [36] Soon James F. Byrnes, now director of Economic Stabilization, did indeed ask Kennedy to become head of the Office of Civilian Supply. "But," Kennedy wrote Frank Kent (as FDR had predicted), "I told him to stick it, because I feel that, in the first place, if I were of any use at all from an executive point of view, I was entitled to something I could really get my teeth into, and if it was merely an excuse for putting me under cover, I resented that equally, so . . . I am still in the leper colony." [37]

His disaffection grew. "Dissatisfaction is rife and lack of confidence in the leaders and in Congress is definitely high," he wrote Beaverbrook in August 1942, "and there is a great undercurrent of dissatisfaction with the appointment of so many Jews in high places in Washington. . . . If you want everybody yelling for our team, Mr. Roosevelt is certainly not attaining that end."* Kennedy rejoiced to Beaverbrook that the August primaries had thus far "returned, with one or two exceptions, men who were accused of being isolationists

---

* JPK to Beaverbrook, August 12, 1942, Beaverbrook Papers. Kennedy gently deflected an evident attempt by Beaverbrook to enlist him in the campaign for a second front in Europe. They had become stout friends and, as outsiders who had made it against the establishment, they had much in common. In a later letter (October 23, 1944) Kennedy spoke of "the natural cynicism that I know you and I share about

previous to Pearl Harbor." He mentioned the fight that James A. Farley was organizing in New York on behalf of the conservative Democrat John Bennett against the New Deal senator James Mead for the gubernatorial nomination. He added casually, "I am taking a slight interest in the politics of Massachusetts and that is giving me something to work on."[38]

The issue in the Massachusetts primary was the senatorial nomination. The administration was backing a debonair liberal congressman from Clinton named Joseph E. Casey, a personal favorite of FDR's whom William V. Shannon has accurately described as "the brightest of the young Irish New Dealers in the House."[39] Where the incumbent Republican senator, Henry Cabot Lodge, Jr.,* in the tradition of the Lodges, had taken an anti-English position from 1939 to 1941, Casey, against the tradition of the Massachusetts Irish, had vigorously supported aid to Britain.

Kennedy wanted to vindicate pre–Pearl Harbor isolationism. He wanted to warn Roosevelt not to disregard the conservative Democrats. In addition — or so Casey came to believe — Kennedy had a precautionary eye on the future. Once elected to the Senate, an Irish Democrat with Casey's ability and broad appeal would very likely stay there indefinitely. Since in those days the Yankees had a lien on the other Senate seat, this would place a considerable impediment in the political future of Joseph P. Kennedy, Jr.[40] The elder Kennedy even found a candidate within the family. John F. Fitzgerald, who had nearly beaten Cabot Lodge's grandfather for the Senate a quarter century before, was only seventy-nine and delighted at the prospect of a last hurrah.

"In Washington," wrote William E. Mullins of the *Boston Herald*, "this contest is regarded as a fight between Joseph P. Kennedy and the Roosevelt administration."[41] Of Fitzgerald's reported expenditures, contributions from Joseph P. Kennedy, Joseph P. Kennedy, Jr., and John F. Kennedy covered more than half.† The elder Kennedy took an active role in the campaign. Once, submitted an especially nasty anti-Casey cartoon for his approval, Kennedy rejected it summarily: "I have two boys now in the war. Should they go into

---

a great many things." Kennedy, it is true, could in a moment of exasperation call Beaverbrook a "treacherous little bastard" (as reported in *The London Journal of Raymond E. Lee, 1940–1941*, ed. James Leutze [Boston, 1971], 92). But then all of Beaverbrook's best friends thought this about him at one time or another.

* So he somewhat inaccurately styled himself. His father was the poet George Cabot Lodge, about whom Henry Adams wrote a moving memoir. The original Henry Cabot Lodge was his grandfather; but Henry Cabot Lodge II doubtless seemed an impractical name for practical politics.

† Fitzgerald filed his financial report, claiming a total expenditure of $5575, with the Secretary of the Commonwealth in December (*New York Herald Tribune*, December 8, 1942).

public life, their heritage must be high — if they come back." As
Edward B. Hanify, later a prominent Boston lawyer, remembered
the scene:

> With those words, "if they come back," he turned away, his head
> bowed. The summer day seemed suddenly dark and cold. The crum-
> pled cartoon lay at his feet like a withered leaf blown by a wintry wind.[42]

Kennedy was surely under no illusion that his father-in-law could
beat Casey. But no doubt the thought passed through his mind that
Honey Fitz's more fervent supporters might move on to Lodge in the
general election. Casey took the primary comfortably by 30,000
votes, carrying eighteen out of twenty-two wards in Boston itself. In
November, Lodge, who used recordings of Fitzgerald's radio attacks
on Casey in his own campaign, won 53 percent of the vote. There
may have been more specific assistance too. According to Arthur
Krock, Lodge was "substantially aided by Joe Kennedy."[43] Casey, his
political career at an end, went on to a lucrative law practice. One of
his first important clients was Aristotle Onassis.

IV

"I am so damn well fed up with everything," Joe Kennedy wrote in
March 1943, "and so disgusted at sitting on my fanny at Cape Cod
and Palm Beach when I really believe I could do something in this
war effort that it is better I don't see anybody."[44] As his older
children began to go to the war fronts, he was haunted by a mixture
of emotions — pride in his sons, fear for them, fear for the world.

He was proud that Joe Jr. and John had enlisted before Pearl
Harbor. "I often smile to myself when I think of the many groups
that suggested America should go to war. . . . Too many of them are
storekeepers, in the intelligence service or other types of work belying
their former firm determination to fight, but that is life and it is
amusing at times."* Not so amusing perhaps. John was the first to
go overseas. By March 1943 he was commanding his own PT boat
in the South Pacific. His parents waited eagerly for his letters, du-
plicated them and sent them on to the other children.

> Going out every other night for patrol. On good nights it's beautiful
> — the water is amazingly phosphorescent — flying fishes which shine
> like lights are zooming around and you usually get two or three por-
> poises who lodge right under the bow and no matter how fast the boat
> goes keep just about six inches ahead. . . .
> I was interested in what you said about MacArthur's popularity. Here

---

* JPK to John Griffin, August 24, 1943, JPK Papers. As a pre–Pearl Harbor interventionist who ended in the intelligence service, I fully accept the justice of Mr.
Kennedy's strictures.

he has none — is, in fact, very, very unpopular. His nickname is "Doug-Out-Doug." . . . There is no doubt that as men start to come back that "Doug-Out-Doug" will spread — and I think would probably kill him off. No one out here has the slightest interest in politics — they just want to get home — morning, noon and night. . . . I didn't mean to use "They" — I meant "WE." . . .

As far as Joe wanting to get out here, I know it is futile to say so, but if I were he I would take as much time about it as I could. . . . As regards Bobby, he ought to do what he wants. You can't estimate risks, some cooks are in more danger out here than a lot of flyers.[45]

In August, the Japanese destroyer rammed John Kennedy's PT-109 in Blackett Strait west of New Georgia, plunging Kennedy and his crew into the waters of Ferguson Passage, now suddenly aflame with burning gasoline. With calm bravery he led most of the crew to refuge on a tiny island, towing one man through the water by gripping the end of the life jacket belt in his teeth. "We have just come back from a fairly trying time," he wrote a friend. "They believed us lost for a week — but luckily thank God — they did not send the telegrams — but unfortunately some telegrams will have to be sent."[46]

He wrote his parents:

On the bright side of an otherwise completely black time was the way that everyone stood up to it. Previous to that I had become somewhat cynical about the American as a fighting man. I had seen too much bellyaching and laying off. But with the chips down — that all faded away. I can now believe — which I never would have before — the stories of Bataan and Wake. For an American it's got to be awfully easy or awfully tough. When it's in the middle, then there's trouble.

It was a terrible thing though, losing those two men. One had ridden with me for as long as I had been out here. He had been somewhat shocked by a bomb that had landed near the boat about two weeks before. He never really got over it; he always seemed to have the feeling that something was going to happen to him. He never said anything about being put ashore — he didn't want to go — but the next time we came down the line I was going to let him work on the base force. When a fellow gets the feeling that he's in for it, the only thing to do is to let him get off the boat because strangely enough, they always seem to be the ones that do get it. . . . He had a wife and three kids. The other fellow had just come aboard. He was only a kid himself. . . .

John Kennedy was an old man of twenty-six.

When I read that we will fight the Japs for years if necessary and will sacrifice hundreds of thousands if we must — I always like to check from where he is talking — it's seldom out here. People get too used to talking about billions of dollars and millions of soldiers that thousands of dead sounds like a drop in the bucket. But if those thousands want to live as much as the ten I saw — they should measure their words with great, great care.[47]

In November 1943 he wrote that he would not be home by Christmas. "Am now the Executive Officer of our squadron. . . . I only hope I don't stay here long enough to make the next step — Supreme Commander — or I'll never get out."[48] In fact, however, the ramming of PT-109 had revived his old back injury, and he was invalided home in December 1943.

In the meantime, Kathleen Kennedy, now twenty-three, had gone to London in June to work in a Red Cross canteen. She was witty and enchanting, "full of charm and love of life and uncertainty," Susan Mary Patten wrote, and, as an English friend put it, with the transcendent quality of making life gay when things were desperate.[49] Kick, as everyone called her, resumed her English friendships with zest and entertained the family with lively accounts of wartime London. Joseph Jr. had been chafing in the United States while his younger brother was becoming a hero in the Pacific. It was "the first time," their mother wrote, "Jack had won such an 'advantage' by such a clear margin. . . . It cheered Jack and must have rankled Joe Jr."[50] In September, Joe finally received overseas orders, going to England in the first naval squadron to fly B-24s with British Coastal Command. Here he distinguished himself in dangerous winter missions over the Bay of Biscay.

V

Robert Kennedy had been chafing too, and for some time. "He is begging and pleading to be permitted to enlist," his father wrote a friend in October 1942, "so that he can get into the Navy Aviation."[51] BOTH UNCLE SAM AND I EXPRESS OUR HEARTIEST CONGRATULATIONS TO YOU ON THE OCCASION OF YOUR BIRTHDAY, his brother John wired when he turned seventeen in November 1942, AS DOES GENERAL HERSHEY.[52] Robert importuned his older brothers for news about life in the Navy. "Thanks for your letter[s]," John replied early in 1943. "What makes them so particularly interesting is that you can't read them. . . . I got my orders yesterday, and am on my way to war. Am taking my boat to Florida. . . . For a guy who never could find Gardiners Rock, Florida seems like a hell of a distance. Am typing this as we b⁰ᵤnᵤe along."[53] In the spring his father went to Washington to see about getting Robert into Navy V-12 training. "He still wants to follow in yours and Joe's footsteps," he wrote John, now on his PT boat in the Pacific.[54]

For a moment, PT boats seized Robert's imagination. His brother disapproved and wrote their father:

In regard to Bobby, he said something about wanting to get in the boats. I don't think he should. By the time he's ready to get out here the usefulness of the boats will be done, I think. I think he should go into that V-12 and stay there for a good long time. He's too young to be out here for a while. When he has at least a year of college he'll be more effective and he will be better able to handle a job. The war will last long enough for him. . . . To try to come steaming out here at 18 is no good. . . .

It's just that the fun goes out of war in a fairly short time and I don't think Bobby is ready yet to come out.[55]

In October 1943, six weeks before his eighteenth birthday, Robert enlisted in the Naval Reserve as apprentice seaman, released from active duty until the following March when (after some behind-the-scenes intervention by his father) he was scheduled to enter the V-12 training unit at Harvard.* Meanwhile he returned to Milton. His photograph appeared in the newspapers — "Kennedy's Third Son in Navy" — with Robert a pensive and exceedingly youthful seventeen and his proud father standing by his side.[56] The picture provoked much merriment in the family. "He certainly didn't look too enthusiastic," Joe Jr. wrote his parents.[57] John Kennedy wrote his younger brother:

The folks sent me a clipping of you taking the oath. The sight of you up there, just as a boy, was really moving particularly as a close examination showed that you had my checked London coat on. I'd like to know what the hell I'm doing out here while you go stroking around in my drape coat, but I suppose that [is] what we are out here for. . . .

In that picture you looked as if you were going to step outside the room, grab your gun, and knock off several of the houseboys before lunch. After reading Dad's letter, I gathered that the cold vicious look in your eyes was due to the thought of that big blocking back from Groton.[58]

There would be more about that "cold vicious look" in years to come.

In March 1944 Robert reported for active duty at the V-12 headquarters at Harvard. He was assigned to a room and roommate in Eliot House. "His name is Hans Anguelmueller so you can see he is not very Irish."[59] He sent his course of studies to his father, who responded with sage paternal counsel:

I think you have developed a maturity in the last year that is surprisingly good. Previous to that time I don't think any real strides had been made. I don't think I'd be thinking too much about what's going to happen three, four, five, or six months from today. The important

---

* "I was more than happy to be of service to you in accomplishing a change in the college assignment of Bobby to permit him to go to Harvard" (Captain J. B. Lynch to JPK, April 28, 1944, JPK Papers).

thing is to do a good job now and every day. After all, you've got your life ahead of you and heaven knows it's going to require all the brains anyone has to work out a satisfactory existence, and the only way to do that is to be well prepared, so stick in there and work as hard as you possibly can."[60]

Robert wrote Joe Jr. in England that "at this point you are about the only one of us many that has the good graces of the head of the family but I give you about four days after your arrival in this country before you'll join the rest of us. I am constantly hacking things but the hope still remains in the family that the Navy will make a man of me." He added, "We haven't really had too much action here in Harvard Square but we're on the alert at every moment for an attack and I'm sure when that time comes we will conduct ourselves according to Navy standards."[61]

It was one more unsettling shift in a highly unsettled education. To his parents: "I'm certainly not hitting the honors like my older brothers."[62] "I am now sitting in my room by myself with no friends," he wrote David Hackett, who was still at Milton, "wishing I were back at Milton where although I might not have had any friends I had a radio I could listen to and had pictures of my family I could look at, neither of which I have here at present but c'est la guerre."[63] The tone was mock self-pity, but vulnerability showed. "I don't like this place," he soon wrote, "and I'm homesick and need someone to cheer me up. I also feel sorry for myself. . . . I wish I could be throwing furniture out of windows or having a fight with you or making you a hero with some superb blocking."[64] In time his spirits revived. "Say hello to all the Irish Catholics for me, and tell em that next to John F. Fitzgerald and J. P. Kennedy I'm the toughest Irishman that lives which makes me the toughest man that lives."[65]

VI

The spring of 1944 brought a new threat to the integrity of the family. Chief among the prewar friends whom Kathleen saw again in London was William John Robert Hartington, the eldest son of the Duke of Devonshire, a descendant of Cavendishes and Cecils, now an officer in the Coldstream Guards. They fell in love. But the Devonshires were as ineradicably Anglican as the Kennedys were invincibly Catholic. History as well as religion argued against the match. The Duke of Devonshire was a leading Freemason, and the Cavendish family had played a notorious role, at least in the Irish view, as British viceroys in Ireland. When Honey Fitz had taken his daughter Rose to Dublin long ago, he had shown her the spot in

Dublin where Irish revolutionaries had assassinated Billy Hartington's granduncle.[66]

It all looked hopeless. "Of course I know he would never give in about the religion," Kathleen wrote her brother John in the summer of 1943, "and he knows I never would. . . . It's really too bad because I would be a most efficient Duchess of Devonshire in the post-war world."[67] As time went on, and with the knowledge that Hartington would go into combat when the second front was opened in France, the lovers grew increasingly determined to circumvent the obstacles. There were complicated negotiations. The apostolic delegate in London was brought in, Cardinal Spellman, even the Vatican. But the rules were firm. Kathleen could be married within the church to a non-Catholic only if Hartington agreed to have the children raised as Catholics, and this the son of the Duke of Devonshire could not do.

In this troubling time Joe Jr. became Kathleen's closest confidant. "I went to see the Bishop who was a nice fellow," he wrote his father in April 1944.

> His attitude seemed to be that if they loved each other a lot, then marry outside the church. He didn't seem to be disturbed about its creating a bad example. He of course said there would be a danger of her drifting away from the church, as she would not be able to receive the sacraments.
>
> I told her that she would have to decide whether she thought her life would really be broken up if she didn't marry him. . . . If this is the love of her life, and she will not be happy with anyone else, then she will probably be unhappy the rest of her life for not having done it.
>
> I think that Billy should make some kind of concession on the girls, for it wouldn't affect his family line. He on the other hand thinks it would be difficult to have half the family one thing, and the rest the other. It is a hell of a problem, and I certainly sympathize with her. . . .
>
> She was quite worried as to what Mother would think of the whole business.[68]

Kathleen was right to worry. Her mother was much distressed by the prospect of her daughter marrying outside the church.

Her father was at his best. He cabled her: I FEEL TERRIBLY UNHAPPY YOU HAVE TO FACE YOUR BIGGEST CRISIS WITHOUT MOTHER OR ME. YOUR CONFIDENTIAL MEMORANDUM WORTHY OF CHESTERTON MAGNIFICENT. WITH YOUR FAITH IN GOD YOU CAN'T MAKE A MISTAKE. REMEMBER YOU ARE STILL AND ALWAYS WILL BE TOPS WITH ME. LOVE DAD.[69] A long letter of reassurance followed.

> I want you to know that I feel for you very deeply. It is very difficult for a person to get a true perspective three thousand miles away from the source of action. I can remember when I was in the picture business

that when I was in New York and sent wires to California every time I arrived in California and read the wires I sent I was sure that I was crazy and so I would send wires back to New York outlining a plan of action and when I got back to New York and read them I know that nobody lived crazier than I, so unless you are on the ground you can't understand what goes on in people's minds. . . .

You will be interested to know that as Mother and I walked down the street last Saturday she said that of all the children, and she made no exceptions whatsoever, she felt that basically your judgment was the soundest.

Now none of this means that I am attempting to tell you how you should handle your life. You are the one that has to live it and it is a long one and also quite a difficult one, but as I have said to you before you are "tops" with me and you always will be.

He added a budget of family gossip. "Jack, much to my amazement because I am not particularly impressed with the depth of his Catholic faith, feels that some kind of concession should be made on the part of Billy." Bobby was playing rugby at Harvard. "He now weighs 165 and is easily the strongest of all the boys." Cardinal O'Connell had died, and they had been having masses, processions and parades. "I think it is a disgraceful exhibition but it seems to be what they want up there. I, of course, will have to go up and be an honorary pallbearer because he married Mother and me." [70]

On May 6, 1944, Kathleen Kennedy and William Hartington were married at the Chelsea Registry Office in London. Joe Jr. gave his sister away. "Never did anyone have such a pillar of strength," she wrote later, "as I had in Joe in those difficult days before my marriage. . . . He constantly reassured me and gave me renewed confidence in my own decision." [71] In Washington, Franklin Roosevelt, hearing the news, "laughed and laughed. . . . The President said he had known the Duke of Devonshire for a long time and that his family were rabid Protestants." [72] In Boston, Rose Kennedy, ill in the hospital, asked nuns to offer special prayers. Jean Kennedy was in a convent; and her father, concerned that the nuns would upset her by condemning Kathleen's decision, came to see her, explaining that this was her sister's conscientious choice and that the main thing was to keep on loving her. [73] "I see now that I've lost one of my daughters to England," Joseph Kennedy wrote Beaverbrook later. "She was the apple of my eye and I feel the loss because I won't have her near me all the time, but I'm sure she's going to be wonderfully happy and I can assure you that England is getting a great girl." [74]

He added that young Joe had just volunteered for an extra month's duty, which would give him forty missions. "Although he's had a large number of casualties in his squadron, I'm still hoping and praying we'll see him around the first of July." But when the first of

July came, young Joe, his second tour safely over, volunteered to fly
an experimental Navy plane designed to strike German flying bomb
sites on the French coast. On August 12, 1944, his plane exploded
in midair over England.* His body was never recovered.[75]

It was the most terrible moment of his father's life, the annihilation
of the hope and dream of nearly thirty years. John Kennedy defined
the impact: young Joe's "success was so assured and inevitable that
his death seems to have cut into the natural order of things."[76]
Desolate and inconsolable, Joseph Kennedy withdrew in silent anguish, spending long hours sitting alone and listening to his beloved
symphonies.[77] "He was the oldest boy," Joseph Kennedy wrote Grace
Tully, FDR's secretary, "and I have spent a great deal of time making
what I thought were plans for his future. I have learned, however,
that God takes quite a hand in the lives of His souls on earth."[78]

The pain was inextinguishable. A few weeks later, Harry S. Truman, the Democratic candidate for Vice President, came to Boston.
Kennedy, still in agony, said to him, "Harry, what the hell are you
doing campaigning for that crippled son of a bitch that killed my son
Joe?"† The wound never healed. When the reporter Bob Considine
mentioned young Joe thirteen years later, Joseph Kennedy "suddenly
and terribly burst into tears at the lunch table and for five full minutes
was racked with grief that cannot be described."[79]

Then, a month after the Kennedys had received word of Joe's
death, they heard in early September that Lord Hartington had been
killed leading an infantry patrol in Belgium.

"For a fellow who didn't want this war to touch your country or
mine," Joseph Kennedy wrote Max Beaverbrook in October, "I have
had rather a bad dose — Joe dead, Billy Hartington dead, my son
Jack in the Naval Hospital. I have had brought home to me very
personally what I saw for all the mothers and fathers of the world."[80]

VII

It was a black time for the Kennedys. With one brother killed and
another nearly killed, with one sister married outside the faith and

---

* Among the witnesses of the midair explosion was Colonel Elliott Roosevelt, accompanying the mission in a Mosquito; the lives of the two families were curiously
intertwined.

† Or so Truman recalled over a quarter century later in a not very reliable set of
conversations with Merle Miller. Truman claimed to have told Kennedy, "If you say
another word about Roosevelt, I'm going to throw you out the window." He nonetheless accepted a sizable campaign contribution. He added to Miller, "Old Joe Kennedy is as big a crook as we've got anywhere in the country" (Merle Miller, ed., *Plain
Speaking* [New York: Berkley reprint, 1974], 199).

the country and then her husband killed, the global war had ravaged the family itself. In the meantime, another sister retired from the family; it had become sadly clear that Rosemary, falling farther and farther behind, would live more tranquilly with the nuns in a Wisconsin convent. Robert Kennedy in his nineteenth year was receiving abrupt instruction in the incertitudes of life.

During the summer of 1944 he was still in V-12 training at Harvard. It has been written that he objected to Kathleen's marriage;[81] but this finds no corroboration in evidence or memory. More probably he followed the lead of his father and oldest brother. He was at Hyannis Port when word came of Joe's death. His father, summoning the children, told them they must be brave; that was what their brother would want from them. He even insisted they go ahead with their plans to race that day. Robert was probably among those who "obediently" did so. Only "Jack could not. Instead he walked for a long time on the beach in front of our house."[82]

No one knows how Robert Kennedy, in his solitary life, dealt with the inner turmoil incited by these events. He paid a visit to Milton and reported to David Hackett, who was now with the paratroops: "School was kind of queer and I certainly wouldn't want to go out there too much. It all seems so inconsequential changing to blue suits and being on time for Chapel and all that sort of thing. I am very glad I am finished with it although we did have such good fun the last year."[83] He got to Hyannis Port over Labor Day. Kathleen, husband at the front, was home for a visit. John had invited PT boat friends from the South Pacific. A rollicking lot, they bridled under the ambassador's strict one-cocktail-before-dinner rule and took to foraging in the kitchen liquor closet after the cooks had departed. One of their wives recalled Robert, "a scrawny little guy in a white sailor suit ... very upset that we were sneaking booze in the kitchen. He was afraid his father might catch us and he knew his father's wrath. But Kathleen handled him. She told him to get lost."[84]

In November Robert was scheduled to move on to the V-12 unit at Bates College in Lewiston, Maine. "Due to the fact that I'm off to Maine for the winter and after that off to beat the Japs, and you're off to the paratroopers and after that?" Robert wrote Hackett, "then we might as well let the fair sex see more of *me* and you."* His particular girl was Betty Blackburn of Pittsburgh, whom he had met in summers on the Cape and who was now at Sarah Lawrence.

---

* RFK to Hackett, postmarked October 4, 1944, Hackett Papers. He added a postscript: "Just read this letter over and can't understand any of it. I should join the Code Department of the Navy."

Lewiston turned out to be far away — from family, from friends, from the war, even from the faith; for the local church held mass for its French Canadian constituency in French, a language that continued to elude him. Finally he located an English-speaking church; he wrote his parents that by the end of the week his soul should be "as white as snow. Hope everyone is impressed."[85]

He wrote endlessly. "Bobby is the best letter writer in the family," Kathleen, who had gone back to live in London, told her family in February. "Practically every week a little billet doux arrives from him reporting on the latest activities in the Maine woods."[86] He bragged to Hackett about his skiing prowess: "but you know how I am in sports, natural athlete and everything."[87] He described himself as "my usual moody self. I get very sad at times. . . . Did you know that Barry Turner has gone overseas. If I don't get the hell out of here soon I'll die."[88] Going overseas was an obsessive motif. "Still can't get over Tom Clark [?] saying he'll be fighting before me and the possibility makes me feel more & more like a Draft Dodger or something. But I suppose somebody has to be in V-12 but the attitudes of some of these guys really makes me mad especially after Joe being killed."[89]

The big thing was to get overseas before the war ended. His family regarded this idea with limited enthusiasm. The decision was that he should go to officers' training school. "I wish to hell," he wrote Hackett, ". . . people would have let me alone to do as I wished, but I suppose I simply must be an officer. . . . I guess the rest of this war is going to be won by you and your brothers for I think the Kennedys have all about shot their bolt."[90] His marks, he told Hackett, were pretty good, and he had been appointed platoon leader. "Something I've learned from the first year in my country's service is that you certainly have to be more independent and you have to depend much more on yourself than before." He kept trying to wriggle out of officers' training. "I talked to Papa about it and told him I didn't feature the R.O.T.C. and would try to get back in aviation right away. . . . I am not sure, between you and me, just how much I go for flying but I guess that's the best thing to do."[91] In May he reported, doubtless with some inner relief: "Am not going to fly as I found out yesterday I failed a test in Flight aptitude. This, of course, hurts me, but everything bad that happens now people say it's God's will so I guess I will chalk that up to it also."[92] At the end of June he was transferred back to Harvard to complete his officers' training.

His father was more than ever in these years the dominating figure in Robert's life; the struggle to command his respect more than ever the dominating theme. "I congratulate you on your marks," his father had written him in January 1945. ". . . By the look of some of those subjects of yours, I know that I would never be able to get even an E." He concluded: "Let me know if there is anything I can do for you."[93] Robert responded at once. "I wish, Dad," he wrote from Maine, "that you would write me a letter as you used to Joe & Jack about what you think about the different political events and the war as I'd like to understand what's going on better than I now do."[94] Joseph Kennedy sent back a two-page, single-spaced letter expatiating upon FDR, military developments in Europe and in the Far East, Henry Wallace and his new book *Sixty Million Jobs.* "There will be great confusion finding jobs for everybody. There will be a cry that the only way to solve all these problems is to let the Government do it, and that will be the beginning of some form of socialism in this country. Now, it sounds like a terrific prophecy to make, but I can't see anything except the Government mixing itself more and more into everybody's lives." His father concluded: "This is a quick summary of the conditions as I see them, and I will write you about it all the time if this is what you want. Just tell me what problems seem to be sticking in your mind."[95] "Thanks very much for your letter Dad," Robert replied, "which is just what I wanted."[96] Joseph Kennedy's letters continued through the winter and spring:

> I was very much interested in your letter about the Morgenthau Plan. I don't think anybody really had an idea of what the plan was because it was never fully explained. Of course, it was never his plan — he wouldn't know enough to have a plan. It was conceived by a Jew named Bernstein, who is now Assistant Secretary of the Treasury.

> This morning I read of the discussion about three votes for Russia [in the United Nations], and one can begin to see plenty of disputes for the future.

> [enclosing a document] This is about as good an outline of the OPA situation as you can find.[97]

His pessimism about the war so sadly verified, Joseph Kennedy looked on the postwar world with unremitting apprehension. "I see that the problems of living standards, economics, stability, and national pride," he wrote Beaverbrook, "are still standing on shaky ground." He could not share Beaverbrook's hope about postwar collaboration with the Soviet Union; the Russians would play the game, Kennedy thought, only so long as it was going their way.[98] Roosevelt's death in April 1945 found him in a bitter mood. There was "real sorrow" for two or three days, he wrote Kathleen, "but there is also no doubt that it was a great thing for the country." Roosevelt had stirred up so much hatred that "no matter whether he proposed anything good or bad, half the country would be against it." As for Truman, "I know him very well"; his great attributes were his common sense, his choice of advisers and "his advocating Americanism as the Mid West knows it and not as the elements in New York want it. . . . I'm seriously considering . . . whether I might not say to him that I'd like to help any way I can; but if he's going to give me a job, I'd rather have him give it to Jack and maybe make him minister to some country or Assistant Secretary of State or Assistant Secretary of the Navy."[99]

VIII

In the autumn of 1945 the Navy commissioned a new destroyer and named it the *Joseph P. Kennedy, Jr.* Without telling his family, Robert went to Washington and asked the Secretary of the Navy, James V. Forrestal, to release him from officers' training and assign him to the ship as a seaman.[100] On February 1, 1946, Robert Francis Kennedy, Apprentice Seaman, Class V-12, U.S. Naval Reserve, having passed satisfactorily courses in elementary and advanced seamanship, damage control and engineering, piloting and celestial navigation, left the Naval Reserve Officers' Training Corps program to join the new boat on its shakedown cruise. "During his stay in the NROTC unit," the executive officer wrote, "his conduct was without question, and his loyalty and cooperation were above average. He is leaving . . . at his own request so that he may acquire sea experience. It is recommended that when possible he be given a Quartermaster rating because of his intense interest in navigation. We are sorry that the above-mentioned man is leaving this Unit."[101]

The *Joseph P. Kennedy, Jr.*, was under orders for Guantanamo, the American base in Cuba. The voyage down, Bobby told his parents, was uneventful after the first day, which "was rather rough & it was a very sea sick crew, but your No. 7 passed the test in true Kennedy fashion & held on to his stomach."[102] He wrote Lem Billings, "It's not much better than a bum's life particularly for a seaman." Most of the crew was from the south and had spent even less time in the Navy than he had. "I get along with them very well," he told Billings, "for they have a lot of something, a lot of those guys at Harvard lacked."[103] He wrote his parents, "I certainly am meeting people with a different outlook & interests in life. . . . P.S. The Franklin Delano Roosevelt just moved into the bay so we're all together once more."[104]

After six weeks he was transferred — "from the lowest grade of chippers, painters & scrubbers, the 2nd Division, up into one of the highest grades of chippers, painters & scrubbers, the I division. . . .

The fact that I stands for Intelligence must make you all realize instantaneously without further discussion that I am in exactly the right place for me. I am sure if the fact that Robert Kennedy was in the Intelligence Division of one of the U.S. Navy's fighting ships was made known to the public women & children would sleep more comfortably in their beds." He noted that international news, as summarized in the ship's newspaper, often gave "the wrong idea & to us all it seemed that we would all certainly be at war by the end of the month." (This was the time of Stalin's hard speech of February 9, which William O. Douglas privately described as "the Declaration of World War III," and of Churchill's Iron Curtain rejoinder a month later at Fulton, Missouri.)[105] "If they are going to have a war again," Robert wrote his parents, "I can tell you they are going to have a very tough time getting this group of men to fight it."[106] In the spring he was back in the United States. On May 30, 1946, he received his honorable discharge.

The war had hardly brushed him, but it had altered things nevertheless. With young Joe dead and John confirmed by war in attitudes of detachment and independence, Joseph Kennedy was paying more attention than ever before to his third son. Robert responded with alacrity. The transformations and griefs of war intensified a process of change that had begun much earlier. The sweetness and defenselessness of the small boy had gone underground. The code of the Kennedys discouraged undue emotion, unless disguised as humor. "You didn't express your emotions," said Charles Spalding, "in front of anybody."[107] Robert Kennedy was, in effect, remaking his personality, or at least his *persona*, in his determination to overcome doubts of his own worth and to win the love of the most important person in his life. Unlike John, who imperturbably went his casual and cool way, Bobby felt an overpowering need to please his father; to demonstrate his sense of family, always more sustained than John's; to demonstrate that, like his father, he was not soft but tough.

Emotionalism remained, both as concern and as vulnerability, expressed in moodiness, taciturnity or a certain brusque harshness, expressed too in persisting sympathy for other underdogs. But a protective covering formed over it. He was learning not to expose himself too much, not to seem to care too much. "His idea of a man," said Joseph Dolan, a close associate in later years, "was that you never asked for any quarter."[108] He became in these years the scrappy adolescent, the relentless kidder, the ferocious competitor, the combative mick. Only those who knew him very well perceived the gentleness within.

There was perhaps something very Irish about it all — the loyalty to family, the irony and self-mockery, the mingling of romantic defiance with a deep sadness; something very Irish American too, for the Irish legacy in its Kennedy form had to accommodate itself to the puritan ethic, the belief in discipline, work and achievement. Being an Irish American, Henry James might have said, was a complex fate.

(4)

The Third Son

Robert Kennedy came out of the Navy just in time for a new family undertaking: the election of his brother John to the House of Representatives from Massachusetts. Apart from summers on the Cape and four winters at Harvard, John Kennedy had not lived in the state for nearly twenty years. Though a Massachusetts voter, he was in effect a carpetbagger. His residence during the campaign was Boston's famous political hotel, the Bellevue, across the street from Bulfinch's State House. But the Eleventh District, which wandered from East Boston and the North End into Charlestown, Somerville and Cambridge, was rich in Fitzgerald and Kennedy memories. The young candidate worked the neighborhoods diligently. Robert spent a day during a naval leave accompanying John on his rounds. They met, he wrote their mother, "quite a divergent group of people from Irish bartenders to Negro members of the V.F.W. [Veterans of Foreign Wars]. Everyone seems to agree he is doing very well."[1]

I

Once discharged, Robert offered to join up full-time. "It's damn nice of Bobby wanting to help," John Kennedy told Paul Fay, a Navy friend who had crossed the continent from San Francisco, "but I can't see that sober, silent face breathing new vigor into the ranks. The best plan is to make it known to the press. One picture of the two brothers together will show that we're all in this for Jack. Then you take Bobby out to [the] movies." When Robert arrived, Fay knew he was present, he later wrote, "only because I could see him. Words came out of his mouth as if each one spoken depleted an already severely limited supply." Off they went to the movies, Fay trying to make conversation, Robert Kennedy responding with a terse yes or no. When the stage show began, a comedian reduced Fay to helpless laughter. "I looked over to make sure he was enjoying this as much as I was. From his expression he might have been paying his last respects to his closest friend."[2]

Robert was eventually turned over to LeMoyne Billings, who was in charge of Cambridge, and Billings assigned-him three Italian wards in East Cambridge. There was already a generational split in the campaign. Joseph Kennedy had set up headquarters at the Ritz Hotel, where he telephoned politicians of his own age, briefed the press and dispensed campaign funds. Honey Fitz, a still sprightly eighty-three, and other venerable politicos crowded John Kennedy's rooms at the Bellevue, filling the air with cigar smoke and ancient tales about forgotten elections. The campaign, however, went off in another direction. John Kennedy summoned his contemporaries — schoolmates like Billings, Navy pals like Fay, young Irish veterans like his campaign manager Mark Dalton. His dashing sisters enlisted friends of their own. Jean Kennedy brought along her Manhattanville roommate, a spirited girl from Greenwich, Connecticut, named Ethel Skakel.

Robert Kennedy observed the tug of generations. Grandpa, he said, "felt very strongly about [the campaign], and he felt very close to my brother. But I think that his effectiveness, in some of these areas, was not overwhelming. He had some important introductions and contacts which were significant. But the appeal that John Kennedy had was to an entirely different group. It was to young people and it was to people who had not been involved in politics and a lot of servicemen."[3]

In East Cambridge Robert himself set grimly to work, not bothering to hide his disgust with the old pols sitting around the Bellevue. East Cambridge was home territory for his brother's leading opponent. "I was in a tough area," Robert recalled, "so there was a good deal of antagonism toward . . . me and the operation."[4] He disarmed at least the young by playing softball in a park across the street from his headquarters. "All that propaganda that the Kennedys were the high-hat kind," said a Cambridge politician, "was dissipated in my area by Bobby playing with those kids."[5] Robert lost his wards on primary day, but the margin was smaller than anticipated. John Kennedy easily carried the district. The primary result in June determined the election in the fall.

For both brothers it was an initiation into politics; also, in a sense, into their Irish legacy. Though traditional Boston politicians left them skeptical (in John's case) or exasperated (in Robert's), the young Kennedys relished the stories and characters of the past. David

Powers, whose friendship with John Kennedy began in this campaign, said, "It took Jack three months before he found out that Mother Galvin wasn't a woman" — Mother Galvin so called because of the maternal solicitude he displayed for the interests of his constituents. They were enchanted by the discovery that Jim Curley's henchman Up-Up Kelly received his name by running into political meetings shouting, "Up! up! Everybody up for Governor Curley!"[6] The baroque politics of Boston was not their style. John Kennedy benefited indeed from the contrast. But it remained part of their inheritance.

With the primary out of the way, Robert joined Lem Billings on a trip to Latin America. It was Robert's first long journey away from the family. They began at Rio de Janeiro, where Robert appalled his friend by insisting on plunging into the treacherous surf of Copacabana Beach in the pitch dark of midnight. They went on to Montevideo and Buenos Aires, to Chile and Peru, ending in Panama and Mexico. Apart from Robert's athletic excesses, it was, somewhat to Billings's dismay, an exceedingly seemly trip. Robert was determined to win the $2000 prize offered by Joseph Kennedy to all his children who did not drink or smoke till their twenty-first birthday. When they went to bars, he ordered Coca-Colas and demanded to move on as speedily as possible to dinner.[7] The travelers were occasionally asked what they had done in the war. "Bobby didn't have anything to say," Billings remembered later. "I used to kid him about it. He didn't think it especially funny."[8] In September 1946 Robert was back at Cambridge and Harvard.

II

Harvard after the war was a confusion of ages. Graying veterans of Anzio and Okinawa mingled with boys fresh from secondary school. Robert Kennedy was chronologically in between. He had received academic credit for his work in V-12 and, now almost twenty-one, was entering his junior year. His marks had been mediocre, except for a B in International Law in the winter term of 1945–46. They did not improve now. He was promptly put on probation, from which he was relieved in the spring of 1947.[9]

His field of academic concentration was government. This was nominal. He attended the classes of eminent scholars and popular lecturers — C. H. McIlwain, Roscoe Pound, Payson S. Wild, William Yandell Elliott, Rupert Emerson, Bruce Hopper. No one can remember a professor who particularly interested him; no professor can particularly remember him. Wild, who had known his brothers well, had no impression of Robert at all.[10] "I didn't go to class very much," Robert said later.[11] His grades attested to that. A string of C's was broken only by D's in Anglo-American Law and in the Principles of Economics.

Like his older brothers, he was elected to the Spee Club, but he got mad when another Irish Catholic was blackballed, and never went to the club again. Or so at least he pretended: "I think he just used it for an excuse," said Sam Adams, his Milton friend and now his fellow clubman. "He just wasn't particularly happy around those guys; he was rather uncomfortable."[12] He was rather uncomfortable with most of his college contemporaries, including some who later became good friends. George Plimpton remembers seeing him once at the house of K. K. Hannon, a beauty of the day. "It always struck me; there was a party going on. There was this one fellow working on a paper or something [on a table] in the kitchen."[13] John Knowles — not the novelist but the physician and later head of the Rockefeller Foundation — found himself next to Kennedy in classes where seating was arranged in alphabetical order. "Intellectually I think he was very bright. . . . I found him very serious in the classes I was in with him . . . nor did I in any way find him so rigid or dogmatic. He was tough on himself and tough on the people around. . . . He was tougher than hell. He stood up for what the hell he believed and kept going at it." Knowles added that he was looked on as "kind of a nasty, brutal, humorless little fellow when he got going"; and that was no doubt how many classmates regarded him.[14] Thus Anthony Lewis, later of the *New York Times*: "Hardly knew him and didn't like him and thought he was callow and tough. . . . He was a remote figure . . . sort of a football player."[15]

For a while he went with other Catholic students to the St. Benedict Center where Father Leonard Feeney inveighed against the infidel. Feeney expounded a rigid version of the doctrine *extra ecclesiam nulla salus* — no salvation outside the church — and anathematized Harvard as "a pest-hole" of atheists and Jews. All this vastly irritated Robert Kennedy, who put up an argument, angrily abandoned the center and shocked his mother by denouncing Father Feeney at the Kennedy dinner table. Feeney's suspension by Cardinal Cushing in 1949 and his excommunication by the Vatican in 1953 no doubt gave Rose Kennedy retrospective comfort.[16]

He was still diffident about girls, though his sisters did their best to encourage flirtations. Over the Christmas holiday of 1945–46 Jean had brought Ethel Skakel along on a skiing expedition at Mont Tremblant in Canada. Robert liked Ethel and, after the 1946 congressional campaign, invited her occasionally to Cambridge. But he took out her older sister Patricia rather more. In his senior year, when Robert invited Pat Skakel to Palm Beach, Ethel managed an invitation for herself from Jean. Ethel attracted him by her high spirits and her proficiency in sports. But all this was on the margin of his life.

His real field of concentration at Harvard was football. Perhaps this seemed the most direct way to prove to his family, and to himself, that he was approaching paternal standards of toughness. An early football friend was Kenneth O'Donnell of Worcester, Massachusetts, whose father was football coach at Holy Cross and whose older brother was captain of the 1946 Harvard team. Ken O'Donnell himself was to be captain later. "I can't think of anyone," he said afterward, "who had less right to make the varsity squad than Bobby." Veterans from across the country were flocking to Harvard under the GI bill. When Robert Kennedy, 5′ 10″, 155 pounds, reported for practice as an end in 1946, he was faced "with the biggest and most talented group of ends that the University had ever gathered." Kennedy was relegated to the lowest squad. He would not give up. He arrived on the field an hour early every afternoon and stayed an hour late. In scrimmage "he'd come in from his end like a wild Indian. If you were blocking Bobby, you'd knock him down, but he'd be up again going after the play. He never let up," O'Donnell said. ". . . He just *made* himself better."[17] "His intensity about football," said a classmate and fellow fanatic, William J. Brady, Jr., of Philadelphia, "was really an all-consuming feeling. . . . I don't think he gave a darn about anything else."[18]

In 1947, his senior year, he started the first game as varsity end and scored the only touchdown of his college career. A few days later he broke his leg in practice but characteristically told no one and kept on trying to play till he collapsed on the field, showing once again, as his brother John had said, either a lot of guts or no sense at all, depending on how you looked at it. The injury knocked him out for the rest of the season. But the sympathetic coach sent him in for his letter during the closing moments of the game with Yale. His leg was still in an enormous cast, and he somehow got mixed up in the final scrimmage but emerged unscathed. It was a small triumph. His father had won a Harvard letter (in baseball), but neither of his older brothers had done so.

It was among athletes that Robert Kennedy felt truly comfortable. They were not traditional Harvard types. Most would not have been at Cambridge except for the GI bill. "They were," said Kenneth O'Donnell,

a very irreverent, disinterested group of fellows just out of the service . . . and weren't very much interested in who the hell Bob Kennedy was. . . . They were all two, three, four, sometimes five years older than he was. At first they didn't like him very much. They didn't want to pay any attention to him; they didn't think he was that good a football player, and they thought he was just a rich kid who was hanging around.[19]

Most of them, O'Donnell said, "couldn't scrape twenty-five cents together."[20] Bob exasperated them especially by his negligence about money. "We'd go on trips," said Bill Brady, "and he never had any money and whatever it was I'd pay. . . . It always amazed me that he literally never had any money."[21] But they forgave him for it. "After about six months, he became one of the group," remembers O'Donnell.[22] "He was rather lonely when he was young," observed Eunice, ". . . so I think he picked friends who were sort of different. They weren't in the usual social stream at all; they were tough and rough . . . all big and bulky and very unsophisticated"; very different, Eunice thought, from John's friends, so smooth and assured and adept with girls and parties. "My father was always very immaculate and well dressed. . . . Bobby thought it was rather amusing to bring home people that were neither."[23]

His Harvard life was the football field, the training table, the Varsity Club. But, if his college years were spent with athletes, they were not altogether consumed by athletics. "I used to talk and argue a lot, mostly about sports and politics," he said later. "I began thinking about issues about the time I went to college." Sometimes football itself posed the issue. In 1947 Harvard had a black tackle; and the team was scheduled to meet the University of Virginia at Charlottesville. No Negro had ever played against a white university in the south. When word came of southern objection, it was, said O'Donnell, "a shock to everybody on the team. . . . Everybody agreed that if he couldn't stay in the same hotel with us, we wouldn't play."[24] Robert Kennedy shared the general outrage, and Harvard made its point.

Politics, as one of the water boys recalled twenty years later, was "the favorite topic of conversation outside of football shop talk."[25] The veterans had wide experience, strong opinions and little respect for authority. "We were," said O'Donnell, "the dozen most know-it-alls in the history of the world."[26] The talk Robert Kennedy heard at the training table could very well have been better than the talk he missed at the Spee Club.

III

Robert Kennedy's opinions were still substantially his father's. Joseph Kennedy in the years after the war lived according to quiet routine at the Cape or in Palm Beach, plugged into the world of affairs by the omnipresent telephone. "I really began to make money," he told James M. Landis, who had remained a Kennedy intimate and became, after he left government in 1948, a Kennedy lawyer, "when I came down here to sit on my butt and think."[27] He seemed to drop out of politics, although, according to recurrent rumors (which Felix Frankfurter regularly passed on to his brethren on the Supreme Court), Kennedy was "actively promoting" William O. Douglas for the presidential or vice presidential nomination.[28] While there is no evidence of great activity, that was certainly his preference. Arthur Krock wrote in 1952 that Douglas was always Kennedy's first choice for the Presidency.[29] Douglas remembers Kennedy saying to him at Palm Beach in the late 1940s, "I must be nuts. The two men in public life that I love most are Jack and you. And I disagree with you guys more than anyone else. What's wrong with me?"[30]

His consuming public concern was as it had been before the war: that the United States might be deluded into mounting a world crusade for democracy. He was perfectly consistent. He saw communism in the forties as he had seen Nazism in the thirties — as a detestable system but not as a mortal threat to American security. "Russia does not want a major war now or in the near future," he wrote in *Life* in March 1946. The Soviet Union no doubt had an ideological hope for the worldwide victory of communism, but in practice, he said, Communist ambitions would be subordinate to Russian national interests.

The United States, as Joseph Kennedy saw it, had the choice of two foreign policies. The idea of "pressing steadily and persistently to establish liberal democracy throughout the world" attracted supporters who were "ardent, sincere, active and optimistic. It appeals to the best in the best of men and the best of nations." But it had, Kennedy thought, signal defects. By its very nature it involved the United States "in a program of minding other people's business on a global scale." He warned against "the war-breeding risks of worldwide meddling under the guise of benign interest. . . . *The basic world policy for the U.S. should be to prevent World War III.*"[31]

The Truman Doctrine of 1947 brought the elder Kennedy into open opposition to the Cold War. "If it could be demonstrated that giving dollars to stop the spread of Communism in the Balkans could accomplish that objective," he wrote in an article for the Hearst newspapers, "I would make no comment. . . . But it is obvious to anyone with even limited experience in world politics that a few hundred million dollars is but a beginning." As for aid to Greece or Turkey, "I suspect that the Russian people resent that as our people would resent Russian aid to a Mexican Pancho Villa or a Communist adventurer in Cuba." What was the panic about? Kennedy's recommendation was to "permit Communism to have its trial outside the Soviet Union if that shall be the fate or will of certain people. In most of these countries a few years will demonstrate the inability of Communism to achieve its promises, while through this period the disillusioned experimenters will be observing the benefits of the American way of life, and most of them will emulate it. . . . The dangers at home are far more real to me" — budget deficits, high taxes, inflation, depression, socialism.* "I insist that, when public officials advocate the policy of underwriting the salvation of the rest of the world from Communism, they are morally bound to show the American people just where the money can come from — OUT OF THE POCKET OF THE AMERICAN TAXPAYER."[32]

As the Cold War deepened, Kennedy insisted more and more on parallels between 1938 and 1948. After the policy of "cooperation" failed, he wrote, Britain had shifted in 1939 "to a policy that parallels the one we have been pursuing under the so-called Truman doctrine." As the British course had rashly committed British prestige to the defense of vulnerable European states, so "the Truman doctrine has committed us in situations all over the world. . . . It assumes that our nation of 145 million people will control the destiny of Europe's 245 million, not to speak of Asia's hundreds of millions." He added that we were now told that the Marshall Plan — "the grandiose subsidization of the world" — would bring about European recovery in a few years and thereby halt the rising tide of communism. "In my judgment it will do neither."[33]

Kennedy was willing to send $5 billion to Europe for humanitarian purposes. But Europe's real difficulties, he said, were not physical but moral — especially the historic selfishness of European capitalism in declining to share its gains with the workers. "The seeds that

---

* Interview with Arthur Krock, *New York Times*, March 21, 1947. I am bound to quote my own response to that interview at the time: "Even in America, the capitalist fatherland, the death wish of the business community appears to go beyond the normal limits of political incompetence. . . . The foreign policy of the business community is characteristically one of cowardice rationalized in terms of high morality. The great refusal to take on the Russians today is perfectly typical. That *doyen* of American capitalists, Joseph P. Kennedy, recently argued that the United States should not seek to resist the spread of Communism . . ." (*Partisan Review*, May–June 1947). In retrospect, I believe there was more force in Kennedy's position than I understood in 1947. On the other hand, I still think he gravely overestimated the freedom that "disillusioned experimenters" would have to abandon totalitarian experiments. Given the economic chaos and political confusion in western Europe in the year before the Marshall Plan, the Kennedy prescription still seems to me unduly complacent, if my response to it seems unduly shrill.

make for communism in Europe are too deep to be reached by a recovery plan." Nor could he see that the United States had any great stake in the balance of power in Europe.

> We cannot base our foreign policy on the thesis that the United States cannot tolerate the attainment by any nation of a predominant position in Europe. If we do this, we shall either have to fill the vacuum ourselves or fight war after war to prevent some other nation from achieving that paramount position. This, however, is what we are asked to do when we are urged to make defensive and offensive alliances with the sixteen or more nations to whom we contemplate extending aid.

As for the larger design for which the Marshall Plan might be only the prelude: "No matter what we try to do, 145 million people cannot reform, police and feed the globe. Russia cannot do it either. No nation that has sought to assume that burden has been able to stand the strain. And the atom bomb will not alter that basic fact, for wars as such solve nothing." [34]

In supposing that a nation, once it had tried communism, could readily overthrow a police state, Kennedy was unwontedly optimistic. And in disclaiming the idea that the United States had no stake in the European balance, he was rejecting an ancient theme in American foreign policy. "It cannot be to our interest," Jefferson had written in 1814, "that all Europe should be reduced to a single monarchy"; it would be better to fight "than see the whole force of Europe wielded by a single hand." [35] If the European balance was deemed essential to the safety of the United States in the age of the frigate, it would conceivably seem even more essential in the age of the bomber.

For Kennedy the answer lay in hemisphere solidarity. "We cannot dominate the world. But we can make ourselves too expensive to conquer. This is not a plea for isolation, but it is a plea against imperialism." He visualized "the Western Hemisphere as a self-sufficient economic unit" under North American protection. The pursuit of hemisphere self-sufficiency, he believed, would be far less costly than the world war to which global meddling would eventually lead. "War, at any time, would be hideous. At present or in the foreseeable future war would wreck our present civilization." [36]

"People are talking war," he wrote Beaverbrook early in 1948. ". . . I shudder to think what the end of the next one would be. There are so many more places now where war could break out, as compared with the last World War, that I haven't much confidence in the situation." If Henry Wallace stayed in the presidential contest, Joseph Kennedy added, "any Republican can beat Truman. . . . Wallace is surprisingly strong because he is advocating peace." [37]

IV

His views were sharply at odds with the orthodoxies of the day. They did not prevail even in his own family. Congressman John F. Kennedy strongly backed both the Truman Doctrine and the Marshall Plan. The disagreement between his father and his brother may have been one explanation of Robert Kennedy's remark that he began thinking about issues when he went to college. These were certainly among the questions vigorously debated when the know-it-alls gathered in the evening at the Varsity Club.

Of the Varsity Club group O'Donnell had the most lively interest in politics. He was an all-out New Deal Democrat. Nothing impressed him less than the discovery that his new friend was the son of the celebrated Ambassador Kennedy. "I didn't really care very much who Ambassador Kennedy was," O'Donnell said later, "and I don't know anybody in the city of Worcester that cared.... The Kennedys were famous people. They were not famous people in Massachusetts." Moreover, O'Donnell had never thought much of the ambassador; "he was an isolationist and I was not ... I had been very pro-Roosevelt and he was not."[38]

"Bobby espoused, naturally, his father's views," O'Donnell recalled. "... I had just spent three years in Europe and I saw it totally differently." The football players sat around and listened to O'Donnell and Kennedy argue; "we had a nightly dissertation on politics in which there'd be twenty or thirty members of the squad and those who came over to listen just for the fun.... We were the acknowledged leading debaters in that segment of the University."[39]

Robert used to bring O'Donnell to Hyannis Port on weekends, perhaps in order to expose him to his father, perhaps in order to expose his father to him, perhaps out of sheer mischief. On an early visit, Robert outlined an argument and provoked his father into vehement rejoinder. "I didn't say anything," O'Donnell remembered, "because I was a guest in the house and wasn't that rude as yet." After the ambassador finished, Robert said, "Dad, now that you're all done, I want you to know that Kenny absolutely, totally disagrees with you, don't you, Kenny?" O'Donnell laughed nervously, "and the old man laughed like hell. We had our confrontations, but he was always a gentleman."[40] O'Donnell said later, "I totally disagreed with Ambassador Kennedy throughout our whole association, and he never cared too much for me.... His ideas and mine were very, very contrary."[41] What bothered the ambassador, however, was not contrariety in ideas but moral delinquency. Once the two boys joked at dinner about finishing last in a sailing race. Mr. Kennedy listened with mounting disgust. "What kind of guys are you to think that's funny?" he said and left the table.[42]

v

In March 1948 Robert Kennedy received his A.B. degree. His father regarded him with increasing hope but still evidently with a certain dubiety. "He is just starting off," he wrote Beaverbrook shortly after Robert's graduation, "and he has the difficulty of trying to follow two brilliant brothers, Joe and Jack. That in itself is quite a handicap, and he is making a good battle against it."[43]

The next step in the battle was a trip abroad. As usual, Joseph Kennedy had everything in hand. He arranged for the *Boston Post* to accredit his son as a correspondent and for Cardinal Spellman and James M. Landis to supply letters of introduction. Armed also with a financial letter from the National City Bank of New York and certification from the Cambridge police department that he had no criminal record, Robert set forth on the *Queen Mary* on March 5, accompanied by a college friend, George Terrien.[44] When his father discovered that Beaverbrook was crossing on the same boat, he enjoined Robert to go and speak to him. This perhaps was one of the last things Robert, diffident as ever, wished to do; but, as his mother said later, he knew that his father would be disappointed if he didn't do it, so he did it.[45]

They dined together on the second night out. Robert wrote in his diary:

> We discussed the differences in his and Dad's pre-war views. Beaverbrook believed that war was not inevitable. Turn Germany toward East. To hell with Czech[s]. Attacked Chamberlain's policies not [for] the reason Churchill did but because they did not go far enough. Dad believed War between Germany and Russia would bring on a Universal War.[46]

They saw the evening movie and then went to the lounge where Beaverbrook ruminated at length.

> Beaverbrook believes we are living in potentially the most happy of times, because it is the most exciting of times. Happiness comes only from the facing and the attempting of answers for problems. The Victorian age could not achieve happiness except thru the pursuit of women.

Robert remarked that he was going to the Middle East. Beaverbrook said that the United States was a "subjugated nation to a Jewish minority." As for Britain, it had become "a satellite to the United States." He said he did not like the British (he was, of course, a Canadian), but he gave them full credit for "the two great attributes of tolerance and bravery." Roman Catholicism, Beaverbrook added, was "the great hope of mankind because of its organizational ability and its great leaders . . . Spellman and Griffiths [he probably meant Cardinal Bernard Griffin] for example." Arriving in London, Beaverbrook wrote Joseph Kennedy: "He is a remarkable boy. He is clever, has good character, energy, a clear understanding, and fine philosophy. You are sure to hear a great deal of him if you live long enough."[47] At least Robert Kennedy was a good listener.

After a few days in London, Robert and his companion went on to Cairo. The fervor of Egyptian nationalism impressed him. "Everyone hates the British . . . who, they believed, kept them subjugated without advance in order to keep their own prominence." The poverty impressed him too. "No middle class at all. The lower class are absolute peons. . . . Poor children have a terrible existence with flies crawling in eyes & nose & it seems to bother them not one iota. They know no better. . . . Beggars everywhere & all want American Dollars. Will not take English hard currency. . . . Honking of horns incessant. . . . Men go around holding hands like mad."

He met a young Jew from Tel Aviv who gave him letters to leaders of Haganah, the Jewish defense organization, and then a representative of the Arab League who "became most vehement & said he doubted very much if the people to whom I have the letters would still be alive when I arrived. 'We will cut them out by the roots.'"[48] The RKO Radio Pictures representative in Cairo, to whom Joseph Kennedy had entrusted the boys, told them they should not go on to so agitated a land.[49]

This had the usual effect. On March 26, Good Friday, they flew to Lydda airport and traveled to Tel Aviv by armored car, with Haganah escort, along roads menaced by Arab guerrillas. In Tel Aviv, soldiers took them to Haganah headquarters for an examination of their credentials. After release, "went for a walk, but picked up by the police & as large crowd gathered were blindfolded & put in car & again taken to headquarters to be interrogated by Security Police. Advised us to stay off streets unless with someone." They wanted to push on to Jerusalem but it was difficult until a British major who had heard of Ambassador Kennedy offered them a ride in an armored car. "Very fortunate we arrived as we did, as the next Jewish convoy that left Tel Aviv if reports are correct is being cut to ribbons in its attempt to get through."

Remembering his assignment for the *Boston Post*, Robert talked to everyone he could find — to Haganah soldiers who held the British responsible for everything; to a British soldier who had been there for two years and was sympathetic toward the Jews. "Admitted British had been responsible for much of terrorism but said officers had tough time controling [sic] men who had seen some of their buddies blown to hell." He talked to members of Irgun, the resistance organization that had recently dynamited a British train and the King David Hotel. "They'll fight any soldiers no matter what uniform they are wearing if they attempt to administer their homeland." He talked to a girl in the propaganda service.

Very nice & only 23 years old. I never saw anyone so tired. . . . It was through her that I really began to see the imense [sic] tragedy & horribleness of it all. She was 6th generation & has four brothers in the Haganah starting at 13 years of age. Before the youngest was finally taken her mother used to make him behave by threatening him [that] the Haganah wouldn't take him unless he behaved. Told one story of a girl who carried on with the British police until finally the Stern Gang caught up with her and shaved her head. When her brother returned from the front and saw her, he shot and killed her. We had a very good time as she laughed a great deal but —

He visited a kibbutz through the kindness of a Jew who, forty years before, had made speeches in Boston for Honey Fitz. "It was self supporting & very impressive but several things I did not like, i.e., taking children away from their parents at age of one year so Mother and Father can concentrate whole of their time on work of farm & not be bothered with children the whole day. Children & parents see each other at night when both are at their best & in much more love & respect supposedly." He talked to a former major in the Russian army who believed that the Russian leaders and people "as a whole are more anti-Semitic than the Germans." Nor did he find much enthusiasm about Americans. "We are certainly not the good little saints," he noted, "we imagine ourselves in the eyes of these people." He summed up Jerusalem: "Firing is going on at all times. Still impressed by vehemence of all parties. More & more horrible stories pouring in. . . . Correspondents all very jumpy. As much shooting is going on things very nervous."

They went on to Lebanon. A letter arrived in Beirut from his mother bringing news of Ethel Skakel's prowess on the tennis court. Rose Kennedy concluded: "If anyone does a lot for you, such as luncheon at the Embassy, you might send the lady flowers afterward. At any rate, try to write a note. As you are young, it need not be a big gesture."[50] He wrote his parents about Palestine early in April: "the only way to describe the situation is 'very very horrible.'

That and a universal hatred for the British seems to be the only points agreed on by all." He was considerably impressed by the Jews. "They are different from any Jews I have ever know[n] or seen." As for the Arabs, "I just wish they didn't have that oil."[51]

On May 14 the British mandate came to an end. A Jewish provisional government proclaimed the republic of Israel, and the Arab League soon launched an attack on the new state. The *Boston Post* ran four articles from its "Special Writer" in the Middle East on June 3–6. The first bore a headline guaranteed to sell papers in Boston: BRITISH HATED BY BOTH SIDES. The author, after commenting on the ambiguity of the Balfour Declaration, gave objective statements of the Arab and Jewish viewpoints. "It is an unfortunate fact that because there are such well-founded arguments on either side each grows more bitter toward the other."

His second piece revealed his own commitment. The Jews in Palestine, he wrote, "have become an immensely proud and determined people. It is already a truly great modern example of the birth of a nation with the primary ingredients of dignity and self-respect." His original draft added at this point: "Many of the leading Jewish spokesmen for the Zionist cause in the United States are doing immeasurable harm for that cause because they have not spent any or sufficient time with their people to absorb the spirit." On reflection he deleted this thought and simply praised the Jews in Palestine as "hardy and tough," their "spirit and determination" created not only by their desire for a homeland but by "the remembrance of the brutal inhuman treatment received by the Jews in the countries of Europe." He gave a lyrical account of his kibbutz visit, omitting the reservations expressed in his diary. The Jews, he said, had "an undying spirit" the Arabs could never have. "They will fight and they will fight with unparalleled courage."

The third piece was sharply critical of British policy for its "bitterness toward the Jews." As for the United States, he noticed himself becoming more and more conscious, as he traveled abroad, "of the great heritage and birthright to which we as United States citizens are heirs and which we have the duty to preserve." A prime motive in writing his dispatches, he said, "is that I believe we have failed in this duty or are in great jeopardy of doing so." We had failed because we had been taken in by the British.

Our government first decided that justice was on the Jewish side in their desire for a homeland, and then it reversed its decision temporarily. Because of this action I believe we have burdened ourselves with a great responsibility in our own eyes and in the eyes of the world. We fail to live up to that responsibility if we knowingly support the British government who behind the skirts of their official position attempt to crush a cause with which they are not in accord. If the American people knew the facts, I am certain a more honest and forthright policy would be substituted for the benefit of all.

The final piece dismissed the notion, then prevalent, that a Jewish state might go Communist. "That communism could exist in Palestine," Robert Kennedy said, "is fantastically absurd. Communism thrives on static discontent as sin thrives on idleness. With the type of issues and people involved, that state of affairs is nonexistent. I am as certain of that as of my name." He recalled how Robert Emmet met the accusation that he was in league with Napoleon by saying he had not fought one foreign oppressor in order to replace him by another.

His original draft concluded: "I think the word 'hate' sums up the situation in Palestine better than anything I can think of. There will be a bloody war because there must be a bloody war. There is no other alternative." He was marginally more hopeful in the *Post*. He had seen Jews and Arabs working side by side in the fields and orange groves outside Tel Aviv; perhaps these would make a greater contribution than those who carried guns. If a Jewish state were formed, it might be the "only stabilizing factor" in the Middle East. "The United States and Great Britain before too long a time might well be looking to a Jewish state to preserve a toehold in that part of the world." In any case, peace-loving nations could not stand by and watch people kill each other. "The United States through the United Nations must take the lead in bringing about peace in the Holy Land."

His father, with his dread of American meddling, conceivably winced a little at this last proposition. Not too much, probably: he cared less what his sons said than whether they said it persuasively. On these grounds he must have admired Robert's series. The pieces showed a maturity, cogency and, from time to time, literary finish creditable for a football player of twenty-two hardly out of college.

VI

From the Middle East the wayfarers went on to Istanbul, Athens and Rome, where they arrived on April 15, three days before the election of 1948. They rushed to hear Togliatti, the Communist leader, denounce the United States before a crowd of 200,000 (or so Robert estimated it). "The fellow who was translating for us was very frightened by the crowd & the fact he had to speak English. He was very anti-Commie but took to clapping when the crowd did so as not to

be noticeable." The crowd, Robert reported, sang "The Internation-
ale" again and again but the speech did not appear to arouse uncontrollable enthusiasm. Christian Democratic planes, no doubt subsidized by the CIA, circled overhead dropping leaflets.

With the Christian Democrats returned by a safe majority on April
18, Kennedy and Terrien headed south, meditating — two young
men who had missed the war — over the smashed monastery at
Monte Cassino; then to Naples and afterward the breathtaking
Amalfi road — the "most beautiful drive that I have ever seen."
Suddenly tragic news: Kathleen Hartington had flown from London
in a private plane for a weekend in the south of France; the plane
crashed in fog near Lyons; the ravishing girl was dead. It was one
more appalling blow for the Kennedy family, already sorely tried by
the hand of God. Joseph Kennedy, who was in Paris, brought his
beloved daughter's body back to England. "I can still see the stricken
face of old Joe Kennedy," Alastair Forbes recalled nearly thirty years
later, "as he stood alone, unloved and despised, behind the coffin of
his eldest daughter amid the hundreds of British friends who had
adored her and now mourned her." She was, Forbes said, "the best
and brightest-eyed of all Kennedys . . . upon whose gravestone at
Chatsworth is most truthfully written 'Joy she gave.' "52

Robert Kennedy, unable because of jaundice to attend Kathleen's
funeral, went to London immediately thereafter. "I never saw Dad
here in England," he wrote his sister Patricia, "but everyone said he
looked terrible. I talked to him on the phone & he was no laughs to
use an old expression. All these people loved Kick so much it's really
impressive." He added: "Between you & me except for a few individuals you can have the bunch."53

The fun had gone out of the grand tour. They pressed on to
Belgium.

> Went to Billy's grave. People around the area where war has passed
> through so many times take it all rather objectively which is hard for
> me to understand. They went thru all the motions of how Billy was
> killed & how he fell etc. which was a little much for me. The farm
> house he was killed attacking & into which he threw his grenade now
> houses a couple of little children & parents with the man having diffi-
> culty keeping his pants up. The people all tell their stories of the war
> as if they just came out of a wild western movie.

In Holland they encountered widespread criticism of the English.
"Came to the conclusion today that if I were an Englishman I would
be very proud of so being." They entered Germany by way of
Aachen, not far from the convent that Rose Kennedy had attended
forty years before. Aachen "was far worse than anything I imag-

ined. Streets not cleared of the rubble which at places reached heights of 10 feet or more. . . . Passed blocks and blocks where only the shells of houses were left standing. It was even more eerie than Cassino." They went to Cologne, where tourists were forbidden to go into the cathedral, "but as I looked like a workman dressed as I was I went through inside. Very sad. Statues with missing heads or limbs, rubble all over floor, piled up high on sides."

They went to Heidelberg and to Frankfurt. "The people in all these cities," he wrote his parents, "walk around in dazed silence. . . . If one blows one's horn, they jump wildly as if you woke them out of a dream, but that's the only reaction for they don't give you more than a fleeting glance." He reported it as "the universal opinion that the majority of the Germans don't feel the least sorry for their 'sins.'" He asked a German veteran where he thought Hitler was. The answer was "in Heaven." He was also critical of the American occupation.

17 & 18 year olds & bitter displaced nationals & Jews because they could speak German were placed in prominent positions where justice was anything but imparted. If you were in a concentration camp you automatically received without investigation an administrative post with the result that people who had been placed in the camps because they were sex perverts or such became mayors.

Many people, he added, "expect war by July tenth. . . . Nobody believes the Russians want it but as the backing down by either side would be a great loss of face almost anything could happen."[54] He was soon writing from Vienna, where he plunged into the atmosphere of *The Third Man.* "The intrigues, plots, and counterplots make even a visitor feel that he is part of something unreal. It is Alfred Hitchcock at his best. The city is a primary focal point in the struggles of the so-called East and West." There was more worrying talk about the future:

Everybody in Embassy wants to go to war with their comprehension of results built not on history but on own rather idealistic beliefs. I'm afraid they might sweep us right into war. The same may well apply for military.

Many here believe we want war now as we have A Bomb and Russia is not yet back on her feet. Many also evidently believe we are imperialistic ranging from the extreme that we actually want to take over the territory till that we want to impose such business ties on the countries that we are supposedly helping that they will virtually be our colony. Many think we are using Marshall Plan because we have to have an export market. Military Men & Civs always shooting mouths off that US is best in everything alienates them, particularly when they refuse to take criticism on own country. Belief infused by the fact we are supporting these countries & have a right to say what we wish. Of course ideally we should be just the opposite.

In Vienna as in Frankfurt he heard much complaint from the American military about the wartime and postwar agreements with the Soviet Union. In early July, he decided to see eastern Europe for himself. They crossed the Russian zone of Austria into Czechoslovakia. "Were kept for an hour at border while the authorities went minutely thru our bags reading as best they could *all* our letters." Czechoslovakia — this was four months after the death of Jan Masaryk and a month after the resignation of President Beneš — was depressing. "All mumble now — What can we do? My question is what are we all going to do?"

He failed to get into Hungary. He had applied for a visa in London; but, after being assured there would be no problem, he was stalled, and no visa ever arrived. Someone told him later that the Hungarians feared he might be carrying secret messages from Cardinal Spellman to Cardinal Mindszenty. Instead the travelers went back to London and then back again to West Germany, where the great airlift, by which the western allies hoped to circumvent the Soviet blockade of ground traffic into West Berlin, was now, in mid-July, two and a half weeks old. Planes were taking off and landing every three or four minutes at the Frankfurt airport. "It is a very moving and disturbing sight to see plane after plane take off amidst a torrent of rain, particularly when I was aboard one."

West Berlin was filled with gloom. "Everybody seems to agree the blockade can't possibly hold out & we will have to ignominiously pack up or attempt to run in with armed force." The airlift "can't possibly" provide sufficient coal, he was told, and, "as coal gets shorter, unemployment will increase & although Germans strongly behind us new resentment is bound to grow & we will not have right to subject them to such increased hardships."

The travelers journeyed on to meet a couple of Kennedy sisters in Copenhagen. "In their thirst for culture they included George and I [the Kennedys could never get this right] so every morning bright and early we hustled off to a different museum to study Denmark's art. I never even knew there was any."[55] Then to Stockholm and to Dublin and the horse show. The summer was drawing to an end. "Would you please make the necessary reservations for George & I to return to the US," Robert wrote his father's office. "We would like to go tourist if possible."[56]

His diary notes for the six months abroad showed sharp and generally dispassionate observation, relentless curiosity, laconic courage, wry humor and a certain sobriety of judgment. He did not evidently share his father's view that the United States should stop minding other people's business. On the other hand, he was appalled at the talk of war and was not, at least in the field, an ideological crusader. This changed somewhat when he got home. At the end of the year the Hungarian government arrested Mindszenty and put him on trial. Robert Kennedy wrote an emotional article in the *Boston Advertiser*:

> We can look back over the last four or five years and see the colossal mistakes that we have made. Every day we hear someone groaning about it. Every day someone is heard to say, "How could anyone be so stupid as to act like that." Yalta, Potsdam, how ignominious!
>
> Here is a great opportunity for forceful action. . . . All eyes will now be turned our way to see if [Mindszenty] will be betrayed, or if resistance to evil can expect support. This is our second chance. Let us not now once again grovel in uncertainty.
>
> For if we fail, the fault as with Julius Caesar's Romans will be "not in our stars but in ourselves."
>
> LET US NOT FAIL![57]

VII

"I just didn't know anything when I got out of college," he said later. "I wanted to do graduate work, but I didn't know whether to go to law school or business school. I had no attraction to business, so I entered law school."[58] His marks were not good enough for Harvard or Yale. The University of Virginia told him that "in view of your record at Harvard College, you are unlikely to be admitted to this Law School, unless you do well on the [Law] Aptitude Test."[59] As a fallback, Robert applied to the Harvard Business School. "They let football players in with very little trouble," he informed his father's office.[60] He wrote his sister Patricia:

> I suppose you have heard it rumored about that there is a small question of what college is going to have my services for the next couple of years. Virginia law school has rather insulted me as well as my whole family [of] which when I looked around you were a member by saying that I had a "far from outstanding record at Harvard." We must stick together on this sort of thing. . . .
>
> There has been some mention of business school . . . I think the idea of me at business school has more comical than serious aspects. The result I'm afraid is that we should all suck around Dr. Cloney [the family dentist] & get old Bobby into Georgetown Dental School.[61]

But Charlottesville finally decided to admit him, adding that "unless he does better work than he did at Harvard, he is most unlikely to succeed in this Law School."[62]

He returned on the *Queen Mary* in August and arrived at Charlottesville in September. As usual, nothing came easy, but, as usual, will was hardened by difficulty. He did not talk much in the class-room but listened carefully and applied himself diligently. "It was like everything he had done," said Charles Spalding; "it was like trying to make the football team at Harvard all over again."[63]

For Professor Charles O. Gregory's noted seminar on labor relations he wrote a paper (submitted with a handwritten note: "Here it is — what you have been waiting for these many weeks!") analyzing the application of seniority rules to layoffs and finding in each case for the union.[64] His great effort in constitutional law was a discourse entitled "The Reserve Powers of the Constitution." This was an encomium on the Ninth and Tenth amendments, which had been designed, he wrote, "to prevent the [national] government from entering into fields which have not been specifically covered by the rest of the Constitution and in that way claiming the prerogative by default." The republic had started to go wrong, he suggested, when John Marshall, "the leading proponent of an all powerful government," took the supremacy clause "as his bible." Fortunately, from Marshall's death to 1936, the power of the states had been "comparatively well protected." In recent years, however, the Supreme Court had been looking at the Constitution "with a magnifying glass and if it could discover there any express or implied power for the federal government to act, then that act was constitutional, the 10th Amendment ineffective." His peroration, no doubt pleasing to his father, called for "effective control of the great leviathan, the federal government."[65]

His major success — at least in the eyes of the Law School, which deposited the paper in the Law Library's Treasure Trove — was an essay on Yalta, written in his last year, for Hardy Dillard's seminar on international law and relations. The Korean War had begun; the Cold War was growing colder; and Robert Kennedy shared enthusiastically in the conservative revulsion of the time against the diplomacy of Franklin Roosevelt. "By Feb. 11, 1945," he declared in his paper, "it can be said that the peace of the world was lost." If the United States had gone to Yalta knowing the real problems and "the correct philosophy," it would not have been too late to rectify earlier errors. But "we were there seeking solutions to the wrong problems and we were there with a bankrupt philosophy. By Feb. 11 we had taken the final step from which there was no salvation."*

---

* He added in a footnote: "I do not believe that because we have lost the peace, that war is inevitable. We have lost the peace but it does not follow that we have lost also the opportunity to prevent a war. That fortunately lies before us" (Robert F. Kennedy, "A Critical Analysis of the Conference at Yalta, February 4–11, 1945," 33, University of Virginia Law Library).

The Far Eastern agreement at Yalta, he argued, was the "most amoral of acts whose potential disaster has long since become for us present-day catastrophe" — a reference to Chinese intervention in Korea. It had not been in our interest, he wrote, to beg for Russia's entry into the Pacific war. "For this mistake we are paying in blood." As for the agreement on liberated Europe, this was

> a clear violation of the basic tenets of International Law. With a stroke of the pen we made forever farcial [sic] the principles for which we had repeatedly told the world we were fighting. England violated her guarantee of Poland of March 1939. . . . The U.S. and G.B. violated the Atlantic Charter.

He drew a fancy conclusion: "Human history has demonstrated too dramatically that terms such as these only give birth to new conflicts. The God Mars smiled and rubbed his hands."

Why these monumental errors? "President Roosevelt felt that the way to beat the Common Enemy as well as to have future peace was to stay friendly with Russia. . . . This was the philosophy that he and his lieutenants, Hopkins, Harriman, Winant, etc. kept as their guiding star throughout the entire war — a philosophy that reached its epitomy [sic] at Yalta, and a philosophy that spelled death and disaster for the world."* The United States had failed to place itself on the right side morally, "and it will be to our everlasting dishonor. One lesson we can learn from all this is that the Bible concept of 'It is better to give than to receive' does not always apply."[66]

The paper was a forceful statement of the right-wing case against Roosevelt's Russian policy. It was written, of course, long before later fashions in revisionism condemned Roosevelt as an imperialist who ignored legitimate Soviet security concerns and proposed to dominate eastern Europe in the interests of expanding American capitalism. It was revealing for the arguments it did not make. Though written after Alger Hiss's conviction for perjury and after Senator Joseph McCarthy's discovery that Communists had been running United States foreign policy, it attributed the errors of Yalta to bad judgment, not to sinister intent. Hiss, for example, had been at Yalta. Kennedy did not mention his name. In this regard it differed from the right-wing boilerplate of the day.

---

* In a draft he said, "In criticizing the Yalta Agreement, I recognize that Franklin Roosevelt and his advisors were faced with immensely complicated problems, and that they were dealing with a dictator who did not share their sincere and generous hopes for the future. But even though this may be true, I believe that mistakes were made and that the American people can only plan for the future by knowing these mistakes."

VIII

Robert Kennedy came into his own at Charlottesville in the extracurricular life — especially when in his last year he became president
of the somewhat moribund Student Legal Forum, an institution set
up to bring outside speakers to the Law School. Drawing on his
father's stock of friends, he produced in rapid succession James M.
Landis, William O. Douglas, Arthur Krock and Joe McCarthy. Thur-
man Arnold and Ralph Bunche came too, as well as Congressman
John F. Kennedy and, on December 12, 1950, Joseph P. Kennedy
himself.

The elder Kennedy spoke six months after the Truman administration had brought the United States into the Korean War and three
weeks after MacArthur's insistence on invading North Korea had
provoked massive Chinese intervention. It was a melancholy December. In Kennedy's view the Korean debacle vindicated every apprehension he had felt about postwar American policy. "From the start
I had no patience with a policy — what has become known as the
'Truman Doctrine' — that, without due regard to our resources —
human and material — would make commitments abroad that we
could not fulfill." Beginning with our intervention in the Italian
elections and our aid to Greece and Turkey, "we have expanded our
political and financial programs on an almost unbelievably wide
scale." To what end? "What business is it of ours to support French
colonial policy in Indo-China or to achieve Mr. Syngman Rhee's
concepts of democracy in Korea?" The interventionist policy, Kennedy said, was "suicidal." It won no dependable friends; it scattered
military power over the globe; it thrust us into battlefields impossibly
removed from our sources of supply. Far from containing communism, "it has solidified Communism, where otherwise Communism
might have bred within itself internal dissensions."

The only hope was "to wash up this policy" and return to fundamentals. The first step, Kennedy said, "is to get out of Korea —
indeed, to get out of every point in Asia which we do not plan
realistically to hold in our own defense." The second step was to
apply the same principle to Europe. "What have we gained by staying
in Berlin?" The Russians could, he believed, push us out whenever
they chose.

> It may be that Europe for a decade or a generation or more will turn
> Communistic. But in doing so, it may break of itself as a unified
> force. Communism still has to prove itself to its people as a government
> that will achieve for them a better way of living. The more people that

it will have to govern, the more necessary it becomes for those who govern to justify themselves to those being governed. The more people that are under its yoke, the greater are the possibilities of revolt.

Moreover, it seems certain that Communism spread over Europe will not rest content with being governed by a handful of men in the Kremlin. French or Italian Communists will soon develop splinter organizations that will destroy the singleness that today characterizes Russian Communism. Tito in Yugoslavia is already demonstrating this fact.

He added, with impressive foresight, "Mao in China is not likely to take his orders too long from Stalin, especially when the only non-Asiatics left upon Asiatic soil to fight are the Russians." His policy, Kennedy said, would be denounced as appeasement. But was it appeasement to withdraw from unwise commitments, arm yourself to the teeth and make clear for what you will fight? He would gladly, he said, choose a Munich to escape a Dunkirk.[67]

It was in retrospect a tantalizing mixture of prescience and perversity. *Pravda* happily reprinted the full text. Congressman John Rankin of Mississippi, the most rabid reactionary in Congress, inserted it with high praise into the *Congressional Record*. Walter Lippmann saw merit in it.[68] Joseph and Stewart Alsop saw none, writing that ten years before Kennedy had advocated giving the world to Nazi Germany; now he "has just crawled out of his richly upholstered burrow to advocate giving the world to the Soviet Union." He might not put it quite that way, they conceded, but his program "insures total triumph for Stalin."[69]

Kennedy's Charlottesville speech, followed by a speech on similar lines from Herbert Hoover a week later, opened the so-called Great Debate of 1951. The debate, however, was speedily derailed, once Senator Robert A. Taft entered it, into an argument, with merit of its own, over the President's constitutional power to commit troops abroad. Kennedy's substantive propositions — especially his opposition to American armies on the Asian mainland and his conviction that communism would not be a monolith to the end of time — were ignored. His contention that the United States should not commit itself to the defense of untenable positions also deserved more attentive hearing. Unfortunately he defined tenable as penuriously as the globalists defined it extravagantly. Western Europe, even South Korea, were not in fact beyond the reach of American military power. Kennedy's perversity in the end defeated his prescience. His critics, fixing on the excesses in his argument, found it easy to dismiss him as an incorrigible appeaser repeating in 1950 his folly of 1940.

If Joseph Kennedy at the Student Legal Forum was a national event, Ralph Bunche caused considerably more excitement locally. Remembering Charlottesville's agitation over the black tackle on the

Harvard football team, Robert Kennedy must have anticipated trouble. When he proposed Bunche, there was, he later recalled, "tremendous opposition." Opposition mounted when a courteous letter arrived from Bunche: "May I ask whether the seating of the audience will be on a strictly unsegregated and unseparated basis? . . . As a matter of firm principle, I never appear before a segregated audience."[70]

Virginia law, however, forbade the mingling of blacks and whites in meeting halls. Kennedy decided to ask the student government to call for an integrated audience. A classmate, Endicott Peabody Davison, described ensuing developments:

> There was a meeting with about ten people from all the classes. Bobby said we must adopt a resolution. Everyone agreed until he asked them to sign it. Then the Southern boys began to say, "I've got to go home to Alabama later — I can't sign it. I'm for it, but I can't put my name to it." Bobby blew his stack. He was so mad he could hardly talk. He had a lack of understanding of the problems these people faced; to him it seemed illogical to support something but be unwilling to sign for it. It's his black-and-white view of things. The resolution failed.

It may have failed there, but the Student Legal Forum adopted it, and a ringing statement went to the university president under Kennedy's signature. "We would like to register the strong conviction," the statement said, "reenforced by our belief in the issues presented by the last war in which most of us fought, and by our belief in the principles to which this country is committed in the Bill of Rights and the United Nations Charter, that action which would result in the cancellation of Dr. Bunche's lecture appears to us morally indefensible."[71]

In the past, the university had evaded the state law by posting one section of the hall for blacks and then letting people sit where they pleased. When Dillard proposed this as the way out, Kennedy rejected the compromise; he felt the principle was bad and had to be confronted. He carried the case to the Law School's governing body, the Board of Visitors, where, according to Davison, "the madder he got, the worse he got at talking. Very little came out." Finally, Professors Gregory and Dillard, impressed by his concern, accompanied the Law School dean to see the university president, Colgate Darden. Darden, who came from an old Virginia family and had been governor of the state, was personally opposed to this form of segregation. When the Law School professors noted that the Supreme Court in the recent case of *Sweatt* v. *Painter* had required the University of Texas Law School to admit a Negro, Darden seized the point, defined the Bunche lecture as an educational meeting and declared that it would be unsegregated.

Bunche found an integrated audience, one third black, and was visibly affected. In his talk, three months after Joseph Kennedy, he vigorously condemned the new isolationism that, he said, had "come out of hiding and has even become respectable again." [72] The Bunche episode may well have given Robert Kennedy second thoughts about the virtue of concentrating governmental power on the state level. In any case he had struck a blow for decency and, after some instruction in the limitations of rage, had learned the efficacy of law.

He graduated in June 1951, fifty-sixth in a class of 124. His best course was Constitutional Law, where he was marked 3.5 out of 4.00, followed by Labor Law (3.0); his average for all courses was 2.54. The impression among the faculty was that he had not worked hard academically because native ability enabled him to do sufficiently well with a minimum of effort. [73] If he had not distinguished himself, he had shown that he could handle the material. He was now the first Kennedy of his generation to have a profession.

IX

He was seeing Ethel Skakel with increasing frequency. He called on her at Manhattanville College; she visited him at Charlottesville; and by 1949 they were enthusiastically in love. [74]

Ethel, born in Chicago on April 11, 1928, was two and a half years younger. The Skakel and Kennedy families had superficial resemblances. George Skakel, a self-made millionaire, had built a small company into the Great Lakes Carbon Corporation and now sold his petroleum coke around the world. He was a quiet man, withdrawn from the family hurly-burly. A Dutch Protestant, he had married Ann Brannack, an Irish Catholic, fat, warm-hearted, eccentric and fervently religious. They had seven children; Ethel was next to last. The Skakels moved east in the mid-thirties, into a stupendous brick mansion in the midst of a large estate outside Greenwich, Connecticut. The Skakel family, like the Kennedy family, was athletic, boisterous, competitive and filled with the will to win.

There were differences. The Kennedys had absorbed enough of the Boston ethos to have a certain reserve, discipline and frugality as well as a large measure of civic and intellectual aspiration. The Skakels, more in the style of Chicago and Fairfield County, did not conceal their wealth or restrain their children. The two oldest boys were always swinging from trees, shooting air rifles from bedroom windows, driving cars into the pond. Ethel's friends often found them scary. The Skakels, in so far as they cared about public affairs, were conservative Republicans. George Skakel wrote Joseph Kennedy after his Charlottesville speech, "It was a fine, constructive and courageous thing for you to do."[75]

Ethel was a sparkling girl, bubbling over with jokes and charm. The Manhattanville yearbook caught her rather well: "An excited hoarse voice, a shriek, a peal of screaming laughter, the flash of shirttails, a tousled brown head — Ethel! Her face is at one moment a picture of utter guilelessness and at the next alive with mischief."[76] An excellent athlete, she hated losing as much as any Kennedy. She was bright and responsive, not, at this point, much of a reader but indefatigably curious. Her gregariousness, teasing humor and endless vitality made her the jolliest of companions. Existence around her was an unceasing swirl of activity. Her ideas of life were simple and passionate. Her sympathy was inexhaustible, which gave her the capacity to grow. Her spirit was indomitable. Her piety rivaled Rose Kennedy's. She loved Robert Kennedy with all her heart.

An obstacle arose: Ethel's sudden conviction that instead of marrying she should become a nun. Robert, walking along the beach at Hyannis Port, said disconsolately to his sister Jean, "How can I fight God?"[77] This passed. They decided to marry with Robert still in law school. "I had misgivings about that," said Rose Kennedy, though none about Ethel.[78] "My financee [sic] followed me down here," Robert wrote his sister Pat from Charlottesville in the spring of 1950, "and wouldn't let me alone for a minute — kept running her toes through my hair & things like that. You know how engaged couples are."[79]

The Greenwich wedding in June was preceded by a memorable bachelors' dinner at the Harvard Club in New York, where the redoubtable Skakel brothers encountered Robert's rowdy football friends. The Harvard Club later sent the Kennedys a bill for the wreckage. Most of his ushers were football players, so big that they could hardly get down the aisle side by side. Mrs. Skakel, said Kenneth O'Donnell, "couldn't understand where Bobby got these characters that were around him, who all weighed 250 pounds and were Greeks, Armenians, Italians."[80] Someone dropped a few coins when the wedding procession was forming, and, as Lem Billings, an usher, politely bent to help pick them up, George Skakel, Jr., kicked him hard in the pants; they had not yet been introduced.[81]

The seventh child thus produced the second marriage. Most of the Kennedys married late. "That family," said Lem Billings, "stayed home longer than any family in history."[82] Except for Kathleen, Robert was the only one of the children to marry before their late twenties, and, with Kathleen dead, he was the only married child until Eunice married R. Sargent Shriver, Jr., three years later. Perhaps he felt a particular need for a family of his own.

The young couple — Robert was twenty-four, Ethel, twenty-two — escaped to Hawaii for their honeymoon and settled, in the autumn, in Charlottesville for Robert's last year at law school. Ethel had technical defects as a wife. She was a flop at cooking and sewing, detested cleaning and spent money carelessly. In the circumstances the sins were venial. They could afford cooks, maids and extravagance. In fundamental ways she was a superb wife, at least in the old sense of being her husband's helpmate. For Ethel Kennedy, reared as she was and believing as she did, ministering to her husband was not self-sacrifice or self-betrayal but self-fulfillment. It did not result in any manifest repression of an irrepressible personality.

For Robert Kennedy it was the best thing that could have happened. Her enthusiasm and spontaneity delighted him. Her jokes diverted him. Her social gifts offset his abiding shyness. Her inextinguishable gaiety lightened his times of moodiness and pessimism. Her passion moved him. Her devotion offered him reassurance and security. Robert, his sister Eunice said, was more dependent emotionally than his brothers, "more anxious for affection and approval and love, and therefore when he got a lot of that, which I think he did when he married Ethel, he blossomed." She awakened his sympathy and his humor and brought him out emotionally.[83] He never had to prove himself to her. Ethel gave him unquestioning confidence, unquenchable admiration, unstinted love.

The Brothers: I

THIRTEEN MONTHS after the wedding Robert Kennedy became the father of the first Kennedy grandchild. They named her Kathleen Hartington. In another year Joseph Patrick Kennedy III was born; early in 1954 Robert Francis Kennedy, Jr.; in 1955 David Anthony Kennedy; in 1956 Mary Courtney Kennedy.[1] In the meantime, after rejecting the idea of running for Congress from Connecticut, Robert took a job with the Internal Security Division of the Department of Justice at $4200 a year. Ethel found a house at 3214 S Street in Georgetown.

I

In September 1951, before he went to work, the *Boston Post* sent him to cover the conference in San Francisco where fifty nations had gathered to conclude a peace treaty with Japan. What impressed him most was the strength of Asian nationalism. "It is a new Asia with which we are dealing," he wrote in his final piece. By tradition Americans should be "in complete sympathy with the revolutionary aspirations." But, with "communism rampant throughout the world," the United States had to act warily. If revolutionary forces were "subservient to Russia," then should not the United States intervene against them? Still,

> if we adopt the policy of intervention, a strong temptation is presented to those in power to picture all opposition forces as communistic in order to gain American aid, remain in control and incidentally save their own lives. The United States if it wishes to keep its friends in Asia cannot afford to be duped.[2]

This was an early formulation of the problem that harassed American policy for the next generation. "Bobbie is really growing up and thinking hard in this last article of his," James Landis wrote his father.[3]

He learned more about Asian nationalism when Congressman John F. Kennedy invited him and their sister Patricia to come along on a trip from Israel to Japan in early October. In Israel an old friend, Congressman Franklin D. Roosevelt, Jr., received all the attention. "I have finally solved the problem," Robert wrote home, "as to why the Jews did not accept Jesus Christ. F.D.R., Jr. is what they have been waiting for."[4] He added in his diary, "He has been very nice to us, though his statements sometimes irritate me." One Arab with whom they talked

> suggested U.S. join itself with the liberal elements of the countries of Asia & prepare for the nationalistic revolutions that are sweeping over the land. FDR, Jr. & rest of State Dept. that was present said this impossible to do as U.S. not that kind of nation.[5]

"He summed up his feeling to an Arab leader," Robert wrote his father, "by saying, 'Don't you people realize the important question is the struggle between U.S. and Russia. Why do you get so upset about these smaller things?' I almost got sick." FDR, Jr., he thought, had "simply completely missed the whole point of the nationalist revolution that is sweeping Asia." As for the State Department, Robert would get rid of the whole crowd "and replace these 'diplomats' with W. O. Douglas and James Landis. Then we might have a chance."[6]

In another week they were in India. Nehru had them to dinner but gazed, bored, at the ceiling as if he did not know why he was thus wasting his time. Only their pretty sister roused him from languid condescension. "He hardly spoke during dinner," Robert noted in his journal, "except *to Pat.*" Afterward John asked Nehru about Vietnam. It was, Nehru said, a classic example of colonialism fighting a doomed battle; "said twice 'poor French' in patronizing way." The west, he added, was "pouring money & arms down a bottomless hole." As for communism, it thrived on discontent. "Might beat communists in a war but will create a situation which will give birth to other communism by destroying all sources of human wealth." (Writing this in his journal, Robert penciled on the margin: "Same as *Taft!*") Communism, Nehru told the boys, was "something to die for. Must give the same aura to democracy. . . . We only have status quo to offer these people. Commies can offer a change." Robert appended a family note: "Dad would have a hayday [sic] here. Nehru's posit. on India is same as his on U.S."

On October 19 they were in Saigon, where the French were fighting Ho Chi Minh and his Viet Minh. "Cannot go outside city because
of guerrillas," Robert observed. "Could hear shooting as evening
wore on." General de Lattre de Tassigny assured them the French
army could not be beaten, adding that, if the west lost the Mekong
Delta, it would lose Asia. They met Bao Dai, the former emperor of
Annam whom the French had installed as token chief of state. John
Kennedy noted that he seemed "in [S. J.] Perelman's words — 'fried
in Crisco.'" Robert thought that, if a plebiscite were held, 70 percent
of the people throughout Vietnam would vote for Ho Chi Minh.

The French, Robert wrote his father, were "greatly hated. Our aid
had made us also "quite unpopular." He wondered how to persuade
the "masses that we are not back of the French in keeping control of
this country." Had the Voice of America made enough of the fact
that the United States had freed the Philippines? "Also do we stress
fact that we do not care what kind of Gov [sic] a country has as long
as that Govt [sic] is not controlled by a foreign nation i.e. Russia?"
He told his father: "Our mistake has been not to insist on definite
political reforms by the French toward the natives as prerequisites to
any aid. As it stands now we are becoming more & more involved in
the war to a point where we can't back out." He concluded: "It
doesn't seem to be a picture with a very bright future."[7]

II

The trip in a real sense introduced Robert Kennedy to his older
brother. Eight years had been a large gap. "The first time I remember meeting Bobby," Jack once said, "was when he was three and a
half, one summer on the Cape."[8] Soon John left home, for school,
for college, for war. Afterward Robert was away at college and law
school. Apart from family gatherings they had seen little of each
other. Seven weeks of arduous travel made them closer than ever
before.

They had flown 25,000 miles, "eastward, always eastward," John
Kennedy said in a radio speech after their return, over lands "in
which the fires of nationalism so long dormant have been kindled
and are now ablaze . . . an area of revolution." If there was one thing
the trip had taught, he continued, it was that "Communism cannot
be met effectively by merely the force of arms." The United States
appeared "too ready to buttress an inequitable status quo." Vietnam
produced a severe paragraph:

We have allied ourselves to the desperate effort of a French regime to
hang on to the remnants of empire. . . . To check the southern drive

of Communism makes sense but not only through reliance on the force of arms. The task is rather to build strong native non-Communist sentiment within these areas and rely on that as a spearhead of defense rather than upon the legions of General de Lattre, brilliant though he may be, and to do this apart from and in defiance of innately nationalistic aims spells foredoomed failure.

He was severe too on American diplomats. "One finds too many of our representatives toadying to the shorter aims of other Western nations, with no eagerness to understand the real hopes and desires of the people to whom they are accredited, too often aligning themselves too definitely with the 'haves' and regarding the action of the 'have-nots' as not merely the effort to cure injustice but as something sinister and subversive."[9] These sons of American capitalism displayed no interest in integrating the underdeveloped world into the American economic system. In their view, the more a nation did for its have-nots the better, so long as its foreign policy was not made in Moscow.

The trip, Robert Kennedy said later, made a great impression on his brother, "a very very major impression . . . these countries from the Mediterranean to the South China Sea all . . . searching for a future; what their relationship was going to be to the United States; what we were going to do in our relationship to them; the importance of the right kind of representation; the importance of associating ourselves with the people rather than just the governments, which might be transitional, transitory; the mistake of the war in Indochina; the mistake of the French policy; the failure of the United States to back the people."[10]

A month after John Kennedy's radio report, his father addressed the same subject on the same network but to very different effect. By contracting unnecessary alliances, the ambassador warned, "we have delegated to others the power to determine our own fate. . . . Perhaps our next effort will be to ally to ourselves the Eskimos of the North Pole and the Penguins of the Antarctic." The great need was, not to assume new burdens, but "to disentangle ourselves from the far-flung commitments . . . recently made." Nehru was right to question American policy in the Cold War: "were I of his race, I could easily join in his hesitation."[11]

"I couldn't possibly have a worse argument with anyone about foreign policy than I have had with my son," Joseph Kennedy said the next year.[12] At the end of the decade John Kennedy said, "We don't even discuss [foreign policy] any more because we're just so far apart, there's just no point in it. I've given up arguing with him."[13]

III

At the end of 1951, Robert Kennedy began work for the Department of Justice. The Internal Security Division was engaged in investigations of Soviet agents, real and fancied. After a short time Kennedy transferred to the Criminal Division, where his first important assignment was to help present before a Brooklyn grand jury a case against two former Truman officials charged with corruption. The work absorbed him — so much so that when his older brother decided to run for the Senate in Massachusetts, Robert, Kenneth O'Donnell later said, was really "out of touch with Jack and unaware of Jack's problems."[14]

He was sufficiently aware to ask O'Donnell to join his brother's campaign. Challenging the popular Republican senator Henry Cabot Lodge, Jr., was hardly a project without risk. As in 1946 the ambassador moved to take personal charge. But the generational problems had intensified. Joseph Kennedy, O'Donnell complained, associated with "elder statesmen who knew nothing about the politics in this day and age, and he assumed, despite the fact that he had not been in Massachusetts for twenty years, that it was the same thing . . . and he's such a strong personality that nobody could — nobody *dared* — fight back."[15] The campaign, O'Donnell thought, was headed toward "absolute catastrophic disaster."[16] He pleaded with Robert Kennedy in New York to come up and take over. Robert was "very angry," O'Donnell remembered. He loved what he was doing in the Department of Justice. He did not look forward to a summer arguing with his father. In any case, he knew nothing about Massachusetts politics: "I'll screw it up and . . . I just don't want to come." O'Donnell said, "Unless you come, I don't think it's going to get done."[17]

He came and took over. "Bobby could handle the father," said O'Donnell, "and no one else could have."[18] The Kennedys spent money freely, setting up committees in a multitude of ethnic and professional categories. "We couldn't win relying on the Democratic political machine," Robert said later, "so we had to build our own machine."[19] Polly Fitzgerald, a cousin, and Helen Keyes, daughter of the Kennedy dentist in Brookline days, organized the famous teas for the candidate's mother and sisters. Franklin D. Roosevelt, Jr., campaigned in Jewish areas where John Kennedy, because of his father's reputation, was supposed to be weak. Robert worked eighteen hours a day, losing a dozen pounds in the process. "I didn't become involved in what words should go in a speech, what should be said on a poster or bill-board, what should be done in television. . . . I was so busy with my part of it that I didn't see any of that."[20]

His part of it — organization — was in its nature an affront to the old-line politics of the state; and Robert himself, twenty-six years old, cocky, unknown except as a millionaire's son, pressing, always pressing, was an affront to old-line politicians. Nor did he do much to assuage wounded feelings. Perhaps, in addition, he felt that a tough exterior was the only means by which a novice could hope to keep the campaign under control. Perhaps, too, he was proving things to his father, and to himself.

The "ruthless Bobby" idea seems to have been born in 1952. One day when a leading labor figure sat chatting in the headquarters, as Massachusetts politicians were wont to do, Robert brusquely ushered him out: "If you're not going to work, don't hang around here."[21] A local pol, celebrated for singing "Danny Boy" from a sound truck, was incredulous to discover that Robert Kennedy had never heard of him. "And you call this a political headquarters?" he said with a burst of profanity, thereby inciting Kennedy, who disapproved of swearing in the presence of women, to throw him out.[22] "Politicians do nothing but hold meetings," he said later, reflecting on the 1952 campaign. ". . . You can't get any work out of a politician." The interviewer noted that he spat out the word "politician" as if it were something "unclean and unwanted."[23]

As Paul Dever, the veteran Democratic governor running for re-election, saw himself slipping behind, he proposed to John Kennedy that they merge organizations and run in tandem. The ambassador thought a joint campaign a fine idea. The candidate disagreed. "Don't give in to them," John told Robert, "but don't get me involved in it."[24] The younger Kennedy broke the bad news to Dever in a stormy meeting. "He got furious at me and almost broke off any kind of relationship with us, except it was patched up really by my father, who [sic] he liked. It was rather a big debacle." Dever sent word to keep that "young kid" out of his office, saying that hereafter he would deal only with the ambassador. "In any case," Robert reflected afterward, "we ran our own campaign, which is all that I wanted."[25] The ambassador accepted it philosophically. "He had confidence in any Kennedy, whether they're right, wrong or indifferent," said O'Donnell. "They really couldn't do anything wrong."[26]

The elder Kennedy was meanwhile cheered by national developments. Eisenhower, he concluded (prematurely) in September, was "the most complete dud we have had for a candidate in my fifty years of presidential elections." As for the Democratic nominee, Adlai Stevenson of Illinois, "he has made a sensational start," Kennedy wrote Beaverbrook. ". . . the hottest thing we have in America on television, [in his speeches] he runs Churchill an even race." Unless the newspapers succeeded in sticking a leftist tag on him, "I would

think that he would murder Eisenhower — and I don't think it would
even be close. . . . Unquestionably he would be a great man for the
United States." Best of all: "He is a great friend of mine and of the
family."*

Eisenhower carried Massachusetts by 200,000 votes. Dever lost by
15,000. Kennedy won by 70,000. It was a triumph for Robert as
well. But it cannot be said that he enjoyed politics, Massachusetts
style. "I never heard Bob Kennedy ask anyone 'How's the family?'"
said Lawrence O'Brien, a master organizer from Springfield who
played a key role in the victory. "The entire hand-shaking, small-
talking side of politics was repugnant to him; he often said to me,
'Larry, I don't know how you stand it.'"[27] Robert had a job to do
for his brother, and he meant to do it. He did not care whether or
not people liked him along the way.

IV

His brother was considerably impressed by the job Robert had
done. "I don't think [Jack] was aware," Lem Billings said, "that Bobby
had all this tremendous ability." John understood, too, the supreme
importance in politics of a partner on whom he could count absolutely, for loyalty and for results. Their intimacy grew in the senatorial campaign. It was still, for a time, limited to politics. Along
with the difference in age there remained differences in temperament and outlook. "All this business about Jack and Bobby being
blood brothers has been exaggerated," their sister Eunice once said.
". . . They had different tastes in men, different tastes in women.
They didn't really become close until 1952, and it was politics that
brought them together."[28] "Bobby and Jack," said Lem Billings, who
loved them both, "had very different personalities and really different interests outside of politics. . . . Their closeness was really based
on their political work."[29]

John Kennedy was the more secure, the freer, of the two — freer
of his father, of his family, of his faith, of the entire Irish American
predicament. "Jack tended to be a little bit fey, a little offhand
around his father," Stephen Smith, whom Jean married in 1956, later
said.[30] Robert Kennedy, on the other hand, accepted the tradition.
Seeking the paternal approval for which he had longed as a small

---

* JPK to Beaverbrook, September 6, 1952, Beaverbrook Papers. As a member of
Governor Stevenson's campaign staff that autumn, I can attest to Mr. Kennedy's eager
interest in Stevenson. He knew me as a friend of his older son's and called me from
time to time in Springfield to pass on his ideas about the tactics and strategy of the
campaign.

boy, he somberly took on the obligations from which his older brother had casually disengaged himself. His father appreciated that. "It was Bobby with whom he would discuss family matters," said O'Donnell.[31] "Bobby's a tough one," Joseph Kennedy himself told Bob Considine, the reporter, in 1957; "he'll keep the Kennedys together, you can bet."[32]

The two brothers had moved in different directions in adolescence and young manhood, John establishing his intellectuality, Robert his toughness, John his independence, Robert his commitment to the family. John had got along better with his father, Robert with his mother; and the result was not unexpectedly paradoxical. "Bobby was more like his father," said William O. Douglas, who saw both brothers a good deal in the 1950s, "and Jack was more like his mother. . . . Bobby was more direct, dynamic, energizing; Jack was more thoughtful, more scholarly, more reflective."[33] They emerged as contrasting figures. "Jack Kennedy," said Paul Dever after the campaign, "is the first Irish Brahmin. Bobby is the last Irish Puritan."[34] But intimacy had begun and would increase.

The head of the family was equally impressed by Robert's work in the campaign. It was, Lem Billings later said, as if "the father suddenly found he had another able son, which I don't think he realized."[35] In 1948 he had told Beaverbrook that Robert was making a good battle against his handicaps. In 1952 the battle was largely won. Robert was proving himself at last. He persuaded his father that he was truly his father's son.

His father was supposed to have said in later years that Robert was more like him than any of the other children because "he hates like me." In 1960 he denied to John Seigenthaler of the *Nashville Tennessean* that he had ever said this. But he did call Robert "the most determined person" he had ever known.[36] To another reporter he said proudly, "Bobby's as hard as nails."[37] Chatting with Charles Spalding, he remarked of John Kennedy, then well on his way to the Presidency, "I don't understand it. He's not like me at all. I never could have done it. But Bobby is like me."[38]

In part it was an imposture. The Irish were always good role players. The gentle self was never extinguished, and the tension between Robert Kennedy's soul and his *persona* took the form of dialectic rather than war. Friendships — Hackett, O'Donnell, Billings — nourished the self; marriage nourished it even more. The inner tension, instead of stultifying, eventually enlarged and enhanced sensibility and character. It did not destroy the ultimate unity of personality. Arthur Krock described Robert Kennedy at fifteen: "bullied by his elder brothers but still managing to keep a kind of savage individuality."[39] That individuality became confused for a season, out of admiration and adoration of his father. It was never crushed.

It took long years, a family of his own, experience with life, experience with death, to reintegrate the elements of personality loosened by the shock, in so many respects the salutary shock, of his upbringing. But it happened — in spite of his father, because of his father. Robert Kennedy was indeed a most determined person. His father made him what he appeared to be. Yet Jacqueline Kennedy, who adored both Robert and his father, once said that, of the brothers, he was "least like his father."[40]

(6)

The First Investigating Committee:
Joe McCarthy

VICTORY brought euphoria. Robert and Ethel, with Kenneth
O'Donnell and Lawrence O'Brien and their wives, joined the
new senator in Hyannis Port for the family celebration. John Kennedy was the only Democrat running statewide in Massachusetts to
have survived the Eisenhower sweep. His brother had managed the
campaign. The younger generation was feeling proud of itself. But
Joseph Kennedy, as usual, represented the reality principle. Late
one November afternoon he said to Robert, "What are you going to
do now?" Robert said, "What do you mean?" "You haven't been
elected to anything," his father replied. "Are you going to sit on
your tail end and do nothing now for the rest of your life? You'd
better go out and get a *job*." [1]

I

Early in December 1952 Joseph Kennedy telephoned to Senator
Joseph McCarthy, Republican, of Wisconsin, whom Eisenhower's victory had made chairman of the Permanent Subcommittee on Investigations of the Senate Government Operations Committee. Joe
McCarthy was a recent Kennedy acquaintance. He claimed to have
met John Kennedy fleetingly in the Solomon Islands in 1943;[2] and
in the late 1940s, after he slipped into the Washington scene as
senator in 1947, he had come to know other members of the family. He took Pat and Jean out amiably in the evening and spoke for
Robert's forum in Charlottesville. Eunice liked him, as did Sargent
Shriver, who was the ambassador's staff assistant and whom Eunice

married in May 1953. Joseph Kennedy liked him most of all and invited him from time to time to Hyannis Port.

He was a black Irishman, lowbrow and roughneck, with a certain animal vitality, coarse charm, broad humor and amusing impudence. Something about him, perhaps this instinctive insolence toward the establishment, may have reminded Joseph Kennedy agreeably of his own youth. In the 1940s he was a man of marked personality without known political principles. Writing in 1947 about "The Senate's Remarkable Upstart," in the *Saturday Evening Post*, Jack Alexander noted that McCarthy was studying Russian. "His friends insist that he has hopes ultimately of . . . charming Stalin in his own language."[3] In the late fall of 1949, suddenly discovering the threat of communism, he forswore this earlier ambition.[4] The crusade against Communists, which he launched inadvertently at Wheeling, West Virginia, in February 1950, quickly became the vehicle for his free-floating destructiveness. The Korean War began four months after Wheeling. Growing casualty lists gave him an ever more attentive audience. If Reds were killing American boys in Korea, why should they be free to walk the streets of America? McCarthy explored the frustrations of limited war with cunning and skill. A natural demagogue, he had rare ability to implant suspicions, to confuse issues and, when challenged, to change subjects. He had a bully's tactile instinct for weakness or fear. No politician of the age, wrote Richard Rovere, had "surer, swifter access to the dark places of the American mind."[5]

All this was amply evident when Joseph Kennedy called him in December 1952. None of it worried the ambassador, even though on occasion he himself criticized the practice of smearing all leftists as Communists. "There is a strong liberal feeling throughout Europe that is a valuable asset in the fight against communism," he said in an interview in 1951, fifteen months after McCarthy had begun his jihad. "By terming this as a communistic movement we are only convincing the people over there that we are driving them into a war."[6] Nevertheless he was fond of McCarthy and now asked him to appoint Robert Kennedy chief counsel of the Permanent Subcommittee on Investigations.[7]

His older son was not enthusiastic. McCarthy's vulgarity was hardly to John Kennedy's Brahmin taste. He understood perfectly that the adoration McCarthy aroused in one section of his constituents was matched by the loathing of another section, less numerous but more influential and one whose approval meant a good deal to him. He told his bright new senatorial assistant Theodore Sorensen that he hoped Bobby would not take the job. "His reasons," Sorensen wrote later, "were political, not ideological."[8]

Joseph Kennedy's request was embarrassing for McCarthy too. He had already decided to name as chief counsel a twenty-five-year-old New York lawyer, notable for anti-Communist zeal, named Roy Cohn. According to Cohn, McCarthy feared that Robert Kennedy's appointment might be taken as a payoff for his father's help in McCarthy's campaigns. One day Cohn found McCarthy sitting irritably in his office listening to Joseph Kennedy's voice crackling over the telephone. After a time McCarthy laid the receiver on his desk, the ambassador's voice still sounding away, scribbled a note and passed it to Cohn: "Remind me to check the size of his campaign contribution. I'm not sure it's worth it."[9]

He named young Kennedy an assistant counsel. McCarthy and Cohn set noisily to work to unearth subversives in the State Department and the Voice of America. Kennedy, assisted by LaVern Duffy, a young investigator, undertook a more prosaic inquiry into the trade carried on by American allies with Communist China. Duffy later recalled his first sight of Robert Kennedy — sleeves rolled past the elbows, collar open, necktie pulled loose, and, "most remarkable" of all, his white, woolen athletic sweat socks. These last, Duffy said, he invariably wore until Ethel or someone persuaded him that they did not go with business suits.[10]

Kennedy and Duffy pored over the Lloyds of London shipping index, British parliamentary debates and reports from the Maritime Commission and the Central Intelligence Agency. In a couple of months they were able to demonstrate that, since the outbreak of the Korean War, 75 percent of all ships carrying goods to mainland China had sailed under western flags — most of this while Chinese troops were shooting at Americans in Korea. Kennedy also discovered that some of the shippers were simultaneously collecting large sums from the United States government for delivering defense materials to western Europe. The British role in the China trade appeared particularly reprehensible. "Great Britain's rubber shipments went up sixty per cent last year over the previous year," Kennedy told one newspaperman. "Natural rubber is needed for the tires of jet planes." He said, typically, that he could not "appreciate" the English position.[11] Joe McCarthy appreciated the English position even less. One day he wrathfully showed the Senate a copy of Beaverbrook's *Sunday Express* with the headline BIG BOOST IN OUR TRADE TO CHINA. "We should perhaps keep in mind," said McCarthy,

the American boys and the few British boys, too, who had their hands wired behind their backs and their faces shot off with machine guns — Communist machine guns, Mr. President, if you please — supplied by those flag vessels of our allies. . . . Let us sink every accursed ship carrying materials to the enemy and resulting in the death of American boys, regardless of what flag those ships may fly. . . . We can go it alone.

He finished with compliments to "Comrade Attlee," the British Prime Minister, and offered the Senate a photograph of Attlee reviewing the International Brigade in Madrid, his hand raised, McCarthy asserted, in the Communist salute.[12]

Kennedy's statistics also showed a large Greek role in the China trade. McCarthy and the staff opened negotiations with Greek ship owners. As a result the owners of 327 vessels promised to stop carrying to Communist China. This infuriated the State Department, which in more circumspect fashion had been trying to persuade allied carriers to abandon the China trade and regarded McCarthy's action as a flagrant violation of the separation of powers. Harold Stassen, the director for Mutual Security, told McCarthy that his intervention "undermines the effort we are carrying on."[13] McCarthy was now the infuriated party. Fearing a brawl, Vice President Richard M. Nixon arranged a summit meeting between McCarthy and John Foster Dulles, the Secretary of State. Dulles cut the ground out from under Stassen, endorsing McCarthy's action as "in the national interest."

II

Robert Kennedy was at McCarthy's side when he announced the Greek arrangement. The *Boston Post* reported him "alarmed and shocked at the manner in which the McCarthy haters hopped all over the deal. . . . 'I don't know,' said Kennedy. 'I really don't. I had supposed that it just didn't make sense to anybody in this country that our major allies, whom we're aiding financially, should trade with the communists who are killing GIs.'" The *Post* added, "Kennedy, contrary to general belief, apparently, has not taken part in any of the McCarthy committee's probes into subversives in the State Department, the Voice of America, and other government agencies. What he has done is work on ways to shut off strategic materials of war to countries in the Russian zone, including North Korea. He'll continue to do so."[14] His work pleased the committee. On April 2 his salary was increased from $4952.20 to $5334.57. "This raise," Flanagan informed him, "is given in recognition of the very excellent job you have been doing in connection with our investigation into East-West trade. You are to be commended on the energy, judgment and imagination which you have exercised in the conduct of this inquiry."[15]

In May, Robert Kennedy, with Ethel and a sister or two looking on, laid out the damning statistics in a public hearing. It was his debut. LaVern Duffy noticed his trembling hands and general nervousness.[16] The issue was not perhaps so simple as the assistant counsel supposed. John Leddy, the acting deputy assistant secretary of state for economic affairs, set forth the complexities. England had interests in Hong Kong and Malaya that would be damaged by a cutoff of trade with mainland China. Japan had been shipping seaweed to China and receiving iron ore in return. "It would be very hard to come to a conclusion that this exchange aids Red China." American efforts to force allies to stop trading with China might use up diplomatic bargaining power needed for more vital purposes. In any case, Leddy said, the effects of blacklisting allied ships engaged in the China trade would be "negligible." There would be "some slight reshuffling of ships' availability, but the overall effect on the 8,000 or 10,000 tramp streamers available in the world would be barely visible." In answer to a direct question, Leddy forthrightly told McCarthy, "We are not seeking at this time the discontinuation of all trade . . . with Red China."[17]

McCarthy shut off Leddy in a pretense of anger and stalked from the room. "No diplomatic subtleties can excuse this condition to the American people," said the *New York Journal-American* irately. "We may be thankful to the McCarthy committee and to its hard-digging assistant counsel Kennedy (brother of Sen. John F. Kennedy, D.-Mass.) for bringing it to light."[18] A few days later McCarthy harangued the Senate. He expressed his appreciation to the ranking Democratic member of the committee "for consenting to lend me, for the day, the very able minority counsel, Mr. Kennedy . . . so that he can furnish me facts and figures." With Kennedy sitting at his desk, McCarthy restated his thesis:

> Of 81 million tons of shipping available to the entire world, non-Communist nations control 78 million tons. If we deny that shipping to the Communists, it will be a death blow to their war-making power. . . . We can do it by simply telling our allies that they will not get one American dollar while any of their flag ships are plying this immoral, dishonest, indecent trade.[19]

The next day, when testimony made it clear that the Defense Department tended to agree with McCarthy, Senator Stuart Symington of Missouri, a Democratic member of the committee, proposed that the chairman ask Eisenhower what the American policy was.[20] Robert Kennedy drafted the letter. It inquired whether State or Defense was speaking for the President, whether the White House thought it a breach of security to disclose the volume of strategic materials shipped to China and whether the government would refuse contracts to allied shipping firms trading with Communist China.

McCarthy had retired for a moment to the Bethesda Naval Hospital, where the letter went for his signature. Robert Kennedy then carried it to the White House. The document was respectful in tone, and Congress had every claim to the information sought. But the request exasperated a White House staff already sorely tried by McCarthy. As so often, the executive branch began talking about national security and the need to protect secret information. This was reaching rather far. The information was no secret from the Chinese Communists, who were after all the recipients of the strategic materials. Still it was deemed important to keep it from the American Congress. Sherman Adams, Eisenhower's chief of staff, noted that, if Eisenhower pronounced on allied trade with China, "he could avoid neither antagonizing the British nor stirring up criticism from anti-Communist groups at home. . . . Like most of the calamities built up by McCarthy, whatever was done about it seemed likely only to worsen the situation."[21] Nixon once again got them out of it. He persuaded McCarthy that he had fallen into a Democratic trap. McCarthy thereupon withdrew the letter. When a newspaperman asked Kennedy about his White House visit, "he appeared flustered and asked, 'Did somebody see me go in there?'"[22]

In the meantime, Kennedy worked on an interim report, filed with the Senate on July 1. The report rested on two unexamined premises: that the United States could compel its allies to conduct their trade in ways contrary to their sense of their own interests; and that total cessation of trade with mainland China would make a substantial difference to the Korean War. The animating tone was moral outrage. "This shocking policy of fighting the enemy on the one hand and trading with him on the other," the report said, "cannot be condoned." Even if official action had little practical effect, the subcommittee believed "that on moral grounds these shipping firms should be made to choose."[23]

Within the limits of its assumptions it was an able job, its facts well marshaled, its argument well organized, its tone cool. At no point did the report hint, in the McCarthy manner, that traitors were making American policy. Arthur Krock praised it as "congressional investigation at its highest level with documentation given for each statement represented as a fact and with conclusions and opinions expressed dispassionately."[24] The tart liberal columnist Doris Fleeson called it "that rara avis" in McCarthy's stormy history — "a documented and sober story. It was thoroughly prepared by Robert Kennedy. . . . It received much more credence, therefore, than normal for anything to which Senator McCarthy's name is attached."[25] Kennedy's salary was raised to $7342.[26]

This did not offset his growing dissatisfaction with the committee. While he had been tediously analyzing commercial statistics, McCarthy and Roy Cohn had been loudly hunting Communists in the American government. McCarthy's interest in foreign policy was at best fitful. Though he paused now and again to bellow about the "blood trade" with China, he had juicier fish to fry at home. Kennedy's shipping report differed from the characteristic McCarthy investigation in two particulars: no one questioned Kennedy's facts, and Kennedy did not question the loyalty of those under investigation. Roy Cohn's investigatory style was directly opposite: nearly everyone questioned his facts, and he questioned nearly everyone's loyalty. McCarthy had been right in making him chief counsel, since this was McCarthy's style too.

The twenty-seven-year-old Kennedy was no doubt irritated from the start by the fact that he had been passed over in favor of a man eighteen months younger. As Cohn proceeded to cater to McCarthy's hit-and-run inquisitorial instincts, personal dislike was compounded by genuine concern. Cohn attached to the committee staff as "chief consultant" a young man named G. David Schine who had qualified himself as an expert by writing a hair-brained tract called "Definition of Communism," copies of which had been placed next to the Gideon Bible in hotels owned by Schine's father. After pursuing subversives in the Voice of America in February and March 1953, Cohn and Schine left for a magical mystery tour of western Europe, where they careened through embassy libraries, plucked offending books from the shelves and terrorized hapless employees.

Kennedy considered it all madness. He told McCarthy that he "disagreed with the way that the Committee was being run," and that "the way they were proceeding I thought it was headed for disaster. . . . I told him I thought he was out of his mind and was going to destroy himself." Later he commented, "Cohn and Schine took him up the mountain and showed him all those wonderful things. He destroyed himself for that — for publicity. He had to get his name in the paper. . . . He was on a toboggan. It was so exciting and exhilarating as he went downhill that it didn't matter to him if he hit a tree at the bottom." As he summed up the McCarthy committee after he had run a major investigation of his own:

> With two exceptions no real research was ever done. Most of the investigations were instituted on the basis of some preconceived notion by the chief counsel or his staff members and not on the basis of any

information that had been developed. Cohn and Schine claimed they knew from the outset what was wrong; and they were not going to allow the facts to interfere. Therefore no real spade work that might have destroyed some of their pet theories was ever undertaken. I thought Senator McCarthy made a mistake in allowing the Committee to operate in such a fashion, told him so and resigned.[27]

Thus ended, after six months, Robert Kennedy's partnership with Joe McCarthy. By May 1953 he had sufficiently distanced himself so that McCarthy introduced him to the Senate as the minority counsel; and in July, after he completed the only solid report ever issued by the committee, he got out. In later years he bore criticism over the McCarthy relationship stoically. He rarely tried to explain, partly because this was not his way, but also because he had initially shared McCarthy's concern. A decade later the writer Peter Maas said to him, "I just cannot understand how you could ever have had anything to do with Joe McCarthy." Kennedy replied, "Well, at that time, I thought there was a serious internal security threat to the United States; I felt at that time that Joe McCarthy seemed to be the only one who was doing anything about it." After a moment, he said, "I was wrong."[28]

Most of all, he retained a fondness for McCarthy. McCarthy was an affable fellow who put on a tremendous show of beastliness and somehow did not expect his victims to take it personally. "He was no kind of fanatic," wrote Richard Rovere in the most perceptive book about him; he was "as incapable of true rancor, spite, and animosity as a eunuch is of marriage. He just did not have the equipment for it. He faked it all and could not understand anyone who didn't."[29] The record is filled with anecdotes of McCarthy's astonishment when people he vilified in the hearings cut him in the corridors. "He was a very complicated character," Robert Kennedy said a few years later. "His whole method of operation was complicated because he would get a guilty feeling and get hurt after he had blasted somebody. He wanted so desperately to be liked.... He was sensitive and yet insensitive. He didn't anticipate the results of what he was doing. He was very thoughtful of his friends, and yet he could be so cruel to others."[30]

Kenneth O'Donnell, who described himself as "one of the few Irish Democrats that were violently anti-McCarthy in the state of Massachusetts," used to call Robert Kennedy and fulminate about McCarthy. "There'd be silence.... After a while he'd say, 'He isn't that bad, but I agree, you know — but ...' And he kept defending his pasha. Bobby had the greatest capacity for the underdog."[31] He spent much time defending his pasha. Once, at a Kennedy dinner, someone asked William McCormick Blair, Jr., Adlai Stevenson's personal aide in 1952 and now his law partner, his view of McCarthy. Blair said exactly what he thought. Robert was loud in disagreement. "Mr. Kennedy, Sr., of all people," Blair later recalled, "had to come over and tell us to quiet down." [32]

Given this personal feeling, resignation from McCarthy's staff could not have been easy. There is no record of his father's reaction. When Robert Lovett, the former Secretary of Defense, had observed to Robert's father, "That's pretty tough company he's travelling with," the ambassador replied, "Put your mind to rest about that. Bobby is just as tough as a bootheel." [33] One assumes that he respected the toughness of his son's decision. For Robert the experience was another step in education. He was learning in particular that patriotic declarations did not make due process of law superfluous and that he owed a debt to his own inner standards.

"With the filing in the Senate of the Subcommittee report on Trade With the Soviet Bloc," Kennedy wrote McCarthy formally, "the task to which I have devoted my time since coming with the Subcommittee has been completed. I am submitting my resignation at this time as it is my intention to enter the private practice of law." The liberal Democrats on the committee, Stuart Symington and Henry Jackson, sent glowing letters:

> Your investigation into the details of shipping to communist China during the war in Korea, along with your subsequent testimony before the Committee and your work on the final report, in my opinion, is one of the two outstanding jobs I have seen done in committee work during the some eight years I have been in Washington. (Symington)

> I was particularly impressed with the thorough, impartial and fair way in which you handled all matters coming to your attention. I noted that in all your investigations you were most diligent in adhering to the facts in order to obtain the best judicial determination in each instance. (Jackson) [34]

IV

One finds no indication that Robert Kennedy seriously intended to enter private practice. His bent more than ever was toward public service. His father, who had been a member of the Commission on Reorganization of the Executive Branch, headed by Herbert Hoover in 1947–49, was now a member of a second Hoover commission appointed by Eisenhower. In August 1953 young Kennedy joined the staff as his father's assistant.

In the twenty years since he had opposed Hoover's reelection, Joseph Kennedy had come to have inordinate admiration for the old man.[35] Robert Kennedy, in principle, shared that admiration. Hoover, who had his eightieth birthday the month young Kennedy came aboard, presided over the commission in a fairly arbitrary way. Thus when Clarence Brown, a conservative Republican congressman from Ohio, proposed a task force on military procurement, Hoover flatly turned the proposal down. Brown took Robert Kennedy aside to complain.

"He says," Robert reported to his father, "that he told Hoover he was dissatisfied, that the people who had opposed this Commission originally, Arthur Flemming and [Nelson] Rockefeller, are now Hoover's big pals, while he is shunted aside, but that he did not intend to be shunted aside any more. . . . He told Hoover that all Hoover thought of him was an irascible old man with no judgment or sense, but he would show him that he had been around a long time too and knew what he wanted." Hoover had replied by telling Brown that, if these disputes continued, he might resign. "Brown said that in view of the circumstances and the way he had been treated, he was not at all sure that it was a bad idea."[36] "My impression of the whole thing," Robert wrote after three months, "was that Hoover was very impatient with those who disagreed with him, and felt that setting up these task forces with eighty-five men was a major accomplishment and that he shouldn't be criticized on some of these smaller matters."[37]

Robert was soon exasperated by the squabbling among the septuagenarians. His own job seemed inconsequential. Probably it was difficult (or so Lem Billings thought) "to be working under the old man who still considered him a child." Most of all, the uncertainty of his own future oppressed him. "He had decided to dedicate himself to the government," Billings said later, ". . . and felt he was accomplishing nothing and getting nowhere. He just didn't see any future."[38]

This was a low time in his life — a time, as Billings said, of "bad doldrums." Lawrence O'Brien recalled Kennedy touch football one afternoon in a Georgetown park. Georgetown University students playing baseball nearby kept hitting flies into the middle of the game. Ted Kennedy shouted to them to stop. They persisted. Finally Ted squared off for a fight with one of the larger among the enemy. Robert, though he gave away about thirty pounds, pushed his younger brother aside. "It was an uneven match," said O'Brien, "which was no doubt what Bob wanted." No one tried to stop it. "It became a bloody brawl, with each man determined to score a knockout. . . . Finally the fight stopped when they simply couldn't raise their arms any longer." That night they all had dinner at John Kennedy's house. Ethel, looking at her husband's bruised face, said,

"That must have been a rough game of touch." "Yeah," Robert mumbled and went on eating.[39]

When they went out, evenings often ended in embarrassing argument. Robert was a "really very cross, unhappy, angry young man," Billings remembered, always telling people off and getting into fights.[40] My own first brush with him came in these months. In February 1954 he sent a letter to the *New York Times* condemning the Yalta agreement.* I dispatched a testy reply, describing his letter as "an astonishing mixture of distortion and error."[41] The exchange amused John Kennedy who told me, "My sisters are very mad at you because of the letter you wrote about Bobby." Bobby was mad too and dashed off a second letter to the *Times*, concluding that he did not "wish to appear critical of Mr. Schlesinger's scholarship for his polemics cover such a wide variety of subjects that he is not always able to read all of the documents he so vigorously discusses." This letter, he wrote me crisply, ought to "clarify the record sufficiently for you to make the necessary public apology."[42] I replied in like spirit.

Many people formed their first impression of him in this period. He was, thought Theodore Sorensen, "militant, aggressive, intolerant, opinionated, somewhat shallow in his convictions ... more like his father than his brother."[43] The view was widely shared.

v

In July 1953 the three Democratic members of the Investigations subcommittee — Symington, Jackson and John L. McClellan of Arkansas — had walked out when the Republican majority voted to give McCarthy sole control over the hiring and firing of committee staff. Then in January 1954, having forced McCarthy to relinquish control over appointments, the Democrats ended their boycott. As part of the settlement, they received the authority to appoint a minority counsel. The job was promptly offered to Kennedy, who as promptly accepted. In a kindly response to his letter of resignation from the Hoover commission, Herbert Hoover said he realized that "a restless soul like you wants to work."[44] In February, McCarthy requested an FBI investigation of Kennedy, but none was made since Kennedy had already been appointed.[45]

McCarthy made these concessions only because his budget was

---

* *New York Times*, February 3, 1954. He was commenting on a *Times* editorial of January 26 attacking the Bricker Amendment. Without endorsing the amendment, Kennedy argued that "the United States Senate, if this matter had been put up to them, would have done a more thorough job" than Roosevelt, with Averell Harriman and Harry Hopkins as his "sole advisers," had done at Yalta.

coming before the Senate. A few days later J. William Fulbright of Arkansas was the single senator to vote against his hectoring demand for a larger appropriation than ever. The Wisconsin senator seemed at his zenith. This was illusory. The Korean War had made McCarthyism possible. But that war was coming to an end. The fears and frustrations it had generated were subsiding. Not only was McCarthyism losing its emotional base, but McCarthy himself, in his increasingly erratic course, had become embroiled with a formidable antagonist, the United States Army. In the fall of 1953 he had launched a sensational, if unproductive, search for Communists and spies in a Signal Corps installation at Fort Monmouth, New Jersey. In the meantime, G. David Schine had been threatened by the draft; and Roy Cohn showed quite singular zeal, first in demanding an Army commission or a soft job in the CIA for his sidekick, and then, after Schine was drafted, in getting him treatment and privileges never before accorded buck privates. An Army document later recorded forty-four specific instances of improper pressure. A Republican member of the committee, Senator Charles E. Potter of Michigan, observed of Cohn, "His campaign over David Schine bordered on lunacy."[46]

Enraged by the Army's callousness, Cohn urged McCarthy on to battle. McCarthy, who lived by headlines, needed no encouragement. "The key," he soon announced, "to the deliberate Communist infiltration of our armed forces" lay in the mysterious promotion of a fellow-traveling dentist named Irving Peress from captain to major.[47] The "Who Promoted Peress" inquisition absorbed him for several pleasurable weeks. In a high point he told one much decorated general he was not fit to wear his uniform. The Army, at last tormented beyond endurance, retaliated in March 1954 by releasing a chronological record of Cohn's interventions on behalf of Schine. Feelings were running high within the committee. "I think I shall enjoy my new job," Kennedy wrote Herbert Hoover. "However every night when I come home from work I feel my neck to see if my head is still attached."[48]

One day McCarthy summoned a hapless black woman named Annie Lee Moss, a teletype operator in the Signal Corps, as a further example of the way the Army coddled Communists. Mrs. Moss timidly but firmly denied Communist connections. Her demeanor impressed the Democrats on the committee. McCarthy, seeing things were going badly, excused himself, leaving the questioning to Cohn.[49] Cohn proceeded to harry Mrs. Moss about her statement that a black man named Robert Hall had left a copy of the *Daily Worker* in her house, insinuating that Hall was a Communist organizer. A newspaperman informed Robert Kennedy that the Communist Hall was white. Kennedy tried to pass on the point to Cohn. Cohn ignored him and persisted in harsh interrogation. A sharp exchange ensued:

SENATOR JACKSON. Mr. Kennedy has something. . . .

MR. KENNEDY. Was Mr. Hall a colored gentleman, or —

MRS. MOSS. Yes, sir.

MR. KENNEDY. There is some confusion about it, is there not, Mr. Cohn? Is the Rob Hall we are talking about, the union organizer, was he a white man or colored man?

MR. COHN. I never inquired into his race. I am not sure. We can check that, though.

MR. KENNEDY. I thought I just spoke to you about it.

MR. COHN. My assumption has been that he is a white man, but we can check that.

SENATOR SYMINGTON. Let us ask this: The Bob Hall that you knew, was he a white man?

MRS. MOSS. He was colored, the one I knew of.

SENATOR SYMINGTON. Let's decide which Robert Hall we want to talk about.

MR. KENNEDY. When you spoke about the union organizer, you spoke about Rob Hall and I think we all felt that was the colored gentleman?

MR. COHN. I was not talking about a union organizer, Bob. I was talking about a Communist organizer who at that time, according to the public record, was in charge of subscriptions for the Daily Worker in the District of Columbia area.

MR. KENNEDY. Evidently it is a different Rob Hall.

MR. COHN. I don't know that it is. Our information is that it was the same Rob Hall.

SENATOR MCCLELLAN. If one is black and the other is white, there is a difference.

Cohn, retreating, said that perhaps this was something the committee should "get some more exact information on." "I think so too," said Kennedy.[50]

Symington asked Mrs. Moss whether she had heard of Karl Marx. "Who is that?" she replied. At the end of her testimony, he told her he thought she was telling the truth ("I certainly am," Mrs. Moss said), and added that if the Signal Corps did not take her back, he would get her a job himself.[51] Later Cohn told Symington that he was making a mistake; "our information is totally accurate." Symington said he would check with the Federal Bureau of Investigation. He then sent Kennedy to see whether the Bureau had anything. According to Cohn, not an unbiased source, Kennedy "stormed in and demanded" the Moss dossier. The assistant to the director, Louis Nichols, told him that no outsider could see FBI files. Kennedy then asked for a decision from J. Edgar Hoover personally. "The attitude of Kennedy in this matter," Hoover noted, "clearly shows need for absolute circumspection in any conversation with him."[52] Roy Cohn later wrote, "Kennedy was not one of the Director's favorites. He regarded him as an arrogant whipper-snapper, who pushed around his family's money and power." Hoover, on the other hand, was fond of McCarthy, with whom he used to dine, and fond enough of Cohn to offer the ultimate accolade in 1954 and insist that Cohn call him "Edgar."* Kennedy recalled the incident later as "a rather major dispute [with the FBI] because they lied to me about some documents that they made available to the Committee. . . . So I had a fight with them, and that was really my first."[53]

Early in April the once loyal *Boston Post* ran an article from its Washington correspondent saying that "no chairman of a communist-hunting Congressional committee has ever escaped smearing" and that it was now McCarthy's turn. A chief villain, it appeared, was the minority counsel, who "represents the subcommittee members who have been vigorously hostile to McCarthy." It was "no longer even an open secret that Robert F. Kennedy . . . has no use for Roy Cohn." But "McCarthy, of course, will stand by his chief counsel to the bitter end. . . . He is not expected to attempt to rebuke Kennedy even though the younger brother of the Massachusetts Senator has made comments to at least one person that were considered to be highly out of order."[54]

VI

The accumulation of charges and countercharges involving Mc-Carthy, Cohn and the Department of the Army produced a general cry for an investigation. The Army-McCarthy hearings began on April 22, 1954. Joseph Welch, a soft-spoken and elaborately ironic Boston lawyer, represented the Army. His assistant was another Boston lawyer named James St. Clair. Ray Jenkins of Tennessee was counsel for the subcommittee majority. A fascinated television audience, amounting at times to 20 million people, watched the hearings.

* Roy Cohn, "Could He Walk on Water?" *Esquire*, November 1972. In 1958 the Subversive Activities Control Board claimed that, according to Communist party records, Mrs. Moss had been a party member in the mid-1940s. Her listing on the rolls, if true, is hardly conclusive, given what we know now about the disposition both of CPUSA organizers and of FBI informants to pad membership lists in order to increase their prestige. Nothing specifically has come to light indicating that Mrs. Moss was anything but the bewildered and harmless woman that Symington and Kennedy supposed her to be.

If they looked hard, they could see Robert Kennedy, sitting quietly behind the Democratic senators and passing them occasional notes as the witnesses testified. His role was inconspicuous. There was one exception. On June 11, Jackson was interrogating McCarthy about an anti-Communist plan proposed by G. David Schine. Let us begin with Roy Cohn's account.

> Bobby Kennedy's job was to write out pertinent questions for the Democratic senators to ask at the hearing. In this Schine plan he saw an opportunity to gibe at us. He fed his questions to Senator Jackson, who used them to fire a barrage of ridicule. . . . They picked at point after point in Dave's plan, finding something hilarious in each. And every time Kennedy handed him something, Senator Jackson would go into fits of laughter. I became angrier and angrier.[55]

The hearing recessed. Cohn had sent out for a file marked "Jackson's record." One reporter wrote, "Carrying it like a weapon he headed for the committee table."[56] There he found Kennedy. "I want you to tell Jackson," Cohn said furiously, "that we are going to get to him on Monday." Cohn threatened, Kennedy told reporters later, to accuse Jackson of having "written something favorably inclined toward Communists." "His blue eyes snapping" (according to the *New York Daily News*), Kennedy said, "You can't get away with it, Cohn." Cohn told reporters he had accused Kennedy of "having a personal hatred" for the McCarthy side. Kennedy said, "I told him he had a ———— nerve threatening me. I told him he had been threatening all the Democrats and threatening the Army and not to try it on me." When he said to Cohn, "Don't you make any warnings to us about Democratic senators," Cohn replied, "I'll make any warnings to you that I want to — any time, anywhere." Then: "Do you want to fight right here?" He started to swing at Kennedy. Two men separated them, and Kennedy, with a small, contemptuous smile, turned away. The banner headline in the *Daily News* the next day was: COHN, KENNEDY NEAR BLOWS IN 'HATE' CLASH.[57]

Walking out with LaVern Duffy, Kennedy said, "I lost my temper." "It isn't like you," Duffy said. "You shouldn't have done it." In visible distress Kennedy went to the Capitol in search of his brother, finally finding him in the Senate cloakroom. For a few moments they had a serious and evidently consoling talk.[58]

The hearings ended a week later. The Republican majority roundly supported McCarthy. The Democratic minority, in a report written by Kennedy, concluded that for "inexcusable actions" in the Schine case "Senator McCarthy and Mr. Cohn merit severe criticism," that McCarthy deserved further criticism for his conduct "throughout the hearings" in "attacking the character and impugning the loyalty of individuals who were in no way associated with or involved in the proceedings" and that McCarthy's "countercharges against Secretary Stevens [of the Department of the Army] and Mr. Adams [general counsel of the Department of the Army] which impugned their patriotism and loyalty, were totally unsubstantiated and unfounded."[59] In a section of the draft the minority did not adopt, Kennedy had written of McCarthy, "The Senator cannot escape responsibility for the misconduct of Cohn. Nor can he excuse the irresponsibility attaching to too many of his charges. The Senate should take action to correct this situation."[60]

The minority report registered popular sentiment. The national audience, well trained by TV westerns to distinguish between white hats and black hats, had little trouble deciding to which category McCarthy belonged. After thirty-five days of the grating voice, the sarcastic condescension, the weird giggle, the irrelevant interruption ("point of order, Mr. Chairman, point of order") and the unsupported accusation, McCarthy had wrought his own downfall. On July 30, Ralph Flanders of Vermont submitted a resolution of censure against McCarthy. At the end of August a committee under the chairmanship of Arthur Watkins of Utah, and including such respected conservatives as John Stennis of Mississippi and Sam Ervin of North Carolina, began hearings. On September 27 the Watkins committee recommended censure. On December 2 the full Senate voted censure.

The Kennedys viewed all this with mixed feelings. John Kennedy prepared a speech in support of censure, but he made his case on exceedingly narrow grounds. He declined to base his vote on "long-past misconduct of Senator McCarthy to which I registered no public objection at the time." His judgment, he said, involved "purely a question of the dignity of the United States Senate" — a dignity impeached by Cohn, acting with McCarthy's full support.[61] Robert Kennedy, much closer to McCarthy, evidently saw his downfall as a mixture of public necessity and personal tragedy. Three years later, after a member of Symington's staff asserted that Kennedy's friendship with McCarthy had prejudiced his conduct in the hearings, Kennedy wrote Symington: "As you know, I always liked Senator McCarthy. However, after the very close relationship that I enjoyed with you during the Army-McCarthy hearings, I am sure that you would not feel that the statement that I allowed my likes and dislikes to interfere with my work was justified."[62] No one thought him opposed to censure. Two weeks after the vote, John Stennis — who had said in the debate that McCarthy had poured "slush and slime" on the Senate and that if censure failed "something big and fine will have gone out of this chamber" — wrote Kennedy: "One of the best parts of my year has been my work and association with you and I look forward to our work together in 1955." [63]

Still his Irish conception of loyalty turned him against some he felt had treated McCarthy unfairly. In March 1954, Edward R. Murrow of CBS had castigated McCarthy in a brilliant television 'documentary.' The show had great impact, though even anti-McCarthy commentators, like Gilbert Seldes in the *Saturday Review* and John Cogley in *Commonweal,* criticized the technique of offering as a "report" what was in fact a superbly calculated attack. In January 1955, Murrow spoke at the banquet honoring those, Kennedy among them, who had been selected by the Junior Chamber of Commerce as the Ten Outstanding Young Men of 1954. Kennedy grimly walked out.*

On the other hand, Joseph Welch, who far more than Murrow — far more than anyone, indeed, except for McCarthy and Cohn — had brought about McCarthy's ruin, congratulated Kennedy on the Chamber of Commerce award. Kennedy replied with utmost cordiality. "You were very kind to think of me," he wrote,

> and I greatly appreciate it. I hope you will give my best wishes to Jim St. Clair. I saw him briefly at a football game last fall and he looked so relaxed I hardly recognized him. I hope the same thing has happened to you. I don't get to Boston very often, but I hope on one of my visits there we can get together for a few minutes anyway.

Welch as promptly responded:

> What you say about being in Boston makes me say that it would be a delight to see you and that I shall be sad indeed if you come this way and don't come in. [64]

Other reactions to the award were less enthusiastic. "Things have come to a pretty pass," one woman wrote Robert Kennedy, "when a man is commended for doing everything he can to stop a fellow countryman from fighting the worst enemy this country has ever known. . . . You know that God and communism do not go together. I have chosen God." Kennedy wrote back, "I am sorry that I have already turned out to be such a disappointment to you." [65]

VII

The midterm elections in 1954 produced a Democratic majority in the Senate. John McClellan became chairman of the Investigations Subcommittee, and Robert Kennedy chief counsel. The committee, Theodore Sorensen advised Robert Kennedy in December 1954,

---

* The resentment was transient: two and a half years later Murrow visited the Kennedys at home to do an installment of his popular program *Person to Person.*

ought now to revise its rules to show that it was "off to a fresh and thoughtful beginning, and intended to safeguard the rights of individuals." This was done, and the *Louisville Courier-Journal* reported in January, when Kennedy arrived for the Junior Chamber of Commerce dinner, that he was "rather proud of the new set of rules." [66]

"Hardly anyone recognized boyish Robert F. Kennedy," the *Courier-Journal* added, "when he strode bare-headed and casual into the Kentucky Hotel last night, carrying a couple of books under his arm." His reading habits were beginning to improve. He said that he thought the Investigations Subcommittee would gradually return to its old function — looking into waste, fraud, corruption and mismanagement in government. But that did not mean, he continued, that it was "a whit less interested in rooting Communists out of government than it was when Senator McCarthy ran the show." [67] Recommending subjects for investigation, Kennedy wrote McClellan that it seemed only fair to all concerned to answer the unanswered questions in the Peress case. "If there is a rotten situation in the Army it should be exposed; if there is not, the bogeyman should be put to rest." [68]

The committee held a new series of Peress hearings in March 1955 and rendered its report in July. The question was whether "the original appointment of Peress, his subsequent change of orders, and his promotion had been, in fact, inspired by subversive interests." The committee found nothing to substantiate McCarthy's insinuations. The report went on to criticize the Army for forty-eight errors "of more than minor importance" and concluded that Peress's promotion and honorable discharge resulted from "individual errors in judgment, lack of proper coordination, ineffective administrative procedures, inconsistent application of existing regulations, and excessive delays, not from treason." [69] Though this conclusion put McCarthy's bogeyman to rest, he extracted what solace he could and blandly commended the committee staff.

Kennedy also reviewed McCarthy's files on Communist infiltration into defense plants. "The derogatory information," he informed McClellan, ". . . is not extensive, ranging anywhere from a couple of sentences to a page or so. The charges vary from allegations that an individual was seen reading the Daily Worker, to the charge that the individual is an active member of the Communist Party." Most of the persons accused did not work at plants handling classified material; those who worked at such plants mostly did not do classified work. "I mention these facts so there will be some understanding of this investigation and to place it in correct perspective. However, there is no intention of minimizing its importance for it is recognized that even if one security risk is removed from a strategically important spot, a distinct service has been performed."[70]

Kennedy was in fact more involved in Communist investigations than he had ever been when McCarthy was chairman, no doubt because the Democrats now in charge did not wish to seem less zealous in saving the Republic than their Republican predecessors. His approach to the professional witnesses, however, was skeptical. "Being anti-Communist," as he observed of Roy Cohn in January 1955, "does not automatically excuse a lack of integrity in every other facet of life."[71] Instead of seeking out voluble and imaginative informers like J. B. Matthews and Louis Budenz as his experts, he turned to scholars like Professor Philip Mosely of Columbia.[72] He even went to see Earl Browder, who had been till 1945 the leader of the American Communist Party. "I appreciate your kindness in talking to me last Saturday morning," he subsequently wrote Browder. "It was of tremendous interest to me to receive your ideas and thoughts on the history of the Communist Party in this country. It is only with an understanding of the past that we can proceed to map out our future with confidence."[73] Browder found Kennedy "pleasant and eager" and was impressed that anyone from a congressional committee should try to learn what communism was all about.[74]

More and more, however, as Kennedy had forecast in Louisville, the committee was returning to its old functions. It now concentrated on such subjects as fraud in the government purchase of military uniforms (an inquiry that resulted in six convictions); contracts let by the Foreign Operations Administration for the erection of grain storage elevators in Pakistan; inefficiencies in the Department of Agriculture's grain storage bin program; and similar issues of waste or graft in federal operations. Kennedy proposed other areas, such as wiretapping in Washington. This prospect agitated the FBI. "Kennedy's attitude," it was reported within the Bureau, "was such that impression could be gained he might not be friendly disposed toward the FBI."[75] The counsel for the Republican minority protested, and the investigation stopped.[76] His relations with the Bureau remained prickly. When he began an investigation into the manufacture of military uniforms, Hoover noted acidly, "Kennedy was completely uncooperative until after he had squeezed all the publicity out of the matter he could."[77]

The committee dropped out of the headlines. One 1956 investigation did, however, attract attention. Harold Talbott, Eisenhower's Secretary of the Air Force, had retained half-ownership in Paul B. Mulligan & Company, a firm of so-called efficiency engineers. Early in 1955 John Kennedy's friend Charles Bartlett heard that Talbott was hustling business for the firm out of his Pentagon office. Bartlett thereupon proposed a joint inquiry to Robert Kennedy, an alliance of the newspaperman and the Senate investigator, each tapping his own sources and pooling the results.

Talbott was a genial fellow. His vivacious wife was even more popular. They were Palm Beach acquaintances of the elder Kennedys and had many friends in the Senate. But Bartlett soon published letters by Talbott, on Air Force stationery, soliciting contracts for Mulligan from firms doing business with the Pentagon. Called before the committee, Talbott denied wrongdoing. Committee members responded admiringly. Joe McCarthy said virtuously that he deemed it "extremely important . . . to make sure that by innuendo, or otherwise, that the Secretary does not get an unfair deal and a smear." "None of the members of McClellan's committee," Bartlett wrote later, "would give Kennedy any support, and the young counsel had to pursue the case in a hearing in which he was the only one asking any questions."

Kennedy, aged thirty, began a tough cross-examination. He asked Talbott, aged sixty-seven, whether he had ever instructed anyone in the Air Force to make representations to RCA for not renewing a Mulligan contract. Talbott: "Absolutely not." Kennedy broke down his story. Talbott, wrote Arthur Krock, was not a crook but rather a man whose "excessive egotism" led him to suppose that no one could question his motives. Eisenhower requested Talbott's resignation. Bartlett won a Pulitzer Prize. Kennedy confirmed a reputation as a relentless prosecutor.[78]

VIII

These were good years for the young Kennedys. Robert, after the bad doldrums of the Hoover Commission, was working hard and cared deeply about what he was doing. Ethel was raising children — five by 1956 — and rushing to cheer Bobby on in hearing rooms at the Capitol. They lived unpretentiously on S Street, simple in their tastes, unsophisticated in their ways. For a time, there was neither alcohol nor even ashtrays in the house. David Ormsby-Gore* stayed with them for a week in 1955. "No doors were ever shut," he recalled, "and everybody wandered through every room all the time. . . . It was great fun but it was quite unlike living in any other house I'd ever lived in before." These were the days when, as Ormsby-Gore later said, the dogs still outnumbered the children. "It was the happiest time," said David Hackett, who had moved to Montreal but

* Later, Lord Harlech.

visited Robert and Ethel whenever he could. "There was not yet a world to solve."[79]

Robert remained private, intense, with an exacting sense of individual responsibility. Once he picked up Charles Spalding at the airport. As they drove back, a dog raced in front of the car. There was no way of stopping, and the dog was killed. "It upset him so," Spalding said later. ". . . We must have spent three hours going up and down the road. . . . We went into every single house for ten miles, I suppose, until we finally found . . . the owner. And Bobby explained what had happened and said how terribly sorry he was and asked about the dog — could he replace it or was there anything possible that could be done."[80] When Ethel's father and mother were killed in a plane crash in October 1955, Robert, not unused to sudden death, drew on deep wells of sympathy and strength for Ethel.

He was still a prickly figure. "He was just always very sensitive," said Eunice, "and he got hurt easily. . . . He just either would look mad or he'd be a little sarcastic and talk back to you." His friends understood this and knew it did not last. As Eunice said, "Bobby never sulked."[81] Acquaintances did not understand it, misconceived his laconic humor, and sometimes found him curt and contentious. He did not like to be put on the defensive and avoided this, in the family style, by asking his own questions before others could question him first. This was partly, too, the result of his incessant curiosity. But, where John Kennedy asked questions impersonally, Bobby asked them with curious intensity. This added to his reputation as an inquisitor.

Politically he was accounted a conservative Democrat, his father's son, though more interested in investigations than in politics. Arthur Krock, seconding his nomination in 1955 for that most respectable of institutions, the Metropolitan Club of Washington, called him in some respects "the number one member of that remarkable tribe . . . distinguished for both scholarship and athletic ability, in the military service for courage . . . of unsurpassed integrity," etc.[82] No doubt the "number one member" crack was induced by Krock's apprehensions about John Kennedy, whose politics had begun to display what he doubtless saw as wayward tendencies.

Nineteen fifty-four was the year when Secretary of State Dulles and Vice President Nixon advocated American military intervention in Vietnam. "In Dienbienphu the French are fighting heroically the battle of the free," as the *New York Times* resoundingly said. The world's biggest democracy could not "stand idly by" while communism swallowed small countries and toppled the dominoes of Asia.[83] John Kennedy opposed intervention. "No amount of American military assistance in Indochina," he told the Senate, "can conquer an enemy which is everywhere and at the same time nowhere, 'an enemy of the people' which has the sympathy and covert support of the people."[84] Later he condemned the administration's refusal to share information and decisions in foreign affairs. Documents were withheld "whose publication could do no more harm than the release of the Secretary of State's laundry ticket." The makers of foreign policy looked with "condescending disdain on those who challenge their omniscience," supposing the layman unable to comprehend the "mystic intricacies" of diplomacy. They would have to "give up the luxury of their exclusive control over our role in world affairs."[85]

Robert Kennedy doubtless agreed. But his interests lay elsewhere. His brother, though now a member of the Government Operations Committee, was not on the Investigations Subcommittee. They saw each other mainly in the evenings. John Kennedy was taking out the exquisite Jacqueline Bouvier. The courtship was on occasion conducted at the younger Kennedys' house in Georgetown, where the four often met at the end of the day for dinner and a movie. At one big Kennedy weekend at Hyannis Port, the boys were playing touch football on the lawn while Jacqueline read on the veranda. John caught a pass and called to her: "Did you see that?" She answered indifferently. In a few moments Robert came over and said that he could well understand why she might think football barbaric, but it did mean a lot to Jack. He said it with such sweet concern that, Jacqueline said later, "I felt at once that I had a friend who would always explain things to me."[86]

John and Jacqueline were married in September 1953 at Newport. The Kennedys gathered in advance at Hyannis Port. Spirits were high; and Bobby and Teddy, completing their assignment in a scavenger hunt, stole a policeman's hat. The policeman complained in person the next morning. An outraged father called the family together and delivered a polemic on the enormity of this offense; it would get into the columns and make the family look ridiculous. When Bobby tried to explain, the ambassador fixed him with a baleful eye and said, "No. You keep quiet and listen to me. This is childish behavior, and I don't want anything more like it."[87]

After the John Kennedys came back to Washington, they moved from Georgetown into an ante-bellum mansion in McLean, Virginia, called Hickory Hill. John was increasingly drawn toward foreign policy, Robert increasingly to the exposure of corruption. For a season the two brothers, their fondness undiminished, had separate preoccupations.

Interlude: William O. Douglas
and Adlai Stevenson

THE TALBOTT investigation ended at 1:30 P.M. on July 28, 1955. Two hours later, Robert Kennedy, in the same rumpled, drip-dry seersucker suit he had worn at the last hearing, was on a transatlantic plane. He was in bad humor. His mother, who flew with him to Paris, gathered that making the trip was about the last thing he wanted to do.[1] But it was his father's project, and the simplest thing was to accept his fate. After a long midnight wait at Orly, he boarded another plane. At 7:30 the next morning, dead tired and harassed, he arrived in Teheran. Here he met his father's friend William O. Douglas. The two now set off on a long-meditated tour of Soviet central Asia.

I

Joseph Kennedy was the only man in the country who could preserve simultaneous friendships with Senator McCarthy, the deadly enemy of the Bill of Rights, and Justice Douglas, its devoted champion. Off the bench, Douglas was an incorrigible traveler. He preferred high mountains, primitive lands and places where few men had traveled before. He had already traversed non-Soviet central Asia and in 1950 conceived the plan of visiting the contiguous Soviet republics to find out whether communism had made a difference.

Joseph Kennedy, never neglecting a chance to widen a son's experience, asked Douglas to take Robert along. Douglas recalled only the "very much subdued" little boy he had seen years before; but he liked the young law student he soon encountered when he spoke at the Charlottesville Forum.[2] The two made their first visa applications in February 1951. These were the last paranoid years of Stalin. The Soviet press had unmasked Douglas as an American spy when he climbed Mt. Ararat in Turkey on an earlier trip. The applications were turned down then and in succeeding years. Robert Kennedy went to work for McCarthy. Mercedes Douglas asked her husband why he still wanted to take this rigid and hostile young man to Russia. Douglas replied, "I have to do it. Joe wants me to do it. Joe thinks it would do him some good."[3]

In 1955 they applied for a fifth time. Stalin was dead, the line had changed, and the Kremlin was preparing for a conference with Eisenhower at Geneva. They received visas within a week.[4] Kennedy also received indignant letters. "I am sorry that you so heartily disapprove," he replied to one Irish American critic. "However, I would like to call your attention to the fact that up until the past several weeks, there has been an American Catholic priest in Russia since 1933."[5]

The American embassy in Teheran had assigned a young Foreign Service officer named Frederick W. Flott to Douglas. When Kennedy arrived, Flott's impression was of a high-strung, take-charge young man. Douglas had suggested that Flott come along as interpreter. Kennedy told Flott loftily, "I haven't decided whether or not you are going." Douglas had also wangled a miscellany of equipment from the Army. The bush jackets came in three sizes and were distributed according to seniority. Douglas took 'large' and Kennedy 'medium,' leaving 'small' to Flott. 'Small' turned out far too small, so Flott ruefully gave his bush jacket to a girl in Teheran.[6]

After an audience with the Shah, they crossed the Elburz Mountains by automobile and drove along the steaming Caspian Sea to Pahlevi. "Pahlevi and the hotel we stayed [in] smelled," Kennedy wrote. "It was terribly, terribly hot and I was never so glad to leave any place in my life."[7] Kennedy and Douglas embarked for Baku the next morning, carrying their Army gear and a supply of ball-point pens with which to edify the natives. Flott, who had not yet received his Soviet visa, parted company with them at Pahlevi.

"The boat is very nice and clean, newly painted," Kennedy reported. "Nobody speaks English but we are treated in the highest style. The cabins are very good and the food never stops coming. . . . They put this terrific spread before us of soup, all kinds of fruit, watermelon, honeydew, apples, cherries, chicken, etc. We both agreed that we would give in easily. I slept on the deck all afternoon sunning myself." They smelled the oil of Baku fifteen miles away. As they disembarked, Douglas pressed a ball-point pen on the captain, who at first declined, till, as Kennedy wrote, "the crew raised a storm and he finally took it."

The Intourist agent in Baku told them no interpreters were available. Douglas said that the Soviet embassy in Washington had promised them one. The Intourist man was "singularly unimpressed." They went to the hotel, holding on the way a whispered conversation on strategy. Once in their rooms, Douglas said loudly to the concealed microphones that he would call Khrushchev and tell him of their wretched treatment. Within an hour an interpreter arrived.

They liked the bustle of the city. At a music festival, "the dancers showed boundless energy and the music for the most part was very catchy." At a collective farm near Baku, they were served a twenty-one-course meal. Their hosts "were disappointed in me that I did not eat more. Douglas made a valiant effort." Someone toasted Douglas and his family and then Ethel and the four children. "One of our hosts said it was very proper that they should drink to us as we had already fulfilled the plan."

From Baku they flew on to Ashkhabad, the capital of Turkmenistan, where Kennedy, an ardent photographer, was arrested for taking snapshots. The Intourist guide got him off. In Bokhara, once the holy city of central Asia, it was 145 degrees in the sun. They inspected the notorious Bug Pit, where, a century before, the Emir Nasullah had consigned two adventurous Englishmen for refusing to acknowledge Mohammedanism as the true faith; after two months among reptiles and vermin, they were beheaded. The American travelers, Kennedy wrote later, "were fortunately received somewhat more warmly by the local officials." [8]

In Samarkand they saw the tomb of Tamerlane, the great blue-domed mosque of Shir Dar, the graceful square of the Registan. Kennedy wrote his children from Stalinabad, the capital of Tadzhikistan: "Say your prayers and get to bed on time and you won't grow up to be a juvenile delinquent like your mother." They watched Tadzhik dancing — "everyone by himself — jumping up and down — moving in rhythm, swaying and moving heads back and forth. Sensuous, weird and interesting." Cultural notes alternated in Kennedy's diary with data about wages, prices, working conditions, housing, child care, courts, factories, and rather especially churches.

Douglas and Kennedy disagreed on the tactics of information gathering. Douglas, the judge, was all suavity. His technique was to appear interested and accommodating, drawing out the people with whom he talked until, having won their confidence, he could hit them with a factual issue and hope for a frank response. Kennedy, the investigator, was exasperated and hostile, as if determined to find

nothing good. Fred Flott, who finally got his visa, followed their trail
through central Asia. At each airport, he heard much the same story:
"We liked Justice Douglas. We can see he is a wise man. But [great
sigh, looking at the ground] with him there is Mr. Kennedy. He
seems always to be saying bad things about our country." Kennedy
carried mistrust to inordinate extremes. The Intourist guide brought
along a footlocker filled with caviar. Douglas could not eat caviar
because of a liver complaint. Kennedy announced that Russian food
was dirty and refused the caviar. The fortunate Flott therefore had
three jars of caviar every meal. Throughout the trip Kennedy ate
very little and drank nothing, subsisting mainly on watermelon and
nervous energy.

II

Kennedy's automatic contentiousness irritated Douglas. He explained to Flott that he hardly knew the boy; his father, to whom he
owed his Supreme Court seat, had said that Bobby had never gone
on a he-man trip and asked that he be taken along. "You can never
argue with these fellows," Douglas finally told Kennedy, "so why
don't we just forget about it, and spend an evening doing something
else rather than wasting it trying to convert some guy who will never
be converted." This worked. "We never had any problems then."[9]
   Even Douglas could not always avoid argument. At the University
of Tashkent Kennedy asked an English-speaking professor why no
one ever criticized the Soviet government. Douglas joined the discussion, pointing out, Kennedy wrote in his journal, "that we in the
United States had been brought up on the fundamental [right] we
possessed to find fault with government where we found things going
wrong. The professor said that in U.S. people criticized just to criticize and that was because Democrats always wanted to criticize Republicans and vice-versa so that they could be elected." Douglas
brought up academic freedom, suggesting that the professors would
be in trouble if they criticized communism in their classrooms. This
was not well received.
   The travelers here and elsewhere inquired especially about the
indigenous peoples of central Asia, whose high cheekbones and dark
complexions, language, schools and customs set them so sharply apart
from their Russian overlords. They were as different, the travelers
thought, as the Moroccan from the Frenchman or the Malay from
the Englishman, and quite as subject to imperial rule. When Douglas
mentioned that one third of the Kazakh population had apparently
been liquidated during the collectivization of the farms, the professor
dismissed the figure as the invention of British intelligence. Douglas

and Kennedy pointed out that Soviet census statistics showed a decline of almost one million Kazakhs between 1926 and 1939. The professor was displeased and, as Kennedy noted, "became rather vehement."

"I hope," Kennedy wrote the children from Tashkent in mid-August, "you are all . . . doing what your mother and Ena [the nanny] tell you to do. (I admit that this gets a little difficult when they tell you to do opposite things but play along with them.)" It was now a rather jolly party. Cables of congratulation to Kennedy over the Talbott affair pursued them; and his companions took pleasure in introducing him as the People's Prosecutor, the American Vyshinsky. Kennedy was forever sending postcards, when he could find any — a portrait of Lenin, for example, to Cardinal Spellman with a cryptic message scratched on the back.

They went to the opera at Tashkent. "Beautiful opera house," Kennedy noted. "Tremendously large chandelier. Kirghesé design of white figures on gold background. One of the prettiest I have ever seen. . . . Live horses on the stage. . . . A few good looking girls in the chorus and the heroine looked like a Metropolitan Opera star." They visited a Baptist church with a congregation of more than a thousand. "Douglas made a speech which was good and caused some of them to cry." At a Tashkent park, Douglas got into a discussion with a crowd. Someone asked about the Sacco-Vanzetti case. "Douglas told them that a worker's payroll had been stolen. I borrowed a fellow's bike and had myself a ride around town."

They visited a collective farm near Frunze in Kirgizia. Coming back, Kennedy fell off his seat when the jeep went over a bump. "The two or three Russians, who were riding with me, grabbed hold of me for every bump afterwards. It did not speak very well for the strength and resilience of American youth." In Alma Ata, to the east, they visited more collective farms and were served more gargantuan meals. At one "there were two young things who kept staring at Douglas and vice versa." At another, there were "many toasts, the host got looped, spilled champagne over Douglas. We sang the Whiffenpoof Song, Home on the Range, which didn't go over, and America the Beautiful."

Kennedy's natural curiosity was overcoming his initial hostility. "He worked hard on the trip," Douglas said later, "and came back with some considerable insight."[10] "People have no idea how our system of government works," Kennedy wrote in his diary. ". . . We, I think, are equally ignorant of their system." In the university at Alma Ata, the rector, a biochemist, observed that, while Soviet scientists read American scientific journals, American scientists, judging from the absence of citation, did not read Soviet scientific journals.

Was this because of a lack of interest? "Douglas explained he was sure it was not and he was surprised to hear about it. I am afraid it is a result of the McCarran Act" — the Internal Security Act, passed over Truman's veto in 1951.

Douglas insisted that the travelers stop for a breather in the Tien Shan range. "We'd get up to ten, eleven thousand feet I think," Douglas recalled. "I wouldn't call it real mountain climbing." He was struck by the wholehearted way Kennedy threw himself into mountaineering. "Bobby was an all-out. . . . If he was climbing a peak or hiking a trail there was nothing else in the world to do but that."[11] The Siberian weather was already turning cold. On August 29 they flew west from Novosobirsk to Omsk. Kennedy developed a chill. By the time they arrived in Omsk, "I was shaking all over and was freezing cold except for my head which felt hot." Douglas decided to summon a doctor: "No communist is going to doctor me," Kennedy said. "I promised your daddy I would take care of you," said Douglas. "I'm going to call a doctor." It took three hours for a large, heavy-boned woman physician to appear. Robert's temperature was now well over 100 and rising steadily. There were signs of delirium. He was shot full of penicillin and rushed to bed, while the others went on, as scheduled, by train.

"Slept most of the next day," Kennedy recorded in his diary. ". . . Doctor stayed with me all yesterday and two nights. Mayor came to see me yesterday." In this laconic fashion he glided over both an intense illness and the devotion of the Soviet doctor. Perhaps he was disinclined to yield too much credit either to his own weakness or to Communist medical care. In another twenty-four hours he had recovered sufficiently to take the plane west. On September 2, Ethel Kennedy, accompanied by Jean and Pat, met the party in Moscow. Looking at Bobby, pale and thin, Ethel cried to Douglas, "What have you done to my husband?"[12]

Charles E. Bohlen, the American ambassador, had them to dinner. They saw *Swan Lake* at the ballet. Then the Kennedys left Moscow to inspect the beauties of Leningrad. A few hours later, a hotel maid knocked at the door of Fred Flott's room and handed him, with a note from Robert Kennedy, an Army bush jacket, 'medium,' freshly laundered.

III

In the aftermath of the Geneva summit, the travelers encountered a confusion of moods in the Soviet Union. Douglas offered the more optimistic reading. While doubting that the Kremlin had changed its objectives, he thought the Russian people "in the mood for friendship with the American. . . . They want peace so that they can enjoy the dividends of the society they have created."[13] As for Kennedy, his views, he wrote David Lawrence, editor of *U.S. News & World Report,* differed "from the reports of some of the other people who visited Russia in that I was far more distrustful of Russia's intentions.[14] "We all want peace," he told an interviewer in Paris on his way home, "but are apt to be blinded by the bear's smile."[15] "My feeling," he said on arrival in New York, "is that before we make a great number of concessions we have to have something more concrete from the Russians, such as the dissolution of the Comintern and a workable disarmament plan."[16]

He developed this theme in October at Georgetown University. He had called Angie Novello, the most efficient stenographer on the Investigations Subcommittee staff, to Hyannis Port where he dictated a text. Slides selected from the more than a thousand photographs he had taken on the trip illustrated the lecture.[17] They were magnificent. But Angie, sitting in the middle of the auditorium, could hardly hear the speaker. "His voice was very high-pitched and he'd slur his words and go down to nothing." Afterward Kennedy asked, "How did you think it went?" She said, "The pictures were great, but I think you could have spoken better than that." Ethel Kennedy said quickly, "I thought he was very good!"[18] Lem Billings agreed with Angie Novello. He said later, "Bobby was the worst speaker that I *ever* . . . It was just horrifying to hear him."[19]

His primary point, after a meticulous survey of life and labor in the Soviet Union, was that history had shown it to be "suicidal for the United States or its allies to make major concessions to the Soviet Union without a quid pro quo." He had a ringing Cold War peroration:

> All I ask is that before we take any more drastic steps that we receive something from the Soviet Union other than a smile and a promise — a smile that could be as crooked and a promise that could be as empty as they have been in the past. We must have peaceful coexistence with Russia but if we and our allies are weak, there will be no peace — there will be no coexistence.[20]

There was a subsidiary theme which grew in the next months until he transmuted it into a forceful article for the *New York Times Magazine.* This was the question that so exercised Douglas of the Russian treatment of ethnic nationalities in central Asia. Kennedy gave the Soviet government full credit for economic development. "Every time I have spoken on my trip through Central Asia," he wrote the director of the United States Information Agency, "I have said that

I felt that the standard of living of the people had improved." But he also insisted on the price the native peoples had to pay. "Western Colonialism hasn't been made acceptable by arguing that material gains have been made in Colonial areas. Why then should Soviet Colonialism?"[21] In fact, Kennedy argued, the Russian record itself reproduced "all the evils of colonialism in its crudest form" — denial of self-government, segregation of Western masters from native subjects, mass executions of local peoples in order to preserve imperial rule.[22] By every standard used to judge Western colonialism, he told the Virginia State Bar Association, the Soviet performance "qualifies as colonialism of a peculiarly harsh and intractable kind."[23]

His anticolonialism was severely consistent. Unlike more obdurate Cold Warriors, he did not see western empire as a bulwark against communism nor suppose that anticolonialism in Asia and Africa was organized in Moscow. It was difficult for Americans to understand the power of anticolonialism in the underdeveloped world, he explained to one correspondent, "because we think that the uppermost thought in all people's minds is Communism."[24] He wrote in the *New York Times Magazine,* "We are still too often doing too little too late to recognize and assist the irresistible movements for independence that are sweeping one dependent territory after another."[25] "The fatal error" in United States foreign policy, he told a communion breakfast at Fordham University, was its commitment to colonialism. "If we are going to win the present conflict with the Soviet Union, we can no longer support the exploitation of native people by Western nations. We supported the French in Indochina far too long."[26]

In the short run, the journey confirmed Kennedy's mistrust of the Soviet government, which was reasonable enough, and his mistrust of negotiations and contracts, which he carried a little far. The ultimate impact was more complicated. The regime was detestable, and he detested it; but he came back with a considerably more differentiated sense of the Soviet Union. His notes include a paper entitled "Good Things":

> More school construction work.
> Great deal of hustle — people seem generally better off than people in comparable areas of Middle East.
> If you agree with principal [sic] of nurseries certainly the ones we saw were well managed.
> Rent very little though housing is admittedly a problem.
> Government loosening up
>   1) people allowed in Kremlin
>   2) Amnesty for criminals
>   3) Sentences being reviewed
>   4) MVD power cut

    5) Farmers allowed more freedom
    6) Visitors allowed in country
People enthusiastic about their work
Music good — Azerbaijan music group in Baku as good as anything
we have seen — music catchy and dancers excellent.[27]

Before the trip he saw Russians as soulless fanatics. Afterward
he saw them, Douglas said later, as human beings and achievers,
"people with problems."[28] When Kennedy displayed his Russian
photographs one evening, David Ormsby-Gore noticed that "he'd
developed very much of a sympathy with the Russian people irre-
spective of the regime."[29] In *U.S. News* Kennedy questioned the
efficacy of Communist propaganda, "because, if they had been told
for 9 or 10 years that we were really their enemies and believed it,
I don't think they could turn that off and have people as genuinely
friendly as they were."[30] The Russian experience took time to penetrate, but Douglas and his wife later thought that it meant for
Kennedy "the final undoing of McCarthyism."[31]

IV

Back in Washington, Kennedy returned to the Investigations Subcommittee and his old concern that east-west trade might build the
military power of Communist nations. The ensuing investigation led
to a prolonged wrangle in which the administration claimed the right
to deny information about trade policy to Congress — both classified
information, on the ground that it was a breach of national security;
and "working-level papers," on the ground that exposing the internal
policy process was contrary to the public interest.[32] The term "executive privilege" had not yet been coined to give such denial specious
constitutional status. But the idea had emerged in full force and with
the support of right-minded people during the Army-McCarthy
hearings. It was now adorned with sweeping historical assertions and
buttressed by an irrelevant anthology of judicial decisions (none bearing on the question of the executive power to withhold information
from the Congress).

The administration, half persuaded it was defending some lofty
constitutional principle, righteously declined to answer questions
about items removed from the embargo list in 1954, even though, as
Kennedy said, "some of our allies have published the lists of strategic
materials which can be shipped to the Communists, and certainly the
Soviet Union knows what it has bought and can buy."[33] The administration declined, in addition, to explain the reasons behind American policy on east-west trade. One Department of Commerce official

said he only wished he could tell the committee why the executive branch had removed petroleum from the embargo: "I think it makes at least a reasonable record." [34]

This attitude infuriated the committee. "It is not national security," said Jackson, ". . . it is just sort of a national coverup." "Can you tell me," said Sam Ervin, "anything in the world that a witness is privileged . . . to tell this committee about what this committee is investigating?" "The real crux of this thing, as I see it," said McClellan, "is whether the Congress is entitled to know what is going on." [35] "I am afraid if the Senate does not wholeheartedly support Senator Mc-Clellan," Robert Kennedy wrote William Loeb, the publisher of the *Manchester Union-Leader*, "we will all be in difficulty in the years to come."* "If the position of the executive department were to prevail," the committee concluded, "it would mean that they would have an absolute right to deny to Congress or the public any information from the executive departments except in such matters as are required to be public by statute." [36]

As Robert Kennedy took leave from the committee to go as a Massachusetts delegate to the Democratic convention in Chicago, Wayne Morse, the old incorruptible from Oregon, wrote him, "In your unwavering dedication to the principle that the facts must speak for themselves . . . you have earned for yourself a very fine reputation among the members of the Senate." [37]

v

John Kennedy had endorsed Adlai Stevenson the previous October. "My brother . . . admired him a good deal," Robert said later; there had never been a question of John's not supporting him for renomination.† As for Robert himself, "In 1952 I had been crazy about him. I was excited in 1956." [38]

Their father, however, had soured on Stevenson and looked with extreme disapproval on John's determination to depose the anti-Stevenson leadership of the Democratic state committee in Massa-

---

* RFK to William Loeb, April 4, 1956, RFK Papers. A prescient remark — though by the time, fifteen or so years later, the difficulty threatened the republic itself, Loeb was on the other side.

† RFK, in recorded interview by John Stewart, July 20, 1967, 36; August 1, 1967, 51, JFK Oral History Program. In 1971 Lyndon Johnson described a phone call from Joseph Kennedy in October 1955 saying that he and John wanted to support Johnson for President in 1956; if he was not interested, they would go for Stevenson. Johnson said he was not interested (Lyndon B. Johnson, *The Vantage Point*, New York, 1971, 3). This incident reflected the father. The son came out for Stevenson that same month.

chusetts. Staying out of local fights, Robert recalled, was a lesson his father had taught them "when we were very, very young." Local politics was "an endless morass from which it is very difficult to extricate oneself." As Robert summed it up, "You're either going to get into the problems of Algeria or you're going to get into the problems of Worcester."[39] Nevertheless John Kennedy and his Massachusetts agents, O'Donnell and O'Brien, went ahead and won their battle.

The ambassador was equally disapproving of his older son's growing interest in the vice presidential nomination. "If you are chosen," he wrote John in May, quoting Clare Boothe Luce with approval, "it will be because you are a Catholic and not because you are big enough to do a good job. She feels that a defeat would be a devastating blow to your prestige."[40] By August the ambassador must have thought the idea had passed, for he went off to the Riviera. But Theodore Sorensen was busy lining up support for Kennedy as Vice President. As for Kennedy himself: "I think," Robert Kennedy said later, "he just wanted to ... put his foot in the water and see how cold it was, but he hadn't made up his mind to swim."[41]

The convention, Robert Kennedy's first, began on August 13. He had brought O'Donnell along. Peter Lawford, the actor, who had married Patricia Kennedy in 1954, offered them the penthouse suite he rented by the year in the Ambassador East. Robert's puritanism was offended by the idea of keeping so expensive an apartment and using it so rarely. To O'Donnell's dismay, he insisted that they move to a double room several floors below.[42]

Stevenson was nominated on August 16. He now faced the conundrum of the Vice Presidency. His preferences were unclear. On July 26 he had recounted to me his discussions with Harry Truman, James A. Farley, Sam Rayburn and other party elders about the contenders — Kennedy, Estes Kefauver of Tennessee and Hubert Humphrey of Minnesota:

> HST thinks Humphrey "too radical"; dismisses Kennedy as a Catholic; has no use for Kefauver. Mentions [Albert] Gore [of Tennessee] and [George] Leader [of Pennsylvania].
> Farley told AES, "America is not ready for a Catholic yet."
> Rayburn: "Well, if we have to have a Catholic, I hope we don't have to take that little piss-ant Kennedy. How about John McCormack?"[43]

A few days later Stevenson gave me the impression that he was "strongly for Humphrey on the merits, and also as a slam-bang campaigner."[44] But he liked Kennedy and had obligations to Kefauver, who, after contending with him in the primaries, had assured his nomination by declaring for him before the convention. In the

end, Stevenson made his decision, or nondecision, to throw the vice
presidential nomination to the convention.

John Kennedy, who had been testing the water, decided to swim.
"Call Dad," he said to Robert, "and tell him I'm going for it." He
then intelligently departed while Robert put the call through to Cap
d'Antibes. "The Ambassador's blue language flashed all over the
room," said O'Donnell, who remained. "The connection was broken
before he was finished denouncing Jack as an idiot who was ruining
his political career. Bobby quickly hung up.... 'Whew!' Bobby
said. 'Is he mad!'"[45]

Twelve hours of frenzy followed. Abe Ribicoff of Connecticut
nominated Kennedy. John McCormack, "literally propelled toward
the platform at the last minute by Bob Kennedy," as Sorensen described it, seconded the nomination.[46] "Bobby and I ran around the
floor like a couple of nuts," said O'Donnell. ". . . We didn't know two
people in the place." Bobby did know Senator McClellan; but
McClellan told him, "Just get one thing through your head. . . .
Senators have no votes; I'm lucky to be a delegate; Orval Faubus is
the governor of Arkansas, and that's it; and where he goes the
Arkansas delegation goes."[47] In the end, Faubus went for Kennedy. So did much of the south in Confederate resentment over
Kefauver's courage on civil rights.[48] Pat Lynch, whom John Kennedy
had made Massachusetts state chairman, told James A. Farley that
he had "one hell of a nerve" opposing Jack Kennedy after young Joe
Kennedy had stuck with him in 1940.[49] Sorensen asked Hubert
Humphrey's campaign manager, Congressman Eugene McCarthy,
whether, if Humphrey dropped out, Minnesota would go for Kennedy. McCarthy replied concisely, "All we have are Protestants and
farmers," later complaining that Kennedy had sent a boy to talk to
him. Lyndon Johnson announced that his state was switching to
Kennedy: "Texas proudly casts its votes for the fighting sailor who
wears the scars of battle." Then Tennessee shifted from Albert Gore
to Kefauver, and the fight was over.[50]

Kennedy in a graceful speech asked that Kefauver be named by
acclamation. Privately he was depressed. No Kennedy liked losing.
Robert said to him, "You're better off than you ever were in your
life, and you made a great fight, and they're not going to win and
you're going to be the candidate the next time."[51] "This morning all
of you were telling me to get into this thing," John Kennedy said,
"and now you're telling me I should feel happy because I lost it."[52]
In a day or two his natural equanimity reasserted itself.

Bobby was angry longer. "He was bitter," said the Massachusetts
delegate who sat next to him on the plane back to Boston. "He said
they should have won and somebody had pulled something fishy and

he wanted to know who did it."[53] But he was learning: the importance of communications at a convention; the importance of an accurate delegate count; the importance of the rules ("If we had known parliamentary procedures, we could have stopped things after Tennessee switched");[54] the importance of friendship ("Kefauver had visited people and sent out cards and this is what paid off");[55] the unimportance of celebrated senators; the importance of uncelebrated party professionals — all points filed away in that retentive mind.

John Kennedy went off to Europe to rest up from the convention. Jacqueline, now pregnant (when she told Robert and Ethel she wanted a baby, they had said they would pray for her every morning), was with her mother in Newport. Suddenly she was taken sick and rushed to the hospital. When she recovered consciousness around two in the morning, her baby lost, the first person she saw, sitting by her bed, was Robert. Her mother had phoned him, and he had driven over at once from the Cape. Many years later Jacqueline learned that Robert had arranged for the burial of the dead child. "You knew that, if you were in trouble, he'd always be there."[56]

VI

Robert Kennedy's education in national politics was only beginning. Stevenson now asked him to join the campaign party. From the Stevenson viewpoint, Robert Kennedy's presence would signal Catholics, conservatives and friends of his father that all was not lost. From the Kennedy viewpoint, here was a chance to find out how to run, or not to run, a presidential campaign. From my own viewpoint on Stevenson's staff, Robert Kennedy seemed an alien presence, sullen and rather ominous, saying little, looking grim and exuding an atmosphere of bleak disapproval.

My reaction was colored by recollections of our altercation two years before. William McCormick Blair, Jr., back again as Stevenson's aide, said later, "I know he felt that we made lots of mistakes, and, of course, we did, but he was always nice about it and never critical." He was, Blair said, "very helpful and did a lot of things on his own initiative." Blair may have had in mind the whistle stop in Lewiston, Pennsylvania, when a group of children made so much noise that Stevenson could not begin his speech. After a moment of confusion, Kennedy climbed down from the train, shouted, "I can beat anyone from here to the lamppost," and took off with the children in hot pursuit.[57]

Blair was a family friend and therefore all right. Robert Kennedy seemed uncomfortable with the rest of us. One's memory was of Bobby making notes, always making notes, whether huddled by the

window in the rear of the bus or plane or even sitting on a railroad
track while the candidate spoke from the rear of the train. Occasionally he revealed himself, but in a rather solitary way. When we
came into Springfield, Illinois, Kennedy said to Newton N. Minow,
Stevenson's law partner, "I have heard the Governor speak before,
but I have never been in Springfield before. Do you know where
the Lincoln house is?" They skipped the speech, and Minow took
him to the Lincoln house. Kennedy grew even more silent than
usual, and Minow saw that he was deeply moved. "For the first time,"
Minow said later, "I understood there were depths of feeling in him
I had never suspected."[58]

One afternoon in October the campaign party found itself in Morgantown, West Virginia. Stevenson was scheduled to speak that evening in New York City. Fog and rain set in, and a single plane was
available to fly the candidate out of Morgantown. Eventually buses
arrived to take the rest of us to Pittsburgh. Under pelting rain we
scrambled aboard, groping for seats in the gathering darkness. I
turned to look at my seatmate and found — I am sure to our mutual
dismay — Robert Kennedy. For the next several hours we rode
through the storm to Pittsburgh. Having no alternative, we fell into
reluctant conversation. To my astonishment he was altogether pleasant, reasonable and amusing. We became friends at once and remained so for the rest of his life. I could never take seriously thereafter the picture of Robert Kennedy as a bearer of grudges, vindictive
and unforgiving.

As for Stevenson, the candidate "sparkles," Kennedy wrote, ". . .
in small groups discussing matters in which he is interested." When
Kennedy took him to see the bishop of Denver, he talked about the
Middle East "and really handled himself very well." Kennedy also
liked Stevenson's humor. In Arizona they coincided with a parade
for the Feast of St. Anthony. Back on the plane, Mrs. Ernest Ives,
Stevenson's sister, asked Kennedy, "Isn't St. Anthony the one who
keeps travelers safe?" Kennedy said, "Mrs. Ives, St. Christopher
keeps travelers safe — St. Anthony finds things." At this moment
Stevenson, passing by, said, "I hope he can find me some votes."[59]
For the rest, the campaign was a deeply disillusioning experience.
He considered it, he told O'Donnell, "the most disastrous operation"
he had ever seen.[60]

Harrison Salisbury of the *New York Times* recalled Kennedy, sitting
as so often on a railroad track during a whistle stop, talking about
what a candidate should do in a campaign, "and he should do practically all the things that Stevenson didn't do."[61] Stevenson, it is true,
spent too little time cultivating politicians and too much embellishing

speeches. Kennedy also correctly objected, in a memorandum written after the campaign, to Stevenson's practice of reading prepared texts even on whistle stops:

> Because of his looks and his accent and his known background, this gave an appearance of insincerity. He was talking about the price of goods and the fact that big business was controlling the U.S. and I am sure the people felt that if he really believed all of this, that he could give it off the cuff and from the heart and would not have to read it. ... He could never get it clear that it was not so much what he said but how he said it. I spoke again and again to people around him to try to get him to change his technique but he never would do it.[62]

He also spoke to Stevenson himself. On the way to Boston: "I said again that I felt his speeches should be extemporaneous and that he should concentrate on the women there to win or lose the election. He seemed interested." As for his Boston speech, an attack on big business, Kennedy pronounced it "incredibly bad."

> I told him I didn't think it would go over in Boston but Schlesinger and Galbraith were of a different opinion although a compromise was ultimately worked out. I told him that I felt he should talk about something he was interested in so he would appear sincere and that necessarily would be foreign policy. I told him I thought it was a great mistake to talk about matters he was obviously not interested in because he did not make an impression on television.[63]

Kennedy was even more distressed by the candidate's disorganization. He joined us one Sunday at Stevenson's house in Libertyville for an exasperating session on campaign planning in general and on Stevenson's controversial proposal for a ban on nuclear testing in particular. "It was probably the most enlightening discussion I ever had with him," Kennedy noted later,

> and it was a real eye-opener for me. It was a conference that really disturbed me as far as thinking that he should be president of the United States. He spent all day long discussing matters that should have taken, at the most, a half hour. For instance, for four hours, with 8 or 12 people there, we discussed who should go on his television program — Senator Anderson, Senator Jackson or Senator Symington. For another couple of hours we discussed whether the hydrogen bomb was fission or fusion and whether it was safe to say hydrogen bomb tests should be stopped at one mega-ton.
>
> At the first discussion, I thought one of his clerks could have made that decision and contacted Anderson, Jackson or Symington and made the arrangements. On the second question, I didn't think it made much sense for 8 or 12 men to discuss scientific matters about which no one knew anything.
>
> Stevenson just did not seem to be able to make any kind of decision.

... The ostensible purpose of the meeting was to plan how the campaign for the last three weeks would be handled and that was hardly discussed at all. The subject of Nixon came up and I was strongly against making the campaign built around an attack on him. . . . People who didn't like Nixon were with Stevenson and they were not going to convince anybody by making any statements about him. . . .

These matters were discussed rather pleadingly, and again no decisions were made. Stevenson was just not a man of action at all.[64]

In the end, Robert Kennedy found his candidate a profound disappointment. "People around Stevenson lost confidence in him. There was no sort of enthusiasm about Stevenson personally. In fact, to the contrary, many of the people around him were openly critical, which amazed me."[65] "It was a terrible shock," Kennedy said later. ". . . I came out of our first conversation with a very high opinion of him. . . . Then I spent six weeks with him on the campaign and he destroyed it all." On election day Kennedy quietly marked his ballot for Eisenhower.[66]

Stevenson wrote him afterward, "It was good to have you along and I have seldom heard sounder, more sensible and thoughtful remarks stated with better verbal conservation. I hope we'll see more of each other as time goes on, even if my public role is sharply diminished. And my affectionate regards to your charming wife."[67]

The campaign was another stage in education. "What struck me about him at that time, in spite of his great youth and seeming detachment from what was going on," said Harrison Salisbury, "was his great analytical ability . . . I had the feeling after the Stevenson campaign that Bobby knew every single thing there was to know about a campaign. He just squeezed all that absolutely dry."[68]

(8)

The Second Investigating Committee:
Jimmy Hoffa

EIGHT DAYS after the 1956 election, Robert Kennedy, back as counsel for the Senate Permanent Subcommittee on Investigations, arrived in Los Angeles. He had embarked on a new inquiry. An investigation earlier in the year into government procurement of uniforms had turned the committee's attention to a fresh subject — labor racketeering. Hoodlums and mobsters, attracted by the spectacular growth of trade-union welfare and pension funds, attracted also by the convenience of unions as a cover for exercises in bribery and shakedown, were, it appeared, engaged in a systematic invasion of the American labor movement.

I

The labor movement itself was in a time of stagnation. The idealism generated in the great organizing battles of the 1930s was largely spent, except in the United Automobile Workers and a few other unions. The American Federation of Labor and the Congress of Industrial Organizations, mending their historic split in 1955, had chosen George Meany of the AFL Plumbers as president. But the new federation was doing little to police its own ranks. Nor, for that matter, did state or federal law provide effective standards for the management of the burgeoning welfare and pension funds.[1]

Congress had looked into the misuse of the funds, but its investigation was diffident, its glance quickly averted. In the meantime, newspapermen across the country — Edwin Guthman in Seattle, John Seigenthaler in Nashville, William Lambert and Wallace Turner in Portland, Pierre Salinger in San Francisco, Clark Mollenhoff in

Washington — dug into the problem. They found the International Brotherhood of Teamsters an especially rewarding field. Mollenhoff, a lawyer as well as a newspaperman, was a hectoring and fearless reporter in the fashion of the old muckrakers. Early in 1956 he began to importune Kennedy about setting the Investigations Subcommittee on the trail of labor racketeers.

Kennedy was skeptical. He doubted whether the Government Operations Committee had jurisdiction; trade union affairs belonged to the labor committees. Mollenhoff replied that the labor committees were afraid of labor's political power. If unions were misusing tax-exempt funds and rendering dishonest financial reports to the Labor Department, this involved government operations. "Occasionally, I taunted him by questioning his courage to take on such an investigation."[2] Kennedy talked to people involved in earlier congressional inquiries. In August he got Senator McClellan to authorize a preliminary check into labor racketeering.

Kennedy remembered something about labor law from Charlottesville, but he knew little about the labor movement and was especially innocent about the Teamsters. This was understandable. The Teamsters were still in good repute in lay circles. Daniel J. Tobin, president for nearly half a century, had been everybody's friend, especially Franklin Roosevelt's. It was to the Teamsters that FDR had given the famous speech in the 1944 campaign about his little dog Fala. David Beck, Tobin's successor in 1952, was vain and pompous, but he had diligently used modern management methods to build the Western Conference of Teamsters. "I run this place just like a business," he said, " — just like the Standard Oil Company. . . . Our business is selling labor."[3] Beck, Murray Kempton of the *New York Post* wrote, was "the first national labor leader who unashamedly talks the language of a Chamber of Commerce secretary."[4] With his Rotary Club orations and his entrée to the Eisenhower White House, Beck was saluted in business circles as the labor statesman of his generation. In a welcome to the convention that anointed Beck, Governor Earl Warren of California declared his own "increasing admiration" for the Teamsters and pronounced the union as "not only something great of itself, but splendidly representative of the entire labor movement."[5]

Through his new friend, the rising young criminal lawyer Edward Bennett Williams,* Kennedy met a Williams associate, Edward Cheyfitz; and Cheyfitz invited him early in 1956 to visit the Teamsters' new building in Washington, erected by Beck as a monument to

---

* Kennedy had met Williams as Joe McCarthy's counsel during the censure proceedings. Williams had become McCarthy's lawyer a year earlier, when Drew Pearson,

himself and popularly known as the 'marble palace.' Beck was not there, but Kennedy met other Teamster grandees. Over lunch in the Teamsters' cafeteria, Cheyfitz, an ex-Communist, confided to Kennedy that he had seen Walter and Victor Reuther in Russia in 1933 and was still not sure the Reuther brothers had ever broken with communism. At Cheyfitz's suggestion, Robert asked his brother John to send Beck an inscribed copy of his new book, *Profiles in Courage.* "I don't know if he ever did," Robert said later. "I have never dared inquire."[6]

II

The Teamsters had begun in 1903 as an organization of city delivery crafts — essentially drivers of coal, milk, ice and bread trucks. In 1933 the union membership was only about 75,000, hardly three times what it had been when Dan Tobin became president in 1907.[7] Tobin, a craft unionist of the old school, had no use for the mass unionism of the 1930s. "We do not want," he liked to say, "to charter the riffraff."[8]

Younger leaders, however, embraced the new gospel of industrial organization. In Minneapolis a group of Trotskyites, led by Farrell Dobbs and the Dunne brothers, controlled the Teamster locals. They were all for chartering the riffraff. Farrell Dobbs's particular goal was to bring in the "over-the-road" drivers — the long-haul interstate truckers — whom the old leadership had disdained. He also wanted to shift from citywide to regional bargaining. His instrumentality was the new Central States Drivers Council. One of his ardent disciples was a young organizer from Detroit named James Riddle Hoffa. "I was studying," Jimmy Hoffa said many years later, "at the knee of a master."[9]

In 1939 Dobbs decided to concentrate on the Socialist Workers party, whose presidential candidate he later became. Hoffa, who moved to the fore after Dobbs's departure, disavowed his mentor's Trotskyism. But he never disavowed his admiration for Dobbs as an organizer and later offered large salaries to lure him back to the union.[10] Even Dobb's Trotskyism may have had more influence than Hoffa admitted. Dobbs, Hoffa told the writer John Bartlow Martin, "was a very far-seeing man. He used to talk by the hour about what was going to happen in the country, and we'd listen."[11] Ralph and Estelle James, after long conversations with Hoffa in the 1960s, concluded that Dobbs's ruminations had left a deep imprint. Hoffa, they

---

the columnist, sued McCarthy after a brawl at the Sulgrave Club; McCarthy's Washington lawyer, a stalwart Italian American named John J. Sirica, decided he had a conflict of interests and passed McCarthy on to Williams (Edward Bennett Williams, in interview by author, January 23, 1976).

wrote, was "in the incongruous position of one who likes the present system, but does not believe it can work."[12] He plainly regarded capitalism as a racket that the strong manipulated to their own advantage — a system in which everyone was on the take, morality was bullshit and no holds were barred.

In any case Hoffa had come out of a hard school. The thirties was a savage decade for labor. Hoffa was designed for fighting — five feet, five and a half inches tall, with broad shoulders, big hands, thick legs; built, as John Bartlow Martin observed, like one of his own trucks.[13] He had an incisive intelligence, an unparalleled knowledge of the trucking industry, a stinging tongue, unlimited energy and, for his followers, unlimited charm. Unlike Beck, Hoffa could not care less about respectability. The rank and file saw in him their own aggressiveness, primitive prejudice and profane disrespect writ large. His achievements in raising wages, reducing hours and improving working conditions for the drivers were beyond dispute. Even the trucking associations were soon grateful to him for stabilizing labor relations in an industry notoriously vulnerable to unexpected work stoppages. He was also devious and perverse, had a raging temper, a bullying and sadistic disposition, a total absence of scruple and an impressive capacity for hatred. Paul Jacobs, an ex-Trotskyite himself and far from unsympathetic to Hoffa, wrote, "When Hoffa becomes really angry, his gray-green eyes get incredibly cold and menacing. It's then that his ruthlessness, his obvious belief in physical violence as an instrument of power, shows through as an important element in his personality."[14]

The Dobbs-Hoffa campaign to build a mass union in the middle west paralleled Beck's drive on the Pacific coast. By 1956 the Teamsters, with a million and a third members and welfare and pension funds rising toward a quarter of a billion dollars, were the largest and richest union in the country.

There was one other exception to the prevailing labor stagnation — the United Auto Workers. The two unions had a symbiotic relationship: the Auto Workers built the machines the Teamsters drove. If the Teamsters owed their vitality in great part to the organizing energy of Jimmy Hoffa, the Auto Workers owed theirs to the intellectual drive of another Detroit trade unionist, Walter Reuther. Like Hoffa, Reuther was a veteran of the bloody thirties. Both men had been baptized in political radicalism: Reuther by the Socialists as Hoffa by the Trotskyites. Both lived in Detroit. On one occasion, when John L. Lewis conducted a raid against the Detroit Teamsters, Reuther opposed Lewis and saved Hoffa.[15] "We knew each other in the rough and tumble days," Hoffa once said to John Bartlow Martin. "For some reason, it's kind of inexplicable, we drifted apart. There's no reason today we should have any arguments. He depends on us and we depend on him.... If Walter would have stayed in the labor movement rather than getting into politics, we'd never had any trouble. We still see each other. He's all right. What the hell, he's got his problems, I got mine. At least he knows what the union business is all about. George Meany don't."[16]

Despite this nostalgia for the old days, Reuther and Hoffa stood by 1956 at opposite poles. Reuther, with his commitment to union democracy, his modest salary, his personal frugality, his activity in the Democratic party, his spacious liberalism on every question of domestic and foreign policy, represented the afterglow of the labor idealism of the thirties. Hoffa, with his dictatorial control, his high salary, his expensive suits, his Cadillacs, his manifold deals on the side, his sweetheart relations with employers, his sympathy for the Republican party, was the epitome of business unionism. As he told Paul Jacobs, "I don't want to change the world."[17]

III

The west coast trip was Robert Kennedy's initiation into labor racketeering. An accountant named Carmine Bellino accompanied him. Bellino, who had run the accounting unit in the FBI before becoming a consultant to congressional committees, was an artist in his profession. With ledgers and canceled checks spread out before him, he could reconstruct the most carefully disguised financial transactions. He had watched earlier congressional investigations into labor racketeering come to nothing. "Unless you are prepared to go all the way," Bellino had told Kennedy, "don't start it." "We're going all the way," Kennedy replied.[18]

In Los Angeles they learned of collusive arrangements between the Teamsters Union and trucker associations. They heard reports of extortion and beatings. They met a Los Angeles union organizer who, as Kennedy wrote later, "had traveled to San Diego to organize juke-box operators. He was told to stay out of San Diego or he would be killed. But he returned to San Diego. He was knocked unconscious. When he regained consciousness the next morning he was covered with blood and had terrible pains in his stomach. The pains were so intense that he was unable to drive back to his home in Los Angeles and stopped at a hospital. There was an emergency operation. The doctors removed from his backside a large cucumber. Later he was told that if he ever returned to San Diego it would be a watermelon. He never went back."[19]

Kennedy flew north to Portland. There William Lambert and Wallace Turner of the *Oregonian* explained to him the complex local relations among politicians, Teamster officials and the underworld. His next stop was Seattle, Dave Beck's home town. Clark Mollenhoff had phoned Edwin Guthman of the *Seattle Times*, a Beck specialist, to say that Kennedy was coming. Guthman had won the Pulitzer Prize for national reporting in 1950. He had never heard of Kennedy and told Mollenhoff he had no interest in being used by the Senate committee and then left high and dry. "Can you trust him?" he asked. Mollenhoff vouched for Kennedy. Guthman, still dubious, met Kennedy at the airport. When Kennedy assured him that the committee was serious and that informants would be protected, Guthman decided to cooperate.[20]

He put Kennedy in touch with a group of dissident Teamsters. "I noticed almost immediately," Guthman wrote later, "that . . . he began to identify with the rank and file members of the union."[21] Kennedy soon uncovered information about Beck's raids into the union treasury. He learned, for example, that Beck, who had sold his lakefront Seattle house to the union for $163,000 two years before while receiving rent-free tax-free occupancy, had actually used union funds to build the house in the first place. Kennedy's informant also told him that Nathan W. Shefferman, head of a Chicago labor relations (i.e., as it later developed, union-busting) firm, held the key to Beck's financial operations.

Two weeks later Kennedy and Bellino were back in Seattle. This time they saw the waggish Frank Brewster, head of the Western Conference of Teamsters, who owned racehorses and used union funds to transport his jockey and trainer and maintain a box at the track. Then they went to Chicago a few days before Christmas to interview Shefferman, who had been making purchases for Beck and receiving payment from union funds. Shefferman yielded up his records, and they carried them back through a blizzard to the hotel for examination. "In an hour," Kennedy wrote later, "we had come to the startling but inescapable conclusion that Dave Beck, the president of America's largest and most powerful labor union, the Teamsters, was a crook."[22]

He went back to Hyannis Port for Christmas in a state of exhilaration. The atmosphere was unsympathetic. His father thought the inquiry into labor racketeering a terrible idea. An investigation, he said, would not produce reform; it would only turn the labor movement against the Kennedys. He was deeply, emotionally opposed, and father and son had an unprecedentedly furious argument.[23] The ambassador, who did not give up easily, asked Justice Douglas to intervene. Douglas told his wife that his mission to Robert had failed: "He feels it is too great an opportunity."[24]

The top labor leadership did not want an investigation. Then Kennedy, as George Meany's biographer put it, "jolted" the AFL-CIO president by telling him that Beck was a crook. "It's going to be rough, Mr. Meany," Kennedy said. "The evidence is here, and there's no way Beck can answer it." CIO leaders like Reuther meanwhile reminded Meany that they had dealt with their Communists; now the AFL leaders had to deal with their criminals. Moreover, the new AFL-CIO constitution, breaking with hallowed traditions of union autonomy, authorized the executive council to suspend or expel unions involved in corrupt practices. Late in January 1957 Meany got the executive council, over Beck's bellowing opposition, to pledge full cooperation with a congressional investigating committee.[25]

But which committee? Meany thought John McClellan "an anti-labor nut" and objected to the Investigations Subcommittee.[26] McClellan objected to what he called the "traditionally gentle" Labor Committee.[27] The Senate compromised by establishing a select committee with four members drawn from Investigations and four from Labor. This led to much soul-searching and maneuver among the senators. McClellan told Kennedy he deserved the chairmanship; "he had taken a lot of bad jobs from the Senate . . . and if they were going to deprive him of this one, he was going to be very upset."[28] Neither Symington nor Jackson, the Democrats next in seniority on Investigations, wanted to go on the select committee. Jackson, declining the assignment, explained to Kennedy that Beck had "a great deal of power [in the state of Washington] and . . . could make things most difficult" in 1958 when he was up for reelection.[29] After some vacillation Symington, who had presidential ambitions, followed Jackson's example. This made Sam Ervin, a conservative on labor issues, the second Democrat from Investigations on the select committee.

Senator Patrick McNamara of Michigan, a former trade unionist, agreed to be one of the two Democrats from the Labor Committee. John Kennedy was next in seniority. He felt at first that one Kennedy investigating the labor movement was quite enough. Joseph Kennedy undoubtedly agreed. Lyndon Johnson, the majority leader, told Kennedy it was madness for him to think he could take on labor and survive as a serious contender in 1960.[30] Still, if Kennedy refused, his place would go to Strom Thurmond of South Carolina, a Democrat in 1957 and noted then, as later, for regressive views on all questions. "Bobby wanted me on that committee," John Kennedy subsequently said, "in order to keep it more balanced. He asked me to come and give him some support or the committee would be too conservative and would have seemed anti-labor. You see, McClellan and Ervin went on, and if I hadn't, then Thurmond would have, and you can imagine what would have happened with Thurmond thrown in with McClellan and Ervin."[31]

The Republican members were Joe McCarthy and Karl Mundt from Investigations; Irving Ives of New York, a liberal, and Barry Goldwater of Arizona, a conservative, from Labor. On January 30, 1957, the Senate unanimously approved the Select Committee on Improper Activities in the Labor or Management Field with an appropriation of $350,000 and McClellan as chairman.

IV

Kennedy set up headquarters in room 101 of the old Senate Office Building. The office, cluttered with documents stacked in piles, had, John Bartlow Martin thought, the air of a secondhand bookshop.[32] Kennedy asked the tireless and cheerful Angie Novello, who happened also to be Carmine Bellino's sister-in-law, to be his secretary. When she said she would think about it, he said, "You're not going to think about it. I'm telling you. You are going to be my secretary." She said with a laugh, "Oh, yes, sir," saluted and clicked her heels.[33] He drafted his old friend Kenneth O'Donnell as his administrative assistant. "If the investigation flops," Robert Kennedy told him, "it will hurt Jack in 1958 and in 1960, too. . . . A lot of people think he's the Kennedy running the investigation, not me. As far as the public is concerned, one Kennedy is the same as another Kennedy." Assuming that the job had been cleared with John Kennedy, O'Donnell came to Washington. When the senator heard about it, he had words with his brother and later told O'Donnell irritably, "I need you up in Massachusetts. Nobody tells me anything."[34] O'Donnell became the point of coordination for the investigations as well as the guardian of the door to Robert Kennedy's inner office. Within, while they discussed investigatory strategy, they used to pass a football back and forth as if they were still at Harvard.[35]

Carmine Bellino was of course a necessity. Kennedy took the reliable LaVern Duffy from the Investigations Subcommittee. In February he broke from the usual committee pattern by hiring a portly, cigar-smoking newspaperman, Pierre Salinger. The previous summer, *Collier's* had asked John Bartlow Martin to do an exposé of the Teamsters. When Martin joined the Stevenson campaign, the assignment fell to Salinger, an experienced reporter in San Francisco. In the autumn, Salinger heard about the Senate investigation. He thereupon arranged an interview with Kennedy. "In five minutes," Salinger wrote later, "the roles were reversed. I was not interviewing him; he was interviewing me. . . . I was surprised at the information I gave him." In December, as the Salinger series was awaiting publication, *Collier's* expired as a magazine. In the gloomy aftermath Salinger received two phone calls. Einar Mohn, a Teamsters vice president whom Salinger knew from California, asked him to be public relations director for the Teamsters. Robert Kennedy asked him for his Teamsters research. Salinger, who suspected the Teamster offer as a device to insure his silence, turned his material over to Kennedy. The job followed.

McClellan, who disapproved of newspapermen, was not enthusiastic. Glaring at Salinger across his desk, he said, "If I ever hear of you talking to the press, I will deal with you in the harshest manner possible." On Salinger's first day in the office Kennedy directed him to serve a subpoena; it turned out to be for Einar Mohn. Salinger walked over to Mohn's office in the marble palace and said, "Einar, you know you offered me a job as public relations director for the Teamsters. I want you to know I was also offered another job at the same time and that was investigator for the Senate Labor Rackets Committee. I'm here to tell which job I've accepted. Here," and handed him the subpoena. That evening Kennedy took him home for dinner. When they reached the main course, Ethel asked, "What would you like to drink?" Salinger said, "I'd like a glass of red wine." Ethel said, "You know, we've never had any wine in this house." Salinger said, "Well, Mrs. Kennedy, don't worry about it. I happen to have a bottle in my suitcase."[36]

The next morning Salinger was on his way to Seattle. He spent most of the next year on the road. Kennedy later wrote that he "could grasp the importance of a document better than almost anyone on the staff."[37] He was also a man of courage, spirit and irrepressible wit and soon became a Kennedy intimate. Kennedy kidded him about his cigars, devised contests to bring down his weight, without much success, and took the Salingers along on Christmas skiing expeditions.[38]

In April another investigator joined the staff, a lawyer and former FBI agent named Walter Sheridan. Kennedy, rushing to a hearing, hired Sheridan going up the stairs from his office. "It was so typical," Sheridan said later, "because most of my conversations with him since then were walking up steps or riding in cars or up and down elevators or going to airplanes because he was always on the move and the only time he had to talk was while he was getting there." Later they discovered they were both born on the same day in the same year. Soon after Sheridan was taken on, Kennedy ordered him to Chicago to join Salinger on the investigation of Nathan Shefferman. Sheridan made a reservation on the noon plane. The next morning, coming upon Sheridan going through the files in the office, Kennedy "gave me this soft look and said, 'I never go any place at noon.' So, after that, I never went any place at noon." [39]

Leaving at noon wasted a day. They all worked, Sheridan said, "at an unbelievable pace." No one worked harder than Kennedy, jacket off, shirtsleeves rolled up, necktie pulled loose under his button-down collar, horn-rimmed glasses pushed to the top of his head, sitting under photographs (improbably) of Franklin Roosevelt and Douglas MacArthur. [40] For all his intensity, Kennedy, as John Bartlow Martin observed him, "infused his office and staff with a certain youthful light-hearted atmosphere." [41] The staff, numbering more than a hundred by 1958, was devoted to him, and he to them.

The investigators, Kennedy liked to say, "are the backbone of the hearings, and without their work we'd have nothing. I think we have the best investigators and staff of any congressional committee." [42] He was demanding but fair. "I'd worked in the FBI," said Sheridan, "and if an agent makes a mistake in the FBI, as Bob Kennedy would say, 'It's his ass.' But with him, you always knew that if you made an honest mistake, he was going to be behind you and he always was. . . . He was aggressive and he was direct and he was terribly honest in his approach to people and to life." [43] Years later, LaVern Duffy, reflecting after nearly a quarter century on the Government Operations Committee, said, "I've seen a lot of counsels here. There was no one like him. He had an uncanny ability to get people to do more than they thought they could do. He didn't do this by bringing pressure on them. It was because they wanted to please him. He gave people a sense of personal interest in themselves, their work, their families. This was his secret. He never got mad except when someone lied to him. He couldn't stand that." [44]

As the committee got under way, letters flowed into the office, a thousand a week. People wrote about union officials who declined to call union meetings or to audit union books. They wrote about rigged elections and theft of union funds, about racial discrimination, goons, intimidation, extortion, about beatings, acid throwing, murder. "I have been a union member for 25 years," someone wrote from Akron, "and until lately, have never been ashamed." "Please come in and restore our union rights and privileges to us. If we can get this exposé going we can kick them all out, with open elections" (Philadelphia). Letters ended: "I regret that I do not feel secure enough to sign my name, but my children also play in the streets"; or, "Please do not use my name — please — I'll be a gone goose if you do"; or simply, "I cannot sign my name." [45] About a third contained information that warranted at least preliminary investigation. [46] "There's something very curious I've learned," Kennedy mused to Murray Kempton one day. "If you get a letter typed on stationery, seven paragraphs in length, and signed by somebody, you can be absolutely sure it's a lie. But if you get a letter which says, 'I saw Jimmy Hoffa take $300 from somebody in a bar in 1947,' signed 'A Workingman,' it's always true."[47]

He kept in close touch with newspapermen. He was "under the illusion," Kempton said, "that journalists knew something about crime." Even the sardonic Kempton was impressed by his "way of asking direct, pertinent questions . . . just cutting to the heart of it."[48] In return Kennedy leveled with the press. "As a reporter," Edwin Guthman said, "I had never encountered a person in public life who answered my questions as candidly and completely as he did."[49] "Most reporters who covered Mr. Kennedy during this period liked him," wrote Joseph A. Loftus, the labor reporter for the *New York Times*. ". . . He was both available and truthful; there is no more effective way to disarm a reporter who by training and experience is a skeptic."[50]

John Seigenthaler of the *Nashville Tennessean* was not disarmed. Distressed by Teamster corruption and violence in Tennessee, Seigenthaler tried to give the story to Kennedy but caught him on a busy day in New York, was sent summarily to the assistant chief counsel, the able Jerome Adlerman, and told to put it all in a memorandum. A second attempt in Washington produced the same result; so did a third approach through Senator McClellan. As Seigenthaler later liked to tell it, Kennedy "kept saying, 'See Mr. Adlerman,' and . . . Mr. Adlerman kept saying, 'Send me a memo on it.' " Seigenthaler decided that Kennedy was a rich snob and a phony.

Around eleven o'clock at night some weeks later, Kennedy called Seigenthaler in Nashville and said he was sending down two investigators. The investigation led eventually to the impeachment and conviction of a judge, the dismissal of a district attorney, the defeat of a sheriff and the conviction of a number of police officials. Kennedy himself went down to testify against the judge. There were two planes back to Washington at the end of the afternoon, two hours apart, and, as Seigenthaler was driving Kennedy to the earlier flight, Kennedy said, "You know, I've never seen the Hermitage. . . . I'm a great admirer of Jackson and I'd like to go out there." They went to the Hermitage, where Robert called Ethel to say where he was and spent so much time that he nearly missed the second plane.[51]

v

The countdown on Dave Beck had begun even before the formation of the select committee. Early in January 1957 Kennedy met the

Teamster president in New York. Beck began with a forceful denunciation of racketeering. "His face," Kennedy wrote later, "grew
red and florid and his voice began to climb to a higher pitch until he
was almost shouting." It was as if he felt "he could yell us into
believing in the righteousness" of his cause.

Kennedy listened politely and then began to ask questions, derived
from the Shefferman papers, about Beck's financial life. Beck said
that, though he had nothing to hide, he could not indulge in a
discussion of private matters. Kennedy asked about his house in
Seattle. Beck said it was a private matter. Watching Beck, Kennedy
was struck by his eyes — at first appearing lost in his large, moonlike
face, but "soon you realized it was Dave Beck's eyes more than anything else that attracted your attention. They seemed like tiny pin-
points of light that were constantly moving back and forth." [52]

The investigation went forward. Many of Beck's vital files had
been destroyed. But the committee staff, drawing on independent
records and ancillary witnesses, was able, with the accounting wizardry of Carmine Bellino, to reconstruct much of the story —
$370,000 taken by Beck from the Western Conference of Teamsters;
$85,000 turned over to Shefferman to pay for personal purchases;
loans from trucking companies; even thefts from a trust fund set up
for a friend's widow. When Beck appeared before the committee in
the Senate caucus room in March 1957, Kennedy felt momentarily
sorry at the spectacle of a man who had grossly betrayed his union
brothers, stolen their money and was now "about to be utterly and
completely destroyed before our eyes." [53]

But Beck, exuding confidence, dominated the early hours of the
hearing. Mundt and Goldwater fed him general questions to which
he returned the pontifical answers of a certified labor statesman.
Toward the end of the day, McClellan called on Kennedy. The chief
counsel asked about Beck's use of union funds. Faltering before the
onslaught of detail, Beck retreated to the Fifth Amendment, invoking
it sixty-five times before the hearing was over. "The closing half
hour on Wednesday," Edwin A. Lahey, the seasoned correspondent
of the *Chicago Daily News,* wrote Kennedy, "was about the finest job
I've ever seen done on Capitol Hill. If ever Providential justice was
ladled out in the Caucus room, you did it that day." [54]

In subsequent appearances Beck's story, confidence and reputation
steadily disintegrated. "This is the case," Meany soon observed, "of
a very, very wealthy individual spending every waking moment, if
you can credit the testimony, to find some way to use his union as an
instrumentality for his own personal profit." [55] In due course Beck
was convicted and imprisoned for larceny and income tax evasion.

(In May 1975 he received a full pardon from President Gerald Ford.) The abrupt downfall of one of the most powerful and respected labor leaders in the country made the Rackets Committee overnight a national phenomenon.

The spotlight played particularly on the chief counsel. "Even with a couple of veteran inquisitors" like McClellan and Mundt on hand, wrote Harry Taylor in the *Cleveland Press*, "it is plain that this is Bob Kennedy's show so far."[56] Not all television viewers liked what they saw. An assistant professor of political science at Berkeley wrote Kennedy, declaring himself

> sickened by the radio and television accounts of your activities, . . . your humiliating, badgering questioning of the witnesses. . . . Perhaps another Kennedy will contribute . . . *Profiles in Bullying.* . . .
> To those of us who see your brother, John Kennedy, as probably the most dramatic liberal personality on the national scene today, the unfortunate possibility that he might be a victim of "guilt by association" and suffer from your current juvenile antics is depressing.[57]

Another correspondent complained to John Kennedy:

> Being wholeheartedly in sympathy with your committee, I was therefore greatly disturbed to observe in the televised segments that the chief interrogator . . . engaged in brow-beating, badgering and attempts to bulldoze the witnesses. . . . It was this same insolent and overbearing manner which was the downfall of Senator McCarthy.[58]

From Chicago, Sargent Shriver relayed to Joseph Kennedy (who promptly attended the hearings to see for himself) the observations of Richard Walsh, the executive editor of Hearst's *Chicago American*. "Words used to describe Bobby's manner," Walsh said, "ranged all the way from 'overeagerness' or 'intolerance' to 'viciousness.' " Even his collaborator in the Talbott case, Charles Bartlett, spoke of "the solemn moral tone with which he is wont to suggest the evil of a witness's ways."[59] He lacked, wrote Paul Healy of the *New York Daily News*, "the cool inward poise of his brother."[60]

Some reviews were more favorable. The *New York Herald Tribune* thought him "calm and polite, never baiting witnesses, but his questions are so penetrating that a witness is often caught off base."[61] The *Toledo Blade* remarked on his "innocent-looking blue eyes" and his habit of twisting a paper clip as he questioned bull-necked men about payoffs and call girls; "you expect him to blush any moment."[62] John R. Cauley of the *Kansas City Star* found him "stern in his interrogation of witnesses, but never abusive and seldom sarcastic"; all in all, "a remarkable personality even for Washington," who had "by his own intelligence, initiative and courage established himself as a real 'comer' and at only 31."[63] Robert Kennedy, Robert

Riggs wrote in the *Louisville Courier-Journal*, was winning "a reputation as one of the most competent investigators any Congressional committee ever had."[64]

"My biggest problem as counsel," Robert Kennedy himself conceded, "is to keep my temper. I think we all feel that when a witness comes before the United States Senate he has an obligation to speak frankly and tell the truth. To see people sit in front of us and lie and evade makes me boil inside. But you can't lose your temper — if you do, the witness has gotten the best of you."[65]

VI

For all the exasperations in the caucus room, this was the most fulfilling time Robert Kennedy had ever known. He was committed to something that mattered; he was convinced of the rectitude of his cause; and he knew he was doing a good job. "He wasn't frustrated during that period," Lem Billings recalled. "This is when he blossomed. . . . He wasn't the angry man any more and he was much more pleasant to be around because he hadn't this terrible feeling that he wasn't contributing."[66]

In October 1957, when the Soviet Union beat the United States into space, Lyndon Johnson said that "a successful investigation of Sputnik could only take place if [it] had someone like young Kennedy handling it." Recording this remark, Kennedy added: "Am very pleased with myself."[67] He must have been even more pleased by a Christmas letter from Clark Mollenhoff. "You have carried a candle," Mollenhoff wrote, "that has been a beacon to hundreds of reporters and editors, thousands of politicians and labor leaders, and literally millions of the rank-and-file labor union members and their families. . . . You may go ahead to higher office than committee counsel, but it is doubtful if anything you do will have greater force for good government and clean labor than what you have done this year."[68]

The job was hard on family life. Early in 1957 he bought from his brother John — for the same price, $125,000, that John had paid three years earlier — the fine old manor in McLean, Virginia, known as Hickory Hill, once the Civil War headquarters of General George B. McClellan, latterly owned by Justice Robert H. Jackson of the Supreme Court. After Jacqueline's miscarriage in 1956, the great house, complete with its new nursery, was too sad and echoing for the senator and his wife. It was ideal for Bobby and Ethel, whose five lively children were overflowing their small house in Georgetown. More children came along: Michael LeMoyne on February 27,

1958, Mary Kerry on September 8, 1959. A handsome white quasi-Georgian mansion, rich in historical association, with high ceilings, white woodwork and crystal chandeliers, surrounded by five and a half acres of rolling green land, Hickory Hill happily absorbed the Kennedys, their children and assorted dogs, horses, rabbits, chickens, goats and pigs. The Civil War association did not, however, excite the new owner. "McClellan," he would say sharply, "didn't press!"[69]

Robert Kennedy pressed. Often he did not get out to Hickory Hill till nine or ten at night. "It's nice Ethel has a big appetite," he once observed, "because she eats with the children at 6:30 and then with me again when I come home." ("I get reacquainted with the youngsters," he added, "every weekend.")[70] No doubt Ethel needed two dinners after a day of driving children back and forth to school and, in between, rushing to the Capitol to lend support at the hearings. But nothing discouraged her, and it took a good deal to rein in her high spirits. On an August night in 1957 one of her Skakel sisters-in-law gave a party in Stamford, Connecticut, at which Ethel was photographed standing cheerfully next to a couple identified by the hostess as Mr. and Mrs. Dave Beck, Jr. The local newspaper reported that Dave Beck himself had been at the party too. "Although our politics differ," he was quoted as saying, "socially we get along famously." After this bulletin got on the wires and attracted a certain attention, the "Becks" turned out to be New York friends of the young Skakels. The editor of the *Stamford Advocate*, taken in by the hoax, sourly informed his readers, "If you have any thoughts on adult delinquency, I wish you'd drop me a note." Nor did Robert Kennedy, who had missed the party, think it very funny.[71]

There were other irritations. Eddie Cheyfitz, he heard, was saying "that he had already established information that I had homosexual tendencies and that while in college, I ran across the campus in a girl's dress."[72] Some problems were more worrying. "Mr. Kennedy exposed himself and his family," wrote Joseph Loftus of the *Times* in 1960, "to terrible danger for three years."* In 1959 the *New York Herald Tribune* reported that Kennedy had received anonymous phone calls warning that acid would be thrown in the eyes of his children. He told newspapermen, "I don't care to discuss it."[73] He declined to take special precautions. "Asked by newspaperman if I

---

* Joseph A. Loftus, review of *The Enemy Within*, in New York Times Book Review, February 28, 1960. This statement enraged Hoffa, who, Loftus said, "accosted me with great fury.... I just sneered at him, and he walked away.... The tone that he used at me was a threatening kind of tone. Knowing his background and his ability to get even, why, I was somewhat concerned" (Joseph A. Loftus, in recorded interview by Jean Stein, March 26, 1970, 8, Stein Papers).

had 4 bodyguards," he noted at one point. "His paper had that
report & that they had been hired by my father to protect me. I
replied that the Kennedys have been taking care of themselves for
years & bodyguards unnecessary."[74] Fame, however, was incomplete. "Taxicab driver in N.Y.C. said he recognized me," Robert
wrote in one of his scribbled memoranda. "... 'I know who you
are. You're Roy Cohn.'"[75]

VII

Among those who watched the downfall of Dave Beck with equanimity was Jimmy Hoffa. Walter Sheridan's conclusion was that
Hoffa had "decided to try to use Kennedy to eliminate Beck."[76]
Certainly Eddie Cheyfitz, a Hoffa ally, gave Kennedy leads about
Beck while trying at the same time, as Kennedy said, "to implant the
thought that after a wild and reckless youth during which he had
perhaps committed some evil deeds, Hoffa had reformed."[77] In due
course Cheyfitz, suggesting that Hoffa could be a force for good,
urged Kennedy to sit down with him. "More out of curiosity than
anything else," Kennedy agreed to dine with Hoffa at Cheyfitz's
house.[78]

A week before the dinner, a lawyer named John Cye Cheasty told
Kennedy he had information that would make Kennedy's hair stand
on end. "In those days there were few people I talked with who did
not claim to have information that would make my hair stand on
end, and I tried to see them all."[79] Cheasty's information turned out
to be adequately hair-raising. Hoffa had hired him, he said, as a spy,
charged with penetrating the McClellan committee and reporting
back on the progress of the Teamsters investigation.

Cheasty, an Irish Catholic, had served in the Navy, Secret Service
and Internal Revenue Service. "He was the kind of fellow," Kenneth
O'Donnell said, "[who] thinks Reuther is a Communist and thought
Joe McCarthy was great."[80] As a Catholic, he later told Kennedy,
"I believe in right and wrong and that I must answer to God for my
conduct. All my life I have regarded good as something to work for,
and evil as something to be fought."[81] He produced bribe money,
ticket stubs and other evidence to corroborate his story. Kennedy
took him to McClellan. Cheasty agreed to work with the FBI as a
double agent. He was already handing selected information to Hoffa
when Kennedy went to Cheyfitz's house on the snowy evening of
February 19, 1957.

Hoffa's enthusiasm about the meeting had been even less than
Kennedy's. "I got nothing to talk to him about," he told Cheyfitz

crossly, but finally acquiesced.[82] Kennedy was impressed by Hoffa's "strong, firm" handshake. Hoffa was less impressed by Kennedy's handshake: "I can tell by how he shakes hands what kind of fellow I got. I said to myself, 'Here's a fella thinks he's doing me a favor by talking to me.'"[83] Cheyfitz offered his guests a drink. Hoffa "made rather a point of the fact that he did not drink," Kennedy noted in a memorandum two days later.[84] Kennedy, who rarely drank, declined also. Cheyfitz reluctantly followed suit.

They talked for a while about the labor movement. Kennedy, as was his custom, asked direct and personal questions: how Hoffa had got into the union, how much a labor leader earned and so on. He meant nothing by it; he was genuinely curious. Hoffa thought the questions condescending. "It was as though he was asking, with my limited education what right did I have to run a union like this?" He answered them in his own way. He told of battles against employers and cops. He said people were always trying to destroy him and therefore he had to destroy them. Kennedy tried to joke about Hoffa's toughness — "maybe I should have worn my bullet proof vest" — but "it seemed to go over his head." Hoffa kept repeating how tough he was; Kennedy remembered this as "the theme of all his conversation." ("I thought to myself that anyone who kept talking about how tough they are is not really tough.") "I do to others what they do to me," Hoffa said to him, "only worse."

During dinner Cheyfitz rambled on about how Dave Beck had ruined his son by choosing his friends, ordering his meals and turning him into a "jellyfish." Kennedy agreed; of all Beck's sins, he later wrote, "his attitude toward his son was his worst." After dinner, Cheyfitz said something about getting down to business. "If you don't ask a few of the embarrassing questions," he said to Kennedy, "then I will." Cheyfitz proceeded to bring up the "paper locals" in New York City. Paper locals are locals that exist only on paper: locals with charters and officers but no members. In an effort to get control of the New York Joint Council of Teamsters, Hoffa had arranged for the chartering of seven such locals just before the council election in 1956. Many officers in these paper organizations were hoodlums who had worked with Johnny Dio (Dioguardi), Tony (Ducks) Corallo and other eminent racketeers. Hoffa freely admitted organizing the phony locals but claimed it was done to circumvent the no-raiding pact scheduled to go into effect after the AFL-CIO merger.

Around half-past nine the phone rang: it was Ethel Kennedy. Bobby said, "I'm still alive, dear. If you hear a big explosion, I probably won't be." Ethel said that a driver, skidding on the icy road, had banged into a tree at Hickory Hill and was now sitting, hysterical, in their living room. Kennedy made apologies and prepared to go. Hoffa said, "Tell your wife I'm not as bad as everyone thinks I am."

Kennedy, perhaps parodying Hoffa's own line about doing to others what they did to him, but worse, noted later, "Personally, I thought he was, but worse." Hoffa was equally unadmiring. After Kennedy left he told Cheyfitz, "He's a damn spoiled jerk." Cheyfitz said, "You gotta get to know him." Hoffa said, "You know 'um, I don't wanta know 'um." It was a foredoomed encounter. Each represented what the other detested most. Hoffa saw Kennedy as the arrogant rich kid for whom everything in life had been easy. Kennedy saw Hoffa as the cynic who betrayed honest workingmen and had no object in life beyond money and power for himself.

VIII

Their next meeting took place at midnight three weeks later in the federal courthouse. John Cye Cheasty had continued to supply Hoffa with material, especially Carmine Bellino's analyses of Beck's financial sprees. "If that's the kind of stuff they have on Beck," Hoffa told Cheasty, "it looks as though his goose is cooked." The government decided to spring its trap on the evening of March 13. Cheasty handed Hoffa a large manila envelope on Dupont Circle. Hoffa, shaking hands with him, stuffed a wadded two thousand dollars in his palm. As Hoffa entered the Dupont Plaza Hotel, the FBI arrested him on charges of bribery and conspiracy.

At the courthouse an hour later, Hoffa and Kennedy sat in uneasy silence. "He stared at me for three minutes," Kennedy wrote, "with complete hatred in his eyes." After an interval they fell into an almost friendly conversation about who could do the more pushups. Kennedy claimed the higher figure. Then or later Hoffa said that no doubt Kennedy could do more pushups; "what the hell weight does it take to lift a feather?" Hoffa remembered saying at the time, "Listen, Bobby, you run your business and I'll run mine. You go on home and go to bed, I'll take care of things. Let's don't have no problems." He told John Bartlow Martin afterward, "He was very unhappy because I called him Bobby."[85]

It looked now as though it were Hoffa's goose that was to be cooked. When a reporter asked what he would do if Hoffa were acquitted, Kennedy said lightheartedly, "I'll jump off the Capitol." The trial began at the end of June. Hoffa's lawyer was Kennedy's friend Edward Bennett Williams. Kennedy was a prosecution witness. Hoffa, in his usual style, baited Williams: "Kennedy is a friend of yours so you will go easy on him." Williams, overcompensating, was therefore especially tough.[86] Kennedy lacked "courtroom experience," Clark Mollenhoff wrote, "and Williams tried to take advantage of this fact with a needling cross-examination. Kennedy had a momentary unsteadiness, but then became amazingly firm."[87] Hoffa, to the surprise of the prosecution, took the stand in his own defense. Employing all his persuasiveness, he contended that he had hired Cheasty as a lawyer and as a matter of course accepted any documents his lawyer gave him. The federal attorney, caught off guard by Hoffa's appearance, was feeble in cross-examination.[88]

The jury was two-thirds black; and the sight one day of Joe Louis embracing Hoffa in the courtroom no doubt had its effect. The *Washington Afro-American* published what was in effect a pro-Hoffa issue, in which columns and advertisements portrayed the defendant as a lifetime friend of the Negro people — he had indeed opposed segregation in the Teamsters — and denounced the judge, Cheasty and Senator McClellan as enemies of civil rights. This had obviously been done, said Edward Bennett Williams, who had not known about it, "in an effort to influence the jury in Hoffa's favor."[89]

Williams's summation was masterly; the prosecutor's was not. The verdict was returned on July 19. Cheasty called Angie Novello: "Angie, you won't believe this, but Hoffa was acquitted." Angie went to the hearing room, where Kennedy was interrogating a witness, and passed him a note. He gasped and looked, Clark Mollenhoff reported, "deathly sick." Mollenhoff himself rushed over to the courthouse in time to hear Williams remind the press of Kennedy's remark about jumping off the Capitol and add, "I'm going to send Bobby Kennedy a parachute." Mollenhoff said to Hoffa, "Jimmy, you are a lucky bastard." "I just live right," Hoffa replied with a wink.[90]

Kennedy, still shaken, returned to his office. Observing downcast faces, he suppressed his own disappointment. "Come on now," he said. ". . . We have a lot of work to do."[91]

IX

A fortnight after Hoffa's acquittal, Kennedy announced a new series of hearings. Racketeers and hoodlums were bad enough anywhere in the labor movement, he said, but worst of all in the Teamsters, for there was "no organization, union or business, that has a greater effect on the community life in this country."[92]

A few days later Pierre Salinger slipped into Detroit. He arrived, it need hardly be said, on an early morning plane, and by midafternoon he had served subpoenas on officials in five Teamster locals.

When he returned to his office in the federal building, the phone rang. It was Frank Fitzsimmons from Hoffa's own local. He said, "Jimmy wants to see you." The next morning Salinger, reinforced by Carmine Bellino, went to Teamster headquarters.

After more than an hour's wait, they were ushered into Hoffa's office and seated in low chairs in front of his desk. Hoffa, his own chair placed, like Mussolini's, on a slightly raised platform, stared down at them. Five union officials, clutching subpoenas, stood uncomfortably in a semicircle behind. In an apparent rage, Hoffa rose, grabbed the subpoenas from their hands and waved them furiously at the men from the Rackets Committee: "You can tell Bobby Kennedy for me that he's not going to make his brother President over Hoffa's dead body." Salinger and Bellino persisted. Hoffa's mood suddenly changed. He became affable and agreed to let them into his records. Given the subpoenas, he had no choice. The great Hoffa investigation had begun.[93]

The records were in poor shape, with names torn off checks and hotel bills. Potential witnesses fell silent or left town. When the hearings began on August 20, 1957, Hoffa appeared confident. "We'd rehearse [with the lawyers] what Kennedy would do," he told John Bartlow Martin. "He isn't the brightest fella in the world. And he's got to investigate for weeks and weeks to find out what we already know. . . . I know what I done wrong and what I didn't. I know what they'll uncover and what they won't. And that's two-thirds of the worrying. All my life I been under investigation."[94]

Hoffa began strong, exchanging deep thoughts about the labor movement with the senators, patronizing the counsel, whom he called "Bob," and responding aggressively to the occasional questions Kennedy got in about his private business dealings. The senatorial meanderings were largely surplusage:

SENATOR MUNDT. Basically, do you believe in socialism?
MR. HOFFA. I positively do not.
SENATOR MUNDT. Do you believe in our private-enterprise system?
MR. HOFFA. I certainly do.
SENATOR MUNDT. Is there some difference in political philosophy between you and some other prominent labor leaders in this country?
MR. HOFFA. Well, there certainly is, and it is going to remain so. . . . I believe that management and labor very definitely must at all times have something in common because one without the other cannot survive.
SENATOR MUNDT. I agree with you on that statement 100 per cent.[95]

Finally Kennedy was able to control the questioning. "Mr. Hoffa," he asked, "do you know Johnny Dio?"[96] Dio, a great chum of Hoffa's, was a particularly unsavory figure. The committee in its 1958 interim report described him as a "three-time convicted labor racketeer and suspected instigator of the blinding of Columnist Victor Riesel."* As Kennedy now dug methodically into Hoffa's relations with mobsters, Hoffa's cockiness began to wilt. Determined not to plead the Fifth Amendment, he chose instead to plead the imperfections of memory. Kennedy pursued him relentlessly, playing tapes of phone conversations recorded — legally in those days — by Frank Hogan, the district attorney of New York County. Under the pressure, the newspapers reported, Hoffa's eyes narrowed, his lips tightened, his facial muscles twitched.[97] At times he lapsed almost into gibberish. In answer to one question:

MR. HOFFA. To the best of my recollection, I must recall on my memory, I cannot remember.

MR. KENNEDY. "To the best of my recollection I must recall on my memory that I cannot remember," is that your answer?

. . . .

MR. HOFFA. I cannot recall the substance of this telephone call, nor place the facts together concerning what it pertains to.

THE CHAIRMAN. But if these things [the tapes] do not refresh your memory, it would take the power of God to do it. The instrumentalities of mankind, obviously, are not adequate. . . .

MR. KENNEDY. You have had the worst case of amnesia in the last 2 days I have ever heard of.[98]

John Bartlow Martin later asked Hoffa why he had performed so poorly. "Maybe I did look silly," Hoffa said mysteriously, "but there was a reason for it." He did not say what the reason was. Kennedy thought that the kind of proof made the difference. "He can say very forcefully someone's a liar — that's easy. But here we had his own voice on the tapes. He couldn't deny it."[99] "Talked with Hoffa after hearing," Kennedy noted.

Told him I did not want him to talk to our investigators as he had been doing. Told him if he wanted to hate anyone to hate me. He agreed. — His eyes were bloodshot. The last two days of hearing he was a beaten man compared to the beginning. His tone was subdued & no longer did he give the hate looks that he enjoyed so much at the beginning.[100]

A little later Kennedy wrote that things were much more relaxed after the Hoffa hearings; the committee, he thought, might be able to finish its job in another year. "Talked to Tommy Corcoran re doing legislative work. Said he had left Roosevelt Administration

* Riesel, a labor reporter, had been exposing underworld penetration of the labor movement. Senate Select Committee, *Investigation of Improper Activities in the Labor or Management Field: Interim Report*, 85 Cong., 2 Sess. (March 24, 1958), 253.

voluntarily — only one in Washington that did — Fight with Hopkins — Mistrusted Dad & him because 'We were too Irish to trust the English & too Catholic to trust the ?'"[101] The question mark was Kennedy's; he could not remember Corcoran's word, which must have been "Russians." Communism was a remote issue now.

For a moment Hoffa seemed to be on the run. The AFL-CIO Ethical Practices Committee reported that he used union funds for personal enrichment, that he worked with racketeers and that his union was "dominated or substantially influenced by corrupt influences."[102] He had already been indicted in New York on charges of illegal wiretapping. Now came another New York indictment, this time for perjury before a grand jury. But he was a hard man to keep down. The Teamsters rallied behind him, electing him their new president in September by a vast majority and defiantly voting Beck an annual pension of $50,000. At a highly emotional AFL-CIO convention in December, Meany succeeded in expelling the Teamsters. Hoffa remained unrepentant. In the same month his trial on the wiretapping charge ended in a hung jury — 11–1 for conviction. Soon the perjury indictment was dismissed.

The wiretapping case was retried in the spring of 1958. One day Kennedy and Hoffa ran into each other in the elevator in the federal courthouse in New York: "Hello, Jimmy" — "Hello, Bobby" — then silence until Kennedy asked how the trial was going. "You never can tell with a jury," Hoffa said. "Like shooting fish in a barrel."[103] As Hoffa remembered the encounter, "I said, 'Hello, Bob, how you getting along?' He'd give me that silly smile and go on about his business."[104] Hoffa was again acquitted; evidently fish could be shot in a barrel. He traveled around the country in triumph, proclaiming new organizing drives and haranguing the faithful.

x

Meany told Kennedy in April 1958 that "he expected great problems and trouble from the Teamsters and that Jimmy Hoffa was a genuinely evil man and that if we did not take care of him by investigation nothing would be done to deal with him."[105] Salinger, Bellino, Sheridan and others sought out witnesses and pored through files. In July a new set of hearings began. After a parade of mobsters marched across the stand, Hoffa himself returned grimly to the Senate caucus room.

The fun had gone out of the hearings. "Kennedy no longer seemed light-hearted," John Bartlow Martin wrote, "Hoffa no longer bantering."[106] The committee had tantalizing leads, but nothing conclusive, like the tapes of the year before. Hoffa maintained an effective defense of evasion and denial. Sometimes, for an hour or so, he appeared, Kennedy wrote in his journal, morose and discouraged. But he had "great stamina and he would bounce back again as forceful as ever." It went on, day after day. The strain told on both men. At the end of one grueling afternoon McClellan proposed a night session. "I was bushed," Kennedy recalled. "I looked at [McClellan] and smiled wearily." Hoffa, standing across the table, snickered, "Look at him, look at him! He's too tired. He just doesn't want to go on."[107] Kennedy himself noted in mid-September:

> I am mentally fatigued — more than during any other hearing. We have been going on for a long time without a break & I have about had it. I shall be happy when Hoffa is finished next week. McClellan also very tired. This year seems to have been tougher than last. Plodding grind. . . . I feel like we're in a major fight. We have to keep going, keep the pressure on or we'll go under.[108]

When Martin asked what Kennedy pinned his hopes on, Kennedy said, "I felt a little like 1864, when Grant took the Union Army to go back into the Wilderness — to go back slugging it out. I didn't pin my hopes on any one thing. Just all these things."[109] On the last day, Martin observed, Hoffa was "confident, almost flip. He seemed to know he was winning." Just before lunch, Kennedy accused him of threatening to kill Sol Lippmann, the general counsel of the Retail Clerks Union. "If I did it," Hoffa had said to Lippmann (or so a frightened Lippmann had told Kennedy the same day), "no jury would even convict me. I have a special way with juries." Hoffa, denying the story, said menacingly, "Bring Lippmann around here."

During the luncheon recess, Kennedy slumped at his desk before a sandwich and a glass of milk while Pierre Salinger called Lippmann. Taking the phone, Kennedy said, "Sol, can I get you to come over here and testify? It would make a real difference." There was a long silence. Then: "Are you going to let him stand up here and kick everybody around. Just because you can make a deal with him for six months or a year? . . . The fact he threatened you and made these statements to a leading union official like you — everybody would know you're telling the truth. Hell, I talked to you within 20 minutes of the time and I knew it was the truth." Finally: "O.K., Sol," and he slammed the phone down. He turned to Salinger and said, "He said, 'The Retail Clerks depend on the Teamsters in many ways. We have to get along with them.' . . . Shit." He drank his milk somberly.[110] Ethel Kennedy said, "This was the committee's worst day."[111] Hoffa, Kennedy wrote in his journal, had been alternately "arrogant, angry, pleasant, antagonistic. Tuesday of last week he looked awful. Thursday morning he did well for himself. He is difficult to figure out unless he's slightly mad which I think he might very well be." After the hearings Hoffa was more cocky than ever. "The noble traditions of Congress," he wrote in the *International Teamster*, "will continue to be abused as Senator McClellan of Arkansas and his millionaire counsel, Robert Kennedy, bend to their anti-labor bias and lend fuel to the raging fire of class hatred. . . . America is not the exclusive domain of the wealthy or the privileged. It belongs to all of us."[112]

In December Kennedy noted, "Jimmy has announced that he is going to organize Police & Firemen. This has thrown everyone into a tizzy. . . . His tactics are evidentally [sic] like those of Joe McCarthy — always take the offensive."[113] Then Hoffa called for an alliance of the Teamsters with Joe Ryan's International Longshoremen's Association, expelled from the AFL-CIO because its leaders were corrupt, and Harry Bridges's International Longshoremen's and Warehousemen's Union, expelled as Communist-dominated. This threw even Kennedy into a tizzy. He called it "an unholy alliance that could dominate the United States within three to five years . . . a subversive force of unequaled power in this country."[*]

Impregnable in his union, a victor thus far in the courts, Hoffa gleefully reminded Clark Mollenhoff that the Rackets Committee would not be around forever. "They'll be going for another nine months, and Bob Kennedy will be running off to try to elect his brother President."[114] "If it is a question, as Kennedy has said, that he will break Hoffa," Hoffa told a cheering audience of Teamsters, "then I say to him, he should live so long."[115]

XI

For all the impassable differences in values and objectives, the two men were not without superficial similarities. Both were blunt, candid and commanding. Both drove themselves and their staffs relentlessly. Both, in separate ways, had an instinct for the underdog; both had the instinct of underdogs themselves. Both had devoted friends and deadly enemies. Both had strong veins of sardonic humor. Both prided themselves on physical fitness. Neither smoked. Kennedy drank sparingly; Hoffa never. Neither wore hats. Both loved their wives. Both were risk-takers: "Hoffa," said Edward Bennett Williams, "is the kind of man who will jump out of a sixth floor

---

* Robert F. Kennedy, "Hoffa's Unholy Alliance," *Look*, September 2, 1958. Hoffa described Beck's longtime Pacific coast campaign against Bridges as "the worst mistake ever made by the Teamsters." Bridges, appearing at the Teamsters convention in 1961, called Hoffa "one of the greatest, outspoken, and fighting labor leaders of this country" and the Teamsters "one of the greatest hopes of our country." Romer, *Teamsters*, 138.

window one day. If he survives, he will jump out of an eighth floor window the next day."[116] There was that aspect to Kennedy.

The same image recurred when journalists wrote about them: "the coiled spring of a man constitutionally unrelaxed . . . like the impatient discipline of the hungry fighter, completely confident but truly relaxed only in the action he craves."[117] This was Hal Clancy of Boston writing about Kennedy; but John Bartlow Martin, writing about Hoffa, said, "It is as though the core of Hoffa's personality were all a tightly-wound steel spring. . . . He is always stripped for a fight, for action."[118] Martin summarized the resemblances in the *Saturday Evening Post* in 1959. They were both, he wrote, "aggressive, competitive, hard-driving, authoritarian, suspicious, temperate, at times congenial and at others curt."[119] Nothing bothered Kennedy more in Martin's manuscript than this passage. "He was amazed and simply could not understand; it had never occurred to him; he had thought of himself as good and Hoffa as evil." Martin retained the passage; his relations with Kennedy were not impaired.[120]

It was an intensely personal duel. Going home, Kennedy had to drive past the Teamsters' marble palace. One night, smiling wryly, he said to Martin, "My first love is Jimmy Hoffa."[121] Another night Kennedy and Salinger left the office at one in the morning. As they drove along, Kennedy noticed a light burning in Hoffa's office. "If he's still at work, we ought to be," Kennedy said, and they went back for two more hours.[122] Hoffa often did indeed work late, but he also, according to Williams, heard of Kennedy's reaction and thereafter, when his day's work was done, took special pleasure in leaving his office lights blazing away.[123]

Hoffa knew what respectable people thought of him and did not care. "I'm no damn angel," he told John Martin. ". . . I don't apologize. You take any industry and look at the problems they ran into while they were building up — how they did it, who they associated with, how they cut corners. The best example is Kennedy's old man."[124] As for the son, "Hell, I hated the bastard," said Hoffa at the end of his life. ". . . He was a vicious bastard and he had developed a psychotic mania to 'get' me at any cost."[125]

During the hearings Hoffa used to fix Kennedy with, as Kennedy perceived it, "a deep, strange, penetrating expression of intense hatred . . . the look of a man obsessed by his enmity, and it came particularly from his eyes. There were times when his face seemed completely transfixed with this stare of absolute evilness." The look sometimes went on for five minutes or so, as if he thought that by staring long and hard enough he could destroy his adversary. "Now and then, after a protracted, particularly evil glower, he did a most peculiar thing: he would wink at me."[126] Hoffa told Victor Lasky years later, "I used to love to bug the little bastard. Whenever Bobby would get tangled up in one of his involved questions, I would wink at him. That invariably got him." Hoffa laughed and laughed.[127]

Kennedy vacillated between thinking Hoffa evil and thinking him mad. It was widely supposed he hated Hoffa; but some who worked most closely with him did not think so. Angie Novello said later, "Deep down he had great respect for Hoffa." Carmine Bellino and LaVern Duffy both said that he "liked" Hoffa.[128] In some obscure personal sense he did, though liking was far overborne by moral disapproval. In later times, when his self-awareness and sense of the complexity of life increased, Kennedy might have seen Hoffa in another context, appreciated his vitality, his impudence, his struggle and the material benefits he had won for the Teamsters. But, just as Hoffa never forgave Kennedy for having been born rich and for believing in the possibility of justice under capitalism, so Kennedy could never forgive Hoffa for his vindictiveness toward dissenting members of his own union, for his ties with the underworld, for his conviction that American society was irremediably corrupt. "We were like flint and steel," said Hoffa. "Every time we came to grips the sparks flew."[129] Hoffa thought that Kennedy "got his jollies playing God."[130] Kennedy, Murray Kempton said, perceived "the devil" in Hoffa, "something absolutely insatiable and wildly vindictive. . . . He recognized in Hoffa a general fanaticism for evil that could be thought of as the opposite side of his own fanaticism for good."[131]

Fanaticism for good was better than fanaticism for evil, but it contained dangers of its own. Nor was Kennedy unaware of the perils of obsession. On January 15, 1959, he scribbled a note after a conversation with Eddie Cheyfitz: "Chafitz [sic] said that I was too e[n]meshed with Hoffa — that it was like Nixon on Alger Hiss."[132]

XII

The testimony before the committee had convinced Robert Kennedy that Jimmy Hoffa had stifled democratic procedures within the union, had ordered the beating and very possibly the murder of union rebels, had misused union funds to the amount of at least $9.5 million, had taken money and other favors from employers to promote personal business deals,* had brought in gangsters to consolidate his control and had tampered with the judicial process in order to escape prosecution.[133]

---

* Asked by Jerry Stanecki in a 1975 interview whether he was a millionaire, Hoffa replied, "I would say" (*Playboy*, December 1975).

Kennedy's critics were quick to reduce his pursuit of Hoffa to base motives: frustration over the failure to jail him; exasperation over his success at the hearings; vengefulness and vindictiveness. A great deal more was involved. Kennedy's exposure to the labor movement had given him impassioned sympathy for those in the Teamsters who, at tremendous personal risk to themselves, opposed the Hoffa dictatorship. He came to identify himself with workers who were expelled, beaten, murdered. Hoffa, as Kennedy saw it, had corrupted a movement of the oppressed. He was running not a bona fide trade union but "a conspiracy of evil," a conspiracy that had seized control of "the most powerful institution in this country — aside from the United States Government itself." As he concluded his case,

> Between birth and burial, the Teamsters drive the trucks that clothe and feed us and provide the vital necessities of life. They control the pickup and deliveries of milk, frozen meat, fresh fruit, department store merchandise, newspapers, railroad express, air freight, and of cargo to and from the sea docks. Quite literally your life — the life of every person in the United States — is in the hands of Hoffa and his Teamsters.[134]

He exaggerated Hoffa's importance. The Teamsters no doubt occupied a strategic position in the economy; but their power to bring the nation to a stop was theoretical. It depended on their failure to exercise it. Indeed, the Teamsters were traditionally cautious about calling strikes. It was true also, as Paul Jacobs insisted, that many of Hoffa's sins were practiced by his brethren in the labor movement. But even Jacobs conceded that "Hoffa's underworld ties do mark him off from many other union leaders with whom he otherwise has much more in common than is generally admitted."[135] Dave Beck, for example, had nothing to do with organized crime. It was the mob that roused Kennedy's deepest concern. "We are always going to have stealing and payoffs," he told John Martin. "As long as we have bank tellers we'll have that. And strike violence we'll always have. But this is far more sinister."[136]

There was a Damon Runyon aspect to it — Barney Baker, for example, six feet four inches tall, over three hundred pounds, a Hoffa organizer who had been a "collector" for a New York gang, had palled around with Bugsy Siegel, Meyer Lansky, Joe Adonis and other edifying figures and had served two terms in prison before becoming a Teamster official. In his testimony Baker mentioned "a Mr. Dunn."

MR. KENNEDY. Cockeyed Dunn?
MR. BAKER. I don't know him as Cockeyed Dunn. I knew him as John Dunn.

MR. KENNEDY.   Where is he now?
MR. BAKER.   He has met his Maker.
MR. KENNEDY.   How did he do that?
MR. BAKER.   I believe through electrocution in the city of New York. . . .
MR. KENNEDY.   What about Squint Sheridan? Did you know him?
MR. BAKER.   Andrew Sheridan, sir?
MR. KENNEDY.   Yes.
MR. BAKER.   He has also met his Maker.
MR. KENNEDY.   How did he die?
MR. BAKER.   With Mr. John Dunn.[137]

There was Joey Gallo, who ran an 'enforcer' union in New York and appeared one day in Kennedy's outer office, "dressed like a Hollywood Grade B gangster (black shirt, black pants, black coat, long curls down the back of the neck)," knelt down, felt the rug and said, "It would be nice for a crap game." Someone entered the office on an errand of his own. Gallo pounced on him, frisked him and explained to the bemused Kennedy staff, "No one is going to see Mr. Kennedy with a gun on him. If Kennedy gets killed now everybody will say I did it. And I am not going to take that rap." The visitor fled in consternation. On the stand, Gallo took the Fifth on all questions. Afterward he told Kennedy, "I'll line up my people for your brother in 1960." It was not really very funny. Joey Gallo had been a suspect in a murder case where the victim had been shot so many times in the face that identification was impossible. Kennedy asked him about this and received a giggle and shrug in reply.[138]

There was Joey Glimco, a Chicago hood who gained control of Teamsters Local 777 by strong-arm methods and thereafter enriched himself by misappropriating union funds and shaking down lesser Teamster officials, taxicab drivers and poultry dealers.[139] There was Paul (Red) Dorfman who took over the Waste Handlers local union in Chicago when the president was murdered in 1939. (Among those the police picked up for questioning after the murder was the local's secretary, Jack Rubenstein, who later changed his name to Jack Ruby and moved to Dallas, where in November 1963 he was in frequent touch with Red Dorfman and Barney Baker.)[140] Dorfman served as middleman between Hoffa and the Chicago mob, gaining in exchange the insurance for the Teamsters' Central States Welfare Fund for his stepson. Young Allen Dorfman, who did not even have an insurance brokerage license at the start, became a rich man in short order and a convict in the longer run.[141]

There was Momo Salvatore (Sam) Giancana, whom Kennedy described in the hearings as "chief gunman for the group that succeeded the Capone mob"[142] and other sources identified as don of the Chicago *consiglio* (i.e., boss of organized crime in Chicago).[143] Giancana, who had spent seven years of his youth in penitentiaries, used the Brotherhood of Electrical Workers (AFL) to take over the juke-box, coin-machine and cigarette-vending field in Chicago. He evaded a committee subpoena until investigators finally tracked him down in Las Vegas. "They couldn't catch me for a year," he boasted to the *Chicago Tribune*. "I was in Chicago all the time. . . . It was fun." He added: "What's wrong with the syndicate? Two or three of us get together on a deal and everybody says it is a bad thing. Businessmen do it all the time and nobody squawks." Called before the committee, he took the Fifth Amendment thirty-three times while Pierre Salinger set forth the more colorful features of his record. Finally:

> MR. KENNEDY.   Would you tell us if you have opposition from anybody that you dispose of them by having them stuffed in a trunk? Is that what you do, Mr. Giancana?
>
> MR. GIANCANA.   I decline to answer because I honestly believe my answer might tend to incriminate me.
>
> MR. KENNEDY.   Would you tell us anything about any of your operations or will you just giggle every time I ask you a question?
>
> MR. GIANCANA.   I decline to answer because I honestly believe my answer might tend to incriminate me.
>
> MR. KENNEDY.   I thought only little girls giggled, Mr. Giancana.[144]

Giancana was flashier than most. He was, as sociologists put it, upwardly mobile. One son-in-law ended up on the payroll of an Illinois congressman; the congressman promptly emerged as a doughty defender of Jimmy Hoffa.[145] A dapper fifty in 1958, Giancana liked to hang around show people and attracted show people who liked to hang around gangsters. The singer Frank Sinatra became his buddy;[146] another singer, Phyllis McGuire, his mistress. After a time Giancana's party life disturbed his associates. In 1966, the *Chicago Tribune* reported that he had been haled before a syndicate tribunal in order to "defend himself against charges of having permitted his international playboy activities to interfere with more serious, day-to-day business of running the underworld organization.[147] Giancana's activities were far more international than anyone could have supposed. Fifteen months after he giggled before the Rackets Committee, a fellow hood named John Rosselli approached him on behalf of the Central Intelligence Agency to solicit his help in a plot to murder Fidel Castro. The syndicates were penetrating corridors of power that not even McClellan or Kennedy in their worst imaginings would have suspected.

Then there was Tony Provenzano of New Jersey, who controlled Local 360 of the Teamsters, had close relations with the mob — New

Jersey law enforcement officials listed him as a "capo" in the Vito Genovese "family" — and spent his spare time shaking down businessmen. Before the Rackets Committee he took the Fifth forty-four times. In 1963 he was convicted of extortion and sentenced to seven years in prison. In the meantime, he rose to an international vice presidency in the Teamsters. In 1976 he was indicted for the kidnap-murder of a union opponent. His longtime alliance with Hoffa came to a bitter end in the early 1970s. It was Tony Pro with whom Jimmy Hoffa may have thought he had a rendezvous when he himself disappeared on July 30, 1975.[148]

XIII

With his quasi-Trotskyite views of capitalism, Hoffa probably considered gangsters no more immoral than businessmen or politicians. "Twenty years ago," he told an interviewer, "the employers had all the hoodlums working for them as strike-breakers. Now we've got a few and everybody's screaming."[149] He believed, Ralph and Estelle James reported, "that underworld forces lie just beneath the surface in most American cities, uncontrolled by the ineffective and/or cooperative police. He is fond of stating that a professional killer can be hired 'just like that' (with snap of fingers) for a mere $2000."[150] When Kennedy asked whether Barney Baker's lurid record disturbed him, Hoffa cracked back, "It doesn't disturb me one iota."[151]

It disturbed Kennedy a great deal. He loathed the labor racketeers. "They are sleek," he wrote, "often bilious and fat, or lean and cold and hard. They have the smooth faces and cruel eyes of gangsters; they wear the same rich clothes, the diamond ring, the jeweled watch, the strong, sickly-sweet-smelling perfume."[152] They preyed particularly on decentralized industries where employers were small and weak — trucking, coin machines, garbage removal, laundry and dry cleaning, newspaper distribution. The hearings documented in grim detail the brutality with which they punished those who would not go along.

As Kennedy saw it, Hoffa had begun by hiring hoods as business agents in Detroit because he needed strong-arm men. Once in, he allowed them to exploit workers and employers. Then they told colleagues in other cities that Hoffa was a "right guy" and they should let him in. "Hoffa wants ex-convicts, I think," Kennedy told Martin,

> because, first, he does a favor for them and they'll stick by him; second, by hiring an ex-convict he's doing a favor for another gangster, so then the gangster will do Hoffa a favor; and third, he uses them to move in on a city. To get into Chicago he needed the okay of Joey Glimco and

Paul Dorfman, and he gave Dorfman the Teamsters' insurance business. To get into Philadelphia he used Shorty Feldman. . . . To get into New York he was using Johnny Dio.[153]

He was becoming a vital link between the underworld and respectable society. Kennedy was learning from testimony what Hoffa had learned from life — that organized crime was a great unseen power. The Eighteenth Amendment, by driving the liquor business underground, had first changed gangsters into businessmen. The word *racket* appeared in the 1920s to describe the passage of crime from the free enterprise stage into the era of cartels and syndicates. The Twenty-first Amendment ended prohibition but not the criminal organizations prohibition had spawned. The 'gangster-outlaw,' like John Dillinger and Pretty Boy Floyd, gave way to the 'gangster-businessman,'[154] whose specialty was not simple and reckless deeds, like robbing banks, stealing automobiles, kidnapping, but the provision of services. "The criminal as such," Walter Lippmann wrote in 1931, "is wholly predatory, whereas the underworld offers something in return to the respectable members of society. Thus, for example, burglars are lawbreakers who, if they could be abolished miraculously, would not be missed; but bootleggers, panderers, fixers and many racketeers have a social function . . . for which there is some kind of public demand."[155]

The nature of these criminal organizations remained a mystery. In 1950–51 a Senate committee headed by Estes Kefauver, after communing with a number of racketeers of Italian extraction, suggested that "a loose-kni organization" called the Mafia provided the connection among the major criminal syndicates in the United States.[156] The Federal Narcotics Bureau, which was having troubles of its own in stopping the smuggling of heroin, welcomed the alibi of an international criminal cartel. The theory of the Mafia, moreover, plucked chords of romantic memory going back not only to the ancient secret society in Sicily but to the Camorra and Black Hand of America at the turn of the century. It hardly did credit, however, to the notable contributions that Irish and Jewish gangsters had made to the rise of organized crime. Nor did it find sustenance in the experience of the thirties. Thomas E. Dewey conducted his campaign against organized crime, including Lucky Luciano, without ever mentioning the Mafia. After breaking up Murder, Inc., Burton Turkus wrote in 1951, "As a factor of power in national crime, the Mafia has been virtually extinct for two decades."[157]

The theory of the Mafia gained new currency when high chiefs of the mob held a conclave at Apalachin in upstate New York in November 1957. A sergeant of the state police, his curiosity piqued by

the influx of black limousines, frightened the gangsters into igno-
minious flight into the surrounding forests. Fifty-eight conferees, including Vito Genovese himself, were picked up; twenty-two of these
were involved in labor or labor-management relations.[158] Sam Giancana made it into the underbrush. "There exists in America today,"
McClellan said, "what appears to be a close-knit, clandestine, criminal
syndicate. This group has made fortunes in the illegal liquor traffic
during prohibition, and later in narcotics, vice, and gambling. These
illicit profits present the syndicate with a financial problem, which
they solve through investment in legitimate business. These legitimate businesses also provide convenient cover for their continued
illegal activities."[159]

What had seemed in 1951 a loose-knit organization was beginning
to appear as something far more menacing: a monolithic, formally
organized, ethnically homogeneous, strictly disciplined secret society,
based on esoteric rituals, commanding the absolute obedience of its
members and controlling all organized crime — in short, an invisible
government of the underworld. The existence of the Mafia in this
sense has never been demonstrated.[160] No one could doubt, however,
the existence of interstate criminal cartels, sometimes working together, sometimes waging savage war against each other. The word
Mafia did not appear in Robert Kennedy's book about the Rackets
Committee, *The Enemy Within.* But he took from the hearings an
appalled sense, defined more carefully than McClellan had done, that
"the gangsters of today work in a highly organized fashion and are
far more powerful now than at any time in the history of the country. They control political figures and threaten whole communities.
They have stretched their tentacles of corruption and fear into industries both large and small. They grow stronger every day."[161]

The Kennedys doubted whether existing methods were adequate
against organized crime. After Apalachin, Robert Kennedy had
asked for the FBI files on the conventioneers. "The FBI didn't know
anything, really, about these people who were the major gangsters in
the United States," he said later. ". . . That was rather a shock to me.
. . . I sent the same request to the Bureau of Narcotics, and they had
something on every one of them."[162] The Kennedys now called for
the establishment of a national crime commission as a clearing-house
for criminal intelligence. The FBI, Robert Kennedy said, "won't keep
these fellows under surveillance. They do it for suspected spies, but
they won't do it for gangsters and racketeers."[163]

He was growing disillusioned about the FBI. Angie Novello recalls
that he generally mistrusted and sometimes disliked investigators who
had once worked for the Bureau; Bellino and Sheridan were excep-

tions.[164] The Bureau seemed suspiciously wary of unpromising cases. "They want the publicity from their press releases," Kennedy noted in May 1958, "but where they have a natural opportunity to get into these things, particularly where it affects us, they are very reluctant to do so."[165] Kenneth O'Donnell expressed Kennedy's views bluntly in 1959: "The FBI has never been aggressive on big crime. It went after Communists and stayed there." O'Donnell did not spare the then sacrosanct J. Edgar Hoover. "The FBI chief is afraid to step on the toes of local officials — afraid he will lose some popularity and get involved in local politics. But local authorities can't handle nation-wide syndicates."[166]

"If we do not on a national scale attack organized criminals with weapons and techniques as effective as their own," Robert Kennedy concluded, "they will destroy us."[167] They might even destroy Hoffa: asked in 1958 whether Hoffa could clean up his union if he wanted to, Kennedy replied, "He can't say, 'You're out.' He wouldn't live."[168] In the end, Kennedy was indisputably right. By 1975 Hoffa must surely have agreed with him.

The Second Investigating Committee:
Walter Reuther

Not all the Rackets Committee was enthusiastic about the pursuit of Jimmy Hoffa. Senators Goldwater and Mundt did not condone Hoffa's excesses. But they declined to see him as the primary threat. He was, after all, a supporter of Republican candidates, a model of cozy relations with employers, a declared believer in private enterprise. His purpose, he explained to the committee, was to get a better deal for his Teamsters, not to transform the system. "Mr. Hoffa," Goldwater observed, "we have labor leaders in this country today, labor leaders who are not particularly friendly to you, labor leaders . . . who do not think like that." Seizing the cue, Hoffa obligingly condemned the use of union power to bring about political and economic change. "We both recognize," Goldwater said, "that in the writing in the clouds today there is an individual who would like to see that happen in this country. . . . For the good of the union movement I am very hopeful that your philosophy prevails."[1]

I

There was no mystery about the individual whom Goldwater saw in the clouds and whose philosophy he hoped Hoffa would defeat. Walter Reuther had become the supreme bogeyman for American conservatives. Hoffa, for all his sentimental memories of Reuther, craftily nourished this fear. He called Reuther "the leader of Soviet America," insinuated Reuther was protecting organized gambling in the auto plants and hinted that Robert Kennedy was going easy on him in order to get the United Auto Workers behind his brother in

1960. He bragged to Clark Mollenhoff that he had passed this "information" to the Republicans on the Rackets Committee.[2]

The Republicans hardly needed prompting. "Mr. Reuther and the UAW," said Goldwater, "have done more damage and violence to freedom than was accomplished by all the peculiar financial transactions of Mr. Dave Beck."[3] Joe McCarthy promised that, if and when the committee got around to Reuther, the activities uncovered would leave Beck smelling like a rose. Asked about this on *Meet the Press*, Kennedy said, "We have no information at the present time showing that Mr. Walter Reuther personally misused any union funds. There have been certain complaints about . . . the Kohler strike."[4] The Kohler strike, which the UAW had been conducting in Sheboygan, Wisconsin, since April 1954, had excited national attention because of its bitterness and violence. The Republicans now set up a cry for an investigation of the strike, the UAW and Reuther.

At this point, they had to confront the rather private way Senator McClellan and Robert Kennedy ran the committee. When asked in July whether the committee as a whole decided what to investigate, the counsel said candidly, "No, I pretty well decide it and I have consultations with the Chairman . . . and he is aware of where we're sending investigators and what we're looking for."[5] McClellan was exactly twice Kennedy's age, but after several years together on the committee a surprising fondness had developed between the dour Baptist from the public schools of Sheridan, Arkansas, and the Irish Catholic from the Ivy League. "His practice of treading where McClellan feared to walk," wrote Charles Bartlett, "earned him the senator's deep devotion, broad license, and increasing support. It is almost certain that no hired official of the Senate has ever wielded more committee authority or operated with so free a hand."[6] Kennedy for his part always deferred to McClellan, teased him, LaVern Duffy said, "with respect"[7] and admired him as "the most devastating cross-examiner I ever heard."

Kennedy was also fond of Sam Ervin. He was exasperated by Pat McNamara, who had both Hoffa and Reuther among his constituents and, an old trade unionist himself, regarded the investigation of unions with extreme discomfort. When McNamara resigned from the committee in 1958, Frank Church of Idaho took his place. "He was absolutely fearless," Kennedy wrote later of Church, "asking some of the most astute and penetrating questions that I heard put. I envied his articulateness." Kennedy also liked Irving Ives, a former dean of the New York State School of Industrial and Labor Relations, and valued his special knowledge of labor problems. Despite their disagreements, he liked Goldwater. "He could cut you to ribbons, slit your throat, but always in such a pleasant manner that you would have to like him."[8]

But he considered Karl Mundt a sanctimonious hypocrite who played up to Hoffa and Reuther when they were on the stand and "attacked them viciously when they were not present to defend themselves."[9] Mundt's leaks to the press were a perennial irritation. Once Kennedy and Salinger, in order to verify their suspicions, sent the eight senators a draft report in which the key page was written in eight different ways. The version that promptly appeared in the newspapers was the one sent to Mundt.[10]

II

In the spring of 1957 the Rackets Committee underwent its first change in membership. Joe McCarthy was in marked decline, a specter haunting the scenes where he had so recently swaggered. There were still flashes of the old brutality. Walking past Arthur Watkins's seat in the Senate, he liked to lean over and whisper, "How is the little coward from Utah?"[11] But something had broken within him. Because he was a sensationalist rather than a true believer — his was the glory drive, Edward Bennett Williams used to say, not the power drive — he had no core of inner certitude to sustain him in adversity. By the time of the Rackets Committee, he was a wreck. "He would come into the Caucus Room late," wrote Richard Rovere, "interrupt a line of questioning with questions of his own, some of which were incoherent, and after twenty minutes or so wander out in an almost trance-like state."[12] "He has been drunk the last three times he has been to the hearing," Robert Kennedy noted on March 5, 1957.[13]

One day McCarthy demanded that Kennedy fire an investigator named McGillicuddy on the ground that he was anti-Catholic and had been indicted. After some talk Kennedy ascertained that McCarthy was talking about a man named Robert Greene.

> I told him the Bob Greene indictment had been dismissed and denounced as a political indictment because of the work Greene was doing investigating Mayor Kenny in Jersey City, and furthermore, that he was an Irish Catholic and could not be considered anti-Catholic. The next thing I heard was that he made a speech on the Senate Floor stating that the Committee had hired an investigator who had been indicted. . . . I told the press that afternoon that I had great confidence in Greene and also told the facts surrounding his indictment back in 1951.

McCarthy persisted, explaining to the committee that anyone who wallowed with hogs — Greene had worked for the anti-McCarthy

Long Island paper *Newsday* — must be a hog himself. Finally Senators Mundt and McNamara were designated to look into the matter and promptly cleared Greene.[14]

Kennedy watched McCarthy's disintegration with sadness. "I felt sorry for him," he said later, "particularly in the last year, when he was such a beaten, destroyed person particularly since many of his so-called friends, realizing he was finished, ran away from him and left him with virtually no one."[15] The incorruptible Edwin Guthman of Seattle, though impressed by Kennedy's exposure of Beck, still felt, he later said, that "I had to get to the bottom of the Kennedy-McCarthy relationship before I could fully trust Bob." They discussed McCarthy. Guthman decided that Kennedy's chief feeling about McCarthy was pity: "Pity for an acquaintance who could be a pleasant companion but who had made a ruin of his career; pity for the men and women and their families whom McCarthy had forced needlessly and unfairly to live under a cloud."[16] Once Kennedy took Lem Billings over to McCarthy's house. Mrs. McCarthy did most of the talking. McCarthy sat by in a stupor. Billings, somewhat mystified by the occasion, said later, "It was typical of Bobby to go in and see somebody who was in trouble."[17] "Joe has been looking so terribly," Kennedy himself noted. ". . . Couldn't stand straight & often appeared in a trance. His conversation was not intelligible."[18] At the end of April McCarthy entered the Bethesda Naval Hospital. There he died on May 2, 1957.

"Very upsetting for me," Robert Kennedy wrote in his intermittent journal.

> I am only happy that I had a friendly conversation with him last week.... Fred Perkins of Scripps Howard after the Bob Green [sic] incident had evidentally [sic] gone to Joe & told him I had called him an S.O.B. — Evidentally a Wisconsin newspaperman had heard the same thing because latter told me he was going to print it.... I am sure that Joe McCarthy knew of this & I felt a little strain about him. I am glad that we talked in the manner we did before he died.
>
> I dismissed the office for the day. It was all very difficult for me as I feel that I have lost an important part of my life — even though it is in the past.[19]

"The death of so young a Senator of such strong convictions," said Senator Jacob K. Javits of New York, "is truly sad. It should be our hope and aim that in the national interest the warnings of Communist subtlety and duplicity uttered by Senator McCarthy will long survive the controversy about his methods of investigation."[20] There was — for the first time since the death of William E. Borah — a state funeral in the Senate chamber. Robert Kennedy also attended the ultimate funeral in Appleton, Wisconsin.

Senator Carl Curtis of Nebraska replaced McCarthy on the committee. The Republicans, except for Irving Ives, were now more
deeply conservative than ever. Kennedy detested Curtis, as he later
did Homer Capehart of Indiana, who replaced Ives in 1959.

III

As the Rackets Committee set out after Hoffa, the Republicans increased their demand that Reuther and the Kohler strike receive
equal time. The National Labor Relations Board had recently concluded two years of hearings on Kohler. Kennedy doubted that the
committee would turn up anything new. Newspaper stories quoted
unidentified Republican senators as saying that the Kennedys were
covering up for Reuther. Kennedy believed Mundt to be the
source.[21] In fact, John Kennedy hardly knew Reuther, and Robert
had never met him. As for Reuther, he had supported Kefauver
against John Kennedy during the vice presidential contest in 1956
and distrusted Robert because of the McCarthy committee.

In July *Newsweek* cited Republican complaints that Robert Kennedy
had "ignored continual demands for an investigation of Reuther"
and had concentrated instead on the harassment of the union leaders
"who had stood in the way of Reuther's domination of American
labor." They complained too that Kennedy never told them what he
was doing.[22] At the next meeting of the committee, John Kennedy,
flourishing *Newsweek,* said that, if this were the way Republican senators felt, let them speak up.

Mundt proposed a reorganization designed to reduce Robert Kennedy's control of the investigations.[23] This was rejected, but the
counsel agreed to give the committee a full briefing on investigative
developments every Monday morning. As for the Reuther problem,
Robert Kennedy suggested that Jack McGovern, a Republican hired
at Goldwater's request ("the only appointment made to the staff,"
Kennedy said later, "because of political considerations"), be assigned
to the UAW-Kohler investigation.[24] Goldwater said he was "as happy
as a squirrel in a little cage." Mundt said he was "as happy as a South
Dakota pheasant in a South Dakota cornfield." Curtis, Kennedy
wrote later, "made no comparison between himself and animals or
birds" but expressed general happiness. McClellan said ironically,
"We shall now adjourn this love feast."[25]

In the next months Kennedy heard nothing from Jack Mc-
Govern. The Republican senators, however, received regular re-
ports, and in October McGovern sent a long memorandum to Mc-
Clellan. While purporting to be the product of McGovern's own

investigation, this document, Kennedy noted, was mostly "copied from NLRB files with none of the derogatory info on the company — if he was not a political appointee I would fire him."[26] Kennedy told Goldwater that the full NLRB report seemed essentially correct "and that therefore a new investigation was unlikely to accomplish what he wanted and expected, namely to destroy Walter Reuther."[27]

Goldwater persisted. In January 1958 he called Reuther and the UAW "a more dangerous menace than the Sputnik or anything Soviet Russia might do to America."[28] (When Reuther objected, Mundt disarmingly defended his colleague: "That would not make you very dangerous. Sputniks are not dangerous. They are weather beeping things going around.")[29] "Corruption, misappropriation of funds, bribery, extortion and collusion with the underworld," according to the Republican argument, existed in the UAW "as in the Teamsters." Reuther was an enemy of the capitalist system who used violence and intimidation as "essential parts" of his "formula for power."[30]

The UAW decided to take the offensive. Reuther placed his administrative assistant, Jack Conway, in charge of the operation. Conway, an Irish ex-Catholic with a degree from the University of Chicago, dumped on Kennedy's desk a huge stack of documents covering, he said, everything the UAW knew about the Kohler strike. "I made it very clear that we were going to be open and above board and direct and that if he was the same way that we'd have no trouble but if he wasn't we'd have trouble." Kennedy questioned the UAW charges against Kohler: things couldn't be that bad in Sheboygan. Conway said, "If you believe that, why don't you go out and take a look yourself?"[31] Kennedy said, "I will," and, to the astonishment of Conway, who thought he might go in a month or so, flew to Wisconsin the same night.[32]

He found Sheboygan infinitely depressing. The effects of the long strike "still hung over it like a shroud, touching everything, affecting everyone." It reminded him of the Middle East. "The hatred [between Jew and Arab] I had felt on that visit to Palestine was just as livid here. Unless you can see and feel for yourself the agony that has shattered this Wisconsin community, it is difficult to believe that such a concentration of hatred can exist in this country. . . . It was difficult to believe that this was America 1958."[33]

The strike was nearing the end of its third year. Kohler, a family-owned corporation operating a company town, was the second largest manufacturer of plumbing ware in the United States. It had a history of labor violence. In 1934 it had brought in armed guards to break a strike, killing two strikers and wounding nearly fifty more. In 1950 the company union, feeling that band concerts and similar feudal benefits were inadequate substitutes for low wages, sought affiliation with the UAW.*[34] In 1952 Kohler reluctantly signed a contract. In 1954, with the contract up for renewal, the company rejected the union demands. The union struck. It engaged in illegal mass picketing. There were violence and vandalism, condoned if not encouraged by UAW officials. Kohler fought back and soon resumed full operation. After a time the union was ready to concede every point except for a small wage increase. The company declined to bargain in good faith. It was evidently seeking — so the NLRB trial examiner reported in October 1957 — not an agreement with the union but "the reduction of the union to impotency as an effective bargaining representative of their employees."[35]

Kennedy went first to the Kohler plant. He saw the enamel shop, where the heat of the ovens brought the temperature well above a hundred degrees and workers took off face shields for a moment to gobble lunch; the UAW had been asking for a thirty-minute luncheon break. He talked to Lyman Conger, the company attorney, who "made no secret of his deep and abiding hate for the union. It was an all-consuming hate — a thing unpleasant to see." He went to the union office, which had the defeated look of "the campaign headquarters of some lesser political candidate who knew he couldn't whip city hall." The contrast between the UAW organizers and the Teamsters struck him. "They wore simple clothes, not silk suits and handpainted ties; sported no great globs of jewelry on their fingers or their shirts; there was no smell of the heavy perfume frequently wafted by the men around Hoffa." They explained the ordeal the town had been through since 1954. "These men felt hatred, too. But it was hatred born of anger and frustration — not the insatiable hatred Conger seemed to feel."[36]

Kennedy had started with "no strong convictions . . . one way or another" about the strike.[37] Sheboygan changed that. On his return to Washington he joined members of the committee at a dinner with Meany and other labor leaders. After dinner ("the ice cream was good"), Meany sat on a couch with Ives on one hand and McClellan on the other. "The couch was too small for the three of them," Kennedy noted, "and they made a rather peculiar looking group — Ives looking sickly and Meany looking very fat with his cigar; McClellan on the other side looking like he did not want to get too close to a labor leader." Meany, though no fan of Reuther's, observed that

* Kohler had an engine plant department, which made UAW involvement less incongruous.

the UAW investigation was "a political investigation because we knew as well as he did that there wasn't anything to look for. He said a lot of people disagreed with Walter Reuther politically but that otherwise it was generally known and understood that he was honest."[38]

IV

The time had come for the committee to consider the counsel's draft of a report on its first year. "Mundt, Curtis and Ives," Kennedy wrote in his journal, "were all anxious to defend management — to [delete] references we had to businesses. In my estimation it was a sickening spectacle."[39] When the expurgated report was released, labor spokesmen found it sickening too. It provoked McNamara's resignation. Meany said it raised "grave doubts as to the [committee's] impartiality, objectivity and integrity."[40] Persuaded that the UAW had a case on the merits, Kennedy recommended that the committee hold hearings on the Kohler strike.

This might offer a chance of recovering labor confidence. Nor did Kennedy and O'Donnell neglect the political possibilities in the UAW connection. "Their curiosity about detail and how we worked in different states and areas of the country was more than the usual," Jack Conway recalled. "I was personally convinced that what they were attempting to do was to feel out a potential alliance." One night, at dinner with John Kennedy, Robert Kennedy said bluntly that, if they went into the 1960 convention, they did not want to get caught in the situation of 1956. "We don't want you on the opposite side from us."[41]

The Republicans were counting heavily on a letter Walter Reuther and his brother Victor had written from the Soviet Union in 1934, ending in the spirit of the early depression: "Let us know definitely what is happening to the YPSL [Young People's Socialist League], also if the Social Problems Club at City College is still functioning and what it is doing. Carry on the fight. Vic and Wal." Through the years enemies of the Reuthers had improved this peroration, amending it to read in some versions: "Carry on the fight for a Soviet America." At least five different texts of the letter were in circulation by 1958. The recipient of the letter, Melvin Bishop, had long since turned against the Reuthers. Kennedy brought Bishop to Washington. Bishop said that he had given the original to a schoolteacher in 1935 and had not seen a copy since. Carmine Bellino tracked down the schoolteacher; she said she had never heard of the letter. McClellan spoke contemptuously of Bishop's "established unreliability" and dismissed the whole thing.[42] No one — not even Mundt — brought it up in the hearings.

The Republicans, repenting their enthusiasm for hearings, urged postponement. The Kennedys, knowing they would be blamed for protecting Reuther, insisted that the committee go ahead. The Republicans then concentrated on defeating Robert Kennedy's recommendation that Reuther should be the first witness for the UAW. He would steal the show, Ives said, and everything else would be anticlimactic. "I said," Kennedy wrote in his journal, "that I felt he was the best witness and should have the right to appear." McClellan said grumpily that he could not understand what the Republicans had to fear from letting Reuther lead off. (Kennedy and McClellan lunched together after the meeting. "He didn't speak for 20 minutes," Kennedy noted. "I imagine he is mad because of the fact that he seems to be appearing on the side of the UAW which, of course, kills him.") McClellan, Ervin and John Kennedy argued through four more executive sessions for the early calling of Reuther. "Goldwater, Mundt and Curtis were all very bitter and bitter-looking during these conferences," wrote Robert Kennedy. "Their faces were distorted to what was almost hate." Eventually McClellan yielded. "The proceedings that will be followed here," he said carefully when the hearing opened, "are not the proceedings in keeping with the Chair's views as to how this matter should be presented."[43]

Reuther himself observed that most people hired lawyers in order to stay away from the Rackets Committee; he was hiring a lawyer in order to appear. He had wanted Arthur Goldberg, the special counsel of the AFL-CIO, to represent the UAW. But Jack Conway favored the Washington civil liberties lawyer Joseph L. Rauh, Jr. "Goldberg," Conway said later, "is a broker and Joe Rauh is a battler." Goldberg himself, when approached, felt that his own first obligation was to the Steelworkers, of which he was counsel, and also recommended Rauh.[44] Rauh was indeed a battler. So was Goldwater. The result was a series of long, acrimonious sessions.

Goldwater, for example, defending the arsenal of guns, shells and tear gas built up by the Kohler management, observed that CIO strikes up to 1954 had caused thirty-seven deaths; "this is an example of what this union has been doing in violence ever since its inception."[45] "He did not mention at all," Robert Kennedy wrote angrily in his notes, "that all these people [killed] were strikers."[46] When Goldwater called it "strange" that violence originally associated with the Communist party was now associated with the UAW, John Kennedy observed, "Mr. Chairman, my brother's name is Joe and Stalin's name is Joe. The coincidence may be strange but I don't draw any inference from it."[47]

Lyman Conger was as intransigent in Washington as he had been in Sheboygan. Robert Kennedy pressed him about lunch in the enamel shop.

> MR. KENNEDY. You say that the men can put the [enamel] equipment in the oven, then they can step back and eat their lunch during that period of time. How much time is there then before they have to do some more work?
>
> MR. CONGER. From 2 to 5 minutes, depending on the piece.
>
> MR. KENNEDY. So you feel they can step back from the oven and take off their mask and have their lunch in 2 to 5 minutes?
>
> MR. CONGER. Mr. Kennedy, they have been doing it for 36 years.[48]

This exchange "made me furious," Kennedy later wrote, "madder really than when witnesses took the Fifth Amendment. . . . To hear a reputable American businessman, in 1958, matter-of-factly advocate a two- to five-minute luncheon period — well, until then I had believed that that kind of thinking had long since disappeared from the American scene."[49]

Kennedy regretted in retrospect "that I did not question Conger more. There were those who felt, and I felt at the time, he looked so bad there might be some sympathy created for him. But there were so many things I felt we could have gotten into and established more clearly about the Kohler Company that I should have done it more than I did." However, he had found no support on the committee for bringing out this material, "other than Jack and of course, he and I were both suspect. Neither McClellan, Sam Ervin nor McNamara asked any critical questions of Conger or [Herbert V.] Kohler. I asked all the critical questions of Conger and Jack did all of Kohler."[50]

"Bobby must have had some form of communication with his brother Jack," Joseph Rauh observed later, "or there really is a thing called extrasensory perception. Every time we were getting into trouble, Jack would enter the Hearing Room, take his seat on the Committee dais and help us out. It got to be a joke inside our crowd as Jack walked down the aisle each time Mundt, Goldwater or Curtis was scoring points against us."[51]

Points were occasionally scored. The UAW official in charge of the Kohler strike was the secretary-treasurer, Emil Mazey, a truculent and emotional man who, as Robert Kennedy accurately said, "shouted first and thought later."[52] Mazey had not been a force for restraint in the early months of the strike; nor was he now at the hearings. After one session he called Conger "a despicable ———," as Kennedy primly reported the scene in his notes,

and the following day Conger raised a big stink about it. Joe Rauh then got into a shouting match with McClellan in which Mundt joined in.

The whole thing was most unpleasant. Rauh went in to see McClellan at the noon hour and apologized.[53]

In the afternoon Mazey lost his temper again and raged against the Catholic priests of Sheboygan for signing a pro-Kohler statement. Later in the day Rauh, who stopped by Kennedy's office each afternoon to check the next day's schedule, found Kennedy and O'Donnell passing the inevitable football back and forth. Kennedy, seeing Rauh's woebegone expression, said, "Don't feel so bad, Joe, the Pope hasn't called yet."[54]

Toward the end of March, Goldwater, after his months of righteous complaint about the Reuther cover-up, astounded the committee by announcing that he saw no reason to call Reuther as a witness. "I thought," Kennedy wrote in his journal, "Senator McClellan was going to faint."[55] The chairman, when he recovered, said he would not be a party to denying Reuther an opportunity to testify. Reuther sent along his opening statement which, Kennedy thought, "wasted time on flowery language . . . democracy, God and mankind, etc."[56] That was, of course, Reuther's style. Once on the stand, with his inexhaustible intelligence, confidence and fluency, he dominated the hearing. He freely admitted past UAW error — mass picketing, violence — and then went on the offensive. When he concluded, Goldwater looked, Kennedy thought, "like a man who had taken a very bad beating." "You were right," he told Kennedy. "We never should have gotten into this."[57] McClellan, who had been saying privately that the senators were all making fools of themselves,[58] publicly disclaimed responsibility for the hearings* and brought them to a close. ("I have to say about that old son of a gun," Jack Conway remarked later, ". . . that he did treat us with fairness.")[59] In September 1960 the NLRB found the Kohler Company guilty of unfair labor practices and of failure to bargain in good faith. The long, cruel strike came to an end, with most strikers given their old jobs and the largest back payment award in American history.

v

As for Goldwater, who had wished the philosophy of Hoffa to prevail over the philosophy of Reuther, he found himself trapped in a paradox. "If we want our trade union organizations to be sympathetic to our private enterprise business system," as Professor John Dunlop

---

* "I did not organize these hearings. I did not arrange for the presentation of them, as I have others" (Senate Select Committee on Improper Activities in the Labor or Management Field, *Hearings*, 85 Cong., 2 Sess. [March 29, 1958], 10172).

of Harvard stated the dilemma in 1957, ". . . we should not be surprised that business unionism by and large adopted business morality. It is inconsistent to expect at the same time business unionism and the morality of the ascetic."[60] The Hoffas, who supported the business system, were, alas, crooked. The Reuthers, who criticized it, were, alas, honest.

Certainly the example provided by businessmen was not always edifying. The Rackets Committee was presumably required by its title to inquire into improper activities in management as well as in labor. Here labor leaders charged an absence of zeal. Kennedy's response was that, because of limited jurisdiction, the committee could not go into improper business activities unless there was some direct connection with labor. Certain investigations, however, like Kohler, did involve what Kennedy later called "the respectables."[61]

The trail of Nathan Shefferman thus led not only to Dave Beck but also to the venerated American house of Sears, Roebuck. Shefferman specialized in wrecking union organization drives in the name of labor relations. After a quarter century diddling labor for Sears, Roebuck, he established a firm that, Walter Sheridan said, was a Sears, Roebuck "creation" and engaged in union-busting activities on behalf of Sears, its subsidiaries and suppliers.[62] When Salinger and Sheridan first asked about Shefferman, the Sears vice president "outright, flatly lied" to them for several hours and, indeed, weeks. Sears officials came to Washington to meet with Kennedy. "If you're going to lie to me," he said, "you might as well go back to Chicago." Kennedy persevered; hearings were held; and, faced with the gloomy necessity of testifying under oath, Wallace Tudor of Sears, Roebuck admitted what had been previously denied, called Shefferman's actions "inexcusable, unnecessary and disgraceful" and promised that the company would not permit a repetition of "these mistakes."[63]

This was rather typical of the committee's experience with the respectables. Nothing took place in the hearings to alter Robert Kennedy's bleak view of business. "Little or no accurate information," he wrote, "came to us from the business community. . . . No investigation was touched off by any voluntary help we received from management." And where the labor movement at least took action against corrupt union leaders, "not one management group or association has made a single move to rid itself of members who were found to be involved in collusive deals." Not one firm had fired a vice president for wrongdoing. "These corrupt businessmen are still sitting down to luncheon and dinner meetings with business groups across the country," Kennedy observed, "and they are getting encouragement and admiration — not censure."[64]

As for the unions themselves: "We thought we knew a few things about trade-union corruption," Meany said in 1957, "but we didn't know the half of it, one tenth of it, or the one-hundredth part of it."[65] A. J. Hayes of the Machinists, chairman of the AFL-CIO Ethical Practices Committee, added that only the subpoena power of government could ferret corruption out.[66] Kennedy's investigations led to the exposure and overthrow of James Cross as president of the Bakers, William E. Maloney as president of the Operating Engineers, Anthony Valente as president of the United Textile Workers. The committee released reports on dismal situations in other unions — the Meat Cutters, the Carpenters, the Hotel and Restaurant Workers, the Sheet Metal Workers. At the same time, the Kohler investigation proved that the committee could distinguish between honest and crooked unions. Among thoughtful labor leaders the committee soon lost its initial antilabor reputation. Arthur Goldberg wrote Robert Kennedy in 1960:

> At our first meeting you told me in straight-forward terms that in your capacity as counsel for the committee you intended to carry on the committee's work in a fair and impartial manner and without any bias, fear or favor. I think it appropriate to say that throughout the entire course of the conduct of the committee's activities, you never departed from this statement of your objectives. . . . You have every reason to be proud of your outstanding work . . . and of the contribution that you have made to decent honest democratic American trade unionism.[67]

VI

Nineteen fifty-nine found Robert Kennedy in a mood of discouragement. In March he told McClellan that he planned to leave the committee in August.[68] The shocking disclosures over two years had had singularly little effect. "Candor compels me to say," he wrote in June, "that in the months since the committee began to work, conditions in the labor and management fields have actually grown worse." Hoffa seemed more secure than ever. Communities infected by corruption had given "the merest lip-service" to clean-up efforts. The National Association of Manufacturers and the U.S. Chamber of Commerce had taken no action whatever against businessmen engaged in questionable activities. Bar associations had done nothing to discipline lawyers engaged in unethical practices. "There is appalling public apathy. . . . Members of Congress receive virtually no mail on the subject."[69]

Most disappointing of all, the Department of Justice had obtained convictions of only three persons — and one was a witness who had helped the committee. "McClellan has written several letters to Justice criticizing their tardiness in moving on some of the cases," Kennedy had noted in 1958. ". . . Their reply which evidentally [sic] was handled by [Attorney General William] Rogers himself was bitter & factually in error."[70] William G. Hundley, chief of the Organized Crime Section, explained to the newspapers that the committee's cases required further investigation to produce evidence admissible in courts and that many involved deeds which, however reprehensible morally, were not violations of federal law.[71] Privately Hundley shared the prejudice of Department of Justice lawyers against congressional committees and simply did not think that Kennedy's cases were solid, at least as they came from the Hill. "We'd refer them out to the FBI or the Internal Revenue Service for investigation," he said later, "and I don't think at that time the FBI or the Internal Revenue Service were breaking their back trying to make cases for Bobby Kennedy." Once Kennedy told Hundley, "When you admit to me that there's obvious wrongdoing here and you tell me on top of that that you can't make a case out of it, it makes me sick to my stomach." Hundley said, "I can't be responsible for your gastric juices." Kennedy hung up on him. Twenty minutes later he called back and said he was sorry. Hundley said he was sorry too.[72]

There remained the possibility of new federal law. "What people expect now, in my estimation," Kennedy wrote a friend, "is not more investigation but laws to deal with the problems that we have uncovered."[73] In early 1958, John Kennedy had begun, with the assistance of Professor Archibald Cox of the Harvard Law School, to prepare a bill. He talked with George Meany, who indicated (Robert Kennedy reported) "that possibly matters dealing with finances would be accepted but that the statements about internal democracy were absolutely out of the question." A. J. Hayes told him, "If you start regulating internal union affairs that is the beginning of the destruction of the free union movement."[74] The Kennedys found Walter Reuther "far more liberal" in his approach; "control of union funds, democracy within the unions, limitation on trusteeships, and the rest — he supported all of these in principle."[75]

In due course Kennedy enlisted the support of Irving Ives. The Kennedy-Ives bill required union officials to guarantee regular elections with secret ballots, to file financial reports with the Labor Department and to disclose financial transactions that might involve conflicts of interest. The AFL-CIO endorsed the bill. It passed the Senate 88–1 in June 1958. Two months later a strange coalition of business, the Teamsters and the Eisenhower administration killed the bill in the House.

Kennedy reintroduced the bill in 1959. Again, after amendment,

it passed the Senate. Only Goldwater voted against it. In the mean-
time, the House was considering a far more severe bill proposed by
Robert Griffin of Michigan and Philip Landrum of Georgia. The
Landrum-Griffin restrictions, Robert Kennedy wrote Congressman
Chester Bowles of Connecticut, went "beyond the scope of the
McClellan Committee's findings to affect the economic balance at the
bargaining table by honest and legitimate unions and employers."[76]

Public apathy still appeared to prevail. Robert Kennedy now
launched a personal campaign in support of his brother's bill. He
gave speeches, wrote articles, went on television. "Bob's done a terrific job," said the ranking Republican on the House Education and
Labor Committee. "I've never had so many letters . . . as I received
in response to his TV appearances."[77] Alas, the letters failed to
discriminate between the Kennedy-Ervin and Landrum-Griffin bills;
and the House, responding to a personal appeal by President Eisenhower, passed Landrum-Griffin. Only heroic effort by John Kennedy in conference achieved a tolerable compromise between the
House and Senate bills. Though labor leaders harshly denounced
the new law, the machinery it established for honest and supervised
elections made a vital contribution in later years to union democracy
in the United Mine Workers, the Steelworkers and other unions.

"With the passage of effective labor legislation by the Congress
— the purpose to which we have pointed two years of effort —"
Robert Kennedy wrote McClellan on September 10, 1959, "I submit
my resignation."[78] These had been extraordinarily busy years —
nearly 300 days of public hearings, more than 1500 witnesses, 20,432
pages of testimony in the printed record, 2.5 million miles traveled
in pursuit of evidence.[79] Few congressional investigations in American history had lasted so long, spread so wide a net, raised such
difficult questions — from the rights of a worker in a union to the
rights of a witness before a congressional committee — or drawn so
much attention to the committee counsel. When John Bartlow Martin's absorbing *Saturday Evening Post* series "The Struggle to Get
Hoffa" came out in the summer of 1959, even McClellan seemed a
little annoyed, afraid perhaps that Robert Kennedy was getting all
the glory.[80] No longer merely his father's son or his brother's
brother, Kennedy was emerging as a national figure in his own right.

Goldwater made the point after his resignation. Still on the trail
of Reuther, Goldwater had snapped out a statement accusing Kennedy of running out on the UAW investigation. Kennedy, indignant,
asked if there was anything further Goldwater thought he ought to
do. "No, no," said Goldwater. "I want to get back to Arizona now.
I don't want any more hearings." "Then why did you say it?" asked
Kennedy. "That's politics," Goldwater said. Kennedy said he did not

consider himself in politics; it was a bipartisan committee, and he had tried to run the investigation on nonpolitical lines. Goldwater said crisply, "You're in politics, Bob, whether you like it or not."[81]

VII

As a national figure, Kennedy continued to receive mixed notices. Some observers rejoiced to see a fearless young crusader exposing wickedness. Others saw more than ever a merciless prosecutor, battering witnesses and overriding constitutional rights.

An assessment of Kennedy's performance requires an understanding of the congressional investigative power. In defending congressional inquiries into the Harding scandals thirty years before, Felix Frankfurter had rejected the analogy between a trial in the courtroom and a hearing before a congressional committee. The first, he said, placed a man's liberty in jeopardy and therefore called for stringent rules of evidence and procedure; the second was an "exercise of the informing function of Congress" and had different standards. Frankfurter condemned the idea of trying to introduce the "technical limitations" of a trial "into a field where they have never been resorted to and where they are wholly out of place."[82] Three years later, the Supreme Court in *McGrain* v. *Daugherty*[83] rejected the theory that a Senate committee, by inquiring into the affairs of Harding's Attorney General, was exercising a judicial function. Anything the committee reasonably claimed as necessary to the exercise of a valid legislative function, the Court concluded, was permissible. "Unlike a criminal trial where the prosecution has the burden of proving its essential facts by proof beyond a reasonable doubt," Professor Philip B. Kurland has written, "a congressional investigation may arrive at a judgment that is based on the preponderance of the evidence."[84]

In discharging the informing function, Congress often encountered resistance. When Senator Hugo Black conducted his investigation of the public utility holding company lobby in 1935, he wrote, "There is no power on earth that can tear away the veil behind which powerful and audacious and unscrupulous groups operate save the sovereign legislative power armed with the right of subpoena and search." Newspapers might complain about the "bullying and badgering" of witnesses; but Black, reprinting passages of disingenuous and evasive testimony, observed that, after an investigator

> has tried every technique, politeness, kindness, blandishments, . . . he is sometimes driven in the presence of a witness who is deliberately concealing the facts to attempt to shake it out of him with a more drastic attack.[85]

This was Robert Kennedy's view. He quoted his friend Justice Douglas:

> The right to ask a question and the right to demand an answer are essential to the democratic process.... That is why the embarrassing question from the Senator or Congressman sitting at the end of the table serves a high function. His right to demand an answer spells in a crucial way the difference between the totalitarian and the democratic regime.[86]

Still the McCarthy experience, as Kennedy knew at first hand, had aroused justified concern about collisions between the investigative power and the Bill of Rights. In 1954 a former Communist named John T. Watkins told the House Committee on Un-American Activities that he would answer questions about himself or about persons he believed still to be members of the Communist party but not about people who had left the Communist movement. This refusal led to his conviction for contempt. Joseph Rauh argued the Watkins case before the Supreme Court. In overturning the conviction in 1957, the Court vigorously reaffirmed the right of Congress to conduct inquiries in furtherance of law-making tasks. But there was "no congressional power to expose for the sake of exposure.... Investigations conducted solely for the personal aggrandizement of the investigators or to 'punish' those investigated are indefensible." Moreover, the Court ruled, the investigative power was subject to the restraints of the Bill of Rights.[87]

VIII

In spite of the multitude of investigators — "the largest investigative team," Kennedy said, "ever to operate from Capitol Hill"[88] — and the service of more than eight thousand subpoenas for witnesses and documents, the Rackets Committee, unlike Black's committee twenty years before, was not charged with abusing the subpoena power. Kennedy, Pierre Salinger said, talked "to every investigator before he went on the road about the proper use of the subpoena — the fact that these were not to be used for fishing expeditions. You had to have some legitimate reason to believe that what you were going after was there."[89] Nor, Kennedy said most explicitly, did the committee ever wiretap.[90] Carmine Bellino developed a technique by which, through close analysis of long distance telephone toll tickets, he could find out to whom anyone talked on a given day. The staff indexed thousands of toll tickets and plotted the calls on worksheets.[91] "This is why they always thought we were wiretapping," said Walter Sheridan, "but we weren't."[92]

The chief criticism was directed not at the investigations but at the hearings. The committee rules were in fact designed to avoid the excesses of the McCarthy era. Witnesses were assured the right to counsel and the right to submit questions for cross-examination. One-senator hearings, a device beloved by McCarthy, were banned. When the committee anticipated derogatory testimony about a person not present, Kennedy notified him in advance and offered the opportunity for instant denial.

Remembering the slovenliness of the McCarthy investigations, Kennedy prided himself on the care with which his own hearings were prepared. He checked and rechecked testimony before he put witnesses on the stand; and he rarely asked a question in public hearings to which he did not know the answer. "For every witness who was called before the Committee," he said, "we interviewed at least thirty-five. For every hour a witness testified on the stand before the Committee, he was interviewed on an average of five hours. For every document that was introduced, five thousand were studied. For every hearing held, there were approximately eight months of intensive investigation."[93]

All this was very well. But Kennedy's style as cross-examiner caused serious controversy. There was, it should be said, testimony in his favor. "I did not think he was 'persecuting' Hoffa or infringing his civil liberties," John Bartlow Martin said. "As a crime reporter, I had seen far far worse prosecutions. Indeed, I thought he treated Hoffa fairly, on the whole."[94] Paul Porter of the New Deal law firm of Arnold, Fortas & Porter wrote Kennedy that the 1957 Hoffa hearings were "an outstanding example of preparation, presentation and fairness. . . . I was impressed with the Chairman's observance of the *Watkins* case criteria and am convinced that your magnificent architecture of this proceeding demonstrates that *Watkins* constitutes no serious limitation on proper congressional investigations."[95] Joseph Rauh, the *Watkins* winner, did not think that Kennedy violated the guidelines. "He was trying to be a fair investigator," Rauh said later, "and any abuses were not due to 'vindictiveness,' but to his lack of experience. If it sometimes led to abuse of witnesses, it sometimes led to witnesses like Hoffa getting away with murder. . . . Far from browbeating Hoffa, it was more of a case of Hoffa browbeating *him*."[96]

Goldwater agreed with Rauh on this if on nothing else. Kennedy, he later told Victor Lasky, often became flustered before labor hoods or their lawyers. "When the going would get rough," according to Goldwater, "John McClellan was likely to take over the questioning, thus saving Bobby from embarrassment."[97] Nor, for that matter, did Hoffa think for a moment that he had been persecuted. He considered Kennedy's questioning "painfully amateurish" and said it was "ridiculously simple to get him to lose his temper. It was equally easy to get him to wander from the subject he was pursuing and to tie him up in knots of verbiage and dialectics."[98]

On the other hand, Professor Alexander Bickel of the Yale Law School summed up a widespread impression in a severe article in the *New Republic*. Where Rauh, Frankfurter's first law clerk, viewed congressional investigations in the spirit of Professor Frankfurter, Bickel, the twentieth clerk, viewed them in the spirit of Justice Frankfurter, a more conservative figure. Accusing Kennedy of conducting "purely punitive expeditions," Bickel cited two Fifth Amendment cases — Dave Beck and the Chicago hood Joey Glimco — where Kennedy had persisted in questioning witnesses whom he knew had debarred themselves (by declining on self-incrimination grounds to answer a single question) from answering related questions. In such cases, Bickel argued, Kennedy was not discharging the informing function but rather trying to punish by publicity. "No one since the late Joseph R. McCarthy," wrote Bickel, "has done more than Mr. Kennedy to foster the impression that the plea of self-incrimination is tantamount to a confession of guilt"; under his direction the Rackets Committee "often — though not always — held hearings for the sole purpose of accusing, judging and condemning people."[99]

Part of Bickel's case rested on the fallacy rejected by Frankfurter and by the Supreme Court — that a congressional hearing was the equivalent of a trial, and that a witness had "the right to an impartial judge who is not at the same time also the prosecutor."[100] The theory that the Rackets Committee had abused judicial functions was never taken seriously enough for anyone — even Glimco or Beck — to invoke it as a basis for an appeal to the courts. The committee was plainly dealing with questions on which legislation could be (and was) had; and it is hard to see that Bickel's horrid examples were greatly different from things that Hugo Black and other sorely tried congressional investigators had done in the past.

Nonetheless, he raised a troubling point about the Fifth Amendment — a point to which Kennedy himself was sporadically sensitive. In 1957 Edward Bennett Williams, speaking at the University of Virginia's Student Legal Forum, accused the committee of calling witnesses into executive session, ascertaining that they would take the Fifth and then asking them questions in open session. Edward Kennedy, now at the Virginia Law School, challenged Williams's assertion.[101] He also reported Williams's charge to his brother, who wrote angrily in his journal, "As of yet we have not had a witness in Executive [session] who has taken 5th & whom we have had in open. Old

McCarthy trick & it irritates me that he now an officer of American Civil Liberties Union [Williams was on the ACLU national committee] should relate it to us."[102]

Nearly one fifth of the witnesses who appeared before the committee — 343 out of 1526 — refused questions on grounds of self-incrimination.[103] Kennedy stoutly defended the Fifth Amendment and called it one of the distinguishing marks between democracy and dictatorship. "I would not have it changed." On the other hand, he clearly used it as a polemical weapon, and did so, as he admitted, because, "although the Fifth Amendment is for the innocent as well as the guilty, I can think of very few witnesses who availed themselves of it who in my estimation were free of wrongdoing." Practically, in such cases as Beck and Glimco, he may well have been right. But his attitude weakened what he himself called "one of the rights that a free people possess against the potential abuses of government."

Moreover, this approach encouraged further abuse. It was one thing if the interrogator pursued witnesses, as Kennedy declared he had done, "only on matters I would have asked about if he were answering questions." But he admitted he did not always have "positive proof" on such matters; and he instanced a case where Senator Curtis asked a Fifth Amendment witness a question on which the committee had no information at all and thereby put a wholly fanciful innuendo into circulation.[104] All this suggests that the Fifth Amendment presents problems that Kennedy, and no doubt Bickel too, failed satisfactorily to solve.

But there lay indistinctly behind Bickel's not always forceful contentions a genuinely forceful apprehension — that Kennedy in these matters had displayed an excess of zeal; that he was a man driven by a conviction of righteousness, a fanaticism of virtue, a certitude about guilt that vaulted over gaps in evidence. "Mr. Kennedy knows," Bickel wrote, "and when he knows he is very sure that he knows." The courts might acquit Hoffa, and Kennedy might say he accepted that as "a citizen or a lawyer," but still in some deeper sense, Bickel felt, he *knew* Hoffa to be guilty. "He sees the public interest in terms of ends," Bickel wrote, "with little appreciation of the significance of means."[105]

IX

After resigning as chief counsel, Kennedy set to work on a book about the committee. He asked John Seigenthaler, who was completing a year as a Neiman Fellow at Harvard, to come to Hickory Hill and help on the manuscript. Kennedy wrote the first draft.

Seigenthaler cut and edited. When Seigenthaler went off to bed at night, Kennedy would be "sitting on the edge of the bed writing in long hand," and when Seigenthaler got up in the morning, "he'd be up already writing again." In the afternoon they took walks. "During those walks I got to know Bob Kennedy . . . better than I got to know anybody during my whole life. He was completely candid with me about everything."[106]

The Enemy Within, as he called the book, was written in brisk journalistic prose interspersed with personal passages of a curious eloquence. "It is a detective story which can be put on the same shelf as the best thriller fiction of the day," the master journalist Lord Beaverbrook told him. "I feel rather like one who has been sent an account of the Crusade written by Richard the Lionheart himself."[107] The book was briefly a best seller, the royalties going to the care of retarded children.

Most books and articles in America in the 1950s with titles like *The Enemy Within* had a single subject — communism. But the enemy within, as Robert Kennedy saw it, was organized corruption, spreading from the underworld into labor, business and politics, expressing, he asserted, the moral sickness of a greedy society. American history left no doubt, he had told the students at Notre Dame in 1958, "that the great events of which we are proud were forged by men who put their country above self-interest — their ideals above self-profit."[108] Dangerous changes were taking place in the moral fiber of American life. "It seems to me imperative that we reinstill in ourselves the toughness and idealism that guided the nation in the past."[109]

The Rackets Committee was a significant chapter in his education. For the first time he had been forced to confront the power structures of organized labor, of organized business, of organized crime and of law enforcement. It also had unexpected emotional impact. The wracking experience of these years deeply affected him. He saw men lie, cry, run terrible risks, display great cowardice and great courage. He learned, as in the case of the jolly rogue Frank Brewster, that he could quite like people of whom he morally disapproved. "He had an immense sense of humor about sinners," Murray Kempton said, "which I had really not expected."[110] He was also, Kempton thought, surprisingly perceptive. "Let two thugs with identical records on paper come before the committee and he could tell you how they differed as human beings."[111] He unlocked resources of sympathy within himself about ordinary people — as when, after attending mass (in Lithuanian) in Sheboygan, he wrote with unwonted feeling about "the strong, stern faces of people who have worked hard and who have suffered."[112] Emotion, long dammed by the requirement of toughness, was beginning to flow again. The experience helped strengthen the instinct of sympathy within him — that instinct more or less repressed since childhood — and gave it for the first time social direction.

Moreover, he understood more than ever that toughness and concern were not necessarily enemies. "What impressed Bobby Kennedy about the UAW," Walter Reuther said later, "is that we shared the sense of human compassion. . . . Too often, I think he felt that liberals articulated this sense of compassion, but never did anything about it. . . . I think this is the thing that finally rang the bell with Bobby Kennedy. And Jack said to me, 'You know, you fellows are educating Bobby.' And we said, 'Well, we're doing what comes naturally.'"[113] Other things brought the Reuthers and the Kennedys together: both, for example, were teams of brothers. Walter Reuther, Victor said later, actually felt more rapport with Robert than with John Kennedy: Bobby was less detached, more committed, more activist.[114] John Kennedy himself said later of his younger brother, "He might once have been intolerant of liberals as such because his early experience was with that high-minded, high-speaking kind who never got anything done. That all changed the moment he met a liberal like Walter Reuther."[115]

Many liberals still could not understand him. His concern with organized crime, for example, did not fit into their categories; such a preoccupation appeared unseemly and irrelevant. They found him hard to understand because he was a Catholic. Not many liberals believed in original sin. Expressing confidence in 1957 in the future of the trade union movement, Kennedy had said that "the brightest spot" was the work of the Association of Catholic Trade Unionists.[116] Kempton was almost alone in seeing the clue. Catholic politicians fascinated Kempton. "The only non-Catholic interesting politician I know," he once said, "is President Kennedy."[117] He early perceived the Catholic essence in Robert Kennedy.

Kennedy had proceeded into the jungle of labor racketeering, Kempton wrote in 1960, without guide or experience. "But there are persons so constituted that they can go nowhere without some piece of faith to serve for light. Robert Kennedy is a Catholic; and naturally he sought his faith there. It is the difference between his brother, the Senator, and himself, the difference between those who are only properly oriented and those who are truly involved." Robert Kennedy, Kempton wrote, looked upon labor racketeers as men who had betrayed a priesthood. He had become for this journey "a Catholic radical." In the cynicism of American politics, he represented "the survival of the spirit."[118]

(10)

1960

Except for the jaunt with Stevenson in 1956, Robert Kennedy had stayed away from politics after 1952. Party grandees exasperated him. Party rituals bored him. When John Kennedy came up for reelection in 1958, Robert's main contribution was to fly to Boston and hear the returns. Nonetheless, as John prepared to run for President in 1960, his younger brother remained, in Theodore Sorensen's words, his "first and only choice for campaign manager." Robert Kennedy, as Sorensen said, could be "trusted more implicitly, say 'no' more emphatically and speak for the candidate more authoritatively than any professional politician." [1]

I

Preoccupied with the Rackets Committee and then with *The Enemy Within*, Robert did not enter the campaign full-time until the end of 1959. Sorensen had run things in the early stages. Later Stephen Smith, who had married Jean Kennedy in 1956, took charge in an inconspicuous Washington office. He was thirty-one years old, had received his political initiation in John Kennedy's 1958 campaign, was soft-spoken, sharp and funny. "He made a major difference to the campaign," Robert Kennedy said later, "when he started back in 1959." [2] O'Donnell and O'Brien worked with Smith.

In the course of the year Robert added David Hackett, the great friend of his youth. Hackett, who had attended McGill University after the war and was now editing a Montreal magazine, was so content in Canada that he was contemplating a change in citizenship. Then his old chum told him over the phone that, since he had done so well in English at Milton, he would be ideal to handle Senator Kennedy's political correspondence. Hackett came to Washington.³ Robert also recruited Pierre Salinger to set up the press operation. John Kennedy, Salinger recalled, made it very clear "he was hiring me strictly because Bobby had asked him to." Soon Robert directed him to put out a statement. The next day John asked Salinger with some asperity who had authorized it. "Check those things with me," John Kennedy said. "You're working for me, not for Bob now."⁴

"All right, Jack," Robert said in December at Palm Beach, "what has been done about the campaign? . . . A day lost now can't be picked up on the other end. It's ridiculous that more work hasn't been done."⁵ This speech defined his role. Inside the campaign he was the tireless invigorator and goad, responsible for everything except the speeches. Outside he became the man to do the harsh jobs, saying no, telling people off, whipping the reluctant and the recalcitrant into line.

When reporters began in due course to write about the "well-oiled Kennedy machine," the myth arose that the Kennedys were master political planners. This was a misconception. They were not systematic calculators but brilliant improvisers. Their reactions, observed Fred Dutton of California, one of the shrewd political minds of the period, "jump[ed] all over the place — just an incisive thing here, incisive thing there, and you've got it. . . . More was almost communicated in their silences or things they skipped over than in their questions."⁶ Robert Kennedy's genius as manager lay in his capacity to address a specific situation, to assemble an able staff, to inspire and flog them into exceptional deeds and to prevail through sheer force of momentum. "When their things worked," said Harris Wofford, the campaign's civil rights coordinator, "it was not because of a great machine but because they released energy in a lot of directions and sort of trusted that . . . the general goal was right."⁷

"If you win," Robert Kennedy told me during his senatorial campaign in 1964,

the reporters will always write about well-oiled machines and superplanning. If you lose, they will always write about hopeless incompetence. I have a simple theory: that campaigns come to an absolute end within a given time; that you can recruit good people and ask them to do a lot of things in that brief period; that some things will work and some won't; that, if something is not working, change it, and be prepared to change it again; and that out of all the activity a momentum will develop which will carry you through.

This guaranteed organizational confusion. "And yet," as the lawyer Barrett Prettyman, Jr., noted, "the very frenzy of it somehow brings more out of people than if the whole thing was just calculated to the *n*th degree."[8]

This was the Robert Kennedy the campaign staff saw and generally adored. As an emissary to Democrats, he was less adorable. Governor Michael DiSalle of Ohio, a jovial veteran of Truman days, aspired to be his state's favorite son. Kennedy offered DiSalle this distinction if DiSalle would agree to endorse him forthwith for President. DiSalle knew that Kennedy could probably beat him in the primary but had no desire to be the first governor to come out for Kennedy — among other reasons, perhaps, because Truman was so hostile to the Kennedys. He therefore sought to delay his announcement. Robert Kennedy and John Bailey met DiSalle to put on additional pressure. It was an unpleasant session. Bailey had been in politics a long time and, as Kenneth O'Donnell said, did not shock easily, but, he told O'Donnell later, he was "startled by the going-over that Bobby had given DiSalle."[9] Bailey was still talking about it in March when I encountered him at the Democratic Midwest Conference. According to my journal, "I had a series of drinks with John Bailey, who was most charming and amusing and regaled me with stories about the Kennedy campaign (e.g., Bobby Kennedy's talking to Mike DiSalle in a way which almost led Mike to cancel the Ohio deal)."[10] Still the deal survived. Early in January DiSalle pledged the Ohio delegation to Kennedy.

This left the Kennedys free to concentrate on their first critical test — Wisconsin. "Senator Humphrey is our big rival now," Robert Kennedy said as they prepared in February to run the gantlet of the primaries. He added: "But over the long run I expect Senator Johnson to be the principal opposition."[11]

II

Hubert Humphrey was one of the brightest and most likable men in the Democratic party. Though he and John Kennedy were in broad agreement on issues, their lives could hardly have been more different. Nearly forty-nine, six years older than Kennedy, Humphrey had come of age during the depression, growing up a druggist's son amidst the dust storms of South Dakota. He worked his way through college, taught, entered politics and flashed into national attention as the fiery champion of civil rights in the Democratic convention of 1948. He was a man of swift intelligence, immense sympathy and great gusto, a stem-winding orator in the style of midwestern populism, sentimental, vulnerable, accommodating, a generous and open man but easily hurt and not always easily forgiving.

His implacably loyal wife, Muriel, joined him in Wisconsin, as did a group of devoted supporters from Minnesota, especially Orville Freeman, the able governor, and the witty and somewhat cynical congressman Eugene McCarthy. "I should be the candidate for President," McCarthy was supposed to have said. "I'm twice as liberal as Humphrey, twice as smart as Symington and twice as Catholic as Kennedy." (Joseph Alsop told John Kennedy that he had encountered McCarthy on the plane to Wisconsin, reading an enormous missal all the way to Milwaukee. Kennedy said, "Well, Joe, there's this old saying in Boston politics, never trust a Catholic politician who reads his missal in the trolley car.")[12]

Humphrey's Minnesotans worked hard. But the Kennedys seemed to be everywhere. "They're all over the state," Humphrey said, "and they look alike and sound alike. Teddy or Eunice talks to a crowd, wearing a raccoon coat and a stocking cap, and people think they're listening to Jack. I get reports that Jack is appearing in three or four different places at the same time."[13] He felt, Humphrey said, like an independent merchant competing against a chain store.[14] It was not only the quantity of Kennedys; it was even more their money, their assurance, their beautiful people, their personal pollsters, their private Convair, the *Caroline,* as against Humphrey's limited funds, plain midwestern folk and rented campaign bus. Hearing a plane overhead one night as he jolted along trying to sleep in his bus, Humphrey shouted, "Come down here, Jack, and play fair."[15]

Robert spoke as often as the rest. Patrick Lucey, the state chairman, told him he should decide whether to be manager or campaigner. "But he insisted on doing both and did both quite well. It was amazing to me how he could find the amount of energy that he was able to put into the thing."[16] He was still not fluent on his feet, but he was gaining confidence. A slight figure — he lost a pound a week in Wisconsin — bare headed, his collar turned up against the icy March winds, he drove from town to town along roads piled high with snow, stopping at high schools, Rotary clubs and political rallies to explain why his brother should be President.

He made new friends in Wisconsin. One was an Italian American named Jerry Bruno, whom John Kennedy had persuaded the summer before to leave his job on the staff of Senator William Proxmire and organize Kennedy clubs around the state. Bruno had dropped out of school in the ninth grade; but, as he later wrote, "John and Robert Kennedy sized you up very quickly. If there was something about you they thought O.K., then your background and education and all the rest didn't matter." Bruno displayed uncommon talent in an obscure but indispensable political specialty: 'advancing' appearances by the candidate. It was a risky job. If the meeting was a success, credit went to the candidate; if a failure, everyone blamed the advance man. (Later in Ohio, when Kennedy and DiSalle found themselves hopelessly mired in a traffic jam and far behind schedule Kennedy said, "I wonder how Hannibal ever made it across the Alps?" "It was easy," DiSalle said, looking at Bruno. "He didn't have an advance man.")[17]

Another new friend who became, even more than Bruno, a fixture in Robert Kennedy's political operations was a raffish, entertaining and outrageous rogue named Paul Corbin. Forty-six years old, of Canadian birth, at one period very close to the Communist party, at another a devotee of Joe McCarthy, he had lived by his wits on the fringes of the labor movement and the Democratic party. With gravelly voice, authoritative manner and unlimited effrontery, he was a natural-born con man. He took cheerful delight in causing trouble and in reorganizing the truth. When Helen Keyes of Boston came to help in Wisconsin, Robert Kennedy committed her to Corbin's doubtful protection. Corbin paraded her up and down the streets of small Wisconsin towns; and, if anyone opposed John Kennedy on religious grounds, Corbin would point to Helen Keyes, a devout Catholic, and say, "Well, this lady's a Baptist from Boston, and *she's* for Kennedy." Helen Keyes said later, "I kept expecting to be struck by lightning for denying my faith."[18]

Paul Corbin fell in love with Robert Kennedy. He was prepared thereafter to do anything he conceived to be in Kennedy's interests, whether or not Kennedy, if he had known what Corbin intended, would have approved. In time Corbin even became a Catholic himself in order to have Robert and Ethel as godparents. "You need Paul Corbin in the Catholic Church like you need leprosy," Helen Keyes protested. ". . . He wants to be a Catholic because *you're* a Catholic." Kennedy said, "Well, who cares? If he gets something out of it, and he's been through all the instruction. . . ."[19]

Something about Corbin appealed to Robert Kennedy. His impudence perhaps touched some Irish folk memory; Corbin could have been a character out of Ed O'Connor's Boston. His entirely cynical view of people's motives was joined to wondrous investigative skills. As Joseph Dolan of Colorado put it, "He could find out who was cutting the ticket pretty fast, when he went into a town." And Corbin was unquestionably effective. "If you have a job and you want to get it done," Helen Keyes summed it up, "and you don't care how it's done, send Paul Corbin out to do it — but just understand you can never send him back to the same district."[20]

Stephen Smith and John Seigenthaler liked Corbin. Dolan and

O'Donnell detested him. Robert Kennedy himself was defensive about him and on occasion exasperated by him. In later years, Robert McNamara, hearing of some Corbin mischief, suggested that he be given a lie detector test. "A lie detector test?" said Robert Kennedy. "Why he'd break the machine!" Many were mystified by the attachment. But the attachment remained.

Humphrey saw in the Kennedy campaign "an element of ruthlessness and toughness that I had trouble either accepting or forgetting."[21] He blamed Robert Kennedy for a rumor that the Teamsters were helping him (which it would not have been surprising for the Teamsters to do, not out of love for Humphrey but out of hatred for the Kennedys). And, though he was not sure that the Kennedys were responsible, he was enraged by the anonymous mailing of anti-Catholic tracts to Catholic households across the state. Others blamed this on Paul Corbin.[22] John Kennedy himself was determined to play down the religious issue. But the issue was irrepressible. The Pope did not have many fans among Wisconsin Lutherans. Nor could the press leave religion alone. Whenever Kennedy shook hands with a nun, the picture appeared in the papers. Corbin, inviting a group of seminarians to a Kennedy rally, begged them to wear sport shirts.

The contest grew acrimonious. Two days before the vote, Humphrey told a Jewish audience in Milwaukee, "They say the Humphrey campaign is disorganized — but I want you to know that the most organized thing that ever happened almost destroyed civilization." "Those cutting remarks," his campaign coordinator observed, "were not lost on a meeting in the Jewish Community Center."[23] Kennedy hoped to win decisively enough to drive Humphrey from the race. He won. But, after a harrowing night, it was evident that the margin was too small to be conclusive. Commentators diminished it further by attributing it to the crossover of Catholic Republicans. John Kennedy was disappointed; Robert Kennedy was exhausted and angry. O'Brien noted the "terrible sense of gloom" in Kennedy headquarters.[24] In an election-night interview Walter Cronkite of CBS asked John Kennedy about the Catholic vote. Kennedy answered frigidly. As soon as the interview was over, as Cronkite later recalled it, "Bobby stormed through the studio saying, 'You violated an agreement! We had an agreement that no question be asked about Catholicism and the Catholic vote. . . . I'm going to see that you never get another interview.'"[25] This was the mood of the evening. In fact, the agreement had not been communicated to Cronkite, and he was to have many more interviews with both Kennedys.

An exultant Humphrey claimed a moral victory. A depressed Robert Kennedy took the *Caroline* to West Virginia.

III

"What I would like," Robert Kennedy told the Kennedy organizers, "is to have a cold-blooded appraisal from each one of you on whether we can win the State of West Virginia; whether . . . these ministers will start telling people that they can't vote for a Catholic because the Pope is coming over. . . . I want to go over, county by county, as to what chances we have."[26]

They told him. A man shouted, "There's only one problem. He's a Catholic. That's our God-damned problem." What had happened to a December poll showing Kennedy ahead, seventy to thirty? In December, the reply came, no one had known that Kennedy was a Catholic. A Catholic could not be elected dogcatcher in West Virginia. The room was in a tumult. "I looked at Bobby," O'Donnell later wrote. "He seemed to be in a state of shock. His face was as pale as ashes."[27] He rallied to lay out the campaign strategy. "If we try to meet the religion issue head on — deal with it frankly," he finally said,

> I think we are feasible. . . . Once that is done and in Jack's speeches he gives answers to rational questions, then he must go on to show that there is something more important to those people than anything to do with religion. Got to have something like the feeling they had for FDR. . . . In Mingo County there are six thousand unemployed. They don't care if he is Catholic — if he can get to them that "I care for you.". . .
> It is simply food, family and flag in southern West Virginia.[28]

There was a month to go. John Kennedy addressed the religious issue — "Nobody asked me if I was a Catholic when I joined the United States Navy"[29] — and a state filled with kindly and patriotic people responded to the appeal for tolerance. Having dealt with flag, he proceeded to food and family. This was not difficult. He was genuinely appalled by what he saw as he traveled across the verdant countryside — hungry, hollow-eyed children, dispirited families living on cornmeal and surplus lard, gray, dismal towns, despair about the future. He had never expected to find anything like this in the United States. "West Virginia," Pierre Salinger said later, "brought a real transformation of John F. Kennedy as a person."[30] It upset Robert Kennedy quite as much. Arriving at the tipple of coal mines as the shifts were changing, he would speak to the miners — "My name is Bob Kennedy. My brother is running for President. I want your help" — shake their hands and then drive on, his own hands now black with coal dust, his face grim.[31]

The objective, Robert Kennedy had said, was to rouse among West Virginians "something like the feeling they had for FDR." Joseph Kennedy decided to do this in the most direct possible way. Shortly after Wisconsin he asked Franklin D. Roosevelt, Jr., to meet with his sons in Palm Beach. "You know, Franklin," the ambassador greeted Roosevelt, "if it hadn't been for that guinea [Carmine de Sapio, who had swung the Democratic gubernatorial nomination in New York from Roosevelt to Averell Harriman in 1954], you would have been elected governor, and now we would all be working for you."[32] This was still the season of gloom. The father refused to be downcast. "He was always confident," said Charles Spalding, who was also at Palm Beach, "and always willing to believe it could be done and it would be done and it was just a question of going and doing it."[33] Joseph Kennedy was operating on his familiar principle that, when his children made a decision, no matter how mistaken their father might have thought it, it became at once the best decision in the world. The support he unstintingly gave sometimes helped make it so.

So in July 1957 John Kennedy had made his speech in the Senate calling for the independence of Algeria. The speech shocked the foreign policy establishment. Dean Acheson assailed him with characteristic scorn. Even Adlai Stevenson, whom I saw in Paris a few days after the speech, was appalled: Kennedy was criticizing an ally, imperiling the unity of NATO and so on. Joseph Kennedy privately told James Landis the speech was bad policy and bad politics. But when John Kennedy, disturbed by the criticism, called his father, Mr. Kennedy said, "You lucky mush. You don't know it and neither does anyone else, but within a few months everyone is going to know just how right you were on Algeria."[34] In December 1959, alone with Paul Fay, the ambassador had expressed pessimism about 1960. "Jack is going to have a rough time. . . . I hate to see him and Bobby working so hard if the timing is all wrong." When John entered the room a moment later, his father said emphatically that he would bet a million dollars in Las Vegas on winning the nomination and the election. "He had made," Fay wrote later, "a 180-degree change of position in less than sixty seconds."[35] "The great thing about Dad," John Kennedy said in 1960, "is his optimism and his enthusiasm and how he's always for you. He might not always agree . . . but as soon as I do anything, there's Dad saying: 'Smartest move you ever made. . . . We really got them on the run now.'"[36] "My father would be for me," Kennedy told Dorothy Schiff of the *New York Post*, "if I were running as head of the Communist Party."[37]

Roosevelt, as confident as the ambassador about West Virginia, spoke so eloquently about the prospects that Rose Kennedy thought to herself, "I am glad Franklin's helping Jack and not running against him."[38] Finally Robert said, "There's no question because of the attitude of the people there toward both your mother and your father that you can have a big influence in that state." Would Roosevelt campaign in West Virginia? Roosevelt agreed at once. ("A lot of his kids were there," Roosevelt recalled about Robert Kennedy. "I remember being impressed with the amount of time and attention he devoted to them, and how devoted they were to him.")[39]

IV

Roosevelt's look, his voice, his charm evoked his father; and, with his entry and John Kennedy's head-on attack on the religious issue, the campaign began to turn around. Still West Virginia was tricky terrain. National politics mattered considerably less than local politics. The problem was to acquire the strongest slate in each county. "Bobby was the one," Roosevelt later said, "who had to decide which local county slate we should get Jack on. . . . He would go into a county, analyze it, talk to everybody he could talk to, going directly to the point; and his judgment turned out to be exactly right."[40]

But there was always a quid pro quo. Slating provided a quasi-legal cover under which money flowed to local politicians in the guise of campaign expenses. This was a cherished West Virginia tradition. O'Brien negotiated the payments for the Kennedy organization. "Neither Jack nor Bob Kennedy knew what agreements I made — that was my responsibility."[41] There was no escape from the local folkways. "Our campaign," Humphrey wrote, "had, in fact, already given in to the system. . . . Hoping to get consideration from the party, it had raised money and given it to the party for slating in various counties. . . . Obviously, our highest possible contribution was peanuts compared to what they had received from the Kennedy organization."[42]

"I must tell you," Walter Lippmann had written John Kennedy in January, "that I do not look forward with any pleasure to the contest between Humphrey and you. You both will be doing and saying things that as President you would wish you hadn't had to say and do."[43] "I can't afford to run through this state with a little black bag and a checkbook," Humphrey cried as the tide began to turn. ". . . I can't buy an election. . . . Kennedy is the spoiled candidate and he and that young, emotional, juvenile Bobby are spending with wild abandon. . . . Anyone who gets in the way of . . . papa's pet is going to be destroyed."[44]

For their part the Kennedy people regarded Humphrey as a

spoiler, incapable, after losing Wisconsin, of winning the nomination for himself, yet determined to knock out Kennedy and no doubt financed in this ambition by those who could ultimately benefit — Symington, Stevenson, Lyndon Johnson.* An anonymous Minnesotan sent Lawrence O'Brien material representing Humphrey as a draft dodger in the Second World War. "We decided to do nothing with this for the time being, but we agreed that if Humphrey hit us with some extremely low blow, we might use the material to retaliate." They also decided that the retaliation should come, not directly from the Kennedys, but from Franklin Roosevelt, to whom they passed on the material.[45] What happened thereafter is beyond the capacity of history to reconstruct. Roosevelt's memory is that, as Humphrey became more intemperate, he was under insistent pressure, especially from Robert Kennedy, to bring up Humphrey's war record. According to O'Brien, Roosevelt, exasperated by Humphrey's demagoguery, brought it up on his own.[46] In any case, Roosevelt finally told an audience, "There's another candidate in your primary. He's a good Democrat, but I don't know where he was in World War II."[47] The outcry was immediate and vivid. Humphrey had in fact been turned down by both services during the war because of physical disabilities. He later wrote that his organization "made repeated contacts with the Kennedys demonstrating the untruth of the allegations. They believed me, but never shut F.D.R., Jr. up, as they easily could have."[48] John Kennedy heard about it after a highly successful afternoon of campaigning. "Something always happens at the end of a good day," he said angrily to Theodore White, the campaign historian. "I didn't want that."[49] He promptly disowned the issue. But he praised Roosevelt's contribution to the campaign, which indeed had been great, and kept him on as a campaigner. Roosevelt felt he had been used, blaming it on what he saw as Robert Kennedy's determination to win at any cost; "I don't think that Jack really had anything to do with deciding whether to insist on my going ahead with asking the question."†

The contest concluded in bitterness. Robert Kennedy and O'Donnell did not think their gains had sufficed to overcome the anti-Catholic prejudice. "We felt," O'Donnell recalled, "that our Pres-

---

* Humphrey wrote later, "We began to listen to offers of help from other candidates . . . but when the chips should have been down, Johnson's people refused money absolutely and Symington's were to come up with only a little" (Humphrey, *Education of a Public Man* [Garden City, N.Y., 1976], 213). Humphrey said nothing about Stevenson money, of which he may have received a certain amount.

† It was, Roosevelt said later, "the biggest political mistake" of his career. The next year he went to Humphrey's office to apologize. Fifteen years later Humphrey declared himself still "unforgiving." Roosevelt, in recorded interview by Jean Stein, December 9, 1969, 6–8; Humphrey, *Education of a Public Man*, 475.

idential campaign was over." "Damn that Hubert Humphrey," Robert Kennedy said. That day West Virginia gave Kennedy a stunning
victory. As the votes came in, Robert Kennedy said, "God, this must
be awful for poor Hubert, ending up this way after working so
hard."[50]

In Humphrey's hotel suite his advisers debated what to do next.
Joseph Rauh and James Loeb, who had founded Americans for
Democratic Action and was now a newspaper publisher in Saranac
Lake, New York, argued that Humphrey should endorse Kennedy.
Others urged Humphrey to keep his delegates together till the con-
vention. Rauh said this would only prove that Humphrey was part
of a stop-Kennedy movement. Loeb typed out a statement of with-
drawal without endorsement. Humphrey approved it, "and I called
Bobby Kennedy," Rauh said later, "and read it to him. Then a
strange incident occurred." Let Rauh tell it:

> A few minutes after I called Bobby Kennedy and read him Humphrey's
> generous withdrawal, the telephone rang in the outer room of Hubert's
> suite. . . . Someone picked up the telephone from downstairs and said,
> "Mr. Kennedy is on the way up." Of course, we thought it was Senator
> Kennedy coming over to thank Hubert. Then we looked at the televi-
> sion, and there Senator Kennedy was, live on television. . . . I said to
> myself, "Oh my God, it's Bobby, and he is the devil as far as this camp
> is concerned." He was the one to whom the enmity went. He was the
> one whom all our people were so bitter about. . . .
> The door opened. Bobby walks in. It was like the Red Sea opening
> for Moses. Everybody walked backwards, and there was a path from
> the door to the other side of the room where Hubert and Muriel were
> standing. I'll never forget that walk if I live to be a hundred. Bobby
> walked slowly, deliberately, over to the Humphreys. He leaned in and
> kissed Muriel. I've always wondered whether she had on her mind at
> that moment that she was going to poke him because she was really not
> very happy about the outcome. Anyway, he was very nice and very
> gracious, and it was the right thing for him to have done. But at the
> moment, it was sure something.*

V

Humphrey's withdrawal left Lyndon Johnson, as Robert Kennedy
had predicted in February, the principal opposition. A week after
West Virginia I received a call from John Kennedy. He was lunching

---

* Joseph Rauh, in recorded interview by Charles Morrissey, December 23, 1965,
70–72, JFK Oral History Program. Humphrey writes that for the first time in his
political career he saw his wife angry about an election. "She did not want to see any
Kennedy, much less be touched by one. When Bob arrived in our room, he moved
quickly to her and kissed her on the cheek. Muriel stiffened, stared, and turned in
silent hostility, walking away from him, fighting tears and angry words" (Humphrey,
Education of a Public Man, 221).

on the road in Maryland. His talk was punctuated by bites and swallows. He calculated that he needed between eighty and a hundred delegates. Adlai Stevenson, he felt, could provide them. "He is the essential ingredient in my combination. I don't want to go hat in hand to all those southerners. But I'll have to do that if I can't get votes from the north."[51] But Stevenson, though an avowed noncandidate, wanted the nomination himself. His friends were organizing on his behalf. The shooting down of a CIA U-2 over the Soviet Union and the consequent cancellation of the Paris conference between Khrushchev and Eisenhower now galvanized his candidacy. If foreign affairs were crucial, Stevenson appeared the most qualified Democrat. It seemed to some of us that he might develop enough strength to stop Kennedy without developing enough to nominate himself. The beneficiary would be Johnson.

In mid-May, Johnson, trying to block a rush of delegates to Kennedy, told Stevenson, "If I don't get it, it will be you."[52] A few days later Kennedy called on Stevenson in Libertyville. "I told him," Kennedy told me afterward, "that Lyndon was a chronic liar; that he had been making all sorts of assurances to me for years and has lived up to none of them."[53] (Stevenson later told me the mention of Johnson had provoked the only flash of temper on Kennedy's part. Recalling how fiercely Johnson had spoken against Kennedy, he added, "Obviously the feeling between the two of them is savage."[54] This was not, I think, altogether so. Early in June at Hyannis Port, Kennedy talked about Johnson with a mixture of admiration and despair. "He went into an historical excursus," I noted, "comparing J. to Peel and others who were omnipotent in Parliament but had no popularity in the country. He was quite amusing on J.'s personality, describing him as a 'riverboat gambler.'")[55]

It was a difficult time, especially for those of us who loved Adlai Stevenson and had labored for him in two campaigns. Early in June I joined a group of liberals — Rauh, John Kenneth Galbraith, Henry Steele Commager, Arthur Goldberg, James MacGregor Burns and half a dozen others — in a statement endorsing Kennedy. Stevenson enthusiasts denounced us for not insisting on the best. I noted in my journal:

> This, I suppose, is the real irony. I have come, I think, to the private conclusion that I would rather have K as President than S. S is a much richer, more thoughtful, more creative person; but he has been away from power too long; he gives me an odd sense of unreality.... I find it hard to define this feeling — a certain frivolity, distractedness, over-interest in words and phrases? I don't know; but in contrast K gives a sense of cool, measured, intelligent concern with action and power. I

feel that his administration would be less encumbered than S's with commitments to past ideas or sentimentalities; that he would be more radical; and, though he is less creative personally, he might be more so politically. But I cannot mention this feeling to anyone.[56]

My wife, Marian Cannon, announced that she was standing by Stevenson. A few days later I received a letter on another matter from Robert Kennedy. He scrawled in a postscript: "Can't you control your own wife — or are you like me?"[57]

VI

Kenneth O'Donnell and Lawrence O'Brien have written in their books (1972 and 1974) that they went to Los Angeles quietly confident about the outcome. I do not remember all this quiet confidence at the time. Theodore Sorensen wrote in 1971 that "privately, Kennedy was nervous."[58] Robert Kennedy told me in 1965, "We had to win on the first ballot. We only won by 15 votes. Well, North Dakota, they had the unit system . . . we won it by half a vote. California was falling apart. . . . Carmine de Sapio came to me and said what we'd like to do is make a deal — 30 [New York] votes will go to Lyndon Johnson and then you'll get them all back on the second ballot. I said to hell with that; we're going to win it on the first ballot. So, you know, it was all that kind of business. There wasn't any place that was stable."[59] That was certainly the way it seemed at the time.

If Kennedy did not win at once, there might have been, as O'Donnell later contended, some strength held in reserve, but there were also delegates, as in Pennsylvania, who would have left him after the first ballot. The main threat, the Kennedys had no doubt, was Johnson. Stevenson aroused far more emotion among the people who flocked around the sports arena. But the operation organized on his behalf consisted not just of true believers, like Eleanor Roosevelt, Herbert Lehman, Thomas Finletter, James Doyle, George Ball; but of surprising newcomers, like Hubert Humphrey and Eugene McCarthy, who were thought to be for Johnson but had unexpectedly declared for Stevenson after they arrived in Los Angeles. The Stevenson drive at the convention therefore struck many of us, from Reuther and Rauh to Robert Kennedy, as a Johnson front, organized to prevent a Kennedy nomination on the first ballot. If it succeeded, Johnson, with his southern delegates and his potential support among northern bosses, appeared far more likely to benefit than Stevenson, a two-time loser with few delegates.

The atmosphere in Los Angeles was ill tempered. John Connally of Texas announced in a press conference that Kennedy had Addison's disease and was physically unfit for the Presidency. Johnson himself was even uglier in private. He described Kennedy to Peter Lisagor of the *Chicago Daily News* as a "little scrawny fellow with rickets." "Have you ever seen his ankles?" he asked. "They're about so round," and he traced a minute circle with his finger. Lisagor passed this on to Robert Kennedy.[60] In public Johnson attacked Joseph Kennedy as an appeaser and, by implication, pro-Nazi, bellowing: "I was never any Chamberlain umbrella policy man. I never thought Hitler was right."[61] When Johnson challenged Kennedy to a debate before the Texas delegation early in convention week, the ambassador, who was ensconced in Marion Davies's estate, told John Seigenthaler as they sat around the swimming pool that Jack would be "a damned fool if he goes near him." Jean Kennedy Smith said, "I know, Daddy, but he's challenged him to a debate"; and to Seigenthaler, "You'll see. That's the way they are. He'll debate him." In a few moments the flash came by radio that Kennedy had accepted the challenge. The ambassador did not look back for a minute; "he will absolutely take this fellow apart."[62]

Everyone who attended the Kennedy staff meetings at room 8315 in the Biltmore remembers the electric impression made by Robert Kennedy, a slight and boyish figure, jacket off, tie loose, climbing on a chair to call his troops to order, stating the tasks of the day with a singular combination of authority, incisiveness, humor and charm. Keeping track of delegates was an arduous job. On Tuesday, when Kennedy announced that the group would meet at eight the next morning, one member mildly suggested making it later; "I didn't get back to my hotel until five, and it's the third night in a row." Kennedy, who probably had not slept at all, fixed him with a cold eye and said, "Look, nobody asked you here. You're not getting paid anything. If this is too tough for you, let us know and we'll get somebody else." Yet, when Joseph Tydings of Maryland misjudged the situation in Delaware and, by agreeing to a unit rule, lost the state to Johnson, he went at once to the Biltmore to tell Robert Kennedy he had blown Delaware. Kennedy said, "The important thing is that you told me right away so that we knew it, so that we're not counting on them. Next time you'll be a little bit wiser."[63]

On Wednesday, the critical day, "he ran through the states," I noted in my journal, "in order to get the rockbottom Kennedy tally. Once again, I was impressed with the sharpness of his direction of the campaign. He insisted practically on the name, address and telephone number of every half-vote. 'I don't want generalities or guesses,' he would say. 'There's no point in our fooling ourselves.

I want the cold facts. I want to hear only the votes we are guaranteed to get on the first ballot.'" The count showed 740 — 21 short of a majority. He concluded crisply: "We can't miss a trick in the next twelve hours. If we don't win tonight, we're dead."[64]

VII

They won that night. The next problem was the Vice Presidency. There has been much speculation as to why and how John Kennedy chose Lyndon Johnson. Early in 1965 Robert Kennedy read my account in the first draft of *A Thousand Days* — an account I had based in part on a memorandum written shortly after the event by Philip L. Graham. In 1960 Graham, the publisher of the *Washington Post* and one of the most brilliant and attractive men of my generation, had served as an intermediary between Johnson and Kennedy, both of whom he knew intimately.[65] Bobby shook his head and said that the Graham memorandum was a reasonable version of things as Graham saw them but that Phil didn't know the real story. "If the indication in the Graham memorandum is what I interpreted from it, that the President had the thought of Lyndon Johnson as Vice President prior to his own nomination other than just passing through his mind, that's not true."[66]

John Kennedy's first thought after his own nomination, his brother told me, was "how terrible it was that he only had twenty-four hours to select a Vice President. He really hadn't thought about it at all. He said what a mistake it was when it was such an important position that you only had twenty-four hours to think about it." I asked why it had not been thought about earlier. "Because we didn't know, you know — we wanted to just try and get the nomination. . . . To keep everybody happy and to keep in contact with everybody, as you remember, it was just a hell of an operation, so we weren't even thinking about [the Vice Presidency]."

'Keeping everybody happy,' I take it, was a term of art. Presidential candidates always prefer to keep the vice presidential situation fluid until their own nomination is assured. Key political figures who are led to think they might be under consideration for the Vice Presidency can be helpful in tight spots. Beyond such time-honored tactics, however, John Kennedy unquestionably had given serious, if fragmentary, thought to the second place in the weeks before the convention. He had asked Walter Reuther and Jack Conway to talk to Humphrey about going on the ticket,[67] and he told Galbraith and me at Hyannis Port on June 12 that he was for Humphrey.[68] After Humphrey repeatedly declined to meet with Reuther, Kennedy told Clark Clifford toward the end of June that Symington was his first

choice and repeated this to Clifford in Los Angeles.[69] John Seigenthaler, who saw a good deal of Robert Kennedy during the convention, had the "distinct impression" that Robert was also for Symington
and wrote for the *Nashville Tennessean* that Symington was the probable choice.[70] Early in the convention labor decided to back
Symington.[71]

Actually Robert had reservations about Symington. Edwin Guthman, who came down from Seattle to cover the convention, asked
about his own senator, Henry Jackson, another veteran of the Investigations Subcommittee. "He's my choice," Kennedy said, "and Jack
likes him. I'd say he and Symington have the strongest chance.
Orville Freeman is another possibility. It depends on how things
develop." He added that Jackson would have to convince leaders like
Mayor Daley of Chicago, Governor Lawrence of Pennsylvania and
John Bailey that he could help the ticket the most. "We've told him
that, and he understands it. I hope he can do it."[72] As for Freeman,
Bobby told me in 1965 that John Kennedy had liked and respected
him but did not think he would help the ticket.

Humphrey finally eliminated himself on Tuesday when he went
back on the understanding that he would endorse Kennedy.* This
left Symington, Jackson, Freeman — all men sufficiently acceptable
to the liberals and labor. The one name that no one ever mentioned
was Lyndon Johnson. Quite the contrary: the Kennedy people told
everyone as categorically as possible that he was not in the picture.
"The labor people had warned me repeatedly that they did not want
Johnson on the Kennedy ticket," O'Donnell has written. "I had
promised them that there was no chance of such a choice. We had
given the same assurance, with Kennedy's knowledge, to wavering
liberal delegates during the movement earlier in the week to draft
Stevenson."[73] This expressed the understanding in the Kennedy
headquarters. "There was *never* any talk in the office," wrote Evelyn
Lincoln, Kennedy's secretary, "that Mr. Johnson was to be the run-
ning-mate."[74]

On Monday, Philip Graham and Joseph Alsop had put the case for
Johnson to Kennedy. Kennedy seemed to agree — "so immediately,"
Graham wrote, "as to leave me doubting the easy triumph." When
Graham passed the word on to James Rowe of Johnson's staff, Rowe
dismissed it as a traditional and transparent attempt to coax a rival
out of the race — a view taken by Johnson himself, who said impatiently that he supposed the same message was going out to all the
candidates.[75] Graham in the meantime encouraged the *Post* to report

* The day before, I had sent Kennedy a memorandum, now for some reason in
Theodore H. White's possession, arguing that Humphrey's nomination would be the
surest way of enlisting the Stevensonians in the campaign.

that Kennedy would probably offer the Vice Presidency to Johnson. When the *Post* containing this word was delivered to the District
of Columbia delegation, Joseph Rauh, who had been laboring pro-
digiously to keep black delegates in line for Kennedy, went in some
agitation to see Robert Kennedy. Kennedy assured him, as he had
before, that it would definitely not be Johnson.

On Wednesday evening, after the presidential nomination, John
Kennedy told his brother "that he was going to go offer [the Vice
Presidency] to Lyndon Johnson the first thing in the morning." He
made clear that this was entirely *pro forma.* "The idea that he'd go
down to offer him the nomination in hopes that he'd take the nomination is not true," Robert Kennedy told me in 1965. "The reason
he went down and offered him the nomination is because he thought
that he should offer him the nomination; because there were enough
indications from others that he wanted to be offered the nomination, . . . *but he never dreamt that there was a chance in the world that he
would accept it.*"

On Thursday morning John Kennedy made the offer. "I was up
in his room when he came back," Robert Kennedy continued, "and
he said, 'You just won't believe it.' I said, 'What?' and he said, 'He
wants it,' and I said, 'Oh, my God!' He said, 'Now what do we
do?' . . . We both promised each other that we'd never tell what
happened — but we spent the rest of the day alternating between
thinking it was good and thinking that it wasn't good that he'd offered
him the Vice Presidency, and how could he get out of it."

The offer caused consternation in Kennedy's immediate circle.
When O'Donnell heard the news, "I was so furious that I could
hardly talk. I thought of the promises we had made to the labor
leaders and the civil rights groups. . . . I felt that we had been double-
crossed." He told Robert Kennedy that it was a "disaster" and John
Kennedy that it was "the worst mistake" he ever made; "in your first
move after you get the nomination, you go against all the people who
supported you." Pierre Salinger joined the protest. John Kennedy
became pale with anger, so evidently wounded that it took a moment
before he could collect himself. He said that the nomination had
been offered but not yet accepted. He added: "If we win, it will be
by a small margin and I won't be able to live with Lyndon Johnson
as the leader of a small majority in the Senate. Did it occur to you
that if Lyndon becomes the Vice President, I'll have Mike Mansfield
as the leader in the Senate, somebody I can trust and depend
on?"[76] This was the rationalization that, Robert Kennedy later told
me, the Kennedys found most persuasive — the thought that Johnson
"would be so mean as Majority Leader — that it was much better
having him as Vice President where you could control him."

At eleven o'clock a labor group — Reuther, Arthur Goldberg, Jack Conway, Alex Rose — came to see Kennedy. "When we went in," Conway said later, "Bob was there, and he obviously was very distressed. Ken O'Donnell looked like a ghost. Jack Kennedy was very nervous. It was obvious that they were going to convey some information to us that we weren't going to like." John Kennedy made the argument he had made to O'Donnell, adding that all he had done was to make Johnson an offer: "I don't see any reason in the world why he would want it." Conway said to Robert Kennedy, "We've come a long way. If you do this, you're going to fuck everything up."[77]

VIII

It was, Robert told John Bartlow Martin in 1964, "the most indecisive time we ever had.... We changed our minds eight times during the course of it."[78] "Finally," he told me, "we decided by about two o'clock that we'd try to get him out of there and not have him because Jack thought he would be unpleasant with him, associated with him, and if he could get him to withdraw and still be happy, that would be fine."

Robert Kennedy thereupon saw Johnson twice — first to feel him out; then, after further consultation, to ask him to withdraw. Subsequently he described their talk to me:

> There were just the two of us. He was seated on the couch and I was seated on his right. I remember the whole conversation.... This was the famous conversation in which Phil Graham says I went down by myself and on my own.... Obviously with the close relationship between my brother and I, I wasn't going down to see if he would withdraw just as a lark on my own — "my brother's asleep so I'll go see if I can get rid of his Vice President" — you know, that's what flabbergasted me about Phil Graham's memorandum.... So I said, "There's going to be a lot of opposition" — this was what we worked out, that I'd tell him there was going to be a lot of opposition, that it was going to be unpleasant, that it was going to focus attention, and we were going to have trouble with the liberals ... and the President didn't think that he wanted to go through that kind of an unpleasant fight.

Instead Robert Kennedy proposed that Johnson become chairman of the national committee.

> The President wanted him to play an important role and he could run the party — the idea being that to run the party he could get a lot of his own people in; and then if he wanted to be President after eight years or something, he could have the machinery where he could run

for President or do whatever he wanted. I mean that was sort of the idea at the time. We didn't really know whether he'd want to go through with it, and, in any case, the President wanted to get rid of him.

As Kennedy spoke his lines, Johnson, he said,

> burst into tears. I mean, it was my feeling at that time, although I've seen him afterwards look so sad I don't know whether it was just an act.* . . . He is one of the greatest sad-looking people in the world — you know, he can turn that on. . . . But he just shook and tears came into his eyes, and he said, "I want to be Vice President, and, if the President will have me, I'll join with him in making a fight for it." So it was that kind of a conversation. I said, "Well, then that's fine. He wants you to be Vice President if you want to be Vice President."

Here the Graham and Kennedy recollections diverge. Graham remembered his calling John Kennedy, Kennedy's reassuring Graham ("Bobby's been out of touch and doesn't know what's been happening"), reassuring Johnson, then speaking to his brother. Robert remembered returning to his brother's room. John Kennedy told him, " 'I just got a call from Clark Clifford or somebody saying that "this is disastrous. You've got to take him." I'm going to make an announcement in five minutes.' . . . But during that whole three or four hours, we just vacillated back and forth as to whether we wanted him or not . . . and one time we wanted him and one time we didn't want him and finally we decided not to have him, and we came upon this idea of trying to get rid of him, and it didn't work." "The only people who were involved in the discussion," he told John Bartlow Martin, ". . . were Jack and myself. Nobody else was involved in it." [79]

It was a hard day. According to Bobby Baker, secretary of the Senate and Johnson's *homme de confiance,* the new vice presidential candidate, "as soon as his performance was over . . . spoke of Robert Kennedy in terms of 'that little shitass' and worse." Robert Kennedy said to James Rowe, "My God, this wouldn't have happened except that we were all too tired last night. Don't you think this is a terrible mistake? Don't you think Stuart Symington should be the candidate?"[80] Bobby's old friend Charles Spalding said, "That was a real surprise." "Charlie, we really weren't at our best this time," Robert Kennedy replied. "We were all too tired."[81] "Yesterday was the best

---

* Robert Kennedy's surmise was doubtless correct. Carl Rowan, who served in the Kennedy and Johnson administrations, remembers Johnson crying in his presence as he told a story. "He looked me in the eyes and said, 'You know, a man ain't worth a damn if he can't cry at the right time'" (Carl Rowan, "Accenting the Positive," *New York Post,* June 16, 1976).

day of my life," he said to Charles Bartlett, "and today was the worst day." A year later he said enigmatically to Seigenthaler, "We were right at eight, ten and two and wrong at four."[82]

That evening they dined with their father. The sun was setting, and Robert Kennedy's children were splashing in a fountain in front of Marion Davies's house. Mr. Kennedy, in a velvet smoking jacket, stood grandly in the doorway, his hands clasped behind his back, as John Kennedy drove up. He had had nothing to do with the Johnson decision, despite later stories to the contrary, nor is it easy to imagine why he should have intervened in favor of Johnson on Thursday after having been called pro-Nazi by Johnson on Tuesday. But, as usual, he exuded confidence in his sons. "Don't worry, Jack," he said. "Within two weeks they'll be saying it's the smartest thing you ever did."[83]

IX

Robert Kennedy had no patience with the theory expounded by Richard Nixon, the Republican candidate, of "peaks" and "rhythms" in a campaign. His formula was simple: "run and fight and scramble for ten weeks all the way." Once, when Theodore White was chatting with a couple of Kennedy aides in the Washington headquarters, the manager, tieless and in shirtsleeves, burst into the cramped office. "What are you doing?" he cried. "What are we all doing? Let's get on the road. Let's get on the road tomorrow! I want us all on the road tomorrow!" He slammed the door and disappeared without waiting for an answer.[84]

Action was one principle of his politics. Another was what Seigenthaler, who became his assistant, called "the Kennedy theory . . . embrace your enemies and broaden your base of support."[85] This theory was in my experience rather more characteristic of the Kennedy *modus operandi* than the usual notion that they were a clan of Irish haters. The offer to Johnson was a spectacular example. The morning after Kennedy's nomination, I began urging the importance of bringing the Stevenson people into the campaign. Robert Kennedy listened patiently, then put his hand on my knee and said, "Arthur, human nature requires that you allow us forty-eight hours. Adlai has given us a rough time over the last three days. In forty-eight hours I will do anything you want, but right now I don't want to hear about the Stevensonians. You have to allow for human nature."[86] In a short while he had James Doyle, who ran the Stevenson campaign, Eugene McCarthy and me carrying the message to the Stevenson loyalists of southern California. Eventually most of

the Stevenson leaders and even Johnson's John Connally received
jobs in the Kennedy administration.

Robert's tactics were flexible. Where the state organization seemed
strong, he worked with it. Where it was weak, he worked through
the Citizens for Kennedy Committee, headed by Byron White of
Colorado. White, an all-American football player, PT-boat veteran
and Rhodes Scholar, represented a combination of accomplishments
irresistible to the Kennedys. Robert even wanted to make him chairman of the national committee. ("Bobby Kennedy displayed a naiveté
at times that was downright astonishing," O'Donnell commented with
disapproval.)[87] Henry Jackson, disappointed in his hope for the Vice
Presidency, got the job.

New York, still in 1960 the most populous state, was vital. Robert
harshly told the New York Reform Democrats, in the presence of
such properly revered figures as Herbert Lehman and Thomas K.
Finletter, to drop their feuds and get together with the regulars:
"Gentlemen, I don't give a damn if the state and county organizations
survive after November, and I don't give a damn if *you* survive. I
want to elect John F. Kennedy."[88] The tone was offensive but effective. He sent William Walton, urbane and shrewd, to pull the Kennedy operation together in the city. He sent Paul Corbin to get
things moving upstate. Professor Richard Wade, the Kennedy coordinator in the Rochester area, told Corbin he had to stop promising
the embassy in Ecuador; he had already offered it to three people;
there were nineteen republics down there, and he could at least
spread his offers around.[89]

Kennedy traveled from state to state lashing local Democrats into
activity. Ralph (Rip) Horton, the Kennedy coordinator in Nebraska,
had gone to Choate with John Kennedy (whom he had joined in
inventing an underground society called the Muckers) and had
known Robert as a little boy. Now Robert was appalled by the scan-
tiness of the Nebraska crowds. "Bobby gave me an extremely hard
time as a result of it," Horton later recalled. "I stood at the foot of
his bed as he lay and relaxed one evening before dinner and he said,
'Rip, can you possibly explain what goes on here to me?' I had no
explanation and I felt like I did when I was a Mucker and they were
throwing me out of Choate."[90]

Joseph Tydings, the Florida coordinator, saw Robert Kennedy cas-
tigate the mayor of Jacksonville, telling him in effect "that we knew
he was cutting us, and by God we didn't like it, and that we were
going to hold him accountable if we lost Duval County and not for
him one minute to think that he was getting away with anything.
. . . That shook the hell out of him." Kennedy, said Tydings, was

"absolutely strong, steel-willed, didn't mind telling a mayor of a city or governor or anyone else what he thought if he stepped out of line. He just was blunt and hard and tough and was of course a magnificent campaign manager." Tydings added that Kennedy only got rough at the top; "he was never unkind or inconsiderate of anybody who was down in the lower echelon"; he was "fantastic" with volunteers, particularly with women and with the young.[91]

California was, as ever, chaos. The fight between the regulars, organized by the forceful professional Jesse Unruh, and the liberals, led by Thomas Braden, then a newspaper publisher in Oceanside, was more than usually embittered. Joan Braden thought Kennedy "the most terrific man to work for that I've ever known. His decisions were made [clap] like that. You went in and you said, 'I would like to do boom, boom, boom, boom.' And he'd say, 'Do this, do this. Don't do this. And do this.' Over. . . . If he said 'No' and you wanted to do it, he would listen; and he might change, but he didn't want you to stay in there an hour and go over and over it."[92]

His campaign role crystallized the impression of Robert Kennedy many had taken from Rackets Committee days — hard, unforgiving, 'ruthless.' "Little Brother is Watching You," aggrieved Democrats warned each other. "He has all the patience of a vulture," a Washington journalist said, "without any of the dripping sentimentality."[93] Norman Mailer was reminded of those unreconstructed Irishmen from Kirkland House he had faced in Harvard intramural football games. You said hello as you lined up for scrimmage after kickoff, "and his type would nod and look away, one rock glint of recognition your due for living across the hall from one another all through your Freshman year, and then bang, as the ball was passed back, you'd get a bony king-hell knee in the crotch."[94]

Bobby didn't care. "I'm not running a popularity contest," he told Hugh Sidey of *Time*. "It doesn't matter if they like me or not. Jack can be nice to them. . . . Somebody has to be able to say no. If people are not getting off their behinds and working enough, how do you say that nicely? Every time you make a decision in this business you make somebody mad."[95]

His brother watched all this carefully and admiringly. Once he sent Robert to call on Harry Truman. The former President gave the young campaign manager a genial scolding. "You should learn to moderate your attitude in your dealings with people. I've heard some stories about how you react." Everybody, Truman said, thought Robert Kennedy a son of a bitch. "I understand that," Truman added, "because everybody thinks I'm a son of a bitch." He counseled Kennedy to stop doing things that made people angry. Kennedy

protested that the stories were exaggerated; he was not such a monster.[96] But he did not relent. (Nor did he convince Truman, who
said the next year: "I just don't like that boy, and I never will. He
worked for old Joe McCarthy, you know, and when old Joe was
tearing up the Constitution and the country, that boy couldn't say
enough for him.")[97]

John Kennedy recognized that a campaign required a son of a
bitch — and that it could not be the candidate. Robert was prepared
to do what the candidate should not have done. Stewart Alsop called
it a "sweet-and-sour brother act, Jack uses his charm and waves the
carrot and then Bobby wades in with the big stick."[98] "Every politician in Massachusetts," John Kennedy said, "was mad at Bobby after
1952 but we had the best organization in history."[99]

                                      X

The most celebrated incident in the campaign, next to the debates
with Nixon and the discussion of church and state before the Houston
Ministerial Association, was Kennedy's phone call to Martin Luther
King, Jr.

The initial Kennedy interest in Negro affairs was political. The
brothers had been entirely devoid of racial prejudice but only intermittently sensitive to racial injustice. The debate over the Civil Rights
Act of 1957 had begun John Kennedy's education. Robert Kennedy
was busy elsewhere. His own response to discrimination had always
been in individual terms: the black football player at Harvard; Ralph
Bunche at Charlottesville; a young witness before the Rackets Committee. "If he was a white fellow," Robert had noted in his journal,
"and had a chance for an education he would do extremely well.
McClellan, for some reason, got furious with him and threw the
various union cards in his face. Poor old Sonie's eyes filled up with
tears and sweat appeared on his forehead and his voice got lower
and lower. I felt very sorry for him."[100] But Robert Kennedy had
not generalized the few cases that had come his way into an encom-
passing national problem.

In 1959 John Kennedy was ready to welcome what white southern
support he could get. When Robert Kennedy made a trip into the
Confederacy, his brother charged him to recall that, on two crucial
civil rights roll calls in 1957 where "the friends and foes of the South
split," Kennedy had voted with Lyndon Johnson while Humphrey
and Symington had voted for the north.[101] At the same time, the
Kennedys fully intended to stand for civil rights in the north, and
believed they could do so without undue harm to their southern

prospects. "Obviously they are going to make a fight on the civil rights problem," Robert Kennedy noted after talks with white leaders in Georgia and Virginia in November 1959, "but I think that they have pretty much decided they will stay with the Democratic Party."[102]

The primaries began to alter John Kennedy's attitude, though politics was still the dominant note. He told me in May that he had learned a lot in West Virginia. He had carried the black vote, he said, simply because Senator Robert Byrd, an ex-Klansman, had backed Humphrey. "This proved to me that it's absolutely fatal to have southern support. I want to be nominated by the liberals. I don't want to go screwing around with all those southern bastards."[103] Soon Robert Kennedy, alarmed by what he deemed an excessively antisouthern tone in one of his brother's public statements, sent him an admonitory memorandum: "I think it should be clarified that what you said was that you feel that you will be nominated without Southern votes — not that you are not interested in having votes in the South. Otherwise, this is going to look like a gratuitous insult to the Southern political leaders who have been interested in you. . . . This should be straightened out today if possible."[104]

Southern concern was not dispelled. In May the candidate had asked Sargent Shriver to organize a civil rights section in the campaign. Shriver brought in Harris Wofford, an ardent young law professor, recommended by Chester Bowles and Father Theodore Hesburgh, as civil rights coordinator. "We really don't know much about this whole thing," Robert Kennedy told Wofford. ". . . I haven't known many Negroes in my life . . . . It's up to you. Tell us where we are and go to it."[105] Shriver and Wofford went to it, and so in due course did Robert Kennedy. Discussing civil rights with southern delegates in Los Angeles, the campaign manager finally said, "My, you've got a lot of cantankerous politicians in the south, haven't you?" A southern voice from the back of the room: "What the hell do you think you Boston Irish are?"[106]

At the convention, Bowles, as chairman of the resolutions committee, drafted with Wofford a "maximum" civil rights plank designed to be used for negotiating purposes. In one of those morning meetings at the Biltmore, Robert told the Kennedy organizers that this was "the best civil rights plank the Democratic party has ever had" and directed them to carry the word that the Kennedy forces were "unequivocally" for it. "Those of you who are dealing with southern delegations make it absolutely clear how we stand on civil rights. Don't fuzz it up. Tell the southern states that we hope they will see other reasons why we are united as Democrats and why they should support Kennedy, but don't let there be doubt anywhere as to how the Kennedy people stand on this." Wofford, astonished, felt like crying out to Kennedy that the plank was beyond where they really wanted to go; they had written it expecting it to be cut down. "When I finally got to Bowles to tell him that the Kennedy word was out to support his *maximum* plank, he couldn't believe it."[107] The southern Democratic seigneurs communed gloomily with each other. "Dick Russell [of Georgia] told me today," Harry F. Byrd of Virginia wrote James F. Byrnes of South Carolina a month after the convention, "that Kennedy will implement the Democratic platform and advocate Civil Rights legislation beyond what is contained in the platform."[108]

XI

On October 19, 1960, Dr. Martin Luther King, Jr., the Baptist preacher who had emerged as the most eloquent southern leader in the battle for racial justice, was arrested in Atlanta for sitting in in a department store. Six months before he had received a suspended sentence and probation for a minor traffic transgression in DeKalb County, Georgia. Though the Atlanta authorities released him, the DeKalb County judge seized the opportunity to sentence him to four months of hard labor for violating probation.

The sentence was frightening to all who knew what hard labor in Georgia might mean, especially for uppity blacks. "The death of Martin King in the state public works camp," wrote the black novelist John A. Williams, "was imminent."[109] Nor was lynching, the traditional Georgia remedy, beyond possibility. King's wife, who was five months pregnant, was wild with concern — all the more so when King was brusquely awakened at four-thirty one morning, put in handcuffs and leg chains, removed from the county jail, transported along dark country roads and deposited in a penitentiary deep in rural Georgia.

Wofford drafted a statement of protest for Kennedy. Then Governor Ernest Vandiver of Georgia begged the candidate not to issue it; it would be politically disastrous in the south, and he would get King out himself. But nothing happened. Mrs. King called Wofford in despair; she was, Wofford said, "terribly hysterical." Wofford thought it might be a good idea for John Kennedy to speak to her. Louis Martin, the wise black newspaperman from Chicago, also working in the civil rights section, agreed. At first they could not reach Kennedy. Wofford meanwhile told the story to Bowles. "I'll call her," Bowles said, "and as a matter of fact I'll have Adlai call her.

He's here for dinner." Mrs. King phoned Wofford the next morning to say how wonderful Bowles was to call. She did not mention Stevenson. Bowles later told Wofford, "I tried my best to get Adlai to talk to her but he said he just couldn't possibly, he'd never been introduced to her."

Wofford finally reached Shriver, who tracked down Kennedy in a motel in the Chicago airport. As has often been told, Shriver waited for everyone else to leave the room, fearing that general discussion would kill the idea. Kennedy at once asked Shriver to get Mrs. King on the phone. Later in the day, as the party was arriving in Detroit, he said casually to Salinger, "By the way, I talked to Mrs. King this morning."

Robert Kennedy's initial reaction was very different. He called in Wofford and Martin and gave them hell. "You bombthrowers probably lost the election," he said. Three southern governors had told him that, if his brother were to say a good word for Jimmy Hoffa, N. S. Khrushchev or Martin Luther King, it would end Kennedy hopes in the south. "You've probably lost three states and . . . the civil rights section isn't going to do another damn thing in this campaign."

At the same time he was personally angered by the judge's refusal to grant bail to Dr. King. He wondered whether he ought not call the judge and protest. Might this not, he asked John Seigenthaler of Nashville, now his campaign assistant, draw some of the fire from his brother? Seigenthaler, who had been hearing all day from southern leaders about John Kennedy's phone call, thought it a poor idea. One more Kennedy shock might be too much for the south. Robert was scheduled to speak in New York that afternoon. Seigenthaler drove him to the airport, confident that he had been dissuaded from calling the judge.

Later in the day the wire services began querying the headquarters about a story that a brother of the senator had called the judge and asked for King's release. Edward Kennedy was far away on the west coast. Everyone knew how Robert felt. Seigenthaler told Roger Tubby of the press section to deny the story. Later Robert phoned in. "Guess what that crazy judge says in Georgia?" Seigenthaler said. "He says you called him about King not getting bail." Kennedy (long pause): "Did he say that?" Seigenthaler: "Yes, but don't worry, I had Tubby put out a denial." Kennedy: "Well, you better get him to retract it." Seigenthaler said, "I can't believe it." Kennedy said, "Yes, it just burned me all the way up here on the plane. It grilled me. The more I thought about the injustice of it, the more I thought what a son of a bitch that judge was. I made it clear to him that it was not a political call; that I am a lawyer, one who believes in the right of all defendants to make bond and one who had seen the rights of defendants misused in various ways . . . and I wanted to make it clear that I opposed this. I felt it was disgraceful." The conversation, Kennedy added, was restrained and calm; he had not been angry nor the judge hostile. The next day King was released.[110]

The brothers, in making their phone calls, had acted independently of each other. "The finest strategies," John Kennedy observed to John Kenneth Galbraith in this connection, "are usually the result of accidents."[111] King himself remarked when liberated, "There are moments when the politically expedient is the morally wise," and declined to endorse Kennedy.[112] His father, however, said, "I'll take a Catholic or the Devil himself if he'll wipe the tears from my daughter-in-law's eyes. I've got a suitcase full of votes — my whole church — for . . . Senator Kennedy."[113] (Kennedy remarked later of Daddy King, "He said he was going to vote against me because I was a Catholic, but that since I called his daughter-in-law he will vote for me. . . . That was a hell of a bigoted statement, wasn't it? Imagine Martin Luther King having a bigot for a father." Then he smiled and said lightly, "Well, we all have fathers, don't we.")[114]

After the election, which John Kennedy may well have won because of the black vote, Murray Kempton asked Robert Kennedy whether he was glad he had called King's judge. "Sure I'm glad," Robert Kennedy said, "but I would hope I'm not glad for the reason you think I'm glad."[115]

XII

"Bobby Kennedy holds his head down and looks up through his eyebrows like a coon peering out of a henhouse," wrote Eugene Patterson in the *Atlanta Constitution*. ". . . Throw an arm around those shoulders and the big white teeth might snap at you. . . . A Southerner is a warm-blooded soul who likes his potlikker and his politics hot. The Kennedys are chill dishes indeed. But you feel that they know what to do in a hot fight."[116]

He drove himself relentlessly. George McGovern, running for the Senate in South Dakota, called to say that he was falling behind; could the national committee send someone out to help. Robert Kennedy knew that McGovern was in trouble because of his staunch support of John Kennedy in an anti-Catholic state. To McGovern's astonishment he flew out to South Dakota himself on the Friday before the election. "South Dakota had four electoral votes; it was obvious they wouldn't go to John Kennedy; and it was obvious that I was behind — probably a losing candidate; and yet . . . yet he came out." After a series of speeches, they sent him back to Chicago on a bitterly cold, windy night, in a lightweight plane with an inexperienced pilot, a local doctor. "He didn't complain."[117]

He never complained, never relaxed, rarely slept. "We can rest in November," he said. By October he was thin, weary, deep gray circles under his eyes, his shoulders slumped with fatigue. "He's living on nerves," John Kennedy said. "Jack works as hard as any mortal man can," said their father. "Bobby goes a little further."[118] Once the two brothers ran into each other at a windswept airport. "Hi, Johnny," said Robert Kennedy. "How are you?" "Man, I'm tired," said the candidate. "What the hell are you tired for?" said the campaign manager. "I'm doing all the work."[119] "I don't have to think about organization," John Kennedy said later. "I just show up. . . . He's the hardest worker. He's the greatest organizer. . . . Bobby's easily the best man I've ever seen."[120]

XIII

"We need someone," Henry Kissinger had told me just before Labor Day, "who will take a big jump — not just improve on existing trends but produce a new frame of mind, a new national atmosphere. If Kennedy debates with Nixon on who can best manage the status quo, he is lost. The issue is not one technical program or another. The issue is a new epoch. If we get a new epoch and a new spirit, the technical programs will take care of themselves."[121]

This, as we saw it, was the issue in 1960. Election night found the Kennedys back in Hyannis Port. The evening began cheerfully enough. Then it sank into an endless strain of uncertainty. By midnight people were wandering restlessly, almost grimly, around the communications center in Robert Kennedy's house. Around three in the morning the atmosphere lightened. Pennsylvania and Michigan looked fairly definite. At three-fifteen Robert Kennedy told his brother to go to bed and sleep well: they had won. But there was no word from Nixon. John Kennedy strolled with Cornelius Ryan across the dark lawn between the houses in the compound. Ryan thought he heard Kennedy say, "I'm angry," and supposed that the awful wait had finally snapped the imperturbable composure. Later, when Theodore White put the story in the manuscript of his book, Kennedy told White, "That isn't what I said in the dark. What I said was, 'I'm hungry.'"[122]

After breakfast the Kennedys met informally with some of the staff at the ambassador's house. More than a third of the states were going one way or another by 2 percent of the vote or less. They expected that Nixon would challenge the results. John Kennedy was calm. If he won by a single electoral vote, it would be enough, and he proposed to go forward on the assumption that he had been elected President of the United States.* Finally they heard that Nixon had scheduled a press conference in Los Angeles. Herbert Klein emerged on Nixon's behalf and read a curt concession. Someone at Hyannis Port observed that Nixon might at least have had the decency to appear himself and thank the people who had voted for him. John Kennedy said, "He went out the way he came in — no class."[123]

Victory was accompanied no doubt by apprehension. Earlier in the autumn Walter Reuther and Jack Conway had come to Hyannis Port. Kennedy suggested they take a swim. It was a crisp fall day, but the sun was warm, and the labor men sat pleasantly on the beach talking with Jacqueline Kennedy. "She really was obsessed," Conway said later, "with this whole idea of the change in her life, the change in Kennedy's life, and how do you protect against assassinations." Gunmen had tried to kill Walter Reuther in 1948 and Victor Reuther in 1949. Jacqueline wondered what security precautions did to family

* Pierre Salinger, in recorded interview by Theodore H. White, July 19, 1965, 101–104, JFK Oral History Program. Out of nearly 69 million votes cast, Kennedy had a plurality of 120,000, but he had a lead of 84 votes in the electoral college. A legend subsequently arose that Mayor Daley of Chicago put Kennedy over by stealing the election in Illinois. The margin in Illinois was less than 9000, and the Cook County machine was unquestionably skilled in producing Democratic votes out of vacant lots and graveyards. So were the Republicans downstate. One can assume that in Illinois one party stole as many votes as the other. When the Illinois count was challenged, the state electoral board, which was 4–1 Republican, voted unanimously to certify the Kennedy electors (see the AP story by Don McLeod, "Did the Democrats Steal It in 1960?" *New York Post,* July 12, 1973).

Even if Illinois had gone to Nixon, Kennedy would still have won by 276 to 246 in the electoral college. Nixon would have had to carry Texas as well to get a majority. While there were probably irregularities in Texas, Kennedy's margin of 47,000 votes was too great to be reasonably attributed to fraud or acceptably reversed by a recount.

The more difficult question, applying, however, only to the popular vote, was raised by Alabama. The Republican ticket elector slate received 238,000 votes. The Democratic slate consisted of six unpledged and five Kennedy-pledged electors. The unpledged electors voted for Harry F. Byrd in the electoral college. It is impossible to determine what Kennedy's popular vote in Alabama was. If the Kennedy electors were credited with 5/11 of the popular vote, the remaining 6/11 being allocated to a national unpledged elector vote, then the state totals would read: Nixon, 238,000; unpledged, 177,000; Kennedy, 147,000 — and Nixon would have won the popular vote by 58,000. (For calculations, see *Congressional Quarterly,* February 17, 1961, and Neal R. Pierce, *The People's President* [New York, 1968], 102–104.) Kennedy would still have won in the electoral college. The possibility of the popular-vote winner losing in the electoral college argues for a reform of the present (1978) system, though not, in my judgment, for a shift to direct popular election. For a proposed change retaining a reformed electoral system while attaining the objectives of direct popular election, see the Twentieth Century Fund task force report "Reform of the Presidential Election Process" (New York, 1978).

life. Don't let the precautions run your life, Reuther said; you can't stop living.[124]

Such thoughts were now set aside. In the afternoon the President-elect played touch football. He went down for a pass, and someone ran into him, and he fell to the ground. "That's my brother," said Robert Kennedy, who had called the play, "all guts, no brains."[125] A cable arrived from England: ON THE OCCASION OF YOUR ELECTION TO YOUR GREAT OFFICE, it read, I SALUTE YOU. THE THOUGHTS AND SUPPORT OF THE FREE WORLD WILL BE WITH YOU IN THE CHALLENGING TASKS THAT LIE AHEAD. It was signed, WINSTON S. CHURCHILL.[126]

11

To the Department of Justice

" WE CAME into this administration without a political obligation," Robert Kennedy recalled in 1964. ". . . I don't mean
that people didn't do favors for us and that we wanted to do favors
for them. [But] there weren't any promises made to anyone that they
would get a job. . . . There were certainly no cabinet positions promised. . . . It was really a fresh slate."[1]

I

On February 9, 1961, he dictated a hasty memorandum about the
process of choosing a cabinet. "The toughest decision for all and
probably and undoubtedly the most important," he said, "was the
selection for Secretary of State."

The President-elect had a clear view, derived from the Roosevelt
model, about what he sought in the State Department. "It is the
President alone," John Kennedy had said imperially in January 1960
in a speech on the Presidency at the National Press Club, "who must
make the major decisions of our foreign policy."[2] He did not want,
he now told Robert Lovett, a Secretary of State who would operate
a "one-man show" in the style of John Foster Dulles.[3] He was looking
for someone who would be wise in counsel, persuasive on the Hill
and effective in modernizing what he regarded as an unduly passive
and conservative Foreign Service.

Jack always wanted William Fulbright [Robert Kennedy wrote]. He had
worked with Fulbright, knew him better, was very impressed with the
way he ran the Senate committee. Prior to the time he was appointed
to the Foreign Relations Committee, he thought Fulbright to be rather
pompous, opinionated, and didn't like him too much. But after he went

on the committee he gained a great respect for him. He didn't know any of the others that were being considered. . . . It finally came down, the last week, to David Bruce and Dean Rusk and William Fulbright. I was very strongly against Fulbright. I felt . . . that any time we took a position against any of the nations of Africa, or any time that we would take a position that was unpopular, that the Russians and the others would be able to say, "Well, this is all done because we have a Secretary of State who feels that the white man is superior to the Negro." I said in the last analysis Jack was going to be the Secretary of State; this was a position in which he would be making his own decisions. All he needed was advice and ideas by an expert but the decisions would be made by him and that we'd be unnecessarily burdened by having a man in that job who had this terrible impediment of having signed the Southern Manifesto.*

We had a final discussion about it at the hospital during the christening of young John Kennedy. Charlie Bartlett was there and took the position that Jack could appoint Fulbright. I was extremely strong against it and ultimately prevailed. On the Friday before the selection, Jack was ready to take David Bruce. Arthur Krock was for David Bruce, [as were] Maggie [Marguerite] Higgins [of the *New York Herald Tribune*] and others. Maggie Higgins was for him because she said he was tough with the Russians and would be able to stand up. Nobody knew Dean Rusk particularly. I received word by noon that Robert Lovett was for Dean Rusk so I called up Jack just before he left for Florida and suggested that he call Robert Lovett. It was the conversation with Lovett that convinced him that Dean Rusk should be the selection. The following day he called Dean Rusk and asked him to come down. He came to Florida and the announcement was made the following Sunday afternoon.[4]

Rusk had served in Truman's State Department and was now head of the Rockefeller Foundation. Acheson as well as Lovett backed him. Chester Bowles had recommended him to Kennedy as under secretary. Rusk considered his first meeting with Kennedy a failure — "Kennedy and I could not communicate," he told Bowles immediately afterward[5] — but Kennedy thought him lucid, competent and self-effacing and offered him the job.

In later years Robert Kennedy repented his fight against Fulbright. He also mused over the curious fact that he and his brother had not considered the man who in retrospect would have made the best Secretary of State of all — Averell Harriman. But the Kennedys hardly knew Harriman in 1960. His defeat two years before for

---

* The Southern Manifesto of 1956 had attacked the Supreme Court decision calling for school desegregation as a "clear abuse of judicial power" exploited by "outside agitators" and had demanded "a reversal of this decision which is contrary to the Constitution." One hundred one legislators from the old Confederacy had signed it, Fulbright among them. The southern senators who refused were Estes Kefauver and Albert Gore of Tennessee and Lyndon Johnson.

reelection as governor of New York was supposed to have ended his career. He was now sixty-nine years old, deaf in one ear, his tall figure a little stooped, his mien, now that he was out of power, tinged with melancholy. He was twice Robert Kennedy's age, and even John Kennedy saw him as a figure from the remote Rooseveltian past. Nor was Harriman greatly admired by those of the Roosevelt generation the younger Kennedys knew best. I suppose that Joseph Kennedy thought as little of Harriman (whom Roosevelt had sent to London a few months after Kennedy had departed) as Harriman thought of him. Max Beaverbrook, who had had memorable rows with Harriman during their Moscow mission in 1941,[6] often made cracks about him in subsequent years. After an evening with Beaverbrook in 1958, Robert Kennedy recorded the master's comment on Harriman: "Never has anyone gone so far with so little."*

For such reasons the Kennedys simply did not think of Harriman as a possible Secretary of State. When I mentioned him to John Kennedy on December 1, the President-elect dismissed him as "old hat."[7] A fortnight later I urged him on Robert, who said more sympathetically, "Are you sure that giving Averell a job wouldn't be just an act of sentiment?"[8] Eventually Kennedy made him roving ambassador, considering it a mainly decorative appointment. Both Kennedys changed their minds rapidly, of course, as the old crocodile got his teeth into things. I have often thought that the nation would have escaped much trouble in the 1960s if John Kennedy had chosen Averell Harriman as his Secretary of State.

II

Sargent Shriver [Robert Kennedy wrote in his 1961 memorandum] was responsible for the selection of . . . Robert McNamara.

We were not sure whether we wanted McNamara in the Treasury Department or going in as Secretary of Defense. It was originally thought that maybe [Thomas S.] Gates could stay on as Secretary and possibly I would go in as Under-Secretary. McNamara knew that he had a choice but he wanted to consider it over the weekend. When he came in the following Monday he indicated that he wanted to be Secretary of Defense. By that time Jack had made up his mind that one of the problems in keeping Gates was the fact that the political leaders including David Lawrence of Pennsylvania was concerned that Gates would come and run for Governor in the State of Pennsylvania and so they asked specifically that he not be retained.

---

* RFK, handwritten note, 1958, RFK Papers. Beaverbrook must have been laying all about him that night. He said of John Hay Whitney, then American ambassador to London, "There are rumors of his demise but as of this time I understand he has refused to lie down."

The first conversation that Jack had with McNamara, he spoke to him about the possibility of hiring Franklin Roosevelt, Jr. McNamara objected saying that the reports he had received on Franklin Roosevelt, Jr., were not very complimentary. Jack said that at least he'd like to have him considered. When McNamara came down on Monday Jack and I both met him. He had a letter with him which he presented to Jack, which set forth that McNamara had complete control over the collection [selection] of those in the Department of Defense. Jack readily agreed as he had agreed verbally before. I put in a plug for Franklin Roosevelt, Jr., but it seemed to go on relatively blank ears.[9]

McNamara preserved his control over Defense appointments with a single exception; "but that wasn't the President's doing," Roswell Gilpatric, the under secretary of defense, said later; "it was Bob Kennedy's doing." Robert Kennedy asked them to find a place for John Kennedy's old PT-boat friend Paul Fay in the Navy Department; and, having turned down Roosevelt for the secretaryship, McNamara decided to accept Fay for the under secretaryship.[10]

As far as the Secretary of the Treasury was concerned [Robert Kennedy recalled in 1961], everybody was considered. I don't think there is any question that the head of the Morgan bank — his name slips me right at the moment [Henry C. Alexander] — would have received the job if during the course of the campaign he hadn't come out as a political supporter of Richard Nixon just after Jack had gone to see him and had spent more than an hour with him. Jack felt that this was a personal insult.

Joe Alsop was a tremendous booster of Douglas Dillon. In view of all the favors that Alsop had done I don't think there's any question that this was a factor. However, obviously if Dillon hadn't shaped up and hadn't impressed Jack the decision would have been quite different.

"My father was opposed," he added in 1964. ". . . He felt he didn't know enough about finances. That's what slowed the President down selecting Dillon."[11]

"There is no question," Robert Kennedy dictated in the 1961 memorandum, "that there was a good deal of pressure on Dillon by Nixon and others — there is no question that they were very angry that he took the job. Politically, it was extremely helpful, because taking Dillon, a Republican, a man of considerable responsibility, . . . buried the idea that the Kennedy Administration was going to pay no attention to their responsibility in the economic field."[12]

III

Stewart Udall [as Secretary of the Interior] was the first position agreed upon. I was a great booster of him as was Teddy. I was a booster despite the introduction Stewart Udall gave me when I spoke in Arizona. There were some rumors around at the time his appointment was being considered that [Senator] Clinton Anderson [of New Mexico, who had been Truman's Secretary of Agriculture] himself was personally interested. Jack called Clinton Anderson after it was virtually agreed that Stewart Udall would be the appointee and Clinton Anderson indicated that he was in fact interested. We were in rather an embarrassing position. We talked it over and the dangers of Clinton Anderson getting in there and putting in cronies and associates, a number of whom we had seen during the course of the campaign who were rather unsavory, unquestionably, and certainly of different viewpoints, different generation, ideas and thoughts as we have. Finally, Jack called him and told him that he just thought that he had to go with Stewart Udall, that he didn't want to make Clinton Anderson give up his Senate position in seniority — it would be too dangerous. Clinton Anderson bought it without any problem or difficulty so Stewart Udall was the first one agreed upon.[13]

Agriculture proved troublesome. The Kennedys were urban types for whom agricultural policy was a mystery. ("I don't want to hear about agricultural policy from any one except you," John Kennedy once said to John Kenneth Galbraith, who had started out as a farm economist, "and I don't want to hear about it from you.") Hubert Humphrey and Stuart Symington pushed a Missouri farm leader named Fred V. Heinkel.

Jack asked me to go in and interview him for a few moments to try to size him up. I asked him if he thought there was any solution to the problem in agriculture and he said that he thought there was, then mentioned as one of the major solutions to sell our surpluses abroad. I then asked what was the answer to the criticism of that made by those who said that it would "glut" the market and then therefore destroy the markets of countries such as the Argentine and Canada who were trying to sell wheat or meat to the same areas. He thought for a moment and said, "Well, that is a problem but I expect there are experts in the Department of Agriculture who had thought about that.". . . He was singularly unimpressive.

We had always considered as the second choice George McGovern but there was a good deal of opposition to him by the members of Congress on the Agricultural Committee. They felt that he did not have the seniority which would warrant such an appointment. This was his biggest problem and then the idea of appointing somebody who had just been defeated as a Congressional candidate was also a problem.

Finally . . . we came back to Orville Freeman because of how impressed everybody was out in Los Angeles with him, his integrity and his vigor. An impossible job. I believe from what I heard that originally he was interested in being Attorney General. I believe that afterwards he felt that somebody had gotten to that position first. . . .

As far as George McGovern is concerned he was disappointed but I had told him right at the beginning that if he didn't get this position he could be head of the Food for Peace [the agency distributing food surpluses abroad under Public Law 480]. I also told him I thought that if he was interested in ever running for political life again that the Food for Peace was a far better job. He thought the prestige, however, going with the Department of Agriculture would greatly help him in South Dakota. I said that I thought the problems of the Department of Agriculture were insoluble or virtually insoluble and it was rather an unpopular job. . . . He told me afterwards that it finally ended up that he was happy to be over as head of the Food for Peace.[14]

Health, Education and Welfare went to Abe Ribicoff of Connecticut, to whom John Kennedy had originally offered Justice. "Ribicoff didn't want to be Attorney General because he felt that it wouldn't be proper for a Catholic and a Jew to try to put Negroes in white schools in the South. I think also he wants to be Justice of the Supreme Court and he felt that by being Attorney General of the United States that he would incur the wrath of the Southerners who would have to vote on it."[15]

Robert Kennedy's 1961 memorandum did not go into Arthur Goldberg's appointment as Secretary of Labor. Working with Goldberg during the Rackets Committee had given both Kennedys confidence in his intelligence, resourcefulness and friendship. But Goldberg's name oddly was not on the list of candidates submitted by George Meany on behalf of the AFL-CIO, and Goldberg himself had reservations about accepting the job. John Kennedy said to him, "Okay, you can act like Adlai Stevenson if that's what you want, and you can say you haven't decided whether or not you will accept. . . . I am going to tell George Meany I'm appointing you."[16] Robert Kennedy told me in 1965, "Arthur Goldberg wanted to be a judge, and so he wanted to be assured that if he took this he could be perhaps appointed as a Circuit Court judge out in Illinois. . . . I said I couldn't promise him. I wanted him to decide about the position. The building trades were all against him. . . . We had a lot of opposition from the labor movement."[17] In the end Meany received Goldberg's appointment affably and was present at the announcement.

Edward Day was selected [as Postmaster General] in a rather peculiar manner. Jack wanted to finish the selection of the Cabinet at an early date because it had been delayed so much. Somebody suggested Edward Day as a businessman and without a great deal of thought, without really clearing it through any political figures, Jack called him on the phone and asked him to be Postmaster General and he accepted. It probably would have been better politically to have chosen somebody else but I was sure, at least based on the first month or so, that Edward Day would turn out well. We always thought, looking back on it though, that this was a decision that was made with too much alacrity and that perhaps somebody else would have been more deserving. Edward Day had not been active in the campaign and had been difficult to contact when we needed his help.[18]

Robert was also concerned about the White House. This was much on his mind when I lunched with him in Washington on December 15. Effective as the Senate and campaign staffs had been, he said, the President needed some new people around him. Moreover, the two staffs — the Sorensen group and the O'Donnell group — did not get along too well. Bobby planned to introduce a neutral figure to ease the tension. Fred Dutton had accepted this tricky assignment. He asked me how I proposed to serve my country. I said that two weeks earlier his brother had asked whether I would be interested in an ambassadorship. (I had said that I didn't think so. John Kennedy said, "I think an ambassadorship would be a great job. I'd like to be one myself. Are you sure you wouldn't want one?" I repeated that I didn't think so.)[19] Robert now asked whether I would be interested in becoming a special assistant to the President. I said I would. He turned to other matters.

I heard nothing more and really supposed that this had been a passing courtesy. On January 9, John Kennedy came to Massachusetts to address the legislature. He set up headquarters in my house in Cambridge. In between a succession of visitors, he said casually that Bobby had spoken to him; was I ready to come to the White House? I said, "I am not sure what I would be doing as special assistant, but, if you think I can help, I would like very much to come." He said that he didn't know what he would be doing as President either but guessed there would be enough to keep us both busy.[20]

IV

In 1965 I asked Robert Kennedy whether he had given any thought to working in the White House himself. His response was a candid definition of the fraternal relationship. "I didn't want to do that," he said,

> because if I was going to do anything, I wasn't going to work directly for him — the President. I mean, that would be impossible. I had to do something on my own, or have my own area of responsibility; and it would be impossible with the two of us sitting around an office looking at each other all day. You know, it was like the campaign. He had his role in the campaign and I had my role in the campaign and we would meet and discuss when there was a crisis that came up like the Houston ministers. . . . But he never involved himself in what I was doing — the running of the campaign — and I was not directly associated with what he was doing. . . . So for brothers — or at least for the two of us — we had to have our own areas. I had to be apart from what he was

doing so I wasn't working directly for him and getting orders from him as to what I should do that day. That wouldn't be possible. So I never considered working over at the White House.[21]

In fact, he was extremely perplexed about his future. In June 1960 John Kennedy had said to me, "If I'd been knocked out of the presidential thing, I would have put Bobby into the Massachusetts picture to run for governor. It takes someone with Bobby's nerve and investigative experience to clean up the mess in the Legislature and the Governor's Council."[22] Robert himself wrote in his 1961 memorandum, "I thought I'd enjoy going back to Massachusetts in 1962 and running for Governor."[23] There was also now the possibility of appointment to John Kennedy's vacant Senate seat; but Robert said, "I won't take that. The only way I'll go to the Senate is to run for it."[24]

He had a long luncheon with Justice Douglas to discuss his future — the federal government? would this help or hurt his brother? Massachusetts? a college presidency? head of a foundation? Douglas, Kennedy recalled, "suggested very strongly that I go back to Massachusetts" and that "I should make a name on my own, not just work for my brother."[25] "The things that came out most clearly," Douglas said later, "were that he didn't want to practice law. . . . The first choice would be public service; second choice as an interim thing, would be a college presidency maybe. That's the way his mind was going."[26]

John and Robert Kennedy had talked in Hyannis Port immediately after the election. "He asked me if I wanted to be Attorney General. I said I didn't want to be Attorney General. . . . In the first place, I thought nepotism was a problem. Secondly, I had been chasing bad men for three years and I didn't want to spend the rest of my life doing that."[27] Others reinforced his doubts. "You would make a whale of a good Attorney General," the columnist Drew Pearson wrote him. ". . . But in the same breath, I don't believe that you ought to take this job. You would do such a good job, and you would handle so many controversial questions with such vigor that your brother in the White House would be in hot water all the time."[28] Robert agreed it

would be a very bad mistake. I was very struck with the fact that when Bill Rogers [Eisenhower's last Attorney General; later Nixon's first Secretary of State] went with Nixon to South Carolina [during the campaign] he hid on the plane. . . . He arrived there and never got off. He hid on the plane so they wouldn't know he was on the plane. And [Governor] Fritz Hollings found out he was on the plane and it was a big issue in South Carolina. . . . I thought with the name the same, and brothers, I would be creating so many problems in civil rights. It would be the "Kennedy brothers" by the time a year was up, and the President would be blamed for everything we had to do in civil rights; and it was an unnecessary burden to undertake. . . .

So then we talked about going as Under Secretary of Defense. They were going to keep Gates on . . . and then Gates would retire and I would become Secretary. [But the acquisition of McNamara eliminated that possibility.] Then I thought about being Assistant Secretary or something for Latin America. My father kept saying that in the first place the President is going to have a lot of problems; and second, considering the relationship that is going to exist between the two of you, you can't be working for somebody else because it is going to make it so difficult for the somebody else . . . so you've got to be cabinet.[29]

By "cabinet" Joseph Kennedy meant Attorney General. This had been a subject of family badinage. At Christmas time in 1959 Eunice Shriver, joking about the future, said, "Bobby we'll make Attorney General so he can throw all the people Dad doesn't like into jail. That means we'll have to build more jails."[30] A year later, the President-elect told Clark Clifford, who was helping to manage the transition, that his father was "absolutely determined" that Bobby become Attorney General. John Kennedy, Clifford recalled, had "very serious reservations."[31]

If so, the reservations concerned the public reaction, not his brother's ability to do the job. Having been turned down by Ribicoff, Kennedy decided to test the water. On November 19, William Lawrence of the *New York Times,* after golfing with the President-elect, reported from Palm Beach that Robert Kennedy was under serious consideration for Justice. The *Times* promptly denounced the idea editorially: "It is simply not good enough to name a bright young political manager, no matter how bright or how young or how personally loyal, to a major post in government that by rights (if not by precedent) ought to be kept completely out of the political arena."[32] Precedent, as the *Times* tacitly conceded, did not oppose the appointment. Homer Cummings, Franklin Roosevelt's first Attorney General, had once been chairman of the Democratic National Committee. So had J. Howard McGrath, whom Truman made Attorney General. Eisenhower had given the job to Herbert Brownell, his campaign manager. Still, the *Times* had an argument. And there were other obvious objections: nepotism; youth; inexperience; Robert's own deep reluctance, reiterated on his return from Mexico after Thanksgiving.

When I talked with John Kennedy on December 1, he still seemed undecided. I mentioned Orville Freeman. Kennedy did not think him quite right by temperament or interest; besides Freeman had been beaten for reelection as governor of Minnesota; "he'd be too vulnerable politically." Then he startled me: "I'd like Stevenson for Attorney General and Paul Freund [of Harvard] for Solicitor General." But when he sounded out William Blair that evening, the word returned that Stevenson preferred foreign affairs.[33]

V

So it came back to his brother. Obviously his father had a point: John Kennedy needed someone in the cabinet whom he knew intimately and trusted utterly. The reaction to the trial balloon suggested that criticism would be sharp but transient and, if Robert did as well as his brother expected, quickly forgotten. Robert remained resistant. "We just vacillated back and forth," he said in 1965, "almost like we did on the Vice President."[34]

He made a round of consultations. J. Edgar Hoover told him he should become Attorney General. (Hoover said afterward to William C. Sullivan of the FBI's Domestic Intelligence Division, "I didn't like to tell him that, but what could I say?"[35] Kennedy said afterward to John Seigenthaler that he really didn't think Hoover meant it.)[36] William Rogers, Kennedy noted in the 1961 memorandum, "said there really wasn't much of a challenge because there wasn't a great deal to be done." James McInerney, his first boss at Justice a decade earlier, said he shouldn't do it. David Ormsby-Gore, in Washington for a night, thought it a poor idea. Justice Douglas said that Kennedy would serve with distinction and that the voices of his critics would be drowned out by his achievements. "The question for him and Ethel, I ventured to say, was whether that office at that time fitted the pattern of the professional career he had for himself."[37]

It was a trying time. "I don't know what's wrong with him," his father said. "Jack needs all the good men he can get around him down there. There's none better than Bobby. You know for six years he hasn't told me what he wants to do."[38] Finally in mid-December, after discussing the question a last time with Ethel and John Seigenthaler over dinner at Hickory Hill, Robert made up his mind to say no. As he picked up the telephone to call his brother, he observed wryly, "This will kill my father."[39] "So I called Jack up," he said later, "and said that I'd decided . . . and I'd go back to Massachusetts or do something else."[40]

"Don't tell me now," John Kennedy said. "I want o have breakfast with you in the morning." Taking Seigenthaler with him, Robert set out the next morning for his brother's house in Georgetown. It was very early and very cold. Snow lay deep on the ground. They chatted over bacon and eggs about the formation of the cabinet. Finally Robert said, "Now, Johnny, can we talk about my situation?"

John Kennedy replied at length and with emphasis, addressing his remarks as if to Seigenthaler. "In this cabinet," Seigenthaler later remembered his saying, "there really is no person with whom I have been intimately connected over the years. I need to know that when problems arise I'm going to have somebody who's going to tell me the unvarnished truth, no matter what . . . and Bobby will do that." He described his cabinet. "I believe McNamara will make a great contribution, but I don't know him. And Dean Rusk is going to be my Secretary of State. The truth of the matter is, I [have] had no contact with him. . . . Most of these people I have had some cursory contact with. None of them have I had a long-standing close relationship with." He spoke about civil rights: "I don't want somebody who is going to be fainthearted. I want somebody who is going to be strong; who will join with me in taking whatever risks . . . and who would deal with the problem honestly."

Suddenly he was talking to his brother: "If I can ask Dean Rusk to give up a career; if I can ask Adlai Stevenson to make a sacrifice he does not want to make; if I can ask Bob McNamara to give up a job as head of that company — these men I don't even know . . . certainly I can expect my own brother to give me the same sort of contribution. And I need you in this government." With that, he pushed back his chair, got up and went into the kitchen, as if for more coffee. Seigenthaler said, "Let's go, Bob." Robert Kennedy said, "No, wait. I've got some points I want to make." Seigenthaler said, "There's no point to make." John Kennedy walked back and said, "So that's it. General, let's go."[41]

"I told him," John Kennedy subsequently remarked to Peter Maas, "I had made up my mind and didn't want to talk about it any more. I also reminded him that every danger is an opportunity."[42] Robert Kennedy laconically summed up the conversation in his 1961 memorandum: "He said he thought it would be important to him and that he needed some people around that he could talk to so I decided to accept it."[43] "I did it," he said later, "not so much to become Attorney General as to be around during that time."[44]

He had tried for a moment to escape his brother and confirm an independent identity. But, if his brother wanted him, he would not say no — a sacrifice of himself to which John and Jacqueline Kennedy responded in their separate styles. For Christmas they gave him a copy of *The Enemy Within* handsomely rebound in red leather by a London bookbinder, title and author inscribed in gold. Inside John Kennedy wrote in his *en famille* vein: "For Bobby — The Brother Within — who made the easy difficult. Jack, Christmas 1960." Above Jacqueline had written: "To Bobby — who made the impossible possible and changed all our lives."

VI

I lunched with him the day he had agreed to become Attorney General. He seemed both rueful and fatalistic. He wrote in the 1961 memorandum, "Once I had made up my mind . . . I was glad it was all behind. Arthur Goldberg was a great support as were some of the labor people generally. Alex Rose and David Dubinsky were very enthusiastic. . . . On the other hand the building trades [unions] met in Washington and sent an emissary to Jack telling him that he shouldn't appoint me as they would consider it with disfavor. This was one of the motivating forces in my taking the job."[45]

Disfavor was not confined to the building trades, as John Kennedy perfectly understood. When Benjamin Bradlee of *Newsweek* asked how he planned to announce the appointment, the President-elect said, "Well, I think I'll open the front door of the Georgetown house some morning about 2:00 A.M., look up and down the street, and, if there's no one there, I'll whisper, 'It's Bobby.'" Robert Kennedy himself recorded the great moment. "He told me to go upstairs and comb my hair to which I said it was the first time the President had ever told the Attorney General to comb his hair before they made an announcement. And then when we were outside he said, 'Don't smile too much or they'll think we're happy about the appointment.'"[46]

Others were decidedly unhappy. "If Robert Kennedy," the *New York Times* noted sarcastically, "was one of the outstanding lawyers of the country, a preeminent legal philosopher, a noted prosecutor or legal officer at Federal or State level, the situation would have been different. But his experience . . . is surely insufficient to warrant his present appointment."* Robert Novak wrote in the *Wall Street Journal* that Robert Kennedy could turn out to be an "unqualified disaster."[47] Alexander M. Bickel summed up the opposition case magistrally in the *New Republic*. More than any executive officer, Bickel wrote, the Attorney General is required to regard means as above ends, process above results. "In his brief but highly visible professional career, Mr. Kennedy has demonstrated specific grounds of

---

* *New York Times*, December 29, 1960. Implicit in the *Times*'s attacks on Robert Kennedy's appointment was the notion that nepotism was a wicked thing. Robert Kennedy took sardonic pleasure in observing the retirement of Arthur Hays Sulzberger as publisher of the *Times* and his replacement by his son-in-law Orville Dryfoos, with Arthur Ochs (Punch) Sulzberger standing next in line. When Punch Sulzberger eventually took over his father's job, Kennedy said, "Well, Punch and I both worked our way to the top through diligence and merit." There were many such jokes. RFK, introducing Adlai E. Stevenson III of Illinois, c. 1966: "If there's one thing I can't stand, it's a fellow running on his family name."

disqualification. . . . On the record, Robert F. Kennedy is not fit for
the office."[48]

Press comment was not wholly adverse, and senators were generally
indulgent. There were, however, rumbles of opposition on the Judiciary Committee. James O. Eastland, the committee chairman, was
a combination of Mississippi demagogue and Washington cynic. Lyndon Johnson once said of him, "Jim Eastland could be standing right
in the middle of the worst Mississippi flood ever known, and he'd say
the niggers caused it, helped out by the Communists — but, he'd say,
we gotta have help from Washington."[49] Most northerners found
Eastland grouchy and obscurantist, a planter rich on federal cotton
subsidies, given to peevish talk about big government and the "mon-
grelization of races" and devoid of any socially redeeming quality,
like the courtly manners of his Mississippi colleague John Stennis.[50]
Robert Kennedy, with his Irish weakness for rogues, liked Eastland,
however, and Eastland evidently liked him too.

When Kennedy called on him before the hearings, Eastland leaned
back in his chair, his face wreathed in cigar smoke, and ruminated
about the Department of Justice. He had got along with Attorney
General Rogers, he said, but added: "You can't have the same confidence in him, of course, as you could in [Lawrence E.] Walsh [the
deputy attorney general]. . . . Did you know that he never brought
a civil rights case in the state of Mississippi?" With that Eastland
winked broadly at Kennedy. Kennedy interpreted the wink as meaning (he put it later) "that Bill Rogers hadn't met his responsibility and
done his duty in Mississippi, and Jim Eastland felt a certain amount
of contempt for him."[51]

Not all the southerners were so equable. "Dick Russell," Lyndon
Johnson told Bobby Baker, "is absolutely shittin' a squealin' worm.
He thinks it's a disgrace for a kid who's never practiced law to be
appointed. . . . I agree with him. . . . But I don't think Jack Kennedy's gonna let a little fart like Bobby lead him around by the
nose." Deciding it was his role to disarm southern opposition, Johnson told Baker: "I want you to lead all our Southern friends in here
by their ying-yangs. . . . I'm gonna put it on the line and tell 'em it's
a matter of my personal survival."[52] "Russell, and I believe [Herman]
Talmadge [of Georgia], were against me," Kennedy himself wrote
in his 1961 memorandum, "and [Allen] Ellender [of Louisiana] had
suggested to Eastland that he ask me some tough questions on Civil
Rights. Actually, they wouldn't have been tough questions because
there wasn't much I was prepared to say. Russell was against me
. . . because he felt that [it] caused him a problem and difficulty in
the South to vote for any Attorney General."[53]

As usual, he prepared carefully for the hearings. "I conferred with Clark Clifford," he noted in 1961, "and then went over and saw Ben Cohen. He is a great gentleman and had some good ideas."[54] James M. Landis helped. On the morning of January 13, 1961, Robert Kennedy rather nervously entered the gleaming marble and walnut-paneled hearing room of the Judiciary Committee. Eastland read into the record messages of support from Whitney North Seymour, president of the American Bar Association, and from J. Edgar Hoover. Everett Dirksen of Illinois, the Republican leader, began the questioning. He saluted Kennedy's "impeccable character" but pointed out that the Attorney General–designate had never practiced law in any state or federal court. Senator Roman Hruska of Nebraska pursued the inexperience point with dogged fervor. Kennedy's old friends from the Rackets Committee, John McClellan and Sam Ervin, intervened from time to time, defending his experience under them. Pressed by Hruska, Kennedy himself finally responded with feeling. "I decided at quite a young age," he told the committee, "that I would . . . work for the Government. . . . I have been working for the Government. In my estimation I think that I have had invaluable experience." He recalled Oliver Wendell Holmes's remark that private practice had hardly prepared him for the broad reach of public law he confronted on the bench. "I would not have given up one year of experience that I have had over the period since I graduated from law school," Kennedy said, ". . . for experience in practicing law in Boston." Hruska persisted: "You have, as I understand it, never negotiated a settlement, for example, of a litigated civil case for damages or the breach of a contract or tort case." Kennedy: "I doubt if I am going to be doing that as Attorney General."[55]

For the rest, it was clear sailing. Asked about his financial holdings, Kennedy said, "I disposed of everything that I could possibly dispose of. . . . The result is now that I do not have any interest in anything other than Government or municipal bonds."[56] Asked about civil rights, he said he would execute the policies laid down by the President. Asked about a national crime commission, he said that he had already worked out an arrangement with the FBI, the Bureau of Narcotics and the Internal Revenue Service for the pooling of intelligence; he planned to give this a try first. Asked about the withholding of information from Congress on the plea of executive privilege, he said, "I feel very strongly about that," and affirmed that Congress, and the public too, had "a right to know many of these matters."[57]

Senator Kenneth Keating, a New York Republican, describing himself as "a longtime personal friend of the nominee and his gracious wife and all the members of his family,"[58] commended the appointment. Except for Hruska, the committee was complaisant. Kennedy had come, James Reston wrote, "expecting to do battle with Jack Dempsey, and instead found himself confronted, most agreeably, by Shirley Temple."[59] The committee, Hruska included, reported the nomination out unanimously. A week later, with only one senator speaking in opposition — Gordon Allott, a conservative Republican from Colorado — the Senate confirmed Robert Kennedy as Attorney General of the United States.

VII

Robert Kennedy turned thirty-five two months before he was sworn in as Attorney General. Only two of his sixty-three predecessors had been younger — Caesar Augustus Rodney, thirty-five by a bare two weeks when Jefferson made him Attorney General in 1807, and Richard Rush, thirty-three when appointed by Madison in 1814. In the days of Rodney and Rush, however, Attorneys General were private lawyers on presidential retainer. They kept up their own law business on the side and did not even live in Washington. It was not until 1853 that an Attorney General — Caleb Cushing — took up full-time residence in the capital, nor until 1870 that the Department of Justice was established. It grew rapidly thereafter. By 1961 it had more than thirty thousand employees. Its annual budget was more than $300 million. The Attorney General, as the government's chief legal officer, had a wide mandate throughout the executive branch. That mandate had been historically construed, in the main, as procedural and protective: to enforce federal law and to assure and defend the legality of executive actions. The Department of Justice was by tradition as well as by function a citadel of the status quo — or at least of *stare decisis*. It had only intermittently been an agency of innovation and reform.

Despite the hostile comment, Robert Kennedy was not wholly unqualified. In one field of importance to law enforcement — organized crime — he knew considerably more than his critics. As for the argument that he lacked courtroom experience, John Kennedy thought this quite missed the point. "The basic requirement for the job," he told an interviewer, "is not that at all. It is the ability to administer a great department."[60] So lightly did the President-elect take the professional complaint that, to Robert's discomfort, he made public jokes about his brother's inadequate preparation. "I can't see that it's wrong," he was soon explaining to the annual banquet of the

Alfalfa Club, "to give him a little legal experience before he goes out
to practice law." (When Robert objected that this was not very funny,
his brother said he had better get used to kidding himself; people
liked it. "Yes, but you weren't kidding yourself," Robert said. "You
were kidding *me*.")[61]

In picking his associates, the new Attorney General sought men
who would make up for his own deficiencies, especially too little law
and too much politics. He wanted first-class lawyers who would be
strong, nonpolitical and young. "It's ridiculous to wait until a man
is 40 to give him a responsible job," he had said in 1957. "By that
age he may have lost most of his zeal."[62] The key appointment was
deputy attorney general. After the election, Robert, on his brother's
instruction, had offered their admired Colorado friend Byron White
a choice of jobs: Secretary of the Army, Secretary of the Air Force,
deputy attorney general.[63] White, who had not gone to Yale Law
School or clerked for a Chief Justice for nothing, decided for Justice. He was a man of stern and circumspect intelligence, realistic
judgment, dry humor and solid legal proficiency. He also was well
acquainted among the lawyers of his generation, especially where the
writ of the Yale Law School ran, and he now recruited some of the
best for his new Department.

White's contribution to the roll of assistant attorneys general in-
cluded Nicholas Katzenbach (Exeter, U.S. Air Force, Yale Law
School, Balliol, Chicago Law School faculty; now Office of Legal
Counsel); Burke Marshall (Exeter, U.S. Army, Yale Law School, Cov-
ington and Burling; now Civil Rights Division); Louis Oberdorfer
(Dartmouth, U.S. Army, Yale Law School, clerk to Justice Black; now
Tax Division); William Orrick (Yale, California Law School, U.S.
Army; now Civil Division). Later Katzenbach brought in Norbert
Schlei (U.S. Navy, Yale Law School, clerk to Justice Harlan; succeed-
ing Katzenbach in the Office of Legal Counsel) and John Douglas
(son of Senator Paul Douglas, U.S. Navy, Yale Law School, Oxford,
clerk to Justice Burton, Covington and Burling; succeeding Orrick
in the Civil Division).

Their overlapping experience — the Ivy League, the war, the *Yale
Law Journal*, Rhodes scholarships, Supreme Court clerkships, established law firms — stamped these men with a common moral outlook. They cared deeply and professionally about the law. Military
service had trained them to assume responsibilities. They were all
Democrats who saw themselves as heirs of the great legal generation
of the New Deal, but they were without partisan inclination. Only
White had taken an active part in the campaign. Their qualities —
integrity, judgment, drive, understatement, personal reserve — com-

mended them to Kennedy. There was logic too in the Yale connection. Kennedy was temperamentally more attuned to the result-oriented jurisprudence propagated in New Haven than to the process-oriented jurisprudence of Cambridge; to William O. Douglas, say, than to Felix Frankfurter. (Alexander Bickel, the most formidable critic of his nomination, was a Harvard outrider in the Yale wilderness.) The appointment of Professor Archibald Cox of the Harvard Law School to the post of solicitor general, the third-ranking job in the Department, broke the Yale monopoly. This was John Kennedy's doing. Cox had worked with the President-elect on labor legislation and had headed his research staff in the 1960 campaign. He was already celebrated, at least in the law school world, for his patrician bearing, direct and lucid mind, crisp style, crew-cut hair and bow ties, the last abandoned when he dressed himself magnificently in swallow-tailed coat to argue before the Supreme Court.

Other places were filled from outside the Ivy League. William O. Douglas and Sam Rayburn urged the appointment of the youngest of the assistant attorneys general, Ramsey Clark of Texas, thirty-three years old, to head the Lands Division. Clark's father had been Attorney General under Truman and was now on the Supreme Court, and Clark himself had met Professor Katzenbach at the Chicago Law School. For Antitrust, Sargent Shriver and Hubert Humphrey recommended Lee Loevinger of Minnesota, forty-eight years old and the oldest of the new appointees. The head of the Criminal Division, Herbert J. Miller, Jr., was a Republican whom Kennedy had known as a member of the Board of Monitors appointed by the courts in a vain effort to clean up the Teamsters. Kennedy originally intended to abolish the Internal Security Division.[64] But this threatened complications; so he left it in a separate building under an accommodating Republican holdover, J. Walter Yeagley, who was content to phase out prosecutive action and, as a former FBI agent, able to do so without unduly estranging J. Edgar Hoover.

For his personal staff Kennedy drew on newspaper friends from Rackets Committee days. John Seigenthaler became his administrative assistant, Edwin Guthman his press man. Angie Novello remained his secretary. Byron White brought from Denver as his deputy Joseph Dolan, who had worked for John Kennedy on Senate committees and had served in the Colorado legislature. Seigenthaler, Guthman and Dolan helped segregate the rest of the cast from the importunings of politics.

It was an outstanding group. Even the skeptical Bickel was impelled to second thoughts. "One began to hear of it in a month or so before he took office," Bickel said later. ". . . It was the most brilliantly staffed department we had seen in a long, long time, and that was very impressive. One immediately had the sense of a fellow who wasn't afraid of having able people around him and indeed of a fellow who had a vision of public service that would have done anyone proud."[65]

VIII

The center of action was the Attorney General's office, a vast, somber, walnut-paneled chamber, with arched roof, indirect lighting, red carpet, stately fireplace and, in the past, a general air of solemnity. Over the next years Kennedy filled it with sofas, tables, lamps, children's photographs, a varnished sailfish over the mantlepiece, an old blunderbuss on the wall, a model of a Chinese junk on a table, a stuffed tiger standing near the fireplace. (The Attorney General carefully explained to a foreign newspaperman that he had not shot the tiger himself. If he had shot it himself, the newspaperman reflected, he would not have said anything.)[66] Affixed to the panels by Scotch tape was an ever-changing montage of children's drawings. "There were fish," recalled the astonished Roy Wilkins of the National Association for the Advancement of Colored People, "that only they would describe as fish. There were chimneys sticking up at wrong angles and roads that were going everywhere and airplanes that were flying through what they said was the air. But they were wild, beautiful colors, the kind of thing that would jump out of a child's imagination and fantasy. Robert Kennedy never offered any apology for these pictures, nor did he offer any explanation of them."[67]

The Attorney General leaned casually back in his leather armchair in the midst of genial clutter, foot propped on an open drawer, jacket off, tie yanked aside, trousers rumpled, hair uncombed. "You'd never confuse him with Dean Acheson," said Ethel.[68] Often, in defiance of government regulations, he brought with him his menacing dog Brumus, a large and ill-tempered Labrador to whom Robert and Ethel were devoted, perhaps because most people, even veteran dog-lovers, detested him. On weekends he might import an assortment of children.

A new informality pervaded the staid old Department. Kennedy wandered the corridors, put his head into offices to find out what was going on, talked to people who had worked in the Department for years without ever speaking to an Attorney General. Patricia Collins, a lawyer who had first come to Justice when Robert Kennedy was nine years old and had served fourteen Attorneys General, said, "We never had an Attorney General who had [that] close personal relationship with the staff. He always said he'd been here himself, he'd been an attorney down in the Criminal Division, and nobody had ever paid any darn attention to him, and he was going to fix that."[69] But there was no sacrifice of authority. Early on, calling in the Department's lawyers in a series of informal meetings, he sat on his desk, one of them recalled, "legs crouched beneath him Indian fashion, necktie open, shirtsleeves rolled up, hair askew. . . . Strangely enough, in that posture of utter informality, almost humility, Robert Kennedy lent an air of dignity to the assembly. I felt at one with him and was aware that that was the mood of the entire group."[70] He was a superb question-asker, an excellent listener and an excellent rememberer.

He carried the personal touch to the outlying parts of his domain — federal attorneys through the country, FBI offices, immigration and naturalization offices, prisons. Before he took a trip, he would pass the word to his assistant attorneys general in Washington: "Aren't there some areas I can help you in?"[71] Armed with detail, he prodded regional offices into new effort by his display of knowledge and concern. "He took a very special interest," said Robert Morgenthau, whom he made federal attorney for the Southern District of New York, "in the work being done by young lawyers. . . . This was one of his great gifts: to make people feel that they were part of the team."[72]

As an administrator, he matched his taste in selecting associates by his readiness to leave them alone. He relied on Byron White for day-to-day management of the Department. He expected the assistant attorneys general to keep him informed about anything they were doing that might have public implications, but did not interfere unless they requested his counsel or support. "He really did not second-guess people to whom he gave responsibility," said Louis Oberdorfer.[73] He interrogated them carefully about their legal recommendations, "but I don't believe," said Seigenthaler, "I ever saw him override one."[74] "He never questioned anybody's legal judgment," said Katzenbach; his style was to say, "Everybody here knows more law about this than I do, but I'm the Attorney General and it's my responsibility, so you've got to make it clear to me so that I can make a decision."[75] "Sometimes he would be a little skeptical," said Norbert Schlei, "as to whether the law required some result that seemed to him harsh or foolish or wrong in some way, but basically he would accept the legal assertion that he was confronted with and make his contribution on the level of the policy question."[76]

Archibald Cox summed up the personal characteristics that he felt made Kennedy "a first-rate Attorney General": his "willingness to listen and reconsider his initial reactions"; "his capability to go to the heart of the problem"; his "ability to delegate while reserving his time for what it was most important for him to do"; and, supremely important for his associates, "when a conference in his office ended, one knew what decision had been taken. The decision might be, 'Do what you think best.' In that event he supported you and shared the blame even if what you did turned out to be a mistake. Or the decision might lay down a line of action." [77]

He used his assistant attorneys general not simply as specialists in their own jurisdictions but as problem-tacklers across the board. He never hesitated to put the heads of the Tax, Lands, Legal Counsel and Civil offices (Oberdorfer, Clark, Katzenbach, Douglas) into civil rights, or his civil rights and lands men (Marshall, Clark) into antitrust. [78] "One of the hallmarks of the Kennedys," said William Orrick, "was that they expected you to do everything." [79] But this was done without violation of comradeship. "We never felt we were competing with each other," said Louis Oberdorfer. "There was no feeling that people were being displaced because someone else was called in to help. . . . One always knew where one stood. . . . He had that quality of leadership that made all of us play above our heads." [80]

Above all, said John Douglas, "he treated people decently. . . . He didn't duck or try to shift blame. He didn't say one thing in private and another in public. He never let his subordinates down. He stood up for them, followed their careers and interests, tried to help them, looked for the best in them, and usually found it. . . . He was the most fastidious man in public life I ever met in his personal relationships." [81] He had the quality, said Katzenbach, "of bringing out the very best in everyone who worked for him." [82] "He was as good at delegating power as anybody I ever saw," said Ramsey Clark. ". . . All he really required in return was performance." [83] His staff repaid confidence not only by performance but by protective devotion. They dispatched word about his virtues to the bar associations and the law schools. They took care always to call him Bob, rightly supposing that his detractors used the diminutive in order to emphasize youth and imply unfitness for a serious job. (Nothing, however, could cure his family of calling him Bobby.) They saw themselves, in the romantic title of Edwin Guthman's memoir of the period, as a band of brothers.

IX

Below the top staff, he exerted control, according to mood and need, through humor, through brusqueness, through solicitude. Addressing Justice employees, he sought to inspire them by the example of his own dazzling rise from the ranks: "I started in the Department as a young lawyer in 1950. The salary was only $4000 a year, but I worked hard. (Pause.) I was ambitious. (Pause.) I studied. (Pause.) I applied myself. . . . And then my brother was elected President of the United States."[84] Rueful jocularity, ordinarily at his own expense, was more than ever a conditioned reflex. When Angie Novello admonished him in a plaintive memorandum that "it would be *extremely* helpful if the Attorney General of the United States . . . would notify his immediate staff of his whereabouts at all times," Kennedy returned it with a handwritten scrawl, "What if I'm lost. Love."[85]

He was still on occasion preoccupied, moody and unpredictable. "He has a horror of wasting time," observed one reporter, "and although he can make a monumental effort to restrain his impatience with long-winded argument, the struggle is sometimes dramatic in the extreme. . . . He clasps his hands as if in prayer and regards his clenched knuckles with a cataleptic stare. If the ordeal continues he comes slowly upright again, pats the arms of his chair in slow and feverish rhythm and looks off into the beyond with . . . muted despair."[86] "Sometimes I wondered," said John Douglas, "if he regarded a display of congeniality or ease with someone he didn't respect as a reflection on his own integrity."[87] "He had," said Robert Morgenthau, "an unusual gift of sensing whether people knew what they were talking about . . . or whether they were just trying to string him along."[88] When he encountered what he took to be indolence or evasiveness, disgust was unconcealed. "His most obvious fault," wrote Patrick Anderson, information officer for the Department's Committee on Juvenile Delinquency, "is his rudeness. His face, when it lacks that boyish, photogenic grin, is not a pleasant sight. It has a certain bony harshness and those ice-blue eyes are not the smiling ones that Irishmen sing songs about. At best, he recalls Fitzgerald's description of Gatsby: 'an elegant young roughneck.'"[89] There were, however, few explosions. Joseph Dolan, who worked closely with him for eight years, thought him exceptionally self-contained. "You used to think you could tell when he was angry. . . . The color reflects in his face, and he'd say, 'Well, really, we can do better, can't we?' You would assume that he was furious, but you couldn't say [so] under oath. . . . You could [only] say, 'Well, you know, he was red, and he was speaking in a kind of different tone than he usually did.'"[90]

He could be tough all the same. He also at times struck people as spoiled in the way the rich, powerful and charming are spoiled: in the sense of entitlement to service. Victor Navasky in his able book *Kennedy Justice* quotes a career attorney: "Bob had the failings of a prince. It might be you who had to schlep Brumus out of the office.

If you worked for him he regarded you as working for *all* of him. So if Ethel lost her bags, you might have to track them down through the airport." But he was also unusually considerate. The same witness added: "You knew he would listen, so if you got his attention — it was rewarding."[91]

He genuinely cared about the people who worked for him. Over the resistance of career officials he had a gymnasium built on the roof of the Department. Ethel arranged for a snack bar with tables and umbrellas in the courtyard. When secretaries fell sick, the Attorney General sent them handwritten notes: thus, "I have just learned that you have gone to the hospital. I want you to come back here quickly. The Department of Justice is falling apart without you."[92] When attorneys won difficult cases, they received a telephone call or a letter. He sent periodic circulars to the assistant attorneys general: "Again I would like to ask that you bring to my attention the names of employees in your Division who have been doing outstanding work. Some assistant attorneys general have been doing this. . . . Others I have not yet heard from."[93] A year later: "Once again I want to remind all of you of my continued interest in knowing who in your Division has done so well that I may call or write a note. I have been forced to the conclusion that everyone is doing only mediocre work as I hardly ever hear from you regarding outstanding contributions."[94]

He imbued the Department with his rage for quality. "Robert Kennedy," his discriminating critic Navasky has written, "operated on the simple theory that standards were contagious, that excellence was catching and within the reach of the Justice Department. He expected it at every level and looked for it everywhere."[95] He imbued it too with his conviction of urgency. "I've known the Department of Justice an awfully long time," said Ramsey Clark in 1969. "I used to walk the halls as a kid — preteenage, because my father had joined the Department as a lawyer when I was nine years old — and I'd come to sense the atmosphere and the mood of the place. It was a quiet and sleepy place until January of '61. . . . Then it came *alive.*"[96] The premium was always on action. "Don't tell me what I can't do," Kennedy would say to cautious lawyers, "tell me what I can do."[97] He was determined to transform the Department from a citadel of *stare decisis* into an agency of reform.

He was confronted by massive forces of institutional resistance. "We have about as much chance of changing that place," Byron White observed to Ramsey Clark one day as they were walking up to the Department, "as you and I do of . . . putting our shoulders against the building and moving it."[98] And there was so little time. When

William Orrick had to decide between job offers at Justice and the Pentagon, Kennedy told him, "Take all the time you want but could you let me know by tomorrow morning?"[99] Clark, going to his first interview with Kennedy, found White gulping down a cup of ice cream. "He just all of a sudden said, 'Come on.' He just went dashing out and leaping up the steps about three steps at a time. I thought, 'My God, I can't carry on like this.'"[100] In two months White was down with a duodenal ulcer.

But they all did carry on like this. It was part of the postinaugural euphoria. "Those were the days," Robert Kennedy later said, "when we thought we were succeeding because of all the stories on how hard everybody was working."[101] He was no fool.

12

The Pursuit of Justice:
J. Edgar Hoover

T HE CENTER of institutional resistance was the Federal Bureau of Investigation. The FBI, the largest single unit within Justice, disposed of nearly 50 percent of the Department's thirty thousand employees and over 40 percent of its budget. Its director, technically a subordinate official, had far more power and fame than his nominal chiefs. Attorneys General came and went. J. Edgar Hoover was there, it seemed, forever. "Mr. Hoover became the Director of the Bureau in 1924, the year before the Attorney General was born," the FBI tour guides used to say in 1961, until Robert Kennedy heard about it.[1] No one could remember the Bureau before Hoover. No one could imagine it after Hoover. Rarely had any institution been so totally the lengthened shadow of a single man.

I

Hoover was a professional bureaucrat of remarkable gifts. Appointed director by Attorney General Harlan Stone in 1924 to clean up a corrupt operation, he rapidly won Stone's unaffected admiration as a man "of exceptional intelligence, alertness, and executive ability."[2] Stone, sensitive to the possibility that "a secret police may become a menace to free government," laid down careful specifications for the Bureau of Investigation. It must not, he said, concern itself "with political or other opinions of individuals" but only with *conduct* forbidden by the laws of the United States. "When a police system passes beyond these limits, it is dangerous to the proper administration of justice and to human liberty."[3] In Stone's view Hoover met those standards. In 1933, when Stone, now on the Supreme Court, heard that the newly elected Franklin Roosevelt might replace Hoover, he sent Felix Frankfurter a powerful letter urging the director's retention.*

The Bureau of Investigation (Hoover added Federal to the title in 1935) soon gained a quiet reputation, largely deserved, as an effective, technologically advanced and law-abiding organization manned by fearless and incorruptible agents. The director proved himself adept at modernizing his methods, at dominating his subordinates and at propitiating his superiors. To first-class administrative capacity he joined unexpected talent for showmanship. Perhaps the rumors of 1933 stimulated him to consolidate his position. Beginning in 1934, articles, books and movies began to pour out celebrating the G-men and their victories over crime. The torrent of publicity about the Bureau dismayed Stone, who observed in 1937, "I only hope that the ultimate effect will not be to break down its morale."⁴ But publicity served the director's purpose. It brought national support to the FBI. It won increasing appropriations from Congress. It gave Hoover himself a cloak of infallibility and invincibility.

His investigations and files multiplied. By 1940 Westbrook Pegler, the cranky columnist, could describe the FBI as "the greatest deposit of personal dirt ever amassed."⁵ International crisis meanwhile enlarged the Bureau's mandate. Roosevelt, confident of his ability to control Hoover, charged him with investigations of Nazi and Communist activities in the United States. National security was now propelling Hoover across Stone's line between conduct and opinion. When war broke out in Europe, Hoover was forty-four years old; he had been director since he was twenty-nine. Pearl Harbor increased his conviction that he alone knew how to protect the nation from subversive forces. He established a catchall "Security Index" listing people the FBI thought should be locked up in time of national emergency. In 1943 Attorney General Francis Biddle ordered him to destroy the list. Hoover simply disobeyed. He instructed his agents to change the label on the files, to "continue to prepare and maintain security index cards" and, above all, not to mention it to anyone outside the Bureau.⁶ All this — the enemies list, the covert defiance of his superior, the secrecy — set a pattern for the future.

The Cold War gave the Bureau new power and the director new glory. Hoover's dossiers grew. So did his command of Congress, his manipulation and intimidation of the press, his stature in the country. When Truman wanted the Civil Service Commission to run

---

* A. T. Mason, *Harlan Fiske Stone: Pillar of the Law* (New York, 1956), 152. Frankfurter passed the letter on to Roosevelt, who sent a reassuring message back to Stone via Frankfurter (Max Freedman, ed., *Roosevelt and Frankfurter: Their Correspondence, 1928–1945* [Boston, 1967], 129).

loyalty-security investigations of government employees, Hoover insisted that this was a job for the FBI. Truman told his staff: "J. Edgar will in all probability get this backward looking Congress to give him what he wants. It's dangerous."[7] Truman was right on both counts.

Truman's comment suggested the widening gap between the inside perception of the director and the popular myth. Those who dealt closely with the great patriot came to know his authoritarian instincts, his profoundly right-wing prejudices, his egomania and his vindictiveness. They knew his agitation over "bleeding hearts," "professional do-gooders" and other malefactors. "Swaydo-liberals," as he called them, was a favorite term of denunciation. This implied no affection for true liberals; all liberals (except for the New York lawyer Morris Ernst) were evidently pseudoliberals. "He refused," said George Allen — self-appointed chum of notables, author of the faintly amusing book *Presidents Who Have Known Me*, the director's closest friend outside the FBI — "to read either the *Washington Post* or the *New York Times* . . . and he was suspicious even of the *Wall Street Journal*."[8]

The insiders knew too that Joe McCarthy could not have sustained his rambling crusade without help from Hoover. George Allen: "Hoover was crazy about McCarthy."[9] "You had us preparing material for him regularly, kept furnishing it to him," William C. Sullivan, who rose to number three man in the Bureau, wrote Hoover in later years, "while you denied publicly that we were helping him."[*] They knew above all that the FBI had long since forgotten Harlan Stone's stipulation that it should concern itself only with conduct and not with opinion. They felt that Hoover had too much power and too little accountability.

The FBI was far from a contemptible organization. Its standards

---

[*] Sullivan to Hoover, October 6, 1971, Sullivan Papers. Sullivan told me on July 26, 1976, that the FBI aid program for McCarthy began very shortly after McCarthy's Wheeling speech in 1950 when McCarthy called Hoover in search of support. Ralph de Toledano confirms this in his fervently pro-Hoover biography: "Hoover spent many hours with Joe McCarthy talking 'business' — lecturing him on Communist strategy and tactics, giving him leads and insights into the Communist apparatus in the United States, and pointing him in the direction of suspect individuals. It is also undoubtedly a fact that though Hoover never gave McCarthy access to FBI files, he did allow him to see reports written to various high-ranking government officials on Communist infiltration" (Ralph de Toledano, *J. Edgar Hoover: The Man in His Time* [New York: Manor Books reprint, 1974], 280). Note also the testimony of William W. Turner, then an FBI agent: "It is very doubtful whether Senator Joseph McCarthy would have long survived in his reckless smearing campaign without the ammunition provided him by Hoover. FBI agents put in long hours poring over Bureau security files and abstracting them for Roy Kohn" (W. W. Turner, *Hoover's FBI* [New York: Dell reprint, 1971], 176). Only the egregious Cohn persists in denying that Hoover gave information to McCarthy (Roy Cohn, "Could He Walk on Water?" *Esquire*, November 1972).

compared favorably to those of most police forces at home and abroad. A decade before the *Miranda* decision, Hoover had instructed FBI agents to advise persons in custody of their rights to silence and a lawyer. Agents did not use the third degree. They were under instructions to fire only in self-defense or the defense of others; there was no shooting down of people in flight. Esprit de corps was high; discipline unremitting. This had bad as well as good aspects. "We were sealed off from the outside world and the experiences and thinking of others from the very beginning," said William Sullivan, "and we . . . steadily became inbred for thirty years."[10] Still the impression eventually arose even among agents that their director was a rather eccentric man.

Hoover was a bachelor who had lived for years with his mother. After her death his cherished companion was an FBI associate, Clyde Tolson. He had a brother and sister, but George Allen could not recollect his ever mentioning them.[11] His extracurricular passion was to go to the races with Tolson and Allen. He was obsessed by what he saw as a decline in sexual morality and was exercised even over the personal appearance of FBI agents. He detested long hair and beards. Jackets were to be worn. Suits were to be dark, shirts white, ties conservative. Red ties were evidence of insincerity. He penalized agents who were overweight or who had moist palms.[12] He feared germs, hated flies and could not abide people, especially agents, stepping on his shadow. Subordinates who displeased him were exiled to remote places like Butte, Montana.

"In the FBI under Mr. Hoover," wrote one special agent after twenty-three years, "you had to work on the premise that the Director was infallible. If you did not really believe this — and of course most employees certainly did not — you nevertheless had to pay lip service to it to survive. . . . If you were ambitious and desired to rise in the organization, you had to pay a still higher toll in the form of exaggerated sycophantic respect and adulation for him." In time Hoover's Bureau began to take on the internal lineaments of the Communist party he so fervently hated. "You just listen to what our Leader says about Commonist [pronounced in the Hoover style] Party discipline," one agent remarked to another, "and see if it reminds you of any other organization." Survivors developed great skill in sending Hoover worshipful letters and collecting funds to give him elaborate presents on Christmas and his birthday. As the director grew older, no flattery became too gross for him to swallow.[13]

He regularly misused Bureau employees, equipment and funds for entirely personal purposes. Subordinates rushed to do whatever he told them, zany as his orders might sometimes sound. "You claim you do not want 'yes men,'" William Sullivan protested, "but you become furious at any employee who says 'no' to you."[14] When his limousine had an accident making a left turn, Hoover magnificently forbade his chauffeur to make left turns in the future. After inspection of a new agent training class, Hoover told the instructor, "One of them is a pinhead. Get rid of him!" The instructor dared ask no more but surreptitiously looked at the class. Alas, no head was conspicuously smaller than the rest. Finally he examined hats. Three members of the class wore smaller hats than the others. Taking no chances, the instructor fired all three.[15] Hoover liked to scribble comments on the borders of memoranda; once, when an agent neglected to leave enough space, Hoover wrote irritably, "Watch the borders." For days thereafter special agents patrolled the Canadian and Mexican borders looking for they knew not what.[16] "Most agents," one FBI veteran wrote, "have ambivalent feelings toward the FBI. They are dedicated to the concept of the organization . . . but at the same time they seethe in frustration."[17] Even the worshipful Roy Cohn reported: "A popular joke with Bureau agents was: 'Did you hear what Tolson said to Hoover on Miami Beach? He said, "Nobody's looking now, John, you can take your walk on the water."'"[18]

"We did all possible to build up your legend," Sullivan told Hoover toward the end. ". . . No government Bureau should spend the vast sums of money on public relations as we have done and still do." Sullivan's staff wrote books for Hoover's signature — *Masters of Deceit* (1958), *A Study in Communism* (1962). FBI field offices promoted the books. "We even wrote reviews here at the Headquarters which were sent to the field to have printed by different papers." Since the work was done on government time, Sullivan recommended that the royalties go to charity. Only recently, he told Hoover in 1971, "did I learn that you put some thousands of dollars in your own pocket and Tolson likewise got a share."[19] The propaganda was a huge success. Support for the FBI became a patriotic duty, Hoover himself a national idol. "He was," said Sullivan, "one of the greatest con men the country ever produced."[20]

The Hoover personality cult constrained those who knew better to silence. There were additional inducements: for agents, apprehension about their FBI careers or, should they resign, about blacklisting in private industry; for Presidents, Hoover's assiduity as a courtier; for legislators and journalists, the ominous sweep of his ever-expanding files, crammed with information about the sexual vagaries of public figures. Hoover, William Sullivan said in 1973, was

a master blackmailer. . . . The moment he would get something on a Senator, he would . . . advise the Senator that "We're in the course of an investigation and by chance happened to come upon this data on your daughter. But . . . don't have any concern — no one will ever learn about it." Well, Jesus, what does that tell the Senator? From that time on the Senator's right in his pocket.[21]

Hoover was nowhere regarded with more veneration, or fear, than in Congress. The House Appropriations Committee always gave him all the money he requested and, on occasion, more. The Senate Appropriations Committee hardly ever asked him to appear at all.[22] In Congress criticism of Hoover was a form of lèse majesté — a fact forgotten in later years by congressional committees when they piously condemned executive officials for reluctantly yielding the FBI a latitude that every Congress for a generation had demanded that they accord him enthusiastically and reverently.

Nor was the press much better. A few, very few, intrepid newspapermen defied the personality cult — James Wechsler of the *New York Post,* I. F. Stone in his *Weekly,* Fred Cook in the *Nation.* But one would search the *New York Times* and *Washington Post, Time* and *Newsweek,* of the 1950s in vain for the exposure and disapproval such journals lavished on Hoover twenty years later when it at last became safe to take on the FBI. Roosevelt might have contemplated getting rid of Hoover, but there is no compelling evidence even of that. None of his successors seriously considered taking so drastic a step — a step that would have enraged Congress and the country.

The FBI, as Dean Rusk, a cautious man, said in 1974, had "developed into an extraordinarily independent agency within our Government. It is hard to exaggerate that. Mr. Hoover, in effect, took orders only from himself, sometimes from an Attorney General, usually from a President, and that was it. He had created a kind of kingdom . . . almost unparalleled in the administrative branch of our Government, a combination of professional performance on the job, some element of fear, very astute relations with the Congress, and very effective public relations."[23] By 1961 the FBI was a runaway agency, and its director sacrosanct and untouchable.

II

Through the years, the FBI had depended on a variety of investigative methods. Most important, from the viewpoint of obtaining convictions, were 'informants' — members of or infiltrators into groups under suspicion. The advantage of informants was that their testimony could be used in court. Other techniques were employed to gain 'strategic' intelligence rather than to build legal cases. There was 'electronic surveillance.' This had two legally and technologically distinct forms: 'wiretapping' — the physical tapping into telephone wires in order to record calls; and 'bugging' — the planting of concealed microphones near the person under surveillance. Bugging required 'trespass' if the bug was planted inside a person's room; nontrespassory bugs were also possible. Then there was 'surreptitious entry' or 'black bag' jobs — breaking into an office or dwelling in order to copy papers or for other purposes, such as, very often, the installation of a bug. And there was 'mail opening' — intercepting and photographing personal letters.

Some of these methods violated federal law. But anything was all right so long as the director approved. Clarence M. Kelley, an FBI agent for twenty years and later one of Hoover's successors as director, remarked in 1976 that some agents still clung "to the idea that you can do anything you want so long as you follow certain ideas that were embedded in them throughout the many years of the leadership of Mr. Hoover. . . . It was an authoritarian type of leadership."[24] It was indeed. Hoover took care to keep his superiors in ignorance about his methods. "We do not obtain authorization," William Sullivan noted in an internal FBI memorandum in 1966, "for 'black bag' jobs from outside the Bureau."[25] Sullivan later told me, "We never, never told any Attorney General about surreptitious entry or about the opening of mail."[26] FBI intelligence reports, even when distributed to the Attorney General, methodically camouflaged the sources. "T-1, a confidential informant of known reliability" might mean a tap, a bug, a burglary, a mail intercept or even a confidential informant of known reliability. Hoover actually invented a "Do Not File" procedure designed to hide what the Bureau was doing.[27] Since some of the things it was doing were patently illegal means to allegedly patriotic ends, it was not the habit of Attorneys General to inquire too closely.

Of the FBI methods, only one — wiretapping (as distinct from bugging) — required the Attorney General's specific authorization. This was for particular historical reasons. When Harlan Stone reorganized the Bureau of Investigation in 1924, he had banned wiretapping. The Treasury Department's Prohibition Bureau, however, tapped freely. A Seattle bootlegger named Olmstead, protested that wiretapping was a form of unreasonable search and therefore forbidden by the Fourth Amendment. In 1928 the Supreme Court ruled in the *Olmstead* case that wiretapping did not breach the Constitution. It was a 5–4 decision. Justice Brandeis in dissent warned that "the progress of science in furnishing the government with means of espionage is not likely to stop with wiretapping." Justice

Holmes contemptuously described wiretapping as "dirty business" and said: "I think it is a less evil that some criminals should escape than that the government should play an ignoble part."[28]

In 1930 the Prohibition Bureau was transferred to Justice. Faced with the anomaly that one of his bureaus was tapping while another was forbidden to do so, Herbert Hoover's Attorney General removed Stone's ban on tapping by the Bureau of Investigation. Congress responded by including in the Communications Act of 1934 a section making it a crime for any person not authorized by the sender to "intercept any communication and divulge or publish the . . . contents." Though J. Edgar Hoover expressed public doubts about tapping till 1939, FBI agents tapped nonetheless while Section 605 was awaiting clarification by the judiciary. The Supreme Court soon ruled that wiretap evidence and evidence derived from wiretap leads were inadmissible in federal courts.[29] In 1940 Attorney General Robert H. Jackson reinstituted Stone's ban. He added his hope that Congress would authorize tapping under safeguards.

Something should be said about the atmosphere of the time. Wiretapping was not the black-and-white issue it became in later years. Ardent New Dealers like Henry Morgenthau, Jr., Harold Ickes and Fiorello La Guardia favored tapping for their own presumably benign purposes. Moreover, the rise of Hitler and fears of his Fifth Column argued for vigilance. "In general," said Franklin D. Roosevelt, "my own personal point of view is close to that of Justice Holmes in his famous dissent in the *Olmstead Case*."[30] But, with Nazi armies poised to overrun western Europe, FDR was willing to make exceptions.

Three months after Jackson's order and under pressure from Morgenthau,[31] Roosevelt gave Justice qualified permission to wiretap "persons suspected of subversive activities against the United States." He directed the Attorney General "to limit these investigations so conducted to a minimum and to limit them insofar as possible to aliens." He added that "under ordinary and normal circumstances wiretapping by Government officials should not be carried on for the excellent reason that it is almost bound to lead to abuse of civil rights."[32] He would also require, he added in 1941, that "the Attorney General be acquainted with the necessity for wire tapping in every single case and that he himself sign a certificate indicating such necessity."[33]

Jackson, worried about reconciling Roosevelt's directive with the Federal Communications Act, resolved the problem by construing Section 605 to require *both* interception and divulgence to become operative and by construing divulgence as disclosure *outside* the executive branch.[34] This strained construction licensed the FBI to tap wires and disseminate the results within the government. Jackson, who regarded the whole business with distaste, thereafter left the problem to Hoover without himself passing on each case. Francis Biddle, his successor in 1941, established the practice of case-by-case authorization by the Attorney General, sometimes turning down Hoover's requests.[35]

Roosevelt saw wiretapping primarily as a wartime measure of counterespionage. After his death and the war's end the FBI continued to tap. The Cold War offered new justification. In 1946 Attorney General Tom Clark, transmitting a draft prepared in the Bureau, asked Truman to renew Roosevelt's 1940 authorization.[36] The FBI version of the directive, however, broadened it spectacularly, omitting Roosevelt's instruction to confine tapping to aliens, espionage and sabotage. After Truman approved the Hoover-Clark draft, George Elsey of his White House staff discovered the distortion and called the new language "obviously a very far cry from the 1940 directive."[37] Nothing was done about it. Thereafter Hoover claimed that the Roosevelt memorandum authorized the FBI to tap in a wide variety of cases, though still with case-by-case approval by the Attorney General. State and local police were tapping happily in at least half the states. In most of these states wiretap evidence was admissible in state courts.

"The progress of science," as Brandeis had foreseen, was meanwhile improving the technology. The rise of the unwired microphone — the 'bug' — gave the FBI a means of electronic surveillance they used without application to the Attorney General. In 1952, however, Attorney General Howard McGrath caught on and forbade the Bureau to install bugs involving trespass. But in 1954 Hoover procured a memorandum from Attorney General Herbert Brownell envisaging "the unrestricted use" of bugs "in the national interest" and omitting to require case-by-case authorization from the Attorney General.[38] Hoover took the Brownell memorandum as a general license for bugging, even with trespass. He could now switch back and forth, using bugs when he wanted to minimize an Attorney General's knowledge, taps when he wanted to maximize an Attorney General's complicity.

In 1961, as Kennedy became Attorney General, the situation was as follows: (1) the FBI could wiretap only on the written authorization of the Attorney General; (2) state and local police taps, often stimulated by the FBI, went freely along as state law permitted; (3) the FBI bugged ad lib at the director's discretion and without notice to Attorneys General; (4) the FBI indulged ad lib in surreptitious entry in order to install bugs or for other purposes, and did so without the knowledge of any Attorney General.*

III

The FBI leadership was pleased neither by John Kennedy's election nor by his brother's appointment. "We'll be stuck with the Kennedy clan," Clyde Tolson observed sourly, "till the year 2000."[39] But they were realists. The Kennedys were realists too. With so minuscule an advantage in the popular vote, the President-elect was not about to take liberties with a national icon. In any case, as Robert Kennedy said in 1964, "Father was a good friend" of Hoover's and his brother "didn't have anything against Hoover, and he thought it was well that Hoover stayed."[40] Immediately after the election Kennedy announced that he would keep Hoover on along with Allen W. Dulles of the CIA.

In the 1950s Robert had had more direct dealings than his brother with the director. His early impulse no doubt had been to place Hoover in the Great American Patriot niche along with Herbert Hoover and Douglas MacArthur. Then came his fight with the Bureau over the Annie Lee Moss case. Later, relations seemed to improve. "I saw [Hoover] on various occasions where he was very friendly and nice." Once in Justice, "I really deferred to him," Robert told Anthony Lewis in 1964.

> I mean, I made a real effort because I recognized the fact that I was young and coming in there and all the rest of the business; and I was making so many changes, not only in organized crime but what we wanted to do in civil rights. . . . I made arrangements with my brother that he would call him every two or three months and then have him arrange to have J. Edgar Hoover, just by himself, for lunch. . . . It's what kept Hoover happy for three years because he had the idea that he had direct contact with the President, and he used to tell me that every month, I guess, Franklin Roosevelt used to have him over for lunch.[41]

Amenities were carefully observed between the Attorney General and the director. Katzenbach thought the relationship "very distant and formal. Bobby always felt embarrassed calling him Edgar."[42] Bobby himself believed that his effort, on the whole, had succeeded. On the surface, he said later, relations were "smooth. I knew that he

---

* "There is no indication that any Attorney General was informed of FBI 'black bag' jobs, and a 'Do Not File' procedure was designed to preclude outside discovery of the FBI's use of the technique" (Church committee, *Final Report*, bk. II: 61; see n. 7).

didn't like me much. . . . But the relationship that we had was not difficult. I mean, it was not an impossible relationship."[43]

John Kennedy regarded Hoover with skepticism. "The three masters of public relations in the last half century," the President told me in October 1961, "have been Baruch, Hoover and Allen Dulles."[44] Once he summoned Theodore Sorensen and me to the residential part of the White House to discuss a speech. We arrived just as Hoover was leaving the lunch table. Kennedy conspicuously omitted to introduce us, remarking later that he had not wanted to upset Edgar too much.[45] One of his favorite lines, the last phrase of which he adapted to the occasion, went: "The three most overrated things in the world are the state of Texas, the FBI and" whatever was exasperating him most at the moment.[46] Still, getting on with Hoover seemed an elemental political necessity. As 1964 approached, Kennedy told Ben Bradlee of *Newsweek* "how FDR used to have Hoover over regularly [a thought Hoover constantly pressed in slightly exaggerated form on both Kennedy brothers], and said he felt it was wise for him to start doing the same thing, with rumors flying and every indication of a dirty campaign coming up. 'Boy, the dirt he has on those senators,' Kennedy said, shaking his head."[47] But Kennedy, as William Sullivan observed, "distrusted Hoover and wouldn't have dared to make a political request of him."[48]

Some Presidents, and Attorneys General, were beguiled by Hoover's gossip. "Edgar was not above relishing a story derogatory to an occupant of one of the seats of the mighty," Francis Biddle recalled of his own days as Attorney General. ". . . I confess that, within limits, I enjoyed hearing it."[49] "I didn't get the feeling that President Kennedy was that kind," observed Clark Clifford. "He wasn't a gossipy man. Some presidents . . . love to get little tidbits, particularly on their enemies and all. I didn't find that quality in the President."[50] According to Kenneth O'Donnell, the designated White House recipient of FBI reports about government officials, Kennedy "didn't like to hear about their sex life, he didn't want to hear whether they drank, he didn't want to hear anything personal about anybody." His reactions were certainly not always those that Hoover intended. Once the prim-prurient director sent over an FBI report describing a mettlesome American ambassador who, caught in a woman's bedroom, hastily departed through a window without his trousers. When the White House failed to react, Hoover queried O'Donnell. O'Donnell, who had had little use for Hoover since Rackets Committee days, said that, yes, the President had seen the report. Asked what he had said, O'Donnell replied: "He said that from now on he's going to hire faster ambassadors."[51]

Hoover, said Robert Kennedy, "was really of no particular importance to him." As O'Donnell put it, "We had a hell of a lot more
problems than J. Edgar Hoover. . . . He was so far down the list. I
don't think the President and Bobby even discussed it. . . . Bobby
took care of his own problems pretty good. He never went crying to
anybody."[52] Robert Kennedy later summed it up from his viewpoint:
"It was important, as far as we were concerned, that [Hoover] remained happy and that he remain in his position because he was a
symbol and the President had won by such a narrow margin and it
was a hell of an investigative body and he got a lot of things done
and it was much better for what we wanted to do in the south, what
we wanted to do in organized crime, if we had him on our side."[53]

IV

For Hoover, the very existence of an Attorney General who was the
brother of the President represented an unprecedented threat. It
was not, of course, just the family tie. Brothership worked no political
magic for Sam Johnson, for example, or Donald Nixon or Billy
Carter. But Robert Kennedy was an aggressive executive in his own
right. Moreover, if Roy Cohn can be believed, Hoover had cordially
disliked him ever since the Annie Lee Moss incident.

Kennedy's informality certainly infuriated the director now. He
was upset by Kennedy's excursions around the Department, and
FBI memos of the period reflect a continuing surveillance by the
Bureau of the Attorney General's movement. "This certainly proves
the point we have been stressing," the director noted on the border
of one, " — our employees should always be busy; engage in no horse
play; and be properly attired. No one knows when & where
A. G. may appear."[54] Once Hoover and Tolson came to the Attorney
General's office. As the three men talked, Kennedy, to Hoover's
repressed rage, idly tossed darts at a board on the wall. "In other
words," explained a Tolson assistant, "he did not give the Director
undivided attention." Hoover became even more upset when Kennedy missed the board, and darts lodged in the paneling. "It was
pure desecration," the director said later. "Desecration of government property." Hoover and Tolson cut their visit short and later
agreed, as the director told another subordinate, that it was "the most
deplorably undignified conduct they had ever witnessed on the part
of a Cabinet member."[55]

"I was in Bobby's office once when he summoned Hoover," William
Hundley of the Organized Crime Section recalled. "I couldn't get

over it." Hundley had been reporting on a political case where Hoover had been, as he saw it, dragging his feet. Kennedy asked whether Hundley wanted to hear Hoover's explanation. "That was too much, so I said 'Yes.' So he hit a goddam buzzer and within sixty seconds, the old man came in with a red face, and he and Bobby jawed at each other for about ten minutes. And Hoover kept looking at me."[56] "Nobody had ever buzzed for Hoover!" said Walter Sheridan, who as an ex-FBI agent observed the relationship with a certain relish. "And Hoover came. But I'm sure it griped him very much to come. And Bob would go in his office in his shirt-sleeves; nobody had ever done this before."[57] After one such call, Hoover, deeply disturbed, said to Assistant Director William Sullivan, "It is ridiculous to have the Attorney General walking around the building in his shirtsleeves. Suppose I had had a visitor waiting in my anteroom. How could I have introduced him?"[58]

Worst of all, the Attorney General did not sufficiently respect the Bureau's highly prized autonomy. Robert Kennedy's deference, which he evidently thought abundant, was from Hoover's viewpoint strictly limited. While the Attorney General was happy to appease the director with occasional White House luncheons, he insisted (with imperfect success) that FBI communications to the White House pass through his own office. No Attorney General had tried to do this since Biddle fifteen years before. He instructed the Bureau to send its press releases to Guthman's office instead of putting them out, as it had done for years, on its own letterhead. Instead of clearing Hoover's speeches and statements automatically, as his predecessors had done, Kennedy sometimes returned them with requests for revision "in the interest of accuracy or Administration policy."[59] He installed a direct line from his own office to the director's desk. Once when Hoover's secretary answered a Kennedy call, the Attorney General informed her curtly that hereafter he expected the director to pick up the phone himself.[60]

Hoover had always demanded that communications from the Attorney General to the FBI go through him to the agents. Kennedy did not hesitate to call FBI men working on cases without speaking to the director first.[61] Hoover was further irritated by Kennedy's habit of dropping in on regional FBI offices and asking questions about current cases. The Attorney General even persuaded FBI offices to place a photograph of John Kennedy next to the obligatory photograph of Hoover, though he affected to believe that it was always the same photograph, circulated by the Bureau to be put up ten minutes before his arrival.[62] In any case, he got around. As special agent William Barry, who often accompanied Kennedy, said later, "Hoover never visited the offices [Kennedy] visited." To Kennedy such visits were part of the Attorney General's responsibility. He wanted, as Barry said, to "let people know that the Attorney General did exist, that they were part of the Justice Department."[63] To Hoover the visits were an inexcusable intrusion.

Over the years the director had successfully withdrawn the Bureau from the oversight of the Department of Justice. The arrival at Justice of a President's brother, a brash young man determined to annex the Bureau to the Department, radically altered his position. It also fed his incipient paranoia. "For the first time in his career," as Ralph de Toledano described Hoover's view in his reverent biography, "Hoover was forced to deal with an Administration that was overtly hostile, that wanted him out."[64] The idea that the Kennedys were planning his dismissal seized possession of his mind. Byron White was occasionally mentioned as a possible successor. For a while Hoover got it into his head that Adam Yarmolinsky, a blameless and scholarly lawyer working for McNamara in Defense, was scheming to get his job. In his usual manner, he leaked items to the press designed to discredit Yarmolinsky.[65] Hoover was no more enthusiastic about the idea of a successor from inside the Bureau and harbored dark suspicions that one or another subordinate was aspiring to the directorship. "What the hell's wrong with wanting to be director?" said one of them (John P. Mohr). "Does the President get mad at every kid that wants to be President? . . . I never said that to Hoover, though."[66] To be in the line of succession was nearly as perilous in Hoover's FBI as in Mao Tse-tung's China.

From time to time — or so it seemed from the Attorney General's office — someone would organize a keep-Hoover campaign. Letters poured into Justice, many, Seigenthaler said later, "identical to other letters."[67] Actually Hoover had little to fear during Kennedy's first term. The political risks appeared too great. But on January 1, 1965, the director would have his seventieth birthday. Hundley was convinced "that the thing that finally destroyed their relationship was that Bobby mentioned to too many people who complained to him about Hoover that, 'Look, just wait,' and we all got the message that they were going to retire him after Jack got reelected and Hoover hit seventy. And it got back to him."*

Convinced that the Kennedys were out to get him, Hoover resorted to his traditional techniques of self-defense. He tried ingratiation.

* Ovid Demaris, *The Director* (New York, 1975), 147; William Hundley, in interview by author, August 4, 1976. RFK in December 1964: "President Kennedy would have gotten a replacement before he left office" (RFK, in interview by Anthony Lewis, December 4, 1964, IV, 35, JFK Oral History Program).

One finds in Robert Kennedy's papers the following letter from Hoover:

> After observing your car in the Department garage, I would like to thank you for coming to work on February 22nd, a national holiday. . . . The spirit you demonstrated — the spirit of Valley Forge and Monte Cassino — will, we hope, spread through the entire Department of Justice. Keep up the good work.[68]

What in the world was this? Flattery? Sarcasm? Kidding? In other moods he was more menacing. "Every month or so," Robert Kennedy said later,

> he'd send somebody up or a memo . . . to give information on somebody I knew or a member of my family or allegations in connection with myself, so that it would be clear whether it was right or wrong that he was on top of all of these things and received all of this information. He would do this also, I think, to find out what my reaction to it would be. . . . I remember on one occasion that somebody said that my brother and I had a group of girls on the 12th floor of the LaSalle Hotel and that, I think, the President used to go over there once a week and have the place surrounded by Secret Service people and then go up and have assignations. . . . A lot of it was so far fetched that even on the face of it it didn't make any sense. I mean, if you were going to do that kind of thing, you wouldn't go on over to the LaSalle Hotel with the Secret Service surrounding the place. But I think the idea was just so that you would know that they were continuously getting into this information.*

In public, the Attorney General meticulously defended the director. One disaffected agent, having tried in vain to persuade Kennedy to act on his grievances, concluded: "In retrospect, it would seem that the Kennedys, having made a political decision to retain Hoover, temporarily felt that they had to put on a public face of supporting him."[69] Privately it was very different. If Robert Kennedy generally contained himself, his irrepressible wife could not. The director, who was perennially feuding with city police chiefs, particularly disliked Chief William H. Parker of Los Angeles. Once at a Department of Justice party Ethel Kennedy had some words with Hoover. "There

---

* RFK, in Lewis interview, V, 2–3. Another example of Hoover's admonitory operation was provided by the disclosure in December 1977 of a memorandum from the FBI director to Robert Kennedy on June 5, 1963, containing assertions that Robert Kennedy in 1960 had settled out of court a breach-of-promise suit, going back to 1951, for $500,000. The story was improbable on the face, for a decade-old breach-of-promise suit against a President-elect would have been generally discounted as publicity seeking. In any event, the woman named, Alicia Darr Purdom Clark, promptly denied that she and John Kennedy had been lovers or that she had ever received a cent from the Kennedys. She added not implausibly: "In fact Jack owed me money. . . . He used to borrow dollars from me for cab fares. He never paid them back" (see *New York Times,* December 15, 1977; *Washington Post,* December 16, 1977; *New York Post,* December 17, 1977; "Walter Scott's Personality Parade," *Parade,* April 2, 1978).

was a little electricity between the two of them," said William Barry, who observed the scene. "On the way out she put in the FBI suggestion box [a note saying] 'Chief Parker in Los Angeles for Director.'"[70]

When a police system passes beyond the limit of dealing only with such conduct as is forbidden by the laws, Harlan Stone had said in 1924, "it is dangerous to the proper administration of justice and to human liberty." Forty years later Anthony Lewis asked Robert Kennedy whether he thought Hoover just a nasty person or truly dangerous. "No," Kennedy replied. "I think he's dangerous."[71] "He's rather a psycho," he told John Bartlow Martin, ". . . I think it's a very dangerous organization . . . and I think he's . . . become senile and rather . . . frightening."[72]

The word in the FBI, said William Sullivan, was that the two men in the world J. Edgar Hoover hated most were Martin Luther King second and Robert Kennedy first.*

---

* Sullivan, in interview by author, July 26, 1976. Jack Anderson reported (*New York Post*, October 9, 1975) that Hoover told Herbert Jenkins, the Atlanta police chief and then president of the International Association of Police Chiefs, that the three enemies he hated most in the world were Kennedy, King and Quinn Tamm, a former FBI assistant director who had become executive director of the International Association.

13

The Pursuit of Justice:
The Mob

THE ENCOUNTER between Robert Kennedy and J. Edgar Hoover
was, within its peculiar limits, a fascinating case history in the
politics of administration: the aggressive executive versus the supreme bureaucrat; the presidential brother versus the national
hero. The limits were imposed on one side by the fact that even a
presidential brother dared not provoke a national hero to resignation,[1] on the other by the fact that even a national hero dared not
break with the President's brother. At stake in this first round was
Robert Kennedy's determination to complete the unfinished business
of the Rackets Committee and bring the men who ran the mob and
the labor rackets to the bar of justice.

I

Hoover's obsession in 1961 remained the Communist movement.
For years he had warned his fellow countrymen against the diabolic
machinations of these godless masters of deceit. Despite the patriotic
endeavor of the FBI, however, the threat, at least as measured by
the director's rhetoric, apparently continued to grow. In an agitated
memorandum a fortnight before John Kennedy's inauguration, Hoover informed the new Attorney General that the American Communist party presented "a greater menace to the internal security of
our Nation today than it ever has since it was first founded in this
country in 1919."[2]

Robert Kennedy was unmoved. Party membership had in fact
drastically declined from its 1944 peak of 80,000. But Hoover
needed the threat to preserve his mystique. "When the membership

figures kept dropping lower and lower," William C. Sullivan reminded Hoover in 1971, "you instructed us not to give them out
. . . not even to the Justice Department."[3] The Attorney General,
undeceived, saw communism as a danger *to* but hardly *in* the republic. "It is such nonsense to have to waste time prosecuting the Communist Party," he told Henry Brandon of the *London Sunday Times* in
December 1961. "It couldn't be more feeble and less of a threat, and
besides its membership consists largely of FBI agents."[4] "As far as
having any real influence as a party in the United States," he observed
in an interview with representatives of the Knight newspapers, "it's
zero."[5] When I wrote a piece for the *Saturday Evening Post* in 1962
entitled "The Failure of World Communism," Hoover, who had described me in an internal FBI memorandum as "a stinker" as early
as 1950 and had forbidden FBI agents to talk to me in 1954, minuted
to an associate, "I understand better now the views of the A.G. since
the latter is a great admirer of Schlesinger."[6] (As usual, the director
exaggerated.)

The enemy within, in Kennedy's view, was something very different. The Rackets Committee had instructed him in the covert realities of power and corruption in American life. "Organized crime,"
he said succinctly in his first speech after becoming Attorney General,
"has become big business."[7] What appalled him, as he told a congressional committee in 1963, was the "private government of organized
crime, resting on a base of human suffering and moral corrosion,"
infiltrating business, labor, politics and government. The racketeer
was no longer "someone dressed in a black shirt, white tie and diamond stickpin, whose activities affect only a remote underworld circle. He is more likely to be outfitted in a grey flannel suit and his
influence is more likely to be as far-reaching as that of an important
industrialist."[8]

In 1961 Jerry Wald, the Hollywood producer, proposed to Kennedy that *The Enemy Within* be made into a movie. Of the writers
Wald mentioned, Kennedy promptly chose Budd Schulberg, whom
he remembered as the author of *On the Waterfront*, the powerful Sam
Spiegel–Elia Kazan film of 1954 about labor racketeers. Schulberg
discussed *The Enemy Within* one night at Hickory Hill. What attracted
him to the book, Schulberg said, was the theme that "something at
the core of our society was beginning to rot." Responding, Schulberg
recalled, "simply and modestly; in fact, diffidently," Kennedy said,
"Good. . . . I wrote those last pages very carefully — I didn't want
the book to seem to be aimed against a single man or a single union.
It is the society that produces a Beck or a Hoffa or a Johnny Dio. I
don't know how you are able to bring that out in a picture, but that

seems to me the only real reason for making the picture." The next ten years, he told Schulberg, might mark a turning point in American history — either an America infected with corruption or the rebirth of the idealism with which the nation had begun.*

Kennedy was determined to stop the drain of power in America to obscure forces beyond moral and legal accountability. In insisting on the spreading threat of organized crime, he offended J. Edgar Hoover doubly — by dismissing the cherished Red menace and by raising a question the director had done his best for forty years to ignore.

Here one encountered another aspect of the gap between inside and public perceptions of the FBI. For the G-men had made their reputation as gangbusters. In 1934, after Roosevelt's anticrime legislation established federal jurisdiction over a variety of interstate felonies (including flight across state lines to avoid prosecution or even to avoid giving testimony in court), the FBI ended the careers of notorious outlaws like Dillinger. Nor had the Bureau let anyone forget these triumphs. They were not, however, triumphs over organized crime. The outlaws were petty entrepreneurs. Often, like Bonnie and Clyde, they were amateurs engaged in sporadic and desperate acts. They were leftovers from an individualist past, irrelevant to the new criminal world of the syndicates, the world characterized, as Daniel Patrick Moynihan wrote in 1961, by "a large, wealthy, and firmly entrenched criminal class."[9]

The much advertised lists of Public Enemies, Top Fugitives and Most Wanted Men kept up the illusion that the FBI was unrelenting in its war on criminals. In fact, of the hundreds of crooks thus honored, very few had any connection with organized crime.[10] While the Bureau had chased Bonnie and Clyde, it left Vito Genovese and Meyer Lansky of the syndicates to flourish in peace. In the meantime, Elmer Irey, chief of the Treasury Department's Intelligence Unit, quietly brought down Al Capone, Johnny Torrio, Waxie Gordon and Willie Bioff. Thomas E. Dewey in New York broke up the Lucky Luciano ring. Harry Anslinger of the Bureau of Narcotics was spinning tales about the Mafia. Organized crime was not precisely a

---

* Budd Schulberg, "RFK — Harbinger of Hope," *Playboy*, January 1969. Kennedy liked Schulberg's script, but Wald, who, Schulberg said, "alone had the courage to produce it," died suddenly. A "labor tough" warned Twentieth Century–Fox that, if the picture were made, the Teamsters would refuse to deliver the print to the theaters. When Columbia, the distributor of *On the Waterfront*, showed a brief interest, William Bufalino, a Hoffa attorney, wrote a letter to the studio pointing out that Twentieth Century–Fox had wisely abandoned the project as soon as the possible consequences had been laid before it and that he felt confident Columbia would be intelligent enough to do likewise. (See Schulberg's introduction to Walter Sheridan, *The Fall and Rise of Jimmy Hoffa* [New York, 1972].) As for Kennedy, Schulberg wrote in *Playboy*, "I understood that it was not his role to ask any studio to produce his book."

secret even in the 1930s, except evidently to J. Edgar Hoover. None of the mobsters above, save for a passing reference to Capone, was even mentioned in Don Whitehead's authorized history, *The FBI Story*, written with Hoover's cooperation in 1956.

Robert Kennedy had discovered Hoover's indifference to organized crime when he requested FBI dossiers on the Apalachin conferees in 1957. Apalachin shook the Bureau itself for a moment. William Sullivan of the Intelligence Division persuaded Hoover to assign the best FBI crime researcher, Charles Peck, to prepare a report on organized crime. The Bureau also started up some electronic surveillance. But, when Hoover read the Peck report, he denounced it as "baloney" and ordered it suppressed.[11] In 1959 the FBI New York office had over four hundred agents working on communism, four on organized crime.[12]

The shock to the Department of Justice lasted a little longer. Attorney General Rogers brought William Hundley over from the Internal Security Division to head up the Organized Crime Section of the Criminal Division. In Internal Security Hundley had found "cooperation from the FBI that was unbelievable."[13] In Organized Crime, Hundley found the FBI's noncooperation equally unbelievable. "It was like pulling teeth to get anything done in this field at all. . . . I had come out of Internal Security where you had agents coming out of your ears, and get over into Organized Crime and you couldn't find an agent."[14] Rogers also set up the Special Group on Organized Crime under Milton Wessell, a New York lawyer. When the group asked the FBI about Apalachin, "the G-men," said one of Wessell's assistants, "acted as if they had never heard of Apalachin."[15] Hoover denied FBI assistance to Wessell, who eventually retired, defeated, to New York. The director explained to the worshipful House Appropriations Subcommittee in 1961 that too many people were seeking "fantastic panaceas as to how to solve *local crimes*." (Agent W. W. Turner noted that, while Hoover was turning down the Organized Crime Group, FBI agents were busy copying number plates of out-of-state cars parked in airports in the hope of finding a stolen car.)[16]

II

Why did Hoover veer away from organized crime? His own favorite explanation was legalistic. "The truth of the matter," he said in 1964 (and on many other occasions), "is the FBI had very little jurisdiction in the field of organized crime prior to September 1961."[17] This proposition was a transparent fraud. The 1934 legislation had given the FBI ample jurisdiction. Eminent gangsters were crossing state lines every day, for example, to avoid prosecution or the giving of testimony. When mobs put out contracts on their enemies, they tended as a matter of standard operating procedure to import gunmen from another state. Many federal statutes covered syndicate activities.* Hoover had carved out his spacious jurisdiction in internal security with virtually no statutory basis at all. As James V. Bennett, director of the Bureau of Prisons, put it, "If they want to do it, it's within the jurisdiction of the FBI. If they don't want to do something, they'll tell you it's outside the jurisdiction of the FBI."[18]

Why then? Former FBI agents laid great stress on Hoover's infatuation with statistics. He liked to regale Congress with box scores of crimes committed, subjects apprehended, crimes solved. Organized crime did not lend itself to statistical display. It required a heavy investment of agents in long, tedious investigations that might or might not produce convictions at the end. The statistical preoccupation steered Hoover toward the easy cases — bank robbers, car thieves, kidnappers and other one-shot offenses.†

The statistics race — each year had to be better than the year preceding — also increased the Bureau's dependence on local law enforcement. The FBI very often inflated its statistics by claiming credit for work performed by local police, as in the recovery of stolen cars.‡ This dependence argued further against the pursuit of organized crime. For the mobs often had their own arrangements with local police, politicians and businessmen. Hoover did not want to get into that. Even the admiring Toledano admitted that Hoover considered it "futile to take on the crime which is entangled with legitimate business or local politics until the community — national or otherwise — was ready for it."[19]

He also, according to Toledano, feared that "to make the attempt

---

* "Among the statutes under the FBI's wing were acts relating to the kickback rackets, extortion, anti-racketeering, labor-management relations, unauthorized publication or use of communications, obstruction of justice, bankruptcy (the scam has long been a favorite Cosa Nostra fraud), the white slave traffic, perjury, misprision of felony, trust formation, interstate transportation of stolen property, interstate transportation of gambling devices, lottery tickets, and wagering data, hijacking, bribery, fraud by wire, and federal firearms control" (William W. Turner, *Hoover's FBI* [New York: Dell reprint, 1971], 162–163).

† William C. Sullivan, in interview by author, July 26, 1976; Turner, *Hoover's FBI*, 4, 16, 159, 246–253. Note William Hundley: "I concluded that the inability to build statistics in organized crime was a factor in the Bureau's reluctance to enter the area" (Pat Watters and Stephen Gillers, eds., *Investigating the FBI* [New York, 1973], 165).

‡ Warren Olney III, head of the Criminal Division in the years 1953–57, called Hoover's statistics "hogwash" and said of auto theft cases, "In at least half, and possibly more of these, the thieves are arrested and the cars returned by local officials" (*Newsweek*, December 7, 1964). Olney's cavalier attitude toward the Bureau doubtless explains Hoover's fierce opposition when Chief Justice Warren proposed Olney as chief counsel of the Warren Commission (see *Clandestine America*, January–February 1978).

would expose the FBI to all the temptations of corruption and big money, returning it to its sorry state under the Wilson and Harding administrations"[20] — an unnecessary slur, it would seem, on the probity of his agents. Still there was little chance of their being bribed by the American Communist party. In addition, the great proponent of the Mafia thesis was Harry Anslinger, the director's ancient rival and a leading figure on his hate list. Beyond all this, one cannot omit the possibility that Hoover, who was among other things a right-wing ideologue, honestly believed that Communists were far more dangerous to the republic than organized criminals.

For whatever reason,* the director of the FBI had stayed out of a battle that the former counsel of the Rackets Committee considered crucial to the national future. After his experience with the Teamsters and the mobs, Robert Kennedy now found himself unexpectedly embroiled with a third secret organization, almost as impenetrable and self-hallucinated as the others, but this time operating under the color of law.

III

The story is told of an early exchange between Kennedy and J. F. Malone, head of the FBI office in New York. "Mr. Malone," Kennedy said, "could you please bring me up to date on what's been happening with organized crime?" "To tell you the truth, Mr. Attorney General," Malone replied, "I'm sorry, but I can't, because we've been having a newspaper strike here."[21]

The national government had twenty-six investigative units at work on the domestic scene.[22] But none knew what the others had discovered about the underworld, and the FBI seemed to know least of

---

* More sinister reasons have been proposed and will only be noted here. I have no reason to believe that they are true — or false. According to *Time* (December 22, 1975): "Hoover was reluctant to move against organized crime. Some FBI agents think they know why. They tell stories of Hoover sometimes traveling to Manhattan to meet one of the Mafia's top figures, Frank Costello. The two would meet in Central Park. Costello apparently convinced Hoover that there was no organized Mafia — merely a loose collection of independent racketeers." The fact that Hoover met with Costello is confirmed by William Hundley and Edward Bennett Williams.

Hank Messick, author of several books on organized crime, points out ("The Schenley Chapter," *Nation*, April 5, 1971) that, according to the testimony of his fourth wife, Lewis S. Rosenstiel of Schenley Industries and a good friend of Hoover's was also a longtime friend of Meyer Lansky. Other testimony suggests that Rosenstiel and Lansky had been in business together during prohibition. Rosenstiel subsequently employed Louis B. Nichols, formerly an assistant director of the FBI, and in 1969 set up the J. Edgar Hoover Foundation. Messick raised the not uninteresting question: "How . . . can a man be a friend of J. Edgar Hoover on the one hand and a chum of Meyer Lansky on the other?"

all. This situation had led to Kennedy's call from the Rackets Committee for the pooling of intelligence through a national crime commission. Hoover hated to share FBI intelligence on the pretext that other agencies could not be trusted and that somewhere an informant would be betrayed. He hated even more to be caught without information to share. He therefore fought the national crime commission to the end.

Luther Huston, the departing information officer for Justice, described what may have been the climactic moment. Scheduled to pay his formal farewell to Hoover, Huston had to wait because the new Attorney General was already in the director's office. "He hadn't called or made an appointment. He had just barged in. You don't do that with Mr. Hoover. Then my turn came and I'll tell you the maddest man I ever talked to was J. Edgar Hoover. . . . Apparently Kennedy wanted to set up some kind of supplementary or overlapping group to take over some of the investigative work that the FBI had been doing. My surmise is that Mr. Hoover told Bobby, 'If you're going to do that, I can retire tomorrow. My pension is waiting.'"[23]

That evidently ended the national crime commission. Frustrated here, the Attorney General approached the problem circuitously. Two key agencies in the campaign against organized crime were the Internal Revenue Service and the Narcotics Bureau, both in the Treasury Department. Kennedy brought Secretary Dillon and Mortimer Caplin, who had been one of his law professors at Charlottesville and was now commissioner of internal revenue, together with Hoover and talked about the importance of sharing criminal intelligence.

> I said that this was something that I knew that President Kennedy was interested in (this was the advantage of being the President's brother), and it was something that I was interested in. . . . I said, "This is what we're going to concentrate on, and this is what we're going to do. I want it done." . . . Then I got the Bureau of Narcotics. . . . There was no question that I could do it [only] because of my relationship. They wouldn't have paid any attention to me otherwise.[24]

Hoover, having prevailed on the national crime commission, had to go along on this more modest proposal. For the first time all the federal groups that knew anything about the underworld came together in the same room and exchanged information.

Kennedy gave the job of coordination not to the FBI but to the Department's Organized Crime Section. There researchers compiled a roster of leading racketeers, circulated the list among the investigative agencies and used the result as the basis for prosecutions. At the same time, Kennedy quadrupled the Organized Crime Section in size and encouraged its lawyers to conduct investigations and try cases. Recalling his problems on the Rackets Committee with Hundley, Kennedy replaced him by Edwyn Silberling, who had worked for District Attorney Frank Hogan in New York. In time he had problems with Silberling too. After two years he brought back Hundley, who had stayed on as one of his special assistants and had handled special assignments with dash and discretion.

Kennedy was resolved to force the FBI into action against the mob. If lack of jurisdiction was Hoover's alibi, then Kennedy would make FBI jurisdiction explicit; at least this would weaken the alibi. In April 1961 he submitted a legislative program. In September Congress enacted five of his eight proposals, the most important of which covered interstate travel in support of racketeering, interstate transportation of gambling equipment and interstate transmission of gambling information. Even with this enabling legislation, however, it remained a battle to get Hoover to acknowledge the existence of organized crime.

Then in 1962 Joseph Valachi, a member of the Genovese gang now in the Atlanta penitentiary, killed another prisoner whom he mistook for a fellow mobster commissioned to kill him. Faced with the death penalty, Valachi agreed to talk in exchange for life imprisonment. He knew a great deal, apparently had total recall and poured out voluble recollections. His account of Italian criminal "families" gave the disparate intelligence accumulating in Justice a unifying theory. "What he did," Hundley said, "is beyond measure. . . . In the past we've heard that so-and-so was a syndicate man, and that was about all. Frankly, I always thought a lot of it was hogwash. But Valachi named names. He revealed what the structure was and how it operates."[25]

Among the names Valachi provided was a new one for the phenomenon that Harry Anslinger had so long called the Mafia — La Cosa Nostra ("our thing"). Since Hoover had always scoffed at the idea of the Mafia, the new phrase allowed him to discover organized crime without appearing to admit that the detested Anslinger had been right. In early 1963 an article signed by Hoover and intended for the *Reader's Digest* reached Edwin Guthman's desk for clearance. Casually imbedded in the middle was a quick reference to the Cosa Nostra. The rest of the piece was routine. Hoover "just wanted," Guthman said later, "to get those two paragraphs printed so that he could say, 'The FBI knew all along.'"[26] Guthman demanded their deletion. Hoover withdrew the article.

This hardly seemed the way to make the most of Valachi's testimony. "We talked about it," recalled Robert Kennedy, "and said that if it was going to come out, it should come out in a more organized fashion."[27] This led to a piece by Peter Maas in the *Saturday Evening Post*. The McClellan committee then wanted Valachi to testify. Kennedy, Hundley said, was "really very reluctant"; Hoover even more so. Hundley warned that, if Justice did not cooperate, the committee would subpoena Valachi. After intricate negotiation, the FBI seemed to agree that Valachi's appearance would not compromise other underworld sources. "But we lost control," said Hundley. The committee, without consulting the Department, put the hearings on television. Valachi was unimpressive on the stand. Italian Americans raised a fuss, claiming a libel on their people.[28]

Hoover lost no time in expressing public displeasure. Valachi, he wrote, had only "corroborated and embellished the facts developed by the FBI as early as 1961." His public appearances, Hoover added, "magnify the enormous task which lies ahead."[29] All this implied that Kennedy was making his job harder. When newspapers thus construed Hoover, "he refused," said Kennedy, "to deny the story that he was upset about Valachi, which irritated me."[30] After all, the FBI had presumably agreed to this use of Valachi. "We didn't understand," Guthman observed later of this episode, "that they never really said what they were thinking."[31]

Hoover was engaged in a rearguard action. He had lost the battle. The importance of Valachi, Robert Kennedy said later, was that "for the first time the FBI changed their whole concept of crime in the United States."[32]

IV

Subversion was out. Organized crime was in. Hoover grudgingly went along. But he exacted his price. He got more money as well as more explicit and possibly wider jurisdiction, and he preserved the autonomy and secrecy of FBI operations.

Wiretapping remained the single FBI investigative technique requiring the specific approval, case by case, of the Attorney General. Kennedy, who had watched Jimmy Hoffa wilt before Frank Hogan's wiretaps in 1957, believed that tapping had a significant if restricted role to play in law enforcement. "My feeling," he said shortly after assuming office,

> is that the use of legal wiretaps should be limited to major crimes such as treason, kidnaping, and murder. In each instance it should only be done with the authority of a Federal judge. . . . At the same time, there is a great deal of indiscriminate wiretapping that now goes unpun-

ished. Therefore, it is essential that the penalty for indiscriminate wiretapping be greatly stiffened. . . . Legalized wiretapping is a two-edged
sword that requires the most scrupulous use. For that reason, I would
not be in favor of its use under any circumstances — even with the
court's permission — except in certain capital offenses.[33]

On a cold day in March 1961, he convened his advisers on the
stone terrace at Hickory Hill to discuss the wiretap problem. Byron
White and Herbert Miller were strongly for legislation. Archibald
Cox and Burke Marshall thought a reasonable statute could be
drafted. Ramsey Clark, Joseph Dolan and William Orrick demurred,
Orrick most emphatically of all. Kennedy said to Orrick, "What if
your child was kidnapped?" Orrick replied, "Well, Bob, hard cases
make bad law." Kennedy, who could not imagine letting legal niceties
halt the search for a kidnapped child, grew heated. "I wanted to get
off that," Orrick said later. "He just stayed with it. He said, 'Well,
I just don't understand your attitude.' I got a little irritated. I said,
'You asked us to come up with our own opinions and I did.'"[34]

Kennedy decided to go ahead. His goal was a comprehensive
system by which all wiretapping, federal, state or local, would be
made a federal crime, except when specifically authorized by a federal judge. In effect, he wished to move toward the British system,
where tapping was forbidden without a warrant, though his proposal
that the warrant be issued not by the executive (the Home Secretary)
but by the judiciary improved on the British model. The FBI, however, demanded that the bill skip court orders for national security
cases, a position he reluctantly accepted.[35] In the end, instead of
seeking new legislation that year, Kennedy carelessly (and without
White House clearance) decided to support a bill already introduced
by Senator Kenneth Keating of New York.

The Keating bill, while purporting to limit federal wiretaps, would
have permitted state and local cops to tap wires in any way authorized
by state law. It was a poor bill. Civil libertarians attacked it, and
Kennedy soon abandoned it. In 1962 Katzenbach drafted a bill[36]
designed to clear up what Kennedy described as the "chaotic" situation. The new bill proposed to outlaw warrantless wiretapping,
whether conducted federally or locally, publicly or privately, except
in cases involving serious threats to national security. It defined cases
where taps might be permitted under judicial supervision, limited
the officials who could apply for court orders, required that each
application state grounds in detail and provided for automatic termination of all judicial warrants.[37]

Some observers considered the 1962 bill a responsible effort to
strike a balance between the competing claims of privacy and law

enforcement. Francis Biddle, who had clerked for Holmes and defended civil liberties for a lifetime, testified for it. The bill "struck me at the time," said Alexander Bickel, "as by far the best solution to the wiretapping problem that had yet been put forward."[38] But the bill also roused a storm of criticism, led by the American Civil Liberties Union and by Kennedy's friend from Rackets Committee days Joseph Rauh.

The critics believed that a wiretap system was inherently uncontrollable and that the dangers to personal freedom far outweighed the gains to law enforcement. They denied the possibility of an American compromise along British lines. They found the categories of tappable crimes dangerously vague. Recalling the familiar phenomenon of "judge shopping," they questioned whether judicial warrants could provide effective control. In any case, where a search warrant authorized only a single search for specified objectives, a wiretap warrant was a dragnet. Nor did they believe that wiretap evidence made all that much difference. They quoted Justice Frankfurter: "My deepest feeling against giving legal sanction to such 'dirty business' . . . is that it makes for lazy and not alert law enforcement."[39] Where Kennedy sought to put an end to warrantless wiretapping, his critics demanded the flat prohibition of all forms of electronic surveillance.

The bill failed in Congress in 1962 and, despite expectation in 1963 that the Valachi testimony would help, it failed thereafter.[40] The chaotic situation continued. State and city cops went their own way. So did private tappers. Federal wiretaps remained under the control, not of the federal courts, but of the Attorney General.

Robert Kennedy authorized 140 wiretaps in 1961, 198 in 1962 and 244 in 1963 (as against 519 authorized in 1945, 214 in 1955 and 115 in 1960).[41] Nearly all were 'national security' (as distinct from criminal) cases. Nearly all were directed against aliens. In three and a half years as Attorney General, Kennedy authorized taps on about sixteen American citizens involved in eight separate cases. (This figure omits an undisclosed number tapped in 1964 at the request of the Warren Commission.) Five of these cases had national security implications: defense leaks to newspapermen (Lloyd Norman of *Newsweek* and Hanson Baldwin and his secretary of the *New York Times*);* attempts by the Dominican government to influence sugar

---

* Baldwin had written a story about photographic surveillance of Soviet missiles by United States satellites. As a result of the publication of the story, the Soviet Union took steps to conceal their missiles, thereby causing great difficulties for American intelligence (see RFK, in interview by John Bartlow Martin, April 13, 1964, II, 9–10, JFK Oral History Program).

quota legislation (the clerk of the House Agriculture Committee; three officials of the Agriculture Department; an American national registered as a Dominican lobbyist; and, most inexcusably, the switchboard of a law office where one of the lawyers was working for the Dominicans); a high CIA official accused (wrongly) of having given classified information to a Yugoslav diplomat; and Dr. Martin Luther King, Jr., and two associates, tapped in order to test Hoover's assertion that King was involved with Communist agents. The three remaining taps on American citizens were urgently requested by the FBI without evident international pretext: on a former FBI agent whom Hoover claimed had revealed "confidential" Bureau information; on an Alabama Klan leader; and on Malcolm X, the black nationalist. In some cases — Baldwin and his secretary, Norman, the erring ex-FBI agent — the Bureau tapped first and got Kennedy's written approval later.*

In the main, Kennedy consented to Hoover's wiretap requests. Nor did he do much to tighten the controls he had inherited from his predecessors. Always the improviser rather than the planner, he handled wiretap authorizations with surprising casualness and inattention. "You never knew," Katzenbach said later, "whether [a tap] was disconnected or still going on, and you never knew what developed out of it, whether it was fruitful or not. So at no given time did any Attorney General know how many wiretaps there were and who they were on."[42] Kennedy could have done something administratively about that, as his successors, Katzenbach and Clark, tried to do thereafter.

But Kennedy did not promote wiretapping. For all his strong feelings about organized crime and Jimmy Hoffa, he authorized no wiretaps against racketeers or Teamsters. He continued to deplore the power given him as Attorney General to decide whose phone should be tapped; "there's no check on what I do whatsoever." In 1964 he condemned the continuing failure of Congress to transfer that power to the federal courts, where "you would . . . have to keep continuously making reports as to what you were doing."[43] He thought wiretapping a valuable tool in a very few very particular circumstances. He wished it, if used at all, to be hedged around by statutory safeguards. As Attorney General, he opposed its total abolition.

---

* Senate Select Committee to Study Governmental Operations with respect to Intelligence Activities, *Final Report*, bk. II, *Intelligence Activities and the Rights of Americans*, 94 Cong., 2 Sess. (1976), 63–65. This tabulation does not include wiretaps on foreign nationals. Such wiretaps sometimes caught Americans, including members of Congress — e.g., Senator Thomas J. Dodd, when the FBI tapped Michael Struelens, a Belgian who was operating in the United States as the agent for Moise Tshombe's secessionist regime in Katanga.

V

Kennedy evidently supposed that the power to authorize wiretaps was the power to contain the FBI's electronic surveillance. He could not have been more mistaken. The wild card was bugging. The war against organized crime had put Hoover on his mettle. The director was under exasperating pressure from the Attorney General. His reputation was on the line. He well knew how much ground the FBI had to make up. The quickest way was to do what he had long done in internal security: to bug the enemy and thereby draw in the information that would lay the groundwork for penetration.* He had already started bugging the mob in a small way after Apalachin. In 1961 he began to bug it in a big way. He did this, as he had always done, on his own. At no time did he seek bugging authorization from the Attorney General. In later years, however, when the public began to catch on to the extent of FBI bugging, he claimed vigorously that Kennedy had known what the FBI was doing — a claim that Kennedy with comparable vigor denied.

In December 1966 Hoover wrote to Congressman H. R. Gross of Iowa, "Your impression that the FBI engaged in the usage of wiretaps and microphones only upon the authority of the Attorney General of the United States is absolutely correct."[44] Hoover's suggestion that taps and bugs were on an equal basis was false. Courtney Evans, who had served as FBI liaison with Kennedy in Rackets Committee days, continued in that role in Justice. "It was my understanding at the time," Evans said in 1975, "that any tap required the written authorization of the Attorney General, but that no such authorization was required for the use of microphone surveillance."[45] Accordingly, when Evans briefed Kennedy in January 1961 on wiretap procedures, he said nothing about bugging.[46] "The Attorney General's personal approval was not sought [for bugging]," said Katzenbach, who succeeded White as deputy attorney general in April 1962 and Kennedy as Attorney General in September 1964, "nor was he even directly advised of any microphone surveillance despite their increased use through the late 1950's and early 1960's."[47]

Hoover nonetheless argued forcibly that Kennedy *knew*. When Harold Cooley, chairman of the House Agriculture Committee, met secretly with Dominican agents in a New York hotel room in 1961,

---

* See Victor Navasky, *Kennedy Justice* (New York, 1971), 67. William Hundley said in 1975: "I think somewhere internally a decision was made that organized crime was heating up and they didn't have any informants so they'd better do something to catch up" (Ovid Demaris, *The Director* [New York, 1975], 144).

Hoover, according to his own memorandum of a talk with Kennedy, told him "we are trying to cover" the meeting. Later he gave Kennedy a summary of what was said. Kennedy no doubt supposed that the meeting had been covered by an informant. Actually the FBI had bugged the room. An FBI review of the case in 1966 concluded, "Our files contain no clear indication that the Attorney General was specifically advised that a microphone surveillance was being utilized."[48]

Hoover claimed to Gross that Kennedy had "in different metropolitan areas . . . listened to the results of microphone surveillances."[49] The director had particularly in mind a session at the Chicago FBI office where FBI agents, obviously hoping to please the Attorney General, played a tape on which a Chicago hood complained to a crooked police captain that, after Kennedy had taken office, they couldn't fix cops any more.[50] In 1966 Hoover demanded that the Chicago agents prepare affidavits saying that in the light of the discussion Kennedy would have had to conclude that the Bureau was engaged in bugging. Some agents complied with Hoover's demand; others refused.[51] William Hundley, who was present, denied that Kennedy had any reason to believe it was an FBI tape.[52] Edwin Guthman, who was present, said, "The clear impression we received was that the FBI had obtained the recordings from local law enforcement sources."[53]

Hoover also produced two documents. One was a letter from Herbert Miller to Senator Ervin in 1961 reporting that the FBI had sixty-seven bugs in operation. Miller assumed, he said later, that the bugs had been legally installed — as, with the owner's consent, in a restaurant or apartment, or on an informant's person. "I had no knowledge then, and am firmly convinced that no one in the non-FBI section of the Department of Justice had knowledge that the FBI was in fact trespassing to place such devices. Had this been known to the Department, I am convinced that immediate steps would have been taken to stop the practice."[54]

Hoover's second document was an FBI request, initialed by Kennedy himself, for authority to lease a special telephone line in order to monitor microphone surveillance in New York City. Kennedy, observing that the request had to do with telephones, evidently thought that he was approving the leasing of a line to be used for wiretaps, and that the Bureau would come back to him for authorization in each specific case.[55] The FBI request was in any event disingenuous, since the leasing never took place. Navasky's conclusion is persuasive: "The FBI's purpose . . . was apparently not to lease a telephone line but to get the Attorney General's participation on the record. Whether or not Robert Kennedy realized the implications of what he was signing is something we will never know."[56]

Courtney Evans may have, with the best of motives, contributed to misunderstanding. He had been in the FBI for a quarter century and had great loyalty to the organization. He also had great respect for Kennedy. He saw his task as that of keeping the Attorney General and the director together. It was an impossible job. As William Hundley said, "Evans walked a very tight rope."[57] Still, "it worked very well for one basic reason," observed Katzenbach, "and that is Courtney would explain something to Bobby one way and explain something to Hoover another way. And I don't think anybody could have done the job in any other way."[58]

In July 1961 the Attorney General held a meeting on organized crime. According to Edwyn Silberling in 1966, Kennedy told Evans that he wanted the FBI to use more "technical equipment" against the syndicates.[59] G. Robert Blakey, a lawyer in the Organized Crime Section, recalled "quite vividly" in 1966 "when you turned to Courtney Evans and told him more or less, 'Let's get some of this security equipment on the street. You tell me all the things it can do.' . . . You were telling the Bureau to stop making excuses for their failure to develop adequate information and to start using electronic equipment. . . . It now looks to an awful lot of good guys like myself that you are cutting out when the going is getting rough. . . . Your present posture indicates you are a liar or a fool. I know you to be neither."[60] After the meeting, Courtney Evans reported to Hoover:

> It was pointed out to the Attorney General that we had taken action with regard to microphone surveillance in these cases. . . . The Attorney General stated he recognized the reasons why telephone taps should be restricted to national-defense type cases and he was pleased we had been using microphone surveillances, where these objections do not apply, wherever possible in organized crime matters.[61]

On the other hand, Evans told the Church committee in 1975 that he had been "purposely vague" in talking about electronic surveillance because Hoover had instructed him not to discuss technical devices outside the Bureau — an admonition, Evans said, that applied even to the Attorney General.[62] And Evans had written Kennedy in February 1966, "I did not discuss the use of these devices with you . . . nor do I know of any written material that was sent to you at any time concerning procedure, or concerning the use, specific location, or other details as to installation of such devices."[63] "Judas Iscariot," Hoover said of Evans when he learned of this letter. "Judas Iscariot." ("Judas Iscariot?" exclaimed one agent. "Does the Director think he's Jesus, for Christ's sake?")[64]

VI

Who outside the Bureau knew about the bugging? "I knew for years that the FBI was making widespread use of electronic bugs in organized crime investigations," Hundley said later. "But the top brass of the Bureau would flat-out lie to me, denying it, whenever I officially asked about it. . . . I figured that this was obviously so delicate that it must be something that was just between the AG and the Director."[65] So he never discussed FBI bugging with Kennedy, any more than he had with Rogers before him; nor did he discuss it with Miller, his immediate superior, who in any case supposed it to be nontrespassory.[66]

Byron White did not know the FBI was bugging. In May 1961 Hoover had sent him one of his memoranda-for-the-record entitled "Technical and Microphone Surveillance." There is no evidence that White ever saw it, nor did Hoover release it in his 1966 barrage against Kennedy.* Katzenbach did not know, save for a single incident. Once an FBI report on Las Vegas casinos contained internal evidence of FBI bugging. Katzenbach brought the report to Kennedy. "We both thought," Katzenbach said later, "this is something that had just happened in Las Vegas and no place else. . . . In hindsight that was stupid on my part and on his. He said to me, 'Make sure this is stopped.' And I said to Courtney [Evans], 'Am I absolutely assured that this is stopped?' and Courtney said, 'Yes.' . . . It was stopped in Las Vegas. That was exactly what we'd asked for and that's what we got."[67]

"You look back," said Henry Ruth of the Organized Crime Section, "and you feel stupid. We'd get furious because we'd propose an investigation and they'd say, 'There's absolutely nothing there!' *Now* I know what this meant — either they knew from bugging that there was nothing there, or they knew that whatever was there was tainted — it couldn't be used as evidence in court because it was the result of a bug."[68] "Until 1965," said Burke Marshall, "I did not know that the bureau engaged in unlawful electronic bugging without authorization or review by anyone outside the bureau."[69] "I believe," John Douglas wrote Kennedy in 1966, "that as Attorney General you did not know of any installation and use by the FBI of electronic listening devices in connection with investigations of organized crime."[70]

* Navasky, *Kennedy Justice*, 94–95, 448–449. Navasky surmises that Hoover did not use this memorandum in 1966 because it misrepresented the bugging authority Hoover claimed to have from the Brownell memorandum of 1954 and perhaps because White by this time was on the Supreme Court.

If the ignorance seemed singular, there were constraints against undue curiosity. "One thing you just don't do in the jungle of law enforcement," said Ramsey Clark, "particularly when you're in a staff meeting and here are twenty-five guys, is say 'Well, who was your informer?' or, 'Where did you get that information?'"[71] Dolan felt in addition that the electronic world was a mystery to the Attorney General. When Kennedy used phrases like technical equipment, "He didn't really know what he was talking about. . . . He had a very bad, poor, mechanical, technical, scientific background. . . . He couldn't tell a spark plug from a generator. . . . I was always surprised at his lack of knowledge of techniques."[72]

Certainly when the facts burst on him in 1966 Kennedy struck everyone as genuinely astonished. Hundley went to see him. "Unless he was the greatest actor in history," he said later, "he really got very upset. I liked Bobby Kennedy, but we weren't that close. What the hell, he fired me once. And he said, 'You knew about it? You knew and you didn't tell me?' . . . I came away convinced, and I'm convinced to this day, that he didn't know. . . . One thing he never was — he wasn't a liar." Hundley, thinking to himself, "I'd rather go down in history as a guy who might have moved around a little than as an idiot," told Kennedy that, even if he hadn't known, it would be better politics to say he had. Kennedy replied, "How can I? I didn't know."[73]

At the same time Hoover might have truly supposed, as a consequence of Courtney Evans's mediation, that Kennedy did know. "I think in part," Katzenbach said later, "the whole wiretap, bugging issue was really that. I really honestly think that Hoover thought that this had all been explained to Bobby by Courtney."[74] This, without malice toward Evans's "unfortunate human trait of trying to please everybody," appeared to be Kennedy's own conclusion.*

Robert Kennedy, in my judgment, had not understood that the FBI was bugging the mob, certainly not with trespass. Nor had he authorized wiretaps against the mob. Where then did he suppose

---

* Frank Mankiewicz, in recorded interview by L. J. Hackman, September 9, 1969, 35, JFK Oral History Program. On the other hand, William Hundley recalled in 1971: "I was talking to Katzenbach one day, and Katzenbach told me that Hoover himself had told him, Katzenbach, that Bobby Kennedy did not know about the bugs. So rather cynically I said to Katzenbach, 'I hope you had a witness.' And he said, 'I had a witness. I had the best witness in the world.' He wouldn't tell me any more than that. Well, I found out later on from talking to Bobby. . . . This was during the Indiana campaign. I told him the story and he just laughed. He said, 'Well, the only mistake that Katzenbach made is when he told you that he had the best witness in the world. He really had the worst witness in the world.' I said, 'Who do you mean?' He said, 'The witness was President Johnson.'" (Hundley, in interview by J. A. Oesterle, February 22, 1971, 38, RFK Oral History Program.)

that, beyond bugging by local cops, the increasingly voluminous FBI intelligence on organized crime was coming from? The answer was penetration. "They started to get informants," he told Anthony Lewis in 1964. "I asked them to go into it like they went into the Communist Party."[75] Hoover sustained this impression at every opportunity. "Infiltration, the same technique which had proven so successful in the FBI's drive against fascism, communism, the Ku Klux Klan and other enemies of freedom," he said, "was brought to bear on the underworld."[76] "We have carried on a very intensive campaign against the Cosa Nostra. We have been able to penetrate it, and we know quite well what its activities are."[77]

The question remains: could Kennedy not have found out about the FBI's bugging? Here was a first-class administrator, an indefatigable question-asker, experienced in the war against crime, surrounded by clues. Granted the extenuating factors — the difficulties of dealing with Hoover, the Bureau's addiction to secrecy, the need to secure the Bureau's cooperation, the easy assumption that the Bureau was exploiting bugs laid down by local police forces, the equal ignorance of his deputies — still, why did it not occur to him to raise the bugging question with Miller or Silberling or Hundley — or Hoover?

In a later time he would readily agree that of course he should have done this. I imagine there were several reasons that he didn't. He was probably the busiest man in Washington, increasingly involved in affairs far outside the Department of Justice. And, though he was coming to understand the importance of the Bill of Rights, his sensitivity to civil liberties in these years was less than his concern about organized crime. So he did not pursue the question as he should have. Perhaps, like all his eminent predecessors, he did not want to know.

VII

Under Kennedy's pressure the national government took on organized crime as it had never done before. In New York, Robert Morgenthau, the federal attorney, successfully prosecuted one syndicate leader after another. The Patriarca gang in Rhode Island and the De Cavalcante gang in New Jersey were smashed. Convictions of racketeers by the Organized Crime Section and the Tax Division steadily increased — 96 in 1961, 101 in 1962, 373 in 1963.[78] So long as John Kennedy sat in the White House, giving his Attorney General absolute backing, the underworld knew that the heat was on.

So did Jimmy Hoffa. After the Rackets Committee reports, Eisenhower's Department of Justice had decided in the summer of 1960 to indict Hoffa for mail frauds in connection with Sun Valley, a Florida real estate venture in which he had a personal financial interest. At the last minute, while James Dowd, the Justice Department lawyer, was making the case for indictment in a Florida grand jury room, the Department changed its mind and summoned Dowd back to Washington. Hoffa was supporting Nixon in the presidential campaign. Some thought this lay behind the cancellation of the indictment. After the election, Attorney General Rogers went ahead with the case. Hoffa claimed he had been double-crossed. "I know for a fact," Allan Oakley Hunter, a former FBI man and California congressman, wrote Hoffa in an effort to divert his rage from Nixon, "that your side of the case was put before the Vice President and that he discussed the case with the Attorney General. . . . The Vice President has been sympathetic toward you and has felt that you were being subjected to undue harassment by certain parties. . . . Mr. Nixon having lost the election, I doubt that he has since been in a position to exercise any decisive degree of influence." [79]

Rogers, observed Paul Healy of the *New York Daily News*, has left the Kennedys "an inaugural present on their doorstep — the indictment — and is in effect saying: 'All right, let's see what you can do!' " [80] Robert Kennedy lost no time in responding to the challenge. He may well have put Herbert Miller in charge of the Criminal Division less because of Miller's criminal law experience, which was scant, than because of his experience as counsel for the anti-Hoffa members of the Teamster Board of Monitors. Kennedy then asked Walter Sheridan to set up a unit on labor rackets in the Organized Crime Section. It was an unconventional appointment. Sheridan was not even a lawyer. But he was a superb investigator, patient and careful; and he surrounded himself with crack lawyers, like James Neal and Nathan Lewin. J. Edgar Hoover took the Sheridan appointment as one more affront. He did not like competing investigative teams and he mistrusted agents who had left the Bureau. "Hoover hates my guts," Sheridan noted later. [81] But, though out of favor in the Washington headquarters, Sheridan, as an ex-agent, had many FBI friends and gained extensive cooperation in the field.

The Sheridan group became known popularly, though never in the Department, as the Get-Hoffa Squad. Hoffa quickly felt its impact. By October 1961 he was complaining to the press that twenty-nine grand juries had been convened to get him, that half the lawyers in the Department and half the agents in the Bureau were enlisted in the effort. "There were, in fact, at the time," Sheridan wrote later, "thirteen grand juries . . . sixteen attorneys . . . and perhaps thirty

FBI agents."[82] Still this was not unimpressive. A number of Teamster leaders, including Anthony Provenzano and Barney Baker, were in due course convicted and sent to prison. But the essential objective was Hoffa himself.

Rogers's Sun Valley indictment was dismissed for technical reasons: the grand jury had not been properly empaneled. Kennedy moved for reindictment. Then an investigation of the Teamsters' Central States Pension Fund in Chicago disclosed a larger case. Soon a new indictment was in preparation, charging Hoffa with conspiring to defraud the trustees of the pension fund. Part of the charge included a misappropriation of more than $100,000 to rescue his Sun Valley investment. Since the new indictment covered Sun Valley, Kennedy decided to drop the Florida case and concentrate on the Chicago prosecution.[83]

In the meantime, a grand jury in Nashville indicted Hoffa on the charge of taking illegal payments from an employer in violation of the Taft-Hartley Act. This, like Sun Valley, was a case Kennedy had opened up in the Rackets Committee. In 1948 Hoffa had intervened to settle a damaging strike against a Detroit trucking firm in the company's favor. Shortly thereafter the Test Fleet Corporation, a truck-leasing company, was incorporated in Tennessee. One of the incorporators, signing only her maiden name, was Mrs. James Hoffa. The Detroit trucking firm became Test Fleet's main customer. Through the years this convenient arrangement had provided Hoffa with hundreds of thousands of dollars.

The Nashville trial began in October 1962. Soon thereafter E. G. Partin, a secretly dissident Teamster still in Hoffa's entourage, made contact with Sheridan. Sheridan was dubious. Partin had a criminal record. His stories were lurid. He reported that Hoffa had once asked him whether he knew anything about plastic bombs. "I've got to do something about that son of a bitch Bobby Kennedy," Partin recalled Hoffa as saying. "He's got to go." Kennedy had guts, Hoffa continued; he drove around by himself in a convertible, swam alone in his Hickory Hill pool and had no guards in the house. Sheridan questioned the story. Partin said he would take a lie detector test. Kennedy said wryly to Sheridan, "What do we do if that fellow passes the test?" For what it was worth, Partin did.[84]

Partin now functioned as a government informant in the Hoffa camp. The trial proceeded. The result was a hung jury. Hoffa was evidently still shooting fish in a barrel. The judge declared a mistrial and called for a grand jury investigation of jury tampering. Sheridan and James Neal, the Nashville prosecutor, were sure that, with Partin's testimony, they had quite enough evidence to indict Hoffa for conspiracy to fix the jury. In April 1963 Kennedy convened his top staff to consider what to do. Douglas, Marshall, Katzenbach, all felt they had a strong case. Cox said he would have no difficulty arguing the case before the Supreme Court. Two advisers were doubtful. Hundley felt that it would come down to Hoffa's word against Partin. Ramsey Clark feared a Sixth Amendment problem — that is, that the claim would be made that Partin, an informer who had infiltrated the defense, had compromised Hoffa's constitutional right to counsel. Kennedy asked what the alternative was. "Should we have told Partin we didn't want to hear about jury tampering? And once we had heard about it, should we have done nothing about it?" It was decided to go ahead.[85]

In May 1963 Hoffa and five others were indicted for jury tampering. William Loeb of the *Manchester Union-Leader,* who had received a $500,000 loan from the Teamsters' pension fund five weeks before, wrote a flaming editorial denouncing the "persecution" of Hoffa. "The fight of James Hoffa against the Kennedys," Loeb said, "is becoming the fight of all Americans who want to stay free men and women."[86] Hoffa, as part of a series of delaying tactics, asked for a change of venue. The trial was scheduled for Chattanooga in January 1964. The Chicago trial was due later in the spring.

The pursuit of Hoffa was an aspect of the war against organized crime. The relations between the Teamsters and the syndicates continued to grow. An FBI electronic microphone, planted from 1961 to 1964 in the office of Anthony Giacalone, a Detroit hood, revealed Hoffa's deep if wary involvement with the local mob.[87] For national purposes a meeting place was the Rancho La Costa Country Club near San Clemente, California, built with $27 million in loans from the Teamsters' pension fund; its proprietor, Morris B. Dalitz, had emerged from the Detroit underworld to become a Las Vegas and Havana gambling figure.[88] Here the Teamsters and the mob golfed and drank together. Here they no doubt reflected that, as long as John Kennedy was President, Robert Kennedy would be unassailable. (And here Richard Nixon in October 1975 made one of his first sallies from his San Clemente fastness to play golf in the company of Frank Fitzsimmons, Hoffa's successor as head of the Teamsters, Anthony Provenzano, a leading suspect in Hoffa's murder, and the ex-convict Allen Dorfman.)[89]

VIII

He was tired, Robert Kennedy had said, of chasing people. Yet he was spending a conspicuous part of his life as Attorney General doing exactly that. All this engraved more deeply the public face he was trying to escape — the face of the implacable prosecutor.

His critics seized on the pursuit of Hoffa as a pretext for new criticism. Kennedy was represented as "single-minded" in that pursuit, as if chasing Hoffa consumed all his days and nights. In fact, Kennedy was one of the least single-minded men in Washington. No one except the President divided his mind among a greater variety of concerns and enterprises. When he thought about Hoffa, he thought hard about him; but thinking about Hoffa occupied a very small part of his time. I saw a great deal of him in these years and can hardly recall his mentioning Hoffa.

Thinking about Kennedy unquestionably occupied a greater part of Hoffa's time. "Between 1957 and 1964," report Ralph and Estelle James, who talked often to Hoffa in this period, "Hoffa's life was dominated by his fear of 'that little monster.' . . . Under Kennedy's direction, Hoffa believes, FBI agents followed him wherever he went, tapped his phone, opened his mail, and beamed electronic listening devices on him from half a mile away, aided by invisible powder they had rubbed onto his clothes." [90] Hoffa's paranoia gave rise to the myth that Kennedy had ordered the tapping and bugging of Hoffa. In 1969, under the Nixon administration — which was sympathetic to Hoffa, later granted him clemency and would have been delighted to dig up something on the Kennedys — the Justice Department was ordered by the Supreme Court to turn over all electronic overhearings related to the Hoffa cases, whether or not relevant to the indictments. After a thorough search, the Department found, as Sheridan summarized it, "that in the entire investigation in connection with the Hoffa cases there had not been one instance of wiretapping or bugging of Hoffa." [91]

The only bugs in the jury-tampering case were both legal. Neither was directed at Hoffa. One was authorized by the judge in order to record a Hoffa lawyer conspiring to fix a juror. The other was installed with Partin's permission in his car for a single ride (the tape turned out to be unintelligible). [92] In 1969 Sheridan learned that the FBI without telling anyone had bugged Moe Dalitz's office and recorded Dalitz's half of a phone conversation with Hoffa; had also monitored several Hoffa conversations in Teamster radio cars in Detroit. "I had not known about these overhearings," Sheridan said, "until long after I left the Department." [93] These tapes contributed nothing to the Hoffa prosecutions. Whether the FBI also opened Hoffa's mail, burgled his office or rubbed invisible powder on his clothes no one outside the FBI can say.

Still, though no illegal methods were used in the prosecutions of Hoffa (a fact twice affirmed by the Warren Court, the most sensitive of all Supreme Courts to prosecutorial improprieties), Kennedy's pursuit of Hoffa raised troubling questions — questions well defined in a famous statement by Robert H. Jackson while still Attorney General:

> The most dangerous power of the prosecutor: that he will pick people that he thinks he should get, rather than pick cases that need to be prosecuted. With the law books filled with a great assortment of crimes, a prosecutor stands a fair chance of pinning at least a technical violation of some act on the part of almost anyone. . . . It is in this realm — in which the prosecutor picks some person he dislikes or desires to be embarrassed, or selects some group of unpopular persons and then looks for an offense, that the greatest danger of abuse of prosecuting power lies.[94]

This was the gravamen of the case against Robert Kennedy: that in singling out Hoffa, or Sam Giancana, or Joseph Aiuppe, the Chicago hood convicted under the Migratory Bird Act, or Louis Gallo, Crazy Joe's brother, convicted for false statements on a Veterans' Administration loan application for a home mortgage, he was abusing the prosecuting power. He was deciding that people were guilty and then looking for something they could be found guilty of. He was convicting them not for their real crimes but for slips that anyone might have made. Beginning with the criminal rather than with the crime led to selective justice. "Never in history," said Navasky about the Test Fleet case, "had the government devoted so much money, manpower and top level brainpower to a misdemeanor case."[95] Would it have done so, would it have bothered to bring the case at all, had the defendant not been Jimmy Hoffa? Hoffa was no doubt a menace to society, and the Get-Hoffa Squad might have a righteous purpose. But the precedents created might beget in other hands Get-NAM Squads, Get-ADA Squads, Get-Enemies-List Squads in any form.[96]

The question of discretionary justice bore with special weight on tax prosecutions. Kennedy and Mortimer Caplin had brought the Internal Revenue Service roaring into the battle against organized crime. After meeting with Kennedy, Caplin instructed the IRS investigators: "The Attorney General has requested the Service to give top priority to the investigation of the tax affairs of major racketeers. Such an undertaking . . . has my wholehearted endorsement." He added that racketeers would be "subjected to the 'saturation type' investigation. . . . Full use will be made of available electronic equipment and other technical aids." The IRS was soon bugging nearly as much as the FBI. In reviewing tax returns, Caplin gave priority to racketeers. In the end, 60 percent of all organized crime cases in these years were tax cases.[97]

This practice raised questions. Some lawyers believed that the purpose of the tax laws was to collect revenue, not criminals. Crooks should not be singled out. Their tax returns should be subject to the same random review as those of any other citizen. Kennedy thought the argument "specious."[98] So long as the same tax standards were applied to crooks as to other citizens, then it did not matter if their cases were moved to the top of the pile. Louis Oberdorfer, the highly scrupulous chief of the Tax Division, accepted a working compromise according to which the Criminal Division could make recommendations to the IRS for investigations (Oberdorfer was not then aware of the electronic methods used by the IRS), and the Tax Division would apply the same criteria to racketeers as to all other cases submitted by the IRS for possible action. "The whole time Kennedy was here," Oberdorfer said later, "never once did he try to push me to indict somebody or to push me not to indict somebody."[99]

The traditional criminal investigation, as G. Robert Blakey pointed out, moved "from the known crime to the unknown criminal." Someone committed a murder, a rape, a robbery, and the job of law enforcement was to find out who did it. The new world of the syndicates was characterized by the opposite problem — "known criminals but unknown crimes."[100] So you had to begin with the criminal. In addition, big shots rarely committed gross crimes themselves. That was what the 'enforcers' and 'soldiers,' like Joe Valachi, were for. Because top racketeers were remote from overt acts, they were vulnerable only on the margin, as in their tax returns. Nor were they forlorn individuals, outmatched by the massed power of the federal government. As Nathan Lewin said of Hoffa, "They had limitless funds and used them, and didn't play by the rules. . . . During the Nashville case Hoffa had six high-priced lawyers sitting in a hotel room to consult with the four other lawyers they had in the courtroom. That's more than the government had working on the trial. He probably spent as much money defending himself as the government spent bringing him to justice."[101]

Robert Kennedy was not the first government official to begin with the criminal rather than the crime. In March 1929 a group of Chicago citizens waited on President Herbert Hoover. They told him that their city was ruled by Al Capone. "At once," Hoover wrote later, "I directed that all the Federal agencies concentrate upon Mr. Capone. . . . The Attorney General set up a special Deputy Attorney General and equipped him with the best men from every bureau of investigation in the government."[102] Later Hoover explained to Elmer Irey of the Treasury Department: "I always wanted to see something done about Capone"; after hearing the Chicago delegation, "I gave the order to put Capone in jail." Hoover and a group of cronies used to begin each day by tossing medicine balls at each other in the White House. "When the exercising starts," Secretary of the Treasury Andrew Mellon informed Irey, "Mr. Hoover says 'Have you got that fellow Capone, yet?' And when exercise is done and everybody is leaving, the last thing Mr. Hoover always says is 'Remember, now; I want that man Capone in jail.'"[103] Hoover's Get-Capone Squad did its job. Irey's T-men put Capone away for income tax evasion. Hoover was not condemned as Kennedy was in later years. Indeed, the conviction of Capone was regarded as one of the few accomplishments of his administration.

It was not as if Kennedy had decided to go after the mob and the Teamsters out of abstract prejudice. The Rackets Committee had dug up a mass of concrete information on lawbreaking. As Attorney General, he felt an obligation to pursue that information. The targets may have received disproportionate attention in the Department of Justice. But selectivity is inescapable in any situation of limited resources, and the suspects received due process in the courts. Was there still unfairness in concentrating on one set of lawbreakers rather than another? If there were, remedy lay in the larger constellation of forces outside the courtroom. The methods that sent Capone and Hoffa to prison achieved useful social results that might not have been achieved otherwise. When those methods are abused by unscrupulous prosecutors, other forces — the courts, the legislature, the press, public opinion — must act, and in the past have acted, to restore the balance. The criticism of his pursuit of Hoffa baffled and dismayed Robert Kennedy; but it was part of the process that kept that pursuit from creating precedents by which his successors could use less defensible methods against less deserving targets.

14

The Pursuit of Justice:
Civil Rights

"I WON'T SAY I stayed awake nights worrying about civil rights before I became Attorney General," Robert Kennedy remarked in these years. He had always acted with impetuous decency when racial discrimination forced itself on his attention. But this was response to individual wrong, not to national shame. "My fundamental belief," he said soon after taking office, is "that all people are created equal. Logically, it follows that integration should take place today everywhere." But, he added, "other people have grown up with totally different backgrounds and mores, which we can't change overnight."[1] In the longer run, it would all work out; when his grandfather arrived in Boston, "the Irish were not wanted there. Now an Irish Catholic is President of the United States. There is no question about it. In the next forty years, a Negro can achieve the same position."[2] He saw civil rights in 1961 as an issue in the middle distance, morally invincible but filled for the moment with operational difficulty. He did not see racial injustice as *the* urgent American problem, as *the* contradiction, now at last intolerable, between the theory and the practice of the republic.

I

Nor did many white Americans. It is hard to recapture the myopia of another age. After the collapse of Reconstruction, the ex-slaves and their descendants had been consigned to legal and social inferiority, forgotten by the courts, by Congress, by Presidents. In the 1930s Franklin Roosevelt did more for black Americans than any President since Lincoln, thereby switching black political allegiance from the Republican to the Democratic party. But he did it initially by general aid to the dispossessed and by government appointments, White House hospitality and other symbolic gestures for blacks. It was not until A. Philip Randolph, the leading black trade unionist, threatened a march on Washington that Roosevelt sponsored a direct if limited attack, through the Fair Employment Practices Committee, on the structure of segregation.

After the war, Congress rejected Truman's plea to continue the FEPC. With the legislative branch hostile and, after 1952, the executive indifferent, the civil rights movement turned to the courts. The National Association for the Advancement of Colored People challenged the constitutionality of segregation in a series of suits, many argued by Thurgood Marshall, a leading black lawyer. Under Chief Justice Earl Warren's leadership, the Supreme Court began to fill the void in the constitutional process.

The fifties were premonitory: in 1954 the Supreme Court decision ordering integration in public schools; in 1955–56 the bus boycott in Montgomery, the twenty-seven-year-old Baptist minister Martin Luther King, Jr., preaching the gospel of nonviolent resistance; in 1957 Governor Faubus of Arkansas defying the Supreme Court and Eisenhower's reluctant dispatch of federal troops to Little Rock to uphold the Constitution; the same year, the enactment of the first civil rights legislation since Reconstruction. The 1957 Civil Rights Act, strengthened by a second act in 1960, was intended to secure southern blacks the right to vote. But the laws were weak, and Eisenhower's Department of Justice applied them with all deliberate lethargy; nor, save for a few determined liberals in the Senate, did white Democratic politicians accord racial justice the highest priority. In the middle fifties Adlai Stevenson, the liberal hero, had actually contemplated an appeal to black leaders for a year's "moratorium" on agitation. No white leader seemed to understand what was going on in the consciousness of black America: the ebbing away of patience; the mounting bitterness; the spreading demand, all other roads blocked, for direct action. In 1958 King founded the Southern Christian Leadership Conference. In February 1960 the sit-ins began. The pace of protest was beginning to accelerate.

The Kennedys were abstractly in favor of equal opportunity. But it took presidential politics to involve them with the movement; and then the prospect of responsibility to make them think intensively about the problem. On an August morning in 1960 John Kennedy picked up Harris Wofford, his campaign man on civil rights, in his red convertible and said, as they drove downtown, "Now, in five minutes, tick off the ten things that a President ought to do to clean up this goddamn civil rights mess." Kennedy needed the black vote. He was also, in Wofford's view, appalled by the goddamn mess. "He considered it irrational. He was not knowledgeable about it. It was alien to most of his experience. He was ready to learn."[3] During the campaign Kennedy made much of the statistics of racial discrimination and promised vigorous presidential leadership in the cause of equal rights. "There is a good deal," he said, "that can be done by the executive branch without legislation.... Many things can be done by a stroke of the Presidential pen" — an executive order ending discrimination in housing, for example.[4]

The narrowness of Kennedy's victory strengthened the argument that the executive route offered the best immediate prospect of progress. Wofford doubted that Congress would pass civil rights legislation in the near future, and he was sure that the legislation already passed had not been fully used. "Although it is heresy in the civil rights camp to say this," he wrote the President-elect during the interregnum, the need now was for "a minimum of civil rights legislation and a maximum of executive action."[5] Nor was this view all that heretical. The liberal Southern Regional Council reminded the administration that the Presidency was "the center of American energy." John Hannah, chairman of the Civil Rights Commission, the independent government agency established under the Civil Rights Act of 1957, gave Kennedy a series of recommendations in February, all relying on executive authority.[6] "If we go into a long fight in Congress," Kennedy himself soon told Martin Luther King, "it will bottleneck everything else" — including measures of vital importance to black Americans, like the increased minimum wage and federal aid to education — "and still get no [civil rights] bill."[7]

With executive action the designated route, the Attorney General became the key actor. "I ... kept [the President] advised," Robert Kennedy said later, "but I think it was just understood by us, which has always been understood, that I have my area of responsibility; and I'd do it.... If I had a problem about alternative steps, I'd call him.... But I wouldn't call him just to be gabby about what was going on in the south."[8] Of all his appointments, the assistant attorney general for civil rights was "the major one that I worried over.... The fellow that should naturally have been appointed was Harris Wofford.... I was reluctant to appoint him because he was so committed to civil rights emotionally and what I wanted was a tough lawyer who could look at things ... objectively."[9] Byron White, who had decided during the campaign that Wofford was too much a crusader, now proposed Burke Marshall, thirty-eight years old, a Yale Law School graduate and a member of the eminent Washington firm of Covington and Burling. Wofford, who had conducted a joint seminar with Marshall at Howard University, also generously recommended him (and himself became special assistant for civil rights in the White House).

Marshall arrived for an interview. He was "terribly nervous," John Seigenthaler recalled. ". . . He kept getting up and walking around."[10] At last he was ushered in to see the Attorney General. Marshall was as shy as Kennedy and as little adept at small talk. They faced each other gloomily and sat through intervals of painful silence. Marshall departed, convinced he had lost his chance. Kennedy wondered whether he would ever be able to communicate with Marshall.[11] But he gave him the job on White's say-so.

The next problem was getting him through the Senate Judiciary Committee. Kennedy went to Chairman Eastland's office and said, "I have Mr. Marshall outside who is going to be head of our Civil Rights Division. I thought you'd like to meet him. He's going to put the Negroes in your white schools in Mississippi." As Kennedy recorded the story in his journal, "Eastland snorted and said. 'I don't want to see him.' I said, 'Come on, see him for a minute.' So I brought Marshall in. They looked each other in the eye and exchanged some words. Afterwards Eastland said he wouldn't hold Marshall's appointment up but he wanted it understood that he was going to vote against him. As he put it, 'I'd vote against Jesus Christ if he was nominated for that position.'"[12]

Almost at once the Department of Justice confronted a crisis in New Orleans, where Judge Skelly Wright had ordered school integration and the Louisiana legislature was determined to forbid compliance by the New Orleans school board. Kennedy said, "We'll have to do whatever is necessary." "He was really mad," Marshall said later. ". . . It was the first time that either one of us had been involved directly . . . with the way that the segregation system worked in practice, and how difficult it was going to be. And they were so irrational about it down there."[13] Marshall instituted contempt proceedings. The state yielded.[14] The experience established a partnership. Kennedy was quickly impressed by the self-effacing Marshall's precise mind, incorruptible character, dry humor and intense moral conviction. There was no one on whose judgment he relied more during the rest of his life.

II

Robert Kennedy was not learned in the intricacies of civil rights — considerably less so than his brother, who had debated these questions in the Senate. Nicholas Katzenbach recalls taking the executive order establishing the President's Committee on Equal Employment

Opportunity out to Hickory Hill one evening in February 1961. To his surprise he found John Kennedy there for dinner. Producing the order, he said to the Attorney General, "You've got to sign this and give it to the President." Robert Kennedy asked what the order meant. As Katzenbach started to explain, the President cut in with questions that Katzenbach could not have answered himself had he not been studying the problem for a week. Finally Robert said to his brother, "I don't know why the hell I sign this thing and give it to you. Why don't you sign it and give it to me?"[15]

Education began at once. When Kennedy and John Seigenthaler returned from an early tour of the Department, Kennedy said, "Did anything occur to you as strange in our visit around to these offices?" Seigenthaler said he was impressed that everyone was working so hard. Kennedy said, "Yes, but did you see any Negroes? . . . Get me a study of how many Negroes are working here."[16] Of the Department's 955 lawyers in Washington, it developed, only ten were black. At his first staff meeting in February, Kennedy, stressing the importance of example, called for "thorough integration" of all Justice's offices, north and south.[17] He wrote leading law schools asking for their best black graduates. In short order the Department had many more black lawyers than ever in history.

The same belief in the necessity of example led him to a private challenge of the Metropolitan Club, which refused to admit blacks even as guests for luncheon. That policy, he wrote the board of governors in April 1961, "does not seem to me logically tenable, humanly honest, or diplomatically sound." His protest unavailing, he wrote again in September, "It is inconceivable to me, in this day and age, that the privileges of this Club which holds such a unique and peculiar position in the Nation's Capital would be denied to anyone merely because of his race. . . . I cannot in good conscience remain a member and herewith submit my resignation."* (When I sent him an editorial from the Pilot, the Boston diocesan weekly, commending his stand, he responded, "The Pilot put it very well. We Catholics are really coming along nicely, don't you think?")[18]

As he set to work on the problems of racial justice in the south, he soon concluded that the right to the vote was the heart of the matter. "From the vote, from participation in the elections, flow all other rights," he told Anthony Lewis in 1964. ". . . I thought it could make a major difference and be of far more help than anything else if they

---

* When Arthur Krock, who had put Robert Kennedy up for the Metropolitan Club, later attacked John Kennedy for refusing a visa to Tshombe, the black secessionist leader in the Congo, the President proposed a deal: "I'll give Tshombe a visa and Arthur can give him dinner at the Metropolitan Club" (Arthur M. Schlesinger, Jr., A Thousand Days [Boston, 1965], 577).

just . . . registered a hundred people a day. . . . That was the key to opening the door to all of what they wanted to accomplish in education, housing, jobs and public accommodation." This was, moreover, the area in which Congress had conferred a clear mandate and "in which we had the greatest authority." And "I felt nobody could really oppose voting."[19]

The Eisenhower administration, except for a few months at the end when Harold Tyler became chief of the Civil Rights Division, had enforced the voting rights act "with about the vigor and imagination," wrote Alexander Bickel, "displayed by William McKinley in enforcing the Sherman Antitrust Act."[20] Only six suits were brought, and none at all in Mississippi, where the exclusion of blacks from voting was most flagrant but also where the chairman of the Senate Judiciary Committee made his home. Kennedy and Marshall were determined to change all that.

III

A major problem in voting rights cases had been the difficulty in finding witnesses willing to testify against the system of exclusion and intimidation. "I understand from Tyler and others," Harris Wofford informed the Attorney General early in 1961, "that the FBI has been the bottleneck here. Apparently the FBI has given little or no priority to requests for investigation by the Civil Rights Division. Moreover, the FBI has, or uses, few if any Negro investigators. White FBI investigators, who often have a southern accent, run into great resistance from skeptical, fearful Negroes."[21]

J. Edgar Hoover had the racist instincts of a white man who had grown up in Washington when it was still a southern city. In 1946 he had complained to Attorney General Tom Clark that the Bureau was expending too much manpower "investigating murders, lynchings and assaults, particularly in the Southern states." The result, he said, was only to enrage "vociferous minority groups" when the Bureau failed to produce immediate results.[22] In 1948 Joseph Rauh asked Clark to have the FBI look into the attempted murder of Walter Reuther. Clark, after conferring with the director, brought back the irrelevant but revealing reply: "Fellows, Edgar says no. He says he's not going to send the FBI in every time some nigger woman gets raped."[23]

In 1957, Hoover gave the Eisenhower cabinet a briefing on racial tensions. The south, the director said, was in a state of "explosive resentment over what they consider an unfair portrayal of their way of life, and what they consider as intermeddling." Southerners simply did not accept "mixed education" as the means "whereby the races can best be served. And behind this stalks the specter of racial intermarriages." The prosegregationist White Citizens Councils were nonviolent; their membership included "some of the leading citizens in the South." The Ku Klux Klan was "pretty much defunct." As for the "crusade for integration," its leaders were threatening violence, and its organizations were major targets for Communist infiltration.[24]

The tradition of cooperation with the local cops further set the Bureau against racial cases in the south. As Truman's Committee on Civil Rights had explained in 1948, it was "difficult for the FBI agent to break this relationship and to work without, or even against, the local police when a civil rights case comes along."[25] And, if local police should stop helping on stolen cars, FBI statistics would suffer. For reasons of policy as well as prejudice, Hoover succeeded in withdrawing the FBI almost completely from civil rights investigations.[26] Internally he preserved it as a lily-white agency.

It was easy enough to bring blacks into the rest of the Department. The FBI, however, was as resistant as the Metropolitan Club. When Kennedy asked Hoover about black agents, the director replied virtuously that the Bureau did not catalogue its employees by race, creed or color. Kennedy said that Hoover's answer was most commendable, but he still wanted to know how many black agents there were. Kennedy's insistence caused, as William Sullivan remembered it, an "uproar" in the Bureau. For there were no genuine black agents. Hoover had inherited a couple when he took over the Bureau in 1924. During the war he had made his chauffeur and office boy special agents in order to protect them from the draft. But none of the so-called black agents conducted real investigations or received, as white agents did, retraining in Washington; they mostly served as drivers. Hoover finally told Kennedy there were five black agents. Professing great astonishment — five black agents out of five thousand — Kennedy instructed Hoover to begin hiring more at once.[27] This order infuriated the director. "He wanted me to lower our qualifications and to hire more Negro agents," Hoover told a reporter in 1970. ". . . I said, 'Bobby, that's not going to be done as long as I'm director of this bureau.' He said, 'I don't think you're being cooperative.' And I said, 'Why don't you get a new director?'"[28] In fact, Hoover gave ground, but even to call his recruitment tokenism — ten black agents by the end of 1962 — would have been wild overstatement.* [29]

Kennedy's next move was to order the FBI to start gathering

---

* By 1977 there were only 139 blacks among the Bureau's 8400 agents (*U.S. News & World Report*, August 29, 1977).

information about black voting patterns in the south.[30] But FBI agents, as John Doar, a Republican holdover whom Burke Marshall put in charge of voting litigation, quickly discovered, were not "geared up to their assignment in any way, shape or form." Many were southerners. All were accustomed to working with the friendly neighborhood sheriff. Doar began to supply them with elaborate questionnaires so that they would have no excuse for not coming back with the needed data. One 1962 memorandum, Doar recalled, "went on in the most minute detail for 174 pages, explaining, anticipating, cautioning and coaching the Bureau agents."[31] In time they learned.

And in time Hoover accepted inevitability on racial justice. "I don't think that he's got a great sympathy for this," Kennedy said in 1964. "I think that he also recognizes where the power is and what he has to do; and, once he reaches that decision, that is paramount. I think he reached the decision that we were going to do things in civil rights and that that's the way it was going to be. . . . So that either you'd have to do it or you'd have to get out. . . . When they did things, frequently they did them damned well."[32]

IV

The Rackets Committee had given Robert Kennedy a favorable reputation in the white south. He knew the welcome would not last. In March 1961 he sent over to Justice Hugo Black at the Supreme Court an editorial torn from an Alabama paper. The editorial lamented that, while his native state ought to salute "a great judge and great man," Black was "so unpopular that were he to visit Birmingham hardly a corporal's guard of personal friends would greet him." Kennedy wrote Black, "My congratulations to you."[33]

Early in May, Kennedy determined to go south himself. The occasion was Law Day at the University of Georgia, where two black students had gained uneasy entry through federal court order. Griffin Bell, who had managed the Kennedy campaign in Georgia, made the advance preparations and accompanied Kennedy from the Atlanta airport to the university at Athens. It was Kennedy's first address as Attorney General. He declared his personal belief in the Supreme Court's decision on school desegregation. "But my belief does not matter — it is now the law." Those who defied judicial rulings challenged "the foundations of our society." "I say to you today that if the orders of the court are circumvented, the Department of Justice will act. We will not stand by and be aloof. We will move."[34] The audience, to his astonishment, gave him an ovation. "Never before, in all of its travail of by-gone years," wrote Ralph

McGill, the liberal editor of the *Atlanta Constitution*, "has the South heard so honest and understandable a speech from any Cabinet member."[35]

In enforcing the law, Kennedy said in Georgia, he hoped "to achieve amicable, voluntary solutions without going to court." He was "very sensitive," Katzenbach remembered, "as to the explosive nature" of the situation.[36] Southerners had first to have a chance at self-correction; "you didn't solve all the South's problems for it."[37] Kennedy and Marshall, as Carl M. Brauer summarized it, believed that Negro rights had "failed to survive after the Civil War because they had been established by an alien federal force; when that force was removed, the Negroes' rights rapidly withered and died. They hoped to see civil rights take root this time and were therefore determined to make the absolute minimal use of federal force."[38] "I don't think," Kennedy told Peter Maas in March 1961, "we would ever come to the point of sending troops. . . . I cannot conceive of this administration's letting such a situation deteriorate to that level."[39]

There were no negotiators more patient than Marshall and Doar. Kennedy also used his two southern assistant attorneys general, Ramsey Clark of Texas and Louis Oberdorfer of Alabama, as well as his southern assistant, John Seigenthaler of Tennessee. Preparing for the new wave of school desegregation in the fall of 1961, these men traveled about conferring with local educational officials. "I'd go in," said Seigenthaler, "my southern accent dripping sorghum and molasses, and warm them up. Burke would tell them what the law was. . . . a pretty effective team."[40] In September schools were desegregated without violence in New Orleans, Atlanta and Memphis.

v

"The hardest problems of all in law enforcement," Kennedy had told the University of Georgia, "are those involving a conflict of law and local custom."[41] The white south's customs were rooted in a bitter history. The Supreme Court's law came on them as a threat to tradition, womanhood and social order. Southern blacks, awakening after so many years to their constitutional rights, were growing as militant for federal law as whites were for local custom. The negotiating approach underestimated both the intractability of the old-school whites and the passion of the new-school blacks.

In December 1960 the Supreme Court had outlawed segregation in the terminals as well as in the trains and buses used in interstate transportation.[42] The white south largely ignored the decision. The following May, James Farmer of the Congress of Racial Equality (CORE), an organization inspired by Gandhi's principles of nonresistance, led a small party, black and white, on a journey to assert constitutional rights in southern bus stations. The Freedom Riders traveled through Virginia, the Carolinas and Georgia with an occasional fracas and arrest but without major trouble. Then, dividing into two groups, one Greyhound, one Trailways, they entered Alabama. In Anniston a screaming white mob burned one bus, boarded the second and beat up civil rights workers. In Birmingham, members of the Ku Klux Klan, promised fifteen minutes of immunity by the police, set on the Riders with baseball bats, lead pipes and bicycle chains. Gary Thomas Rowe, Jr., an FBI informant in the Klan, had alerted his superiors. The FBI neglected to pass on the warning to Kennedy.[43] "Everybody who got off the bus," Rowe wrote later, "was clubbed, kicked or beaten. . . . When people looked up, I couldn't see their faces for blood. . . . I observed several FBI men . . . taking movies of the beatings." No Klansman was arrested.[44] It was Mother's Day, Eugene (Bull) Connor, the Birmingham police commissioner, later explained, and he had let most of his force off to perform filial duties.

"I never knew they were traveling down there," Robert Kennedy said in 1964, ". . . before the bus was burned in Anniston."[45] The first Washington reaction was one of dismay mingled with a certain exasperation. When Peter Maas had asked Kennedy in March about the sit-ins, the Attorney General had replied, "My sympathy is with them morally."[46] But undue militancy threatened the strategy of suasion. In the White House, John Kennedy, preparing for meetings with de Gaulle and Khrushchev in Europe, thought the Riders an unnecessary burden. "Tell them to call it off," he said to Harris Wofford. "Stop them." "I don't think anybody's going to stop them right now," Wofford said.[47]

Plainly the Riders were within their constitutional rights. Plainly the federal government had an obligation to protect travel in interstate commerce. Robert Kennedy sent John Seigenthaler to Birmingham to see what could be done. Seigenthaler found the beleaguered group, heads bloody and bandaged, still in the bus terminal. Simeon Booker, the black newspaperman, told him, "I don't believe we'll ever get out of here." After long negotiations — the airlines had received bomb threats — Seigenthaler took them on a plane to New Orleans. As they disembarked, he ran into Barry Goldwater. "This is horrible," Goldwater said. "Just horrible. Never should have happened. I'm glad you're with them."[48]

With the first group *hors de combat*, more Riders descended on Birmingham to take their place. Acting on the principle of initial resort to local authority, Robert Kennedy asked for assurances of safe passage across the state. Seigenthaler met with Governor John Patterson. Patterson had been a strong Kennedy man in 1960. "He was," Robert Kennedy remembered in 1964, "our great pal in the south."[49] Times had changed. Patterson now launched into a tirade. He assured Seigenthaler that Alabama could keep the peace without the Kennedys. "By God," he said, "I'm telling you if federal marshals come into Alabama, there'll be blood in the streets." Floyd Mann, the state director of public safety and a man of courage, told Seigenthaler his troopers would protect the bus on the highways.[50]

But no one could be found to drive the Greyhound bus to Montgomery and on to Mississippi. Kennedy called the bus company:

GEORGE CRUIT (the Greyhound superintendent). Drivers refuse to drive.

KENNEDY. Do you know how to drive a bus?

CRUIT. No.

KENNEDY. Well, surely somebody in the damn bus company can drive a bus, can't they? . . . I think you should — had better be getting in touch with Mr. Greyhound or whoever Greyhound is and somebody better give us an answer to this question. I am — the Government — is going to be very much upset if this group does not get to continue their trip. . . . Under the law they are entitled to transportation provided by Greyhound. . . . Somebody better get in the damn bus and get it going and get these people on their way. Mr. Cruit, I think that if some of your people would just sit down and think for a few minutes that somebody would be able to drive a bus 80 or 90 miles.*

Mr. Greyhound eventually solved the problem. Floyd Mann's state troopers escorted the group to the outskirts of Montgomery.

A menacing crowd, carrying chains and ax handles, surrounded the Montgomery terminal. "The passengers are coming off," John Doar reported to Kennedy over the telephone. Then: "Oh, there are fists, punching. A bunch of men led by a guy with a bleeding face are beating them. There are no cops. It's terrible. It's terrible. There's not a cop in sight. People are yelling 'Get 'em, get 'em.' It's awful."[51] Seigenthaler, who was circling the terminal in a car, saw a Freedom Rider, a white girl, struggling to escape. A fat woman was chasing her, beating her over the head with a pocketbook; in front a young skinny kid "was facing her and dancing backwards like a boxer and smacking her in the face." Seigenthaler got out of his

---

* Transcript of Kennedy-Cruit conversation, May 15, 1961, RFK Papers. The call was recorded by Cruit, not by Kennedy. One of Kennedy's statements — "We have gone to a lot of trouble to see that they get to this trip and I am most concerned to see that it is accomplished" — was later used in Alabama and elsewhere in an effort to show that Kennedy had put CORE up to the Freedom Rides.

car. The girl said, "Mister, get away. Leave me alone. . . . You're only going to get killed. This is not your fight." Seigenthaler tried to push her into the car. Suddenly he was clubbed from behind. He lay unconscious on the ground for half an hour. FBI agents on the scene took careful notes but raised not a finger.[52] Seigenthaler awakened much later in the hospital with a concussion. In Washington, Peter Maas saw Robert Kennedy immediately after he heard about Seigenthaler. He was, Maas recalled, "possessed by an enormous anger . . . as if he had been down in Montgomery himself and been hit."[53] Kennedy phoned Seigenthaler: "You did what was right." Seigenthaler said from his hospital bed, "Let me give you some advice. . . . Never run for Governor of Alabama. You couldn't get elected."[54]

Local authority had manifestly failed to preserve civil peace. On Sunday, May 21, Robert Kennedy sent in five hundred marshals with Byron White in command. Meanwhile Martin Luther King, Jr., was flying to Montgomery. Governor Patterson informed Kennedy that he and General Henry Graham of the Alabama National Guard could not guarantee King's safety. Kennedy said, "I don't believe that. Have General Graham call me. I want to hear a general of the United States Army say he can't protect Martin Luther King." But Kennedy took no chances. "Fifty marshals," Patterson later said disgustedly, "met King at the airport and escorted him through town . . . just like he was the president of the United States."[55]

Patterson told Byron White rather inconsistently that Alabama could handle the situation. According to Patterson, White thereupon advised Kennedy to pull out the federal force.* Certainly, White explained at a press conference, the marshals were there only to maintain order and would not intervene if the cops arrested the Freedom Riders; "I'm sure they would be represented by competent counsel."[56] But Kennedy rejected the recommendation and told White to stay on in Montgomery.

Fifteen hundred blacks gathered on Sunday evening to hear King at the First Baptist Church. An angry white mob burned an automobile in front of the church and evidently meant to burn the church too. King called Kennedy at the Department of Justice. Kennedy assured him that the marshals were on their way, adding that, so long as he was in church, "he might say a prayer for us." "He didn't think that was very humorous," Kennedy recalled in 1964. Even when the marshals arrived, it was a serious question whether they

---

* A telephone operator at Maxwell Field, where White made his call, was the wife of a state trooper and "monitored" White's calls for Patterson (Howell Raines, *My Soul Is Rested* [New York, 1977], 310).

could hold the church. "At one point," said Burke Marshall, "it really looked as if the mob might overcome the marshals and get in there." Airborne troops were put on alert at Fort Benning nearby. Military intervention was the last thing the Kennedys wanted, except that defiance of federal law and racial war were even worse. But, reinforced by the valiant Floyd Mann and a squad of state troopers and finally by the National Guard, the marshals drove the mob back. The enraged whites did not disperse, however, and the blacks in the church dared not leave.

Through the long night Kennedy was on the phone with Patterson and with King. "Robert showed no knowledge whatsoever of the problem," Patterson complained later, "and he ran the whole affair from his command post in the Justice Department, and making telephone calls at two or three o'clock here in the morning, telling us we didn't have enough policemen on a certain corner. . . . He didn't understand who these people were. . . . They were known Communists, some of 'em."[57] At one point Patterson said that the affair would ruin them both politically. "Now, John, don't tell me that," Kennedy replied evenly. "It's more important that these people survive than for us to survive politically."[58]

Around four in the morning King berated Robert Kennedy for allowing the situation to continue. "Now, Reverend," Kennedy said, "don't tell me that. You know just as well as I do that if it hadn't been for the United States marshals you'd be as dead as Kelsey's nuts right now." There was a long silence. Finally King said, "All right. All right." Perhaps he was baffled by the reference. "Kelsey's nuts" was evidently an Irish expression of unknown provenance.[59] A militant young Rider, John Lewis, later called the battle of the First Baptist Church "a real testing moment . . . without the support of the federal government at that time and a commitment, I think it could have been real bad."[60]

Around five in the morning Byron White sent William Orrick, also down from Washington, to negotiate a truce with the National Guard commander. "I was treated," Orrick recalled, "like I might have been treated in Russia. . . . There wasn't a sign of the American flag . . . just the Confederate flags. I've been four and a half years in the Army, and I know GIs like the back of my hand. They were just outright hostile. . . . The general was a lily white Wasp. He was polishing his boots when I came in and talking to his staff about being clean." After an initial refusal the Confederate leader agreed to get the people in the church safely home. (Orrick was left behind in charge of the marshals and civic tranquility in Birmingham. He called Washington and said, "I don't know how to be a law enforcement officer. I'm a corporation lawyer from San Francisco." Ken-

nedy said, "Yeah." Orrick said, "You remember my background.
I'm in the Civil Division." Kennedy said, "Yeah. Can't you do it?"
Orrick said, "Yes, yes." Kennedy said, "Well, don't bother me. . . .
Do it." He did.)[61]

VI

Kennedy's plan was to move the Freedom Riders out of Alabama as
quickly as possible. "I thought that people were going to be killed,"
he said in 1964, "and they had made their point. What was the
purpose of continuing with it?"[62] He called for a "cooling-off" period. James Farmer told a reporter, "We had been cooling off for
100 years. If we got any cooler we'd be in a deep freeze."[63] Kennedy
said that racial troubles would embarrass the President of the United
States in his meeting with Khrushchev. Ralph Abernathy, King's
number two man in the Southern Christian Leadership Conference,
said, "Doesn't the Attorney General know that we've been embarrassed all our lives?"[64] The Freedom Riders determined to press on
to Mississippi. "I don't think any of us thought that we were going
to get to Jackson," said James Farmer. "Everyone on that bus,"
recalled another Rider, "was prepared to die."[65] Nor was this hysteria. James P. Coleman, the former governor of Mississippi, warned
Burke Marshall that his successor, Ross Barnett, could not be trusted
and that the Riders would be killed before they reached Jackson.[66]

Kennedy now turned to his friend Senator Eastland. "He always
kept his word," Kennedy said later, "and he always was available, and
he always told me exactly where he stood and what he could do and
what he couldn't do. He also told me who I could trust and who I
shouldn't trust in the state of Mississippi." Eastland promised there
would be no violence. "I think Jim Eastland really took a responsibility for it," Kennedy said. Eastland added that the Riders would be
arrested as soon as they arrived in Jackson. "I said to him that my
primary interest was that they weren't beaten up. So I, in fact, I
suppose, concurred [in] the fact that they were going to be arrested,
though I didn't have any control over it."[67]

When Martin Luther King called him from Jackson on the night
of the arrests, Kennedy's preoccupation was with getting the Freedom
Riders — Farmer among them — out of jail. To his surprise, King
said they would stay in.

KING. It's a matter of conscience and morality. They must use their
lives and their bodies to right a wrong.
KENNEDY. That is not going to have the slightest effect on what the
government is going to do in this field or any other. The fact that
they stay in jail is not going to have the slightest effect on me.

KING. Perhaps it would help if students came down here by the hundreds — by the hundreds of thousands.

KENNEDY. The country belongs to you as much as to me. You can determine what's best just as well as I can, but don't make statements that sound like a threat. That's not the way to deal with us. [A silence]

KING. It's difficult to understand the position of oppressed people. Ours is a way out — creative, moral and nonviolent. It is not tied to black supremacy or Communism but to the plight of the oppressed. It can save the soul of America. You must understand that we've made no gains without pressure and I hope that pressure will always be moral, legal and peaceful.

KENNEDY. But the problem won't be settled in Jackson, Mississippi, but by strong federal action.

KING. I'm deeply appreciative of what the administration is doing. I see a ray of hope, but I am different than my father. I feel the need of being free now!

KENNEDY. Well, it all depends on what you and the people in jail decide. If they want to get out, we can get them out.

KING. They'll stay.[68]

King also had his practical side. Even before the Freedom Ride, he had suggested that the Interstate Commerce Commission act against segregation in the terminals. Kennedy had responded that the ICC was an independent commission, not subject to presidential order and notoriously slow-moving on its own.[69] Now Katzenbach's Office of Legal Counsel proposed (a lawyer named Robert Saloschin had the idea) that the Attorney General formally petition the ICC. On May 29, nine days after the Montgomery mob had tried to burn down the First Baptist Church, Kennedy, in a quite novel administrative step, asked the ICC to issue regulations ending segregation in interstate bus terminals. On September 22, 1961, the order was issued. Kennedy's negotiators set to work again, persuading local officials to take down WHITE and NEGRO signs. Though mopping-up operations were required in McComb, Mississippi, and elsewhere, Jim Crow was soon banished from bus travel. The Department then turned to the railroads and airlines. By 1963, the Attorney General could say, "Systematic segregation of Negroes in interstate transportation has disappeared."[70]

VII

Education and transportation were important; but voting rights remained the primary target. Negotiation proved no more successful here, though, as Marshall observed, the effort was essential if only to expose the illusion that white southern communities had "the will and the power to correct wrongs themselves."[71] Mediation failing, the next step was the lawsuit. Voting rights cases had ordinarily been argued by local federal attorneys; but Marshall sent his own lawyers into the field and the courtroom. John Doar, a lanky, reticent man of great personal courage, led many of these expeditions into the south. He even invaded Mississippi. Kennedy may have thought that Eastland's wink licensed this violation of his sanctuary. At any rate, Kennedy said later, Eastland "never made any effort to stop us from doing anything in the state of Mississippi, or try to impede anything that we did in the state."[72]

In conducting interviews, Doar's lawyers concentrated on blacks, who were generally frightened by FBI agents, and the agents concentrated on whites, who were often frightened by Justice Department lawyers.[73] Each suit required thousands of pages of evidence. "The Division," Doar said later, "was not prepared to take the terrible risk of losing a single case because of lack of proof. We faced tough judges. We wanted the proof to be so overwhelming so as to lock up the trial judge . . . and to convince the country as well."[74] In the end, Kennedy's Department of Justice brought fifty-seven voting suits, including thirty in Mississippi and one in Jim Eastland's own Sunflower County.[75]

The voting rights thesis was not universally accepted. The Civil Rights Commission, though all for black enfranchisement, denied that it was a cure-all. Its 1961 report, Burke Marshall complained, "went out of its way" to point to counties in the deep south where blacks registered freely with no discernible effect on the caste system.[76] For that matter blacks voted without constraint in the north, but black inequality remained. The administration did not believe such points invalidated the strategy. "If it does not bring change by itself," Marshall said, "the vote is nonetheless the key to possible change."[77]

Black leaders suspected the administration's motives. "I felt that what they were trying to do," said Lonnie King of the Student Nonviolent Coordinating Committee (SNCC), "was to kill the Movement, but to kill it by rechanneling its energies."[78] Vincent Harding, the black historian, thought it an attempt "'to get the niggers off the street.' [The Kennedys] probably recognized that the constant movement on the streets might well lead to the tearing and the breaking of the fabric. . . . That they were probably deathly afraid of." Harding believed the breaking "absolutely necessary for something real to take place."[79] The Kennedys were determined on peaceful change. In June 1961, meeting with representatives from King's Southern Christian Leadership Conference (SCLC), CORE and SNCC, Robert

Kennedy argued that voter registration would be far more productive than demonstrations. "I said that it wasn't as dramatic; and that perhaps there wasn't going to be as much publicity. . . . But I thought that's where they should go and that's what they should do. I had some conversations with Martin Luther King along those lines. I think that they rather resented it." [80]

He was more persuasive than he supposed. Marshall and Wofford followed up in subsequent meetings. The Taconic, Field and Stern foundations offered financial support. The black leaders understood that the administration would provide protection. One SNCC activist recalled someone, perhaps Wofford, saying that, if it were necessary in the course of protecting people's right to vote, "the Kennedy Administration would fill every jail in the South." [81] In August, after heated debate between "direct action" and "voter registration" wings, SNCC decided to join the campaign. [82] James Forman, SNCC's revolutionary-minded director, saw registration work "as a tool by which consciousness might be aroused, politicized, and organized." [83] In September James Farmer brought in his comrades at CORE. [84] In November Martin Luther King, after noting that the civil rights movement was laboring on many fronts, declared with emphasis: *"The central front, however, we feel is that of suffrage."* The vote would "give us the concrete tool with which we ourselves, can correct injustice. . . . This is the pattern for changing the old South, and with it the nation as a whole." [85] In January 1962 the Voter Education Project came into existence — "Education" because contributions to an educational activity collecting information on registration problems were tax-exempt. (Kennedy said later, "I was able to work out with Mort Caplin for them to receive tax-free status.") [86] The VEP was supposed to last for two and a half years. It was still going strong fifteen years later with John Lewis, the Freedom Rider, as executive director.

Early results, outside the cities, were disappointing. In the rural south the drive bogged down in white intimidation, black fear and apathy. Roy Wilkins of the NAACP thought it futile to work the countryside. But Forman of SNCC, welcoming the hatred of the sheriff and the Klan, believed that violence "would create more exposure and thereby more consciousness." [87] The argument between voter registration and direct action was adjourned. Voter registration in a hostile setting *was* direct action. And it was action, as one CORE worker later put it, with widespread repercussions. "The fact that activity was brought to the smallest towns and rural areas instilled black residents there with enormous hope and pride. . . . [It] left these communities with a legacy of organizational skills, higher expectations, and, most importantly, the sense that they could take positive action to change their own lives."[88]

It also wore out Robert Kennedy's southern welcome. "The Freedom Riders do not seem to have hurt President Kennedy much," wrote Samuel Lubell, the expert political observer. ". . . It is 'brother Bobby' in the Attorney General's office, rather than President Jack, who has been blamed." Lubell asked a Mississippi housewife to name the biggest problem in the country. She replied: "Kennedy has too many brothers."[89]

VIII

At the same time, the invasion of the rural south exposed a grave weakness in the Kennedy strategy. The civil rights activists had interpreted the administration's assurance of cooperation as a promise of protection. They felt bitterly betrayed when white sheriffs, policemen and good old boys struck back with apparent impunity.

In Itta Bena, Mississippi, in the summer of 1963, fifty-seven blacks made a night march to the house of a deputy sheriff to ask police protection for voter registration. The next day all over fourteen years old were peremptorily tried, convicted, fined and sent to prison for months on charges of disturbing the peace. In Clinton, Louisiana, twelve blacks wrote respectful letters to the mayor and district attorney requesting the establishment of a biracial committee to combat racial tensions. They were arrested on charges of intimidating public officials. A black walking down the street in Clarksdale, Mississippi, wearing a T-shirt with CORE on the front and FREEDOM NOW on the back was arrested for parading without a permit.[90]

These were the milder cases. As activists went in increasing numbers into the deep south, as southern blacks asserted their rights with new vigor, local retaliation grew more atrocious. Beating became frequent; murders were not uncommon. The rage of those who believed they were defending the Constitution became harder to contain or to answer. Why did not the federal government use injunctive sanctions? Why did not FBI agents make on-the-spot arrests? Why not bring cases involving constitutional guarantees directly into the federal courts? Was national authority powerless to protect American citizens whose only offense was to use constitutional methods to claim constitutional rights? These questions came to torment the Department of Justice.

Under the federal system police power was reserved for local authorities. The national government could intervene only when, as

Robert Kennedy said, there was a specific and clear "federal responsibility" — as when Alabama failed to protect interstate travel or when
local authorities defied federal court orders.[91] Nor did existing legislation provide sure footing for more general intervention. Sections
332–334 of Title 10 of the United States Code permitted the President to send federal troops in case of a widespread breakdown of
law and order. It was not a remedy for sporadic incidents of brutality
by local police officials. Sections 241–242, Title 18, of the Code had
been enacted during Reconstruction to prevent the deprivation of
civil rights. But the first section required proof of conspiracy, the
second of specific intent. The Supreme Court had long since narrowed their scope and attenuated their force.

Burke Marshall sought to make it easier to initiate criminal prosecutions when local cops or mobs interfered with persons exercising
constitutional rights in the cause of racial justice. He got the FBI to
make investigations of alleged police brutality and encouraged the
filing of criminal informations to get around the reluctance of southern grand juries to return indictments.[92] The Department of Justice
in March 1962 sent up a bill that would have made forms of police
brutality a federal crime; no one pressed the bill in Congress, and
there were no committee hearings.[93] Beyond the infirmities in the
law, the southern system of justice remained the most formidable
obstacle. In civil rights troubles, white police were more likely to
arrest the victims than the leaders of mobs. White juries rarely
convicted white men charged with violating the rights of blacks.
White judges often imposed heavy bail or harsh sentences on civil
rights workers while going easy on segregationists. The rules of
federal-state comity discouraged federal interference in state criminal
proceedings. With only forty attorneys, Marshall felt he could deploy
limited manpower more profitably in voting than in protection cases.

"The point about protection," Marshall wrote to a civil rights
advocate,

is the most difficult and frustrating we have to live with under the
federal system. I say over and over again — hundreds of times a year
— that we do not have a national police force, and *cannot* provide
protection in a physical sense for everyone who is disliked because of
the exercise of his constitutional rights. The marshals we have used are
taken from duty in the service of papers, and the like, in many states.
At most, I can round up maybe 100, for short-term duty. They are not
police officers. They are not hired for that, and we have no budget for
it. Congress would never grant one.... There is no substitute under
the federal system for the failure of local law enforcement responsibility. There is simply a vacuum, which can be filled only rarely, with
extraordinary difficulty, at monumental expense, and in a totally unsatisfactory fashion.[94]

Lawyers understood Marshall's point. Asked privately in 1964 whether he thought the decision against a vast federal police presence in the absence of specific court orders was wise, Thurgood Marshall of the NAACP, victor in the school desegregation case and a dozen other civil rights cases, replied, "I certainly do. . . . The law is quite clear that the federal government is not the policing authority. Policing authority rests with the several states. That," he added, ". . . some of us can understand, but the average layman cannot understand it."[95] Certainly civil rights activists risking their lives in the war for the Constitution did not. "Careful explanations of the historic limitations on the federal government's police powers," as Robert Kennedy said the same year, "are not satisfactory to the parents of students who have vanished in Mississippi or to the widow of a Negro educator shot down without any reason by night riders in Georgia."[96] "If we are murdered in our attempts," cried one SNCC worker to the Kennedys, "our blood will be on your hands; you stand in the judgment of God and our people."[97]

IX

The anguishing dilemma led Burke Marshall in 1964 to write *Federalism and Civil Rights,* a small volume that, beneath lucid prose, was packed with perplexity and near-despair. Under the federal system, as Marshall later put it, "no matter how far reaching the equitable decrees of the federal courts may be, the states and local political institutions are going to retain operating control over all of the institutions involved in the granting or denial of equal rights. . . . There are many things the federal court can do to prevent misuse of state power, but it cannot replace the system of justice — of unfair police, biased juries, and locally elected judges applying a double standard in a segregated courtroom."[98] Or it could do so only by abandoning the federal balance, at least in connection with the police power.

This course seemed to Marshall full of hazard. It would require the creation of a national police force. Only a few years before, Justice Robert H. Jackson had warned somberly that "the establishment of the supremacy of the national over the local police authorities" could be a first step toward a totalitarian state.[99] Nor was there reason to believe that a national police force could easily overcome massive local opposition. "Nothing short of the pervasive presence of armed men will do," wrote Alexander Bickel. ". . . As a regular and more or less permanent device, it is something from which we recoil, deeming it destructive of the values of a free society."[100]

There had been no alternative to sending the Army to Little Rock. But this could not become the routine way of enforcing federal law. The federal balance had historically been the means by which a far-flung, diverse and turbulent nation had preserved a measure of cohesion. It was not to be lightly discarded.

Thus Marshall's bleak choice: keeping the federal balance at the cost of terrible injustice to civil rights workers; or seeking justice for civil rights workers with unforeseeable consequences for the federal balance. Marshall in the end favored the preservation of the federal system. Yet he understood that this could have awful consequences too — above all, in cynicism about the power, or will, of law and government to achieve justice.[101] His pessimism was mitigated only by his conviction that new local perceptions and new federal statutes might give the executive branch new opportunities. "The final cure," he said, "can come only from the political process."[102]

The question remained whether this argument was not put too absolutely — whether, within the confines of federalism, limited federal intervention in critical times and places might not still have made a difference. As pressure rose, the Department of Justice found itself doing things each new year it had declined to do the year preceding. This confirmed Marshall's suggestion that the question was one of political as well as of constitutional possibility. "Notwithstanding all the elaborate answers which the Justice Department produced, most of them in the language of constitutional construction and executive propriety," wrote a pair of thoughtful critics, "the true answer was to be found in the remaining power of southern white supremacy in Washington."[103] Increase in intervention therefore had to await increase in opposition to white supremacy, and this had to await increase in crisis. In the meantime, as John Doar said, "you've just got to keep going back. . . . Laws and court decisions and speeches up in Washington mean nothing. You've got to come *here* [he was musing in Mississippi] and chip away. You'll lose some games and win others — you've got to expect that. You'll think it's taking forever — you've got to expect that too. But the only alternative would be police state methods. . . . So you do the only thing you can do."[104]

In 1964 Anthony Lewis asked Robert Kennedy whether he would alter the system that gave primary responsibility for law enforcement to the states and communities. Kennedy said:

> No . . . I still think it's wise. I think that for periods of time that it's very, very difficult. . . . I just wouldn't want that much more authority in the hands of either the FBI or the Department of Justice or the President of the United States. . . . You would have accomplished much

more if you had had a dictatorship during the period of time that President Kennedy was President because you would have gotten . . . all kinds of domestic legislation. You would have been able to protect our open spaces . . . to do something about congestion in the cities and . . . about the slums. . . . We could have sent perhaps large numbers of people down to Mississippi and be able then to protect that group down there. But I think that it comes back to haunt you at a later time. I think that these matters should be decided over a long range of history, not on a temporary basis or under the stress of a particular crisis. . . . I think it's [best] for the health of the country.[105]

A generation rendered sensitive by Watergate to the case for constitutional process finds this view more persuasive than did the men and women who went south in defense of elemental rights in the early 1960s. For a time Robert Kennedy became almost as unpopular among civil rights workers as he was among segregationists.

x

Federal judges, Burke Marshall wrote, were "a principal factor in efforts to make federal rights for Negroes a reality in the South."[106] Of all forms of federal presence around the country, the judicial presence was the most solemn. Nothing lay more clearly within the field of executive initiative than good judicial appointments. Yet nothing in Robert Kennedy's attorney generalship received more valid criticism than his early recommendations for the southern bench.

Every judicial nomination had to have the approval of the Senate Judiciary Committee. In the spirit of local patriotism Chairman Eastland wished Kennedy's first appointment to come from Mississippi; wished, in addition, the appointee to be his college roommate, William Harold Cox. With many more appointments coming up, the administration thought it essential to keep Eastland reasonably happy. The nomination of a "close friend" of Eastland's as district judge in Mississippi, Roy Wilkins wired John Kennedy, "is not a surprise in the light of the general situation." But Cox on the bench, he added, boded ill "for any litigation not keyed to the mores of 1861. For 986,000 Negro Mississippians Judge Cox will be another strand in their barbed wire fence, another cross over their weary shoulders and another rock in the road up which their young people must struggle."[107]

Wilkins could hardly have been more prescient. Still there was nothing specific against Cox. He had not been associated with the Klan or the White Citizens Council. He had no public record at all.

The FBI found him blameless. The American Bar Association gave him its highest rating — "exceptionally well qualified." Robert Kennedy had Cox in. "We sat on my couch in my own office; and I talked to him. I said that the great reservation I had was the question of whether he'd enforce the law and whether he'd live up to . . . the interpretation of the Constitution by the Supreme Court. He assured me that he would. He was the only judge, I think, that I had that kind of conversation with. He was very gracious; and he said that there wouldn't be any problem. . . . I was convinced that he was honest with me, and he wasn't."[108]

Soon after Cox's nomination, the administration sent three black names to the Judiciary Committee — two, like Cox, for district judgeships, and Thurgood Marshall for the circuit court. This last nomination was exceptional. Up to that point only one Negro, William Hastie, had been appointed to the circuit court. The famous story of Eastland's remark to Robert Kennedy — "tell your brother that if he will give me Harold Cox I will give him the nigger"[109] — is perhaps apocryphal, but was true in some vaguer sense. Cox helped soften Eastland for Marshall. When Congress adjourned without acting on Marshall's nomination, John Kennedy gave him a recess appointment. In May 1962 a Judiciary subcommittee brooded over Marshall but rendered no report to the full committee. "Keating and Javits," Kennedy said later, "raised a fuss continuously and we didn't raise too much because I'd had a conversation with Jim Eastland [who said] before it was over he'd put it through. So that there wasn't any reason to get all excited about it and I told Thurgood Marshall that he'd be appointed. So he was relaxed."[110] On June 8, 1962, Kennedy's list of phone messages contained one from Eastland: "The matter you all discussed on the telephone this morning is all right with him and Senator Stennis." Kennedy noted in his crabbed script: "Don Wine [for federal attorney in Iowa] & Thurgood Marshall."[111] In September the Judiciary Committee approved Marshall. In the end the Kennedys appointed ten black judges. Eastland, said Robert Kennedy, held some up for a time but "never . . . caused us any trouble."[112] They were an odd couple.

Still Cox was a heavy price to pay. He no sooner got on the bench than he denied the Department of Justice the right, granted in the 1960 Civil Rights Act, to inspect public voting records in a Mississippi county where no black had registered for thirty years. No judge was reversed more regularly thereafter on civil rights cases.[113] Reversal no doubt increased Cox's fury. Early in 1963 he sent a letter to a Mississippi county registrar charging that the Attorney General typically preceded voting suits by "a political barrage of propaganda for voter consumption among negroes."[114] Cox was only warming up.

In an October letter to John Doar, who had had the impertinence to ask for a trial date in a voting case: "I spend most of my time fooling with lousy cases brought before me by your department in the Civil Rights field and I do not intend to turn my docket over to your department for your political advancement." Cox went on to call Doar stupid, unless he was trying to impress his "boss man."[115] Kennedy frigidly wrote Cox that he was turning the letter over to the ABA's Standing Committee on Federal Judiciary.[116]

Robert W. Meserve, chairman of the ABA committee, conferred with Cox and thought matters were under control. Then in March 1964, when Doar was trying to expedite black registration in a Mississippi county, Cox roared against the applicants in open court as "a bunch of niggers . . . acting like a bunch of chimpanzees."[117] Leon Jaworski of the ABA committee wrote Burke Marshall, "This latest outburst was very disappointing to Meserve. I have strongly recommended that Committee action be taken recording a strong disapproval of Judge Cox's conduct."[118] (In a few years, Doar and Jaworski joined in even more spectacular disapproval of official misconduct.)

Cox was only one of the sour Kennedy appointments to the district courts of the Fifth Circuit, covering the deep south from Florida to Texas. Judge E. Gordon West, named at the behest of Senator Russell Long of Louisiana, called the school desegregation case from the bench "one of the truly regrettable decisions of all time."[119] Judges Robert Elliott of Georgia and Clarence Allgood of Alabama were nearly as blatant. Their strategy was the ruthless use of judicial discretion and procedural manipulation to protract civil rights cases, turning every action, in Marshall's phrase, into "an endless chain of litigation."[120]

Why such appointments? Administration senators, according to firm tradition, proposed the judges for their own states, especially for the district bench. Southern senators were particularly in need of propitiation in 1961 because they held committee chairmanships vital to Kennedy's program — not only Judiciary itself, but Finance, Armed Services and Foreign Relations. Justice's information about the candidates was inadequate. FBI reports, said Burke Marshall, were "worthless."[121] The ABA generally went along with the legal establishment, hardly a sure guide in the south (or elsewhere). None of the judges appointed was a segregationist on the public record, except for Elliott; and Elliott was endorsed by the courageous liberal judge Elbert H. Tuttle as well as by black leaders in Georgia. (Nor was the public record necessarily an indicator. Alexander A. Lawrence of Georgia, appointed in later years, was a segregationist in politics but an integrator on the bench.) One civil rights official attributed to Kennedy's Justice Department the romantic view that "every redneck who came down the pike was another Hugo Black."[122]

It was evident all too quickly that the administration had not done well. Robert Kennedy wrote Eleanor Roosevelt in May 1962 that, while they had made every effort to make sure the new judges would do their duty, "I must say that we have had some disappointments . . . as I have had in a lot of areas."[123] He learned, as usual. After early 1962, Navasky has written, "he nominated no more judicial disasters."[124] Indeed, the final Kennedy record of southern judicial appointments was, despite public impression to the contrary, comparable to that of the Eisenhower administration. It is well known that Eisenhower, who, as a Republican, had no southern senators he was required to propitiate, appointed to the southern bench wise and fearless men like Tuttle, John M. Wisdom and Frank Johnson (about whom Robert F. Kennedy, Jr., wrote his senior honors essay at Harvard in 1976). He also had his share of duds.* A careful analysis of judges named by Eisenhower and Kennedy in the Fifth Circuit showed that each was responsible for five segregationists, Eisenhower for eight moderates and Kennedy for three, Eisenhower for two integrationists and Kennedy for eight.[125] Still damage was done.

XI

The selection of southern judges was one point at which the practice of the executive-action strategy fell considerably below the theory. Another, for a painful twenty months, was the campaign promise of an order ending discrimination in housing "by a stroke of the Presidential pen." The President in 1961 found the problems involved — legal, political, administrative — more complex than the campaigner had anticipated in 1960.

He almost put out the housing order in the fall. Then, after he and Robert Kennedy walked the beach at Hyannis Port over Thanksgiving pondering the question, he decided to try first for a department of urban affairs. He intended to appoint Robert C. Weaver, an eminent black economist, as secretary. Legislative leaders made it clear that Congress would not create the new department if he issued the housing order. He agreed to postpone the order. The

* One of them, Ben C. Dawkins, Jr., the chief judge in the western district of Louisiana, advised Robert Kennedy that "forced association cannot and will not work" and that civil rights legislation would be as "unworkable" as the Eighteenth Amendment. "I don't believe the Supreme Court is unworkable," Kennedy replied. ". . , I hope that several years from now we both will agree that none of the problems were insurmountable and our country and all of its people are better off because of the work of the three branches of the Federal Government." (Ben C. Dawkins, Jr., to RFK, June 18, 1964; RFK to Dawkins, August 4, 1964, Burke Marshall Papers.)

House Rules Committee, hating the idea of the first black cabinet member in history, killed the reorganization bill anyway. This was the congressional mood — and explains why there appeared no alternative to the executive-action route.

Counterpressure mounted on the civil rights side. An "Ink for Jack" campaign began. Thousands of pens descended on the White House. The President, recalling that the fatal phrase came from Harris Wofford, suggested that they be piled on Wofford's desk. As the issue became a symbolic test of the executive-action strategy, civil rights groups wanted to extend the order beyond housing affected by federal loans and guarantees to cover housing financed by savings and loan associations and even banks. The 1962 elections were drawing near. Democratic congressmen, including northern liberals, were apprehensive about the impact the order might have on the campaign. "There's no question," Robert Kennedy said later, "that we waited until after the election because of the political implications."[126] The order, finally promulgated on Thanksgiving eve, was limited in its reach. It proved neither the calamity predicted by opponents nor the blessing anticipated by advocates.

The housing order raised an interesting question, little discussed at the time. The civil rights leaders, seeking every road to equality, had no fear of the Presidency. "The power inherent in Executive orders," Martin Luther King, Jr., said in early 1961, "has never been exploited; its use in recent years has been microscopic in scope and timid in conception."[127] Opponents disagreed. "An order of this nature," Senator Sam Ervin protested to John Kennedy, "would constitute an invasion of the legislative field. The Constitution vests the legislative powers of the Federal government in the Congress and none in the President."* Indeed Congress had regularly rejected antidiscrimination amendments to housing bills. But the constitutional objection was generally discounted in those years as a neo-Confederate pretext. Perceptions sharpened by the Imperial Presidency found it more impressive later. Archibald Cox, recalling in 1975 that the order was issued "at a time when Congress would have rejected such legislation," noted that, though the pressures had been great and the results commendable, "in retrospect one has doubts about the method."[128]

If there could be doubt whether, except in compelling circumstances, a President should through executive order take a quasi-legislative action that Congress would not take, there could be no doubt

---

* Sam Ervin to JFK, October 6, 1962, JFK Papers. In 1955 Justice Black had said that executive orders not based on legislative command looked like legislation to him and that "the Constitution does not confer lawmaking power on the President" (*Peters vs. Hobby*, 349 U.S. 331, 350 [1955]).

about presidential power to combat racial discrimination by contractors doing business with the federal government. In establishing the President's Committee on Equal Employment Opportunity, John Kennedy charged it "to permanently remove from Government employment and work performed for the Government every trace of discrimination."[129]

Nixon had headed a much weaker committee of this sort under Eisenhower. The Kennedy executive order not only forbade discrimination but imposed on government contractors the requirement of "affirmative action" to assure equality in employment. Kennedy, desperate, like all Presidents, to find things for his Vice President to do, asked Lyndon Johnson to serve as chairman of the new committee. Johnson, his intimate friend William S. White reported, "privately flinched" from the assignment, fearing that as a southerner he would be blamed if the committee were a bust.[130] Kennedy insisted. Finally, Johnson, whose civil rights record before 1957 had been deplorable but who was himself wholly lacking in racial bigotry and perhaps perceived political advantage in proving himself where he had previously been regarded with suspicion, accepted.

As chairman, Johnson was uneasy from the start. One of the public members was Robert Troutman, Jr., an amusing and energetic Georgian who had known young Joe Kennedy years before and had been a Kennedy lieutenant in 1960. Troutman's contribution was Plans for Progress, a campaign to sign up firms in voluntary agreements to employ blacks. Johnson resented Troutman's access to the Kennedys as well as his initiative on a committee of which the Vice President was nominal head.[131] Robert Kennedy lacked Troutman's faith in voluntary action, but respected his organizational drive and was impressed that a southerner, who still pronounced the word 'Nigra,' cared about equal rights. "As aware as he was of the inadequacies of Troutman's approach," said Seigenthaler, "he felt, 'At least we've got something going here.'"[132] Despite opposition from civil rights groups, Plans for Progress had a small success. In 1962, Troutman, discouraged by Johnson's attitude, retired to Atlanta.

After September 1962 the executive vice-chairman was Hobart Taylor, a black lawyer from Detroit, a Texan by birth and a long-time Johnson man. Though the executive order empowered the committee to conduct hearings, terminate contracts and recommend criminal penalties for companies furnishing false information, Johnson and Taylor favored conciliation and cajolery. It was not, Johnson said, "a persecuting or prosecuting committee."[133] No hearings were held, no contracts were canceled. Since the committee lacked statutory authorization, this caution was understandable. It was also unproductive. Robert Kennedy and the successive Secretaries of Labor, Arthur Goldberg and Willard Wirtz, came to feel that the committee was badly run and that it was Johnson's fault. "There wasn't any problem about making it an effective organization," Kennedy said later, ". . . just if the Vice President gave it some direction. It was mostly a public relations operation. I mean, it accomplished a good deal more than it had accomplished under Nixon. But a lot of it was public relations. Secondly, there wasn't any adequate follow-up. Thirdly, the head of the staff . . . Hobart Taylor, whom I have contempt for because I thought he was so ineffective."[134]

XII

"What do you see as the big problem ahead for you?" an interviewer asked Robert Kennedy in May 1962. "Is it Crime or Internal Security?" Kennedy answered, "Civil rights."[135] Yet, for all their good intentions, for all their sensitivity to racial injustice, for all their unpopularity with the segregationists, the Kennedys after a year and a half were proving something of a disappointment to civil rights leaders. The refusal to call for legislation, the failure to protect civil rights workers, the interminable litigation over voting rights, the southern judicial appointments, the prolonged postponement of the stroke of the pen, the leniency with government contractors — all these were beginning to convince the movement that good intentions were not solving many problems.

Within the government itself, the Civil Rights Commission became a strong voice of concern. The commission's authority was limited to investigation, including the holding of hearings, and recommendation. Under Eisenhower it had been a cautious and rather technical body, compiling information and proposing modest remedies. When the new Department of Justice preempted the practical ground, the commission saw itself increasingly as a source of fresh ideas, moving out ahead and educating a larger public.[136] This forward conception turned it into a critic of the Kennedy strategy.

The commission had thus questioned the administration's bet on voting litigation, arguing instead for an across-the-board attack on inequality through legislative as well as executive action. It campaigned vigorously for the housing order and for the attempt to broaden its coverage. And it hit very hard on the question of southern justice. Father Theodore Hesburgh, the president of Notre Dame, was an influential member. On his annual retreat in 1961, Hesburgh read documents on police brutality that "just made my blood run cold.... With these horrible things going on all over, somehow I felt a cry of anguish, or a real loud blast from my moral soul." He therefore wrote an addendum to the 1961 report demanding more protective intervention by the federal government.[137] This question much preoccupied the commission in subsequent years.

When the commission met with the President in 1961, an argument arose over the number of southern state universities that admitted blacks. Kennedy had one figure; the commission's expert another. "He was right and our staff fellow was wrong," Hesburgh said. "... That impressed our office quite a bit." Still, for all the knowledge and incisiveness of the Kennedy brothers, Hesburgh felt that "the civil rights issue really imposed itself upon them, rather than they imposing themselves on civil rights." He had begun, he said, "with great optimism," but after the first year "I felt rather pessimistic."[138]

As the Department of Justice saw it, the commission was always getting in the way. "They had a very unpleasant meeting with Bob Kennedy," Harris Wofford recalled, "which turned all of them against Bob Kennedy, who seemed to be treating them as if he could tell them what to do."[139] The commission did indeed annoy Kennedy. "I didn't have any great feeling that they were accomplishing anything of a positive nature," he said in 1964. "I thought it was almost like the House Un-American Activities Committee investigating communism.... They were not objective investigations.... So I had no confidence in them." Moreover, "they were investigating violations of civil rights in areas in which we were making investigations. I thought that they could do more in the north. I thought that there were ... subjects that they could go into which perhaps would be helpful. But ... they were going over old ground, and they were doing what we were really doing: voting."[140]

Kennedy was overreacting; essentially because as a strong executive he wished to hold as many elements as possible under his own control. Hesburgh, a strong executive himself, understood the Attorney General's exasperation: "I'm sure that there were many times," he said in 1966, "when they thought we were complicating their business because it's always easier if you have one organization attending to business rather than two." But the commission's value, Hesburgh argued, was "precisely in being a kind of national conscience, a kind of burr under the saddle of the administration."[141] He was right. Few agencies have had a better record of recommendations translated into statutes than the Civil Rights Commission. Still, Kennedy found criticism from within the family wounding.

XIII

As for the black leadership, it gave the Kennedys mixed reviews. "This administration," Martin Luther King said in early 1962, "has reached out more creatively than its predecessors to blaze some new trails," notably in voting rights and in government appointments. Its "vigorous young men" had launched "imaginative and bold forays" and displayed "a certain *élan* in the attention they give to civil-rights issues." Still the "melancholy fact" was that the administration was "aggressively driving toward the limited goal of token integration." King demanded much more and so, he asserted, did the American people. "The nation is ready and eager for bold leadership in civil rights."[142]

Even blacks within the administration were somewhat dispirited. The loyal Robert C. Weaver sensed an "overcautiousness" in the President.[143] Roger Wilkins, a young lawyer in the foreign aid program, Roy Wilkins's nephew, contended in a 1962 memorandum (set down for Ralph Dungan to pass on to the Kennedys) "that this Administration seems to be using appointments and the achievement of voting rights as an excuse for doing little or nothing else. . . . Presidential indifference is the unkindest cut of all."[144]

The activists risking their lives on southern back roads were bitter. "The young Negro militants," wrote Loren Miller, the black civil rights attorney, "'don't believe in liberals at all.'" Let liberals shut up, Miller concluded, until they were "ready to enlist as foot-soldiers and subordinates in a Negro-led, Negro-officered army under the banner of Freedom Now."[145] "Kennedy is the smartest politician we have had in a long time," said Bayard Rustin, a founder of CORE in the forties and a special assistant to King in the fifties. "He calls the Negro leaders together and says in effect: 'I want to help you get money so Negroes can vote.' That's when he is bowing toward us. Then he turns and bows to the Dixiecrats and gives them Southern racist judges who make certain that the money the Negro gets will not achieve its purpose."[146] In the summer of 1962 disillusioned CORE leaders even considered "aggressive nonviolent action" against the Kennedys.[147] Forman of SNCC regarded John Kennedy as "quick-talking, double-dealing" and his voter registration drive as preparation for his reelection in 1964.[148]

Those who dealt personally with the Kennedys could not share this bitterness. Something about them — their openness, interest, the alertness of their sympathy — redeemed the hesitancy of their policy and offered hope for the future. "He was always ready to listen,"

said Martin Luther King of John Kennedy.[149] "He doesn't act as if
he knows everything," said Clarence Mitchell, the NAACP's Washington representative, of Robert Kennedy. "He is always willing to
listen. . . . The whole Kennedy family seems to have . . . an attitude
of decency which means a lot to people."[150] "They would listen very
carefully," said Thurgood Marshall, "but they would always be frank
and give their objections to what you were proposing, give their
reasons for it. . . . I think their frankness was the important factor
in dealing with organizations and individuals."[151]

And yet, and yet. "On the one hand," said Martin Luther King's
friend Vincent Harding, King "really hoped that he could believe in
people like Bobby; and on the other hand, a certain black wisdom
told him that he had better not try to believe in them."[152] And they
had one manifest disagreement. King was sure the nation was eager
for bold leadership in civil rights. Discerning no such national eagerness, doubting whether presidential exhortations could create an
emotional urgency the white majority was resolved not to feel, knowing the resistance they would encounter on Capitol Hill, the Kennedys felt they had to row to their destination with muffled oars. To
ardent spirits they seemed friendly but somehow detached; uninvolved? After a talk with John Kennedy, King said to Harris Wofford, "I'm afraid that the fact is he's got the understanding and he's
got the political skill . . . but the moral passion is missing."[153]

This is the way matters stood in the early autumn of 1962.

15

The Pursuit of Justice:
Ross Barnett and George Wallace

THE KENNEDY CIVIL RIGHTS STRATEGY, however appropriate to the congressional mood of 1961, miscalculated the dynamism of a revolutionary movement. "Every civil rights victory," Loren Miller proudly said, "adds to the Negro's intransigence; he becomes ever more impatient and demanding."[1] "It is not always when things are going from bad to worse that revolutions break out," Tocqueville had said long before. ". . . Patiently endured so long as it seemed beyond redress, a grievance comes to appear intolerable once the possibility of removing it crosses men's minds."[2]

I

The possibility of removing grievances crossed many men's minds. So James Meredith, his spirit quickened by John Kennedy's inaugural address, applied on January 21, 1961, for admission to the all-white state university at Oxford, Mississippi. Had Nixon been elected, he said later, he might not have applied.[3] Meredith was a lonely, taciturn and quixotic man of courage and purpose. "Nobody handpicked me," he said later. "I believed, and I believe now, that I have a Divine Responsibility to break White Supremacy in Mississippi, and getting in Ole Miss was only the start."[4]

Everyone recalls the roll of events grimly unfolding: Meredith's rejection by Ole Miss; his suit against the university, filed with the encouragement of Medgar Evers, the state NAACP director; the "calculated campaign of delay, harassment and masterly inactivity" waged by Ole Miss — a campaign, the Fifth Circuit Court added, that would have done credit to Quintus Fabius Maximus; the court's

ruling in June 1962 that Meredith must be admitted forthwith; more
delay, more harassment; on September 10, the ruling handed down
from the Supreme Court by Hugo Black, an Alabaman, commanding
no further interference with the judicial order; on September 13,
Ross Barnett, the governor of Mississippi, on statewide television:
"We will not surrender to the evil and illegal forces of tyranny."[5]

The Justice Department had been in touch with Meredith from
nearly the start and supported him as *amicus curiae* during the appeal
before Justice Black. Hoping to avert trouble, Robert Kennedy now
began a series of phone conversations with Barnett — twenty between
September 15 and September 28. He also considered the advisability
of a presidential statement. On September 17 he asked me to come
to the Department. My journal:

> I went over ... about six thirty and found the legal group in Bobby's
> garish office. Everyone was in shirtsleeves except Nick Katzenbach, who
> sat saturninely in the background and said softly that he saw no reason
> to issue a statement at that time. Guthman had drafted a statement, to
> which I added a quotation from the Mississippi Legislature in 1832
> denouncing the interposition doctrine. Bobby pondered this and then
> engaged in quiet conversation with his people about the best mode of
> protecting the Negro student when he tried to register. The mixture
> of shirtsleeved casualness, soft voices and evident determination was
> impressive. In the end, Bob agreed with Nick about the statement, and
> it was deferred. Instead, he issued quiet orders and his troops dis-
> persed. He told me, by the way, that he regarded Governor Barnett
> ... as genuinely loony — that he had been hit on the head by an
> airplane propellor last summer and had never been the same since.[6]

The conversations continued. Barnett wobbled between bluster
and placation. On September 25:

BARNETT. That's what it's going to boil down to — whether Mississippi
can run its institutions or the federal government is going to run
things. . . .

KENNEDY. I don't understand, Governor. Where do you think this is
going to take your own state?

BARNETT. A lot of states haven't had the guts to take a stand. We are
going to fight this thing. . . . This is like a dictatorship. Forcing him
physically into Ole Miss. General, that might bring on a lot of trouble. You don't want to do that. You don't want to physically force
him in.

KENNEDY. You don't want to physically keep him out. . . . Governor,
you are a part of the United States.

BARNETT. We have been a part of the United States but I don't know
whether we are or not.

KENNEDY. Are you getting out of the Union?

BARNETT. It looks like we're being kicked around — like we don't
belong to it. General, this thing is serious.

KENNEDY. It's serious here.

BARNETT. Must it be over one little boy — backed by communist front — backed by the NAACP which is a communist front? . . . I'm going to treat you with every courtesy but I won't agree to let that boy to get to Ole Miss. I will never agree to that. I would rather spend the rest of my life in a penitentiary than do that.

KENNEDY. I have a responsibility to enforce the laws of the United States. . . . The orders of the court are going to be upheld. As I told you, you are a citizen not only of the State of Mississippi but also of the United States. Could I give you a ring?

BARNETT. You do that. . . . Good to hear from you.[7]

But Barnett's tone was more often plaintive than truculent. Though a doctrinaire segregationist, he really did not wish to take on the federal government, and he liked to be liked. Kennedy later abandoned the idea that he was loony. "An agreeable rogue, and weak," he called him in 1964. ". . . I suppose I thought he must have been sane. I thought that after we got into it a bit that it was in his interest. I'm influenced a lot by whether what a person does or says he's going to do is in their interest in doing it."[8]

The pace accelerated. Federal marshals escorted Meredith to the university. They were several times turned back. Once Barnett read a grandiloquent proclamation rejecting Meredith — and then gave him the embossed document as a souvenir. Eastland suggested that Kennedy might back down a little. Kennedy: "You don't really believe that, Senator. You've been in the Senate too long to believe that." Barnett proposed a show of force, federal marshals with their guns drawn, as a cover for his own retreat. Even Eastland thought this ridiculous, but Kennedy was willing to try anything. When the moment came, the crowd seemed too menacing for the drama. "A lot of people are going to be killed," Barnett told Kennedy. "It would be embarrassing to me." "I don't know if it would be embarrassing," said Kennedy, "— that would not be the feeling." Good old boys from the country were coming by truckload into Oxford. In Washington apprehension grew that federal marshals might not be enough.[9]

"Sending in troops is a hell of a thing for the country," Robert Kennedy said early on.[10] His consuming fear was a mini–civil war with GIs and Mississippians shooting each other down. "What I was trying to avoid, basically," Robert Kennedy said in 1964,

was having to send troops. . . . What [Barnett] was trying to accomplish was the avoidance of integration at the University of Mississippi, number one, and, if he couldn't do that, to be forced to do it by our heavy hand; and his preference was with troops. Because he wants to be an agreeable fellow, he continued to have these conversations with me, where he got [in] deeper and deeper, where he was trying to satisfy and please me and trying to talk me out of James Meredith coming to the University

of Mississippi. While he was doing that, he ... got more and more
involved himself. He had people pulling and pushing at him from so
many different directions that I think he just got himself into a bigger
and bigger box. He eventually pulled me in with him.[11]

II

On September 28, Robert Kennedy met with General Maxwell Taylor, chairman of the Joint Chiefs of Staff, and Cyrus Vance, who had
become Secretary of the Army two months before. Maps of Mississippi were improbably unrolled in the Pentagon War Room. Later
that day the Attorney General told me

> he understood better now how Hitler had taken over in Germany.
> "Everyone in Mississippi is accepting what this fellow is doing," he
> said. "There are no protests anywhere — from the bar or from profes-
> sional men or from professors. [He forgot the valiant historian James
> Silver on the Ole Miss faculty.] I wouldn't have believed it." I asked
> him how his talks with Barnett had been. He said, "Some of them have
> been hard to believe. Today Barnett said to me, 'Why can't you per-
> suade Meredith to go to another college? I could get some money
> together and we could give him a fellowship to any university he wanted
> outside the state. Wouldn't that be the best way to solve the prob-
> lem?'" Bobby couldn't believe it.[12]

Up to this point the President had been held in reserve. The time
had come to see whether his personal intervention could sway Barnett. Early Saturday afternoon, September 29, I was summoned to
the Oval Office. The Attorney General, Burke Marshall and Kenneth
O'Donnell were with the President. "Matters were approaching a
climax, and the President was about to call Barnett. As the phone
rang, the President, with the air of a master of ceremonies, announced, 'And now — Governor Ross Barnett.' Bobby said, mocking
a prize-fight manager, 'Go get him Johnny boy.' The President,
rehearsing to himself, said, 'Governor, this is the President of
the United States — not Bobby, not Teddy, not Princess Radziwill'
[Jacqueline Kennedy's sister, then married to Prince Stanislaw
Radziwill]. Then the call came through, and he got down to
business."[13]

Kennedy was firm, patient and impersonal. "I don't know Mr.
Meredith," he said, "and I didn't put him in the University, but on
the other hand under the Constitution I have to carry out the orders
of the Court. ... I would like to get your help in doing it."[14] He
made this point in one form or another for several minutes. After
the conversation ended, the President said bemusedly, "You know

what that fellow said? He said, 'I want to thank you for your help on the poultry problem.'" Turning to his brother, he said, "You've been fighting a sofa pillow all week." They called Barnett back in another hour. The President reported that Barnett had said, "Just tell 'em to cool off and you can sneak him in. I'll bring everyone up to Oxford — and then you can register him in Jackson." Robert Kennedy said: "What a rogue!"[15]

Early that Saturday evening Barnett proposed sneak registration again. This time the Kennedys reluctantly accepted it. Three hours later Barnett canceled it. At midnight Norbert Schlei of the Office of Legal Counsel came to the White House with a proclamation ordering persons obstructing justice in Mississippi to desist and disperse and an executive order federalizing the Mississippi National Guard. Kennedy took Schlei into a small study on an upper floor, sat down at a table, turned on a lamp and read the documents. After a moment he asked, "Is this pretty much what Ike signed in 1957 with the Little Rock thing?" Schlei pointed to a few refinements. Kennedy signed, snapped off the light and headed into the hall. Then he paused and rapped the table with his hand. "You know," he said, "that's General Grant's table." They said good night. As Schlei went down the stairs to tell waiting reporters what had happened, Kennedy suddenly sprinted to the top balustrade and called down to him: "Don't tell them about General Grant's table."[16]

Sunday morning Robert Kennedy and Barnett had a rough conversation:

KENNEDY. The President is going on TV tonight. He is going through with the statement [he] had with you last night. He will have to say why he called up the National Guard; that you had an agreement to permit Meredith to go to Jackson to register. . . .
BARNETT. That won't do at all.
KENNEDY. You broke your word to him. . . .
BARNETT. Don't say that. Please don't mention it. . . . Why don't you fly him in this afternoon; please let us treat what we say as confidential. . . . We will go on and do it any way. Let's agree to it now and forget it. I don't want the President saying I broke my word. . . . We will cooperate with you.

The theory of registration on Sunday was that few students would be around. "You won't mind," Barnett said, "if I raise cain about it?" Kennedy said, "I don't mind that; just say law and order will be maintained."[17] It seemed finally that Barnett had come aboard. When Robert Kennedy went over that Sunday night to the White House to hear the presidential telecast, "we thought," he said later, "that it had been accomplished."[18]

III

"Oh, God, I hope nothing happens to Meredith," Robert Kennedy had said earlier that day to Budd Schulberg. "I feel responsible for him. I promised we'd back him up all the way — and I'm worried for McShane and the others, too."[19] James McShane, who had been on the Rackets Committee staff, was now chief marshal, and Kennedy had designated him as Meredith's escort. John Doar was with McShane and Meredith at the Millington Naval Air Station near Memphis. There, too, were Louis Oberdorfer and Joseph Dolan, rehearsing the federal marshals.

It was a typical mixture of preparation and improvisation. On Sunday morning Kennedy had asked Katzenbach whether he had any plans for the afternoon. Katzenbach said no. Kennedy said, "Well, I'd like you to go down there and follow through on this plan and be in charge of things." Katzenbach picked up Edwin Guthman and Norbert Schlei. Kennedy added an old football friend, strong as a bull and faithful in a pinch, Dean Markham.[20] Harold Reis, an able lawyer in the Office of Legal Counsel, poked his head into the office. Kennedy said, "Harold, you can help. Can you go too?" It was not till they were halfway to the Andrews Air Force Base that Reis learned where he was going. (He said later, "The next time I'll ask, 'Go where?'")[21] Burke Marshall and Ramsey Clark manned the Justice headquarters. Kennedy bade Katzenbach a characteristic farewell: "If things get rough, don't worry about yourself; the President needs a moral issue."*

In Oxford, Katzenbach found hundreds of people milling about the Lyceum, the university's administrative building. Soon Meredith arrived on campus flanked by McShane and Doar — "two of the bravest men I have ever known," Meredith called them later. "I never saw either of them weaken under any and all conditions of danger."[22] He was spirited into a dormitory, but the crowd supposed him in the Lyceum around which federal marshals, in white helmets and bright orange vests, had stationed themselves. After a time the

---

* Jean Stein and George Plimpton, *American Journey* (New York, 1970), 104. The humorless English editor Anthony Howard, reviewing *American Journey* in the *New Statesman* (June 18, 1971), seized upon this remark as proof of Robert Kennedy's callous determination to sacrifice everyone to the greater glory of the Kennedys. Katzenbach responded with admirable restraint that, while Howard had made "manifestly clear his dislike for humour in tense moments, he does not make explicit the fact that the statement was made in jest to cover up a deep concern and affection. Perhaps this fact will have some significance for those readers less familiar with the context than Mr. Howard, who may possibly have thought the remark serious" (*New Statesman*, July 2, 1971).

state police, despite Katzenbach's protestations, were withdrawn. Hard-faced men from the countryside, many armed, began to appear. The crowd, at first merely noisy and raucous, turned ugly. As it grew dark, people started to throw bricks and bottles. Five minutes before the President went on the air in Washington, Katzenbach told the marshals, some hurt and bleeding, to protect themselves by discharging tear gas.

Kennedy's speech was designed for a different occasion. It assumed that Meredith had arrived on campus in relative tranquility and appealed to white Mississippi's sense of honor, "won on the field of battle and on the gridiron." There was no reason, he concluded, "why the books on this case cannot now be quickly and quietly closed."[23] At Ole Miss the mob howled obscenities and surged forward toward the line of marshals. In a while the crash of bottles was punctuated by the crackle of bullets.

The President had no idea things were falling apart in Oxford. Now the reports from the front grew suddenly alarming. The Attorney General asked Guthman what it was like at the Lyceum. "Pretty rough," said Guthman. "It's getting like the Alamo." Robert Kennedy: "Well, you know what happened to those guys" — a famous response that, as Guthman wrote later (without irony), "raised our morale and helped us through the night."[24] With the marshals under siege — many injured, tear gas running low, gunfire increasing, reports of deaths — and under the President's personal order not to shoot except to protect Meredith's life,* Katzenbach made a bitter decision and asked Washington to send in the Army.

As usual, the Pentagon's plans were impressive. Cyrus Vance, looking at the military timetable, promised the President quick action. But in Memphis the troops had gone off alert after hearing the speech. When the summons arrived, the first group, according to the Army theory of riot control, was armed only with nightsticks. The situation in Oxford had long since passed this point. "So they stood out in the dark at Memphis," Katzenbach said later, "opening crates of guns and opening crates of gas ... by flashlight."[25] After two hours Katzenbach asked the White House where the devil the Army was. John Kennedy, in between conversations with Barnett, called Vance and the Pentagon. Vance, consulting his schedule, assured him the troops were on their way.

Robert Kennedy said later:

> That happened six, eight, ten times during the course of the evening. "They're leaving in twenty minutes." We'd call twenty minutes later, and they hadn't even arrived to get ready to leave. "They're ready to

* Compare with Kent State and Jackson State in 1970.

go now," and they hadn't been called out of their barracks to get into
the helicopters yet. "They're in the helicopters now." They were just
forming up. "The first helicopter's leaving and will be there in forty
minutes." The first helicopter went in the air and then circled and
waited for the rest of the helicopters. You know, all that kind of
business. . . . They said they'd get there in two hours if I gave them the
notice — and they didn't arrive for five hours. In the meantime, our
marshals were being overwhelmed. . . . Two people were killed right at
the beginning, so we had visions of the whole night — and then their
getting in and killing Meredith. . . . It was so frustrating for the Army
to continuously give false, wrong, inaccurate information.

As the hours dragged on, John Kennedy called Vance in the Pentagon, General Creighton Abrams in Memphis — "the worst and harshest conversations with Cy Vance and with the General that I think
I've ever heard him [conduct]," Robert recalled.[26]

Even after the troops arrived in Oxford, there was inexplicable
delay. At the Lyceum, the marshals had finally run out of tear gas.
At the airport, half a mile away, the field commander, operating by
the book, waited for his whole force to disembark and fall in before
moving. Agonizing moments passed. John Kennedy sent the commander a coldly furious message: "People are dying in Oxford. This
is the worst thing I've seen in forty-five years. I want the military
police battalion to enter the action. I want General Billingslea to see
that this is done."[27] In the end, it took the Army over an hour to
make the last half mile.

"It was a nervous time for the President," Robert Kennedy reflected, "because he was torn between, perhaps, an Attorney General
who had botched things up and the fact that the Attorney General
was his brother."[28] Robert Kennedy felt a terrible responsibility. His
brother had placed him in charge. Now everything was going
wrong. The call for troops was itself a confession of failure. Still
worse, the troops had not come. The President, wrote Theodore
Sorensen, who was present, "cursed himself for ever believing Barnett and for not ordering the troops in sooner."[29] Robert Kennedy
recalled: "We could just visualize another great disaster, like the Bay
of Pigs, and a lot of marshals being killed or James Meredith being
strung up."[30]

In retrospect, the Attorney General believed that it had been right
to delay the order to the troops. Barnett had promised to keep
things under control. Burke Marshall considered it "doubtful
whether the President had any constitutional choice about refusing
in advance to accept the word of the governor of a state."[31] In any
event, it would have been hard to justify military intervention before
trouble began. "If we had sent the troops in, and then they had a

riot," said Robert Kennedy, "that would have been disastrous." But, once the shooting started, "the only way to deal with that kind of operation is to have overwhelming force. . . . The question of the evening would have been quite different if the troops had gotten there at the time that they were supposed to have gotten there." Still, "I was the Attorney General. The fact that I said that the troops would arrive and they didn't arrive was my fault."[32]

In Oxford they hung on through the wild night. The Army and daybreak brought a melancholy quiet to the embattled campus. Guthman turned wearily to Katzenbach and said, "What'll we do in the morning?" "Why," Katzenbach replied, "we're going to register Mr. Meredith at 8 o'clock."[33] So they did, and another wall crumbled. In the next months Meredith doggedly pursued his destiny in face of harassment by loutish classmates. He prevailed. "I am a graduate of the University of Mississippi," he was able to write Robert Kennedy in September 1963. "For this I am proud of my Country — the United States of America. The question always arises — was it worth the cost? . . . I believe that I echo the feeling of most Americans when I say that 'no price is too high to pay for freedom of person, equality of opportunity, and human dignity.'"[34]

Robert Kennedy had already replied to Meredith's question in a speech the Saturday night after the battle of Ole Miss: "We live in a time when the individual's opportunity to meet his responsibilities appears circumscribed by impersonal powers beyond his responsibility. . . . But even today there is so much that a single person can do with faith and courage, and we have had a number of outstanding examples just this week. . . . James Meredith brought to a head and lent his name to another chapter in the mightiest internal struggle of our time."[35]

IV

The Oxford crisis, most of all the military intervention, shocked the white south. It also shocked the Kennedys. They had never thought it would come to this. Their assumption in 1961 was that unreasonable problems would yield to reason and law. Ole Miss showed them how stubborn, savage, deeply rooted the problems were. "We lacked," Guthman reflected, "a sense of Southern history."[36] They had been brought up to believe, for example, that Reconstruction was a matter of southern whites rescuing their states from ignorant and incompetent ex-slaves — the view reflected in *Profiles in Courage* where John Kennedy had described Thaddeus Stevens as "the crippled, fanatical personification of the extremes of the Radical Republican movement."[37] Now, after reading the Mississippi legislature's report on Oxford, he told Robert he could never take this view of Reconstruction again. "He said," his brother recalled, "that they can say these things about what the marshals did and what we were doing . . . and believe it. They must have been doing the same thing a hundred years ago."[38] The next year, when Medgar Evers, Meredith's friend and counselor, was murdered in front of his house in Jackson, the President said sadly to me, "I don't understand the South. I'm coming to believe that Thaddeus Stevens was right. I had always been taught to regard him as a man of vicious bias. But when I see this sort of thing, I begin to wonder how else you can treat them."*

After Oxford, the Kennedys began to understand how profoundly the republic had been trapped by its history. They were men endowed by a bitter ancestral history of their own with a conviction of the precariousness of social forms in face of the bloodiness of human existence. "Before my term has ended," John Kennedy had told Congress in his first annual message, "we shall have to test anew whether a nation organized and governed such as ours can endure. The outcome is by no means certain."[39] Nothing tested the nation more severely than the challenge of racial justice. If survival were the issue, far better to use force in support of law than to accept violence in defiance of law.

This was the way that most Americans, including many white southerners, perceived the battle of Ole Miss. Thus a courageous Mississippi legislator, Karl Wiesenburg, argued that Barnett had followed "the road to riot" and that the bloodshed at Oxford was the "price of defiance."[40] "Tell [Wiesenburg] for me," Robert Kennedy wrote the federal attorney in Oxford, "that he is the one who deserves a Pulitzer Prize or more. It is all very well for we in the North to talk about these matters. That takes no courage at all. But people such as you and Mr. Wiesenburg are the ones who really carry the banner."[41] The southern Vice President applauded the action. Lyndon Johnson had not in fact been consulted. "Rightly or wrongly," observed Guthman, "Bob felt his presence would complicate the discussions rather than further them."[42] Johnson had even been permitted to spend the climactic weekend on his Texas ranch and might well have been disaffected; but on his return to Washington he spoke to the southern Secretary of State, who passed his message along to the White House. "The situation in Mississippi," Dean Rusk reported Lyndon Johnson as saying, "had been handled better than he could ever have thought of handling it."[43]

---

* Author's journal, June 20, 1963. Alas, he had been thus taught by the Harvard history department.

Around the world the use of troops dramatized as nothing else could have done the commitment of the administration to the cause of racial justice. Howard P. Jones, the ambassador to Indonesia, wrote the Attorney General, "This was a battle which *had* to be won. . . . What might have been a severe set-back to our prestige in Asia and Africa was turned into a gain."[44] "For one small Negro to go to school," the representative of Upper Volta told the UN General Assembly, the American government "threatens governors and judges with prison. . . . It sends troops to occupy the University of Mississippi."[45]

Only American black leaders lacked enthusiasm for the Kennedy performance. Sitting with Martin Luther King and James Baldwin, the black novelist, in a Boston railway station shortly after Oxford, William Goldsmith, the political scientist, listened to their "despairing reactions. . . . It was not only the clumsiness and hesitancy with which the whole business was handled that depressed them, but the lack of moral conviction in the President's remarks as he spoke patronizingly of war heroes and football stars to the rioting students and yahoos in Oxford that awful Sunday night."[46] The behind-the-scenes dealing with Barnett, King felt, "made Negroes feel like pawns in a white man's political game."[47]

King was despondent. In March 1963 he described 1962 as "the year that civil rights was displaced as the dominant issue in domestic politics. . . . The issue no longer commanded the conscience of the nation." He ascribed this to the readiness to accept token victories — one man in the University of Mississippi — as evidence of serious change. "In fairness," King concluded, "this Administration has outstripped all previous ones in the breadth of its civil-rights activities. Yet the movement, instead of breaking out into the open plains of progress, remains constricted and confined. A sweeping revolutionary force is pressed into a narrow tunnel."[48]

v

On February 23, 1963, John Kennedy finally submitted a legislative program on civil rights. His message to Congress enumerated in ringing language the costs of racial discrimination. "Above all," he said, "it is wrong."[49] His recommendations included, as King agreed, "constructive measures," especially in the voting field.[50] But on the whole the proposals gravely disappointed the civil rights leadership — nothing serious about fair employment, school desegregation, public accommodations or federal protection for civil rights workers.

Burke Marshall was convinced that school desegregation legislation could not possibly pass and that "problems of law enforcement in some places, before the free exercise of the franchise, are almost insurmountable at present under the federal system."[51] So voting was still paramount. Could even voting reform pass? "We thought that it was very difficult," Robert Kennedy said in 1964, "but that it was such a basic right, that we had so much on our side, that we might be able to get it through." The bill, however, fell into a vacuum of apathy. "There wasn't any interest in it," said Kennedy. "There was no public demand for it. There was no demand by the newspapers or radio or television. There was no interest by people coming to watch the hearings. . . . Nobody came. Nobody paid any attention."[52]

In the meantime, King and the SCLC leadership held a two-day strategy review in Dorchester, Georgia. A tacit alliance, they decided, had developed between the administration and the movement. If the movement could create a public demand, the administration would respond. They decided also to make their next move in Birmingham, Alabama. If that brutal city, so raw in its emotion and so grim in its segregation, could be turned around, "the results," as Stanley Levison, King's close associate, put it, "would radiate across the South."[53] Early in April 1963 King launched a campaign to end discrimination in the department stores and lunch counters of downtown Birmingham, where the police commissioner Bull Connor had boasted a year earlier he could solve the racial problem with "two policemen and a dog."[54]

A moderate city administration was about to take over in Birmingham. Robert Kennedy urged King not to force the issue while the horrible Connor was still in power. King, who had agreed to an earlier postponement, declined this one. "For years now I have heard the word 'Wait,'" he soon wrote. ". . . This 'Wait' has almost always meant 'Never.'"[55] He wrote this from the jail where Bull Connor thrust him after he defied a local injunction and led a protest march on Good Friday. King was held incommunicado. His wife, unable to reach him, telephoned John Kennedy. She could not reach him either. Then Robert Kennedy returned her call: "The President wasn't able to talk to you because he's with my father who is quite ill. He wanted me . . . to find out what we can do for you." Coretta King expressed her fears for her husband. "We have a difficult problem with the local officials," the Attorney General replied. "Bull Connor is very hard to deal with. . . . But I promise you I will look into the situation." The FBI in Birmingham checked that night. The next day John Kennedy told Mrs. King that her husband was well and would ring her shortly. In fifteen minutes King was on the phone. She told him about her appeal to the Kennedys. King said, "So that's why everybody is suddenly being so polite."[56]

King had put his time to good use composing his moving Letter from Birmingham Jail. Released on bond after eight days, he led the campaign into a new phase. On May 2 several thousand black children marched. Bull Connor's police arrested almost a thousand. King characteristically thought that participation in the protest would give the children pride in their race, belief in their capacity to influence events, a "sense of their own stake in freedom and justice."[57] Robert Kennedy characteristically thought the children might be hurt.[58] (Malcolm X objected too: "Real men don't put their children on the firing line.")[59] Another march took place on May 3. This time Connor's cops retaliated frightfully with nightsticks, pressure fire hoses and growling police dogs. On May 4 newspapers across the land (and around the world) ran appalling photographs of the dogs, teeth bared, lunging viciously at the marchers.

The Attorney General sent Burke Marshall and Joseph Dolan to Birmingham. They found a desperate situation. The white leaders would not talk to Martin King. "They wouldn't talk to anybody that *would* talk with him," Marshall said later. ". . . Somebody had to bridge that." The blacks had their own suspicions about Marshall. King subsequently acknowledged "initial misgivings," fearing that Marshall had been sent to urge a cooling-off period.[60] Marshall met around the clock with everybody, listening endlessly, talking quietly, forever seeking common ground. He tried to show the whites that King was not intending revolution; all he wanted was recognition of "what was right and what was unfair. And once he got that in a way that was public, then that was enough."[61] He tried to persuade the blacks to accept pledges of future action on desegregation and job discrimination: the important thing was to establish the principle.

A week of marathon negotiations produced an agreement on a measure of desegregation and black employment. Then Bull Connor arrested King again. The Reverend Fred Shuttlesworth, another black leader, prepared to lead the marchers back into the streets. A bloody clash loomed ahead. "Once again," wrote King's best biographer, "the Kennedy touch was needed."[62] Dolan phoned the Attorney General. "It was just a real tense situation," Andrew Young, King's top lieutenant, recalled. ". . . I never knew what [Kennedy] said to Fred Shuttlesworth. But when they got through talking, things, you know, were pretty much resolved. And I guess I began to get an appreciation for that kind of way of working with people, rather than with principles and laws. . . . He was very much the diplomat. And, you know, I began to like that."[63] The march was canceled, and Kennedy secured King's release. On May 10 the Birmingham settlement was announced.

That night segregationists bombed the house of King's brother and the motel where the civil rights leaders had their headquarters. In savage response angry blacks streamed down the streets, throwing rocks, breaking windows, burning shops and even attacking police officers. "Let the whole fucking city burn," some shouted. ". . . This'll show the white motherfuckers!"[64] Nothing like it had ever happened in the south. "The passivity and nonviolence of American Negroes could never again be taken for granted," two southern political scientists wrote later. ". . . The 'rules of the game' in race relations were permanently changed in Birmingham."[65]

Despite the renewal of violence then and in months to come, the Birmingham settlement held together. In that summer and that city it was a remarkable achievement. Marshall "did an invaluable job," King finally conceded.[66] "Bobby Kennedy, Burke Marshall and Joe Dolan . . . did some phenomenal things," Andrew Young said. ". . . You began to get a kind of unofficial, personal reconciliation with both whites and blacks, which was very new. . . . Birmingham was in such a state then that either the black or white community could have gotten out of hand any minute. . . . The country in 1963 could have gone either way."[67]

Not everyone was so pleased. Arthur Hanes, the retiring mayor, cursed out Robert Kennedy. "I hope that every drop of blood that's spilled he tastes in his throat," Hanes said, "and I hope he chokes to death."[68]

VI

On May 24, a few days after the Birmingham settlement, Robert Kennedy met with a group of black writers and artists brought together by the novelist James Baldwin. The meeting came about in the following manner.

Birmingham convinced Kennedy that the next great battlefield for racial justice lay in the cities. James Baldwin was born in Harlem. His extraordinary *New Yorker* piece of November 1962, "Letter from a Region in My Mind," had exposed in searing words the humiliation, despair and rage of Negro Americans. Baldwin and Kennedy had met the year before at the White House dinner for Nobel Prize laureates. They had agreed then that they wanted to talk some more. Now Kennedy invited Baldwin to breakfast at Hickory Hill.

Baldwin was a brilliant, passionate, sensitive, dramatic man imbued with a conviction of utter hopelessness about the black fate in white society. "The Negro's experience of the white world," he had written in the *New Yorker*, "cannot possibly create in him any respect for the standards in which the white world claims to live."[69] His belief, one felt, was that all whites by definition hated all blacks and that white liberals were worst of all because they pretended to deny their innermost feelings. Nonetheless he caught an early plane to Washington. "We had a very nice meeting," Kennedy said later.[70] "I was really quite impressed by him," said Baldwin. ". . . He seemed honest and earnest and truthful."[71] Burke Marshall, who was present, said, "He and Bob Kennedy had a rather good conversation about the cities."[72] Baldwin's plane had been late, however, and Kennedy had to leave for another engagement. He therefore proposed that Baldwin assemble a group with thoughts about the northern ghetto.

This led to the meeting in the Kennedy family apartment at 24 Central Park South in New York. Baldwin, acting on short notice, made an effort to enlist experts on the northern city, like Kenneth B. Clark, the social psychologist, Edwin C. Berry of the Chicago Urban League and Clarence B. Jones, an attorney for Martin Luther King. But what Baldwin called "this sociology and economics jazz" was not his métier.* He also invited the playwright Lorraine Hansberry and the singers Lena Horne and Harry Belafonte — artists concerned, like Baldwin himself, less with solutions than with the anguish of the problem. Then he brought along Jerome Smith, a young civil rights worker, who began as a Gandhian pacifist, became a Freedom Rider and a CORE field worker and, according to the historians of CORE, "had probably spent more months in jail and been beaten more often than any other CORE member."[73] He was now in New York for medical treatment.†

Robert Kennedy and Burke Marshall had spent an unpleasant morning urging owners of chain stores to desegregate their lunch counters in the south. The white executives seemed to feel, Kennedy recalled, that "the Devil Incarnate had arrived in New York, and they were asked to meet with him."[74] They turned out not to be the only people in New York that day who regarded the Attorney General as the devil incarnate.

Clark and Berry arrived armed with statistics and proposals. They never had a chance. Jerome Smith, as Baldwin put it later, "set the

---

* Harold Cruse, *The Crisis of the Negro Intellectual* (New York, 1967), 193–194. Cruse adds: "This failure to discuss the racial conflicts either in terms of possible practical solutions, or in terms of American economic and sociological realities, made Baldwin's assault on white liberals a futile rhetorical exercise; it was further weakened by the intellectual inconsistencies, incoherence and emotionalism of his line of argument."

† Also present were Edward False, James Baldwin's secretary, Baldwin's brother David and Thais Aubrey, a friend of David Baldwin's; and three white men: Rip Torn, the actor, Robert P. Mills and Henry Morgenthau III, a television producer who was to tape an interview of Baldwin by Kenneth Clark later in the afternoon (R. P. Mills to Burke Marshall, June 7, 1963, Marshall Papers).

tone of the meeting because he stammers when he's upset and he
stammered when he talked to Bobby and said that he was nauseated
by the necessity of being in that room. I knew what he meant. It
was not personal at all. . . . Bobby took it personally." This was
perhaps understandable. To say, as the Attorney General heard it,
that being in the same room with Robert Kennedy made him feel
like throwing up seemed a rough way to begin. "Bobby took it
personally," Baldwin continued, "and turned away from him. That
was a mistake because he turned toward us. We were the reasonable,
responsible, mature representatives of the black community. Lorraine Hansberry said, 'You've got a great many very, very accomplished people in this room, Mr. Attorney General. But the only
man who should be listened to is that man over there.'" [75]

Smith talked on with vehement emotion. He told what he had
been through in the south. He said he was not sure how much longer
he could stay nonviolent. He said, "When I pull the trigger, kiss it
good-bye." Baldwin asked him whether he would fight for his country. He said, "Never! Never! Never!" [76] This shocked Kennedy, for
whom patriotism was an absolute. "We were shocked that he was
shocked," said Kenneth Clark. ". . . Bobby got redder and redder
and redder, and in a sense accused Jerome of treason, you know, or
something of that sort. Well, that made everybody move in to protect
Jerome and to confirm his feelings. And it became really an attack!" [77]

"This boy," Lena Horne said afterward, "just put it like it was. He
communicated the plain, basic suffering of being a Negro. The
primeval memory of everyone in that room went to work after
that. . . . He took us back to the common dirt of our existence and
rubbed our noses in it. . . . You could not encompass his anger, his
fury, in a set of statistics, nor could Mr. Belafonte and Dr. Clark and
Miss Horne, the fortunate Negroes, keep up the pretense of being
the mature, responsible spokesmen for the race." [78] Lorraine Hansberry said to Kennedy, "Look, if *you* can't understand what this
young man is saying, then we are without any hope at all because
you and your brother are representatives of the best that a white
America can offer; and if *you* are insensitive to this, then there's no
alternative except our going in the streets . . . and chaos." [79] The
whites, she said, were castrating the Negroes. She talked wildly about
giving guns to Negroes in the street so that they could start killing
white people. [80]

Kennedy said, as he had before, that his grandparents had encountered discrimination and now, two generations later, his brother
was President; a Negro would be President within forty years. Baldwin replied furiously, "Your family has been here for three genera-

tions. My family has been here far longer than that. Why is your brother at the top while we are still so far away? That's the heart of the problem."[81] Kennedy tried to turn the conversation to statistics and legislation. "In that moment, with the situation in Birmingham the way it was," said Lena Horne, "none of us wanted to hear figures and percentages and all that stuff. Nobody even cared about expressions of good will."[82]

Kennedy said he had come in search of ideas. Baldwin suggested that John Kennedy personally escort two black students whom Governor George Wallace was loudly threatening to bar from the University of Alabama. Robert Kennedy thought this theatrical posturing. (Asked about this proposal two days later, James Meredith said, "I don't think this would be the proper thing for a President of the United States to do.")[83] They denounced the FBI. Kennedy, who privately agreed, passed the question to Burke Marshall, who said that "special men" — that is, lawyers from the Civil Rights Division — went into areas where the Bureau seemed delinquent. This answer, said Clark, "produced almost hysterical laughter."[84] Birmingham came up. When Kennedy explained how closely he had worked with Dr. King, they jeered and cried, "That's not true."[85]

Kenneth Clark: "It became really one of the most violent, emotional verbal assaults . . . that I had ever witnessed before or since." Baldwin: "Bobby didn't understand what we were trying to tell him . . . didn't understand our urgency." Kennedy: "They seemed possessed. They reacted as a unit. It was impossible to make contact with any of them." Clark: "Bobby became more silent and tense, and he sat immobile in the chair. He no longer continued to defend himself. He just sat, and you could see the tension and the pressure building in him."

It went on for three hours; then suddenly stopped, out of sheer exhaustion. Two incidents as the meeting broke up increased Kennedy's shock. Clarence Jones, King's lawyer, drew Kennedy aside and said, "I just want to say that Dr. King deeply appreciates the way you handled the Birmingham affair." Kennedy said, "You watched these people attack me over Birmingham for forty minutes, and you didn't say a word. There is no point in your saying this to me now." Harry Belafonte had tried to smooth things over earlier by mentioning to the group the hospitality he had enjoyed at Hickory Hill; after a time, he had fallen into uncomfortable silence. Now he said to Kennedy, "Of course you have done more for civil rights than any one else." Kennedy said, "Why do you say this to me? Why didn't you say this to the others?" Belafonte said, "I couldn't say this to the others. It would affect my position with these people, and I have a chance to influence them along a reasonable way. . . . If I sided with you on these matters then I would become suspect."[86]

The meeting shattered them all. Clark and Baldwin talked it over that night. "Our considered judgment," Clark said, "was that the whole thing was hopeless; that there was no chance that Bobby heard anything that we said. . . . Kennedy was not unimpressive. He didn't minimize or condescend. But he just didn't seem to get it."* Clark told James Wechsler in a day or two, "The fact that Bobby Kennedy sat through such an ordeal for three hours proves that he is among the best the white power structure has to offer. . . . There were no villains in that room — only the past of our society."[87] That is what made it all seem so hopeless.

Kennedy said to me on his return to Washington, his voice filled with despair: "They don't know what the laws are — they don't know what the facts are — they don't know what we've been doing or what we're trying to do. You can't talk to them the way you can talk to Martin Luther King or Roy Wilkins. They didn't want to talk that way. It was all emotion, hysteria — they stood up and orated — they cursed — some of them wept and left the room." I worried (as I noted) "that his final reaction would be a sense of the futility rather than of the urgency of trying to bridge the gap. He may have felt himself that I might have had this fear, because he called me later in the evening on another matter and seemed thoroughly calm."[88]

In subsequent months he tried to explain to himself why there had been this outburst against him. Of course the meeting was misconceived. He was interested in policy, the blacks, Clark and Berry apart, in witness. What Clark called the "excruciating sense of impasse"[89] seemed inevitable. Guilt, Kennedy thought, was also a factor. "There was a complex about the fact that they had not been involved personally themselves, they were not suffering like Negroes were suffering. . . . You've seen it with white people . . . children of wealthy parents . . . who've got some social problem about where they stand in life . . . and therefore become extreme and difficult emotionally."[90]

---

* Kenneth Clark, in recorded interview by Jean Stein, January 30, 1970, 5, Stein Papers; Clark, in interview by author, September 9, 1976. Clarence Jones on June 7 sent a letter to the *New York Times*, in which he said: "The Attorney General and Mr. Burke Marshall have been more vigorous in their prosecution of actions in behalf of civil rights than any previous Administration. . . . [But] when the chief legal officer of the United States displays a certain shock over the sentiments expressed by the participants at that meeting, that fact . . . is indicative of how the Administration underestimates the explosive ingredients inherent in the continued existence of racial discrimination." He sent a copy to Robert Kennedy, who sent it along, on June 12, to Marshall with the wry note: "He is a nice fellow & you have swell friends."

But none of this could quite explain away the violent jolt he received that spring afternoon. He began, I believe, to grasp as from the inside the nature of black anguish. He resented the experience, but it pierced him all the same. His tormentors made no sense; but in a way they made all sense. It was another stage in education. Thirteen years later it remained for Kenneth Clark "the most intense, traumatic meeting in which I've ever taken part . . . the most unrestrained interchange among adults, head-to-head, no holds barred . . . *the* most dramatic experience I have ever had."[91]

VII

But education came in the main from the rush of events. Racial justice was no longer an issue in the middle distance. Robert Kennedy now saw it face to face, and he was on fire. To complete the Birmingham settlement, he had to get King's demonstrators out of jail. He called on his old UAW friends, Walter Reuther, Jack Conway and Joseph Rauh, and $160,000 in bail money was rapidly on its way. He began to prod the federal government across the board. For two years he had heard from Lyndon Johnson's Committee on Equal Employment Opportunity bland reassurances about the steady increase of blacks in government jobs. But, when the committee met on May 29 — its first meeting since the previous November — he discovered that, aside from the Post Office and a Veterans' Administration hospital, only 15 of the 2000 federal employees in Birmingham were black — less than 1 percent in a city that was 37 percent black.[92] He was furious.

He remained furious. At the committee's next meeting, twenty days later, the Attorney General was on the warpath. Jack Conway was there representing the federal housing agency. "Within a matter of three or four minutes," Conway recalled, "the Vice President found himself on the defense because Bob just tore in . . . and asked for facts and statistics." Kennedy turned on James Webb, head of the National Aeronautics and Space Administration, an amiable and competent executive and a Johnson protégé. NASA had contracts of $3.5 billion and, Kennedy discovered, only one and a half men in Washington working on equal employment. A stinging colloquy followed:

KENNEDY. I asked how many people you have got in your program and you said one and a half men.
WEBB. I am not sure. I can furnish you that, but with nine centers throughout the country we are putting a great deal of effort on it. . . .
KENNEDY. What is the experience and background of the people you have selected for these jobs? . . . Do they have other responsibilities?

WEBB.   Oh, yes. They have the responsibility, if they are in the Personnel Department, of operating a total personnel system of which
this is a part. . . .
KENNEDY.   Mr. Webb, I just raised a question of whether you can do
this job and run a center and administer its $3.9 billion worth of
contracts and make sure that Negroes and non-whites have jobs. . . .
I think that unless we can get down into the specifics, Mr. Webb,
unless you get down to the specifics and the particular individuals
and find out what they are doing and have them understand that
. . . this Committee and the President of the United States are interested in this program, I don't see that the job will be done. . . .
WEBB.   I would like to have you take enough time to see precisely what
we do.
KENNEDY.   I am trying to ask some questions. I don't think I am able
to get the answers, to tell you the truth.

Johnson, obviously angry, slumped grimly in his chair, his eyes half
closed. "It was," said Conway, "a pretty brutal performance, very
sharp. It brought tensions between Johnson and Kennedy right out
on the table and very hard. Everybody was sweating under the
armpits. . . . And then finally, after completely humiliating Webb
and making the Vice President look like a fraud and shutting Hobart
Taylor up completely, he got up. He walked around the table . . .
shook my hand . . . and then he went on out."[93]

Subsequently Willard Wirtz, the Secretary of Labor, charged that,
in spite of the committee's self-congratulatory publicity, two thirds of
the companies holding government contracts still did not employ
blacks. Kennedy backed Wirtz vigorously. It was not enough, Kennedy told the committee, to make "generalized statements." "Why
hasn't somebody followed it up, what companies have you actually
gotten in touch with, is anybody going to see any of the
companies? . . . You have to go out and look for these people. I
thought that that should be the policy and that we should push
it."[94] In July, Robert Kennedy told his brother that they

> were having a difficult time with Johnson and the committee because of
> his oversensitivity of any outside criticism. One of the great problems
> has been the fact that this committee has not really accomplished what
> . . . the press releases would indicate that it had, the result being a
> stepped up effort which Burke, Bill Wirtz and I have had to take with
> a great deal of care. Johnson was so furious several weeks ago with Bill
> Wirtz and for a while it appeared that he would never talk to him. This
> was all because Bill Wirtz had given some statistics on the number of
> government contractors who had no Negro employees.[95]

These were, Kennedy recalled, "the sharpest disputes I had with
Vice President Johnson." When he took Wirtz's statistics over to his
brother, the President "almost had a fit. . . . He said, 'That man can't
run this Committee. Can you think of anything more deplorable

than him trying to run the United States? That's why he can't ever
be President.'"[96] This was the autumn of 1963.

VIII

A new crisis was building in Alabama. An ambitious and cocky
politician named George Wallace had been elected governor. During
his campaign he had promised to place himself in the doorway of
any schoolhouse under court order to admit blacks. On his inauguration in January 1963 he noted that he was standing where Jefferson
Davis had stood when he became president of the Confederate States
of America; and now, "from this Cradle of the Confederacy, this very
heart of the great Anglo-Saxon Southland, . . . I draw the line in the
dust and toss the gauntlet before the feet of tyranny. And I say,
Segregation now! Segregation tomorrow! Segregation forever!"[97]
Martin Luther King considered Wallace "perhaps the most dangerous racist in America today. . . . I am not sure that he believes all the
poison he preaches, but he is artful enough to convince others that
he does."[98]

King was right. Privately Wallace was less intransigent than his
rhetoric. Aware that black students were applying to the University
of Alabama at Tuscaloosa, he had asked Senator John Sparkman to
say to the White House, "Please give Alabama as much breathing
time as possible," adding that he was under no misapprehension
about his ability to stop anything.[99] "Perhaps if we put this off until
next September," John Kennedy had told his brother in November
1962, Ole Miss etched in both their minds, "we might . . . get a
favorable solution."[100] But delay was impossible. On May 21 the
federal district court ruled that the university must admit the black
applicant to its summer session.

The Attorney General had already begun preparations to prevent
Tuscaloosa from becoming another Oxford. The university's president and board of trustees were ready to comply with the court
order. The administration now sought to rally Alabama business
leaders in a campaign to dissuade Wallace from defiance. "We wrote
down into a book," Robert Kennedy recalled, "the names of every
company with more than one hundred employees, I think, in the
whole state of Alabama. All those names were distributed at a cabinet
meeting. . . . A cabinet member . . . called, I guess, every one of
them."[101] He tried to talk to Wallace himself. The governor refused
his calls — when asked his opinion of Robert Kennedy, his reply,
according to an observer, was "too revoltingly obscene" for repetition[102] — but finally consented to a meeting.

On April 25, Kennedy, accompanied by Marshall and Guthman,

arrived at the state capitol. State troopers, their steel helmets painted
with Confederate flags, surrounded the building. "The point," recalled Kennedy, "was to try to show that my life was in danger in
coming to Alabama because the people hated me so much." Actually,
"they were all very friendly, the people. I shook hands with everybody." But Wallace "had the biggest State Troopers you ever saw all
guarding the way in, and they all had big sticks. . . . One of them
took his stick and put it into my stomach . . . and belted me with the
stick. . . . Not for laughs at all. They were most unfriendly." [103]

Wallace began by asking Kennedy whether he had any objection
to his taping their conversation. "Whatever you like, Governor,"
Kennedy replied, realizing at once that Wallace planned to use the
meeting for political effect. After desultory chat, Kennedy said, "I
didn't know whether we might discuss the problem we are perhaps
facing here in the state." "I don't hear good," said Wallace. Kennedy
said, "I don't think that the Union or the country means anything
unless the orders of the land are enforced." He continued:

> KENNEDY. Do you think it is so horrifying to have a Negro attend the
> University of Alabama?
> WALLACE. I think it is horrifying for the federal courts and the central
> government to rewrite all the law and force upon people that which
> they don't want. . . . For a hundred years we operated under what
> the Supreme Court said was legal . . . and spent millions and millions
> of dollars building a beautiful educational system. . . . I will never
> submit myself voluntarily to any integration in a school system in
> Alabama. . . . There is no time in my judgment when we would be
> ready for it in my lifetime.

They continued talking across each other, Kennedy reiterating his
duty to uphold the Constitution, Wallace reiterating his determination not to submit to federal tyranny and hoping as he went to trap
Kennedy into damaging admission. After a time Wallace said it
looked as if the matter would wind up in court. "As long as it winds
up in Court," Kennedy said, "I will be happy, Governor. That is all
I ask. . . . I just don't want it to get in the streets. I don't want to
have another Oxford, Mississippi." This gave Wallace his opening:
"You folks are the ones that will control that matter, because you
have control of the troops."

> KENNEDY. We have a responsibility, Governor, to insure that the integ-
> rity of the courts is maintained . . . and all of the force behind the
> Federal Government will be used to that end.
> WALLACE. . . . That is what you are telling me today, if it is necessary
> you are going to bring troops into Alabama.
> KENNEDY. No, I didn't say that, Governor.
> WALLACE. Well, you said all the force of the Federal Government.

KENNEDY. To make sure that the orders of the court are obeyed.

WALLACE. I know, but all the force includes the troops, doesn't it?

KENNEDY. Well, I would hope . . . that we would be able to litigate a settlement.

WALLACE. But it does involve troops if the law is disobeyed.

KENNEDY. Well, I am planning and hoping that the law will be obeyed. . . . Maybe somebody wants us to use troops, but we are not anxious to.

Wallace caught the personal thrust of Kennedy's last remark and fell into a trap himself. "I can assure you this," he said, "that I do not want you to use troops, and I can assure you that there is not any effort on my part to make a show of resistance and to be overcome." Kennedy said dryly, "I am glad to hear that." As they talked on, Kennedy appeared increasingly incredulous, speaking, wrote Marshall Frady, "with a curious low, dull uncertainty, as if not quite sure that he was actually engaging with the governor of Alabama about the kind of question he seemed to be engaged in." Toward the end Wallace tried to retrieve his point about the troops. "According to you all's attitude," he said, "you in the background say we may have to send troops and jail you as a governor of a state." "We never said that, Governor," Kennedy said sharply. In a moment, Wallace: "Let us get it straight. You would not use —" Kennedy replied that he expected the court order would be obeyed. "I have no plans to use troops. You seem to want me to use troops." Wallace: "Well, then we could say you are not going to use troops." Kennedy objected, rallied, evaded Wallace's snare and returned the conversation to the question of the supremacy of law. They circled around each other for a few more moments.[104] Then Kennedy departed in a state of bemusement.

Three weeks later John Kennedy himself went south. He spoke first in Tennessee. "For one man to defy a law or court order he does not like," the President said, "is to invite others to defy those which they do not like, leading to a breakdown of all justice and all order." He added that any citizen who sought "to subject other human beings to acts that are less than human, degrades his heritage."[105] He went on to Alabama, where Wallace accompanied him on the helicopter from Muscle Shoals to Huntsville. They talked about Birmingham. Kennedy observed that the very people who objected to Negroes working in downtown stores had Negroes serving their tables at home. Wallace said that the problem in Birmingham was that "faker" Martin Luther King. King and Fred Shuttlesworth, Wallace elegantly continued, vied with each other to see "who could go to bed with the most nigger women, and white and red women too."[106]

IX

The Kennedys could not make out what Wallace was going to do. It looked, Burke Marshall told the cabinet on May 21, as if he meant "personally to prevent the entry of a Negro student into the University. He has intended to require his own arrest or removal by force, and to require the federal government to use troops to enforce the court order."[107] Bulletins from the president of the university and from anti-Wallace trustees were not illuminating. "They'd report," Robert Kennedy recalled, "that he was acting like a raving maniac. So the result was that when you're trying to make plans . . . should you make plans on the basis of what a reasonable man would do? . . . You couldn't do that. The great question for us was not having to arrest the Governor, not having to charge him with contempt, and not having to send troops into the state of Alabama — and yet getting the students into the University."[108]

Did Wallace really want to be arrested? "We changed our minds a number of times about that," Kennedy recalled.

> I don't think that he wanted to go to jail, and there were some tough federal judges in Alabama who might have sent him to jail. . . . The other major question for us, if he was going to stand in the door, how many troops did we need? We didn't want to go through the same thing that we went through at the University of Mississippi. If you start having violence and you only have 300 troops and the police weren't going to help, how many troops did you need? . . . Dr. [Frank] Rose [the university president] made it quite clear that [Wallace] wouldn't step aside . . . for marshals. We had to have troops to get him to step aside. . . . So the final plans for the University of Alabama were really not made until that morning.[109]

June 11 was the day the black students planned to register. The federal judge warned Wallace's lawyers that, if Wallace violated the court order in a substantial way, he would receive a substantial sentence. The Attorney General also contemplated the immediate federalization of the Alabama National Guard. "That was before the first confrontation, so the first step would have been taken by us rather than by them. The President said, 'Let's wait,' . . . and that was the correct thing, as it turned out."[110]

"It is June 11th," Robert Kennedy wrote a Virginia friend, "and I am in the midst of trying to register two of your fellow countrymen at the University of Alabama and for one reason or another am having some difficulty."[111] Someone had proposed bringing the students onto the campus with troops. Katzenbach, who was in charge in Tuscaloosa, disagreed: "I think we should let Wallace have his show." In the early morning, Kennedy and Katzenbach worked out a final plan over the telephone. Katzenbach would escort the students to the university ("Dress as if you were going to church," John Doar told them), leave them in his automobile and, if Wallace blocked the doorway to registration, simply declare them registered and take them to dormitories to which they had already been assigned. This would permit Wallace to have his show. It would also permit the Justice Department to avoid arresting him for contempt because the students would not be physically present. "Dismiss George Wallace as a sort of second-rate figure," Robert Kennedy told Katzenbach, "wasting your time, wasting the students' time, causing a great fuss down there." If Wallace persisted in obstruction, then federalize the Guard; but leave the onus for that on Wallace. (And have the regular army on the ready at Fort Benning across the border in Georgia, troops sitting in helicopters; the Attorney General was not about to repeat the dismal Oxford experience with a dilatory army.)[112]

Shortly before 11 A.M. Katzenbach arrived and began the long walk under the broiling Alabama sun to the registration building. Inside the building a highly nervous Wallace repeatedly asked a state detective, "Ben, do you think they'll actually arrest me?" He sent word to Katzenbach through the National Guard commander that there would be no problem; he just wanted a chance to say something.[113] Now, standing incongruously behind a lectern, a microphone draped around his neck, he awaited Katzenbach. White semicircles painted in the doorway indicated where Wallace should stand in order to look his best on television; this irritated Katzenbach, who felt the situation was sufficiently theatrical already. Katzenbach began by saying he had a presidential proclamation commanding Wallace to cease and desist from unlawful obstructions. Wallace, holding up his hand like a traffic cop, stopped Katzenbach and read a proclamation of his own denouncing the "Central Government." They stood together, the rangy Katzenbach towering over the bantam governor, "exchanging vaguely irritable and exasperated phrases," as Frady saw it from a distance, "like a short, idle, haphazard argument on some street corner; Katzenbach with arms folded tightly and a faint expression of pained sufferance, beginning to glisten a little with sweat." Katzenbach again asked Wallace to step aside: "If you do not, I'm going to assure you that the orders . . . will be enforced. From the outset, Governor, all of us have known that the final chapter of this history will be the admission of these students." "I stand according to my statement," said Wallace. Katzenbach went back to the car. Wallace retired inside the auditorium.[114]

The President federalized the National Guard. Katzenbach took the students to their dormitories. No one stood in the doorway there. When the students appeared again for registration in the afternoon, Wallace stepped aside. Nor did he block the enrollment two days later of a third black student at the university's branch in Huntsville.

Wallace had lost — essentially, Robert Kennedy thought later, because of the appeal to respectable opinion in the state. "He was getting fifteen to twenty calls every day from business people," said Burke Marshall. "He'd have to guarantee them that there wouldn't be any repeat of the Oxford incident, that there wouldn't be any violence." Kennedy said: "Wallace really didn't know what he was going to do, I'm convinced, up until at least a couple of days before and, maybe, a few hours before. . . . If he had received great popular support in the state . . . standing in the door, continuing to stand in the door, I think that's what he would have done. But he was trying to get off the hook." [115]

Wallace explained his defeat: "I can't fight bayonets with my bare hands." Skeptics noted that there had been no bayonets to fight. [116] Still, as Robert F. Kennedy, Jr., wrote perceptively in 1978, Wallace "did not fear defeat as a political pitfall." "A good fight and a noble defeat" were never disqualifications in the south. "In the past, defeat had been imposed upon Alabama, and in a way it had a redemptive quality." [117] The tableau remained: the defiant Wallace standing in the schoolhouse door; the sweating Katzenbach making the lonely walk back to the limousine.

The encounter did not extinguish Wallace. But black students were now in the University of Alabama. At the end of the long day, Robert Kennedy scrawled notes to his children. To David: "Shortly after you left my office the two negroes attempted to register. . . . They were opposed by Governor Wallace but then Jack called up the National Guard and tonight at 6:00 they are registered and preparing to attend classes tomorrow." To Michael: "I hope when you are Attorney General these kind[s] of things will not go on." [118]

16

The Pursuit of Justice:
Martin Luther King

O N THE NIGHT of Wallace's capitulation John Kennedy delivered a television address to the nation. It was a response not alone to Birmingham and Tuscaloosa but to an entire springtime of gathering outrage and atrocity — the white postman from Baltimore murdered on a 'freedom walk' in Alabama; black demonstrators in Mississippi slugged, stomped on and then arrested for disturbing the peace; protests, riots; hardening resolve, hardening resistance. The President, deeply disturbed, decided the time had come to assert leadership and appeal to the nation. His advisers disagreed. O'Donnell and O'Brien did not think he should get personally involved. Sorensen was dubious. Burke Marshall later recalled to Robert Kennedy that, at the White House strategy meeting, "you were the only one who urged him to do it." "He just decided that day," Robert himself said. ". . . He called me up on the phone and said that he was going to go on that night."[1]

I

It was a busy week for John Kennedy. On Monday he had given a notable speech on foreign policy at American University. For most of the next day he and Robert were on the telephone discussing developments in Tuscaloosa. They could not know until late Tuesday afternoon whether the President might not have to go on television, as his brother later put it, to announce "the occupation of Tuscaloosa or the arrest of the Governor." At last the good news from the campus: "a pleasant moment," said Robert, "after working so hard."

An hour before the telecast the Attorney General and Burke Marshall met the President and Sorensen in the cabinet room. The President, Robert said later, "thought that he was going to have to do it extemporaneously, so that the two of us talked about what he'd say . . . outlining and organizing" while JFK made notes "on the back of an envelope or something." Sorensen made notes too; then left the room. John Kennedy's casualness stunned Marshall: "I remember him sitting there, putting down these notes, without any speech, when he was about to go on television, on what I thought was [an issue] of momentous importance." Five minutes before the President went on the air Sorensen returned with most of a draft; no peroration, however. John Kennedy, his brother recalled, "went over it quickly, decided it said what he wanted to say. . . . Then the President used a little of the off-the-cuff stuff at the end. . . . The speech was good. I think that it, . . . if he had given it extemporaneously, would have been as good or better."[2]

It was, in fact, a great speech, its power deriving from a passionate declaration on racial justice never before uttered by an American President: "We are confronted primarily with a moral issue . . . as old as the scriptures and . . . as clear as the American Constitution." Its power came too from its unusual sense of identification with the victims of segregation. Let the white American, Kennedy said, contemplate the plight of his black fellow citizen. "Who among us would be content to have the color of his skin changed and stand in his place? Who among us would be content with the counsels of patience and delay?" The time had come, he said, "for this Nation to fulfill its promise" of freedom.[3]

That night the white killer shot down Medgar Evers in front of his house in Mississippi. A few days later, Roy Wilkins spoke at the funeral in Jackson. One man pulled the trigger, Wilkins said, but "the Southern political system put him behind the rifle." A great crowd gathered behind the cortège, chanting, "We want the killer! We want the killer! . . . We want equality! We want freedom!" Some headed toward the business district. Policemen with growling dogs massed to block them off. The mourners began to throw rocks and Coke bottles. Cops took out their riot guns. Then John Doar in shirtsleeves stepped between the groups. He said to the blacks, "I'm your friend. You can't win this way. . . . Don't throw bottles, that's what they want you to do." James Meredith, who was there, said Doar "restored order in an act of bravery and judgment seldom matched in human history."[4]

Medgar Evers, a war veteran, was buried in the Arlington Cemetery in Washington. Charles Evers, Medgar's brother: "Bobby sat with me all during the funeral, and he consoled me. President Kennedy carried Mrs. Medgar and I and the children back to the White

House. We stayed there for the rest of the day."[5] Robert Kennedy gave Charles Evers his personal phone numbers. "He said I could call him any time, day or night, if Negroes were being harassed and intimidated. . . . Whenever I had the need to call him, I never found it too late or too early."[6]

It was a stormy time. There was trouble in Cambridge, Maryland, where a black protest movement and the local authorities were in bloody collision. Kennedy called a meeting in his office. "I'll never forget it," said John Lewis of SNCC. ". . . Somehow he was able to bring the people together. . . . He asked questions and he also did a great deal of explaining, trying to convince some of the officials of Cambridge — some of them were very stubborn. . . . Before the meeting was over, people were talking to one another in fellowship, and we were getting pictures made together."[7]

In Prince Edward County in Virginia, there had been no public education at all since 1959, the local board of supervisors preferring to close schools rather than to integrate them. White parents had set up 'private' schools for their children. Seventeen hundred black children had no place to go. In 1961 the Department of Justice had sought a judicial order requiring the county to reopen the schools; but the case had progressed glacially in the courts. In February 1963 John Kennedy told his brother that some way must be found in the meantime to educate the black children. Kennedy called in a bright and resourceful New York lawyer, William J. vanden Heuvel. Vanden Heuvel worked out a plan for a private operation called the Prince Edward Free School Association. "We would raise the money privately; we would lease the public schools from the County and teach these children in the most effective way until the courts reopened the schools. . . . Although Bobby was fully aware that there were many pitfalls . . . he never hesitated once in giving the go-ahead." With highly qualified teachers coming in from outside, the Free Schools even attracted a handful of white children. In May 1964 the Supreme Court ordered Prince Edward County to reopen its public schools.[8]

Such activity was mostly behind the scenes. The militants of racial justice continued to condemn the Attorney General for not doing enough. A few days after the murder of Medgar Evers, Kennedy spoke to an angry demonstration in front of the Department of Justice. Jack Newfield, a radical reporter from the *Village Voice* of New York, was in the crowd. "I had, by then, been jailed twice in civil rights demonstrations," he wrote later, "and with pure fury stared at Kennedy's crew-cut face. It was, I remember, a hard Irish face; alert, but without much character, a little like the faces that used to follow me home from Hebrew school, taunting, 'Christ killer.'" Someone shouted that they hadn't seen many Negroes come out of the building. "Kennedy tensed up even more. His skin seemed to draw even tighter around his sharp features, and the hostility radiating from his blue eyes became even more intense." He shouted back, "Individuals will be hired according to their ability, not their color." Newfield thought it "exactly the sort of impersonal, legalistic response, blind to the larger moral implications of our protest, that we felt made Kennedy such an inadequate Attorney General."[9]

On September 15, fifteen sticks of dynamite blasted the Sixteenth Street Baptist Church in Birmingham, killing four small girls who were changing into choir robes in the basement. Robert Kennedy said the next year:

> The responsibility for the bombing might have been those two Klansmen. But in the last analysis . . . it's Governor Barnett and Jim Eastland and John Stennis and the business community. . . . It's George Wallace and political and business leaders and newspapers. They can all deplore it . . . [but] they're the ones that created the climate that made those kinds of action possible. . . . With all the smiles and all the graciousness of Dick Russell and Herman Talmadge and Jim Eastland and George Smathers and [Spessard] Holland . . . none of them really made any effort to counter this. And they're the ones, really, who have the major responsibility, rather than these stupid figures who think that they become national heroes by taking on these tasks.[10]

II

"I shall ask the Congress," John Kennedy had said in his speech on the night of Tuscaloosa, ". . . to make a commitment it has not fully made in this century to the proposition that race has no place in American life or law."[11]

Kennedy's minimal February program had died before an indifferent nation. James Baldwin had not even been aware of its existence. Now Bull Connor and George Wallace had at last made civil rights a legislative possibility. The blacks were in the streets, moreover, and the President felt he would lose control over an increasingly dangerous situation unless he exerted leadership. He also was sensitive to the verdict of posterity. "There comes a time," he told Luther Hodges, his Secretary of Commerce, "when a man has to take a stand and history will record that he has to meet these tough situations and ultimately make a decision."[12]

Hodges, a southerner, was predictably doubtful. But O'Donnell and O'Brien also opposed civil rights legislation, fearing it would wreck the rest of the administration's program. "Every single person who spoke about it in the White House," Burke Marshall said later, "— every single one of them — was against President Kennedy's sending up that bill." Only Robert Kennedy wanted it. "He urged it, he felt it, he understood it. And he prevailed. I don't think there was anybody in the Cabinet — except the President himself — who felt that way on these issues, and the President got it from his brother."[13]

Robert Kennedy was much irritated in later years by the theory that his brother "only realized there was a civil rights crisis after Birmingham in 1963; or otherwise he would have tried to obtain ... legislation in 1961 or 1962. ... That was just a lot of hogwash. There wasn't anything he could do then. ... Nobody was ready." Birmingham had not only created the mood but demonstrated the need. The discrimination in department stores and lunch counters, against which Martin Luther King had launched his protest, violated no federal statute. The moment had arrived, the Kennedys felt, to make equal access to places of public accommodation a matter of law. "For the first time," Robert Kennedy said, "people were enough concerned about it, and there was enough demand about it that we could get to the heart of the problem and have some chance of success."[14]

The administration bill also revived the voting rights proposals fruitlessly submitted to Congress in February. As for education, the drafters considered and ultimately abandoned the idea of deadlines for total school integration and settled instead on a provision authorizing the Attorney General to sue for school desegregation. As for federal funds used in discriminatory state or local activities, they originally supposed that withholding would be too much for Congress. Finding to their surprise general support for such a provision, they added it to the bill. The question of employment gave great difficulty. John Kennedy thought Truman had made a mistake fifteen years earlier "by prematurely throwing FEPC at the Congress." To include it now, he feared, might jeopardize the bill as a whole. The civil rights leadership and the labor movement protested. Hubert Humphrey carried their protest to the White House. Kennedy finally decided to endorse fair-employment-practices legislation in his message, but not put it in the bill.[15]

Doubts remained within the administration. The Vice President was a particular puzzle. Unable to decipher his warily delphic utterances in meetings, John Kennedy sent Burke Marshall to see him. Johnson at first filibustered, reminiscing at length about the way he had found jobs for Negroes when he worked for the National Youth Administration in Texas during the depression. Jobs, he said, were what mattered. Civil rights legislation "wouldn't get passed and was impossible and would cause a lot of trouble and was the wrong thing to do."[16] Robert Kennedy sent Norbert Schlei, who was in charge of drafting the bill, on a similar mission. Johnson, a little disgruntled, said he had asked for fifteen minutes alone with the President on the bill. Schlei was obviously an unsatisfactory substitute. The bill, Johnson said, "would be disastrous for the President's program and would not be enacted if submitted now." He said this, he explained, not because he opposed civil rights legislation — no one in the country wanted to do more for the Negroes — but because so little had been done to prepare the country for it.[17] In a few days Bobby Baker, the secretary of the Senate, took a poll that, as he reported to Mike Mansfield, the majority leader, "shows conclusively that it is virtually impossible to secure 51 Senators who will vote for the President's Bill."[18] Even the faithful Mansfield thought the bill had no chance.[19]

John Kennedy himself somewhat shared this pessimism, but saw no alternative except to go ahead. "We always felt that maybe this was going to be his political swansong," Robert Kennedy recalled the next year. ". . . He would ask me every four days, 'Do you think we did the right thing by sending the legislation up? Look at the trouble it's got us in.' But always in a semi-jocular way. It always seemed to me quite clear that was what we needed to do."[20] The President agreed. The reason, Robert said, was

> not only the passage of legislation, but what in my judgment was even more important, to obtain the confidence of the Negro population in their government and in the white majority. I thought there was a great danger in losing that unless we took a very significant step such as the passage of legislation. . . . There's obviously a revolution within a revolution in the Negro leadership. We could see the direction of Martin Luther King going away from him to some of these younger people, who had no belief or confidence in the system of government . . . and thought . . . that the way to deal with the problem is to start arming the young Negroes and sending them into the streets, which I didn't think was a very satisfactory solution because, as I explained to them, there are more white people than Negroes and although it might be bloody, I thought that the white people would do better.[21]

III

The bill went to Congress on June 19, 1963. "In the way he had of getting everything organized at once," said Burke Marshall, Robert Kennedy took command.[22] Together he and Marshall talked to all the members of Congress they could find. He proposed that the President invite influential groups to the White House — lawyers, educators, clergymen, businessmen, labor leaders, hotel, restaurant and theater owners. Some in the White House continued to think John Kennedy should not get so personally identified with the fight. The Attorney General disagreed. (So did his brother.) The whole point, Robert Kennedy said later, was to display "the active involvement of the President" as well as to show local leaders "why it was in their interest to do something rather than to wait until violence occurred."[23] Robert planned the effort, "saw that the meetings took place, did the groundwork for them, did the follow-up on them" (Burke Marshall).[24] The two Kennedys and Johnson addressed group after group in their contrasting styles — the President crisp and businesslike, the Vice President emotional and evangelical, the Attorney General blunt and passionate. Vincent Harding recalled Robert Kennedy at one meeting: "He said something about how *he* felt he would respond if he were a black person at this particular time; on a certain level, I had the feeling that he was serious in trying to say how he would feel as a black person, even though I thought that at certain levels it was really impossible for him to do more than *want* to feel that."[25]

The civil rights leaders thought that the bill, in Joseph Rauh's words, "contained the Administration's best estimates of what could be enacted, rather than what was needed."[26] Still, as Rauh told them, "It is the most comprehensive civil rights bill ever to receive serious consideration from the Congress of the United States."[27] Their major disagreement with the administration now came over tactics.

In the previous December, Bayard Rustin, the talented black organizer, had called on A. Philip Randolph, the revered president of the Brotherhood of Sleeping Car Porters. Recalling Randolph's 1941 victory in threatening a march on Washington and getting the wartime FEPC in exchange, Rustin proposed a new march.[28] The Kennedys were about as enthusiastic over a march on Washington as Roosevelt had been twenty-two years before. "We want success in Congress," the President told the black leaders in the cabinet room on June 22, "not just a big show at the Capitol. Some of these people are looking for an excuse to be against us; and I don't want to give any of them a chance to say, 'Yes, I'm for the bill, but I am damned if I will vote for it at the point of a gun.' It seemed to me a great mistake to announce a march on Washington before the bill was even in committee. The only effect is to create an atmosphere of intimidation — and this may give some members of Congress an out."[29]

Speaking with dignity and force, A. Philip Randolph, James Farmer and Martin Luther King defended the march.[30] The blacks were already in the streets; better that they march under nonviolent leadership; if the march were called off, they might turn to new and desperate leaders. "It may seem ill-timed," King said. "Frankly, I have never engaged in any direct action movement which did not seem ill-timed. Some people thought Birmingham ill-timed." The President said, "Including the Attorney General."

"This is a very serious fight," John Kennedy said. "The Vice President and I know what it will mean if we fail. . . . A good many programs I care about may go down the drain as a result of this — so we are putting a lot on the line. What is important is that we preserve confidence in the good faith of each other. I have my problems with Congress; you have yours with your own groups. We will undoubtedly disagree from time to time on tactics. But the important thing is to keep in touch."[31]

The civil rights leaders came away impressed by the President's commitment. "Joe," Kennedy had said to Rauh, "this is going to be a long, tough fight, but we have to do it."[32] "I liked the way he talked about what *we* are getting," said King. "It wasn't something that he was getting for you Negroes. You knew you had an ally."[33]

IV

If the march could not be headed off, the Kennedys decided that it would have to be made a success. John Kennedy talked to Walter Reuther, who shared the concern that an all-black march might lose votes on the Hill. Reuther proceeded to bring in white leaders, the churches and the labor movement. Jack Conway, now back with the UAW, served as his representative in the planning sessions.[34] Within Justice, Robert Kennedy, following his principle that a good man could do anything, gave the charge to John Douglas, the newly appointed assistant attorney general for the Civil Division. For the next five weeks Douglas worked full-time on the march. "Good administration," he told his assistant Alan Raywid, "presumes what can go wrong will go wrong," and he directed Raywid to draw up a list of everything that might conceivably go wrong with a march. The list, extending finally to nearly seventy-five items, covered everything from route and timetable, the sound system and security forces, to food, soft drinks (in paper cups, so there would be no bottles to throw) and toilet facilities.

The Douglas group, meeting twice a day in Justice, kept its hand as invisible as possible. Working through Conway, they persuaded the planners in New York to forgo the idea of a demonstration on Capitol Hill in favor of a march from the Washington Monument to the Lincoln Memorial. Remembering Birmingham, they forbade the District police to include police-dog units in their riot-control precautions. They brought in Jerry Bruno, the king of advance men, to work out the platform arrangements. In every way they sought to carry out the Attorney General's instruction that, as Douglas put it later, "the government not give the impression that they were worried about violence ... [but rather] an impression of organization and confidence."[35]

By August 15, President Kennedy was concerned that the march might not be large enough. If less than 100,000 came out, he told me, it might persuade legislators that the demand for civil rights legislation was exaggerated.[36] He need not have worried. A quarter of a million came to Washington on August 28, 1963, to hear Martin Luther King's dream: "Free at last! Free at last! Thank God Almighty, we are free at last!"

Others wished to seize the occasion for more drastic statement. John Lewis of SNCC forced the issue by preparing a tough speech. "We cannot support the Administration's civil rights bill," he planned to say. "There's not one thing in the bill that will protect our people from police brutality.... I want to know — which side is the federal government on?" The next time, Lewis's text concluded, "we won't march on Washington, but we will march through the South, through the Heart of Dixie, the way Sherman did."[37] All this upset the religious leaders, especially Archbishop Patrick O'Boyle, who said he would withdraw unless Lewis moderated his views. Burke Marshall spoke to Reuther, and there were tense last-minute negotiations in the small office behind Lincoln's statue in the Memorial. "I looked up at Abe," Reuther recalled, "and I said, 'Abe, I need your blessing and your help.'"[38] Even James Forman of SNCC was sufficiently carried away by the mood to aid in the revision of Lewis's speech, though he repented later.[39]

Had the Kennedys, in the language of the time, co-opted the March? What they did effectuated rather than betrayed the intentions of the planners — of Randolph, Rustin and King, of Roy Wilkins and Whitney Young. The revolutionaries of course wanted to go further than the respectables. "Behind our backs," Forman said afterward, "they and the Kennedy administration developed plans of their own as to how the March on Washington could be used to advance their own respective interests."[40] "There wasn't a single logistics aspect uncontrolled," wrote Malcolm X. "... Who ever heard of angry revolutionists swinging their bare feet together with their oppressor in lily-pad park pools, with gospels and guitars and 'I Have a Dream' speeches?" He called it the "Farce on Washington."[41] As King gave his famous oration, some in the crowd, joining hands, swaying back and forth, cried almost ecstatically, "Dream some more." But one black man shouted furiously: "Fuck that dream, Martin. Now, goddamit, NOW!"[42]

It remained a beautiful day — a day that we saw then as the beginning of a new era but that came to seem in retrospect the end of an old one. No demonstration, Bayard Rustin wrote ten years after, "has influenced the course of social legislation and determined the shape of institutional reform to the degree that the March did."[43] Perhaps so; certainly that afternoon in August had a purity that no one present can ever forget. Yet it also marked the effective, and no doubt inevitable, end of white and black collaboration in the mass movement for racial justice. Thereafter the movement took its own course under black leadership, assumed new forms and passions, suffered vicissitudes and disenchantments, but bravely persevered in the necessary fight.

v

Among those who viewed the March on Washington with particular loathing was J. Edgar Hoover. Though the FBI under Robert Kennedy's pressure had finally entered the struggle for racial justice, Hoover's personal feelings about blacks had not abated. As if in compensation for the ground he had been forced to give on civil rights, he redoubled his determination to discredit the movement — discredit the Kennedy administration too, if it persisted in its alliance with Martin Luther King. In so doing, the director fell back on the cry that had never failed him in the past: the ineluctable threat, evidently undiminished despite all his effort, of Communist infiltration into American institutions.

He first had to extirpate heresy in his own ranks. Before the march, William C. Sullivan, the assistant director in charge of the FBI's Domestic Intelligence Division, sent Hoover a memorandum asserting that the Communist party had failed "dismally" in its forty-year effort to influence the American Negro. A subsequent memorandum pointed out that less than two hundred Communists were planning to take part in the march. These memoranda infuriated the director. They suggested, he noted sarcastically, that Communists were having an "infinitesimal" effect on the blacks. They reminded him "vividly," he wrote Sullivan, "of those I received when Castro took over Cuba. You contended then that Castro and his cohorts were not Communists. . . . Time alone proved you wrong."[44]

The assistant director was in trouble. Hoover stopped speaking to him, and, luckily for the historical record, the two men confined themselves to written exchanges for several weeks. Sullivan had no doubt he must provide the director the words he wanted to hear. He said later: "We had to engage in a lot of nonsense which we ourselves really did not believe in. We either had to do that or we would be finished."[45] Certainly Sullivan, who hoped to succeed Hoover, would have been finished. Sounding for all the world like a Communist caught in some wretched deviation, Sullivan confessed error and affirmed his leader's infallibility. "The Director is correct. We were completely wrong. . . . It may be unrealistic to limit ourselves as we have been doing to legalistic proof or definitely conclusive evidence that would stand up in testimony in court or before Congressional committees [a little dangerous irony here?]. . . . It is obvious to us now that we did not put the proper interpretation upon the facts which we gave to the Director." And, on Martin Luther King: "We must mark him now, if we have not done so before, as the most dangerous Negro of the future in this Nation from the standpoint of communism, the Negro and national security."[46]

These last were above all the words that Hoover wanted to hear. He had long since made up his mind about the iniquity of Dr. King. In May 1961, reading in an FBI memorandum that King had not been investigated, Hoover wrote on the border, "Why not?" In January 1962 he told the Attorney General that a leading King adviser was a Communist.[47] In February he scribbled on an FBI memorandum: "King is no good." In May he put King on the list of people to be rounded up in event of a national emergency. In October the FBI opened a formal investigation of King and the SCLC. In November, after King told the *Atlanta Constitution* that FBI agents were often southern-born and pals of the local police, an enraged Hoover directed Cartha (Deke) DeLoach, head of the Crime Records Division, to call King and set him straight. When King neglected to return DeLoach's call, DeLoach described him as "a vicious liar" who "constantly associates with and takes instructions from [a] . . . member of the Communist Party."[48]

In the meantime Hoover bombarded Robert Kennedy with baleful memoranda about King's allegedly sinister associations. The following, of April 1962, is typical:

> This Bureau has recently received additional information showing the influence of Stanley David Levison, a secret member of the Communist Party, upon Martin Luther King, Jr. You will recall that I have furnished you during the past few months substantial information concerning the close relationship between King and Levison.
>
> A confidential source who has furnished reliable information in the past advised on April 16, 1962, that he had learned that Levison is forming in King's name an organization to be known as the Ghandi [sic] Society for Human Rights. Levison contemplates sending invitations signed by King to approximately twenty prominent people to attend a

luncheon on May 17, 1962, in Washington. . . . You as well as the
President, Senator Clifford Case, Senator Eugene McCarthy and former
Attorney General William P. Rogers are among those being considered
to be invited. . . . The informant said that he is under the impression
that Theodore Kheel . . . Harry Belafonte . . . and A. Philip Randolph
are involved in the formation of the organization.

A handwritten note on the document put the Attorney General on
notice that he was not to file and forget: "This is a very sensitive
source. I assume the White House knows about Levison." [49]

If it was hard to regard inviting William P. Rogers and Clifford
Case to a Gandhi Society luncheon as a major threat to the republic,
Stanley Levison did present the Attorney General a more difficult
problem. A few days later, the Senate Internal Security Subcommittee, no doubt tipped off by Hoover, subpoenaed Levison and interrogated him in executive session. "To dispose of a question causing
current apprehension," Levison told the committee, "I am a loyal
American and I am not now and never have been a member of the
Communist Party." He went on, however, with the leftist attorney
William Kunstler at his side, to challenge the committee's right to
inquire into political beliefs and thereafter took the Fifth Amendment. He did so, he later explained, because he feared the session
would turn into an attack on King.[50] Actually no one brought up
King. Instead Eastland, who sounded as if he had been briefed by
the Bureau, asked such questions as: "Isn't it true that you are a spy
for the Communist apparatus in this country? . . . Isn't it true that
you have gotten funds from the Soviet Union and given them to the
Communist Party of the United States?" Levison declined to answer. Finally, after much exasperated quizzing and effective stone-
walling, McClellan terminated the hearing, calling it "one of the
shabbiest performances I have ever heard before a Senate Committee
by any witness." [51]

Levison was a New York lawyer of independent means and benevolent disposition who, with introductions from A. Philip Randolph
and Bayard Rustin (both resolutely anti-Communist), had met King
in Montgomery in 1956. King was not skilled in organizational matters. SCLC affairs were in chronic disorder. Tactful and self-effacing, Levison helped supply King's administrative deficiencies. Soon
he was donating much of his time to the SCLC, raising money, trying
to straighten out the internal finances, writing King's formal addresses, contributing legal passages to his articles and books, providing King professional counsel and personal understanding.[52] King
came to rely heavily on him. Coretta Scott King described him as
one of her husband's "most devoted and trusted friends." [53]

VI

Stanley Levison, Hoover informed Robert Kennedy, was a secret member of the CPUSA national committee charged with controlling King. The Bureau came up with a second supposed Communist around King. This was one of his executive assistants, a young black named Hunter Pitts (Jack) O'Dell, whom Hoover also declared to be a member of the national committee. In October 1962 the Bureau leaked incriminating material about O'Dell to southern newspapers. Soon King placed him on temporary leave.[54] No incriminating material was leaked on Levison, except the general allegations to the Internal Security Subcommittee. Probably there was none to leak.

The allegations greatly troubled Kennedy and Marshall. They simply took Hoover's word that Levison was a secret party mucka-muck. However hopeless the Bureau was in organized crime and civil rights, internal subversion was its undisputed field of expertise. The CPUSA was swarming with FBI informants. Levison's appear-ance before the Senate committee was hardly reassuring. At Hoover's request, Kennedy authorized a national-security wiretap on Levison's office telephone in March 1962 and on his home telephone in No-vember. The Bureau also planted a microphone in Levison's office, though as usual without telling the Attorney General.[55] The wiretaps on Levison's phones produced no evidence of Communist deviltry; but this only showed, the Bureau replied, what a truly secret muck-amuck he was.[56]

Kennedy and Marshall admired King's courage and leadership. They thought it essential that his gospel of nonviolence prevail. Ac-cordingly they decided that he must be warned for his own protection and for the protection of the movement. Nothing would play more effectively into the hands of white supremacists than proof that Mar-tin Luther King was under Communist control.

The warning process began early. Kennedy asked Harry Belafonte about Levison. Belafonte, supposing that Levison was under consid-eration for a government appointment, gave a warm endorsement. Kennedy shook his head and said, "No. Martin ought to get rid of him."[57] Harris Wofford, King's old friend, was asked to speak di-rectly to King about Levison. King was incredulous.[58] John Seigen-thaler spoke to him. King said he did not like to accuse people who volunteered to help the cause; in any case, his associates could change neither his will nor the movement's policies.[59] Burke Marshall spoke to him. Robert Kennedy spoke to him. King remained incredulous.

Once the Kennedys decided to go all-out for civil rights legislation, Levison was no longer just a King problem. If word of his role reached the Hill, "not only would it damage [King]," Kennedy observed, "but it would also damage all of our efforts and damage any possible chance of the passage of legislation."[60] The Kennedys had committed themselves to King. "The basic identity of constitutional and political interest," wrote Katzenbach, ". . . is the necessary predicate of all subsequent events. Anything which discredited Dr. King, or his movement, would have been a disaster to the Kennedy administration. . . . More importantly, it would have been a disaster to the country."[61] Kennedy directed Burke Marshall to find out what the hard evidence against Levison was.

Levison had indeed known Communists in New York. He had worked with them in the Henry Wallace and anti-McCarthy campaigns. He was, in fact, a goodhearted and undiscriminating liberal of the type for whom there was no enemy to the left. His twin brother, Roy Bennett, the New York correspondent for Aneurin Bevan's paper, the *Tribune* of London, was a democratic socialist and an outspoken anti-Stalinist. Levison had fallen under the suspicion of the Bureau because, he eventually came to believe, a business acquaintance under FBI investigation had tried to ingratiate himself with the Bureau by denouncing Levison.[62] Nor did the Bureau ever offer any proof of party membership. Early in 1963, it had even reported that Levison had criticized the CPUSA on the ground of Communist neglect of civil rights.[63] In 1964 the New York FBI office told Hoover: "Mr. Levison is not now under CP discipline in the civil rights field. There has been no indication, however, that Mr. Levison has not continued his ideological adherence to communism."[64] William C. Sullivan told me in 1976 that he had never seen any evidence that Levison was a party member.[65] Andrew Young considered him "a conservative influence . . . very cautious about becoming involved with [i.e., against] the war in Vietnam, for example."[66]

When Marshall now applied to the FBI for the hard evidence, he was told that Levison was something far more baleful than a secret member of the CPUSA. The Bureau advised him in the most formal and emphatic manner and for the first time (though Hoover must have muttered something of the sort to Eastland, judging by Eastland's questions before the Senate committee) that Levison was in fact a key figure in the Soviet intelligence apparatus in the United States. Marshall was further advised that disclosure of this information to King would endanger an FBI agent in place inside the KGB.[67]

It is hard to know what to make of this. It would have been an act of gross bureaucratic imperialism for the FBI to invade CIA territory and run an agent in Moscow. Possibly Hoover meant a New York agent known as Fedora, "officially a Soviet diplomat with the United Nations," Edward Jay Epstein has written, who began in March 1962 to supply the Bureau with information about Soviet espionage.[68] Hoover's first warnings about King's Communist adviser had been pre-Fedora; but perhaps Hoover checked Levison with his new hot source, and Fedora obligingly confirmed the director's suspicions. The Levison wiretap was placed in the same month that Fedora started talking to the FBI. CIA counterintelligence later came to regard Fedora as a double agent under KGB control. If this were so, then the KGB may have collaborated with the FBI in setting up Levison, both agencies wishing to discredit King and destroy his alliance with Kennedy, the KGB receiving the added benefit of heightened racial bitterness in America.

Or perhaps Hoover had simply got it into his head that Levison was a Communist, as he had got it into his head that the Communists dominated the civil rights movement, and was prepared to say anything to make others accept what he so urgently believed the truth. Certainly the story the Bureau told Marshall was difficult to reconcile with Hoover's earlier portrait of Levison as a top party functionary. When people entered the Soviet intelligence service, they ordinarily abandoned all connection with the CPUSA (note Whittaker Chambers). But Marshall, whose specialty was civil rights, not Communist espionage, understandably believed the FBI story.

In June 1963 all anyone had was the certification by the high priest of national security that Stanley Levison, Martin Luther King's intimate adviser, was a top Soviet intelligence operative — and this at a time when the Kennedy administration was staking its future on the integrity of the civil rights movement.

The Attorney General's next move was to bring in the President. On June 22, after his meeting with the civil rights leaders, John Kennedy invited King to stroll in the Rose Garden outside the Oval Office. Kennedy began (so King later told Andrew Young) by saying that Eastland and others were denouncing the March on Washington as a Communist trick. The FBI, Kennedy said, was greatly concerned; "I assume you know you're under very close surveillance." He warned King against discussing significant matters over the phone with Levison, whom Hoover regarded, Kennedy said, as "a conscious agent of the Soviet conspiracy."* (King, noting that Kennedy had

---

* Kennedy's wiretapping reference was obviously to Levison's phone. King's phone was not tapped for another four months.

walked him out into the garden, reflected, "The President is afraid
of Hoover himself. . . . I guess Hoover must be buggin' him too.")[69]

Then Kennedy said (so King later told Levison), "You've read
about Profumo in the papers?" The fall of the British Secretary of
State for War had dominated the spring headlines. Kennedy: "That
was an example of friendship and loyalty carried too far. Macmillan
is likely to lose his government because he has been loyal to a
friend. You must take care not to lose your cause for the same
reason." He named Levison and O'Dell. "They're Communists.
You've got to get rid of them. . . . If they [the opponents of civil
rights] shoot *you* down, they'll shoot *us* down too — so we're asking
you to be careful." King finally said, "I know Stanley and I can't
believe this. You will have to prove it." Kennedy said that Burke
Marshall would show the proof to Andrew Young.[70]

King wrote O'Dell in ten days making his temporary resignation
permanent. He reminded O'Dell of the attacks on the movement as
Communist-inspired; "in these critical times we cannot afford to risk
any such impressions."* In O'Dell's case there was indication of at
least past Communist connections. As for Levison, King awaited the
Marshall-Young meeting.

They met at the federal courthouse in New Orleans, where Marshall was discussing a civil rights case. Marshall walked Young
around the corridors and told him that the FBI considered Levison
a Soviet spy. "Burke never said anything about any evidence he had,"
recalled Young; "he always quoted what the Bureau said it had. I
didn't feel this was conclusive. They were all scared to death of the
Bureau; they really were." Young reported back to King. King was
still reluctant to break with Levison. "I induced him to break," Levison told me. "I said it would not be in the interests of the movement
to hold on to me if the Kennedys had doubts. I said I was sure it
would not last long."[71]

VII

It did not last long. In the meantime, however, the Kennedys went
bail for the movement. The President in his July 17 press conference
said there was "no evidence" to show that any civil rights leaders were
Communists or that the demonstrations were Communist inspired.
"I think it is a convenient scapegoat to suggest that all the difficulties

* King wrote O'Dell that, after a "thorough inquiry," the SCLC was "unable to
discover any present connections with the Communist party on your part" (Martin
Luther King, Jr., to H. P. [Jack] O'Dell, July 3, 1963; in Victor Navasky, *Kennedy Justice*
[New York, 1971], 143-144; emphasis added).

are Communist and if the Communist movement would only disappear that we would end this."[72] "Based on all available information from the FBI," Robert Kennedy wrote senators, ". . . we have no evidence that any of the top leaders . . . are Communists, or Communist controlled. This is true as to Dr. Martin Luther King, Jr., about whom particular accusations were made, as well as other leaders."[73]

Could they be sure? There was one way to find out, the Bureau said. That was to tap King's telephone. This possibility had been on Hoover's agenda for some time. He had directed Sullivan to push the Attorney General's office on wiretaps for King, and Sullivan had been faithfully pushing.[74] Kennedy himself could hardly escape the question. If Levison was indeed a top Soviet agent, then, as Marshall put it later, "it would not be responsible for . . . the Attorney General to refuse a tap" on the American leader with whom Levison had established so intimate a relationship.[75] They were all "scared to death" of the FBI, Andrew Young had said. They were not scared without reason. The still sacrosanct Hoover had vast power to do damage to King, to the civil rights bill, to the Kennedys. In the short run, Hoover had only to leak to the Hill and the right-wing press. In the longer run, he could do to John Kennedy what he had done to Truman ten years before over the case of Harry Dexter White; he could tell a congressional committee that a President had deliberately ignored his solemn warnings and had thereby endangered the republic. If Robert Kennedy refused a tap on King and anything went wrong, Hoover would have a field day. On the other hand, a tap might end the matter by demonstrating King's entire innocence, even to the satisfaction of the FBI.

Such considerations undoubtedly led Kennedy to raise with Courtney Evans on July 16, 1963, the possibility of "technical coverage" on King. This was the meeting that Hoover cited half a dozen years later to support his claim that Kennedy was pressing the idea on a reluctant FBI. Asked whether Kennedy had in fact originated the idea, Evans testified in 1975, "No, this is not clear in my mind at all. . . . There were pressures . . . from both sides, the Bureau wanted to get more specific information, and the Department wanted resolved the rather indefinite information that had been received."[76] Asked where the idea came from, Sullivan said, "Not from Bobby. It came from Hoover. . . . The whole impetus came from us."[77] "To say or imply that this tap was the original conception of Robert Kennedy — that he was the moving force in this situation — or that he had any doubt whatsoever about Dr. King's integrity or loyalty is false," said Katzenbach.[78]

In any case, on July 23 Hoover made a formal request for wiretap authorization. On July 24 Kennedy rejected the request. The Bureau kept up its pressure. It put out a report on King's "Affiliations with the Communist Movement." Kennedy commented acidly that, if that report reached the Hill, he would be impeached.[79] Then, though King and Levison did not meet for another year or so, King decided that his promise to drop Levison had been a mistake,[80] and they resumed telephonic communication — a fact Hoover joyously passed on to Kennedy. Kennedy had a natural desire to find out whether this renewal of association would hurt the common cause. The civil rights bill was in trouble on the Hill. On October 7 Hoover again requested authorization for taps on King's home .telephone and on the New York SCLC office. Kennedy was worried but agreed to go ahead "on a trial basis, and to continue it if productive results were forthcoming." A few days later Hoover requested a tap on the Atlanta SCLC office. "The Attorney General is apparently still vacillating in his position as to technical coverage," Evans reported back. Kennedy finally consented, stipulating that all the King taps be evaluated "at the end of thirty days" and that the question of continuing the surveillance be decided then. Thirty days from October 21, 1963, was November 20, 1963 . . . On its own, the FBI maintained taps on King's home phone till mid-1965, on the SCLC phones for even longer, on Levison and at least one other King associate, not to speak of at least twenty-one microphone installations in King's hotel and motel rooms.*

The Kennedys authorized the taps on King for defensive purposes — in order to protect King, to protect the civil rights bill, to protect themselves. "Bobby Kennedy resisted, resisted, and resisted tapping King," said Sullivan. "Finally we twisted the arm of the Attorney General to the point where he had to go. I guess he feared we would let that stuff go in the press if he said no."[81] Installing the taps shifted responsibility for King to Hoover; and the Kennedys were confident that King was clean and that the taps would yield nothing of interest. What was less excusable, if understandable for a period after November 22, 1963, was the failure of the Attorney General to terminate the taps.

Though outsiders have condemned the taps as indefensible in any context, neither King nor, for that matter, Levison held the taps

---

* Sixteen reported in Church committee, *Final Report*, bk. III, 81, 103, 115–116 (see n. 44); five more in Department of Justice, "Report of Task Force to Review King," 127–128 (see n. 46). One of these was installed by the New York field office without Hoover's authorization and remained despite instructions from Clyde Tolson that it be removed "at once."

against the Kennedys in this particular context. The SCLC people did not feel all that threatened by wiretapping. After all, John Kennedy had personally warned King about FBI surveillance. "We knew," Andrew Young said in 1976, "they were ... bugging our phones; but that was never really a problem for us. It's hard to remember back then, but when you were anxious about your life, civil liberties seemed a tertiary consideration. In our conduct of a nonviolent movement there was nothing that we did not want them to know anyway."[82] Beyond this, they specifically and sympathetically recognized the dilemma in which Hoover had placed the Kennedys and therefore were not alienated by the decision to take protective action. King had declined to endorse John Kennedy in 1960 and remained opposed in principle to the idea of endorsing presidential candidates. In 1964 — after the tapping had begun — he wrote in *Why We Can't Wait,* "And yet, had President Kennedy lived, I would probably have endorsed him in the forthcoming election."[83] What is even more fascinating is that these lines were actually written for King by Stanley Levison.[84]

In 1976, Levison said: "I was neither an international agent nor even a domestic party member, but at that time Hoover's credibility was unchallenged: it would have diverted the movement at a critical stage to take on so difficult a civil liberties battle." As for Robert Kennedy: "I really understand his position. You have to recall the time. We weren't too far away from the McCarthy period. They were so committed to our movement, they couldn't possibly risk what could have been a terrible political scandal. When I realized how hard Hoover was pressing them and how simultaneously they were giving Martin such essential support, I didn't feel any enmity about their attitude toward me."[85]

VIII

The wiretap authorization did not content Hoover. The Bureau had expanded its dissertation on King, and in October Hoover directed that the resulting top-secret "monograph" be sent around the government. One of his assistant directors warned him that it could be taken "as a personal attack on Martin Luther King. . . . This memorandum may startle the Attorney General, particularly in view of his past association with King, and the fact that we are disseminating this outside the Department. He may resent this." Hoover on the border: "We must do our duty."[86] The Attorney General was more than startled. "Bobby got furious," said Katzenbach,[87] and demanded that the Bureau retrieve all copies at once. Hoover, sensing

Kennedy's rage, quickly agreed that the document be recalled. Afterward he complained bitterly to his associates;[88] then reissued the monograph a year later in an amplified edition with authorization from a new White House (granted by Bill Moyers of all people) to distribute to "responsible officials in the Executive Branch."[89]

Kennedy had been furious because the document was one-sided and "very, very unfair." It expatiated on King's tie with Levison but contained none of the abundant "offsetting material . . . which indicated that he didn't want to have anything to do with the Communists."[90] An added reason for his outrage was that the monograph was indeed a highly personal attack, including an indictment, in Burke Marshall's words, of "the moral character and person of Dr. Martin Luther King."[91]

King was not only a great and honorable citizen and a noble leader: he was also, like other distinguished leaders, a passionate man. This was well known to close associates and did not disturb them. Some white friends explained it in psychological terms as an unconscious attempt to prove himself as virile as his father. Black friends stressed the importance of machismo, the affirmation of masculinity, among subjugated peoples; they recalled the historic obligations of evangelical preachers in the south to the women of the congregations. In any case, as one told me, "Martin really believed in the gospel of love." Those who worked with King accepted this as an expression of both a cultural tradition and a demonstrative temperament.

After the exertions and triumphs of the March on Washington, King had relaxed in his suite at the Willard Hotel with friends of both sexes. Further developments were recorded on a hidden microphone. It is not clear who bugged the room. Kennedy, Marshall and Katzenbach, when they heard about it later, assumed it had been done by the Washington Police Department, with the tape then passed on to the FBI. The FBI was quite capable of planting the bug itself, but this has not been proved.* In any event, the results of the August séance accounted for the zeal with which the Bureau bugged King's hotel rooms in the future.

Robert Kennedy was genuinely shocked by the revelations but did not feel that King's private life had anything to do with his probity as a public leader. Hoover abominated sex in general and regarded interracial sex in particular as evidence of moral degeneracy. King, it appeared, was not only subversive but depraved, a more fright-

---

* The first FBI bug on King established by the Church committee was planted on January 5, 1964, also at the Willard (Church committee, *Final Report*, bk. III, 120 [see n. 44]).

ening menace than ever. Within the Bureau, Sullivan, still trying to redeem himself, called a meeting on the King problem. A nine-hour session on December 23, 1963, resulted in a plan, as the penitent Sullivan explained in a memorandum to the front office, to reveal King to the country "as being what he actually is — a fraud, demagogue and scoundrel" — and thereby "to take him off his pedestal and to reduce him completely in influence." Hoover noted on the memorandum: "I am glad to see that 'light' has finally, though dismally delayed, come to the Domestic Int. Div. I struggled for months to get over the fact that the communists were taking over the racial movement but our experts here couldn't or wouldn't see it."[92]

There followed a weird crusade. The Bureau on Hoover's orders stalked Martin Luther King much as allied psychological warfare had stalked Nazi *Gauleiter* during the Second World War, calling on all the resources of deception, counterespionage and black propaganda to "neutralize," as a Hoover memorandum put it in April 1964, "or completely discredit the effectiveness of Martin Luther King, Jr. as a Negro leader."[93] If supposed Communist influence on King was Hoover's real concern, his effort would obviously have been to neutralize and discredit Stanley Levison. But it was King himself whom Hoover unfathomably hated. When *Time* made King Man of the Year in 1963, Hoover scribbled on the press release, "They had to dig deep in the garbage to come up with this."[94] When King received the Nobel Peace Prize in 1964, Hoover said he was the "last one in the world" who should have got it.[95]

The Bureau expended much of its energy, it must be said, on singularly inane objectives — trying to stop King's honorary degrees, the publication of his articles, his meetings with British leaders after he collected his Nobel Prize. The vendetta reached its climax on November 18, 1964, when the director, in a rare press conference, denounced King as "the most notorious liar in the country." King responded that Hoover must be "faltering" under his burdens. On November 24, in a speech in Chicago, Hoover carried on wildly about "zealots of pressure groups" who indulged in "carping, lying and exaggerating" and were "spearheaded by Communists and moral degenerates."[96]

The hotel tapes had given Hoover a weapon against King, but not an easy one to use. Goaded by King's defiance, Hoover ordered Sullivan to prepare a composite tape, culling high spots from Washington, San Francisco and Los Angeles, and to mail it to Mrs. King. If the marriage broke up, Hoover reasoned, this would diminish King's stature. The package also contained a brutal letter: "King, there is only one thing left for you to do. You know what it is. You have just 34 days. . . . You better take it before your filthy fraudulent self is bared to the nation."[97]

By the time the Kings got around to opening the package, the deadline had long passed. Since her husband often spoke extemporaneously, Mrs. King had asked for tapes of his speeches; and the FBI tape had been added to the pile. They listened to the tape. It was, said Andrew Young, "basically just a bunch of preachers that were relaxin' after a meeting. . . . Toward the end there was a recording of somebody moanin' and groanin' as though they were in the act of sexual intercourse, but it didn't sound like anybody I knew, and certainly not Martin."[98] When they saw the letter, "we began to realize that this was the most serious kind of intimidation we had so far experienced." They had no doubt that the tape and letter came from Hoover. King said, "What does he think I am?"[99]

Hoover, who seems to have taken leave of his senses that November, also dispatched DeLoach to offer newspapermen the dossier on King's private life. When Ben Bradlee of *Newsweek* reported this to Katzenbach, now Acting Attorney General, "I was shocked," Katzenbach testified, ". . . and felt that the President should be advised immediately." Katzenbach and Marshall thereupon flew to President Johnson's Texas ranch on November 28. "I told the President," Katzenbach said, "my view that it should be stopped immediately and that he should personally contact Mr. Hoover. I received the impression that President Johnson took the matter very seriously and that he would do as I recommended."[100] Johnson, according to Hugh Sidey, a conscientious reporter, had enjoyed listening to the tapes, even to the noise of the bedsprings;[101] and what in fact he did, so far as the record shows, was to instruct Bill Moyers to warn the Bureau that Bradlee was an enemy.[102]

The refusal of newspapers to touch the King tapes attests to the standards of decency the press observed in those forgotten days. Legislators were equally scrupulous. In the midst of the civil rights debate Senator Richard Russell of Georgia asked the Justice Department what the rumors were all about. When Katzenbach briefed him, he "just wasn't interested; wondered why I was wasting his time. He's a pretty good fellow on hitting below the belt. He said to me, 'You and I are a mile apart on civil rights, but I'll tell you, I'm a hundred miles away from George Wallace.'"[103]

The tapes presented a dilemma, however, to King himself. Summoning a group of advisers to New York, he outlined alternatives: either he could extend the hand of peace and conciliation to Hoover, or he could dare Hoover to do his damndest. Kenneth Clark and Clarence Jones urged the second course: "Let Hoover reveal his damned tapes"; it would hurt Hoover more than it would King. Others felt that disclosure would destroy King's "image," especially among his white supporters. At the end, King said, "I'm going to extend the hand of peace." Clark said angrily, "Damn it, you may be Christ-like, but you're not Christ." The others laughed, but Clark felt the time had come when love and conciliation were not enough.[104]

The summit meeting took place on December 1, 1964. Contrary to mythology, Hoover did not bring up King's private life. Most of the meeting consisted of a monologue by Hoover on the Bureau's marvelous record in the south. DeLoach, who was present, described it as "more or less of a love feast." Andrew Young, who was also present, thought it more like "a mutual admiration society."[105] The session quieted the public feud but altered no private feelings. King told someone over the phone later, "The old man talks too damned much." The FBI excised this sentiment before they passed the transcript on to Hoover, fearing his rage.[106] Six years later, Hoover, in an otherwise fictitious account of the meeting, said of King, "I held him in complete contempt."[107] This was a sentiment that King certainly reciprocated.

IX

"The United States," Robert Kennedy told the Senate Commerce Committee on July 1, 1963, "is dominated by white people, politically and economically. The question is whether we, in this position of dominance, are going to have not the charity but the wisdom to stop penalizing our fellow citizens whose only fault or sin is that they were born. That, Mr. Chairman, is why Congress should enact this bill, and should do it in this session."[108]

The House of Representatives proved more receptive than the administration had anticipated. The Judiciary Committee added a fair-employment-practices provision and strengthened the bill in other particulars; and, on the day of the March on Washington, John McCormack, now Speaker after Rayburn's death, predicted that an FEPC could pass the House. Katzenbach and O'Brien regarded this thought as "a disaster," Katzenbach said later, "but, if you were to interview Speaker McCormack, he would say this was probably the key to achieving the civil rights bill as it finally came out, and he might be right. . . . I'm delighted that he did it, although at the time I could have shot him."[109]

Determined to bring out of the House a bill the Senate could accept without amendment, Robert Kennedy resisted the strengthening process. The Kennedys thought they faced a hard choice: either to accept the liberal Democratic bill, lose the vote and gain an issue; or to seek Republican support for a modified version of the administration bill. John Kennedy doubted whether Republicans would go along with any bill that contained what he wanted. But he decided to make the try. He had a crucial ally in William McCulloch of Ohio, a man of principle and ranking Republican on the House Judiciary Committee.

In mid-October the President called the House leaders to the White House. He was never more persuasive. To his astonishment, not only McCulloch but Charles Halleck, the conservative Republican leader, agreed to support a bill acceptable to the civil rights leadership, even with the FEPC. The Attorney General thought that the President's eloquence had swept Halleck along, "that finally he nodded his head in assent, and he had given his word, so he didn't want to go back on his word. But the President could never understand."[110] Robert Kennedy now saluted the House bill as superior to the original administration draft.

The battle promised to be more difficult in the Senate. "States' rights, as our forefathers conceived it," Kennedy said in his opening statement to Eastland's Judiciary Committee, "was a protection of the right of the individual citizen. Those who preach most frequently about states' rights today are not seeking the protection of the individual citizen, but his exploitation. . . . The time is long past — if indeed it ever existed — when we should permit the noble concept of States' Rights to be betrayed and corrupted into a slogan to hide the bald denial of American rights, of civil rights, and of human rights."[111] But the Judiciary Committee had never brought a civil rights bill to the floor and was not ready to break this tradition. (The administration planned to evade this by routing the bill also through the more cooperative Commerce Committee.) The Judiciary Committee's constitutional sage was the charming and formidable Sam Ervin — an authentic expert, save that his Constitution stopped with the ten amendments adopted in 1789; he appeared to regard the rest, especially the Fourteenth and Fifteenth, as apocrypha. A stack of law books piled on his desk, Ervin called the bill "as drastic and indefensible a proposal as has ever been submitted to this Congress."

For nine days between July and September, the Attorney General was the administration witness. A courteous but implacable duel took place. It was an exhausting business. One day Ethel Kennedy, encountering Ervin at a White House dinner, said, "What have you been doing to Bobby? He came home and went straight to bed." On the last day, Kennedy made a personal appeal to Ervin, telling him that, if he came out against discrimination and in support of the bill, he could make "a major difference in ending these kinds of practices, as well as bringing the country through a very difficult period of transition."

> ERVIN. The only thing you have a right to ask of me is that I stand and fight for the Constitution, and that is what I am doing now. You are not correct in saying that I have never spoken out against discrimination.... As a citizen, lawyer and judge in North Carolina, I have always stood for the right of all men to stand equal before the law....
>
> KENNEDY. I asked you whether, in your judgment, any discrimination existed in the State of Mississippi against Negroes. You said to your knowledge it did not. I don't see how you can ...
>
> ERVIN. I told you the truth, Mr. Attorney General. I have never been in the State of Mississippi in my life, except on one occasion when I was asleep in a Pullman on a train which passed through the northern portion of Mississippi on the way to Memphis. I said I could not corroborate by personal knowledge any conditions in Mississippi....
>
> KENNEDY. I would say, Senator, I have a lawyer here who could take you down into some of these places and maybe your eyes would be opened.
>
> ERVIN. I have got to stay here and fight the destruction of Constitutional government which this bill contemplates, and haven't time in this session of Congress....

Kennedy made a final try. "Senator," he said, "it would only take forty-eight hours. Would you be willing to do that?" Ervin said, "I do not have even forty-eight hours to spare from my fight to preserve Constitutional principles and the individual freedoms of all citizens of the United States."[112]

Ervin sought always to bring the colloquy around to abstract principles; Kennedy, to concrete conditions. For the Attorney General, life had taken racial justice out of the middle distance. Civil rights had displaced organized crime as the issue of law enforcement closest to his heart. "The more he saw," Burke Marshall said, "... the more he understood. The more you learned about how Negroes were treated in the South, ... the madder you became. You know he always talked about the hypocrisy. That's what got him. By the end of a year he was so mad about that kind of thing it overrode everything else." When Victor Navasky asked Marshall to measure the rise in Kennedy's civil rights consciousness, Marshall shot his right arm up toward the sky.[113]

(17)

The Politics of Justice

"**I** AM GETTING OUT of all political activities as fast as I can," Robert Kennedy said in his first interview as Attorney General. "Partisan politics played no role when I was chief counsel for the McClellan Labor-Rackets Committee. They will play no part now."[1] He told the assistant attorneys general at his first staff meeting: "No politics — period. You don't attend political functions, you don't speak on political matters, you don't get involved in any way."* He was, Ramsey Clark said later, "incredibly sensitive about excluding political influence from legal decisions."[2] To keep politics out of prosecutions, he put a Republican, Herbert J. Miller, Jr., in charge of the Criminal Division and soon gave William Hundley back the job he had held in the Eisenhower administration as chief of the Organized Crime Section.†

I

Still the campaign manager of 1960 could hardly expect to take the veil altogether in 1961, especially when his brother was President. Though he shielded his department from partisan politics, Robert

* As recalled by Ramsey Clark, in Richard Harris, "The Department of Justice," *New Yorker*, November 15, 1969. Note Clark's testimony before the Senate Judiciary Committee, Subcommittee on Separation of Powers, March 28, 1974, 6 (mimeographed): "Presidential appointees and their personal assistants in the Department of Justice should be prohibited from giving, receiving or soliciting political contributions, from managing or advising a political party or campaign while in office ... from attending political meetings, public or private; from political speeches and endorsements. Robert F. Kennedy asked this of his assistants when he became Attorney General."

† It is a comment on both the caliber and the nonpartisanship of the Kennedy Department of Justice that the Watergate crowd turned to so many of its members in

Kennedy himself kept a vigilant eye on party affairs. He installed Paul Corbin in the Democratic National Committee. This inevitably caused wide offense. It offended J. Edgar Hoover, who assured the press and the House Un-American Activities Committee that Corbin had been a Communist. The Attorney General, Courtney Evans told the Bureau, "seems to have gone completely overboard in trying to defend Corbin. He has suppressed any and all references to our report detailing Corbin's communist activity."[3] It offended Hubert Humphrey. "Who is this fellow Corbin?" Humphrey wrote irritably to John Bailey, the national chairman. "... This man has no place at the DNC unless we are trying to make enemies and lose friends.... To say the least, I am shocked and to be frank about it, I am damned good and mad."[4] It doubtless offended the long-suffering Bailey. It certainly was not popular in the White House. John Kennedy could never understand why his brother valued Corbin, and Kenneth O'Donnell detested him. But Corbin, as ever, had his uses. He sniffed truculently around, smelled graft, some fancied, much real, and eventually uncovered highly dubious goings-on lower down in the national committee itself, thereby confirming Robert Kennedy's regard for him and producing a schism in the Irish Mafia.

The Attorney General also took a lively interest in patronage. "He passed on applicants for the top, appointive positions," J. Edward Day observed of the Post Office Department. "... He telephoned in person or sent word through his staff about certain appointments of postmasters and rural letter carriers. One afternoon I talked to him three times by telephone about a single rural letter carrier who was to be appointed in a small town in Mississippi in which Bobby was intensely interested."* I don't know whether this last was an example; but Robert Kennedy particularly wanted to increase the number of blacks in government service. Among those whom he nagged relentlessly was Adlai Stevenson, not one of his favorites at best. After a time, Stevenson finally offered a job to Frank Montero, a well-qualified Negro but a Bronx Reform Democrat and for that reason objectionable to the ancient Bronx boss, Charlie Buckley, whom the Kennedys remembered with gratitude as their original supporter in New York. A member of Lawrence O'Brien's staff called me (I served as liaison between the President and Stevenson) and said the Montero appointment had to be stopped. I disagreed

---

their time of trouble — Nixon and Richard Kleindienst to Miller, John Mitchell to Hundley, John Dean to Charles Shaffer, Howard Hunt to William Bittmann.

* J. Edward Day, *My Appointed Round: 929 Days as Postmaster General* (New York, 1965), 9. This does not imply a great upheaval. Day wrote: "During my first year in office only 70 of 35,000 postmasters were removed. As many of them were Democrats as Republicans and the charges against each were of an extremely serious nature" (ibid., 13).

but consented to look into the matter. To my delight, I discovered the UN mission was hiring Montero on the insistent and repeated request of Robert Kennedy.[5]

On another occasion the White House politicos opposed an appointment for James MacGregor Burns, the Williams political scientist, on the ground of his seditious activities as an antiorganization Democrat in Massachusetts. My notes at the time:

> I brooded about this for a while and finally sent a memorandum to the President, with a copy to Bobby, asking whether he knew there was a White House veto on Burns. The President instantly took the matter up, called in Kenny [O'Donnell] and said he could see no reason why Burns should not receive an appointment. Ken said that Jim had been fighting the organization. The President said that this did not matter; Jim was a good fellow, he liked him, and the State Committee was terrible anyway.... The same day I received a memo from Bobby saying that there was no tenable objection to Jim's appointment.[6]

Robert Kennedy followed New York politics with particular interest. He had been outraged by the cavalier treatment given to Herbert Lehman and Eleanor Roosevelt, in the 1960 campaign, by Carmine de Sapio, the Tammany leader, and his protégé Mike Pendergast, the state chairman. "I understand," he wrote John Bailey early in 1961, "the recommendations of Mike Pendergast in New York are being accepted. This seems to me most inadvisable if true. I would appreciate a report on it."[7] For all their personal loyalty to men of the past like Charlie Buckley, the Kennedys wanted to put a new face on the party in what was still the nation's most populous state.

Thus early in 1962 John Kennedy astonished Edward R. Murrow, whom he had brought into the administration as head of the United States Information Agency, by suggesting that he run for the senate in New York. "It was one of the few times," observed Donald Wilson, Murrow's deputy, "I've seen Ed totally at a loss for words. He didn't really say anything so the matter was dropped. I was informally asked by Bob Kennedy a few days later whether I thought Murrow would be interested or not and I said . . . he had told me very frankly that he didn't have any interest in it, and that was the end of it."[8]

In the same spirit Robert Kennedy sought a fresh candidate for governor. In 1961 he had, over Buckley's opposition, named Robert Morgenthau federal attorney in New York City. No doubt he took dynastic satisfaction in appointing the son of the man who had reintroduced Joseph Kennedy to Franklin D. Roosevelt thirty years before. In 1962 he thought that Morgenthau, who had rapidly proved himself a first-class public servant, might be a good candidate for governor.[9] When the embattled Democrats gathered in Syracuse, Kennedy persuaded Buckley to switch from Frank O'Connor, the candidate of the old school, to Morgenthau. O'Donnell told me dryly the day after Morgenthau's nomination, "Charlie Buckley says this morning that, when he put over [Frank] Hogan in Buffalo in 1958 [as the nominee for senator], he was a boss, but now, when he puts over Morgenthau in Syracuse, he becomes a statesman."[10] Morgenthau proved a colorless candidate, however, and Nelson Rockefeller won reelection.

The major family political enterprise in 1962 was the election of Edward Kennedy to the Senate from Massachusetts. "The person who was primarily interested in having him run was my father," Robert Kennedy told John Bartlow Martin in 1964, ". . . just as I would never have been Attorney General if it hadn't been for him. . . . He just felt that Teddy had worked all this time during the campaign and sacrificed himself for the older brother and we had our positions and so that he should have the right to run, that it was a mistake to run for any position lower than that. . . . If it was left up to [Teddy] he probably would have run for Attorney General," an elective office in Massachusetts. Robert added: "I think that some of the people at the White House mumbled about the fact that . . . it would hurt the President but as far as the President and I were concerned, I was pleased that he was running and I think the President was pleased that he was."[11]

The Attorney General went up to Hyannis Port in August to help prepare Ted for a debate with his opponent in the Democratic primary, Edward McCormack, the Speaker's nephew. "Have no feah, we are heah," Robert announced on arrival. The two brothers discussed the debate as they tossed a football on the lawn. Ted wondered how best to explain why he should be in the Senate. Robert answered, "If you get that question, tell them about public service. Tell them why you don't want to be sitting on your ass in some office in New York." He suggested that Ted ask McCormack what he thought about the situation in West Irian; "he'll think it's West Iran and he'll talk about the Middle East and he'll look like a fool."[12] Edward did well in the debate, without benefit of West Irian.

It was an angry primary. Ancient resentment of the Kennedys among Irish pols mingled with wrath at Edward's presumption in the intellectual community. The admirable Mark DeWolfe Howe of the Harvard Law School spoke of the candidacy as "reckless" and "childishly irresponsible." This was wounding. "What bothered [Ted] most," said Gerald Doherty, who helped Stephen Smith run the campaign, "was the almost total boycott by the liberal establishment."* But the voters gave him the nomination by a wide margin.

---

* Burton Hersh, *The Education of Edward Kennedy* (New York, 1972), 160, 175. Mark Howe, before his untimely death, became a friend and supporter of Edward Kennedy.

My journal on primary night: "I received two phone calls. . . . The first was from Bobby in Hyannis Port to say that, in view of Eddie McCormack's performance in the primaries, he wished that I would persuade Mark Howe to endorse George Lodge [the Republican candidate] in similar terms for November. No sooner had he hung up than a call came [in the same vein] from Ethel in McLean."[13]

Though Robert Kennedy was by now taking strong liberal positions on most issues, he continued to suspect the liberal establishment. He thought that "professional liberals" had a "sort of death wish, really wanting to go down in flames. . . . Action or success make them suspicious; and they almost lose interest. I think that's why so many of them think that Adlai Stevenson is the second coming. But he never quite arrives there; he never quite accomplishes anything. . . . They like it much better to have a cause than to have a course of action that's been successful."[14] John Kennedy, it must be said, sometimes felt the same way. "The only one [of them] I care about," the President once remarked to Ben Bradlee, "is Joe Rauh. He's great."[15] Robert wasn't always too sure about Rauh.

He did in fact make exceptions. One was George McGovern, who, after an effective interlude as Food for Peace director, was thinking of trying again for the Senate. He consulted with Robert Kennedy. "His question to me was . . . can you win? You know, he didn't believe . . . in these long-range hopeless efforts." When McGovern came down with hepatitis in the midst of his campaign, Robert sent Ethel out to South Dakota. "I was very ill then," McGovern recalled. "I could hardly hold my head up. . . . The publicity was fantastic. We had maximum coverage on television, radio and the press, and she was gay and hopeful. . . . I never knew until later that she was four or five months pregnant at that time. She has held a special place in my heart ever since. Indeed, I probably could not have made it to the Senate without her."[16]

II

In May 1961 Congress passed a bill creating seventy-one new judge-ships. This, along with normal attrition among older judges, meant that the Kennedys made 128 judicial nominations by mid-October 1962. For a year they were sending names to the Hill at the rate of nearly ten a month.*

* Harold W. Chase, *Federal Judges: The Appointing Process* (Minneapolis, Minn., 1972), 49. Professor Chase wanted to make a study of the way judges were appointed and asked Katzenbach, a college classmate, whether he could observe the Kennedy process

The Attorney General traditionally recommended judicial nominations to the President. Robert Kennedy took his responsibility
conscientiously but had one small problem. He really could not
understand why anyone wished to be a judge. Going on the bench
took people out of the fight; "it's so boring, isn't it?" "Anybody he
really liked, he didn't want to make a judge," noted Joseph Dolan,
who handled the initial explorations in Justice.[17]

Still people inexplicably yearned for the bench. Appointments had
to be made, a crossfire of pressures endured. The party organization
had little influence in the process. "We never cleared, we never
checked, we never called the Democratic National Committee about
judicial appointments," Dolan said.[18] In New York, Buckley and De
Sapio forwarded lists of names. "None of these recommendations
was accepted," Robert Morgenthau recalled in 1971.[19] The White
House politicos, O'Donnell especially, argued tenaciously for deserving Democrats. "The President was more understanding," recalled
Robert Kennedy. ". . . We appointed some [Republicans], and that
didn't go over well [with O'Donnell], and then some of the time I'd
turn [Democrats] down because I didn't think they were qualified."*

The American Bar Association's Standing Committee on the Federal Judiciary spoke for the legal establishment. Justice regarded the
ABA imprimatur as desirable but not essential. "I would be very
surprised," Katzenbach told the ABA House of Delegates in 1962,
"if this committee were omniscient and infallible. . . . At least some
of the judges found by this committee to be qualified will . . . prove
to have been bad appointments."[20] After all, the ABA had given the
egregious Judge W. Harold Cox its highest rating. Nor did the ABA,
having insinuated itself into the appointing process during the Eisenhower administration, want to risk its position now by giving too

---

firsthand. "Before we could go ahead, we needed the approval of Robert Kennedy.
His reaction was typical for him. Judicial selection was the people's business as well as
his. He was willing for me to do the study and call the shots as I saw them but he
wanted to be reassured (by Katzenbach) that I would do a responsible job" (ibid., x).
Chase was set up in a conference room between the offices of Katzenbach and Joseph
Dolan, who did the preliminary screening of judicial possibilities. His book is interesting and illuminating. See also Victor Navasky, *Kennedy Justice* (New York, 1971),
259.

* RFK, in recorded interview by John Bartlow Martin, April 13, 1964, I, 41–42, JFK
Oral History Program. Of Kennedy's 130 judicial appointments, 11 — 8.5 percent
— were Republicans; Eisenhower's proportion of Democrats, 15 out of 201 judges,
had been even smaller — 7.5 percent. Kennedy and Eisenhower were both following
precedent. The only President in more than half a century to appoint less than 90
percent of his judges from his own party was Hoover (82.7 percent). J. C. Goulden,
The Benchwarmers: The Private World of the Powerful Federal Judges (New York: Ballantine
reprint, 1976), 66; Chase, *Federal Judges*, 72, 77.

much offense. "The Kennedy people weren't the corporate types," one member of the committee said; "in a lot of areas we just didn't synchronize. I had the idea that if we turned down too many judges, Bob Kennedy and Katzenbach would have told us to go to hell." In the end, 6.3 percent of Kennedy's judges, as against 5.7 percent of Eisenhower's, failed the ABA test.[21]

By far the strongest pressure on the process was the Senate. "Basically," Robert Kennedy said in 1964, "it's grown up as a senatorial appointment with the advice and consent of the President." Sometimes senators had interests contrary to the administration's interests; "and this is where the struggle takes place. The President of the United States is attempting to obtain the passage of important legislation in many, many fields; and the appointment of a judge recommended by the chairman of a committee or a key figure on a committee can make the whole difference on his legislative program."[22]

Robert Kerr of Oklahoma was the second ranking Democrat on the Senate Finance Committee. Since the doughtily reactionary Harry Byrd of Virginia was chairman, Kerr was essential to the success of the administration's tax proposals. He was an imperious figure, generally accounted, after the departure of Lyndon Johnson, the most powerful and ruthless man in the Senate. Now he demanded that Luther L. Bohanon, an Oklahoma friend, be named district judge. Both Justice and the ABA found Bohanon unqualified. Byron White and William Geoghegan so informed Kerr. "Young men," said Kerr, "I was here a long time before you came. I'm going to be here a long time after you go. I stand by my recommendation."[23] Meanwhile tax legislation stalled. The President asked the secretary to the Senate, Bobby Baker, to find out from Kerr what the problem was. "Tell him," the senator said, "to get his dumb fuckin' brother to quit opposing my friend."[24] Robert Kennedy was determined not to yield; "I struggled and fought about that." John Kennedy had other concerns. "He just [felt] we weren't going to get any legislation through that committee unless we had Bob Kerr." It was the only judicial appointment, Robert said later, in which "we really . . . were on opposite sides."[25]

The dilemma was genuine enough. "It was the judge that I felt strongest about, and [Kerr] was really so blatant about it that I really disliked him. I might say that President Kennedy rather liked him. . . . He was so effective — I mean, the way he operated." In the end the Attorney General capitulated. "You really have to balance off. It sounds terrible. . . . [But] you stand fast on principle, and Kerr doesn't get his judge, and you don't get any tax legislation. They play it as tough and as mean as that. . . . He liked the

President; and through his efforts we were able to obtain the passage of a good number of bills that would have never gotten by. . . . He knew his stuff, and he was the only person who could handle Byrd." So Bohanon became a judge — and turned out, to general surprise, not too bad, at least on civil rights.[26]

One ABA plank was opposition to the appointment of persons over sixty. (They also opposed judges who had not had at least fifteen years of practice, including "a substantial amount of trial experience" — a standard that did not excite the Attorney General.)[27] A united front of Texans, from Vice President Johnson to Senator Ralph Yarborough, were calling for the appointment to the federal bench of Sarah T. Hughes, a respected state judge who became sixty-five in 1961. The Attorney General mentioned the ABA objection. "Sonny," Sam Rayburn told him, "in your eyes everybody seems too old." Kennedy asked Ramsey Clark, another Hughes backer, to explain in three minutes why she should be nominated. Clark replied: "There are three reasons. First, we need women judges. She has served twenty-six years on the bench and has been a good judge. Second, women live longer than men, so if you want to be scientific about it, give her the benefit of the doubt. Third, a gentleman never asks a lady her age." "Bob was standing there," Clark recalled, "and he just whirled around and said something like, 'Let's do it.' . . . Bob was a joyous person in so many ways."[28] Judge Hughes retired in 1975 after fourteen years of distinguished service on the federal bench.

Robert Kennedy was also determined to appoint a Mexican American; "there had never been one," he told Clark, "and it was way past due." The ABA disapproved three names submitted by the Department. After a talk with Kennedy, the committee agreed to give Reynaldo B. Garza a "qualified" rating. The Vice President later exhorted Garza: "Reynaldo, I want you to be such a great judge that when you walk down the streets of Brownsville, all the little Mexican-American boys will think you're the second coming of the Lord."[29]

John Kennedy, I noted in my journal in July 1961, "was quite funny on the problem of appointing judges — there were so many appointments to be made, he said, that every senator thought he had a God-given right to have his own bum appointed. 'Even I have mine,' he said."[30] His bum was Francis X. Morrissey of Boston, Joseph Kennedy's loyal retainer across the years and head of John Kennedy's Boston office in his congressional period. The ambassador hated to harass his sons to find out what they were up to; instead, he had come to rely on reports from Morrissey. A cheerful, round-faced little man, Morrissey was not without his subtleties and decencies; but the sons had come to see him almost as their father's spy, nor did they suppose for a moment that his appointment would elevate the federal judiciary.

A few days later (my journal again): "The President complained of the press concentration on the allegedly prospective appointment of Frank Morrissey . . . and the simultaneous neglect of the excellent judicial appointments under way in the rest of the country." Mc-George Bundy, who was present, remarked that this was one more sign Morrissey's appointment would cause unnecessary trouble. The President said, "Look: my father has come to me and said that he has never asked me for anything, that he wants to ask me only this one thing — to make Frank Morrissey a federal judge. What can I do?" Bundy answered forthrightly that his father knew he shouldn't ask this, that the President knew he should reject the request, and that his father would know that he was right to reject the request. "The President looked gloomy and said that Morrissey couldn't have worked for him for eleven years without picking up something in the way of a sense of public service and that in any case he would be greatly superior to his predecessor Windy Bill McCarthy."[31] (This conversation should be noted, among other reasons, for the candor John Kennedy invited from his staff.)

The ABA had meanwhile rendered a devastating report on Morrissey. Anthony Lewis of the *New York Times* asked the Attorney General whether he meant to appoint him. "He really was furious," Lewis recalled, "and could be extremely gruff and angry. He said, escorting me to the door, 'We are going to appoint Frank Morrissey just to show that the *New York Times* isn't making the judicial appointments around here.'"[32] But they didn't appoint Morrissey (though they did leave the judgeship vacant).

III

Except for the disastrous southern judges, the Kennedy-Kennedy record in the circuit and district courts was in the main excellent. The Supreme Court, of course, mattered most of all. These were, said Katzenbach, speaking of John Kennedy, "almost the only judicial appointments he took seriously. . . . The others he was perfectly satisfied to leave to other people's judgment, and he would ask only occasional questions about them. The Supreme Court [appointments] he did feel were personal."[33] His first opportunity came with the resignation of Justice Charles E. Whittaker in the spring of 1962.

"I had originally been for putting a Negro on," Robert Kennedy said in 1964. ". . . My first recommendation was [Judge William H.] Hastie." Hastie, a Harvard Law School graduate, appointed by Truman in 1949 as the first black circuit court judge, had proved himself steady, able and hard working. The Attorney General was dismayed by the resistance he encountered. "I went up and saw [Chief Justice Earl] Warren about Hastie. He was violently opposed to having Hastie on the Court. . . . He said, 'He's not a liberal, and he'd be opposed to all the measures that we are interested in, and he would just be completely unsatisfactory.'" Robert's old friend Justice Douglas agreed with Warren. Hastie, he told Kennedy, would be "just one more vote for Frankfurter."[34]

"A lot of people in the White House," Robert Kennedy recalled, "were opposed to having a Negro" — not on racial grounds, but because it seemed too political, "too obvious." The President called in Clark Clifford, who after observing that the judiciary had been Truman's great failure, exhorted Kennedy to make only distinguished appointments. When Kennedy mentioned Hastie, "I said I thought it would be a mistake for him to reach out just to put a Negro on the bench" for the sake of having a Negro; Hastie's record was "reasonably good" but not great. Robert Kennedy argued back that obviousness was no reason for rejection, that Hastie "was a good judge, and that it was an appropriate time. . . . If you were going to consider the five or six best judges of the Circuit Court, you'd put him there. So that was the basis of it. . . . I didn't know when another vacancy would come, and I thought that it would mean so much overseas." Finally Katzenbach, after an analysis of Hastie's opinions, assessed them as "competent and pedestrian. . . . When [I said] I thought his opinions were rather pedestrian, this just killed him with President Kennedy."[35]

Clifford was high on Paul Freund of Harvard. So was Katzenbach, though he asked the President whether there would be a problem if he named a second Jew with Frankfurter still on the Court. "He blew up at that and . . . said, 'Why the hell shouldn't I?'" Naming a Harvard professor was more difficult. "We've taken so many Harvard men," Robert Kennedy said to me, "that it's damn hard to appoint another."[36] Warren and Douglas were cool on Freund, no doubt supposing he too would be one more vote for Frankfurter.[37] Other names were considered: two eminent state judges, Roger Traynor of California and Walter Schaefer of Illinois; the president of the University of Chicago, Edward Levi; the Texas lawyer Leon Jaworski; Arthur Goldberg, the Secretary of Labor (the President crossed Goldberg off the list, saying that would come later; he was needed in the cabinet now).[38]

Senator Russell soon threatened to bring a delegation over to ask the President for a conservative. The President, wishing to forestall the request, now proposed appointing Byron White at once. O'Donnell objected, reminding him of the hard time his brother had given them "whenever we tried to reward a nice young man, who had helped us politically, with a Federal judgeship." White, O'Donnell said, "doesn't have any of that Oliver Wendell Holmes background that the Justice Department is always demanding when we try to give somebody a judge's job."[39] The Attorney General himself was dubious — not on the merits but because he could not conceive why White, whom he saw as a man of action, would want to go on the bench, and because he felt he needed him in the Department. Katzenbach dealt with the second point by remarking that, since Kennedy had just lost John Seigenthaler, who had left to become editor of the *Nashville Tennessean,* it would be hard on him to lose White too; "you ought to consider it before recommending Byron." Kennedy, bridling, said, "I'm not going to stand in Byron's way. I can handle the Justice Department without Byron."[40]

On March 30, 1962, I was with the President when the White House telephone operators tracked down White somewhere in the west. "Well, Byron," Kennedy said, "we've decided to go ahead on you." There was a moment's silence, and the President said, "We want to get the announcement out in twenty minutes, so we need an answer right away." Another silence, and the President said, "All right, we'll go ahead." Turning to me: "Could you write a statement right away explaining why I am choosing White." He gave some reasons, and I called Bobby, who gave some more. Make the point, the Attorney General characteristically said, that White was no mere professor or scholar but had actually seen 'life' — in the Navy, in private practice, in politics, even on the football field. So White left Justice, with Katzenbach taking his place as deputy attorney general. As for the Court, the President said to me, "I figure that I will have several more appointments before I am through, and I mean to appoint Freund, Arthur Goldberg and Bill Hastie."[41]

Frankfurter, now seventy-nine years old, was beginning to show the infirmities of age. John Kennedy at one point sympathetically asked Max Freedman of the *Manchester Guardian,* Frankfurter's designated biographer, whether the Justice would be willing to resign if he and the President could agree on a successor. I later asked Kennedy what Frankfurter had replied. Kennedy said, "I guess he decided that he was indispensable to the court."[42] Over the summer, however, Frankfurter recognized he could not continue.

"The President called me in the middle of the week," Robert Kennedy wrote in a memorandum in early September 1962,

> and said that he had heard from Warren that Frankfurter was not going to return to the Court. We discussed who might replace him and he thought about putting Paul Freund on. I made the point that Archie Cox had come to work for the Administration and had done a fine job

while Paul Freund had refused the position of Solicitor General. Therefore, I thought that Archie Cox deserved the appointment more than Paul Freund.

We talked about the necessity of putting a Jew on. I said I thought if a Jew was placed on the Court it should be Arthur Goldberg although I thought it was very difficult to part with him. We discussed it a couple of times again and the only reservation about Arthur Goldberg was the fact that he was so valuable to the Administration and could handle labor and management in a way that could hardly be equalled by anyone else.

Finally, however, the President called and said he had talked to Arthur Goldberg and that Goldberg was anxious to get the appointment.... Certainly a major factor in giving up the services of Arthur Goldberg was the necessity of replacing Frankfurter with a Jew. The President said that Mike Feldman [of the White House staff] indicated clearly that the various Jewish organizations would be upset if this appointment did not go to a person of the Jewish faith.[43]

Frankfurter must have regretted his refusal of Kennedy's earlier proposal. As Dorothy Goldberg later put it, "Arthur and Justice Frankfurter were at opposite poles in their thinking."[44] Nor was the Justice a great admirer of the Attorney General. "What does Bobby understand about the Supreme Court?" he told an interviewer in 1964. "He understands it about as much as you understand about the undiscovered 76th star in the galaxy."[45]

The Attorney General was kinder about Frankfurter. I will never forget one of the Justice's last public appearances. It was at the swearing-in of Carl McGowan as judge of the District of Columbia Court of Appeals. Frankfurter had asked to say a few words.

He sat in his wheelchair and spoke fluently and coherently, though not very relevantly, for about twenty minutes. Since everyone expected a short tribute to McGowan, and since the Justice did not mention McGowan and showed every sign of being able to go on for hours, a restiveness crept over the courtroom. People passed notes, and finally Joe Rauh, who had accompanied the Justice, leaned over and suggested that perhaps he had talked long enough. The Justice said cheerfully — and loudly — "I may be Joe Rauh's judge, but he is not my lawyer" — and continued. Finally something happened to his microphone and, though he went on talking, no one could hear him. Judge [David] Bazelon said, "Mr. Justice, Mr. Justice, your power has gone" — a terrifying metaphor, given the occasion. Finally the Justice stopped, and Joe wheeled him out.... There were mixed reactions afterward. Tony Lewis considered it a great tragedy for the Justice to expose himself in his rambling and garrulous old age. Bobby Kennedy, however, with whom I rode away, said, "If that experience gave the old man half an hour of pleasure, no one in the room had such pressing business that he couldn't stay for a few extra minutes."[46]

After the Goldberg appointment, John Kennedy mused about Cox and Freund. "Bobby thinks it would be impossible to appoint them both," he said, "but I don't see why. After Archie has been down here a little while, he will seem more Washington than Cambridge. I think we'll have appointments enough for everybody."[47]

IV

"I shall withhold from neither the Congress nor the people," John F. Kennedy said in his first annual message, "any fact or report, past, present, or future, which is necessary for an informed judgment of our conduct and hazards."[48] Imperfect as the execution of this pledge later was in foreign affairs, Kennedy's statement was a deliberate repudiation of the theory devised by the Eisenhower administration of an absolute presidential right to deny information to Congress. This supposed right, born in response to the irresponsible requisitions of Joe McCarthy, received in 1957 the name of 'executive privilege' and thereafter acquired with mysterious rapidity the status of ancient and hallowed constitutional doctrine.

Toward the end of the Eisenhower years Congressman Porter Hardy, chairman of a House Subcommittee on Foreign Operations and Monetary Affairs, tried to get records relating to the foreign aid program in Peru. Eisenhower promptly ordered that the documents be withheld. When Kennedy came in, the State Department renewed the Eisenhower directive.[49] On March 10, 1961, the frustrated Hardy dropped a letter of complaint at the White House. Before he reached home, the new President had called him. Three days later Secretary of State Rusk promised delivery of the documents. At the next hearing, however, State Department witnesses showed up with a letter from Rusk forbidding them to testify. Hardy complained again to Kennedy. The Department's legal adviser hastily withdrew the Rusk letter, asking Hardy to "treat it as though it had not been sent," and transmitted the documents.[50]

"You seem more willing than previous Presidents and Administrations," James Reston of the *New York Times* said to Robert Kennedy on *Meet the Press* in September 1961, "to give information sought by Congress." The Attorney General replied:

> I was associated with a Congressional committee for five or six years and had battles with the Executive Branch of the government regarding obtaining information. I think it is terribly important [in order] to insure that the Executive Branch of the government is not corrupt and that they are efficient, that the Legislative Branch of the government has this ability to check on what we are doing in the Executive Branch of the government.
>
> So in every instance that has been brought to our attention in the

Department of Justice so far by various departments of the Executive Branch where this question has been raised, we have suggested and recommended that they make the information available to Congress. We will continue to do that.

I don't say that there might not be an instance where Executive privilege might have to be used, but I think it is terribly important, with the Executive Branch of the government as powerful and strong as it is, that there be some check and balance on it. And in the last analysis the group that can best check and insure that it is handling its affairs properly is the Congress of the United States. So we will lean over backwards to make sure that they get the information they request.[51]

There were cases where withholding had a serious point; but these, Robert Kennedy felt, were the rarest exceptions. John Kennedy invoked executive privilege only once. In February 1962 a Senate committee, on an expedition of harassment, demanded to know by name which Pentagon officials had deleted warlike passages from speeches by fire-eating generals and admirals. Byron White, as Acting Attorney General, sent the President a cautionary memorandum, which, he said, "accurately reflects the views of the Attorney General." No doubt executive privilege had to be asserted in this particular case, White wrote, but Justice was concerned about "internal pressure in State and Defense, particularly the former, to use this invocation as precedent for future situations." The President had to make clear "whether or not this Administration is adopting the Eisenhower principle and will apply it automatically as was done in the past Administration."

The memorandum acknowledged the "great appeal" of the Eisenhower principle for the executive branch.

[But] it is not enough to assert "managerial responsibility" without at the same time weighing the needs of Congress for the information withheld. From the viewpoint of Congress, automatic application of this principle would make meaningful investigation almost impossible in many situations. . . . The Executive branch has some tendency to hide its errors and many Congressional investigations have disclosed errors which would not otherwise have come to the attention of an agency head. . . . Furthermore, I am inclined to believe that the internal advice rule, however honestly designed to protect subordinate officials from Congressional criticism, as often as not has the effect of protecting those whom we claim to be responsible, for frequently the information withheld would show that an agency head got good advice and ignored it.

Justice therefore recommended an explicit presidential declaration that the use of the privilege in this single case "is *not* to be regarded as precedent for its invocation in other circumstances where Congressional questioning of subordinate officials has more justification in bringing out relevant information. Each case will be looked at on its own merits."[52]

The President followed this recommendation. That he really believed in an executive responsibility to supply information to Congress was shown in a few weeks when Porter Hardy again sought foreign-aid documents, this time about Angola, and the State Department again withheld them. Within an hour after the question was raised in a press conference, Hardy received the records.[53] Later, John Kennedy wrote Hardy that, because he was convinced "your investigation will be both constructive and careful and of real help to both the Congress and the Executive Branch," he was directing the State Department to give the committee everything pertinent except personnel files; even these would be turned over "for your eyes only as Chairman."[54]

Where Kennedy's pledge of full disclosure failed was in areas, like Cuba and Vietnam, about which Congress made no requests for information. Where Congress requested, Kennedy, in the words of Raoul Berger, "sharply limited resort to executive privilege."[55] Congressman John Moss observed of executive privilege in 1963, "The powerful genie . . . momentarily is confined." He added prophetically: "but can be uncorked by future Presidents."[56]

v

In the field of law enforcement, Robert Kennedy confounded those who had predicted he would play politics with justice. "The top political animals in the Kennedy administration," recalled Ronald Goldfarb of the Criminal Division, "helplessly watched Robert Kennedy's organized crime drive knock off politician after politician around the country." At one meeting, lawyers reported on recent indictments. The Attorney General asked questions, including inevitably, "Was he Republican or Democrat?" After several answered "Democrat," Kennedy observed with a smile, "If you guys keep this up my brother is going to have to take me out of here and send me to the Supreme Court."[57] His Department of Justice indicted two Democratic congressmen, three Democratic judges, five Democratic mayors, divers smaller fry.

Bringing some of these cases was difficult. There was the case of Judge J. Vincent Keogh of the New York State Supreme Court. The Kennedys owed nothing to Judge Keogh, but they owed a great deal to his brother Congressman Eugene Keogh, a distinguished Irishman of the old school, a long-time friend of Joseph Kennedy's, influential both in Congress and in New York politics and, in Robert Kennedy's judgment, one of the "five people who were most helpful to the President in the election."[58] But Gene Keogh's brother, it began to appear, had taken a bribe to induce a colleague on the bench to go easy in sentencing a man convicted of bankruptcy fraud. The matter arose in Robert Kennedy's first year. The evidence against Keogh was persuasive but not conclusive. Department of Justice lawyers watched to see what the Attorney General would do.

The White House was watching too. One day the President asked Seigenthaler whether his brother planned to indict Vincent Keogh. Seigenthaler said there were problems. John Kennedy said, "My God, I hope that he doesn't. Gene Keogh was my friend, and, if there's any way I can honestly help him, I'd want to help him."[59] O'Donnell suggested that they drop the case and let Judge Keogh resign. Perhaps to make it clear that politics would not affect his decision, Robert Kennedy put the Eisenhower holdover, William Hundley, in charge. "It was not an open-and-shut case," Hundley recalled. ". . . I don't have any doubt that Kennedy was hoping that the grand jury investigation would show that Keogh didn't do it." But the investigation convinced Hundley that the government's evidence was credible and that Keogh had in fact taken the money.[60] "You've *got* to prosecute this," Byron White said to Robert Kennedy. The Attorney General buried his face in his hands and said, "Goddam it, I told my brother I didn't want this job."[61] Then he said, "If the evidence is there, as difficult as it would be for the President of the United States to hurt his friend, and as difficult as it would be for me, we'll proceed."[62] "He didn't just sit back and say, 'I disqualify myself,'" Hundley recalled. "That's the easy way out: 'I disqualify myself.' Then he could go running back to Gene Keogh and say, 'Well, it was that son-of-a-bitch Hundley.' . . . He never did that. It was his decision. . . . He took the heat personally."[63]

When Hundley said he thought they just had to go ahead, Kennedy said, "Remember, if we lose this case, you and I are going to look like a couple of real shits." "That," Hundley observed later, "was his way of saying, 'Make damn sure you win.'"[64] Vincent Keogh was convicted and sentenced to two years in prison. "It was a very painful case for Bob Kennedy," said Hundley. ". . . But we were closer after that than we had ever been before."[65] Someone at the Democratic National Committee later said to Walter Sheridan, "I can't believe that you couldn't have fixed that case." Sheridan said, "Of course we could have, but that's the difference [between] you and us."[66]

Another painful case involved George Chacharis, the mayor of Gary, Indiana. It appeared that Chacharis had received over

$200,000 in payoffs and failed to report them in his tax returns. There was no evidence that Chacharis had used the money for himself; he had faithfully plowed it back into his political organization and evidently did not suppose he had done anything wrong. Congressman Ray Madden of Gary, a powerful figure in the House, brought Chacharis to see the Attorney General. Madden reminded Kennedy that Chacharis, a Greek immigrant who had come up the hard way, had been the driving force for John Kennedy in Indiana when Kennedy needed help. As the conversation proceeded, Chacharis, an emotional man who had actually been proposed as ambassador to Greece, broke down and cried. His attorney was the Democratic National Committeeman from Indiana. The prosecutor was aggressive, theatrical and, in the view of some in Justice, a grabber of headlines. But the pressure of evidence overcame the pressure of politics. Chacharis was indicted, convicted and sent to prison.[67]

Another case involved Matthew H. McCloskey of Philadelphia, a veteran Democratic fund raiser and Kennedy supporter whom John Kennedy named ambassador to Ireland in 1962. The McCloskey Construction Company had built — very badly — a Veterans' Administration hospital in Massachusetts. The Civil Division proposed a suit against the McCloskey firm for faulty construction and breach of contract. "We made that recommendation to the Attorney General," said John Douglas, "and he didn't bat an eyelid; he just said, 'Go ahead.'"[68]

A couple of cases complicated the Attorney General's prickly relations with the Vice President. Fred Korth, a Johnson protégé, had succeeded John Connally as Secretary of the Navy. Korth, it was discovered in 1963, was using department stationery, the Navy yacht *Sequoia* and other official facilities on behalf of his bank back in Fort Worth. Robert Kennedy, over Johnson's protest, recommended Korth's dismissal. The President agreed. "Clark Clifford," Kennedy reminded Johnson in 1964, "was also for retaining Korth but after reviewing the material (and also, incidentally, arguing with us at the Department of Justice) he came to the conclusion that Korth should be relieved."[69]

Then there was the case of Bobby Baker, so long Johnson's *homme de confiance*. "I have two daughters," Johnson used to say. "If I had a son, this would be the boy."[70] Congressional investigations began to turn up evidence that eventually led to Baker's conviction for tax evasion and fraud. The Vice President was sure Robert Kennedy was pursuing Baker in order to force Johnson off the ticket in 1964. Actually both Kennedy and Johnson were, in different ways, on the spot. The Attorney General, Katzenbach said later, "never got into much of the details of the case.... I don't really think he wanted to know a great deal about it.... He was sure that Johnson would believe that somehow he was behind this, and [that] it was designed to embarrass Lyndon Johnson, which is the way President Johnson often thought."*

Ironically, as it was in some sense easier to prosecute Kennedy loyalists than Johnson loyalists, so, in at least one major case, it was easier to prosecute Democrats than Republicans. Soon after he became Attorney General, Kennedy was faced with the problem of the relationship between the Boston textile manufacturer Bernard Goldfine and Sherman Adams, a governor of New Hampshire and subsequently President Eisenhower's chief of staff. The disclosure that he had accepted favors from Goldfine had forced Adams to resign in 1958. The favors, it now appeared, went far beyond hotel rooms and the famous vicuña coat. By 1961 Goldfine, old and sick, was ready to talk. His lawyer was Robert Kennedy's one-time friend and later antagonist Edward Bennett Williams. Goldfine claimed, Robert Kennedy noted in February 1961, that Adams had received "more than $150,000 in cash over the period of about five years." Goldfine also implicated a former Democratic governor of Massachusetts and a couple of Republican senators, among them Styles Bridges of New Hampshire. "Jack called me the other day and said that he understood Styles Bridges had cancer and, obviously, we did not want to develop this information about him as he was dying."[71]

One problem was that Goldfine himself was so ill with arteriosclerosis that he could not always remember things coherently. "Sunday of last week," Kennedy wrote, "[he] burst into tears and said he couldn't go on because his mind was so befuddled." A trail of cashier's checks Adams had given his Washington landlady seemed to

* Nicholas Katzenbach, in recorded interview by L. J. Hackman, October 8, 1969, 9, RFK Oral History Program. When Kennedy and Johnson discussed Bobby Baker in 1964, Johnson emphasized "that Bobby Baker got into all this difficulty after he, Lyndon, had left the Senate. He said Bobby Baker wanted to go to Puerto Rico on one occasion when he was Vice President, when he, Johnson, was going down to the inauguration of Bosch. [He meant the Dominican Republic.] He said for some reason Bobby's request concerned him and he didn't take him. Later on, he said, he found out he went anyway. But he said he could have gotten into so much difficulty if he had taken him as part of his official party. Secondly, he said God must have been watching over him because he did not have financial dealings with Bobby. He could very well have done so because if Bobby asked him to loan him $10,000 for the purchase of some stock he certainly would have done so. He said the only business deal he had with Bobby Baker was in connection with the purchase of some insurance. The insurance company then placed $1500 (I think that was the figure) worth of advertising with his television station. He said this really didn't mean anything because the competitor company was willing to place twice as much" (RFK, memorandum, August 4, 1964, 3–4, RFK Papers).

corroborate Goldfine's allegations in general; but Goldfine himself would obviously not be a strong witness. "I am not optimistic," said Kennedy; "however, I think there is probably a fifty-fifty chance. He's got everything to gain and nothing to lose." [72]

Kennedy turned the case over to the indispensable Hundley. Hundley interviewed Adams, who seemed extremely nervous — as if he were ready to jump out the window, Hundley thought — and, confronted by the cashier's checks, admitted receiving money from unremembered donors, but not, he insisted, from Goldfine. In the end, gaps remained in the evidence. Obviously the Kennedys could not afford to bring a case against Adams and lose it. This would look like a bungled attempt at political vengeance. Because it would not be hard to confect a defense — people might always arise to swear they had given the money to Adams out of disinterested patriotism — Louis Oberdorfer, as head of the Tax Division, recommended against prosecution.[*]

It is said, though I have been unable to verify the story, that John Kennedy sent the incriminating data on Adams over to Eisenhower at Gettysburg.[73] The new President was keenly aware of his predecessor's continuing popularity and might have thought it useful to put the old general under obligation. The relations between the two men were cool. On inauguration day, as they rode together to the Capitol, Kennedy had asked what Eisenhower thought of Cornelius Ryan's vivid account of D-day, *The Longest Day*. The new President "was rather fascinated," Robert Kennedy recalled in 1964,

> that Eisenhower never read the book and in fact hadn't seemed to have read anything. He thought he had a rather fascinating personality, could understand talking to him why he was President of the United States . . . [but] just felt that he . . . hadn't done any homework, didn't know a great deal about areas that he should know. I think he always felt that Eisenhower was unhappy with him — that he was so young and that he was elected President and so he always, feeling that Eisenhower was important and his election was so close, . . . went out of his way to make sure that Eisenhower was brought in on more matters and that Eisenhower couldn't hurt the Administration by going off on a tangent and that's why he made such an effort over Eisenhower, not that Eisenhower ever gave him any advice that was very helpful.[74]

According to the story, Eisenhower, after looking at the Adams dossier, sent back a message to Kennedy expressing fervent hope that

---

* William Hundley, in recorded interview by J. A. Oesterle, December 9, 1970, 21, RFK Oral History Program; Hundley, in interview by author, August 4, 1976; Louis Oberdorfer, in interview by author, December 29, 1976. The Department of Justice had recently lost a case where J. Truman Bidwell, chairman of the board of governors of the New York Stock Exchange, escaped conviction by persuading a jury that he had come upon money in other ways than those charged in the indictment.

Adams be spared further humiliation. This was not the reason the Department of Justice failed to indict Adams; but Eisenhower presumably did not know this, and the apparent favor may well have, for a period, restrained his criticism of his successor.*

VI

The most painful cases of all involved friends of the family. Igor Cassini — Cholly Knickerbocker, the gossip columnist of Hearst's *New York Journal-American* — was no friend of Robert Kennedy's. But his brother Oleg made dresses for Jacqueline Kennedy and attended White House parties. And Igor — Ghighi to his intimates — was an acquaintance of Joseph P. Kennedy's in Florida. He was also a friend of the Dominican dictator Trujillo, with whose bloody regime the United States had recently broken relations; and he claimed one Palm Beach evening that Trujillo would reform if the United States restored recognition; adding darkly that, if Washington did not alter its policy, the Dominican Republic would fall to communism.[75] The old ambassador was sufficiently impressed to tell the President. In April 1961 the State Department sent the veteran diplomat Robert Murphy to take a look. Cassini was at Murphy's right hand.

After their return Robert Kennedy heard rumors that Cassini's interest in Trujillo was more than academic. He thereupon instructed the FBI to investigate reports that Cassini, who had not registered as a foreign agent, was taking money from the Dominican tyrant. Cassini protested. "I've performed this function with Murphy," he told the Attorney General, "and then I come back, and somebody starts a rumor, and you start an investigation." He swore to God that he was clean. The investigation, he pleaded, was hurting him with his newspapers; "the Department of Justice is just finishing me." The FBI, after several weeks, came up with nothing. Impressed

---

* Another version of what may be the Sherman Adams story appears in Bobby Baker with Larry King, *Wheeling and Dealing: Confessions of a Capitol Hill Operator* (New York, 1978), 82–84. According to this account, a federal grand jury was on the verge of indicting Eisenhower's "former aide, 'Mr. Jones'" for income tax evasion. Mrs. Eisenhower feared that "Jones might commit suicide." The former President asked Senator Dirksen to ask President Kennedy "as a personal favor to me, to put the Jones indictment in the deep freeze. . . . He'll have a blank check in my bank if he will grant me this favor." According to Dirksen, as rendered by Baker, Robert Kennedy strongly opposed dropping the indictment, and the President told his brother, "If you can't comply with my request then your resignation will be accepted." JFK later cashed in his blank check by insisting that Eisenhower and Dirksen publicly endorse the test ban treaty of 1963.

by Cassini's vehemence and by his invocation of the Almighty, Kennedy told the Bureau to discontinue; it was, he said later, "the only investigation I called off since I've been Attorney General." Back in New York, Cassini airily informed an interviewer that his Dominican visit had been "well known to the White House. . . . But Trujillo has such a reputation for having paid off everybody that anybody who ever went there was supposedly paid off. . . . I was not among the lucky ones."[76]

He was lying through his teeth. At an early point the Dominican consul general in New York reported to El Supremo that Cassini "entertains the idea of arranging a meeting for you with Mr. Joseph P. Kennedy, father of the President, or with the President himself, both of whom, according to Cassini, are his intimate friends and with whom he maintains close and frequent relations." Cassini assertedly shared in payments of nearly $200,000 stashed away in a Geneva bank account.*

For a time the matter lay dormant. Then Trujillo was killed. A new democratic regime opened the Trujillo files to newspapermen and congressional investigators. The journalist Peter Maas got onto the story. Maas remarked to someone in this period how much he liked Robert Kennedy. He was told, "Forget it if you run the Cassini story, because the lead . . . involves Bobby's father; 'ruthless' Bobby'll get you for that." Maas said to himself, "If that's the way it is and Bobby is going to get me, the hell with it. I'm not giving up on the story." His wife thought that at least he was obligated, as a friend, to tell Kennedy what he had discovered.

Maas went to see the Attorney General. Kennedy asked him what it was all about. Maas: "I've just been doing a long piece on Ghighi Cassini." Kennedy, almost half-rising out of his chair: "No. No. We've looked through all that. . . . There's nothing to it." Maas: "There is a lot to it." Maas told Kennedy in detail what he had found. "He sat there," Maas said later, "and didn't say anything at all. He sat there kind of hunched over in that big chair in that big room with his chin on his fingers. . . . I finished. All he said was, 'Thank you very much,' and I left."[77]

Kennedy sent the FBI back to uncover what it had missed. More pleas from Cassini and his friends: "There was an awful lot of pressure . . . not to prosecute," Kennedy said later.[78] Cassini's wife wrote the President expressing shock at "Bobby's harsh and punitive atti-

---

* Brock Brower, *Other Loyalties* (New York, 1968), 101. For another view of the case, see the account by Cassini's lawyer, Louis Nizer, in the chapter "Foreign Agent," in *Reflections without Mirrors* (Garden City, N.Y., 1978).

tude." Her husband, she said, could not understand "why the son of a man whom he considered one of his closest friends for 17 years ... should now be determined to bring him down to total ruin."[79] The President, according to Ben Bradlee, was "obviously upset, and wondered out loud about the virtue of prosecuting Cassini."[80] The Attorney General went ahead. Cassini pleaded nolo contendere, paid a $10,000 fine. "If there was one thing that turned me completely pro-Bobby," said Maas, it was the Cassini case. ". . . I was so used to fixes in covering politics that this was such an extraordinary thing. There were fifteen different ways his people could have fouled up the investigation or done something to have it thrown out of court. But it obviously never entered his mind."[81] Robert Kennedy told Maas, "God knows this was the last case in the world I was looking to prosecute, but what's right is right."[82]

The saddest case involved someone far closer to the Kennedy family than Igor Cassini. James M. Landis had been a faithful friend of Joseph Kennedy and all his children for nearly thirty years. He had performed legal work, written articles, helped the father on his memoirs, drafted legislation and speeches for the sons. He was a singular and in some ways eccentric man. Gruff taciturnity concealed a certain inner perversity. He never, for example, despite protestations from Joseph Kennedy and from his own law partners, sent Kennedy bills for his legal services. He was terribly negligent about his own affairs. In 1956, under business pressures, he had delayed filing an income tax return, applied for one extension, forgot to apply for another, put the whole thing out of his mind. The next year he filled out his tax form but could not figure out how to answer the question whether he had filed a return the year before; so he deposited the amount owed the government in a separate bank account and put the return in a drawer. He did the same thing for the next three years.

After the 1960 election, John Kennedy asked Landis, who, among other things, was an expert on the administrative process, to prepare a report on the regulatory agencies. This assignment led to a routine security check and in turn to the discovery that the Internal Revenue Service had no recent tax returns from Landis. With the help of Joseph Kennedy's accountant, new tax forms were filled out; and, with the help of a loan from Joseph Kennedy (the money Landis had set aside covered taxes and interest but not penalties), he filed delinquent returns, paying $48,000 in 1961 and another $46,000 in 1962. The critical issue now was whether the late filing was voluntary. If deemed voluntary, the IRS would not recommend prosecution. When the IRS asked Landis why he decided to file, Landis, in his way, said shortly, "Because Joe Kennedy told me I'd damn well better."[83] Was it voluntary when the father of a President directed a citizen to pay his back taxes?

Had Landis not been a friend of the Kennedys, his setting aside the money would have been taken as evidence of good faith; his inability to file would have been laid to not uncommon psychological blocks (Landis had in fact been seeing a psychiatrist); his late returns would have been accepted, and the case closed. But, with a friend of the President involved, no one was willing to declare the action voluntary. Commissioner Mortimer Caplin of the IRS passed the buck to the Justice Department. On Katzenbach's insistence, and for self-evident reasons, Robert Kennedy disqualified himself. "Bobby had nothing to do with it," said Katzenbach later. "I wouldn't let him. I said, 'You cannot make a decision not to prosecute Landis. If you can't make that decision, don't make any.'"[84] Louis Oberdorfer, the head of the Tax Division, disqualified himself on the ground of possible conflict of interest; he had consulted Landis before his confirmation hearing, Landis had never billed him, therefore he might be thought to owe something to Landis. This left the case to Katzenbach, who made the decision to prosecute. No doubt he felt that otherwise the Republicans would say that Robert Kennedy had fixed the case to protect his father's friend.

Landis was not a well man. He was undergoing neurological treatment at the Columbia Presbyterian Hospital. He pled guilty. Everyone supposed, in view of the circumstances, that, like the far worse malefactor Igor Cassini, he would receive a fine and a suspended sentence. He came, however, before a self-righteous judge who, after a lecture, sentenced him to thirty days in prison. Robert Morgenthau, the prosecutor, was furious. This was not a fraud case; a sentence so harsh for a man so ailing made no sense in terms of rehabilitation, deterrence or anything else. Justin Feldman, Landis's partner, said to Morgenthau: "He emotionally can't survive it. What can we do?" "There's only one thing we can do," Morgenthau replied. "The Attorney General has the right to determine that he should serve the sentence in a hospital, and not in a jail."[85]

Robert Kennedy at once certified that Landis should serve his sentence in a hospital. The unyielding judge, rejecting the plea of Landis's physician that he needed the facilities at Columbia's Neurological Institute, sent him to the Public Health Service Center on Staten Island, where he was flung into a ward largely populated by screaming psychotics. Feldman went to Washington to ask for Landis's transfer to a hospital. Katzenbach agreed that the Attorney General had the authority to make the transfer but opposed his doing it: "There'll be a whole series of stories, and we're back where we started from — cronyism and favors to friends." Then Feldman saw Kennedy:

KENNEDY.   What does Nick say?

FELDMAN.   Nick says that you shouldn't do it.

KENNEDY.   Does he say I shouldn't do it, or I can't do it?

FELDMAN.   He says you shouldn't do it.

KENNEDY.   Oh, I can't believe that. . . . [punching the intercom to Katzenbach] Is Justin's legal opinion correct that I can do it? . . . You know, Nick, I don't understand you sometimes. . . . Sure we have to worry about the reaction of people in the press, and so forth, but if any goddam reporter wants to say that the Kennedy Administration, having prosecuted one of the best friends they've ever had — somebody who's been practically a father to me, who helped me get through my law school exams, and who's been close to Ted and close to the whole family and close to my father, and has done us so many favors — if they want to say that, having prosecuted him, having exposed him to the public through this, that we are now soft on criminals by having him serve that stupid thirty days in some degree of comfort, they can go to hell! [86]

That settled that. It was an unhappy business. Reading a sympathetic piece by David Lawrence in *U.S. News & World Report,* Robert Kennedy scribbled a note to Lawrence: "I just wanted to let you know how understanding and kind I thought your article was on Jim Landis. He has suffered terribly, as a sensitive proud person would. Your piece will help him to stay alive mentally and physically." [87]

Reflecting about the parade of cases — Landis, Chacharis, Keogh and the rest — Kennedy said in 1964, "It's one of the reasons I was rather reluctant to become Attorney General because I knew I'd get into these kinds of cases as well as somebody that wanted to be a judge and all that other business. That's why I think it's . . . got some dangers to it for a person involved in politics to be Attorney General. . . . It is a position that should really be removed from politics or as completely as it can be." [88]

(18)

Justice and Poverty

H E WAS TIRED, Robert Kennedy had said, of chasing people. The Rackets Committee had persuaded him that the American malaise was a problem not just of crooked individuals but of a corrupt system. The Department of Justice now gave the problem a larger context. He increasingly perceived a nation stained with inequities — in the distribution of power, in the dispensation of justice, in access to opportunity. The "essential" trouble, he wrote in 1964, was "poverty — poverty of goods and poverty of understanding."[1] The New York World's Fair might salute a shining future; "but less than an hour away in Harlem, people live in squalor and despair more closely resembling the nineteenth century. A few hundred miles away, in the remote hovels of Appalachia, the life of the people is, if anything, worse than it was a hundred years ago."[2] Such disparities were unacceptable. "Government belongs where evil needs an adversary and there are people in distress who cannot help themselves." It was his ambition, he said, to make the department over which he presided "more than a Department of Prosecution and . . . in fact, the Department of *Justice*."[3]

I

"The law, especially in criminal cases," he said in his first interview as Attorney General, "favors the rich man over the poor in such matters as bail, the cost of defense counsel, the cost of appeals, and so on."[4] In his first press conference he announced a committee to explore the plight of the poor in federal courts. Early in 1963 the committee, headed by Professor Francis Allen of the University of

Chicago Law School, submitted a searing report, "Poverty and the Administration of Federal Criminal Justice." The findings documented the conviction Kennedy had taken from the Rackets Committee that the United States had in fact two systems of criminal law. Most guarantees of due process — the rights to counsel, to present evidence and call witnesses, to trial by jury, indeed to the very presumption of innocence itself — were effective only for those who could pay for them. The poor lost out at every point along the line.

The Allen committee devoted particular attention to bail, the most visible example of the power of money in the system of justice. Every year thousands of poor people remained behind bars for months after arrest. "They are not proven guilty," Kennedy said. "They may be innocent. They may be no more likely to flee than you or I. But they must stay in jail because, bluntly, they cannot afford to pay for the freedom."[5] In March 1963 Kennedy instructed federal attorneys to release defendants in specified categories without bail. (In the first year the default rate, 2.5 percent, turned out to be slightly less than that for those released on bail, and a thousand man-years of liberty were saved.)[6] In the spring of 1964 he convened the National Conference on Bail and Criminal Justice. In due course Congress passed the federal Bail Reform Act.

He took up the problem of counsel for indigent defendants. "Any person haled into court, who is too poor to hire a lawyer," Justice Black wrote for a unanimous court in the 1963 case of *Gideon* v. *Wainwright*, "cannot be assured a fair trial unless counsel is provided for him." *Gideon* applied to state courts. But within ten days of the decision the Department of Justice drafted the bill that, as the Criminal Justice Act of 1964, extended guarantees of competent defense to every stage of federal judicial proceedings.[7] Kennedy then established the Office of Criminal Justice to deal with social issues affecting the criminal process from arrest to rehabilitation.

He gave strong support to James V. Bennett, the notably humane director of the Bureau of Prisons. "Let us reject the spirit of retribution," he said in one speech, "and attempt coolly to balance the needs of deterrence and detention with the possibilities of rehabilitation." J. Edgar Hoover noted on the border, "Sounds like some of Bennett's philosophy," and thereafter fretted about Kennedy's "soft treatment approach to criminals."[8] But, over Hoover's opposition, Kennedy and Bennett closed down the gloomy dungeons of Alcatraz.[9] They set up halfway houses to help ease the return of convicts to civilian life. Kennedy worried especially about youthful offenders, too often incarcerated with, and thereafter indoctrinated by, their hardened seniors. In 1963 they began planning youth centers for persons under twenty convicted of federal crimes.*

He was much concerned, as were Bennett and the President, with inequality in sentencing. "On pardons and commutations," Robert Kennedy said later, "we made a real major effort and really major breakthrough. President Eisenhower didn't believe in them because he thought it was interfering with judicial prerogatives."[10] John Kennedy, Bennett wrote in his memoirs, "whose compassion exceeded that of most men, struck out against the injustices that he perceived in our sentencing procedures," reducing the sentences of more than one hundred prisoners and granting unconditional pardons to five hundred more who had demonstrated good citizenship after release. "Without exception," said Bennett, "Kennedy approved my recommendations for executive clemency, and he invariably reminded me to let him know when I came across a glaring example of excessive sentencing."

Robert Kennedy's "contribution to even-handed justice," Bennett added, "was also immense." Bennett instanced the case of a destitute young man who robbed a bank in quest of money to support his family and then, overcome with remorse, turned himself in to the FBI. He got forty years in the penitentiary. The Attorney General had it reduced to fifteen years, the more customary sentence for bank robbers.[11] (Kennedy was fond of Jim Bennett. At the surprise party marking the director's twenty-fifth anniversary as head of the prison system, the Attorney General gave him a cake with a file baked inside.)[12]

The case of Junius Scales aroused particular attention. Scales had been the only Communist sent to prison under the membership clause of the Smith Act. There was no suggestion of espionage or other criminal deeds. He had received a longer sentence than top Communists convicted under other sections of the act. After the Red Army invaded Hungary in 1956, he had broken with the party and was thus serving his sentence as an ex-Communist. In 1962 anti-Communists of the left, led by Norman Thomas and Joseph Rauh, urged his release. But Scales, even as an ex-Communist, flatly refused on grounds of principle to identify people he had known in the party. This infuriated J. Edgar Hoover. The release of Scales,

---

* Dr. Robert Coles attended the dedication in 1968 of the Robert F. Kennedy Youth Center in Morgantown, West Virginia. An old miner said to him, "I'm sure they'll take a lot better care of those boys than they did with us when we went into the mines. I was 15, like they must be, going into that Center, and there was no Robert Kennedy looking over us, no sir, there wasn't — just a lot of greedy coal operators that didn't care if we lived or died" (Robert Coles, "Youthful Offenders," New Republic, October 4, 1969).

he told the Attorney General, would make it impossible for the FBI to insist that Communists or ex-Communists name names in the future. Byron White and Nicholas Katzenbach successively recommended against commutation of Scales's sentence.[13]

"There was some fuss about it," Kennedy told John Bartlow Martin in 1964,

> and I guess Norman Thomas and a lot of other people came to see me at various times or wrote me letters and I said that . . . I would take a look at it but I wasn't going to take any look at it if they picketed or made it appear that it was going to be because of political reasons. . . . I just asked them if nobody would get in touch with me for a period of time that I would take a look at it and if I thought it should be done that I'd do it. And nobody did do anything and I never heard for a period of time and I looked at it. . . . I just thought that . . . based on the penalties for others, the amount of time of his sentence and . . . the problems he had at home [a sick wife] that . . . I had his sentence commuted.
>
> I never discussed it with the President.[14]

Anthony Lewis recalls the Attorney General mumbling late in 1962, "We're going to let your friend Scales out." He said it, Lewis continued, "embarrassedly, almost grumpily; he did not like to look soft."[15]

Scales was back with his family on Christmas Eve. Rauh summed up the significance of Kennedy's decision: "The act of granting clemency to Scales without his 'cooperation' by 'naming names' is a repudiation of the loyalty test long used by the FBI. . . . For the first time since the rise of McCarthyism an Attorney General has refused to treat a man's unwillingness to inform on others as a ground for withholding favorable government action in his case."[16]

John Stewart Service, the old China hand, who had been dismissed from the State Department on security grounds in Eisenhower days and subsequently won reinstatement in the courts, sued State for his back pay. John Douglas of the Civil Division believed that Service's suit had merit and that Justice should settle but warned the Attorney General there would be objection on the Hill. Word promptly came back to go ahead.[17] A well-known scholar on Sino-Soviet affairs applied for the validation of his passport to permit travel to mainland China. The CIA and the FBI scented espionage, because the scholar, who had once worked for the research side of CIA, had, in violation of regulations, taken copies of his own studies with him when he moved on to the RAND Corporation. Might he not have taken other documents and be planning to defect to Peking? Kennedy told Abba Schwartz, head of the Bureau of Security and Consular Affairs at State, he was sure any violation was technical; but the FBI was pressing for criminal action. It would be a tragedy to ruin a man on flimsy evidence. Would Schwartz look into the matter? After talking to the scholar, Schwartz agreed with Kennedy. Together they persuaded the security agencies that the scholar was at most subject to administrative reprimand. "The decent instinct of RFK," Schwartz said later, "saved an individual and an important career."[18]

There were other such cases. "I should have been happy, looking back," his most distinguished living predecessor wrote him early in 1962, "to think that I had done as much for civil liberties when I was Attorney General as you have done in your first year." "Such praise, from you especially," Kennedy replied to Francis Biddle, "is tremendously appreciated."[19]

II

Argument before the Supreme Court was the responsibility of the solicitor general. Few solicitors general had discharged this duty with more style than Archibald Cox. He knew the law, and he knew the Court; and, like all the best solicitors general, he entered so intimately into decisions as to become, at times, a sort of tenth justice. The Attorney General had not known Cox well before 1961. Their most extended contact had come in 1958–59 when Cox was working with John Kennedy on labor legislation and Robert Kennedy had argued for more stringent provisions than Cox, or John Kennedy, thought wise.[20] But the two men quickly established relations of mutual confidence.

It had become a custom for Attorneys General to argue one case in person at the summit. Even had there been no such custom, Kennedy would have wished to appear before the Court. He wanted, as Burke Marshall said, "to show to himself and everybody else that he could do it."[21] The obvious field was civil rights. But "I had done so much in civil rights," Kennedy recalled. "I was up to my ears in civil rights . . . so I selected an apportionment case."[22]

In a number of states the overrepresentation of rural areas, whether in state legislatures or in party primaries, had given the countryside disproportionate power against cities and suburbs. "The yokels hang on," as H. L. Mencken had written in 1928, "because old apportionments give them unfair advantages. The vote of a malarious peasant on the lower Eastern Shore counts as much as the votes of 12 Baltimoreans."[23] By 1962, despite large population shifts, apportionment lines in more than half the states had not been redrawn for over a quarter century.[24] City mayors found the situation intolerable. So did political scientists. Malapportionment was held the cause of a host of troubles — especially the neglect by state government of the cities and the urban poor, the consequent encroachment of the national government and the enfeeblement of federalism. Still there were, as Marshall observed later, "lots of reasons" for Robert Kennedy not to get into it. Situations of malapportionment had historic sanction. Reform threatened vested political interests, especially in the south, where the Attorney General was sufficiently unpopular anyway. Nevertheless Kennedy insisted. He saw apportionment as an issue of deep concern and wide consequence. The case he chose at the end of 1962 involved the county-unit system in Georgia.

Earlier in the year the Court had changed its mind about its power to deal with inequality in voting. In 1946 it had pronounced unequal representation a political question, beyond judicial remedy. "Courts," Justice Frankfurter warned in *Colegrove* v. *Green*, "ought not to enter this political thicket."* But did citizens really have no recourse when unequal representation paralyzed the very processes of political self-correction? With the increase in urban and suburban population in the fifties, protest rose against the impasse created by judicial restraint. In 1958 Senator John Kennedy, adapting Lincoln Steffens's phrase, called discrimination against city dwellers in the apportionment of state legislatures "the shame of the states."[25] In the same year, Anthony Lewis, Robert Kennedy's Harvard classmate and after 1961 his increasingly close friend, made a tightly reasoned argument for judicial intervention in the *Harvard Law Review*.[26]

Tennessee presented a case in point. A law passed in 1901 still determined the composition of the state assembly. Legislators showed no inclination to change the law against their own interest. Local remedies were exhausted. As the Kennedy administration took over, a Tennessee lawsuit — *Baker* v. *Carr* — was making a slow ascent toward the Supreme Court.[27] Robert Kennedy's Tennessee pals — Seigenthaler, John Jay Hooker, Jr., E. William Henry — asked that the Department of Justice enter the case as *amicus curiae*. The solicitor general's intervention in fact rescued the case. Accepting Cox's proposition that gross and unreasonable malapportionment was in conflict with the equal protection clause of the Fourteenth Amendment, the Court held in *Baker* v. *Carr* in March 1962 that apportionment was a justiciable issue and that courts could provide relief. Though it declined to reach the question of substantive standards, it reversed its 1946 decision in effect if not in form. Robert Kennedy promptly saluted the decision as "a landmark in the development of representative government."[28]

"The right to fair representation and to have each vote count

---

* *Colegrove* v. *Green*, 328 U.S. 549, 556 (1946). Though Frankfurter wrote the opinion of the Court, actually only two other justices agreed with him that representation was a political question and not justiciable. Justice Rutledge concurred in the result on other grounds and was responsible for the 4–3 decision.

equally," President Kennedy said in his next press conference, "is, it seems to me, basic to the successful operation of a democracy."[29] Senator Richard Russell, Georgia on his mind, observed grimly that the Court had "set out to destroy . . . the system of checks and balances."[30]

III

The Georgia county-unit system had survived four earlier challenges before the Supreme Court.[31] The issue it presented was not, strictly, apportionment but suffrage. The system heavily weighted the Democratic primaries in favor of low-population rural counties. One vote in Echols County counted as much as 99 in Fulton County. The issue was whether or not 16 percent of the electorate had the right to decide statewide nominations (indeed, elections; for Georgia was then a one-party state).

This, Cox said in a memorandum to Kennedy, was an "easier" question than legislative apportionment. "The distribution of seats in the legislature almost inevitably requires some choice between different theories of representation. There is no reason not to give everyone an equal vote in a statewide election." However, he warned the Attorney General against adopting the position "that all votes must be given equal weight. In my opinion, we should not argue for this dogmatic interpretation even in the case of statewide elections, for although Justices Black and Douglas might buy it, the argument is very unlikely to command the assent of a majority. Furthermore, we do not need to go so far in the present case." Cox recommended the narrower argument that any departure from equality of voting power had to have "rational foundation in terms of a permissible State policy." Kennedy's problem of advocacy, Cox said, was to knock down all conceivable justifications for discriminating in favor of the country voter. "The best argument for you to make will be whatever you, yourself, find most persuasive. . . . The burden of convincing the Court that it is just wrong-*wrong*-WRONG will come in oral argument, for the point is just as much emotional as rational."[32]

Kennedy retired over the Christmas holiday in 1962 with Cox's memorandum, a lengthy brief from the solicitor general's office and a yellow legal-sized pad, on which he scratched page after page of notes.[33] On January 17, 1963, improbably clad in morning coat and striped trousers, looking, said one reporter, "like a nervous and uncomfortable bridegroom,"[34] Kennedy confronted the Court in the case of *Gray* v. *Sanders*. It was his first, and last, appearance as a lawyer in a courtroom. His mother, his wife, two sisters, his brother

Edward, his sister-in-law Jacqueline and four of his children were in the audience. The Kennedys, it was observed, outnumbered the justices.

The Attorney General spoke without notes. His argument partook of Cox's constitutional caution. Kennedy thus asked the Court only to determine the validity of the existing Georgia system, not "of some other county-unit system which has not yet been enacted." The "ideal" was "to have one's vote . . . given equal weight with the votes of other citizens," but he was not, he said, "arguing that in all cases all votes must be identical in value." Still any deviation from the ideal must be for "a full, good and complete reason." Ultimately at stake, he concluded, was the capacity of state government to deal with the problems of an increasingly urban and suburban nation.[35] Overcoming his early unease, he finished with a flourish by quoting "that old Massachusetts saying, 'Vote early and vote often,'" and concluding: "With Georgia's county-unit system, all you have to do is vote early."[36] He handled questions from the bench with dispatch and was adjudged to have performed his task in style. (Success did not go to his head. A few days later, a group of students came in to see him. "Do you know what they asked me?" he said to Joseph Dolan, and he quoted a question about reapportionment. Looking at Dolan "with that look he has," Kennedy said, "They got a lot of nerve asking the Attorney General of the United States questions like that. They should ask a lawyer.")[37]

On March 18 the Court, speaking through Justice Douglas, struck down the county-unit system. Douglas distinguished the issue of suffrage from that of legislative representation. Then, sweeping beyond Kennedy's argument, he proposed a conclusion applicable to both issues: "The conception of political equality from the Declaration of Independence, to Lincoln's Gettysburg Address, to the Fifteenth, Seventeenth, and Nineteenth Amendments can mean only one thing — one person, one vote."[38]

This was Robert Kennedy's destination too. He was all for the courts plunging into the political thicket. "He wanted, instinctively," said Marshall, "to see the democratic process work right. It was the same instinctive reaction he had about the Negro vote. If you set up a political structure dependent on the vote and rig it so that the vote doesn't count, that's no good."[39] Cox wished the same result but by a different route. A strong liberal in objectives, he was conservative about the role courts should play in attaining them. "In a state constitutional convention," he wrote the Attorney General, "my vote would go to apportion both houses of a bicameral legislature in accordance with population, but I cannot agree that the Supreme

Court should be advised to impose that rule upon all 50 states by judicial decree. . . . The court's ability to make the system work depends upon exercising enough self-restraint not to get too far ahead of the country on too many fronts." [40]

A flood of apportionment cases was now rising toward the Court. The problem of substantive standards, avoided in *Baker* v. *Carr*, could no longer be escaped. In August 1963 Cox recommended that representation on the basis of population be required in one house of every state legislature but that deviations — substantial but not extreme — be permitted in the other house. "We should not argue that both houses must be apportioned on the basis of the principle — one man, one vote." [41] Kennedy wanted to go faster and farther. But he deferred to Cox's knowledge of the Constitution and the Court. And, as Anthony Lewis, whose expertise on the issue gave him a lively role behind the scenes, said later, "Bobby was not in a position to order Archie to say something he did not believe; to have a fight with his Solicitor General, moreover, would not help the case because the court . . . might see any brief filed over Archie's head as a political, not a legal, document." [42]

Still Kennedy kept pushing Cox in the gentlest way, suggesting, directly and through intermediaries, that the solicitor general's briefs, if they did not affirm the one-man, one-vote standard, must at least not discourage plaintiffs who wished to move toward it. A Colorado case, where malapportionment had been sustained by a popular referendum, gave Cox particular difficulties. As Cox and Marshall were taking an elevator to Kennedy's office to discuss the case, Marshall said, "Look: Bob won't order you to file a brief you really think you shouldn't file. He won't file any brief without your name. But you must recognize his problem. At your urging we have supported the plaintiffs in all these cases. Suddenly to turn around and oppose the Colorado plaintiffs puts him in an impossible position." Cox said, "I'll work out something." [43]

Cox still believed that the Court would not buy one-man, one-vote; nor did he think it would be good for the Court if it did. But, recognizing the problem, he drafted language in the Colorado case, that, as Kennedy said, "would satisfy his own conscience and would cover what I thought should be the position of the United States Government." [44] Cox, as he freely confessed thereafter, was wrong in his forecast. In June 1964 Chief Justice Warren, speaking for a nearly unanimous Court, ruled that both houses of a state legislature had to be apportioned on the basis of population. When the Court was first asked to treat malapportionment as a justiciable issue, Cox said a decade later, "the counsel of caution was ardently advocated

... [but] the Court overrode the objections, entered the political thicket, and emerged stronger than ever before."[45]

On retirement, Warren called the reapportionment cases the most important decisions of his Court. This seems an overstatement. The reapportionment revolution did not have the results predicted in the reinvigoration of representative government, the salvation of the cities or the strengthening of federalism.[46] Evidently malapportionment was not the root of all evil. Nonetheless, by outlawing manifest distortions and irrationalities, the reapportionment decisions increased the chance of finding solutions within the political process.

As for Kennedy, Cox remembered two salient qualities. One was his willingness to listen. "He often had strong immediate reactions; but, if you said, 'Now, wait Bob . . .' and outlined other considerations, he would wait, listen, reconsider." The other was his legal intuition. "He did not know a great deal about case law. But he had the quality Hugo Black had so strongly. I would present a technical problem to him in technical terms — very often a problem remote from his training or experience — and he could put his finger at once on the gut issue."[47]

IV

Robert Kennedy's main connection with businessmen was through the Anti-Trust Division. He had long embraced its guiding principle. "Too much power scares me," he said in his first interview, "whether we find it in a trade union or in a corporation."[48] But, as he told Lee Loevinger, the assistant attorney general in charge, antitrust was "a strange field of law" for him.[49]

Loevinger, a fluent Minnesotan, spent too much time for the Attorney General's comfort making speeches around the country and too little taking hold of his organization. "He was not an administrator," said Seigenthaler, "and he never really got that division going."[50] On his part Loevinger evidently felt inhibited by the administration's desire not to hurt business confidence at a time of nascent economic recovery. "It is probably true," he told a reporter, "that we are affected by business uncertainties to the point where we are holding up cases with a novel or uncertain legal approach. We are sticking pretty much to the predictable, to the established lines."[51]

In 1962 Kennedy asked Ramsey Clark and Joseph Dolan to survey the Anti-Trust Division. Their report confirmed Kennedy's feeling that, as Clark put it, most antitrust cases arose from the complaints of competitors; there were "no priorities or purpose; you just sat there and whatever happened, happened."[52] The next year Ken-

nedy moved a reluctant Loevinger over to the Federal Communications Commission. William Orrick took his place.

In antitrust, as elsewhere in Justice, the Attorney General was
determined to keep politics out. "The Anti-Trust Division during
the period I was there," Loevinger said later, "was run on a strictly
merit basis, and I never had one hint from either Bob Kennedy or
Byron White or anyone else that it should be run any other way.
. . . There was simply no pressure from either of the Kennedys."[53]
Orrick said: "He would sign the complaint, and he'd ask, 'Is it
right? Must you always sue our largest contributors?' And he
laughed, but then he'd sign it. There was no question with him about
ducking."[54]

In spite of his nondescript antitrust record, Robert Kennedy was
regarded with gloomy mistrust by businessmen. In November 1961
he spoke to the Economic Club of New York. Joseph Kennedy
dropped by his suite at the Waldorf while his son was changing into
a dinner jacket. "I see you're living it up, really in the tall cotton,
aren't you?" said the ambassador. "My father's a millionaire," said
the Attorney General. The title of his speech — "Vigorous Anti-
Trust Enforcement Assists Business" — was evidently thought up by
someone who supposed that businessmen really believed in free competition. Kennedy began with a trail of jokes, all of which fell flat,
and went downhill from there. He had hoped to repeat the success
of his Georgia speech where he candidly told a southern audience he
intended to enforce civil rights decisions. Candor about enforcing
the antitrust law had less success. When he finished, the applause,
reported Edwin Guthman, was "scarcely perfunctory."[55]

Then came the great steel battle of 1962. This episode had its
origin in the administration's effort to hold down prices and wages
in the bellwether automotive and steel industries and thereby to
break, in Walt Rostow's phrase, "the institutional basis for creeping
inflation."[56] This policy, fully understood by the union leadership,
was, as Roger Blough, U.S. Steel's executive officer, mystifyingly said
a dozen years later, "simply not comprehended by those negotiating
the steel contract on the company side." He added that this belated
illumination might "help explain President Kennedy's reaction to the
steel price increase and make his reaction more understandable."[57]
Indeed it might.

The administration had thought its wage-price guideposts secure
when it induced the Auto Workers and then the Steelworkers to
accept noninflationary contracts. A few days after the steel wage
agreement, on a Tuesday afternoon in April, Blough made his celebrated call on the President to announce U.S. Steel's decision to

raise prices six dollars a ton. Five other companies promptly followed suit. Blough later justified his *démarche* by condemning the idea of a President setting out "to substitute his own action for the action of the marketplace by trying to set prices for any competitive product."[58] Kennedy agreed that, if Blough's challenge succeeded, it meant the end of the government's stabilization effort. He felt, moreover, that Blough, who had said nothing to him, to Secretary of Labor Goldberg or to David McDonald of the Steelworkers about raising prices, had personally double-crossed them all. "You kept silent," Goldberg rebuked Blough, "and silence is consent. One thing you owe a President is candor."[59] Kennedy, coldly furious, called McDonald: "Dave, you've been screwed and I've been screwed."[60]

The details of the battle for price stability are amply recorded elsewhere; but Robert Kennedy's participation is relevant here. "The President called me in my office around 6:00 P.M.," he noted in a memorandum dictated a few days later, ". . . to tell me that Blough had been in to see him. . . . Obviously he was outraged." The next day Sorensen read the Attorney General the statement the President planned to make in his press conference. Robert Kennedy:

> I thought the statement was (1) too personal; and (2) that the President's statement should be one of sorrow for the country and the betrayal of the steel executives rather than on any personal feeling. I suggested, therefore, that he bring in about the Sergeants in Vietnam, the reservists who had been called up, the unions that had kept their wage demands down, and the other sacrifices that were being made. . . . It was agreed that Sorensen would write it up and the President will give it along those lines.[61]

So far, so good. But the President had also told his brother on the first day that "he was particularly anxious that we should check on the newspaper report of the interview of the President of Bethlehem Steel." Bethlehem was the second largest steel company; and the press had quoted Edmund F. Martin, its president, as saying in a stockholders' meeting shortly before, "There shouldn't be any price rise. We shouldn't do anything to increase our costs if we are to survive."[62] When Bethlehem imitated U.S. Steel and put up its prices, the question arose of price fixing in violation of the antitrust laws. The Attorney General accordingly asked the FBI to check Martin's exact words with the reporters who wrote the stories. Preliminary investigation in antitrust cases was a routine mission for the Bureau. This one was executed with unbecoming alacrity.

Katzenbach instructed Courtney Evans to have the agents interview the steel executives at their offices. He gave no specific directions about interviewing the newspapermen.[63] With its knack for the strict construction of orders, the Bureau evidently felt free to question the reporters at any place or time. FBI agents thereupon visited them at their homes in Philadelphia and Wilmington before dawn the next morning. This caused great excitement in the newspapers, with much scholarly reference to police-state tactics, the Gestapo and the dreaded 'knock on the door.'

The Attorney General, as usual, took full responsibility for the Bureau. In actuality, beyond his general request that the FBI find out what Martin had said, he had nothing to do with the midnight calls. "That was a decision the FBI made," he said privately, ". . . nor did they discuss it with me, nor did I even know who they were interviewing."[64] "We were the ones," said William Sullivan of the FBI, "who made the decision to interview [them] at night, not Kennedy."[65] The Attorney General often wondered about the bizarre timing. "He was sometimes suspicious," observed Douglas Dillon, "that things like this might have been done on purpose by Edgar Hoover to embarrass him. I can't have any judgment on that, but I know that that was his feeling."[66]

v

If Robert Kennedy must be acquitted on the Gestapo charge, there can be no question that he was prepared to play against U.S. Steel, as he himself said later, with a "hard ball." He convened a grand jury to see whether the steel companies had violated criminal law. "Then we looked over all of them as individuals. . . . We were going to go for broke — their expense accounts and where they'd been and what they were doing. I picked up all their records. . . . I told the FBI to interview them all — march into their offices the next day. . . . We weren't going to go slowly. . . . So all of them were hit with meetings the next morning by agents. All of them were subpoenaed for their personal records and . . . their company records."[67]

At the same time, there was strong public reaction against the companies. The price increase, said William Scranton, Republican candidate for governor in Pennsylvania, was "wrong for Pennsylvania, wrong for America, wrong for the free world."[68] The combination of national indignation and hard ball had its effect. On Friday morning Clark Clifford, who had been brought in to help Goldberg in negotiations with Blough, reported that U.S. Steel was ready to roll back halfway; what would the President say? "Not a damn thing," said Kennedy. "It's the whole way."[69] Late Friday afternoon Blough surrendered. It had been an intense seventy-two hours.

"Jack was very anxious to make peace with industry," Robert wrote a few days later. "His political strength rested with staying in the middle of the road. That was why it was so difficult to attack him, he felt. Therefore, it was important to make up to business so that they would not consider him or the Administration anti-business." Clifford and Charles Bartlett, who had also been active behind the scenes in arranging talks with steel executives, "were anxious that the Grand Jury be delayed or forgotten. I was very anxious that I continue although in low key." Clifford "expressed himself as very anxious that we do not take any action against the companies. . . . I had a strong exchange with him that caused Jack to say later to Charlie Bartlett, joking I think: 'What we need in this Administration is a good Attorney General that can be fixed.'" [70] One likes that "joking I think."

The Attorney General finally accepted the line. On Monday night, according to Blough, he assured the general counsel of U.S. Steel that the administration was not vindictive and there would be no further problems with subpoenas and grand juries. [71] The Attorney General was talking, however, about inquiries related to the recent price rise. The Anti-Trust Division had been preparing other indictments against the steel industry before Blough exploded his bomb. "Those were subsequently delayed slightly," said Loevinger, "in order to avoid any impression that they were the result of the steel pricing matter." [72] In these matters the steel companies eventually pleaded nolo contendere.

The President's quest for peace was not helped when word got around that, like his father, he thought that all steelmen were sons-of-bitches. Robert Kennedy, reflecting later on his brother's attitude toward businessmen, said, "Well, he never liked them. I mean, he just always felt . . . that you couldn't do anything with them and there's no way to influence them. I suppose the fact [was] we were brought up thinking that [way]. . . . My father thought businessmen didn't have any public responsibility and we just found out that they were antagonistic." [73] Robert Kennedy concluded for himself the January after the steel episode: "The business community always has a greater mistrust of any Democratic Administration than of a Republican Administration. It is an ideological reflex — obsolete, in my opinion — but that's one of the facts of life. I don't know that businessmen, the big ones, anyway, no matter what we do, will ever be in love with us." [74]

The Kennedys kept on hoping, however, and ended in the worst of both worlds. The steel battle was something of a shock for the President as well as for Blough. Kennedy did not like to use that much power and realized it could not be easily used again; and he did not like to excite that much enmity. His policy toward business thereafter was one of mild appeasement. Archibald Cox, who had been through earlier struggles as chairman of Truman's Wage Stabilization Board, suggested in June that the President might be given authority to delay vital price or wage increases until a public commission rendered a report on economic impact. Kennedy was disturbed, not, as he told me, at the content of Cox's speech but at the timing.[75] The timing was indeed unlucky. The bottom fell out of the stock market the same day, and the White House hastily disclaimed responsibility for Cox's ruminations. (Cox, a little wounded, spoke to Robert Kennedy about the White House reaction. The Attorney General said: "That is an outrage. I read your speech in advance and approved it. You're not supposed to have any political sense. If anyone is to blame for that speech, it's me" — whereupon he called for his car and went over to the White House to set everybody straight.[76]

John Kennedy, Cox said in retrospect, "thought of power as a bank account. You could only draw out so much. He didn't understand sufficiently that power can also be a snowball, gathering momentum as it rolls."[77] In any event, antitrust policy remained in the doldrums, nor did the administration support with much enthusiasm the valiant efforts of Estes Kefauver and his antimonopoly committee on the Hill. Appeasement was in vain. Most businessmen, perceiving that the Kennedys did not like them, responded to their attitudes rather than to their actions and credited them with antibusiness designs they did not have.

The question remained whether the steel crisis had led, as was charged then and later, to a gross misuse of executive authority, especially by Robert Kennedy. "After this display of naked power," wrote Erwin D. Canham in the *Christian Science Monitor*, "whatever its provocation or justification, how free will the American economy be?"[78] "The Government set the price," cried the *Wall Street Journal*. "And it did this by the pressure of fear — by naked power, by threats, by agents of the state security police."[79] In the *New Republic*, Professor Charles Reich of the Yale Law School, noting that the government had demobilized as soon as the price increase was rescinded, asked: "Can acts or conditions that are criminal on Tuesday become less criminal by the following Monday? Are crimes by steel and other industries permitted so long as the criminals 'co-operate' with the administration?" Reich concluded, "It was dangerously wrong for an angry President to loose his terrible arsenal of power for the purposes of intimidation and coercing private companies. . . . Congress has given the President no power to fix the price of steel."[80]

It could be responded that Congress had not given U.S. Steel power to fix the cost of living. Weighty national interests were involved. When Blough made clear his determination to wreck the stabilization effort, Kennedy had to choose between stopping him or letting him get away with it. American Presidents had tried to defend the general welfare against private corporations since the age of Jackson, and Kennedy was in a respectable tradition. Once he had decided to stop Blough, he could no more risk losing than Andrew Jackson could have risked losing to Nicholas Biddle. "Supposing we had tried and made a speech about it, and then failed," John Kennedy said later. "I would have thought that would have been an awful setback. . . . There is no sense in putting the office of the Presidency on the line on an issue, and then being defeated." [81]

So they played with a hard ball, and won. There was no transgression of law or the Constitution. There was, however, selective deployment of executive power. When asked in 1964 whether the possession of so much discretionary authority was not alarming, Robert Kennedy replied, "Yes, it is, and rather scary, there's no question about that; and it can be abused and misused." Still the alternative, he thought, "would have been bad for the country . . . and it would have been bad all around the world because it would have indicated that the country was run by a few manufacturers, and I don't think we would ever have reestablished ourselves. . . . I agree it was a tough way to operate but under the circumstances we couldn't afford to lose it." [82]

As Professor William Goldsmith observed in his magistral study of presidential power, executive aggression sometimes reflects "significant dislocations in the political process itself, which have created conditions and circumstances where it becomes clear that nothing short of an overwhelming mustering of presidential power and influence can correct the balance of public and private interests." In the steel case Goldsmith concluded that "the administration did not overstep the limits of executive power . . . and acted quite properly to protect the common good." [83] The guideposts were preserved. Inflation was contained. Under Kennedy the rate of increase in consumer prices was what another decade would regard as unimaginably small: 1.2 percent in 1962; 1.6 percent in 1963.

Nor did the Kennedys emerge intoxicated with their use of power. Success did not even embolden them to seek improvements in the stabilization policy beyond the guideposts. They simply did not want another major fight with business. In view of this aftermath, it was not perhaps such a famous victory as it seemed at the time. As the circumstances faded from the public mind, what lingered, especially in the business community, was the fantasy of the Attorney

General ordering FBI agents to pound on doors in the dead of night. The story quickly expanded to include businessmen as well as reporters; in a few years Lyndon Johnson was reminiscing about "the FBI interviewing Barry Goldwater at dawn."[84] The incident, suitably embellished, consolidated the image of Robert Kennedy as ruthless prosecutor riding roughshod over individual rights, any means to an end.

VI

The Kennedys were more resolute in fighting for the poor and dispossessed than they were in fighting against the rich and powerful. Justice for the Negro was only the most dramatic commitment. There were the Spanish-speaking Americans — half of whom, Robert Kennedy pointed out in August 1963, earned less than the average per capita income, far too many "still the victims of poverty and of social and economic discrimination." When he named a delegation of jurists to represent the United States at the Inter-American Congress of Attorneys General, a majority "for the first time in history" were Spanish-speaking Americans.[85]

Then there were the American Indians. Ramsey Clark's office in Justice, the Lands Division, disposed of more money than any other in the Department; but the Attorney General was not deeply moved by land titles and condemnations. "He thought it was a great bore," Joseph Dolan recalled. ". . . One day he made a great discovery: Ramsey had Indians. And then the friendship began to blossom."[86] Ramsey had Indians because his division dealt with Indian land claims. The division's traditional role had been to oppose Indian claims. Clark and Kennedy now perceived the opportunity to try the novel policy of settling the claims. This approach was not altogether popular. Professional Indian lawyers and witnesses, who had been living off the litigation, did not like to see the cases resolved. Indians themselves, with exalted ideas about the value of their land, were often disappointed by the results. Still Clark, with Kennedy's backing, persisted.[87]

Congress, which had to pass appropriations in each particular case, was especially resistant to big payoffs to Indians. But political pressure on Indian issues, Clark recalled, "would never come directly to me. Bob would never pass it on, just never." When the Oklahoma delegation, led by the redoubtable Robert Kerr, objected to the size of a proposed settlement with the Cherokee Nation, the Attorney General intervened personally. Clark: "Bob didn't have much patience with that because his primary instincts were not legal but humanitarian. . . . 'For God's sake we owe it to the Cherokees, and don't worry about these niceties.' . . . He felt we pushed them around and said so."[88] He said so in speeches too, describing the red man as "the victim of racial discrimination in his own land," pointing out that "the infant mortality rate among Indians is nearly twice that of any other racial group in the country; and their overall life expectancy is twenty years less."[89]

Clark used to wonder why Kennedy cared so much. "He may have had some specific experience with Indians that I don't know about," Clark said, but finally decided that "he just had this burning passion to help people who had been denied justice."[90]

VII

The settlement of land claims and the enforcement of civil rights statutes could go only so far in the redress of grievances. If government belonged where evil needed an adversary and people in distress could not help themselves, then government must devise new forms of action.

It all began mildly enough when Robert Kennedy in his first press conference spoke about the "alarming increase" in juvenile delinquency. The predicament of the young always concerned him; in part, no doubt, because of the contrast between his own economically secure childhood and the scramble for survival among kids in the slums. So much depended on where one began. Jack Newfield once asked him what he might have become if he had not been born a Kennedy. He replied, "Perhaps a juvenile delinquent or a revolutionary."[91] Widespread delinquency seemed to him a self-evident symptom of derangement in the social order.

This had not been the prevalent view in his Department. In the January 1961 issue of the FBI's *Law Enforcement Bulletin,* J. Edgar Hoover laid about at those "muddle-headed sentimentalists who would wrap teenage brigands in the protective cocoon of the term 'juvenile delinquency.'" The very phrase, he said, should be "banished forever from our language." He continued this barrage throughout Kennedy's attorney generalship, fulminating periodically against all who would "coddle" those "beastly punks."[92] The Attorney General, undeterred, insisted that juvenile delinquency must receive "priority attention." He named as his special assistant charged with coordinating the federal effort his intimate friend of so many years David Hackett.[93]

Hackett knew nothing about juvenile delinquency. But he was an intelligent man with organizational drive. "Though hardly an idea

man himself," said one historian, "the square-jawed former hockey
star excelled at recruiting original thinkers, persuading them to join
the administration, and then sitting around a table to knock heads.
. . . He batted around ideas as he had once shot hockey pucks, never
holding onto them for detached or deep speculation, but outpacing
many other Kennedy aides in shrewdness, dexterity and guts."[94]

Ignorance may well have been an advantage in a field beset by
competing diagnoses. "Everyone I talked to," Hackett said later, "had
a conflicting area, and I just listened."[95] The traditional social work
approach had seen juvenile delinquency as an expression of individual maladjustment, rooted in the disordered slum family and requiring educational or psychiatric remedy. But some sociologists had
long doubted whether case treatment was in fact the answer to delinquency. Hackett encountered the ideas set forth by two professors
at the Columbia School of Social Work, Lloyd Ohlin and Richard A.
Cloward, in their 1960 book *Delinquency and Opportunity*. Ohlin and
Cloward argued that delinquency was more a product of social than
of individual pathology. It was what happened when a culture excited middle-class aspirations and then maintained barriers — of
status, income, education, race — that made these aspirations unat-
tainable for the poor by legitimate means. The disparity between
ideals and opportunities left the poor with the choice of accepting
inferior status, thereby emasculating themselves, or of asserting identity through defiance and lawlessness.

Delinquency was thus not the function of individuals nor of groups
but of "the social systems in which these individuals and groups are
enmeshed." Its cure, Ohlin and Cloward suggested, lay not in changing the client but in changing the structure of opportunity. "The
major effort of those who wish to eliminate delinquency should be
directed to the reorganization of slum communities."[96] Not everyone
accepted the 'opportunity theory.' Those who traced delinquency to
the disturbances of early childhood, for example, considered it excessively optimistic in its intimations of rapid social improvement.[97]
Yet its very optimism made it attractive to activists who wanted to *do*
something.

While Hackett was pursuing ideas in New York, he was also promoting organization in Washington. The Children's Bureau in the
Department of Labor hoped to dominate any federal antidelinquency
effort. Hackett, who was greatly impressed by the opportunity theorists, feared that the Children's Bureau was more interested in treatment than in prevention. Robert Kennedy thought the antidelinquency program needed visibility. In May 1961 John Kennedy
established the President's Committee on Juvenile Delinquency and

Youth Crime, with Robert Kennedy as chairman and Hackett as executive director. A poor title: as it worked out, it might better have been named the Attorney General's Committee on Poverty and Urban Reorganization.[98] Congress passed the Juvenile Delinquency Act in September.

Hackett and Ohlin, who became head of the Office of Juvenile Delinquency, enlisted a number of ardent spirits. In time they were called 'Hackett's guerrillas' and lived, as one said, "off the countryside of the government and foundations."[99] The deeper they got into the subject, the more they saw delinquency as a cover word for poverty. "We made no distinction between the two," Hackett said later.[100] Soon they began to see it also as a cover word for racial discrimination. The attack on delinquency required, they believed, comprehensive planning to coordinate federal effort at the local level, where services were actually delivered. Instead of working through a collection of uncoordinated agencies, they wished to create "new local structures through which public and private agencies cooperate."[101]

The Committee on Juvenile Delinquency thus joined the National Institute of Mental Health and New York City in funding a program on New York's Lower East Side called Mobilization for Youth. Mobilization for Youth experimented with a diversity of new approaches — preschool education, vocational training and placement, employment of people from the neighborhood, decentralization in welfare departments, day care and community service centers, legal services for the poor. In 1962 planning began for Harlem Youth Opportunities Unlimited (HARYOU), which had come into being under the direction of Dr. Kenneth Clark when Harlem residents protested the city's decision to import an out-of-neighborhood agency to work with black youth. The Juvenile Delinquency Committee, Clark said later, gave HARYOU "its basic financial and psychological support"; David Hackett's "involvement was such that it could be accurately described as that of an adjunct member of the HARYOU staff."[102]

Hackett meanwhile sent out advance men to stimulate other communities to prepare their own plans. As the committee staff roamed the country seeking promising sites for projects, "we were dismayed," Ohlin said later, "by the absence of demonstrated indigenous leadership in the slum communities."[103] This argued for a planning period to permit the development of local institutions capable of handling outside grants. By the end of 1963, with its initial appropriation coming to an end, the committee had made seventeen grants for demonstration projects in sixteen cities. Along with their insistence on comprehensive planning (followed by careful evaluation of results), the guerrillas had reached certain political conclusions: that the orthodox welfare bureaucracies had a vested interest in regarding delinquency as a problem distinct from poverty and race; that the poor themselves offered a basis from which to challenge the established social service agencies; that involving the poor in their own salvation through 'community action' would enhance their confidence and mobilize them as a constituency. The watchword was "institutional change." [104]

VIII

"Bob Kennedy gave me complete responsibility for it," said Hackett, "and we only asked him to do things that we needed his help on. He gave us in essence no direction." But, Hackett continued, he was eager to be involved. He interrogated the experts, visited projects, chatted with youth gangs in the streets of East Harlem, [105] talked up the program in speeches.

He helped particularly in dealing with Congress, first getting the legislation through and then keeping Edith Green happy thereafter. Mrs. Green, a comely and strong-minded woman who had led the Kennedy forces in Oregon before Los Angeles, was chairwoman of the House subcommittee handling the program. Her ties were with the Children's Bureau. The Juvenile Delinquency Act meant one thing to Hackett's guerrillas, something quite different to Edith Green. She had expected a scattering of small pilot projects flung into operation as fast as possible and administered by agencies already in the field. She profoundly disliked Hackett's conception of a long planning period followed by comprehensive efforts, as she put it, "to bring about any social reforms either city-wide or nation-wide." Mobilization for Youth was her bête noire, dismissed with contempt as that "two million dollar Cloward-Ohlin experimentation program." She finally summoned Hackett to her office for an accounting. The discussion was heated. Thereafter it seemed better for Hackett to stay out of Mrs. Green's way.

Finally Robert Kennedy, of whom Mrs. Green was fond, came himself. Wesley Barthelmes, her administrative assistant, accustomed to seeing legislators call on cabinet members, was surprised by Kennedy's readiness to go to the Hill and spend time on what was, compared to civil rights and organized crime, a relatively small program. The Attorney General contended quietly that projects like Mobilization for Youth were not inconsistent with the aims of the legislation. Phone calls followed: "Well, Mrs. Green, is it going to your satisfaction? I hope we can reach an agreement. It's a very important program." Mrs. Green, only a bit mollified, did not relent in her determination to keep the guerrillas from running wild. Her attitude toward the Attorney General, recalled Barthelmes, was that "of a sorrowful mother for an errant son, which is an attitude Mrs. Green had toward a good number of things. . . . [But] it wasn't sorrowing mother toward an errant son in respect to Dave Hackett."[106]

The Attorney General made another congressional intervention in 1963 at the behest of his sister Eunice. Fifteen years before, Eunice had served in the Department of Justice as executive secretary to the National Conference on Juvenile Delinquency. After marrying Sargent Shriver she had worked with teenage delinquents in Chicago. Her consuming interest now was in a domestic peace corps modeled on Franklin Roosevelt's Civilian Conservation Corps as well as on the undertaking her husband was running with such élan overseas. She worried the President about it until he finally said he had heard enough: "Why don't you call Bobby? See if Bobby could get it going. It's not a bad idea."[107]

She called Bobby, who put Hackett on it. Hackett suggested a cabinet-level committee with the Attorney General as chairman and Richard Boone of the Ford Foundation as staff director. Boone, who had begun as a community organizer in Chicago, has recorded his impressions of the period: "The most important thing about Robert Kennedy then and later was his ability as a listener. He did not pontificate. . . . The thing I remember most was his anxiousness to learn, his ability to put proper questions so everyone could learn, his conservatism in that he did not believe the federal government could do all things for all people, his belief that . . . one of the principal functions of the federal government was to support citizen participation at the local level and to build wherever possible from the bottom up rather than from the top down."[108]

Boone organized a feasibility study, the Attorney General brought his cabinet colleagues together for three Saturday-morning meetings, and in February 1963 the President recommended the establishment of a national service corps designed to apply the Peace Corps principle to the domestic scene. "We visited a state hospital for the mentally retarded on a bright April day when you would have expected all the children to be playing outside," Robert Kennedy said as he portrayed the need to a House committee.

> The children were inside, standing in a room which was bare but for a few benches. The floor was covered with urine. Severely retarded patients were left naked in cubicles, which suggested kennels. . . . Patients were washed by a device resembling a car wash. . . . The only toilets for the approximately seventy patients in a large ward were

located in the middle of the room, permitting no privacy. The hospital's
hard-working but inadequate staff could provide at best only custodial
care.

There is not even custodial care for great numbers of migratory farm
workers, who live in almost unbelievable squalor. . . . One family had
been living in their car for three months. . . . The mother was seriously
ill. The children were suffering from malnutrition, and were unbelievably dirty because of the lack of sanitary facilities. . . . They had no
money and virtually no hope; they did not know where to turn for help.

The Attorney General mentioned other national shames: the Indians, the Alaskans, the illiterates. As thousands of Americans had
already volunteered to serve in far corners of the world, so "we are
convinced that Americans are equally willing to take on the toughest
jobs in this country, whether in a city slum, an Indian reservation or
a mining town. . . . Every sixth citizen in the United States needs our
help; there are five of us who should help him." The program, he
concluded, asked Americans "to invest a year of their lives, at no
salary and under Spartan conditions, to help millions of their fellow
citizens who, through no fault of their own, are denied the essentials
of a decent life. I am proud to admit that this concept is idealistic."[109] The legislation passed the Senate in 1963 but failed in the
House. The idea survived.

Kennedy meanwhile took the District of Columbia as his personal
responsibility. In 1962 he began paying visits, without publicity and
often without notice, to Washington public schools. He organized an
annual benefit — Ella Fitzgerald performing one year, Sammy Davis,
Jr., the next — to raise a fund to prevent teenage blacks from having
to drop out of school for economic reasons. He also arranged for
playgrounds, recreation centers, sports programs, summer job programs. He knew these were no more than palliatives, but they "were
at least intended," as he later explained to his brother's successor,
"to show the children and juveniles that their government cared
about their problems."[110]

Typically, he asked his assistant on transportation matters, Barrett
Prettyman, Jr., an able young lawyer who had clerked for three
Supreme Court justices, to take on juvenile delinquency in Washington. The second precinct led all the rest in juvenile crime. Kennedy
discovered that the Dunbar High swimming pool had been closed for
eight years for want of $30,000 to make necessary repairs. "He
couldn't believe it," Prettyman said. They raised the money in the
next two weeks.[111] Patrick Anderson, the information officer for the
Committee on Juvenile Delinquency, accompanied the Attorney General on an inspection of the pool, an occasion, Anderson said later,
when "I saw him, within the space of ten minutes, at his very best
and very worst." An official from the city recreation department,

perhaps hoping for commendation, approached him. The Attorney General pointed out curtly that the shallow end of the pool was too deep for small children and asked him what he was going to do about it. "Grant that the man was a bumbler," Anderson wrote later. "A Southern politician would have thrown his arms around his shoulders, listened to his problems — and in five minutes had his promise. ... But Kennedy had no time for this." Then, moments later, as he began to leave the area, hundreds of black children ran after him, calling his name. "Kennedy moved among them slowly, smiling, rubbing their heads, squeezing their hands, reaching out to the smaller ones who could not get near him. This was not done for show — there were no reporters or photographers along — but because he loved those slum children, loved them as much as he disdained the fool of a bureaucrat who could not give him the answers he wanted."[112]

IX

"That the Kennedys were serious in this approach to juvenile delinquency," wrote Kenneth Clark, "was reflected in the tremendous amount of time and energy which Robert Kennedy devoted to this responsibility."[113] Those involved in the poverty wars often speculated about the reasons for this intense personal concern.

Richard Cloward and Frances Fox Pliven suggested that one motive was political — the desire "to strengthen the allegiance of urban blacks" to the Democratic party.* Certainly, because the Juvenile Delinquency Committee had a three years' supply of funds and kept the lowest of profiles, it provided a means of getting money to blacks in the inner city without provoking further strife with Confederates on Capitol Hill. But this had more to do with preserving urban peace than with attracting urban votes. By 1963 the civil rights struggle in the south assured the overwhelming popularity of the Kennedys in the ghettos, even if they had done nothing about juvenile delinquency. Many blacks, in addition, did not vote; and helping the black underclass carried with it the backlash risk of estranging the white working class.

* Frances Fox Pliven, "The Great Society as Political Strategy," *Columbia Forum*, Summer 1970; Frances Fox Pliven and Richard A. Cloward, *Regulating the Poor: The Functions of Public Welfare* (New York: Random House, Vintage reprint, 1971), ch. 9. David Hackett could recall later only two grants made for political reasons (and both met the committee's criteria). One, made to appease the ravenous appetite of Congressman Adam Clayton Powell in Harlem, involved a grant for his own welfare organization, set up as a rival of HARYOU, an action followed later by Powell's capture of HARYOU and explusion of the stubbornly independent Kenneth Clark; the other was a program requested by Congressman Albert Thomas of Houston. (David C. Hackett, in interview by author, January 27, 1975.)

The committee's problem was the opposite: not that of responding
to an impassioned constituency but that of creating one. The President could never understand why the poor were not angrier. "In
England," he remarked to me one spring day in 1963, "the unemployment rate goes to two per cent, and they march on Parliament.
Here it moves up toward six, and no one seems to mind."[114] Hackett's guerrillas were in fact far out in front of both those their programs were designed to benefit and the Democratic National Committee. "The war on poverty," Moynihan accurately observed, "was
not declared at the behest of the poor."* He could have added: nor
at the behest of the politicians.

Robert Kennedy was of course determined to secure his brother's
reelection; but the Committee on Juvenile Delinquency was not an
important factor in his calculations for 1964, except as the failure to
redeem implied promises might cause trouble. His concern about
poverty had deeper sources. Richard Boone attributed it to his compulsive exploration of social reality. "Bobby best understood things
by feeling and touching them," he said. "He did feel and he did
touch and he came to believe and to understand."[115] Dr. Leonard
Duhl of the National Institute of Mental Health, one of Hackett's
guerrillas, thought the clue lay in Robert Kennedy's personal history
— in his "notion of what an extended family is, and how a family
works, and the mutual relationships between a family. When Dave
Hackett started to take him around . . . to visit the ghettos," he
responded both because he felt so keenly the misfortune of children
denied the nourishment of family and because his extended family
embraced all children in distress.[116]

All this was true enough. But the question was never a mystery to
David Hackett. This was simply the Robert Kennedy who twenty
years before at Milton rushed to the side of the underdog and never
worried about embarrassing his friends. Life had only enlarged his
knowledge of underdogs — first the Rackets Committee, then the
civil rights movement, then the poverty wars. His convictions about
participation, his readiness to bypass established bureaucracies, his
impulse to experiment with new institutional forms, above all, his
instinct for sympathy: these were the key to his growing identification
with the minorities of the republic — and theirs with him.

---

* D. P. Moynihan, *Maximum Feasible Misunderstanding: Community Action in the War on
Poverty* (New York, 1969), 25. Even Cloward and Pliven concede that the programs
were "thrust upon a Congress and a public that were at the outset virtually indifferent
to the specifics of the legislation" (Pliven and Cloward, *Regulating the Poor*, 266).

The Kennedys and the Cold War

THE LAST SEVEN CHAPTERS record an unusually preoccupied life. Few Attorneys General had ever been so busy in so many diverse directions. Yet the Justice Department occupied only a portion of Robert Kennedy's days. In foreign affairs, his brother was navigating tricky waters with a crew of strangers. When storms blew up, there was all the more reason to summon a shipmate of longer standing. On that awful evening in April 1961, as the bad news began to trickle in from the Cuban beaches, the President said to Kenneth O'Donnell, "I should have had Bobby in on this from the start."¹ After the Bay of Pigs he took care to have Bobby in on the critical international decisions.

I

A son of the Cold War, John Kennedy saw it not as a religious but as a power conflict. He did not believe in holy wars. In 1958 he had rejected the proposition that "we should enter every military conflict as a moral crusade requiring the unconditional surrender of the enemy."² He was an incisive and persevering critic of John Foster Dulles. He opposed Dulles's idea of responding to local aggression by massive nuclear retaliation at times and places of American choosing. He opposed Dulles's excommunication of the neutralist states of the Third World. He opposed Dulles's insistence on regarding the islands of Quemoy and Matsu off the Chinese coast as casus belli. He worked for closer economic ties with Poland and Yugoslavia, for increased assistance to India and other underdeveloped countries, for the independence of Algeria and of Indochina. He described Castro as "part of the legacy of Bolívar" and his revolution as the result "of the frustration of that earlier revolution which won its war against Spain but left largely untouched the indigenous feudal order." "The world-wide struggle against imperialism, the sweep of nationalism," he wrote, "is the most potent factor in foreign affairs today."[3] When Chester Bowles prepared an article for *Foreign Affairs* advocating the admission of Communist China to the United Nations under a two-China policy, "I showed it to Jack Kennedy, and he enthusiastically endorsed it."[4] Within the Democratic Advisory Council Kennedy sympathized with Harriman, Bowles and Stevenson as against the imperious Dean Acheson and the dedicated Cold Warriors.

Kennedy believed in coexistence because he knew what war was like. "The romanticist point of view of war," said his intimate friend Charles Spalding, "was just stupid as far as he was concerned. He had a high regard for courage and what you inescapably have to do, but he was no romantic about war."[5] His father, moreover, continued to view the Cold War as hopeless folly. "America comes with dollars and talks of economics," the ambassador wrote in 1956, "while the communists come with ideas and talk of justice." If communism were expanding, it was because of the "vacuum of the spirit created by our own pre-occupation with economic and military problems alone." America's pretension of world leadership was full of dangers for the United States: above all, Joseph Kennedy pointed to "an imperialism of the mind, the attempt to make every state into a copy of America."[6] Though the son supported the containment policy, the paternal doubts could not have been entirely without impact.

Kennedy's interest in negotiation with the Soviet Union went back at least to 1953.* His concern grew as the decade wore on. "For over two years," he said publicly in 1960, "I have been urging that we should negotiate with the Russians at the summit or anywhere else.... It is far better that we meet at the summit than at the brink." Negotiation would not be easy. But the superpowers had "certain basic interests" in common, and progress in areas of shared concern "might well end the current phase — the frozen, belligerent, brink-of-war phase — of the long Cold War."[7]

One such area was arms control. In 1959 Kennedy argued against the resumption of nuclear testing (as recommended by Governor Nelson Rockefeller), and called for a test-ban agreement.[8] Here Kennedy's old friend from his father's ambassadorial days in London,

---

* That year he sent me for comment a memorandum by Marshall Shulman of Columbia entitled "Starting Point for a Discussion of Possibilities for Negotiation with Soviet Union under Present Circumstances," October 19, 1953; Evelyn Lincoln to author, November 5, 1953, Schlesinger Papers.

David Ormsby-Gore, British delegate to the nuclear-weapons talks at Geneva and the United Nations, played an important role. Robert Kennedy later called him "the motivating force to have the President make this a major area of interest." Ormsby-Gore actually provided the statistics John Kennedy used so effectively in campaign speeches describing how few people in the American government were working on disarmament.[9]

Another area might be central Europe. When George Kennan's heretical disengagement proposals in his BBC lectures of 1958 brought the foreign policy establishment, led by Acheson, down on his head, Kennan, brooding in his Oxford study, was astonished to receive a letter from John Kennedy about his lectures:

> I should like to convey to you my respect for their brilliance and stimulation and to commend you for the service you have performed by delivering them. . . . Needless to say, there is nothing in these lectures or in your career of public service which justifies the personal criticisms that have been made. I myself take a differing attitude toward several of the matters which you raised in these lectures — especially as regards the underdeveloped world [Kennan had expressed skepticism about economic aid to poor countries] — but it is most satisfying that there is at least one member of the "opposition" who is not only performing his critical duty but also providing a carefully formulated, comprehensive and brilliantly written set of alternative proposals and perspectives.[10]

After Kennan defended his disengagement proposal in a subsequent article, Kennedy wrote: "You have disposed of the extreme rigidity of Mr. Acheson's position with great effectiveness and without the kind of *ad hominem* irrelevancies in which Mr. Acheson unfortunately indulged last year."[11]

In 1960 Kennedy took a restrained line over the collapse of the Paris summit meeting after the shooting down of the American U-2 spy plane over the Soviet Union. Adlai Stevenson told me he thought Kennedy's postsummit remarks had been "perfect."[12] Kennedy himself was privately critical of a statement put out by the Democratic Advisory Council. He told me "he could not go along with the passage condemning Eisenhower for having gone to the summit without assurance (as a result of preliminary low-level negotiations) that there would be positive results."[13] He took his negotiating credo from Liddell Hart, whose book *Deterrent or Defense* he reviewed that year: "Keep strong, if possible. In any case, keep cool. Have unlimited patience. Never corner an opponent, and always assist him to save his face. Put yourself in his shoes — so as to see things through his eyes. Avoid self-righteousness like the devil — nothing is so self-blinding."[14] "We should be ready to take risks," he wrote in 1960, "to bring about a thaw in the Cold War."[15]

II

Kennedy joined his call for negotiation with growing concern that an apparent decline in the relative American position might turn the terms of negotiation against the United States. He charged in the campaign that the United States was in trouble — that the Soviet Union had been far more successful in identifying itself with nationalism in the Third World; that the massive-retaliation strategy left the United States helpless in face of aggression too limited or local to justify the use of nuclear weapons; even that Russia was well on its way to nuclear superiority.

The first two points in the indictment were not much disputed. Nor, for the most part, was the third. Legend has it that in 1960 Kennedy and the Democrats knowingly foisted on the American voter the false notion of a 'missile gap.' In fact, far from being a cynical Democratic fabrication, the idea had been first put forward in 1957 by the so-called Gaither committee, appointed by Eisenhower to examine the American defense position and including in its panels such experts as Jerome Wiesner of the Massachusetts Institute of Technology, soon to become Kennedy's science adviser, William C. Foster, soon to become head of the Arms Control and Disarmament Administration, I. I. Rabi of Columbia and James Killian of MIT. This was not a gang of ideological Cold Warriors. "The USSR," the Gaither report said, "will probably achieve a significant ICBM delivery capability with megaton warheads by 1959. . . . U.S. will probably not have achieved such a capability. . . . This appears to be a very critical period for the U.S."[16]

Eisenhower's Secretary of Defense subsequently confirmed the theory of the gap.[17] The *New York Times* and the *Washington Post* issued grave warnings about it. Retired generals — James M. Gavin in *War and Peace in the Space Age,* Maxwell Taylor in *The Uncertain Trumpet* — enthusiastically expounded it. "Fifteen more years of a deterioration of our [military] position," Henry Kissinger wrote in *The Necessity for Choice* in 1960, ". . . would find us reduced to Fortress America in a world in which we had become largely irrelevant."[18] It is hardly surprising that the Democrats accepted it in good faith in their campaign.

Even critics of the hard line perceived ominous changes in the power balance. In August 1960 George Kennan sent Kennedy a stern letter. "We may, by January," he concluded, "be faced with an extremely disturbing if not calamitous situation. It will in any case be an unfavorable one and in urgent need of improvement." Kennan called for "a strengthening of our defense posture, particularly in the conventional weapons." He urged Kennedy to avoid commitments about summit meetings but hoped that some alleviation might come through reciprocal concessions worked out in private discussions with Moscow. Kennedy replied that he was "very much in accord with the main thrust of your argument, and with most of your particular recommendations." [19]

After the election, Adlai Stevenson, whom Kennedy had named head of a task force on foreign policy, warned the President-elect: "This revolutionary age confronts us with a potential shift in the world power balance of a magnitude hitherto unknown." [20] At the end of December, asked for his thoughts about Kennedy's inaugural address, Stevenson recommended first of all "a frank acknowledgment of the changing equilibrium in the world and the grave dangers and difficulties which the West faces for the first time." [21]

III

In the view from Moscow any capitalist election was, as Khrushchev said, "a circus wrestling match." Still marginal distinctions were worth noting. When the Eisenhower administration in the heat of the canvass asked the Russians to release two American fliers shot down over the Arctic Circle, Khrushchev put it to the Politburo that this would help Nixon. "We thought," he later recalled, "we would have more hope of improving Soviet-American relations if John Kennedy were in the White House." The release of the fliers was deferred. [22]

Khrushchev's remark was not necessarily a compliment. It may have meant simply that he supposed Kennedy would be easier to hornswoggle. In any case amiable signals flashed from Moscow after the election to Harriman, Bowles and the President-elect's brother. On December 12, 1960, Mikhail Menshikov, the Soviet ambassador, invited Robert Kennedy to luncheon. "Khrushchev," Menshikov said, "admired Senator Kennedy's intelligence and vigor and believed that it was now possible to have clear and friendly understanding between the USSR and the US." The ambassador blamed past trouble on middle-level American officials who had distorted the Soviet position on major issues. "He said," Robert Kennedy reported to Dean Rusk, "that he foresaw no barrier to resolving the difficulties between the two countries; that the Soviet Union had no objection to Berlin remaining free under the United Nations, and that he felt that this was another case in which the position of the Soviet Union had been distorted."[23]

The Soviet ambassador continued to hustle members of the incoming administration till John Kennedy, wanting to end the indirect approaches, asked David Bruce to find out specifically what Menshikov had in mind. Menshikov handed Bruce a formal document. Kennan, to whom Kennedy showed the paper, found it "considerably stiffer and more offensive" than Menshikov's own remarks and believed it had been drafted in the Kremlin. He advised Kennedy that "these people had no right whatsoever to rush him in this way. . . . In any case, it was difficult to see how an American President could conceivably meet with people who were putting their signatures to the sort of anti-American propaganda which had recently been emanating from Moscow and Peking."[24]

As the incoming administration struggled to decide from conflicting evidence whether Moscow intended to cooperate or to make trouble, Khrushchev himself appeared to answer the riddle. In a pugnacious speech on January 6, 1961, the Soviet leader proclaimed the irresistible triumph of communism. This would take place neither through world war, he said, because nuclear war would destroy civilization, nor through local war, which could so easily grow into nuclear war. It would come, he explained, through "national liberation wars" in the Third World — the "only one way of bringing imperialism to heel." Citing Vietnam, Algeria and Cuba as promising examples, Khrushchev called Asia, Africa and Latin America "the most important centers of revolutionary struggle against imperialism" and pledged full Soviet support to the insurgencies. "Communists are revolutionaries," he concluded, "and it would be a bad thing if they did not take advantage of new opportunities."[25]

In retrospect it is possible to surmise that Khrushchev's belligerence was intended less as a challenge to the United States than as part of a complex maneuver involving Communist China. In reaffirming his rejection of nuclear war and his belief in "peaceful coexistence," he insisted on positions that Peking found abhorrent. By imbedding them in an otherwise truculent context, Khrushchev may have been trying to prove to the world Communist movement that nuclear coexistence was not incompatible with revolutionary militancy. Perhaps he thought that his bellicosity would bemuse the Chinese while his softer words would gratify the west. Inevitably Peking and Washington heard only the passages written for the other.

Certainly no one in Washington then recognized the extent of Russian-Chinese differences. Much of the State Department continued for three or four more years to go on about "the Sino-Soviet bloc" as if it had been a marriage made in heaven. Nor, recalling

Khrushchev as the man who had banged his shoe on a United Nations desk a few months before, did many conceive him as notably subtle. So Washington missed his subtleties, if indeed they were there, and heard only boasts and threats. The State Department so read the Moscow mood in a report to the next President: "There is confidence that the bloc's growing economic and military power will enable it to operate increasingly from a position of strength in dealing with the West. . . . This self-confident posture indicates that vigorous Soviet actions on the international scene are to be expected."[26] The Khrushchev speech appeared to substantiate Hans Morgenthau's proposition: "While Stalin conducted a Cold War of position, Khrushchev was the champion of a Cold War of movement."[27]

The speech, Robert Kennedy said later, was "a major factor. . . . The President made everybody on the National Security Council read the speech and he also made the military people all read the speech."[28] John Kennedy himself took it not only as the authoritative exposition of Soviet policy, especially in its emphasis on national liberation wars, but as a calculated personal test. He responded by devoting his inaugural address a fortnight later almost exclusively to foreign affairs.

This was, he hyperbolically announced, an "hour of maximum danger. . . . Let every nation know, whether it wishes us well or ill," he added with a flourish, "that we shall pay any price, bear any burden, meet any hardship, support any friend, oppose any foe, in order to assure the survival and the success of liberty." But, like Khrushchev, Kennedy mingled hard with soft. He condemned the arms race, asked to "bring the absolute power to destroy other nations under the absolute control of all nations," observed that "civility is not a sign of weakness" and declared: "Let us never negotiate out of fear. But let us never fear to negotiate."[29]

Latter-day critics have concentrated on Kennedy's grandiloquent response — paying any price, bearing any burden, etc. — to what he saw as Khrushchev's flung gauntlet. That passage attracted no attention at the time. It was the cant of the age. The fresh note struck that day was very different. Surveying reactions to the address, the *New York Times* reported as widely "singled out . . . for special mention" the sentence about never fearing to negotiate.[30] This was appropriate. The passage on negotiation reflected Kennedy's deeper purpose. He did not intend the inaugural as a trumpet blast summoning Americans to war, nor was it so read in 1961. "I wanted to tell you," Eleanor Roosevelt said in a handwritten letter to the young President, "how I felt about your Inaugural address. I think 'gratitude' best describes the kind of liberation and lift to the listener which you gave. I have re-read your words several times and I am filled with thankfulness. May we all respond to your leadership and make your task easier."[31]

IV

For all the inaugural rodomontade, Kennedy had no illusions about a pax Americana. He felt that previous Presidents, especially Eisenhower through his network of pacts, had overcommitted the United States, as in Southeast Asia. He made no new security commitments during his own Presidency. His essential view, declared in a speech nine months after the inauguration, was "that the United States is neither omnipotent nor omniscient — that we are only 6 percent of the world's population — that we cannot impose our will upon the other 94 percent of mankind — that we cannot right every wrong or reverse each adversity — and that therefore there cannot be an American solution to every world problem."[32]

He looked ahead to what he called in 1962 "a world of diversity" — a world of nations, various in institutions and ideologies, "where, within a framework of international cooperation, every country can solve its own problems according to its own traditions and ideals." The "revolution of national independence," as he saw it, was the source and guarantee of the world of diversity. Communism was one element in this pluralistic world. But diversity, he argued, was ultimately incompatible with the Communist dogma that all societies passed through the same stages and all roads had a single destination. "The great currents of history," he said, "are carrying the world away from the monolithic and toward the pluralist idea — away from communism and toward national independence and freedom."[33] In his American University speech of June 1963, he summed up his policy in a deliberate revision of Wilson's famous line: "If we cannot now end our differences, at least we can help make the world safe for diversity."[34]

This was his abiding objective. But, if he did not think there could be an American solution to every world problem, he did not think there could be a Soviet solution either. That was the immediate issue. Khrushchev's speech left no doubt in Washington that the Cold War of movement was on full blast. Lest this drive the world toward the ultimate war, Kennedy considered it imperative to restore the military balance now tilting so alarmingly (it was believed) in the Soviet favor. This meant both overcoming the alleged gap in nuclear missiles and creating a new capacity to fight nonnuclear war.

Eisenhower had stubbornly declined to do these things.[35] As a general, he was far less impressed by the Joint Chiefs of Staff, about whose business he knew everything, than he was by the Secretary of the Treasury, about whose business he knew nothing. He did not think the Russians crazy enough to start a war. He discerned a self-serving budgetary design behind the wailing at the Pentagon wall. He believed that an increase in defense spending could wreck the sacred free-enterprise system and turn America into a garrison state. So he was damned if he would move. In a not unusual irony, Eisenhower was right for the wrong reasons; Kennedy, who did move, was wrong for the right reasons.

Kennedy moved on two fronts, increasing both nuclear and conventional force. After becoming President, he discovered that the CIA questioned the missile gap.[36] McNamara shared this skepticism. But the professional military remained obdurate. No one could be quite sure where the truth lay. Kennedy, in addition, was reluctant to get into a fight with the Joint Chiefs at a time when McNamara was struggling to gain control of the military establishment. General Curtis LeMay, who spoke for the Air Force in the JCS, thought there should be at least 2400 Minutemen missiles; General Thomas Powers, head of the Strategic Air Command, talked to Kennedy of 10,000.[37] When Kennedy and McNamara imposed a limit of 1000, it almost provoked, Under Secretary of Defense Roswell Gilpatric recalled, "a major cleavage with the Chiefs."[38]

The decision to build only 1000 was itself an overreaction, as some in the administration felt at the time and most felt eventually. By autumn 1961 the new reconnaissance satellite persuaded even the Pentagon that the Russians had produced only a tiny proportion of the missiles claimed by the gap-mongers. "If we had had more accurate information about planned Soviet strategic forces," McNamara said later, "we simply would not have needed to build as large a nuclear arsenal as we have today."[39] The build-up had unhappy consequences. It sent the wrong message to Moscow. It ended any hope of freezing the rival strategic forces at lower levels. It compelled Khrushchev to start worrying about the Soviet missile gap.

At the same time, Kennedy called for an increase in conventional force. "We must regain the ability," he had said in June 1960, "to intervene effectively and swiftly in any limited war anywhere in the world."[40] This policy was less controversial at the time. It caused even more trouble in the end. For a dozen years men of good will had sought ways of saving the world from nuclear suicide. The best way to make sure nuclear weapons would never be used, it appeared, was to abandon the Eisenhower-Dulles strategy of predominant reliance on the bomb. 'Flexible response,' through the multiplication of 'options' — a pet word of the day — would make the punishment fit the crime and thereby, it was supposed, more effectively discourage criminality. The nuclear arsenal should be only the shield against a Soviet first strike. Limited-war forces should become the active element in American military strategy. The purpose of all this was precisely to insure against nuclear catastrophe. This view had been elaborated by theorists at MIT, Harvard, Princeton and RAND. It was urged with persuasive force by Maxwell Taylor in *The Uncertain Trumpet*. Kennedy and McNamara embraced it wholeheartedly. It was supported by most of the scientists, economists, political scientists and historians in the Democratic party.*

'Flexible response' was adopted for NATO reasons. It was designed to enable the west to respond to Soviet aggression in Europe without immediate resort to nuclear weapons. But a fateful side effect was to create forces that could be thrown into local conflicts in the Third World. Taylor's *Uncertain Trumpet* had forecast this possibility in an ominous comment on Dulles's appeal for American intervention to save Dien Bien Phu in 1954: "Unfortunately, such [conventional] forces did not then exist in sufficient strength . . . to offer any hope of success."[41] Not only would the creation of capabilities now make such intervention feasible; the escalation process, imperceptibly raising the level, step by small step, could make commitment irreversible. "One effective way of keeping out of trouble," Bernard Brodie, a penetrating critic of flexible response, later wrote, "is to lack the means of getting into it."[42]

We missed these fatal implications at the time. The diversification of the American defense posture, along with new initiatives in foreign aid, seemed the best way to stabilize that changing power equilibrium about which even men like Kennan and Stevenson had warned Kennedy and about which Kennedy, through a long campaign, had

---

* The major dissent in the party's intellectual ranks came from Thomas K. Finletter and Stuart Symington — both of whom served as Secretary of the Air Force under Truman — and Roswell Gilpatric, who had served as under secretary. Their protest was unwisely discounted as the view of otherwise intelligent men who had spent too much time with the air marshals. Gilpatric summed up the opposition case in a prescient letter to the *New York Times* in 1954. The survival of communism in North Korea, North Vietnam and China, Gilpatric said, suggested that "regardless of the amount of United States or United Nations force committed to a local war in the Far East, the odds are against the free world's winning it by conventional arms. A foreign policy based on a willingness and ability on the part of the United States to engage in local wars does not make much sense. . . . There is less likelihood of a 'brush fire' war becoming a major conflict if United States forces are not brought into the combat theatre. . . . Direct intervention by military means should be minimized or wholly avoided. In the Far East the chances of heading off local wars or sealing them off once started hinge largely on non-military measures. . . . By no longer trying to block Communist China's admission to the United Nations the United States might be able to bring about a reduction of tension in southeast Asia that would lessen the chances of further Communist 'nibbling' or 'brush-fire' type of aggression" (*New York Times*, December 29, 1954; see also Thomas K. Finletter, *Power and Policy* [New York, 1954]).

warned the nation. We all saw it as a prelude not to war but to negotiation. Kennedy never tired of quoting Winston Churchill: "We arm to parley."

v

In June 1961 Kennedy met Khrushchev in Vienna. The President's hope was to make a standstill agreement with the Soviet Union. Each superpower, he proposed, should abstain from doing things that, by upsetting the rough balance into which the postwar world had settled, might invite miscalculation and compel reaction by the other. As everyone knows, Khrushchev brusquely rejected Kennedy's general idea (apart from Laos) and threatened specifically to conclude a treaty with East Germany that would extinguish western rights in West Berlin. Kennedy replied that so drastic an alteration in the global balance was unacceptable. Khrushchev himself, he observed, would not submit to a comparable shift in favor of the west. Khrushchev rejoined that, if America wanted war over Berlin, there was nothing the Soviet Union could do about it; he was going to sign the treaty by December 31.*

Earlier in the year Kennedy had asked Dean Acheson to undertake special studies of the problems of NATO and Germany. Probably he did this in order to hear the 'worst case.' The result was to define the issue of Berlin in the bleakest terms. In Acheson's view the way to answer Khrushchev was to make the most conspicuous preparations for a military showdown, including the proclamation of a national emergency. The west, he argued, could only lose by negotiation. Merely to propose negotiations would be an admission of weakness. After Acheson alarmed Harold Macmillan by expounding this line during the British Prime Minister's visit to Washington in early April, Averell Harriman said to me, "How long is our policy to be dominated by that frustrated and rigid man? He is leading us down the road to war."[43]

The President listened to Acheson but, as Presidents do, kept his own counsel. "I was never quite sure," Acheson said later, "whether Jack Kennedy was completely sold on the conclusions."[44] His doubts

---

* The legend has arisen that Khrushchev browbeat and bullied Kennedy at Vienna. A reading of the record dispels that impression. Each man made his points lucidly and vigorously. Each held his own. Neither yielded ground. Kennedy, said Khrushchev, "impressed me as a better statesman than Eisenhower. ... He felt perfectly confident to answer questions and make points on his own. This was to his credit, and he rose in my estimation at once. ... He was my partner whom I treated with great respect. It was my judgment that ... he wouldn't make any hasty decisions which might lead to military conflict. ... I think that Kennedy was more intelligent than any of the Presidents before him" (N. S. Khrushchev, *Khrushchev Remembers* [Boston, 1970], 458; *Khrushchev Remembers: The Last Testament* [Boston, 1974], 497–500).

were justified. Meditating on the two influential columnists of the day, the Achesonian Joseph Alsop and the anti-Achesonian Walter Lippmann, Kennedy observed to Bundy and me in May 1961, "I like them both, but I agree more with Lippmann than I do with Alsop." (Bundy said, "That's exactly what Joe is afraid of.")[45] Asked later what the President thought of Acheson, Robert Kennedy said in 1965: "He respected him, and found him helpful, found him irritating; and he thought his advice was worth listening to, although not accepted. On many occasions, his advice was worthless."[46]

In fact, John Kennedy had long advocated talking with the Russians about Berlin. "I do think it may be possible for us to reach a *modus vivendi* with them," he had said in December 1959, "particularly if they feel that any real attack on our position in Berlin would bring war."[47] His policy after Vienna was clear enough: to increase military readiness in order to lay the basis for negotiation. In July the President requested more defense appropriations, called out 150,000 reservists and announced a program, which he soon regretted, of fallout shelters for protection against nuclear attack. At the same time, he rejected Acheson's idea of a national emergency, rejected also the notion of dramatizing the crisis by a tax increase and, explicitly acknowledging Russia's security interests in central Europe, declared in a speech on Berlin, "We believe arrangements can be worked out to meet those concerns."[48]

Robert Kennedy disagreed in minor detail. The "great battle" for five days before his brother's speech, he noted,

> was whether we should ask for a tax increase. I was the one who fought for that policy and was supported by all of the Cabinet with the possible exception of Dillon, plus Edward R. Murrow. On the other side was Sorensen and Walt[er] Heller and the financial people [the Council of Economic Advisers]. They felt it would not do any good financially and, in fact, might harm the economy of the country. I felt first that it would be helpful psychologically. It would give everybody the sense of participation which they were lacking at the present time.
>
> The day before the speech we heard some rumblings that Congress might cut down on the foreign aid program if we asked for the tax increase. I think it was this more than anything else that finally determined Jack not to go for it. The following weekend he asked me if I didn't think he was right. I said, "Why don't you admit that you were wrong about it?"[49]

The Attorney General also opposed the fallout-shelter program. "I was a minority of one," he said later. ". . . I thought during that period of time that really you're never going to get it going very well and if we got too far out on a limb on it it was just going to collapse eventually."[50] He was appalled by the horrid emotions the program unleashed. In September, Father L. C. McHugh, a former professor of ethics at Georgetown, argued in *America*, the Jesuit magazine, that shelter owners had the moral right to kill unsheltered neighbors clamoring for admission.[51] Over Thanksgiving, in a Hyannis Port conclave, I presented at the President's direction a critical review of the shelter program. When I discoursed before assembled potentates on the divisive results of the *sauve-qui-peut* ethic, the Attorney General interjected sourly, "There's no problem here. We can just station Father McHugh with a machine gun at every shelter."[52]

In Moscow, meanwhile, Khrushchev directed a group of nuclear scientists, among them Andrei Sakharov, the father of the Soviet hydrogen bomb, to prepare a new series of nuclear tests in order, Khrushchev explained, to bolster the Soviet policy on Berlin. "To resume tests after a three-year moratorium," Sakharov protested, "would undermine the talks on banning tests and on disarmament, and would lead to a new round in the armaments race." Later, at dinner, Khrushchev held forth in characteristic fashion: "Sakharov is a good scientist. But leave it to us, who are specialists in this tricky business, to make foreign policy. Only force [counts] — only the disorientation of the enemy. We can't say aloud that we are carrying out our policy from a position of strength, but that's the way it must be. I would be a slob, and not chairman of the Council of Ministers, if I listened to the likes of Sakharov."[53]

The announcement of the Soviet decision to resume testing produced, Robert Kennedy recorded,

the most gloomy meeting at the White House ... since early in the Berlin crisis. I had talked to Jack previously and he was at a loss to explain Khrushchev's decision to resume testing. It was obviously done to try to intimidate the West and the neutrals.... I offered the explanation that we were as puzzled about this as Khrushchev must have been puzzled about our invasion of Cuba [the Bay of Pigs]. As we discussed at the time, he must have thought there was something sinister and complicated behind it all or otherwise, we would not have done anything quite as stupid.

The meeting considered steps to be taken if the Soviet Union tried to shoot down allied planes flying to West Berlin. "The major point of contention raised by Jack was if anti-aircraft guns fired on our planes whether we should fire back in that case. General [Lyman] Lemnitzer and General [Maxwell] Taylor felt very strongly that we should fire back. Jack raised the question as to whether we should not have some control over that action.... McNamara finally came out for not firing back at land targets without permission."[54]

The atmosphere was grim. "It is obvious," Robert Kennedy continued,

> that as we get closer to D-day the situation becomes more difficult. The people are less and less anxious to stand firm. It is my feeling that the Russians feel this, feel strongly that if they can break our will in Berlin that we will never be able to be good for anything else and they will have won the battle in 1961. They speak with one voice — they do not have to worry about opposition or what other countries think. Their plan is obviously not to be the most popular but to be the most fearsome and to terrorize the world into submission.
>
> My feeling is they do not want war but they will carry us to the brink. . . . I remember Chip Bohlen said in the beginning of the year that 1961 was going to be a fatal one on decision. This was the year the Russians were going to come the closest to nuclear war. I don't think there is any question but that that is true.
>
> After the meeting Jack asked me what I thought. I said I wanted to get off. He said, "Get off what?" and I said, "Get off the planet." I said to all of them that it was much more cheerful at the Department of Justice and that I was not going to come into that room again or have anything to do with them. . . . I told him I wasn't going to take Paul Corbin's advice and run against him in 1964 — I didn't want the job.[55]

VI

"I want to take a stronger lead on Berlin negotiations," the President informed Rusk on August 21.[56] "All wars start from stupidity," he said to Kenneth O'Donnell. "God knows I'm not an isolationist, but it seems particularly stupid to risk killing a million Americans over an argument about access rights on an Autobahn."[57] "I am almost a 'peace-at-any-price' President," he told his friend the painter William Walton.[58]

He grew increasingly frustrated in his dealings with the State Department. Those who supported negotiation, like Harriman, Bohlen and Stevenson, were excluded from the Berlin discussions. The pronegotiation cables from Ambassador Llewellyn Thompson in Moscow were discounted. The head of the Berlin working group, Foy Kohler, was an all-out Achesonian. "If we make it a choice between the status quo and atomic war," Harriman said to me somberly at the end of August, "we will get atomic war."[59] Since the Berlin crisis of 1958–59, the Department had not come up with a fresh thought on Berlin. "The only person who had had any innovations or any ideas or even questions," Robert Kennedy observed, "was Jack himself."[60]

In a White House meeting on September 4, 1961, the President laid out a Berlin negotiating program: the submission of the legal dispute to the World Court; the transfer of the United Nations to

West Berlin; the internationalization, under UN control, of the autobahn to Berlin; a UN plebiscite in Berlin; a nonaggression pact between the NATO and Warsaw Pact powers. Perhaps, he said, he should offer these points in the speech he was about to give to the UN General Assembly. Stevenson said they were all excellent points, except the one about moving the UN to Berlin. Kennedy said, "I don't think enough of the UN not to be prepared to trade it for a nuclear war."[61]

The prospect of negotiation encountered wrathful opposition in the country. "Kennedy," said James L. Buckley, "has chosen to identify himself with that segment of American society which is either unwilling or unable to regard Communism as more than a childish bugaboo."[62] Barry Goldwater, already a contender for the next Republican nomination, inveighed against Kennedy's "no-win policy" and asked, in the title of his popular book, *Why Not Victory?* "Our objective," Goldwater wrote, "must be the destruction of the enemy as an ideological force possessing the means of power. . . . Our effort calls for a basic commitment in the name of victory which says we will never reconcile ourselves to the Communists' possession of power of any kind in any part of the world."[63]

The frenetic atmosphere put the very idea of negotiation in jeopardy. James Wechsler of the *New York Post,* interviewing Kennedy in September, found him "disturbed by the frustrated fury of many of his countrymen who believe our national manhood can be affirmed only by some act of bloody bluster." On the other hand, Kennedy was not prepared to accept humiliation. If the west were pushed out of Berlin, he believed, it would transform the whole European balance and make peace impossible. "Caught in the crossfire of Russian intransigence and domestic Know-nothingism," Wechsler wrote, "he is both keeping his head and sustaining his nerve."[64]

If Professor Robert Slusser's detailed reconstruction of the debate within the Kremlin is correct, the situation may have been even more dangerous than Washington supposed.[65] Khrushchev probably had to contend with his own "why not victory?" crowd. Then Kennedy's military preparations, including even the egregious shelter program, made it clear that the Soviet Union could not win Berlin by psychological warfare. This very likely strengthened Khrushchev against Frol R. Kozlov and the hard-liners. The erection of the Wall in August meanwhile stopped the flight of refugees, thereby reducing the East German pressure for a separate treaty. In October, Khrushchev postponed the treaty deadline. The Berlin crisis, for a moment so frightening to the world, subsided, and was rather quickly forgotten.

Kennedy's desire for a peaceful solution was manifest throughout. "Several attempts have recently been made," Professor Slusser wrote later,

> to depict John F. Kennedy as a dogmatic anti-Communist, a Cold Warmonger whose actions helped create the very crises with which his administration tried to cope. . . . The evidence of Kennedy's patient search for negotiations with the Soviets over Berlin is part of the historical record, and it is only by ignoring the plain facts that the authors in question have managed to construct their image of Kennedy as a Cold Warrior.[66]

Khrushchev clearly agreed. A notable by-product of the Berlin crisis was the extraordinary personal correspondence between the two leaders, initiated by the Russian in order, as he wrote Kennedy, to bypass the foreign office bureaucracies and to eliminate the propaganda obligatory in formal communications.[67]

At home Kennedy went on a speaking tour to vindicate the idea of negotiation against those Americans, like Acheson, who considered it stupidity or, like Goldwater, treason. As long as we understood our vital interests, Kennedy declared in November 1961, "we have nothing to fear from negotiations at the appropriate time, and nothing to gain by refusing to take part in them." He condemned "those who urge upon us what I regard to be the pathway to war: equating negotiations with appeasement, and substituting rigidity for firmness." He concluded with deep feeling: "If their view had prevailed, we would be at war today, and in more than one place."[68]

VII

The Berlin episode confirmed John Kennedy's misgivings about the State Department. The next President, he had said in 1960, "will face a world of revolution and turmoil armed with policies which seek only to freeze the status quo and turn back the inevitable tides of change."[69] Kennedy genuinely sought to move beyond the status quo with policies responsive to a world of revolution. His State Department seemed a stuck record, spinning endlessly in the grooves of the past.

The Kennedy skepticism was of long standing. The boys had doubtless learned it at their father's knee. Their own travels in the early 1950s had verified the ambassador's discontents. In 1960 (as recalled by Peter Lisagor) John and Robert Kennedy discussed their plans to regenerate the Department. Their father listened patiently. Finally he said, "Sons, I want to tell you that I once went to see Franklin D. Roosevelt, who made much the same kind of talk that you're making now. He lamented the State Department. He talked about razing the whole thing and starting from scratch. He didn't do a damn thing about it, and neither are you." [70]

They tried. The first problem was to gain control of the national security machinery. Robert Kennedy later summarized his impressions of his brother's impressions of the crucial appointments:

> McNamara he thought the most highly [of]. . . . Rusk speaks very well, [was] awfully loyal and a very nice man. . . . But as far as focussing attention on what should be done himself and sticking with his guns . . . he didn't have that. . . . He didn't have any strong point of view. . . . The President and I discussed on a number of occasions [the possibility] after the [1964] election of moving Rusk out perhaps to the UN and appointing Bob McNamara Secretary of State.*. . . McNamara, under President Kennedy, would have been a good Secretary of State . . . because President Kennedy knew so much about it that he could make the major decisions. . . . Under Lyndon Johnson I'm not so certain he would be a good Secretary of State. . . .
>
> [Adlai Stevenson] was always complaining [but] he was the best ambassador we could have had there. . . . Averell Harriman, of course, performed the kinds of functions he performed tremendously well and, of course, the President was very pleased with what he did in Laos and the test ban. . . . He thought [George Ball] was able enough but didn't have any rapport with him particularly. . . . Tommy [Llewellyn] Thompson he thought was outstanding. . . . He liked Ed Murrow very, very much. . . . [But] he really felt at the end that the ten or twelve people in the White House who worked under his direction with Mac Bundy . . . performed all the functions of the State Department. [71]

Kennedy to an exceptional degree became his own Secretary of State — more so than FDR, thought Harriman, who had served them both. "Roosevelt selected the things that he wanted to deal with himself and then left to Sumner Welles the handling of the [other] things. . . . Kennedy dealt with every aspect of foreign policy. . . . He knew about everything that was going on." [72] Still, even if the President might try to know everything, he could not do everything. After the Bay of Pigs, Robert became his brother's eyes and ears around the national security establishment. "If you wanted to get a dissenting idea into the White House," Michael Forrestal told David Halberstam, ". . . the best channel — almost the only channel — was Bobby Kennedy." [73] Kennedy ambassadors, frustrated by State Department orthodoxies, made the Attorney General's office a port of call on home leave. "Where I had need to bring pressure on the State Department," said John Kenneth Galbraith (India), ". . . I found a very good ally in Bob." [74] He was, said William Attwood

---

* Rusk, on accepting the appointment, had told Kennedy that for financial reasons he could remain for only four years.

(Guinea, later Kenya), "the most attentive listener in town."[75] The national security community soon recognized him as a major countervailing force on international decisions. "He seemed to have a damn good intelligence collection apparatus," said Roswell Gilpatric of Defense, "that kept him very well informed of what was brewing."[76] "Kennedy, a student of guerrilla warfare," observed Patrick Anderson, "was applying its techniques to intergovernmental relations."[77]

VIII

The Attorney General's dissatisfaction with the State Department was acute. He nagged the Secretary about the Department's dilatoriness in hiring blacks. Sending Rusk, for example, a *Washington Afro-American* editorial entitled "Anti-Colored U.S. State Department," Kennedy wrote, "Confidentially, there is a great deal of feeling that the fellow who is handling the matter over there is an Uncle Tom and is just not getting the job done."[78]

The Kennedys were also exasperated by State's philosophic attitude toward the travel restrictions that mocked America's pretense of being a truly open society. The Immigration and Nationality Act of 1952 excluded from the United States past and present members of proscribed organizations, such as the Communist party and Communist-front organizations, and advocates of proscribed doctrines, such as communism. However, it also empowered the Attorney General, when State so recommended, to grant waivers for reasons of compassion or of the national interest. The Eisenhower administration administered the proscriptive sections of the act with enthusiasm. Dulles selected Scott McLeod, a Joe McCarthy protégé, as head of the Bureau of Security and Consular Affairs. In the Dulles atmosphere, consuls abroad automatically denied visas to all remotely suspicious persons.

The Kennedys thought it all ridiculous. Exchange agreements with Communist countries permitted the entry of party members from behind the Iron Curtain; but artists, writers and scientists from western Europe or Latin America whom someone had charged with committing an offense against American ideas of political propriety were regularly turned away. Their admission, the President and the Attorney General believed, was unlikely to destroy the Republic. Exposure to the United States might even have a salutary effect. In 1962 President Kennedy appointed Abba Schwartz administrator of the bureau. No greater contrast with McLeod could have been imagined. Schwartz, a liberal, a law partner of James Landis's, a close friend of Eleanor Roosevelt's, had long experience in refugee and immigration problems. He had served as liaison between Mrs. Roosevelt and the Kennedys in the 1960 campaign. "We've lost a good deal of time," the President told Schwartz,

> and I'm fed up with the image we have as a police state. I keep seeing reports about excluding visitors because of their political views. We act like a closed society. I want you to change that. . . . You will be my personal appointee. I expect you to assert your authority and act for me. You know what I want — an Open Society — with the freest possible movement in and out of the country. And a new immigration law. Keep in touch with the Attorney General. I don't know what you'll find in the State Department but you'll just have to work that out.[79]

What Schwartz found in the Department, he wrote later, was total unawareness of "the President's conception of what the Bureau should and should not do." The Secretary appeared indifferent; indeed Kennedy had never mentioned Rusk in his talks with Schwartz. The Passport and Visa offices contained what Schwartz described as "the hard-core remnants of the McCarthy era."[80] But, as directed, Schwartz kept in touch with the Attorney General. Together they vigorously used Justice's waiver authority to reopen the United States to those formerly excluded.

When, for example, Professor Kaoru Yasui, a Japanese winner of the Lenin Peace Prize, was invited to the United States in the autumn of 1963, his visa application was at first turned down. Schwartz got on the case, discussed it with the Kennedys, the Attorney General issued the waiver and Yasui made his visit. The republic survived, though in a few months Schwartz had to justify the action before seven closed sessions of the House Un-American Activities Committee.[81] Carlos Fuentes, the Mexican novelist, had been refused a visa in 1962 before Schwartz came to the Department. He applied again in 1964. Angie Novello sent a note to Harold Reis of Justice: "It seems he got into a fight with Ambassador [Thomas] Mann who called him a Communist which Fuentes has denied. Would you check with Abba Schwartz about this? The Attorney General is interested in this and thinks he should be granted a visa."[82] Schwartz at once recommended to Rusk that State request a waiver from Justice. Mann continued to object. Rusk stalled. Finally George Ball, Acting Secretary on a day when Rusk was out of town, sent the request over to Justice. Kennedy promptly gave Fuentes his visa.[83]

He favored admitting foreigners to the United States — and permitting them to stay when it was dangerous for them to return home. Once the Iranian ambassador told Rusk that the Shah wanted twenty Iranian students in American colleges sent back to Iran. The

Secretary explained to the Attorney General that they were Communists, and the United States had better get rid of them. Kennedy consulted his old friend William O. Douglas in the justice's capacity as an expert on Iran. "I told him," said Douglas, "that that meant the Shah was making up lists for the firing squad"; the students should not be sent home "unless there was a real interest of the national security of this country because these students were lawfully here." "Don't do it," he said to Kennedy, "unless the FBI can prove conclusively that they actually are Communists." In a few weeks Kennedy called back: "The FBI report is in, and not a bloody one of these kids is a Communist, so I just told Rusk to go chase himself."[84]

The freedom of Americans to travel abroad had also suffered attrition under the Cold War. In the forties and fifties the Department grew accustomed to refusing passports to politically suspect citizens. In 1958 the Supreme Court outlawed this practice in the absence of statutory authority.[85] Eisenhower asked Congress to grant the authority. Senator Kenneth Keating of New York introduced a bill giving the Secretary of State power to withhold passports. Nothing happened. The Kennedy administration declined to support the Keating bill. In 1961, however, the Court, by upholding the constitutionality of the Internal Security Act of 1950, brought into play provisions in that law forbidding passports to members of organizations required to register under the act. Robert Kennedy promptly construed the statute as entitling a spurned applicant "to be confronted with the precise information against him and to be apprised of the source of such information."[86] This interpretation outraged Frances Knight, the director of the Passport Office, and she refused to accept it. (Later, the security-obsessed Keating signed a senatorial démarche to the Department designed to entrench Knight in her job.) Robert Kennedy initiated a test case, which resulted in a Court decision in 1964 to invalidate the passport provision in the Internal Security Act.[87]

The most irksome constraint on travel came from the practice of placing the presumably more offensive Communist countries — China, North Vietnam, North Korea, Albania, Cuba — off limits. The President, Schwartz said, "abhorred" travel control. "Everyone should be able to go to China, or any other place, if they can get in," John Kennedy said. "We know too little about China. The bans should be removed."[88] Backed by Robert Kennedy, Ball, Harriman and Edward R. Murrow, Schwartz repeatedly recommended to the Secretary of State that the restrictions be lifted. Rusk, fearing an outcry on the Hill, declined to act. Schwartz concluded that the President, facing so many difficulties with State, "must have decided that others than himself should deal with the Secretary on this problem rather than for him to direct that the restrictions be removed over the Secretary's spoken or silent objections."* His brother's activity in this field, otherwise laudable, was not calculated to endear the Attorney General to the Department.

IX

The essential Kennedy problem with State was how to annex the Department to the New Frontier. It was difficult to do anything about Rusk; so the presidential eye moved on to the under secretary. Ironically, Chester Bowles was the most devoted New Frontiersman in the Department. The President liked him and shared most of his outlook on the world. But, save for his first-class ambassadorial selections, Bowles had had small impact on the entrenched bureaucracy. One "trouble" with that generous-hearted, intelligent, imaginative man, as his friend Galbraith conceded from New Delhi, was the "uncontrollable instinct for persuasion which he brings to bear on the persuaded, the unpersuaded and the totally irredeemable alike." [89]

At Kennedy's suggestion Rusk asked Bowles in July 1961 whether he would not like to become a roving ambassador. A disturbed Bowles called the President, who promptly invited him to luncheon. "This is a luncheon which I shall never forget," Bowles wrote Stevenson (in a thirteen-page, single-spaced letter).

> . . . I have never heard the President talk as thoughtfully and passionately on foreign affairs. He said precisely the things you and I have been advocating for years, and he said them well.
> He then expressed his deep disappointment at the failure of the State Department effectively to translate his views into action. He went on to say that changes must be made, and since it was generally assumed that the Under Secretary was in charge of administration, I was the logical one to change. [90]

An affectionate Bowles lobby swung into action and saved him for a moment. Three weeks later Robert Kennedy sent his brother a glowing account by Ambassador Philip Kaiser of Bowles's performance at a conference of American diplomats in Lagos. "This is the kind of thing Bowles should be doing all the time," the Attorney General wrote.

> I am sure conferences such as these will be going on around the world continuously and he could be our best and most forceful emissary. I would think that working out something like this with Bowles and

---

* Abba Schwartz, *The Open Society* (New York, 1968), ch. 5. Restrictions on travel to Cuba, Vietnam, North Korea and Cambodia were finally removed fourteen years later by President Jimmy Carter, in March 1977.

putting George Ball or Roswell Gilpatric as Under Secretary would be the answer. I know it is difficult but there is no question that with Rusk and Bowles in the two top positions things are going to continue to flounder at the State Department. Rusk is no detail man and with both of them in those top jobs things are going to get steadily worse.[91]

George Ball, who succeeded Bowles in November 1961, brought a refreshing crispness to the job. But Ball's interest lay in policy, not in management. The administrative responsibility fell to George McGhee, under secretary for political affairs and Rusk's best friend in the Department. Nothing improved. "In every conversation you had with [McGhee]," Robert Kennedy said later, "you couldn't possibly understand what he was saying. I was involved with him a good deal in 1962, and it was just impossible." Finally the President told his brother, "If you feel so strongly . . . why don't you go see Dean Rusk and ask him to get rid of him." Accompanied by the monstrous Brumus, the Attorney General called on the Secretary of State one Saturday morning "and said I thought it was discouraging and gave him some examples of the fact that George McGhee didn't know what he was doing." Rusk asked whether this was something personal. Kennedy said no, it wasn't personal. Rusk said wearily he would take it under consideration.[92]

McGhee was soon shipped off as ambassador to Bonn, and the seventy-year-old Harriman took his place. Harriman had the signal advantages of wisdom, confidence and the knowledge of how to deal with Presidents. Presidents, Harriman said, always look around the room to see who agrees with a proposed decision and who does not. "I've learned . . . that you have a split second. . . . You can't wait." Harriman seized those split seconds and, when he disagreed, explained why in a few blunt sentences — a talent that won him the White House nickname of 'the crocodile.' "Mouth open, and I bite, you know," said Harriman.[93] He also had a singular capacity to get things done within the Department.

But even Harriman could not take care of everything. In an effort to tighten the internal workings, Robert Kennedy sent William Orrick over from Justice in late 1962 to become deputy under secretary for administration. Orrick later recorded his marching orders from the President. Two things in particular bothered John Kennedy: "Number one, he could not understand the Foreign Service. He wanted me, if I could, to interpret . . . the Foreign Service to him by whatever means I could. And, secondly, it just annoyed him beyond measure that he couldn't get answers out of the Department."[94] State regarded Orrick with suspicion as the agent of the Attorney General. When Orrick overcame that, he discovered both that the problems were deep and that the Kennedys to some extent misconceived them. With relief he seized the chance to come back to Justice in 1963 as chief of antitrust. "After the Orrick experience," John Seigenthaler said later, "there just was the feeling that what the hell can you do about it?"[95]

All this activity insured the Attorney General's unpopularity with the professionals — and theirs with him. "They all had problems with Bob," said Orrick. ". . . When I'd say something nice about them, Bob would say, 'Well, you don't really know them.'"[96] They saw him, said Donald Wilson of the United States Information Agency — who, like Orrick, admired the Attorney General — "as a headstrong, unreliable upstart in the field of foreign affairs. . . . He was regarded as quite a menace by most of the people in the State Department.. . . . He'd be pretty outrageous sometimes in some of the things he'd talk about or get excited about. You weren't ever sure whether he was just speaking for himself or for the President. Actually, I think in most cases he was speaking for himself, and if he thought what he elicited . . . was of significance, he would pass it on to the President. If he thought it wasn't, then he wouldn't."[97]

x

Robert Kennedy summed up his own views in a memorandum to his brother in April 1962 proposing themes for a presidential speech to the Foreign Service Association (views rendered by the President soon after in less provocative language). "Innovations, imagination," he wanted his brother to tell the diplomats, "yes, even revolutionary concepts are essential. . . . *It is your responsibility not just to carry out policies that have been established but to suggest and come forth with new ones.*" Foreign policy, the Attorney General thought the President should point out, was no longer a matter of transactions among chancelleries. "More and more the people themselves are determining their country's future and policies. Therefore, greater attention has to be given than has been true in the past to these new and sometimes revolutionary elements" — labor leaders, intellectuals, students.

> A major effort must be made to meet with representatives of all kinds of groups. . . . For those discussions our Foreign Service must know what we are doing here in this country; understand internal problems that we are facing — unemployment and racial difficulties — and know also what remedial steps we are taking. They must understand what this government has accomplished for its people in a social way such as social security, minimum wage, housing, and what we are trying to do in the future. They must know our history and our culture so that they can discuss these matters freely and with candor.[98]

The Kennedys had a romantic view of the possibilities of diplomacy. They wanted to replace protocol-minded, striped-pants officials by reform-minded missionaries of democracy who mixed with the people, spoke the native dialects, ate the food, and involved themselves in local struggles against ignorance and want. This view had its most genial expression in the Peace Corps, its most corrupt in the mystique of counterinsurgency. The gospel of activism became the New Frontier's challenge to the cautious, painstaking, spectatorial methods of the old diplomacy. Of all the New Frontiersmen, no one was a greater enthusiast for action diplomacy than Robert Kennedy.

What was sometimes later described as Kennedy activism was in fact its opposite. So in Italy the White House after long and painful argument induced the Department to remove the veto imposed by the Eisenhower administration on a center-left government in Rome. This withdrawal of American intervention in Italian politics was peculiarly criticized by State Department people as itself an act of intervention.[99] And in special circumstances activism was useful — especially so in places where the United States in the preceding decade had intervened *against* social change. In India, Africa and Latin America activist ambassadors gave valuable counsel to governments struggling against massive difficulties. But action diplomacy was easier to advocate than to execute. John Paton Davies, Jr., the old China hand, inspected the scene with dismay. "Bold new ideas and quick decisions," he wrote, "were asked of men who had learned from long, disillusioning experience that there were few or no new ideas, bold or otherwise, that would solidly produce the dramatic changes then sought." Davies was not surprised that the Foreign Service did not win the confidence of the New Frontier. "Crusading activism touched with naiveté seldom welcomes warnings of pitfalls and entanglements."[100]

The issues were ably canvassed in a letter from Philip Bonsal, one of the best and most liberal-minded of the old school, to Robert Kennedy's friend the journalist John Bartlow Martin, who served with distinction as ambassador to the Dominican Republic. Acknowledging Martin's achievements, Bonsal observed that they "involved your direct and forceful participation in many of the top level decisions with which these [Dominican] governments were confronted. . . . You functioned much as would the authoritative coach of a rather backward football team." Still Bonsal could not help wondering

> whether if we had let the Dominicans stew in their own juice politically while assisting them economically they might not have cooked up a stew better adapted to their palates and digestions than what was in the event

served up to them.... A process involving the constant participation of the American Ambassador, no matter how high his motives or how great his political skill, in the domestic affairs of the country to which he is accredited cannot produce constructive results.[101]

Perhaps there was more to be said for the old diplomacy than Robert Kennedy, at least, realized in 1961. Philip Graham of the *Washington Post* made the case after the President spoke to him late in 1962 about superseding Rusk by McNamara. "Defense," Graham wrote John Kennedy, "needs action, movement, and above all a competence and courage to take decision after decision after decision.... In State, energy and dynamism can be specious qualities." Apologizing for the "effrontery" of his letter, Graham concluded: "It is just that I — who have a ghastly weakness for action, movement, and go — have painfully learned the value of wisdom, slowness and the kind of calm sense owned by a handful of men such as David Bruce."[102]

The New Frontier style suffered too often from this ghastly weakness — as the State Department style suffered from passivity, complacency, careless identification with the possessing classes and the traumas of the Dulles-McCarthy years. The Kennedys, observed Orrick, "held no brief for the time-honored traditions of diplomacy," according to which "you must be patient and listen and do anything but get involved.... They wanted to get involved every place.... For a few underdeveloped countries that might be desirable. But mostly it isn't. It just isn't. And they never understood that."[103]

Action diplomacy, carried too far, entangled the United States in the fortunes of nations beyond the area of American interest or understanding. It encouraged the illusion that an internal crisis in a remote land might affect the peace of the world. It led American diplomats to try to do things in foreign countries that the people of the country ought to have done for themselves. It nourished the faith that American 'know-how' (odious word) could master anything; the 'when in doubt, do something' approach to an intractable world; the officious solicitude that degenerated into cynical manipulation and finally into bloody slaughter.

The professionals were right to question New Frontier activism in the execution of policy. They were equally wrong to resist, as so many did, all departures in the substance of policy from the hallowed doctrine as revealed to Acheson and Dulles years before. Dulles had purged the Department of independent-minded Foreign Service officers and converted it into a mighty temple of Cold War orthodoxy. Rusk was no heretic. The failure of State to contribute to the President's drive for Berlin negotiations showed the grip of Cold War dogmas. "There would never be any initiative that would come from the State Department," Robert Kennedy said later with only minor exaggeration. ". . . I would think 60 per cent of this sort of new concepts or new ideas or plans really came from the President — oh, no, maybe 80 per cent came from the President and the White House, of which probably 60 of the 80 per cent came from the President himself."[104]

Much of John Kennedy's initiative was devoted to moving beyond the Acheson-Dulles theory of the Cold War. In subsequent years critics sought oddly to portray the President himself as a rigid and zealous Cold Warrior. An English historian spoke of "that most dangerous and megalomaniac of Administrations — the late John Kennedy's."[105] "He prosecuted the Cold War more vigorously, and thus more dangerously," wrote an American author, "than did Eisenhower and Dulles."[106] Khrushchev, perhaps a better authority, disagreed. "I had no cause for regret once Kennedy became President," the Soviet leader wrote in his memoirs. "It quickly became clear he understood better than Eisenhower that an improvement in relations was the only rational course. . . . He seemed to have a better grasp of the idea of peaceful coexistence than Eisenhower had. . . . From the beginning, he tried to establish closer contacts with the Soviet Union with an eye to reaching an agreement on disarmament and to avoiding any incidents which might set off a military conflict."[107]

(20)

The CIA and Counterinsurgency

T HE STATE DEPARTMENT remained the citadel of caution, professionalism and entrenched Cold War ways. But diplomats were
no longer in control of foreign policy. The Pentagon and the CIA
had grown far more than State at the pap of the Cold War. Almost
at once the Kennedys were forced to confront the power of the
military-intelligence complex. The occasion was the invasion of Cuba
in April 1961 by the so-called Cuban Brigade, a band of twelve
hundred anti-Castro exiles, recruited, trained and launched by the
CIA on a plan approved by the Joint Chiefs of Staff.

I

"This will just be some thoughts on Cuba and the effect it has had
on the Administration and the President," said Robert Kennedy in
a memorandum dictated on June 1, six weeks after the disaster.

> I was brought in on the Cuban situation about four or five days prior
> to the actual event — I think about the twelfth of April. Dick Bissell
> [the CIA's deputy director of plans] at Jack's instructions came over to
> the Department of Justice and briefed me. He told me at that time that
> the chances of success were about two out of three and that failure was
> almost impossible because . . . even if the force was not successful in its
> initial objective of establishing a beachhead, the men could become
> guerrillas and, therefore, couldn't be wiped out and would be a major
> force and thorn in the side of Castro. This, as it later turned out, was
> completely incorrect. I think that the President might very well not
> have approved of the operation if he had known that the chances of
> these men becoming guerrillas was practically nil.
> Subsequently, I attended some meetings at the White House. After-
> wards I said to Jack that I thought that . . . based on the information

that had been given to him by Dick Bissell, Allen Dulles, the Joint Chiefs of Staff, Mac Bundy and the others . . . there really wasn't any alternative to accepting it. These men had to be gotten out of Guatemala and Nicaragua; and if we brought them back to the United States and turned them loose, it could be a tremendous problem both here in this country and abroad. The chances of success as outlined by the CIA and outlined by the Joint Chiefs of Staff were extremely good.[1]

In 1964 he gave John Bartlow Martin additional reasons for his brother's decision. The advocates of the operation — Dulles, Bissell, General Lyman Lemnitzer, chairman of the Joint Chiefs of Staff — had, the new President supposed, been "trusted by his predecessor; so he thought that he could trust them and when they said it was much more apt to succeed than Guatemala [where the CIA had overthrown a leftist regime in 1954], when the military looked it over and said it was a good plan, then he went ahead." President Kennedy also had to reckon the political repercussions if he were to disband his predecessor's Cuban liberation corps in the face of expert assurance that the project would prevail. "If he hadn't gone ahead with it, everybody would have said it showed he had no courage because . . . it was Eisenhower's plan, Eisenhower's people all said it would succeed, and you turned it down."[*] In the last days before the landing, Robert Kennedy added, a highly decorated Marine colonel brought in an ecstatic report on the Brigade's military prowess. This, Robert Kennedy thought, "was the most instrumental paper in convincing the President to go ahead."[†]

Back to Robert's June 1 memorandum:

Jack on the first day realized there was going to be difficulty. He called me down in Williamsburg where I was speaking and said we had run into some trouble. He asked me to come to the White House. When

---

[*] RFK, in recorded interview by John Bartlow Martin, March 1, 1964, II, 17, JFK Oral History Program. When Eisenhower and Kennedy met the day before the inauguration, "President Eisenhower said with reference to the guerrilla forces which are opposed to Castro," as Clark Clifford, who accompanied Kennedy, recorded his words, "that it was the policy of this government to help such forces *to the utmost*. At the present time we are helping train anti-Castro forces in Guatemala. It was his recommendation that this effort be continued and accelerated" (Clifford to JFK, January 24, 1961, attaching "Memorandum on Conference between President Eisenhower and President-elect Kennedy and their Chief Advisers on January 19, 1961," JFK Papers; emphasis added).

[†] RFK, in Martin interview, II, 4–5. Kennedy added (6–7): "The one person who was strongly against it was Arthur Schlesinger. And Arthur Schlesinger came to my house . . . and I remember having a conversation with him in which he said that he was opposed to it and I said I thought that everybody had made up their mind and that I thought he was performing a disservice to bring it back to the President. . . . I remember at the house telling him that once the President had made up his mind, once it seemed to have gone this far, that we should make all efforts to support him, and he should remain quiet. Which he did." For an extended account of the Bay of Pigs, see Arthur M. Schlesinger, Jr., *A Thousand Days* (Boston, 1965), chs. 10–11.

I arrived there there was a meeting going on and the situation already looked dark.

He agreed to permit the Navy to come in closer to shore and ended the meeting by saying he wanted to be kept closely advised, that he'd rather be called an aggressor than a bum. In other words we didn't want to have this thing planned and the United States support [it] and have it go wrong.

That afternoon Dick Bissell called me at the Department of Justice and said that troops had been able to capture the air fields and the situation looked better. We had continuous conferences and conversations that day. The situation didn't seem to improve. There was also reports of shortages of ammunition.

Jack told them to bring in the [ammunition] ships and gave his permission. It then turned out that the ship the CIA had in mind was the ship that was supposed to arrive there two or three days later and they couldn't get it there any earlier and they took no cognizance of the ammunition ships that had run out on the action. It was complete lack of communication. Nobody was able to determine what was going on really there. At D plus one the information was again very gloomy. There was talk of trying to evacuate the men. We kept asking when the uprisings were going to take place. Dick Bissell said it was going to take place during the night. Of course, no uprisings did take place.

A critical time was on D plus one when the CIA asked for air cover. Jack was in favor of giving it. However, Dean Rusk was strongly against it. He said that we had made a commitment that no American forces would be used and that the President shouldn't appear in the light of being a liar. It was pointed out of course that the alternative was possibly destruction. I took the position that we didn't really have enough information to know whether the air cover would make any difference or not. If the air cover should save the operation, perhaps we should take that step. On the other hand if it just meant the delay of a few hours before absolute and ultimate collapse, there wasn't any sense in it. It was, therefore, decided that the Navy would make a reconnaissance over the beach and try to determine where the fighting was taking place and whether there was any possibility or chance of these men holding out. Admiral [Arleigh] Burke was given that assignment. He explained that it would take six or seven hours to get a reply and that meant that there couldn't be any air cover that day because we didn't get the request until about noon or one o'clock. Anyway we went ahead with that and they came back about seven o'clock that night and said they couldn't see any fighting going on and it would appear that the beach[head] was extended to about ten miles.

The next report came about 11:45 the following morning which was that the beach[head] was about to collapse.... The President for the previous three hours was battling hard to make arrangements to evacuate. By the time we received this information, however, the men were in the water. It was too late to send the destroyers in because they could be destroyed by artillery fire. He ordered them, however, to run up and down the beaches as close as they could and try to pick up any survivors. Over a period of the next six or seven days they picked up 25 or 30 men. Once again lack of communications had caused incalculable harm. Ultimately from the dispatches that I read and studied the CIA should have known that this was in bad shape and that we should take action to evacuate these men at a much earlier time than we did.

I think it was incompetency and lack of communications and nervousness. Dick Bissell always twists his hands and he was twisting them even more by the time this was over and for the following week. Allen Dulles for the next six or seven days looked like living death. He had the gout and had trouble walking and he was always putting his head in his hands. It even had a great physical effect on Jack. I noticed him particularly a week from the following Saturday when we had a talk for a half hour before the meeting began and even during the meeting he kept shaking his head, rubbing his hands over his eyes. [Robert Kennedy later told Martin, "We'd been through a lot of things together, and he was more upset this time than he was any other."]

He felt very strongly that the Cuba operation had materially affected the standing — his standing as President and the standing of the United States in public opinion throughout the world. We were going to have a much harder role in providing leadership. The United States couldn't be trusted. The United States had blundered. During the midst of this conversation about the difficulty in re-establishing a position and the fact that he felt that he had been badly undermined, Mrs. Lincoln [his secretary] walked in, heard him say this, and said, "Oh, no, everybody thinks you are doing wonderfully."[2]

II

During the long gray hours of the Bay of Pigs, John Kennedy decided he had better find out why, after all the expert assurance, the operation had ended as that rarity, an immaculate failure. He therefore summoned General Maxwell Taylor out of retirement to perform an autopsy. Robert Kennedy, Allen Dulles and Admiral Arleigh Burke joined Taylor on a board of inquiry. The Cuba Study Group met for six weeks in April and May and interrogated some fifty witnesses. The Attorney General told his associates to report during this period to Byron White, observing dryly: "It will be terrible if the Department improves while I'm gone."[3] May 1961 — the month of the Freedom Rides — was not an easy month in which to go. After spending the day on the Bay of Pigs, Kennedy generally worked in his office on racial justice and other matters till late at night.

The transcript shows that the Attorney General took the lead in examining the witnesses. His unsparing questions disclosed how pathetically ill considered the adventure was. His June 1 memorandum recorded some of the things he learned:

What comes out of this whole Cuban matter is that a good deal of thought has to go into whether you are going to accept the ideas, advice and even the facts that are presented by your subordinates. Before you can rely on an individual you have to know him pretty well.... Does he exaggerate? does he push his project too much because he is interested in it? does he therefore alter or inflate the facts that are involved because he wants to sell the idea? If you work with a person for a long period of time, you take that into consideration when you are considering his ideas. But here the President didn't have a chance to work with these people. He was taking them at face value and when they told him that this was guerrilla country, that chances of success were good, that there would be uprisings, that the people would support this project, he accepted it. And that is going to be the difference between the President before Cuba and after Cuba. . . .

The fact that we have gone through this experience in Cuba has made the President a different man regarding his advisers, and the Joint Chiefs of Staff are well aware of this. They realize far better now what their responsibilities and obligations are in this field. This evidently they did not realize before. Their study of the Cuba matter was disgraceful. The President was looking to them for their military evaluation and they really didn't give it the attention that was necessary and study that was essential and didn't analyze the facts. For instance they didn't make any study whether this country was proper guerrilla country. It was guerrilla country, but it was guerrilla country between 1890 and 1900. Now, with helicopters, it is no longer guerrilla country and there was no way these men fighting in here in this swampy area could possibly supply themselves. This should have been realized by the military.

The military had realized all too little. Their assessment of an earlier version of the invasion plan, Robert Kennedy scribbled in private notes on the Study Group meetings, showed "what really bad work the Joint Chiefs of Staff did on this whole matter. The plan as they approved it would have been even more catastrophic than the one that finally went into effect." The Chiefs were supposed to pass on military feasibility. "There is considerable evidence that initially at least they did not give the plan adequate study and attention and that at no time did they individually or as a body properly examine all of the military ramifications." There was "no explanation as to how C.E.F. [Cuban Expeditionary Force] could possibly hold this beachhead for a long period of time. How would they get all the ammunition ashore by daybreak. How would they keep their airport from being knocked out. How would they keep soldiers from coming through the swamp after them."[4] One reason, Admiral Burke explained to his colleagues on the board, was that the military review had to be done "on a personal study basis by the Chiefs themselves. The emphasis on utmost secrecy did not permit the usual staffing."[5] This plea that they should be excused because they had had to do the work themselves no doubt struck the Attorney General as further proof of their incompetence.

The board of inquiry was oddly constituted. Dulles and Burke were in effect investigating themselves. Their real function was to preserve the CIA and JCS from feelings of persecution.* The Attorney General was guarding the presidential interest. Taylor, as chairman and draftsman, concentrated on securing a unanimous report. "I was far more critical . . . of the military," Robert Kennedy said later, "than perhaps Maxwell Taylor wanted to be."⁶ For his part, Taylor, caught in a crossfire, had, he said later, "more difficulty with the Chiefs than with the others."⁷

The Eisenhower administration, the report concluded, ought to have recognized by November 1960 the impossibility of keeping the American hand secret and should thereupon have assigned the project to Defense as an overt American operation. "Failing such a reorientation, the project should have been abandoned." If the national interest required that it go ahead, "restrictions designed to protect its covert character should have been accepted only if they did not impair the chance of success."⁸ Taylor did not think, however, that President Kennedy's much criticized cancellation of a D-day air strike "materially" affected the outcome.⁹ Actually Castro had dispersed his planes and, he told John Nolan of the Justice Department in 1963, had moved fifty antiaircraft guns to the area shortly after the invasion began. As Nolan reported: "It was for this reason that he said that air cover for the Brigade would not have been decisive; his antiaircraft was so strong that it would have been able to knock out even much stronger air support."† Even if, against these precautions, the strike had been tactically successful, it would still have left 1200 CIA Cubans on the beachhead, surrounded by an avid enemy force two hundred times its size.

The board of inquiry gave Robert Kennedy a new friend. Maxwell Taylor was fifty-nine years old, articulate, dashing, urbane, both a consummate Army politician and an authentic war hero. "I was terribly impressed with him," the Attorney General said later, "— his intellectual ability, his judgment, his ideas."¹⁰ For his part, Taylor was greatly taken by young Kennedy; he told him — it was Taylor's highest praise — that the Attorney General could have made his old 101st Airborne Division; "you're the kind of guy we wanted around to take a hill or hold a trench."¹¹ He was impressed by Kennedy as an interrogator, "always on the lookout for a 'snow job,' impatient

* According to the egregious Howard Hunt, Colonel J. C. King of the CIA was the board's "general factotum," and Hunt himself was assigned to Dulles's office to provide "documentary and other answers to questions — or rather charges — made by the . . . inquisitors" (E. Howard Hunt, *Give Us This Day* [New York: Popular Library reprint, 1973], 213).

† Nolan was in Cuba with James B. Donovan to negotiate the exchange of Bay of Pigs prisoners (Nolan to RFK, n.d. [April 1963], RFK Papers).

with evasion and imprecision, and relentless in his determination to get at the truth." He was impressed too by Kennedy's attitude toward his brother — "a reversal of the normal fraternal relationship of a big brother looking after a younger one. In this case, Bobby, the younger brother, seemed to take a protective view of the President."[12] And he welcomed Robert Kennedy's instinct for organization — something Taylor found conspicuously wanting in the President. "You'll have to remember my brother doesn't think the way you do," the Attorney General would say. "He thinks about issues and people," not about structure.[13] A warm friendship, solidified on the tennis court, survived subsequent policy disagreements and lasted to the end of Robert Kennedy's life.

III

The collapse of the 'missile gap' had given the Kennedys an early skepticism about the Joint Chiefs. The President would sometimes look up and down the cabinet table during military discussions and ask sardonically, "Who ever believed in the missile gap?" (Only Maxwell Taylor would raise his hand.)[14] After the Bay of Pigs, the Kennedys began to question the Chiefs' professional competence. They also resented their public relations tactics. The new President was not one for men in uniform with pointers reading aloud sentences off flip charts he could read much faster for himself. In the spring of 1961 Kennedy received the Net Evaluation, an annual doomsday briefing analyzing the chances of nuclear war. An Air Force general presented it, said Roswell Gilpatric, the deputy secretary of defense, "as though it were for a kindergarten class. . . . Finally Kennedy got up and walked right out in the middle of it, and that was the end of it. And we never had another one."[15] Nor was the President any fonder of the public displays so dear to the armed forces. "John Kennedy," wrote Jerry Bruno — who, as advance man, had to deal with these matters — "hated military pageants — with a passion. And his brother Robert hated them even more."[16] In 1962 the battle of Ole Miss completed their disenchantment. "The Army had botched it up," Robert Kennedy said, adding, "but we didn't have an exercise with the Army in which they didn't screw it up."[17]

Of the inherited Joint Chiefs, they liked best David Shoup, the commandant of the Marine Corps, who believed that the job of the Marines was "to teach fighting, but not hate."[18] They liked least Curtis LeMay of the Air Force. Every time the President had to see LeMay, said Gilpatric, "he ended up in a sort of fit. I mean he would just be frantic at the end of a session with LeMay because, you know,

LeMay couldn't listen or wouldn't take in, and he would make what Kennedy considered . . . outrageous proposals that bore no relation to the state of affairs in the 1960s." But LeMay's popularity in the ranks and on the Hill gave him immunity. "We would have had a major revolt on our hands," said Gilpatric, "if we hadn't promoted LeMay."[19]

Except for Shoup, the military leaders were Cold War zealots. They had sedulously cultivated relations with powerful conservative legislators — John Stennis, Mendel Rivers, Strom Thurmond, Barry Goldwater. They hunted and fished with right-wing politicians, supplied them aircraft for trips home and showed up at their receptions. The alliance between the military and the right disturbed the Kennedys. This was why the President backed McNamara so vigorously in the effort to stop warmongering speeches by generals and admirals.

Direct military excursions into politics were still more disturbing. Radicals of the right were conducting an impassioned crusade not only against negotiation with the Soviet Union but also against social measures at home (the fluoridation of water was deemed a particular threat) and against the Warren Court. The John Birch Society was flourishing. The Minutemen were training their members in guerrilla warfare. Major General Edwin A. Walker, the right-wing demagogue who had made a menacing appearance at Ole Miss during the Meredith affair, was in a sense the heir of Major General George Van Horn Moseley, who had gone, in the thirties, from high Army commands to the domestic fascist movement. When the popular thriller *Seven Days in May* depicted a Pentagon attempt to take over the government, the President remarked, "It's possible. It could happen in this country, but the conditions would have to be just right. If, for example, the country had a young President, and he had a Bay of Pigs, there would be a certain uneasiness." If there were a second Bay of Pigs, the military would begin to feel it their patriotic obligation to preserve the nation. "Then, if there were a third Bay of Pigs, it could happen. . . . But it won't happen on my watch."[20] Still, he took no chances and, to the dismay of the Pentagon, encouraged John Frankenheimer to film the novel as a warning to the republic.*

In November 1961 the President assailed the extremists in a series

* Frankenheimer: "Those were the days of General Walker and so on. . . . President Kennedy wanted *Seven Days in May* made. Pierre Salinger conveyed this to us. The Pentagon didn't want it done. Kennedy said that when we wanted to shoot at the White House he would conveniently go to Hyannis Port that weekend" (Charles Higham and Joel Greenberg, *The Celluloid Muse: Hollywood Directors Speak* [New York: New American Library, Signet reprint, 1972], 92).

of speeches on the west coast. "They call for 'a man on horseback,'" Kennedy said. ". . . They find treason in our churches, in our highest court, in our treatment of water. . . . They object quite rightly to politics intruding on the military — but they are very anxious for the military to engage in their kind of politics."[21] On the President's return, Walter Reuther, a cherished target of the right himself, congratulated Kennedy on the speeches. Kennedy asked Reuther what he thought the federal government could do, within the First Amendment, to curb right-wing extremism. Reuther, who never liked to be at a loss for a program, said he would make some recommendations. Kennedy said, "Fine. Give them to Bobby. He's greatly concerned, and I've asked him to work on it."[22]

Reuther handed the assignment to his brother Victor and Joseph Rauh. In December 1961 they gave the Attorney General a memorandum addressed particularly to the problem of the radical right and the armed services, and recommending, among other things, withdrawal of tax exemptions from right-wing foundations and reexamination of free radio and television time for radical-right commentators.[23] The memorandum, which received fairly wide distribution around the government, soon leaked, producing an aggrieved counterattack on the Attorney General.[24] For this and other reasons, the administration moved cautiously on the Reuther recommendations.

But the Kennedys continued to worry about the problem. The President tried to interest Senator John Pastore of Rhode Island, chairman of the Senate Communications Subcommittee, in doing something about right-wing harangues on the airwaves. Eventually, as Charles Daly of the White House congressional staff put it, "he abandoned the idea of pushing Pastore because they agreed it would look too much like a purge."[25] Instead, in October 1963, Kenneth O'Donnell asked Wayne Phillips, a former reporter now in the administration, to see whether the 'fairness doctrine' — the rule that a radio or television licensee must afford a reasonable opportunity for contrasting viewpoints on controversial issues — could be used to counter extremist commentary. Phillips moved to the Democratic National Committee and encouraged persons and organizations attacked on the air to demand reply time. The effort resulted in over five hundred replies in the next few months and doubtless had an inhibiting effect on right-wing broadcasts.[26]

In the meantime, the President counted on Robert McNamara to contain the military. Robert Kennedy shared his brother's admiration of and affection for the Secretary of Defense. He became another new friend, closer even than Maxwell Taylor. The Kennedys also thought highly of Gilpatric, Cyrus Vance and the other civilian officials. Still, as civilians, the McNamara people remained outsiders; so, after the Cuba board of inquiry finished its work, Kennedy kept Taylor on as military representative in the White House and a year later, when Lyman Lemnitzer's term expired, made him chairman of the Joint Chiefs of Staff. With Taylor reinforcing McNamara, Kennedy believed that the Joint Chiefs of Staff, in the absence of further Bays of Pigs, were under control.

IV

But the Joint Chiefs of Staff had only approved the Bay of Pigs. The CIA had invented it. "I made a mistake," the President said to me while the Cubans were still fighting on the beachhead, "in putting Bobby in the Justice Department. He is wasted there. Byron White could do that job perfectly well. Bobby should be in CIA. . . . It's a hell of a way to learn things, but I have learned one thing from this business — that is, that we will have to deal with CIA. McNamara has dealt with Defense; Rusk had done a lot with State; but no one has dealt with CIA."[27]

Dealing with CIA was more of a problem than any of us imagined. The Bay of Pigs raised acute questions about the Agency's responsiveness to higher authority. "We found out later," Robert Kennedy said in 1964, "that, despite the President's orders that no American forces would be used, the first two people who landed on the Bay of Pigs were Americans — CIA sent them in." Haynes Johnson's excellent book *The Bay of Pigs* reported CIA operatives telling the Cuban Brigade that, if the President tried to call off the invasion, they should 'arrest' the operatives and go ahead anyway — "virtually treason," Robert Kennedy thought.[28]

Dulles and Bissell, it was evident, had misled the White House about vital features of the plan — especially the uprising behind the lines that, they had given the White House to understand, would follow the landings. "It was believed," as Rusk told the board of inquiry, "that the uprising was utterly essential to success." So too Bundy: "Success in this operation was always understood to be dependent upon an internal Cuban reaction."* It turned out that

---

* In the May 4 meeting of the Cuba Study Group. Lemnitzer on May 18: "The ultimate effect centered upon the uprisings that would be generated throughout the island and the reinforcements which would be gravitating toward this particular beachhead. . . . I remember Dick Bissell, evaluating this for the President, indicated there was sabotage, bombings, and there were also various groups . . . asking or begging for arms." Robert Kennedy noted after the April 26 meeting: "Evidently no probability of uprisings written up and put in memo form. No formal statement of opinion was given or asked for (RFK Papers)."

Dulles and Bissell had no reason to expect an uprising. A CIA Special National Intelligence Estimate in December 1960 foresaw no development "likely to bring about a critical shift of popular opinion away from Castro" and predicted that any disaffection would be "offset by the growing effectiveness of the state's instrumentalities of control."* The White House had been misled too about Castro's army. The Attorney General, in his personal jottings, observed that we "never would have tried this operation if [we] knew that Cuba forces were as good as they were and would fight." And the CIA had given a totally false impression about the ease with which the invaders could slip away from the beachhead to become guerrillas in the back country.[29]

In addition to misinformation, one surmises that there were equally vital thoughts withheld. It is hard to believe that Dulles and Bissell assumed the operation would succeed within the limits imposed. They may well have assumed that, if it faltered, the President, having gone so far, would have no choice but to follow the road to the end and send in American troops. CIA operatives actually assured the Brigade leaders American military assistance would be there if they needed it.[30] Howard Hunt, the original chief of political action for the project, had no doubt about this. When Kennedy publicly excluded the use of American troops before the invasion, "I did not take him seriously. The statement was, we thought, a superb effort in misdirection."[31]

Better "an aggressor than a bum" may have been Kennedy's first reaction. But, despite great pressure, he held the line. The CIA planners did not reckon with his inherent prudence and his ability to refuse escalation. Reflecting, that summer at Hyannis Port, on those who had wanted to go in with full force, Kennedy said, "We're not going to plunge into an irresponsible action just because a fanatical fringe . . . puts so-called national pride above national reason. Do you think I'm going to carry on my conscience the responsibility for the wanton maiming and killing of children like our children we saw here this evening?"[32] In 1965, Dulles reminisced with Tom Wicker of the *New York Times* about the Bay of Pigs. Kennedy, he said, obviously had not liked the idea. In order to get him to go along, Dulles had to suggest repeatedly, without ever saying so explicitly, that, if Kennedy canceled the project, he would appear less zealous than Eisenhower against communism in the hemisphere.

* CIA Special National Intelligence Estimate, December 8, 1960. Richard Goodwin asked Bissell for CIA's evidence on the extent of disaffection in Cuba. Bissell said to General Cabell, Dulles's deputy, "Don't we have an NIE on it?" Cabell: "Yes." The NIE, for obvious reasons, was not shown to the White House (Goodwin, in interview by author, June 11, 1977).

"This was a mistake," Dulles said. "I should have realized that, if he had no enthusiasm about the idea in the first place, he would drop it at the first opportunity rather than do the things necessary to make it succeed."[33]

The second withheld thought was a plan to assassinate Castro. That plan made no sense in isolation. The elimination of Castro in a vacuum would only have led to a more violent Cuban regime under Fidel's radical younger brother Raul or under the Argentine revolutionist Che Guevara. In the context of invasion, however, the murder of Castro might produce enough shock and disorganization to put the regime in jeopardy. As early as December 1959, Colonel J. C. King, chief of CIA's Western Hemisphere Division, recommended that "thorough consideration be given to the elimination of Fidel Castro" in order to "accelerate the fall" of his government.* Dulles and Bissell approved King's inquiry.[34] In May 1960, Howard Hunt, offering his recommendations to Bissell, listed first: "Assassinate Castro *before* or coincident with the invasion. . . . Without Castro to inspire them the Rebel Army and *milicia* would collapse in leaderless confusion."[35] Though Dulles and Bissell said not a word about it at the Bay of Pigs meetings or thereafter to the board of inquiry, the CIA had pursued, if with spectacular ineptitude, its plan to murder Castro before or during the invasion.

v

Intelligence agencies, sealed off by walls of secrecy from the rest of the community, tend to form societies of their own. Prolonged immersion in the self-contained, self-justifying, ultimately hallucinatory world of clandestinity and deception erodes the reality principle. So intelligence operatives, in the CIA as well as the FBI, had begun to see themselves as the appointed guardians of the Republic, infinitely more devoted and knowledgeable than transient elected officials, morally authorized to do on their own whatever they believed the nation's security demanded. Let others interfere at their peril. J. D. Esterline, the CIA's supervisor of planning for the Bay of Pigs, bitterly told the board of inquiry, "As long as decisions by professionals can be set aside by people who know not whereof they speak, you won't succeed."[36]

The CIA had struck out on its own years before. Congress had liberated it from normal budgetary restraints in 1949. Through most

---

* Actually the American ambassador to Cuba, Arthur Gardner, had proposed to Batista in 1957 that the CIA or FBI murder Castro. Batista said, "No, no, we couldn't do that: we're Cubans" (as told by Gardner to Hugh Thomas, *Cuba or the Pursuit of Freedom* [London, 1971], 947).

of the 1950s the fact that the Secretary of State and the director of Central Intelligence were brothers gave the Agency unusual freedom. "A word from one to the other," wrote Howard Hunt, "substituted for weeks of inter- and intra-agency debate."[37] Eisenhower, reluctant to commit conventional armed force, used the CIA as the routine instrument of American intervention abroad. Covert-action operators, working on relatively small budgets, helped overthrow governments deemed pro-Communist in Iran (1953) and Guatemala (1954), failed to do so in Indonesia (1958), helped install supposedly prowestern governments in Egypt (1954) and Laos (1959) and planned the overthrow and murder of Castro in 1960.

Congress and the press looked on these activities, insofar as they knew about them, with complacency. Only one group had grave misgivings and informed criticism; expressed, however, in the deepest secrecy. This, improbably, was the President's Board of Consultants on Foreign Intelligence Activities, created by Eisenhower in 1956 and composed of unimpeachably respectable private citizens.

Almost at once the board had appointed a panel, led by Robert Lovett and David Bruce, to take a look at CIA's covert operations. "Bruce was very much disturbed," Lovett told the Cuba board of inquiry in 1961. "He approached it from the standpoint of 'what right have we to go barging around into other countries buying newspapers and handing money to opposition parties or supporting a candidate for this, that or the other office?' He felt this was an outrageous interference with friendly countries.... He got me alarmed, so instead of completing the report in thirty days we took two months or more."[38]

The 1956 report, written in Bruce's spirited style, condemned

> the increased mingling in the internal affairs of other nations of bright, highly graded young men who must be doing something all the time to justify their reason for being.... Busy, moneyed and privileged, [the CIA] likes its "King Making" responsibility (the intrigue is fascinating — considerable self-satisfaction, sometimes with applause, derives from "successes" — no charge is made for "failures" — and the whole business is very much simpler than collecting covert intelligence on the USSR through the usual CIA methods!).

Bruce and Lovett could discover no reliable system of control. "There are always, of course, on record the twin, well-born purposes of 'frustrating the Soviets' and keeping others 'pro-western' oriented. Under these almost any [covert] action can be and is being justified.... Once having been conceived, the final approval given to any project (at informal lunch meetings of the OCB [Operations Coordinating Board] inner group) can, at best, be described as pro forma." One consequence was that "no one, other than those in the

CIA immediately concerned with their day to day operation, has any detailed knowledge of what is going on." With "a horde of CIA representatives" swarming around the planet, CIA covert action was exerting "significant, almost unilateral influences . . . on the actual formulation of our foreign policies . . . sometimes completely unknown" to the local American ambassador. "We are sure," the report added, "that the supporters of the 1948 decision to launch this government on a positive [covert] program could not possibly have foreseen the ramifications of the operations which have resulted from it." Bruce and Lovett concluded with an exasperated plea:

> Should not someone, somewhere in an authoritative position in our government, on a continuing basis, be . . . calculating . . . the long-range wisdom of activities which have entailed our virtual abandonment of the international "golden rule," and which, if successful to the degree claimed for them, are responsible in a great measure for stirring up the turmoil and raising the doubts about us that exist in many countries of the world today? . . . Where will we be tomorrow? [39]

In December 1956 the full board passed on to Eisenhower its concern about "the extremely informal and somewhat exclusive methods" used in the handling of clandestine projects.[40] (Among those signing this statement was another board member, Joseph P. Kennedy. "I know that outfit," the ambassador said after the Bay of Pigs, "and I wouldn't pay them a hundred bucks a week. It's a lucky thing they were found out early.")[41] In February 1957 the board pointed out to the White House that clandestine operations absorbed more than 80 percent of the CIA budget and that few of the projects received the formal approval of the so-called 5412 Special Group, the National Security Council's review mechanism. The CIA's Directorate of Plans (i.e., covert action), the board said, "is operating for the most part on an autonomous and free-wheeling basis in highly critical areas." All too often the State Department knew "little or nothing" of what the CIA was doing. "In some quarters this leads to situations which are almost unbelievable because the operations being carried out by the Deputy Director of Plans are sometimes in direct conflict with the normal operations being carried out by the Department of State."[42]

When this happened, State did not always prevail. Indeed, State was often itself a CIA target. In 1957 the CIA station in Indonesia decided that the United States should back a military revolt against Sukarno. "We began to feed the State and Defense departments intelligence," one CIA man said later. ". . . When they read enough alarming reports, we planned to spring the suggestion that we should support the colonels." The spooks moved on many fronts, even fabricating a Sukarno mask for an actor to wear in a pornographic film to be ground out in the blue movie mills of Los Angeles and used thereafter to discredit the Indonesian leader. John Allison, the ambassador in Djakarta, opposed the idea of overthrowing the regime. "We handled this problem by getting Allen Dulles to have his brother relieve Allison of his post within a year of his arrival in Indonesia."[43]

The Indonesian adventure — a far more exact model for the Bay of Pigs than Guatemala — was a total failure. The colonels rebelled in February 1958 after Allison's departure. Though Allen Dulles had personally promised to keep the new ambassador, Howard P. Jones, apprised of CIA activity in Indonesia, Jones was not informed that CIA was giving military assistance to the rebels.[44] Eisenhower, fully aware of the CIA role, piously described American policy as one of "not to be taking sides where it is none of our business." As the rebellion collapsed in May, a CIA pilot, Allen L. Pope, was shot down and captured.[45] "There was no proper estimate of the situation," the board told Eisenhower in a meeting in December 1958, "nor proper prior planning on the part of anyone, and in its active phases the operation was directed, not by the Director of Central Intelligence, but personally by the Secretary of State, who, ten thousand miles away from the scene of operation, undertook to make practically all decisions down to and including even the tactical military decisions." The chief result, the board said, had been to strengthen the Communists in Indonesia. There were "no present provisions," it added, "for any regular external review of Clandestine Cold War programs and no formal accounting of them." It begged Eisenhower once more to reconsider these "programs which find us involved covertly in the internal affairs of practically every country to which we have access."[46]

The board pressed its campaign in 1959 and 1960. Allen Dulles made minor organizational changes. In 1959 the 5412 Special Group began for the first time to meet regularly.[47] The board was not satisfied, then or later. When Dulles, Bissell and J. D. Esterline briefed the board late in 1960 on the Cuban project, its members, Lovett particularly, registered dismay, especially over the manner in which the planning was being administered.[48] In its last written report to Eisenhower, in January 1961, the board said grimly: "We have been unable to conclude that, on balance, all of the covert action programs undertaken by CIA up to this time have been worth the risk or the great expenditure of manpower, money and other resources involved. In addition, we believe that CIA's concentration on political, psychological and related covert action activities have tended to detract substantially from the execution of its primary intelligence-gathering mission. We suggest, accordingly, that there should be a total reassessment of our covert action policies."[49] "I have never felt," Lovett told the Cuba board of inquiry, "that the Congress of the United States ever intended to give the United States Intelligence Agency authority to conduct operations all over the earth."[50]

The Board of Consultants had no visible impact. Allen Dulles ignored its recommendations. Eisenhower gave it no support. But its testimony demolishes the myth that the CIA was a punctilious and docile organization, acting only in response to express instruction from higher authority. Like the FBI, it was a runaway agency, in this case endowed with men professionally trained in deception, a wide choice of weapons, reckless purposes, a global charter, maximum funds and minimum accountability.

VI

The Bay of Pigs dramatically verified the board's forebodings. The President called in Lovett. Lovett told him that CIA was badly organized, dangerously amateurish and excessively costly. It ought to be run, he said, in a hard-boiled fashion. This, he said, had not been possible with Dulles as director and Eisenhower as President.[51]

Determined to deal with CIA, Kennedy began by reviving the Board of Consultants. He called it the President's Foreign Intelligence Advisory Board and named James R. Killian, Jr., of MIT as chairman and Clark Clifford as a member. "I've made a terrible mistake," Kennedy told Clifford. A second Bay of Pigs would destroy his administration. "I have to have the best possible intelligence," he said. ". . . This has come early. I would hope that I could live it down, but it's going to be difficult."[52] Between May and November 1961, the Foreign Intelligence Advisory Board met twenty-five times — more than in its previous five years of existence.[53] Because the board enjoyed greater presidential support under Kennedy than before (or since), most of its recommendations were adopted. Clifford became chairman in 1963. But the board was an imperfect watchdog. It was not organized to find out things an agency adept at concealment wished to conceal. "I was not conscious," Clifford said later, "of any activity in the Kennedy administration in which the CIA was being used improperly."[54]

As for CIA itself, Kennedy began by redefining its mandate. In National Security Action memoranda 55 and 57 he transferred paramilitary operations from CIA to Defense and sought to restrict the size of the covert operations remaining to CIA. Obviously the Agency had to have new leadership. "Allen Dulles handled himself awfully well, with a great deal of dignity," Robert Kennedy said of the period after the Bay of Pigs, "and never tried to shift the blame. The President was very fond of him, as I was." Still, all recognized he had to go. "The President spoke to me about becoming head of CIA, and I said I didn't want to become head of the CIA and I thought it was a bad idea to be head of CIA, in addition, because I was a Democrat, and brother."[55] In time they settled upon John McCone, neither a brother nor a Democrat but a properly hard-boiled Republican industrialist with government experience. Bissell was replaced as chief of the clandestine services by Richard Helms, an OSS-CIA veteran whose background lay in secret intelligence rather than in dirty tricks. The 5412 Special Group was reconstituted with new authority over CIA covert action and with Maxwell Taylor as chairman.

In the months after the Bay of Pigs Kennedy thought he had put together a foolproof combination to keep CIA under control: tough McCone, the hard-driving executive, as director; experienced Helms, the perfect professional, as deputy director for plans; tough Taylor as chairman of the new Special Group; tough Clifford on the Foreign Intelligence Advisory Board. To complete the chain the President quietly gave Robert Kennedy an informal watching brief over the intelligence community.

VII

Taylor wanted to go much farther. His Cuba report concluded with a proposal for an extraordinary reorganization of the entire government in order to fight the Cold War. "We are in a life and death struggle," the report said, asking the President to consider the proclamation of a national emergency, to review treaties restraining "the full use of our resources in the Cold War" and to reexamine the adequacy of his "emergency powers." Cold War strategy, Taylor recommended, should be decided by an interdepartmental strategic resources group in charge of all situations around the planet certified as "critical" by a Cold War indications center.[56]

Kennedy rejected nearly all of this. He had uses for Taylor but not for his organizational fantasies. Rusk understandably saw the scheme as Taylor's takeover bid for control of foreign policy. Robert Kennedy, filled with inchoate urgencies after the Bay of Pigs and also beguiled by Taylor, had endorsed the grandiose plan but does not appear to have objected to its disregard thereafter. The President did, however, pluck from the exercise not only Taylor himself, whom he saw as a handy instrument for managing a benighted JCS, but also fragments of Taylor's bureaucratic dream. The Cold War indications center thus emerged briefly as the Crisis Center in the State Department (where Robert Kennedy installed his brother-in-law Stephen Smith, who intelligently concluded it was a waste of time and soon departed).

One aspect of the Taylor approach — the idea of concerted action in critical countries — appealed to a President who had believed ever since his visit to Vietnam in 1951 that the characteristic mode of Communist military pressure was not direct confrontation but indirect aggression and especially guerrilla warfare. Subsequent guerrilla successes in Cuba and Laos, accompanied now by rising guerrilla threats in Latin America and Africa, provided the background from which the famous Khrushchev speech derived its plausibility and menace. With nuclear war inconceivable and limited war improbable, guerrilla war had evidently become the chosen Soviet weapon. If Kennedy was deceived about the efficacy of this threat, so no doubt was Khrushchev. In any event Kennedy in 1961 saw counterguerrilla action as the way to plug the great gaping hole in the fabric of peace.

The impetus behind counterinsurgency did not come from the military.[57] Taylor had advocated conventional force for limited war, not special forces for guerrilla war, though he was delighted to add counterinsurgency to the menu of 'options' available in the name of flexible response. The impetus came rather from Kennedy himself and from social scientists he brought to Washington, notably the economic historian Walt Rostow (in the White House in 1961 and thereafter in the State Department) and the political scientist Roger Hilsman (director of the Bureau of Intelligence and Research in State; a veteran of guerrilla fighting in Burma during the Second World War).

The theorists did not see counterinsurgency as any permanent solution to the turmoil of the Third World. In a speech to the Special Forces at Fort Bragg in June 1961, Rostow argued that communism could best be understood "as a disease of the transition to modernization." Guerrilla warfare was the means by which "scavengers of the modernization process" sought to exploit emergent revolutionary aspirations.[58] Counterguerrilla action was the shield that would protect the development process from Communist disruption until the nation gained the social strength to control the situation itself. If counterinsurgency meanwhile persuaded Moscow to drop the weapon of national-liberation wars, the world could look forward to an era of tranquility.

At his first National Security Council meeting, on February 1, 1961, the President directed McNamara to instruct the Pentagon to place more emphasis on counterinsurgency.[59] In March, Kennedy told Congress that guerrilla warfare had been since 1945 "the most active and constant threat to Free World security."[60] Following a study by a committee under Bissell of the CIA, Kennedy created in January 1962 a new NSC committee designated Special Group (Counterinsurgency). Where the Special Group was intended to control CIA covert action, Special Group (CI) was supposed to coordinate overt American efforts to help foreign governments threatened by guerrilla insurrection. Maxwell Taylor was named chairman. Robert Kennedy was an ardent member.

The model of successful counterinsurgency lay at hand in Ramón Magsaysay's victory over the Hukbalahaps in the Philippines in the early 1950s. "All out friendship or all out force" had been Magsaysay's theory: military action combined with land reform, honest elections and generous amnesty, the counterguerrilla forces acting as "brothers of the people."[61] And the Magsaysay formula had an indefatigable American proponent in Brigadier General Edward G. Lansdale, already legendary in 1961 at the age of fifty-three. After working with Magsaysay, the attractive and fervent Lansdale had moved on to Saigon as CIA station chief. Graham Greene portrayed him in *The Quiet American* (1956) as Alden Pyle, the CIA agent who, "with his gangly legs and his crew cut and his wide campus gaze . . . seemed incapable of harm." Pyle's dream was to establish a Third Force in Vietnam, opposed equally to the colonialist French and the Communist Viet Minh. The result, as rendered by Greene, showed the horrible consequences of ignorant good intentions. "He comes blundering in, and people have to die for his mistakes. . . . God save us always from the innocent and the good."[*] (Asked about *The Quiet American* in later years, Lansdale said innocently, "Well, Graham Greene calls his books entertainments. I found it very entertaining.")[62] His next literary incarnation was in William J. Lederer's and Eugene Burdick's *The Ugly American* (1958) as Colonel Edwin B. Hillendale, the officer recommended by Magsaysay to the American ambassador in Sarkhan (Vietnam) because of his ability "to go off into the countryside and show the idea of America to the

---

[*] Graham Greene, *The Quiet American* (New York: Bantam reprint, 1957), 10, 12, 186. The plot turns on Pyle's support of General Trinh Minh Thé, the dissident chief of staff who was fighting both the French and the Viet Minh. The Greene character tells Pyle: "General Thé's only a bandit with a few thousand men; he's not a national democracy. . . . You can't trust men like Thé. They aren't going to save the East from communism" (151). For Lansdale's account of his relations with Thé, see his *In the Midst of Wars* (New York, 1972), esp. 189 ff., 308 ff.

people." "Every person and every nation," opined Colonel Hillendale, "has a key which will open their hearts. If you use the right key, you can maneuver any person or any nation any way you want." *

There was something to both portraits. "I took my American beliefs with me into these Asian struggles," Lansdale wrote later, "as Tom Paine would have done." He combined confident ideological naiveté and a disregard of consequences with genuine affection for alien lands and a certain political realism. As a student of Sun Tzu and Mao Tse-tung, he seized upon a basic point about "people's war" — hat any struggle between a government and Communist insurgency is "over which side will have the allegiance of the people." This meant that "social injustice, bullying by military or police, and corruption must be seen as grave weaknesses in the defense of a country." Washington policy makers, he believed, were too infatuated with "the brute usages of our physical and material means" to understand that government 'of the people, by the people, for the people' was "the strongest defense any country could have against the Communists." He had his own equivalent of the sayings of Chairman Mao: "The harder a Communist attack on a weak point in the social fabric, the more honest we must become in strengthening it. . . . The poorest view of an insurgency is from an office desk." [63] The lesson of the Philippines, he wrote, "was that there must be a heartfelt *cause* to which the legitimate government is pledged, a cause which makes a stronger appeal to the people than the Communist cause. . . . When the right cause is identified and used correctly, the *anti*-Communist fight becomes a *pro*-people fight." [64]

His trouble was that he identified the "right cause" in terms of American beliefs brought to Asian struggles. In the end, innocence was not enough. Still, his emphasis on the political character of people's war — on preempting the revolution and thereby taking the human base away from the guerrillas — was at the heart of the Kennedy theory of counterinsurgency.

VIII

The mission of Special Group (CI) was less the waging of guerrilla war than its prevention. It was essentially an ad hoc, trouble-shooting, expediting body, established, not to create policy, but to make a slow-moving bureaucracy act quickly. It spent the most time on Latin

---

* William J. Lederer and Eugene Burdick, *The Ugly American* (New York: Fawcett reprint, 1960), 153, 233. Hillendale was *not* the Ugly American (nor was that term used invidiously; it was applied to Homer Atkins, who was physically ugly but devoted himself to building dams and roads in backward countries).

America, then on Southeast Asia — Cambodia, Laos, Siam; not very much, however, on Vietnam.[65]

Counterinsurgency was not intended as a gospel of counterrevolution. "The United States does not wish to assume a stance against revolution, per se, as an historical means of change," the CI Group said in a 1962 policy statement. "The right of peoples to change their governments, economic systems and social structures by revolution is recognized in international law. Moreover, the use of force to overthrow certain types of government is not always contrary to U.S. interests. A change brought about through force by non-communist elements may be preferable . . . to a continuation of a situation where increasing discontent and repression interact."[66]

Nor was counterinsurgency conceived in military terms. Robert Kennedy defined it as a political strategy. "A military answer," he wrote,

> is the failure of counterinsurgency. . . . Insurgency aims not at the conquest of territory but at the allegiance of man. . . . That allegiance can be won only by positive programs: by land reform, by schools, by honest administration, by roads and clinics and labor unions and even-handed justice, and a share for all men in the decisions that shape their lives. *Counterinsurgency might best be described as social reform under pressure.*
>
> Any effort that disregards the base of social reform, and becomes preoccupied with gadgets and techniques and force, is doomed to failure and should not be supported by the United States.[67]

"It wasn't just a case of getting out and shooting guerrillas, by any manner or means," Taylor said later. "For the first time, I, at least, sensed the tremendous political and social aspects to this problem." Even the Joint Chiefs of Staff got religion, at least on paper, proclaiming improbably that the United States must "align itself, whenever possible, with the dynamic forces of social and economic progress."[68] The official doctrine placed great store on 'civic action.' The Special Forces were instructed in harvesting crops, digging wells, curing diseases, delivering babies and other benign undertakings.[69] The goal was 'winning the hearts and minds' of the people.

The CI Group was more effective, however, in other areas — especially in training Latin Americans to cope with insurgencies. Academies in Washington and Panama instructed police officers in riot control, a practice much criticized later. But, as John Bartlow Martin observed in the Dominican Republic, "the police, trained under Trujillo, knew nothing but to shoot to kill. Teaching them how to control riots with means short of deadly force seemed humanitarian."[70] Latins also learned wider arts of counterinsurgency.

Nor were they fighting phantoms. "Of course we engage in subversion, the training of guerrillas, propaganda! Why not?" Castro told Herbert Matthews of the *New York Times* in 1963. "This is exactly what you are doing to us."[71] Actually Castro had begun his covert action program against Latin America in the spring and summer of 1959, some months before the United States went into the business against him.[72]

By 1963 the progressive democratic regime led by the dauntless Romulo Betancourt in Venezuela was Castro's main target. Castro plainly saw Betancourt as a greater threat than, say, Perez Jimenez, the Venezuelan dictator in the fifties, would ever have been. "If the Communists have been so hostile to my regime," as Betancourt said in 1963, "it is . . . because we are carrying out the type of social action that strips the Communists of support and followers."[73] The protection of Venezuelan democracy became a primary American concern. Latin America, we thought in the early sixties, might be facing an ultimate choice between the Castro way and the Betancourt way. "You represent," John Kennedy told Betancourt when he visited the White House, "all that we admire in a political leader." Betancourt's fight for democracy, Kennedy continued, had made him "a symbol of what we wish for our own country and for our sister republics. And the same reasons have made you the great enemy of the Communists in this hemisphere. . . . You and your country have been marked number one in their efforts to eliminate you and what you stand for."[74]

The Havana radio beamed a steady barrage in support of the Communist-controlled insurgency. "When the people of Venezuela are victorious," cried Blas Roca, the veteran Cuban Communist, in January 1963, ". . . then all America will take fire."[75] In November 1963 the Cuban government staged a "Week of Solidarity with the Venezuelan Revolution" to cheer on the disruption of the presidential election scheduled for December. Later in the month the Venezuelans discovered four tons of arms shipped by Castro to a cache on their northwest coast.[76] Castro's campaign was a flop. Despite the threat that voters would be picked off by revolutionary snipers, 90 percent went to the polls. United States technical assistance to the Venezuelan police unquestionably helped the Betancourt government to put down the guerrillas, though no amount of police efficiency would have availed had not Betancourt's social policy won hearts and minds. For the first time in Venezuelan history one elected president peacefully succeeded another. The insurgency faded away.

Police training had its risks. There is no evidence that the American schools ever offered courses in torture. Still the instruction was much concerned with the psychology of interrogation; and too many of the students added torture when they got home. The Costa-Gavras film *State of Siege* was much exaggerated but had its point. There was also the danger that police training might be used for the repression of progressive democrats seeking constitutional change as well as of terrorists sent in from abroad. This was much on Robert Kennedy's mind. He kept stressing, Gilpatric recalled, the importance of distinguishing "subversion from outside the country" from "revolutionary situations developing within the country."[77]

IX

For a season in 1961–62 counterinsurgency was the rage in Washington. The President took a highly publicized personal interest, enlarging the Special Forces, sending to Fort Bragg for their literature on guerrilla war, putting them in green berets as a mark of distinction.[78] By June 1963 the Special Forces numbered nearly 12,000 men. Counterinsurgency training had been given to 114,000 American officers and nearly 7000 from foreign countries.[79] Foreign Service officers, even ambassadors, had to attend counterinsurgency seminars before assignment to the Third World.

The Counterinsurgency Group met regularly. Robert Kennedy was its goad. The activity and the era put him in an inordinately gung-ho mood. After the Bay of Pigs and the Berlin Wall, he did not feel his brother could afford another setback. "The great events of our nation's past," he had written in *The Enemy Within*, "were forged by men of toughness."[80] The guerrilla, brave and hardy, fighting in the shadows, evoked memories of a more heroic America — Francis Marion, the Swamp Fox; Mosby's Raiders; Merrill's Marauders. The Attorney General cross-examined those called before the CI Group as caustically as, in the same period — before the Committee on Equal Employment Opportunity — he cross-examined officials whose agencies had not hired enough blacks. "If the man knew what he was after," said Averell Harriman, who became chairman of the CI Group in 1963, "he got very sympathetic questioning from Bobby. If Bobby got the idea he did not know, . . . he treated him as if he was a witness on the stand. . . . It was not a very pleasant experience."[81] After a notably harsh session, a man from the State Department told a reporter, "If one of you guys writes one more time about his looking like a choirboy, I'll kill you. A choirboy is sweet, soft, cherubic. Take a look at that bony little face, those hard opaque eyes, and then listen to him bawl somebody out. Some choirboy!"[82] Such pressure was valuable when the question was minority employment. In foreign affairs, it could lead to ill-judged activism.

Fortunately, however, the new gospel had inherent limits. Despite the President's benediction and the Attorney General's zeal, counterinsurgency retained strong opponents within the government. The Army was institutionally hostile to freewheeling officers like Lansdale and irregular formations like the Green Berets. Guerrilla action was simply beyond the conception of an opulent military establishment dedicated to high-technology war. The State Department, except for Harriman, who thought counterinsurgency had its uses in Latin America, was even less enthusiastic. "The amount of effort and theology with which that whole business was invested," George Ball said later, "was totally incommensurate with anything we ever got out of it. I think it led us into messing into a lot of situations where we would have done much better to have stayed out." [83]

Counterinsurgency faced still deeper problems. The theory the Kennedys had taken from Lansdale — winning the people by "social reform under pressure" — had a superficial plausibility. But it was not carefully thought through. Very few governments under guerrilla attack cared much about their own dispossessed. That was precisely why they were vulnerable. Magsaysay and Betancourt were the exception, not the rule. The regimes most severely challenged were those most suicidally determined to hold on to despotic power and privilege. Modernization and reform might well undermine whatever stability the regime had without eliciting the active loyalty of a long-suffering populace. And since the American ambassador was by definition debarred from using the ultimate sanction — withdrawal of American support — he lacked leverage to force apprehensive oligarchs to take measures they believed threatening to their own survival.

This was the fatal fallacy in the liberal theory of counterinsurgency.[84] With the United States so often obliged to work through repressive local leadership, the reform component dwindled into ineffectual exhortation. The professional military, American and native, lacked ability or desire to make social revolutions. Armies inevitably did what came naturally. "While the policy purports to cover the full spectrum of counter-insurgency," wrote Charles Maechling, Jr., the CI Group's staff director, "it in fact only fully addresses itself to the equipment and training aspects." [85] "It is simply much easier," Robert Komer of the White House reported to McGeorge Bundy, "to get people moving on active problems like Laos, Vietnam, or Thailand than to get them to focus on the preventive medicine (e.g. civic action or police programs) needed to forestall such problems arising elsewhere." [86]

In military hands, counterinsurgency shifted from winning the hearts and minds of the people to controlling their behavior, if necessary by brutal coercion. "When the going gets tough," as Knute Rockne had said, "the tough get going." "People tend to be motivated," wrote Morton H. Halperin, a counterinsurgency theorist of the new school, "not by abstract appeals, but rather by their perception of the course of action that is most likely to lead to their own personal security."[87] Or, as it was formulated by the Green Berets, "When you've got 'em by the balls, their hearts and minds will follow." The Special Forces themselves tended by a self-selecting process to attract, with notable exceptions, men whose taste was less for good works than for violence. The 'dirty dozen' complex compounded trouble. As Peter De Vries rewrote Rockne, "When the tough get going, the going gets tough."[88]

Despite the original intent, counterinsurgency moved irresistibly toward counterrevolution. Robert Kennedy watched this development with dismay. "He was, in all fairness to him," said Michael Forrestal, who sat in on the meetings, "as worried about dictatorships of the right as he was about dictatorships of the left."[89] To John Bartlow Martin, Kennedy spoke in 1964 about the "conflict" within the American government. The idea that "it doesn't matter what kind of system you have so long as it's anti-Communist . . . was against the philosophy of the President and against mine."[90] "As time went on," Maxwell Taylor recalled, "the business diminished and his interest also diminished."[91] Ralph Dungan of the White House wrote the Attorney General in January 1963 urging the CI Group's dissolution.[92] It was decided to let it linger, but its heyday had passed. As a weapon, counterinsurgency was turning cruelly in the hand of the user. In the end it was a ghastly illusion. Its primary consequence was to keep alive the American belief in the capacity and right to intervene in foreign lands. God save us always from the innocent and the good. The failure of counterinsurgency was another phase in the education of Robert Kennedy.

21

The Cuban Connection: I

THERE REMAINED, after the Bay of Pigs, the exasperations of Cuba itself: of an exultant Castro, more galling and dangerous than before; of an elated Kremlin with ever closer ties to Havana; of eager guerrillas dispatched by Cuba against democratic Latin governments; of right-wing clamor in the United States over that putatively terrible threat ninety miles from Key West; of brave Cuban exiles, sent to the beaches by American folly and now herded miserably into Castro's prisons.

I

The night the Bay of Pigs collapsed, Maxwell Taylor was struck by John Kennedy's already "deep remorse" about the Cubans captured by Castro.[1] "During the months after the invasion," Kenneth O'Donnell wrote later of the President, "he thought of those prisoners constantly. . . . Within the privacy of his office, he made no effort to hide the distress and the guilt that he felt." He had had trouble sleeping, he told O'Donnell one morning; "I was thinking about those poor guys in prison down in Cuba. I'm willing to make any kind of a deal with Castro to get them out."[2] As Robert Kennedy said in 1964: "The one thing that really hung over from the Bay of Pigs was the fact that 1150 or so prisoners remained in jail and that they were going to die and not only did we have the responsibility for carrying out the Bay of Pigs in an ineffective way but also cost people's lives. . . . So we wanted to do whatever was necessary, whatever we could, to get them out. I felt strongly about it. The President felt strongly about it."[3]

Consequently when, a month after the invasion, Castro offered to exchange the prisoners for an indemnity of five hundred bulldozers, the President asked Mrs. Roosevelt, the always reliable Walter Reuther and two Republicans — Milton Eisenhower, president of Johns Hopkins University, and George Romney, president of American Motors — to head a fund-raising committee. Eisenhower accepted, though with "an odd sense of foreboding."[4] Romney declined. Most Republican leaders promptly denounced the proposal. Barry Goldwater fulminated against "surrender to blackmail." "Human lives," said Richard Nixon in one of his communiqués on public morality, "are not something to be bartered."

There were other problems. From start to finish Castro spoke with rare consistency of an "indemnity" of five hundred bulldozers. The committee gagged at the implications of indemnity, preferring to talk of an "exchange." Castro gagged at that. Moreover, the machines Castro specified — D-8 Super Caterpillar bulldozers — seemed better suited for the construction of airfields and missile bases than for agricultural work. The committee offered farm tractors. Castro finally said he would settle for the dollar value of the bulldozers, which he estimated at $28 million. By now Republican criticism was reaching a crescendo. Eisenhower wrote bitterly to Kennedy and threatened to resign. The committee expired at the end of June 1961 in a burst of recriminatory statements to and from Castro. "Well, boys," Castro told a committee of the prisoners, "the Americans are not going to give you those tractors." Offering them drinks, he explained that Mrs. Roosevelt was a "*vieja chocha,*" a silly old lady, and Reuther, a "labor baron." In Key West, exiles formed the Cuban Families Committee for the Liberation of the Prisoners of War.

Nine months later, at the end of March 1962, the prisoners were brought to trial. Roberto San Román, whose brother Pepe was in the dock but who had himself escaped from the beachhead, flew from Florida to Washington. Brushed off by the State Department, he went to the Attorney General. "This man was completely different," San Román said later. "This was like talking to a Brigade man. He was very worried about the Brigade." Finally the Attorney General said, "All right, Roberto, . . . I give you my word we will do everything possible to keep them from being shot." He invited San Román to telephone "ten times a day" and sent him to see Richard Goodwin at the White House, who, among other things, got President João Goulart of Brazil to ask Castro to spare the prisoners.[5]

Robert Kennedy now proposed an exchange of $28 million worth of agricultural products for the prisoners. This was peculiarly his initiative. On the evening of April 5, I noted in my journal, "the President called me at 7 o'clock to find out what the Food-for-Prisoners deal is all about. This is an operation which had been approved

by the Special Group and is strongly backed by Bobby. . . . We had
all assumed that the President knew about it, but apparently no one
had told him. This morning Adlai Stevenson had mentioned it to
him, and JFK had not known what he had been talking about. The
President asked me to check into the matter and let him know how
far things had gone. I called Bobby, who seemed a little surprised
and chagrined that his brother did not know about it, and agreed to
call him and bring him up to date. JFK's mood over this lapse was
less one of indignation than of wistfulness."[6]

On April 8 the Revolutionary Tribunal gave the prisoners thirty
years at hard labor — or, it added, Washington, if it cared so much,
could buy their freedom by paying fines of $62 million. Two days
later, representatives of the Cuban Families Committee, led by Alvaro
Sanchez, Jr., arrived in Havana with pledges of $28 million in agri-
cultural products. Castro stuck to his new figure. To prove good
faith, however, he said he would release sixty sick and wounded
prisoners to the United States. For this first installment he expected
an interim payment of $2.9 million.[7]

The sixty men arrived in Miami exactly one year after they had
left Nicaragua for the Bay of Pigs. Sanchez called the Attorney
General. "I saw you on TV," Kennedy said. "I watched everything." Explaining that the government could not pay the ransom,
he advised the exiles to form a fund-raising committee. But the idea
of ransom drew a bad press. Most Americans wanted to forget the
Bay of Pigs. Public relations sages advised the Cubans to sign up an
eminent chairman for a sponsoring committee. Kennedy said to
forget about a chairman; "what you need is a man who knows how
to deal with Castro."[8] He recommended James B. Donovan, who
had been general counsel of OSS during the war and who, the
preceding February, after meticulous negotiation, had effected the
exchange of the Soviet spy Colonel Rudolf Abel for Francis Gary
Powers, the U-2 pilot.[9]

Donovan took the case without knowledge of the Attorney General's interest[10] (the two men had never met) and without fee. Reports from the prisons gave an alarming picture of hunger, sickness
and degradation. At the end of August, Donovan went to Cuba.
Castro asked about the $2.9 million due for the prisoners released in
April. Donovan assured him it would be taken care of. After discussion they agreed, in effect, that Castro could call the ransom
indemnification and in return would take it in medicine and food.
There were new complications. Soviet arms shipments to Cuba were
frightening some Americans. In September, Robert Kennedy noted
that it seemed "virtually impossible for us to send $60 million worth
of equipment to Cuba with part of our population and a number of

political leaders calling for an invasion of Cuba"; $25 million perhaps, but not $60 million.[11] Donovan became the Democratic senatorial candidate against Jacob Javits (Robert Kennedy thought he should retire from the race; other Democrats thought he should spend more time campaigning). In early October, though suffering from acute bursitis, Donovan flew again to see Castro. On October 20 he received Castro's shopping list. Two days later John Kennedy announced the discovery of nuclear missile bases in Cuba.

II

Redeeming the prisoners was only part of the unfinished business with Castro. There was also the regime itself, now by Castro's own statement a Marxist-Leninist outpost in the western hemisphere. On April 19, 1961, as resistance was ending on the beaches, Robert Kennedy added a prescient final sentence to a memorandum: "If we don't want Russia to set up missile bases in Cuba," he wrote the President, "we had better decide now what we are willing to do to stop it."

He identified three possible courses: (1) sending American troops into Cuba — which "you have rejected . . . for good and sufficient reasons (although this might have to be reconsidered)"; (2) placing a strict blockade around Cuba — which as an act of war "has the same inherent problems as Number (1)" and as "a drawn-out affair . . . would lead to a good deal of world-wide bitterness"; or (3) calling on the Organization of American States to prohibit the shipment to Cuba of arms from any outside source, including both the Soviet Union and the United States. "At the same time they would guarantee the territorial integrity of Cuba so that the Cuban government could not say they would be at the mercy of the United States." Perhaps the OAS might be persuaded to do this, he ruminated, "if it was reported that one or two of Castro's MIGS attacked Guantanamo and the United States made noises . . . that we might very well have to take armed action ourselves. . . . Maybe this is not the way to carry it out but something forceful and determined must be done. . . . The time has come for a showdown for in a year or two years the situation will be vastly worse."[12]

On the same day — D-day plus two — there was a meeting at the White House. Robert Kennedy recalled it in his memorandum of June 1, 1961:

> It was a very gloomy meeting. Lemnitzer, Burke, Rusk, Mac Bundy, Bissell, Allen Dulles, several others present. Everybody really seemed to fall apart. I said in rather strong terms that I thought we should pick ourselves up and figure what we were going to do that would be

best for the country and the President over the period of the next six
months or a year. . . . Jack and I then left the meeting and took a walk
and they worked for another three or four hours. I came back and
visited them around three o'clock. There was discussion about whether
we should send American troops into Cuba. Chip Bohlen was strongly
in favor of doing something like this. Dean Rusk advised against it.
There was other discussion about establishing a blockade. Ultimately,
it ended up in not doing a great deal.[13]

Chester Bowles also remembered this meeting. The "consensus,"
he said later, "was to get tough with Castro. Some of the people at
the meeting were very outspoken and took an attitude of 'He can't
do this to us. We've got to teach him a lesson.' If the President had
at that point . . . decided to send in troops or drop bombs or whatever,
I think he would have had the affirmative votes of at least 90 percent
of the people."[14] "We were hysterical about Castro at the time of
the Bay of Pigs and thereafter," Robert McNamara said later.[15]

Bowles stoutly opposed the prevailing militancy. His own thoughts
were deemed insufficiently militant. "Later on," Robert Kennedy
said in his June 1 memorandum:

a plan came in from a task force that was studying Cuba which was just
God awful. It was presented by Chester Bowles at the next National
Security meeting. I said that I thought it was a disgrace; and a new task
force was formed which came in with some suggestions and ideas which
made far more sense. . . .
     Chester Bowles hasn't contributed a great deal. He talks in such
general language. It is very difficult to pin him down as to specific
action. He started complaining after one of the meetings on D plus two
that he was against Cuba. I told him that he might have been against
it before but that he was for it from then on. However, ultimately it got
in *Time* magazine that he was complaining around town that he was
against the project and his position has been badly undermined with
the President. . . . The other great question is getting an Assistant Secretary of State for Latin America. I thought John Connally would do
the job. Not an expert in the field but at least he is a man of action. . . .
     I had a slight flare-up with Lyndon Johnson at one of these meet-
ings. At D plus two we started to go around the table to try to find out
who was in favor of the project. We had the impression he was just
trying to get off it himself. I had a talk with Dean Rusk after D plus
two and I believe with McNamara to make sure that everybody in their
Departments understood that the important matter was to stand firm
and to not try and pass the blame off to somebody else. This could lead
only to bitter recriminations and even more disaster. They all accepted
that and I think we had very little difficulty along those lines.[16]

I had gone to Europe shortly after the Bay of Pigs. When I
returned early in May, I attended meetings of the Cuban task force
on postdebacle policy. "There was complete agreement," I noted,

"against any thought of direct intervention. There was general opposition, the Pentagon and CIA dissenting, to the idea of enlisting a Cuban Freedom Brigade in the United States Army." At the National Security Council meeting the next day, McNamara expressed Department of Defense opposition to this idea too. "He also said that both the Army and the Navy rejected an Air Force plan for an assault against the Castro regime."[17] Ah, the Air Force!

"The Cuba matter is being allowed to slide," Robert Kennedy said on June 1. "Mostly because nobody really has the answer to Castro. Not many are really prepared to send American troops in there at the present time but maybe that is the answer. Only time will tell."[18] The Attorney General was unquestionably right on one point: nobody had the answer to Castro.

John Kennedy politely asked Richard Nixon, "What would you do now in Cuba?" "I would find a proper legal cover," Nixon replied (by his own account), "and I would go in."[19] In mid-June the Cuba Study Group chimed in. "There can be no long-term living with Castro as a neighbor," Maxwell Taylor, Robert Kennedy and their associates solemnly concluded. The group saw two possible policies: either to hope that time and internal discontent would eventually end the threat, "or to take active measures to force its removal. . . . Neither alternative is attractive. . . . While inclining personally to a positive course of action against Castro without delay, we recognize the danger of dealing with the Cuban problem outside the context of the world Cold War situation."*

Senator Mike Mansfield, who was regularly wiser on questions of foreign policy than most members of the National Security Council, offered the best answer. "If we yield to the temptation to give vent to our anger at our own failure," he wrote the President, "we will, ironically, strengthen Castro's position." Mansfield sensibly recommended "gradual disengagement of the U.S. government from anti-Castro revolutionary groups, . . . a taciturn resistance to the political blandishments or provocations from those at home who would urge that we act directly in Cuba, . . . a cessation of violent verbal attacks on Castro by officials of the government" and full steam ahead on the Alliance for Progress because without economic progress "*Castro-ism is likely to spread elsewhere in Latin America whether or not Castro remains in power in Cuba.*"[20] Robert Kennedy agreed over the long run. "If the Alliance for Progress goes into operation fully," he said

---

* Cuba Study Group, "Conclusions," Recommendation 6, June 13, 1961, Schlesinger Papers. The context makes it entirely clear that "Castro" was used throughout the document as shorthand for "the Castro regime." "Removal," for example, meant overthrow of the regime, not the assassination of Castro.

in 1963, "if reforms, social, economic and political, are put into effect, then Communism and Castroism will collapse in Latin America."*

III

But neither the Alliance for Progress nor the diplomatic and economic isolation of Cuba promised immediate results. With his brother under Communist harassment in Berlin and Southeast Asia and under Republican harassment in the United States, with Castro's operatives plotting against democratic regimes in Latin America, Robert Kennedy was determined to find quicker ways of striking back. In the interstices of his other battles of 1962 — with the segregationists and with the mob, with Jimmy Hoffa and with Edgar Hoover — he now began to expend much time and concern on the overthrow of the Castro regime. He vividly felt, as Maxwell Taylor said later, "that the United States had suffered a great humiliation in Cuba, as we had, and hoped it would be possible to reverse that in some way by utilizing the Cuban exile resources, assisting as we could on the American side."[21]

He turned to the CIA, which, as the Justice Department's Office of Legal Counsel extravagantly contended in January 1962, had through the Presidency "the constitutional power . . . *by any means necessary* to combat the measures taken by the Communist bloc" and to do so "without express statutory authorization."[22] The Attorney General had had a hand in John McCone's appointment, and, despite nearly a quarter century's difference in age, the two men became good friends — a friendship sadly consolidated when Ethel Kennedy provided daily solace during the fatal sickness of McCone's wife. McCone proved, as expected, an able executive and, as not expected by some of us, a detached analyst of intelligence, even when the intelligence might clash with his own robustly conservative views. Robert Kennedy used to kid him about his conservatism and try to nudge the Agency to the left. Thus he wrote McCone early in 1962 suggesting he "talk to Jack Conway, Arthur Goldberg and perhaps Walter Reuther. . . . You might pick up some good ideas from these people. . . . Arthur is a good judge on who is effective in the labor movement. Jack Conway is a bright articulate fellow, who feels very strongly but makes a good deal of sense."[23] (The Attorney General said to me more than once that Jack Conway was exactly the sort of man who might be able to make some sense out of CIA.)

With sure bureaucratic instinct, the Agency seized on the Cuban

---

* He added: "If effort is not made and reforms are not forthcoming, we will have problems in South America even if there is no Cuba or Castro" ("Robert Kennedy Speaks His Mind," *U.S. News & World Report*, January 28, 1963).

problem as the way of making its comeback from the Bay of Pigs. "We wanted to earn our spurs with the President," as Richard Helms said.[24] But as usual it thought it alone knew how to do the job. The CIA wished to organize Castro's overthrow itself from *outside* Cuba, as against those in the White House, the Attorney General's office and State who wished to support an anti-Castro movement *inside* Cuba. The CIA's idea was to fight a war; the others hoped to promote a revolution. Any successful anti-Castro movement inside Cuba would have to draw on disenchanted Castroites and aim to rescue the revolution from the Communists. This approach, stigmatized as *Fidelismo sin Fidel*, was opposed by businessmen, both Cuban and American, who dreamed of the restoration of nationalized properties. But the CIA alternative was probably dictated less by business interests than by the Agency's preference for operations it could completely control — especially strong in this case because of the Cuban reputation for total inability to keep anything secret.

As I wrote in July 1961 to Richard Goodwin, the White House liaison for Cuba, the CIA's Cuban Covert Plan contemplated "a *CIA* underground formed on criteria of operational convenience rather than a *Cuban* underground formed on criteria of building political strength sufficient to overthrow Castro." The CIA specifications favored those Cubans "most willing to accept CIA identification and control" and discriminated against Cubans who insisted on running their own show. I had in mind the anti-Castro radical Manuel Ray, who was believed to have the most effective network on the island. "The practical effect," I concluded, "is to invest our resources in the people least capable of generating broad support within Cuba."[25]

These disagreements were temporarily papered over at the end of August 1961 when the Cuba Task Force — which included George Ball and some Latin Americanists from State, Goodwin from the White House and a CIA delegation led by Richard Bissell — came up with a formula. They all agreed, as Goodwin reported to the President, on a campaign directed "toward the destruction of targets important to the economy, e.g., refineries, plants using U.S. equipment, etc." This was the CIA plank; but Goodwin also won nominal CIA acceptance of "the principle that para-military activities ought to be carried out through Cuban revolutionary groups which have a potential for establishing an effective political opposition to Castro within Cuba."[26]

But the Goodwin plank was against the CIA's operational code. The Agency could not bring itself to trust those it could not control. In the early fall Bissell was told, perhaps in a meeting with the President and the Attorney General, to "get off your ass about Cuba."[27] After a meeting with Bissell, Esterline and other Agency people on October 23, I noted: "At bottom, there is a conflict between operational interests and diplomatic interests. CIA wants to subordinate everything else to tidy and manageable operations; hence it prefers compliant people, like [Joaquin] Sanjenis [head of Operation 40, a right-wing clandestine group funded by the CIA] to proud and independent people, like Miro [Cardona, the head of the Cuban Revolutionary Council and Castro's first prime minister]."* Hoping to bring the CIA into line, Goodwin proposed to the President in early November that Robert Kennedy "would be the most effective commander of the anti-Castro campaign."[28] At just this point, the famed General Lansdale returned from a trip to Saigon. Robert Kennedy, remembering Lansdale's doctrine of preempting the revolution and aware that the Army did not want him in Vietnam, thought that the savior of Magsaysay might have the answer to Castro.

On November 4, 1961, Cuba was the subject of a White House meeting. Present, according to Robert Kennedy's handwritten notes, were

> McNamara, Dick Bissell, Alexis Johnson [the deputy under secretary of state for political affairs], Paul Nitze, Ed Lansdale (the Ugly American). McN said he would make latter available for me — I assigned him to make survey of situation in Cuba — the problems & our assets.
> My idea is to stir things up on island with espionage, sabotage, general disorder, run & operated by Cubans themselves with every group but Batistaites & Communists. Do not know if we will be successful in overthrowing Castro but we have nothing to lose in my estimate.[29]

Lansdale made his survey. He recommended "a very different course" from the CIA "harassment" operations of the summer, conceived and led as they were by Americans. Instead, the United States, Lansdale argued, should seek out Cubans who had opposed Batista and then had become disillusioned with Castro. His theory was to work within Cuba, taking care not to "arouse premature actions, not to bring great reprisals on the people there." The objective was to depose Castro in the same way Castro had deposed Batista — to have "the people themselves overthrow the Castro regime rather than U.S. engineered efforts from outside Cuba."[30]

The President decided in favor of the Goodwin-Lansdale thesis and against the CIA. At the end of November he put out a top secret instruction "to use our available assets . . . to help Cuba overthrow

---

* Author's journal, October 23, 1961. The next day Miro told me, "I can fight against Castro — against the corruption of the past, against Prio [Socarras, a right-wing Cuban politician who was trying to launch a new exile movement] — against Batista — but, if I have to fight the CIA on top of all this, I declare myself vanquished" (ibid., October 24, 1961).

the Communist regime."[31] Lansdale was appointed chief of operations, reporting to a new review committee known as the Special Group (Augmented).* Operation Mongoose was born.

IV

Mongoose, like the baby in the old story, was attended at the cradle by good fairies with divergent wishes. The President's wish was that, as Taylor described it, "all actions should be kept in a low key." Since "anything big was going to be charged to the United States," Mongoose had to be kept small, functioning at what was known in intelligence circles as a low noise level.[32] The American hand was to be concealed. Nor did the President wish undue activity to jeopardize the lives of the Bay of Pigs prisoners.

The Attorney General had a separate wish, as rendered in CIA notes of a meeting at Justice with Mongoose planners in January 1962: that "no time, money, effort — or manpower . . . be spared." Mongoose was "top priority."[33] But he was never clear how the time, money, etc., were to be used.

Lansdale's wish was activity — a lot of it — leading to internal revolution. He had a multitude of ideas: nonlethal chemicals to incapacitate sugar workers; "gangster elements" to attack police officials; defections "from the top echelon of the Communist gang"; even (at least according to one witness before the Church committee; Lansdale later disclaimed the project) spreading word that Castro was anti-Christ and that the Second Coming was imminent — an event to be verified by star shells sent up from an American submarine off the Cuban coast ("elimination by illumination," a waspish critic called it). In February, Lansdale presented a six-phase plan designed to culminate the next October in an "open revolt and overthrow of the Communist regime."[34] All this was a little rich for the Special Group (Augmented), which directed him instead to make the collection of intelligence the "immediate priority objective of U.S. efforts in the coming months." While the group was willing to condone a little concurrent sabotage, the acts "must be inconspicuous," it told Lansdale, and on a scale "short of those reasonably calculated to inspire a revolt." It further insisted that all "sensitive" operations,

---

* The regular Special Group — Taylor, McGeorge Bundy, Alexis Johnson, Gilpatric, Lemnitzer and McCone — would meet at two o'clock every Thursday afternoon. When its business was finished, Robert Kennedy would arrive, and it would expand into the Special Group (CI). At the end of the day, Cuba would become the subject, and the group, with most of the same people, would metamorphose into Special Group (Augmented).

"sabotage, for example, will have to be presented in more detail on a case by case basis."[35]

The Augmented Group could prescribe policy to Lansdale. It was harder for Lansdale to prescribe operations, which remained in the hands of CIA. Task Force W, the CIA unit for Mongoose, soon had four hundred American employees in Washington and Miami, over fifty proprietary fronts, its own navy of fast boats, a rudimentary air force and two thousand Cuban agents. The Miami headquarters became for a season the largest CIA station in the world. All this cost over $50 million a year.[36] The CIA had its special wish for Mongoose too. Whereas the Special Group (Augmented) had accepted the presidential decision that "the one thing that was off limits was military invasion," the Agency persisted in seeing the objective as the creation of "internal dissension and resistance leading to eventual U.S. intervention" (October 1962).[37] The Agency, in short, was more bent than ever on fighting a war. It proved this by the men to whom it offered command of Task Force W.

Bissell's first choice, if Howard Hunt can be believed, was Hunt himself, an operative notorious in the Cuban community for his division of the exiles into (as one anti-Castro exile put it) the "good guys," who did his bidding, and the "bad guys," who "refused to be coerced into accepting his standard operating procedures."[38] Hunt, however, declined the job on the ground that "it was obvious there was no serious interest in overthrowing Castro, and I was reluctant to conduct operations for their own sake, to give the appearance of activity."[39]

The next CIA candidate was no more amenable to the Lansdale-Goodwin political strategy. William King Harvey had begun as an FBI counterespionage agent, renowned for having turned the Nazi spy William Sebold into a double agent during the Second World War. Fired by Hoover for drunkenness in 1947, he caught on with the CIA and became one of its celebrated operators. His great coup, the 'Berlin tunnel,' had enabled the CIA for many months in 1955–56 to intercept communications between East Berlin and Moscow. He was a histrionic fellow who always packed a gun, even in CIA headquarters in Langley. "If you ever know as many secrets as I do," he would say mysteriously, "then you'll know why I carry a gun."[40] Far from wanting the independent Cuban movement envisaged by Lansdale, Harvey was determined to reduce his Cuban operatives to abject dependence. "Your CO [case officer] was like your priest," one of Harvey's Cubans said later. ". . . You learned to tell him everything, your complete life."*

* Taylor Branch and George Crile III, "The Kennedy Vendetta," *Harper's*, August 1975. The speaker was Eugenio Rolando Martinez, who learned the lesson of obe-

Lansdale found Harvey intensely secretive, almost paranoid. Lansdale finally said to him, "I'm not the enemy. You can talk to me." But the momentum of the overblown Miami establishment generated its own excesses. "Mostly the things we needed were not the things they wanted to do," Lansdale said later. "Still, if the equipment exists, the temptation to use it becomes irresistible." Lansdale would ask what they expected random hit-and-run raids to accomplish. They had no good answer. A bridge would be blown up. "Why did you do it?" Lansdale would say. "What communications were you trying to destroy?" Harvey would reply, "You never told us not to blow up the bridge." Lansdale made his directives increasingly precise in the hope of stopping aimless sabotage and saving courageous Cubans from pointless death.[41]

Harvey protested to McCone about the "tight controls exercised by the Special Group" and the "excruciating detail" he was expected to provide.[42] The controls must have had some effect. "They never let us fight as much as we wanted to," lamented Ramón Orozco, a Cuban commando, "and most of the operations were infiltrations and weapons drops."[43] In October 1962 Robert Kennedy pointed out that, after almost a year of Mongoose, "there had been no acts of sabotage and that even the one which had been attempted [against the Matahambre copper mines] had failed twice."[44] CIA itself complained that same month, "Policymakers not only shied away from the military intervention aspect but were generally apprehensive of sabotage proposals."[45]

Still Harvey's Cubans evidently did more than Washington imagined. "The difficulties of control were so great," Taylor Branch and George Crile have written, "that the Agency [itself] often didn't know which missions were leaving in which directions." Harvey's people included soldiers of fortune like William (Rip) Robertson, a flamboyant figure who, in defiance of presidential orders, had landed on the beach at the Bay of Pigs. "When we didn't go [on missions]," said Orozco, "Rip would feel sick and get very mad." Once Robertson told Orozco, "I'll give you $50 if you bring me back an ear." Orozco brought him two, and "he laughed and said, 'You're crazy,' but he paid me $100, and he took us to his house for a turkey dinner."[46] They all sounded crazy. It was the 'dirty dozen' spirit, action for action's sake, without concern for the safety of the Cubans involved or for the reprisal effect or for political follow-up.

Lansdale, Robert Kennedy said in 1964, "came to cross purposes with CIA and they didn't like his interferences."[47] "Revolutions have

---

dience so completely that, when Howard Hunt asked him a decade later to break into the Watergate, Martinez, still on a CIA retainer and supposing this to be a CIA operation, obliged without second thought.

to be indigenous," Lansdale said in retrospect. ". . . We have a tendency as a people to want to see things done right — and, if they aren't, we step in and try to do them ourselves. That is fatal to a revolution." Harvey, however, had the decisive advantage of operational control. The political strategy fell by the wayside.[48] As for Robert Kennedy, he found Harvey and his meaningless melodrama detestable. "Too much 'Gunsmoke stance' for the Attorney General," said Howard Hunt.[49] "Why lose lives," Kennedy used to say to Taylor and to Lansdale, "if the return isn't clearly, clearly worth it?"[50] As for Harvey, he "hated Bobby Kennedy's guts," said a CIA colleague, "with a purple passion."[51]

The Attorney General was always dissatisfied with Mongoose. He wanted it to do more, the terrors of the earth, but what they were he knew not. He was wildly busy in 1962 — a trip around the world in February, the steel fight in April, civil rights always, Ole Miss in September, organized crime, Hoffa, apportionment, fighting with Lyndon Johnson about minority employment, fighting with Edgar Hoover about everything. Castro was high on his list of emotions, much lower on his list of informed concerns. When he was able to come to meetings of Special Group (Augmented), as he did his best to do, he made up in pressure for what he lacked in knowledge. His style there, as everywhere, was to needle the bureaucracy. If there was a problem, there had to be a solution. He conveyed acute impatience and urgency. With the intelligence-collection phase ending in August 1962, the Mongoose group meditated a "stepped up Course B Plus" intended to inspire open revolt. Kennedy, who was away on the west coast, endorsed B Plus in a message to Taylor. "I do not feel," he said, "that we know yet what reaction would be created in Cuba of an intensified program. Therefore, I am in favor of pushing ahead rather than taking any step backward."[52]

In October he urged the Augmented Group, in the name of the President, to give more priority to sabotage. The group responded by calling for "new and imaginative approaches with the possibility of getting rid of the Castro regime."[53] The State Department recommended the use of Manuel Ray, the anti-Castro radical. "I do believe," Lansdale wrote Robert Kennedy on October 15, "we should make a real hard try at this — since helping Cubans to help themselves was the original concept of Mongoose and still has validity." But CIA had always rejected Ray as too independent. Lansdale added: "I believe you will have to hit CIA over the head personally. I can then follow through, to get the action desired."[54] . . . On the same day, CIA experts were poring over photographs just taken by a U-2 plane over western Cuba.

v

The CIA had yet another program against Castro, the one initiated in 1960 and dedicated to his death. The Agency had considered assassination within its purview since the early Eisenhower years when it gave a special unit responsibility for, among other things, kidnapping and murder. The Church committee found no evidence that this special unit tried to kill any foreign leaders.[55] There are indications, however, that CIA operatives abroad tried, or at least wished, to kill Chou En-lai in 1955.* But the fever seems not to have struck in full force until the last Eisenhower year. In August 1960 the CIA decided to kill Patrice Lumumba, the pro-Soviet Premier of the newly independent Congo, and delivered a bag of poison to Leopoldville for that purpose. (Lumumba's actual murder the following January was, however, the work of his fellow countrymen and not of the CIA.) In August also, the CIA set in motion the plot to kill Castro.

The early planning was evidently confided to a team of humorists. One idea was to dust Castro's shoes, if he chanced to leave them outside his hotel room, with thallium salts in the expectation that this

---

* An Air India plane on which Chou En-lai had booked passage to Bandung, Indonesia, for the Afro-Asian Conference in April 1955 blew up in midair. Chou En-lai, who at the last minute flew instead to Rangoon to meet Nasser, was not on board. A detonating mechanism was recovered from the wreckage. The Hong Kong police described the incident as "carefully planned mass murder." On November 21, 1967, an alleged American defector named John Discoe Smith claimed in *Literturnaya Gazeta* (Moscow) that he had delivered the explosive mechanism on behalf of CIA to a Chinese Nationalist in Hong Kong (see Brian Urquhart, *Hammarskjold* [New York, 1972], 121; and *New York Times*, November 22, 1967). The deputy chief of the special unit told the Church committee that the recommendation to murder Chou was disapproved in Washington and the CIA station was "strongly censured" (Church committee, *Final Report*, bk. IV, 133; see n. 55). According to W. R. Corson's intermittently well-informed but too often undocumented book, Allen Dulles canceled the assassination of Chou lest General Lucian Truscott, whom Eisenhower had inserted into CIA as guardian of the presidential interest, inform the White House (W. R. Corson, *The Armies of Ignorance* [New York, 1977], 365–366). Despite all this, though, someone apparently blew up a plane on which Chou would have traveled except for his last-minute change of plan.

The year before, according to documents made public in 1978 under the Freedom of Information Act, the CIA designed a mind control experiment under the code name Artichoke to assess this "hypothetical" problem: "Can an individual of [deleted] descent be made to perform an act of attempted assassination involuntarily under the influence of Artichoke?" Sheffield Edwards, Artichoke's security officer, was the man who in 1960 proposed bringing in the mob to murder Castro. (See Nicholas M. Horrock, "C.I.A. Documents Tell of 1954 Project to Create Involuntary Assassins," *New York Times*, February 9, 1978.)

would cause his beard to fall out and destroy his charisma. Another was to lace a box of cigars with a chemical that produced temporary disorientation and get Castro to smoke one before delivering a speech. Since Castro's speeches often gave an impression of disorientation anyway, it is not clear how much difference the toxic cigar would have made. By August 1960 the Agency had progressed to the project of impregnating cigars with botulinum, a poison so deadly that a man would die after putting one in his mouth. The cigars were ready in October and were delivered to an unidentified person the following February.[56]

The Agency also may have recruited Marie Lorenz, a pretty German girl whom Castro had taken as a mistress in 1959 and later cast off. She fell in with Frank Fiorini, an adventurer who had fought beside Castro, but turned against him and went to work for CIA; under the name of Frank Sturgis, Fiorini figured in the Bay of Pigs and later in Watergate. The CIA, Lorenz claimed in 1976, appealed to her patriotism, promised her money for her old age and asked her to kill Castro. Sturgis gave her two poison capsules, which she secreted in a jar of cold cream. When she arrived in Havana, Castro greeted her warmly, took his phone off the hook, ordered food and coffee and then fell asleep on the bed with a cigar, evidently not a CIA model, in his mouth. Lorenz went to the bathroom and opened the cold cream. "I couldn't find the capsules," she said in 1976. "They had melted. It was like an omen. . . . I thought, 'To hell with it. Let history take its course.'"[57]

The Agency decided to turn to experts. When Castro's revolution swept into Havana at the end of 1958, it swept out not only Batista but the dictator's partners in the North American mob. Batista and the criminal entrepreneur Meyer Lansky fled on the same day. The closing of the casinos, the bordellos and the drug traffic cost the mob perhaps $100 million a year. Back in God's country, Lansky was reputed to have persuaded his associates to join in a pledge of a million dollars for Castro's head. Frank Sturgis claimed in 1975 that the mob had offered him $100,000 in 1959 to kill Castro. The masterminds of CIA now decided the syndicate possessed uniquely both the motives and skills required to rid the world of the Cuban revolutionary. Since the mob had its own grudge against Castro, a gangland hit would be less likely to lead back to the American government.[58]

The CIA commissioned Robert Maheu — a former FBI agent, later a private eye involved for a time with Hoffa's friend Eddie Cheyfitz and now working for Howard Hughes — to put out the contract. Maheu subsequently said that assassination was presented to him as "a necessary ingredient ... of the overall invasion plan."[59] Maheu brought in John Rosselli, a minor crook-about-town he had known in Las Vegas. Rosselli, who had no illusions about his middling rank in the underworld, brought in the Chicago big shot Sam Giancana, who was a don and could make the right connections. Giancana's vital connection for this purpose was Santos Trafficante, Jr., the boss of organized crime in Florida; not so long before, in Havana.

CIA case officers made their contact with the underworld notables before they informed their own chief, Allen Dulles.[60] One would have supposed that an alliance between the CIA and the mob might have required prior approval at least by the CIA director, if not by the Special Group. That this was not the case suggests the liberties casually taken even by lesser CIA officials in the Agency's golden age.

In the fall of 1960, CIA installed Giancana and Rosselli in a Miami Beach hotel. The Agency stuck staunchly by Giancana even after J. Edgar Hoover sent Bissell in October a report that, "during recent conversations with several friends, Giancana stated that Fidel Castro was to be done away with shortly." For his part Giancana, plotting away in Miami and jealously apprehensive that his girl, the singer Phyllis McGuire, might have her own plots in Las Vegas, asked the CIA to arrange an illegal wiretap in the room of his putative rival, the comedian Dan Rowan. The Agency obliged: anything to keep Giancana happy. When the tap was discovered, the Agency did its best to stop prosecution.[61]

As for Trafficante, he was a leader in his field. He had been a prime suspect in the barbershop murder of Albert Anastasia in 1957. Later that year he had been picked up at the Apalachin seminar. Havana was his base, and he was the only syndicate boss to stay on after the revolution. Castro imprisoned him in 1959. For an enemy of the people, Trafficante lived behind bars in surprising comfort. Chums from the mainland visited him, among them the Dallas hood Jack Ruby.[62] On release in September 1959, Trafficante went to Tampa. With his Cuban business presumably destroyed, but with part of the gang still in Havana, Trafficante had both motives and men for the CIA job.

It may have been more complicated than that. Why, after all, had Castro let Trafficante go? A Federal Narcotics Bureau document in July 1961 reported rumors in the exile community that Castro had "kept Santo [sic] Trafficante, Jr., in jail to make it appear that he had a personal dislike for Trafficante, when in fact Trafficante is an agent of Castro. Trafficante is allegedly Castro's outlet for illegal contraband in the country."[63] While banning drugs at home, Castro might have wished to earn foreign exchange by permitting them to flow through Cuba to the United States, as they had done so lucratively under Batista. He might also have hoped to promote drug addiction in the United States — even as Chou En-lai organized the export of Chinese opium in order to demoralize American troops in Vietnam.* If Trafficante was indeed a double agent, one can see why Castro survived so comfortably the ministrations of the CIA.

With Giancana shot seven times in the throat and mouth as he was frying sausages in his Chicago kitchen in June 1975, with Rosselli hacked up, stuffed into an oil drum and dumped into the sea near Miami in July 1976, it is impossible to know how they viewed their CIA mission. Rosselli, according to some accounts, was in his way a patriot, persuaded he was serving his nation. But he had also entered the country illegally and may have looked on patriotic duty as insurance against deportation (as it turned out to be).† The cynical Giancana, a draft dodger during the Second World War, very likely entered his country's service now exclusively in the expectation that he would thereafter be immune to prosecution. For the rest, he evidently did not take it all too seriously. He told one friend that Maheu was "conning the hell out of the CIA." According to Rosselli, when Giancana heard that local police had discovered the CIA wiretap in Las Vegas, he almost swallowed his cigar "laughing about it."[64] One has the strong impression that the hoods, if they did not regard the whole thing as a big joke, at most only went through the motions of carrying out their assignment.

Their first idea was to slip some pills into Castro's drink shortly before the Bay of Pigs. After one batch of CIA pills failed to dissolve, a more soluble batch killed some innocent monkeys and was deemed suitable. The obliging Trafficante produced a Cuban who, he said, had access to one of Castro's favorite restaurants. Maheu gave the Cuban the pills and $10,000. Castro survived. A second attempt was made in April immediately before the Bay of Pigs. Castro survived.[65]

After the Bay of Pigs, the Castro project was, in the CIA patois, "stood down" — i.e., suspended. In the meantime William K. Harvey had entered the picture. In January 1961, before Kennedy's inauguration, Bissell discussed with Harvey, and soon directed him to establish, an "executive action capability" for the disabling of foreign

---

* Or so, according to the Egyptian editor Mohammed Heikal, Nasser's great confidant, Chou told Nasser in 1965, adding, "The effect this demoralization is going to have on the United States will be far greater than anyone realizes" (see Heikal, *The Cairo Documents* [Garden City, N.Y., 1973], 306–307).

† In 1971 the CIA intervened with the Immigration and Naturalization Service to discourage Rosselli's deportation (Church committee, *Interim Report*, 85; see n. 15).

leaders, including assassination as a "last resort." No one has been able to discover that the executive-action crowd did much apart from Cuba.[66] In April 1962, Richard Helms, who had now succeeded Bissell, ordered Harvey to reactivate the Castro project. Whatever else Harvey was, he was a professional, and he found the six-link operation — CIA to Maheu to Rosselli to Giancana to a Cuban exile to a Cuban assassin — intolerably loose. He thereupon instructed Rosselli to cut out Maheu and Giancana. Now it all started again: poison pills in April 1962; a three-man assassination team in June; talk of a new team in September; Castro always in perfect health. Harvey began to doubt whether the operation was going anyplace. He finally terminated it in February 1963.[67] The mob's conning of CIA came to an end.

CIA's stalking of Castro did not. Some genius proposed that James Donovan, during his negotiations for the Bay of Pigs prisoners, give Castro a scuba diving suit contaminated by a tubercle bacillus and dusted inside with a fungus designed to produce skin disease. Donovan, who knew nothing of this, innocently foiled the Agency by giving Castro a clean diving suit on his own. Then Desmond Fitz-Gerald, who replaced Harvey as head of Task Force W in January 1963, suggested depositing a rare seashell, rigged to explode, in a place where Castro might skin-dive and pick it up. This inspiration proved beyond the Agency's technical capacity.[68]

VI

Who authorized the CIA to try to murder foreign leaders? The answer remains obscure. In the Lumumba case, Eisenhower said something in a meeting of the National Security Council that at least one person present understood as an assassination order. A week later Gordon Gray, Eisenhower's special assistant for national security affairs, passed on to the Special Group the President's "extremely strong feelings on the necessity for very straightforward action"; not intending, however, to imply assassination. The next day Allen Dulles sent the assassination order to the CIA station in Leopoldville.[69]

One doubts whether this is really what Eisenhower had in mind. Assassination seems out of character for him. It was contrary, moreover, to what John Eisenhower described as his father's philosophy that "no man is indispensable."[70] But a mist of ambiguity envelops such situations. Intelligence people might fear to say too much on the ground that Presidents do not, or should not, want to know too much. Certainly words like "removal" and "elimination" caused trouble, then and later. To most people around a table, they meant removal from political office. To specialists in dirty tricks, they could mean removal from life.

The Castro authorization is even less documented. Richard Bissell assumed that Allen Dulles had cleared the idea with Eisenhower, using a "circumlocutious approach" in order to maintain "plausible deniability." But, according to Thomas Parrott, the CIA officer who served as secretary of the Special Group, Dulles's practice was to insist on specific orders rather than "tacit approval." Parrott found the theory of the "circumlocutious" approach "hard to believe." As for the Special Group, while it approved the plan to train Cuban exiles for an invasion of their homeland, it never approved — and probably, though the evidence is murky, never considered — any plot to kill Castro.[71]

Yet that plot was an integral part of the invasion plan. And it is hard to suppose that even the runaway agency mordantly portrayed in the reports of the President's Board of Consultants would have decided entirely on its own to kill the chief of a neighboring state. Dulles must have glimpsed a green light somewhere in 1960. Could it have been flashed by the Vice President of the United States? "I had been," Richard Nixon said in 1964 of the invasion project, "the strongest and most persistent advocate for setting up and supporting such a program."[72] Philip Bonsal, American ambassador to Cuba in 1959–60, called Nixon "the father of the operation."[73] Brigadier General Robert Cushman, Nixon's military aide and later his deputy director of the CIA, described him to Howard Hunt in 1960 as "the project's action officer in the White House."[74] On the other hand, Gordon Gray told the Church committee that such descriptions greatly exaggerated Nixon's role.

Nixon's apprehensive allusions to the Bay of Pigs in the privacy of his own Presidency remain enigmatic. On September 18, 1971, according to John D. Ehrlichman's notes, Nixon instructed him to tell CIA to turn over "the *full* file [on the Bay of Pigs] *or else.* ... π [Ehrlichman's shorthand for the President] was involved in Bay of Pigs. π must have the file — deeply involved."[75] On July 23, 1972 — six days after Hunt, Martinez and the Cubans were arrested in the Watergate building — Nixon brooded to H. R. Haldeman:

Of course, this Hunt, that will uncover a lot of things. You open that scab, there's a hell of a lot of things, and we just feel that it would be very detrimental to have this thing go any further.... [Tell people] "Look, the problem is that this will open the whole, the whole Bay of Pigs thing, and the President just feels that ah, without going into the details" — don't, don't lie to them to the extent to say there is no involvement, but just say ". . . the President believes that it is going to open the whole Bay of Pigs thing up again."

And later that same day, again to Haldeman:

If it gets out that this is all involved, the Cuba thing, it would be a fiasco. . . . It is likely to blow the whole Bay of Pigs thing, which we think would be very unfortunate — both for the CIA, and for the country.

The same day, later still, this time Haldeman speaking:

The problem is it tracks back to the Bay of Pigs . . . The leads run out to people who had no involvement in this, except by contracts and connection, but it gets into areas that are likely to be raised.*

Whether or not it had been so commanded by someone in the

---

* Staff of the *New York Times, The End of a Presidency* (New York, 1974), 330, 335-336, 347-349.

In a notarized statement received by the Church committee on March 9, 1976, Nixon, while asserting that there were "circumstances in which presidents may lawfully authorize actions in the interests of the security of this country, which if undertaken by other persons, or even by the president himself under different circumstances, would be illegal," claimed (for what it is worth) that "assassination of a foreign leader" was "an act I never had cause to consider and which under most circumstances would be abhorrent to any president" (Church committee, *Final Report*, bk. IV, 157-158; see n. 55). The committee did not submit specific interrogatories about the Bay of Pigs.

In August 1976, Congressman Thomas Downing of Virginia released affidavits from Robert D. Morrow, who said he was a CIA contract agent in 1960, and Mario Kohly, Jr., the son of a right-wing Cuban exile, now deceased. The affidavits claimed that the elder Kohly had told them of an agreement between himself and Nixon (in the words of Morrow's affidavit) "for the elimination of Miro Cardona and all the leftist Cuban Revolutionary Front leaders in order that Kohly could immediately take over the reins of power in Cuba, once a successful invasion of exiles being trained by the CIA had been accomplished. . . . [Manuel] Artime [commander of the Cuban Brigade] and his followers were to be assassinated by [Kohly's] force once a successful landing had been accomplished."

In October 1963, Robert Morgenthau procured the indictment of the elder Kohly on charges of conspiracy to counterfeit Cuban pesos. Kohly was convicted and sentenced to a year in prison. On March 9, 1965, Nixon wrote Judge Edward Weinfeld asking for a suspension or reduction of Kohly's sentence on the ground that "the patriotism, courage and energy of the exiles in attempting to mount a counterrevolution have been in the past, and may in the future again be regarded as advantageous to the interests of the United States." (See *Richmond Times-Dispatch*, August 1, 3, 1976; *Newport News* [Virginia] *Daily Press*, August 1, 1976. I am indebted to Eston Melton of the *Times-Dispatch* for assembling the documents on this interesting but inconclusive matter.)

In a weird book entitled *Betrayal* (New York: Warner reprint, 1976), Morrow claimed that he had discovered nuclear missile sites in Cuba in 1961 and that the Kennedys deliberately suppressed the evidence. He described the CIA leadership as bitterly antagonistic toward the Kennedys. Robert Kennedy was depicted as the nemesis of Kohly and his anti-Castro counterfeiting scheme. The book's accuracy may be judged by its account of Owen Brewster, who had retired in 1952, as sitting senator from Maine in 1962 (ibid., 95).

White House, the CIA in the Eisenhower years had confidently taken unto itself the authority to kill foreign leaders.

VII

What of the Kennedy years? Did John and Robert Kennedy authorize the CIA to continue its attempts to murder Castro in 1961–63? Though the assassination plan was confided to Robert Maheu, though it became a staple of Sam Giancana's table talk and a joke among the mob, there is *no* evidence that any Agency official ever mentioned it to any President — Eisenhower in 1960, Kennedy after 1960, Johnson after 1963 — except (in Johnson's case) for operations already terminated. The practice of "plausible denial" had, as Colonel William R. Corson, a veteran intelligence officer, put it, "degenerated to the point where the cover stories of presidential ignorance really are fact, not fiction." [76]

Nor was the Castro murder plan submitted to the Special Group, the supposed control mechanism for covert action, either in 1960 or thereafter; nor to the Special Group (Augmented), sitting on top of Mongoose. Nor was it disclosed after the Bay of Pigs to Maxwell Taylor and his review board. [77] Rusk, McNamara, Bundy, Taylor, Gilpatric, Goodwin, Rostow, all testified under oath that they had never heard of it (nor had Kenneth O'Donnell, [78] nor, for that matter, had I). On every occasion in the Kennedy years when CIA officials might naturally have brought it up — Bissell's briefings of his old Yale friend Mac Bundy after the inauguration, the Bay of Pigs meetings, the Taylor board, the missile crisis — they studiously refrained from saying a word.

The argument that the Kennedys knew and approved of the assassination plan comes down, in the end, to the argument that they *must* have known — an argument that, of course, applies with equal force to Eisenhower and Johnson. The CIA, it is said, would never have undertaken so fearful a task without presidential authorization.

In John Kennedy's case, Bissell and Helms offered contradictory theories about the nature of that authorization. Bissell assumed the project had been cleared with Kennedy, as with Eisenhower, in ways that were tacit, "circumlocutious," camouflaged, leaving no record. Helms thought CIA's authority was derived, not from supposed "circumlocutious" talks, but from the "intense" pressure the Kennedys radiated against Castro. The CIA operators were undoubtedly misled by the urgency with which the Kennedys, especially Robert, pursued Mongoose. "We cannot overemphasize the extent," the CIA inspector general said in 1967, "to which responsible Agency officers felt themselves subject to the Kennedy Administration's severe pressure to do something about Castro and his regime." "It was the policy at the time to get rid of Castro," said Helms, "and if killing him was one of the things that was to be done in this connection, that was within what was expected." Having been asked "to get rid of Castro," Helms added, "... there were no limitations put on the means, and we felt we were acting well within the guidelines." No member of the administration told Helms to kill Castro, but no one had ever specifically ruled it out, so the Agency, he believed, could work for Castro's overthrow as it deemed best. Moreover, as Helms said, "Nobody wants to embarrass a President ... by discussing the assassination of foreign leaders in his presence." [79]

Still it seems singular that, even if the CIA people believed they had an original authorization of some yet undiscovered sort from Eisenhower, they never inquired of Kennedy whether he wished these risky and disagreeable adventures to continue. The project after all had begun as part of the Cuban invasion plan. In this context the murder of Castro had an arguably rational, if wholly repellent, function. After the Bay of Pigs and the abandonment of invasion fantasies, the radical change in context ought surely to have compelled reconsideration both of the project itself and of the alleged authorization. The Mongoose committee was demanding at this time that all "sensitive" operations be presented in "excruciating" detail, case by case. But the Agency, without consulting superior authority, resuscitated the assassination project on its own, the murder of Castro now becoming, without the invasion, an end in itself — a quite pointless end, too, in the unconsulted judgment of the CIA's Intelligence Branch. In October 1961 the Special Group asked for a contingency plan in the event Castro died from whatever cause. The CIA Board of Estimates responded that Castro's death "by assassination or by natural causes ... would almost certainly not prove fatal to the regime." Its main effect would probably be to strengthen the Communist position in Cuba.[80] Unfortunately the Intelligence Branch and the Clandestine Services were hardly on speaking terms.

It appears that the CIA regarded whatever authorization it thought it had acquired in 1960 as permanent, not requiring review and reconfirmation by new Presidents or, even more astonishingly, by new CIA directors. For neither Bissell nor Helms even told John McCone what his own Agency was up to. Harvey, after supplying Rosselli with a new batch of poison pills in 1962, briefed Helms and, according to the 1967 report by the CIA inspector general, "obtained Helms' approval not to brief the Director." "For a variety of reasons which were tossed back and forth," Harvey said later, "we agreed that it was not necessary or advisable to brief him." One reason was that the project had "arisen with full authority insofar as either of us knew long before I knew anything about it, and before the then-Director became Director." Harvey also knew that McCone, a devout Catholic, would fiercely disapprove. When assassination arose in another connection, McCone told Harvey, "If I got myself involved in something like this, I might end up getting myself excommunicated." Much later, in August 1963, McCone read in the *Chicago Sun-Times* that Giancana had a relationship with CIA. He asked at once for an explanation. Helms gave him a memorandum that described the operation as having been closed out before McCone took over. "Well," said McCone, "this did not happen during my tenure," unaware of what was still happening during his tenure.[*]

VIII

There is evidence, in other connections, of John Kennedy's attitude toward assassination. Lumumba had not been killed during his tenure; but Nasser of Egypt wrote the new President soon afterward reproaching the American government for the part he was sure it had had in Lumumba's murder. "Political assassination," Kennedy replied, while his own CIA was hustling the mob to poison Castro, ". . . should be . . . vigorously investigated and condemned."[81]

Then, on May 30, 1961, a group of Dominican patriots gunned down Rafael Trujillo, the most sadistic and squalid of Latin American tyrants. To its credit, the Eisenhower administration had established contact with anti-Trujillo Dominicans, and in December 1960 the Special Group approved a program of covert support for the dissidents. Henry Dearborn, the able American consul in the Dominican Republic, was in charge of the operation. The dissidents believed — and Dearborn agreed — that successful action against the regime required the assassination of Trujillo. On January 12, 1961, eight days before Kennedy took over, the Special Group approved the transfer of three revolvers and three carbines to the conspirators. Since the leading conspirators were generals, they had sufficient access to arms of their own; their point was evidently to obtain symbolic support from the United States. The Kennedy White House

* Church committee, *Interim Report*, 99–108; see n. 15. According to Corson, McCone was "fired for not knowing what was going on in the CIA with respect to the Castro 'hit'" (Corson, *Armies of Ignorance*, 35). This seems improbable. McCone left in April 1965. Johnson apparently first learned of the CIA assassination plots from a column by Drew Pearson in the spring of 1967. Johnson's demand for an investigation then led to the Inspector General's Report of May 23, 1967 (Church committee, *Interim Report*, 179).

knew in a general way about the contacts, and approved. It did not know about the transfer of arms.

In May rumors began to circulate that Trujillo might be assassinated. "I looked into the matter," Robert Kennedy said in notes dictated two days after Trujillo's death, "after I had a discussion with Allen Dulles over at the Pentagon. I called the President. He had known nothing about it and it was decided at that time that we'd put a task force on the problem and try to work out some kind of an alternative course of action in case this event did occur. Dick Goodwin from the White House worked on it."[82] Goodwin promptly got from CIA the details of its covert program in the Dominican Republic. When he read that the dissidents had received weapons "attendant to their project efforts to neutralize TRUJILLO," he circled the paragraph, underlined "neutralize" and took the document in to Kennedy.[83]

This was the first the White House knew about weapons. Kennedy reacted at once. "If Trujillo goes, he goes," he said, "but why are we pushing that?" In fact, Kennedy did not want Trujillo's overthrow until he knew what would come thereafter. He detested Trujillo and favored the OAS sanctions against him, but for international reasons, preferred the Trujillo regime to a Communist takeover.[84] He told Goodwin to advise Dearborn that the United States was not going to get involved in assassination and that we were not going to give any more weapons to anyone. Dearborn replied that, after a year of dealing with the conspirators, it was "too late to consider whether United States will initiate overthrow of Trujillo." Goodwin drafted a rejoinder intended to preserve friendly relations with the rebels while avoiding American implication in their plot. Kennedy personally strengthened the draft by adding: "US as a matter of general policy cannot condone assassination. This last principle is overriding."[85] The cable was sent on May 29. The Dominicans who killed Trujillo the next day did so for Dominican reasons and with Dominican weapons.

Neither Eisenhower nor the Kennedys nor the CIA ordered Trujillo's assassination. At most, they were sympathetic to Dominicans who saw no other way to liberate their country. When Trujillo was killed, the President was in Paris, where Pierre Salinger, running ahead of hard information, told reporters that Trujillo was dead, a statement that some thought indicated American foreknowledge. In Washington, however, there was only confusion. "When Allen Dulles called me yesterday morning," Robert Kennedy said in his June 1 notes, "the reports still had not been verified. A half hour later he called me back and said that it looked like it was probably 99 chances

out of 100 that Trujillo had in fact been assassinated." Kennedy's
observations indicated a degree of ignorance among top government
officials incompatible with an authorized CIA operation:

> We had been in contact with some of the generals who were anti-
> Trujillo. They had been asking for some arms and it had been decided
> in the last week or so that we wouldn't furnish them. I was quite
> surprised that the extent of their requests were for sub-machine guns.
> It didn't seem to me very extensive. . . .
> We still don't know how much popular support this anti-Trujillo
> group has. We don't know who is involved in it; we have at the present
> time no contact of any kind with them; we haven't received a report
> from the consul indicating that he has any contact with them. . . . The
> great problem now is that we don't know what to do because we don't
> know what the situation is.[86]

IX

As for Castro: in November 1961, after recommending Tad Szulc,
the Latin American correspondent of the *New York Times,* for a job in
the administration, Goodwin brought him in to see John Kennedy.
According to notes Szulc made immediately after the talk, the President asked what the United States might do about Cuba, "either [in]
a hostile way or in establishing some kind of dialogue." Then Kennedy said, "What would you think if I ordered Castro to be assassinated?" Szulc said the murder of Castro would not necessarily
change things and it was not anything the United States should be
doing. Szulc's notes: "JFK then said he was testing me, that he felt
the same way . . . because indeed U.S. morally must not be part[y] to
assassinations. JFK said he raised question because he was under
terrific pressure from advisers (think he said intelligence people, but
not positive) to okay a Castro murder." Szulc remembered Kennedy
going on "for a few minutes to make the point how strongly he and
his brother felt that the United States for moral reasons should never
be in a situation of having recourse to assassination."[87] If Kennedy
had ever planned to give such an order, he would hardly have begun
by telling the *New York Times* about it. A day or two later, Goodwin,
out of curiosity, raised the question with Kennedy, observing that it
sounded like a crazy idea to him. "If we get into that kind of thing,"
Kennedy said, "we'll all be targets."[88] That same month Kennedy
said in a speech, "We cannot, as a free nation, compete with our
adversaries in tactics of terror [and] assassination."[89]

Kennedy's reference to pressure from the intelligence people is
puzzling, since the relevant CIA officials still alive in 1975 (Dulles
died in 1969) all denied ever having discussed assassination with
him. But there was evidently something in the air. Kennedy's old

friend Senator George Smathers of Florida used to press him in those days to take a tougher stand against Castro. "My feeling," Smathers said in 1964, "was that he was not sufficiently aware as to the importance of Cuba, and I used to talk with him rather insistently about it. . . . He always identified me with pushing, pushing, pushing." On one occasion they speculated about Castro's assassination. Kennedy, Smathers recalled, was certain it could be done but doubted whether it would accomplish what he wanted. Smathers thought Kennedy "horrified" at the idea. Smathers continued to push. At last, one summer evening in the White House, "I remember that he took his fork and just hit his plate and it cracked and he said, 'Now, dammit, I wish you wouldn't do that. Let's quit talking about this subject. . . . Do me a favor. . . . I don't want you to talk to me any more about Cuba.' . . . And I never did."[90]

The only evidence that any Kennedy was told about the CIA operations came in the spring of 1962. CIA officials, anxious to protect Sam Giancana from the legal consequences of the wiretap he had demanded on Dan Rowan's Las Vegas telephone, brought insistent pressure on the Justice Department to refrain from prosecution. At first they spoke vaguely to Justice about Giancana's part in "CIA's clandestine efforts against the Castro government." This meant, they explained, that CIA was utilizing underworld sources in Havana for intelligence purposes. Robert Kennedy, unimpressed, scribbled on the FBI report of the CIA representations: "Courtney I hope this will be followed up vigorously." It was. The Agency continued to object, saying in April 1962 that prosecution "would lead to exposure of most sensitive information relating to the abortive Cuban invasion in April 1961." This was still not enough.[91]

Finally, on May 7, 1962, Lawrence Houston, the CIA's general counsel, explained why the case was so momentous. The CIA, he told Robert Kennedy, had recruited Giancana to murder Castro; therefore the Las Vegas prosecution involved national security. It was Houston's understanding, as he later testified, that "the assassination plan aimed at Castro had been terminated completely," and he so informed the Attorney General. He did not know that it had been resurrected and that Harvey had already passed on a new vial of CIA poison pills to Rosselli. Houston recalled that Kennedy was "upset" by CIA's use of Giancana. "If you have seen Mr. Kennedy's eyes get steely and his voice get low and precise, you get a definite feeling of unhappiness." Kennedy told Houston in sarcastic understatement, "I trust that if you ever try to do business with organized crime again — with gangsters — you will let the Attorney General know." Kennedy subsequently told J. Edgar Hoover about the CIA plot against Castro in terms that led Hoover to describe it as a situation "which had considerably disturbed him."[92] Plainly Kennedy had known nothing about assassination plots.

Houston did not recall that Kennedy expressed himself on the principle of assassinating Castro. Kennedy thought he had. In 1965, pressed by a group of Peruvian intellectuals — one person insisting, "Castro! Castro! What have you done about Castro?" — Kennedy said to his interpreter, "Tell them that I saved his life."[93] In 1967 Jack Anderson, breaking the story about the CIA and the mob, wrote that "Bobby, eager to avenge the Bay of Pigs fiasco, played a key role in the planning."[94] Kennedy, outraged, said to his assistants Peter Edelman and Adam Walinsky, "I didn't start it. I stopped it. . . . I found out that some people were going to try an attempt on Castro's life and I turned it off."[*]

x

Had the Attorney General approved the assassination of Castro, he would have been more appreciative of Giancana's patriotic zeal. Moreover, the gangster no doubt thought he had another, and equally esoteric, hold on the Kennedys. For, through his friend Frank Sinatra, Giancana had met in March 1960 a pretty girl named Judith Campbell whom Sinatra had introduced to John Kennedy the month before. If Campbell can be believed, she devotedly cultivated her relations with both men, the don and the President, over the next two years.[95]

White House logs show that Campbell telephoned some seventy times after January 1961 in pursuit of her friendship with John Kennedy.[96] One cannot conclude whether she was doing it out of a weakness for Presidents or whether she was put up to it by Giancana so that the underworld might get something on Kennedy. It seems improbable that she had a more complicated role. "I couldn't imagine her being privy to any kind of secret information," said William Campbell, her ex-husband. "She wouldn't understand it anyway. I mean . . . they weren't dealing with some kind of Phi Beta Kappa."[†] In any case, Hoover, in February 1962, informed the Attorney General about Judith Campbell's associations with Giancana. On March

---

[*] Seymour M. Hersh, "Aides Say Robert Kennedy Told of CIA Castro Plot," *New York Times*, March 10, 1975. See also the testimony of another Kennedy senatorial assistant, Frank Mankiewicz: "He told me that there was some crazy CIA plan at one time for sending some Cubans in to get Castro which he called off" (Mankiewicz, in recorded interview by L. J. Hackman, October 20, 1969, 70, RFK Oral History Program).

[†] *New York Post*, December 22, 1975. Campbell herself thought it "possible that I was used almost from the beginning" (Judith Exner, *My Story* [New York, 1977], 142).

22, Hoover brought a memorandum about her to one of his luncheons with the President. The Campbell calls ceased.[97]

Between CIA protection and the blackmail potentiality, Giancana expected immunity from federal prosecution. He was badly disappointed. The Attorney General remembered him well as the powerful Chicago racketeer who had giggled before the Rackets Committee. "Robert Kennedy was always out to get him," William Hundley said later. "He would raise hell about him. . . ." "Bobby pushed to get Giancana at any cost."[98] "They can't do this to me," Giancana said to Edward Bennett Williams, whom he tried to hire as his lawyer. "I'm working for the government."* But they did. He was put under heavy surveillance. FBI cars clogged the street near his house. FBI agents followed him everywhere — to the Armory Lounge, his favorite bar; to church; to the golf course, where four agents played the hole behind him. When he missed a putt, they would all boo (or so he claimed). Giancana sent a message to the Attorney General: "If Bobby Kennedy wants to talk to me, I'll be glad to talk to him, and he knows who to go through." Lest this allusion be unclear, the intermediary explained how close Giancana was to Sinatra.[99] Such a message was not calculated to impress Kennedy. Receiving no response, Giancana then sued in federal court for relief. The judge obligingly instructed the FBI agents to play golf at least two foursomes back. For unknown reasons — perhaps the CIA again — the federal attorney declined cross-examination. Later Giancana, given immunity from prosecution by a grand jury, refused to testify, was held in contempt and went to prison for the first time since 1942.[100] On his release, after a year, he went into gilded exile in Cuernavaca.

The Sinatra connection raised problems. Sinatra had been an enthusiastic Kennedy fund raiser in the campaign. He had sung at the inaugural gala. Peter and Patricia Lawford were in his intimate circle. But "it seemed he almost took a pleasure," Peter Lawford said later, "in flaunting his friendship with Giancana. . . . Giancana often stayed at Frank's home in Palm Springs."[101] Sinatra was financially involved, moreover, in dubious casinos and racetracks. A report of August 13, 1962, prepared by the Organized Crime Section at Robert Kennedy's request, set forth in detail Sinatra's associations with the mob.[102] Six months earlier, the Kennedys had already decided to draw away. Sinatra, expecting to entertain Kennedy in his Palm Springs house, had put in a helicopter pad and made other prepa-

---

* Williams repeated this statement to Hundley, who said it was ludicrous. Then Williams asked Maheu, an old schoolmate. Maheu verified Giancana's claim (Edward Bennett Williams, in interview by author, January 23, 1976).

rations. Shortly before a California trip in the spring of 1962, Kennedy called Peter Lawford. "I can't stay there," he said. "You know as much as I like Frank I can't go there, not while Bobby is handling this [Giancana] investigation."[103] Kennedy then compounded the humiliation by staying at the Palm Desert house of Bing Crosby, not only the rival troubadour of the day but a Republican. The President's action deeply wounded Sinatra. The lovely actress Angie Dickinson told me the next year that she had heard Sinatra say, "If he would only pick up the telephone and call me and say that it was politically difficult to have me around, I would understand. I don't want to hurt him. But he has never called me." I asked Angie about Lawford and Sinatra. She said, "Well, the President excluded Frank, so Frank excluded Peter."[104]

John Kennedy appreciated Sinatra's services in 1960 and liked him personally; no one, when the mood was on him, could be more charming than Sinatra. I think the President may have felt a little badly about it. For a moment in late 1963 Sinatra was on the guest list for a White House state dinner; at the last minute, on a pretext, the invitation was withdrawn. It was subsequently said that Robert Kennedy shared this regret and ignored recommendations from the Organized Crime Section for an investigation of Sinatra.[105] According to Hundley, however, his people investigated Sinatra with care and found nothing incriminating beyond associations.[106] Sinatra himself said in 1976: "I read the report about how Bobby Kennedy protected me from investigation by Government agents and now realize that after five grand-jury subpoenas, two [Internal Revenue Service] investigations (which probably utilized about 30 men) and a couple of subpoenas to Congressional committees, if you have a close friend in high office, you don't need any enemies."[107]

XI

The record shows that the only assassination plot disclosed to Robert Kennedy involved Sam Giancana — and that Kennedy did his best thereafter to put Giancana behind bars. As the CIA's inspector general put it in the 1967 report, "The Attorney General was not told that the gambling syndicate operation had already been reactivated, nor, as far as we know, was he ever told that CIA had a continuing involvement with U.S. gangster elements."[108] Nor did any of these arresting facts emerge in the course of Operation Mongoose. Helms testified that he never told Robert Kennedy about any assassination activity and never received any assassination order from him. Harvey testified that he never informed the Special Group (Augmented) or any of its members of the ongoing plots. Lansdale testified that Harvey had never told him about any such activities and that he had never discussed assassination with Robert Kennedy.[109]

The subject of assassination came up at one of the Augmented Group's meetings, on August 10, 1962. Robert Kennedy was not present. McNamara raised the question — one can only assume, in view of his ready dismissal of the idea a few days later, either in an academic survey of options or in a moment of acute frustration. McCone and Murrow took sharp exception. The "consensus," Lansdale said, "was ... hell no on this and there was a very violent reaction." Nevertheless Lansdale inexplicably produced a memorandum three days later that, among other things, asked Harvey to look into the question of "liquidation of leaders." (No one knew that Harvey had been looking into this question for some time.) On reading the memorandum, McCone blew up, called McNamara and demanded that it be withdrawn. "I agreed with Mr. McCone," McNamara said, "that no such planning should be undertaken." When McCone's stern disapproval of assassination was later communicated to Helms and Harvey, neither thought it necessary to tell their boss about Rosselli.[110]

Nor is it correct to say that John Kennedy did not communicate his own disapproval of assassination. He took care to send CIA his statement of May 29, 1961, that the United States could not as a matter of general policy condone assassination. It was a statement he never thereafter withdrew, qualified or amended. No doubt the CIA people took it as another of those superb efforts in misdirection to which they were addicted in their own daily work. Perhaps the failure of the Kennedys, McCone and others at the top specifically to forbid the assassination of Castro may have reinforced the CIA's delusion that killing foreign leaders was permissible. On the other hand, since none of them, including McCone, knew that Castro's assassination remained so heavily on CIA's mind, they would probably have regarded such an order as superfluous.

Several circumstantial points strengthen the conclusion that the Kennedys knew nothing about the continuing assassination policy. There is the McCone problem. If the Kennedys had been informed, the informer would have had to stipulate that they not mention the plots to McCone, who was deliberately not informed. Given John Kennedy's reliance on McCone to bring the Agency under control and Robert Kennedy's close personal friendship with McCone, it strains credulity to suppose that either would have accepted such a stipulation from some subordinate CIA official. In any case, both Bissell and Helms denied ever talking about assassinations with either Kennedy.

Then there is the problem of the Bay of Pigs prisoners. If the Kennedys had any obsessive feeling about Cuba, it was to save the lives of the prisoners. Nothing would more surely have doomed the prisoners than the assassination of Castro, whether attributed to anti-Castro Cubans or to the CIA.

Some have written that the Kennedys had got their Irish up and were determined to get even with Castro at any cost.[111] This over-looks the fact that the CIA had organized the plots well before the Irish moved into the White House. The Kennedys no doubt felt a personal animus toward Castro; but, if they were consumed by a vendetta, they would have had the perfect pretext when the Soviet Union sent nuclear missiles to Cuba in October 1962. Many advisers urged them then to seize the opportunity, invade Cuba and smash Castro forever. Robert Kennedy led the fight against this course. John Kennedy made the decision against it.

Then there is the fact that Allen Dulles withheld from the Warren Commission all information regarding the assassination plots. One obvious reason might have been his fear that, if Robert Kennedy now learned about the extent of the plots, he might hold the CIA responsible for his brother's death.

The available evidence clearly leads to the conclusion that the Kennedys did not know about the Castro assassination plots before the Bay of Pigs or about the pursuit of those plots by the CIA after the Bay of Pigs. There is a final consideration. No one who knew John and Robert Kennedy well believed they would conceivably countenance a program of assassination. Like McCone, they were Catholics. Robert, at least, was quite as devout as McCone. Theodore Sorensen's statement about John Kennedy applied to them both: assassination "was totally foreign to his character and conscience, foreign to his fundamental reverence for human life and his respect for his adversaries, foreign to his insistence upon a moral dimension in U.S. foreign policy and his concern for this country's reputation abroad and foreign to his pragmatic recognition that so horrendous but inevitably counterproductive a precedent committed by a country whose own chief of state was inevitably vulnerable could only provoke reprisals." "I find," said McGeorge Bundy, "the notion that they separately, privately encouraged, ordered, or arranged efforts at assassination totally inconsistent with what I knew of both of them, . . . their character, their purposes, and their nature and the way they confronted international affairs." McNamara said it would have been "totally inconsistent with everything I knew about the two men."[112] I too find the idea incredible that these two men, so filled with love of life and so conscious of the ironies of history, could thus deny all the values and purposes that animated their existence.

22

Robert Kennedy and
the Missile Crisis

EARLY IN 1961 Robert Kennedy had become acquainted with a Soviet official named Georgi Bolshakov. Nominally the embassy press attaché, Bolshakov conveyed the impression that he was in fact Khrushchev's personal representative in Washington. That may well have been the case. Certainly the CIA considered him a top KGB agent.[1] "If so," said Ben Bradlee, "he was a gregarious spy, could drink up a storm, and liked to hand wrestle."[2] "He was Georgi to everyone," recalled James Symington, who succeeded Seigenthaler as the Attorney General's assistant, "and he seemed to find satisfaction in being kidded. With self-deprecating nods, smiles, and circus English, he enjoyed Bob's predilection for harmless buffoons, and had almost unlimited access to the inner sanctum."[3]

I

Bolshakov was full of chaff and badinage. In November 1961 Kennedy noted, "Russian told me K[hrushchev] is Kennedyizing USSR Gov. — bringing in young people with new vitality, new ideas — I told this to Jack who laughed & said we should be Khrushchevizing the Am. Gov."[4] After the FBI men woke up the reporters during the steel imbroglio, Bolshakov observed to the Attorney General that "he saw in the papers that we were learning a few things from the Russians i.e., sending the police around in the middle of the night."[5] Later: "Georgi said that what had happened in the steel dispute had made a big impression in the Soviet Union but that Khrushchev still honestly believes the country is run by the Rockefellers, the Morgans and the big financiers on Wall Street. He said Khrushchev had his eyes opened tremendously on his visit to the

United States but still many of those who report to him report the old fashioned ideas and thoughts and do not realize what the United States is like today."[6]

They talked every fortnight or so. "I met with him about all kinds of things," Robert Kennedy said in 1964. ". . . Most of the major matters dealing with the Soviet Union and the United States were discussed. Arrangements were made really between Georgi Bolshakov and myself. . . . Unfortunately — stupidly — I didn't write many of the things down. I just delivered the messages verbally to my brother and he'd act on them and I think sometimes he'd tell the State Department and sometimes perhaps he didn't."[7]

They reviewed the preparations for the Vienna meeting. They went through the Berlin crisis of 1961 together. "I kept telling Georgi Bolshakov we would fight. He said he kept reporting it."[8] In the last days of the crisis there was a frightening confrontation of American and Soviet tanks at the Brandenburg Gate. "I got in touch with Bolshakov and said the President would like them to take their tanks out of there in twenty-four hours. He said he'd speak to Khrushchev, and they took their tanks out in twenty-four hours. So he delivered effectively when it was a matter of importance."[9]

In the spring of 1962, when the Kennedy administration was laboring to establish a neutralist Laos under Prince Souvanna Phouma, the Communist Pathet Lao resumed the military offensive and derailed the negotiations. Robert Kennedy, summoning Bolshakov, said that his brother had relied on Soviet assurances there would be no new hostilities and "felt now he had been double-crossed." This, the Attorney General emphasized, was a personal message to Khrushchev. Several days passed before Bolshakov returned with an answer.

> He said this was personal from Khrushchev to the President. He said there would not be any further armed action in Laos but they were anxious that the whole matter be resolved peacefully. . . . Georgi came back to see me the following day and I told him that the President was pleased with the message and again, that we were going to do all that we could with Souvanna Phouma but Khrushchev also had to work on his people on the other side.[10]

Khrushchev was "very pleased with the settlement in Laos," Bolshakov soon reported, but worried about American troops in Thailand. "Georgi said," Robert wrote his brother, "that Khrushchev understood that the reason the troops were sent in was because of the possibility of an outbreak in Laos. Therefore, now that the agreement had been made in Laos . . . Mr. Khrushchev hoped that it was possible for the United States to withdraw its troops from Thailand."[11] The President told Robert to reply that withdrawal would begin in ten days.[12] "Khrushchev sent back a message that it meant a great deal. So some of these crises which were building up and which perhaps [were] due to misunderstanding were discussed and an effort made to resolve them through these kinds of discussions."[13]

The Attorney General was not the only unconventional channel. On September 30, 1961, Bolshakov delivered to Pierre Salinger the first in the extraordinary exchange of letters initiated by Khrushchev as the Berlin crisis was subsiding. Thereafter the letters were passed along in cloak-and-dagger fashion, Salinger meeting Bolshakov at Washington street corners or bars, receiving documents enclosed in folded newspapers or placed surreptitiously in the pocket of his trench coat. On one rainy night, Bolshakov, the transaction completed, clapped Salinger on the back and said, jovially, "Every man has his Russian, and I'm yours."[14]

He was above all Robert Kennedy's Russian. An authentic friendship grew. Officially Robert Kennedy remained on Moscow's enemies list. *Za Rubezhom*, the weekly political journal, described him in June 1962 as an enemy of the working class, defender of monopolies, organizer of guerrilla warfare against Cuba, champion of preventive war.[15] But Bolshakov, on behalf of the Kremlin, invited him to visit the Soviet Union. The President thought not, feeling a visit by his brother would "ruffle a lot of feathers over in State."[16] Then the Russians indicated they would welcome the Attorney General as successor to Llewellyn Thompson, now about to leave Moscow after five years as ambassador. "The President spoke to me once about sending me," Robert Kennedy recalled. ". . . In the first place I couldn't possibly learn Russian. I spent ten years learning French. And secondly . . . for the first couple of months I might have done something but, after that, I don't think . . . the amount of good it would have done would have remained."[17] President Kennedy finally decided, as he told Thompson, that Bobby "was too greatly needed in Washington."[18]

In the meantime, the Foreign Service was "strongly" urging the appointment of Foy Kohler, a rigid Cold Warrior. "I had been involved in a lot of conferences with Foy Kohler," Robert Kennedy said later. ". . . I wasn't impressed with him at all. He gave me rather the creeps and I don't think he'd be the kind of a person who could really get anything done with the Russians. . . . Finally, my brother said well, . . . if I could come up with a decent candidate he'd send him. . . . [I] was not able to offer any alternative so he was finally appointed."[19]

The Georgi-Bobby exchanges and the Khrushchev-Kennedy cor-

respondence provide the counterpoint against which the public declarations of the period must be read. Behind the obligatory Cold
War rhetoric, the two leaders were trying to forestall conflict and to
seek accommodation. All this built in the course of 1962 a measure
of mutual understanding, even of mutual confidence.

II

Having established a measure of confidence, Khrushchev proceeded
to abuse it. He had begun to ship military equipment in growing
amounts to Castro. The administration was disturbed. It would have
been far more disturbed had it known that Khrushchev was sending
nuclear missiles too. But in early September, Anatoly Dobrynin, the
new ambassador, gave Robert Kennedy a reassuring message from
Khrushchev for his brother: not to worry, no offensive weapons, no
dirty tricks to embarrass the administration during the impending
congressional election. Shortly afterward Bolshakov returned from
Moscow with further reassurance. He had met with Khrushchev and
Anastas Mikoyan, the first deputy chairman. This was the time to
lower tensions, Khrushchev had said; Castro was getting only defensive weapons. Here Mikoyan had interjected: short-range, surface-
to-air missiles for use against planes; in no circumstances would long-
range, surface-to-surface missiles go to Cuba.*

Of course they were lying. But had the American secret war
against Castro left the Russians no alternative but to send nuclear
missiles secretly to Cuba? Certainly Castro had the best grounds for
feeling under siege. Even if double agents had not told him the CIA
was trying to kill him, the Mongoose campaign left little doubt that
the American government was trying to overthrow him. It would
hardly have been unreasonable for him to request Soviet protection. But did he request nuclear missiles?

The best evidence is that he did not. Castro's aim was to deter
American aggression by convincing Washington that an attack on
Cuba would be the same as an attack on the Soviet Union. This did
not require nuclear missiles. "We thought," he told Jean Daniel of
L'Express in November 1963, "of a proclamation, an alliance, conven-

---

* Bolshakov read the notes of his Moscow talk to Charles Bartlett on October 24.
The Russian, Bartlett reported to Robert Kennedy, was "clearly puzzled and disturbed"
but insisted "he could not believe there were long-range missiles on the island. . . . He
agreed with me that if things were as they appeared, he was in the position of having
been deceived by his own officials" (Bartlett to RFK, October 26, 1962, RFK Papers;
see also Robert F. Kennedy, *Thirteen Days: A Memoir of the Cuban Missile Crisis* [New
York, 1971 ed., with afterword by Richard E. Neustadt and Graham T. Allison], 5).
Bolshakov's reaction contradicts the claim of some revisionist historians that Khrushchev was innocent of deception, that the Americans just misunderstood his messages.

tional military aid." It was Khrushchev who thought of nuclear missiles. "The initial idea," Castro said, "originated with the Russians and with them alone." Khrushchev's proposal, Castro said to Daniel with emphasis, "*surprised us at first and gave us great pause.*" No doubt he worried (rightly) that the Soviet missile bases in Cuba would be an intolerable provocation to Washington. So he resisted the missiles. "When Castro and I talked about the problem," Khrushchev recalled, "we argued and argued. Our argument was very heated. But, in the end, Fidel agreed with me." "We felt that we could not get out of it," Castro told Claude Julien of *Le Monde* the next March. "This is why we accepted them. It was not in order to ensure our own defense, but primarily to strengthen socialism on the international plane." "We finally went along," Castro told Daniel, "because on the one hand the Russians convinced us that the United States would not let itself be intimidated by conventional weapons and secondly because it was impossible for us not to share the risks which the Soviet Union was taking to save us. . . . It was, in the final analysis, a question of honor."[*][20]

Why did Khrushchev wish to force nuclear weapons on Castro? In his memoirs he alleged the protection of Cuba as his primary motive. The thought, he wrote, was "hammering away at my brain; what will happen if we lose Cuba?" — much as in American brains the thought was starting to hammer away about the dire consequences if Washington 'lost' Vietnam. Losing Cuba, Khrushchev believed, would be a "terrible" blow, gravely diminishing Soviet influence throughout the Third World, especially in Latin America[21] — the domino theory, I expect. It would also, though Khrushchev did not say this, expose the Soviet Union to devastating ideological attack by Peking.

Still, if the protection of Cuba had been the only point, this could have been done far more simply, as Castro had proposed, through a proclamation or an alliance or the stationing of Soviet troops on the island. If nuclear weapons were to be used, tactical weapons would have been easier to install, harder to detect and less likely to provoke. As Graham Allison later wrote in his careful study of the

---

[*] In other moods Castro told other stories. Thus he insisted to Herbert Matthews of the *New York Times* in October 1963 that the missile idea was his alone, "not the Russians," and reiterated this after Matthews read the Daniel interview (Herbert L. Matthews, *Return to Cuba* [Palo Alto, Calif., 1964], 16). Possibly he wanted to reassure the romantic Matthews that Cuba was its own master. In 1967, he gave Matthews substantially the same account he had given to Julien and Daniel in 1963 (Matthews, *Fidel Castro* [New York, 1969], 225). In later years he told K. S. Karol and Frank Mankiewicz that the missile idea had come from Khrushchev (see K. S. Karol, *Guerrillas in Power: The Course of the Cuban Revolution* [New York, 1970], 261–262; Frank Mankiewicz and Kirby Jones, *With Fidel: A Portrait of Castro and Cuba* [Chicago, 1975], 174).

crisis: "It is difficult to conceive of a Soviet deployment of weapons less suited to the purpose of Cuban defense than the one the Soviets made."[22] Long-range nuclear missiles, in short, served Russian, not Cuban, purposes. Khrushchev, as Castro told Herbert Matthews in 1967, "was acting solely in Russian interests and not in Cuban interests."[23] The secret war against Castro was therefore *not* the cause of the Soviet attempt to make Cuba a nuclear missile base.

By 1962 Khrushchev was in a state of acute frustration — blocked in Berlin, at odds with Peking, stalled in the Third World, left badly behind by the inordinate American missile build-up. Inside the Soviet Union industrial growth was slowing down; agriculture was in its usual trouble; generals were demanding a larger share of limited resources for the military budget; old-line Stalinists were grumbling against internal liberalization. Khrushchev was desperate for a change of fortune.

The emplacement of nuclear missiles in Cuba would prove the Soviet ability to act with impunity in the very heart of the American zone of vital interest — a victory of high significance for the Kremlin, which saw the world in terms of spheres of influence and always inflexibly guarded its own. It would go far to close the Soviet Union's own missile gap, increasing by half again Soviet first-strike capacity against American targets.* It would do so without the long wait and awful budgetary strain attendant on an intensified missile production program. Secrecy would conceal the missiles until they were deployed. Once they were operational, Kennedy, as a rational man, would not go to nuclear war in order to expel them. If they were discovered prematurely, the congressional elections, not to speak of Kennedy's past irresolution on Cuban matters, would delay the American response. With one roll of the nuclear dice, Khrushchev might redress the strategic imbalance, humiliate the Americans, rescue the Cubans, silence the Stalinists and the generals, confound the Chinese and acquire a potent bargaining counter when he chose to replay Berlin. The risks seemed medium; the rewards colossal.

A plunger, he plunged. He sent 42 medium-range (1100-mile) nuclear missiles, 24 intermediate-range (2200-mile) missiles (which never arrived), 42 IL-28 nuclear bombers, 24 antiaircraft missile sites (SAMs) and 22,000 Soviet troops and technicians. This was something, he bragged to Castro, Stalin would never have done.[24] He explained his objective in his memoirs: "Our missiles would have equalized . . . 'the balance of power.'"[25] As Mikoyan put it in a secret meeting with Communist diplomats in Washington on November 30,

---

* According to Herbert Dinerstein, it would have advanced "the Soviet attainment of nuclear parity by almost a decade" (H. S. Dinerstein, *The Making of a Missile Crisis* [Baltimore, Md., 1976], 156).

1962, the purpose of the deployment was to achieve "a definite shift in the power relationship between the socialist and the capitalist worlds."*

III

Of Khrushchev's dream the Americans knew nothing. But Cuba remained a problem; and, for a diversity of reasons — Mongoose, the inertia at Foggy Bottom, the Republican exploitation of the Castro issue in the fall campaign — Robert Kennedy had seized the lead in Cuban policy. Early in September he privately recorded his concern that "Cuba obtaining [nuclear] missiles from the Soviet Union would create a major political problem here." The defensive surface-to-air missiles identified by American overflights, he feared, were only the first step. "We would have to anticipate there would be surface to surface missiles established in Cuba. What was going to be our reaction then?"[26]

He had already directed Norbert Schlei, a new assistant attorney general, to see what the United States could do under international law to forestall missile bases in Cuba. Schlei produced a brief that justified intervention on the principle of self-defense, ideally with the sanction of the Organization of American States. Missile shipments, he concluded, might be countered by "total blockade" or by "'visit and search' procedures."† The Schlei memorandum led to a White House meeting on September 4. The Attorney General urgently recommended a presidential statement warning the Russians that long-range missiles would raise the gravest issues. "My point is that I think it is much more difficult for them to take steps like that after you have made that statement." Rusk demurred, saying that the

---

* According to the testimony of Janos Radvanyi, then the Hungarian chargé in Washington (see Chalmers Roberts, *First Rough Draft* [New York, 1973], 205). It should be added that this too was the conclusion of Michel Tatu, *Le Monde*'s Moscow correspondent from 1957 to 1964, writing from Soviet sources in his book *Power in the Kremlin: From Khrushchev to Kosygin* (New York, 1969). He called Khrushchev's adventure "a truly dynamic move . . . undertaken in order to alter the world balance of power in their favour . . . an inexpensive means of altering the strategic balance" (229, 231).

† When Schlei suggested that the Monroe Doctrine gave the United States special rights in the hemisphere, John Kennedy snapped: "The Monroe Doctrine. What the hell is that?" — meaning: what the hell standing did it have in international law? The answer was, of course, none; and Kennedy, who knew how much Latin Americans resented unilateral declarations from Washington, never mentioned the doctrine throughout the subsequent crisis. (The Schlei memorandum, completed about August 29, 1962, is reprinted in appendix 1 of Abram Chayes, *The Cuban Missile Crisis: International Crises and the Role of Law* [New York, 1974], 108–132. See also Schlei to Chayes, May 22, 1968, in the same book, 132–134; and Schlei, in recorded interview by John Stewart, February 20–21, 1968, 6–12, JFK Oral History Program.)

problem with the Soviet Union was worldwide, not just in Cuba.[27] The President accepted his brother's proposal. With Katzenbach and Schlei, Robert Kennedy drafted a statement issued, with modifications, the same day.

Despite such precautions, however, almost no one in the administration believed Khrushchev mad enough to do what in fact he was doing. Though the Attorney General had apprehensions, his attention was soon diverted by James Meredith and the Battle of Ole Miss. Though John McCone had presentiments, he never communicated them to the Kennedys* and went off to Europe on a honeymoon. But the Republicans, led by Keating of New York and Capehart of Indiana, jeered at the administration view that Russia was only sending defensive weapons. At the end of of September, a House Republican caucus called Cuba "the biggest Republican asset" in the approaching election. On October 5, Clare Boothe Luce wrote in *Life:* "What is now at stake in the decision for intervention or nonintervention in Cuba is the question not only of American prestige but of American survival." On October 10, Keating claimed "100 percent reliable" evidence from exiles that nuclear missile bases were under construction. In the *New Yorker,* Richard Rovere reported a war party in Washington as active as the party that had produced the war over Cuba in 1898.[28]

On October 15, a U-2 camera discovered the sites. Studying the blown-up photographs, Robert Kennedy, as he wrote in his excellent posthumous book *Thirteen Days,* saw only what seemed the clearing of a field, as for a farm or the basement of a house.[29] The CIA photo-interpreter traced out the missile sites for him. Reverting to his preoccupations of a few days before, Kennedy asked, "Can they hit Oxford, Mississippi?"[30] John Kennedy was both infuriated by the Soviet deception and alarmed by the Soviet design. He saw the project precisely as Khrushchev saw it. It was, Kennedy said, "an effort to materially change the balance of power," "a deliberately provocative and unjustified change in the status quo"[31] — just what he had counseled Khrushchev against in Vienna. The consequences of so swift an alteration in the power equilibrium might be incalculable. If the United States accepted it, where might the plunger, dizzy with success, plunge next? How long could a fragile peace survive such shocks? Therefore, the President decided, the missiles had to be removed.

But how? The catchwords *hawk* and *dove* made their debut in the

* Robert Kennedy said in 1965, "I never even heard about it, and I used to see him all the time.... It was certainly never communicated to the President" (RFK, in recorded interview by author, February 27, 1965, 22, JFK Oral History Program).

missile crisis.* Within the closed meetings of the so-called Executive Committee of the National Security Council, Robert Kennedy was a dove from the start. If you bomb the missile sites and the airports, he said on the first day, "you are covering most of Cuba. You are going to kill an awful lot of people and take an awful lot of heat on it." If the Americans said they were bombing because of the missiles, "it would be almost [incumbent] upon the Russians to say that we are going to send them in again and, if you do it again, we are going to do the same thing in Turkey."[32] Listening to the war cries of the hawks, he sent his famous note to Sorensen: "I now know how Tojo felt when he was planning Pearl Harbor."[33] His steadfast allies were McNamara, Gilpatric, Ball, Llewellyn Thompson, Sorensen, Stevenson and Lovett.

The most formidable of the hawks was Dean Acheson, backed by Maxwell Taylor ("a twofold hawk from start to finish," he later described himself),[34] the Joint Chiefs (save for Shoup), McCloy, Nitze, and, initially, Dillon and McCone. Others shifted their views as the debate progressed. Bundy, Kennedy wrote at the time, "did some strange flipflops. First he was for a strike, then a blockade, then for doing nothing because it would upset the situation in Berlin, and then, finally, he led the group which was in favor of a strike — and a strike without prior notification, along the lines of Pearl Harbor."[35] Rusk was equally perplexing. At first he was for a strike; later he was silent or absent. He had, Robert Kennedy wrote laconically in *Thirteen Days*, "other duties during this period of time and frequently could not attend our meetings."† Privately, Kennedy was less circumspect. Rusk, he thought in 1965, "had a virtually complete breakdown mentally and physically."[36] After the crisis, Rusk explained that his function, as he saw it, was to keep the group from moving too far or too fast. "He said," Kennedy noted, "he had been playing the role of the 'dumb dodo' for this reason. I thought it was a strange way of putting it."[37]

While the deliberations went on, the deliberators maintained the fiction of business as usual. During the week of secrecy an unknowing Joseph Dolan brought the Attorney General a paper to sign. Robert Kennedy seemed more than usually preoccupied. "The room looked

---

* I do not recall their use at the time; but Stewart Alsop and Charles Bartlett thrust them into circulation soon thereafter in their article "In Time of Crisis" (*Saturday Evening Post*, December 8, 1962). Of course Jefferson had invented the term "war hawk" in 1798.

† RFK, *Thirteen Days*, 24. Acheson commented: "One wonders what those 'other duties and responsibilities' were, to have been half so important as those they displaced" (Acheson, "Dean Acheson's Version of Robert Kennedy's Version of the Cuban Missile Affair," *Esquire*, February 1969).

different," Dolan recalled, "and I was just trying to remind him that
he had a human being in the room. I says, 'Something looks different
here,' and he just looks at me, and he says, 'I'm older.' And so I
stuck the thing in front of him, and he signed it without a word."[38]

IV

"I could not accept the idea," Robert Kennedy wrote later, "that the
United States would rain bombs on Cuba, killing thousands and
thousands of civilians in a surprise attack."[39] The moral issue was
paramount in his mind. "We spent more time on this moral question
during the first five days," he said, "than on any single matter."[40]

For a while he listened more than he talked, exerting what Gilpatric
called a "disciplinary presence"[41] on the Executive Committee, seeking to raise hard questions, to elicit alternatives, to force his colleagues
to consider the implications of possible decisions. But he could not
conceal moral anguish. Acheson detested moral anguish. He
thought Robert Kennedy "moved by emotional or intuitive responses
more than by the trained lawyer's analysis." In particular, he could
not abide Robert Kennedy's talk about a "Pearl Harbor in reverse."
On Thursday, October 18, the imperious old man had a private talk
with the President. When John Kennedy brought up Pearl Harbor,
Acheson decided that the President was simply repeating "his
brother's clichés." He told the President — or so he later claimed
— that this was a "silly" way to analyze the problem: "It is unworthy
of you to talk that way."[42] Subsequently he denounced Pearl-Harbor-
in-reverse as "a thoroughly false and pejorative analogy."[43]

Later that Thursday afternoon, however, the Executive Committee, without Acheson, substantially agreed — Taylor and the Chiefs
dissenting — on a visit-and-search blockade, the idea broached by
Schlei weeks before, arrived at independently and advocated persuasively by McNamara and Gilpatric and soon to be rechristened "quarantine," in memory of Franklin Roosevelt. They so reported to the
President. But the next day Bundy reopened the issue of the
strike. Acheson was present. John Kennedy was away campaigning. He had spoken with the President that morning, Bundy said,
and, now speaking for himself, he favored "decisive action with its
advantages of surprise and confronting the world with a *fait accom-
pli*."[44] Sorensen intervened (I am quoting the record made at the
time by Leonard C. Meeker, the State Department's deputy legal
adviser), saying it was not fair to the President to reconsider a matter
on which they had all reached a decision the day before. Robert

Kennedy, according to his own notes dictated twelve days later: "I said I thought this was such a vital and important matter that obviously no solution was going to be completely correct and that people should still express themselves and voice and discuss their ideas."

Acheson then spoke, Kennedy said, "eloquently and at some length."[45] According to Meeker:

> Mr. Acheson said that Khrushchev had presented the United States with a direct challenge, we were involved in a test of wills, and the sooner we got to a showdown the better. He favored cleaning the missile bases out decisively with an air strike. . . .
>
> Secretary Dillon said he agreed there should be a quick air strike. Mr. McCone was of the same opinion. General Taylor said . . . it was now or never for an air strike. He favored a strike. If it were to take place Sunday morning, a decision would have to be made at once. . . . For a Monday morning strike, a decision would have to be reached tomorrow. Forty-eight hours' notice was required. Secretary McNamara said he would give orders for the necessary military dispositions. . . . He did not, however, advocate an air strike.[46]

When he had heard Acheson lay down the militant case during the Berlin crisis, Robert Kennedy thought that he "would never wish to be on the other side of an argument with him."[47] But with the blockade decision on the brink of reversal, he was not to be intimidated. Meeker summarizing his remarks:

> The Attorney General said with a grin that he too had had a talk with the President, indeed very recently this morning. There seemed to be three main possibilities . . . one was do nothing, and that was unthinkable; another was an air strike; the third was a blockade. He thought it would be very, very difficult indeed for the President if the decision were to be for an air strike, with all the memory of Pearl Harbor and with all the implications this would have for us in whatever world there would be afterward. For 175 years we had not been that kind of country. A sneak attack was not in our traditions. Thousands of Cubans would be killed without warning, and a lot of Russians too. He favored *action,* to make known unmistakably the seriousness of United States determination to get the missiles out of Cuba, but he thought the action should allow the Soviets some room for maneuver to pull back from their over-extended position in Cuba.[48]

As Kennedy himself recalled a dozen days later: "I said I just did not believe the President of the United States could order such a military operation. I said we were fighting for something more than just survival and that all our heritage and our ideals would be repugnant to such a sneak military attack."[49]

He talked, said Douglas Dillon, "with an intense but quiet passion. As he spoke, I felt that I was at a real turning point in his-

tory. . . . The way Bob Kennedy spoke was totally convincing to me.
I knew then that we should not undertake a strike without warning. . . . With only one or two possible exceptions, all the members
of the Excom were convinced by Bob's argument."[50] Said Alexis
Johnson of the State Department: "Bobby Kennedy's good sense and
his moral character were perhaps decisive."[51]

It was a compelling moment. Everyone in the room remembered
it for a long time. But the record suggests that its impact was not
instantaneous. On Saturday afternoon the group met with the President. McNamara made a forceful case for blockade as "the only
military course of action compatible with our position as a leader of
the free world." The air strike planned by the Joint Chiefs of Staff,
he said, would involve eight hundred sorties. "Such a strike would
result in several thousand Russians being killed." McNamara
doubted that Moscow would take it "without resorting to a very major
response. In such an event, the U.S. would lose control of the situation which could escalate to general war."

Maxwell Taylor made the opposing case, Acheson having disgustedly retired to his country seat. "Now was the time to act," Taylor
said, "because this would be the last chance we would have to destroy
these missiles. If we did not act now, the missiles would be camouflaged in such a way as to make it impossible for us to find them."
Taylor scouted McNamara's "fear that if we used nuclear weapons in
Cuba, nuclear weapons would be used against us" — a mysterious
remark, for there is no other reference to an American nuclear attack
on Cuba. The "Air Strike Scenario" prepared for the meeting specified that the missiles and sites were to be destroyed "by conventional
means." Probably Taylor was seeking to counter McNamara's point
about the risk of escalation to general war, and the rapporteur became confused.

The Attorney General informed the President that the air strike
was supported by Taylor, the Chiefs, Bundy and, with minor varia-
tions, by Dillon and McCone (the last two, however, disclaimed this
support before the meeting was over; Robert Kennedy had been
more persuasive the day before than he knew). As for himself, he
believed it essential to begin with a blockade. If the Russians did not
stop work on the missile bases, it would always be possible to move
on to the strike.[52] For a moment there was silence. Then Gilpatric
said: "Essentially, Mr. President, this is a choice between limited
action and unlimited action; and most of us here think that it's better
to start with limited action."[53]

John Kennedy said that the IL-28 bombers Moscow had given
Castro did not bother him. "In his view the existence of fifty planes

in Cuba did not affect the balance of power, but the missiles already in Cuba were an entirely different matter." He gave orders to begin with the blockade; also to make contingent preparations for an air strike and invasion. "The President acknowledged that the domestic political heat following his television appearance [to announce the crisis] would be terrific. He said he had opposed an invasion of Cuba but that now we were confronted with the possibility that by December there would be fifty strategic missiles deployed there. In explanation as to why we have not acted sooner . . . he pointed out that only now do we have the kind of evidence which we can make available to our allies in order to convince them of the necessity of acting."[54]

McNamara later thought that Robert Kennedy's arguments "did much to sway the President" to the quarantine.[55] Robert Kennedy gave the credit to McNamara. No one knows when John Kennedy reached his decision. "Most of those who attended the meeting on Saturday afternoon," Robert Kennedy wrote a month later, "felt he had made up his mind then. I think it is true that he had pretty much decided but he . . . held out his interest in a quick strike at the missile bases until Sunday morning." Then the head of the Tactical Air Command told him that even a very considerable strike might leave missiles capable of being fired back at the United States.[56]

Perhaps John Kennedy did not really make up his mind till Saturday or Sunday; or perhaps, having made his mind up earlier, he had simply, in the presidential manner, reserved his views. The advisers he trusted most — his brother, Sorensen, McNamara — had all advocated quarantine since at least Wednesday. He had provisionally endorsed it himself on Thursday. The hawks never considered him an ally. "It had been apparent after the first day or two," wrote Taylor, "that President Kennedy was inclining strongly to the side of the quarantine."[57] Acheson decided on Thursday that the President had been subverted by Robert's clichés. One imagines that the President's insistence on further military consultation expressed, not indecision, but a desire to establish a record. Well aware of the Chiefs' capacity for political reprisal, he may well have concluded he had better make a show of canvassing the military alternative to the end. He evidently found this exercise more than a little nerve-racking. Gilpatric saw Kennedy after LeMay had boomed forth on the beauties of an air attack. "He was just choleric. He was just beside himself, as close as he ever got."[58] To Kenneth O'Donnell, Kennedy said savagely, "These brass hats have one great advantage in their favor. If we . . . do what they want us to do, none of us will be alive later to tell them that they were wrong."[59]

V

Critics have called it a needless crisis. Kennedy's conduct, according to R. J. Walton, was "irresponsible and reckless to a supreme degree." Driven by "the *machismo* quality" in his character,[60] by rage over deception, by egotistical concern over personal prestige, by fear of setbacks in the elections, by a spurious notion of international credibility, Kennedy exaggerated the danger, rejected a diplomatic resolution and instead insisted on a public showdown. His purpose was the conspicuous humiliation of Khrushchev — eyeball to eyeball. Pursuing so insensate a course, he risked the incineration of the world in order to satisfy his own psychic and political needs.[61]

If Kennedy's idea had been to seek a showdown with Khrushchev, he went about it in a strange way. Instead of making a big issue of Soviet military aid to Castro during the summer of 1962, he minimized and, in the increasingly agitated Republican view, excused this aid. As late as the day before the missiles were discovered, Bundy said on national television that there was "no present evidence" and in his judgment "no present likelihood" of a "major offensive capability" in Cuba.[62] Indeed, according to some of the same critics, Kennedy's warnings about the arms shipments were so weak that the Soviet Union honestly misinterpreted them as a green light for nuclear missiles[63] — a view obviously incompatible with their theory he was on the prowl for confrontation.

Most critics have left the impression that, once the nuclear missiles were in Cuba, Kennedy should have done nothing at all. But suppose he had not forced the missiles out. Khrushchev by his own statement had sent the missiles in order to alter the balance of world power. American acquiescence would have been a stunning vindication of the Soviet 'Cold War of movement.' In a secret message via Bolshakov in June, Khrushchev had already warned the Kennedys of his intention to begin again in Berlin.* Had the missiles remained, the sixties would have been the most dangerous of decades. And acquiescence would have produced a shattering reaction in the United States. "If you hadn't acted," Robert Kennedy told his brother, "you

---

* On June 18, 1962, Bolshakov gave Robert Kennedy a message from Khrushchev. Khrushchev said that, if the United States continued to insist on occupation rights in West Berlin, "the Soviet Union will face the necessity of signing a peace treaty with the GDR [German Democratic Republic] and the question of liquidation of war remnants will be solved and on this basis the situation in West Berlin — a free demilitarized city — would be normalized" (RFK Papers).

would have been impeached." The President said, "That's what I think — I would have been impeached."[64]

R. J. Walton was an exception among the revisionists in conceding that, "given the political realities, Kennedy had to get the missiles removed."[65] But he and others have asserted that, instead of negotiating them out, Kennedy manufactured a public crisis. Even if he had not sought a showdown in August, he definitely sought one in October: both for personal reasons — obsession with his image, Irish temper, machismo, etc. — and for political reasons — the November election. He should, say, have discussed the missiles when Andrei Gromyko, the Soviet foreign minister, called on him at the White House three days after their discovery. If the Executive Committee had not yet figured out what to do, why did he not call Gromyko back as soon as the decision was made and before instituting the quarantine? Why did he not yield to Khrushchev at once the things that would enable the Soviet leader to remove the missiles with dignity: a no-invasion guarantee for Cuba and the removal of American Jupiter missiles from Turkey?

Charles Bohlen, that brilliant aficionado of Soviet policy, attended the meetings on the first two days. Departing on the third day, to take up new duties as ambassador to France, he left Kennedy a valedictory memorandum. The missiles, Bohlen agreed, had to be eliminated; and "no one can guarantee that this can be achieved by diplomatic action — but it seems to me essential that this channel should be tested out before military action is employed."[66] On the same day, Sorensen, summing up the discussion thus far, defined a choice between what he called "the Rusk or the Bohlen approaches." Rusk, he said, favored a limited strike without prior warning. Bohlen and "all blockade advocates" favored a "prompt letter to Khrushchev, deciding after the response whether we use air strike or blockade. . . . If you accept the Bohlen plan, we can then consider the nature of the letter to K[hrushchev]."[67]

Critics claim that Kennedy rejected the "Bohlen plan." In fact Kennedy did his best to pursue the plan. For two days, at his direction, Sorensen and others worked on letters to the Kremlin. On October 20, Sorensen reported to the President: "No one has been able to devise a satisfactory message to Khrushchev to which his reply could not outmaneuver us"[68] — by, for example, demanding submission of the dispute to the UN or to a summit meeting, thereby plunging the whole affair into a protracted diplomatic wrangle and making other forms of American reaction difficult while the missile bases were rushed to completion. The not unreasonable decision was therefore to announce the quarantine *before* beginning talks with the Russians.

After the announcement, the President began to lay the foundation for a peaceful resolution. On Tuesday, October 23, he sent his brother to make clear to Dobrynin the implications of Khrushchev's course. Robert Kennedy's memorandum of the meeting contains details omitted from *Thirteen Days*. The Attorney General reminded the ambassador that his brother, on the basis of Soviet assurances, had taken a "far less belligerent position [on the arms shipments] than people like Senators Keating and Capehart, and he had assured the American people that there was nothing to be concerned about. I pointed out, in addition, that the President felt he had a very helpful personal relationship with Mr. Khrushchev. Obviously, they did not agree on many issues, but he did feel that there was a mutual trust and confidence between them on which he could rely." The Attorney General noted that the American President had met Khrushchev's request for the withdrawal of American troops from Thailand. Sneaking nuclear missiles into Cuba now, he said, displayed the Soviet leaders as "hypocritical, misleading and false." So far as he knew, Dobrynin stoutly replied, there were no nuclear missiles in Cuba. "Dobrynin seemed extremely concerned."[69]

Everyone was extremely concerned. Plans were circulated for the evacuation of top officials to underground installations outside Washington. "I'm not going," Robert Kennedy told Edwin Guthman. "If it comes to that, there'll be sixty million Americans killed and as many Russians or more. I'll be at Hickory Hill."[70]

VI

"I sat across from the President," Robert Kennedy scribbled on a pad. It was Wednesday night, October 24. Two Soviet ships had approached to a few miles of the quarantine line.

> This was the moment we had prepared for, which we hoped would never come. The danger and concern that we all felt hung like a cloud over us all. . . . These few minutes were the time of greatest worry by the President. His hand went up to his face & covered his mouth and he closed his fist. His eyes were tense, almost gray, and we just stared at each other across the table. Was the world on the brink of a holocaust and had we done something wrong? Isn't there some way we can avoid having our first exchange be with a Russian submarine — almost anything but that, he said. . . . We had come to the edge of final decision — & the President agreed. I felt we were on the edge of a precipice and it was as if there were no way off.[71]

Robert thought inexplicably of Joe's death, of Jack's brushes with death. Then word came that the Russian ships had stopped in the water.[72] So far so good; but the quarantine met only part of the problem. It could stop the delivery of additional missiles to Cuba. It could neither remove those already there nor prevent the completion of the bases. From the start, the Kennedys and their chief advisers had known that this was the task of negotiation, and they were prepared to negotiate. When they were trying to execute the Bohlen plan, one of Sorensen's trial drafts to Khrushchev had the President say that, if the missiles were removed under effective surveillance, "I would be glad to meet with you . . . and to discuss other problems on our agenda, including, if you wish, the NATO bases in Turkey and Italy."[73] On Friday, October 19, according to the Meeker record, "More than once during the afternoon Secretary McNamara voiced the opinion that the US would have to pay a price to get the Soviet missiles out of Cuba. He thought we would at least have to give up our missile bases in Italy and Turkey and would probably have to pay more besides."[74] The next day McNamara again said, "We would have to be prepared to accept the withdrawal of US strategic missiles from Turkey and Italy and possibly agreement to limit our use of Guantanamo to a specified limited time. He added that we could obtain the removal of the missiles . . . only if we were prepared to offer something in return."[75]

It was at this Saturday meeting that Adlai Stevenson, arriving late from the UN in New York, proposed that the quarantine, which he supported, be accompanied by an immediate statement of the American negotiating position. He specifically suggested, according to the NSC minutes, "a settlement involving the withdrawal of our missiles from Turkey and our evacuation of Guantanamo." Most present felt that to begin with concessions would legitimize Khrushchev's action and give him an easy triumph. It was Stevenson's timing and his Guantanamo proposal, Robert Kennedy said later, not his Turkish thoughts, that created his reputation as the supreme dove.[76] In the general frustration and fatigue, everyone jumped on Stevenson.

The President emphatically disagreed that the initial presentation to the UN should include our notion of an eventual political settlement. He also "sharply rejected the thought of surrendering our [Cuban] base. . . . He felt that such action would convey to the world that we had been frightened into abandoning our position." However, "he agreed that at an appropriate time we would have to acknowledge that we were willing to take strategic missiles out of Turkey and Italy if this issue was raised by the Russians. . . . But he was firm in saying we should only make such a proposal in the future."[77] Stevenson, in what O'Donnell, who was present, called "an impressive show of lonely courage,"[78] held his ground. He reiterated "his belief that we must be more forthcoming . . . that the present situation required that we offer to give up such bases in order to induce the Russians to remove the strategic missiles."[79]

The President said afterward to O'Donnell, "I admire [Stevenson] for saying what he did."[80] Robert Kennedy, not a Stevenson fan at best, did not like it. "We had a rather strong argument with him," he wrote in his contemporaneous notes. ". . . I suggested [privately to the President] that we get someone else up at the United Nations to replace Stevenson as he was such a weak man in these kinds of negotiations."[81] In the end, John McCloy was summoned to sit at Stevenson's right hand. The President sent me to New York also to help on the UN presentation. (After reading my draft speech for Stevenson, the President called for the deletion of the more truculent passages, especially the threat of an American strike if the build-up persisted.) The Attorney General took me aside as I was rushing for the plane. "We're counting on you to watch things in New York," he said. "That fellow is ready to give everything away. We will have to make a deal in the end; but we must stand firm now. Our concessions must come at the end of negotiation, not at the start."[82] In *Thirteen Days* he made an *amende honorable:*

> Stevenson has since been criticized publicly for the position he took. . . . I think it should be emphasized that he was presenting a different perspective than the others, one which was therefore important for the President to consider. Although I disagreed strongly with his recommendations, I thought he was courageous to make them, and I might add they made as much sense as some others considered during that period of time.[83]

The last line was pure Robert Kennedy. And the disagreement, it should be emphasized, was not over the necessity for but the timing of negotiations.

In short, the advocates of quarantine assumed from the start that their course implied diplomacy somewhere down the road. The questions were when the trading should begin, how long it should take and what was to be put on the block. Critics later accused Kennedy of waiting too long, imposing too peremptory a deadline and refusing the most obvious deal — Cuban missiles for Turkish missiles.

VII

"I would sum it up, Prime Minister," John Kennedy told Harold Macmillan on October 25, two days after the Soviet ships turned back, "by saying that by tomorrow . . . we should be in a position to know whether there is some political proposal that we could agree to" — perhaps, he had explained earlier, withdrawing the missiles in exchange for an international guarantee of Cuba — "and whether the Russians are interested in it or not. . . . On the other hand, if at the end of 48 hours we are getting no place and the missile sites continue to be constructed then we are going to be faced with some hard decisions." [84]

Critics have detected ulterior motives behind the forty-eight-hour deadline. Unless Kennedy forced Khrushchev to retreat before the election on November 6, the journalist I. F. Stone argued in 1966, "the nuclear menace from Cuba would certainly have cost the Democrats control of the House of Representatives. . . . If Kennedy was so concerned [about peace] he might have sacrificed his chances in the election to try and negotiate. . . . Kennedy could not wait. But the country and the world could." [85] The argument lacked Stone's customary cogency. What chance would negotiations have had if the administration were trounced by a Republican party shouting for the invasion of Cuba? Repudiation of a dovish President by a hawkish electorate would have doomed any hope of a peaceful resolution. And in the meantime some of the nuclear bases would have become operational.

If negotiations were protracted, in short, their success required a strong Kennedy showing in the election. This perception only complicated Kennedy's problem. He thought the missiles an unmitigated political disaster. "We've just elected Capehart in Indiana," he said when he handed the U-2 photographs to O'Donnell, "and Ken Keating will probably be the next President." [86] If he had wished, as charged, to exploit the crisis for political advantage, the one course that would infallibly have insured a Democratic triumph in 1962 was the air strike. Ironically, this was in the main advocated by Republicans and opposed by Democrats on the Executive Committee — which suggests how little party politics mattered during the thirteen days.

The problem remains of the validity of the administration's claim that the bases were about to become operational. Critics have challenged this argument on diverging grounds. Ronald Steel has contended, correctly, that the long-range missile sites would not have been operational until mid-November; therefore Kennedy had plenty of time for negotiation, and the deadline must have derived from the election. Steel, however, overlooked the medium-range missile sites, four out of six of which, Barton J. Bernstein has argued, citing a CIA estimate, were already operational by Tuesday, October 23. Therefore, according to Bernstein, the deadline had already passed, and Kennedy must have had some base motive for pretending otherwise.[87]

The CIA estimate did indeed list four medium-range missile "sites" as "fully operational." But it added, "We are unable to confirm the presence of nuclear warheads."* "Fully operational," in short, had an exceedingly narrow significance. Obviously the presence of nuclear warheads, not the technical operability of the sites, determined the deadline. No one could be sure when the warheads would arrive. McCone thought that the missiles would be operational by the end of the week.[88] "The last 24 hours' film," Kennedy told Macmillan on October 24, "show that they are continuing to build those rockets."[89] "What was clear," scrawled Robert Kennedy in his notes on that evening, "was that the work in Cuba was proceeding and several of the launching pads would within a few days be ready for use."[90]

The reason for the negotiating deadline was simply the conviction that, once nuclear warheads were in place and pointed at the United States, the terms of trade would undergo drastic change. "It is obvious," wrote Michel Tatu of Le Monde, "that if this work could be completed under cover of diplomatic negotiations over broader issues ... Khrushchev would have won all he wanted: not only strategic reinforcement as a result of his newly built Cuban base, but also the diplomatic initiative resulting from his greater strength."[91]

Forty-eight hours remained. Was there a "political proposal" on which both sides could agree? On Thursday, October 25, Kennedy had cabled Khrushchev his hope that the Soviet Union would "permit a restoration of the earlier situation." Khrushchev's famous reply, arriving Friday evening, was a rather marvelous letter in the annals of correspondence among statesmen. While rearguing the Soviet case with due vehemence, Khrushchev set forth in highly personal language his fears about the slide toward catastrophe, ending with his knot-of-war image ("the more the two of us pull, the tighter that knot will be tied"). He concluded with a proposal: if the United States gave Cuba a no-invasion pledge, "this would immediately change everything" — no more arms sent to Cuba, no further need for Soviet military specialists in Cuba.[92]

Kennedy had no problem about what he had already described to Macmillan as a "trade of these missiles for some guarantee for Cuba."[93] He had never wanted an American invasion of Cuba, during the Bay of Pigs or thereafter. On Friday night the Executive Committee went to bed happy for the first time in eleven days. Relief

---

* CIA, Office of Current Intelligence, "Readiness Status of Soviet Missiles in Cuba," October 23, 1962, JFK Papers. Needless to say, Bernstein, in the manner of revisionist scholarship, suppressed the passage about nuclear warheads.

was premature. On Saturday a second letter, signed by Khrushchev but having the ring of a declaration by the Presidium, added the Turkish missiles to the list of Soviet conditions. We do not know what had gone on in Moscow. Khrushchev well understood that the Turkish missiles — Jupiters of an ancient vintage — were obsolete. Nonetheless their renunciation, he later claimed and may have thought at the time, would have "symbolic" value.[94] It would be something to show the Stalinists and the Chinese, and it might cause difficulties for the United States in NATO. On the other hand, Khrushchev may have temporarily lost control of the Presidium. There were hawks in the Kremlin too.

The Executive Committee, Robert Kennedy wrote, decided that the change in tone "indicated confusion within the Soviet Union." He dryly added: "but there was confusion among us as well."[95] For the Turkish missiles raised problems. It was not that anyone except General LeMay supposed them of much use. The President had ordered the forever dilatory State Department to get them out of Turkey months before and was incensed to find them still there. When McNamara and Stevenson had argued in the first week that Turkish missiles were a reasonable trade for Cuban missiles, Kennedy had agreed. Still, for the United States by unilateral decision to sacrifice the defense of a NATO ally would understandably affront the Turks, whose objections had delayed the missile withdrawal, and might well upset NATO.

Dubious as Harold Macmillan had been at the start about the American reaction, and fearful as he remained throughout of the dangers of escalation, the Prime Minister never wavered in his resistance to the bargaining away of NATO assets. "He may *never* get rid of Cuban rockets except by trading them for Turkish, Italian or other bases," he mused on the day of Kennedy's television speech. "Thus Khrushchev will have won his point." Toward the end of the crisis: "Anything like this deal would do great injury to NATO." Ten years later: "To make an exchange would have been fatal. If the President had done that, all credibility in the American protection of Europe would have gone."[96] Henry Kissinger, later no mean bargainer himself, took the same line outside the government.[97] Llewellyn Thompson, whose opinion carried particular weight because of his knowledge of the Soviet leadership, warned that a Turkish deal would be seen in Moscow as "proof of weakness" in Washington.[98] Nevertheless, as Robert Kennedy said soon after, "it was a rather reasonable offer," and "we would have difficulty explaining [refusal] if discussions were prolonged."[99] On the climactic Saturday the President therefore instructed Gilpatric to draw up a scenario for the

early removal of the Turkish missiles — this regarded still as an item
for subsequent negotiations rather than an immediate response to
the second Khrushchev letter.[100]

Then word arrived that a U-2 plane had been shot down over
Cuba, its pilot killed. Four days before, the Executive Committee
had recommended, in such an eventuality, "immediate retaliation
upon the most likely SAM site involved."[101] Washington's reaction
now was very macho — "almost unanimous agreement," in Robert
Kennedy's words, "that we had to attack early the next morning . . .
and destroy the SAM sites." The Joint Chiefs joyfully reminded the
President "that they had always felt the blockade to be far too weak
a course and that military steps were the only ones the Soviet Union
would understand." "There was the feeling," said Robert Kennedy,
"that the noose was tightening on all of us." The President pulled
everyone back. "It isn't the first step that concerns me," he said, "but
both sides escalating to the fourth and fifth step — and we don't go
for a sixth because there is no one around to do so."[102] According
to Graham Allison, Kennedy's order to call off the Air Force reprisal
was "received with disbelief" in the Pentagon. Soon he increased the
dismay of the hawks by ordering the defusing of the Turkish missiles. Allison: "For some members of the group, this was the darkest
moment."[103]

VIII

In the early afternoon the State Department submitted a draft reply
to Khrushchev flatly rejecting the Turkish deal. "I disagreed with
the content and tenor of the letter," Robert Kennedy said later.[104]
Everyone remembers his own inspired suggestion that the President
forget the second Khrushchev letter and answer the first. "There
were sharp disagreements," Robert Kennedy recalled. The hawks,
dreaming of a Monday morning war, rallied behind the hard line.
The Attorney General persisted in his criticism of the State Department letter. The President asked him to try a draft himself. In forty-
five minutes Robert Kennedy and Sorensen produced a new letter
offering a no-invasion guarantee in exchange for the removal of all
offensive weapons systems from Cuba, all this to be accompanied by
termination of the quarantine and UN inspection. The President
dispatched it without delay.[105]

The deftness with which this letter elided the Turkish problem has
obscured the dénouement. For, soon after the letter went off, the
President instructed his brother to talk once more to Dobrynin. The
two men met half an hour later at the Department of Justice. There
are two accounts of this Saturday night meeting: one written three

days afterward by Robert Kennedy, and one written in 1971 on the basis of Dobrynin's dispatches by Anatoly Gromyko, a Soviet historian and son of the foreign minister.[106] Kennedy began the conversation by emphasizing the gravity of the situation. If Cubans were shooting down American planes, he said, Americans were going to shoot back. "We had to have a commitment by at least tomorrow that those bases would be removed. This was not an ultimatum, I said, but just a statement of fact." Dobrynin recalled Kennedy as very blunt and very tough.[107] According to the Gromyko account, Kennedy said that, "should war break out, millions of Americans would die. The U.S. government, he said, was trying to avoid this. Later he gave his opinion that he was sure that the Soviet Union adhered to the same view. Any delay in finding ways out of the crisis was fraught with great danger."

Kennedy went on to say, according to the Soviet version, that "the Pentagon was exerting strong pressure on his brother" in connection with the shooting down of the U-2. "Kennedy noted that, in the United States, among the highly placed generals there were many stupid heads who were always eager for a fight. He did not exclude the possibility that the situation could get out of control and lead to irreparable consequences." These remarks reflected the Attorney General's opinion, but did not appear in his own memorandum of the conversation. Later Khrushchev, dictating his memoirs, recalled Dobrynin's sketch of an exhausted Robert Kennedy — "one could see from his eyes that he had not slept for days" — saying: "Even though the President himself is very much against starting a war over Cuba, an irreversible chain of events could occur against his will. That is why the President is appealing directly to Chairman Khrushchev for his help in liquidating this conflict. If the situation continues much longer, the President is not sure that the military will not overthrow him and seize power. The American army could get out of control."[108] The escalation from the Dobrynin-Gromyko version of the *situation* getting out of control to the Khrushchev version of the *U.S. Army* getting out of control attests to Khrushchev's melodramatic temperament, spirited memory and desire to portray himself as the savior of American democracy from a military coup. While the Kennedys did not altogether dismiss *Seven Days in May*, it is unlikely that Robert Kennedy would have confided such forebodings to the Soviet ambassador.*

Dobrynin then asked what the Americans proposed. Once the Russians agreed to remove the missiles and dismantle the bases,

---

* Elie Abel has suggested that Khrushchev may have confused the Kennedy-Dobrynin meeting with the postcrisis negotiations over the removal of IL-28 bombers. The Russians, he wrote, "had been given reason to believe that if a fire-fight started

Kennedy said, "in return, if Cuba and Castro and the Communists ended their subversive activities in other Central and Latin American countries, we would agree to keep peace in the Caribbean and not permit an invasion from American soil." Both accounts say that Dobrynin was the first to mention the Turkish missiles. "I replied," Kennedy reported to Rusk, "that there could be no quid pro quo — no deal of this kind could be made.... It was up to NATO to make the decision. I said it was completely impossible for NATO to take such a step under the present threatening position of the Soviet Union. If some time elapsed — and per your instructions, I mentioned four to five months — I said I was sure that these matters could be resolved satisfactorily." This understanding, the Attorney General added, would be canceled at once if the Soviet government tried to claim public credit for it. They parted. Robert Kennedy's last words, Dobrynin cabled Moscow, were: "Time will not wait, we must not let it slip away."

So concluded a singular exercise in secret diplomacy — secret not only in process, as most diplomacy must be, but in result, which is generally inadmissible. Macmillan and NATO were told nothing, nor was the American Congress, nor even the Executive Committee. "They were very private," Douglas Dillon said of the Kennedy-Dobrynin talks, "and were not always necessarily reported in great detail, even to those of us on the ExComm.... All of us felt that we just couldn't take [the Turkish missiles] out as part of the deal, even though we knew that they were coming out anyway." [109] Probably no one except the Kennedys, McNamara, Rusk, Ball and Bundy knew what Robert Kennedy had told Dobrynin. The American people certainly did not know until Robert Kennedy himself described the meeting in *Thirteen Days*, a half dozen years later — and then his account was so muted that Harold Macmillan believed to the end of the day that no such bargain was ever struck, and critics continued to portray John Kennedy as recklessly preferring confrontation to negotiation.*

The Russians made one attempt to put the deal on the record.

---

the Washington 'hawks' might overpower the 'doves,' invade Cuba and get rid of Castro.... 'We were happy to leave them believing that could happen,' one of the American negotiators recalled." (Abel, *The Missile Crisis* [New York: Bantam reprint, 1966], 191. See also Abel, "The Khrushchev Version of the Cuban Missile Crisis," *Washington Post*, January 4, 1971.)

* See, e.g., FitzSimons: "In face of this admittedly reasonable proposal, President Kennedy persisted in refusing to negotiate, preferring to force the Soviet retreat at the risk of war rather than even discuss a 'not unreasonable' proposal" (Louise Fitz-Simons, *The Kennedy Doctrine* [New York, 1972], 167). Also Bernstein: "Why, on Saturday, the 27th, when the only issue blocking settlement was American withdrawal of our missiles from Turkey, did the administration reject this condition and risk ... moving toward nuclear war?" (Barton J. Bernstein, "'Courage and Commitment': The

Two days after their decisive conversation, Dobrynin brought Robert Kennedy an unsigned letter from Khrushchev to the President spelling everything out. After brooding over the letter, the Attorney General called the Soviet ambassador to his office. Robert Kennedy's scribbled notes for the interview define the American position:

> Read letter — Studied it over night.
> No quid pro quo as I told you.
> The letter makes it appear that there was.
> You asked me about missile bases in Turkey. I told you we would be out of them — 4-5 months. That still holds. . . . You have my word on this & that is sufficient.
> Take back your letter — Reconsider it & if you feel it is necessary to write letters then we will also write one which you cannot enjoy.
> Also if you should publish any document indicating a deal then it is off & also if done afterward will further affect the relationship.*

Dobrynin withdrew the letter. Nothing was heard of it thereafter.†

In short, the Kennedys made a personal, but not official, pledge that the Turkish missiles would go. On October 29, the day after the settlement, John McNaughton, the general counsel of the Defense Department, told an interdepartmental task force: "Those missiles are going to be out of there by April 1 if we have to shoot them out."[110] With NATO assent, they went. On April 25, 1963, almost exactly five months after Robert Kennedy's talk with Dobrynin, McNamara sent John Kennedy a handwritten note: "The last Jupiter missiles in Turkey came down yesterday. The last Jupiter warhead will be flown out of Turkey on Saturday."[111]

Was this secret diplomacy justified? — a testing question for those who think that no President should ever make a secret commitment. It is extremely doubtful that any Turkish agreement could have been made publicly in the forty-eight hours before the bases acquired

---

Missiles of October," *Foreign Service Journal*, December 1975). Also Horowitz: "Khrushchev's offer . . . to exchange the missile bases in Turkey for the bases in Cuba was turned down" (David Horowitz, *The Free World Colossus* [New York, 1965], 386).

* RFK, handwritten notes, n.d. [October 30, 1962], RFK Papers. At one point Dobrynin said that of course the Soviet government would not publish the correspondence. Kennedy said, "Speaking quite frankly, you also told me your government never intended to put missiles in Cuba" (RFK to Dean Rusk, reporting on the interview, October 30, 1962, RFK Papers).

† Later the bargain was mentioned by Khrushchev and Anatoly Gromyko in their accounts of the crisis, though Gromyko, faithful to the agreement, based his account on *Thirteen Days* rather than on Dobrynin's cables. (See Khrushchev, *Khrushchev Remembers: Last Testament* [Boston, 1974], 512; Gromyko, "Diplomatic Efforts of the USSR to Liquidate the Crisis," *Voprosy Istorii*, August 1971.) The Russians evidently briefed Castro but swore him to silence too. He told Lee Lockwood in December 1966, "One day perhaps it will be known that the United States made some other concessions in relation to the October Crisis besides those which are made public" (Lockwood, *Castro's Cuba, Cuba's Fidel* [New York, 1967], 204). Castro told Herbert Matthews in October 1967: "Kennedy was willing to give up the Turkish and Greek [sic] bases" (Matthews, *Fidel Castro*, 225).

nuclear warheads. If the deal appalled Harold Macmillan, it would have been even more appalling to NATO governments less determined to get along with Washington. At home the deal would have caused an uproar in Congress, in the Pentagon, among the voters; very likely within the Executive Committee itself. "I did not accept the explanation that the [Turkish] missiles had become obsolete," said General LeMay afterward, ". . . nor did any other military man I know."[112] Only secret diplomacy could have assured Khrushchev the Turkish withdrawal that, along with the Cuban no-invasion guarantee, covered his retreat before his opponents on the Presidium. More important, the willingness to make this additional concession doubtless helped persuade the Russians that the American government was truly bent on peace (and accounts for the indulgent treatment the Kennedys later received in Soviet accounts of the crisis). Perhaps there may be a place for secret diplomacy, at least when nuclear war is involved and when no vital interests of a nation or ally are bartered away.

As the world knows, on that golden Sunday morning, the day after Robert Kennedy and Dobrynin had their chat about the Turkish missiles, Khrushchev announced his nuclear retreat from Cuba. In Havana, Castro — who had not been consulted — cursed, kicked the wall in a rage and broke a mirror.[113] "The possibility that the Soviet Union would withdraw [the missiles]," he said later, "was an alternative that had never entered our minds."[114] Having begun his misbegotten adventure by lying to Kennedy, Khrushchev ended by lying to Castro.[115]

In Washington most of us felt a sense of limitless relief. Not all the Chiefs, however: "Admiral [George] Anderson's reaction to the news," Robert Kennedy noted, "was, 'We have been had.' General LeMay said, 'Why don't we go in and make a strike on Monday anyway.'" "The military are mad," the President told me the next morning. "They wanted to do this. It's lucky for us that we have McNamara over there."[116] "The first advice I'm going to give my successor," he said two weeks later, "is to watch the generals and to avoid feeling that just because they were military men their opinions on military matters were worth a damn."[117] Still, for the moment, even the original hawk, Dean Acheson, was confounded. On October 29, he congratulated Kennedy on his "leadership, firmness and judgment," adding that the dénouement "amply shows the wisdom of the course you chose — and stuck to."*

Contrary to the latter-day supposition, the President did not think

---

* A month later he wrote again, praising Kennedy's policy as "wisely conceived and vigorously executed" (Dean Acheson to JFK, October 29 and November 30, 1962, JFK Papers).

the resolution would help him in the election. It was too bad, he said to me, that this had happened in the midst of the campaign. The Republicans would now feel compelled to denounce the settlement. They will say, he said, "that we had a chance to get rid of Castro and, instead of doing so, ended up by guaranteeing him."[118]

On Sunday he had said to his brother that this was the night he should go to the theater, like Lincoln. Robert Kennedy:

> I said if he was going to the theater, I would go too, having witnessed the inability of [Vice President] Johnson to make any contribution of any kind during all the conversations. Frequently, after the meetings were finished, he would circulate and whine and complain about our being weak but he never made ... any suggestions or recommendations. On Saturday night he started to move around saying he heard that the people felt we were backing down. When I asked him for specifics he had none.*

"He was against our policy on Cuba," Robert Kennedy said the next year. "... I never quite knew what he was for but he was against it. ... He was shaking his head, mad."[119]

"The 10 or 12 people who had participated in all these discussions," Robert Kennedy wrote that November, "were bright and energetic people. We had perhaps amongst the most able in the country and if any one of half a dozen of them were President the world would have been very likely plunged in a catastrophic war."[120]

IX

Loose ends remained. Castro in his rage refused to permit UN verification of the missile removal. He also refused to yield the IL-28 bombers. These, unlike the missiles, could be reasonably claimed as Cuban property. Khrushchev sent Mikoyan to Havana. Castro kept him waiting so long that Adlai Stevenson said it might be necessary to send James Donovan down to get him out.[121] Llewellyn Thompson advised Kennedy to insist on the IL-28s; Khrushchev, having swallowed a camel, would not strain at a gnat.[122] "Wondering at times," Sorensen said, "whether his stand was necessary," the President insisted.[123]

Castro was obdurate. Stevenson and McCloy, conducting the negotiations in New York, were prepared to drop UN inspection, give

---

* RFK, memorandum, November 15, 1962, RFK Papers. Robert Kennedy told me in 1965: "[Johnson] was displeased with what we were doing. ... He said that he had the feeling that we were being too weak ... and that we should be stronger. We discussed it afterwards and the President knew it was because he was talking to perhaps some of his congressional friends" (RFK, in interview by author, February 27, 1965, 24, also 38, JFK Oral History Program).

the Russians time to deal with the Cubans over the bombers, make the no-invasion pledge anyway and lift the quarantine. Robert Kennedy, as usual, thought Stevenson excessively weak. His own platform was no inspection, no pledge. He was, however, prepared to lift the quarantine at once if the Russians would withdraw the bombers in thirty days. Thompson supported an offer along these lines, deeming it important that "Khrushchev have something to show to his colleagues in the Kremlin."[124]

The Attorney General carried this proposal to Dobrynin on an evening when the Soviet ambassador was entertaining the Bolshoi Ballet. After the two men talked in private, Dobrynin introduced the President's brother to the dancers. The prima ballerina, the enchanting Maya Plisetskaya, said, "You and I were born on the same day and same year." "We gave each other a little kiss," Kennedy wrote in his journal. "I asked her where she would be on November 20th and she said, 'In Boston.' I said I would send her a present." Then he said a few words to the troupe, telling them that "I felt a little awkward speaking to them as I had trouble doing the two-step; that although I had read in the papers that there were some differences between their country and mine (to which they laughed) that all of us in both countries recognized excellence and that it was a great honor to have this group visit the United States."[125]

But the Russians rejected his proposal. Three days after the Bolshoi evening, Robert Kennedy held "the most unpleasant conversation I have had with Dobrynin. It made it seem that perhaps I had seen him too frequently. I said this to the President the next day. I said I thought it was good that I could do this but certainly not in every day exchanges nor developing a familiarity which made statements by me at critical times less effective."[126] Debate continued within the administration. "President reluctant to send in low level flights," Robert Kennedy noted (on the back of an envelope) regarding a meeting on Monday, November 19. ". . . How far can we push K[hrushchev]."[127] That afternoon the Attorney General warned Georgi Bolshakov that low-level reconnaissance, with all its attendant dangers, would start again unless the Russians promised to remove the bombers. He needed an answer, Kennedy said, before the President's press conference the next day.[128]

The next day — November 20 — was Robert Kennedy's thirty-seventh birthday. Dobrynin came to his office in the morning. Smiling broadly, he said, "I have a birthday present for you," and handed him a letter from Khrushchev agreeing to remove the IL-28s in thirty days. "Reading the letter I noticed a number of rather humorous references"; also "that the letter was rather disorganized. Dobrynin

said these letters were dictated by Khrushchev himself ... walking
around the room, never looking at the girl, ... [who] just wrote as
he talked." That afternoon John Kennedy announced the agreement. Bolshakov was in the Attorney General's office. "We had a
drink together and listened to the President's news conference. He
pronounced it very good. I then had him sing 'Happy Birthday' in
Russian to Maya Plisetskaya who was in Boston. I don't know how
he reported that back to the Kremlin."[129]

This was Bolshakov's farewell performance. The State Department
had never liked his special relationship with the Attorney General;
nor presumably had Dobrynin, who had by now established a special
relationship of his own with the Attorney General. After Joseph
Alsop exposed Bolshakov's role in a column, the Kremlin decided
that his usefulness in Washington was over and recalled him. In
March 1963, Robert Kennedy sent him a handwritten note in
Moscow:

> There is still peace even though you have been gone from the United
> States for more than two months. I would not have thought that
> possible.
>
> Anyway we all miss you. I hope you are telling all your Communist
> friends what nice people we are over here — and that they believe
> you.... Give my best wishes to my friend Maya. Why don't you two
> jump into one of those brand new luxurious jet liners and fly over and
> see us? She could dance, I could sing and you could make a speech.
> Best wishes from your friend,
>
> > Robert F. Kennedy[130]

X

Why had Khrushchev given up? "This is a mystery," Castro said
bitterly the next March. "Maybe historians will be able to clarify this
twenty or thirty years hence. I don't know."[131] The assurances about
the Cuban guarantee and the Turkish missiles undoubtedly helped
sweeten his retreat.* But the political concessions were face savers.

---

* On October 30, U Thant told Castro in Havana that the United States had
proposed a UN inspection team "composed of persons whose nationalities would be
acceptable to the Cuban government.... The United States told me [U Thant con-
tinued] that as soon as this system is set up they would make a public declaration, and
if necessary in the Security Council, that they would relinquish any aggressive intentions against the Cuban government and would guarantee the territorial integrity of
the nation. They asked me to tell you this." As Maurice Halperin well says, this offer
"might have appeared to anyone except Fidel, as a real turning point in revolutionary
Cuba's uncertain and extremely costly struggle for survival" (Maurice Halperin, *The
Rise and Decline of Fidel Castro* [Berkeley, 1972], 196). However, since Castro chose to
resist inspection to the end, the guarantee never went into formal effect. In his
November 20 statement on the IL-28s, John Kennedy said only: "For our part, if all

The real reason Khrushchev pulled out was his hopeless military situation. In his explanation to the Supreme Soviet in December 1962, he emphasized that Cuba was to be attacked in two or three days. An American invasion of Cuba would have been a disaster for Khrushchev personally and for the Soviet claim to world revolutionary leadership — especially when Khrushchev's own recklessness had handed America the pretext. What Communist state would trust Soviet promises thereafter? And, if Kennedy were serious about invasion, Khrushchev could do nothing about it, short of nuclear war against a stronger nuclear power. He was not suicidal. "It would have been preposterous," Khrushchev said later, "for us to unleash a war against the United States from Cuba. Cuba was 11,000 kilometers from the Soviet Union. Our sea and air communications were so precarious that an attack against the US was unthinkable."* Lacking conventional superiority in the Caribbean, he could neither break the blockade nor protect Cuba against invasion. Lacking strategic superiority, Khrushchev could not safely retaliate elsewhere in the world.

"What would have happened," Barton J. Bernstein later asked for the critics, "if Khrushchev . . . had refused to back down, and had chosen war instead of humiliation?"[132] But Khrushchev had never for a moment considered choosing war; and Kennedy was prepared to help him escape humiliation. The missile crisis was a triumph, perhaps the only triumph, of flexible response. Had conventional strength remained at January 1961 levels, McNamara later informed Kennedy, "we would not have had enough forces to meet the minimum requirements stated by the Joint Chiefs of Staff to carry out a successful invasion of Cuba . . . even if we had put every available combat-ready unit not committed to an overseas theatre into the operation, leaving no reserve for other contingencies."[133] When Kennedy mobilized, Khrushchev had no choice but to cash in his chips and get out of the game. Kennedy's genius lay in making sure he had chips to cash in.

As for Kennedy, he had thought the odds on war, according to Sorensen, "between one out of three and even."[134] McNamara, re-

---

offensive weapons are removed from Cuba and kept out of the hemisphere in the future, under adequate verification and safeguards, and if Cuba is not used for the export of aggressive Communist purposes, there will be peace in the Caribbean" (*Public Papers . . . 1962* [Washington, 1963], 831). In practice, aerial reconnaissance took the place of inspection, and the United States without formal commitment refrained from invasion (which it had never intended in any case).

* Khrushchev, *Last Testament*, 511. According to Hervé Alphand, the French ambassador to Washington, "De Gaulle said there was really no risk. Because he was sure of what the Russian reaction would be. It was a sure decision" (interview in *W*, October 28–November 4, 1977).

calling the magnificent sunset over the Potomac on Saturday evening, October 27, said later he had wondered how many more sunsets he was destined to see.[135] Did these highly intelligent men really believe we were on the edge of catastrophe? Recalling one's own tumult of emotion, and subtracting a certain relish in crisis not unknown on the New Frontier (not much shared, however, by the rather prudent Kennedys or by the analytical McNamara), I would say: one lobe of the brain had to recognize the ghastly possibility; another found it quite inconceivable. The President "believed from the start," his brother said, "that the Soviet Chairman was a rational, intelligent man who, if given sufficient time and shown our determination, would alter his position."[136] The two men had taken each other's measure in Vienna. A year's secret correspondence had enlarged each one's sense of the other. Kennedy's grim odds were based on fear, not of Khrushchev's intention, but of human error, of something going terribly wrong down the line. This is why he took such care to maintain tight control over American reactions.

In retrospect, I am bound to say that the risk of war in October 1962 seems to me to have been exaggerated — pardonably at the time and less pardonably in historical recapitulation. Still Khrushchev, no less than Kennedy, might have been betrayed down the line — and perhaps he was more at the mercy of his military than Kennedy was of the Pentagon.* Even with the justified assumption of reciprocal rationality a terrible risk remained. Robert Kennedy had hoped to include in *Thirteen Days* a discussion of the ultimate ethical question: "What, if any, circumstance or justification gives ... any government the moral right to bring its people and possibly all people under the shadow of nuclear destruction?"[137] How many statesmen would raise such a question? How many philosophers answer it?

Had the Americans, as Soviet theology supposed, been looking for an excuse to smash Castro or, for that matter, the Soviet Union, they could hardly have found a better one than the Soviet attempt to make Cuba a nuclear outpost. Kennedy's restraint now persuaded Khrushchev of the point the Soviet leader had rejected in Vienna — that neither side dare tamper carelessly with the explosive international equilibrium. The Kremlin fell silent about Berlin and for

* Khrushchev told Norman Cousins in 1963, "When I asked the military advisers if they could assure me that holding fast would not result in the death of five hundred million human beings, they looked at me as though I was out of my mind or, what was worse, a traitor. The biggest tragedy, as they saw it, was not that our country might be devastated and everything lost, but that the Chinese or the Albanians would accuse us of appeasement or weakness" (Cousins, "The Cuban Missile Crisis: An Anniversary," *Saturday Review*, October 15, 1977).

some years thereafter showed a certain respect for existing world frontiers. On the other hand, the Russians, determined never to be caught again in a position of gross nuclear inferiority, stepped up their production of missiles. Within a decade they came close to parity with the United States. Khrushchev meanwhile disappeared from the scene, in part a casualty of Cuba. In dismissing him in October 1964, the Presidium spoke harshly of his "harebrained scheming, hasty conclusions, rash decisions and actions based on wishful thinking."[138] His downfall moderated the Cold War of movement. It also ended the de-Stalinization campaign, the loosening in the joints of a despotic society, the gusts of fresh air that this rollicking, erratic, passionate, indelibly human, curiously attractive figure brought to the nation he loved so much.

Once Khrushchev accepted the Vienna principle, it was possible to move toward détente. If nuclear missiles had remained in Cuba, there would have been no American University speech, no test ban treaty, no 'hot line' between the White House and the Kremlin, no relief from the intolerable pressures of the Cold War. After Cuba, relief was not only thought to be possible but felt to be imperative. During the crisis, talking with Ormsby-Gore, his friend of so many years, the President had suddenly burst out: "You know, it really is an *intolerable* state of affairs when nations can threaten each other with nuclear weapons. This is just so totally irrational. A world in which there are large quantities of nuclear weapons is an impossible world to handle. We really must try to get on with disarmament if we get through this crisis . . . because this is just too much."[139] This was Khrushchev's feeling too, the theme of his impassioned letter; and the crisis deepened both their urgencies. "If thou gaze too long into the abyss," said Nietzsche, "the abyss will gaze into thee."

XI

The record demands the revision of the conventional portraits of Kennedy during the crisis: both the popular view, at the time, of the unflinching leader fearlessly staring down the Russians until they blinked; and the later left-wing view of a man driven by psychic and political compulsions to demand unconditional surrender at whatever risk to mankind.[140] Far from rejecting diplomacy in favor of confrontation, Kennedy in fact took the diplomatic path after arranging the military setting that would make diplomacy effective.

The hard-liners thought him fatally soft. Dean Acheson in another year: "So long as we had the thumbscrew on Khrushchev, we should have given it another turn every day. We were too eager to make an agreement with the Russians."[141] Richard Nixon inevitably thought that Kennedy's doves had "enabled the United States to pull defeat out of the jaws of victory."[142] Or Daniel Patrick Moynihan in 1977: "The Cuban Missile Crisis was actually a *defeat*. . . . When anybody puts missiles into a situation like that, he should expect to have a lot of trouble with the United States, and real trouble — and all that happened was the agreement: 'O.K., you can have your man down there permanently.'"[143]

The revisionists, on the other hand, portrayed Kennedy as reckless and irresponsible. This was not the view of those in the best position to judge — neither of Khrushchev nor, in the end, of Castro himself. In a time of "serious confrontation," Khrushchev said, ". . . one must have an intelligent, sober-minded counterpart with whom to deal. . . . I believe [Kennedy] was a man who understood the situation correctly and who genuinely did not want war. . . . Kennedy was also someone we could trust. . . . He showed great flexibility and, together, we avoided disaster. . . . He didn't let himself become frightened, nor did he become reckless. . . . He showed real wisdom and statesmanship."[144] In 1967 Castro told Herbert Matthews that he thought Kennedy had "acted as he did partly to save Khrushchev, out of fear that any successor would be tougher."[145] And in 1975 Castro told George McGovern: "I would have taken a harder line than Khrushchev. I was furious when he compromised. But Khrushchev was older and wiser. I realize in retrospect that he reached the proper settlement with Kennedy. If my position had prevailed, there might have been a terrible war. I was wrong."[146]

In all this, Robert Kennedy was the indispensable partner. Without him, John Kennedy would have found it far more difficult to overcome the demand for military action. Even Senator Fulbright, in Kennedy's meeting with congressional leaders before his television speech, had advocated the invasion of Cuba as a "wiser course" than the quarantine.* It was Robert Kennedy who oversaw the Executive Committee, stopped the air-strike madness in its tracks, wrote the reply to the Khrushchev letter, conducted the secret negotiations with Dobrynin. "Throughout the entire period of the crisis," Mc-

---

* Fulbright's point was that a quarantine, by involving a direct confrontation with Soviet ships, would be more likely to provoke a nuclear war than an invasion that would pit Americans against Cubans. He evidently forgot about the 22,000 Russian soldiers in Cuba. He said later, "We had not gone through the days of considering various alternatives, advantages and disadvantages. . . . Had I been able to formulate my views on the basis of facts since made public rather than on a guess as to the nature of the situation, I might have made a different recommendation" (*Congressional Record*, December 10, 1973, S22289; see also interview with R. W. Howe and Sarah Trott, *Saturday Review*, January 11, 1975).

Namara said in 1968, "a period of the most intense strain I have ever operated under, he remained calm and cool, firm but restrained, never nettled and never rattled."[147] "For this happy outcome to such long and agonizing negotiations," Adlai Stevenson wrote him, "I think you are entitled to our gratitude."[148] Khrushchev, recalling the discussions during the crisis, said the Americans "had, on the whole, been open and candid with us, especially Robert Kennedy."[149] "Looking back on it," said Harold Macmillan, "the way that Bobby and his brother played this hand was absolutely masterly. . . . What they did that week convinced me that they were both great men."[150]

John Kennedy himself, in the cruelest hour of the crisis, on the black Saturday before the golden Sunday, said fervently, "Thank God for Bobby."[151] As for Robert Kennedy, he sent a letter, "in that strange, little cramped handwriting," to Robert Lovett thanking him for his counsel and especially for a quotation Lovett had produced during the ordeal.

The quotation was: "Good judgment is usually the result of experience. And experience is frequently the result of bad judgment."[152]

23

The Cuban Connection: II

T HE DAILY LIFE of an Attorney General went on. The missiles of October were an interruption, framed between Ole Miss in September, the housing order in November, the preparation for the Georgia county-unit case in December. Still Robert Kennedy could not completely disengage from Cuba. There remained Operation Mongoose. There remained the Bay of Pigs prisoners.

I

The crisis finished Mongoose. "We had a terrible experience," Robert Kennedy recalled in 1964. CIA's Task Force W was "going to send sixty people into Cuba right during the missile crisis." One of them sent word to the Attorney General that they did not mind going but wanted to make sure he thought it worthwhile. "I checked into it, and nobody knew about it. . . . The CIA didn't and the top officials didn't." The ineffable William K. Harvey, it developed, had conceived on his own the project of dispatching ten commando teams to Cuba. Three had already departed. Kennedy called a meeting at the Pentagon. As Harvey later put it, the Attorney General took "a great deal of exception." "I was furious," Kennedy remembered, "because I said you were dealing with people's lives . . . and then you're going to go off with a half-assed operation such as this. . . . I've never seen [Harvey] since."[1]

On October 30 the Executive Committee canceled all "sabotage or militant operations during negotiations with Soviets" and sent General Lansdale to Florida to make sure that Task Force W obeyed.[2] Shortly thereafter both Mongoose and the Special Group (Augmented) were abolished. The CIA, taking care of its own, made

Harvey station chief in Rome, where he was soon sodden with drink. Lansdale moved on to other matters, then retired in the autumn of 1963. Rip Robertson went off to Vietnam.

Asked for his retrospective assessment, Maxwell Taylor said Mongoose "didn't work well at all."[3] Lansdale thought it worse than that. CIA's mindless hit-and-run tactics, he said, far from creating a political movement against the regime, stiffened the "national resolve" behind Castro.[4] The effects beyond Cuba were no better. Had sabotage been more successful, the result could have been only to increase Cuba's economic dependence on the Soviet Union. The secret war, not unreasonably seen by Castro as preparation for a new and better invasion, intensified the Cuban desire for Soviet protection.

The program Robert Kennedy and Lansdale had intended was different from the program CIA carried out. But the political base for their anti-Castro uprising simply did not exist inside Cuba, nor did the CIA wish to create such a base. The Bay of Pigs ought to have made it sufficiently clear how covert action degenerated as directives passed from immaculate conference rooms at CIA headquarters to embattled officers in the field and from there to war lovers behind the lines. Lansdale tried to control Harvey; but a Lansdale in Washington implied a Harvey in Miami, as a Harvey implied a Rip Robertson, and a Rip Robertson implied a Ramón Orozco, grinning as he brought his two severed ears back from Cuba.

The problem was why they had not called off Mongoose long before. The answer perhaps was a driven sense in the administration that someone ought to be doing *something* to make life difficult for Castro. Mongoose was poorly conceived and wretchedly executed. It deserved greatly to fail. It was Robert Kennedy's most conspicuous folly.

II

Mongoose had always weighed less on Robert Kennedy's mind than his other Cuban commitment of 1962: the Bay of Pigs prisoners.

Castro had given his list of drugs and medicines — ten thousand items, specified in detail — to the unflagging James Donovan in mid-October, two days before the world knew of the Soviet missiles. The crisis behind him, Castro was almost madder at the Russians than at the Americans. He was looking for a way to reestablish himself as an independent actor on the world scene. He needed the medical supplies. And he had no easy alternative for the prisoners. "You can't shoot them," Donovan had said in his frank, semikidding way.

"Maybe you could have done that at one time, but you can't do it now. . . . If you want to get rid of them, if you're going to sell them, you've got to sell them to me. There's no world market for prisoners."[5]

In mid-November, Castro let Alvaro Sanchez of the Cuban Families Committee visit the prisons. Appalled, Sanchez rushed back to the United States. "I'm a cattleman, Mr. Attorney General," Sanchez told Robert Kennedy on November 24, "and these men look like animals who are going to die." He could tell, he said, by looking at the back of their necks. "If you are going to rescue these men, this is the time because if you wait you will be liberating corpses." Kennedy said, "You are right. I think this is the moment." "We put them there," he told Edwin Guthman, "and we're going to get them out — by Christmas!" Guthman said it was not possible. "We will," Kennedy said.[6] This meant, under the terms of the Castro-Donovan agreement, that within a month Castro would have to receive 20 percent of the items on his list and acceptable guarantees for the delivery of the rest within six to nine months; also the $2.9 million owed for the sick and wounded prisoners sent to the United States the previous April.

There followed a classic Robert Kennedy operation. Nicholas Katzenbach, who had spent two years himself in a prisoner-of-war camp, became coordinator; Louis Oberdorfer of the Tax Division, field commander. Several Washington lawyers were enlisted: John Nolan to work with Donovan; Barrett Prettyman to solve the transportation problems; John Douglas as a general trouble-shooter. (All three joined the Department in 1963.) Mitchell Rogovin of the Internal Revenue Service handed down quick rulings on the tax deductibility of corporate contributions. Lawrence Houston of the CIA operated indistinctly in the background.

"None of us knew anything about any of the problems," Joseph Dolan said later. "Nobody knew anything about transportation, drugs, or baby food." Dolan, tying up loose ends at Ole Miss, was summoned back to Washington by Oberdorfer. He arrived to find a meeting in progress.

> Oberdorfer looked up and said to me, "Bob has a project going; we are going to try to get the Cuban prisoners out." I said, "Oh, damn." Everybody looked at me in a surprised way. I sat down and was there for about a minute when all of a sudden I said, "Oh." Lou said, "What is the matter?" I said, "You mean we are going to negotiate?" . . . I really thought that with Bob Kennedy we were going to get something going with some boats and we were going down to get the prisoners out. I had just returned from Oxford, Mississippi, and expected anything.[7]

The first problem was the drug industry — under congressional investigation for price markups and other sharp practices, filled with self-pity over what it regarded as bureaucratic persecution, distrustful of the Kennedy administration, even less fond of Castro and not disposed to do favors for either by donating drugs to ransom prisoners. Speaking with uncommon eloquence before industry leaders, the Attorney General argued the American responsibility for brave men who, at American instigation, had risked their lives for freedom and now, unless freed themselves, would die miserably in Castro's jails. His presentation, according to Oberdorfer, "had a tremendous impact on those businessmen. They came back to my office with red eyes . . . and they really got busy."[8]

The Attorney General had to tread a difficult line. Many of the sixty-three companies that made donations were under antitrust or Federal Trade Commission investigation.* He repeatedly said that cooperation would bring no favors, refusal no reprisals. In fact, the drug industry made no great sacrifice. Tax rulings based on wholesale prices produced windfall profits for some companies because of the high markups. A few companies even tried to unload inventories of obsolescent items. And, as Lloyd Cutler, the counsel for the Pharmaceutical Manufacturers Association, said after it was all over:

> The action of the drug industry in responding to the Attorney General's request was not followed by any visible change in the attitude or policies of any division in the Department of Justice. . . . Both antitrust and criminal prosecutions have been just as vigorous, and their legislative attitude is essentially the same and not entirely what we ourselves would think was the correct government policy.[9]

Supplies rapidly accumulated. So did problems. The Pentagon, for example, worried about items of alleged military value on Castro's list, such as retractable steel rulers. The Attorney General said tersely, "Are they bombs?" and they stayed in.[10] The biggest problem was getting the goods to Havana by Christmas. "Experts in the field," said Barrett Prettyman, "told us flatly that this operation was impossible — $11 million worth of goods could not be solicited from all over the country, prepared for shipment, transported by rail, air and truck, reloaded at a commmon point for transporting by sea and air, and unloaded in Cuba, all within less than two weeks" — and especially in the Christmas season.[11] But they all charged ahead. Prettyman has left a description of Oberdorfer's headquarters:

> . . . constant calls (often with four or five people waiting for each of us on incoming lines), a steady barrage of incoming and outgoing memo-

---

* At the time, thirteen firms were actually defendants in antitrust actions, twelve in FTC actions (Victor Navasky, *Kennedy Justice* [New York, 1971], 338, 453–457).

randa and files, quick and pointed conferences between from two to a dozen persons, sandwiches and coffee snatched at odd moments, the influx of businessmen, railroad men, airline executives, shippers and government personnel, the large donations chart that was changed and watched and worried over, the sudden appearance of the Attorney General with words of advice and some hurried decisions, the tension when a big donation hung in the balance, the irritation at red tape and confusion, the relief when a donor or a shipment developed out of nowhere, the laughter when a company tried to unload an absurd product, the concern when the direct line to Donovan remained silent too long.[12]

On December 18, Donovan flew to Havana, where Nolan and Prettyman soon joined him. Nolan: Donovan and Castro had established a "very cordial, bantering" relationship; "Castro, I think, regarded Jim as kind of a character, a role that Donovan played to the hilt."[13] Prettyman: Castro "leaped at the chance to take a group of us to Hemingway's home," a wild ride with two cars abreast along narrow country roads.[14] The drug shipments were arriving, but, with all the jollity, a difficulty remained: the ransom left unpaid for the April prisoner release. After half the remaining prisoners had been loaded into planes, the Cubans made it clear that this was it until they received the $2.9 million.

Robert Kennedy turned to Cardinal Cushing of Boston, a sponsor for the Cuban Families Committee. "I remembered a talk I had with Jack about the Bay of Pigs prisoners," Cushing said later. "It was the first time I ever saw tears in his eyes."[15] Such a reaction from the least sentimental Kennedy impressed the highly sentimental cardinal, who raised $1 million in a few hours from Latin American friends ("I promised them it would be repaid within three months and it was").[16] General Lucius Clay, another committee sponsor, valiantly raised the rest on his personal note. Word flashed to Havana. The last members of the Brigade boarded the planes.

It was now the day before Christmas. In Miami wives and children waited in a tumult of emotion and relief. When the last of 1113 prisoners had disembarked, Donovan called Katzenbach in Washington, and Katzenbach called Kennedy at Hickory Hill, where he was surrounded by his children on Christmas Eve: "Bob, they are all in; it's over." Katzenbach added in his ironic way, "Bob, I don't think I will come in tomorrow." "Why not?" said Kennedy. "No reason at all, Bob," said Katzenbach. "I'm just not coming in." "All right, you guys," said Kennedy, "what about Hoffa?" He was being ironic too, but, reflected Dolan, this was the essence of Robert Kennedy: "always on to the next hill, on to the next hill, on to the next hill."[17]

III

Four days after Christmas John Kennedy went to the Orange Bowl
in Miami and addressed the reunited Bay of Pigs Brigade. Rusk and
Bundy advised against his going. So did O'Donnell, who said: "It will
look as though you're planning to back them in another invasion of
Cuba." But Robert encouraged him, believing, O'Donnell thought,
that the appearance "would ease the President's sense of guilt."[18]
In the emotion of the day, Kennedy made a promise, not in the
script, that the Brigade's battle flag would be returned "in a free
Havana."[19] The exiles, taking this as O'Donnell had feared, chanted
"Guerra! Guerra! Guerra!"[20] Actually Kennedy's script was designed
to signal Havana that Washington's objection was to Cuba's external
alliances and aspirations, not to its revolution. "We support the right
of every free people," he also said, "to transform the economic and
political institutions of society so that they may serve the welfare of
all." Both the Brigade and Castro received the wrong message.

In Washington the Executive Committee, now diminished in size
and rebaptized the Standing Group, wrestled with Cuba policy. Its
members were McNamara, McCone, Bundy, Sorensen and Robert
Kennedy. The administration was no closer to an answer to Castro
than in earlier years. On one extreme was Bundy, who proposed on
January 4, 1963, an exploration of the possibility of communicating
with Castro. Thereafter what Bundy called in April the "gradual
development of some form of accommodation with Castro" became
a standard item in lists of policy alternatives.[21] On the other extreme
was the Defense Department, which, as Sterling Cottrell of State
reported to Bundy late in January, still favored "increasing degrees
of political, economic, psychological and military pressures" to bring
about "the overthrow of the Castro-Communist regime." If anti-
Castro groups in Cuba requested assistance, Defense thought the
United States "should be in a position to respond with open military
support . . . up to the full range of military forces."[22] The Pentagon
felt more than ever, as one of State's Cuban specialists put it, that
"we had missed the big bus" in not invading Cuba during the missile
crisis.[23]

As for Robert Kennedy, he was particularly worried by Castro's
own secret war against Latin America, especially when the Cubans in
February organized a guerrilla front against Betancourt in Venezuela. On March 14, following what he evidently found a most
unsatisfactory NSC meeting, he sent his brother a testy memorandum

urging new efforts to counter Cuban-trained operatives in South America. As for Cuba itself,

> John McCone spoke at the meeting today about revolt amongst the military. He described the possibilities in rather optimistic terms. What is the basis for that appraisal? What can and should we do to increase the likelihood of this kind of action? . . . Do we have evidence of any break amongst the top Cuban leaders and if so, is the CIA or USIA attempting to cultivate that feeling? I would not like it said a year from now that we could have had this internal breakup in Cuba but we just did not set the stage for it.[24]

The President, as usual, felt less tragically about Cuba. Though he did tell Bundy to send his brother's questions to CIA, he did not respond directly to the Attorney General, who wrote him plaintively nearly two weeks later: "Did you feel there was any merit to my last memo? . . . In any case, is there anything further on this matter?"[25]

IV

Most of the 200,000 Cuban refugees in the United States were hard-working, law-abiding people, happy at the chance of a new life in a land of relative political and intellectual freedom.* A minority were violent men, consumed with a single hope and a single hatred, living for the day when they could return to Havana with Fidel Castro's head on a pike. Embittered by the peaceful resolution of the missile crisis, the anti-Castro zealots were now determined to force the administration into action against its will — by rumors and raids and exile manipulation of American domestic politics. "European embassies," wrote the English ambassador to Havana, "began to be plagued . . . with anonymous phone calls and letters and mysterious visitors bringing with them sketches and plans of caves all over the country where the Cuban Government was alleged to have hidden away a proportion of their stock of rockets. . . . It seemed 90 percent certain to most of us that this was another attempt by would-be counter-revolutionaries to bring about a United States invasion in a last desperate gamble to bring down the Castro regime."[26] The rumors raced on to Miami and thence to the United States Senate.

On January 31, 1963, the indefatigable Kenneth Keating declared that the Soviet Union had cunningly filled Cuban caves with nuclear missiles. Soon Keating shifted his target to the Soviet soldiers still in

---

* Of these refugees, 153,634 entered legally from January 1961 to the missile crisis; 29,962 from the missile crisis to November 1965; others arrived 'illegally' on boats and rafts (L. D. Bender, *The Politics of Hostility: Castro's Revolution and United States Policy* [Hato Rey, Puerto Rico, 1975], 118).

Cuba. Richard Nixon denounced Kennedy's Cuban policy — "we have goofed an invasion, paid tribute to Castro for the prisoners, then given the Soviets squatters' rights in our backyard" — and demanded a "command decision" to get the Russians out.[27] John Kennedy muttered privately that 17,000 Soviet troops in Cuba were not so bad compared to 27,000 American troops in Turkey, but added wearily, "It isn't wise politically, to understand Khrushchev's problems in quite this way."[28] It wasn't wise politically because the visible pressure was all for drastic action. In April, when James Reston of the *New York Times* wrote in some detail about the "subversive war" conducted by the American government against Cuba,* no one in Congress or the press saw this as a matter for investigation or criticism.

Actually, by the time Reston wrote, the secret war was almost at an end. The exile raids had become too much. Under such designations as Alpha 66 and Commandos L-66, daring men in small fast boats, setting out from the Bahamian keys, sometimes from Florida itself, had been landing saboteurs in Cuba and firing torpedoes at Cuban ships; even, as on March 18 and again on March 26–27, at Soviet ships. These raids, John Kennedy told Marquis Childs, the columnist, "made everything worse."[29] The NSC met gloomily on the subject at the end of March. Robert Kennedy noted:

> McCone presentation — Gave facts & then said felt less criticism by newspapers & Congress if we do not stand down the raids.
> Rusk: If these raids are going to be carried out we have to accept responsibility — better if they are going to be done that we do it. . . .
> Decided to proceed & work up plan to prevent attacks from continuing.[30]

The next day the administration announced it would "take every step necessary" to stop raids from the United States. As the CIA station chief in Miami recalled it, "The whole apparatus of government, Coast Guard, Customs, Immigration and Naturalization, FBI, CIA, were working together to try to keep these operations from going to Cuba."[31] The FBI entered exile camps and seized caches of dynamite and bomb casings. A number of Cubans were indicted. Early in April, Kennedy terminated CIA financial support for Miro Cardona and the Cuban Revolutionary Council. The Standing Group meanwhile decided that the CIA sabotage program was "not worth the effort expended on it." On April 3, Bundy informed a Cuba meeting that no further sabotage operations were under way.[32] When Robert Kennedy brought up the subject again the next

* James Reston, "Kennedy and His Critics on Cuba," *New York Times*, April 21, 1963. The Reston column proves that the "secret war" was not much of a secret.

month, Bundy ordered another review and then told the Attorney General: "The sum and substance of it is that useful organized sabotage is still very hard to get. . . . Proposals which do more good than harm are rare."[33]

V

Early in April 1963, Donovan and Nolan had returned to Havana to wrap up the prisoner exchange negotiations. This time Castro took them to the Bay of Pigs and delivered an amiable battlefield lecture. To demonstrate the impassability of the area beyond the beachhead, he strode several paces into the salt marsh until mud oozed to the top of his boots. They lunched on a boat in the bay and spent the sun-drenched afternoon fishing and skin-diving, guided by a Russian PT boat.[34]

Donovan's mission now was to secure the release of a number of Americans, including CIA men, from Cuban prisons. (When this was first broached to him, Donovan had said, "Jesus Christ, I've already done the loaves and the fishes, and now they want me to walk on the water.")[35] Castro asked about future United States policy. Donovan pointed to the restrictions placed on exile groups but added that the prisoners remained "a stumbling block." Castro pursued the question. "His ideal government, he emphasized, was not to be Soviet oriented." He asked how diplomatic ties might be resumed. Donovan replied: the way porcupines make love — very carefully. At a minimum Donovan thought there must be assurances that Cuba would leave other Latin countries alone. Later, Dr. René Vallejo, Castro's intimate friend and personal physician, who had interned in Boston and served with the U.S. Army in the war,[36] took Donovan aside and said that Fidel "wanted to officially establish such relationships . . . even though certain Communist officials in the Cuban government were unalterably opposed."*

Donovan considered Castro "a most intelligent, shrewd and relatively stable political leader."[37] "Throughout the prisoner exchange negotiations," Nolan reported to Robert Kennedy, "he had been both reasonable and reliable and has not been difficult to deal with."[38] "Our impressions," Nolan said later, "would not square with the commonly accepted image. . . . Castro was never irrational, never

---

* This account of the Castro-Donovan talks is drawn from Donovan's debriefing by M. C. Miskovsky of the CIA and Miskovsky's memorandum of April 13, 1963, to John McCone. Donovan later wrote of Castro: "A handsome man and witty conversationalist, he little resembles the caricature which we see in the United States. His grooming in personal life is impeccable" (James B. Donovan, *Challenges* [New York, 1967], 100–101).

drunk, never dirty." "What do you think?" Robert Kennedy asked. "Can we do business with that fellow?"[39]

Donovan's efforts secured the release of nearly 10,000 Cubans and Americans from Cuban detention by July 4.[40] After Donovan's visit and on his recommendation,[41] Castro gave Lisa Howard of the American Broadcasting Company ten hours of interview in late April. She concluded that Castro was "looking for a way to reach a rapprochement," probably for economic reasons. The "U.S. limitations on exile raids," Castro had said, were "a proper step toward accommodation." Che Guevara and Raul Castro, Lisa Howard thought, opposed accommodation; Vallejo and Raul Roa, the foreign minister, favored it. Castro himself, Howard felt, was ready to discuss rapprochement "with proper progressive spokesmen." He indicated, however, "that if a rapprochement was wanted President John F. Kennedy would have to make the first move."*

But a few days after his talks with Lisa Howard, Castro was on his way to the Soviet Union. Returning after five weeks in a glow of

---

* CIA debriefing of Lisa Howard, May 1, 1963, RFK Papers. This had not always been Che Guevara's position. On August 17, 1961, after the Punta del Este conference, he had held a conversation in Montevideo with Richard Goodwin. He then said, as Goodwin reported to President Kennedy, "that they didn't want an understanding with the United States, because they knew that was impossible. They would like a *modus vivendi* — at least an interim *modus vivendi*. . . . He thought we should put forth such a formula because we had public opinion to worry about whereas he could accept anything without worrying about public opinion. I said nothing, and he waited and then said . . . (1) That they could not give back the expropriated properties . . . but they could pay for them in trade. [Payment for expropriated properties, Goodwin said later, was "the thing that meant least to JFK." Che's idea that it was of primary importance revealed his own Marxist dogmatism about democratic leaders.] (2) They could agree not to make any political alliance with the East — although this would not affect their natural sympathies. (3) They would have free elections — but only after a period of institutionalizing the revolution had been completed. . . . This included the establishment of a one-party system. (4) Of course, they would not attack Guantanamo. . . . (5) He indicated, very obliquely and with evident reluctance because of the company in which we were talking, that they could also discuss the activities of the Cuban revolution in other countries. . . . He said they could discuss no formula that would mean giving up the type of society to which they were dedicated." He also thanked Goodwin for the Bay of Pigs: "It had been a great political victory for them — enabled them to consolidate — and transformed them from an aggrieved little country to an equal" (Goodwin to JFK, August 22, 1961, Schlesinger Papers).

Perhaps Guevara was following instructions from Castro, rationalizing this ("an interim *modus vivendi*") as a Leninist tactic in the tradition of Brest-Litovsk, or perhaps he was less radical then than he became by 1963. There was no follow-up in Washington because, as Goodwin reminded me in 1977, the timing and psychology were wrong. It was too soon after the Bay of Pigs humiliation. More important, Betancourt, Haya de la Torre, Frei and the democratic left in Latin America would have been appalled. A modus vivendi would have legitimized a Marxist regime in the hemisphere and therefore given a color of legitimacy to Communist actions against democratic regimes in Venezuela and elsewhere (Richard Goodwin, in interview by author, June 11, 1977).

enthusiasm, he did his best to dispel the anti-Soviet doubts he had himself fostered after the missile crisis. The collapse of the 1963 sugar crop made him more dependent than ever on Soviet aid. On the other hand, he may have also come back impressed by Khrushchev's postcrisis desire for détente.

The American government was no less vacillating. When the Standing Group met in Washington at the end of May, bafflement prevailed. McCone argued for sabotage in order to "create a situation ... in which it would be possible to subvert military leaders to the point of their acting to overthrow Castro." McNamara questioned the utility of sabotage and preferred overt economic pressures. The Attorney General, ever the activist, said the United States "must do something against Castro, even though we do not believe our actions would bring him down."[42] On June 3 the Special Group recommended the exploration of "various possibilities of establishing channels of communication to Castro." The American government appeared, like Castro, to be facing in two directions. For on June 19 the Special Group suddenly approved a new sabotage program directed at major segments of the Cuban economy.[43]

VI

The June 19 decision meant the resurrection, in a highly qualified way, of the secret war. The object was no longer to overthrow the regime; only to "nourish a spirit of resistance and disaffection which could lead to significant defections and other byproducts of unrest."[44] The new campaign, while less ambitious than Mongoose, seems, after the failure of Mongoose, even more pointless. Given the imperfect documentation, one must surmise what lay behind the June revival: the desire to divert Castro from Venezuela, where his secret war against Betancourt was building to a climax; a desire also to intensify his economic stringencies; beyond this, a relief from frustration, a feeling that Castro deserved harassment, a hope that what the White House saw as no more than "pinpricks"[45] would, without interrupting détente, reassure Latin American governments, Cuban exiles, Republican critics and the CIA that the administration was not faltering in its opposition to Castro.

The imperfect record also makes it hard to discover what had actually been going on since Mongoose. When Desmond FitzGerald, who had replaced Harvey as the CIA's man for Cuba, was asked by the Special Group on June 19 whether the Cubans might retaliate, he replied that they always had this capability "but that they have not retaliated to date, in spite of a number of publicized exile raids."

The phrase "exile raids" implied that the Agency had been doing nothing itself. Certainly the Special Group, according to the Church committee, had authorized "little, if any," CIA sabotage in the first six months of 1963.[46] But had CIA been up to its old tricks? Had it indeed been responsible for the March attacks on Soviet shipping — the attacks the President and the National Security Council had blamed on Alpha 66 and Commandos L-66? Captain Bradley Earl Ayers, a paratrooper assigned to CIA, recalled General Victor Krulak, the JCS counterinsurgency specialist, telling him in the spring of 1963 that the operations attributed to exile groups were mostly "planned and conducted under the supervision of the CIA . . . from bases in southern Florida." When Ayers himself went to southern Florida to train Cuban commando units, he learned that "customarily, either by prearrangement through exile operatives or because of their own wish to capitalize on the political impact of such incidents, one of the splinter, independent Cuban exile groups, such as Alpha 66, would publicly take credit for the [CIA] raids."[47] Despite the death of Mongoose and the lack of Special Group authorization, CIA/Miami evidently continued, under exile cover, to wage its private war against Castro. "When the target diminishes," as James Angleton, CIA's counterespionage chief, said later, "it's very difficult for a bureaucracy to adjust. What do you do with your personnel? We owed a deep obligation to the men in Miami."[48] It was easier to crank up a large clandestine operation than to wind it down.

The FBI continued raiding exile camps after June 19,[49] but perhaps these were camps not under CIA control. CIA/Miami, according to Ayers, meanwhile received orders to "increase the effectiveness and frequency of hit-and-run raids by exile commando groups."* Robert Kennedy thought Desmond FitzGerald an improvement over the detested Harvey. Sabotage was "better organized than it had been before and was having quite an effect." What kind of projects? asked John Bartlow Martin. Kennedy replied, a little inconsistently, "Well, just going in blowing up a mine . . . a bridge. Some of them ended in disaster, people were captured, tried and confessed. It wasn't very helpful." "Any direct assassination attempt on Castro?" "No." "None tried?" "No." "Contemplated?" "No."[50] Alas, Kennedy did not know all the projects FitzGerald was organizing.

Other efforts were mounted against the Castro regime from Cen-

---

* B. E. Ayers, *The War That Never Was* (Indianapolis, Ind., 1976), 100. Ayers also claimed that Robert Kennedy himself paid two visits to CIA installations in Florida — one after it had been decided to hit a major Cuban oil refinery (ibid., 76, 147–148). Kennedy's appointment books show no trips to Florida between April 27–28 (when he went to see his father at Palm Beach) and November 28–December 3 (at Palm Beach and Hobe Sound). He made two earlier weekend trips that year to Palm Beach: January 25–28 and March 15–18.

tral America. Here the record is unusually murky. At the end of June, Luis Somoza, son of the thieving Nicaraguan dictator and a former president of Nicaragua himself, asked to see the Attorney General. The State Department advised Kennedy that Somoza wanted to know the American attitude toward an anti-Castro base about to be set up in Nicaragua by Manuel Artime, the Bay of Pigs leader. "We recommend that you limit your reply to . . . general terms," State said; Kennedy might express sympathy with the exiles but no particular knowledge of Artime.[51] There was, indeed, an Artime operation and Hal Hendrix of the *Miami News* supposed it managed either by CIA or, "on a hip pocket basis," by the Attorney General himself.[52]

Somoza was soon telling Caribbean notables that he had received a "green light" from Robert Kennedy to mount anti-Castro raids from Nicaraguan bases.[53] In another month he claimed he was "leading a movement of the five Central American countries to overthrow Castro and that he had the blessing of the United States government." When a Central American repeated this remark to the State Department, the coordinator of Cuban affairs responded that Somoza had received neither a green nor even an amber light from Washington and that his claims should be treated with "extreme reserve."[54] In 1964, the Somozas alleged that Robert Kennedy and McNamara had assured them of full U.S. support for Artime. The State Department promptly cabled the American ambassador in Managua, "No high USG officials have made statements about USG support for Artime such as those attributed to them," though some may have said they considered Artime "a responsible, dedicated Cuban leader."[55]

My guess is that CIA was financing Artime,* that State disapproved and may not even have known and that Robert Kennedy, if he knew, thought Artime a brave man who had earned an opportunity to show what he could do. But the whole episode remains perplexing. Robert Kennedy understood so lucidly the enormity of a Pearl Harbor air strike on a small country. It is odd that he did not see that the same principle applied to the secret war. Still, in the May of Birmingham and Bull Connor, in the June when George Wallace stood in the schoolhouse door, in the summer when Medgar Evers died and Martin Luther King dreamed, in the autumn of the civil rights bill, Cuba was not a subject to which the Attorney General devoted sustained attention.

---

* It certainly was by 1964–65. See Senate Select Committee to Study Governmental Operations with respect to Intelligence Activities (hereafter cited as Church committee), *Interim Report: Alleged Assassination Plots Involving Foreign Leaders*, 94 Cong., 1 Sess. (1975), 89–90. (Artime is referred to under the code designation B-1.)

VII

It deserved sustained attention. The White House saw the resumed secret war as a way of keeping Castro off balance and neutralizing bureaucratic discontent while deciding whether it would be possible to risk a try at accommodation. But the CIA, under its distended theory of authorization, saw it as a license to renew its attempts to kill the Maximum Leader. These attempts made even less sense in 1963 than they had before. The notion of invading Cuba had been dead for years. I suppose that, in 1961 and 1962, CIA might still have regarded assassination as the ultimate logic of a frenetic over-throw-Castro policy. But in 1963, with invasion absolutely excluded, with the anti-Castro policy drastically modified and with the White House drifting toward accommodation, assassination had no logic at all.

In the spring of 1963, moreover, the Standing Group had asked the CIA's Intelligence Branch to assess possible developments in the event of Castro's death. This had nothing to do with assassination plots; estimating the consequences of the death of a national leader, from whatever cause, was a favorite intelligence exercise, applied indifferently to de Gaulle, Khrushchev, Salazar or Castro. The CIA Office of National Estimates decided that Castro's death, far from benefiting the United States, would probably mean that "his brother Raul or some other figure in the regime would, with Soviet backing and help, take over." And, if Castro were by any chance assassinated, "the U.S. would be widely charged with complicity."[56]

So far as the Church committee could discover, the Clandestine Service's assassinatory enthusiasm had lain dormant after January 1963, when its technicians failed to perfect the explosive seashell designed to blow up the Maximum Skindiver. The June 19 decision to revive the secret war now rekindled that old feeling. CIA turned to an 'agent in place' in Havana — Rolando Cubela Secades, a revolutionary zealot who had killed Batista's military intelligence chief in 1956, fought beside Castro in the Escambray Mountains, seized the Presidential Palace in advance of Castro's own arrival in Havana in 1959 and in 1960 was Castro's instrument in destroying the ancient freedoms of the University of Havana. Thereafter Cubela claimed disillusionment. The Agency recruited him in 1961 and gave him the code name of Am/Lash. A heavy drinker and a psychiatric patient, Cubela was not the ideal operative; better, however, in the CIA view, than no man in Havana at all.[57]

The Agency, which had dropped contact with Cubela after the

missile crisis, got in touch again. In early September he met a CIA man in São Paulo, Brazil, and said he was prepared to attempt an "inside job" against Castro's life.[58] On September 7, the very day that CIA/Washington received the report on the Brazil meeting, Castro attended a party at the Brazilian embassy in Havana. There he told Daniel Harker of the Associated Press, "Kennedy is the Batista of our time, and the most opportunistic President of all time." He warned against "terrorist plans to eliminate Cuban leaders." He said: "We are prepared to . . . answer in kind. United States leaders should think that if they assist in terrorist plans to eliminate Cuban leaders, they themselves will not be safe."[59]

The fact that Castro chose the Brazilian embassy for this interesting disquisition alarmed CIA counterespionage experts. Was he signaling his knowledge of the São Paulo meeting? Their suspicions increased when Cubela requested murder weapons and a meeting with Robert Kennedy. Was Cubela an *agent provocateur*? Even if not, was it safe, given his known instability, to bring him into direct contact with high American officials? "My disapproval of it was very strong," one counterespionage officer testified. "Des FitzGerald knew it . . . and preferred not to discuss it any more with me." Richard Helms solved the Robert Kennedy request by deciding, according to the CIA inspector general's report, that "it was not necessary to seek approval from Robert Kennedy for FitzGerald to speak in his name."[60]

It was in this period that McCone read about Sam Giancana and the CIA in the *Chicago Sun-Times* and demanded an explanation. His subordinates, on the principle of admitting only defunct operations, explained Giancana away as a historical incident. They told him nothing about Am/Lash. Nor, so far as the record shows, was anyone outside CIA — in the White House, State, Defense or Justice — told about Am/Lash.*

Presenting himself falsely as Robert Kennedy's "personal representative," Desmond FitzGerald met Cubela on October 29, 1963. Cubela asked anxiously for some means of killing Castro without being killed himself. FitzGerald claimed to have told Cubela that the United States would support a successful coup but would have "no part of an attempt on Castro's life." The case officer, who served as interpreter, did not remember any such disclaimer. In any event,

---

* Church committee, *Assassination Plots*, esp. 161–166, 175; Church committee, *Final Report*, bk. V, *The Investigation of the Assassination of President John F. Kennedy*, 94 Cong., 2 Sess. (1976), 27, 69. In 1966 CIA said in a memorandum to the Secretary of State, "The Agency was not involved with [Am/Lash] in a plot to assassinate Fidel Castro." Richard Helms later told the Church committee that this memorandum was "inaccurate" (*Assassination Plots*, 178).

when the three met again three weeks later, FitzGerald, according to the CIA record, promised Cubela "everything he needed (telescopic sight, silencer, all the money he wanted)." In addition, FitzGerald presented Cubela with a ball-point pen containing a hypodermic needle so fine that the victim would allegedly not notice its insertion. Cubela grumbled that CIA could surely produce "something more sophisticated than that." No one remembers whether he took the poison pen with him or threw it away. The meeting took place in Paris on November 22, 1963.[61]

The Agency plied Cubela with weapons and encouragement for another year and a half. In the spring of 1965, it brought him together with Manuel Artime in Central America. The two men planned to revive the Bay of Pigs formula: an exile invasion combined with the murder of Castro. But Castro, as the CIA belatedly recognized, had penetrated the operation.[62] Cubela was arrested. In 1966 he confessed plotting with Artime and begged for the firing squad. Castro interceded on his behalf. Cubela received thirty years.[63]

The Church committee found evidence of at least eight CIA attempts to kill Castro from 1960 through 1965. In 1975 Castro gave George McGovern a list of 24 supposed CIA attempts over the same period, some against himself, some against other Cuban leaders. Oddly there was little duplication. The CIA denied involvement in fifteen of Castro's cases. In the other nine it admitted relationships with people mentioned but not for the purpose of assassination.[64] From all this mighty effort Castro emerged unscathed. Either the CIA repeatedly bungled or else Castro knew in advance — in the Rosselli case, perhaps through Trafficante; in the Am/Lash case, perhaps through Cubela himself, either because he was a double agent (how else to explain his freedom to travel abroad?) or simply because he was emotional, suggestible and careless.

In the course of his talks with Donovan and Nolan in April 1963, Castro had got on the subject of assassination. Asked later whether Castro thought the United States government was trying to kill him, Nolan replied, "If he did, he didn't let on. . . . But we talked for, as I recall, an hour or so riding in the car about the possibility of somebody, a disaffected Cuban, shooting Castro."* In the spring Castro may not have been sure about an official American hand in

---

* In 1966 Castro gave Lee Lockwood a detailed account of CIA activities against Cuba but said not a word about assassination (Lockwood, *Castro's Cuba, Cuba's Fidel* [New York, 1967], 202–203). Donovan and Nolan were much impressed by Castro's security — the handling of the cars, for example, that made it impossible for any other car on the road to get into a shooting position. "They did this without apparent effort . . . like they'd run a lot of drills on it" (Nolan, in recorded interview by Frank DeRosa, April 25, 1967, 17, JFK Oral History Program).

the assassination attempts. After all, the underworld had reasons of its own to rub him out, and who would have supposed that gangsters were working for the CIA? Am/Lash was another matter. When Cubela met with what he believed to be Robert Kennedy's "personal representative," Castro may have had for the first time what seemed conclusive evidence tying the Kennedys directly to the plots against his own life.

At this point, total murkiness takes over. In 1967 John Rosselli, now fighting deportation as well as charges of gambling fraud, disclosed to his lawyer, Edward Morgan of Washington, his role in the CIA assassination plots. He had subsequently learned, he said, from "sources in places close to Castro," that Castro had found out about the plots and decided that, "if that was the way President Kennedy wanted it, he too could engage in the same tactics." So, Rosselli claimed, Castro "despatched teams ... to the United States for the purpose of assassinating President Kennedy." [65] Rosselli later suggested to the columnist Jack Anderson that Castro, with Latin irony, intended to use against Kennedy the same members of the old Santos Trafficante gang in Havana that the CIA had intended to use against him. [66]

Trafficante and Rosselli were friends, at least in the fashion of the underworld. They dined together at Fort Lauderdale twelve days before Rosselli was hacked up and stuffed into the oil drum. "Authorities believe," according to the *New York Times*, "it was a member of the Trafficante organization who was able to lure Mr. Rosselli to his death." [67] Like all members of his trade, Trafficante regarded Robert Kennedy as Public Enemy Number One. No Attorney General in history had pursued the syndicates so relentlessly. There was every indication, after the Valachi testimony in the autumn of 1963, that the worst was yet to come. Trafficante, in addition, was a friend of Jimmy Hoffa's. A year before, he had discussed the President with a Cuban acquaintance. "Have you seen," the gangster said, "how his brother is hitting Hoffa, a man who is a worker, who is not a millionaire? ... Mark my words, this man [John] Kennedy is in trouble, and he will get what is coming to him. ... He is going to be hit." [68]

VIII

Yet it was more complicated than that. Both Castro and Kennedy had been pursuing dual policies since the missile crisis: at one moment, reaffirming impassable ideological antagonism; at the next, squinting toward accommodation.

Castro's goal was to secure independence for his country. After Khrushchev's willfulness during the crisis, he no doubt saw a need to assert Cuban sovereignty against the Soviet Union as well as against the United States. This argued for placing his country in the position where it could play one superpower off against the other. On Kennedy's side, his objection had never been to the Cuban revolution per se. It was to a Soviet-aligned Cuba, a repository for Soviet missiles, a base for the subversion of the Alliance for Progress. The rudiments of a deal were there: Cuba retaining its revolution and sovereignty but no longer a satellite or a subverter. Each leader, however, surrounded by Cold War doctrinaires in his own camp, had to proceed in stealth. Neither trusted the other. Neither disdained the other.

Asked in April 1964 about a deal with Castro, Robert Kennedy said, "We always discussed that as a possibility, and it was a question of trying to work it out."[69] His own attitude toward the Cuban revolution, for all his recurrent needling about sabotage, notably relaxed in the course of 1963. The beguiling Donovan-Nolan portrait of Castro probably impressed him. His continuing campaign to lift the travel restrictions stamped in American passports now focused on Cuba. He thought it preposterous to prosecute American students who wanted to inspect the Castro revolution. "What's wrong with that?" he said to me one day. "If I were twenty-two years old, that is certainly the place I would want to visit. . . . I think our people should go anywhere they want."[70]

The State Department objected, asking how Latin American countries could be expected to keep their citizens out of Cuba if the United States let theirs in. Katzenbach, on Kennedy's behalf, responded that their travel controls were for their own self-protection and that their security problems should not govern the rights of United States citizens to wander as they chose.[71] A memorandum from Ball to Rusk summed up the situation at the end of 1963:

> At meeting this afternoon concerning Attorney General's proposal to remove travel restrictions on Cuba, [Edwin] Martin [the assistant secretary for inter-American affairs] and others made strong presentation that removal or easing of restrictions . . . would make more difficult our policy of getting Latin Americans to maintain travel restrictions and would erode U.S. policy of isolation of Cuba in hemisphere. . . . On the other hand, there was general agreement that travel restrictions are contrary to American tradition and put U.S. in unfavorable light around the world. Accordingly our recommendation is that we give consideration to abolishing travel restrictions to all areas except Cuba.[72]

Even this, however, went too far for a cautious Secretary of State confronted by a Cold War Congress.

The summer of 1963 saw the negotiation of the limited test ban treaty in Moscow, followed by John Kennedy's unexpectedly successful western swing in defense of the treaty and then, on September 24, by its comfortable ratification in the Senate. The people, it appeared, were considerably ahead of the politicians on Cold War issues. This enlarged possibilities within the hemisphere and increased the interest of the Kennedys in what Robert described a few months later as "some tentative feelers that were put out by [Castro] which were accepted by us . . . through Bill Attwood."[73]

Attwood, a gifted journalist, formerly an editor of *Look*, was American ambassador to Guinea, on home leave to convalesce from an attack of polio. While recovering, he had been seconded to work with Adlai Stevenson at the United Nations. The Guinean ambassador to Cuba had told him that Castro, unlike the Communists around him, disliked Cuba's satellite status and was looking for a way of escape. Attwood, who had interviewed Castro in 1959, found this reasonable. The Cuban leader had seemed, like Sekou Touré of Guinea, sufficiently naive to be swayed by Communist advisers but too idiosyncratic to endure Communist discipline.[74] Then, in early September, Carlos Lechuga, the Cuban ambassador to the UN, raised with Attwood the possibility of exploratory talks.[75]

On September 18 Attwood sent the State Department a "Memorandum on Cuba." The policy of isolating Cuba, he argued, not only intensified Castro's desire to cause trouble but froze the United States before the world "in the unattractive posture of a big country trying to bully a small country."

According to neutral diplomats and others I have talked to at the UN and in Guinea, there is reason to believe that Castro is unhappy about his present dependence on the Soviet bloc; that he does not enjoy in effect being a satellite; that the trade embargo is hurting him — though not enough to endanger his position; and that he would like to establish some official contact with the United States and go to some length to obtain normalization of relations with us — even though this would not be welcomed by most of his hard-core Communist entourage, such as Che Guevara.

All of this may or may not be true. But it would seem that we have something to gain and nothing to lose by finding out whether in fact Castro does want to talk and what concessions he would be prepared to make. . . .

[Attwood proposed] a discreet inquiry into the possibility of neutralizing Cuba on our terms. It is based on the assumption that, short of a change in regime, our principal political objectives in Cuba are:

  a. The evacuation of all Soviet bloc military personnel.
  b. An end to subversive activities by Cuba in Latin America.
  c. Adoption by Cuba of a policy of non-alignment.

The time and place for the inquiry, Attwood suggested, were the current session of the UN General Assembly. As a journalist who had interviewed Castro, he could plausibly meet Lechuga and chat about old times. If Castro were interested, one thing might lead to another. If Attwood were invited to go to Cuba, he would travel "as an individual but would of course report to the President before and after the visit. . . . For the moment, all I would like is the authority to make contact with Lechuga. We'll see what happens then."[76]

IX

This was a bold proposition for 1963. With a presidential election coming up the next year, it was a course filled with extraordinary risk. If Keating or Goldwater (or Nixon or Rockefeller), if the Cuban exiles, caught a whiff of it, there would be hell to pay. Averell Harriman, now under secretary of state for political affairs, was the first man in Washington to receive the memorandum. Responding the next day, he told Attwood that he was "adventuresome" enough to favor the plan but that, because of its political implications, Attwood should discuss it with Robert Kennedy.[77] "Bill Attwood got in touch with me," Robert Kennedy said the next year, "and I had him get in touch with Mac Bundy." The Attorney General thought the effort "worth pursuing." Bundy, who had long favored a look at accommodation, reported back to Attwood that the President was in favor of "pushing towards an opening toward Cuba" to take Castro "out of the Soviet fold and perhaps wiping out the Bay of Pigs and maybe getting back to normal."[78] Attwood talked to Lechuga. Lisa Howard, brought in because of her Cuban friendships, talked to Dr. René Vallejo. On October 31, Vallejo told Howard that Castro was ready for a meeting and that, understanding the importance of secrecy, he would send a plane to fly a designated American official to a private airport near Havana.[79] "The President gave the go-ahead," as Robert Kennedy summed it up in 1964, "and [Attwood] was to go to Havana, I don't know, December last year or January of this year and perhaps see Castro and see what could be done [to effect a] normalization of relationship."[80]

In the meantime, President Kennedy was pressing the matter on an entirely separate channel. Jean Daniel of L'Express was in Washington. Attwood and Ben Bradlee, both old friends of Daniel's, urged the President to see him.[81] Nothing happened. Bradlee called again, mentioning that Daniel was on his way to Havana. "Have him come tomorrow at 5:30," Kennedy said at once. They began — it was October 24 — with the obligatory chat about de Gaulle. Daniel brought up Indochina. "We haven't enough time to talk about Vietnam," said Kennedy, "but I'd like to talk to you about Cuba."

He had thought about few subjects, Kennedy said, with greater care. "There is no country in the world . . . where economic colonization, humiliation and exploitation were worse than in Cuba, in part owing to my country's policies during the Batista regime. I believe that we created, built and manufactured the Castro movement out of whole cloth and without realizing it." He assured Daniel "that I have understood the Cubans. I approved the proclamation which Fidel Castro made in the Sierra Maestra. . . . I will go even further: to some extent it is as though Batista was the incarnation of a number of sins on the part of the United States. Now we shall have to pay for those sins. In the matter of the Batista regime, I am in agreement with the first Cuban revolutionaries. That is perfectly clear."

Unfortunately, Kennedy continued, it had ceased to be a purely Cuban problem. It had become a Soviet problem. Castro had betrayed his promises of the Sierra Maestra and "has agreed to be a Soviet agent in Latin America." In so doing, he had brought the world to the verge of nuclear war. The Russians understood this; "I don't know whether [Castro] realizes this, or even if he cares about it." Kennedy rose; the interview was over. Daniel detained him for two quick questions. Did the American President see an incompatibility between American liberalism and socialist collectivism? Kennedy said, "We get along very well with Tito and Sekou Touré." What about the economic blockade of Cuba? "The continuation of the blockade," Kennedy said, "depends on the continuation of subversive activities." Then: "Come and see me on your return from Cuba. Castro's reactions interest me." [82]

Daniel went on to Havana. On November 5, Bundy told Attwood that the President was more interested than the State Department in exploring the Cuban overtures. A State Department memorandum of November 7 certainly took a much harder line than the White House. "Before the United States could enter into even minimum relations with any Cuban Government," the Department opined, Havana would not only have to end political, economic and military dependency on "the Sino-Soviet bloc"* and cease its subversion within the hemisphere, but would have to "renounce Marxism-Leninism as its ideology, remove Communists from positions of influence, provide compensation for expropriated properties and restore private enterprise in manufacturing, mining, oil and distribution.[83] This insistence on repealing most of the revolution was not part of

* It seems incredible, but the troubles between Moscow and Peking had evidently not come to the notice of the State Department, at least not of the American Republics Division.

Kennedy's thinking, nor did it appear in a Bundy memorandum of November 12, prepared as guidance for Attwood. The only "flatly unacceptable" points in Castro's policy, Bundy said, were Cuba's submission to external Communist influence and his subversion directed at the rest of the hemisphere.[84]

Havana was doubtless conducting its own internal debate. Vallejo, the stout proponent of accommodation, kept calling Lisa Howard to promise that Attwood would be brought to Cuba secretly, that only he and Castro would meet him and that Che Guevara specifically would not be present. Vallejo added that he could not come to New York and that Attwood should await word from Lechuga.[85]

On November 18, Kennedy himself flashed a new message to Castro. Speaking before the Inter-American Press Association in Miami, he emphasized once again that the Alliance for Progress did "not dictate to any nation how to organize its economic life. Every nation is free to shape its own economic institutions in accordance with its own national needs and will." As for Cuba, a "small band of conspirators" had made it "a weapon in an effort dictated by external powers to subvert the other American republics. This, and this alone, divides us. As long as this is true, nothing is possible. Without it, everything is possible. . . . Once Cuban sovereignty has been restored we will extend the hand of friendship and assistance to a Cuba whose political and economic institutions have been shaped by the will of the Cuban people."* [86] The next day Bundy told Attwood that the President wanted to see him as soon as he had spoken to Lechuga. The President, Bundy said, would not be leaving Washington again, except for a brief trip to Dallas.[87]

X

In the meantime, Jean Daniel had been spending three fruitless weeks in Havana. He passed on word about his White House meeting. Castro declined to see him — perhaps estopped by his own hard-liners; more likely wishing to postpone bargaining until he knew

---

* The CIA told Cubela that Desmond FitzGerald had helped write the speech and that the passage about the "small band of conspirators" was meant as a green light for an anti-Castro coup (Church committee, *Assassination of President Kennedy*, 20). Edward Jay Epstein repeats this CIA claim as a fact (Epstein, *Legend: The Secret World of Lee Harvey Oswald* [New York, 1978], 240). On its face the passage was obviously directed against Castro's extracontinental ties and signaled that, if these were ended, normalization was possible; it was meant in short as assistance to Attwood, not to FitzGerald. This was the signal that Richard Goodwin, the chief author of the speech, intended to convey. A search of the JFK Papers shows that Goodwin, Ralph Dungan, Bundy, Gordon Chase of Bundy's staff and I were involved in discussions about the speech. No evidence was uncovered of any contribution from FitzGerald and the CIA (W. W. Moss to author, March 30, 1978).

the outcome of the attempt to disrupt Venezuela's December presidential election. This was the month when Castro, with great fanfare, was staging his Week of Solidarity with the Venezuelan revolution. Finally Daniel booked passage to Mexico City for November 20. At ten o'clock on the evening of the nineteenth, Castro, accompanied by Vallejo, unexpectedly came to his hotel room.

They talked till four in the morning. Daniel described his conversation with Kennedy. Castro listened, Daniel thought, with "devouring and passionate interest," stroking his beard, tugging on his paratrooper's beret, "making me the target of a thousand malicious sparks cast by his deep-sunk lively eyes." He made Daniel repeat three times Kennedy's indictment of Batista; three times also Kennedy's remark that Castro himself, in defiance of the superpowers, had almost brought the world to nuclear war. Khrushchev, Castro recalled, had called Kennedy "a capitalist with whom one could talk." Daniel felt that Castro saw Kennedy as an "intimate enemy."

"I believe Kennedy is sincere," Castro finally said. "I also believe that today the expression of this sincerity could have political significance. . . . He inherited a difficult situation. . . . I also think he is a realist." But, instead of addressing himself to the particularities of negotiation, Castro rambled on about the iniquities of American policy. He blandly denied to Daniel the Cuban revolutionary role in Latin America of which he had bragged earlier that month to Herbert Matthews. He refused to discuss Cuban relations with Russia; "I find this indecent." As for relations with Washington, "We have forgotten the United States. We feel neither hatred nor resentment anymore, we simply don't think about the US." Still he could not understand why Washington would not accept Cuba as it was. "Why am I not Tito or Sekou Touré?"

He continued to ramble. Someday, he thought, there would appear in the United States a man capable of understanding the explosive reality of Latin America. "Kennedy could still be this man," he mused. "He still has the possibility of being, in the eyes of history, the greatest President of the United States, the leader who may at last understand that there can be coexistence between capitalists and socialists. . . . He would then be an even greater President than Lincoln. . . . Personally, I consider him responsible for everything, but I will say this: he has come to understand many things over the past few months; and then too, in the last analysis, I'm convinced that anyone else would be worse. . . . You can tell him that I'm willing to declare Goldwater my friend if that will guarantee Kennedy's re-election!"

Castro was in no hurry about negotiation. He supposed he had all the time in the world. He made no comment on Kennedy's Miami hints of the day before. Holding out for the Venezuelan prize, he stalled, putting off Daniel with pieties: "Since you are going to see Kennedy again, be an emissary of peace. . . . I don't want anything. I don't expect anything. But there are positive elements in what you report."[88] That was November 20. Two days later, lunching in his villa at the beach with Daniel, he heard the news from Dallas.

"*Es una mala noticia*," he muttered again and again; "this is bad news." To Daniel: "*Voilà*, there is the end to your mission of peace." Later: "I'll tell you one thing: at least Kennedy was an enemy to whom we had become accustomed. This is a serious matter, an extremely serious matter." In the afternoon he had a fusillade of questions about Lyndon Johnson. Finally, as if well aware that political leaders might not be able to control their intelligence agencies: "What authority does he exercise over the CIA?"[89]

The next day at the UN, Lechuga at last received instructions from Castro to begin talks with Attwood. On December 4, Attwood told me that his secret explorations were, he believed, reaching a climax; that Castro might be trying to get out from under Guevara and the Communists and strike a deal with the United States.[90] But on December 23, Bundy observed at his morning staff meeting that, because Lyndon Johnson expected to run against Nixon, he did not want to give him any openings; "i.e., he does not want to appear 'soft' on anything, especially Cuba."[91] Bundy informed Attwood that "the Cuban exercise would probably be put on ice for a while — which it was," Attwood wrote in 1967, "and where it has been ever since."[92] Castro, looking back a decade later, reflecting perhaps on an opportunity he had squandered because he overestimated his capacity to make a revolution in Venezuela, said of Kennedy, "He was one of the few men who had enough courage to question a policy and to change it. . . . We would have preferred that he continue in the presidency."[93] On balance, it must be judged unlikely that Fidel Castro was conspiring to kill the American President with whom he was striving to come to terms — and whose successor offered no promise of more favorable policies.* Of course, he might not have exercised complete authority over his own CIA.

XI

No one can say how far these explorations might have gone. Castro was possibly prepared, even perhaps pleased, to move a Tito stride

* Donald E. Schulz also emphasizes Castro's public threats in his September interview with Daniel Harker: "If Castro were going to have Kennedy assassinated would he broadcast it to the world? . . . The demand for retaliation would have been overwhelming" ("Kennedy and the Cuban Connection," *Foreign Policy*, Spring 1977).

away from the Soviet Union in exchange for the resumption of trade with the United States. But he was not yet prepared in November 1963 to throw in his revolutionary hand in Latin America. Not until June 1964, after his operatives had flopped irremediably in Venezuela, did he make clear any readiness, in the words of Maurice Halperin, "to cut off all aid to revolutionary movements in Latin America, in return for a normalization of relations with the United States and the other estranged republics."[94] Nor could he be sure how violent the opposition in his own ranks might be to a policy of rapprochement — from Che Guevara, his personal Leon Trotsky; from the Communists; from his intelligence people, who, like the CIA, might have unleashed projects of their own; from allies, if he had such, in the American underworld, who had independent reasons for wishing to end the power of the Kennedys.

Kennedy also faced formidable resistance to a change in course. The State Department, apart from Harriman, was not happy with Kennedy's willingness to accept the social changes wrought by the revolution; nor, it must be supposed, did Kennedy's Miami speech of November 18 delight former owners of plantations, mines and mills for whom the 'liberation' of Cuba meant primarily the restoration of lost wealth and privilege. And, though the Attwood plan was closely held, it seems inconceivable that the CIA knew nothing about it. American intelligence had Cuban UN diplomats under incessant surveillance. It followed their movements, tapped their telephone calls, read their letters, intercepted their cables. Suspecting, as it must have, that Attwood and Lechuga were doing something more than exchanging daiquiri recipes, the CIA, in pursuing the Am/Lash operation, must be convicted either of abysmal incompetence, which is by no means to be excluded, or else of a studied attempt to wreck Kennedy's search for normalization.

But the deepest rage of all against John Kennedy was among the anti-Castro *fanaticos* in the Cuban exile community. In their obsessed view normalization would complete the perfidious course begun at the Bay of Pigs and carried farther during the missile crisis. No one had a greater interest in putting the Kennedy-Castro explorations on ice than those Cubans who had committed their lives to the destruction of the Castro regime.

Robert Kennedy had stayed in touch with his friends in the Brigade. But the administration made no effort to keep the Brigade alive as an entity or to make it a political or military force.[95] Instead those of its members not recruited by CIA were offered only the opportunity to enlist in the United States Army as individuals. The Brigade leaders protested to the CIA in June that commando raids would not overthrow Castro, that the only hope was "a massive U.S.

intervention." They were, Helms reported to the Attorney General, "disheartened in that they do not foresee such an invasion."[96]

Most Bay of Pigs prisoners were constrained for a season by their loyalty to the Kennedys, to whom, after all, they owed their freedom. Other exiles were less constrained. Miro Cardona resigned as chairman of the Cuban Revolutionary Council in an outburst of wild accusations against the Kennedy administration. A broadside of unknown origin told Cuban exiles in Miami that "only one development" would return them to their homeland — "if an inspired Act of God should place in the White House within weeks a Texan known to be a friend of all Latin Americans."[97] The Task Force W Cubans, trained by the CIA in clandestinity and violence, brooded darkly over what one of them, Dr. Orlando Bosch, denounced in an angry letter to the President as their betrayal by the Kennedys.[98] Thereafter the compliant ones, like Barker, Sturgis, Martinez, huddled around the CIA case officers and went down the trail of espionage, deception and dirty tricks that ended in Watergate. The crazy ones, like Orlando Bosch, became terrorists. In 1968 Bosch's exploits won him ten years in the federal penitentiary. Paroled in 1972, he went underground, resumed murderous activities of diverse sorts and in 1977 was in a Venezuelan prison charged with organizing the bombing of an Air Cubana plane and killing all seventy-three aboard. Other CIA Cubans were implicated in a wave of bombings and killings in Miami. Even the once sober Miro Cardona cried in 1973, "We are alone, absolutely alone. . . . There is only one route left to follow and we will follow it: violence."[99]

The men in Washington who in 1960 had planned the murder of Castro and the invasion of Cuba poured a stream of malignant emotion, pro-Castro and anti-Castro, into the very wellsprings of American life.

(24)

Missions to the Third World

EVER SINCE THEIR JOURNEY across Asia in 1951, John and Robert Kennedy had believed that nationalism was the most vital political emotion in the developing world. They instinctively sympathized with new nations struggling for survival; and, after Khrushchev's 1961 prediction of Communist world victory through national liberation wars, they saw the Third World as the crucial battleground between communism and democracy. As President, John Kennedy cultivated the new leaders, welcomed them to Washington and, to emphasize his personal concern, confided the Peace Corps to his brother-in-law and dispatched his brother on Third World missions.

I

Africa was a natural place to begin. The State Department had traditionally regarded Africa as a European preserve. The point of American policy, besides protecting American investments, had been to avoid actions that might offend European allies and, in the jargon of the day, 'weaken NATO.' John Kennedy himself had been accused of doing this when he gave his notorious speech in 1957 on Algeria. In the White House he proposed to deal with African questions on African merits.

In August 1961 the Ivory Coast celebrated its first anniversary of independence. The Attorney General headed the American delegation. He chafed under the official briefings until the State Department produced a young officer named Brandon Grove, just back from three years in the Ivory Coast. Impressed by Grove's intelligence and candor, Kennedy added him to the delegation. Once in the Ivory Coast, Kennedy bridled at the embassy schedule of official receptions and demanded to talk to students and labor leaders. "The embassy," recalled David Halberstam, who covered the trip for the *New York Times*, "felt he ran roughshod over it, pushed its members around, made unnecessary and unfair demands, and insulted the good Ivorien friends. Almost everyone else loved it." [1]

Felix Houphouët-Boigny, the astute and durable president of the new republic, seized the opportunity to give the President's brother an indoctrination in the problems of West Africa. Kwame Nkrumah of Ghana, Houphouët-Boigny warned, "had surrounded himself with sycophantic advisers who had so inflated his existing ego and messianic complex that he was now thoroughly convinced that it was his divinely ordained mission to lead all of Africa." The Russians were encouraging him in these delusions. Robert Kennedy asked whether the United States should help Ghana build the Volta Dam. Houphouët-Boigny's reply was a tepid yes. [2] The Attorney General was not convinced.

The Volta Dam propelled Robert Kennedy into one of his few sharp disagreements with his brother. His suspicions were emphatically reinforced when his old friend from Rackets Committee days, Clark Mollenhoff, returned from a year's fellowship in Africa. Muckraking as usual, Mollenhoff denounced the dam simultaneously as a capitalist plot pushed by American business interests and as a project whose main beneficiary would be the Communists. [3] In September 1961, after Nkrumah inveighed against the United States at a meeting of neutralist leaders in Belgrade, the Attorney General wrote the President: "We are limited to the amount of money we are going to spend in Africa and it would be better perhaps to spend it on our friends rather than those who have come out against us." [4]

On December 5, 1961, the National Security Council faced the decision. Robert Kennedy, an NSC regular after the Bay of Pigs, used to sit modestly along the wall with the staff people rather than at the cabinet table with the statutory members. The President resolved to go ahead. His grounds were that commitments had been made, that the dam was of long-run benefit for the people of Ghana and Africa, that it made economic sense and that, presumably, Nkrumah would not be there forever. As the discussion proceeded, John Kennedy said, "The Attorney General has not yet spoken but I can feel the cold wind of his disapproval on the back of my neck." [5]

On December 7, Robert Kennedy fired a parting shot:

> I would like to state why I think such a project is not in the interests of the United States. First, I think it is very clear that Nkrumah is growing closer and closer to the Soviet Bloc. . . .
> Another strong reason . . . is the effect it will have on the other

African states. Friends of ours such as the Ivory Coast, Togo, Upper Volta, have received anywhere from one to five or six million dollars in aid. Ghana which is bitterly opposed to us and playing "footsies" with the Soviet Union is receiving this amount of money with no strings attached. If I were one of these other countries I would certainly, from now on, figure the best way to get money for my country would be to play the Soviet Union off against the United States. . . .

I think also, if we give this money without any strings attached it will encourage Nkrumah to be more stringent with his repressive measures. He has already placed several hundred of his opposition in jail without trial and within the last month has passed some measures which will give him dictatorial control over life and liberty.[6]

"The President finally decided to put the dam in Ghana," Robert Kennedy reflected in 1964, "and we had some spirited arguments about it . . . and I think probably looking at it in retrospect that it was the correct decision."[7] In time, Nkrumah surrendered unconditionally to his delusions of grandeur, was overthrown, fled to Guinea where he evidently infected Sekou Touré with similar delusions and died. The Volta Dam was a success.

II

Robert Kennedy saw his opposition to Nkrumah as in the interests of true African nationalism. The African Bureau found him its strongest ally in its perennial argument with the Pentagon and with the State Department's Europeanists.

Thus in the summer of 1963 a dispute erupted over impending UN votes on African issues. The Pentagon, arguing the alleged military indispensability of naval bases in the Azores and of tracking stations in South Africa, wanted to support Portugal and South Africa against black Africa. On July 1, I sent Robert Kennedy a memorandum suggesting that Defense "make a much more rigorous examination than it seems yet to have made of the alternatives to Azores and the tracking stations." This was the bitter summer of the civil rights war, and the Attorney General had more urgent matters on his mind and desk. Nevertheless (as I discovered in going through his papers) he underlined a sentence in the memorandum asking whether "these military facilities are so indispensable to us that they must determine our African policy" and scribbled a note to Angie Novello: "Have me speak to Bob M about this Wed." Ten days later McNamara sent Rusk a cost-benefit analysis discounting the supposedly decisive strategic considerations and concluding: "I believe the decisions on these issues should be based on general considerations of foreign policy. I hope that you share the views I have expressed."[8]

Angola and Mozambique were the only European colonies of consequence left in Africa. Robert Kennedy thought that their independence was both right and inevitable, that it was foolish for the
United States to identify itself with Portuguese suzerainty and that
the American interest lay in winning the confidence of the anticolon-
ial leaders. Rusk considered it improper, as by the book it doubtless
was, for men engaged in rebellion against NATO allies to be received
officially in Washington. When Eduardo Mondlane, the head of the
Mozambique Liberation Front, came to the United States, Wayne
Fredericks, the deputy assistant secretary of state for African affairs,
told Kennedy that Mondlane would almost certainly be the leader of
an independent Mozambique (as he would have been, had he not
been assassinated in 1969). Perhaps, Fredericks suggested, the Attorney General might care to meet Mondlane in a neutral spot —
dinner at someone's house, for example. Kennedy said, "I will see
him in the office of the Attorney General of the United States. Bring
him here."[9] Kennedy helped arrange a CIA subsidy to cover Mondlane's travel and other expenses and also got CIA money for Holden
Roberto in Angola — to be spent not on weapons but on aid to
refugees fleeing from villages burned down by the Portuguese and
on the education of promising young Angolans in European and
American universities.

On November 20, 1963, Robert Kennedy wrote Bundy saying that
the Standing Group simply must discuss the "policy of the United
States toward the individuals and organizations which are attempting
to gain independence in Mozambique, South Africa, Angola and
Rhodesia. . . . I gather that we really don't have much of a
policy. . . . These areas are going to be extremely important to us in
the future. . . . Personally I feel if we could take steps now, either
through the CIA and/or making a concentrated effort with students
and intellectuals, we could head off some of the problems that are
undoubtedly going to appear on the horizon in the next year or
so."[10]

"I felt very strongly," he told John Bartlow Martin in April 1964,
"that we should become more involved." At that point he was exercised over Zanzibar, where a "people's republic" of obscure but impassioned views had recently been established. For the last four
months, he said, he had been trying "to get somebody to do something about Zanzibar. . . . Averell Harriman wrote a memorandum
. . . and George Ball wrote back and said it was foolish to waste our
time, it was such a small country and added that, if God could take
care of the little swallows in the skies, He could certainly take care of
a little country like [Zanzibar]. . . . Imagine that!"[11]

Three days after Kennedy said this, the United Republic of Tanzania was formed, and Zanzibar became a problem for Julius Nyerere (and for God). Ball plainly had the better of that argument. Martin asked later how John Kennedy felt about involvement in Africa. Robert said, "Well, I don't know, really. I think he wanted to get involved. I don't know — I never had the discussion about that particular problem with him, but it came up before our Counterinsurgency committee as to what we would do in some of these areas and I was always in favor of becoming more involved."*

III

John Kenneth Galbraith, the ambassador to India, convalescing in a Honolulu hospital in February 1962, recorded in his journal a visit from Robert and Ethel Kennedy. "They were marvelously vital, just having gone sailing outside Pearl Harbor and tipped over their dinghy. I proposed they say they had been cut down by an enemy destroyer with stress on the fact that Bob had saved the whole crew. He could then be President."[12]

This time the Kennedys were going around the world. Edwin Reischauer, the leading American historian of Japan and newly appointed ambassador to Tokyo, had urged the Attorney General to accept an invitation to Japan, where anti-American violence had forced Eisenhower to cancel a visit in 1960. The State Department had jobs for him in Indonesia. Mayor Willy Brandt had invited him to give the Ernst Reuter Lecture at the Free University of Berlin. The Department again provided Brandon Grove as an escort. John Seigenthaler was also in the party, along with a few newspapermen, among them Anthony Lewis of the *New York Times,* Kennedy's Harvard classmate but still a rather wary acquaintance. The uninhibited and sometimes unruly company tried Brandon Grove's Foreign Service sense of propriety. Once, when Ethel Kennedy forgot to turn off the water in their Tokyo hotel and her bath overflowed, Grove said with dignified wrath, "Some day they will learn that this is a *real world* with *real people* in it."[13]

Kennedy set forth the theme of his trip in a speech at Tokyo's

---

* RFK, in interview by John Bartlow Martin, May 14, 1964, I, 5, JFK Oral History Program. He had not always felt that way. Earlier he had sensibly told a questioner at Howard University, "Do you want us to intervene all over the world? Telling this government what to do and that one what not to do? A lot of countries in Africa have one-man governments. If we protested, they'd tell us to mind our own business. We stand for democracy and we try to show people by example, but we can't go running all over the globe, telling people what to do" (Fletcher Knebel, "Bobby Kennedy: He Hates to Be Second," *Look,* May 21, 1963).

Nihon University. This was, he said, the century of "the awakening of peoples in Asia and Africa and Latin America — peoples stirring from centuries of stagnation, suppression and dependency." The resources of the earth and the ingenuity of man could provide abundance for all —

> so long as we are prepared to recognize the diversity of mankind and the variety of ways in which people will seek national fulfillment. . . . We do not condemn others for their differences in economic and political structures. We understand that newer nations have not had time, even if they wished, to build institutions relying primarily on private enterprise as we have done. . . .
>
> We have no intention of trying to remake the world in our image but we have no intention either of permitting any other state to remake the world in its image. . . . We call to the young men and women of all nations of the world to join with us in a concerted attack on the evils which have so long beset mankind — poverty, illness, illiteracy, intolerance, oppression, war.[14]

He had begun at Nihon with a few sentences in what he supposed, after laborious rehearsal by Edwin Reischauer and his Japanese wife, might resemble the native tongue. "The audience thought I was speaking English and waited for the translation. Simultaneously we all realized what had happened and to everyone's relief I restarted in English."[15] The audience laughed, and all seemed well. But ominous reports were coming in from the next stop, Waseda University, where the Zengakuren, the Marxist student clubs, were on the offensive. "We . . . were told we shouldn't go," Reischauer said later, "but we decided at the last minute that it would look bad to back out."[16]

At Waseda three thousand students jammed a hall built for half that number. Several hundred had come to revile the brother of the American President. "It is possible," Kennedy began dryly, over a barrage of cries and hisses, "that there are those here today who will disagree with what I say."[17] In the front row a young man yelled hysterically — "a skinny little Japanese boy," as Seigenthaler remembered him, "tense, shouting, shouting, shouting, screaming at the top of his lungs, really red-faced; just completely wrought up emotionally."[18] After a moment, Kennedy reached out his hand and pulled his heckler to the platform. "Perhaps you could make your statement," he said, "and ask me a question and then give me an opportunity to answer." There was no question, but a ten-minute tirade. When the critic, exhausted, came to an end, Kennedy, enlightened by instant translation, commenced a reply. At this point the public address system, and every light in the house, went dead.

There followed a quarter hour of chaos. Someone from the embassy, anticipating the worst, had brought a battery-operated bullhorn. Reischauer, speaking fluent Japanese, quieted the audience. Kennedy, "without loss of temper or composure," according to the embassy report,[19] took the bullhorn and said: "We in America believe that we should have divergencies of views. We believe that everyone has the right to express himself. We believe that young people have the right to speak out."[20] After a few moments in this vein, he invited more questions from the audience, which by this time was turning in his favor. He concluded with some personal words: "My brother, the President, entered politics at a young age. Although now President, he is still young, and all those who held key positions in his campaign are young. He believes that the future of the world belongs, not just to the younger generation of my country, but to the younger generation of all countries."[21]

Someone bellowed in the rear of the hall; whether in rage or agreement, Kennedy could not tell. The interpreter explained that this was the school cheerleader, desiring to make apology. In a moment, the cheerleader was on the platform gesticulating wildly (inadvertently punching Ethel Kennedy in the pit of the stomach; "she doubled up, stood up immediately, smiled and went on," observed Seigenthaler)[22] and leading the student body in the Waseda song. The audience responded, as if at a football rally before the homecoming game. The Kennedy party, alas, learned the song — "Miyako no seihoku" — and later inflicted it mercilessly on parties at Hickory Hill.[23]

There was, said the Tokyo newspaper *Yomiuri* the next day, "no question who won." Another paper, *Sankei*, said Kennedy was "superb." "In his frankness, simplicity and courage," said a *Yomiuri* columnist, "Mr. Kennedy reminds us of [the] pioneer of old."[24] Moreover, through television, "the Japanese nation," as Reischauer said later, "saw this whole unrehearsed play, as it were, with the ranting young radical student and the calm and reasonable young political leader from America, and it made a tremendous impression. And after that, you know, it was just a smashing success."[25]

IV

Instructions had gone forth to American embassies: the Kennedys could be scheduled to the hilt, but with a minimum of official ceremony and a maximum of informal mingling with students, intellectuals, trade unionists and factory workers. "We drew him up a tremendous program," said Reischauer, "every minute occupied till 10 o'clock at night, and he wired back, 'Program looks fine. What do I do from 6 to 8 A.M.?' So we put on that, too. He skated with workers, 6 to 8 A.M."[26] The embassy, noting Kennedy's "easy rapport with laborers and union representatives," recalled the indifference shown to Japanese workers by Mikoyan a short while before. "The Attorney General, by contrast, started talking knowledgeably about working conditions and labor union activities whenever he came into speaking distance of a laborer."[27]

The combination of bluntness, humor, courtesy and charm was effective. Even the Zengakuren lost their ardor. Pickets still yelled, "Go home, Kennedy. Go home, Kennedy." "Then," Seigenthaler reported, "one of them would yell, 'Where's Ethel, Bobby?' so that we had the feeling we were even getting through to them."[28] When the party went on to Osaka, Nara and Kyoto, crowds lined country roads to watch them drive by. They slept on floors, ate snails and seaweed for breakfast and whale meat for lunch, visited ancient temples and modern plants. "When he went off the beaten track, as he did constantly," said the embassy, "it was to mingle with workers, labor union people, students or intellectuals in their own environment. The 'one of us' reaction was most pronounced among these people."[29] They met, wrote one Japanese journalist, "all classes of the people, boys and girls, workers, students, housewives, and farmers, and saw their actual conditions. . . . My impression is that Kennedy's personal magnetism was overwhelming."[30]

"The overwhelming consensus of opinion," the deputy chief of the embassy's translation services branch reported, "is that the Attorney General's 'whirlwind' visit was the most successful accomplishment by the United States in its postwar relations with Japan."[31] The visit, the embassy concluded, "commanded the attention of more people and elicited a more positive response from the Japanese public than any good will visit in Japanese history."[32]

The schedule was so crowded that Reischauer hardly had a moment alone with Kennedy. They finally reserved fifteen minutes before the group departed for Indonesia. "During those fifteen minutes," Reischauer recalled, "I outlined the Okinawa problem, what we would have to begin doing about it, and one or two other problems. . . . He had hardly gotten back to Washington before action began on all of these. . . . And if I really had something that I just had to get to the President, you know, and that we had to get to work on, I could always do it that way. . . . It was the most important channel."[33]

v

Now sixty years old, the flamboyant and vainglorious Achmed Sukarno had been active in the Indonesian independence movement
before Robert Kennedy was born. In the Eisenhower years the State
Department had looked on him as a sort of communist. John Kennedy, who thought it easy to understand why Sukarno might be anti-
American after what the CIA had done to him in 1958,[34] had him
pegged as a clever nationalist politician and charmed him with notable success when Sukarno visited Washington in 1961.

While the Robert Kennedy party was still in Tokyo, Communists
had stoned the American embassy in Djakarta and painted the city
with KENNEDY GO HOME signs. Sukarno, taking no chances with the
safety of his guests, installed them in the presidential palace (which
Ethel plunged into darkness one evening by plugging in her hair
dryer and blowing the fuses) and surrounded them by soldiers wherever they went. There was one failure of security. When Robert
Kennedy spoke at the University of Indonesia, "suddenly this little
wraith of a kid," as Seigenthaler described it, "a tall, slender boy,
broke through the lines of people and, standing at a distance of about
fifteen feet, took a full windup and let fly a piece of hard shelled
fruit, which hit Bob on the end of the nose. It was thrown with full
force — I mean, he looked like Bob Feller." Kennedy, Seigenthaler
said, "never flinched. The only time he took a backward step was
when it hit him; it knocked him back just a step. He turned around
to me and said, 'Did you see that little s.o.b.?' . . . and he sort of
put his eyes up in the top of his head as he said, 'Whew.'" The
police grabbed the thrower, twisted both arms behind him, turned
him upside down and rushed him out to a car. "We never saw
or heard of him again," Seigenthaler said. "Bob told Gunawan,
the Attorney General, that he hoped they wouldn't do anything to
harm him."*

In the question period, students asked Kennedy insistently why the
United States was not backing Indonesia in its quarrel with the
Netherlands, now threatening to explode into hostilities, over the
future of West New Guinea. Kennedy replied that his country had
"vigorously supported Indonesia's struggle for independence" and
declared his hope that these "two close friends of the United States"

---

* John Seigenthaler, in recorded interview by R. J. Grele, February 23, 1966, 397–
398, JFK Oral History Program. Bob Feller was the Cleveland Indians' pitcher, renowned for his fast ball.

would be able to work out West New Guinea. The audience objected
to his agnosticism. Kennedy went into one of his famous exercises
in bluntness. "We have allies throughout the world," he said,

> and we don't agree with everything that they do, and they don't agree
> with everything we do. I don't know any of you who agree with every
> one of your fellow students. . . . I suppose a day doesn't go by when we
> don't disagree with the English on some matter or other. We are going
> to disagree with Indonesia, and you are going to disagree with us.
>
> But we are both democratic countries. We both should have a foun-
> dation of friendship, so that every time an incident comes up and we
> don't do exactly what you want us to do, you don't say, "To hell with
> the United States."

There was an ovation. "Strangely enough," Kennedy wrote later, "it
was this answer more than any other that seemed to establish a
friendly rapport." [35]
What got him into trouble that night was not West New Guinea
but Texas. A student who had done his homework in American
history brought up the Mexican War. "I would say," said Kennedy,
"that as far as the war with Mexico — although there might be some
from Texas that might disagree — I would say that we were unjustified. I don't think that this is a very bright page in American
history." This response went down well in Indonesia; less well in
Texas. Soon a cable arrived from Salinger: YOUR REMARKS ABOUT
TEXAS APPEAR TO BE CAUSING SOME FUSS. IF YOU ARE PRESSED ON THIS
MATTER, PRESIDENT SUGGESTS YOU ATTEMPT TO MAKE SOME HUMOROUS
REMARK. GOOD LUCK. Later the Attorney General told the press that
he had been instructed to clear any future Texas speeches with the
Vice President. He noted privately, however, that most of the critics
denounced him for supposed aspersions cast on the battle of the
Alamo; they "did not even have the correct war." [36]
Kennedy, the embassy cabled the State Department, made his
"most notable impression on Indonesian students. . . . Young audiences which were initially cold or even hostile rapidly warmed to his
frank, man-to-man approach, admission that not all was perfect in
US, demand for grown-up approach to inevitable differences of
opinion, and assurances that US did not seek to impose its system on
world." Soon, when his car passed, people waved and shouted,
"Hello, Kennedy." When he walked the streets, "people poured out
of nearby stalls or buildings, elbowed each other aside to indulge in
un-Indonesian practice of shaking hands." The embassy concluded:
"No short-term visitor to Indonesia has had impact on public that
Attorney General Kennedy did during February 12–18 visit." [37] The
implied comparison was with Khrushchev, who in 1960 had been put
off by Sukarno's "scandalous" appetite for women, was visibly bored

by Javanese ritual and, when shown artisans printing fabric by hand, had said only, "They could do it faster with machines."[38]

VI

"I didn't like him from what I had heard of him," Robert Kennedy said privately of Sukarno in 1964, "and I didn't like him when I was there and I haven't liked him since." Like Khrushchev, he was offended by Sukarno's satyriasis and by the gallery of nudes that adorned the presidential palace. "I don't have respect for him. I think that he's bright. I think he's completely immoral, that he's untrustworthy. . . . I think he's a liar. I think he's got very few redeeming features. He speaks like hell, he's a demagogue. He's not a Communist. I think he's anti-white but I think he liked the President, . . . he admires the United States and I think he liked me and he liked Ethel."[39] This was indeed the case. "I like people with flame in their eyes," he told Howard Jones, the American ambassador. "He and Bobby hit it off beautifully," Jones said later.[40]

Robert Kennedy's specific mission in Djakarta was to head off war between Indonesia and the Netherlands over West New Guinea, a territory whose status was left unresolved when Indonesia became independent in 1949. Jones had long advocated a forthcoming policy on West Irian, as the Indonesians called it. He saw no future in the United States taking the losing side in a colonial war. He also warned that the Soviet Union was exploiting the issue. True enough: Moscow had already suborned several of Sukarno's generals. "Unbeknownst to him," Khrushchev said later, "some of these generals were even Communist Party members."[41] Nevertheless Jones's recommendations found a dusty answer in Washington. "If Sukarno starts aggressing against his neighbors," Dean Rusk told Jones, "he'll find us on the other side. We learned our lesson with Hitler. The time to stop him was at the beginning." It was not, Jones wrote, till Averell Harriman took over as assistant secretary for Far Eastern affairs in November 1961 that "our recommendations from Djakarta began to receive a sympathetic hearing."[42] "It could have developed," Harriman said later, "into a very difficult situation in which the Dutch would be spending a lot of money and a lot of people would be killed . . . a conflict in which there could only be one end, namely the withdrawal of the Dutch."[43]

Kennedy's particular task was to persuade Sukarno to talk to the Dutch without preconditions. Except for Jones's indispensable counsel, he was on his own. "Your relation to the President," Bundy cabled, "will give ample authority to your own words, and the experience of men like Harriman is that in such a situation it is better not to be bound by a canned message."[44] The talks began. Kennedy, Jones reported to Washington, "stressed that as a major world statesman Sukarno must appreciate utter absurdity of permitting question of peace or war on so important an issue to turn on procedural rather than substantive point." The Indonesians had everything to gain and nothing to lose from negotiation. "In our opinion," Kennedy said, "an acceptable solution would result from talks but if this did not happen and GOI [government of Indonesia] felt it must take military action its position before world would be strengthened." The United States, he added, recognized that the territory was going to Indonesia "in some way at some time. . . . If we did not think talks had chance of success we also would not be urging them." Only there could be no preconditions; "this was a matter of prestige with Dutch and they could not be moved on this."[45]

"In my judgment," Kennedy warned Washington, "Indos will fight unless this issue is resolved, and this would be full of dire implications for the free world in Asia."[46] But Sukarno dropped his preconditions after the visit; and, when Kennedy went to Holland later in the month, he assured the Dutch that, "based on my conversations with Sukarno, Subandrio [the foreign minister] and others that I was sure they were interested in resolving this whole matter without armed conflict."[47] "It is abundantly clear," Jones wrote Kennedy in May, "that you were able to start the ball rolling toward a negotiated settlement."[48] Eventually, interminable meetings between the Dutch and the Indonesians, with Ellsworth Bunker, a veteran American diplomat, serving as mediator, led to an agreement in August 1962 under which West New Guinea, after an interim UN administration, would go to Indonesia. Sukarno told Jones, "It could not have been done without your government."[49]

Critics denounced the settlement as capitulation. In 1963 Sir Robert Menzies, the Australian Prime Minister, complained bitterly to the wife of the American ambassador about Robert Kennedy's role. "It is not as bad as they think," Kennedy told her, "although I guess I must accept some responsibility. I think, however, any alternative action on our part would have brought on a war in that part of the world which would have been no help to Australia or the United States."[50]

On the larger arena it was a master stroke. Moscow sent Mikoyan to Djakarta in July 1962 in a last-minute effort to block the settlement. Khrushchev's irritated summation made the case for the Harriman-Jones-Kennedy policy:

> Sukarno cleverly utilized both the Soviet Union and the United States to achieve his goal of getting Holland to back down. . . . We felt it was wrong of him not to inform us of his intentions in advance. In any

event, while continuing to support the Dutch publicly [not true], the Americans obviously put pressure on them behind the scenes. As a result, Holland submitted to the negotiations and agreed to hand over West Irian.... Since armed conflict had been avoided, our advisors who had been training the Indonesians were no longer needed, so they came home.[51]

VII

The novice diplomat had, in addition, a more secret mission: to secure the release of Allen Pope, the CIA pilot who had been captured in 1958 and condemned to death in 1960. The death sentence had not been carried out. When Sukarno visited Washington in 1961, John Kennedy had asked for Pope's release. Sukarno said it would happen in due course. Mrs. Pope, a comely airline stewardess, flew to Djakarta. "She cried bitterly," Sukarno later said, "and begged me to pardon him. When it comes to women I am weak. I cannot stand even a strange woman's tears. Then his mother and sister visited me and those two sobbing was more than I could bear."[52] However, he managed to restrain his emotion. Pope remained in prison. Mrs. Pope called on Robert Kennedy. "I was tremendously impressed with her and then she talked about the children and the fact that the children were fighting with one another and one child had never seen her father. So I was really very taken up with it."[53]

He brought up Pope at their first meeting. Sukarno said he would have to think about it. He was still thinking at their last meeting. Kennedy concluded that he was trying to use Pope as a bargaining counter over West New Guinea. "I explained to him that ... he could take Pope out and shoot him and it wouldn't affect what we did on West New Guinea because what we were going to do on West New Guinea was ... in the interests of the United States and the interests of justice." There was a long pause. Then Sukarno said, "Well, let me tell you, Mr. Attorney General, you're just going to have to let me handle this in my own way." Kennedy said that was fine; "but could you give me some indication so that I might tell President Kennedy that you do have in mind standing by what you've already told him, that you are going to let this fellow out." Another long pause; then Sukarno: "Mr. Attorney General, you're going to have to leave this to me to handle in my own way." Kennedy, his voice growing testy: "Could you tell me whether you're going to stand by your promise to the President of the United States?" Sukarno said again he would handle the matter in his own way.

"I never really had the feeling that Bob was angry because I know him so well," recalled Seigenthaler, "but I had the distinct impression that Sukarno thought he was angry." Kennedy said, "Am I to go back to the President and say you will not tell us that you'll stand by your word?" He stood up. "I can't understand that," he said. "Everybody tells me that you're a man who stands by his word . . . and the President of the United States believes that, and now you won't say that. I'm his brother. He sent me here as his representative. I'm speaking for him when I'm asking you what you're going to do about this . . . and you won't tell me." He walked out on the balcony.

"We were sitting there in a circle," recalled Seigenthaler. "I looked at Howard Jones, and I thought he was going to fall off his chair." After a moment, Seigenthaler joined Kennedy. Kennedy said, "Do you think that what I said to him is going to have any impression on him?" Seigenthaler thought it was very definitely having an impression. Kennedy said he wanted "to make sure that when I leave here he knows how strongly we feel about this. That kid belongs out." Jones came onto the balcony. "Look," he said to Kennedy, "you feel strongly about this. He also feels strongly about it. . . . I remember riding with him on a boat one time, and he showed me out under the water where a ship had been sunk, and he said, 'That's the work of your CIA agent Pope.' You can't leave him in there like this. You've got to go back in." Kennedy said, "Oh, I'm going back in. Don't misunderstand, Mr. Ambassador. I'm not upset at all, but I want him to know that this is a matter that holds some potential danger."

They returned to the room. Kennedy went at once to Sukarno and said, "If I have offended you by my bluntness . . . I certainly apologize for that. You've been extremely nice to me here. You've been very nice to my wife. . . . But I could not leave without saying to you what I feel about this, what my brother feels about it, and what we will continue to feel so long as Mr. Pope is in prison." Another long silence; then Sukarno: "Mr. Attorney General, you are forceful and you are young, and I am forceful and I am old. . . . I am hopeful that I can do something that will be proper. But I simply am not in a position at this time to say any more than that." Then, said Seigenthaler, they went out together where the press was waiting, "and they were like two foxes. I mean, they just smiled and shook hands."[54]

Anger, or the simulation thereof, is a not unknown diplomatic weapon. Sukarno, the old fox, saw what the young fox was up to and evidently admired his technique. Asked later whether Kennedy had been abrasive, Jones reflected, "He was a very positive guy. But abrasive? His Irish wit kept him from being abrasive. There were one or two points where . . . the discussions became very, very heated;

but then they both calmed down, and their personal relations were cordial and friendly. Evidence of that was that Sukarno welcomed Bobby back two years later";[55] but that is another story.

Kennedy departed the next day. In June, four months after the visit, Sukarno sent word to Pope: "You are pardoned. But I do so silently. . . . Just go home, hide yourself, get lost, and we'll forget the whole thing."[56] "I am back in the United States a free man," Pope wrote Kennedy in July, "and for this . . . my wife and I shall always be eternally grateful."[57] He later came to Washington to express his gratitude in person — "a good-looking fellow," the Attorney General thought, ". . . the soldier of fortune type." He said he was going back to the Far East. Kennedy asked about his wife. "She's not going," Pope said. "We don't live together. We've been separated for some time."*

VIII

On February 19, 1962, the Kennedys arrived in Calcutta. I was in India on a Food for Peace mission with George McGovern, and the Attorney General had asked me to travel with him to Berlin. Galbraith, back from Hawaii, noted in his journal that Robert "looked very tired," but that "Ethel seemed unchanged by the round-the-world campaigning. She spoke appreciatively of the art and culture of Japan, Indonesia and Bangkok with special reference to President Sukarno's collection of nude paintings." Galbraith added:

> Arthur had written a speech for Bob to give in Berlin. I had thought it all right. Bob immediately pinpointed its faults: strictly conventional praise of the bravery of Berliners, strictly conventional damnation of the Communists. On second thought, I was forced to conclude, as did Arthur, that the criticism was sound.[58]

Kennedy was evidently less of a Cold War rhetorician in 1962 than Schlesinger or even Galbraith.

On the plane to Europe Bobby was filled with reflections about his trip. What was clearest of all, he told me, was that, in countries like Japan and Indonesia, America could make contact with the youth and the intellectuals *only* as a progressive nation. "I kept asking myself," he said, "what a conservative could possibly say to these people. I could talk all the time about social welfare and trade unions and reform; but what could someone say who didn't believe in these things? Can you imagine — Barry Goldwater in Indonesia?"[59]

---

* RFK, in Martin interview, April 13, 1964, II, 37. Pope was going to work for a CIA subsidiary, Southern Air Transport (J. B. Smith, *Portrait of a Cold Warrior* [New York, 1976], 205).

In Rome we lunched at Alfredo's. The press party, which by this time had fallen in love with Bobby and Ethel, sent over champagne from their neighboring table, procured an accordionist and started an impromptu dance that went on till four-thirty in the afternoon. I somehow remember through vinous mists the newspapermen bringing in a motor scooter and Ethel driving it precariously between the tables, though that may have been another occasion. The next day the Kennedys concentrated on matters spiritual, such as an audience with the Pope, while I stuck to the temporal side, meeting with Nenni, La Malfa and other political leaders and clearing the Attorney General's German speeches, now amply revised, with Washington.

Entry in my journal, February 22:

> We arrived in Berlin on a cold, blowy, snowy day. Willy Brandt, General Clay and Al Lightner (head of the US Mission in West Berlin) met us at the airport. As we got off the plane, the band played "When the Crimson in triumph flashing, mid the strains of victory." After a brief airport ceremony and gallant but incompetent tries at German by Bobby and Ethel, we set forth on a motorcade to the *Rathaus*. The streets were lined with cheering people, who had waited for hours in the bitter cold. It was all deeply moving until one remembered that a good many of them were cheering just as hard twenty years ago for Hitler.

Edward Kennedy joined us in Berlin. February 22 was his thirtieth birthday; also George Washington's two hundred thirtieth. Both events received due recognition at Mayor Brandt's dinner that evening. After Bobby saluted the two "notable Americans" whose birthday it was, Brandt proposed a toast to "the President, government and people of the United States." The Attorney General, responding, said, "That's the three of us — the President, that's my brother; the government, that's me; and [looking hard at Teddy] you're the people." The otherwise sensible Brandt did not appreciate Kennedy humor and wrote gravely in his memoirs that, though he admired the Kennedys, this event made him "regard the family's political expansion with disquiet."[60]

Robert Kennedy gave the Ernst Reuter Lecture that evening at the Free University. "We do not stand here in Berlin," he said, "just because we are against Communism. We stand here because we have a positive and progressive vision of the possibilities of free society." Marx's indictment of "the heartless laissez-faire capitalism of the early 19th century" now applied precisely, Kennedy suggested, to twentieth-century communism: this had become the contemporary means "of disciplining the masses, repressing consumption and denying the workers the full produce of their labor." And, by "historical para-

dox," it was free society that now seemed most likely "to realize
Marx's old hope of the emancipation of man."[61]

The next morning Lightner assembled a group of editors, ministers
and politicians for breakfast. I had never heard Robert Kennedy in
a foreign policy discussion of this sort before and was considerably
impressed. It was six months after the erection of the Wall. Emotions
were intense in West Berlin; the temptation for visitors to respond
to them was intense too. I noted: "Bobby, I thought, handled himself
exceptionally well. He was frank and direct in discussion; he made
no effort to gratify his audience by saying the things we all knew they
so desperately wanted to hear, and instead spoke with great honesty
and realism about the remoteness of German unification, the impossibility of doing much about the Wall, etc."[62]

In Bonn, Konrad Adenauer, almost half a century older than his
visitor, won Robert Kennedy's heart for a moment by giving him a
scrapbook about the convent that Rose Kennedy had attended in
Germany nearly sixty years before; then lost ground by remarking
later that, of all the world leaders he met, the one he most admired
was John Foster Dulles.[63] Thence Kennedy went to The Hague to
wrap up the Indonesian matter; then to Paris and de Gaulle: "He
wasn't nearly as warm as Adenauer. He was cold and tough [and]
opposed to meeting with the Russians."[64] On February 28, 1962,
they were back in Washington — 14 countries, 30,000 miles in 28
days; exhilaration, exhaustion. The wary Anthony Lewis now concluded that Robert Kennedy was "a very unusual person, a person
who in a kind of tormented way was struggling to do everything on
the merits, which was, I thought, all one could ask. Certainly, nothing
was ever the same after that trip. I mean, he certainly became the
most important person for me in the government."[65] Ethel Kennedy
gave Brandon Grove a pair of cuff links: one engraved REAL WORLD,
the other REAL PEOPLE.

IX

Even the Foreign Service conceded that, for all his strange hobbies,
like youth and counterinsurgency, the Attorney General had peculiar
skills as a propounder of home truths to world leaders. So in December 1962 he was prevailed upon to undertake a mission to Brazil.
This was about the last thing he wished to do. After the Battle of
Ole Miss in September, the missile crisis in October, the housing
order in November, he was now in December liberating Bay of Pigs
prisoners, as well as, in spare moments, preparing for his Supreme
Court debut in January. But Brazil was the largest country in Latin

America, and the Alliance for Progress needed its full collaboration.

In later years the Alliance for Progress was harshly criticized as, at best, a classic example of liberal good intentions overpackaged, overpromised, oversold; at worst, an "arrogant" North American effort to change Latin American countries into "mirror images" of the United States and "to make the region perpetually safe for private U.S. investment." The Alliance unquestionably had its share of illusion and error. The criticism also had its share of myth.[66]

The first myth was that the Alliance was a North American design imposed arbitrarily on Latin America. In fact, the contribution of the Kennedy administration was to give ideas long proposed by progressive Latin American economists and political leaders comprehensive form, collective endorsement and public dollars. The democratic left in Latin America embraced the Alliance. Eduardo Frei of Chile enumerated the objectives of "the Latin American revolution": destruction of the oligarchies; breaking up semifeudal estates and redistributing the land; assuring equal access to education and political power; sharing the gains of economic development; utilizing international capital for the benefit of the national economy. "These," he said, "are precisely the same objectives as those of the Alliance."[67] After John Kennedy received his astonishing welcome in the crowded streets of Bogotá in 1963, Alberto Lleras Camargo of Colombia said to him, "Do you know why these people were cheering you? It's because they think you're on their side against the oligarchs."[68]

The oligarchs, with a few distinguished exceptions, detested the Alliance. Kennedy warned them: "Those who make peaceful revolution impossible will make violent revolution inevitable."[69] They disagreed. Ellis Briggs, an American diplomat of the old school, spoke for them when he denounced the Alliance as a "blueprint for upheaval" and expressed sympathy for "hard-pressed" Latin leaders to whom Kennedy's exhortation "sounded suspiciously like the Communist Manifesto."[70]

Nor was the Alliance either fostered or applauded by American business. If the Kennedy plan had been to integrate the southern hemisphere more firmly than ever in the American corporate economy, presidential speeches stimulating and legitimizing social change in Latin America were an odd way of going about it. It did not even occur to us to invite United States businessmen to the founding meeting at Punta del Este, Uruguay, in August 1961, till the week before. Nor did we follow up on their recommendation that a permanent committee be established to promote foreign private investment.[71] A. F. Lowenthal, a discriminating critic of United States hemisphere policy, has written, "Far from reflecting big business domination of United States foreign policy, the Alliance for Progress commitment emerged in part because of the unusual (and temporary) reduction of corporate influence in the foreign policy-making process."[72]

The strongest witness against the idea that the Alliance was an instrument of Wall Street imperialism was Fidel Castro himself. "In a way," he remarked to Jean Daniel in 1963, "it was a good idea, it marked progress of a sort. Even if it can be said that it was overdue, timid, conceived on the spur of the moment, under constraint, . . . despite all that I am willing to agree that the idea in itself constituted an effort to adapt to the extraordinarily rapid course of events in Latin America."[73] This remained his view. "The goal of the Alliance," he told Frank Mankiewicz a dozen years later, "was to effect social reform which would improve the condition of the masses in Latin America. . . . It was a politically wise concept put forth to hold back the time of revolution, . . . a very intelligent strategy. . . . Basically one has to admit that the idea of the Alliance for Progress was an intelligent one; however, an utopian one."[74]

Utopian because, as Castro told Herbert Matthews in 1963, "you can put wings on a horse, but it won't fly."[75] "Kennedy's good ideas aren't going to yield any results," Castro assured Jean Daniel. Historically the United States had been committed to the Latin oligarchs. "Suddenly a President arrives on the scene who tries to support the interests of another class (which has no access to the levers of power)." What happens then? "The trusts see that their interests are being a little compromised; . . . the Pentagon thinks the strategic bases are in danger; the powerful oligarchies in all the Latin American countries alert their American friends; they sabotage the new policy; and in short, Kennedy has everyone against him."[76]

x

Not, in retrospect, a bad analysis. Kennedy in fact overestimated the possibilities of peaceful revolution in Latin America — as Castro overestimated the possibilities of violent revolution. Both Washington and Havana underestimated the rigidity of the old structures, the inertia of the masses and especially the tenacity of the oligarchs, who simply denied that those who made peaceful revolution impossible made violent revolution inevitable and were well used to calling out the army to suppress either.

The Alliance for Progress required a continent of Betancourts for success. But the progressive democratic parties had their problems.

Governor Luis Muñoz-Marin of Puerto Rico put forward an urgent
one in 1962:

> One terrible disadvantage of political parties of the Democratic Left all
> over Latin America is that they are poor, while their enemies are well-
> heeled. The totalitarian left never seems to lack for funds . . . and the
> totalitarian right is wealthy, per se. . . . In many a country where the
> Alliance may be pouring in tens of *millions*, the ultimate political battle
> may be lost because the parties of the Democratic Left lack a few
> *thousands* of dollars for desperately needed, legitimate, democratic
> action.[77]

This at least Washington could remedy. The CIA had already
helped José Figueres and Norman Thomas establish in Costa Rica
the Institute of Political Education, which Muñoz-Marin called "perhaps the most important seed-bed for democratic leaders in Latin
America."[78] With Robert Kennedy's vigorous support, the Agency
passed funds to parties of the democratic left in Chile and elsewhere. Was this all so base? A decade later found it so. It seemed
useful at the time — why should we not help the friends of peaceful
revolution as much as the Russians and Cubans helped the friends
of violent revolution? — but it established bad habits and contained
an awful potentiality for abuse. The deeper trouble was the ultimate
artificiality of the conditions CIA intervention sought to create. The
progressive democratic leaders, while the finest people in Latin
America, were falling out of touch with fresh currents in politics and
with the militants of the younger generation. If they could not
sustain themselves, it was vain to try to sustain them from without.

President Kennedy never pretended the Alliance would be easy.
In June 1962 he warned against the expectation "that suddenly the
problems of Latin America, which have been with us and with them
for so many years, can suddenly be solved overnight." "We face
extremely serious problems in implementing the principles of the
Alliance for Progress," he said in December. ". . . It's trying to accomplish a social revolution under freedom under the greatest obstacles. . . . It's probably the most difficult assignment the United
States has ever taken on."[79] The Alliance, he said again on November
18, 1963, was "a far greater task than any we have ever undertaken
in our history."

He ended in 1963, as he had begun in 1961, by saying that this
was no miracle to be passed in Washington; it depended in the end
on the people of Latin America. "They and they alone," he had said
at the start, "can mobilize their resources, enlist the energies of their
people, and modify their social patterns so that all, and not just a
privileged few, share in the fruits of growth."[80] It was the Latin

Americans, he said on November 18, 1962, who must "modify the traditions of centuries" and undergo "the agonizing process of reshaping institutions. . . . Privilege is not easily yielded up. But until the interests of the few yield to the needs of the Nation, the promise and modernization of our society will remain a mockery."[81]

XI

Brazil was the largest nation in Latin America. It overflowed with energy, talent, resources, inequities and problems. Its role could be pivotal in the Alliance. It was in a condition of political and economic disarray.

The intelligent but mysterious Janio Quadros had resigned after a few months as President in August 1961. The Vice President, João Goulart, succeeded to the office at the price, exacted by the army and other conservative interests, of having to accept a prime minister and a quasi-parliamentary regime. Goulart, an inept pupil of the renowned Getulio Vargas, was a wealthy landowner of ostentatious populist sentiment. He was not, as some in the Pentagon and the CIA supposed, a Communist. He was rather a left-wing political boss who made deals with Communists or anyone else. An incompetent administrator, he "liked power," as Roberto Campos, the witty Brazilian ambassador to Washington, put it, "but . . . detested government."[82] Most of all he was a *gaucho,* happiest with the cowboys on his ranch.[83]

Goulart paid a visit to Kennedy in April 1962. There was not much on the agenda. Kennedy wished Brazil would do more to support the Alliance. Goulart agreed. Pressed by George Meany to bring up Communist infiltration into the Brazilian labor movement, Kennedy recalled his own experiences on the Senate Labor Committee and introduced the subject "delicately," Lincoln Gordon, now the American ambassador to Brazil, thought. Roberto Campos thought Goulart "a bit irked."[84] Congress was huffing and puffing over the expropriation of International Telephone and Telegraph properties by Goulart's radical brother-in-law, Leonel Brizola, governor of the state of Rio Grande do Sul. Harold Geneen of ITT, making an early foray into foreign policy, was urging what soon became the Hickenlooper Amendment — cutting off foreign aid to any country nationalizing American property without full and speedy compensation. Kennedy made it clear, Campos recalled, that he was not "defending the interests of big business" and saw no future for foreign ownership in fields like telephones and other utilities where "every time a rate had to be raised . . . it became a diplomatic problem." Still, Brazil

must recognize that expropriation without compensation would only
drive Congress to restrictive legislation. His interest, Kennedy emphasized, was "in getting some procedure that would enable us to
engage in . . . frictionless nationalization"[85] — a striking change from
previous American Presidents whose interest was in stopping nationalization per se.*

Kennedy's purpose, Campos recognized, was "to convert Goulart
into a liberal leader of the Alliance for Progress." For a moment he
seemed to have some success. Campos thought Goulart "struck by
Kennedy's personality and liberal posture" and "pleasantly surprised"
by the warmth of his reception.[86] He invited Kennedy to come to
Brazil in July. On his return, however, Goulart plunged into a campaign for the restoration of full presidential powers. Soon he forced
the parliament to agree to a plebiscite on the question in January
1963. In view of the political uncertainty, Gordon recommended
that Kennedy postpone his visit.

Goulart did little about the Alliance, named a Brizola man as prime

---

* I suppose it is impossible to convince the Eric Hobsbawms of this world of it,
though the Harold Geneens had no problem believing it; but preventing nationalization was for John Kennedy an issue of small interest. Lincoln Gordon recalled Kennedy's reaction at a time, shortly before Goulart's visit, when an ITT subsidiary in
Porto Alegre was expropriated. A State Department official put out a statement
condemning the action and talking about the bad effects it would have on private
investment and American aid. Kennedy was furious when he read this and, calling
the assistant secretary for inter-American affairs, asked, "Who is trying to undermine
my Alliance?" (Lincoln Gordon, in interview by author, October 17, 1974).
He was not greatly concerned with internal economic polices. When Jeddi Jagan,
the Marxist Prime Minister of British Guiana, argued that only "socialism" could break
the development bottlenecks in his country, Kennedy told him, "We are not engaged
in a crusade to force private enterprise on parts of the world where it is not relevant.
If we are engaged in a crusade for anything, it is national independence. That is the
primary purpose of our aid. The secondary purpose is to encourage individual
freedom and political freedom. But we can't always get that; and we have often helped
countries which have little personal freedom, like Yugoslavia, if they maintain their
national independence. That is the basic thing. So long as you do that, we don't care
whether you are socialist, capitalist, pragmatist or whatever" (author's journal, October
25, 1962). The difference between Jagan and his rival Forbes Burnham, apart from
the contest between Indians and Africans within Guyana, was not that Jagan was
socialist and Burnham was not but that Jagan was pro-Soviet and Burnham was not.
The British government supported Jagan, though Hugh Gaitskell, the leader of the
Labour opposition, preferred Burnham. The CIA grew considerably exercised over
Jagan, and the British felt — quite rightly, I think, in retrospect — that it was much
ado about nothing. The President's old friend Hugh Fraser was handling the matter
as a junior minister in the Tory government. "American policy," he said later, "was
making this really far more important than it should have been because [British
Guiana] was really nothing but a mudbank" (Hugh Fraser, in recorded interview by
Joe O'Connor, September 17, 1966, 9, JFK Oral History Program). Washington finally
persuaded the British to change the electoral system to proportional representation.
This brought a coalition government under Burnham to power in 1964. As Prime
Minister of the Cooperative Republic of Guyana, Burnham subsequently pursued,
without United States retaliation, a policy of nationalization, hostility to foreign in-
vestment and neutralism.

minister and watched economic confusion grow. Still the *gaucho* in him responded to Kennedy's handling of the missile crisis. "You know we have been defending Cuba's right to be free from invasion," he told Gordon, ". . . but, if they have offensive Russian missile bases there, that's a threat not only to the United States but to the whole western hemisphere, including us, and we are with you." When the missiles were withdrawn, he poured out two huge whiskeys and proposed to Gordon that they drink a toast. As the American ambassador formulated an innocuous speech to world peace, Goulart looked at him and smiled — "he has a very charming smile" — and said in effect, "To hell with that. Let's drink to the American victory."[87]

But Goulart's associates were more and more vocally anti-Yanqui. With political turmoil increasing, the ambassador considered a visit by President Kennedy more untimely than ever. Still he felt that something ought to be done to strengthen the personal relationship established in the spring. In late November 1962 Gordon proposed that Robert Kennedy come to Brazil and express the President's concerns. The Attorney General's reaction, Gordon recalled, was "wry — 'that's a hell of a mission to ask me to take on' — but he was a good soldier."[88]

Robert Kennedy arrived in Brasília on December 17, 1962. A teasing cable from Ethel about Maya Plisetskaya greeted him: IS IT TRUE THE BOLSHOI IS IN BRASILIA? TRES INTERESSANT. KINDLY REMOVE THE MISTLETOE FROM OVER THE TOP OF YOUR HEADS. MME. GOULART DOES NOT UNDERSTAND. . . . YANKEE COME HOME.[89] He began the talks with Goulart by observing that a new government was to be formed and a new program adopted. "This could be a major turning point in relations between Brazil and the United States and in the whole future of Latin America." President Kennedy thought their meeting in the spring had laid the foundation "for the same type of cordial personal relationship that had existed between Vargas and Roosevelt." But recent developments had created "the gravest doubts." It was hard to envisage collaboration if Brazil showed no "spirit of active participation or leadership" in the Alliance and if the Brazilian government and trade unions were "systematically and resolutely anti-American." As for American business, the Attorney General himself "had struggled with some of the same companies with which the Brazilian Government was sometimes in conflict," but, while "business abuse should be combatted . . . business should be treated fairly."

The Americans, Goulart responded, had to understand the contest of "the popular classes against the old dominant elites." If the Brazilian people felt that Washington was allied with the elites, anti-Americanism was inevitable. Some in the government, he conceded, had "fixed hostility to the United States — although not toward

President Kennedy"; but that was a consequence of the continuing political crisis and would change after the plebiscite. As Goulart went on and on, Kennedy scribbled a note to Gordon: "We seem to be getting no place."

Kennedy, when he got the floor after more than an hour, wondered whether Goulart really understood President Kennedy's concern. The Kennedys, after all, had their own problems with American business. Brizola mattered no more than Barry Goldwater. But "a policy to prove Brazil's independence by systematic hostility to the United States cannot be reconciled with good Brazilian-American relations." Goulart asked sharply what Kennedy had in mind. Lincoln Gordon identified government agencies where he thought there had been serious left-wing penetration. There was a moment of tension. Then everyone relaxed, and the meeting ended agreeably. Washington could be sure, Goulart said, that he would not play the Communist game. "In any showdown, there was no doubt that Brazil stood on the side of the United States."[90]

Campos later characterized the results of the meeting as "negligible, if not negative." Goulart, he thought, resented Kennedy's emphasis on Communist infiltration in his government and on the treatment of American corporations.[91] Gordon felt that Kennedy had been too casual in manner, had allowed the conversation to wander, had not seemed sufficiently in earnest — almost as if he was simply performing a chore.[92] Kennedy observed of Goulart later, "I didn't like him. He looks and acts a good deal . . . like a Brazilian Jimmy Hoffa. . . . I didn't dislike him as much as I disliked Sukarno. But . . . I didn't think he could be trusted."[93]

Actually the talks may have had some transient impact. Goulart's cabinet after the plebiscite was not bad. Its key figure, San Tiago Dantas, was intelligent and effective. Unfortunately Dantas was dying of cancer. Goulart was unpopular in Congress, and Dantas failed to get the United States assistance he hoped for. Later in the year Goulart fired his moderate ministers and headed down the Vargas-Perón road of demagogic populist nationalism.

The mission to Brazil lasted only twenty-four hours and therefore gave Kennedy no opportunity for his usual sessions with students and labor. It was less successful than the mission to Indonesia, though in the end neither Sukarno nor Goulart could be 'saved' — either for the democratic or the Marxist worlds; both succumbed to fantasy and were overthrown by their own armies. The Latin American trip, however, left Robert Kennedy with a consuming interest in the Alliance for Progress as offering the most promising formula for democratic development in the Third World. "I might have gotten more involved in Latin America," he mused in 1964, ". . . in the second term, anyway."[94]

His Third World missions persuaded him not only of the folly of the slapdash, self-intoxicated, rhetorical left in the style of Nkrumah and Sukarno but, even more than before, of the hopelessness of supposing the oligarchies could stop the course of history. "Far too often, for narrow, tactical reasons," he wrote after his return, "this country has associated itself with tyrannical and unpopular regimes that had no following and no future. Over the past twenty years we have paid dearly because of support given to colonial rulers, cruel dictators or ruling cliques void of social purpose. This was one of President Kennedy's gravest concerns."[95] It was increasingly one of Robert Kennedy's strongest convictions.

25

The Brothers: II

THE YEAR 1961 was a damned long time ago. For perhaps the last time in their history it was possible for Americans to feel as if all the world were young and all dreams within grasp. I exaggerate of course. But many of us who came to Washington with the Kennedys did suppose that reason could serve as an instrument for social change and that we were moving in the grain of history. When the President on his first day directed Richard Goodwin to find out why the Coast Guard unit in the inaugural parade had no black faces, Goodwin called the Treasury, he said later, with "a rush of energy bordering on elation" and thought, "Why, with a telephone like this we can change the world."[1] Alas, it proved more difficult to change the world than to desegregate the Coast Guard. Euphoria crashed on the Cuban beachhead. Asked in December 1962 how experience had matched expectation, Kennedy replied: "In the first place the problems are more difficult than I had imagined they were. Secondly, there is a limitation upon the ability of the United States to solve these problems."[2] The next year he quoted Franklin Roosevelt on Lincoln — "a sad man because he couldn't get it all at once. And nobody can."[3]

I

For Robert Kennedy these years — his thirty-fifth to thirty-eighth — were fantastically crowded. Unrelenting pressure was etching lines in his face. His rumpled, sand-brown hair was now flecked with gray. Amidst the cascade of public responsibilities, the center of his life remained Hickory Hill, that graceful white mansion — the green lawn rolling down to the tennis court and swimming pool, a pandemonium of children, dogs, ponies, rabbits, cockatoos, jokes, games — where Ethel presided with inexhaustible high spirits.

Hickory Hill was the most spirited social center in Washington. It was hard to resist the raffish, unpredictable, sometimes uncontrollable Kennedy parties. One night at a dinner for the Duchess of Devonshire, with thirty guests crowded into a small dining room, Ethel, who serenely said grace before every meal, finished with a codicil: "And please, dear God, make Bobby buy me a bigger dining-room table." Soon there was a bigger table, in time a new wing, and the parties expanded accordingly. When weather permitted, they were held on the terrace, with the small children "standing around peeping over the hedge, barefooted, in those little nightgowns" (Theodore H. White) or watching "owl-like, with grave, proprietary eyes" (George Plimpton) from a tree house in the tall hickory. One party gave me undesired notoriety when Lee Udall, the wife of the Secretary of the Interior, pushed me fully clothed into the swimming pool as I stood on the edge, contemplating whether I should not jump in anyway to help Ethel, who had been tumbled in a few seconds before when her chair slipped from a platform suspended over the pool. André Malraux, after his exposure to Brumus, the black dog, and Meegan, the white dog, and the children in their red pajamas, was moved to break into English, a language he spoke as rarely as possible, and say, "This house is 'hellzapopping.'"[4]

Their eighth child, Christopher George, was born on Kathleen's twelfth birthday, July 4, 1963; the seventh, Kerry, was now almost four. Robert Kennedy loved children, his own and most others. Children dissolved his reticences, released his humor and his affection, brought him, one felt, more fully out of himself and therefore perhaps more fully into himself. Children in neglect, privation, distress wounded him, like an arrow into the heart. Photographs of the period show him playing with his own children on the beach in summer or in the snow in winter, walking with them hand in hand, carrying them on his shoulders, holding them pensively in his arms. He ruled by encouragement and humor rather than by strict discipline. He was a natural instructor, patient in answering their questions and in teaching them to throw and catch and swim. "There wasn't a problem that the kids didn't have," said Art Buchwald, who became a Hickory Hill familiar, "that he wouldn't interrupt whatever he was doing to solve."[5]

Physical fitness was his accompanying mania. Whenever he could, he swam, played tennis, sailed, skated, skied, pressed more or less unwilling friends into touch football. In February 1963 John Kennedy happened on a 1908 letter from Theodore Roosevelt, another physical fitness maniac, laying down the proposition that Marines ought to be able to hike fifty miles in twenty hours. The President sent the letter over to General Shoup, asking whether present-day

Marines were as fit as their predecessors. "I, in turn," he added, "will ask Mr. Salinger for a report on the fitness of the White House staff."

He then told Salinger that someone from the White House — preferably one whose sacrifice would inspire millions of other out-of-shape Americans — must march the fifty miles with the Marines. The portly Salinger, who had studiously avoided all forms of exercise except cigar-smoking for twenty years, tried for several days to dodge the challenge, finally resorting to the desperate expedient of insisting on step-by-step coverage by the White House press corps. At Camp Lejeune, Marine officers went into training. Around the country athletic citizens started off on fifty-mile walks of their own. On the brink Salinger extracted a statement from the President's Council on Physical Fitness saying that only persons in appropriate physical condition should attempt the hike — a standard, he pointed out, that clearly disqualified him. Carleton Kent of the *Chicago Sun-Times* spoke for the White House press: "Thank God, the Press Secretary is a coward."

On a Friday afternoon at the Department of Justice, Robert Kennedy, who always had to test himself against everything, decided that he too would make the hike. To Edwin Guthman, Louis Oberdorfer and James Symington, who had the bad luck to be with him, he said, "You're all going with me, aren't you?" At five o'clock on Saturday morning the four, joined by David Hackett, the best athlete at Milton, met by the old Chesapeake & Ohio Canal. Their plan was to walk to, or toward, Camp David in Maryland. The thermometer stood at twenty degrees. The towpath was covered by snow and ice. As the miles passed, Hackett, Oberdorfer and Symington dropped out. A helicopter flew low; the pilot waved. "Maybe there's a national emergency and I'll have to go back," Kennedy said hopefully. It was a *Life* photographer. After thirty-five miles Guthman gave up. The Attorney General whispered, "You're lucky, your brother isn't President of the United States."[6] On he doggedly went to the finish. David Brinkley of NBC wrote him later: "As an aroused and concerned citizen, I feel it my duty to call to your attention certain pertinent facts, as follows: President Theodore Roosevelt, who is now credited with originating all this hiking and exercising therapy, died young, at sixty."[7]

II

A great sorrow marred the familial felicities of these years. On December 19, 1961, while playing golf in Palm Beach, Joseph P. Kennedy, now seventy-three years old, fell suddenly sick. His niece

Ann Gargan drove him back to his house. He said brusquely, "Don't call any doctors."[8] For once his command was ignored. They rushed him to the hospital. It was an intracranial thrombosis, paralyzing his right leg, arm and face, paralyzing the hemisphere of his brain that governed his ability to speak.

He had stayed resolutely in the background from the moment his son became a presidential candidate. "My day is done," he had told the reporter Bob Considine in 1957. "Now it's their day."[9] "I don't want my enemies to be my son's enemies," he told John Seigenthaler in 1960, "or my wars to be my son's wars."[10] He had meticulously avoided public appearances in the election year, declining to speak at the banquet celebrating the twenty-fifth anniversary of the Securities and Exchange Commission,[11] even declining to attend his son's acceptance speech in Los Angeles. The proud old man practiced self-abnegation with sensitivity and grace. His pride in his children had always exceeded his pride in himself.

After the election, the ambassador said that "the saddest thing about all this" was that "Jack doesn't belong any more to just a family. . . . The family can be there. But there is not much they can do sometimes for the President of the United States."[12] After the inauguration he called Stephen Smith. "I want to help," he said, "but I don't want to be a nuisance. Can you tell me: do they want me or don't they want me?" Smith reported the conversation to Robert Kennedy, who listened carefully and thanked him.[13] Soon afterward the ambassador left for Europe. When he returned, he did not visit his son's White House. The President called him frequently at Hyannis Port and Palm Beach. "His father used to wait for the call," Charles Spalding remembers. "And it was really touching if you knew Mr. Kennedy, who was a terribly aggressive individual, the way he would hold himself in check and the way he would make his recommendations . . . in the restraint and in the manner in which he did it."[14] It was almost as if life had moved beyond him, as if his sons, who never stopped loving him and who were his life, needed him no longer.

John and Robert flew at once to Palm Beach after the stroke. Their father developed pneumonia, pulled through, then began a long invalidism. He could not control his speech but did not know it and, since he continued to understand what was said to him, continued to speak. Only his "no" came through clearly, though sometimes it meant "yes." Most words poured out in an unintelligible garble. His family nodded as if they understood. "Thanks, Dad," the President would say, "I'll take care of it. I'll do it your way."[15]

For a time he underwent therapy at Dr. Howard Rusk's Institute of Physical Medicine and Rehabilitation in New York. Ann Gargan became his eternal companion, in perennial contention with the professional nurse, Rita Dallas, over the best mode of treatment. His wife and children enveloped him in affection. "There was never a moment," Rita Dallas later wrote,

> when the love each child felt for Mr. Kennedy was not evident. . . . Time and again I would see them stand outside his door and actually seem to screw up courage from deep inside before entering the sick-room. They went in to him with shoulders squared, but when they left his presence, they would often sag against the wall in despair. . . . Each one radiated confidence to him, and always, when his children were around, Mr. Kennedy took on a new vigor.[16]

No one among them was more attentive than Jacqueline Kennedy.

He was indomitable. "I never saw a man," said his nurse, "fight so hard to stay vital and alive. . . . He would fail and fall back, but his drive for survival invariably forced him forward again."[17] At last he came to his son's White House. Ben Bradlee saw him there in the spring of 1963. "The old man is bent all out of shape," he wrote,

> his right side paralyzed from head to toe, unable to say anything but meaningless sounds and "no, no, no, no," over and over again. But the evening was movingly gay, because the old man's gallantry shows in his eyes and his crooked smile and the steel of his left hand. And because his children involve him in their every thought and action. They talk to him all the time. They ask him "Don't you think so, Dad?" or "Isn't that right, Dad?" And before he has a chance to embarrass himself or the guests by not being able to answer, they are off on the next subject. . . . When he eats, he drools out of the right side of his mouth, but Jackie was wiping it off quickly, and by the middle of dinner there really is no embarrassment left. . . . The Kennedys are at their best . . . when they are family, and forthright and demonstrative, and they were at their best tonight.[18]

At times the frustration became too much to bear. Robert Kennedy took the brunt of it, as he had when he was a small boy so many years before. Once the presence of the President and the Attorney General stimulated the ambassador to rise from his wheelchair and try to walk without a brace. He stood erect for a moment, then commenced to stagger. "In a lightning move," reported his nurse, "Bobby grabbed his father. Mr. Kennedy tried to struggle loose and began swatting at him with his cane. . . . Even though he was ducking blows, Bobby gradually succeeded in easing the tension by starting to laugh and tease his father." The doctor intervened, persuading the old man to sit down. "He was screaming and shaking his fist at his son." Robert leaned over, kissed his father and said, "Dad, if you want to get up,

give me your arm and I'll hold you till you get your balance. . . .
That's what I'm here for, Dad. Just to give you a hand when you
need it. You've done that for me all my life, so why can't I do the
same for you now."

In summers at Hyannis Port, Rita Dallas wrote, "it hurt me whenever Mr. Kennedy treated him roughly. It seemed that he yelled at
him constantly, but as time went on, I realized it was almost as though
he were trying to urge him on to accomplish greater and greater
things." Robert understood, never complained, cherished his father,
exercised every day with him in the swimming pool. "During the
years that followed, I watched Bobby strengthen his father, laughing
with him, praising him, then he would swim away. His eyes would
fill with tears, and a look of deep sorrow would cloud his face, but he
would quickly compose himself, and begin once more doing what he
could to assist him in therapy."[19] Robert himself characteristically
said it was Jack who was "the best with my father because he really
made him laugh and said outrageous things to him."[20] One more
shadow had fallen over the Kennedy family.

III

Among strangers the Attorney General was still diffident and often
uneasy. One day the clerks of the Supreme Court justices invited
him to lunch. "His mastery of the questions — and you can imagine
the Supreme Court law clerks were a very snotty crew — was quite,
quite impressive," said Peter Edelman, Arthur Goldberg's clerk.
"[But] all the time that this tough guy was answering questions . . . his
hands were shaking under the table and were knotted up with one
another."[21]

Shyness made Kennedy appear abrupt or preoccupied. On occasion people saw a veil fall over his eyes; they did not know where
they stood with him or whether they were getting through or not.
On occasion the old combativeness flared up. On occasion he seemed
imperious, too ready to override people in the determination to get
things done, expecting too much too easily in the way of loyalty or
service. He now radiated a sense of power, derived from his own
gifts and intensities as well as from his relationship to the President.
I remember Marie Harriman's birthday party in April 1963. Randolph Churchill, who had come to Washington to accept his father's
honorary American citizenship, was there; also Franklin D. Roosevelt,
Jr., and Robert Kennedy — sons of the three great Anglo-American
political dynasties of the twentieth century. "The dominating figure

of the three," I noted in my journal, ". . . was Bobby, who kidded the others mercilessly."[22] Randolph Churchill described the evening in a London newspaper: "He has a wonderful gift of quick repartee, a delightful smile and a most engaging personality."* Robert Kennedy thanked him for "something nice you wrote about me," adding, "This does not happen with overwhelming frequency here in the United States so I am having your piece made into leaflets and have instructed the Air Force to take them and drop them all across this country so that people will come to realize that they have a fine fellow as Attorney General. Our U-2 pilots will also drop some in Cuba."[23]

Achievers fascinated him, whether James Baldwin or General MacArthur or Richard Daley, whether scholars or astronauts or film stars. I remember a luncheon that William Walton gave for Marlene Dietrich in September 1963:

> Bobby, with his innocent audacity, asked Marlene who was the most attractive man she had ever met. She answered promptly: "Jean Gabin," and then added that the years she had spent with him were the happiest of her life. Bobby asked why she had left him. She said, "Because he wanted to marry me." When Bobby expressed surprise, she said, "I hate marriage. It is an immoral institution. I told him that, if I stayed with him, it was because I was in love with him, and that is all that mattered." Bobby asked her whether she still saw him. She said, "No, he won't see me any more. He has married, and has grown terribly fat, and thinks he is no longer attractive, and does not want me to see him." Bobby said, "Does he still love you?" Marlene: "Of course."[24]

The same audacity brought him to Marilyn Monroe. We both met her the same night after she had sung "Happy Birthday, Mr. President" at a Madison Square Garden celebration of John Kennedy's forty-fifth birthday. It was May 19, 1962, at a small party given by that loyal Democrat Arthur Krim of United Artists. Adlai Stevenson wrote a friend about his "perilous encounters" that evening with Marilyn, "dressed in what she calls 'skin and beads.' I didn't see the beads! My encounters, however, were only after breaking through the strong defenses established by Robert Kennedy, who was dodging around her like a moth around the flame."[25] We were all moths around the flame that night. I wrote:

> I do not think I have seen anyone so beautiful; I was enchanted by her manner and her wit, at once so masked, so ingenuous and so penetrating. But one felt a terrible unreality about her — as if talking to

---

* Randolph Churchill in *News of the World* (London), April 14, 1963. Churchill had had his ups and downs with the Kennedys, as he had had with everybody else. On September 26, 1948, the *New York Times* had published a letter from Joseph P. Kennedy attacking Winston Churchill's war memoirs on the ground of "misquotations" and other "inaccuracies." On October 17, the *Times* carried a long and characteristically truculent reply from Randolph Churchill.

someone under water. Bobby and I engaged in mock competition for her; she was most agreeable to him and pleasant to me — but then she receded into her own glittering mist.[26]

There was something at once magical and desperate about her. Robert Kennedy, with his curiosity, his sympathy, his absolute directness of response to distress, in some way got through the glittering mist as few did. He met her again at Patricia Lawford's house in Los Angeles. She called him thereafter in Washington, using an assumed name. She was very often distraught. Angie Novello talked to her more often than the Attorney General did. One feels that Robert Kennedy came to inhabit the fantasies of her last summer. She dreamily told her friend W. J. Weatherby of the *Manchester Guardian* that she might get married again; someone in politics, in Washington; no name vouchsafed. Another friend, Robert Slatzer, claims she said Robert Kennedy had promised to marry her. As Weatherby commented, "Could she possibly believe that Kennedy would ruin himself politically for her?" Given the desperation of her life, this idea, Norman Mailer suggested, perhaps became "absolutely indispensable to her need for a fantasy in which she could begin to believe." In other moods she spoke more reasonably. She once mentioned the rumors about Robert Kennedy to her masseur Ralph Roberts, with whom according to Mailer she had "a psychic communion that is obviously not ordinary." "It's not true," she said to Roberts. "I like him, but not physically."* On the weekend of August 4 Kennedy was in San Francisco at a meeting of the American Bar Association. She killed herself through an overdose of sleeping pills, probably by accident, perhaps by intent, on the night of August 4. This was less than three months after they met. I doubt whether they had seen each other more than half a dozen times.

He always wanted to learn new things. "I was an orphan," said Art Buchwald, "and I was raised in foster homes, and I was in a Jewish upbringing. And he was very fascinated with this because this is something that he had never experienced. . . . He was very interested in other people's lives, particularly if they weren't Harvard, Yale or Princeton, which he knew about."[27] When he met John Glenn, the first American to enter space orbit, "he was interested," Glenn recalled, "in exactly the personal experience of what it's like. What

---

* W. J. Weatherby, *Conversations with Marilyn* (New York: Ballantine reprint, 1977), 154–155, 164; Norman Mailer, *Marilyn: A Biography* (New York, 1973), 229, 232; Robert F. Slatzer, *The Life and Curious Death of Marilyn Monroe* (New York, 1974), 1, 3–4. Even by Slatzer's highly conspiratorial account, her long-distance phone bill from May 27 to August 5 amounted only to $209 — not much for a Hollywood actress (Slatzer, 141). Peter Lawford subsequently described talk of an affair with either Kennedy as "garbage" (*Star*, February 24, 1976).

does it feel like to be weightless? What did you think about just before the booster lit off? . . . What did the sunset look like? . . . What did you think about during reentry? . . . a thousand and one questions."[28]

He read more than ever, mostly history and biography.* He even listened to long-playing records of Shakespeare's plays while shaving in the morning. At last he was fulfilling his mother's dream of self-improvement — and perhaps exciting new maternal dreams. Rose Kennedy wrote Robert semi-jocularly from Paris in 1961:

> I think you should work hard: and become President after Jack —
>   It will be good for the country
>   And for you
>   And especially good for you know who,
>   Ever your affectionate and peripatetic mother.[29]

Returning from a fortnight of Aspen seminars in the summer of 1961, Robert and Ethel asked me whether I would organize a series of evening meetings in Washington at which heavy thinkers might remind leading members of the administration that a world of ideas existed beyond government. I stalled for a time, trusting they would forget. Not a chance; and the first Hickory Hill seminar convened on November 27, 1961. The lecturers ranged from Isaiah Berlin to Al Capp; the audience from McNamara and Dillon to Alice Roosevelt Longworth. Except for the summers, we managed, under the Attorney General's gentle pressure, a meeting every month or so. (Thus a note in July 1963: "I hope you are working on a schedule for our seminars starting mid-September. You said you would and if you don't I'll tell J. Edgar ——— ."[30] Even Mrs. Longworth, her cynicism well honed by sixty years of life in Washington, became a devotee. "They sound rather precious," she said later, "but there was nothing precious about these lectures. It was all sorts of fun." John Kenneth Galbraith, waylaid on a visit from New Delhi to hold forth on economics, recorded his impression:

> Bobby confined his role to interrogation. But he was a very rapt and eager prosecutor of the positions; you had the feeling that if you were shabby on any important point you could pretty well count on Bobby

---

* Between Christmas and Easter 1962–63, he read E. S. Creasy's *Fifteen Decisive Battles of the World*; S. F. Bemis's *John Quincy Adams and the Foundations of American Foreign Policy*; Irving Stone's *They Also Ran*; Alan Moorhead's *The White Nile*; Barbara Ward's *The Rich Nations and the Poor Nations*; Herbert Agar's *The Price of Union*; Barbara Tuchman's *The Guns of August*; Cecil Woodham-Smith's *The Great Hunger*; Paul Horgan's *Conquistadors in North American History*; and *Seven Days in May* by Fletcher Knebel and Charles Bailey (see Fletcher Knebel, "Bobby Kennedy: He Hates to Be Second," *Look*, May 21, 1963).

to come in and press you on it. This was matched in some degree by the eagerness of the questioning that Ethel put. It stands in my mind as a bright, lively, and professional evening.[31]

"The most striking thing about Bob," Ramsey Clark said, "was his desire and capacity for growth."[32] "He continually embraced new things," noted the writer Peter Maas, "and he didn't reject something just because it didn't fit in with an earlier period."[33] "Most people," said Anthony Lewis, "acquire certainties as they grow older; he *lost* his. He changed — he grew — more than anyone I have known."[34]

IV

John Kennedy's close friends, as Kenneth O'Donnell perceptively noted, were all people he had met in the 1930s and 1940s. Robert kept "making close friends up until the time of his death."[35] Among his intimates Bobby was usually a competitor and sometimes a needler but essentially easy, considerate and so often unexpectedly sweet. "When I first met him," said Michael Forrestal, son of FDR's Secretary of the Navy and himself an able lawyer who had come to the White House to work on Southeast Asia, "I damned near punched him in the nose. I found him very offensive, rude and obnoxious. . . . But within about a month, after seeing a little bit more of him, I became more or less a captive of his, . . . the *most* astounding person I've ever met."[36]

It was in these years that I got to know him well. When he recruited me for the White House in 1960, we had little more than an amiable acquaintance. Nor would I — a liberal, intellectual, professor, writer, agnostic; a non-hiker, non-skier, non-touch football player, non-mountain climber, whose only athletic pleasure was indifferent tennis — have seemed especially his type.

We got along. I served some purpose in the White House in his considerably more synoptic view of the administration. He called me from time to time to ask me to look into this or that problem; no special pattern emerged. Sometimes he asked me to bring up things with his brother. I wondered why he did not call his brother directly but rarely questioned and supposed I had some minor role in a larger stratagem. When I was blocked in something I thought important, I went to him. He always listened and often acted. I found him excellent company. He was very funny in a sardonic throwaway style, and also keenly appreciative of humor in others. When a remark entertained him, he had the most engaging habit of clapping two or three times. Jocular letters arrived in one's mail. Once Senator Hugh

Scott of Pennsylvania attacked me; I replied suitably. Soon I received a handwritten note:

> In connection with your letter to Senator Scott on whether you're a communist or something. You looked like a subversive with Marilyn Monroe in New York and Mark DeWolfe Howe likes you so don't expect any support from over here.
>
> Teddy's brother[37]

Underneath the teasing, one felt that he was deeply protective of friends.

"He could be at times," Kenneth O'Donnell and David Powers wrote in their book, "incredibly naive, too impressed by celebrities, too impulsive or too unrealistic. At most other times, especially when his older brother was depending on his firm support, he was wise, calm, restrained, full of courage and understanding. . . . Always he was the kindest man we ever knew."[38] Most who knew him well would warmly endorse the concluding thought. This was hardly the prevailing public idea in 1963. The impressions of the fifties — his father, the McCarthy committee, the Rackets Committee, the pursuit of Hoffa — had not died. They now received influential restatement at the hands of the writer Gore Vidal. "His obvious characteristics," Vidal wrote in *Esquire* in March 1963, "are energy, vindictiveness and a simple-mindedness about human motives which may yet bring him down. . . . He has none of his brother's human ease; or charity. . . . He would be a dangerously authoritarian-minded President."[39] A dozen years later Vidal summed up his theory of Robert Kennedy: "really a child of Joe McCarthy, a little Torquemada."[40]

I suppose that the historian must suggest a background for Vidal's attack. Vidal was a talented man, a writer of lucid and graceful prose, an authentic wit, excellent historical novelist, adept playwright, brilliant essayist, litigious citizen, always filled with charm and malice. He and Jacqueline Kennedy both had had, in successive periods, Hugh D. Auchincloss as a stepfather. His ambition had long extended to politics. "He tells me he will one day be president of the United States," Anaïs Nin wrote in 1945; adding soon, "He is insatiable for power."[41] "I certainly wanted to be president," Vidal said in 1975. ". . . Before *The City and the Pillar* [his novel about homosexuality], I thought I had a very good crack at it."[42] He ran for Congress in 1960. Robert Kennedy came into Dutchess County to speak on his behalf but, when they met, failed to recognize him. As John English, a New York Democratic leader, later told the story, "Vidal got very uptight about it. And Kennedy was embarrassed because [Vidal] was related to Jackie and so forth, and he was really [hard] on himself for not having a better memory."[43] This no doubt introduced an early edginess into their relationship.

Nevertheless Vidal was delighted to have friends in the White House. In April 1961 he wrote an adoring piece about John Kennedy for the *London Sunday Telegraph*: "the most accessible and least ceremonious of recent Presidents . . . withdrawn, observant, icily objective in crisis. . . . His wit is pleasingly sardonic. Most important, until Kennedy it was impossible for anyone under fifty (or for an intellectual of any age) to identify himself with the President."*

On November 11, 1961, the Kennedys gave a dance at the White House for Lee Radziwill. Vidal was there. As the evening passed, he gave the strong impression of having drunk far too much. Someone, I forget who, perhaps Jacqueline Kennedy, asked me whether I would get him out of there. I enlisted Kenneth Galbraith and George Plimpton. We took Vidal back to his hotel. The next day Vidal came to see me. He often did in those days, but he later told someone that, assuming I was the White House chronicler (not in fact my job or expectation), he wanted in this case to explain particularly what had happened the night before. I noted in my journal:

> Gore Vidal got into violent fights, first with Lem Billings, then with Bobby. According to Gore, Bobby found him crouching by Jackie and steadying himself by putting his arm on her shoulder. Bobby stepped up and quietly moved the arm. Gore then went over to Bobby and said, "Never do anything like that to me again." Bobby started to step away when Gore added, "I have always thought that you were a god-damned impertinent son of a bitch." At this, according to Gore, Bobby said, "Why don't you get yourself lost." Gore replied, "If that is your level of dialogue, I can only respond by saying: Drop dead." At this Bobby turned his back and went away.[44]

Vidal's enemies delightedly embellished the story. I can testify that he was not forcibly ejected, nor cast bodily into Pennsylvania Avenue, nor was there any kind of major scene. Jacqueline Kennedy, however, was irritated by his behavior and resolved not to have him in the White House again.

Vidal and Robert Kennedy were not in any event made for each other. Vidal genuinely admired the President — "an ironist in a profession where the prize usually goes to the apparent cornball," he wrote as late as 1967.† But he could not abide the younger brother's puritanism and zeal. For his part, Robert Kennedy, I imagine, regarded bisexuality with disapproval — he became more understand-

---

* The piece appeared in the *Sunday Telegraph*, April 9, 1961. Reprinting it in 1962 in *Rocking the Boat* (Boston, 1962), 3–14, Vidal noted (282) that the Bay of Pigs had rather undercut the portrait he had drawn for the British. "Nevertheless, I take back nothing."

† Gore Vidal, "The Holy Family," *Esquire*, April 1967. I must perhaps declare an interest, because Vidal, whose company I continue to find entertaining, if problematic, described *A Thousand Days* in the same piece as "the best political novel since *Coningsby*" — though, as an admirer of *Coningsby*, I take this as a high compliment.

ing of the varieties of human experience in later years — and mistrusted Gore's destructive propensities. I surmise that, as awareness
of his excommunication sank in, Vidal's resentment concentrated on
the Attorney General and boiled over in *Esquire.*

Shortly after the piece came out, Robert Kennedy was in the family
apartment in New York. He had invited Budd Schulberg for breakfast along with Edwin Guthman. Just back from mass, he asked his
guests what they would like. They said bacon and eggs. "A nice
Catholic boy like me," Kennedy said, "has to cook bacon and eggs on
Friday for a couple of backsliding Jewish boys." As he prepared the
breakfast, he said, "If only Gore Vidal could see me now — the
lovable Bobby — standing over a hot stove to see that his friends get
a good nutritious start on the day." Behind the wry humor, Schulberg said later, "I felt a real hurt, even a sense of bafflement in Bob
that his public image was so much closer to Vidal's caricature than to
the actual, intensely human being we knew."[45] Yet Robert Kennedy
well understood the function he served. "The President," he said in
an interview, "has to take so much responsibility that others should
move forward to take the blame. People want someone higher to
appeal to. . . . It is better for ire and anger to be directed somewhere
else."[46]

v

The fraternal partnership did not cover every aspect of policy. As
Sorensen said later, the President "made important decisions on
which the Attorney General was not his most influential adviser and
may not have been consulted at all."[47] Except for poverty, Robert
Kennedy was not much involved in economic issues. As civil rights
increasingly absorbed his time and passion, he withdrew from foreign
affairs. In April 1963, when Dobrynin handed him a hectoring Soviet
paper about American foreign policy, the Attorney General told him:
"I thought it was so insulting and rude . . . that I would neither accept
it nor transmit its message." If the Kremlin wanted to send messages
like this, deliver them formally through the State Department, "not
through me."[48] Thereafter, the Attorney General dropped out of
the Soviet relationship.

The President's struggle for détente continued. "One of the ironic
things," the President remarked that spring to Norman Cousins of
the *Saturday Review,* ". . . is that Mr. Khrushchev and I occupy approximately the same political positions inside our governments. He
would like to prevent a nuclear war but is under severe pressure
from his hard-line crowd, which interprets every move in that direction as appeasement. I've got similar problems. . . . The hard-liners

in the Soviet Union and the United States feed on one another."[49] Each leader persevered. Kennedy's American University speech in June 1963 led to the Harriman mission to Moscow, the test ban treaty, the ratification by the Senate.

Opposition to détente persisted in both capitals. Early in October, Kennedy authorized the sale of wheat to the Soviet Union. Lyndon Johnson was unhappy. "The Vice President," John Kennedy told me on October 11, "thinks that this is the worst foreign policy mistake we have made in this administration."[50] Khrushchev had comparable troubles. While he was away from Moscow on a hunting trip in late October, the KGB arrested Professor Frederick Barghoorn of Yale as a spy. Kennedy immediately demanded Barghoorn's release. "Khrushchev returned home," said Yuri Ivanovich Nosenko of the KGB, "and he was so mad. 'Who allowed this operation? What fools have done it?'" (The fool turned out to be Brezhnev.)[51]

Robert Kennedy strongly supported the Soviet wheat deal: "I felt it was difficult to turn down the request for the purchase of food [for hungry people]. . . . Also I thought that . . . the Cold War had diminished somewhat — the problems hadn't but the cold war had."[52] Above all, his brother imparted to him the intense belief that the control of nuclear weapons was the world's supreme issue. John Kennedy, Robert said in 1964, "felt stronger about that question almost than anything else."[53] He told Robert late in 1963 that "his greatest disappointment was that he had not accomplished more on disarmament."[54]

The realities of Kennedy's policy bear little resemblance to revisionist caricature. On October 22, 1963, George Kennan, who had resigned as ambassador to Yugoslavia and from public life, wrote Kennedy: "I don't think we have seen a better standard of statesmanship in the White House in the present century."[55] Mike Mansfield, the Senate's most steadfast voice for restraint and humanity in foreign policy, had been in Congress since 1943. Asked on his retirement in 1976 to discuss the seven Presidents he had served, from Franklin Roosevelt through Gerald Ford, Mansfield described Kennedy in his laconic fashion as "the best of the lot."[56]

VI

Despite the fraternal tie, Robert suffered frustrations familiar to every cabinet member. "Far from being influenced easily by Bobby," O'Donnell thought, "the President was quick to point out an error or weakness in one of Bobby's proposals."[57] The Attorney General, though not the most faithful attendant at cabinet meetings, came to feel that the President would do better if he paid more attention to the cabinet. He wrote his brother in March 1963:

> The best minds* in Government should be utilized in finding solutions to . . . major problems. They should be available in times other than deep crisis and emergencies as is now the case. You talk to McNamara but mostly on Defense matters, you talk to Dillon but primarily on financial questions, Dave Bell on AID matters, etc. These men should be sitting down and thinking of some of the problems facing us in a broader context. I think you could get a good deal more out of what is available in Government than you are at the present time.[58]
>
>   * ME [handwritten footnote]

More lightheartedly in November, he recalled Daniel Webster's reminder to John Tyler that William Henry Harrison's practice had been to let the cabinet make the decisions, "each member . . . and the President having but one vote." "I wonder," the Attorney General said, "if this gives you any ideas about the role that Stew [Udall], Orville [Freeman], Luther [Hodges] and John [Gronouski] and some of the rest of us should be playing at future meetings."[59]

Kidding remained the favorite Kennedy form of communication. Early in the administration, Robert noticed that "Jack had on his desk, 'Profiles in Courage' and the Bible. I asked him . . . if these were the books of the world's two great authors."[60] When *Life* described the Attorney General as the number two man in town, the older brother said darkly, "That means there's only one way for you to go, and it ain't up!"[61] Robert was delighted by a photograph showing two of his children peering from under the presidential desk; the President's inscription was: "Dear Bobby: They *told* me you had your people placed throughout the government." They talked kiddingly even about matters on which they felt seriously; and always talked in the cryptic half sentences that bespoke perfect understanding.

In 1963 Ben Bradlee asked John Kennedy why — "never mind the brother bit" — he thought Robert was so great. The President replied: "First, his high moral standards, strict personal ethics. He's a puritan, absolutely incorruptible. Then he has this terrific executive energy. We've got more guys around here with ideas. The problem is to get things done. Bobby's the best organizer I've ever seen." ("Management in Jack Kennedy's mind," Chester Bowles once commented, ". . . consisted largely of calling Bob on the telephone and saying, 'Here are ten things I want to get done.'")[62] John Kennedy added to Bradlee: "He's got compassion, a real sense of compassion. Those Cuban prisoners . . . weighed on his mind for eighteen

months.... His loyalty comes next. It wasn't the easiest thing for
him to go to [Joe] McCarthy's funeral."[63]

John Kennedy used Robert in part as Franklin Roosevelt used
Eleanor — as a lightning rod, as a scout on far frontiers, as a more
militant and somewhat discountable alter ego, expressing the President's own idealistic side while leaving the President room to maneuver and to mediate. At the same time, the Attorney General was
John Kennedy's Harry Hopkins, Lord Root of the Matter, the man
on whom the President relied for penetrating questions, for follow-
up, for the protection of the presidential interest and objectives.
Robert Kennedy, McNamara said, "recognized that his greatest contribution to the President would be to speak candidly, to contradict
him if he felt he was wrong and to move him to the right course if
he felt he was not on that course. The President never hesitated
to turn down Bobby's advice, but many, many times he took it
when, initially, he, the President, was in favor of an opposite course.
They had an *extraordinarily* close relationship: affection, respect,
admiration."[64]

I do not think Richard Goodwin overstated the case when he wrote
years later that "President Kennedy's most impolitic appointment
... also made the greatest contribution to the success and historical
reputation of his Administration."[65]

VII

The relationship was closest in working hours. For the President,
evenings and weekends were times of respite. Jacqueline Kennedy
once told me that he preferred not to see at night people who insisted
on the problems of the day.[66] He wanted distraction, a change of
subject, easy conversation with old friends.

Robert understood this well enough in the abstract. Explaining
Jacqueline's singular charms for his brother, he told Pearl Buck,
"What husband wants to come home at night and talk to another
version of himself? Jack knows she'll never greet him with 'What's
new in Laos?'"[67] But in practice Robert himself often could not
resist asking what was new in Laos. "Bobby got his relaxation out of
talking about those things," observed Spalding. "... He never seemed
to need any release from it at all, whereas the President obviously
did."[68] For John Kennedy, at the end of a long day, Robert was
often too demanding, too involved in issues, too much another version of himself. Teddy made the President laugh. Bobby was his
conscience, reminding him of perplexities he wished for a moment

to put aside. So Robert and Ethel were not often at the White House on purely informal occasions.

Alike in so many ways, united by so many indestructible bonds, the two brothers were still different men. John Kennedy remained, as Paul Dever had said, the Brahmin; Robert, the Puritan. In English terms, one was a Whig, the other a Radical. John Kennedy was urbane, objective, analytical, controlled, contained, masterful, a man of perspective; Robert, while very bright and increasingly reflective, was more open, exposed, emotional, subjective, intense, a man of commitment. One was a man for whom everything seemed easy; the other a man for whom everything had been difficult. One was always graceful, the other often graceless. Meeting Robert for the first time in 1963, Roy Jenkins of England thought him "staccato, inarticulate . . . much less rounded, much less widely informed, much less at ease with the world of power than his brother."[69] John Kennedy, while taking part in things, seemed, as Tom Wicker observed, almost to watch himself take part and to criticize his own performance; Robert "lost himself in the event."[70]

John Kennedy was a life enhancer. His very presence was exhilarating — more so than anyone I have ever known. "It was like a lot of flags on a ship with Jack, easy and bright," said Spalding. "Gaiety was the key to his nature. . . . He was always the greatest, greatest company; so bright and so restless and so determined to wring every last minute. . . . [He] gave you that heightened sense of being."[71] He was, Robert told John Bartlow Martin, "really an optimist . . . a little bit like my father, always saw the bright side of things."[72] Robert himself was variable, moody; "the pendulum just swings wider for him than it does for most people," as Lawrence O'Brien said.[73] Underneath the action and jokes there was a streak of brooding melancholy. John Kennedy, one felt, was at bottom a happy man; Robert, a sad man.

John Kennedy seemed invulnerable; Robert, desperately vulnerable. Friends wanted to protect the younger brother; they never thought the older brother required protection. One felt liked by John Kennedy, needed by Robert Kennedy. Robert had the reputation for toughness; but, as Kenneth O'Donnell said, John "was much the toughest of the Kennedy brothers."[74] "Robert Kennedy," said Pierre Salinger, "gave the impression of a very tough man when he was in fact very gentle. John Kennedy, under his perfect manners, was one of the toughest men that ever was."*

---

* Pierre Salinger, *Je suis un Américain* (Paris, 1975), 180. Note also Jerry Bruno: "I always got a laugh out of the notion that it was Bobby Kennedy who was the ruthless one while John Kennedy was nice and relaxed" (Jerry Bruno and Jeff Greenfield, *The Advance Man* [New York: Bantam reprint, 1972], 40).

John Kennedy was a man of cerebration; "a man who mistrusted passion," as Richard Neustadt said. Robert trusted his passions.[75] Reason was John Kennedy's medium; experience was Robert's. Both men deeply cared about injustice, but the President had an "intellectual understanding" of the great social problems, said Ben Bradlee; Robert "had it in his gut."[76] One attacked injustices because he found them irrational; the other because he found them unbearable. The sight of people living in squalor appalled John Kennedy but, like FDR before him, he saw it from without. Robert had a growing intensity of personal identification with the victims of the social order.

John Kennedy was an ardent liberal reformer. But he was, in his own phrase, an idealist without illusions. He accepted reality. "There is always inequity in life," he observed in 1962. "Some men are killed in a war and some men are wounded, and some men never leave the country. . . . Life is unfair."[77] He was, said Neustadt, "much more resigned to the restraints of institutional life." He was also responding to a calmer time. Had he been President in the later 1960s, he might well have been as radical as his brother, though with the composure of FDR rather than with Robert's rushing passion, and very likely more effective. Robert Kennedy, in a more turbulent time, rebelled against institutional restraints. He was ready, said Neustadt, "to leap outside established institutions to get things done. . . . He could come to terms with institutions but he always hated to." He had, more than his brother, "this drive to the direct approach." His notion was that "every wrong . . . somehow . . . if you can't right it, you've got to bust in the attempt."[78]

Charles Spalding said in 1963, "Jack has traveled in that speculative area where doubt lives. Bobby does not travel there."[79] The President's mind was witty and meditative; the Attorney General's, direct and practical. Roy Wilkins of the NAACP felt that the President "invited you to commune with him. . . . I never got the impression you're communing with Robert Kennedy. You're talking to him; you're arguing with him; and you're dealing with . . . a hard, clear-thinking, determined public servant who has, in addition to a conviction, a moral concern."[80]

The differences in intellectual outlook came out in attitudes toward ultimate things. John Kennedy was a practicing but conventional Catholic. Lord Longford, an English Catholic, once remarked to Eunice Shriver that a book should be written about President Kennedy and his faith. Eunice replied, "It will be an awfully slim volume."[81] Robert's volume would have been thicker. The President's ethos was more Greek than Catholic. He took his definition of happiness from Aristotle — the full use of your powers along lines of

excellence. He seemed to imply that man achieved salvation by meeting his own best standards rather than by receiving the grace of
God. Life, he appeared to feel, was absurd. Its meaning was the
meaning men gave it through the way they lived. There was also a
Hindu touch. He had these lines incribed on a silver beer mug he
gave David Powers on his birthday in 1963:

> *There are three things which are real:*
> *God, human folly and laughter.*
> *The first two are beyond our comprehension*
> *So we must do what we can with the third.*

"No one else at the White House, then or a year later," wrote Tom
Wicker, "knew that the source of those lines was Aubrey Menen's
version of *The Ramayana*. I could find the words in no book of
quotations. The Library of Congress was not able to tell me who
wrote them. . . . But Ted Clifton, Kennedy's military aide, recalls
him writing down those words one spring morning, quickly and
without reference to any book. He had them by heart."[82]

Robert Kennedy began as a true believer. He acquired his perceptions of the complexity of things partly because his beloved older
brother led him to broader views of society and life and partly because
he himself possessed to an exceptional degree an experiencing nature. John Kennedy was a realist brilliantly disguised as a romantic;
Robert Kennedy, a romantic stubbornly disguised as a realist.

26

Corridors of Grief

I<small>N THE YEARS</small> 1961 and 1962, opponents of the administration
tended to separate the Kennedy brothers. Robert was the villain
who persecuted steel barons and enforced racial integration. The
President, as Robert wryly observed, "wasn't such a bad fellow." But
in 1963, "instead of talking about Robert Kennedy, they started
talking about the Kennedy brothers, which he used to point out to
me frequently." The lowering atmosphere, Robert feared, foretold
a mean contest for reelection. "We saw all this literature that was
coming out and the letters that poured in — you know, just real
hatred." In the fall Robert asked his brother "on what basis I could
get out as Attorney General because I thought it was such a burden
to carry in the 1964 election." The President thought this "a bad
idea, because it would appear . . . that we were running out on civil
rights." They decided he should stay for the time being. They would
look at the situation again in the spring.[1]

I

They speculated about the probable Republican opponent. The man
they yearned for was their old friend from the Rackets Committee,
Barry Goldwater. "We had worked with Goldwater," Robert Kennedy
said later, "and we just knew he was not a very smart man, and he's
going to destroy himself. [The President] was concerned that he
would destroy himself too early and not get the nomination."[2] John
Kennedy told Ben Bradlee, "I really like him, and if we're licked at
least it will be on the issues. At least the people will have a clear

choice."* The Republican the President thought would be "the most difficult to beat" was Governor George Romney of Michigan. "He was always for God," Robert Kennedy said, "and he was always for [motherhood] and against big government and against big labor." The President "was very concerned that Romney was going to win [the Republican nomination], and he never discussed it with anyone, I think, other than perhaps myself." He did not mention Romney as a possible opponent lest this build him up. He mentioned Goldwater at every opportunity.[3]

The Kennedys had their own preparations to make for 1964. The Attorney General was dissatisfied with the Democratic National Committee. He thought John Bailey, the chairman, had become "almost a figurehead,"[4] and he feared, from evidence adduced by Paul Corbin, that other committee officials were misusing their party positions. There was always a family solution. Starting in the spring, Stephen Smith came to Washington several days each week to set up the campaign. Planning proceeded through the summer and early autumn. On November 13 the campaign group — the Attorney General, Smith, Theodore Sorensen, Kenneth O'Donnell, Lawrence O'Brien, John Bailey and Richard Maguire from the National Committee and Richard Scammon, director of the Census Bureau and an eminent psephologist — met with the President. "I was anxious to have it clear through the campaign," Robert Kennedy recalled, "that Steve Smith was going to be running things and people should be reporting to him. So that was one of the major purposes [of the meeting] for the President to say that."[5] The rest of the meeting was devoted to registration, the convention site (it now looked like Atlantic City), funds, films and other campaign detail.[6] They wondered how to give interest to a convention dedicated to the renomination of an incumbent. The President said that, if he and the Attorney General got into a public fight, it would make things more lively.[7]

The press soon heard about the meeting. The absence of the Vice President or any representative, the activity of the Department of Justice in the Bobby Baker case, the desire for a good story in a dull week and, perhaps, the morbid suspicions of the Vice President himself led to a burst of talk that the Kennedys were planning to dump Johnson in 1964. Johnson was chronically nervous about his prospects. In 1962 when Ted Lewis wrote lightly in the *New York Daily News* that Eunice Shriver was talking about a Kennedy-Kennedy

---

* Benjamin C. Bradlee, *Conversations with Kennedy* (New York, 1975), 232. Goldwater himself said later: "I looked forward to running against Jack. And we used to talk about it. We had a hell of a good idea that I think would have helped American politics.... We would travel together as much as possible and appear on the same platform and express our views" (Roy Reed, "The Liberals Love Barry Goldwater Now," *New York Times Magazine*, April 7, 1974).

ticket in 1964, Johnson had sent Walter Jenkins, his assistant, to ask O'Donnell whether this was a campaign to deny him renomination. O'Donnell protested incredulously that, if Eunice had said anything at all, it was obviously a joke. The President later suggested that his sisters reserve their jokes for less sensitive matters.[8]

The idea of dumping Johnson, the President said to Ben Bradlee on October 22, 1963, was "preposterous on the face of it. We've got to carry Texas in '64, and maybe Georgia."[9] When George Smathers mentioned it: "George, you must be the dumbest man in the world. If I drop Lyndon, it will make it look as if we have a really bad and serious scandal on our hands in the Bobby Baker case, which we haven't and that will reflect on me."[10] "There was no plan to dump Lyndon Johnson," Robert Kennedy told John Bartlow Martin the next spring. "It didn't make any sense. . . . And there was never any discussion about dropping him."[11] Nor was there anything resembling evidence for such a plan — at least until 1968 when Evelyn Lincoln, the President's secretary, in a book called *Kennedy and Johnson,* claimed to remember a conversation on November 19, 1963, in which Kennedy told her he was thinking about Governor Terry Sanford of North Carolina as his running mate; in any case, "it will not be Lyndon."[12] When I saw an advance copy of the Lincoln book in February 1968, I alerted Robert Kennedy, who was off skiing in Vermont. He said again there was never any intention to dump Johnson, adding, "Can you imagine the President ever having a talk with Evelyn about a subject like this?"* Johnson himself informed his own brother in 1963, "Jack Kennedy has personally told me that he wants me to stay on the team. Some of the people around him are bastards, but I think he's treated me all right."[13]

The ticket was definitely to be the same. The campaign theme, the President told the November meeting, was to be "peace and prosperity."[14] Looking back at the first three years, John Kennedy was dissatisfied with the amount of time he had had to give to foreign affairs, though he had seen no choice: "each day was a new crisis." "He thought a good deal more needed to be done domestically," his brother said in 1964; ". . . We really had to begin to make a major effort to deal with the unemployed and the poor in the United States," even at the cost of "several billions of dollars each year."[15]

The civil rights struggle had driven Kennedy down in the polls from an overwhelming 76–13 approval in January 1963 to 59–28 in November. This was still a comfortable margin, however, and in a Gallup trial heat against Goldwater in October, Kennedy won easily 55–39.[16] But the President himself felt, his brother said the next

---

* Author's journal, February 10, 1968. The nonexistence of any dump-Johnson plan is fully and emphatically confirmed by Stephen Smith.

spring, "that he had not gotten himself across as a person with much compassion and that people . . . didn't feel personally involved with him."[17] His doubts may have reflected a conversation with Marquis Childs, back from a trip through the south and southwest. Childs had been "startled" by the hostility toward the President. Kennedy was soft on the blacks, soft on the Commies. "From smug editorial writers to filling station attendants, I heard hatred" reminiscent of the hatred of Franklin Roosevelt a generation before. "The rich and privileged were again excoriating that son of a bitch in Washington, and not only the rich and privileged this time. It was for me a deeply disturbing phenomenon." He told Kennedy the time had come for a book like his own *They Hate Roosevelt* of 1936. "I don't believe it," the President said. "I just don't believe that's true. . . . I don't think they feel toward me the way they felt toward Roosevelt. I can't believe that." Childs had never seen him so tense. This was October 31, 1963.[18]

II

On November 20, 1963, Robert Kennedy had his thirty-eighth birthday. There was a party in his office. "All office parties are bad," said Patrick Anderson of the Committee on Juvenile Delinquency, "but this one was miserable."[19] John Douglas thought the Attorney General "glum" and "depressed."[20] After a time Robert Kennedy climbed on his desk and delivered an ironic disquisition. It was great, he said, to have achieved so much in so short a life: to have elected a President; to have become the great asset in the administration; now to have assured the President's reelection by the popularity of his policies on civil rights, Hoffa, wiretapping . . . Ramsey Clark and John Douglas walked out together wondering about the Attorney General's melancholy. He clearly thought himself a terrible political liability. Douglas said, "I guess Bob won't be here by Christmas."[21]

Robert Kennedy went on to the annual White House reception for the judiciary. These parties were attended traditionally by judges and the top people at Justice. The Attorney General had begun inviting long-time Justice employees — clerks, telephone operators, elevator operators. "There were an awful lot of people there," the Attorney General recalled. "But it was a terrific thrill for people who'd . . . been in the Government for long periods of time and never been in the White House. . . . I stayed longer than the President or Jackie. . . . Then I don't know why I left but I went upstairs and I talked to Jackie." She was looking forward, she said, to the impending trip to Texas.

The President joined them. He was looking forward to the trip too. The political problems in Texas — an angry feud between the populist Democrats led by Senator Ralph Yarborough and the oil Democrats led by the Vice President's protégé Governor John Connally — made it the more interesting. The President, his brother thought, "liked John Connally and liked Yarborough and he's so used to people fighting and not liking each other in politics, it didn't surprise him. He always thought those things could be worked out. . . . He said how irritated he was with Lyndon Johnson who wouldn't help at all in trying to iron out any of the problems in Texas, and that he was an s.o.b. . . . because this was his state and he just wasn't available to help out or just wouldn't lift a finger to try to assist." Soon Ethel bounded up from the reception and reminded her husband that he had a birthday party of his own at Hickory Hill.[22]

It was a characteristic Hickory Hill party — large, loud, happy, with satiric toasts and heckling guests. Only Ethel departed from the prevailing mood. Instead of making her usual chaffing toast about her husband, she gravely asked the party to drink to the President of the United States. Robert Kennedy sat up talking with Gene Kelly, the actor, in the library till two-thirty in the morning. Ethel forgot to tell him about her birthday present — a sauna in the basement — till it was three and they were in bed.

III

The next day he presided over an organized crime meeting. Federal attorneys came from across the country to consider the next, phase in the campaign against the syndicates. The meeting continued on Friday, November 22. Shortly after noon the Attorney General suggested a recess. He took Robert Morgenthau and the chief of Morgenthau's criminal division, Silvio Mollo, out to Hickory Hill for a swim and luncheon.

It was a sunlit day, unseasonably warm for November. The three men sat with Ethel Kennedy around a table by the swimming pool, eating clam chowder and tuna fish sandwiches. Shortly before quarter to two, the Attorney General said they had better get back. Fifty yards away workmen were painting the new wing of the house. Morgenthau abstractedly noticed one of them, a painter's hat jammed over his ears, a transistor radio in his hand, run abruptly toward the pool. He was shouting. No one understood what he said. Then a telephone extension rang across the pool. Ethel went to answer it. "She said J. Edgar Hoover was calling," Robert Kennedy remembered later, "so I thought something was wrong because he wouldn't be ... calling me here." Suddenly Robert Morgenthau realized what the workman had cried: "They say the President is shot." On the phone Hoover said,

> "I have news for you. The President's been shot"; or "I have news for you" and I might have said, "What?" and he said, "The President's been shot." And — well, I don't know what I said — probably "Oh" or something — and I don't know whether he then — I asked him or got into whether it was serious, and I think he said, "I think it's serious." ... He said, "I'll call you back ... when I find out more." I don't remember anything more of that conversation.

Morgenthau saw Robert Kennedy turn away and clap his hand to his mouth. There was a look of "shock and horror" on his face. Ethel saw too and rushed to his side. For a few seconds Robert could not speak. Then he almost forced out the words: "Jack's been shot. It may be fatal."[23]

They walked dazedly back to the house. Ethel led Morgenthau and Mollo to a television set in the living room, then accompanied Robert upstairs. Morgenthau wanted to leave but felt he could not until others arrived. "I went off," Robert remembered,

> and called Kenny [O'Donnell, with the party in Dallas], I think. I never got through to him. ... Then I talked to the Secret Service and I think I talked to Clint Hill [a Secret Service man with the party] ... but I don't know who it was ... in the hospital down there and they said that ... it was very serious. And I asked if he was conscious and they said he wasn't, and I asked if they'd gotten a priest, and they said they had.... Then, I said, will you call me back, and he said, yes, and then he — Clint Hill called me back, and I think it was about thirty minutes after I talked to Hoover ... and he said, "The President's dead."

Robert walked downstairs, put his head in the living room and told Morgenthau, "He's dead."

Robert Kennedy: "So then Hoover called ... I don't remember what [his words] were.... He was not a very warm or sympathetic figure." William Manchester later asked him whether Hoover sounded excited. "No, not a bit. No, nor upset ... not quite as excited as if he was reporting the fact that he found a Communist on the faculty of Howard University."* There were many calls.

> I think I called Teddy, and then John McCone called me and said, "I'll come out," and he came out.... I called Teddy and all my sisters, and I tried to get hold of Jean, and she wasn't there.... But I talked to Steve or talked to somebody to tell her to come down to Washington to stay with Jackie.... I talked to Sarge about trying to organize the funeral arrangements and he took charge of that and did a terrific job.

---

* Howard University was the black university in Washington, D.C. Conceivably Hoover might have said Harvard University. In either case, it was one of Hoover's pet hates.

I talked to Teddy and asked him to go up to tell my father and my mother. . . . I talked to my mother and we agreed to wait — not telling my father until the next morning. . . . I talked to Eunice and . . . she decided that she would go up to my mother. She's the closest really to my mother so that made sense. Jean was closest to Jackie so that made sense. . . . I had to stay here, so Teddy went up to tell my father.

John McCone arrived; the CIA was a short distance from Hickory Hill. Kennedy and McCone went out to the lawn. In a moment Lyndon Johnson was on the phone. Kennedy took the call by the pool.

First he expressed his condolences. Then he said . . . this might be part of a worldwide plot, which I didn't understand, and he said a lot of people down here think I should be sworn in right away. Do you have any objection to it? And — well, I was sort of taken aback at the moment because it was just an hour after . . . the President had been shot and I didn't think — see what the rush was. And I thought, I suppose, at the time, at least, I thought it would be nice if the President came back to Washington — President Kennedy. . . . But I suppose that was all personal. . . . He said, who could swear me in? I said, I'd be glad to find out and I'll call you back.

He called Katzenbach, who said anyone could do it, including a district court judge. "So I called Johnson back and said anybody can."*

Dean Markham, Robert's football friend, arrived. David Hackett was soon there, and Byron White, and Edwin Guthman, and others. "How are you doing?" Kennedy said to Guthman. "I've seen better days," Guthman replied. "Don't be so gloomy," said Kennedy. "That's one thing I don't need now." "There's so much bitterness," he soon said as he paced the lawn with Guthman. "I thought they'd get one of us, but Jack, after all he'd been through, never worried about it. . . . I thought it would be me."[24]

Ethel left to pick up the children at their schools and break the dreadful news. Robert walked back and forth between the tennis court and the swimming pool. Brumus trailed along at his heels. The children came, and he embraced them, comforting them as he had tried to comfort his friends. He told them, "He had the most wonderful life." He seemed controlled. But Ethel, noticing his eyes rimmed with red, handed him a pair of dark glasses and joined his restless walk. As the shadows lengthened, he prepared to drive to the Pentagon. From there he would go by helicopter to Andrews Field and meet his brother's body. On the elevator to McNamara's

* In fact this discussion was supererogatory. Johnson became President on his predecessor's death, not on the administration of the oath. As Van Buren said long ago, "The Presidency under our system like the king in a monarchy, never dies" (Martin Van Buren, *Inquiry into the Origin and Course of Political Parties in the United States* [New York, 1867], 290).

office he said to Guthman, "People just don't realize how conservative Lyndon really is. There are going to be a lot of changes."[25]

He did not want a crowd at Andrews. He told Sargent Shriver, "The last thing that Jackie, everybody wants [is] to see a lot of — I wasn't very realistic about it. . . . [Justice] Arthur Goldberg got on the phone and said that this was something more than personal, this is the President of the United States, and I think we should all go. And I said, if you want to go, go. I wasn't going to get into an argument about it. So they all came. I didn't see them, but they were all there, and it was nice."

He arrived at Andrews half an hour before the plane was due from Dallas. He took a solitary walk in the enveloping night. "There were all those people out there, and I didn't want to see any people." He sat for a few minutes in the back of an Army truck. Then he made arrangements to get on the plane without running the gantlet of television cameras. "As the plane came in, I walked around. I don't think anybody saw me, and I went up where the pilot is — the front entrance. And everybody's eyes were on the back entrance." He boarded the plane by himself, hurried past the Johnson party and hugged Jackie.

They went to Bethesda Hospital. There were so many details. The funeral home wanted to know how grand the coffin should be. "I was influenced by . . . that girl's book on [burial] expenses . . . Jessica Mitford [*The American Way of Death*]. . . . I remember making the decision based on Jessica Mitford's book. . . . I remember thinking about it afterward, about whether I was cheap or what I was, and I remembered thinking about how difficult it must be for everybody making that kind of decision." Jessica Mitford's sister, in the circularity of life and death, was the sister-in-law of Kathleen Kennedy's husband Billy Hartington.

The question arose whether the coffin should be open or closed. The casket arrived at the White House early in the morning of the twenty-third. After a brief service in the East Room, "I asked everybody to leave and I asked them to open it. . . . When I saw it, I'd made my mind up. I didn't want it open and I think I might have talked to somebody to ask them to look also." He had indeed asked McNamara to come into the East Room. "After a time," I recorded in my journal, "he came out and asked Nancy Tuckerman [Jacqueline Kennedy's social secretary] and me to go in, look at the bier and give our opinion whether the casket should be open or shut. And so I went in, with the candles fitfully burning, three priests on their knees praying in the background. . . . For a moment, I was shattered. But . . . it was too waxen, too made up. It did not really look like him.

Nancy and I told this to Bobby and voted to keep the casket closed. When Bill Walton agreed, Bobby gave instructions."[26]

He spent the night in the Lincoln bedroom. Charles Spalding went with him and said, "There's a sleeping pill around somewhere." Spalding found a pill. Robert Kennedy said, "God, it's so awful. Everything was really beginning to run so well." He was still controlled. Spalding closed the door. "Then I just heard him break down. . . . I heard him sob and say, "Why, God?"[27]

IV

He lay fitfully for an hour or two. Soon it was daylight. He walked down the hall and came in on Jacqueline, sitting on her bed in a dressing gown, talking to the children. Young John Kennedy said that a bad man had shot his father. His older sister, Caroline, said that Daddy was too big for his coffin. Later in the morning "there was a dispute about where the President would be buried. . . . Kenny [O'Donnell] and Larry [O'Brien] and Dave Powers were all for him going to Boston, and I was for him being buried out here. . . . They were rather strong. So that made it rather difficult for me. . . . They were going either to turn the Boston Common over and build something in the middle or . . . they were going to set some other place aside . . . but I said the place . . . in Boston that he was going to be buried, where Patrick [the baby who died at birth a few months before] was buried, was unsatisfactory." In the afternoon he met Robert McNamara at Arlington Cemetery. It was raining "like hell." They looked first at a site toward the bottom of the hill, next to Oliver Wendell Holmes. "I said that it would make a major difference if we could have it higher."[28]

He went to the services at the Rotunda of the Capitol. Maude Shaw, the governess, had made John wear gloves. Robert told him to take them off. In the limousine Jackie said, "Where are John's gloves?" Robert said that boys didn't wear gloves. They looked out the car's windows. "That was the first time I saw that horse" — Blackjack, the great, black, restless funeral horse. "That was what really kept your mind off it. You didn't know whether he was going to run away. . . . I was so nervous about the fellow who was holding him. . . . It was nice as you look back on it that he was so restless." He listened to the speeches. "I thought they were nice; just [Mike] Mansfield's such a nice man, liked the President so much. I didn't care much what he said. I thought . . . the repetitious business . . . was awkward. And [Chief Justice] Warren, I thought, was inappropriate, to talk about hate."[29]

Then the funeral. He put his PT-boat tie clip, a silver rosary Ethel had given him and a cutting of his hair into the coffin.[30] I wrote in my journal, "The ceremony at Arlington, against a background of wildly twittering birds, was solemn and heartrending. De Gaulle was there, and Eisenhower, and Truman, looking shattered. Evelyn Lincoln said to me, 'The thing he hated most of all was fanatics.' The day was sunny, crisp and cold. I have never felt so depressed."[31] Robert Kennedy sent a letter to each of his children and told his sisters to do likewise. He wrote his son Joe:

> On the day of the burial
> of your Godfather
> John Fitzgerald Kennedy

THE WHITE HOUSE
WASHINGTON

Nov. 24, 1963

Dear Joe,
    You are the oldest of all the male grandchildren. You have a special and particular responsibility now which I know you will fulfill.
    Remember all the things that Jack started — be kind to others that are less fortunate than we — and love our country.

Love to you
Daddy[32]

He appeared, I noted the day after Dallas, "composed, withdrawn and resolute."[33] Ben Bradlee the same day saw him "clearly emerging as the strongest of the stricken."[34] Discipline and duty summoned him to the occasion. Within he was demolished. "It was much harder for him than anybody," said LeMoyne Billings, his friend of so many years. He had put "his brother's career absolutely first; and not anything about his own career whatsoever. And I think that the shock of losing what he'd built everything around . . . aside from losing the loved figure . . . was just absolutely [devastating] — he didn't know where he was. . . . Everything was just pulled out from under him."[35] They had been years of fulfillment, but of derivative fulfillment: fulfillment not of himself but of a brother and a family. Now in a crazed flash all was wiped out. *"Why, God?"*

v

Robert Kennedy was a desperately wounded man. "I just had the feeling," said John Seigenthaler, "that it was physically painful, almost as if he were on the rack or that he had a toothache or that he had a heart attack. I mean it was pain and it showed itself as being pain. . . . It was very obvious to me, almost when he got up to walk that it hurt to get up to walk." Everything he did was done through a "haze of pain."[36] "He was the most shattered man I had ever seen in my life," said Pierre Salinger. "He was virtually non-functioning. He would walk for hours by himself." Douglas Dillon offered him his house in Hobe Sound, Florida, where Robert and Ethel went with a few friends at the end of the month. They played touch football — "really vicious games," Salinger recalled. ". . . It seemed to me the way he was getting his feelings out was in, you know, knocking people down."[37]

Sardonic withdrawal seemed to distance the anguish. Seigenthaler went out to Hickory Hill after the funeral. "Obviously in pain, he opened the door and said something like this, 'Come on in, somebody shot my brother, and we're watching his funeral on television.'"[38] When Helen Keyes arrived from Boston to help with his mail, "I didn't want to see him; I just figured I'd dissolve; and I walked in and he said, 'Come in.' I said, 'All right.' And he said to me, 'Been to any good funerals lately?' Oh, I almost died, and yet once he said that it was out in the open, and, you know, we just picked up and went on from there."[39] Senator Herbert Lehman of New York died early in December. Robert Kennedy, in New York for the services, said to his Milton friend Mary Bailey Gimbel, "I don't like to let too many days go by without a funeral."[40]

Friends did their best. John Bartlow Martin, retiring as ambassador to the Dominican Republic, went to say goodbye. "How his face had aged in the years I'd known him." Martin attempted a few words of comfort. "With that odd tentative half-smile, so well known to his friends, so little to others, he murmured . . . 'Well, three years is better than nothing.'"[41] Peter Maas arrived from New York on the first day the Attorney General went out publicly — to a Christmas party arranged by Mary McGrory of the *Washington Star* for an orphanage.

> The moment he walked in the room, all these little children — screaming and playing — there was just suddenly silence. . . . Bob stepped into the middle of the room and just then a little [black] boy — I don't suppose he was more than six or seven years old — suddenly darted forward, and stopped in front of him, and said, "Your brother's dead! Your brother's dead!" . . . The adults, all of us, we just kind of turned away. . . . The little boy knew he had done something wrong, but he didn't know *what*; so he started to cry. Bobby stepped forward and picked him up, in kind of one motion, and held him very close for a moment, and he said, "That's all right. I have another brother."[42]

He gave his own annual Christmas party for poor children. A three-man clown band played by the Christmas tree in his office. The children received presents, then trooped downstairs to the auditorium for a show. Someone thought the clowns should go down too to entertain them as they waited. The chief clown told Patrick Anderson of the Juvenile Delinquency Committee that they weren't going anywhere until they had a smoke. Anderson: "Kennedy returned and spoke to me: 'The clowns should be where the children are.' Our eyes met for a long moment and it seemed, incredibly, as if he wanted my agreement. 'Yes, sir,' I said, 'they should be,' and I herded the reluctant clowns downstairs."[43]

It went on and on. He went skiing with Charles Spalding. "You almost prolong the pain not to lose the person," Spalding thought. ". . . It just hurts so bad. Then you figure, if it doesn't hurt I'll be further away from what I've lost. So it just seemed that those nights would go on forever."[44]

VI

He refused to involve himself in the problem of who had murdered his brother. He "never really wanted any investigation," Nicholas Katzenbach thought.[45] Nothing would bring John Kennedy back to life. Investigation would only protract the unbearable pain. Almost better, Robert Kennedy seemed at times to feel, to close the book. He left to Katzenbach all dealings with the Warren Commission, appointed by the new President on November 29 to ascertain the truth about Dallas.

The Chief Justice and his colleagues had perforce to depend greatly on the intelligence agencies. They did not know that the agencies had their own secret reasons to fear a thorough inquiry. If it came out that the putative killer might have had intelligence connections, domestic or foreign, that FBI agents should have had him under close surveillance, that CIA assassins might have provoked him to the terrible deed, the agencies would be in the deepest trouble. But if Lee Harvey Oswald could be portrayed as a crazed loner acting on some solitary impulse of his own, they would be in the clear.

In CIA, James J. Angleton, the counterespionage chief and CIA liaison with the Warren Commission, compiled a dragnet of names and called for information from all branches of the Agency. One name on his list was Rolando Cubela Secades. Desmond FitzGerald decided to withhold from Angleton the story of CIA's role in Cubela's plot to murder Castro. He even ordered any mention of the poison pen deleted from the report of the November 22 meeting.* Nor did

---

* My source is the informative book by Edward Jay Epstein, *Legend: The Secret World of Lee Harvey Oswald* (New York, 1968), 253–254. Epstein's source was evidently Angleton. I know no reason to doubt this particular story. But, since Angleton in his quest, necessary but maniacal, for Soviet 'moles' (penetration agents burrowing their

Allen Dulles, a member of the Warren Commission, repair the ignorance of his colleagues. (He may not have known about Cubela, but he certainly knew about the 1960–61 assassination attempts.)

The FBI succumbed equally to the bureaucratic imperative. As Edward Jay Epstein has persuasively argued, the Bureau might well have suspected that Oswald had been involved with the KGB and actually believed he had met with a Soviet intelligence officer in Mexico City two months before Dallas. But Oswald's name was not in the FBI's voluminous Security Index. Hoover at once called for an internal inquiry into the "investigative deficiencies in the Oswald case." After reading the report, he noted despairingly that the findings "have resulted in forever destroying the Bureau as the top level investigative organization." Early in December he secretly censured seventeen FBI officials.[46] Externally he was desperate to avert any suspicion that the Bureau had failed. "The thing I am most concerned about," he told Walter Jenkins of the new White House, ". . . is having something issued so we can convince the public that Oswald is the real assassin." Katzenbach, no doubt reflecting Hoover, wrote Bill Moyers, another of the new President's special assistants:

> 1. The public must be satisfied that Oswald was the assassin; that he did not have confederates who are still at large; and that the evidence was such that he would have been convicted at trial.
> 2. Speculation about Oswald's motivation ought to be cut off, and we should have some basis for rebutting thought that this was a Communist conspiracy or (as the Iron Curtain press is saying) a right-wing conspiracy to blame it on the Communists.[47]

Robert Kennedy, Katzenbach said later, knew nothing about this memorandum.[48]

For reasons of bureaucratic self-preservation, the CIA and the FBI thus found themselves in the ironic position of denying any possibility of Cuban or Soviet implication. Nor did the new administration wish to think about the unthinkable problems that would arise if there were indication of international conspiracy. All the pressures in Washington were toward a quick and uncomplicated verdict. Robert Kennedy, I believe, had his own thoughts. We spent the evening of December 9 together. "I asked him, perhaps tactlessly, about Oswald. He said that there could be no serious doubt that Oswald was guilty, but there was still argument if he had done it by himself or as part of a larger plot, whether organized by Castro or by gangsters.

---

way in the adversary system) at one time thought I might be a Soviet agent in the White House (after a Soviet official in Caracas came up with the date of the Bay of Pigs landing), I may perhaps be pardoned if I do not regard him with the reverence that pervades the Epstein book.

The FBI thought he had done it by himself, but McCone thought there were two people involved in the shooting."[49]

At about the same time, Kennedy asked Walter Sheridan how Jimmy Hoffa had taken the news. "I didn't want to tell him," Sheridan said later, "but he made me tell him." Hoffa in Miami, hearing that Harold Gibbons and top Teamsters in Washington had lowered the flag over the marble palace to half-mast, "flew into a rage." He yelled at his secretary for crying. A reporter asked him about the Attorney General. Hoffa spat out: "Bobby Kennedy is just another lawyer now." A Teamster leader in Puerto Rico soon wrote Robert Kennedy that he planned to solicit donations from union brothers to "clean, beautify and supply with flowers the grave of Lee Harvey Oswald. You can rest assured contributions will be unanimous."[50]

Robert Kennedy perceived so much hatred about, so many enemies: the Teamsters; the gangsters; the pro-Castro Cubans; the anti-Castro Cubans; the racists; the right-wing fanatics; the lonely deluded nuts mumbling to themselves in the night. I do not know whether he suspected how much vital information both the FBI and the CIA deliberately denied the Warren Commission or whether he ever read its report. But on October 30, 1966, as we talked till two-thirty in the morning in P. J. Clarke's saloon in New York City, "RFK wondered how long he could continue to avoid comment on the report. It is evident that he believes that it was a poor job and will not endorse it, but that he is unwilling to criticize it and thereby reopen the whole tragic business."[51]

The next year Jim Garrison, the New Orleans district attorney, started making sensational charges about a conspiracy. I asked Kennedy what he made of them. He thought Garrison might be onto something; NBC, he added, had sent Walter Sheridan to New Orleans to find out what Garrison had. Garrison's villain turned out to be the CIA. Kennedy said to Sheridan something like: "You know, at the time I asked McCone . . . if they had killed my brother, and I asked him in a way that he couldn't lie to me, and they hadn't."* Kennedy asked Frank Mankiewicz of his Senate staff whether he thought Garrison had anything. "And I started to tell him, and he said, 'Well, I don't think I want to know.'"[52] Kennedy told me later: "Walter Sheridan is satisfied that Garrison is a fraud."

I cannot say what his essential feeling was. He came to believe the Warren Commission had done an inadequate job; but he had no

---

* Walter Sheridan, in recorded interview by Roberta Greene, June 12, 1970, 19, RFK Oral History Program. In 1967 Marvin Watson of Lyndon Johnson's White House staff told Cartha DeLoach of the FBI that Johnson "was now convinced there was a plot in connection with the assassination. Watson stated the President felt that CIA had had something to do with this plot" (*Washington Post*, December 13, 1977).

conviction — though his mind was not sealed against the idea of conspiracy — that an adequate inquiry would necessarily have reached a different conclusion. At times his view was, I believe, close to that expressed by his friend Anthony Lewis in 1975. "The search for conspiracy," Lewis wrote, "only increases the elements of morbidity and paranoia and fantasy in this country.... It obscures our necessary understanding, all of us, that in this life there is often tragedy without reason."[53]

VII

Tragedy without reason? But was there anything in the universe without reason? The question echoed: *"Why, God?"* For an agnostic the murder of John Kennedy seemed one more expression of the ultimate fortuity of things. But for those who believed in a universe infused by the Almighty with pattern and purpose — as the Kennedys did — Dallas brought on a philosophical as well as an emotional crisis. Robert Kennedy in particular had to come to terms with his brother's death before he could truly resume his own existence.

In these dark weeks and months, on solitary walks across wintry fields, in long reverie at his desk in the Department of Justice, in the late afternoon before the fire in Jacqueline Kennedy's Georgetown drawing room, in his reading — now more intense than ever before, as if each next page might contain the essential clue — he was struggling with that fundamental perplexity: whether there was, after all, any sense to the universe. His faith had taught him there was. His experience now raised the searching and terrible doubt. If it were a universe of pattern, what divine purpose had the murder of a beloved brother served? An old Irish ballad haunted him.

> *Sheep without a shepherd;*
> *When the snow shuts out the sky —*
> *Oh, why did you leave us, Owen?*
> *Why did you die?* [54]

He scrawled on a yellow sheet:

> The innocent suffer — how can that be possible and God be just.

and

> All things are to be examined & called into question —
> There are no limits set to thought.[55]

Over Easter in 1964 he went with Jacqueline, her sister and brother-in-law, the Radziwills, and Charles Spalding to Paul Mellon's house in Antigua. Jacqueline, who had been seeking her own con-

solation, showed him Edith Hamilton's *The Greek Way.* "I'd read it
quite a lot before and I brought it with me. So I gave it to him and
I remember he'd disappear. He'd be in his room an awful lot of the
time . . . reading that and underlining things."[56] Edith Hamilton's
small classic, then more than thirty years old, opened up for him a
world of suffering and exaltation — a world in which man's destiny
was to set himself against the gods and, even while knowing the
futility of the quest, to press on to meet his tragic fate.

Robert Kennedy's underlinings suggest themes that spoke to his
anguish. He understood with Aeschylus "the antagonism at the heart
of the world," mankind fast bound to calamity, life a perilous adventure; but then "men are not made for safe havens. The fullness of
life is in the hazards of life. . . . To the heroic, desperate odds fling
a challenge." This was not swashbuckling defiance; rather it was the
perception that the mystery of suffering underlay the knowledge of
life. "Having done what men could," Thucydides had written of the
brilliant youths, who, pledging the sea in wine from golden goblets,
sailed to conquer Sicily and died miserably in the mines of Syracuse,
"they suffered what men must." Robert Kennedy memorized the
great lines from the *Agamemnon* of Aeschylus: "He who learns must
suffer. And even in our sleep pain that cannot forget, falls drop by
drop upon the heart, and in our own despair, against our will, comes
wisdom to us by the awful grace of God."

Suffering was the common badge of humanity. Euripides wrote
of "the giant agony of the world" — a world made up of a myriad of
individuals, each endowed with a terrible power to suffer, an "awful
sum of pain" to which no individual could be indifferent save at the
price of his humanity. "Know you are bound to help all who are
wronged," the mother of Theseus cried in the *Suppliants.* As John
Kennedy's sense of the Greeks was colored by his own innate joy in
existence, Robert's was directed by an abiding melancholy. He un-
derscored a line from Herodotus: "Brief as life is there never yet was
or will be a man who does not wish more than once to die rather
than to live." In later years, at the end of an evening, he would
sometimes quote the *Oedipus Tyrannus* of Sophocles:

> The long days store up many things nearer to grief than joy
> . . . Death at the last, the deliverer.
> Not to be born is past all prizing best.
> Next best by far when one has seen the light
> Is to go thither swiftly whence he came.[57]

He read much more of Edith Hamilton — *The Echo of Greece, Three
Greek Plays, The Ever-Present Past* (from which he took the quotation
he used so much subsequently — "to tame the savageness of man and

make gentle the life of the world"). In August 1966, we were chatting in Hyannis Port. "Almost shyly, he pulled out of the briefcase ... *Three Greek Plays* — a well-thumbed volume, with pages loose and falling out — and asked me to read two passages from 'The Trojan Women,' one describing the horrors of war, the other the importance of friendship and loyalty. They are both powerful passages and clearly had great meaning for him."[58] A scribble to Angie Novello after Richard Goodwin had sent him the Arrowsmith translation of *Thetis and Achilles*: "Angie. What other translations are there of William Arrowsmith of Greek plays."[59] "He knew the Greeks cold," recalled one of the bright young men on his Senate staff. ". . . He'd cite some play, and say, 'You know that?' We didn't at all. I think he got some delight out of that."[60]

The fact that he found primary solace in Greek impressions of character and fate did not make him less faithful a Catholic. Still, at the time of truth, Catholic writers did not give him precisely what he needed. And his tragic sense was, to use Auden's distinction, Greek rather than Christian — the tragedy of necessity rather than the tragedy of possibility; "What a pity it had to be this way," rather than, "What a pity it was this way when it might have been otherwise."[61] Next to the Greeks, he read Albert Camus most intently — *The Notebooks; Resistance, Rebellion and Death; The Myth of Sisyphus; The Plague.* His commonplace book of these years is filled with Camus:

> But sometimes in the middle of the night their wound would open afresh. And suddenly awakened, they would finger its painful edges, they would recover their suffering anew and with it the stricken face of their love.

> Smiling despair. No solution, but constantly exercising an authority over myself that I know is useless. The essential thing is not to lose oneself, and not to lose that part of oneself that lies sleeping in the world.

> We are faced with evil. I feel rather like Augustine did before becoming a Christian when he said, "I tried to find the source of evil and I got nowhere." But it is also true that I and few others know what must be done. . . . Perhaps we cannot prevent this world from being a world in which children are tortured. But we can reduce the number of tortured children. And if you believers don't help us, who else in the world can help us do this?[62]

With all he had striven for smashed in a single afternoon, he had an overwhelming sense of the fragility and contingency of life. He had never taken plans very seriously in the past. He could not believe in them at all now. "Who knows if we will be here next week?" If things were worth doing, they were not to be deferred to the precarious future: so he would protect his family, his friendships, his reading from the exactions of public affairs. At the same time he knew he was a child of fortune as well as of fatality — that he had enjoyed far more freedom and happiness than most people, far more opportunity to enlarge his choices and control his existence. This made his obligation to help all who had been wronged the more acute and poignant.

Robert Kennedy at last traveled in that speculative area where doubt lived. He returned from the dangerous journey, his faith intact, but deepened, enriched. From Aeschylus and Camus he drew a sort of Christian stoicism and fatalism: a conviction that man could not escape his destiny, but that this did not relieve him of the responsibility of fulfilling his own best self. He supplemented the Greek image of man against fate with the existentialist proposition that man, defining himself by his choices, remakes himself each day and therefore can never rest. Life was a sequence of risks. To fail to meet them was to destroy a part of oneself.

He made his way through the haze of pain — and in doing so brought other sufferers insight and relief. "For the next two and a half years," wrote Rita Dallas, his father's nurse, "Robert Kennedy became the central focus of strength and hope for the family. . . . Despite his own grief and loneliness, he radiated an inner strength that I have never seen before in any other man. . . . Bobby was the one who welded the pieces back together."[63] As his father had said so long before, he would keep the Kennedys together, you could bet.

He was now the head of the family. With his father stricken, his older brothers dead, he was accountable to himself. The qualities he had so long subordinated in the interest of others — the concern under the combativeness, the gentleness under the carapace, the idealism, at once wistful and passionate, under the toughness — could rise freely to the surface. He could be himself at last.

27

Stranger in a Strange Land

JOHN KENNEDY ALWAYS HAD a certain fondness for Lyndon Johnson. He saw his Vice President, with perhaps the merest touch of condescension, as an American original, a figure out of Mark Twain, not as a threat but as a character. The President, Ben Bradlee observed in 1963, "really likes his roguish qualities, respects him enormously as a political operator . . . and he thinks Lady Bird [Mrs. Johnson] is 'neat.'" But, with the best will in the world, the relationship between President and Vice President is doomed — a generalization almost without historical exception. For the Vice President has no serious duty save to wait around for the President to die. This is not the basis for cordial and enduring friendships. In spite of all Kennedy's indulgence for Johnson, there were times, Bradlee noted, "when LBJ's simple presence seems to bug him."[1] "Every time I came into John Kennedy's presence," said Johnson, "I felt like a goddamn raven hovering over his shoulder."[2]

I

Kennedy nevertheless appreciated the frustrations and humiliations inseparable from that most hopeless of constitutional offices. He once told Johnson, "I don't know how you're able to contain yourself."[3] He tolerated no cracks about Johnson from his staff (nor, so far as I could see, was there much inclination to make cracks, though Johnson stoutly believed otherwise). Johnson himself told Dean Rusk that "he had been better treated than any other Vice President in history and knew it."[4] "He's done much better by me," Johnson used to muse (truthfully), "than I would have done by him under the same circumstances."[5] Shortly before his death he told Bobby Baker, "Jack

Kennedy always treated me fairly and considerately."[6] I once remarked in the company of George Reedy, a Johnson special assistant in these years and afterward an astute philosopher of the Presidency, that Kennedy treated Johnson with a consideration exceedingly rare in the history of the White House. Reedy commented, "Historically, Arthur is absolutely correct. I thought, myself, that President Kennedy was rather generous to Vice-President Johnson. But that didn't mean that Vice-President Johnson appreciated it in the slightest."[7]

For Johnson was insatiable. No amount of consideration would have been enough. He requested an office next to the President's. "I have never heard of such a thing," said Kennedy, and gave him an office in the Executive Office Building across the way. He nagged Evelyn Lincoln to make sure he would be invited to all White House meetings. Failing to find his name on the list for a private White House dinner, he even asked Mrs. Lincoln whether there had not been a mistake. "Tell him," said Kennedy, "that you have checked and you found there was no mistake."[8] Johnson made it a practice to descend on the President with personal complaints, "often," said Kenneth O'Donnell, "about Bobby." In time Kennedy and O'Donnell worked out a routine. "The President would first hear him out alone, and then call me into his office and denounce me in front of Johnson for whatever the Vice-President was beefing about. I would humbly take the blame . . . and the Vice-President would go away somewhat happier."[9]

Johnson was a man notably larger than life in ambitions, energies, needs and insecurities. In the Senate he had been accustomed to the exercise of personality and power. His role was suddenly one of self-abnegation, and he played it with impeccable public loyalty. He had his disagreements with Kennedy, but there was never a whiff of discord, complaint or self-pity to newspapermen or even to old friends on the Hill. This unprecedented self-discipline exacted a growing psychic cost. By 1963 the Vice President faded astonishingly into the background. Evelyn Lincoln calculated that in 1961 he had spent ten hours and nineteen minutes in private conferences with the President, in 1963 one hour and fifty-three minutes. At meetings in the Cabinet Room he became an almost spectral presence. As the President's "sureness and independence increased," Mrs. Lincoln noted, "the Vice President became more apprehensive and anxious to please."[10] Bill Moyers, who had worked for him in the Senate and was now deputy director of the Peace Corps, thought that his self-confidence was trickling away. He was, said Moyers, "a man without a purpose . . . a great horse in a very small corral." Ranching similes were irresistible. Daniel Patrick Moynihan remembered looking into the Vice President's eyes and thinking, "This is a bull castrated very late in life."[11] Ruminating on his Vice Presidency, Johnson said later: "I detested every minute of it."[12]

II

"Johnson," wrote O'Donnell, "blamed his fallen prestige on Bobby Kennedy." "His complaints against Bobby Kennedy," said Bobby Baker, "were frequent and may have bordered on the paranoic."[13] No affection contaminated the relationship between the Vice President and the Attorney General. It was a pure case of mutual dislike. "Maybe it was just a matter of chemistry," Johnson said in retrospect.[14]

Chemistry was certainly part of it. The Vice President and the Attorney General were immiscible. Johnson was seventeen years older, six inches taller, expansive in manner, coarse in language, emotions near the surface. It was southwestern exaggeration against Yankee understatement; frontier tall tales, marvelously but lengthily told, against laconic irony. Robert Kennedy, in the New England manner, liked people to keep their physical distance. Johnson, in the Texas manner, was all over everybody — always the grip on the shoulder, tug at the lapel, nudge in the ribs, squeeze of the knee. He was a crowder, who set his great face within a few inches of the object of his attention and, as the more diffident retreated, backed them across the room in the course of monologue. Robert Kennedy baffled Johnson. Johnson repelled Robert Kennedy.

One night, after a White House dance for General Gavin, they scrambled eggs in an upstairs kitchen. Johnson said, "Bobby, you do not like me. Your brother likes me. Your sister-in-law likes me. Your Daddy likes me. But *you don't like me*. Now, why? Why don't you like me?" "This went on and on for hours," said Charles Spalding, who was there, "like two kids in the sixth form. . . . Johnson kept pursuing him, saying he didn't like him. . . . It was rather persistent questioning and a little difficult for everybody. . . . He was trying to rake up this relationship and find out what it was that was rankling Bobby."[15]

Finally Johnson said, "I know why you don't like me. You think I attacked your father [at the 1960 convention]. But I never said that. Those reports were all false. . . . I never did attack your father and I wouldn't and I always liked you and admired you. But you're angry with me and you've always been upset with me." The next morning Kennedy repeated the conversation to John Seigenthaler. "Bob hardly remembered" the attack on his father and asked Seigenthaler to look up the facts. Seigenthaler soon found the damning story in the *New York Times*. "There can't be much doubt in anybody's mind," he reported to the Attorney General, ". . . that he was vicious."[16] Later, Robert Kennedy recalled with incredulity Johnson's claim that night that he was not interested in the Presidency himself, that he had been for John Kennedy from the start, that he simply could not call off his supporters, that he had never heard anyone saying anything bad about the Kennedys. Robert Kennedy: "My experience with him since then — he lies all the time, I'm telling you. . . . He lies even when he doesn't have to lie."[17]

Still, he admired Johnson's restraint as Vice President. Ramsey Clark said he "never heard Bob Kennedy . . . say anything unkind or political about the Vice President the whole time I was there."[18] "I never heard him criticize Johnson," said Katzenbach, "in terms of any lack of loyalty to President Kennedy."[19] Robert Kennedy told me in 1965 that Johnson "was very loyal and never spoke against the President," adding, "but he wasn't very helpful at times that he might have been helpful."[20]

For Johnson, on the other hand, Robert Kennedy was nemesis. The younger brother had begun by trying to deny him the vice presidential nomination. "He repeated that to me over a period of weeks," Pierre Salinger recalled of Johnson's first months in the White House.[21] After the inauguration, said O'Donnell, Johnson felt Robert Kennedy "had taken over his rightful position as the number two man in the government."[22] The Attorney General was the man who humiliated the Vice President at the Committee on Equal Employment Opportunities; who, Johnson assured Hugh Sidey, Time's White House correspondent, "bugged him all during the time he was Vice President";* who, in the autumn of 1963, Johnson believed, was fomenting the Bobby Baker case in order to deny him renomination.[23] "President Kennedy worked so hard at making a place for me, always saying nice things, gave me dignity and standing," Johnson said to Helen Thomas of United Press International after the 1968 election. "But back in the back room they were quoting Bobby, saying I was going to be taken off the ticket."[24]

Johnson may have had deeper reasons for his obsession with Robert Kennedy. In Doris Kearns's view, he had been torn in childhood between competing self-images — the masculinity demanded by his father, the intellectuality dreamed of by his mother. Despairing of satisfying both ideals, he chose masculinity and told himself that

---

* Hugh Sidey, "L.B.J., Hoover and Domestic Spying," Time, February 10, 1975. This rumor persisted, later taking the form that the Attorney General had tapped the Johnson White House. "There is absolutely no truth to the statement . . . that the White House had been wired by Robert Kennedy and the Justice Department," Ivan Sinclair of the White House staff wrote one inquirer (Sinclair to Glen Mann, September 14, 1964). "The report which you read," George Reedy wrote another, "is untrue" (Reedy to James Tilson, September 9, 1964; both letters in Johnson Papers).

intellectuals were destined to impotence; a man had to be virile to be effective. But Robert Kennedy, it appeared, combined the qualities that Johnson had convinced himself were incompatible. Johnson, Kearns noted, liked to impose tests of manhood, of which the most notorious was bringing politicians to his ranch and insisting that they kill deer.[25] John Kennedy, filled with deep distaste, had killed his deer after the 1960 election. I never heard that Robert Kennedy killed deer at the LBJ ranch.* But Johnson imagined he had, or at least said so for the purpose of tormenting Hubert Humphrey: "Bobby Kennedy got two of them. You're not going to let Bobby get the best of you, are you?"[26] At the same time Kennedy read books, quoted poetry and was (so Johnson thought) the darling of the intellectuals.

Above all, Robert Kennedy became the outlet for the unconscious resentment Johnson felt toward the President — a resentment he would not acknowledge, for he really liked John Kennedy, but could not altogether repress. In the fall of 1963 the Vice President lunched at the *New York Post* with Dorothy Schiff, James Wechsler and Joseph Lash. To the surprise of the *Post* people, a Johnson aide set up a recording machine in the corner. Johnson, Wechsler thought, seemed in a "strange and abstracted state." He expressed great admiration for John Kennedy. When, Johnson said, the President went around the table with the question "What would you do?" he prayed Kennedy would not turn to him first. Wechsler had the impression of "a rather beaten man whose only solace was in contemplating the burdens he had escaped by failing to achieve the presidency." The talk turned to Vietnam. Johnson, explaining the degeneration of the Diem regime, said the situation was much like that in Washington — a President with a "very strong" brother. "The inescapable overtone," Wechsler thought, "was that . . . Bobby Kennedy was running things, and in view of the Vietnam analogy, . . . [it] seemed to be an extremely bitter thrust."[27] This was October 16, 1963.

III

"I am Vice President," said the first man to hold the job. "In this I am nothing, but *I may be everything*."[28] After Dallas, Lyndon Johnson was suddenly everything. For all his towering ego, his devastating instinct for the weaknesses of others, his unlimited capacity for self-

---

* Nor has Jean Kennedy Smith any recollection of RFK's killing deer. He did go deer hunting at the ranch and was disgusted by Johnson's practice of shooting deer from the comfort of an elevated concrete structure. "This isn't hunting," he said on his return, "it's slaughter" (William vanden Heuvel and Milton Gwirtzman, *On His Own: Robert F. Kennedy, 1964–1968* [Garden City, N.Y., 1970], 246).

pity, he was at the same time a man of brilliant intelligence, authentic social passion and deep seriousness about his new responsibilities. He was now filled with apprehension about the nation, grief for the Kennedys and a consuming determination to restore confidence in the American government in the shortest possible time.

His insistence on presidential prerogative during that ghastly flight from Dallas came, not from unseemly haste to grab the perquisites of office, but from compelling desire to calm a frantic people and reassure a shocked world. For the stricken Kennedys, however, and for Robert in particular, it was too much too soon. Stunned and despairing, they thought only of the dead President; the living President had all the future in which to establish himself. And, though Johnson comprehended, even to a degree shared, their misery and sought in his sometimes awkward way to alleviate it, his inordinate sensitivity compounded misunderstandings. He had taken the presidential oath, but, as he sourly recalled the harsh time, "for millions of Americans, I was still illegitimate, a naked man with no presidential covering, a pretender to the throne, an illegal usurper."[29]

When the plane arrived at Andrews Field and Robert Kennedy hurried by the Johnson party to Jacqueline's side, Johnson took it as a deliberate snub: "He ran [Johnson later told Jim Bishop] so that he would not have to pause and recognize the new President."[30] Perhaps some such thought contributed to Robert Kennedy's haste. But a man more secure than Johnson would have sympathized with the terrible urgency carrying him to his murdered brother's wife. Driving from Andrews Field to the Bethesda Naval Hospital, Godfrey McHugh, President Kennedy's loyal and emotional Air Force aide, described to the Attorney General the inexplicable delay before the plane took off from Dallas and the sad confusion once they were airborne. Robert Kennedy recalled: "McHugh said that Lyndon Johnson had been — and I remember the word that he used — . . . obscene. There wasn't any other word to use and it was the worst performance he'd ever witnessed."[31] Then or later Jacqueline Kennedy told Robert of John Kennedy's last comment on Lyndon Johnson. He had said on Thursday night, November 21, probably in reference to Texas politics, that Lyndon Johnson was "incapable of telling the truth."[32]

"Four or five matters," Robert Kennedy told John Bartlow Martin the next spring, ". . . arose during the period of November 22 to November 27 . . . which made me bitterer, unhappy at least, with Lyndon Johnson."[33] He had gone to the Oval Office the day after his brother died. "I wanted to make sure the desk was gotten out of there, and I wanted to make sure all his papers were out." In the outer office he encountered Evelyn Lincoln in evident distress. The new President, she said, was already in the Oval Office and had told her, "I have a meeting at 9:30 and would like you to clear your things out of your office by then so my own girls can come in." At that point Johnson appeared and asked to speak to Kennedy.

Their conversation was brief. "He said how much he needed me — 'I need you more than the President needed you.'" As Kennedy observed to William Manchester in the spring, that soon became "a familiar refrain." "I said I don't want to discuss it and . . . that it was going to take us a period of time to move out of here, and I think, maybe, can't you wait? He said, well, of course, and then he started to explain. . . . McNamara had told him he had to move in. Dean Rusk told him he had to move in because the world would fall apart. . . . He didn't want to move in and everybody told him he should move in." Manchester later asked whether these excuses were true. "I never inquired into it, and then Mac Bundy said to me afterwards it was just a mix-up." Johnson extended Evelyn Lincoln's period of grace till noon. She was out by eleven-thirty.[34]

The first cabinet meeting was scheduled for two-thirty. Kennedy: "I was upset about what had happened on the plane and [by] the fact that he came into the office. So by this time I was rather fed up with him. . . . But I went by and Mac Bundy said it was very important that I come to the cabinet meeting." He arrived late. When he entered the room, some in the cabinet rose to their feet. Johnson did not move. Robert Kennedy:

> Rusk spoke sort of for the cabinet members and made a nice little statement. . . . Adlai Stevenson had written out a page and a half or two pages which he then read on how nice Lyndon Johnson was. . . . It didn't offend me. . . . I felt it was fine. It just struck me that he had to read the damn thing. . . . Afterwards somebody told me . . . how impressed Lyndon Johnson was with Dean Rusk because he's the only one who spoke up at the cabinet meeting. So I thought . . . what he wanted is declarations of loyalty, fidelity from all of us.[35]

The meeting was spiritless and quickly adjourned. A few moments later, Johnson told a cabinet member with "real bitterness" in his voice (so recorded in the cabinet member's diary) that Kennedy had deliberately arrived late in order to spoil the meeting's effect. Kennedy had told an aide, Johnson roundly asserted, "We won't go in until he has already sat down." When Manchester later passed on this story, Kennedy expressed amazement first, then amusement.[36]

This was Saturday. On Sunday the question arose of Johnson's first address to Congress. He wanted to make it on Tuesday. Kennedy: "I didn't like that. I thought we should just wait one day — at least one day after the funeral." When Bundy reiterated that Johnson wanted Tuesday, "I said, well, the hell with it. Why do you ask me about it? Don't ask me about what you want done — you'll tell me what it's going to be anyway. So go ahead and do it."[37] In the end, Johnson waited till Wednesday. It was an excellent address, delivered with dignity and force. I watched from Mrs. Johnson's box. Robert Kennedy "was pale, somber and inscrutable, applauding faithfully, but face set and his lips compressed."[38]

Johnson asked Kennedy to come and see him after the address. The meeting was brief. Again the new President attempted explanations: why the plane had been so long delayed in the departure from Dallas; why on Rusk's and McNamara's insistence he had moved so quickly into the Oval Office. "People around you are saying things about me," he told Kennedy. ". . . You can't let your people talk about me and I won't talk about you and I need you more [than your brother did]." Kennedy later: "I didn't get into an argument about it. . . . I don't know quite what I did say."[39]

Kennedy and Johnson did not see each other in December. On New Year's Day Johnson wired Kennedy from the ranch:

> I KNOW HOW HARD THE PAST SIX WEEKS HAVE BEEN FOR YOU. UNDER THE MOST TRYING CIRCUMSTANCES YOUR FIRST THOUGHTS HAVE BEEN FOR YOUR COUNTRY. YOUR BROTHER WOULD HAVE BEEN VERY PROUD OF THE STRENGTH YOU HAVE SHOWN. AS THE NEW YEAR BEGINS, I RESOLVE TO DO MY BEST TO FULFILL HIS TRUST IN ME. I WILL NEED YOUR COUNSEL AND SUPPORT.
>
> LYNDON B. JOHNSON

Kennedy replied from Aspen, Colorado, two days later:

> GREATLY APPRECIATE THE THOUGHTFULNESS OF YOUR TELEGRAM. I AM LOOKING FORWARD TO VISITING YOU IN WASHINGTON AT YOUR CONVENIENCE. RESPECTFULLY
>
> ROBERT F. KENNEDY[40]

IV

No one acted more expeditiously to emphasize Robert Kennedy's change of fortune than J. Edgar Hoover.

Hoover and Johnson had been Washington neighbors and friends for many years. They understood each other. Shortly after the assassination, the direct line from the Attorney General that Kennedy in another age had instructed Hoover to answer personally rang on the director's desk. "Mr. Hoover," an FBI agent recalled, "didn't answer it, so everyone tried to ignore it. When it finally stopped, Mr.

Hoover said, 'Put that damn thing back on Miss Gandy's desk where it belongs.'"[41] The director replaced Kennedy's friend Courtney Evans by the obsequious Cartha DeLoach as the FBI liaison with the White House. Walter Jenkins, a Johnson special assistant (whose brother was an FBI agent), replaced O'Donnell as the FBI contact. The Attorney General's office was cut out of the chain. When Dolan warned Kennedy that the Bureau was back to its old trick of dealing directly with the White House, Kennedy smiled wryly and said, "Those people don't work for us any more."[42] He spoke to me in early December about the "revolt of the FBI," adding grimly that this was something he would gladly occupy himself with in the next eleven months.[43]

Kennedy never forgot the coldness with which Hoover had broken the news from Dallas. "It wasn't the way that, under the circumstances, I would have thought an individual would talk. That was one thing. Then I knew that within a few days he was over to the White House giving dossiers on everyone that President Kennedy had appointed, in the White House particularly . . . with the idea that President Kennedy had appointed a lot of . . . rather questionable figures." In his talk with Johnson after the address to Congress, Kennedy had warned that the FBI was operating once again as an independent agency. "I said I thought it was a major mistake because I thought that they should have some control over the FBI. . . . His response was mostly that it wasn't going on and that he wanted me to control the Department of Justice." Kennedy rejoined that they both "knew what was going on and that . . . if I had just been appointed Attorney General, I would resign." However, if he did, "it would be considered that I was getting out for a different reason. So that wouldn't do any good; and I was going to accept that relationship through the year; and then I'd get out." Johnson repeated that "that wasn't really the situation that existed, that J. Edgar Hoover never came to him directly. But that wasn't the truth."[44]

Kennedy's friends kept him informed about the reports Hoover pressed on the President; as, for example,

> that the Attorney General's people at the Department of Justice are holding secret meetings to try to play up the Bobby Baker case "in order to cause you embarrassment so that you'll have to take the Attorney General on the ticket." . . . [Ethel Kennedy] is alleged to have said that the FBI is out there [at Hickory Hill] not checking the phones [for wiretaps] but are putting wiretaps in so Lyndon Johnson can listen.[45]

> McNamara used to tell me that Hoover used to send over all this material on me and that Lyndon Johnson would read it to him. Lyndon Johnson told me that he never received an adverse report from J. Edgar

Hoover on me. One time McNamara had a dinner at Nick Katzenbach's house to talk about their children taking a bicycle ride through Cape Cod. . . . Hoover sent a report in to Lyndon Johnson that there was a meeting — I think I was supposed to have been there — at this house in which we were discussing the overthrow of Lyndon Johnson, to take the nomination from him. Or like Abba Schwartz reported to have said at some party, "We've got to get rid of Lyndon Johnson so that Robert Kennedy can become President." . . . Lyndon Johnson says that he never received such a report.[46]

While thus feeding the new President's suspicions, the director at the same time protested to him about, as Kennedy heard it, "a conspiracy, led by me, to get rid of J. Edgar Hoover."[47] Johnson's response to Kennedy's criticisms of the FBI, or so Hoover told an interviewer in 1970, was: "Stand by your guns." As for Hoover himself: "I didn't speak to Bobby Kennedy the last six months he was in office."[48] In mid-1964, Johnson, standing in the Rose Garden with Hoover beaming at his side, announced he was exempting the director indefinitely from the law requiring retirement at seventy. "The nation cannot afford to lose you," the new President grandly said. ". . . No other American, now or in our past, has served the cause of justice so faithfully and so well."[49]

v

On the late afternoon of December 13, 1963, Richard Goodwin and I met Kennedy in his office at Justice. Earlier in the day I had found Averell Harriman "in a state of considerable distress and wrath" over the new President's rumored decision to place Thomas Mann in charge of western hemisphere affairs. Mann, assistant secretary when Kennedy took over and thereafter ambassador to Mexico, was able, hard-boiled, opinionated, a colonialist by mentality and a free enterprise zealot. He was also a Texan, though not, I believe, a particular friend of Johnson's. But they both had the Tex-Mex attitude toward Latin America, well delineated by Johnson himself soon after assuming the Presidency. "I know these Latin Americans," Johnson told a group of reporters.

. . . They'll come right into your yard and take it over if you let them. And the next day they'll be right on your porch, barefoot and weighing one hundred and thirty pounds and they'll take that too. But if you say to 'em right at the start, "hold on, just wait a minute," they'll know they're dealing with somebody who'll stand up. And after that you can get along fine.[50]

Mann's designation, Harriman thought, would "reverse the whole direction of Latin American policy."[51] The Attorney General had also spoken with Harriman. "I don't want to see Averell Harriman get hurt, or anyone else," he now observed to Goodwin and me. "Harriman's got his faults. I've got my faults. We've all got faults. The important thing for us to do now is to stick together."

He talked on in the fading December light. "Our power will last for just eleven months. It will disappear the day of the election. After November 5th we'll all be dead." We must

> use that power in these months to the best possible advantage. . . . There are a hundred men scattered through the government who are devoted to the purposes for which we came to Washington. We must all stay in close touch and not let them pick us off one by one. I haven't the answer in detail yet, but I am sure that the fundamental principle now is collective action.

He got up, stood tensely by his desk, his hands at his side.

> Sure, I've lost a brother. Other people lose wives. . . . But that's not what's important. What's important is what we were trying to do for this country. We worked hard to get where we are, and we can't let it all go to waste. My brother barely had a chance to get started — and there is so much now to be done — for the Negroes and the unemployed and the school kids and everyone else who is not getting a decent break in our society. This is what counts. The new fellow doesn't get this. He knows all about politics and nothing about human beings. . . . I haven't talked to him yet. I don't feel mentally or physically prepared to do so yet. When I talk to him, I am ready to be tough about what we must have.

He continued:

> There are a lot of people in this town. They didn't come here just to work for John Kennedy, an individual, but for ideas, things we wanted to do. It's one thing if you've got personal reasons for leaving, like you may want to leave, Arthur. But I don't think people should run off. . . . Remember, after November 5th we're all done. We won't be wanted or needed.[52]

The next day Johnson announced Mann's appointment. It was, I wrote in my journal, "a declaration of independence, even perhaps a declaration of aggression, against the Kennedys."[53] I wrote the Attorney General:

> Johnson has won the first round. He has shown his power to move in a field of special concern to the Kennedys without consulting the Kennedys. . . . We have supposed that Johnson so badly needed the Kennedy people for the election that we would retain a measure of power for eleven months. . . . We have underestimated the power of the Presidency. The President has nearly all the cards in this contest. . . . So long as he maintains an ostensibly liberal position on issues, it will be very difficult to do anything about it. . . . We are weaker — a good deal weaker — than we had supposed. He has understood that the only

sanctions we have are resignation and/or revolt — and that both sanctions are meaningless, and will seem sour grapes, unless they are provoked by a really understandable issue — and this LBJ will do his best to deny us.[54]

Kennedy thought my analysis unduly pessimistic. Johnson would have to consult the Kennedy people about the Vice Presidency, he said, and we would exert some influence on the choice of Secretary of State in the next cabinet. I said it was hard to impose a deal on Presidents. Kennedy said, "It is not so hard. I will be perfectly willing to ask President Johnson what his plans are for the State Department before we decide the role we are going to play in the campaign." I said, "We will all have to play a role in the campaign, or we will be finished forever in the party." He said, "Yes, but there is a considerable difference between a nominal role and a real role. We can go through the motions or we can go all out."[55]

So the Kennedy loyalists indulged in impotent speculation. Lyndon Johnson unquestionably surmised, if he did not know, that such discussions were going on. He could hardly be blamed for resentment and mistrust. As for myself, believing the new President had the right to his own men in the White House, I had submitted my resignation the day after Dallas, received the Johnson treatment ("I need you more than President Kennedy needed you"), agreed to stay for the transition and in January 1964 resigned again. This time it was accepted with entirely understandable alacrity.

VI

Walter Sheridan asked Robert Kennedy when he was coming back to the Justice Department: "We need you, you know." Kennedy said: "I know, but I don't have the heart for it right now."[56] His world was radically altered. Jacqueline Kennedy had seen it as the end of Camelot* — a romantic fancy that I do not recall on Robert's lips and that I am sure would have provoked John Kennedy to profane disclaimer. For Robert his brother's Washington evoked rather Periclean Athens and high ideals of civic commitment — ideals he feared might now fall before a rabble of wheeler-dealers.

His friends wondered what might recall him to action. Then his old acquaintance Sukarno provided a new crisis in Southeast Asia.

* A few days after Dallas, Jacqueline Kennedy had recalled to Theodore H. White the lyrics of the song "Camelot" written by Alan Jay Lerner, John Kennedy's boyhood friend at Choate and Harvard, for the 1960 musical show of the same name (see Theodore H. White, "For President Kennedy an Epilogue," *Life*, December 6, 1963). The image was not perhaps, on analysis, all that romantic. King Arthur's Camelot concluded in betrayal and death.

The formation under British sponsorship of the Federation of Malaysia in September 1963 had given deep offense to the Indonesian leader. "I have been duped and humiliated by the British," he told Ambassador Howard Jones, his face contorted with rage. "I will not take it."[57] He proclaimed the policy of Konfrontasi — confrontation — sneaked a few guerrillas into Malaysian territory and fulminated against the new state. "We'll crunch up Malaysia," he said, "and spit out the pieces."[58]

Washington, though sympathetic to Malaysia, did not want to drive Indonesia into closer connection with Peking, which after 1962 had displaced the Soviet Union as Sukarno's Communist friend. And, if the British Commonwealth went to war, the United States might be dragged in as a result of obligations under the ANZUS treaty of 1951. All this argued for American initiative in moderating the crisis. When Jones returned to see President Kennedy in mid-November 1963, he had requested authorization to say that, if Sukarno settled the Malaysian dispute, the United States would resume economic aid, now somewhat cut back, and that President Kennedy himself would visit Indonesia the next spring. Harriman, Roger Hilsman and Michael Forrestal backed this recommendation. The President agreed. Two days later he left for Texas.[59]

Indonesia, it thereafter developed, had been one more point of foreign policy disagreement between Kennedy and Johnson. The new President refused to sign a determination that aid to Indonesia was in the national interest. He agreed, however, to one more try at getting negotiations under way between Sukarno and Tunku Abdul Rahman, the Malaysian Prime Minister. Harriman proposed Robert Kennedy as a presidential emissary — both on the merits and in the hope that the mission might rouse him from his depression. Howard Jones thought Kennedy "ideal. . . . He had established a warm rapport with Sukarno, knew many of the people involved, was familiar with the issues, and had proved his skill in delicate negotiation."[60] In the White House Bundy cordially agreed. The new President, however, disliked the idea. Johnson, Bundy said, "felt that he had been sort of maneuvered into approving by staff people who weren't thinking about the Johnson interest."[61] Nor did Kennedy himself much want to go. After the New Year he dubiously committed himself to the trip.

The condition precedent to negotiation was a cease-fire. "[The] Tunku cannot negotiate with a pistol at his head," Forrestal wrote the Attorney General, "and it is up to Sukarno to remove the pistol."[62] Tell Sukarno, Kennedy's instructions read, that, "if the confrontation is stopped, the natural course of history . . . will inevitably lead to a gradual reduction of the British presence. Thus it is Sukarno's own policy which is producing the British reactions to which he is objecting. This pathway is bound to lead to a bloody nose for Sukarno, and when we are forced to choose sides we will not be able to choose his."[63]

Ethel accompanied him, along with Forrestal, Guthman and Peter Maas. The atmosphere on the plane across the Pacific, with Robert and Ethel answering letters of condolence, was melancholy.[64] The first talks with Sukarno were to take place in Tokyo. Kennedy, however, had an earlier stop to make. At what the American embassy called "the insistent urging of the students and faculty," he returned to Waseda University, where students had raged at him only two years before. The auditorium was packed with an enthusiastic crowd. Thousands more, standing outside in a soft rain, listened over loudspeakers. He talked about his brother: "He was not only President of one nation; he was president of young people around the world. . . . If President Kennedy's life and death and his relationship to all in our age group mean anything, it means we young people must work harder for a better life for all the people in the world." The crowd cheered wildly. Then he led the students in singing the Waseda song.[65]

He met Sukarno in Tokyo on January 16. Sukarno said, "Did you come here to threaten me?" Kennedy smiled and said, "No, I've come to help get you out of trouble."[66] Forrestal reported to Washington his impression that Sukarno had been "quite capable of reacting irrationally if the Attorney General had not knocked the idea down."[67] Their talks went surprisingly well. Responding to Kennedy's frank presentation, Sukarno said he would agree to a cease-fire if the Malaysians agreed also. Kennedy flew on to Malaysia. The Tunku protested that Sukarno would use a cease-fire to build up his guerrilla forces but eventually yielded to the Attorney General's persuasion.

On January 22 Kennedy arrived in Indonesia, where the Communists had unleashed a clamor against a cease-fire. But Sukarno kept the word he had given in Tokyo. That afternoon, looking "utterly beat," he unburdened himself to his American biographer, Cindy Adams. "I am tired. So tired," he said. ". . . I have to make a speech at the stadium where my people are waiting for me to shout again 'Ganjang Malaysia . . . Crush Malaysia.' And I cannot. I have agreed to a cease-fire." She asked him how he liked Kennedy. "Bob is very warm," he said. "He is like his brother. I loved his brother."[68]

"He was very, very frank with Sukarno," Forrestal said later; "told him things that nobody else would have dared to say to him." Dealing with two opinionated, angry men, Sukarno and the Tunku, "he never told a lie; he never misquoted anybody; he never concealed. . . . He was always absolutely impersonal, spoke with great directness softened by humor; the truth aroused no antagonism as he delivered it."[69] "I am most grateful," the Tunku wrote Kennedy, "for everything that you have done to bring about a cease-fire. . . . You have done more than anybody else has managed to do so far."[70]

The return to Washington was an anticlimax. Instead of reporting to the new President in private, Kennedy found himself on January 28 at a mass meeting with Johnson and ranking members of the Senate Foreign Relations and Armed Services Committee. The President displayed little interest then and none thereafter in Kennedy's trip.[71] Though the conferences Kennedy had set up between the Indonesians and the Malaysians took place, the parties failed to agree on the terms in which the cease-fire should continue. He tried to follow up. He sent messages to the Tunku and to Sukarno. His feeling grew that the President did not, and the State Department dared not, give a damn about his effort. The episode left him, said Guthman, with "a bitter taste."*

VII

Still, it restored him to activity. "I hadn't wanted to go on that trip," Robert Kennedy told Murray Kempton, "but afterwards I was glad I had."[72] It was, however, almost his last involvement as Attorney General in world affairs. Johnson's first foreign crisis had erupted before the trip — over the Panama Canal Zone. No one had consulted Kennedy. "It's really the worst matter involving an international problem that I have not been in," he said in March. He thought it "very badly handled."[73] He could hardly be kept out of the next crisis, however, for it involved Cuba. The available record does not clarify Kennedy's participation in Cuban policy after Dallas, save for an allusion in a February memorandum to "the Attorney General's statement that there is no point in discussing courses of action . . . until the fundamental decision is made as to whether or

* Guthman, *We Band of Brothers* (New York, 1971), 253. Harriman and Jones continued to argue that military aid to Malaysia would destroy what little influence the United States had left in Indonesia. In July Johnson invited the Tunku to Washington, where they jointly announced a program of American military assistance to Malaysia. At the National Press Club the Tunku compared Sukarno to Hitler. Sukarno would no doubt have succumbed to irrevocable anti-Americanism anyway as he pursued his increasingly megalomaniac course. But Johnson's July decision left him no alternative. (Howard P. Jones, *Indonesia: The Impossible Dream* [New York, 1971], 312, 342–343; W. Averell Harriman, in recorded interview by author, June 6, 1965, 28–30, JFK Oral History Program.)

not it is possible for the United States to live with Castro."[74] One deduces a desire to keep the Attwood initiative alive.

The Cuban crisis of February arose when the Coast Guard seized four Cuban fishing boats two miles off the Florida Keys. Thomas Mann encouraged Johnson to view the incident as a critical Soviet "test" of the new American President. So far as Kennedy could see, it was no more than a traffic violation. The Attorney General had what he subsequently described as a "rather violent" argument with Mann at a National Security Council meeting. Kennedy's demand for information to substantiate the Mann-Johnson thesis caused an adjournment. When the NSC reassembled that afternoon, McCone conceded that "maybe the fishing vessels just arrived there. So then I said that this whole business about the fishing vessels was foolish . . . like a speeding, parking ticket. Why don't they just tell them to get out of there and go home? and, if you wanted to fine them a couple of hundred bucks, fine them, but the idea of locking them up and creating a major crisis about it was foolish."[75]

Kennedy prevailed on the problem of the hapless fishermen, who were fined and sent home. But Johnson still wanted his crisis. Castro, protesting the seizure of the fishing boats, had turned off the water supply to the American naval base at Guantanamo. Mann now proposed to discharge two thousand Cubans who worked on the base. They were, he said, security risks; they brought dollars into the Cuban economy; most important of all, "The only thing the South Americans understood was money and, when you took this money away from Castro, it would be a sign to the rest of the countries of Latin America that this was a new administration which was going to stand up." Latinos, Mann said, were not like North Americans. This exasperated Kennedy. "I said I thought he sounded like Barry Goldwater making a speech at the Economic Club and that this policy of the United States had gone fifty years before." If the Cuban workers were security risks, why had they not created problems during the missile crisis? "I didn't think that made any sense." He lost this one. The employees were fired. This was, Kennedy said later, "the last meeting on any substantive matter" of foreign affairs to which he was invited.[76]

He made a final gesture about Cuba in the summer. A. J. Muste, the aged pacifist, began a peace march from Quebec to Guantanamo. The White House, unduly agitated, sought an injunction to prevent Muste from getting on a boat in Florida. The letter of authorization required the signature of the Attorney General. Kennedy asked what it was all about. He was told, "The White House wants it signed." He listened with incredulity. "Let me get this straight," he said. "You mean you want me to sign that piece of paper to tell an 84 year old man he can't walk 800 miles? . . . I don't think the security of the United States is going to be endangered by an 84 year old man. I'm not going to sign that piece of paper." [77]

He remained unchallenged in his own domain. "I don't think Lyndon Johnson paid one iota of attention to the Justice Department," Katzenbach said later. "Bobby Kennedy had license to do what he wanted to do." [78] The Attorney General was at last beginning to have the heart again for Justice. His collar still seemed a little large, his cuffs a little too close to the knuckles, "not as though he had wasted," Murray Kempton thought, "but as though he had withdrawn." The office looked the same. "I have this feeling," Kennedy mused, "that I go on doing the same old things I always did. . . . A month can go by before I need to take a problem to the White House. After all even my brother never asked me about the Lands Division." [79]

He took up the threads of the past. Jimmy Hoffa went on trial in Chattanooga in January 1964 on the charge of conspiring to fix the jury in the Test Fleet case. In March he was found guilty, receiving an eight-year sentence. In April he went on trial again, this time in Chicago on charges of diverting a million dollars from the Teamster pension funds for his own use. In July he was found guilty once more. This time it was a five-year sentence.

Hoffa's downfall brought Kennedy no pleasure. "I think he lost all interest in Hoffa," Kempton said. ". . . I never heard him say anything about Hoffa that really indicated much more than boredom with the subject in the last years of his life." [80] He went to a Justice Department party after Hoffa's Chattanooga conviction but told Angie Novello that there was "nothing to celebrate." [81] Kenneth O'Donnell thought Kennedy "unhappy that he was going to jail. . . . He had enough tragedy of his own now." [82] "He didn't like the idea of Hoffa having eight years in Lewisburg," said William Hundley. "We used to talk about Jimmy on plane trips. He would say, 'How's Hoffa doing?' If he had ever become President, the first person he would have let out was Hoffa." [83]

VIII

He was through with chasing people. His brother's last wish had been a war against poverty. He now unexpectedly found himself in a position to help carry forward that wish. The Council of Economic Advisers and the Bureau of the Budget, in trying to put an antipoverty program together, had been "bewildered," as William Capron of the council said later, "by the complete disarray of the nominal professionals in the field of poverty."[84] In the search for a "unifying principle,"[85] William B. Cannon of the Bureau of the Budget bethought himself of the President's Committee on Juvenile Delinquency.

David Hackett now proposed that the antipoverty program base itself on a community-action approach, in which local institutions would coordinate federal programs within the community and be accountable to (and funded by) a single responsive instrument in Washington. Communities were to have a year to prepare their programs. By establishing a method rather than imposing substantive solutions, the community-action approach both acknowledged that no one knew how to cure poverty and guaranteed adaptability to local circumstance.

Strongly backed by the Bureau of the Budget and the Council of Economic Advisers, the community-action approach won the initial sympathy of the new President.[86] But it was vigorously opposed by the old-line departments, most of all by Labor, where Willard Wirtz and Daniel Patrick Moynihan forcefully argued that a job program ought to have first priority. Johnson, moreover, wanted visible results well before the 1964 election. "What we said," Hackett recalled, "was, 'Go stage by stage, don't rush into legislation.' But Johnson just said, 'Go.'"[87] After Dallas, the influence of Hackett and the Department of Justice group evaporated. "Few things that Robert Kennedy had touched," Moynihan wrote later, "were not thereafter viewed with suspicion, fear, and distaste by the staff of the Johnson White House, and of course most of all by the President himself."*

In December, Sorensen had given the antipoverty planners the impression that Robert Kennedy might "seriously consider heading" the war against poverty.[88] Whether or not this was so, Johnson decided, while Kennedy was in the Far East, to offer the job to Sargent Shriver. Like Johnson, Shriver wanted immediate visibility and rapid results. As for community action: "It'll never fly," he said. ". . . Where you need the money worst, you'll have the worst plans."[89] Instead he put together an eclectic program more or less combining everybody's ideas. On Kennedy's intercession Shriver reluctantly added community action, though now divested of the emphasis on advance planning.† Still, even if Hackett's guerrillas were

---

* Moynihan, *Maximum Feasible Misunderstanding* (New York, 1969), 80. See William B. Cannon: "The assassination also meant that [the] Hackett-Kennedy group as the agent for translating irresistible social forces and ideas into action were themselves something less than irresistible political forces" ("The Dangerous Abuse of the Middle Class," ms., ch. 6, 51).

† We turned to David Hackett and Richard Boone too for help when it looked like

defeated in their hope of making community action the centerpiece, so much survived from their labors of 1961–63 — the national service corps, reappearing as Volunteers in Service to America (VISTA); legal aid; mental health centers; youth development projects; neighborhood services — that, as Kenneth Clark wrote in 1969, the Juvenile Delinquency Committee had in large part laid "the foundation for the national anti-poverty program."[90]

The community-action title of the Economic Opportunity Act of 1964 was, Hackett said later, "a rearguard action."[91] Its point was to vindicate popular participation against both government and the social welfare establishment. So Hackett, Cannon and Richard Boone inserted the mystic words — "maximum feasible participation of the residents of the areas" — into the law. Their belief was that "poor people themselves could render most effective help to the poor"; their hope, that in the process communities could develop their own competencies.[92]

When the bill came before Congress in the spring of 1964, the single administration witness to speak about community action was Robert Kennedy. His testimony summed up the lessons he had drawn from the Juvenile Delinquency Committee:

> The institutions which affect the poor — education, welfare, recreation, business, labor — are huge, complex structures, operating far outside their control. They plan programs *for* the poor, not *with* them. Part of the sense of helplessness and futility comes from the feeling of powerlessness to affect the operation of these organizations.
>
> The community action programs must basically change these organizations by building into the program real representation for the poor ... giving them a real voice in their institutions.[93]

Beck and Hoffa had taught him long before how essential it was for union members to have the power to challenge hierarchies. Community action, he hoped, would give the poor a way of exerting leverage on the structures that ruled their lives. The ideal of participation remained close to his heart for the rest of his days.

IX

The racial question was more than ever Robert Kennedy's primary concern. Congress was still debating his brother's civil rights bill. In the spring of 1964 the Attorney General carried the campaign to Georgia. A black southerner could die for his country, Kennedy told

the Community Action Program was going to be jettisoned by the Shriver group. I firmly believe that the major reason Community Action survived was because David Hackett prevailed upon Attorney General Kennedy to intercede with Mr. Shriver" (Cannon, "Dangerous Abuse," ch. 6, 44).

them at West Georgia College; but, if his widow wanted to travel north to visit his grave at Arlington Cemetery, "she doesn't know what hotel she can stay at. She doesn't know what restaurant she can possibly stop in ... what restrooms [she can use]. ... And yet her husband ... was killed for all of us." Legislation to end this "continuous insult" was, he said, "long, long overdue. ... How would any of us like it if we were in that situation?" Questioners brought up states' rights. Mississippi and Alabama, Kennedy replied, had not dealt justly on racial matters. Negroes "are entitled to the federal government affording them some protection." His answers, wrote Eugene Patterson in the *Atlanta Constitution,* "were direct and rough-edge. ... And the students responded to what he said with profound, rolling applause that visibly startled him." "There is only one word for Attorney General Robert Kennedy's reception at West Georgia College Tuesday," wrote Reg Murphy, the *Constitution*'s political editor: "Fantastic."[94]

The bill, however, left out one point of supreme importance to the civil rights movement — federal protection of civil rights workers. The question John Lewis had tried to ask during the March on Washington — "Which side is the federal government on?" — still echoed. Nineteen sixty-three had been bad enough. Nineteen sixty-four promised to be worse. Young people, black and white, were flocking south to work for racial justice over Freedom Summer. The Klan organized klaverns in twenty-nine Mississippi counties between February and June.[95] The national government seemed helpless before impending violence. "There is no answer," said Katzenbach bleakly, "which embraces both compassion and law."[96] "We simply can't wetnurse everybody who goes down to try to reform or re-educate the Negro population of the South," snapped J. Edgar Hoover, blaming the trouble on the "harsh approach to the Mississippi situation by the authorities here in Washington."[97]

With the administration sticking to the view that the federal system tied its hands, the movement advanced its own proposals. Federal law-enforcement agencies, Martin Luther King pointed out early in 1964, were long accustomed to "working within secret groups and obtaining effective results. ... There is a need to know what is going on in conspiratorial racist circles. Many of the shocking bombings might have been avoided if such knowledge had been available."[98] "We need the FBI before the fact," said Bob Moses, a brave young SNCC leader after James Chaney, Michael Schwerner and Andrew Goodman disappeared in Neshoba County in June 1964. "We have them now after the fact."[99] Why, asked James Wechsler, has "the FBI so completely and totally failed to engage in any effective form of infiltration of the forces of Southern violence — the Ku Klux

Klan, the White Citizens Councils, and all their front organizations?"[100]

Was not this the answer? Why should the FBI, so successful by its own account in penetrating the CPUSA and the Mafia, not now penetrate the Klan? "President Kennedy spoke to [Hoover] about that," Robert Kennedy said later, "to try to do it on the same basis."[101] In June 1964 the Attorney General formally recommended to President Johnson the use against white supremacy groups of "techniques followed . . . in the infiltration of Communist groups."[102]

What no one outside the Bureau knew was that the FBI's Intelligence Division had long since institutionalized and elaborated such techniques in a project called the Counterintelligence Program — COINTELPRO in the trade. COINTELPRO's mission was to penetrate organizations not, however, merely to find out what they were doing but to disrupt and destroy them. Its weapons were rumor, forgery, denunciation, provocation. "No holds were barred," William C. Sullivan, the assistant director in charge of the Intelligence Division, later testified. ". . . We have used [these techniques] against Soviet agents. They have used [them] against us. . . . [The same methods were] brought home against any organization against which we were targeted. This is a rough, tough business."[103]

COINTELPRO had started against the Communist party in 1956. Two years later Hoover described it to Attorney General Rogers in highly general terms as "a program designed to promote disruption" within the CPUSA. This was about as near as he came to telling any Attorney General about COINTELPRO. The very name was unknown outside the Bureau. In January 1961 Hoover informed Robert Kennedy even more generally about a "program of counterattack against the CPUSA which keeps it off balance. Our primary purpose in this program is to bring about disillusionment on the part of individual members." The only techniques mentioned were the use of informants and the exposure of Communist infiltration of other organizations.[104] When the Kennedy White House requested a formal report on "Internal Security Programs," Hoover mentioned the Bureau's "investigative programs" and said nothing at all of disruptive activities.[105] As Richard D. Cotter put it after twenty-six years' service as research chief of the Bureau's Intelligence Division, COINTELPRO demonstrated "the astonishing degree of independence the FBI had gained in the domestic security area. The Bureau was able to initiate an extensive, risky, and highly questionable operation without consulting the Attorney General. The COINTELPRO project required only the recommendation of the Assistant Director in charge of Intelligence Division and the approval of the FBI Director."[106]

In 1964 Sullivan persuaded Hoover to transfer the Klan from the

General Investigative Division, where "nothing was being done," to COINTELPRO. Sullivan had seen fiery crosses burning on the hills of central Massachusetts as a boy. His father had fought the Klan in the twenties. "I had a real interest," he said later, "in breaking up the Klan."[107] The General Investigative Division, it is true, had Klan informants. Gary Thomas Rowe, Jr., for example, had been recruited in 1959. But, after the Klan COINTELPRO began in September 1964, Rowe's FBI case officers told him, "The bureau is declaring war on the Klan. You can do anything you want to get your information." Their first inspiration was that he sleep with the wives of fellow Klansmen. Their second was that he stir up personal trouble inside the Klan, spread rumors of other infidelities, for example. As for strong-arm activities, "Well, you'll have to do what you have to do. No holds barred."[108]

x

"We have penetrated the Klan very effectively," Hoover declared publicly in December 1964, ". . . particularly in the States of Alabama, Georgia and Mississippi."[109] Nine months later he reported to the Attorney General that, of the Klan's fourteen state organizations, "we have penetrated every one of them through informants and currently are operating informants in top-level positions of leadership in seven of them." In one state the top Klansman was an FBI operative. "As a result, we have been successful to date in holding Klan violence in the entire state to an absolute minimum." In another state the FBI subsidized a runaway klavern critical of the regular Klan. At one point the Bureau even considered installing an informer as the Klan's imperial wizard. In the first year of the Klan COINTELPRO, informers accounted for 70 percent of the new membership and, at the end of the year, for about one sixth of the total — 2000 out of 12,000–13,000. (Never missing a chance to extract bigger appropriations, Hoover told the House Subcommittee on Appropriations ominously, "During the past year there has been a marked increase in Klan membership.")[110]

FBI infiltration led to the discovery of the bodies of Chaney, Schwerner and Goodman, the murdered civil rights workers, and the arrest of their killers. Gary Rowe ended his long masquerade by testifying against the murderers of another civil rights worker, Viola Liuzzo. Informers were responsible for numerous other convictions.[111] The Klan was soon in rout. "When we took it over," Sullivan said, "the Klan had more than fourteen thousand very active members, and when I left in 1971 it had been reduced to forty-three hundred completely disorganized and impotent individuals."[112]

Exhilarated by its triumph, the FBI extended COINTELPRO in 1967 to so-called Black Nationalist–Hate Groups, including, incongruously, the pro-integration, nonviolent SCLC, and in 1968 to New Left groups. Flagrant excesses led some civil libertarians to attack counterespionage per se as a violation of the Bill of Rights. When the Klan bombed a school bus in Michigan, the testimony of an FBI informer convicted the bombers. "My judgment," said Professor Vern Countryman, "would be that if the only way to detect that bombing is to have the FBI infiltrate political organizations, I would rather the bombing go undetected. . . . There are worse things than having people killed." Critics dwelt on the moral corruption of informing, the psychological instability of persons predisposed to inform, the tendency of informers to become *agents provocateurs*, the historical role of informers as agents of despotism.[113]

Conceivably the question was not that simple. It was, after all, possible to inform *against* despotism too. Robert Kennedy had seen this repeatedly on the Rackets Committee. He admired the courage of rank-and-file trade unionists who, at enormous risk and small benefit, 'informed' against corruption and tyranny — men like Terry Malloy as written by Budd Schulberg and played by Marlon Brando in *On the Waterfront*. There was the veteran labor organizer Sam Baron, who had started out with the International Ladies' Garment Workers, fought with the Loyalists in Spain and organized Sears, Roebuck and Montgomery Ward for the Teamsters. Baron, who detested the labor racketeers, provided evidence against Hoffa. When someone called Baron a "fink" and "stool pigeon," Kennedy observed that he regarded Baron as "a citizen who was reporting information and evidence in connection with illegal activities."[114]

"Courts have countenanced the use of informers from time immemorial," Judge Learned Hand had said in sustaining the conviction of Communist leaders under the Smith Act.[115] The Supreme Court, upholding Partin's testimony against Hoffa, ruled that neither the Fourth nor Fifth Amendment limited the use of informers: "The established safeguards of the Anglo-American legal system leave the veracity of a witness to be tested by cross-examination, and the credibility of his testimony to be determined by a properly instructed jury."[116] As a practical matter, many crimes in such areas as racial violence, racketeering, and political corruption would not be solved without informers.

The court decisions, however, assumed that the work of informers was informing. But informers in organizations engaged in criminal activity, like the Klan or the mob, often behaved criminally themselves in order to maintain cover. The FBI had instructed Gary Rowe to avoid violence. Yet "to gather information," as Rowe's FBI case officer told the Church committee, "you have to be there. . . . If [Rowe] happened to be with some Klansmen and they decided to do something, he couldn't be an angel and be a good informant."[117] On FBI instructions,[118] Rowe went along on the expedition that ended in the murder of Viola Liuzzo. There is something appalling about the FBI informant riding in the Klan car on FBI orders, impotently watching, seconds before the shots, "the most awful expression [on the victim's face] . . . that God knows I have ever seen in my life."[119]

Still, Rowe's presence brought the killers to justice. King and other black leaders advocated FBI penetration of the Klan for urgent reasons. The Kennedy and Johnson administrations turned to counterespionage as the best means of federal protection at their command. Refusal of FBI infiltration would have left civil rights workers in the deep south at the mercy of hooded terrorists. It was not necessarily a wicked choice. Burke Marshall reflected a decade later on the way the Klan was broken in a terrible time:

> It was done . . . by bribery, by payments to informers, by whatever eavesdropping was then permitted under the bureau's rules, by the sowing of suspicion among Klan members so that none knew who was an informer and who not, by infiltration and deception, and in at least one incident by the participation of a bureau informer in the planning and attempted execution of a murder.
>
> It did not appear to those involved at the time, and it does not appear to me now, that the criminal conspiracy of violence that existed in the State of Mississippi then could have been handled by less drastic measures.[120]

XI

From the start the new President had left no doubt about his vigorous support of the civil rights bill. "I knew," Lyndon Johnson told Doris Kearns, "that if I didn't get out in front on this issue, [the liberals] would get me. . . . I had to produce a civil rights bill that was even stronger than the one they'd have gotten if Kennedy had lived. Without this, I'd be dead before I could even begin."[121]

He left the day-to-day strategy to the Department of Justice. "I'll do on the bill just what you think is best," he told Robert Kennedy in January 1964. ". . . We won't do anything that you don't want to do on the legislation, and I'll do everything you want me to do in order to obtain passage." Anthony Lewis later asked how Kennedy explained Johnson's blank check. Johnson, Kennedy replied, did not see how the bill could pass the Senate. "If he said, 'Where are you going to get the votes?' as he used to say to me — well, President Kennedy used to say the same thing [in November] — you couldn't tell him where you were going to get the votes. The person who you're going to get the votes from, really, in the last analysis, was Everett Dirksen." And, if the bill failed, Johnson, Kennedy thought, did not want sole responsibility. "If he did what the Department of Justice did, said, recommended, suggested — and particularly me — then he could always say that he did what we suggested.... He had a particular problem being a southerner.... So I think that for political reasons it made a good deal of sense. Secondly, our relationship was so sensitive at the time that I think that he probably did it to pacify me."[122]

The bill, after passing the House by a better than 2–1 vote in February, fell into a Senate filibuster in March. Dirksen was more than ever the key. By late April the minority leader, no one quite knew why, appeared ready to accept the bill with minor modifications. "Whatever Dirksen and the AG agree on," Johnson said, "I am for."[123] Dirksen's collaboration brought along enough Republican votes to end the filibuster. On July 2, Johnson signed the Civil Rights Act of 1964, thereby outlawing discrimination in voting, employment, public education and public accommodations and facilities.

It was a stronger and better bill than the one the Kennedy administration had given Congress a year earlier. The signing took place with the usual nonsense in the Oval Office. The Reuther brothers noticed Robert Kennedy in the rear of the room, staring at the floor. "Surely no one," Victor Reuther thought, "had contributed more to this moment than he." Roy Reuther seized Kennedy by the arm and firmly propelled him across the room to Johnson, saying, "Mr. President, I know you have reserved a pen for your Attorney General." Johnson blandly gave Kennedy a handful of pens.[124] The Reverend Walter Fauntroy, watching the ceremony with Martin Luther King, said later, "Our enthusiasm — that of Dr. King and myself — was sort of dampened by the sadness that we saw in Bobby's eyes and the coldness with which the President obviously treated him.... We commented afterwards at the coldness on the part of Johnson toward Kennedy on that important day."[125]

28

The Vice Presidency

THE SCENE AT THE SIGNING of the civil rights bill was one more grim moment in a grim spring. I do not think that either man can be held responsible for the failure of the relationship. Each in his own way made an attempt. Neither in his heart wanted the attempt to succeed. The record of Johnson-Kennedy meetings shows, apart from the cabinet (which Kennedy infrequently attended), only two meetings in January, one before Kennedy left for the Far East and the other on his return; two meetings on February 7 about Cuba; a meeting on February 11 about civil rights; telephone calls on January 31, February 1, 4, 10 and 26 — and then, if the compilation is complete, no further calls until April 22, no further meetings until the late spring.[1]

I

Johnson, haunted by the circumstances that had brought him to power, was both sorrowing and vulnerable. He felt an immense need, psychological as well as political, for the Kennedys. But he could not control his diverse resentments — over popular comparisons with his predecessor, over glib journalistic comment about his alleged lack of culture and polish, over the failure of the Kennedys to fall in with his plans; at bottom over their reluctance to love him. "Lyndon Johnson had come on stage before a black curtain," wrote Liz Carpenter, Lady Bird Johnson's devoted press secretary, "and the Kennedys made no move to lift the darkness for him, or for the country. He had gone to the well for them so many times. 'Lyndon would like to take all the stars in the sky and string them on a necklace for Mrs. Kennedy,' his wife said softly in the dreadful days that followed Dallas."[2] He was "bitter," recalled Pierre Salinger, who stayed on for a few distracted months as Johnson's press secretary, "about his inability to get Mrs. Kennedy to come to the White House and participate in various ceremonies. . . . He couldn't understand, after all his kindnesses to her, why she wouldn't come down."[3]

Most of all he resented Robert Kennedy: "I'd given three years of loyal service to Jack Kennedy. During all that time I'd willingly stayed in the background. . . . Then Kennedy was killed and I became the custodian of his will. I became the President. But none of this seemed to register with Bobby Kennedy, who acted like *he* was the custodian of the Kennedy dream, some kind of rightful heir to the throne."[4] "That upstart's come too far and too fast," he told Eric Goldman, the Princeton historian who had a short and unhappy time in the Johnson White House. "He skipped the grades where you learn the rules of life. He never liked me, and that's nothing compared to what I think of him."[5] "If Bobby Kennedy's name came up even by accident," recalled Kenneth O'Donnell, who stayed on for a season in the White House, ". . . he'd launch into a tirade about what a son of a bitch Bobby Kennedy was. Ninety-nine percent of the things were untrue. And it'd get back to Bobby Kennedy, and Bobby'd say something about Lyndon Johnson. . . . These two men just didn't know each other; and they built up this picture of each other which was just incredible."[6]

For Kennedy, staying in the administration at all was a major effort. Yet he saw himself as yoked with Johnson in the execution of a legacy and the preservation of a party. Others might go; "but if I go," he told Murray Kempton, pausing and searching for a word, "if I should, uh, *desert*, that would be harmful."[7] In public he was as meticulously loyal to Johnson as Johnson had been to John Kennedy. But everything was different. "What makes me sad is that I see a problem . . . and I can't do anything about it. There was this time when if people had something and couldn't see my brother, they could always see me and I could pick up the phone and call him. . . . It's strange to think that you can't just pick up the phone."[8]

Nor could he, any more than Johnson, control his resentments. If the new President, enveloped in insecurity, was preternaturally sensitive to comparisons with his predecessor, Robert Kennedy, enveloped in grief, was preternaturally sensitive to slights to his brother. Walter Heller sent him a copy of a memorandum to Johnson saying that "under your budget and tax program" the net fiscal stimulus to the economy would be the greatest in peacetime history. Kennedy circled "your" and scribbled in the margin: "I thought this was Pres. K. program."[9] "I thought that an awful lot of things that were going on that President Kennedy did," he said in May, "[Johnson] was getting the credit for and wasn't saying enough about the fact that President Kennedy was responsible."[10] He encouraged others to stay — "If any one of us is in a position to keep him from blowing up Costa Rica or something," he told Richard Goodwin in March, "I guess he had better do it"[11] — but he became unhappy when they seemed to embrace the principle "the king is dead, long live the king."[12] "He really felt," McGeorge Bundy thought, "that . . . if you were fully in the Kennedy administration you had, a continuing allegiance that should, in certain circumstances, be more important to you than your allegiance to the existing President. And I couldn't feel that way."[13]

For once it was not primarily the case, so familiar in politics, of the staffs egging their principals on to combat. Johnson retained Bundy, O'Donnell, O'Brien, Dungan and brought Goodwin back to the White House. His own people were first-rate. Though Eric Goldman recalls unnamed Johnson men calling Kennedy "the perfect model of the liberal fascist" and the like,[14] my own clear impression is that Johnson's top assistants liked Robert Kennedy. I know he liked them. "I think all of them are good," he told John Bartlow Martin in May. ". . . Walter Jenkins seems like a good fellow. . . . Bill Moyers is a smart fellow and an honorable man. . . . George Reedy's fine. They're all scared, of course, of Lyndon. I guess [Jack] Valenti's a nice fellow. And they've all treated me very well. . . . The people that he has immediately around him look good."[15] It was finally a situation where the two principals just could not abide each other.

But they had to go through the motions. The new President would say to O'Donnell and O'Brien that, as Kennedy rendered it, "he thought I hated him and what he could do to get me to like him, and why did I dislike him, and whether he should have me over for a drink."[16] Johnson grew Byzantine in his endeavor to establish at least a record of benevolent intent. In March I met in New York with Alex Rose of the Liberal party, an avid dabbler in the higher politics. Ostensibly we were to discuss the vice presidential situation. But the meeting had in fact been instigated by Johnson, who had charged Rose with a message for me to pass on to Kennedy. The message, I wrote Kennedy (who was away from Washington), was "in effect, that President Johnson loves you, wants to be friends with you, that the door at the White House is always open to you."[17]

After Kennedy got back to Washington, he asked Bundy to find out whether the President wanted him to call, saying he did not wish to stand on ceremony. Bundy, Kennedy told me, "sounded out LBJ and told Bobby that the President did not want a call from him, that he was quite happy with the situation as it was, that he did not want to give [James] Reston the satisfaction of supposing that a summit meeting between the two was the result of his column of some days back, and that they could get together later." When I pressed Kennedy to call Johnson nonetheless, thinking at least to establish the record on his side, he said he felt he had gone as far as he could properly go — that, if he persisted in calling, after receiving Bundy's message, it would look as if he were trying to curry favor; "the White House reaction would be, 'That fellow will do anything to become Vice President.'"[18]

In retrospect this elaborate tiptoeing among grown men seems odd. But, though neither wanted to see the other, neither wanted the responsibility for a break. In the end nothing could have brought them together. What made the gulf ultimately impassable was a remark Johnson made that spring to Salinger — a remark so guileful a man could hardly have made without intending it to get back to Kennedy. "You know the worst thing that Johnson has said?" Kennedy said to me. He then incredulously repeated Johnson's remark: "When I was young in Texas, I used to know a cross-eyed boy. His eyes were crossed, and so was his character. . . . That was God's retribution for people who were bad and so you should be careful of cross-eyed people because God put his mark on them. . . . Sometimes I think that, when you remember the assassination of Trujillo and the assassination of Diem, what happened to Kennedy may have been divine retribution."*

* Author's journal, July 23, 1964. I have combined this story as I noted it down with a slightly more detailed version told by RFK to John Bartlow Martin; Martin interview, April 13, 1964, II, 57–58. Johnson's divine retribution thesis found new ammunition in March 1967 when he learned for the first time of the CIA's attempts to murder Castro (Senate Select Committee to Study Governmental Operations with respect to Intelligence Activities [Church committee] *Final Report*, bk. V, *The Investigation of the Assassination of President John F. Kennedy: Performance of the Intelligence Agencies*, 94 Cong., 2 Sess. [1976], 85–86). The Kennedy administration, Johnson later told Leo Janos of *Time*, "had been operating a damned Murder, Inc. in the Caribbean" (Leo Janos, "The Last Days of the President," *Atlantic Monthly*, July 1973). "Kennedy was trying to get Castro," he told Howard K. Smith of ABC, "but Castro got to him first" (Howard K. Smith, ABC News broadcast, June 24, 1976 [see also *Family Weekly*, September 12, 1976]). Johnson assumed that, because the CIA had tried to murder Castro during the Kennedy administration, Kennedy must have authorized the plots and therefore had only himself to blame for the Dallas horror. In fact, the same reasoning convicted Johnson as well. The CIA had continued to try to murder Castro during the first two years of Johnson's own administration; only he did not know this, because the CIA did not tell him any more than it had told Kennedy (Church committee, *Investigation of the Assassination*, 86). Castro himself in a communication to Senator George McGovern claimed ten assassination plots during the Johnson Presidency (McGovern press release, July 30, 1975).

II

The diverse resentments came to focus in 1964 on a specific question: the Vice Presidency, vacant now, to be filled at the Democratic convention in August.* Calculations on each side were filled with uncertainty. In some moods Johnson was not even sure he wished to run. Passing periodically from self-confidence through self-pity to self-doubt, worried about his health, worried, as he confided to James Reston in the spring, that the nation was "not far enough from Appomattox" to accept him,[19] he had, in his own words, "decidedly mixed feelings about whether I wanted to seek a four-year term."[20] Salinger later declared himself "absolutely convinced that, if [Johnson] thought he could have left the field without yielding the nomination to Robert Kennedy, whom he hated, he would not have run."†

In other moods, however, Johnson was determined to run and win. To O'Donnell, who never heard him express the slightest doubt about running, he candidly defined his intention on the Vice Presidency three weeks after John Kennedy's death. "I don't want history to say I was elected to this office because I had Bobby on the ticket with me," he said. "But I'll take him if I need him."[21] This was as far as he would go toward admitting the possibility; and he was talking, of course, to a Kennedy friend. His brother Sam Houston Johnson thought that the President never "for a single moment" considered Kennedy as his running mate. "The reasons were quite obvious: a) Lyndon hated Bobby. b) Bobby hated Lyndon."[22]

As for Kennedy, he assumed Johnson would run and intermittently wondered whether he ought to seek the second place for himself. As early as December 3, 1963, Joseph Dolan told him that the Democratic National Committeewoman from Michigan and, oddly, Senator Thomas J. Dodd of Connecticut favored him for Vice President.[23] A couple of days later Kennedy asked me what I thought. "My first reaction," I noted, "was negative, though, when he asked me why, I found it hard to give clear reasons. I think, first, that it seems to me a little too artificial and calculated; second, that Bobby should develop his own independent political base; and third, that LBJ might well prefer Shriver on the ground that Shriver would bring along Bobby's

---

* The Twenty-fifth Amendment, empowering the President to fill a vice presidential vacancy, was not ratified until February 10, 1967.

† Salinger added, "I had the very profound impression that, the moment Lyndon became President, he regretted it" (Pierre Salinger, *Je suis un Américain* [Paris, 1975], 294–295).

friends without bringing along his enemies. Bobby added that he did not like the idea of taking a job which was really based on the premise of waiting around for someone to die."[24]

In January the Democratic organization in Buffalo endorsed him for Vice President. "I have received letters from all over the country approving our action," Peter Crotty, the Buffalo leader, informed Kennedy in mid-February. ". . . I don't think there is any President who would place his own election in jeopardy by ignoring the will of the country."[25] Kennedy friends thought that the New Hampshire primary in March might further affirm the country's will. William Dunfey, the New Hampshire national committeeman, encouraged a Kennedy-for-Vice-President write-in, but did so quietly, in the New Hampshire manner: "People kind of ran ads on their own and that kind of thing. . . . Once it got moving, Bob Kennedy called me once or twice, just, you know, how did I look at this thing and was it going to be something that he would get seriously embarrassed on." Dunfey told him there wasn't much he could do about it.[26]

At this point the incorrigible Paul Corbin, still hanging on at the national committee, appeared in New Hampshire. It was impossible for anyone who did not know Corbin to believe it; but, as Edwin Guthman said later, "the truth was that Corbin was acting on his own."[27] "He had a strange way about him," John Seigenthaler, who liked Corbin, said. "When he made up his mind to do something, he'd do it."[28] Dolan, who disliked Corbin, described him as "so determined to do everything that he sincerely thought was in Robert Kennedy's interest that he would sweep all else aside," including Robert Kennedy's own wishes in the matter.[29]

"Corbin in New Hampshire," said Dolan. "That's like trying to conceal Mount Everest in Manhattan Island." After ascertaining that Kennedy had not sent him, Dunfey chased him out of the state.[30] Dunfey's annoyance was nothing, however, compared to Lyndon Johnson's rage. The President had no doubt that Kennedy had sent Corbin. Sam Houston Johnson was surely registering his brother's skepticism when he wrote, "I have never in my life seen a campaign that couldn't be stopped by a candidate who didn't want it." Lyndon Johnson himself told Kennedy after a cabinet meeting that he wanted Corbin out of both New Hampshire and the national committee. According to O'Donnell, Johnson told Kennedy, "If he is such a good fellow, you pay him. He's around town knocking my head off . . . and has been for three years, and I never met the bum in my life. Why should I have him on my payroll?" Kennedy said he did not know that Corbin was in New Hampshire, adding, "He was loyal to President Kennedy. He'll be loyal to you." Johnson responded, "I know who he's loyal to. Get him out of there." Kennedy recalled it as "a bitter, mean conversation. The meanest tone that I heard anybody take. . . . I said . . . I don't want to have that kind of conversation with you." [31]

On February 10 Johnson called Kennedy to say that Corbin had been fired; he wanted no one at the committee pledged to a particular vice presidential candidate. He had sent Kennedy to Indonesia, he explained, only because he "had to keep things equal" among those under consideration for the second place. Kennedy, who thought he had been sent to Indonesia because he had proven he could negotiate with Sukarno, said curtly, "Don't ever do a favor for me again!" The conversation ended. Kennedy stared gloomily out the window for a few moments, then said to Guthman, "I'll tell you one thing, this relationship can't last much longer." [32]

As the voting day approached, it seemed that Kennedy might actually outpoll Johnson. This would have been embarrassing all around. A few days before the primary the Attorney General put out a disclaimer. Early drafts contained statements that he was not seeking the vice presidential nomination. He regularly crossed them out. The final version said simply that the President "should be free to select his own running mate" and therefore Kennedy wanted to discourage efforts on his own behalf. [33] Johnson received 29,600 votes, Kennedy 25,900.

III

"Take it! Take it!" General Douglas MacArthur told him. "He won't live. He gambled on your brother and won. You gamble on him and you'll win." Ethel and Jacqueline urged him to go ahead. [34] Political leaders besought him to signal his availability. Walter Reuther told me, as I noted at the time, that he was "waiting for Bobby to say the word — that he would not move until he knew what Bobby wanted. I mentioned Hubert's claims. Walter said that he loved Hubert but that he could not forget Los Angeles [in 1960]. 'Hubert had a secret deal with Lyndon.'" [35] In April the Gallup poll reported Kennedy the vice presidential choice of 47 percent of Democratic voters as against 18 percent for Adlai Stevenson and 10 percent for Humphrey. [36]

His own motives were indecipherable. No one could quite tell how much, or even if, he wanted the nomination; or whether he was simply demonstrating popularity and keeping options open. The argument *for* was that, as Vice President, he might assure the Kennedy outlook and constituency a role in national policy; in any event, the Vice Presidency had the political advantage, not to be prudently relinquished to someone else, of establishing its occupant as a quasi-heir apparent. The argument *against* lay in the singular stupidity of the job itself. As the historian in his circle, I should have warned far more stringently than I did that the idea was hopeless; but I had not yet studied the Vice Presidency as carefully as I would later.[37] Nonetheless, Kennedy, having observed Lyndon Johnson, well understood the impotence of the Vice Presidency. And on top of the incurable deficiencies of the office, there were the extraordinary demands that Johnson *redivivus* was bound to make on its holder. "I heard Johnson say one time," Congressman Emanuel Celler once said, "that he wanted men around him who were loyal enough to kiss his ass in Macy's window and say it smelled like a rose."[38] Johnson's 1964 aphorism on the Vice Presidency was widely repeated in Washington that spring: "Whoever he is, I want his pecker to be in my pocket."

Searching through the oral histories, one finds that Nicholas Katzenbach, McGeorge Bundy, Richard Goodwin, Robert McNamara, Donald Wilson, Joseph Alsop, Rowland Evans, Joseph Kraft, Ben Bradlee, supposed Robert Kennedy wanted the nomination. Kenneth O'Donnell, Charles Spalding and Charles Bartlett thought he did not. I was never sure either way. The most extended statement I have found by Kennedy himself came in an oral history interview with John Bartlow Martin in mid-May. The one thing Johnson did not want, Kennedy said, "is me as Vice President. . . . I think he's hysterical about how he's going to try to avoid having me. . . . I'm just trying to make up my mind what I'm going to do." If his nomination were forced on Johnson, "it would be an unpleasant relationship. I would lose all ability to ever take any independent position. . . . Johnson's explained quite clearly that it's not the Democratic party any more, it's an all-American party, and the businessmen like it. All the people who were opposed to the President like it. I don't like it very much."

Martin suggested that Kennedy, with his powerful constituency, would be in a different position from other Vice Presidents. "I don't think you can have any influence," Kennedy repeated. "Lyndon Johnson didn't have any influence. . . . He's not going to have to pay any attention to me whatsoever." Martin said Johnson would have to pay attention; he could not afford to break with the Kennedys. Kennedy responded that Johnson was "not doing anything for the Alliance for Progress and he's not paying proper attention to Panama or Brazil. . . . If I was in the United States Senate I would have raised a fuss about Panama." Martin said there were a hundred senators. Kennedy: "Yeah, but I'm not just a senator. [Suppose] I'm senator from New York and I'm head of the Kennedy wing of the Democratic party." He mused on. "I think it's possible to be Vice President. . . . But he's not going to pay any attention to me. I can't go out and speak against . . . what he's doing in South America, what he's doing in Southeast Asia, as Vice President. . . . I make one speech like that and you've broken the — " Martin, interrupting, thought he could do it quietly, short of a speech. Kennedy: "If I was Vice President and that Cuba thing came up again, for instance, I'd be running up to the Foreign Relations Committee and saying . . . why don't they [do something about it]. . . . That's a disloyal operation."

And there was Johnson himself. Charles Spalding thought the MacArthur argument just too "macabre" for Kennedy: "I mean to sit and figure out whether Johnson was going to live or not on top of everything else, that was just too morbid."[39] Johnson alive posed problems almost as morbid: "The fact," as Kennedy put it, "that he's able to eat people up, and even people who are considered rather strong figures — I mean Mac Bundy or Bob McNamara. . . . He's mean, bitter, vicious — [an] animal in many ways. . . . I think his reactions on a lot of things are correct. . . . but I think he's got this other side of him and his relationship with human beings which makes it difficult unless you want to 'kiss his behind' all the time. That's what Bob McNamara suggested to me a couple of weeks ago if I wanted to get along."[40]

IV

On June 19 he was in Hyannis Port. That night Edward Kennedy, flying to a political dinner in Springfield, Massachusetts, crashed in thunderstorms over the Connecticut Valley. The pilot and a Kennedy aide were killed. Edward Kennedy suffered horribly — damaged vertebrae in his lower back, a punctured lung, broken ribs, uncontrollable internal hemorrhaging. Robert Kennedy raced across the state by automobile to the hospital. Stirring from unconsciousness, Ted Kennedy made out his brother through the mist and said, "Is it true that you are ruthless?"[41]

In relief Robert Kennedy and Walter Sheridan left the hospital for a walk in the summer night. "We just lay down in the grass," Sheridan said later, "and he said, 'Somebody up there doesn't like me.' Then he asked me if I thought he ought to go for the Senate or go for the Vice Presidency. I told him I thought he ought to go for the Vice Presidency."[42] The next morning someone wondered whether the Kennedys now intended to retire from politics. "The Kennedys intend to stay in public life," Robert Kennedy said crisply. "Good luck is something you make and bad luck is something you endure." "Is it ever going to end for you people?" "I was just thinking out there — if my mother hadn't had any more children after her first four, she would have nothing now. . . . I guess the only reason we've survived is that there are more of us than there is trouble."[43] William Shannon later asked whether the family disasters were affecting his religious faith. "No, they do not," he said; then, with that smile, "Of course, we do occasionally think that someone up in heaven is out to lunch when he ought to be attending to business."[44]

A week later, Robert and Ethel were on their way to Berlin at Willy Brandt's invitation to dedicate a memorial to John Kennedy. Mrs. Donald Wilson, who was with them, described what they did on disembarkation: "In rapid fire order, [Kennedy] reviewed an honor guard, delivered an arrival statement, visited a school, inspected a factory, held a press conference, went to a luncheon for 300 only long enough to eat, rode in a tank, spoke to 50,000 at the dedication of the John F. Kennedy Platz, received an honorary degree, made a second major address and attended a reception and buffet dinner for 700."[45] They asked him inevitably about German reunification. "We will stand and fight for principle," he said. ". . . But we also know what war is going to bring. . . . I think the best opportunity of breaking down the artificial barriers . . . is by a greater understanding between the peoples that live in the eastern part of Europe and the Soviet Union, and those of us who live elsewhere."[46] In this spirit the Kennedy party pressed on to Poland.

The Polish government had declined to invite Kennedy, consenting only to a "nonofficial" visit. Kennedy's particular interest, as usual, was to meet the young. "Disaffection of youth," Ambassador John Moors Cabot warned Washington, "is not just headache but nightmare to Polish Communist leadership."[47] Once Kennedy arrived, the government did its best to keep his presence secret. But word got out. When the Kennedys went to mass the day after their arrival, a crowd surged around the Cabot limousine. After a moment, Kennedy lifted Ethel to the car's roof, climbed up himself and began waving. The ambassador, an able diplomat of the old school, called to a Kennedy aide: "I say there, would you tell the Attorney General that the roof is caving in."[48] "We had an awful time fixing it," he said afterward. "This is the way we always come home from mass," said Kennedy.[49]

The Poles were determined to frustrate the intruders. Bringing presents to an orphanage, Robert and Ethel discovered that the children had been spirited away. That evening Kennedy lectured

Foreign Minister Adam Rapacki about the incident. "It sounds as if
we took them all to Siberia," Rapacki grumbled later.[50] The next
morning they went to Cracow, where another throng besieged their
car in the Cloth Market Square. The Poles sang "Sto Lat" — "May
you live a hundred years" — and the Kennedys sang back in hurried
improvisation "When Polish Eyes Are Smiling." Asked at a press
conference if he planned to run for President, he said, "No . . . I
think I'll run for mayor of Cracow." An hour late in getting back
for an embassy dinner, Kennedy climbed on the cartop and spoke to
still another crowd in front of Cabot's residence. Finally he said, "I
have to go into dinner now. Would you like to come in with me?"
When they shouted yes, he pretended to consult with Cabot over the
wall and then said, "The ambassador says you can't come" — a
lighthearted moment that amused Kennedy more than it did Cabot.[51]

Jozef Winiewicz, the deputy foreign minister and ranking guest,
was furious. "Now, Mr. Kennedy," he said sarcastically, "you've addressed the crowd outside. Why don't you get on this table and
address us?"[52] "He was very agitated," Kennedy noted later, "that
I had been talking frequently to groups of people." Gomulka, Winiewicz said, would never have done anything like that. Kennedy said,
"Maybe that's what's wrong with Gomulka."[53] The Poles especially
resented Kennedy's call on Stefan Cardinal Wyszynski. Wyszynski
told Kennedy that "the best thing that had happened to the Catholic
Church in Poland was that it had been deprived of its wealth. This
had brought the priests and bishops much closer to the people. The
government officials . . . have become the capitalists and the church
has become the proletariat." Kennedy: "Without question, he is the
most impressive Catholic clergyman I have ever met."[54]

The Polish newspaper *Polityka* described Kennedy as an "indefatigable" politician "resorting to tricks used in the political life of the
United States, where popularity is the most important thing."[55] The
ambassador too: "Frankly, I was very annoyed about the whole business. It had its good point. Kennedy was followed by huge crowds
wherever he went, but he got the Polish government practically livid
with some of the things he did. And he got the embassy so mad that
it could scarcely sputter."[56] Still, Cabot reported the visit to Washington as an "undoubted success" though "it would be naive to underestimate anger and shock of Polish leadership."[57] Another embassy officer called it "an ill-prepared and even dangerous exercise
which came off very well" largely because of Kennedy's "skill in public
relations, grasp of the larger picture, and receptivity to advice" and
because of his "Irish luck."[58] When Khrushchev visited Warsaw three
weeks later, Cabot, noting the "pitiful" turnout on the streets in spite

of vast government publicity, told the department, "The contrast to the welcome given Attorney General Kennedy was particularly striking."[59]

V

"Every day," Lyndon Johnson recalled, "as soon as I opened the paper or turned on the television there was something about Bobby Kennedy; there was some person or group talking about what a great Vice President he'd make. . . . It just didn't seem fair. I'd waited my turn. Bobby should've waited for his. But he and the Kennedy people wanted it now. . . . I simply couldn't let it happen. With Bobby on the ticket, I'd never know if I could be elected on my own."[60]

"My brother's only concern," wrote Sam Houston Johnson, "was how and when to squelch the campaign."[61] The President explored diversionary action — mollifying the Kennedys and the Catholic vote, for example, by putting Sargent Shriver on the ticket. Moyers reported to the President that Shriver was willing and that Bobby would not complain if his brother-in-law were Johnson's running mate. O'Donnell commented, "The hell he wouldn't." In due course, Shriver gave way to Eugene McCarthy in Johnson's shell game. O'Donnell said, "There's only one Catholic in the country. That's Robert Kennedy. If you're just looking for a substitute, [you're] saying we're a bunch of cattle who run around and vote for people because of their religion." Besides, was it defensible to prefer the second senator from Minnesota, however witty and estimable a man, over the first? The Kennedy people believed that, if it was not to be Bobby, the nomination should go to Hubert Humphrey.[62]

As for Kennedy, Ben Bradlee, after going with him to Kansas City in June, had no doubt he wanted the nomination and so wrote in *Newsweek*. (They visited a home for old people. Kennedy went over to a dying woman, rubbed her hand, talked softly to her. "It just brought tears to my eyes," the hard-bitten Bradlee said later. "There was absolutely no reason for him to do it. . . . It was the most poignant thing I've ever seen in my life, and I don't move easy.")[63] In *Newsweek* Bradlee also quoted Kennedy as calling himself "the last man in the world" Johnson would want — "because my name is Kennedy, because he wants a Johnson Administration with no Kennedys in it, because we travel different paths, because I suppose some businessmen would object, and because I'd cost them a few votes in the South."[64]

By the time Kennedy returned from Europe in early July, Johnson was playing a vast and enjoyable game with everybody. In mid-July he was still talking to O'Donnell and O'Brien about putting a Catholic on the ticket.[65] He sent James Rowe to sound out Humphrey and then tormented Humphrey by discussing possible candidates in his presence without mentioning him.[66] Concerned no doubt to head off an independent Kennedy movement, he assured Salinger and Galbraith that Kennedy was very much under consideration.[67]

But Goldwater's victory over Rockefeller in the California Republican primary in early June had extinguished any need Johnson might ever have felt for Kennedy. The prospective nomination of the conservative hero gave Johnson a free ride in the north, where Kennedy was strong, and created difficulties in the south and southwest, where Kennedy would be a handicap. "The antagonism through the South against Bobby Kennedy," Jack Tarver of the *Atlanta Journal* and *Constitution* wrote Johnson, "may be unreasoning and emotional but, believe me, it is real. With him on the ticket, there is serious doubt you could carry a single state south of Ohio or east of Texas." There would be no problem, Tarver added, about Humphrey or McCarthy.[68] In these months, moreover, Johnson had solidly established himself as President in his own right. By June his polls showed that no running mate, Kennedy included, would make more than 2 percent difference to his presidential vote. "Look't here," he said to his brother one night, waving a poll across his dinner coffee. "I don't need that little runt to win. I can take anybody I damn please."[69]

All this was evident enough to Kennedy. In late July I noted: "Bobby says that he feels wholly out of it, that he is certain LBJ does not want the sense of 'a cross little fellow looking over his shoulder,' etc. He was very hostile to the idea of McCarthy and seemed distinctly to prefer Hubert."[70] One night he summoned his brother, O'Donnell, O'Brien, Fred Dutton and Stephen Smith to Hickory Hill. Goldwater was going to be the nominee, he told them, and he had decided to forget about the Vice Presidency, resign from the cabinet and run for the Senate in New York. Edward Kennedy and Smith urged him to move at once. O'Donnell begged him to wait; once Kennedy was out of the running, liberal Democrats would lose their power to force Johnson to accept Humphrey. Kennedy agreed to delay his New York announcement — "one of the finest and most unselfish decisions," O'Donnell thought, "of Bobby's controversial career." As they left, he called after O'Donnell in the darkness, "When they start that 'ruthless brat' stuff, O'Donnell, you'd better be there to bleed along with me."[71]

O'Donnell set to work with Reuther and others to build support for Humphrey. On July 15 Goldwater won the Republican nomination. A few days later Johnson asked Kennedy to come to the White House. "He's going to tell me I'm not going to be Vice President," Kennedy said when he put down the phone. "I wondered when he'd get around to it." [72]

VI

The summit meeting took place on July 29 at 1 P.M. "He's going to record every word you say," O'Donnell warned Kennedy. [73] Instead of taking his visitor in the usual manner to the sofa by the fireplace, Johnson sat formally at the presidential desk. A memorandum written by Clark Clifford lay in front of him. Reading aloud, Johnson said that, in view of Goldwater's nomination and the need for a running mate with appeal to southern and border states, he had concluded "that it would be unwise for our party . . . to select you." The memorandum went on to mention Kennedy's "unique and promising future," to solicit his aid in the campaign and to raise the possibility of his succeeding Adlai Stevenson at the United Nations. According to Johnson, the discussion was "frank . . . but there was no unpleasantness." He remembered Kennedy saying at the end: "I'm sorry that you've reached this conclusion, because I think I could have been of help to you." [74]

Kennedy dictated his own account:

> I sat down in the chair next to his desk. He looked off from me to the wall and to the floor and then said, over a three-minute span, that he was thinking a good deal about the Vice Presidency but he wanted a Vice President who could help the country, help the Party, and be of assistance to him. He repeated these thoughts in different words and concluded them by saying that that person wasn't me. I said that was fine, that I would be glad to campaign for the ticket and whatever he felt was helpful I would be glad to do.
>
> I stayed for about 45 minutes and during the course of the conversation the following points were made: first, that he thought that I had high qualifications to be President; that he wanted to work toward that end [and] although he could not commit himself definitely at that time, he wanted to help me in whatever way he could. He wanted me to know that if I wished to go around the country and speak he would never be jealous of me. . . . If I was interested in another Cabinet position or in going to Paris, London or Moscow, he would be glad to help. He could not give me a job that someone else held right now but he was sure something could be arranged. I said that I wasn't interested in any other position. He said that if he were me he would stay over in the Department of Justice. He said I had an outstanding staff. He thought I should remain there.
>
> He said in comparison to the staff he had at the White House the

group I had selected were unusually competent. The people that contributed in the White House, he said, were the group that was selected by President Kennedy. He said he really could not count on his own people — Jenkins, Valenti, Reedy. He said Bill Moyers was good but his most useful function was rewriting what other people did. I was shocked to hear him being so critical to me of people who had been so loyal to him. Just as much as anything else, it convinced me that I could not have worked closely with him. . . .

He spoke to me about running the campaign. He said not much had been done in the organizational part and asked me if I could run it from the Department of Justice. I said I had made an arrangement with my brother that I would not participate while I was Attorney General. I had done that for two reasons: Number 1, because I felt an Attorney General should be above active political life, and secondly, because I had thought, coming just out of the campaign, I would be doing nothing else once I got started. Therefore, I said, I was reluctant to take on both of these jobs. . . .

I then brought up the subject of Bobby Baker. I said that we had an active case and we had to make a judgment how to proceed. I said I felt he should be advised and I would like to have someone's outside judgment on what we proposed to do. I suggested Jim Rowe. He countered that with Abe Fortas and Clark Clifford. I said I didn't think Abe Fortas would be proper because he had been Bobby Baker's lawyer. He said these are the two people in his judgment on whom he really relied.

He brought up the Fred Korth case. He said he was against President Kennedy's action in dismissing Fred Korth. I said I knew that because I was on the other side. But Clark Clifford was also for retaining Korth but after reviewing the material (and also, incidentally, arguing with us at the Department of Justice) he came to the conclusion that Korth should be relieved. He said Clark finally convinced him of this also.

He then talked approximately 10 or 15 minutes about the fact that Bobby Baker got into all this difficulty after he, Lyndon, had left the Senate. . . . He said God must have been watching over him because he did not have financial dealings with Bobby. He could very well have done so because if Bobby had asked him to loan him $10,000 for the purchase of some stock he certainly would have done so.* He said the only business deal that he had had with Bobby Baker was in connection with the purchase of some insurance. The insurance company then placed $1500 (I think that was the figure) worth of advertising with his television station. He said this really didn't mean anything because the competitor company was willing to place twice as much. He said Bobby Baker went with power and when he left the Senate he latched on to other people. In any case, he said, if there was any corruption in this case it was the Republicans.

He then spent 10 or 15 minutes citing to me the facts in those cases.

---

* According to Baker, he did ask Johnson for a loan. Johnson thereupon called Senator Robert Kerr and said, "*Our* boy Bobby's in trouble and he needs *your* help." "I got nothing directly from Lyndon Johnson," Baker wrote later. ". . . LBJ simply was not a man to share" (Bobby Baker with Larry King, *Wheeling and Dealing: Confessions of a Capitol Hill Operator* [New York, 1978], 67–68).

[Libelous material deleted.] It was obvious . . . that he was receiving detailed reports from the FBI on the activities of several of the Congressmen and Senators. . . . Incidentally, the whole conversation with Johnson was recorded. I saw the buttons on while I was sitting talking to him and later on in a conversation with Kenny and Larry he inadvertently and virtually admitted as much.[75]

Both principals were calm when they parted;[76] both, I suspect, relieved. Kennedy seemed in excellent humor when he got back to his office. He told his associates what had happened. There was a painful silence. "Aw, what the hell," Kennedy finally said with a laugh, "let's go form our own country."[77] Johnson, his creative imagination already at work, informed O'Donnell that Kennedy had offered to manage his campaign. O'Donnell, who had lunched with Kennedy directly after the meeting, said Kennedy had told him just the opposite.[78] One wondered whether Johnson's offer was not repayment, conscious or unconscious, for Kennedy's suggestion in the heat of Los Angeles four years before that Johnson forgo the Vice Presidency and become chairman of the national committee.

A question remained. "The fact that I had not volunteered to say that I would withdraw," Kennedy noted, "was obviously going to cause him some difficulty. How he could say that he had gotten rid of me was his greatest problem."[79] Johnson did not want to take public responsibility for vetoing Kennedy. His solution was to direct Bundy to get Kennedy to announce that he had withdrawn on his own. The President used Bundy partly because Bundy had served from time to time as a liaison with Kennedy, partly as a brutal and characteristic Johnson loyalty test. Kennedy vigorously rejected the proposal, telling Bundy "first, it wasn't true; secondly . . . a large part of the reason that I was at all interested in the Vice Presidency was because of the interest of others. If I suddenly withdraw my name without discussing it with them it would be impossible for them to understand. Mac disagreed and said this would be very helpful to Johnson and helpful to me." Bundy honestly believed this; Kennedy as honestly believed that Bundy's intervention was an act of inexplicable disloyalty.[80] For a moment an old association became embittered. (A month later Kennedy wrote Bundy: "These have been a difficult nine months for all of us. But I wanted you to know, as I am sure you must, that you were a great help to my brother, made a major difference for him, and for your friendship to me, which, even with the misunderstandings, I know was genuine, I am most grateful.")[81]

The day after the Bundy effort, Johnson "told Kenny that he had to put out a statement by six o'clock that afternoon saying I was out but what about a statement from me saying that I had asked my name to be withdrawn. Kenny said he couldn't see how I could say that. Then he said, then why doesn't the Attorney General make up a statement? . . . I made the point that he could put out any statement that he wanted as long as it was in accordance with the facts."[82] That evening Johnson announced that he was excluding from consideration all cabinet members and all who met regularly with the cabinet — the last clause disposing of Adlai Stevenson.

The day following, Johnson had Tom Wicker of the *New York Times,* Edward T. Folliard of the *Washington Post* and Douglas Kiker of the *New York Herald Tribune* to luncheon. He provided an elaborate and caustic account of the Kennedy meeting. He had watched Kennedy, he said, "like a hawk watching chickens," and claimed that Kennedy had given a vast gulp at the bad news, "his Adam's apple going up and down like a Yo-yo." Johnson, who had an unexpected gift for mimicry, thereupon rendered his impression of the Kennedy gulp, "like a fat fish," his admiring brother said, "pulling in a mouthful of air." Within forty-eight hours, word of Johnson's table talk was all over Washington.[83] When Kennedy protested, Johnson denied he had discussed their meeting with anyone. Kennedy said the President could not be telling the truth. Johnson said he would check his schedule to see whether there might have been some conversation he had forgotten. "He tells so many lies," Kennedy noted on August 6, "that he convinces himself after a while he's telling the truth. He just doesn't recognize truth or falsehood."[84]

I spent that weekend at Hyannis Port. Kennedy was, I thought, "very matter-of-fact" about the Johnson meeting and "quite funny" about the subsequent stages. He set forth a number of possible comments if reporters asked about the exclusion of the cabinet: "I am sorry that I had to take so many nice fellows down with me" or "It seems to me premature to stop the vice presidential boom for Dean Rusk" or "I swear to the best of my knowledge I am not now and have never been a member of the cabinet on the ground that it might tend to eliminate me."[85]

VII

For Johnson the crisis was still not over. "He was afraid," Clark Clifford said later, "that at the convention in Atlantic City, because they had planned a tribute to John F. Kennedy and that Bobby was to deliver the tribute, that he might very well stampede the convention and end up being . . . the vice presidential nominee. . . . We talked about it at very considerable length."[86]

The tribute to the fallen President was originally scheduled for

Tuesday night. Johnson had it moved back to Thursday, by which time the nominations would be completed. He took other precautions, the most extraordinary of which was to send Cartha DeLoach and an FBI team of thirty snoops and wiretappers to Atlantic City. The ostensible purpose was to gather intelligence "concerning matters of strife, violence, etc." The real purpose, according to William Sullivan of the FBI, was to gather political information useful to President Johnson, "particularly in bottling up Robert Kennedy — that is, in reporting on the activities of Bobby Kennedy."*[87] The FBI's senior resident agent in Atlantic City later confirmed to the Watergate Committee that "Robert Kennedy's activities were of special interest, including his contacts with [Martin Luther] King."[88] DeLoach instructed him, the agent testified, that the presence of the FBI squad was not to be disclosed to the Secret Service and especially not to the Attorney General.[89]

When Bill Barry of the Bureau drove Kennedy, as he customarily did, to the airport in New York, Kennedy asked him to come along to Atlantic City. "Ethel's pregnant and I have nobody in Atlantic City, and I really want you to go." Although Atlantic City lay outside his division, Barry went nevertheless. On arrival DeLoach told him, "There's some thought that Attorney General Kennedy might try to stampede the convention.... If he does that and he's got an FBI agent by his side, the President will not be too happy with the FBI organization, so you're to immediately leave Atlantic City."[90]

"By means of informant coverage, by the use of various confidential techniques, by infiltration of key groups through use of undercover agents, and through utilization of agents using appropriate cover as reporters," DeLoach later boasted to his superiors, "we were able to keep the White House fully apprised of all major developments during the Convention's course." The "confidential techniques" included tapping Dr. King's hotel room and bugging the SNCC and CORE headquarters. "We disseminated 44 pages of intelligence data to Walter Jenkins," DeLoach continued. "... I kept Jenkins and Moyers constantly advised by telephone of minute to minute developments." When Moyers sent a letter of thanks, DeLoach replied, "Dear 'Bishop': ... I'm certainly glad that we were able to come through with vital tidbits from time to time which were

---

* See also William C. Sullivan to J. Edgar Hoover, October 6, 1971: "I saw clearly at last that the FBI, always presented to the American public as non-political ... was just the contrary. It was immersed in politics and even went so far as to conduct purely political investigations and inquiries. At times it seemed that when we were not asked to perform politically we sought opportunities to do so. I was so concerned about this under Mr. Johnson's administration that I wrote you a letter ... and urged that the FBI not be used politically." I thank Mr. Sullivan for making this letter available to me.

of assistance to you and Walter." Johnson ordered Jenkins to tell
Hoover that "the job the Bureau had done in Atlantic City was one
of the finest the President had ever seen."[91]

This unprecedented effort — no previous chief executive had ever
used the FBI at a party convention — was hardly necessary. Kennedy
had no intention of storming the convention. It is true that his
appearance excited the emotions of delegates and the dreams of
friends. "The applause hit like thunder," said John Seigenthaler
after Kennedy's visit to the West Virginia delegation. ". . . I knew
that the electricity was there, and it could have been put together."
People around him discussed a coup, "but he really didn't want to do
it." He was there to thank Democrats for supporting his brother, not
to win them for himself. "If something had happened, that would
have been fine, but it had to happen in the heart of Lyndon
Johnson."[92]

The heart of Lyndon Johnson remained, understandably enough,
obdurate. With no signs of a Kennedy insurrection, Johnson sustained his vice presidential tease to the end. On the second day of
the convention, he summoned Humphrey — along, improbably, with
Thomas J. Dodd — to Washington. After keeping poor Humphrey
waiting in a limousine outside the south entrance to the White House,
Johnson finally relented and had him in. "Most Presidents and Vice
Presidents," he told Humphrey, "just don't hit it off. . . . This is like
a marriage with no chance of divorce. I need complete and unswerving loyalty. . . . Do you think that you're that man?" "I think I am,"
replied Humphrey, hopeful and guileless.[93] "If you didn't know that
I had you picked a month ago," Johnson cruelly concluded, "maybe
you haven't got brains enough to be the Vice President."[94]

Robert Kennedy was glad about Humphrey. But his mind was on
other matters. On Thursday night he was to introduce a film about
his brother. Seigenthaler accompanied him to the auditorium. They
were led through the darkness into a small dressing room under the
platform. Kennedy said, "Would you check on the program? We
can't hear anything back here. . . . I think Lyndon may just have put
us back here with orders to forget us. They'll probably let us out
day after tomorrow." He pulled out his manuscript and started
reading it once again. Jackie, he told Seigenthaler, had given him a
quotation from *Romeo and Juliet*. (Had either consciously noted the
thrust of the last line? I never knew.)

Soon they were taken to a runway behind the platform where other
dignitaries had already assembled. "He was really sort of a bastard
at the family reunion," Seigenthaler thought. "Nobody was enthusiastic about coming and saying 'hello.' Lyndon had picked the

whole goddamn platform.... [James A.] Farley, for once in his life, was decent and tried to make small conversation." Kennedy looked through his manuscript again, took out a pencil, made small changes. Finally Senator Henry Jackson, who was presiding, motioned him to the rostrum. "When Scoop introduced him, it hit. I mean, it really hit.... It just went on and on and on."[95]

I stood on the floor in the midst of the thunderous ovation. I had never seen anything like it. Ordinarily an organ in the background controls the pandemonium of a convention. This time they stopped the organ after a moment or so. But the demonstration roared on, reaching a new intensity every time that Robert Kennedy, standing with a wistful half-smile on his face, tried to bring it to an end. As Kennedy once more raised his hand to still the uproar, Jackson whispered to him, "Let it go on.... Just let them do it, Bob.... Let them get it out of their system."[96]

He repressed his tears. Many in the audience did not. He seemed slight, almost frail, as the crowd screamed itself hoarse. It went on for twenty-two minutes. Finally he began to speak. At the end, the quotation:

> When he shall die
> Take him and cut him out in little stars
> And he will make the face of heaven so fine
> That all the world will be in love with night,
> And pay no worship to the garish sun.

29

To the Senate

O N AUGUST 22, 1964, two days before the Atlantic City convention, Robert Kennedy announced his candidacy for the Democratic senatorial nomination in New York. "I think," he had scribbled
to a newspaperman a few days before, "I shall respond to the spontaneous draft of my brother-in-law." [1] The possibility had been in his
mind since early spring. The New York residential requirement was
nominal, and Stephen Smith had begun quiet contingency preparations for Robert's entry. Then, after Edward Kennedy's plane crash
in June, Robert briefly considered leaving politics altogether. His
reception later that month in Europe persuaded him he could not
thus abandon John Kennedy's legacy.

I

At the end of July Averell Harriman, David Hackett and I were at
Hyannis Port for the weekend. The most recent New York poll had
shown Kennedy leading Kenneth Keating, the Republican incumbent, by a healthy 52 to 35 percent and running significantly ahead
of the Democratic alternatives — Adlai Stevenson, Robert F. Wagner,
serving his third term as mayor of New York City, and Samuel
Stratton, an upstate congressman. [2]

The Senate plainly tempted him. "He has a great desire," I noted,

to be independent of Johnson and the administration and to have his
own base. . . . He thinks that he might be able to organize a bloc of New
Frontier senators — Joe Tydings, George McGovern, Birch Bayh, Pierre
Salinger [recently appointed to fill a Senate vacancy from California]
and some others — and have some impact on the administration. But
he is quite clear in his mind that he does not want to go into New York
unless there is a summons by a broad and significant group within the
party. Averell is prepared to lead the way on this. The real question
is Alex Rose and the Liberals. [3]

Rose was my assignment. We had breakfast a couple of days later at his usual table in the old Astor Hotel on Times Square. Kennedy had problems, Rose said: mistrust among reform Democrats; prejudice among middle-class Jews; concern among the regulars that a pro-Kennedy movement would have an anti-Johnson thrust. So long as Stevenson was in the picture, Rose added, he could not commit himself to Kennedy. After a long talk, he gave me to understand in his sibylline fashion that he thought he might manage to bring the Liberal party around.[4] Harriman, Stephen Smith and Jack English, the Democratic leader in Nassau County, were meanwhile tackling the regulars. (One county leader told English: "I'll be for Kennedy if you give me one promise: if Corbin doesn't come into the state.")[5] Ronnie Eldridge and Albert Blumenthal, West Side reform leaders, along with William Haddad and others, were working on the reform Democrats.

Wagner turned out to be more of a question than Rose. The mayor was not a candidate himself, but he was the top Democratic officeholder, and, as Harriman reported to Johnson, did not wish "any other stars in the orbit."[6] His preferred candidate was Stevenson, whom he saw as an elder statesman, not as a competitor for party leadership. He was resistant to Kennedy. But Kennedy regarded Wagner's blessing as indispensable. The newcomer had to appear to be responding to an invitation, not conducting an invasion.

Stevenson turned out not to be a candidate either. He watched developments with rueful malice. "Bobby is mad to run now," he wrote a friend on August 10, "and Steve Smith is making unctuous calls daily about getting together. 'He doesn't want to do anything if I'm interested' etc. Bob Wagner is holding out but the K's have unleashed the mafia. . . . The avarice of the K's really makes me sick. I'd almost like to do it to challenge him."[7] But "almost" was not a challenge. Lacking a candidate, Wagner endorsed Kennedy. Kennedy's nomination was now certain. A trail of bitterness remained. In mid-August Stevenson happily cited "the rising protest against the Kennedy invasion from the Liberal Party, the Roosevelt-Lehman Reform groups, Labor, et al. . . . Of course, he can get the nomination, but winning is another matter what with the widespread disaffection."[8]

II

Kenneth Keating was a silver-haired, pink-faced gentleman of sixty-four. A rather conservative congressman from Rochester in the

1950s, he had as senator after 1958 gained a vaguely liberal reputation, enhanced now by his refusal to back Barry Goldwater. His benign appearance, ingratiating personality and passable record made him a tough opponent, especially for a contender vulnerable on grounds of carpetbagging.

In fact, Robert Kennedy was less a carpetbagger than his older brother had been in Massachusetts in 1946. He had lived in New York from a few months after his birth till 1942, indeed, had lived nowhere longer. Except for schools and summers, he had really never lived in Massachusetts. But he had always voted in Hyannis Port and was indelibly identified with his brother's state. Still New York had historically been hospitable to carpetbaggers. Its first senator, Rufus King, had begun by representing Massachusetts in the Continental Congress — a point I tried, as an historian, to make during the campaign to absolutely no effect. Stevenson would have been a carpetbagger. Of the two senators succeeding to Robert Kennedy's seat, the first lived in Connecticut, the second in Massachusetts.*

The carpetbagger problem at first gave rise to jocularity. Ethel suggested as his campaign slogan: "There is only so much you can do for Massachusetts."[9] A reporter, explaining he could not accompany Kennedy on an upstate tour, said, "I'm going to be in Boston, of all places." "Never heard of it," said Kennedy.[10] The novelist Richard Condon proposed that Kennedy meet the issue by ending his speeches, "Ich bin ein New Yorker."[11] It was more serious than that. What gave the carpetbagger charge its edge was the idea that the invasion of New York was one more expression of Kennedy's well-advertised ruthlessness — a power grab by a madly ambitious, arrogant, opportunistic, primitive and dangerous young man, the son of Joe Kennedy, the aide of Joe McCarthy, the confrere of J. Edgar Hoover, the oppressor of Jimmy Hoffa, who stopped at nothing to advance himself and now was cynically using New York as a launching pad for the Presidency.

The *New York Times* denounced "The Kennedy Blitzkrieg" in a series of editorials. "Does the *Times* think," Francis Biddle finally wrote the paper, "Mr. Kennedy should seek office with more gentility, affording that pleasant touch of hypocrisy which is still highly prized by many 'respectable' citizens?" Roosevelt's Attorney General concluded: "As Attorney General [Kennedy] has furthered civil liberties

---

* James Buckley and Daniel Patrick Moynihan. In 1972, in addition, the Democrats nominated Ramsey Clark, who had spent most of his life in Texas. In none of these campaigns was carpetbaggerism an issue.

for Negroes more than any of his predecessors, enforcing the law up to the hilt, but always with a cool and wise judgment."[12] This was to small avail. Kennedy's actual role in Washington — his contributions, for example, to the Cuban missile crisis or his real relations with J. Edgar Hoover — was hardly known in New York, and disbelieved in anti-Kennedy circles when described.

Gore Vidal and Lisa Howard organized a Democrats for Keating Committee. Other liberals — I. F. Stone, Carey McWilliams of the *Nation*, James Baldwin, Richard Hofstadter, Paul Newman, Joseph Mankiewicz, Barbara Tuchman, Nat Hentoff[13] — forgot Keating's ignoble part on such matters as wiretapping, the restriction of passports and the introduction of bills to reverse civil liberties decisions by the Supreme Court, not to mention those Soviet nuclear missiles cunningly stowed away in Cuban caves, and joined the anti-Kennedy crusade. In due course the *Times* endorsed Keating, explaining mystically that Kennedy aroused "an uneasiness that is no less real because it is elusive and difficult to define."[14] "At least," said Kennedy, "they can never say I got my job through the *New York Times*."* As the barrage intensified, he started to decline in the polls. "It's a strange state you fellows brought me into," Kennedy told one supporter. "I haven't made my first campaign speech, and I'm already losing ground."[15]

Under Stephen Smith's direction the organization began to take shape. Justin Feldman of New York worked with John Nolan on scheduling. Two eager young lawyers, Adam Walinsky and Peter Edelman, resigned from the Justice Department and inserted themselves into the campaign. Bill Barry spent his vacation from the FBI traveling with the candidate in the last fortnight. (Later, four days before his sixth child was born, Hoover ordered him to Mobile, Alabama. Barry resigned from the Bureau.) Paul Corbin installed himself in a hotel across the river in New Jersey until he demanded so many telephones that they ejected him, suspecting he was a bookie. Thereafter he secreted himself in an apartment in New York.[16] Ethel Kennedy was pregnant again but traveled with her husband and even on occasion nerved herself to make speeches.

At the start, enthusiasm, especially among the young, was unrestrained. "If I had my way," Kennedy would say to the screaming kids, "I'd lower the voting age to six." The week after Labor Day he concluded an upstate tour by flying to the placid little town of Glens Falls. Due at eight in the evening, he arrived at one in the morn-

---

* The reference was to a series in which the *Times* touted its want ads by portraying people who claimed they had gotten their jobs through the *New York Times* (Gerald Gardner, *Robert Kennedy in New York* [New York, 1965], 12).

ing. "Do you think anybody waited?" he asked before disembarking. To his astonishment a band struck up, and a thousand people surged around the plane. "Here we are, five hours late," he told them. "That's the smooth, hard-driving, well-oiled Kennedy machine for you." As they drove into town, men, women and children in pajamas and bathrobes waved and cheered along the road. Four thousand people, a quarter of the population, were patiently waiting in the town square. Kennedy told them: "I'd like to make my very first commitment of the campaign. I promise that, win or lose, the day after the election I'm coming back to Glens Falls!"[17]

He entertained audiences with local allusions. "The Catskills," he said, "were immortalized by Washington Irving. He wrote of a man who fell asleep and awoke in another era. The only other area that can boast such a man is Phoenix, Arizona." He begged the attention of Johnson City and its shoe industry for two reasons: "First of all, eight small children need a lot of shoes. And second, I'm the one who popularized those fifty-mile hikes."[18] But he felt he knew the real reason that the crowds were pouring out and quoted John Kennedy in every speech. "They're for him," he told Guthman, "— they're for him."[19] Soon he sank into melancholy memories. County chairmen found him moody and uncommunicative. His staff thought him unwontedly irascible. "He hated himself, in my view, for being here," Justin Feldman thought. ". . . He was being purely and simply President Kennedy's brother, and he just wasn't coming through."[20] One day Corbin, observing that he was retracing his brother's 1960 schedule, called him and said, "Get out of this mysticism. Get out of your daze. . . . God damn, Bob, be yourself. Get hold of yourself. You're real. Your brother's dead."[21]

Kennedy had decided that a frontal attack on Keating would only accentuate the impression of a ruthless young tough out to mug a kindly old man. But he continued to drop in the polls. "I was appalled," John Douglas wrote him in early October, "to learn you said yesterday that Keating's voting record is good. Every time you make a statement like that you are cutting your own throat. Keating's voting record is not good."[22] Douglas enumerated the areas. Kennedy irritably called it a "snotty" letter. If he questioned Keating's record, he responded it would look like "a personal vendetta." "In other words," said Douglas, "he thought that if he went after Keating on a perfectly valid basis, that somehow people would view it as a personal attack."[23] The "ruthless" charge had bitten deep.

On October 5 in Rochester, Richard M. Nixon, righteously denouncing what he called "a cloud of corruption" over the Johnson White House, forecast a big Keating victory.[24] On October 6, John

F. Kraft, the Kennedy pollster, reported that "it ain't good" in New York City; on October 8, that Keating was leading 51–49 statewide while Johnson led Goldwater 60–40.[25] The same day Samuel Lubell, the master diagnostician of the American voter, wrote in the *New York World-Telegram*: "Robert Kennedy is running well behind. . . . Keating's lead mainly reflects a heavy break against Kennedy among normally Democratic voters in New York City. . . . In some New York City precincts a third or more of the voters going for President Johnson say, 'I'll split my ticket' on the Senate race."[26]

III

Buoyed by the sympathy flowing to underdogs, Keating had been conducting a careful campaign. But his chief political adviser was Herbert Brownell, an aggressive veteran of political wars. As Kennedy fell in the polls, Brownell smelled an opportunity to administer the coup de grâce. He urged his candidate to go hard after traditional Democratic groups now breaking from Kennedy: the blacks; the Italians; above all, the Jews.

Keating had an excellent civil rights record and solid relations with black leadership. In a speech prepared for the state NAACP convention on October 2, he accused Kennedy of having run out on the civil rights movement: "He abandoned his post at the Department of Justice with an unfinished task before him." Charles Evers, who had taken his brother Medgar's place as head of the NAACP in Mississippi, addressed the convention the same day. The state NAACP president, as Evers fumed at Keating, sent him a note: "I know Bobby's your man, but not here." "God, I was so mad," Evers said later. "I just bit my tongue." When his time came to speak, Evers threw away his text. "You're fortunate to have a man of Kennedy's caliber even visit New York, much less running for Senator," Evers told the audience. Turning to Keating: "Bobby Kennedy means more to us in Mississippi than any white man I know, including yourself, Senator." The NAACP officials refused to speak to Evers when he finished.[27] But his intervention averted an NAACP endorsement. "We in Mississippi cannot vote for [Kennedy]," Evers wrote in the *New York Post* later in the month, "but we sure can urge you to vote for this man who . . . has done more for the Negroes not only in Mississippi but throughout this country, than any other single individual."[28]

Bidding for the Italian vote, Keating accused Kennedy of smearing all Italian Americans by unveiling Joe Valachi and of doing so in order to divert attention from the crimes of Bobby Baker. Mario Biaggi, a police lieutenant and president of the Grand Council of Columbia Associations, denounced the Valachi hearing as a "one-man show produced, staged, directed and prompted" by Kennedy.[29] "He's a crumb," an Italian barber observed of Kennedy to a *Times* reporter. "He made every Italian look like a gangster."[30]

The major Republican effort, however, was directed at the Jews. Kennedy, William Haddad warned Stephen Smith, was "in trouble with the Jewish community. . . . Underlying this problem is a great uneasiness, an unsureness, a fear, as one person put it, of the tough Irish kid that beat him up on the way to school each day."[31] As for Keating, he "goes around with a yarmulke in his pocket," one rabbi told Neil Sheehan of the *Times*, "and every time you can put three Jews together he's there to give a speech."[32] At the end of September, Keating, striking hard for the Jewish vote, accused Kennedy of having made a deal with a "huge Nazi cartel . . . the chemical arsenal for Nazi Germany" when the Justice Department settled the case of the General Aniline and Film Corporation in 1963.[33]

The case, left over from the Second World War, involved the claim of a Swiss corporation to the assets of General Aniline, a company seized as enemy property in 1942. It was a singular case for Keating of all people to bring up, for Keating himself had introduced the bill permitting the sale of General Aniline. "I am gratified," he had told the Senate on March 4, 1963, "that as a result of the General Aniline & Film sale bill we have moved very close to the day when this company can become part of the mainstream of our free enterprise system."[34] He had been "fully informed as to everything that was going on," said William Orrick, who had handled the settlement, "and I say that because I did it. . . . He just lied in the campaign."[35]

Kennedy was outraged. "The charge that I made a deal with Nazis can't help but have an adverse effect on how Jewish people feel about me," he told a press conference. ". . . If this kind of charge were true, I wouldn't deserve to be elected to any public office. The charge isn't true. . . . I lost my brother and brother-in-law to the Germans. The idea that I would turn over money to the Nazis is ridiculous."[36] "I expected this campaign to be on a higher level," he ruminated to James Stevenson of the *New Yorker*. "I found myself being charged with making deals with Nazis and running out on Negroes and being anti-Italian and being a crook, covering up for Bobby Baker, and I just didn't expect it would be as bad as that."[37] But the General Aniline fiasco both hurt Keating and legitimized a shift in Kennedy's tactics. "It gives you the justification for attacking *him* for a change,"

Professor Richard Neustadt told him.[38] "If he hadn't attacked me," Kennedy said later, "I would have had a hell of a time."[39]

IV

Free at last, Kennedy in October struck back on the issues between Keating and himself. Old friends deployed on his behalf: not only Charles Evers and others among the blacks, but Harry Golden from North Carolina in Jewish neighborhoods, while Galbraith, Rauh, Neustadt, Adam Yarmolinsky and I spoke evening after evening to disaffected liberals. I could not forget that four years earlier I had found the same people in the same mistrustful mood when I campaigned for John Kennedy. Now they mourned him and mistrusted his brother.

Norman Mailer came unexpectedly to our assistance. Mailer had never been a fan of Robert Kennedy's. The dead President had been "the existential hero"; "Bobby! — the Irish equivalent of Roy Cohn on the good old McCarthy team; Bobby! — with the face of a Widmark gunsel, that prep-school arrogance which makes good manual laborers think of smashing a fist through a wall; Bobby! —" etc., etc. But Keating, with his face "like the plastic dough children play with," seemed even worse. As between "a neutron" and "an active principle," Mailer announced for the active principle. Besides, "I have affection for Bobby Kennedy. I think something came into him with the death of his brother. . . . Something compassionate, something witty, had come into his face. Something of sinew." Besides, "I think Bobby Kennedy may be the only liberal about, early or late, who could be a popular general in a defense against the future powers of the Right Wing."[40] Like all of Mailer's political journalism, it was brilliant and rather dotty. It helped.

Most important, Kennedy was beginning to find himself as a campaigner. He still read prepared speeches in a monotone but was increasingly effective in extemporaneous delivery. He stumped the state morning to night five or six days a week. When he called, as he regularly did, on his father, he was, Rita Dallas noted, "red-eyed from lack of sleep, and hoarse from hours of speechmaking." The old man gave him indecipherable words of counsel. "I'll make changes, Dad," Bobby would say. "You know I'll make changes. Millions of people need help. My God, they need help."[41]

In mid-October his advertising blitz began. Kennedy mistrusted the Madison Avenue magicians. They showed him a brochure with a photograph in which he was displayed shaking hands with a labor leader. Kennedy frowned. "What's wrong with this one?" he was asked. "That fellow's in jail," Kennedy said. As for television, the aim, he was told, "will be to present you as a warm, sincere individual." Kennedy said, "You going to use a double?"[42] The experts were puzzled how to project him on the tiny screen. He stiffened in the studio. Clips of the candidate surrounded by apparently demented teenagers did not give the desired impression of statesmanship. But he was forceful and persuasive dealing with questions. His strongest television piece emerged from a session with antagonistic students at Columbia.[43]

He also finally agreed to take advantage of the fact that Johnson was heading for a landslide in New York. Steve Smith's new posters read: "Get on the Johnson-Humphrey-Kennedy Team." In mid-October Johnson and Kennedy campaigned together for two days from Buffalo to Brooklyn. Both played their roles to the hilt. "You don't often find a man," said Johnson, "who has the understanding, the heart and the compassion that Bobby Kennedy has." Johnson, said Kennedy, was "already one of the great Presidents of the United States."[44] Underneath, older feelings persisted. Richard Wade, who was running the Kennedy campaign in northwestern New York, thought "it drove Johnson crazy" to see hands outstretched to shake the senatorial candidate's hand and not his own. Then "Johnson would get up on the platform, put his arm around him and say, 'this is ma boy, I want you to elect ma boy,' you know, six feet four and . . . you could see the whites of Bobby's knuckles."[45] At the end of two crowded days, Kennedy asked Johnson whether he had enjoyed it. Johnson looked at him earnestly and said, "Of all the things in life, this is what I most enjoy." Kennedy, telling me this, added incredulously, "Imagine saying that, of all the things in life, this is what you like most."[46]

On October 19 the *Daily News* poll showed Kennedy leading by 59.5 percent to 37.3.[47] Blacks and Puerto Ricans were flocking to him; Jews and Italians were coming back.[48] Keating, increasingly frantic, assailed this "self-seeking outsider" who was conducting "the most arrogant campaign New York State has ever witnessed" and was "temperamentally and intellectually unfit" for the Senate.[49] ("He's on to you," said Steve Smith.)[50] Keating then filed a complaint with the Fair Campaign Practices Committee over Kennedy's claim in a Syracuse speech on October 20 that Keating had "ridiculed" the test ban treaty.

Kennedy had been wrong. A speechwriter, revising a draft from the research staff, had sharpened a considerably more qualified statement portraying Keating's early misgivings about the test ban. In the end, Keating had voted for the treaty's ratification. The FCPC executive director sent Kennedy a letter accusing him of "dishonest and unfair distortion" if not "deliberate and cynical misrepresentation." The Keating staff then leaked the FCPC letter to the *Herald Tribune*. Brownell said that Kennedy had been caught "red-handed with a spectacular distortion." Kennedy, indignant over what he regarded as unwarranted overkill, submitted documentation of Keating's test ban equivocations and personally telephoned members of the FCPC board of directors. Ralph McGill of the *Atlanta Constitution* and others resigned. The FCPC withdrew its letter, saying it "should not have been written, and any accusations in it were necessarily unfair to Mr. Kennedy at that stage." The issue evaporated almost as quickly as it had arisen.[51]

In September, Keating, when he was running behind, had proposed a debate. Kennedy, after evading the challenge, accepted it in early October when *he* was behind. Keating then stalled. Now, behind again, he revived the challenge. It was once more Kennedy's turn to stall. He saw no advantage in debating Keating. "He looks like your grandfather," Kennedy would say, "and there's no way you can win it." At the end of October, Keating bought a half-hour of prime television time and called on Kennedy to appear with him. Harry Golden urged Kennedy to do it. Kennedy said, "Harry, you've got white hair, you debate Keating."[52] With Javits at his side, Keating arranged to debate an empty chair marked "Robert F. Kennedy" — the symbol, he said, of Kennedy's "ruthless contempt" for New York voters.

This was an ancient device. Kennedy, the innovator, suddenly resolved to fill the chair. "I was sure I would get in," he said later. "I couldn't think of any way they could keep me out." When he arrived without warning just before the telecast, studio guards turned him back. He vigorously protested his exclusion to newspapermen. Keating meanwhile fled the reporters, leaving twin photographs for next day's papers — Keating threatening his empty chair; Kennedy looking in frustration at a sign PLEASE KEEP OUT. Kennedy told an audience later that night, "There was Javits and Keating on television really giving it to this empty chair. . . . They kicked that chair all over the room. And there was I outside trying to get in."[53]

He won by 719,693 votes, greatly helped by Johnson, who carried the state by 2.7 million. Kennedy neglected the President in his victory statement. Far away in Texas, Johnson said, and repeated, "I wonder why he doesn't mention me?"[54] Bill Barry congratulated the senator-elect in the midst of election night pandemonium. "If my brother was alive," Kennedy said, "I wouldn't be here. I'd rather have it that way."[55] He quoted Tennyson to the cheering campaign workers:

> Come my friends,
> 'Tis not too late to seek a newer world.

He had an hour's sleep and next morning took the plane to Glens Falls.

v

On January 4, 1965, Robert Kennedy was sworn in as the junior senator from New York. Edward Kennedy, leaning heavily on a cane, his smashed back enclosed in a steel brace, gaunt after months in the hospital, took the oath with him.* Not since 1803 had two brothers — Dwight Foster of Massachusetts and Theodore Foster of Rhode Island — served together in the Senate.† That evening the Women's National Press Club threw its annual soiree for new senators. "First of all," said Robert Kennedy, when presented to the audience, "I want to say how delighted I am to be here representing the great state of . . . ah . . . ah . . ." — a long wait while he made a show of searching through his notes. Having ascertained the name of the state, he concluded: "I have absolutely no presidential ambitions, and neither does my wife — Ethel Bird."‡

His Senate office was small in comparison to the cavernous chamber he had occupied at Justice. The indispensable Angie Novello came along as executive secretary. When Edwin Guthman left in April to resume his newspaper career, Wes Barthelmes, a former paratrooper and reporter who had been working for Congresswoman Edith Green, became press secretary. Guthman gave Barthelmes a single piece of advice, "You've got to learn to read the pauses, these long silences."[56] Joseph Dolan, as administrative assistant, knew how to read the silences. Underneath his close-mouthed Irish manner,

---

* Edward Kennedy's election to the Senate in 1962 was to complete an unfinished term. In 1964 he was elected for the full term.

† Robert was, of course, the third Kennedy brother in the Senate in five years. The Kennedys were still behind the Washburn brothers, three of whom served together in the 34th, 35th and 36th Congresses representing three different states while a fourth later represented still another state both in the House and Senate. These were Israel Washburn, 32nd–36th Congresses from Massachusetts; E. B. Washburn, 33rd–41st Congresses from Illinois; C. C. Washburn, 34th–36th and 40th–41st Congresses from Wisconsin; and W. D. Washburn, 46th–48th Congresses and later senator (1889–1895) from Minnesota.

‡ *Washington Post*, January 5, 1965. Like most contemporary politicians, Kennedy often called on professionals for jokes in speeches. The always helpful Alan King provided the "ah . . . ah . . . ah" gag; but, according to King, it was the Ethel-bird joke "which just broke up the place. That was his own" (Alan King, in recorded interview by Jean Stein, May 1970, 2, Stein Papers).

he was, Barthelmes noted, "a well-disguised intellectual."[57] Dolan and Kennedy communicated in a monosyllabic style that baffled the rest of the staff. Frank Mankiewicz, who replaced an exhausted Barthelmes as press secretary after a year, recalled them grunting and snorting at each other: "This guy . . ." "No, not this fellow because you remember . . ." "Oh yes, that other guy, that's right . . ." "I never was quite sure what they were talking about," Mankiewicz said, "but they seemed to know."[58]

The legislative assistants, Adam Walinsky and Peter Edelman, were in their late twenties. Walinsky had gone to Yale Law School, Edelman to Harvard; Walinsky had clerked for a judge in the Court of Appeals, Edelman for Arthur Goldberg on the Supreme Court; both had been junior attorneys at Justice, who, when they forced their way into the senatorial campaign, brought fresh ideas and enthusiasm at a midpoint of fatigue. They were one of those teams, like Tom Corcoran and Ben Cohen in the New Deal, that materialize around secure patrician reformers. Only a secure man could absorb so much needling brilliance without feeling threatened thereby; only a reformer would wish it. Like Corcoran and Cohen, both did everything, but each inclined to his specialty: Walinsky writing speeches, Edelman drafting bills; one the eloquent salesman, the other the careful researcher; one ostentatiously arrogant, the other deceptively gentle; both passionate, obstinate, hero-worshiping, exasperating, engaging. Kennedy listened to them, argued with them, learned from them. The sixties were breaking open, and they, far more than his New Frontier friends, could tell him what it was all about. He valued their testimony and could handle their indignation. When Walinsky kept putting things in speeches that Kennedy did not wish to say: "Now, Adam, get that out. That's yours, not mine. I don't want that." "Adam," said Mankiewicz, "had a marvelous relationship with him because he was always pushing him. And that's what he wanted."[59]

It was an exceptionally busy office. The mail averaged twelve hundred letters a day. At the start he intended to live within the official allowance, but by the end, in order to keep abreast of business, he was adding a hundred thousand dollars of his own for salaries.[60] As at Justice, the staff had great discretion. "He didn't spend a lot of time telling you the fourteen steps to follow," Walinsky recalled. ". . . He just sort of indicated that he thought something should be done in an area and then he expected you to grab hold and go do it. And then maybe a week or so later . . . he'd say, 'Well, where is it?' or 'How are we doing on that?'"[61] Once Walinsky failed to produce a poverty speech. "He was rather upset about it, but all he said was, 'You know, I thought that was a very important speech, and

I'd really like to do well with it.' And so I said, 'I'm sorry.' He said, 'Thank you.' That was the strongest it ever got. I mean he was not one for chewing people out."[62]

Nor was he effusive in praise, but, as Barthelmes said, his assistants "were mature enough to feel that they were just working for someone who felt everyone was responsible and grown up." The compensation came in his confidence in them. "He always gave me," Barthelmes said, "an extra sense of dimension of myself." And he excited because of the feeling — again Barthelmes — that he carried "a hidden agenda with him. There were always many, many, many, many things to be done, and you were part of it."[63] "He was a very secret man," said Walinsky, ". . . terribly sensitive to people's feelings," but conveying "a certain kind of sadness and a certain sense of loss all the time. . . . There were things he carried around in his head that were unimaginable, things that he just had to live with all by himself."[64]

VI

Robert and Ethel liked New York. They took an apartment high in United Nations Plaza with a startling view of the East River and delighted in the gaieties, the theater, ballet, restaurants, museums, of the most spirited of cities. Walking in New York was fun, Robert Kennedy said, because "everybody's always in such a hurry to get home they don't see me."[65] Unlike most Upper East Siders, he systematically acquainted himself with the darker corners of the metropolis, the ghettos of Harlem and Bedford-Stuyvesant, the joyless stretches of the Bronx and Queens, the decaying housing projects, rat-infested tenements, garbage-strewn alleys, desperate prisons.

From his apartment he could see two huge Consolidated Edison stacks belching smoke into the air across the East River; it infuriated him that no one was doing anything about it. He wished he could swim in the river; but, as he observed of the Hudson, "the water's so bad here, if you fell in, you'd dissolve."[66] He liked the great state rolling out beyond the metropolis, the tranquil villages and the rippling lakes, liked, as he once put it, "the sense of rootedness and the fundamental decency, to be found in our small towns and farms."[67] "He loves travelling around New York," I noted after a talk with him in 1965, "and says that he would really like some day to become governor; 'there is so much that can be done for that state.'"[68] He was surprisingly popular upstate; indeed had done better there, relative to previous Democratic candidates, than in the city. He methodically visited mayors, city councils and regional development groups in places that had never seen a Democratic senator. He worried about the poisoning of Lake Erie, the pulpwood scum on Lake Champlain, the wetlands of Long Island, the preservation of parks and forests.

He worried even more about the preservation of people. In the summer of 1965 he received distraught letters from parents with children in state institutions for the mentally retarded. He made unannounced visits and was appalled. Children, he said, "just rock back and forth. They grunt and gibber and soil themselves. They struggle and quarrel — though great doses of tranquilizers usually keep them quiet and passive. . . . There are no civil liberties for those put in the cells of Willowbrook — living amidst brutality and human excrement and intestinal disease."[69]

The senior senator, Jacob Javits, shared many of these concerns, but their relations began in mutual suspicion. Javits, a sensitive and talented man too liberal for his party, was sometimes driven to bouts of equivocation in order to maintain Republican credentials. He had, in addition, minor vanities, especially on the question of newspaper publicity. The two offices engaged for a time in adolescent competition over which senator was doing more for New York. The principals went on television together. Riding back to his office afterward, Kennedy told Barthelmes, "Javits is so facile. He's computerized for knowledge. . . . I never want to do that again" — which, Barthelmes said later, "was about the nearest one ever got . . . to a reprimand." A few moments later Javits's press secretary phoned Barthelmes and said, "Javits just bitched and bitched at me all the way back. . . . He said, 'Don't ever put me on with Kennedy again. He's glamorous. . . . They won't remember what I said.'"[70]

Kennedy had a certain sympathy for Javits. "It's awfully hard for him, me coming into the whole thing here and sort of upstaging him," he said to Edelman. ". . . If I were in his shoes, I wouldn't like me either."[71] In his first legislative initiative, Kennedy introduced an amendment extending the Appalachia bill to cover thirteen upstate New York counties. Javits, unconsulted, was furious. "I thought it was very rude," he said later, ". . . and would have embarrassed me politically if it were strictly his and not mine also. And I told him so in unmeasured terms and not without heat. And he appreciated that and said that he could understand it perfectly . . . and that he would try to make it clear that we were together in it."[72] Soon they agreed to make joint announcements of federal projects — an unusual arrangement when the senators were from different parties.

Kennedy came to appreciate Javits's solid qualities — his acute intelligence, his breadth of informed interest, the genuine conscientiousness underneath his sensitivities. They became collaborators and in the end friends. I remember chatting with Kennedy in 1967 about Javits's reelection contest in 1968. "I don't regard the defeat of Jack Javits," he said, "as one of the more momentous issues facing the nation." Looking back on Kennedy, Javits said in 1973, "He wanted to conserve human values wherever he found them. . . . He was passionately devoted to whatever was a vivid life expression. . . . His, essentially, was a life force as a senator and as a man."[73]

VII

After his election Robert had visited his brother Edward in the Boston hospital. A photographer asked him to step back a little; "you're casting a shadow on Ted." Ted said promptly, "It'll be the same in Washington."[74]

But in the Senate it was Ted who knew the ropes, who was amiable, courteous and popular, whose return after his accident was greeted by other senators with genuine pleasure as compared to the visible skepticism with which they received the former Attorney General. Edward did his best to instruct his older brother. They called each other "Robbie" and "Eddie" and teased back and forth endlessly. They sat together on the Labor and Public Welfare Committee. At hearings senators asked questions in order of seniority, and Robert was almost at the bottom of the list. Early on, he whispered impatiently to his brother, "Is this the way I become a good senator — sitting here and waiting my turn?" "Yes," Edward said. Robert had an engagement in New York. "How many hours do I have to sit here to be a good senator?" "As long as necessary, Robbie."[75] If Robert in his zeal went on too long before the committee, a note from Edward: "You just lost Lister [Hill, the chairman] — stop talking and let's vote, or you'll lose all the others too."[76] Another member said he would help vote out one of Robert's bills if Robert accepted an invitation to speak in his state. "Is this what you have to do to get votes?" he asked his brother. "You're learning, Robbie," said Ted.[77] When pressing a bill, he would ask Ted, "What should I do now? You know how to handle these fellows. . . . You're the likable one."[78] A scribbled note, Robbie to Eddie, transmitted on the floor:

All I want to know — when I shake my head from side to side and you nod your head up and down. Does this mean I will vote no and you agree? Obviously not. In other words when I wanted to vote no I should have nodded my head up and down — you would have shaken your head from side to side — then I would have known just how to vote.[79]

They consulted constantly, occasionally differing on inconsequential matters, but, for all the playful competition to top the other, Robert remained the head of the family.

Kennedy's daily chums were the three freshmen with whom he shared a new fifth row, installed in the back of the chamber to accommodate Democratic gains in the Goldwater election — Walter Mondale of Minnesota, Fred Harris of Oklahoma, Joseph Tydings of Maryland. Young, liberal and irreverent, the four vastly entertained themselves by sotto voce comment on the scene before them. Kennedy had, of course, old friends in the rows ahead — George McGovern, whom he once described to me as "the most decent man in the Senate," adding in hyperbolic afterthought, "the *only* decent man in the Senate"; Frank Church from Rackets Committee days; Birch Bayh, who had crashed with his brother in the Connecticut Valley; Claiborne Pell, whom he had known from Newport.

"I had not thought at first that I would like Robert Kennedy," said Fred Harris, who had been reared in the populist faith that "rich people were *never* happy and seldom good." But Kennedy "*was* about as impressive a person as I ever met." Harris was drawn to him by his earnestness, his commitment, his "insatiable" curiosity about things, people and issues.[80] McGovern too was struck anew by his continued "hunger to learn" and by his ready cooperation. "I never asked Bob Kennedy in my life for anything that I thought was very important where he failed to follow through."[81] A certain moodiness persisted, however. "One day we would crack jokes for an hour," recalled one of his back-bench colleagues; "the next day he'd chop you off. . . . The next day [after that] your relationship would be just as though it never happened. I'm not sure that he realized how some people were hurt by that."[82]

With conservative senators, some at least, he kept up affable banter. McClellan, of course, was an old patron, though for various reasons — especially the crusty Arkansan's dislike of Robert McNamara — their friendship had somewhat soured; Fred Harris accurately described it as "the remnant of what one would suspect might have been a stronger one — of a kind of father-son relationship."[83] Eastland, though always an opponent, remained a friend. When Kennedy published *To Seek a Newer World* in 1967, he inscribed a copy for the old Confederate: "It is still not too late. Repent now!"[84] Kennedy shared with the aged Harry Byrd an enthusiasm for dogs and the Virginia countryside. Thus a handwritten note in the summer of 1965: "I was sorry to hear that you were not feeling well. We need you back here to keep a check on us young liberals who will get away with everything without you to keep an eye on us."[85] Even a 1966 telegram to George Murphy, the right-wing movie actor who had beaten Pierre Salinger in California: WE ARE APPROPRIATING BILLIONS OF DOLLARS FOR SOCIAL PROGRAMS WHILE YOU ARE OUT OF THE CITY. WE NEED YOU BACK HERE TO PUT US ON THE RIGHT TRACK.[86]

A few, like Edward Long of Missouri, whom he regarded as the Teamsters' senator, and Carl Curtis of Nebraska, whom he had detested since the Rackets Committee, he could not abide. Ralph Nader, beginning his career as consumer ombudsman, tried to testify, over Curtis's interruptions, about the dangers in automobiles. Kennedy challenged Curtis:

> KENNEDY. First, you admit you haven't read the book [*Unsafe at Any Speed*]; and, secondly, you haven't heard his testimony. Why don't you listen to his testimony and then criticize?
> CURTIS. I have no objection to hearing his testimony, but when he loses me with . . .
> KENNEDY. With big words?[87]

He had hoped for the Foreign Relations Committee, but was too low in the seniority ladder. In addition to Labor and Public Welfare, he went on Government Operations, where a decade earlier he had been a subcommittee counsel, and on the District of Columbia Committee. The committees offered ample scope for his concerns with education, civil rights, welfare, poverty, juvenile delinquency. He was generally accounted a more effective senator than his older brother but much less a 'senator's senator' than his younger brother. Critics found too much sudden, isolated attack in his performance, too many eye-catching amendments, too little detailed follow-through.[88] "I would call him a first-act politician," said Senator Eugene McCarthy. "You know it's easy to write the first act. And it's relatively easy to write the third act. Lyndon Johnson was a good third-act man. . . . But it's the second act that is toughest to write. In Congress, that's where the drudgery and hard work come."* Edward Kennedy was a good second-act man.

Robert Kennedy's manner was also on occasion more brusque than senatorial tradition approved. "He was intense, he was furious about the issues and he was rather well prepared," said Fred Harris. Sometimes "he would almost be too mean."[89] His candor, said Tydings, "made him very unpopular among many of his colleagues." The Senate had its courtly idiom of disagreement. "Bobby didn't take the

---

* Dick Schaap, *R.F.K.* (New York: Signet reprint, 1967), 117. Schaap does not identify the source of the quotation. I assume it to have been McCarthy in view of the conclusion of his poem "Lament of an Aging Politician": "I have left Act I, for involution / and Act II. There mired in complexity / I cannot write Act III."

trouble to polish his phraseology very often. To me it was a plus rather than a minus. . . . Many people interpreted it as being blunt and rough and harsh."[90]

But he was a personality. Whether he was in a committee room or on the Senate floor, observed David Burke, his younger brother's chief aide, "you always had the feeling that Robert was ready to explode. Something was going to happen. Watch out! There was always tension, electricity around him."[91] The Senate generally accepted him as sui generis. "When he went after something," said Javits, "he went after it with perseverance and doggedness, and again that quality of aloneness, but of distinction and of passion. . . . He was a very effective speaker . . . and rough-and-tumble debater." Asked whether any of their colleagues hated Kennedy, Javits said, "They feared him. They thought he was a dangerous opponent, or they had a certain reverence about him and his history, or they were fascinated by the loner aspect of him and his boldness and something of a touch of arrogance because of his boldness. But I was not conscious of anybody hating him."[92] One senator complained to a committee chairman that Kennedy was getting preferential treatment for a freshman. "Oh, no," was the reply. "I treat him the same way I'd treat any future President."[93]

VIII

Late in his first session he found himself drawn into a piece of unfinished family business. This was the problem John Kennedy had so long postponed, the only thing his father had ever asked of him, the appointment of the old retainer Francis X. Morrissey to the federal district court in Massachusetts. Morrissey's qualifications were no more apparent in 1965 than they had been in 1961. But, with their father hopelessly disabled, his last request hung on his sons' consciences. Lyndon Johnson seemed surprisingly cooperative. "Before I leave the Department of Justice," Robert Kennedy wrote him on the eve of his resignation, "I want to express to you my appreciation for your statement that, at an appropriate time, you would give sympathetic consideration to Frank Morrissey for appointment to the United States District Court. . . . He bears an excellent reputation as to character and integrity, has judicial temperament, and is, I believe, worthy of appointment."[94] Perhaps, in a spirit of filial piety, Robert Kennedy persuaded himself of this. But Johnson must have noted that the Attorney General had not fought for the appointment himself.

Edward Kennedy now took the initiative. Morrissey and his wife, he explained later, were "very insistent." He added, "My father was sick, . . . and I simply felt it was something that had to be put forward."⁹⁵ At least for the family record: he may not have supposed Johnson would act on his request. Then, with what can only be assumed as unrestrained internal glee, Johnson sent Morrissey's name to the Senate. Robert Kennedy was dumbfounded. "I don't believe it," he told Milton Gwirtzman of his brother's staff. "Johnson's just nominated Frank Morrissey."⁹⁶ Johnson's theory was plain enough. The Kennedys would suffer the obloquy for the nomination; further, it would place them under obligation to him while at the same time exposing their own pious pretensions. Taking no chances, the President called J. Edgar Hoover and directed him to make the most complete investigation of Morrissey; "I don't want him to get that judgeship."⁹⁷

The nomination was greeted with universal disapproval. The *New York Herald Tribune* called it "nauseous." In Boston Judge Charles Wyzanski, who was presiding judge of the district bench, the *Boston Globe* and the respectable bar denounced it.⁹⁸ Faced with a fight, Edward Kennedy set out to win it. He was also exasperated by what seemed to him establishment snobbery toward a poor Irish lad who had not been able to afford an Ivy League law school. For a time he appeared to be succeeding. The Judiciary Committee, if with nearly half its members abstaining, sent the nomination to the floor. But Morrissey's own testimony had shown troubling discrepancies in his account of his legal education — discrepancies soon widened by the investigative reporting of the *Boston Globe*. Senator Dirksen now resolved to make Morrissey a major issue. Senator Tydings, a Kennedy friend but also a guardian of the judiciary, told the Kennedys he would have to vote against confirmation.

Tydings's defection on top of Dirksen's opposition doomed Morrissey. Tydings was plainly right on the principle. But the Kennedys, especially Robert, were bitter. "They could understand my not voting for him," Tydings said later, "but they couldn't understand my actually opposing him."⁹⁹ Tydings, after all, was a Hickory Hill familiar. Robert Kennedy had made him a federal attorney and helped him become a senator. He had violated the Kennedy loyalty system.

In his way Morrissey was violating it too — by insisting on the fight in face of the mounting embarrassment he was causing his patrons. It had been mainly Edward's problem so far, but Robert now became more involved. After counting the votes, the brothers saw they could not win and decided to send the nomination back to the Judiciary Committee. Gwirtzman asked, "How about your commitment to Frank Morrissey?" Robert Kennedy said, "I think we've more than fulfilled our commitment to Frank Morrissey."¹⁰⁰ On October 21,

1965, in an emotional speech, Edward Kennedy defended Morrissey but at the same time moved the nomination's recommitment. A few weeks later Morrissey asked that his name be withdrawn. The only winner in this fracas was Lyndon Johnson.*

IX

The Morrissey episode was typical of the relations between Johnson and Robert Kennedy in 1965. Neither wished an open break. When either sought to embarrass the other, it was done under the cloak of solicitude. Mostly each wanted to keep out of the other's way. They indulged in public praise and private propitiation.

> Dear Mr. President,
>   You were most courteous to me today at the signing of the Voting Bill —
>   I wanted you to know how much I appreciate your thoughtfulness.
> > Respectfully,
> > Bob Kennedy
> > [August 17, 1965]

> Dear Bob:
>   Lady Bird and I join in congratulating you on this, your first birthday in Congress. We hope that this and every day that follows will bring you happiness. . . .
> > [November 20, 1965]

> Dear Bob:
>   The celebration of Chirstmas is a holy time, a time of renewal and hope. It is a time for families and friends, neighbors and colleagues to join in the goodness of God. . . . I could not let this Christmas pass without telling you that Mrs. Johnson, my daughters, and I ask God's blessings on you and yours.
> > [December 20, 1965][101]

Johnson took due precautions within the frame of the armed truce. He saw Kennedy as his only rival within the party and naturally wanted to weaken him as much as possible short of open provocation. He began by insisting on his old friend Edwin Weisl for New York's national committeeman. He did this without consulting Kennedy, who soon rejoiced to discover that Weisl was not even a registered Democrat. Still Weisl rather than Kennedy controlled federal appointments in New York. "Lyndon Johnson did not extend

---

* Also the *Boston Globe*, which received a Pulitzer Prize for its work. Robert Kennedy was at the fiftieth anniversary dinner, where the award was made. When the *Globe* was announced, "those sitting close to Bobby's front-and-center table could not resist stealing furtive looks to catch his reaction. As the scroll was being handed over, Bobby bowed his head, turned toward his dinner companion and, with a boyish grin, winked broadly" (Penn Kimball, *Bobby Kennedy and the New Politics* [Englewood Cliffs, N.J., 1968], 139).

the courtesy of advice and consent to Robert Kennedy," said Dolan.
". . . They used to whiz them by us all the time."[102] Kennedy sponsorship was in fact a handicap. When, for example, Walter Mansfield
was up for a district judgeship, Ramsey Clark, as Attorney General, reassured Marvin Watson, Johnson's political operative, with
Weisl's denial that Mansfield "was in any way close to Senator
Kennedy."[103] The administration actually favored Javits over Kennedy. "The White House was not too anxious to build him up,"
Javits recalled in considerable understatement, "and we had many
indications of that. . . . I think the White House was rather interested
in dealing with me all they could."[104] In his determination to weaken
the New York party, Johnson even repudiated an agreement made
by John Kennedy under which half the Democratic money raised in
New York was returned to the state. "You don't seem to understand,"
Johnson said coldly when Robert Kennedy reminded him of the
agreement. "That was a different President."[105]

Johnson was equally concerned with blocking a Kennedy fifth column in Washington. Vice President Humphrey's right-hand man
wrote Marvin Watson about an ambassadorial nomination, mysteriously delayed: "He is a Johnson man. He is *not* a Kennedy
man. . . . Once again, if anyone has informed you that Rivkin is an
RFK man, you were wrongly informed."* When Roger Wilkins was
named director of the Community Relations Service, he wanted his
commission to read "from New York" so that Kennedy would introduce him to the Senate committee passing on his nomination. The
White House insisted on "from the District of Columbia." Johnson,
Wilkins said, "just didn't want Bob Kennedy up there getting the
credit. . . . The President was really annoyed at my swearing in and
he wouldn't talk to me! Just wouldn't talk to me! . . . Never responded to anything I said to him. Nothing! Just gave me the silent
treatment. . . . The President later sent a message to me through
[Thurgood Marshall, now Solicitor General]. It said, you'd better get
on the straight and narrow."[106]

Johnson could do nothing — nor did he try to do anything —
about McNamara, Harriman, Goodwin and others with an established
Kennedy relationship. But his own people knew the eyes of Texas
were on them if they spent the evening at Hickory Hill. Moyers and
Valenti were particular offenders. Nor were senators exempt. Once,
when Fred Harris and his wife were at Hyannis Port, the phone rang
during dinner. Ethel, answering, said, "It's President Johnson for
you, Fred. He's found you, and you're in big trouble now." Johnson

---

* William Connell to Marvin Watson, June 2, 1965. William Rivkin, an able Chicago
lawyer, had served as ambassador to Luxembourg in the Kennedy administration. He
had close ties to Humphrey and warm relations with Robert Kennedy.

chatted inconsequentially with Harris: "He had wanted nothing in particular," Harris reflected, "except, I figured, to let me know he knew where I was." Later Johnson informed Harris through an Oklahoma oil magnate that "he could do a lot more for you if you weren't so close to those Kennedys."[107]

X

Kennedy took care to respect the armed truce. He was well aware of the widespread impression that he regarded the White House as Kennedy property and Johnson as a usurper. He read every week that he was, in the words of hostile columnists, "high-handed, ruthless, power-grabbing and consumed by an inordinate White House ambition."[108] He recognized that the press was building the Johnson-Kennedy 'feud' as the great human drama of Washington.

Nothing inhibited Kennedy more than the conviction that policy disagreements with Johnson would be inexorably attributed to a personal vendetta. "A lot of what Kennedy said honestly in those days," Walter Mondale recalled, "was dismissed by the press on the grounds that it was politically motivated, that he was trying to get Johnson."[109] "He would hold back in regard to criticism of the President," said Fred Harris. ". . . He was afraid that that would appear to be political . . . that he was speaking out against Johnson out of either personal political ambition or unreasonable animosity."[110] "I think he bent over backwards," said O'Donnell, "to try to give the President whatever benefit there was."[111]

He was circumspect even with his closest senatorial colleagues and his own staff. "I never found Bob Kennedy badmouthing Johnson," said George McGovern.[112] "I don't ever recall the Senator talking in personal terms of the President," said his first press secretary. ". . . I heard him rebuke Walinsky for saying something about 'Lyndon' rather uncomplimentary. You know, it was just simply, 'OK, Adam, that's enough. That'll be enough of that.' So he didn't tolerate it."[113] Before the spring of 1967, said his second press secretary, "I never heard him say anything derogatory about the President."[114] But circumspection did not destroy his feelings. He saw Johnson more than ever as a formidable but flawed man, powerful and dangerous. The President, he confided to Barrett Prettyman, "does not know how to use people's talents, to find the very best in them and put the best to work. But more than any other man, he knows how to ferret out and use people's weaknesses."*

He vigorously supported Johnson's Great Society. Privately he

---

* Handwritten notes, September 1965, Prettyman Papers. Johnson once said to Nancy Dickerson, "I'm just like a fox: I can see the jugular in any man and go for

resented the popular explanations of Johnson's legislative success. The Goldwater debacle in 1964 had given Johnson nearly 40 new Democrats in the House. This made him the first Democratic President since Roosevelt before 1938 with a working progressive majority in Congress. The result was the dazzling legislative record of 1965–66. The newspapers ascribed the result not to parliamentary arithmetic but to presidential genius and contrasted his accomplishments with his predecessor's failures.* Such comment wounded Kennedy. Congressman Donald Riegle of Michigan, watching Kennedy when the President addressed joint sessions of Congress, recalled that he "seldom clapped; he just seemed to smolder."[115]

He thanked heaven every day, he told me, that the Vice Presidency had fallen through; "it would have been a miserable and hopeless relationship, and nothing but trouble could have come from it."[116] The new life was certainly not boring. "Since you left home," he wrote Anthony Lewis, now head of the *New York Times*'s London bureau,

> I have become a beloved figure. The *New York Times*, the liberals in New York City, ~~Lyndon Johnson~~ — all like me.... There have been a few black clouds overhead on occasion. Last week at various hearings on minimum wage, migrant workers and car safety I had slight altercations with the President of the National Association of Manufacturers — the head of the Chamber of Commerce — the head of the Farm Bureau — the President of the G.M. [General Motors] Corporation ($500,000 salary), the chairman of the board of GM ($650,000 or perhaps it is vice versa) and the President of the Chrysler Corporation.[117]

"I didn't really want to be a Senator," Kennedy told an English journalist in 1967, "but I like it more than I expected to."[118]

---

it.... I keep myself on a leash, just as you would an animal" (Nancy Dickerson, *Among Those Present* [New York: Ballantine reprint, 1976], 177).

* Johnson did work harder than John Kennedy at congressional cajolery, bargaining and arm-twisting. But, when Johnson lost 48 Democratic House seats in the 1966 election, he found himself, despite his alleged wizardry, in the same condition of stalemate that had thwarted Kennedy and, indeed, every Democratic President since 1938. Had the sequence been different, had Johnson been elected to the Presidency in 1960 with Kennedy as his Vice President, and had Johnson then offered the 87th Congress the same program actually offered by Kennedy, the probability is that he would have had no more success than Kennedy — perhaps even less because he appealed less effectively to public opinion. And, if Johnson had died in 1963 and Kennedy had beaten Goldwater by a large margin in 1964, then Kennedy would have had those extra votes in the House of Representatives, and the pundits of the press would have contrasted his cool management of Congress with the frenetic and bumbling efforts of his predecessor. In the end, arithmetic is decisive.

30

The Foreign Policy Breach:
Latin America

ROBERT KENNEDY'S FIRST POLICY disagreements with the Johnson administration arose in foreign affairs. He watched the evolution of Latin American policy with alarm. The snuffing out of negotiations with Castro, the Thomas Mann appointment and the controversies over Panama and Cuba were followed in April 1964 by a generals' coup against the Goulart regime in Brazil. Kennedy had no use for Goulart. But he thought that Johnson embraced the new military dictatorship with unbecoming alacrity. He was still in the cabinet, however, if now shut out of foreign policy, and he held his peace.

I

The Alliance for Progress, in the hands of Johnson and Mann, was undergoing a basic transformation — so much so that the historian must talk of two Alliances for Progress. Another program by the same name now struggled on after the political and social components of Kennedy's Alliance — i.e., its heart — had been removed. With the coming of Mann, wrote William D. Rogers, who stayed on as the Alliance's deputy administrator till 1965, "a more dramatic shift in tone and style of U.S. Alliance leadership would have been difficult to imagine."[1]

Mann took over at a propitious time for eviscerating the Alliance. In 1961–63 the fear of Castro had given urgency and legitimacy to pressure from Washington for social change. When this fear diminished — as it had after the missile crisis and after the success of counterinsurgency in Venezuela — interest in social change diminished too. This encouraged Mann to liquidate two of the three goals

of Kennedy's Alliance — structural reform and political democratization — and to convert much of what remained into an instrumentality for North American corporations. "Not until Tom Mann came
back in 1964," said Alphonse de Rosso of Standard Oil of New Jersey,
"did the business community feel that it was 'in' again with the United
States government."[2] So Mann suspended virtually all aid to Peru in
an effort to force the democratic, pro-Alliance Belaunde government
to come to terms with the International Petroleum Company, a Standard Oil of New Jersey subsidiary.

John Kennedy, said Teodoro Moscoso, the Alliance's first chief in
February 1965, had "stood for change — revolutionary change —
and he said so even when it hurt in exalted places. . . . Do we remember [now] that there is a revolution going on?"[3] "The Alliance
for Progress is dead," wrote the caustic Mexican Victor Alba. "What
is left is a bureaucratic structure, mountains of mimeographed paper,
a sarcastic smile on the lips of the oligarchs, and pangs of guilt on
the part of the politicians of the left who did not take advantage of
the Alliance and make it theirs."[4] "The Alliance for Progress, deprived of its leader," wrote Juscelino Kubitschek of Brazil, whose
Operation Pan America had been one of the Alliance's inspirations,
"foundered on a series of acronyms as useless as they were pompous."[5] The "vitality and spirit" with which John Kennedy had imbued the Alliance, said Juan Bosch of the Dominican Republic, "died
with him in Dallas."[6]

II

In the Senate, Robert Kennedy continued to watch in silence. As
usual, he doubted that criticism of Johnson would be received on its
merits. Then, on April 24, 1965, came an uprising in the Dominican
Republic. The target was the ineffectual authoritarian regime installed by the Dominican military in September 1963 after the overthrow of Bosch's ineffectual democratic regime. The rebels were a
mixed group of Boschists, disgruntled officers and romantic leftists.
A panic-stricken American ambassador in Santo Domingo sent out
hysterical cables about bloodbaths and the capture of the revolt by
the Communists. An inexperienced CIA director in Washington,
seven hours in office and zealous to prove his anti-Communist credentials, relayed unverified messages from secret operatives in Santo
Domingo about decapitations in the streets and provided dubious
lists of the Dominican Communists who were supposedly taking
everything over. Johnson sent in 22,000 American troops. He did
this with minimal attention to the Organization of American States,

an institution of which he once said, "It couldn't pour piss out of a boot if the instructions were written on the heel."[7]

I was in Buffalo at the end of a lecture tour on April 30 when Bill Moyers tracked me down and asked me to come at once to Washington. On my arrival, Moyers said that Harriman and Moscoso had gone to various capitals in Latin America to explain the United States intervention; would I do the same in Costa Rica and other Central American countries? Briefed to the teeth by the CIA and foolishly believing its absurd reports, I agreed to go. Then I discovered that Romulo Betancourt was in Washington. I strongly urged on Moyers and Bundy that Johnson call in Betancourt at once. Johnson refused, though in a rambling television speech the next day he erroneously cited Betancourt as a supporter of American intervention.

Since no one in the White House, given their master's attitude, would talk to Betancourt, I decided to see him myself. Ever since Trujillo had sent his thugs to kill Betancourt, the Venezuelans had maintained a top-flight intelligence operation in the Dominican Republic. To my astonishment, I found Venezuelan intelligence entirely skeptical of the CIA thesis. Betancourt told me he had every confidence that the United States could work with the non-Communist leaders among the rebels and swiftly isolate the few Communists.[8] I begged off the Central American mission.

I had several talks during this time with Kennedy. He thought the American military intervention an "outrage" and asked me to put something on paper for him. On May 6 he warned the Senate that the determination to stop Communist revolution in the hemisphere "must not be construed as opposition to popular uprisings against injustice and oppression just because the targets of such popular uprisings say they are Communist-inspired." He criticized the policy of driving "the genuine democrats in the Dominican revolution into association with the Communists by blanket characterizations and condemnations" and called for assurances to "all honest Dominican democrats, including those who took part in the revolution," of a future role in their country. In no case, he added, should the United States "act on our own without regard to our friends and allies in the Organization of American states." Some might dismiss the OAS as weak. "One way to make it stronger is to use it."[9]

The Dominican affair provoked bitter criticism throughout Latin America. In the United States, apprehension spilled over to Johnson's foreign policy in general. "When I consider what the administration did in the Dominican Republic," Adlai Stevenson told me — it was our last talk before he died in July — "I begin to wonder if we know what we are doing in Vietnam." Reluctant as he was to praise John Kennedy, Stevenson nonetheless contrasted Johnson's impulsive and extreme reaction with the methodical exploration of alternatives during the missile crisis.[10] One began to wonder whether Johnson understood the uses of power. John Kennedy had always tried to match threat and response. Johnson, it appeared, was the aficionado of overkill in foreign as well as in personal relations. In any event, the Dominican intervention irretrievably shattered his foreign policy consensus.

III

On June 23, 1965, Robert Kennedy made his formal maiden speech in the Senate. He chose the subject that had most consumed his brother in the last year of his life — the spread of nuclear weapons. This, "not Vietnam, or the Dominican Republic," Kennedy said, was "the most vital issue now facing this Nation and the world." Should nuclear weapons become generally available, the planet would be at the mercy of irresponsible military commanders, internal dissidents, anonymous madmen. Every passing crisis "might well become the last crisis for all mankind." He called on the American government to extend the test ban treaty of 1963 to underground tests and to seek a new treaty covering nonproliferation. Both superpowers must set an example to mankind by cutting back their own nuclear forces or, at the minimum, by freezing these forces at their present level. He also said that "we should vigorously pursue negotiations on this subject with China," then regarded as the most fanatic of Marxist states.[11]

Seventeen senators rose to praise his remarks. The White House restrained its enthusiasm. The speech, though not expressly critical of the administration, was filled with somber quotations from John F. Kennedy about nuclear proliferation and carried the inescapable implication that Johnson was not doing enough to stop it. Johnson was so annoyed that he struck the disarmament proposals from his own pending address to the United Nations lest someone think he was following Robert Kennedy's lead.[12]

The nuclear proliferation speech deepened mutual suspicions over foreign policy. But Latin America especially nagged at Kennedy. He had been there only once — the overnight mission to Goulart in 1962 — since the jaunt with Lem Billings after his discharge from the Navy in 1946. Receiving an invitation to speak in Brazil, he decided to take advantage of the November recess and revisit the subcontinent. In September Jack Valenti asked Johnson for guidance about the Kennedy trip. Jack Hood Vaughn had become assistant secretary for inter-American affairs when in 1965 Mann was promoted to the more powerful job of under secretary for economic affairs; and Bundy had proposed that he brief Kennedy. "Vaughn's question to me," Valenti wrote the President: "Should he discourage Kennedy from going to Argentina, Chile and Venezuela? And what kind of briefing should he give Kennedy?"[13]

Johnson's answer can be safely surmised. When Kennedy arrived for the briefing, he and Walinsky were placed on one side of a long table. The State Department people all sat on the other side. It had the aspect of an adversary proceeding from the start. Vaughn, ordinarily affable, was in an extraordinarily aggressive mood. "The only thing I could think of," said Frank Mankiewicz, who represented the Peace Corps at the briefing, "was that someone had given him orders to be just as hostile and bitter as he could." Kennedy began by wondering what he should say about the Dominican intervention. Vaughn said, "In the first place, nobody will ask you about it because they don't care about that issue. No one asks about that any more." Kennedy said, "Well, you and I don't talk to the same Latins because that's all they ever ask me." He offered to bet that it would be one of his first three questions in Latin America. Vaughn took the bet, adding, "If they *do* ask you, you can always tell them what your brother said about Cuba." "At that," noted Walinsky, ". . . that look comes over Robert Kennedy's face. . . . Those eyes would just go dead flat cold. I mean it was enough to wither tree branches a hundred miles away." Kennedy said frigidly, "What in particular were you suggesting?" Vaughn, after fumbling a moment, mentioned the statement that the United States would not tolerate communism in the western hemisphere. Robert Kennedy said softly, "I just hope you're not using anything that President Kennedy ever said to justify what you did in the Dominican Republic . . . because you know I opposed that."

Despite the withering look, Vaughn persisted in referring to President Kennedy as "your brother." At one point he said that the United States was in trouble in Peru because "your brother" had suspended relations after the military coup in 1963; "we'll never do that again while I'm Assistant Secretary." Kennedy said he thought it was a good thing to have done. Mankiewicz, who had been Peace Corps director in Peru at the time, agreed. The talk turned to the International Petroleum Company dispute. Kennedy questioned the punitive suspension of aid to a democratic regime. "Why," he asked, "should the [American] government get into a contest between the Peruvian government and a private American oil company?" Vaughn defended the oil company.

They moved on. Kennedy said, "What can I say in Brazil, where they've outlawed political parties and closed down the congress and are denying people their rights and so forth?" A Brazil desk officer read off a prepared statement in departmentese expressing regret that a great power had seen fit to curtail certain freedoms temporarily, etc. Kennedy cut him off, saying, "I don't talk like that." Vaughn said, "Well, why don't you just say nothing?" Kennedy said, "Are you kidding?"

"Well, Mr. Vaughn," Kennedy summed up, "let me get this straight. You're saying that what the Alliance for Progress has come down to is that if you have a military takeover, outlaw political parties, close down the congress, put your political opponents in jail, and take away the basic freedoms of the people, you can get all the American aid you want. But, if you mess around with an American oil company, we'll cut you off without a penny. Is that right? Is that what the Alliance for Progress comes down to?" Vaughn said, "That's about the size of it."[14]

IV

The Kennedy party left Miami for Lima on November 10, 1965. Ethel and Angie Novello came along, as did Walinsky, Richard Goodwin, John Seigenthaler, Thomas Johnston, head of Kennedy's New York City office, and assorted friends and reporters. William and Jean vanden Heuvel joined the group in Peru.

The embassy in Lima had wanted Kennedy to go to bullfights, American-owned factories and a white-tie reception for the King of Belgium. "Robert F. Kennedy doesn't even own a white tie," John Nolan, who scheduled the trip, had told them.[15] They now evaded the embassy cordon. Kennedy's first meeting, with university students, struck the keynote. "The responsibility of our time," he told them, "is nothing less than a revolution" — a revolution that would be "peaceful if we are wise enough; humane if we care enough; successful if we are fortunate enough. But a revolution will come whether we will it or not. We can affect its character; we cannot alter its inevitability."[16] The third question from the audience dealt with the Dominican Republic. Kennedy cabled Vaughn: YOU LOSE THE BET.

The questioners went on about American imperialism. Kennedy said that there had been too much American aid to right-wing regimes but pointed out that left-wing countries — Yugoslavia, Poland, Ghana — had received aid too. It was too easy to make the Yankees the universal alibi. "You are the Peruvian leaders of the future. If you think the Alliance for Progress is imperialistic, then don't join it. . . . You have to decide what is in your interests. If you object to

American aid, have the courage to say so. But you are not going to solve your problems by blaming the United States and avoiding your own personal responsibility to do something about them."[17]

This theme was on his mind when he met a group of Peruvian intellectuals at the house of the artist Fernando Seizlo. It reminded Tom Johnston of a meeting of reform Democrats on the West Side of Manhattan. To Kennedy's irritation, they showed no concern about their own problems — the *barriadas* of Lima or the desperately poor Andean Indians — but complained endlessly about American imperialism and the International Petroleum Company. Kennedy finally said, "Well, why don't you just go ahead and nationalize the damn thing. I mean nothing's going to happen.... The United States government isn't going to send destroyers or anything like that. Why don't you just do it?" They were appalled; the Rockefellers ran United States policy; the Rockefellers wouldn't permit it. "Oh, come on," said Kennedy. "In our country we eat Rockefellers for breakfast."[18] A thoughtful guest had hidden a tape recorder under a couch, and a few days later a popular Peruvian magazine put the Rockefeller crack in a headline. A reporter in Buenos Aires asked Kennedy later whether it was true that he had breakfast every morning with the Rockefellers.[19]

At Cuzco a crowd of two thousand burst through a barbed-wire fence to greet him. Kennedy cut his right cheek in the melee but declined a tetanus shot and wore a ripped suit for the rest of the day. He talked to Indians and Peace Corps volunteers and on November 13 flew to Chile, to be inundated by more crowds and more headlines. His old friend Ralph Dungan was now American ambassador. They relaxed in Dungan's residence for a few hours. It was the first day on the trip, someone noted, that Kennedy had failed to visit a slum.[20]

He had luncheon and a meeting of minds the next day with Eduardo Frei, the Christian Democratic President of Chile and, along with Betancourt, the emblem of the Latin American democratic left. The day after, he went to Concepción, Chile's third largest city and a center of Communist enthusiasm. Marxist students at the university threatened a mass protest. A delegation called at his hotel. "Will you disrupt the meeting if I come?" he asked. The student leader said, "We do not condemn you personally, but as a representative of a government whose hands are stained with blood. If it was up to me, I would not let you speak."

"You describe me with blood all over my hands," Kennedy responded. "I haven't eaten you up since I have been here. I haven't had a marine stick a bayonet in you.... The greatest indictment of your position is that you won't let me speak. I think it is a self-confession of the error of your position." He proposed a deal: "I would speak, one of your people would speak, and then I will answer questions. If I don't answer the questions satisfactorily, then, hell, your position is much stronger. Aren't you confident? You don't sound it."[21]

After the students left, Seigenthaler and vanden Heuvel advised Kennedy not to go. Universities were off limits to police by Latin American tradition, and the rector could not guarantee Kennedy's safety. Then two Christian Democratic students knocked on the door. "If you don't come," one said, "it will be a great victory for the Communists." This made up Kennedy's mind. That night he entered a gymnasium packed with students screaming "Kennedy — paredón" ("to the wall") and throwing eggs and garbage. "It was really a frightening goddamned thing," said Martin Arnold of the New York Times.[22] Eggs splattered over vanden Heuvel and others, but Kennedy walked along, never looking back, and was untouched. "If these kids are going to be young revolutionaries," he muttered to Seigenthaler, "they're going to have to improve their aim."[23]

He tried to make himself heard from the platform. After a moment he challenged the Marxists to debate; "Will you test your ideas before the students of this university?" Even as they shouted "Kennedy — paredón," some had leaned across the rails to shake his hand; and now he went toward them and shook hands with the Marxist leader he had met in the afternoon. Another student spat in his face. The leader grabbed the spitter by his shirt and pulled him away. The pandemonium continued. Finally Kennedy left to a thunderous ovation. They went to bed well after midnight. At 4:00 A.M. vanden Heuvel roused the party: "Come on, the senator wants to go to the coal mines."[24]

The miners' union was solidly Communist. The embassy had kept the mine off Kennedy's schedule, and mine officials now begged him not to go down the shaft. He went anyway. "All these coal miners were wild to see him," said Arnold. "It was a miserable dank place with water dropping off and dust coming in. A terrible place! We went to the end of the shaft. It was a mile or two out into the ocean."[25] "If I worked in this mine," Kennedy told an American reporter, "I'd be a Communist too."[26]

He flew the next morning to Argentina for a frenzied forty-eight hours. Over a thousand people awaited him at the airport.[27] Harry Hopkins's son Robert, a Buenos Aires businessman, wrote me later: "Even some of the most fanatic nationalists were won over by his enthusiasm, his desire to understand Argentine problems, his frank answers to provocative questions, his concept of U.S. policy towards Latin America and by his youth. By going straight to the people, he created a surge of pro-U.S. sentiment. . . . If he were an Argentine, he could be elected President of Argentina tomorrow."[28]

On November 20 they left for Brazil. It was his fortieth birthday. Ethel had arranged a party at the São Paulo house of Mildred Sage, who was handling the Brazil schedule. There were the usual birthday poems, skits and jokes. Displaying a toy airplane, Ethel explained that it was a U-2 sent by Lyndon Johnson to report on Bobby's activities. Someone began pulling the ends of party favors to make them explode. The series of quick bangs suddenly sounded like shots. Kennedy sunk his head in his hands and said, "Oh, no. . . . Please don't."[29] On November 22, two years after Dallas, they went to mass in Bahía — "that gorgeous church with all the gold over it in that poor little town," Angie Novello remembered it; ". . . and then the first place he headed for was the *barriada* . . . and all the mud"[30] — mud and open sewers and steaming heat and a stench so vile that it drove the Brazilian security police to refuge in their car. He gathered barefoot children around him in a forlorn community center named for John Kennedy. "President Kennedy was most fond of children," he said. "Can I ask you to do a favor for him? Stay in school, study hard, study as long as you can, and then work for your city and Brazil."[31] On the plane north he sat by himself, his head buried in his arms.[32]

More than a hundred thousand people turned out later in the day in Natal. Exhilarated, he leaped to the top of the truck and shouted to the crowd: "Every child an education! Every family adequate housing! Every man a job!" They spent the night at Recife, in the heart of the horribly poor northeast, and the next morning walked the sugar cane fields where cutters worked for less than a dollar a day. Discovering that the wage was below the legal level, Kennedy became a zealous advocate of the Brazilian minimum wage law. "You're breeding your own destruction," he told a landowner, ". . . if you don't pay people a decent wage."[33] At a university, they asked him about civil rights. Politely challenging the Brazilian myth of racial equality, Kennedy said, "I don't see many dark-skinned faces in this audience."[34]

The party had reached Brazil in a state of exhaustion. Kennedy sent John Nolan a cable: NEXT TIME I SCHEDULE YOU.[35] Some of his entourage appeared petulant and highhanded. They did not like the military regime anyway and were not minding their manners. Mildred Sage found Kennedy himself, however weary, entirely co-operative. He was always considerate in the *favelas* and sugar fields. Fatigue showed when he encountered the establishment. He lectured business leaders, argued angrily with Roberto Campos at the American embassy and harassed Castello Branco, the military President, about the minimum wage. Fatigue showed in other ways. Sitting in a café in front of a Rio hotel on November 24, he heard the crackle of shots. He jumped from his seat, then realized it was the backfire of a car. "Sooner or later," he told Mildred Sage. "Sooner or later." [36]

Goodwin advocated a weekend in the jungle for relaxation. They flew to Manaus, deep in the interior, where they made their way up the Amazon in a paddle-wheel steamboat. Determined to go even farther, Kennedy boarded an ancient single-engine seaplane. Kissing Ethel goodbye, he said, "I must be crazy to get on this thing." An interlude of reckless adventure and misadventure culminated in a dash down foaming rapids in a dugout canoe and swims in piranha-infested waters. "Piranhas," Kennedy said philosophically, "have never been known to bite a U.S. senator." [37] There followed two busy days in Caracas. At the end of the month they were back in Washington.

v

An unsigned report to Lyndon Johnson from an American in Latin America speculated that Kennedy's trip was meant to launch his campaign for the presidency.

> Without actually saying so, he disassociated himself from the Johnson Administration. He portrays himself as the complete liberal. His inference was that Johnson is an old-guard imperialist whose administration is unaware of South American problems.... Embassy aides in the countries he visited shuddered at some of the things he said and did. Nevertheless they waited on him hand and foot.... The newsmen traveling with him from the New York Times, Herald-Tribune, etc., appeared to be under his spell.... Americans closest to the scene ... resented the way he and the group acted. They were cocky and over-bearing and made themselves as much at home as if they were in Massachusetts. [38]

One of the accompanying newspapermen, evidently not fully under Kennedy's spell, described the trip in the *Saturday Evening Post* under the title "The Compulsive Candidate" with the subtitle: "Robert Kennedy Runs for President Every Day." [39] Many found it hard to explain his intervention in the domestic problems of Latin America on other grounds. The theory that he had gone south in order to dish Lyndon Johnson was one more expression of the reductionism that so exasperated Kennedy in these years. In fact, Martin Arnold said, "he did a good job for President Johnson, really, in the sense that he used to tell these people that Lyndon Johnson had come from the soil and that he would understand their type of problems. He was very good that way, even though the two men obviously hated each other."[40] When Castello Branco of Brazil, after enduring Kennedy's complaints about the poverty in the northeast, inquired into the purpose of his trip, Kennedy answered slowly that his only political ambition was reelection to the Senate; Johnson would be the Democratic candidate in 1968. "My brother," he finally said, "cared enormously about Latin America. So I came to get myself educated."[41]

The sense of legacy was the essential reason. His hope was to remind the Latin Americans that there was more to the United States than the International Petroleum Company. In this he was successful. Kennedy, wrote Paulo de Castro in *Correio da Manha* of Rio, was

> the other face of America, the liberal face, which believes in democracy and does not profess the dogma of the West's salvation through the militaristic state, by interventions and the eternity of the oligarchies. . . . Robert Kennedy came to revive the message of liberal revolution that made the United States an exemplary nation, a revolution forgotten by those that today intend to transform a great democratic country into a protector of tyrannies.[42]

Kennedy believed in the American liberal revolution. He felt an obligation to do what one man could to help defenseless people in other parts of the world. He was doing this as an individual, not as a government. He did not in the end suppose it could be done by governments; certainly not by intervention in the Dominican fashion. After all, as Walinsky said, he was "a man who had had several years of experience in trying to affect conditions in a place like Mississippi but not to do it simply by moving in the fiat of the federal government." His idea was to help awaken the people and enlighten the oligarchies in the hope that countries might be moved to solve their own problems. It was a quixotic venture uncommon in history, a premonition perhaps of the world society someday to come. The Latin American trip, the response of the students, the outpouring of crowds, strengthened the feeling that it might be possible to rouse other countries to humanize their policies. "You may think that's a great act of hubris," said Walinsky, "but, indeed, he had the potential to do it and to some degree he probably did."[43]

He hoped even more to humanize the policy of his own country. The big problem, he said on *Meet the Press* the week after his return, was the impression in Latin America that "business determines the internal policy of the United States." The related problem was indiscriminate anticommunism. "If all we do . . . is to associate ourselves with those forces which are against subversion and against Communism . . . then I think it is self-defeating and will be catastrophic."[44] Such talk was not always appreciated. "What we need to preach in Latin America to those young radicals," said one business leader, "is an acceptance of the free-enterprise system and a respect for property rights. And we can't do it if Bobby Kennedy is going down there and upsetting everything we've been saying."[45]

He set to work on a speech designed to recall the Alliance to its original purposes. Walinsky was the main draftsman. Frank Mankiewicz helped too; and, when Barthelmes left the Kennedy office in the spring of 1966, Mankiewicz replaced him. Son of the Herman Mankiewicz who had written *Citizen Kane* and nephew of Joseph L. Mankiewicz, author and director of other notable films, he sprang from an alert and witty tradition and quickly became an intimate. ("Don't worry about Mankiewicz's finances," Bill Moyers wrote Kennedy from the White House. "I have arranged [for] CIA to pay half his salary since he will be working for me part-time.")[46]

Kennedy gave the speech in the Senate on May 9 and 10, 1966. There could be, he said, "no preservation of the status quo in Latin America." The Alliance had been "a pledge of revolutionary change." It proposed an end to the "closed society, a society which reserves all wealth and power and privilege for the same classes, the same families, which have held that wealth and power for the last 300 years." Reform ought to be "a condition of full participation in the Alliance. . . . Aid should not be withheld to force special advantages for U.S. business." As for the threat of communism, "Batista, not Castro," Kennedy said, "was the major cause of communism in Latin America. . . . If we allow communism to carry the banner of reform, then the ignored and the dispossessed, the insulted and injured, will turn to it as the only way out of their misery." He summed up: "There is a revolution now going on down there, and we must identify ourselves with that revolution."[47]

Nothing could have been further from the mood of Washington in the middle sixties with a Latin American policy responsive to American corporations and a President increasingly obsessed with the war in Indochina.

31

Vietnam Legacy

W HEN JOHN AND ROBERT Kennedy visited Vietnam in 1951, they had been greatly "impressed," Robert recalled in 1964, by "the toughness of the French soldiers. A lot of them were Foreign Legion and paratroopers . . . and they had several hundreds of thousands and they were beaten."[1] The memory had abided. "We saw the position the French were in," Robert told Daniel Ellsberg in 1967, "and saw what they were trying to do to the Indochinese. And my brother was determined early that we would never get into that position."[2]

I

The Geneva agreement of 1954 ended the First Indochina War. The French departed. In 1955 South Vietnam became an independent republic. The new leader in Saigon was Ngo Dinh Diem, a robust nationalist opposed to the French and Ho Chi Minh's Communists alike and sponsored in the United States by such exemplary men as William O. Douglas and Mike Mansfield. Having argued for Vietnamese nationhood, John Kennedy felt a responsibility for the new nation. "This is our offspring," he said in 1956, "— we cannot abandon it." Still, much remained to be done "in a country where concepts of free enterprise and capitalism are meaningless, where poverty and hunger are not enemies across the 17th parallel but enemies within their midst. . . . What we must offer them is a revolution — a political, economic and social revolution far superior to anything the Communists can offer."[3]

When Kennedy came to office in 1961, no revolution had been offered or undertaken. Diem had emerged as a rigid and unpopular

despot. The American involvement had deepened, at least rhetori-
cally. "The loss of South Viet-Nam," Eisenhower said in a 1959
speech, "would set in motion a crumbling process that could, as it
progressed, have grave consequences for us."⁴ The wary general did
not, however, enlarge the American military mission in Saigon beyond the 685 advisers permitted under the Geneva agreement. His
South Asian worries were directed elsewhere. In his pre-inaugural
briefing of Kennedy, he ignored Vietnam and called Laos the "present key to the entire area," adding, according to Clark Clifford's
record of the meeting, that "he considered Laos of such importance
that if it reached the stage where we could not persuade others to act
with us, then he would be willing, '*as a last desperate hope, to intervene
unilaterally.*'"⁵ Even more than his advice to Kennedy to plunge
ahead on the Bay of Pigs, this warlike exhortation strikes a peculiar
note. It is hard to believe that the canny Eisenhower would himself
ever have gone it alone with American troops in Laos.

Whatever Eisenhower's motives, Kennedy had no desire to follow
his advice. But he had no desire either to be denounced by the
national military hero. He therefore enlisted Eisenhower's wartime
friend Harold Macmillan to argue the case against intervention. "As
I understand it," the British Prime Minister wrote fraternally to the
former President,

> President Kennedy is under considerable pressure about "appeasement"
> in Laos. . . . I should however be very sorry if our two countries became
> involved in an open-ended commitment on this dangerous and un-
> profitable terrain. So I would hope that in anything which you felt it
> necessary to say about Laos you would not encourage those who think
> that a military solution in Laos is the only way of stopping the Com-
> munists in that area.⁶

The Bay of Pigs finally destroyed any possibility of military intervention in Laos. "I don't think that there is any question," Robert
Kennedy noted in June,

> that if it hadn't been for Cuba, we would have sent troops to Laos. We
> probably would have had them destroyed. Jack has said so himself. . . .
> It was after Cuba that he began inquiring intimately into the situation
> in Laos. It was then that he found out that the Communists could send
> five men into Laos for every one that we sent in. That they could
> destroy the airports and therefore cut off our people after getting only
> a thousand or several thousand into Laos and that the only way really
> that we could win in Laos was drop the atomic bomb. . . . Therefore,
> in order to preserve Laos, for instance, we had to be prepared to engage
> in a major atomic war both with China and with Russia.⁷

After hearing the generals pursue their scenario to its apocalyptic
conclusion, the President instructed Averell Harriman to negotiate

the neutralization of Laos with the Russians. A year later, Harriman told me that the President's policy was still being "systematically sabotaged" within the government by the military and the CIA. "They want to prove that a neutral solution is impossible," he said, "and that the only course is to turn Laos into an American bastion." [8] But Harriman was tougher and shrewder than the opposition and had Kennedy's full support. Though the neutralization accord worked out at Geneva in July 1962 did not stop the fighting, it succeeded at least in removing Laos from the area of great-power conflict.

II

The problem remained, as Robert Kennedy described it in 1961, of "what we were going to do with South Viet Nam which [unlike Laos] was willing to fight and protect itself." [9] Diem had an army of 150,000, a Civil Guard of 68,000 and a Self-Defense Corps of 40,000 — more than 250,000 men in arms against an estimated 12,000 Viet Cong.* "Free world forces," the *New York Times* advised the administration in early April, ". . . still have a chance in South Vietnam, and every effort should be made to save the situation." [10]

The President now made a de facto deal with the national security establishment: if it went along with neutralization in Laos, he would do something for resistance in South Vietnam. The Pentagon called for the immediate dispatch of 3600 combat troops. [11] Kennedy, compromising, sent a hundred military advisers and four hundred Green Berets. "Even so small an increase," recalled Roswell Gilpatric, head of the Vietnam task force, "was greeted by the President with a great deal of impatience. He showed at the very outset an aversion to sending more people out there." [12]

In late April Kennedy discovered an unexpected ally — General Douglas MacArthur, who assured him that it would indeed be a "mistake" to fight in Southeast Asia. "He thinks," the President dictated in a rare *aide-mémoire*, "our line should be Japan, Formosa and the Philippines. . . . He said that the 'chickens are coming home to roost' from Eisenhower's years and I live in the chicken coop." † Soon

---

* "A Program of Action to Prevent Communist Domination of South Vietnam," May 6, 1961, Senate Foreign Relations Committee staff study, "Vietnam Commitments, 1961," 92 Cong., 2 Sess. (1972), 15–16. Substantially the same figures were given to John Kenneth Galbraith when he was briefed by the Saigon embassy, November 19, 1961 (Galbraith, *Ambassador's Diary* [Boston, 1969], 261).

† JFK, memorandum of conversation, April 28, 1961, JFK Papers. It may be well to mention the discussion Kennedy held a few weeks later with another mystical general. Charles de Gaulle claimed in his memoirs that Kennedy "made no secret of the fact that the United States were planning to intervene" in Indochina and ignored de Gaulle's prediction that "you will sink step by step into a bottomless military and

Kennedy brought MacArthur to Washington to lunch with a group of bellicose legislators. "He said," noted the Attorney General, who was also invited, "that we would be foolish to fight on the Asiatic continent and that the future of Southeast Asia should be determined at the diplomatic table."[13] Alexis Johnson of the State Department, present too, thought MacArthur's argument "not correct and not rational. . . . Nevertheless it made a very deep impression on the President. . . . I think that for the rest of the time he was in office this view of General MacArthur's . . . tended to dominate very much the thinking of President Kennedy with respect to Southeast Asia."[14] It made, said Maxwell Taylor, "a hell of an impression on the President . . . so that whenever he'd get this military advice from the Joint Chiefs or from me or anyone else, he'd say, 'Well, now, you gentlemen, you go back and convince General MacArthur, then I'll be convinced.' But none of us undertook the task."[15]

III

In the spring of the Bay of Pigs and the summer of Berlin, Vietnam was a minor concern. For Robert Kennedy it remained so until August 1963. But because he later had to struggle with his brother's Vietnam legacy, it is essential to understand what that legacy was.

The Pentagon kept up its clamor for combat units. American troops, the President observed to Arthur Krock in October 1961, had no business on the Asian mainland; moreover, the United States should not interfere "in civil disturbances created by guerrillas." He was stalling the military, he said, by sending Maxwell Taylor and Walt Rostow to look at South Vietnam.[16] Taylor had backed neutralization in Laos. No doubt Kennedy supposed him to be, along

---

political quagmire, however much you spend in men and money" (de Gaulle, *Memoirs of Hope: Renewal and Endeavor* [New York, 1971], 255-256).

In fact, their discussion concentrated on Laos, not on Vietnam. De Gaulle strongly supported Kennedy's proposal of a neutral government under Souvanna Phouma. The discussion of intervention arose when Kennedy said that the threat of intervention might be necessary to get Moscow to agree to neutralization. De Gaulle said more generally that Southeast Asia did not offer a good terrain for western troops or western politics, and advocated the neutralization of the entire area (see the account, based on the memoranda of conversations, in Arthur M. Schlesinger, Jr., *A Thousand Days* [Boston, 1965], 351). Ambassador James Gavin, who was present, later wrote that he was "startled" to read de Gaulle's account and that it "bore little relationship to what had been said. It all may have been on President de Gaulle's mind at the time, but it certainly was not what he said" (Gavin, reviewing David Halberstam's *The Best and the Brightest*, *Harper's*, December 1972).

It might be noted that on other occasions de Gaulle found it hard to resist the temptation to improve the record. Ambassador Bohlen never heard de Gaulle say, "I am not interested by anything you propose," when they met after Nassau on January 4, 1963, but the phrase appeared in de Gaulle's own minutes of the conversation (see Robert Kleiman, *Atlantic Crisis* [New York, 1964], 22).

with MacArthur, Matthew B. Ridgway and James Gavin, a charter member of the Army's 'never again' club on the question of an American expeditionary force in Asia. Rostow, the high priest of counterinsurgency, would presumably oppose the use of conventional force in a guerrilla war. Kennedy instructed Taylor to "bear in mind" that responsibility for "the independence of South Vietnam rests with the people and government of that country" and that, while the military problem was important, "political, social, and economic elements are equally significant."[17]

"The last thing he wanted," Taylor said later, "was to put in our ground forces. And I knew that. I had the same feeling he had on the subject. But all the way, starting with CINCPAC [headquarters of the commander-in-chief for the Pacific], the feeling was that we'd better get something into South Vietnam."[18] Within a week Taylor recommended that 8000 American combat troops be dispatched at once; "I do not believe that our program to save SVN [South Vietnam] will succeed without it." If infiltration continued from the north, the time might come "when we must declare our intentions to attack the source of guerrilla aggression." Above all, the United States must make a formal and total commitment to prevent the fall of South Vietnam to communism.[19]

Reflecting on the reception of his report, Taylor said later, "I don't recall anyone who was strongly against, except one man and that was the President. The President just didn't want to be convinced that this was the right thing to do.... It was really the President's personal conviction that U.S. ground troops shouldn't go in."[20] Kennedy told me sardonically, "The troops will march in; the crowds will cheer; and in four days everyone will have forgotten. Then we will be told that we have to send in more troops. It is like taking a drink. The effect wears off, and you have to take another."[21]

He also discussed his misgivings with Galbraith, directing him to stop in Saigon on his way back to his post in New Delhi and give the other side (which Galbraith brilliantly did). But what course to follow? "There are limits," he told Galbraith, "to the number of defeats I can defend in one twelve-month period. I've had the Bay of Pigs and pulling out of Laos, and I can't accept a third."[22] Once again, as in the spring, he compromised, rejecting combat units while accepting another increase in military and economic assistance to Diem. But he insisted, as William Bundy of the Defense Department later wrote, on "a stiffer and more *quid pro quo* political program than the *'first* support him, *then* help him to reform' approach of the Taylor-Rostow report."[23] And he absolutely declined in the National Security Action Memorandum of November 15 to make the total commitment to save Vietnam his advisers thought so essential. Nor did he suppose that limited assistance would weaken his control of the situation. When George Ball warned that it would finally lead to 300,000 American troops on the Asian mainland, Kennedy said, "You're crazy as hell; it can't happen" — not so long as he was President.[24]

He was under fire that autumn from Republicans of the "why not victory?" school. The collective security press, led by the *New York Times*, was also in arms. Kennedy feared that, if Congress and the newspapers learned the Pentagon wanted to send ground troops, his own hand might be forced. Accordingly, the White House disingenuously let the impression seep out that Taslor opposed combat units. Though the policy disagreement was thus concealed, the subsequent increase in military assistance was not. On December 11, 1961, the *Times* reported the arrival of two U.S. Army helicopter companies in Saigon — "the first direct military support by the United States for South Vietnam's war against Communist guerrilla forces."[25] On December 20 the *Times* told of uniformed American advisers "operating in battle areas with South Vietnamese forces" and, though not in combat, authorized to fire back if fired on.[26] No newspaper, no member of Congress, raised questions.

Three days before Christmas, Specialist Fourth Class James Thomas Davis of Livingston, Tennessee, became the first American soldier to die in Vietnam.

IV

The *Times*, denouncing Communist "aggression" in Vietnam, had said grimly, "The present situation is one that brooks no further stalling."[27] All the same, Kennedy was determined to stall. But his compromises, minor as they seemed at the time, were storing up trouble for the future. The establishment in February 1962 of the Military Assistance Command Vietnam under General Paul Harkins gave the Pentagon a new stake in the war. The *quid pro quo* stipulations on reform fell by the wayside. Reporters soon described American helicopters ferrying South Vietnamese troops to the front and American pilots strafing Viet Cong territory. By the end of 1962, twenty Americans had lost their lives in battle.[28] Neither Congress nor the press questioned the deepening involvement. All this encouraged the campaign for American combat units. "It is fashionable in some quarters," General Earle G. Wheeler, a later chairman of the JCS, declared publicly in November 1962, "to say that the problems in Southeast Asia are primarily political and economic rather than military. I do not agree. The essence of the problem is military."[29]

The "quarters" in disagreement with General Wheeler were led by Averell Harriman, now assistant secretary of state for Far Eastern affairs, Roger Hilsman, also of State, and Michael Forrestal in the White House. While accepting the strategic importance of South Vietnam, these men saw the problem as one of civil war rather than external aggression. The remedy, they believed, had to be at least as much political and economic as military. "The dasic question of the whole war," Hilsman and Forrestal reported to Kennedy after a trip to South Vietnam in January 1963, was "the attitude of the villagers." The folk in the countryside would remain apathetic "or even become pro-Communist if the government does not show concern for their welfare."[30]

Harriman and his allies advocated a combination of counterinsurgency — guerrillas had to be fought by guerrillas — with political and social reform. The introduction of American troops, they believed, would hand the nationalist issue altogether to the Viet Cong. It would also strengthen Diem's authoritarianism at the expense of the reforms essential for his survival. "In wartime," General Harkins opined, "you have to have someone who is as strong . . . as Diem was — a more or less benevolent dictator."[31] The Harriman group felt on the contrary that Diem was lost unless he relaxed his despotism and broadened his base.

The critique was persuasive. Alas, the remedy was not. "Diem will not reform," Galbraith wrote Kennedy after his visit to Saigon. ". . . He cannot. It is politically naive to expect it. He senses that he cannot let power go because he would be thrown out."[32] The very idea of democratization was unintelligible to Diem, reared to believe, as he once put it, that the "sovereign" was "the mediator between the people and heaven" and therefore entitled to "sacred respect."[33] The oriental mandarin regarded reform talk as typical American simplemindedness. He was right.

As for counterinsurgency, it was never really tried in Vietnam.* Taylor and Rostow, for all their counterinsurgency enthusiasm in Washington, roared home from Saigon dreaming of big battalions. Edward G. Lansdale, the one American with counterinsurgency experience in South Asia, accompanied their mission. But Taylor gave him the task of assessing the feasibility of building a fence along the northern frontier — a concept Lansdale considered preposterous.[34]

---

* It is sometimes said that Vietnam was used as the 'proving ground' of counterinsurgency. That is a myth. Counterinsurgency doctrine contributed to the 'pacification' program of 1967–68; but this was as a supplement to, not as a substitute for, a military strategy resting on large units, high technology and attrition.

Thereafter, though State repeatedly recommended Lansdale's assignment to Vietnam, the Pentagon as repeatedly vetoed him.[35]

The theory that Saigon's guerrillas should fight Hanoi's guerrillas foundered on the Pentagon's irrepressible instinct, as Forrestal put it, "to turn the war into a conventional and American enterprise."[36] The Special Forces were sent to remote regions to help peripheral groups like the Montagnards. At the end of 1963 there were only 100 Green Berets left in South Vietnam.* Hilsman had great hopes that a network of fortified 'strategic hamlets' could cut the Viet Cong off from the food and manpower in the villages. Diem handed the strategic hamlet program over to his creepy brother Ngo Dinh Nhu, who used it as a means of police control. Hilsman, after an inspection, pronounced it "a fraud, a sham."[37]

The military establishment pressed throughout for victory through large units, firepower and attrition. This was the opposite of counterinsurgency. The Counterinsurgency Group, wrote Charles Maechling, Jr., its staff director, never "dreamed that we would attempt to win over the peasantry of Vietnam by burning their villages, strafing them with rockets and machine guns, and then indiscriminately slaughtering the survivors."[38] The dissenters were wholly convincing in their objections to the high-technology war. Their own vision of a counterinsurgency alternative was, however, fantasy.

v

John Kennedy, I believe, belonged to neither school. He thought and said (privately) in 1961 that the United States had been "overcommitted" in Southeast Asia. Still, we were there, and he was willing to give restricted aid to a South Vietnamese government prepared to rally its own people and fight its own war. Diem, if he used the aid well, had the numerical superiority to put down the Viet Cong. It seemed a chance worth taking. But aid was to be restricted. "Resistance was encountered from the President at every stage," Gilpatric said later, "as this total amount of U.S. personnel deployment increased."[39]

Kennedy had no intention of dispatching American ground forces to save South Vietnam. Nor did he accept the Truman-Eisenhower-Pentagon view that a President had inherent authority to send an expeditionary force into battle. If combat troops "in the generally understood sense of the word" — units, not advisers — were required in Vietnam, he told a press conference in March 1962, that would be

* This sounds improbable, but it is so stated in Enclosure A of JCS to Special Group (CI), February 14, 1964, RFK Papers.

"a basic change . . . which calls for a constitutional decision, [and] of course I would go to the Congress."[40] As for counterinsurgency, if he believed it had any chance of licking the Viet Cong, he would surely have pushed it harder than he did. Lansdale, not Harkins, would have been in command in Saigon.

As against both win-the-war factions, Kennedy, I believe, was vaguely searching for a nonmilitary solution — vaguely because Vietnam was still a side show. In April 1962 Galbraith proposed that the United States offer "phased withdrawal" (there were then about 2000 troops in Vietnam) as an element in a larger deal with the Soviet Union, which in return might get Hanoi to call off the Viet Cong. Kennedy asked Defense for comment. After sneers at the idea of a "political solution," the chairman of the JCS, General Lyman Lemnitzer, denounced Galbraith for wishing the United States "to seek disengagement from what is now a well-known commitment to take a forthright stand against Communism in Southeast Asia."[41]

For the Chiefs the commitment may have been "well-known." But they had thus far failed in their efforts to force it on the President. A few days later a George Ball speech stimulated the militant Marguerite Higgins to lead her *Herald Tribune* story: "American retreat or withdrawal from South Viet-Nam is unthinkable, according to Mr. Ball. The American commitment, moreover, is now irrevocable." Ball, making a case, as lawyers do for clients — privately he opposed American involvement — portrayed the possible downfall of South Vietnam as "a loss of tragic significance to the security of free world interests in the whole of Asia and the South Pacific." Kennedy, disturbed, asked Bundy how Ball's disquisition had slipped by. The speech, Bundy informed the President, had "a tone and content that we would not have cleared, simply from the point of view of maintaining a chance of political settlement."[42]

In July 1962, despite the Joint Chiefs' excommunication of Galbraith, Kennedy instructed McNamara to start planning for the phased withdrawal of American military personnel from Vietnam. The assumption was that South Vietnamese troops would take up the slack as their capabilities improved under American training. The target date for complete disengagement was the end of 1965. After several tries, the military produced an acceptable plan in May 1963.[43]

The generals saw the withdrawal plan as a means of putting pressure on Diem to accelerate the training of Vietnamese troops. Kennedy, however, saw it at the very least as a means of setting a limit to demands from the national-security bureaucracy for escalation. He also saw it as the reserve plan for extrication. "McNamara indicated to me," said Gilpatric, "that this was part of a plan the President

asked him to develop to unwind the whole thing." John Mc-
Naughton, assistant secretary of defense for international affairs, told
Daniel Ellsberg, a young Pentagon hawk, in 1964 "that Robert
McNamara had told him of an understanding with President Ken-
nedy that they would close out Vietnam by '65, whether it was in
good shape or bad."[44]

VI

Kennedy's growing misgivings were soon reinforced by the man with
whom he had originally met Diem — Mike Mansfield. Mansfield,
who had served forty years before with the Marines in the Far East
and thirty years before as a professor of Far Eastern history, was no
novice in the area. At Kennedy's request he visited South Vietnam
in November 1962.

Back in Washington, Mansfield warned the Senate Foreign Relations Committee that the United States was nearing "the point at
which the conflict in Viet Nam could become of greater concern and
greater responsibility to the United States than it is to the Government and people of south Viet Nam." If the South Vietnamese
would not do more, Mansfield said, the United States should "reduce
its commitment or abandon it entirely."[45] In private Mansfield told
Kennedy he was wrong to keep on sending advisers; far better to
pull out the Americans and let the South Vietnamese stand on their
own feet. Kennedy, challenged on his policy of small concessions,
was annoyed. "I got angry with Mike for disagreeing with our policy
so completely," he said afterward to Kenneth O'Donnell, "and I got
angry with myself because I found myself agreeing with him."[46]

He was beginning to sense that small concessions were leading him
into a trap. "Kennedy assumed that the war was going well in 1961
and 1962," wrote Henry Brandon, the well-informed Washington
correspondent of the *London Sunday Times,* "but at the end of 1962,
in a conversation with Roswell Gilpatric, he talked in a restless and
impatient way about how the U.S. had been sucked into Vietnam
little by little."[47] Still, if the South Vietnamese were to have a run
for their — and our — money, how could he go public about withdrawal? Nothing would so quickly undermine Diem, or, for that
matter, so surely create trouble on the Hill, in the press, including
the *New York Times* and the *Washington Post,* and among the voters.

No doubt he felt he had to play the hand a little longer. Asked
about the Mansfield report, he told his press conference on March
6, 1963, that the fall of Southeast Asia would inevitably affect "the
security of India" and "begin to run perhaps all the way toward the

Middle East." July 17: "For us to withdraw . . . would mean a collapse not only of South Vietnam, but Southeast Asia. So we are going to stay there." September 2: "I don't agree with those who say we should withdraw. That would be a great mistake."[48] There were other such statements.

Did he really believe them? Or was it all patter intended to give Diem his chance and to keep the Pentagon and Congress quiet on the assumption that, if Diem's incapacity to win on his own became incontestable, the United States would have plenty of time to change course? But his public statements, while they improved morale in Saigon, also misled Americans about the national stake in Vietnam and thereby narrowed choices in Washington. Instead of controlling the situation, the situation was beginning to control him. He grew uncharacteristically snappish about pessimistic newspaper dispatches from Saigon, not, I believe, because he thought they would lead to demands for withdrawal — that was not the national mood in 1963, or for long years after — but because they might lead to demands for escalation. In midsummer, according to the Louis Harris poll, Americans agreed by a 2-1 margin that, "if the Communist threat to South grew worse," they would favor sending U.S. troops "on a large scale."[49] Kennedy also felt that press criticism decreased any possibility, slim as it might be, that Diem could pull it off.

The evidence suggests that, underneath his uncertainties, withdrawal, which had begun for him in 1962 as a precaution, was turning in 1963 into a preference. He asked his good friend Lester Pearson what the United States should do. The Canadian Prime Minister said frankly, "Get out." Kennedy said with equal frankness, "That's a stupid answer. Everybody knows that. The question is: How do we get out?"[50] He now told Mansfield he had been right about total withdrawal. "But I can't do it," he said, "until 1965 — after I'm reelected." Otherwise the Republicans might beat him in 1964 over the 'loss' of Indochina as they had beaten the Democrats in 1952 over the 'loss' of China.* When Mansfield left, Kennedy remarked to O'Donnell, who had been present, "If I tried to pull out completely now from Vietnam, we would have another Joe McCarthy red scare on our hands, but I can do it after I'm reelected. So we had better make damned sure that I *am* reelected."[51] Kennedy "felt we had

---

* Some later condemned this political calculation as immoral. Daniel Ellsberg in his *Rolling Stone* interview (December 6, 1973) thus charged Kennedy with "a willingness to keep on bombing the Vietnamese for a couple of more years in order to get through the election." In fact, there was little bombing in the Kennedy years. In any event, would it have been better to have lost in 1964 to a presidential candidate who agreed with General Curtis LeMay that North Vietnam should be bombed back to the Stone Age? The American people in these years were not of notably dovish spirit.

made an error," Mansfield told the columnist Jack Anderson in
1970. "He was going to order a gradual withdrawal. . . . [He] had
definitely and unequivocally made that decision . . . but he never had
the chance to put the plan into effect."[52]

Kennedy had said all along that the United States could not win
the war for the South Vietnamese. After two years of intensified
American aid, he was starting to conclude that they could not win it
for themselves. He commenced to prepare the American people for
the idea that the United States must limit its involvement. "In the
final analysis," he said on television on September 2, 1963, "it is their
war. They are the ones who have to win it or lose it. We can help
them, we can give them equipment, we can send our men out there
as advisers, but they have to win it, the people of Viet-Nam."[53]

VII

The late summer of 1963: Kennedy still playing out his public hand
while secretly wondering how to get out; his two schools of advisers,
united in perceiving an American responsibility to win the war but
bitterly antagonistic to each other. The issue between them was now
the survival of the Diem regime itself.

The Pentagon backed Diem more fervently than ever. The Harriman group had reached the conclusion that Diem and more immediately his brother Nhu and the serpentine Madame Nhu were
disasters. In the new ambassador to Saigon, Henry Cabot Lodge,
who replaced the complaisant Frederick Nolting, the anti-Diem
school at last had a purposeful, even highhanded, ally in Saigon.
Everyone knows the story of the cable of August 24, drafted by
Harriman, Hilsman and Forrestal and cleared loosely around town
on a summer weekend, authorizing Lodge to tell a camarilla of anti-
Diem generals that, if they overthrew Diem, the United States would
recognize their new regime. The impact in Saigon was negligible.
The generals did nothing. But the impact in Washington was shattering. The President, who was at Hyannis Port, thought the cable,
in the words of Robert Kennedy, "had been approved by McNamara
and Maxwell Taylor and everybody else, which it had not. . . . I
became much more intimately involved in it then."[54]

Up to this point the Attorney General had had little to do with
Vietnam. In February 1962, he had stopped over for two hours in
Saigon on his way from Indonesia to Siam. He did not leave the
heavily guarded airport but held forth boldly on the war. "We are
going to win in Viet-Nam," he said. "We will remain here until we

do win." A British correspondent noted that there had been American casualties; "do the American people understand and approve of what is going on?" "I think the American people understand and fully support this struggle," Robert Kennedy replied. ". . . I think the United States will do what is necessary to help a country that is trying to repel aggression with its own blood, tears and sweat."[55]

He accepted the party line as imparted to him by McNamara and Taylor. After a meeting of the Counterinsurgency Group in November 1962, Forrestal sent him a worried memorandum:

> I became concerned about the kind of information you seem to be getting on South Vietnam. Both Averell and I feel that the war is not going as well out there as one might be led to believe. . . . The reports we get indicate that the political problem is growing relatively worse. There has been very little indication to date that Diem's Government has been able to follow up the military operations with the type of social and economic programs which would convince the people whose security has been assured that they are better off with Diem than with the Viet Cong.[56]

Thereafter the Harriman group kept him abreast of its doubts. But in the spring and summer of 1963 the struggle for racial justice left the Attorney General little time for anything else.

Now American policy in Vietnam was in crisis, and John Kennedy wanted his brother at his side. The President thought the August 24 cable impulsive and precipitate. "He's always said that it was a major mistake on his part," Robert Kennedy recalled the next year. ". . . The result is we started down a road that we never really recovered from." John Kennedy became "very unhappy" with Harriman — so much so that Robert, noting that Harriman "put on about ten years during that period . . . because he was so discouraged," asked his brother "if he couldn't rehabilitate him by just being nice to him . . . because he's a very valuable figure."[57]

The day after the coup cable, Robert Kennedy talked to McNamara and Maxwell Taylor. "Nobody really knew what our policy was; it hadn't been discussed, as everything else had been discussed since the Bay of Pigs, in full detail before we did anything — nothing like that had been done before the decision made on Diem and so by Tuesday we were trying to pull away from that policy."[58] Stormy meetings followed. Harriman, Ball, Hilsman, Forrestal, said that the United States "must decide now to go through to a successful overthrow."[59] Taylor, McNamara, McCone, Lyndon Johnson, strongly opposed a coup. Nolting was brought in on August 27 to give the case for Diem. Gilpatric could not remember when anyone in the presence of a President "took the tongue-lashing that Nolting did from Harriman. And I don't think it would have been tolerated by the President from anybody else."[60] "The government split in two," recalled Robert Kennedy. "It was the only time really, in three years, the government was broken in two in a very disturbing war."[61] "My God!" John Kennedy said one day to Charles Bartlett. "My government's coming apart!"[62]

The Attorney General attended meeting after meeting. "He was beginning," said Forrestal, "to have serious doubts about the *whole* effort. . . . Was the United States capable of achieving even the limited objectives that we then had in Vietnam? Did the United States have the resources, the men and the thinking to have anything useful really to do in a country that was as politically unstable as South Vietnam was? Was it not possible that we had overestimated our own resources and underestimated the problem in South Vietnam? . . . He began forcing people to take a harder look at what it was we were doing there."[63] At the NSC meeting on September 6 Robert Kennedy said that the fundamental issue was what we thought we were doing in Vietnam. "The first question was whether a Communist take-over could be successfully resisted with any government. If it could not, now was the time to get out of Vietnam entirely, rather than waiting."[64] The question hovered for a moment, then died away, a hopelessly alien thought in a field of unexamined assumptions and entrenched convictions.

VIII

In Saigon Lodge was determined to overthrow Diem. THERE IS NO TURNING BACK, he had cabled Washington on August 29.[65] On September 11: TIME HAS ARRIVED FOR THE US TO USE WHAT EFFECTIVE SANCTIONS IT HAS TO BRING ABOUT THE FALL OF THE EXISTING GOVERNMENT AND THE INSTALLATION OF ANOTHER.[66] He had no doubt in his mind and ample Brahmin confidence in his judgment. The American newspapermen said cheerfully: "Our mandarin is going to beat their mandarin."

"The individual that forced our position really at the time of Vietnam," Robert Kennedy said in 1965, "was Henry Cabot Lodge. . . . The President would send out messages and he would never really answer them. . . . My impression was Henry Cabot Lodge didn't pay much attention because he wanted a coup. . . . It was an impossible situation." So impossible indeed that, according to Robert Kennedy, the President decided to call Lodge home in November "and discussed with me in detail how he could be fired, because he wouldn't communicate in any way with us." The Attorney General, who had questioned Lodge's appointment, was not surprised. Robert Kennedy had warned his brother, he later recalled, "that Henry Cabot Lodge in Vietnam would cause him a lot of difficulty in six months, and I brought that up, and he said it was terrific about me because I could always remember when I was right. And he said, do you remember when you were in favor of a tax increase [during the Berlin crisis in 1961]? and that would have had a very adverse effect, and that's when you weren't right."[67]

Lodge was a strong man with the bit between his teeth. On the other hand, his Washington instructions were hardly clear or firm — a reflection both of the continuing quarrel in Washington and of the President's own uncertainties. In the television interview on September 2 the President himself had said that, if Diem did not make changes in personnel and policy, his chances of winning the war would not be good. Thereafter he signed cables laying down a checklist of reforms to which Diem would not conceivably accede — from the exile of the Nhus to a free press and free elections.[68] All this could be interpreted as support for a coup. Robert Kennedy said, however, that the President "was against getting rid of [Diem] until you knew what was going to come along and whether the government that was going to replace it had any stability and in fact [whether it] would be a successful coup."[69] Reflecting the next spring, Robert said,

> Diem was corrupt and a bad leader and it would have been better if we didn't have him. But we inherited him. He came with the job. So what do you do? . . . It's better if you don't have him but you have to have somebody that can win the war, and who is that? . . . It's bad policy to get into for us to run a coup out there and replace somebody we don't like with somebody we do because it would just make every other country nervous as can be that we were running coups in and out.[70]

Here Robert Kennedy adopted the win-the-war perspective shared by supporters as well as by opponents of the coup. But, if John Kennedy had withdrawal in mind, there was all the more reason to take care. For a coup encouraged by Washington would drag the United States ever more irrevocably into responsibility for the fate of South Vietnam. And disengagement was increasingly, I believe, the presidential purpose. "By the autumn of 1963," wrote Henry Brandon, "he seemed sick of it, and frequently asked how to be rid of the commitment."[71]

At the end of September the President dispatched McNamara and Taylor on one more reconnaissance. "The question at the top of McNamara's mind," according to William Bundy, the mission's staff director, was: "Could the U.S. look forward . . . to the withdrawal of at least the bulk of its military advisors by the end of 1965? The insistence on this question shows the degree to which the planning of May [for phased withdrawal] had survived intervening events."[72] On his return McNamara answered the question in the affirmative, recommending in addition the immediate withdrawal of 1000 advisers by the end of 1963. "This is a Vietnamese war," his report said, "and the country and the war must, in the end, be run *solely* by the Vietnamese."[73]

This was the first application of Kennedy's phased withdrawal plan. Taylor went along as a way of bringing pressure on Diem.[74] The political school, however, regarded the recall of 1000 advisers with consternation. William Sullivan, the State Department representative on the McNamara-Taylor mission, threatened a minority report. The sentence about troop withdrawal was deleted. The mission then met with the President. At one point Kennedy took McNamara and Taylor into the Oval Office. When they returned, McNamara directed that the statement be restored. There was much "expostulation," Forrestal recalled. Kennedy listened impatiently, turned on his heel and left the room. The statement remained.[75]

Chester Cooper of McGeorge Bundy's staff, much agitated, rushed to Bundy, who was closeted with his brother. Troop withdrawal, Cooper said, would destroy any credibility the McNamara-Taylor mission could have had. "Both Bundys agreed, but Bill had little elbow room. Finally, in utter exasperation, Bill said, 'Look, I'm under instructions!'" McGeorge Bundy called McNamara but could not persuade him to change his mind. "McNamara seems to have been trapped too," Cooper concluded; "the sentence may have been worked out privately with Kennedy, and therefore imbedded in concrete."* Cooper's suspicion is persuasive. When McNamara left the White House to announce the start of troop withdrawal to the press, Kennedy called after him, "And tell them that means all the helicopter pilots too."[76]

Kennedy was equally insistent on preserving flexibility in declaratory policy. McNamara and Taylor had recommended that the withdrawal announcement affirm "the *overriding objective* of denying this country to Communism and of suppressing the Viet Cong insur-

---

* Chester Cooper, *The Lost Crusade: America in Vietnam* (New York, 1970), 215–216. William Bundy: "I have no recollection of either episode or attitude, but my notes do make clear that the sentence was regarded as settled all through the day, after the meeting between McNamara and the President in the morning" (Bundy ms. on U.S. policy in East Asia during the Kennedy-Johnson years, ch. 9). Bundy in retrospect finds it hard to resolve the contradiction between the report's political pessimism and its withdrawal proposal. No doubt the answer lies in Kennedy's private determination to begin, at whatever cost, a strategy of extrication.

gency." Kennedy had fought off a total commitment for three years and was not about to capitulate now. Instead the statement said mildly that it remained American policy, "in South Viet-Nam as in other parts of the world, to support the efforts of the people of that country to defeat aggression."*

If the South Vietnamese could not win their war for themselves, the United States had to lay the foundation for disengagement. The President called an all-agency conference in Honolulu to consider, as he somewhat confusedly told his press conference on November 14, "how we can intensify the struggle, how we can bring Americans out of there. Now, that is our object, to bring Americans home."[77] On November 20 the conference accepted a speed-up of the withdrawal plan.[78] Kennedy had trapped himself long enough and was looking for the way out.

IX

Did he have an escape route? O'Donnell once inquired how he could pull out without damaging American prestige. "Easy," Kennedy replied. "Put a government in there that will ask us to leave."[79] "We would withdraw the troops," he told his press conference in May 1963, ". . . any time the government of South Vietnam would suggest it."[80] "Remember Laos," he said to Hilsman, making it "abundantly clear," Hilsman later wrote, that if the Saigon government proved incapable of reform, "his intention was to negotiate a settlement along the lines of the 1962 Geneva accord on Laos."[81] His brother, Robert Kennedy told Ellsberg in 1967, would have arranged "a Laotian type solution, some form of coalition government with people who would ask us to leave."[82]

The irony is that Diem and Nhu, without Kennedy's knowledge, were engaged in secret negotiations that could have resulted in somewhat the situation Kennedy had in mind. The Ngo brothers were, in their anachronistic fashion, authentic Vietnamese nationalists. They were reluctant about American troops and resistant to American interference. "Those who knew Diem best," Robert Shaplen of the *New Yorker* wrote after twenty years in Vietnam, "feel that neither

---

* See the McNamara-Taylor recommendation and the White House statement (*Pentagon Papers*, vol. 2, 188, 753, emphasis added). The Pentagon statement would have had Kennedy "adhere to the overriding objective . . ."; but, since Kennedy had steadily vetoed persistent Pentagon attempts to slip by an unconditional commitment, there was nothing for him to adhere to; nor did he. The White House statement said only, "We will adhere to our policy of working with the people and Government of South Viet-Nam to deny this country to communism . . ." (also Peter Dale Scott in *The Pentagon Papers*, Senator Gravel Edition [Boston, 1971], vol. 5, 215-216).

he nor Nhu would ever have invited or allowed 550,000 American soldiers to fight in their country and to permit the devastation caused by air attacks."[83] Diem may also have felt, as Bui Kien Thanh has suggested, that massive American intervention would provoke massive Chinese intervention and deliver Vietnam to its historic enemy.[84] In May 1963 Nhu proposed publicly that the United States start withdrawing its troops. In the summer he told Forrestal in his "hooded" way that the United States did not understand Vietnam; "sooner or later we Vietnamese will settle our differences between us."[85] "Even during the most ferocious battle," Nhu said to Mieczyslaw Maneli, the Polish member of the International Control Commission established in 1954 to supervise the Geneva agreement, "the Vietnamese never forget who is a Vietnamese and who is a foreigner."[86]

Maneli was a political scientist rather than a career diplomat, a survivor of Auschwitz and Maidanek, an admirer of Kennedy, with whom he had discussed Vietnam in Poland in 1959.* In the spring of 1963 Roger Lalouette, the French ambassador in Saigon, asked Maneli to serve as an intermediary between the Ngo brothers and Hanoi. De Gaulle's interest was plain enough — to show up the blundering Americans, to rescue the west from a hopeless predicament, to restore French influence in Indochina. "The next step," Maneli reported to Warsaw in May, "will be neutralization. . . . In this way Vietnam, in addition to neutral Cambodia and Laos, will again become a pearl in the 'grandeur of France.'"[87]

Maneli shuttled between Saigon and Hanoi.† He pressed Ho Chi Minh and Phan Van Dong, the North Vietnamese Premier, about their peace terms. They replied that Hanoi would settle for a coalition government in a neutralized South Vietnam. "I asked," recalled Maneli, "if such a government could be headed by Mr. Diem. In the summer of 1963 the answer was finally yes." The North Vietnamese added that, if the Americans withdrew, Hanoi would give the south guarantees, including perhaps American participation in a supervisory process.[88]

As American pressure mounted against the Diem regime, Hanoi refrained from exploiting the confusion as by a new Viet Cong of-

---

* I thank Dr. Maneli for permitting me to read his account of this meeting in an unpublished paper, "Encounters with John F. Kennedy and Discussions on 'Democratic Socialism,' Polish Policy, and Vietnam," 4–11.

† He was not, he believed, the only channel. The North Vietnamese used him, he thought, "as an additional link to reinforce the dialogue they were having with Saigon through other contacts" (Stanley Karnow, "Lost Chance in Vietnam," *New Republic*, February 2, 1974). Nhu may also have been holding talks with NLF representatives, as stated in Bui Kien Thanh, "Mandarins of Vietnam," *International History Magazine*, January 1974.

fensive. Maneli cabled Warsaw on July 10: "This is certainly because it wishes [Diem and Nhu] to survive for a time yet — long enough to come to an agreement with them behind the Americans' backs." A month later Maneli concluded that there was "a supersecret agreement between Diem-Nhu and Hanoi" and that Hanoi and the National Liberation Front "at the first opportunity will back Diem against the Americans." "In conversation with me and with my deputy in Hanoi," the "Vietnamese comrades" spoke of the possibility of an alliance with Diem-Nhu against the Americans "plainly and without fuss."[89]

On August 29 de Gaulle publicly offered French assistance in transforming the two Vietnams into a unified state, free of all foreign influences. Lodge's evident desire to overturn Diem worried the French. "Only Diem," Lalouette told Maneli, "can conclude peace. . . . If Diem and Nhu are removed, all our plans designed to end the fighting and bring about agreement with the North will come to naught. . . . Any other government will be even more dependent on the Americans, will be obedient to them in all things, and so there will be no chance for peace."[90] Nhu himself encouraged Maneli to carry on. Maneli could not tell how serious the brothers were. "They were carrying on such a complicated and many-sided game that one could not be certain about the direction in which they were heading."[91] Nhu may well have been trying to blackmail the Americans into stopping their campaign against him by the threat of a deal with Hanoi, trying at the same time to see whether such a deal offered the Ngos a better prospect for survival. "He was," Maneli said later, "playing on many instruments at the same time."[92]

His hints about treating with Hanoi were not, according to William Bundy, "taken seriously in either Saigon or Washington."[93] For the pro-Diem faction, the idea of a Diem-Ho deal so flagrantly violated the theory of Diem as a last-ditch anti-Communist that it did not seem worthy of a moment's credit. The anti-Diem faction was quite willing to believe that Nhu and de Gaulle were up to some deviltry. But this only made it the more hostile to the notion of a deal. The State Department was already furious at de Gaulle for European reasons. James M. Gavin, who had served two years as ambassador to Paris, remembered a State Department officer calling de Gaulle "a bastard who is out to get us."[94] He was meddling in Indochina, it was supposed, in order to make sure that the United States could not succeed where France had failed. Neutralization, moreover, was dismissed as a prelude to communization. Even Kennedy, while observing that "anything General de Gaulle says should be listened to, and we listened," added, "It doesn't do us any good to say, 'Well,

why don't we all just go home and leave the world to those who are
our enemies.'"[95] The anti-Diem group saw the secret negotiatioñs,
if true, not as an opportunity for extrication but as a threat to victory. On August 30 a Hilsman memorandum suggested that, if Diem
were in fact negotiating with Ho, Washington should "encourage the
generals to move promptly with a coup."*

No one knew then whether the explorations had any reality. (No
one knows now.) Later Madame Nhu informed a visitor that "the
two brothers had sought to negotiate an end to the conflict, and that
it was to prevent this that the coup was carried out. She declared
emphatically that Mr. Kennedy had activated the brothers in this
effort to achieve peace; he had no part in fomenting the coup."[96]
Conceivably Kennedy's October insistence on starting the troop withdrawal in December could have been taken by Diem and Nhu as
encouragement to seek a negotiated peace. But I found no evidence
that Kennedy was ever adequately informed on the matter. The deal
did not seem real enough to have been a major factor in American
attitudes toward a coup.

It was a tricky business at best. Washington would have been hard
put to save Diem from the generals in any case; nor was there any
guarantee that Diem, once assured of American support, would have
negotiated an end to the war. Had he done so, a Diem-Ho deal
followed by an American withdrawal would not have been applauded
in the United States of 1963. "Fortunately," as the *New York Times*
pontificated after the coup, "the new Vietnamese rulers are dedicated
anti-Communists who reject any idea of neutralism and pledge to
stand with the free world. It is significant that one of their charges
against Mr. Nhu is that he tried to make a deal with Communist
North Vietnam along the lines hinted at by President de Gaulle."[97]

This was the national mood. Still, a Diem-Ho deal could have been
the means of an American exit from Vietnam in 1963, though the
life expectancy of a Diem government in Saigon would have been
minimal thereafter. An opportunity of some sort was perhaps missed
in the autumn of 1963. Whether it might have been seized later, if
Kennedy had never gone to Dallas, is another question. De Gaulle

---

* Hilsman to Rusk, August 30, 1963, JFK Papers. The memorandum, Hilsman
informs me, was a response to a request by Rusk that the Far Eastern Division list
anti-American moves Diem might make and the whole range of *possible* responses. It
should be noted that a document in the RFK Papers entitled "South Vietnam: An
Action Plan," with no date or source but, from internal evidence, September 1963 and
very likely also from Hilsman, listed as an American response to Nhu-Diem tactics: if
"we have sufficient hard evidence of GVN negotiations with any Communists, consider
going to Moscow directly. Sound out the Soviets in a secret demarche as to the
possibility of a neutralization of all Vietnam on the basis of a change of leadership in
South Vietnam."

was scheduled to visit Washington in February 1964. Talking to Gavin in late October 1963, with Vietnam the dominant problem on his mind, Kennedy said, rather happily, Gavin thought, "I am going to see the General in the next few months, and I think that we will be able to get something done together."[98]

X

In Saigon a new group of dissident generals was planning a serious coup. Lodge maintained contact with the conspirators through the CIA. The National Security Council met on October 29. Robert Kennedy thought the situation no different from August, when the generals talked big and did nothing. "To support a coup," he told the group, "would be putting the future of Vietnam and in fact all of Southeast Asia in the hands of one man not now known to us. A failure of a coup risks so much. The reports we have are very thin." The President observed that, since the pro-Diem and anti-Diem forces appeared about equal, any attempt to engineer a coup would be silly.[99]

But Lodge responded to a cautionary cable: "Do not think we have the power to delay or discourage a coup. [General] Don has made it clear many times that this is a Vietnamese affair."[100] When the coup began two days later, Lodge did his best to assure the personal safety of Diem and Nhu. They refused his assistance and were killed. (Lodge did succeed in flying Nhu's children to Bankok. The escort officer was Frederick Flott, with whom Robert Kennedy had traveled around the Soviet Union eight years before.)[101] When the President heard about Diem's murder, he leaped to his feet and, as Taylor recalled it, "rushed from the room with a look of shock and dismay on his face which I had never seen before."[102] I had not seen Kennedy so depressed since the Bay of Pigs. He said that Diem had fought for his country for twenty years and that it should not have ended like this.[103]

"I believe prospects of victory are much improved," an unrepentant and elated Lodge cabled on November 6.

> ... The coup was a Vietnamese and a popular affair, which we could neither manage nor stop after it got started and which we could only have influenced with great difficulty. But it is equally certain that the ground in which the coup seed grew into a robust plant was prepared by us and that the coup would not have happened [as] it did without our preparation....
> All this may be a useful lesson in the use of US power.... Perhaps the USG has here evolved a way of not being everywhere saddled with

responsibility for autocratic governments simply because they are anti-
Communist — a course which can eventually lead many people to believe
that the foreign communist autocracy which they don't know is preferable to the local autocracy which they do know. . . .
    The prospects now are for a shorter war.[104]

Kennedy disagreed. "Just before his death," according to Henry
Brandon, "he gave Mike Forrestal, in private conversation, odds of
a hundred-to-one that the U.S. could not win. But he also knew that
he could not get out of Vietnam before the elections in November,
1964, without inviting his own political eclipse."[105] He still supposed
he had plenty of time for maneuver. He told Forrestal, "I want to
start a complete and very profound review of how we got into this
country, what we thought we were doing and what we now think we
can do. . . . I even want to think about whether or not we should be
there."[106] Wayne Morse, one of the few senatorial critics of the
involvement, saw him in mid-November. Kennedy said, "Wayne, I
want you to know you're absolutely right in your criticism of my
Vietnam policy. Keep this in mind. I'm in the midst of an intensive
study which substantiates your position on Vietnam."[107] "The facts
that he envisaged a troop withdrawal and that he spoke often of the
necessity of ending American participation in the war lest it become
a swamp into which the United States sinks more and more," said
Pierre Salinger, "prove, in my judgment, that he was on the right
track at the time of his death."[108]

Of course he should have asked the searching questions long before. He should have realized the cumulative momentum of the
policy of small concessions. When Kennedy became President, there
were 685 American military advisers in Vietnam. In October 1963
there were 16,732. This was a formidable escalation. Still, Kennedy
sent many fewer troops to Vietnam than Khrushchev sent to Cuba
in 1962 or Johnson to the Dominican Republic in 1965.* The total
number of American soldiers killed in Vietnam as a result of hostile
action from the beginning of 1961 to the end of 1963 was only 73[109]
— 73 too many, but still inconsiderable compared to the tragedy to
come. The process was hardly irreversible. Troops, once sent, could
be withdrawn, as they were in Cuba and the Dominican Republic, as
Kennedy already planned to do in Vietnam.

Kennedy had proved his manhood in the Solomon Islands and did
not have to prove it again. He was a prudent executive, not inclined
to heavy investments in lost causes. His whole Presidency was marked
precisely by his capacity to *refuse* escalation — as in Laos, the Bay of
Pigs, the Berlin Wall, the missile crisis. "Having discussed military

---

* And indeed only 2000 more than Eisenhower sent to Lebanon in 1958.

affairs with him often and in detail for fifteen years," wrote General Gavin, "I know he was totally opposed to the introduction of combat troops in Southeast Asia."[110] After Dallas, Ho Chi Minh said thoughtfully to Maneli that Kennedy would perhaps have contributed to an early end of the war; "his death creates new problems."[111] Kennedy's failure lay in the hopelessly divided legacy he left on November 22, 1963.

32

The Breach Widens:
Vietnam

HOW TO RECAPTURE the way enlightened Americans thought about Vietnam in 1964? Take the vivid book, *The Making of a Quagmire,* written that year by David Halberstam of the *New York Times,* the best and brightest of correspondents. Halberstam was a penetrating critic of the American performance in Vietnam. Yet he did not doubt the premise. Vietnam, he wrote, "is perhaps one of only five or six nations in the world that is truly vital to U.S. interests." Neutralization? "Out of the question": it would only postpone a Communist victory. Withdrawal? It would mean "a drab, lifeless and controlled society for a people who deserve better. . . . Throughout the world the enemies of the West will be encouraged to try insurgencies like the one in Vietnam. . . . We would dishonor ourselves and our allies by pulling out." Americanization of the war? "We should think and prepare for a long, long time before going in with our own troops." We were caught in a quagmire and probably would not learn the lesson in time to save Vietnam. "Perhaps at this moment we are gaining the knowledge necessary to deal with a situation in Thailand, or Angola, or a small republic in South America — some country where it is not yet too late."*

* David Halberstam, *The Making of a Quagmire* (New York, 1965), 315, 317, 319. Note might also be taken of Senator Fulbright's famous "Old Myths and New Realities" speech of March 25, 1964, in which he opposed both a negotiated settlement and the commitment of American troops and called on the United States "to support the South Vietnamese Government by the most effective means available" and "continue to meet its obligations and fulfill its commitments with respect to Vietnam" (Jules Davids, ed., *Documents on American Foreign Relations 1964* [New York, 1965], 29).

I

"We shall continue," the new President said. But Kennedy's Vietnam legacy was dual and contradictory. He had left on the public record the impression of a major national stake in the defense of South Vietnam against communism. He had left steadily enlarging programs of military and economic assistance. He had left national security advisers who for three years had been urging an American expeditionary force and a total commitment to the salvation of South Vietnam. On the other hand, he had consistently refused to send such a force or to make such a commitment. He had left a formal plan, processed successfully through the Pentagon, for the withdrawal of American advisers by the end of 1965. He had left a public campaign, belatedly begun, to instill the idea that American involvement must be limited in a war that only the South Vietnamese could win. And he had left private opposition, repeatedly and emphatically stated, to the dispatch of American ground forces.

Did Kennedy lead the country ineluctably into catastrophe in Southeast Asia? He had not made things easy for his successor. But it is not sufficient to say that Lyndon Johnson was carrying forward Kennedy's policies. The question remains: which of Kennedy's policies? Continuity was an ambiguous command. The new President could choose the continuity of deepening involvement. Or he could choose the continuity of absolute limitations — no combat troops, no heavy bombing — and planned withdrawal by 1965. Each alternative bore the Kennedy stamp. The decision was now Johnson's.

It was Johnson's of course within the framework of public illusion to which Kennedy had made significant contributions. A generation of superpowership had persuaded most Americans that they had the power and the responsibility to work their will around the planet. The strength of this national mood constrained Kennedy to defer disengagement till after the 1964 election. But the mood did not constrain Johnson. He vigorously agreed with it. He had memorized the copybook lesson of the 1930s. Appeasement, he was sure, was an invitation to aggression. In May 1961 Kennedy had sent him to Vietnam. "The basic decision in Southeast Asia is here," Johnson reported on his return. "We must decide whether to help these countries to the best of our ability or throw in the towel in the area and pull back our defenses to San Francisco."[1]

He was a stout supporter of Diem, whom in 1961 he had hailed as the "Winston Churchill" of Southeast Asia. A week after the coup scare of August 1963, he told McNamara and Rusk that "he had never really seen a genuine alternative to Diem ... that from both a practical and a political viewpoint, it would be a disaster to pull out; that we should stop playing cops and robbers and get back to ... winning the war."[2] He watched subsequent developments with resentful disapproval. On the day after Kennedy's funeral, Johnson, showing Hubert Humphrey a portrait of Diem hanging in the hallway of his house, said, "We had a hand in killing him. Now it's happening here."[3]

On November 24 the new President convened a small group to meet with Lodge, just in from Saigon. Robert Kennedy was not invited. Johnson told them, "I am not going to lose Vietnam. I am not going to be the President who saw Southeast Asia go the way China went."[4] Afterward he mused to Bill Moyers, "They'll think with Kennedy dead we've lost heart. So they'll think we're yellow and don't mean what we say.... The Chinese. The fellas in the Kremlin. They'll be taking the measure of us.... I'm not going to let Vietnam go the way of China. I told them to go back and tell those generals in Saigon that Lyndon Johnson intends to stand by our word."[5] Two days later National Security Action Memorandum 273 declared: "It *remains* the *central objective* of the United States in South Vietnam to assist the people and Government of that country to *win* their contest against the *externally directed* and supported communist conspiracy."[6] Here, represented as the continuation of existing policy, were both the total commitment Kennedy had always refused and the diagnosis of the conflict that Kennedy, who had described it in July 1963 as a "civil war which has gone on for ten years,"[7] had never quite accepted.

While reannouncing the December troop withdrawal, NSAM 273 also emphasized that American military programs "should be maintained at levels as high as those in the time of the Diem regime."[8] This nullified Kennedy's extrication intent. One thousand men were nominally withdrawn in December. It was, in the words of the Pentagon history, "essentially an accounting exercise" and achieved no reduction in American strength.[9] Reducing American strength was the last thing Johnson wished to do. "The President," Rusk cabled Lodge, eyes only, on December 6, "has expressed his deep concern that our effort in Viet-Nam be stepped up to highest pitch."[10] On New Year's Day Johnson pledged the Saigon government "the fullest measure of support ... in achieving victory." "By implication," the *New York Times* commented, "the message erased the previous date for withdrawing the bulk of United States forces by the end of 1965."[11] The erasure was made explicit and final, in full Pentagonese, on March 27, 1964.[12]

Johnson's New Year's Day bouquet also denounced neutralism;

"only . . . another name for a Communist take over." De Gaulle meanwhile pressed his campaign. Reports flowed into Washington, according to William Bundy, that the post-Diem junta was "sympathetic to the French position."* Hanoi, Mieczyslaw Maneli later attested, thought it might come to terms with the junta.[13] Then Johnson's worries were unexpectedly eliminated by a second coup on January 30, 1964. General Khanh, the new leader, explained that the junta had been guilty of "paving the way for neutralism and thus selling out the country."[14] This, Maneli said, not the overthrow of Diem, was the "real coup."†

"At all levels concerned with Indochina," recalled William Bundy, Johnson's "personal force and imprint were felt much more strongly, right from the start, than JFK had been at any past times except for May and November of 1961 and September/October of 1963."[15]

II

Robert Kennedy's involvement in Vietnam had been strictly limited before Dallas. John Kennedy had not even found an occasion to discuss his deeper doubts and intentions with his brother; there always must have seemed a world of time to do that later. Robert's own understanding of his brother's position, as he expounded it to John Bartlow Martin in the spring of 1964, was that John Kennedy felt "we should win the war" because the loss of Vietnam would mean the fall of the rest of Southeast Asia; that there had been no consideration given to pulling out and at the same time no disposition to go further in; here Robert Kennedy cited MacArthur, adding, "we couldn't win the war for them and they had to win the war for themselves." Martin persisted: if the South Vietnamese were about to lose, would John Kennedy have sent ground forces? "Well," said Robert Kennedy, "we'd face that when we came to it."[16]

After Dallas, his involvement in Vietnam was nonexistent. Vietnam did not come up when he went to Indonesia and Malaysia in January 1964. It was not on his mind in succeeding months. The fundamental questions he had tried to raise the previous September sig-

---

* William Bundy, ms. on U.S. policy in East Asia during the Kennedy-Johnson years, ch. 12. According to Mieczyslaw Maneli, "the most realistic and responsible generals considered the possibility of neutralizing Vietnam with the active cooperation of France" (Maneli, "Encounters with John F. Kennedy and Discussions on 'Democratic Socialism,' Polish Policy, and Vietnam," unpub., 30).

†Maneli, in interview by author, June 22, 1977. Maneli's conclusion was that neither of the immediate superpowers — America or China — wanted the war to end. The Soviet Union alone, fearing a Hanoi victory would mean either a Chinese dependency or an Asian Titoism, was in favor of negotiated peace but in this period had limited influence in Hanoi.

nified an instinct to get to the root of the matter on his brother's behalf rather than a conclusion of his own that the United States should get out. Insofar as Robert followed Vietnam policy in early 1964, he probably noted that Johnson stayed within the Kennedy guidelines — no combat troops, no heavy bombing — and assumed he was doing what had to be done.

In June he began to focus on Vietnam again. One finds a flurry of Vietnam papers in his files — a CIA document doubting an immediate domino effect from the fall of Vietnam; a memorandum Mansfield had written to Johnson, restating his old doubts and saying that, if the President planned to pursue the war, he had better make a more persuasive case.[17] By this time Lodge had signaled a desire to be relieved; and on June 11 Kennedy sent a handwritten note to Johnson:

> Dear Mr. President:
> I just wanted to make sure you understood that if you wished me to go to Viet Nam in any capacity I would be glad to go. It is obviously the most important problem facing the United States and if you felt I could help I am at your service.
> I have talked to both Bob and Mac about this and I believe they know my feelings. I realize some of the other complications but I am sure that if you reached the conclusion that this was the right thing to do then between us both or us all we could work it out satisfactorily.
> In any case I wished you to know my feelings on this matter.
>   Respectfully,
>   Bob[18]

I have never been sure why he wrote this letter — probably for reasons no more complicated than old-fashioned patriotism. Johnson declined the offer. "I would be accusing myself for the rest of my life," he told Valenti, "if something happened to him out there. He could do the job. He could do it damn well, but I can't trust the security there and someone or some group might want to do him in. I couldn't live with that."[19] Evidently he thought a Vietnamese might want to avenge the murder of Diem.

There was yet no evidence of Kennedy dissent. Vietnam was not an issue in the fall election. Asked for his "exact position" by the students at Columbia, he responded with considerable inexactitude that he had no easy solution, that we should continue, but that he remembered the French in 1951, that the political and social effort was vital and that "we're not going to win that war unless the government has the support of the people. That, in my judgment, is the key to success."[20]

Johnson had no easy solution either. But he escalated the rhetoric, introducing the proposition in March 1964 that Vietnam was the crucial "test case" for Communist strategy[21] and thereafter the proposition, also unknown in the Kennedy years, that the SEATO treaty obligated the United States to intervene with full force in Southeast Asia. The gates of escape were not, however, completely shut. At the end of 1964, according to Chester Cooper, the careful participant-historian of the period, "the get-out option was by no means academic. President Johnson was still pausing warily, still undecided as to which path to follow." In early February, Cooper accompanied Bundy to Saigon. After several days, Cooper wrote later, "'getting out' still seemed to be possible."[22] Then on February 7 the Viet Cong swarmed over the airfield at Pleiku, killing American soldiers. Johnson now made the decision to Americanize the war.

American bombers went for the first time to North Vietnam. American combat units went for the first time to South Vietnam. "Their war," as Kennedy had called it, was becoming "our war." The Washington atmosphere changed radically. "It is important for the public to realize," the former staff director of the Counterinsurgency Group wrote in 1970 (his italics), "that President Kennedy and his brother *regarded Vietnam as a massive source of vexation and concern but not as intrinsically important in itself — only as a counter in a larger game.* As civilized, well-educated Americans they were totally devoid of the obsessive attitudes that characterized President Johnson under the influence of the 'hard-liners.'"[23]

The bombing of North Vietnam disturbed Robert Kennedy. "Late in April," Lyndon Johnson recorded, "Bobby Kennedy came to see me" to urge a bombing pause. It would do no harm, Kennedy said. Maybe something useful would come of it. The President assured him, as Presidents do, that the idea was under consideration.[24] Almost at once, however, Johnson asked Congress for a supplemental $700 million for Vietnam, explaining that this was no "routine appropriation" but a vote of confidence in the whole policy. Kennedy was furious. The appropriation, as he told me, was not necessary; ample funds were already available. What Johnson plainly wanted was congressional approval for further escalation. Still, it would be hard to vote against the request without seeming to let down the boys at the front. "Bobby," I noted in my journal, "was none the less very much inclined to vote against the resolution." This was also the week of the Dominican intervention; and my speech for him touched on Vietnam as well. When Burke Marshall and I discussed the draft with him, "we both said he had to vote for the resolution but should accompany his vote by a clear statement of what he was voting for."[25] His instinct was obviously a good deal

sounder than his advisers' advice. To their eternal honor, Wayne
Morse, Ernest Gruening and Gaylord Nelson voted no.

In his speech on May 6, 1965, Kennedy discussed three possible
courses in Vietnam. Withdrawal would mean "a repudiation of commitments undertaken and confirmed by three administrations" and
would "gravely — perhaps irreparably — weaken the democratic
position in Asia." Escalation would mean "the commitment to Vietnam of hundreds of thousands of American troops" and "might
easily lead to nuclear warfare." There remained the course of negotiation. "This, I take it, is the policy of the administration, the
policy we are endorsing today." He was voting for the resolution, he
said, in the understanding that it was no "blank check" for wider war
but rather the necessary prelude for negotiations. We had erred in
seeing Vietnam "as purely a military problem. . . . I believe that our
efforts for peace should continue with the same intensity as our
efforts in the military field."[26]

III

American involvement deepened. So did Kennedy's apprehension.
He accepted an invitation to give the commencement address at the
International Police Academy in July. The academy, AID's contribution to the counterinsurgency program, was a part of Kennedy's
past. He thought now to suggest how the Americanization of the war
was departing from the theory of counterinsurgency. Adam Walinsky worked with him on the speech, but Walinsky, as he said later,
was not yet on fire about Vietnam. It was Kennedy's "own initiative. Nobody pushed him to do it."[27]

The war, Kennedy said, was ultimately a contest for the allegiance
of men. "The essence of successful counterinsurgency is not to kill,
but to bring the insurgent back into the national life." Destruction
was no solution. "Air attacks by a government on its own villages are
likely to be far more dangerous and costly to the people than is the
individual and selective terrorism of an insurgent movement." The
history of the last twenty years, Kennedy said, had demonstrated
"beyond doubt that our approach to revolutionary war must be political — political first, political last, political always. Victory in a
revolutionary war is won not by escalation, but by de-escalation."[28]

The text, released before delivery, caused great perturbation in
the White House. It could only be read as an indictment of Johnson's
new course. Also, framed as it was by quotations from John F.
Kennedy, it contained more than a hint that Johnson had abandoned
his predecessor's policies. "It will be another Kennedy vs. Johnson

issue," Marvin Watson wrote darkly to the President, adding that government agencies like the Police Academy ought to be instructed to notify the White House "as to who they intend to invite to make speeches."[29] Fearful AID representatives implored Kennedy to add on delivery a passage conceding a military dimension to the problem and to drop the line about deescalation. But the original text was already in the newspapers.

The Police Academy speech had almost an elegiac quality. Counterinsurgency might have done it once, Kennedy seemed to say, but its day had passed. Yet his reluctance to force the issue with Johnson showed the fear of misunderstanding that inhibited him throughout his senatorial years. "He was worried," said Fred Harris, "that people would say that he was just trying to make political capital out of criticizing Johnson."[30] The worry was warranted. Most of the press, as Bill Moyers said later, interpreted his criticism of the Vietnam policy "as a political move on his part rather than a genuine unfolding of his own ideas about Vietnam."[31]

His ideas were plainly unfolding. Walinsky, increasingly seized by the problem, fed Kennedy's concern, stuffing the senatorial briefcase with memoranda, *I. F. Stone's Weekly*, books by Bernard Fall and other incendiary documents. Joseph Kraft, the columnist, interested Kennedy in the case of Gustave Hertz, an AID official kidnapped by the Viet Cong. The National Liberation Front, as the Viet Cong were formally known, had a diplomatic outpost in Algeria. Kraft took Kennedy to Cherif Guellal, the Algerian ambassador, who asked Ben Bella, the Algerian chief of state, to intercede with the NLF and soon went to Algiers himself to press the negotiation.

The NLF agreed to exchange Hertz for one of their own men under sentence of death in Saigon. The NLF representative told Guellal, "We are doing it because of President Ben Bella's intervention; and we know Senator Kennedy and . . . the contribution of his late brother to Algeria's struggle for independence." It was Guellal's impression that the NLF saw an opportunity to open larger contact with the Americans. Once the exchange was made, they would release Hertz — personally to Senator Kennedy in Algiers if he wished. Kennedy took the proposal to Maxwell Taylor, who rejected it on the ground that the arrangement would upset the Saigon government. Kennedy, Guellal thought, was "upset" and "bitter." Hertz died a Viet Cong prisoner.[32]

The effort, Kraft later suggested, made Kennedy understand that the Viet Cong were not roving bands of terrorists but truly a National Liberation Front with a political structure and legitimacy of its own. If counterinsurgency were dead, then negotiations with the NLF might offer the best exit from Vietnam.[33] Guellal thought that the "persistency and consistency" with which Kennedy pursued the Hertz matter was "because he saw there was a possibility, an opening" that might lead to broader negotiations.[34]

The administration preferred victory. The *Time* cover story on October 29, 1965, was entitled "The Turning Point in Viet Nam." "Today," the house dithyrambist wrote, "South Viet Nam throbs with a pride and power, above all an *esprit* scarcely credible against the summer's somber vista.... The remarkable turnabout in the war is the result of one of the swiftest, biggest military buildups in the history of warfare.... All of free Asia ... stands to be ultimately strengthened by the extraordinary — and still burgeoning — commitment of the lives and talent and treasure of America."[35]

Kennedy did not share the euphoria. Professor Eugene Genovese, a Rutgers historian, said that he hoped the Viet Cong would win. Patriots demanded Genovese's dismissal. Kennedy defended his right to heresy. Richard Nixon, denouncing Kennedy, said that a Viet Cong victory "would mean ultimately the destruction of freedom of speech for all time not only in Asia but in the United States as well."[36] But the agitation against the war was spreading. It was the year of teach-ins, the first antiwar turbulence on the campuses, the first bonfires of draft cards. Early in November, Kennedy defended the demonstrations in a press conference at the University of Southern California: "I don't think you can ever discuss these matters enough." As for draft cards,

RFK. If a person feels that strongly and wants to ... burn his draft card.... I don't agree with it personally but I think that obviously [is] the way [chosen by] somebody that feels very strongly about this matter....
PRESS. What about giving blood to the North Vietnamese?
RFK. I think that's a good idea.
PRESS. Is that going too far?
RFK. If we've given all the blood that is needed to the South Vietnamese. I'm in favor of giving [to] anybody who needs blood. I'm in favor of them having blood.
PRESS. Even to the North Vietnamese?
RFK. Yes.[37]

This caused particular excitement. "If you feel strongly enough for the enemy to give him a pint of your blood every ninety days or so," asked the *New York Daily News,* "then why not go the whole hog? Why not light out for the enemy country and join its armed forces? Bobby Kennedy is young, strong and virile, and financially able to provide for his wife and children while he is away at war."[38] The *Chicago Tribune* ran a cartoon showing Kennedy standing on a flag-draped coffin labeled AMERICAN CASUALTIES IN VIETNAM while carrying a placard reading: "I am willing to give my blood to the Communist enemy in Vietnam."[39] It was, said his old friend Barry Goldwater, "closer to treason than to academic freedom. . . . It's appalling to me that the press of the nation hasn't jumped down his throat."[40] Much of it did.

This did not deter him. He lashed out with a sort of reckless impulsiveness on other subjects. "That was the really endearing thing about him," Walinsky said. ". . . No matter how hard he would try and protect himself . . . the things that got him in trouble were the things that were just basically right and decent; they would come out time and time again in these impromptu comments."[41] Thus in January 1966: "We have spoken out against inhuman slaughters perpetrated by the Nazis and the Communists. But will we speak out also against the inhuman slaughter in Indonesia, where over 100,000 alleged Communists have been not perpetrators, but victims?"[42] No other American politician condemned the Indonesian massacre. In the same month the Defense Department denied burial in Arlington Cemetery to the American Communist leader Robert Thompson. Thompson had gone to jail under the Smith Act in 1949. He had also won the Distinguished Service Cross for heroism in New Guinea in 1943. Kennedy, again alone, denounced the action on the floor of the Senate. "To hate and harry the sinner to his grave," he said, quoting a newspaper editorial, "is hardly in the American tradition."[43] He told me later, his reference obvious, "I don't think anyone now buried in Arlington would object to having Thompson buried there, so I don't see why all these living people are objecting."[44]

IV

In mid-November 1965, Amintore Fanfani, the Italian foreign minister, relayed to President Johnson four points for negotiation given to a special Italian mission to Hanoi. Rusk took a fortnight to acknowledge Fanfani's report and then, after various cavils, asked Fanfani to check back with Hanoi for clarification. In the meantime, Fanfani received a new message: if Hanoi or Haiphong were bombed, this would kill negotiations. Kennedy was one of those shown this message.[45] On December 15, before Hanoi responded to Rusk's inquiry, Johnson ordered the bombing of Haiphong.

On December 19 there was a party at Hickory Hill. My son Stephen's notes record Kennedy's mood:

> RFK said: "Why didn't we accept the Fanfani message positively, agreeing to the four points and offering our own interpretation of what the four points meant — plus some points of our own. . . . Then the onus

would be on Hanoi to refuse. This would make us look good whether
the offer was real or not. But to dismiss it out of hand is disastrous.
We lose all credibility. How could the State Department wait for two
weeks? How could they? Why was the State Department message so
bad? If we had acted that way in the Cuban crisis we might have had
war. If, on Friday night, we had asked the Soviet government for an
explanation of their message instead of just agreeing to our own interpretation of it, it might have been chaos.

"I don't believe in pulling out the troops. We've got to show China
we mean to stop them. If we can hold them for about 20 years, maybe
they will change the way Russia has. The next 30 days will be critical
in Vietnam. With the casualty lists growing, the decision to send more
troops will be critical. We may have reached the point of no return
once that decision has been made. I'm upset over our policy in Vietnam. I don't think we've shown an open approach. I really think
Johnson wants negotiations. Ball, Harriman, McNamara too. But Rusk
is against them. Fifty percent of the government is probably for them;
fifty percent against them. We're in a stalemate.

"I'd like to speak out more on Vietnam. I have talked again and
again on my desire for negotiations. But if I broke with the administration it might be disastrous for the country. I'd have accepted the
four points two years ago — at least a year ago. I told LBJ that the last
time I saw him." . . .

He said that if we had accepted the four points a year ago, or even
last week, in a short while, possibly 14 days, a Gallup Poll would show
most of the American people favoring it. Asked how he knew, RFK
said: "I feel it in my gut." He also felt that China would not get involved
in Vietnam.[46]

Hanoi had made it clear that talks were out so long as American
planes were bombing North Vietnam. McNamara, the most ardent
supporter of negotiations in the cabinet, had argued since early November for a bombing pause. Johnson was skeptical. The Joint
Chiefs of Staff were violently opposed.[47] So were Cabot Lodge, who
had returned to Saigon as ambassador, and General William Westmoreland, the new commander in Vietnam. So were Clark Clifford
and Abe Fortas, Johnson's closest unofficial advisers. But Ball supported the pause, and Rusk and Bundy dropped their earlier opposition. On December 18, Johnson, still skeptical, decided to give
it a try. The pause began with a Christmas truce and ran through
the month of January 1966.

About this time McNamara asked me to arrange a private meeting
with New Frontier friends. Galbraith, Richard Goodwin, who had by
now left the Johnson White House, and Carl Kaysen dined at my
house in Georgetown on January 6. McNamara told us that he did
not regard a military solution as possible. His objective, he said, was
"withdrawal with honor." I noted, "He seemed deeply oppressed
and concerned at the prospect of indefinite escalation. Our impression was that he feared the resumption of bombing might well put

us on the slippery slide."[48] The generals were clamoring for resumption; and Kennedy, hoping to stiffen the President against the rising pressure, sent Johnson a copy of Bruce Catton's *Never Call Retreat* with a handwritten note:

> I thought it might give you some comfort to look back at another President, Abraham Lincoln, and some of the identical problems and situations that he faced that you are now meeting. I refer you to pp. 56–63 and 371–381. . . . In closing let me say how impressed I have been with the most recent efforts to find a peaceful solution to Viet Nam.[49]

Johnson replied on January 27 (the letter was written by Valenti):

> Your warm letter arrived at an appropriate time. It was one of those hours when I felt alone, prayerfully alone.
>
> I remembered so well how President Kennedy had to face, by himself, the agony of the Cuba missile crisis. I read the paragraph in Catton's book that you had marked, and then I went to a meeting in the Cabinet Room with the Congressional leaders of both parties. I read them that passage where Lincoln told a friend that all the responsibilities of the administration "belong to that unhappy wretch called Abraham Lincoln." I knew exactly how Lincoln felt.[50]

Four days later Johnson resumed the bombing.

"Bob," George McGovern wrote Kennedy, "I do hope you will continue to raise questions about Vietnam. Your voice is one of the very few that is powerful enough to help steer us away from catastrophe."[51] On January 31 Kennedy arose in the Senate. "If we regard bombing as the answer in Vietnam," he said,

> — we are headed straight for disaster. In the past, bombing has not proved a decisive weapon against a rural economy — or against a guerrilla army. And the temptation will now be to argue that if limited bombing does not produce a solution, that further bombing, more extended military action, is the answer. The danger is that the decision to resume may become the first in a series of steps on a road from which there is no turning back — a road which leads to catastrophe for all mankind.[52]

v

On February 4 the Senate Foreign Relations Committee opened public hearings on Vietnam. Kennedy watched with avid attention and mounting frustration. No one seemed to have thought through the detail of a negotiated settlement. He tried to think it through himself and concluded that the price of settlement was taking the National Liberation Front into the political process. When this point did not emerge before the Foreign Relations Committee, he decided to make it on his own.

He put out his statement on February 19 before going off to ski in Vermont. Each side, he observed, had its irreducible demand. The United States was insisting that South Vietnam should not be turned over to the North. North Vietnam was insisting that South Vietnam should not be a hostile state dedicated to the extermination of Communists. *"A negotiated settlement means that each side must concede matters that are important in order to preserve positions that are essential."* The United States had to recognize that many people in South Vietnam wanted change. "There are three things you can do with such groups: kill or repress them, turn the country over to them, or admit them to a share of power and responsibility." The first two meant a continuation of the war. The last was "at the heart of the hope for a negotiated settlement." [53]

Kennedy meant that the NLF should be admitted to a share in the political process, not necessarily to a place in the government. His statement mentioned "a compromise government" as only one of the possible forms acceptance of the NLF might take. But his language was imprecise, and he was immediately denounced as the advocate of coalition with the Communists. Vice President Humphrey led the counterattack from the far Pacific, where he was touting the war for the administration. "I don't believe in writing a prescription for the ills of Vietnam that includes a dose of arsenic," the old druggist said. "It would be something like putting a fox in the chicken coop." [54] Humphrey later wrote that his doubts about escalation early in 1965 had resulted in exclusion for months from the Vietnam policy councils. On this 1966 tour, he said, he was converted by brainwashing in Saigon, by confidential pleas from Asian leaders (including Indira Gandhi) that the Americans stay in Vietnam and by his own intense desire to be restored to presidential favor.*

In Washington, the day after the statement, George Ball on *Issues and Answers* and McGeorge Bundy on *Meet the Press* assailed Kennedy. To Kennedy's particular irritation, Bundy quoted John Kennedy in opposition to united front governments. Actually, Bundy had planned to quote Robert himself, but Johnson, in one of his flashes of solicitude, warned him not to do so if he valued his friendship with the senator. [55] "I would have obviously appreciated a call," Kennedy wrote Bundy (in a paragraph he deleted after further con-

---

* Hubert Humphrey, *The Education of a Public Man* (New York, 1976), ch. 32, esp. 320–324, 337. George McGovern, however, describes a talk with Humphrey about Vietnam in the spring of 1965 — "the first real argument Hubert and I had had in a decade as neighbors and friends. He replied heatedly that he believed in the policy — that we had to stop the Communists in Vietnam or they would take all of Asia. . . . If I heard others say he wasn't with the President all the way . . . it wasn't true" (McGovern, *Grassroots* [New York, 1978], 105).

sideration), "before you dealt with [my position] and me on Sunday afternoon. Perhaps a call would not have taken any more time than for someone to look up the quote of President Kennedy to use against my position."[56] Later he made the point about advance notice over the telephone. Bundy said that Kennedy might equally have called before he took out against the administration. Kennedy replied that he had in fact discussed the statement with Moyers the day before he put it out.[57]

The uproar was general. John Connally sent word to Johnson that he understood Kennedy to have been "the motivating force behind the Senate hearings and the Saturday statement was only his climax."[58] The *Chicago Tribune* entitled its editorial "Ho Chi Kennedy."[59] Kennedy defended his position on the *Today Show*. Murray Kempton, who accompanied him to the studio, was struck by his "concentrated attempt to be his brother's interior, to talk not as John Kennedy did on the platform, but as he seems to have in private moments of crisis. . . . The most reckless and romantic of the Kennedys was deliberately reshaping himself according to his memory of the coolest and most detached of them." They went out afterward for coffee. The counterman asked Kennedy to autograph the morning's *Daily News* cartoon — a picture of a scrawny Kennedy holding a giant hatchet emblazoned APPEASE VIET CONG. "You want me to autograph *this?*" Kennedy laughed and did so. A White House telephone lineman introduced himself. "We miss him," he said. "You ought to come over and see us." "Do I dare?" said Kennedy.[60]

There was now concern in the White House that the counterattack might go too far. Maxwell Taylor saw no serious divergence, assuming that Kennedy had not implied the establishment of a coalition government in advance of free elections. If the people of South Vietnam themselves brought the fox into the chicken coop, Taylor agreed, the administration would abide by the result. Moyers, who was laboring to keep Johnson and Kennedy together, made this point in a Tuesday press briefing, adding that NLF participation in an interim government would be negotiable. Kennedy accepted the compromise.[61] Two days later, Johnson defended his Vietnam course at a Freedom House dinner. The old New Dealer David Lilienthal noted Kennedy "chewing at a cigar, looking sullen and cynical as Johnson recited all the answers to his critics. . . . 'I have heard all this hokum a thousand times' written on his face."[62]

On February 27 Kennedy appeared on *Face the Nation*. I had been away from Washington during the hullabaloo but was now back, and he asked me to join Fred Dutton in running through questions before the show. Impersonating an interrogator, I challenged him about

John Kennedy's responsibility for the trouble we were in in Vietnam. He thought a moment and said, "Well, I don't know what would be best: to say that he didn't spend much time thinking about Vietnam; or to say that he did and messed it up." Then, in a sudden, surprising gesture, he thrust out his hand to the sky and said, "Which, brother, which?"[63] In a short while he told the television audience that statements "that we will never deal with assassins or murderers make it difficult for [the other side] to believe that they are being asked to come to the negotiating table for anything other than to surrender."[64]

An hour after *Face the Nation*, Humphrey, now back from the Pacific, appeared on *Issues and Answers*. Kennedy and Humphrey had become good friends. When a letter from the mayor of Mobile, Alabama, was misdelivered to Kennedy's office in January, Kennedy had sent it along to the Vice President with a joking note:

> This was sent [to] and opened in our office by mistake. It is not part of any ruthless scheme to open and read all your mail before it goes to your office.
> However I am somewhat shocked and surprised at the close working and social relationships you are establishing with some people.
> Needless to say I shall breathe no word about any of this to Clarence Mitchell or Martin Luther King or anyone like that.[65]

On Humphrey's return three days before, Kennedy had sent another note: "Welcome home. I was taking care of everything back here while you were away. Perhaps you heard. As a matter of fact I felt a little like the fox in the chicken house myself."[66] Kennedy genuinely liked Humphrey and admired his intelligence while still finding him weak; not to be relied on, he said in 1964, to "stand fast on something that's important . . . you know, he just wobbles away."[67] The liking was perhaps one-sided. The unrelenting Humphrey treasured his resentments of 1960. "I never quite forgave him for that period," Humphrey wrote as late as 1976.*

At any event we all now watched Humphrey on the tiny screen. The NLF, he vigorously said, "engage in assassination, murder, pillage, conquest, and I can't for the life of me see why the United States of America would want to propose that such an outfit be made part of any government."[68] "It was a new and different Hubert," I noted sadly as an ancient Humphrey fan, "— hard-faced, except for

---

* Humphrey, *Education*, 374. See also Isabelle Shelton: "I asked him, not long ago, if he still was resentful about that election, and he said he didn't want to talk about it. 'They're all gone now,' he said sadly, meaning Jack and Bobby Kennedy. Then he stood silent for a long minute, lost in thought. . . . Finally he said quietly, a little tight-lipped: 'I haven't changed my mind'" (Shelton, "Memories of Hubert Humphrey," *Washington Star*, January 22, 1978).

some unctuous smiles, and uncharacteristically coarse in language; he sounded like Tom Dodd. His trouble, I fear, is that he cannot say something publicly without deeply believing it privately; and when, as now, he has no choice in his public utterances, he whips up a fervency of private belief. I fear also that someone has persuaded him that this is the issue on which he can knock out RFK." Kennedy himself "was both so irritated and upset by Hubert's performance that, after a few minutes, he went silently away."[69]

For a few weeks Kennedy continued to argue the case for a negotiated settlement. The public remained confused. "I made some mistakes in handling it," Kennedy said later. "I think it was unpopular politically. But I would do it all over again if I had to."[70]

VI

Kennedy brooded about Vietnam but said less in public. The announcement in April 1966 of "no sanctuary" for planes attacking American aircraft from Chinese bases provoked him to rise again in the Senate and warn against escalation.[71] Mostly the old fear that disagreement would be ascribed to a personal vendetta held him back. He took the usual Kennedy refuge and made jokes. "It isn't true that President Johnson and I didn't get along during the time I was Attorney General and he was Vice President," he would say. "We began the Kennedy administration with the best of relations, close, friendly, cordial — and then, as we were leaving the inaugural stand. . . ." Or: "President Johnson and I are very courteous and correct in our correspondence these days. I address my letters to him at the White House and he writes to me at the Senate Office Building. Sometimes he only uses the initials."[72]

One day in the late spring Galbraith, Goodwin and I lunched in New York. "It would be terrible," Goodwin said, "if, when the nuclear bombs begin to drop on Peking or Washington, we had to reflect that all we did in the summer of 1966 was to rest comfortably on one or another beach." Johnson, Goodwin said, was a man possessed, wholly impervious to argument. The only thing he understood was political opposition. We decided to do what little we could to stir public opinion. Goodwin soon wrote his book *Triumph or Tragedy: Reflections on Vietnam*. Galbraith wrote *How to Control the Military*. I wrote *The Bitter Heritage*. We also discussed the formation of a national committee against the widening of the war.

When we raised this latter idea with Kennedy on July 20, he seemed skeptical. A mass movement against escalation, he said, "would have to tie itself to an issue or to a man. The issue of widening the war is too complicated and ambiguous. LBJ could always justify each specific step on the ground that it was necessary to save the lives of American troops and that it would shorten the war. As for tying it to a man —" Here he paused. The polls were already showing him the leader of the Democratic opposition. Goodwin observed that the more an anti-escalation movement seemed a pro-Kennedy movement, the less likely it would be that Johnson would respond to it. Kennedy listened carefully but said nothing.[73]

The next weekend I resumed the discussion with Galbraith and George McGovern at Galbraith's farm in Vermont. "By now," I noted, "I had begun to think for the first time of the possibility of RFK's going for the nomination in 1968." The polls suggested he could win the primaries. He had kept up his relations with the city bosses like Daley of Chicago. When we talked about this in Vermont, McGovern said he was all for it. He added that he thought Johnson had a "great yellow streak" and might conceivably be bluffed out of running again.[74]

A fortnight later I was at Hyannis Port. I did not bring up 1968, but we talked about the establishment of a Vietnam information center in Washington. Kennedy said at once, "I can think of three things right off that such a center could do." One was the problem of Vietnamese elections — "the standards we should demand that they meet." Second, defoliation: "so far as I can tell, this hurts the ordinary people in Vietnam without seriously hurting the Viet Cong, but again we don't know nearly enough about it. Third, there is the question of our turning prisoners over to the South Vietnamese and becoming accomplices in torture. This is wrong in itself, and it is also bad policy, because we can never hope to get them to come over to us if all they can expect is to be tortured."[75]

Publicly he remained silent. In September Dr. Benjamin Spock urged him to open up the fight for negotiation. You are, Spock wrote, "the only person who could lead the country back to peace and decency." Kennedy asked Goodwin to "go see him on my behalf" and tell him "that I would like to see him some time after [the November] election."[76] In October, I. F. Stone scornfully entitled an article "While Others Dodge the Draft, Bobby Dodges the War."[77]

VII

"I knew from the start," Lyndon Johnson told Doris Kearns in 1970, "that I was bound to be crucified either way I moved. If I left the woman I really loved — the Great Society — in order to get involved with that bitch of a war on the other side of the world, then I would lose everything at home.... And I knew that if we let Communist aggression succeed in taking over South Vietnam ... that would shatter my Presidency, kill my administration, and damage our democracy."[78]

He was a man of titanic gyrations of mood and temper. In Vietnam meetings he was controlled and reasonable. George Ball, the most persevering critic of the war among his advisers, found him unfailingly courteous. But Ball, as George Reedy said, was the "official dissenter," whose arguments were "welcomed because they prove for the record that decision was preceded by controversy."[79] In private, as the Vietnam debate grew more bitter, the President became driven, irascible, inflamed by wild suspicions. Chester Bowles, on home leave from India where he was now ambassador, met with him in the summer of 1966. "Literally half our time together was taken up by almost paranoiac references to Bobby Kennedy, Wayne Morse, Bill Fulbright and others." Bowles saw "a man headed for deep trouble, with the probability of an increasing obsession with his 'enemies.'"[80] White House aides, Doris Kearns wrote, "were frightened by what seemed to them signs of paranoia." The President would enter into a compulsive monologue, punctuated by irrelevant laughs:

> Two or three intellectuals started it all you know. They produced all the doubt.... And it spread and it spread.... Then Bobby began taking it up as his cause and with Martin Luther King on his payroll he went around stirring up the Negroes.... Then the Communists stepped in. They control the three major networks, you know, and the forty major outlets of communication. Walter Lippmann is a communist and so is Teddy White. It's all in the FBI reports.... Isn't it funny that I always received a piece of advice from my top advisers right after each of them had been in contact with someone in the Communist world? And isn't it funny that you could always find Dobrynin's car in front of Reston's house the night before Reston delivered a blast on Vietnam?[81]

It was hard to make out. Was this merely an eccentric mode of relaxation? Or did he really believe what he was saying? If the latter, Goodwin and Moyers wondered what could be done. They thought of asking for psychiatric investigation. But, as Goodwin said, he would just talk calmly and rationally to a panel of psychiatrists, "and everyone would think we were the ones who were crazy."[82]

The original Johnson White House began to break up. Walter Jenkins cracked under the strain in October 1964. Reedy left in July 1965, thereafter to write in *The Twilight of the Presidency* the classic account of the way the isolations of power withdrew Presidents from reality. Goodwin left in September 1965, Bundy in February 1966,

Valenti in May. Moyers stayed on till December 1966. Then he left. He told me, "You know, Arthur, I would not be leaving if I thought I could do any good by staying."[83] He told Frank Mankie-wicz, "I thought I could make him more like me, but I've found in the last several months that I'm becoming more like him; so I got out."[84]

Johnson's obsessions centered more than ever on Kennedy. Had there been no escalation in Vietnam,

> there would be Robert Kennedy out in front leading the fight against me, telling everyone that I had betrayed John Kennedy's commitment to South Vietnam. That I had let a democracy fall into the hands of the Communists. That I was a coward. An unmanly man. A man without a spine. Oh, I could see it coming all right. Every night when I fell asleep I would see myself tied to the ground in the middle of a long, open space. In the distance, I could hear the voices of thousands of people. They were all shouting at me and running toward me: "Coward! Traitor! Weakling!"

The language, Doris Kearns noted, betrayed unconscious concerns — "coward," "unmanly," "without a spine" — along with his envy for Kennedy, whose manliness no one questioned. The sense of Kennedy as "the enemy," she believed, only hardened his determination not to change course in Vietnam.[85]

33

The Breach Widens:
South Africa, New York

THE KENNEDY CHALLENGE seemed to be acquiring global propor-
tions. Robert Kennedy had already carried his campaign, as the
exasperated White House saw it, to Latin America. He was meddling
in Vietnam. Now he was proposing to move on to Africa. In the
autumn of 1965 the anti-apartheid National Union of South African
Students (NUSAS) invited him to speak at its annual Day of Affirmation. As Attorney General, Kennedy had supported independence movements in black Africa. He well remembered John Kennedy's concern that South Africa's racist policies "would create a
bitterness and hostility that could not be contained within the country's own borders."[1] Wayne Fredericks, still deputy assistant secretary of state for African affairs, urged him to go; South African
liberals needed all the encouragement the outside world could
give.[2] He accepted the invitation.

I

"Bar Kennedy!" William Loeb begged the South African government
from his New Hampshire eyrie. "Bobby Kennedy is the most vicious
and most dangerous leader in the United States today. It would
make no more sense to us for South Africa to admit Bobby Kennedy
. . . than it would to take a viper into one's bed."[3]

It took South Africa five months to decide to admit him. When
the visa finally arrived, Kennedy was unexpectedly embroiled in a
New York political battle. He told Fredericks he preferred now to
wait until after the fall elections. "Things are moving in South Africa," Fredericks replied. "Ian Robertson [the president of NUSAS]

may not be available in November. Go now. If you postpone, it will confirm the idea that everything takes precedence over Africa, including local politics." In twenty minutes Kennedy called back: all right, he would go on schedule.[4] The week before he left, Lyndon Johnson delivered a speech on Africa, his first (and last) as President. "Cynics will wonder," observed the *New York Times*, "if the attention given to Senator Kennedy's forthcoming visit" did not explain the sudden presidential discovery of the dark continent.[5]

The South African government was even less enthusiastic than the White House. When Kennedy asked the South African embassy for advice on his itinerary, the ambassador responded that he had nothing to say, that his government disapproved of NUSAS and that no ministers would receive him.[6] A fortnight before his arrival, the government placed Ian Robertson under a "ban," excluding him from political and social life for five years. Then it denied visas to forty American newspaper and television correspondents assigned to cover the trip. South Africa, the Department of Information announced, would not allow the visit "to be transformed into a publicity stunt . . . as a build-up for a future presidential election."[7] The omens were all bad. As Thomas Johnston, who advanced the trip, put it, Kennedy was "going into a terribly explosive and delicate situation where the government was completely against him and the largest percentage of the white people were against him; and the blacks were in no position to really have any information, or be for him or against him. . . . The whole trip had the makings of one of the major disasters of his life."[8]

The party — Robert and Ethel, Angie Novello, Adam Walinsky — arrived in Johannesburg just before midnight on June 4, 1966. Johnston, who met them, found Kennedy unwontedly on edge.[9] But more than fifteen hundred people jammed the airport. Hecklers shouted "Chuck him out" and "Yankee go home." They were outroared by the enthusiasts. Kennedy made his way through a screaming crowd, his cuff links torn from his sleeves in the process, and climbed on top of the automobile waiting outside.[10] The reception was at once reassuring and disturbing. It showed both the hunger for a liberal voice and the emotionalism of the family argument into which he was entering.

He spent the next day in Pretoria. The Prime Minister, Dr. Hendrik Verwoerd, declined to see him or to permit other ministers to do so. In the evening he dined with South African businessmen. They could not understand, they told him, why the United States did not embrace South Africa, which was after all the most staunchly anti-Communist of nations. Here and for the rest of his stay he tried to explain that anticommunism was not enough. "What does it mean to be against communism," he asked, "if one's own system denies the value of the individual and gives all power to the government — just as the Communists do?" The South African said, "You don't understand. We are beleaguered." Kennedy wondered who was beleaguered: his dinner companions, talking comfortably over cigars and brandy? or Alan Paton, the novelist? the banned Ian Robertson? Chief Albert Luthuli, the Zulu leader who had won the Nobel Peace Prize in 1961 and was now banished to a remote farm in the back country?[11]

The next day he flew to Cape Town, delighting in the white city glittering in the sun, then chilled as the plane swooped down over Robben Island with its political prisoners. He was two hours late, but three thousand people waited at the airport. He promptly called on Ian Robertson and gave him a copy of *Profiles in Courage,* inscribed by Jacqueline Kennedy. When he emerged from Robertson's apartment, the street was filled with people and resounded with applause.[12]

That evening was the Day of Affirmation speech. Walinsky had written a draft in Washington. Allard Lowenstein, the young liberal activist who had traveled in South Africa, thought it too cautious; you really had to go full tilt, he said. Richard Goodwin was enlisted, and the speech took shape.[13] Going over the text still once more, Kennedy set off for the University of Cape Town. It took half an hour to get through surging crowds into the hall.

He talked first about the long American struggle for racial justice. "Nations, like men, often march to the beat of different drummers, and the precise solutions of the United States can neither be dictated nor transplanted to others." Still, everyone must accept "the full human equality of all our people" and join in "a shared determination to wipe away the unnecessary sufferings of our fellow human beings." The world's cruelty, he said, could not be changed "by those who cling to a present which is already dying, who prefer the illusion of security to the excitement and danger which comes with even the most peaceful progress." For it was "a revolutionary world we live in; and thus, as I have said in Latin America and Asia, in Europe and in the United States, it is young people who must take the lead."

Let no one, he said in his most eloquent passage, be discouraged by

the belief there is nothing one man or one woman can do against the enormous array of the world's ills — against misery and ignorance, injustice and violence.... Few will have the greatness to bend history itself; but each of us can work to change a small portion of events, and in the total of all those acts will be written the history of this generation.

It is from numberless diverse acts of courage and belief that human history is shaped. Each time a man stands up for an ideal, or acts to improve the lot of others, or strikes out against injustice, he sends a tiny ripple of hope, and crossing each other from a million different centres of energy and daring those ripples build a current which can sweep down the mightiest walls of oppression and resistance.[14]

These lines bespoke his profoundest conviction. The next morning the *Cape Times* reprinted the entire text under a nine-column headline: 15,000 ACCLAIM CALL TO YOUTH.[15] Frank Taylor of the *London Daily Telegraph* called it "the most stirring and memorable address ever to come from a foreigner in South Africa."[16] It was Kennedy's greatest speech.

The next day he talked to Afrikaner students at Stellenbosch University. "If the blacks are not 'inferior' to the whites," he remarked in the question period, "why don't they take part in your elections? . . . Why don't you allow them to worship in your churches? What the hell would you do if you found out that God was black?" He received a surprising ovation.[17] On to Durban: DURBAN'S WILD WELCOME FOR R.F.K. was the *Cape Times* headline.[18] He dined that evening with anti-apartheid leaders: Denis Hurley, the Roman Catholic archbishop; two Zulu chiefs; and Alan Paton, whose novels he so much admired.

In the evening he addressed nearly ten thousand people at the University of Natal. "Maybe there is a black man outside this room who is brighter than anyone in this room," he said, "— the chances are that there are many." There was applause.[19] Were South Africa's policies a threat to world peace? "Not at present," he replied. "But if the situation continues as it is, South Africa will place itself in that position, and I do believe that if it continues there will be a crisis."[20] Afterward, on the way back to the hotel, he encountered a group of Africans, and they sang "We Shall Overcome" together.[21]

II

At the beginning, said Walinsky, "his only crowds were the students. . . . But then somewhere as the thing started to get across . . . there were just thousands of ordinary citizens . . . not just Englishmen but Afrikaners coming out there into the streets to cheer and to listen and to shout . . . pulling at him and tugging at him and cheering him and it was really fantastic."[22] "By the end of the third day," said Tom Johnston, "as we moved from Cape Town to Durban, you began to sense this really great change; there were so many, many more people and instead of standing there passive, sort of hostile, why they were like a campaign that begins terribly slowly and then moves to this climax, all condensed into four and a half days!"[23]

The government sensed the change too and acceded to his request to visit Chief Luthuli. He left by helicopter before dawn the next day for Luthuli's farm. "A most impressive man, with a marvelously lined but kind face," Kennedy scribbled in his travel notes. "What did one notice first? The white goatee, perhaps, so familiar in his pictures, but then quickly the smile which lit up his whole presence, the eyes which danced and sparkled, and then, when he talked of the future of his country, of his people, of the relationship between the races, [became] intense and hurt and hard, all at once." "What are they doing to my countrymen, to my country?" Luthuli asked his visitor. "Can't they see that men of all races can work together — and that the alternative is a terrible disaster for us all?"[24]

They spent the last day in Johannesburg. There was a tour of Soweto, the steaming ghetto where half a million blacks lived behind wire fences, passing in and out with identity cards. People swarmed around his car, shouting "Master, Master." "Please don't use that word," Kennedy asked in embarrassment.[25] He made speeches from the roof of the automobile, from the steps of the Catholic cathedral, from a chair in the middle of a school playground. Later the inhabitants papered their shacks with Kennedy photographs cut from the newspapers.

That evening at the University of Witwatersrand he addressed the charge that blacks were too barbarous to be entrusted with power. "It was not the black man of Africa," he said, "who invented and used poison gas and the atomic bomb, who sent six million men and women and children to the gas ovens, and used their bodies as fertilizer." Their common humanity bound white and black inextricably together. Hope lay in "the most fragile and the most powerful of human gifts — the gift of reason. Thus those who cut themselves off from ideas and clashing convictions . . . encourage the forces of violence and passion which are the only alternatives to reason."[26]

Government displeasure did not relent. "This little snip," said one deputy minister, "thinks he can tell us what to do. . . . The next thing NUSAS will invite Nkrumah here and make him its chairman. . . . We are growing militarily stronger every day and we could eat any other African state for breakfast. Kennedy can threaten us as much as he likes, but we will show the world that our policy is the only one for South Africa."[27] After Kennedy departed, *Die Transvaler*, a paper close to the government, sourly offered "deepest sympathy for the American people if Senator Kennedy becomes their future President."[28]

Yet public attacks were rare during an astonishing week. Kennedy, through tact, directness and simplicity, came over to a surprising degree not as meddling intruder but as honest friend, recognizing white South Africa's kinship with as well as its estrangement from the great traditions of western civilization. He departed in a mixture of pessimism and elation. As the plane took off, he said somberly, "If I lived in this country, I would gather up everything I have and get out now." Then, wryly: "If we stayed another two days, we could have taken over the country." It was well to leave, he decided, before he began to have more influence than he wanted to have.[29]

"What caused all that almost frantic enthusiasm?" Anthony Delius wrote in the *Cape Times* after Kennedy's departure.

> ... What, in fact, made the senator a kind of third political force in the country in less than a week? ... Perhaps the most attractive and compelling feature of Senator Kennedy's speeches ... was the insistence that he and all South Africans he spoke to belonged to the same world.... There was a common humanity, he insisted, which went beyond all superficial differences of race and culture. And among all there might be people with extraordinary capabilities — "Why, among the black schoolchildren I saw today there might be an Einstein — or a Dr. Verwoerd." ...
>
> With him a great shaft of outer daylight was let into the air-conditioned comfort and concrete corridors of the bastion that nearly everybody believed, or feared, or hoped was shut tight against such natural illumination.[30]

The *Rand Daily Mail:*

> Senator Robert Kennedy's visit is the best thing that has happened to South Africa for years. It is as if a window has been flung open and a gust of fresh air has swept into a room in which the atmosphere has become stale and foetid. Suddenly it is possible to breathe again without feeling choked....
>
> This younger Kennedy ... has taken the youth of the country by storm through his message of confident, unashamed idealism. That is what so many of the young people of South Africa have been yearning for — some sort of clear and unequivocal endorsement that the hopes and ideals that all decent youngsters feel are indeed part and parcel of the great traditions of the contemporary world and not, as they are being told so often, something alien, unwholesome or worse....
>
> The effects of Senator Kennedy's visit will be felt for a long time to come.... And so, as he and Mrs. Kennedy fly off today, we say to them: Thank you a thousand times for what you have done for us. Come back again. You have a place in our hearts.[31]

"Even in remote areas such as South West Africa," the American embassy soon reported, "the Kennedy speeches were passed from hand to hand in the African townships."[32] The visit, Alan Paton said, could only be described as a phenomenon. "These long waits, this excitement, those outstretched hands, what are they but the signs of a hunger and a thirst, greater than we imagined? And who better able to satisfy them than our visitor?" Kennedy was "like a fresh wind from the wider world." But it would be a mistake, Paton added soberly, "to imagine that the Kennedy visit has made our world anew. . . . He can't fight our battles for us, and it is we who have to live our particular South African tomorrow."[33]

This was for Kennedy himself the ultimate frustration. "I remember one student who spoke at a meeting," he said after his return. "It was very easy for me to go there and speak of principles, and then leave, but that young man was going to stay."[34] He promised to return the next year,[35] but the government made it clear he would never receive another visa.[36] There was, in the end, an awful justice in the verdict of the *Cape Argus*: "Like a meteor, Mr. Kennedy has flashed across the South African sky, and has gone. . . . South Africa remains as it was."[37]

III

The rest of the African trip — Tanzania, Kenya, Ethiopia — was an anticlimax. The black leaders gave him the kind of reception usually accorded visiting chiefs of state. He found Nyerere lucid and forceful; Kenyatta, who gravely toasted the President of the United States, John F. Kennedy, seemed ancient, bemused and out of it; Haile Selassie had presence and dignity but appeared a relic of another age.[38] In Dar es Salaam, Kennedy met with Eduardo Mondlane, still fighting for a free Mozambique. Mondlane's account of the guerrilla war against the Portuguese gave the Americans a chastening sense of how the Vietnam War must look to the Viet Cong.[39] (Mondlane was assassinated in 1969. His movement continued the fight and won full independence in 1975.)

They went on to Rome, where Kennedy saw the Pope. "I told him how important it was the church take a clear position — how effective Bishop Hurley was — how cruel the system was." Black Africans were turning against Christianity, Kennedy said, because, as one had told him, "the Christian God hates the Negroes." He found the Pope "impressive though frequently as I was making a point he would make one of his own that was not completely pertinent." When Kennedy was discussing South Africa, the Pope said, "'We took an interest in African art in North Africa' as if this were a satisfactory response. Nonetheless impressive — & sensitive too. Urged me to continue talking on the moral aspects of [the] problem." His Holiness was also "very concerned about Viet Nam. Appeared solution must come from God." He thought the International Control Commission could be helpful in opening a dialogue between the North and South. "Urged that US cease opposition to China's entry into the UN." Kennedy observed "how harmful internally the war had been in US. [The Pope] said US leaders should keep pounding away at fact that they were interested in peaceful solutions."[40]

There was an intoxication about these trips. But Kennedy well understood their limits. Chester Bowles soon invited him to India. Kennedy sent the letter on to Galbraith with a note: "What do you think? To tell the truth I don't feel much like going off to India in November. I would like to start seeing the developed world like Paris."[41] When talk arose about an autumn journey to eastern Europe, Walinsky sent him an alarmed memorandum saying it's too much, you can't top it, there are too many things to be done in New York. The memo came back with a note in Kennedy's squiggly handwriting: "Adam, I'm not going anywhere this fall. I shall be hand in hand with you while we walk through the ghettoes of New York."[42]

There were indeed things to be done in New York. The trouble Kennedy was encountering in South Africa, wrote Joseph Kraft while he was still away, "is apt to look like child's play when he comes home. For by taking sides in an obscure New York judicial primary, he has set in motion a titanic struggle among the elemental forces of New York politics."[43] On his return on a rainy June afternoon, he opened his press conference by saying, to the delight of the New York reporters, "Everybody in Africa was very interested in Sam Silverman."[44]

IV

Sam Silverman represented Robert Kennedy's most serious foray thus far into New York Democratic politics. It was an impulsive venture. He well remembered his father's warning to his older brother a decade before that local politics was an "endless morass." John Kennedy, apart from one or two lapses, had heeded that warning. "It didn't make a great deal of sense to get into all those . . . fights in a state," the younger Kennedy mused in 1967, "because after a period of time it sucked away all your strength."[45] Nor did he need headlines about ruthless Bobby Kennedy trying to take over the New York party.

He was, in addition, as his perceptive friend John Burns, the state chairman, said, "bored" with politics.[46] His candid correspondent Fred Dutton had pleaded with him in vain to attend to "dull political proprieties" as his brother had.[47] The clubhouse politicians, Kennedy told Richard Goodwin, were not interested in issues; he was not interested in communion breakfasts and county dinners. They felt he neglected them. "I don't suppose I'm the jolliest fellow to have around."[48] Actually, the professionals felt not just neglected but rejected and threatened. "There aren't ten politicians in the whole state I like and trust," he told Jack Newfield of the *Village Voice*.[49] He concealed this feeling inadequately from the rest. "In Massachusetts they steal," he used to say, "but here they steal and lie too."[50]

On the other hand, he was, except for the comptroller, Arthur Levitt, the only Democratic official elected statewide in what had become almost the personal fief of Governor Nelson Rockefeller. This imposed an obligation to try and lead his party out of captivity. He felt an obligation too to the public-spirited young men and women whom the Kennedys had made a business of encouraging to enter politics. He wanted to restore the traditions of Smith, Roosevelt and Lehman, to raise the quality of candidates, to make the party an instrument of progressive government. And he of course wanted to secure his base against the depredations of Lyndon Johnson, who was still withholding patronage and by 1966 would have liked to sneak the state party (and the 1968 delegation to the presidential convention) away from Kennedy. In June, Marvin Watson reported to Johnson discussions with Mike Prendergast, whom Robert Kennedy had driven from the state chairmanship in 1961, about "ways in which your Administration can properly recognize the Democratic Party machinery there and therefore take away the separate organization which Senator Kennedy is planning."[51]

Finding the regulars hopeless, Kennedy was in fact turning increasingly to the reform Democrats and the Liberals. But, except for his West Side friends, like Ronnie Eldridge, the reformers seemed to him to be divided between opportunists trying to get into the organization and emotionalists with what James Wechsler of the *Post* called an "insistent and possibly incurable" will to lose.[52] As for the Liberals, he admired Alex Rose's talents in political maneuver but did not warm to him as a person.

Two weeks before departing for Africa, Kennedy had a drink with Rose. The idea was to renew relations; neither had anything in particular to discuss. "He's not the easiest fellow in the world to talk to, you know," as Rose said later; so, to keep things going, he showed Kennedy a Wechsler column in the afternoon paper about a special election for the surrogate court to be held in June. J. Raymond Jones, the Manhattan Democratic leader and a power in Harlem, had made a deal with the Republican county leader by which Arthur

Klein, the candidate of the Democratic organization, would get Republican endorsement in exchange for Democratic endorsement of a Republican candidate for the state supreme court. Wechsler reprinted excerpts from a twenty-year-old wiretap showing that Klein, wittingly or not, had once nominated a judicial candidate selected by the mobster Frank Costello. As Kennedy read the column, Rose recalled, "I could see his face change." "Can we do something about this?" Kennedy asked. "You can," said Rose. Kennedy: "Do you think we can win?" Rose: "You can't ever lose if you do the right thing — you know what I mean?" The astute Liberal leader threw in the thought that John Lindsay, the dashing Republican congressman elected mayor the previous November, believed nothing could be done.[53]

Kennedy's face changed for several reasons, of which only one was disgust over the intrusion of the underworld into politics. The surrogate court was a scandal. Its function was the probating of wills, in the course of which favored lawyers received large fees out of the estates of deceased citizens. The reform of the surrogate court was a way of protecting poor people who often saw inheritances draining away in surrogate fees. It was also a way of denying the party organizations their most lucrative source of patronage. And it was a way to consolidate his own relations with the Liberals and the reformers; a way to dish Ray Jones, whom Kennedy regarded as an unedifying practitioner of the old politics and who had, in addition, worked for Johnson in Los Angeles in 1960; a way, as Kennedy said, to "stick it to John Lindsay,"[54] who had shunned the fight. He asked Stephen Smith and William vanden Heuvel to come up with a candidate. They proposed Samuel Silverman.

v

Silverman, a justice of the state supreme court, had been a partner in Paul, Weiss, Rifkind, Wharton and Garrison, the city's leading liberal law firm. He had participated in Robert Oppenheimer's defense against security charges brought by the Atomic Energy Commission. He had the highest respect of his profession. Hearing the report on his qualities, Kennedy said, "*Habemus papem.*"[55]

He then telephoned Silverman. New York, Kennedy said, was filled with gifted Democrats who ought to be involved in public affairs. If Silverman would stand for surrogate, this would be an example to others. He added, "We both run certain risks in this." Silverman, who was happy on the supreme court, said he saw the risks Kennedy ran, but the only risk he could see for himself was that he might be elected. Kennedy said he would talk to the reform leaders and, unless they were willing to get off their asses and work, it wasn't worth doing. Silverman finally remarked that he couldn't say no to Kennedy.[56]

On May 22, a few days before leaving for Africa, Kennedy met with the reform leaders. It was, as Jack Newfield described it, "Kennedy at his evangelical best, and the reformers at their nitpicking worst." They were almost resentful to find him urging them on to the barricades. After two hours of bickering, Kennedy said quietly, "I had always been under the impression that you people really were interested in doing something [about judicial reform]. And now that I see that you're not really, I think that I have to reconsider before I ask Judge Silverman to make the commitment that he's willing to make." He started to leave the room. "It all pulled them together," Ronnie Eldridge recalled. "It was like a parent talking to the children."[57] The reformers agreed to join the battle.

Tammany and the old clubs were stoutly behind Arthur Klein. Ray Jones was fighting for political survival, and Kennedy feared that in his desperation he might raise the race issue. "It won't be bad, my being in Africa," he said as he departed, "if, you know, the Jones thing comes up."[58] It came up all right. One statement signed by Harlem district leaders charged that Kennedy's intervention was "aimed solely at J. Raymond Jones, because he is a Negro."[59] Kennedy responded by getting James Meredith and James Farmer to endorse Silverman.

Turning the campaign over to Stephen Smith, who described it as "two unknowns running against each other for a job nobody understood,"[60] Kennedy went off to Africa. Now he was back to find the campaign well organized but uphill. There were ten days to go. He flew to New York from Washington every afternoon to tour the city with Sam Silverman.

It was a rollicking few days. I accompanied the Kennedy motorcade on one suffocatingly hot June evening. At each street-corner rally Silverman would make four crisp points about the surrogate court, and James Farmer would boom away, and then Kennedy would speak. "Don't die in the city of New York," he exhorted his audiences. "Don't die — if you want to leave anything to your wife and children." Pointing to Silverman, "He is an honest man, he will protect you." He liked to tease the children in the crowd. "You all know, of course, what the surrogate is." They would cry, "Yes," and he would call back, "Some in this neighborhood don't tell the truth, I'm afraid. . . . OK, how many of you are going to go home tonight

and tell your mothers and fathers to vote for Judge Silverman?" All
hands shot up. "Let's go over it again. What are you going to tell
your mothers and fathers when you go home?" "Vote," the children
cried. "And vote for whom?" "Kennedy!" He would put his hands
to his head in mock pain and say, "No, no. I don't think you have
the right idea," and the marimba band would strike up and the
motorcade move on to the next rally.[61]

There was always the vein of melancholy. Kennedy sat, as he
customarily did, in the front seat of the car to the right of the
driver. Someone called from another car that his door was not fully
closed. As he slammed the door shut, he said, "I wonder what
proportion of voters in New York would have thought it a good idea
to warn me about this."[62] Silverman found him altogether different
from what he had expected — "thoughtful and gentle. . . . I never
saw any of the, you know, the ruthless reputation. . . . He'd ask what
I'd think and so on, and never attempted to impose his will."[63]

Silverman won easily, even almost carrying Harlem. Kennedy at
the victory party: "I well remember that Sunday when I called him
to see if he would run. And he said in a ruthless kind of way, 'Just
you remember, Mr. Kennedy, Silvermans don't finish second!'" Kennedy introduced the campaign manager — "a beloved figure who is
replacing me on the American horizon, ruthless, mean Stephen
Smith."[64] Murray Kempton titled his column for the Post "Kennedy
Regnant," adding, "The editors of the New York Times have a contempt for the motives of the junior Senator from New York quite as
excessive as the respect they entertain for those of its senior Senator;
quite naturally then there was no mention of Robert Kennedy in the
Times editorial celebrating the newest of our annual deliverances
from the bosses."*

VI

"Kennedy Regnant" — the momentum of the Silverman triumph led
to the next step in the revival of the party: the nomination of a
candidate for governor against Nelson Rockefeller. The regulars
had gathered behind Frank O'Connor, a pleasant and unassuming
Irishman of the pre-Kennedy school who had been district attorney
in Queens and was now president of the New York city council.

* New York Post, June 30, 1966. Kennedy followed up his victory by proposing in
the New York State Constitutional Convention in 1967 the merger of the surrogate
court with the state supreme court and the establishment of a state office of public
guardian. The politicians successfully defended the old system. Silverman served
with distinction on the surrogate court and reformed everything within his power but,
bored and frustrated, resigned in 1971 and returned to the state supreme court.

Kennedy liked him well enough but hardly considered him a model of the new spirit with which he hoped to infuse the party, a feeling that O'Connor, a realist, accepted.

Among other aspirants, Franklin Roosevelt, Jr., had claims on Kennedy's support. John Kennedy had planned to make Roosevelt Secretary of Commerce so that he could build himself up for the New York governorship.[65] But Roosevelt was unpopular with the organization, and Kennedy, though he retained a fondness for him over the years, did not see him as an ideal candidate either. His preference was Eugene Nickerson, county executive of Nassau County, able, informed, honorable, with a tested appeal to the suburbs. Nickerson was little known upstate, however, and not a stirring campaigner. Then the Silverman victory gave Kennedy a moment of maximum influence. Many thought he could now dictate the nomination. O'Connor thought so.[66] Nickerson thought so.[67] The newspapers ran stories about his ruthlessness and predicted he would name the candidate.

Stuck with the reputation, he might as well have been ruthless. Instead he hesitated, vacillated and blew the opportunity. When Jack English, Nickerson's main backer, pressed for an endorsement, Kennedy replied, "If I do that, isn't it going to be boss rule?"[68] "He decided finally," said John Burns, "that he was not going to get himself into that kind of a position where he was going to force Nickerson or anyone else on the party."[69] If Nickerson were clearly a great candidate and O'Connor clearly hopeless, Kennedy told me at the end of July, he would come out for Nickerson; but the gap was not quite that large. "I am weak with the Irish," he added, "and, if I come out for Nickerson, I will get into a lot of trouble there."[70]

English, making a last try, warned Kennedy that O'Connor, if victorious, would back Humphrey for the presidential succession.[71] O'Connor was indeed a friend of Humphrey's and, when asked in 1970 whether the administration had been courting him, replied, "No question about it." "Bob Kennedy — I know now for a fact —" O'Connor said, "believed that I was completely in the Humphrey-Johnson camp, which was not true, but I can understand it, and I don't resent it."[72] At the time, he went down to Hickory Hill and told Kennedy, "If I were you . . . and I was thinking of supporting a candidate for governor of New York, I would want to make darn sure that he was not an enemy of mine. . . . I want to tell you now I'm not out to block you."[73]

Kennedy had lost control of the situation. After O'Connor agreed to let him pick the delegation in 1968 and to keep Burns as state chairman,[74] he put out a statement saying he would abide by the

wishes of the convention. The Johnson White House watched with
extreme pleasure. "The tide is now on the ebb for him," wrote Henry
Hall Wilson of the legislative staff, especially because of "his fiasco in
the New York gubernatorial matter. His problems will be compounded if O'Connor defeats Rockefellow [sic], as now seems highly
probable."[75]

The Democrats assembled in Buffalo on September 7. "It was
Robert Kennedy's burden to introduce O'Connor," wrote Jack Newfield. "The senator is deeply bored by the kaleidoscopic tribalisms of
New York politics. His is the mind of the curious novelist, rather
than the mechanic absorbed by details; his span of concentration is
brief when it becomes focused on tactical intrigues divorced from
substantive issues. . . . Still, the senator knows of O'Connor's dream
of becoming Hubert Humphrey's running mate in 1972. . . . In the
scorebook of presidential politics, Humphrey won in Buffalo and
Robert Kennedy lost." Newfield added: "But the senator is a pro,
and in the last week of October will be in the streets."[76]

VII

When he arrived on the New York streets at the end of October,
Newfield was with him. Three years had passed since the radical
reporter had glared with hatred at the hard-faced Attorney General
shouting to demonstrators before the Justice Department. The previous February Newfield had described Kennedy on a television show
as an existential man, preoccupied by suffering, who could understand himself only in action and to whom Dallas had given a unique
sense of the absurdity of life. Kennedy, chancing to see the show,
had asked Newfield for lunch.

"I was not fully prepared for the changes," Newfield wrote. ". . . Instead of the military crew cut, his graying, ginger hair now lapped
over his earlobes in the shaggy style of the alienated young. His blue
eyes were now sad rather than cold, haunted rather than hostile."[77]
They argued about Kennedy's performance as Attorney General —
whether he had instituted enough voter registration suits and why
the FBI had only taken notes when Newfield saw blacks beaten up in
Mississippi. "I liked him enormously," Newfield said later. ". . . Kennedy was the first national politician I had met who had human
reactions. He wasn't plastic. He wasn't programmed. . . . I felt that
he was bright and that he saw the world from an angle I couldn't
even imagine."* Kennedy liked Newfield enormously too. Newfield

* Victor S. Navasky, "Jack Newfield Talks About R.F.K.," *New Leader*, May 26, 1969;
Clark Whelton, "Memoir of a Hero Who Died at Five," *Village Voice*, June 5, 1969;

had grown up in Bedford-Stuyvesant, the only white kid on the block, helped found Students for a Democratic Society, joined SNCC, went to Mississippi and saw the world from an angle Kennedy wanted more than ever to understand.

Back in New York after speaking for liberal Democrats across the country, Kennedy first wanted to help his favorite member of the New York House delegation, Congressman Hugh Carey of Brooklyn. Shaking hands as he walked with Carey and Newfield along crowded sidewalks, he noticed a ten-year-old girl wearing glasses. He knelt down and said, "You know something? My little girl has glasses just like yours. And I love my little girl very much."[78] Lost for a moment to the world, he rubbed the back of her neck, then turned to shaking hands again. (On another occasion in Bay Ridge, when Kennedy was walking with Carey, someone shouted, "Carey, what are you doing with that Red? Are you a Commie too?" Carey shrugged and said, "That's the way they act around here.")[79]

On the way back to Manhattan to join O'Connor, Kennedy said to Newfield, "Do you still like me, Jack?" Newfield said, "Yes, but why do you have to campaign for a guy like Carey?" "He's such a decent fellow," Kennedy said. "He works very hard and he's very bright. Whenever we meet at least he's concerned about the issues. He's not like [James] Scheuer [a reform Democrat, also in the House] who has such a great reputation, but only asks me about patronage. . . . Also, Carey has trouble with his district. He was telling me the Church was such a problem. I wonder why the kids who come out of parochial school are so conservative."[80]

O'Connor, who had proved a listless campaigner, was behind in the polls; nevertheless, after three rallies in Manhattan, he announced he was cold and tired and would call it a day. "Kennedy's face," observed Newfield, "showed he felt this was a sign of softness." Kennedy completed the schedule, then invited Newfield to his apartment in United Nations Plaza. They had a couple of drinks. Kennedy asked a little shyly whether Newfield liked poetry. Newfield said he liked Yeats and Hart Crane. Kennedy said, "Can I read you some poetry by a poet I like very much?" Newfield, anticipating a con job for the benefit of the rebel press, thought he might come up with something like Allen Ginsberg's "Howl." Instead he read Emerson's "Fame."

Jack Newfield, in recorded interview by Jean Stein, July 21, 1968, 3–5, Stein Papers. Newfield developed his thesis about Kennedy in "Bobby Kennedy: Existential Hero" (*Cavalier*, June 1966). Kennedy told Newfield he had sent a copy of the piece to his father, first ripping it out of the girlie magazine because the rest of the contents were so obscene.

When he finished, Kennedy peered out the window at the barges making their way along the East River. "Look at that!" he said. "There's a ship called *World Justice*, and it's moving away from the United Nations." Then he read Emerson again, this time the "Ode Inscribed to W. H. Channing":

> But who is he that prates
> Of the culture of mankind,
> Of better arts and life?
> Go, blindworm, go,
> Behold the famous States
> Harrying Mexico
> With rifle and with knife![81]

On November 8, O'Connor went down to defeat. Robert Kennedy spent election night with a small group ranging from John Steinbeck and Yevgeny Yevtushenko, the Soviet poet, to Sam Silverman.

34

Time of Troubles

IN THE LAST MONTHS of 1966 Robert Kennedy moved dramatically ahead of Lyndon Johnson when public opinion polls asked whom Democrats and independents favored for 1968. Kennedy's "rise in political appeal," Gallup said in August ". . . has been spectacular."[1] Louis Harris concurred in November. Kennedy himself remained skeptical. The voters, he told me, probably agreed 80 percent with Johnson on the issues between them; it was just that people did not believe Johnson any longer. "He is rueful about his own vogue, which he regarded as transient."[2] He was right.

I

There had been premonitions in the summer of 1966 when Katzenbach as Attorney General told Kennedy the Justice Department was filing a brief with the Supreme Court confessing that the FBI had engaged in illegal electronic eavesdropping in 1963. Kennedy asked Katzenbach to state in the brief that the Attorney General at the time had no knowledge of the FBI's bugging habits either in the case at hand or in general. Katzenbach refused, not because he disbelieved Kennedy but because he thought the addition irrelevant.[3] Kennedy, upset, wrote his old friend: "As you know this is a damn important matter for me. I just don't want to receive a shaft — it's not deserved — and anyway I don't like them deserved or not. I'm getting too old I guess. I can't write you as many memos as J. Edgar Hoover. And there is no sense in talking about it by phone. I feel strongly about it — and I write you just that as there's not much else to say."[4]

The departmental confession initially aroused little attention. Then Senator Edward Long of Missouri, a friend of Roy Cohn and

of the Teamsters, proposed an investigation of wiretapping and bugging. "He is out to get Bobby," Bill Moyers told Goodwin one day.
"Johnson is egging him on."[5] In December 1966 Hoover himself
went public, firing off letters and documents to Congressman H. R.
Gross of Iowa. The press covered the altercation with relish. Either
Kennedy knew about the bugging, Richard Harwood wrote in the
Washington Post, in which case his denials impeached his credibility,
or else he did not know, in which case "his executive competence
could be brought into question."[6] Congressman Wayne Hays sagely
advised the President to "fire J. Edgar Hoover on the strength of
Bobby Kennedy's statement and answer the backlash by pointing at
Kennedy. . . . your best bet for getting rid of both men."[7]

Kennedy took refuge in jokes — "I also want to deny officially the
rumor that J. Edgar Hoover irritated me during that period. He was
no bother at all. Of course, he bugged me now and then"[8] — but he
knew that damage had been done.

II

Greater damage was to come. After Dallas, Jacqueline and Robert
Kennedy had resolved to spare the dead President the indignity of
having a book written about his last days by Jim Bishop, whom
Jacqueline thought a hack and Robert distrusted because of a pro-
Hoffa series Bishop had written in 1959. They therefore approached
— or, as Jacqueline later inaccurately and infelicitously put it —
"hired" William Manchester to do the book.[9] (Bishop did his book
anyway.) Manchester, a skilled writer and a decent but highly emotional man, had written novels, an excellent biography of Mencken
and a deft and admiring sketch of John Kennedy.

On March 28, 1964, Manchester signed a contract providing that
"the final text shall not be published unless and until approved" by
Jacqueline and Robert Kennedy.[10] This was a fundamental error.
A provision requiring family approval of the use of materials made
uniquely available by the family — letters, oral history transcripts and
the like — would have been proper. A provision requiring family
approval of the entire text compromised Manchester's freedom as a
writer and made the Kennedys responsible for Manchester's independent research and interpretation. The Kennedys never should
have proposed such a provision. Manchester never should have
acceded to it. But in the aftermath of the assassination the Kennedys
were in an unduly protective mood. When John Kennedy's old
friend Paul Fay wrote his bouncy memoir *The Pleasure of His Company,*
Robert Kennedy landed on him as if it were the only book through

which posterity would ever know his brother. He wanted Fay to cut nearly two fifths of his jocular text. Galbraith, reading the manuscript at Jacqueline's behest, thought it harmless. Fay published it with some excisions in 1966, and an old friendship lapsed.

In early 1966 Manchester, after prodigious research, turned over a long, deeply felt, greatly overwritten manuscript. Kennedy friends read it with growing dismay. They objected especially to Manchester's treatment of Johnson. The draft opened, for example, with a hunting scene in which a boorish Johnson pressed a reluctant John Kennedy to shoot a deer on the Johnson ranch. This beginning, as I wrote Manchester, had the effect "of defining the book as a conflict between New England and Texas, decency and vulgarity, Kennedy and Johnson." The portrait of Johnson "too often acquires an exaggerated symbolism — so much so that some critics may write that the unconscious argument of the book is that Johnson killed Kennedy (that is, that Johnson is an expression of the forces of violence and irrationality which ran rampant through his native state and were responsible for the tragedy of Dallas)." [11] John Seigenthaler, Edwin Guthman and Pierre Salinger all believed that Manchester's evident hostility to Johnson "would destroy the credibility of the book." [12] Evan Thomas of Harper & Row, the putative publisher, thought the manuscript in part "gratuitously and tastelessly insulting to Johnson." [13]

The Kennedy concern was not, as Manchester supposed, "anxiety over Johnson's reaction." [14] By 1966 Johnson had turned irrevocably against Robert Kennedy. It was rather the impression the manuscript left that the Kennedys had never given Johnson a fair chance. In my own case, I recommended the elimination of the mythodrama depicting Johnson as monster-usurper, but felt that Evan Thomas had gone too far in deleting strictly factual references to Johnson.* Kennedy himself acknowledged the claims of history. He observed to Manchester of one chapter that it "will injure both Johnson and me, but apparently it's factually correct and a contribution to history. I'd like you to change it, but I guess you won't." [15]

Had it not been for the wretched contract, Manchester would have been entirely right in fighting for his reading of the historical record; even within his rights in fighting for his romantic mythodrama. The contract probably should have been rewritten to release the Kennedys from responsibility; but I cannot remember anyone's suggesting

* On April 28, 1966, after consultation with Seigenthaler in Nashville, I wrote Kennedy, "You will be under great pressure to tone down the LBJ passages in the book. Of course, everything that is petty and gratuitous should go; but I hope you will not take out anything which is an essential part of the historical record" (RFK Papers). See also William Manchester, *Controversy* (Boston, 1976), 19.

that. It was probably supposed that no one would believe Manchester was truly a free agent. Or, with the controversy every day more acrimonious, the Kennedys may have considered the contract their only security on a separate issue: Manchester's use of material derived from his interviews with Jacqueline Kennedy.

On this point Manchester was clearly wrong. The proposition that an oral history interview became the private property of the interviewer was indefensible. If the person interviewed lost all rights over the transcript, the oral history program, which promised so much to the historian of the future, would be dead. Manchester had no right to use material from his interviews with Mrs. Kennedy that she did not wish him to use.

III

This became the central issue. With regard to Johnson, Manchester subdued most of his mythodrama and retained most of his history. But he was unwilling to surrender personal detail that Jacqueline Kennedy, increasingly agitated, regarded as an inexcusable violation of her privacy. With *Look* poised to begin serialization, the discussions acquired the hysteric character of a malign farce. Robert Kennedy, who had other things on his mind, became increasingly weary of the business. "It was just a damned nuisance to him," Evan Thomas said. "He hated it all and wished he had nothing to do with it." [16]

But his loyalty to his sister-in-law was unconditional. At one point he impulsively wired Thomas that publication should be canceled. "It just seems to me that rather than struggling with this any longer we should take our chances with Jim Bishop." [17] This last was strong language from Robert Kennedy. When Manchester complained how difficult it all was for him, Kennedy said in amazement, "Do you think you've suffered more than Jackie and me?" [18] By December 1966 a desperate Manchester would make no more changes. "I have reached the point," he informed his agent, "where, if the integrity of my manuscript is violated, I have no wish to go on living. . . . I am ready to die for this book." [19] Jacqueline Kennedy was equally determined to expunge the personal material she found offensive. On December 16 she filed suit to prevent the book's publication. Frank Mankiewicz said to Kennedy, "My God, I think that's a terrible mistake." Kennedy said, "Yes, it's a terrible mistake but nothing can be done about it." [20]

The press had a field day. The columnist William S. White, a Johnson intimate, wrote that the purpose of the Manchester book had been to "gut Johnson" and "that the smile or the frown of the

Kennedy cult has a power over the fortunes of any kind of book that this country has never known before."[21] Editorial writers rode their high horse about the sacred cause of freedom of information. I watched it all with a certain skepticism. Fifteen months before, the press had almost unitedly risen to denounce a writer (me) for reproducing in *A Thousand Days* comment about public officials acting in their public capacities. The historical truth of the comment was not under challenge; but editorial writers then brusquely dismissed the claims of history. Now with almost equal unanimity they invoked historical truth as an absolute in order to rebuke the widow of a murdered President for guarding her grief against a writer who sought to use her confidences without her permission.

The *Washington Post*, which had not in 1965 defended historical truth as a criterion in writing about Secretaries of State, asserted sententiously in 1966, "The lives of public men — the records of their careers, the thoughts of others about them — are not the property of their families, but the property of posterity."[22] Robert Kennedy later wrote to Katharine Graham, the *Post*'s publisher, that Jacqueline was "so upset and really crushed . . . a girl who hadn't committed any great crime but who day after day was being attacked and pilloried in all kinds of scandalous ways."[23]

On December 21 *Look* agreed to remove or modify passages relating to the personal life of Jacqueline Kennedy and her children. "These paragraphs," her lawyer explained, "were the sole reasons for the initiation of her legal action."[24] The suit against Harper & Row was settled in January, the court awarding the oral history tapes to Jacqueline Kennedy. The Kennedys had won their case but, as Robert perfectly understood, at a fearful cost. The stereotype of ruthless, arrogant Bobby Kennedy, riding roughshod over everybody, had received a tremendous infusion of new life. "It's really mostly my fault," he said to Mankiewicz. ". . . There wasn't very much time anyway and I certainly didn't want to spend it on that. We'll just have to move on."[25] Murray Kempton saw him after a visit to wounded men at the Bethesda Naval Hospital. "You know," Kennedy said, "as I was standing there, I thought to myself: there are things more important than who writes a book about somebody."[26]

IV

The men he had visited were back from Vietnam, and one of the things more important was the war. But there was also new trouble, nearer home. One day, as Ethel Kennedy was riding with her children near Hickory Hill, she noticed a cadaverous horse standing miserably in a chicken coop. "It was a bag of bones," she said, "the saddest sight I've ever seen." She told her groom to bring the horse to Hickory Hill where, despite maternal care, it died five days later. The owner sued her for $30,000 as a horse thief. The trial, in January 1967, lasted a long two days, and the jury deliberated for a longer two and a half hours. Ethel began to tease Louis Oberdorfer, her lawyer, about what would happen if they lost when, to popular applause, the jury came back with an acquittal. It was the best publicity the Kennedys had all winter. Someone asked Ethel whether she would steal a starving horse again. She replied, "I don't think I could live with myself if I didn't." As for Robert Kennedy, Oberdorfer recalled him as not too greatly concerned; "he was preoccupied then with the Vietnam business."[27]

He could not escape the Vietnam business. While generally avoiding the war during the autumn campaign, he had called for an end to the bombing in a question period after a speech in California, adding that the people of South Vietnam had a right to determine their own destiny. "Personally, I don't think they want General Ky, just as it is clear they don't want the Communists." This irreverence, at a moment when Johnson was meeting the new leader of South Vietnam in Manila, scandalized the statesmen of the press. It was, said the indignant David Broder, "a hip-shooting display of verbal carelessness, worthy of Barry Goldwater."[28] Three days later at Cam Ranh Bay Johnson exhorted American troops to "come home with that coonskin on the wall." Commentators did not denounce this as hip-shooting. One forgets too quickly the mood of 1966.

In mid-December Jack Newfield pleaded with Kennedy to speak out. If another speech would do any good, Kennedy said, "I would make it tomorrow. But the last time I spoke I didn't have any influence on policy, and I was hurt politically. I'm afraid that by speaking out I just make Lyndon do the opposite. He hates me so much that if I asked for snow, he would make rain, just because it was me. But maybe I will have to say something. The bombing is getting worse all the time now."[29]

At the end of January he went to England to address the Oxford Union. Soon after arrival he received a cable from Ethel:

SUNDAY GALLUP POLL WILL SHOW ... DEMOCRATS REPUBLICANS AND INDEPENDENTS COMBINED KENNEDY 48 JOHNSON 39 NEITHER 13. IF THIS KEEPS UP YOU JUST MAY HAVE TO DUMP OLD HUCKLEBERRY CAPONE LOVE AND KISSES AND GOOD WORK.[30]

Alas, the poll had been taken before the Manchester dénouement. In London he saw old friends. He urged them not to acquiesce in the decline of Britain. "Your history is your power. . . . You have lost faith in your future because you are forgetting how much your past has meant to all of us." As for the preoccupying issue, "Agree or disagree with America's position in Vietnam, but do it because you believe it, not to save the pound." Waiting for Prime Minister Harold Wilson, he wrote his father a letter on 10 Downing Street stationery, recalling the family's stay in London nearly thirty years before.[31]

Then to Oxford, where the first question after his address was about Vietnam. Reluctant to attack his own government beyond the twelve-mile limit, Kennedy confined himself to expressing "grave reservations" about the bombing. The next three weeks, he added, might be crucial. "There are some signs that Hanoi is reconsidering its position." He evidently had in mind the impending visit of Aleksei Kosygin, the Soviet Premier, to London. The questions went on for an hour, ranging from birth control, which he endorsed despite the deplorable example of his own family, to China. He departed to a standing ovation.[32] The next day he left for France.

The French, however much they disagreed on everything else, were united on Vietnam. De Gaulle's position, François Mitterand, the Socialist leader, told Kennedy, had "the support of the majority."[33] Couve de Murville, the foreign minister, said there would be "a Communist government in Saigon. We may not like it but it is inevitable."[34] "The United States cannot do well," André Malraux told him, "when you are involved in a matter of inner contradiction. Vietnam is against American tradition."[35] Henry Kissinger, who was in Paris, took him to see Jean Saintény, once high commissioner in northern Indochina. Saintény, recently back from Hanoi, talked about Ho Chi Minh's determination to be independent of both Moscow and Peking. Kissinger appeared to agree that further escalation would be folly and that negotiation was a possibility to be pursued.[36]

He saw de Gaulle. "As I told your brother," the general began, "the United States is involved in a wrong course in Vietnam." There could be no peace until Washington stopped bombing and declared its intention to withdraw. This would not mean communization. "South Vietnam would not permit the North . . . to run their country. Ho Chi Minh realizes this and would not attempt it." America, the general continued, had always presented the highest ideals to the world. "All of this is being destroyed by your role in Vietnam. . . . History is the force at work in Vietnam, and the United States will not prevail against it." After seventy minutes de Gaulle concluded by offering Kennedy some remarkably bad counsel. "You are a young man," he said, with "a brilliant political future. . . . I am an old man, and I have lived through many battles and wear very many scars, so listen to me closely. . . . Do not become embroiled in this difficulty in Vietnam." Those who opposed the war would be "badly hurt." When they lose their effectiveness, then it will be time for you to step in and "help your country regain its proper course."[37]

Kennedy's most important Paris talk, as it turned out, was with Étienne Manac'h, the director of Far Eastern affairs at the Quai d'Orsay. John Gunther Dean, the Vietnam expert at the American embassy, went with him. Manac'h stressed Hanoi's "complete mistrust" of American peace gestures. "Recent history has shown that, as soon as President Johnson makes offers of peace, there are parallel actions of war." Manac'h felt nevertheless that Hanoi remained ready to negotiate. "The one indispensable condition . . . is a cessation of bombing."[38] This was what Kennedy got out of Manac'h, according to his own memorandum of the conversation. The talk was in French, which he understood imperfectly and which Dean translated on the run.

But Dean, better versed in both the language and the negotiating nuances, got something more specific. Up to this time Hanoi had conditioned negotiation on American acceptance of four prior points. Now Manac'h said, citing a declaration made to him by the head of the North Vietnamese mission in Paris, Hanoi was waiving its preconditions and asking only that the bombing stop. Hanoi, in short, was offering an important concession; and Manac'h evidently felt that, if he passed the message to an American diplomat in Kennedy's presence, it could not be ignored in Washington. Afterward Dean drafted a cable, showed it to Manac'h to make certain he had it right, heard Manac'h say again how important the new Hanoi position was and sent the cable off to the State Department.[39]

Kennedy went on to Bonn, where Chancellor Kiesinger told him that "the majority of people do not understand what and why the United States is doing in Vietnam"[40] and Willy Brandt, now foreign minister, said that German-American relations had "declined as a result of the Vietnam involvement."[41] There was more of the same in Rome. Fanfani complained for an hour about the sabotage of negotiations by American bombing;[42] President Saragat said that the United States had evidently lost all interest in Europe because of Vietnam, and Russia was moving into the vacuum;[43] the Pope cited "extremely reliable sources" indicating that Hanoi had changed its attitude and was somewhat more prepared for discussions.[44] Unable to find any leader on the continent who supported or even understood America's Vietnam policy, Kennedy, his own misgivings redoubled, flew back to the United States.

v

While Kennedy was still abroad, someone in the State Department showed Dean's cable to Edward Weintal, the diplomatic correspondent of *Newsweek*. Kennedy returned on Saturday, February 4. On Sunday evening the *New York Times* received an advance copy of *Newsweek* with Weintal's story about a "significant peace signal . . . unveiled for the benefit of Robert F. Kennedy for reasons best known to the enemy."[45] The Monday *Times* led with the Kennedy "peace feeler."

When Lyndon Johnson read his *Times*, he was enraged. It was another of those recurrent seasons of military optimism. On January 8, Lodge had predicted "sensational military gains."[46] Westmoreland was delivering sanguine pronouncements with imperturbable inanity. "Joe Alsop came in for an hour," Rostow reported to Johnson on January 28. ". . . He thinks we are in the process of winning the war quite rapidly. . . . I cautioned him on some of his exuberance . . . but my personal view is that he is nearer the mark than most of the analysts."[47] Why make concessions when you are ahead? The weekend of Kennedy's return Johnson hardened his negotiating position, demanding that the North stop reinforcing its troops in the South as a condition precedent to the cessation of the bombing.*

It had been infuriating enough in Johnson's view for Robert Kennedy to wander around Europe chatting with heads of state; but for him to try and force the President's hand on negotiation, at a time when the war was going so well, was final proof of malevolence. For Johnson was sure Kennedy had leaked the story to *Newsweek* in Paris — a suspicion presumably verified when the State Department, with customary efficiency, could find no Paris cable on the Kennedy-Manac'h talk. (Dean's account was later discovered to have been misfiled.)[48]

Kennedy, not having understood that Manac'h was relaying a hot tip from Hanoi, was baffled by the *Newsweek* story. He told Mankiewicz he did not have the "foggiest idea" what they were talking about.[49] In any event, the courteous thing, he thought, was to report to the President on his trip. They met Monday afternoon, February 6, the day of the *Times* story. Kennedy left no record of the meeting;

---

* The letter to this effect, dispatched by Johnson to Ho Chi Minh on February 8, was drafted and redrafted February 2–8 (David Kraslow and S. H. Loory, *The Secret Search for Peace in Vietnam* [New York: Random House, Vintage reprint, 1968], 206).

Johnson published none (Kennedy was sure he had the meeting taped); [50] and the others present — Katzenbach, who had become under secretary of state, and Rostow — have held their peace, except to agree that it was most unpleasant. When he got back to his office, Kennedy described it all incredulously to Mankiewicz and Edelman, and the historian must rely on their recollections.

"The President started right in by getting mad at me for leaking the story," Kennedy told Mankiewicz. Kennedy replied that he had not leaked the story; he was not aware there had been a peace feeler and still was not sure there had been one. "I think," Kennedy said, "the leak came from someone in your State Department." "It's not *my* State Department, God damn it," Johnson said angrily. "It's *your* State Department." The President went on about the irrelevance of negotiations. "Those guys are out of their minds," Kennedy told Mankiewicz. "They think they're going to win a military victory in Vietnam by summer. They really believe it." Johnson told him that the war would be over by June or July. "I'll destroy you and every one of your dove friends [he specified Fulbright, Church and a couple of other unfortunates] in six months. You'll be dead politically in six months."

After a time Kennedy asked whether Johnson would like to know what Kennedy thought he should do. Johnson said yes, go ahead. "Say that you'll stop the bombing if they'll come to the negotiating table," Kennedy said, "and then you should be prepared to negotiate." He outlined a series of possibilities — a cease-fire in stages, an expanded International Control Commission to deter further escalation, an international presence gradually replacing American forces, a political settlement allowing all major elements in South Vietnam to participate in the choice of a government. "There just isn't a chance in hell that I will do that," Johnson said, "not the slightest chance." Kennedy and his friends, he said, were responsible for prolonging the war and for killing Americans in Vietnam. Blood was on their hands. Kennedy said, "Look, I don't have to take that from you," and started to leave.

Katzenbach and Rostow tried to compose the situation. They asked Kennedy to tell the waiting press that there had been no peace feelers. Kennedy refused, saying he did not know whether there had been peace feelers or not. "I didn't know what the hell had been said to me." He finally agreed to say that he had not brought home any feelers — true enough, since the message had been transmitted to Washington by Dean. "Well, that wasn't a very pleasant meeting," Kennedy told his aides when he was safely back on the Hill. "He was very abusive," he said later to Newfield. ". . . He was shouting and seemed very unstable. I kept thinking that if he exploded like that with me, how could he ever negotiate with Hanoi."[51]

Versions of the meeting, suitably embellished, rapidly spread about town. *Time,* rising to the occasion, announced that Kennedy had called Johnson a son of a bitch to his face, which he had not.[52] Once again Kennedy was the loser. He appeared a publicity-seeking meddler in international relations, a compounder of trouble for an overburdened President and, as always, ruthless and arrogant. A more legitimate criticism would have been of his failure to spot the negotiation feeler, but the press did not pursue this point. Kennedy, as usual when in doubt, made a joke. Speaking in March at the Gridiron banquet, he offered his own version of the session with Johnson: "We had a long serious talk about the possibilities of a cease-fire, the dangers of escalation and the prospects for negotiations. And he promised me the next time we are going to talk about Vietnam."[53]

In any case, the Manac'h message, when the State Department finally found it, made no difference. Neither did Kosygin's assurance to Harold Wilson in London in the second week of February that unconditional cessation of the bombing would bring negotiations. On February 13 Wilson and Ambassador David Bruce urged Washington to continue the brief bombing pause over Tet, the Buddhist New Year, so that Kosygin might have time to persuade Hanoi to meet Johnson's insistence on reciprocal military deescalation in exchange for a bombing halt. On February 14 Johnson ordered the resumption of the bombing. "Peace was almost within our grasp," said Wilson.* Johnson preferred that coonskin on the wall.

VI

The February 6 meeting marked the end of the road. Kennedy saw no further point, Mankiewicz said, in trying "to mute his criticism of the war."[54] He began on February 8 with a speech in Chicago about China.

In its quest for reasons sufficiently dire to justify the growing fury of American intervention, the administration had evolved the thesis that the real enemy in Vietnam was China. "The threat to world peace," as Hubert Humphrey soon said, "is militant, aggressive Asian communism, with its headquarters in Peking. . . . The aggression of North Vietnam is but the most current and immediate action of militant Asian communism."[55] In Chicago Kennedy flatly rejected

---

* There was much more to the Washington sabotage of the Wilson-Kosygin talks. The bizarre story is well told in Chester L. Cooper, *The Lost Crusade* (New York, 1970), 350–368.

"attempts to portray Viet Nam as a Chinese-inspired conflict." Vietnamese Communism on the contrary was "a native growth, with its own revolutionary traditions and dynamism." As for Communist China, Kennedy called for a new American policy based on "conscious and open recognition that we live in the same world." He looked forward, he said, to the day when an American diplomat would go to Peking bearing the instructions Secretary of State Daniel Webster had given Caleb Cushing in 1843: to tell the people of China "that you are a messenger of peace from the greatest power in America to the greatest in Asia."[56]

The next step was a speech on Vietnam itself. Kennedy took care to talk with McNamara, Taylor and Harriman to get the administration's latest thoughts. I do not know what they told him; he always protected such confidences. McNamara by this time had lost all faith in the efficacy of bombing. I do not imagine he gave Kennedy to understand that victory was imminent. Taylor remained a true believer. Harriman was more complicated. His misgivings about the war were deep but carefully buried because, I always thought, he wanted to conclude his distinguished career by negotiating a Vietnam settlement and feared that any hint of disloyalty would lead Johnson to give the job to someone less determined to make negotiations succeed.

Kennedy also sought out the most extreme opponents of the war. On February 13 he asked Newfield to arrange a meeting with Tom Hayden and Staughton Lynd, two articulate radicals whose book *The Other Side*, an account of a trip to Hanoi, Kennedy had recently read. The New Leftists, to Kennedy's surprise, advocated not unilateral withdrawal but an unconditional bombing halt as a preliminary to negotiations, with which Kennedy of course agreed. Kennedy said he favored a "Laos-type" solution for Vietnam. Afterward Newfield asked his friends what they had made of Kennedy. Lynd said, "Very fair-minded. Sort of detached, and not authoritarian at all. But still very much a liberal." Hayden thought him superior to "Reuther or Rauh or any of those guys" and was reminded of Mendès-France. Kennedy told Newfield that they seemed nice, bright fellows; "but I didn't think they told me everything they felt about immediate withdrawal."[57]

Goodwin, Walinsky and I set to work on drafts. Goodwin, along with Kennedy himself, had the largest hand in the result. In the meantime, public expectancy was growing. On February 17 the *Times* announced in a five-column headline: RFK SETS MAJOR SPEECH ON BOMBING. Other senators, it should be noted, were calling for a halt. The valiant Morse and Gruening had gradually been joined by

Fulbright, McGovern, Nelson, Clark, Church, Javits, Eugene Mc-
Carthy and a dozen more. But Kennedy was the one who alarmed
Lyndon Johnson.

The White House orchestrated a counterattack. Johnson directed
Westmoreland, "in anticipation" of the Kennedy speech, to denounce
a bombing pause.[58] "I don't want to pay one drop of blood," Westmoreland promptly told reporters in Saigon, "for a pig in a
poke."[59] James A. Farley was exhumed to inveigh against Kennedy's
"soaring ambitions." "Insulting, belittling and interfering with the
office of the Presidency," Farley said, "is not the act of a mature
citizen, let alone a United States Senator."[60] Johnson even dispatched
Harriman to request Kennedy "in the national interest" not to encourage Hanoi and undermine the harassed President.[61]

On Sunday, February 26, Kennedy, Goodwin and I spent the
afternoon in my New York apartment putting the speech in what we
hoped was final shape. The Gallup poll that morning had said that
only 24 percent of the country wanted a bombing halt. I had feared
Kennedy might draw back. He muttered that Sorensen had advised
against the speech, but that he was sure it was the right thing to do
and was determined to do it regardless of consequences.[62] About
this time he warned John Burns, the state chairman: "It's going to
hurt me, isn't it?" Burns said it would hurt him for a while, but "I
think that over a period of time you'll be proven correct." "That's
the way I feel about it," Kennedy said, but "no matter whether it
hurts me or whether it doesn't," he would speak anyway. ". . . Somebody has to do something about it."[63]

VII

On March 2 1967, Robert Kennedy came down to breakfast at Hickory Hill. He had been up till three-thirty working with Goodwin,
Walinsky and Mankiewicz to put the speech into final final shape.
Ethel greeted him: "Hail, Caesar." Kennedy said, "I spoke to Teddy
last night. He said to make sure that they announce it's the Kennedy
from New York." Children drifted in. Bobby Jr., the animal collector, brought along a coatimundi which suddenly attacked Ethel,
sinking sharp claws and teeth into her leg. "He's biting me," she
screamed. "Oh, God, he's biting me." The animal was pried away,
and Ethel went off to a doctor. When she returned, her leg in
bandages, she said, "If these are all the scars the Kennedys end up
with by five o'clock, it'll be all right."[64]

Johnson was making his own preparations. In the course of the

day, determined to deny Kennedy the headlines, he made (in the
computation by vanden Heuvel and Gwirtzman)

> two unscheduled speeches in Washington, held an unscheduled news
> conference to announce that Russian Premier Kosygin had agreed to
> talk on reducing the stockpile of nuclear weapons, announced he was
> inviting all the nation's governors to the White House, had Senator
> Henry Jackson of Washington read on the floor of the Senate a predated
> letter from him, explaining why the bombing was necessary, and con-
> firmed the rumor that his daughter Luci was pregnant.[65]

The headlines went to Kennedy all the same.

In the early afternoon Kennedy met with reporters. One asked
about Johnson's charge that speeches like Kennedy's were a disservice
to the boys overseas. "You have to balance that," Kennedy said,
"against what you think does the greatest amount of good. I don't
think we're going to end the war by military action." Did he think
the country was more hawkish than dovish? "Yes," Kennedy said.[66]

He rose in the Senate chamber at twenty minutes to four. His
speech, like all respectable anti-escalation discourse of the time, began
with a brief declaration of American determination to stay in Vietnam
until commitments (undefined) were fulfilled. This had to be said in
order to gain a hearing for what followed. It was also said because
it was still accepted by all except the New Left. "I do not agree with
those who would abandon South Viet-Nam," Chester Bowles, for
example, wrote privately that spring to Hubert Humphrey.[67] Those
of us who hoped for a negotiated peace believed that retention of
American troops in defensive enclaves was essential to give the other
side an incentive to negotiate. Unilateral withdrawal, we thought,
would lead, not to negotiation, but to a Communist Vietnam. In
retrospect, I feel that, while we were right in our forecast of the
consequences of withdrawal, the New Left was right in seeing these
consequences, whatever they might be, as less ghastly than continuing
the war. I wish that unilateral withdrawal rather than negotiation
had been the dove objective in 1967.

But it wasn't; and it would have done even less well than "negotiation now" in a country still a good deal more hawkish than dovish.
In any case, Kennedy rushed through the obligatory ritual in two
sentences. Then he got down to business. He himself had been
involved, he said, in "many" of the decisions that had brought the
United States into Vietnam. Literally this was not true. He meant
that his brother had been involved. "I can testify," he continued,
"that if fault is to be found or responsibility assessed, there is enough
to go around for all — including myself." As late as January 1971,

according to *Facts on File,* Kennedy "was the only major official in either Democratic administration who admitted publicly to being wrong about Vietnam."[68]

The United States had sent more than 400,000 men, he continued, into the "ever-widening war." "The most powerful country the world has known now turns its strength and will upon a small and primitive land." He enjoined his countrymen to visualize the "horror" of this "distant and ferocious" conflict. War was "the vacant moment of amazed fear as a mother and child watch death by fire fall from an improbable machine sent by a country they barely comprehend." It was "the night of death destroying yesterday's promise of family and land and home." It was "a land deafened by the unending crescendo of violence, hatred, and savage fury.... Although the world's imperfections may call forth the act of war, righteousness cannot obscure the agony and pain those acts bring to a single child."

This horror, Kennedy said, was "not just a nation's responsibility, but yours and mine. It is we who live in abundance and send our young men out to die. It is our chemicals that scorch the children and our bombs that level the villages. We are all participants." And now we were "steadily widening the war" at just the time when opportunities for settlement appeared to lie at hand. What was the risk in exploring these opportunities? "No one is going to defeat us, or slaughter our troops, or destroy our prestige because we dare take initiatives for peace." He asked the administration to test Hanoi's sincerity by halting the bombardment of the North and saying the United States was ready to negotiate within the week. If, as the administration objected, conflicting signals came from Hanoi, why not seize on the most favorable message, as President Kennedy had done during the missile crisis? As for the idea that we must punish the North for its iniquity, "We are not in Vietnam to play the role of an avenging angel pouring death and destruction on the roads and factories and homes of a guilty land. We are there to assure the self-determination of South Vietnam.... Can anyone believe this Nation, with all its fantastic power and resources, will be endangered by a wise and magnanimous action toward a small and difficult adversary?"[69]

Henry Jackson, Kennedy's old friend from the McCarthy committee, the party chairman in 1960, read the Senate Johnson's letter defending the bombing. Kennedy's proposals, Jackson said, put the United States into a position of weakness. Fulbright, McGovern, Clark, Tydings, Claiborne Pell, Albert Gore, John Sherman Cooper, supported Kennedy. At the end of the long day, Kennedy returned to his office. An aide suggested that Jackson deserved a gift. "Why not send him the coatimundi?" He caught the shuttle to New York. Phil Ochs, the folk singer, who had come down to Washington to hear the speech, was with him. Kennedy remembered that Bob Dylan was supposed to have changed his name to help his career and asked Ochs whether this was so. Ochs said it was. Kennedy said, "You think it would help me if I changed mine?"[70]

VIII

The inevitable Richard Nixon said that Kennedy's speech "had the effect of prolonging the war by encouraging the enemy. . . . Johnson is right and Kennedy is wrong."[71] Johnson agreed with that. The American people, he said a few days later at a private dinner of the Democratic National Committee, would not stand for "a dishonorable settlement disguised as a bargain for popularity purposes." He went on, said William Dunfey, the national committeeman from New Hampshire, in "an unbelievable tirade," ending by reading aloud a letter to the sister of a soldier killed in Vietnam. "Your brother," the letter said, "was in South Vietnam because the threat to the Vietnamese people is, in the long run, a threat to the free world community." The letter, Johnson revealed with unconcealed relish, was signed John F. Kennedy.[72]

Undeterred, Robert Kennedy pursued his campaign to drive home the meaning of the war. Soon after his Vietnam speech, he went to the University of Oklahoma. The president gave him a cordial introduction. Kennedy said, "That's the nicest thing any president has said about me in a long time." In the question period someone asked about student deferment from the draft. Kennedy said he was against it; he could afford to send his children to college, but others couldn't, and it was unfair. "There was hissing and booing," recalled Senator Fred Harris, who had accompanied Kennedy. Kennedy said to the audience, "Let me ask you a few questions." How many favored student deferment? Resounding cheers from the students. How many favored escalation of the war? The vast majority raised their hands. "Let me ask you one other question," said Kennedy. ". . . How many of you who voted for the escalation of the war also voted for the exemption of students from the draft?" There was, said Fred Harris (who himself still backed the war), a "giant gasp"; then stunned silence; then overwhelming applause.[73] He did this in many colleges. "The poor are carrying the burden of the struggle," he would say, and call for a universal draft determined by lottery.[74] In fact, so long as the Americans dying in Vietnam came from the other side of the tracks, the respectable middle class did not mind the war. Only when the contraction of educational deferments late in 1967 exposed their own sons did community leaders begin to turn against further escalation.

The debate was growing bitter. In mid-April 1967, Joseph Rauh invited Vice President Humphrey to hear the views of long-time liberal friends. Before Rauh's dinner I visited Kennedy at Hickory Hill. He said gloomily that his Vietnam speech "had probably stiffened LBJ's determination to pursue the opposite course. Also he could not get a hearing for anything he had to say on the merits; his every action was always interpreted in terms of political maneuver." At Rauh's house an impassioned Humphrey gave us a defense of the war, at once voluble and pathetic. He talked as if the whole thing were a Chinese Communist plot. He thought a physical barrier across northern South Vietnam — the same old fence Lansdale had derided in 1961 — was a great idea. "I was most depressed of all," I wrote afterward, "by the lack of the sense of the concrete human dimension of problems which had characterized the old Hubert. Not once in his long discourse did he express any dismay over the human wreckage wrought by American policy. . . . This trailing off of humanity is accompanied by an obvious delight in hobnobbing with statesmen — many mentions of the Pope, de Gaulle, Radakrishnan, etc., etc."[75]

A week later Johnson escalated the bombing. Westmoreland denounced the doves for leading the enemy to believe "he can win politically" what he could not win militarily. George McGovern responded bitingly in the Senate. Kennedy arose to praise McGovern for "one of the most courageous speeches delivered in the Senate since I became a Senator." Kennedy went on to warn that American escalation made counterescalation by our adversaries "inevitable." He condemned the policy of seeking peace "through military action which is really going to bring about the destruction of Vietnam and the people."[76]

I spoke to McGovern and Kennedy the next day.

Both were a little melancholy, felt that the senatorial revolt could be no more than a gesture, and wondered what kind of national support they could get. Beyond that, both felt stymied by our constitutional situation. The irony is that all of us for years have been defending the presidential prerogative and regarding the Congress as a drag on policy. It is evident now that this delight in a strong Presidency was based on the fact that, up to now, all strong Presidents in American history have pursued policies of which one had approved. We are now confronted by the anomaly of a strong President using these arguments to pursue a course which so far as I can see, can lead only to disaster.[77]

McGovern and Kennedy were right to doubt their national support. In May 1967, nearly half the college students described themselves as hawks, hardly more than a third as doves.* In October, 53 percent of Americans advocated further escalation.[78] Kennedy's opposition to the war compounded his troubles. His "standing with the American people," Louis Harris had reported at the end of January, "has taken a tumble downward."[79] "The ranks of Kennedy supporters," Gallup added in May, "have steadily declined."†

He still had not figured out how to disagree with Johnson on substance without laying himself open to charges, as Joseph Kraft had warned after the 1966 election, of putting "personal interest above all other things," of acting "to divide his party."[80] In the spring of 1967 even doves like Walter Lippmann urged him not to break with Johnson. "A Johnson-Kennedy fight for the nomination," Lippmann wrote a fortnight after Kennedy's Vietnam speech, "would split the Democrats and ... favor, even if it did not assure, the nomination and election of a right-wing Republican, perhaps Nixon."[81] Old Kennedy hands like Joseph Dolan, Milton Gwirtzman and Fred Dutton ("I suspect that this suggestion goes strongly against your grain") begged him to make conciliatory noises about the President. And, in a way, professions of support licensed a larger measure of opposition on issues by making it harder to charge political ambition or personal resentment.

Returning in June from the London funeral of his beloved Cissy Harlech, the wife of his old friend David Ormsby-Gore, Kennedy introduced Johnson at a Democratic dinner in New York. Sorensen handed him a hyperbolic paragraph as he rushed from the airport to the Americana Hotel. Kennedy spoke about the "height" of Johnson's aim, "the breadth of his achievements, the record of his past, the promises of his future. . . . In 1964 he won the greatest popular victory in modern times, and with our help he will do so again in 1968."[82] Many of us cringed at such extravagance. "How could you say all those things?" Peter Maas asked him the next day. Kennedy replied icily, "If I hadn't said all those things, that would give Lyndon Johnson the opportunity to blame everything that was going wrong. . . . Vietnam, the cities, the race question. . . . on that son of a bitch Bobby Kennedy."[83]

So he avoided charges of ambition and party-splitting by incurring charges of inconsistency and hypocrisy. Privately he was in despair. "An indefinable sense of depression hung over him," I wrote after seeing him in April, "as if he felt cornered by circumstance and did

---

* The breakdown was 49–35 (G. H. Gallup, ed., *The Gallup Poll: Public Opinion 1935–1971* [New York, 1972], vol. 3, 2065).

† He had lost 11 points from the poll Ethel had cabled so cheerily to London. Johnson now led him 49–37 (*Washington Post*, May 10, 1967).

not know how to break out." [84] He had reached the personal conclusion that Johnson was hell-bent on smashing his way to military victory, was deluded as to victory's likelihood and indifferent as to its human consequences and, worst of all, had retreated into some realm beyond the reach of reasoned argument. I dined at Hickory Hill in late May. At the end of the evening, he walked out with me into the soft spring night. Before I drove away, he said, "How can we possibly survive five more years of Lyndon Johnson? Five more years of a crazy man?" [85] But he could not see what he could do to prevent it.

35

Tribune of the Underclass

ROBERT KENNEDY SAW Vietnam not in abstractions but in images — a village smashed, children scorched, the mother clasping the baby while fire rained inexplicably from the sky. This compulsion to be at one with individuals in extreme situations was increasingly the key to his politics. From childhood he had, as an underdog, sympathized with underdogs; then Dallas, at once agony and liberation, had charged sympathy with almost despairing intensity. His own experience of the waste and cruelty of life gave him access to the sufferings of others. He appeared most surely himself among those whom life had left out. Even Gore Vidal once conceded: "I think he had a real affinity for the hurt people of the world: the blacks, the poor, the misunderstood young."* He kept a commonplace book in these years. The epigraph read: "None can usurp the height but those to whom the miseries of the world are a misery and will not let them rest."[1]

I

"In every arena where the poor, the black, and the uneducated suffered indignity and neglect," Jack Valenti said, "President Johnson and Bobby Kennedy thought alike."[2] At the start Lyndon Johnson gave mighty impetus to the twin wars against poverty and racial inequality. So long as the Great Society absorbed the President, there was strong reason for Kennedy, despite differences on foreign policy, not to break with the administration.

---

* Gore Vidal, in recorded interview by Jean Stein, March 25, 1969, 2, Stein Papers. Mr. Vidal wrote me in 1977: "I don't recognize the quote but if I said it I said it but, perhaps, in a different context since I regarded his metamorphosis with my usual suspicion."

As the civil rights struggle moved north, poverty and racial inequality were more than ever intertwined. In 1963 Kennedy had observed that race problems were "at least temporarily ... more easily resolved" in the south. When segregation was based on law, repeal the law, "release a valve," and the oppressed, for a time, were satisfied. "But in the North, in Chicago or Los Angeles, ... what steps would you take to release that valve?" Here segregation had been abolished, political and legal equality affirmed, but inequality remained. "What you have to do is make over some of these cities and really take drastic action."[3]

His last act before he resigned as Attorney General was to send Johnson a memorandum entitled "Racial Violence in Urban Centers." He urged the President to tell a conference of mayors — Kennedy appended a list of cities with large black populations — how the national government would help in meeting the problems "that have created Negro frustrations and hatreds." He suggested palliatives for long hot summers; but "the basic problems of jobs, training, and housing may take more than a generation to resolve."[4] "Problems in the North," he told Anthony Lewis in December 1964, "are not easily susceptible to passage of legislation for solution. You could pass a law to permit a Negro to eat at Howard Johnson's restaurant or stay at the Hilton Hotel. But you can't pass a law that gives him enough money to permit him to eat at that restaurant or stay at that hotel. . . . That's basically the problem of the Negro in the North."[5]

Problems of course remained for the southern Negro, now in the final phase of the drive for political rights. Early in 1965 Martin Luther King began a campaign to register black voters in Selma, Alabama. When local whites responded with tear gas and bull whips, Johnson in his eloquent "We Shall Overcome" speech in March asked Congress for new voting rights legislation. "He's got some guts," Kennedy said admiringly to John Seigenthaler.[6] From all over the country, people came to Selma to march for human rights. The passage in August of the Voting Rights Act of 1965 started a far-reaching transformation of southern attitudes.

Kennedy was asked the next spring to speak both at the University of Mississippi, where four years before the good old boys had hoped to lynch James Meredith, and at the University of Alabama, where three years before George Wallace had stood (briefly) in the schoolhouse door. A Mississippi legislator of the old school compared Kennedy's visit to a murderer returning to the scene of his crime. But five thousand people crowded the Coliseum at Ole Miss to hear him. "We must," Kennedy said, "create a society in which Negroes will be as free as other Americans — free to vote, and to earn their

way, and to share in the decisions of government which will shape
their lives." When in the question period someone asked about Ross
Barnett, Kennedy's deadpan account of Barnett's erratic behavior in
1962 produced, to his astonishment, gales of laughter. "He came
there *persona non grata*," reflected Oscar Carr, a liberal Delta planter,
". . . and I have never seen a politician of any ilk, stature, office at
any time in the state more wildly acclaimed."* The reception in
Alabama, Kennedy told me on his return, was even more cordial.[7]

The trip confirmed his conviction that the next battle for racial
justice lay in the northern city. He remarked that people already
"hate my guts" for the civil rights efforts in the south; but winning
southern blacks the right to public accommodations or to vote was
"an easy job compared to what we face in the North."[8] He had been
appalled by the tragedy of Selma. But he was also concerned when
white northerners rushed off to Selma forgetting their own problems
at home. "Why do they go to Selma?" he said to Richard Rovere.
"Why not to 125th Street?"[9] Thousands had marched against the
brutalities of Alabama, he told the National Council of Christians
and Jews,

> but the many brutalities of the North receive no such attention. I have
> been in tenements in Harlem in the past several weeks where the smell
> of rats was so strong that it was difficult to stay there for five minutes,
> and where children slept with lights turned on their feet to discourage
> attacks. . . . Thousands do not flock to Harlem to protest these condi-
> tions — much less to change them.[10]

Later he said: "All these places — Harlem, Watts, Southside — are
riots waiting to happen."[11]

In August 1965 violence broke out in Watts, the black ghetto in
Los Angeles. Beating, looting, burning, sniping, bombing, went on
for six days, leaving 34 people dead, more than 1000 injured. The
Watts riot, said Dwight D. Eisenhower sternly, "did not occur in a
vacuum. I believe the U.S. as a whole has been becoming atmos-
phered, you might say, in a policy of lawlessness." The former President's solution was "greater respect for law." Kennedy lashed
back. "There is no point in telling Negroes to obey the law," he
said. "To many Negroes the law is the enemy. In Harlem, in Bedford-Stuyvesant it has almost always been used against them."[12] Nor

---

* Oscar Carr, in recorded interview by D. J. O'Brien, May 6–7, 1969, 14, RFK Oral
History Program. W. F. Minor, who had covered Mississippi politics for two decades,
thought the ovation given Kennedy "had more spontaneity and unrestrained enthusiasm than any state politician has ever received" (*New Orleans Times-Picayune*, June 9,
1968). See also Mary McGrory, *Washington Star*, March 19, 1966, Robert E. Baker in
Washington Post, March 19, 1966, and Jack Nelson in *Los Angeles Times*, March 19,
1966. The Kennedy speeches are in his papers.

would new law solve the problem of the ghetto. That problem, Kennedy said in April 1966, would yield "only to other kinds of fundamental change — to the forces created by better education and better housing and better job opportunities. And it will yield only when the people of the ghetto acquire and wisely exercise political power in the community, only when they . . . establish meaningful communication with a society from which they have been excluded."[13]

II

He set out to awaken the north to its responsibilities. His own church, he told the Pope, ought to be "the foremost champion for changing this kind of difficult, poverty-stricken life." But in places like Los Angeles — he was as blunt with Popes as with anyone else — the Catholic Church "was a reactionary force and in New York it was not particularly helpful." The Pope replied laconically, "You cannot judge the Church by its representatives in Los Angeles."[14] Kennedy also reproached black civil rights leaders for their failures in the ghetto. "The army of the resentful and desperate," he said in the week of Watts, "is larger in the North than in the South, but it is an army without generals — without captains — almost without sergeants." The black middle class has "failed to extend their hand and help to their fellows on the rungs below." Demagogues "have often usurped the positions of leadership."[15]

"Religion was the language the south understood," said Andrew Young, "and there was an almost calculated avoidance of any economic questions."[16] But economics was at the heart of the northern problem. The northern black, as Martin Luther King wrote in early 1965, sought "more significant participation in government, and the restructuring of his economic life to end ghetto existence. Very different tactics will be required to achieve these disparate goals."[17] Kennedy for his part understood that King's evangelical nonviolence, deriving from the religious heritage of the old plantation, might not be so compelling in northern slums.[18] Nevertheless King had more influence than anyone else. Kennedy thought he ought to go north.

In January 1966 King came to Chicago — because, as Pat Watters, the historian of the southern civil rights movement has written, "Senator Robert Kennedy had criticized him for not giving more time to northern problems."[19] In Chicago, King encountered something new — black militants quoting not Gandhi but Frantz Fanon and contending that violence alone could bring liberation. "The only time that I have been booed," King said later, "was one night in a Chicago mass meeting by some young members of the Black Power movement."[20] He also encountered White Power in the shape of the Daley machine. His nonviolent demonstrations — the Southern Christian Leadership Conference's first and last campaign in the north — ended in grisly riots. Nonviolence appeared impotent before the despair of the ghetto, the obduracy of the white establishment and the anger of the black militants, who now claimed from King's failure further evidence that "burn, baby, burn" was the road to salvation.

King did not hold Chicago against Kennedy. Still, there remained "a strange attitude of both admiration and caution in Martin's conversation about Bobby," Andrew Young wrote the year after both men died.

> He was extremely impressed with [Kennedy's] capacity to learn, to grow, and to deal creatively in any given situation. . . . Martin tended to feel overly humble about his own accomplishments and somewhat afraid of "power" . . . and saw Bobby as a man of both moral courage and a keen sense of political timing. Martin also talked of this quality in Gandhi. It was one thing that he was always anxious about. He was clear on the moral issues, but anguished over their implementation. He admired Bobby's blend of "crusader" and realistic politician. Closely related to this was the Kennedy "efficiency mystique." "Bobby knows how to get a job done as well as talk about it."

"After the White House years," Young continued, "they met very seldom, if at all." He could remember only a casual chat during the hearings of a Senate committee.

> Perhaps this distance was dictated by the attempts to link the statements and actions of these two great statesmen through some direct financial or political alliance. Neither man could profit by such an overt relationship and both avoided any direct association. Yet they continued down parallel paths of opposition to racism, poverty and war. A distant comradarie [sic] which needed no formal tie or physical link — a genuine spiritual brotherhood which leaped across the widest chasms of our time. . . . If there is an after life, and I have no doubt there is, I am sure they are together — finally able to share the much denied love that could never be fulfilled in a world such as ours.[21]

III

Kennedy had been exploring city slums since the days of the Committee on Juvenile Delinquency and now did so more intently than ever. "I was with him one time in Brooklyn," said Pete Hamill, the writer, "and we went into some *horrible* tenement that was one of the worst I've ever seen; there was a girl with a mangled face all torn up. He said, 'What happened to her?' The Puerto Rican mother explained that the rats had bitten her face off when she was a little baby." Kennedy was outraged: how could such things continue to "happen in the richest city on earth?"[22] He wanted others to know and filled not only speeches but table talk with accounts of life in the *barriadas* of America.

Rats, filth, bad housing, bad schools, unemployment, segregation, powerlessness, alienation, hate, crime, violence, "burn, baby, burn" — all seemed interlocked in an endless chain to which no one had the key. He had rejoiced at Johnson's embrace of the antipoverty program. But, as Vietnam escalated, Johnson began to hold back domestic spending. "In 1966," John C. Donovan, the historian of the antipoverty effort, has written, "the staunchest advocate of Mr. Johnson's war on poverty was Senator Robert Kennedy. He was perhaps the most outspoken of all Senate liberals in his criticisms of the administration's budget policy." Johnson did not like it. When Kennedy persuaded the Labor Committee to increase antipoverty appropriations, the President, as Everett Dirksen reported after a visit to the White House, "fulminated like Hurricane Inez about what we were doing to his budget."[23]

In the meantime, the welfare system was beginning to buckle under the quick expansion of relief rolls. Kennedy disliked welfare. He thought it broke up families (the "man in the house" rule reduced payments for families that stayed together), destroyed self-respect and subjected the poor to a "prying" middle-class welfare bureaucracy.[24] Still, it was manifestly better than nothing. When Wilbur Mills, the powerful fiscal boss of the House, tried to freeze welfare payments in 1967, Kennedy called his bill "the most punitive measure in the history of the country," punishing "the poor because they are there and we have not been able to do anything about them."[25]

Nor did he like the idea of replacing welfare by a guaranteed minimum income. Edelman and Walinsky kept pushing this proposal. Kennedy resisted. He simply believed, Walinsky concluded, that "in the last analysis people had to do whatever they did for themselves.... He did not believe in the government just taking large sums of money and handing it out to people.... He believed in land reform in Latin America because, you know, people ought to have land to work. And he believed devoutly and would have torn the country apart to provide jobs for everybody.... But he would never have proposed large scale government doles."[26]

He feared, in addition, that concentration on the guaranteed annual income would postpone the central need — "a massive effort to create new jobs — an effort that we know is the only real solution." Employment for the poor was the centerpiece of the urban program he first presented in a series of speeches in January 1966. With all the work to be done in rebuilding the country, with wretched housing, crumbling schools, ravaged parklands, "how can we pay men to sit at home? . . . The priority here is jobs. To give priority to income payments would be to admit defeat on the critical battle front."[27] Once the employment effort became effective, Kennedy favored income maintenance for unemployables. But he always disliked the phrase 'guaranteed minimum income.' "He never could get it through his head," said Edelman, "that he really was for it."[28]

By 1967 no one could doubt that Vietnam was swallowing up the Great Society. With the administration otherwise engaged, Kennedy moved out on his own. In July he introduced a bill calling for tax incentives to induce private enterprise to bring plants, shops and jobs into urban poverty areas. His more basic proposal, introduced in the autumn with Joseph Clark of Pennsylvania, was to make the national government itself the employer of last resort. The Kennedy-Clark bill contemplated the creation of two million new jobs through public service employment. Again the White House intervened. The bill, as Gaylord Nelson of Wisconsin reminded the Senate in 1974, was "strongly opposed . . . by the Johnson administration" and "died even in a watered-down version on the floor of the Senate."[29]

Reemployment would only establish the economic preconditions for healthier cities. There remained the gnawing political problems — powerlessness and participation. The crisis of the city, Kennedy believed, came ultimately from "the destruction of the sense, and often the fact, of community, of human dialogue, the thousand invisible strands of common experience and purpose, affection and respect which tie men to their fellows." The history of the human race, "until today," had been "the history of community." Now community was disappearing at the time when its "sustaining strength" was more than ever necessary in a world grown "impersonal and abstract."[30] The child of the ghetto was "a prisoner in an area which is not a community or even a series of communities, but a vast, gray, undifferentiated slum."[31]

The revival of community, he believed, called for something new. The slum clearance and public housing programs of the New and Fair Deals had perpetuated segregation.[32] Massive housing projects, undertaken with the most benign intent, had become "jungles — places of despair and danger for their residents, and for the cities they were designed to save."[33] New programs, even new institutions, were necessary to wipe out the ghettos, reestablish communities and move toward a multiracial society.

But which came first — community or integration? Kennedy understood the Black Power point that black families, economically and psychologically unprepared for white neighborhoods, might better develop pride in their own culture and self before they attempted the adventure of integration. Otherwise the multiracial society would be founded on the white man's values. "The violent youth of the ghetto," he told the National Catholic Conference on Interracial Justice, "is not simply protesting his condition, but making a destructive and self-defeating attempt to assert his worth and dignity as a human being — to tell us that though we may scorn his contribution, we must still respect his power. In some ways it is a cry for love." [34] Blackness, he said, "must be made a badge of pride and honor." [35]

So he was prepared to give first place to economic and moral recovery *within* the ghetto. As this was accomplished, dispersion could begin: "The building of a truly integrated society depends on the development of economic self-sufficiency and security in the communities of poverty, for only then will the residents of these areas have the wherewithal to move freely within the society." [36] And dispersion rather than long-distance busing seemed the best way to resolve the vexed problem of school integration. "My personal opinion," he said in 1964, "is that compulsory transportation of children over long distances, away from the schools in their neighborhoods, doesn't make much sense and I am against it." [37] He said this during his senatorial campaign and mostly avoided the issue thereafter. "True school integration," he said in 1966, "depends on a desegregation of residential patterns." [38]

For the revival of the ghetto he looked not only to jobs but to the ideas of community action fostered by the Committee on Juvenile Delinquency. He proposed in 1966 the establishment of community development corporations, owned and controlled by the residents of the area, mobilizing both local talent and resources and outside capital, private as well as public. The critical element, he emphasized, "should be the full and dominant participation by the residents." Such corporations "could go far to changing perhaps in revolutionary ways our techniques for meeting urban needs." [39]

IV

Speeches and bills did not satisfy him. He wanted to get things done — and in New York City. In 1965 Thomas Johnston, the soft-spoken Kentuckian from the Yale Drama School who headed his office in the metropolis, had worked up with HARYOU, the Urban League and other organizations a program providing some five thousand summer jobs in Harlem. The program was a success, but investigations afterward revealed incompetence or worse in the handling of funds that baffled auditors from the Office of Economic Opportunity for months to come. The experience showed too — as Kenneth Clark had already discovered in HARYOU — how hard it was to get things done in a locality dominated by powerful and selfish personalities like Adam Clayton Powell and J. Raymond Jones.[40]

And Harlem, though far the most celebrated, was not the largest black ghetto in New York City. Across the East River in central Brooklyn lay Bedford-Stuyvesant, where 450,000 people — 84 percent black, 12 percent Puerto Rican — lived, more nonwhites packed together than in any ghetto in the land save for the South Side of Chicago. Where Harlem was filled with large, crumbling tenements, Bedford-Stuyvesant had streets lined with brownstones. Fifteen percent of the people owned their homes as against 2 percent in Harlem. On the other hand, as a New York University study put it in 1967, "Bedford-Stuyvesant is more depressed and impaired than Harlem — i.e., fewer unified families, more unemployment, lower incomes, less job history."[41] Unlike Harlem, it had received almost no federal aid.[42] Its decay seemed almost irreversible.

On February 4, 1966, Kennedy took a long walk through Bedford-Stuyvesant. He saw it all: burned-out buildings, brownstones in abject decay, stripped cars rusting along the streets, vacant lots overflowing with trash and garbage, a pervading stench of filth and defeat. He met with a group of community activists, led by state supreme court judge Thomas R. Jones, the leading black politician in the area. The group was irritated and cynical. One said, "You're another white guy that's out here for the day; you'll be gone and you'll never be seen again. And that's that. We've had enough of that."[43] Judge Jones said, "I'm weary of study, Senator. Weary of speeches, weary of promises that aren't kept. . . . The Negro people are angry, Senator, and, judge that I am, I'm angry, too. No one is helping us."[44]

Kennedy was irritated too. "I could be smoking a cigar down in Palm Beach," he said as they drove back to Manhattan. "I don't really have to take that. Why do I have to go out and get abused for a lot of things that I haven't done?" Then: "Get them a swimming pool." Then, after a time, "Maybe this would be a good place to try and make an effort." What kind of an effort was still obscure. "We didn't have any sort of idea," Johnston said later, "that we could do anything that was very big."[45]

The first thought was to get foundation money. The Taconic Foundation, which had helped Kennedy on the black registration drive when he was Attorney General, was interested. Soon McGeorge Bundy, now head of the Ford Foundation, was drawn in; later Mrs. Vincent Astor and the Astor Foundation were of inestimable help. But money by itself was not a solution and could even create difficulties, as it had in Harlem the previous summer. At this point there were less than a dozen black certified public accountants in the country and none in Bedford-Stuyvesant. In September, Johnston suggested the possibility of enlisting white business leaders who might provide technical and managerial advice as well as capital.

This idea appealed to Kennedy. His relations with the business community were as chilly as ever. But he had some strong business friends, notably Douglas Dillon, Thomas J. Watson of IBM and William Paley of CBS. He believed that it was useful to draw in people who had power. He enjoyed, I believe, arranging incongruous coalitions. And, with Vietnam consuming available federal money, social policy would require a larger infusion of private funds. At the same time, he recognized that he would get nowhere if business saw Bedford-Stuyvesant as a Kennedy promotion. It had to be politically beyond suspicion. So he accompanied his courtship of business leaders with an assuagement of Republican fears. Javits presented no problem. He saw at once the value of the idea, had come by now to like Kennedy and provided wholehearted cooperation. Mayor Lindsay was more uncertain. He saluted the project but suspected the projector. Lindsay and Kennedy were the most striking political figures in New York: competitors today; rivals, it might be tomorrow, for the Presidency itself. Lindsay privately thought Kennedy a publicity-grabber. Kennedy privately thought Lindsay a lightweight. I suspect he was also a little jealous of Lindsay, who was very tall, very handsome and filled with what girls called charm and journalists charisma. But they agreed on most things, especially on the importance of racial justice, and each essentially respected the other. Mary Lindsay and Ethel Kennedy had been schoolmates. Lindsay agreed to give his support.[46]

Armed with nonpartisan credentials, Kennedy approached business leaders. He counted heavily on André Meyer of Lazard Frères, on whom Jacqueline Kennedy had come to rely for counsel. Meyer told Kennedy and Johnston he would help on one condition. Johnston stiffened, wondering what inordinate capitalist demand was about to come. "I will come in," Meyer told Kennedy, "if you will stand up in the Senate and make an even stronger speech on Vietnam than you have. Bedford-Stuyvesant will have no meaning if we don't end that terrible war."[47] Dillon, Watson, Paley, Roswell Gilpatric, J. M. Kaplan of Welch Grape Juice, James Oates of Equitable Life Assurance and George Moore of the National City Bank came readily along. André Meyer also suggested David Lilienthal, the old New Dealer who had carried his developmental genius from the Tennessee Valley to the far corners of the planet. Kennedy met Lilienthal for a drink at the Century Club. "I can't remember ever having my impression of a man change so soon and so suddenly," Lilienthal wrote in his diary. ". . . I asked myself: Could this earnest young man possibly be the same fellow pictured by the press and TV as a cynical, ambitious, ruthless trickster dealing only with political issues that would 'pay off'?"[48]

A later recruit was Benno Schmidt, a managing partner in J. H. Whitney & Company. He too had heard things about Kennedy that, as he said later, "didn't give me a particularly warm feeling toward him at the outset." Schmidt said at once that he had voted for Nixon in 1960 and Keating in 1964. So much the better, said Kennedy, promising that he would "never do anything in connection with this project that you will feel inconsistent with my assurance to you that this thing is non-political and nonpartisan."[49] In time Schmidt succeeded Dillon as chairman of the Bedford-Stuyvesant Development and Services Corporation, becoming as well a close friend of both Robert and Ethel Kennedy.

V

On December 10, 1966, ten months after his walk through the gloomy streets, Kennedy, with Lindsay and Javits by his side, unfolded the development plan to a thousand people gathered in the auditorium of a Bedford-Stuyvesant school. There were two separate corporations: one, representing the people of Bedford-Stuyvesant, to decide on programs; the other, composed of business leaders, to bring in outside investment and to supply managerial assistance. Franklin A. Thomas, an able black lawyer, headed the community group. John Doar was persuaded in 1967 to leave the Justice Department and run the Development and Services Corporation.

Kennedy himself continued to play an active role. Lilienthal observed him at a meeting the next spring, "looking quite handsome in spite of the fact that his tie wasn't fully pulled up, his suit was rumpled, his sox were droopy. He listened with an intentness that was almost painful, occasionally rubbing his eyes with fatigue, his knees drawn up against the edge of the table. . . . He never let his mind wander nor did he relax for a second." As he left, murmuring that he had just flown in from Washington and now had to fly back for an evening engagement, he said, "I'm a yo-yo."[50] "It was his work, his vision, energy, enthusiasm and intelligence," said John Doar, "that kept it going." "If I wasn't the United States senator," Kennedy once said to Doar, "I'd rather be working in Bedford-Stuyvesant than any place I know."[51]

Thomas and Doar, in partnership with the community, managed

in the next years to combine housing and physical renovation with jobs and social services in a way that gave Bedford-Stuyvesant a new life. Revisiting the project in 1978, Michael Harrington of *The Other America* called it "a modest success — which, in the context of so many failures, is to say a remarkable success." And, as Harrington added, if even one American neighborhood had headed toward "economic and sociological hell . . . and then reversed the disastrous trend, that is important news for the nation as a whole."[52]

It was Kennedy's profound hope that Bedford-Stuyvesant would serve as a model for self-regeneration in other ghettos. He had already, with Javits's cosponsorship, amended the Economic Opportunity Act to provide for a "special impact" program intended to channel federal development aid into urban poverty areas. But once more he was having trouble with Lyndon Johnson. Although Congress funded the amendment for two years, the administration, Kennedy said in 1968, "has opposed the program and has spent most of the funds appropriated on other manpower activities."[53] In the meantime, he introduced a bill intended to create housing and jobs in poverty areas through tax incentives and low-interest loans along with safeguards for neighborhood control. Again the President balked. "The Johnson Administration," Robert Semple wrote in the *New York Times* in September 1967, "mounted a concerted attack today on a proposal by Senator Robert F. Kennedy to build more and better low cost housing in the slums through private enterprise."[54] In another month Johnson announced a program of his own, not dissimilar to Kennedy's but more limited in scope. "How can they be so petty?" Kennedy said to Jack Newfield. "I worked on my plan for six months, and we talked to everyone in the Administration in all the relevant agencies. We accepted many of their ideas and put them in our bill. Now they came out with this thing, and the first I hear about it is on television. They didn't even try to work something out together. To them it's all just politics." *

---

* Jack Newfield, *Robert Kennedy* (New York, 1969), 106. Four years later a Twentieth Century Fund Task Force on Community Development Corporations, noting the "deliberate policy decisions made by both the Johnson and Nixon administrations to discourage neighborhood-controlled urban projects," called on the national government to "take immediate steps to create a national system of support for community development corporations" (Report of the Twentieth Century Fund Task Force on Community Development Corporations, *CDCs: New Hope for the Inner City* [New York, 1971], 4, 29, 30). After Kennedy's death Gaylord Nelson introduced a Community Self-Determination Act (*Congressional Record*, July 24, 1968, S9270). Senator Fred Harris observed, "This measure very closely resembles two of the late Senator Robert Kennedy's tremendously innovative bills," adding that the Bedford-Stuyvesant experiment constituted "an important precedent for this bill, and perhaps a demonstration of the likely success of some of the bill's objectives" (S9285). The bill got nowhere. In 1972 Javits and Edward Kennedy expanded the original 'special impact' program by

It was not just politics to the black community. No white leader was more welcome in the ghettos. Black leaders liked Kennedy's openness, his conviction that something could be done, even his occasional abruptness of challenge, proving, as it did, that he was treating them as equals. Dr. Kenneth Clark, whose dismay over Kennedy at the Baldwin meeting in 1963 had been compounded by Kennedy's surrender of HARYOU to Adam Clayton Powell in 1964, met him again in 1967. This time, the longer they talked, "the more I came around to saying, 'You know, it is possible for human beings to grow. This man has grown.' I committed myself to working with him . . . something which I never dreamed that I would after the [Baldwin] thing and the HARYOU thing."[55] "Bobby Kennedy," said the Reverend Channing Phillips, a Washington black leader, "had this fantastic ability to communicate hope to some pretty rejected people. No other white man has this same quality."[56]

VI

His concern was confined neither to cities nor to blacks. His work on the Migratory Labor Subcommittee of the Senate Labor Committee gave him a vivid understanding of the wretched conditions among itinerant farmworkers. When a witness from the American Farm Bureau, after objecting to every specific proposal for the protection of migratory labor, admitted that his group of prosperous farmers had no program of its own, Kennedy was incredulous. He brusquely told the witness, "To be opposed to a minimum wage, to be opposed to legislation which would limit the use of children . . . to be opposed to collective bargaining completely . . . to oppose all that without some alternative makes the rest of the arguments you have senseless."[57]

Soon afterward, in early 1966, Walter Reuther and Jack Conway came to Washington. They were fresh from Delano, California, where the National Farm Workers Association was conducting a strike of migratory grape workers and urging a national boycott against grapes picked by nonunion labor. The leader of the strike was a Mexican American named Cesar Chavez. Reuther strongly recommended that the subcommittee go to California and hold hearings on the situation in the grapefields.[58] It was a poor time for

adding a new Title VII to the Economic Opportunity Act. In an address in 1974 to the Congress of Community Development Corporations, Javits described Bedford-Stuyvesant as the place "where the community economic development corporation idea . . . was born." By this time there were 34 federally funded and 75 privately funded community development corporations (*Congressional Record*, April 9, 1974, S5522).

Kennedy. He was trying to figure how to save the cities; he was digging his way out of the Vietnam chicken coop; and he was reluctant to leave Washington. Yet, as he told Peter Edelman, "if Walter Reuther and Jack Conway want me to do it, I suppose I'll do it."[59] Conway finally said to him, "These people need you."[60] On the plane out to the coast, Kennedy still wondered why he was going.[61] He arrived in time for the second day of the hearings.

The local sheriff came before the committee to explain his manner of keeping the peace. He took photographs, he said, to identify potential troublemakers; he had five thousand, he bragged, in his files. Kennedy said, "Do you take pictures of everyone in the city?" The sheriff: "Well, if he is on strike, or something like that." Kennedy asked why he had arrested forty-four of Chavez's men engaged in lawful picketing. The strikebreakers, the sheriff explained, had said, "'If you don't get them out of here, we're going to cut their hearts out.' So rather than let them get cut, we removed the cause." George Murphy of California, another member of the subcommittee, muttered, "I think it's a shame you weren't there before the Watts riots." Kennedy said caustically, "This is the most interesting concept.... How can you go arrest somebody if they haven't violated the law?... Can I just suggest that the sheriff reconsider his procedures in connection with these matters?... [Can] I suggest during the luncheon period that the sheriff and district attorney read the Constitution of the United States?"[62]

By the end of the day, Kennedy had embraced Chavez and La Causa. "He shouldn't go so far," Chavez whispered to his lieutenant Dolores Huerta, "because it's only going to hurt him." "Instead of that awful feeling against politicians who don't commit themselves," Chavez recalled,

> we felt protective. He said that we had the right to form a union and that he endorsed our right, and not only endorsed us but joined us. I was amazed at how quickly he grasped the whole picture.... He immediately asked very pointed questions of the growers; he had a way of disintegrating their arguments by picking at the very simple questions.... When reporters asked him if we weren't Communists, he said, "No, they are not Communists, they're struggling for their rights." So he really helped us, ... turned it completely around.[63]

"Robert didn't come to us and tell us what was good for us," Dolores Huerta said later. "He came to us and asked two questions ... 'What do you want? And how can I help?' That's why we loved him."[64]

Chavez was two and a half years younger than Kennedy. His heroes were Saint Francis and Gandhi. He had been organizing Mexican Americans since 1952. The two men had in fact met in

1960, when Chavez was running a drive to register Spanish-speaking voters. The Kennedy staff thought he was going about it in the wrong way. "If he's been here for ten years," Robert had said when the problem was brought to him, "why can't he do it the way he wants to do it?"[65] Later *Time* carelessly ascribed the success of the drive to a committee of Mexican-American politicians. Chavez asked Dolores Huerta to protest to Kennedy. "So I had the copy of *Time* magazine in my hand, and I crashed in. And Bobby Kennedy was standing there; he was talking to a lot of people. I think he saw this wild-eyed looking person walking toward him, and he threw up his hands and he said, 'I know. I know.'" The story had been a mistake, he said, and he would correct it, which he did. "So we felt very good about it," said Dolores Huerta.[66]

The meeting in Delano sealed a relationship. "Something had touched a nerve in him," said Peter Edelman, who followed the problem in Kennedy's Senate office. ". . . Always after that, we helped Cesar Chavez in whatever way we could."[67] For all their differences in background, the two men were rather alike: both short, shy, familial, devout, opponents of violence, with strong veins of melancholy and fatalism. Chavez, Kennedy believed, was doing for Spanish-speaking Americans what Martin Luther King had done for black Americans: giving them new convictions of pride and solidarity.

The rural laborer became an abiding concern. Few states treated migrant workers worse than New York. In 1967 Kennedy and Javits visited a work camp upstate. The owner's sign warned: ANYONE ENTERING OR TRESPASSING WITHOUT MY PERMISSION WILL BE SHOT IF CAUGHT. This discouraged most of the party. Kennedy, head down, kept walking. He found three migrant families living in an old bus with the seats ripped out. Inside he saw six small children, their bodies covered with running sores. The stench was overpowering. Kennedy asked an ancient black woman in the bus how much she earned. She said a dollar an hour, picking celery. He made a face, shook his head and held her calloused hand for a moment. Cardboard covered the windows of the next bus, where a child played forlornly on a filthy mattress. "As Kennedy looked down at the child," reported Jack Newfield, "his hand and his head trembled in rage. He seemed like a man going through an exorcism." The owner, as billed, had a gun. "You had no right to go in there," he said. "You're just a do-gooder trying to make some headlines." Kennedy replied in a whisper, "You are something out of the 19th century. I wouldn't let an animal live in those buses." "It's like camping out," the owner said. Once back in the twentieth century, Kennedy demanded that Rockefeller investigate health conditions in the camps and called on labor leaders to organize the migrants.[68]

Rural squalor was not confined to *braceros* and bindlestiffs. Most tragic of all were the Indians. As senator from the state eighth in the country in Indian population, Kennedy visited the remnants of the once mighty Five Nations in their upstate reservations. He talked Indians with LaDonna Harris, the Indian wife of his Oklahoma colleague, and in March 1967 he addressed an organization she had founded called Oklahomans for Indian Opportunity. Answering one question, he said, at once jokingly and seriously, "I wish I had been born an Indian." "It sounded so real and also kind of wistfully funny," said Fred Harris, "that everybody laughed and applauded overwhelmingly."[69]

He learned the grim statistics. "The 'first American,'" he said, "is still the last American in terms of employment, health and education."[70] In 1967 he persuaded the Senate to set up a committee to study Indian education. He went into the schools on the reservations and asked if there were Indian teachers and whether they were teaching Indian culture and history. He looked at library shelves to see what Indians could read about their own past. At the Blackfoot reservation in Fort Hall, Idaho, they turned up only one book on Indians — *Captive of the Delawares*, its jacket showing an Indian scalping a blond child.[71] At one reservation he learned that a baby had died of starvation the same day. He said, and meant, "When that baby died, a little bit of me died too."[72]

On the day Robert Kennedy himself died, a New York Seneca, whose reservation he had visited in 1967, wrote his widow: "We loved him, too, Mrs. Kennedy. Loving a public official for an Indian is almost unheard of, as history bears out. We trusted him. Unheard of, too, for an Indian. We had faith in him."[73] Vine Deloria, Jr., the Standing Rock Sioux who wrote *Custer Died for Your Sins*, observed that Kennedy's intercession had probably discouraged federal action "because of his many political enemies and their outright rejection of causes he advocated." Still, said Deloria in a fine sentence, he was a man "who could move from world to world and never be a stranger anywhere." And Indians thought him "as great a hero as the most famous Indian war chiefs precisely because of his ruthlessness." At last, somewhere, that reputation had its advantages. "Indians," said Deloria, "saw him as a warrior, the white Crazy Horse" — the great war chief of the Oglala Sioux who did, Deloria said, what was best and what was for the people. Kennedy, Deloria concluded, "somehow validated obscure undefined feelings of Indian people which they had been unwilling to admit to themselves. Spiritually, he was an Indian!"[74]

VII

Chicanos, migrant workers, Indians — all presented aspects of the larger shame of poverty. In March 1967 the Senate Labor Committee's Subcommittee on Poverty held new hearings. One of the witnesses was Marian Wright, a twenty-seven-year-old black lawyer from South Carolina and Yale Law School, now in Mississippi for the NAACP's Legal Defense Fund. Marian Wright told the committee how mechanization and the reduction in cotton planting under the federal subsidy program had thrown thousands of blacks out of work in the Mississippi Delta. At this point Mississippi counties had shifted from the program that gave surplus food to the poor to one that required the monthly purchase of food stamps in lump sums the poor did not have. The result, Marian Wright said, was disaster.[75]

Joseph Clark, the chairman, thought the committee should go to Mississippi and see for itself. Kennedy, Javits and Murphy accompanied him. Kennedy sent Edelman down a few days in advance to get a sense of the problem. (Edelman talked particularly to Marian Wright. Fifteen months later they were married.) The committee arrived in Jackson on April 9, 1967. They dined that evening with a spectrum of Mississippians, from Oscar Carr to Charles Evers. Carr thought Kennedy "a very shy man. . . . He continually asked questions."[76] "We talked and talked," said Evers, "and he listened."[77]

The hearings took place the next day. Kennedy, as usual, was not interested in the explanations of officials. He believed, as a reporter on the trip, Nick Kotz of the *Des Moines Register,* said, that "the poor themselves made the best witnesses." Their testimony was appalling. After the hearing Kennedy told Charles Evers, "I want to see it."[78] The next day Kennedy and Clark toured the Delta. They went, said Evers, "into one of the worst places I've ever seen." Kotz described "a dark windowless shack" smelling of "mildew, sickness, and urine." "There was no ceiling hardly," said Evers; "the floor had holes in it, and a bed that looked like the color of my arm — black as my arm — propped up with some kind of bricks to keep it from falling. The odor was so bad you could hardly keep the nausea down. . . . This lady came out with hardly any clothes on, and we spoke to her and told her who he was. She just put her arms out and said 'Thank God' and then she just held his hand."

A small child sat on the floor rubbing grains of rice round and round. "His tummy was sticking way out just like he was pregnant. Bobby looked down at the child, and then he picked him up and sat down on that dirty bed. He was rubbing the child's stomach. He said, 'My God, I didn't know this kind of thing existed. How can a country like this allow it? Maybe they just don't know.'" He tried, said Kotz, to evoke a response from the child, talking, caressing, tickling. The child never looked up, sitting as in a trance. "Tears were running down [Kennedy's] cheek," said Evers, "and he just sat there and held the little child. Roaches and rats were all over the floor. . . . Then he said, 'I'm going back to Washington to do something about this.' No other white man in America would have come into that house." [79]

"Have you ever seen anything like this before?" asked W. F. Minor of the *New Orleans Times-Picayune.* "Yes, I have," Kennedy said. "I've seen it in Southeast Asia and in Harlem." [80] Marian Wright had thought the senators were there for publicity; "and then he came," she later told Roger Wilkins, "and he did things that I wouldn't do. He went into the dirtiest, filthiest, poorest black homes . . . and he would sit with a baby who had open sores and whose belly was bloated from malnutrition, and he'd sit and touch and hold those babies. . . . I wouldn't do that! I didn't do that! But he did. . . . That's why I'm for him." [81]

VIII

The day after they got back, Kennedy and Clark went to see Orville Freeman, the Secretary of Agriculture. Freeman wondered whether conditions were really so bad as they thought. [82] The entire Clark committee then appealed to President Johnson for action to meet the Mississippi emergency: free food stamps for the neediest, cheaper stamps for the poor, investigation of the way local officials distributed federal food. [83] Johnson asked Joseph Califano, his chief assistant on domestic affairs, what it was all about. "Freeman," Califano reported back, "does not want to upset the entire program by either giving free food to these negroes in the delta or by lowering the amount of money they have to pay for food stamps until he has the food stamp program through Congress." [84] Johnson turned the committee letter over to the Office of Economic Opportunity for answer. The reply, as described by one scholar, was "defensive . . . argumentative . . . irrelevant." [85] Daniel Patrick Moynihan surmised that the White House read the committee letter "as an attack by Kennedy on Johnson." [86] "We thought they were exaggerating the extent of the hunger," Califano said later. [87]

The issue would not go away. The Field Foundation sent down a team of doctors headed by Robert Coles, a psychiatrist who combined the concerns of a sociologist with the sensitivity of a literary artist. The Coles group reported in June that children in Mississippi were "living under such primitive conditions that we found it hard to believe we were examining American children in the twentieth century."[88] Back in Washington the doctors waited on a number of officials — Freeman, John Gardner, the Secretary of Health, Education, and Welfare, others — and "pleaded with them," Coles said, to do something for the poor people of Mississippi. "We were not only given the runaround but, in all bluntness, we ourselves were getting so depressed . . . that we were ready to give up." If their efforts offended the southern conservatives on the Agriculture committees, the Johnson officials warned, there would be even less money for commodity distribution and food stamps. "The reason we didn't give up is because several of us had decided that we would go over to see Senator Kennedy." Kennedy told them: "You don't have to take that. This is the beginning, not the end. You don't have to be discouraged."[89]

Clark called a new set of hearings, where Robert Coles and his colleagues testified. In July, John Stennis, the conservative Mississippi senator, introduced an emergency food and medical bill. Clark's committee reported it out at once. The Senate passed it in ten days. But the administration remained hostile, and the Texas chairman of the House Agriculture Committee derailed the bill. In October an administration Nutrition Task Force recommended $300 million more for food programs. Johnson rejected the recommendation. In November Congress at last passed the Stennis bill. The administration stalled on its execution. "Not until April 1968," wrote Kotz, "did interdepartmental haggling finally ebb enough so that someone could begin dispensing what was supposed to be 'emergency' aid to the sick and hungry poor." Later Johnson learned that Freeman was planning to spend $145 million over the 1968 budget on food programs. "I never authorized you to do that," the President said in a rage.[90]

Even Hubert Humphrey joined the campaign. In his only known written criticism of Johnson in these years, he set forth the situation to Mrs. Arthur Krim, who was professionally concerned about the effects of malnutrition on the mental development of small children and whose husband was the Democratic party's chief fund raiser. "There are ways the President could have helped," Humphrey wrote, ". . . in approving some of Orville Freeman's budget requests, in supporting legislation on the Hill, and suggesting administrative change — but he has not." "On at least 12 specific occasions," according to Nick Kotz, "his aides and Cabinet officers had recommended food aid reform and Lyndon Johnson had said 'no.'"[91]

The Great Society was a fading memory. Johnson even banished the phrase from his speeches. He feared to incite inflation by increasing the budget. Perhaps he also feared to incite disorder by encouraging social protest. "Beginning with the 1967 State of the Union message," Daniel Patrick Moynihan wrote in 1968, "civil-rights and poverty issues practically disappeared from Presidential pronouncements, to be replaced by disquisitions on Safe Streets and Crime Control Acts, and other euphemisms for the forcible repression of black violence."[92]

IX

The summer of 1967 was the worst yet in the ghettos. A contagion of riots, marked by arson, looting and sniping, began in the south in May, spread to the north in June and reached an awful climax in July. Twenty-six people were killed in Newark in disorders lasting from July 12 to July 17. On July 23 violence broke out in Spanish Harlem. It was continuing two days later when I had dinner with Robert and Ethel Kennedy, Pete Hamill and José Torres, the gentle (except in the ring) former world's light heavyweight champion and a respected leader of his people. After dinner we piled into Torres's car and drove through the anxious streets of East Harlem, with buildings shuttered and knots of policemen on each corner. At one point Torres stopped the car to show us a message painted in Spanish across Third Avenue: YOU CROSS THIS LINE, YOU BE DEAD.[93]

That same night the greatest violence of all exploded in Detroit, where forty-three people died in the next four days. Governor George Romney called in the National Guard. "They have lost all control in Detroit," J. Edgar Hoover told Johnson on the evening of July 24. "Harlem may break loose within thirty minutes. They plan to tear it to pieces." Later that night Johnson sent tanks and paratroopers into Detroit.[94] He explained his decision over nationwide television in a cold statement, notably lacking any acknowledgment that human despair might possibly lie behind the explosion. Even in the White House Harry McPherson, a Johnson special assistant, thought the presentation "legalistic." There was no point, McPherson protested, in criticizing everyone else, especially Congress, "unless and until we are willing to go before them in joint session and state the case for America's cities."[95]

If McPherson was unhappy over the President's speech, Kennedy was incredulous. "It's over," he told Mankiewicz. "The President is just not going to do anything more. That's it. He's through with domestic problems, with the cities. . . . He's not going to do anything. And he's the only man who can." Mankiewicz asked what

Kennedy would do if he were President. Kennedy said that, first, he would ask the heads of the three television networks to produce as rapidly as possible — and run in prime time — a two-hour documentary showing what it was like to live in a ghetto.

> Let them show the sound, the feel, the hopelessness, and what it's like to think you'll never get out. Show a black teenager, told by some radio jingle to stay in school, looking at his older brother — who stayed in school — and who's out of a job. Show the Mafia pushing narcotics; put a Candid Camera team in a ghetto school and watch what a rotten system of education it really is. Film a mother staying up all night to keep the rats from her baby. . . . Then I'd ask people to watch it — and experience what it means to live in the most affluent society in history — without hope.

Next, Kennedy continued, he would put together the racial data on every major city. He would call meetings at the White House — one a day if necessary — for each city on the danger list. He would invite not only the mayors but the bankers, contractors, real estate men, union officials, ministers; "everybody knows who really has power in a city." "Gentlemen," he would say, "this is your problem, and only you can solve it. If you don't solve it, your city will fall apart in a few years, and it will be your fault." We could do it, he told Thomas Johnston, the way we did Bedford-Stuyvesant. Let the local community define its problems and plans; let the government define the available resources; let us work it out together to save the cities. But it was no use *his* saying these things, he told Mankiewicz. "When I do, it's a political speech. The President of the United States is the only man who has the pulpit. . . . If he leads — if he shows that he cares — people will give him time." [96]

Kennedy had no more indulgence than Johnson for violence. "We must make it unequivocally clear by word and deed," he said in a speech in San Francisco on August 4, "that this wanton killing and burning cannot and will not be tolerated." But repression was not the answer. "If we can spend $24 billion for the freedom and the liberty of the people in Vietnam," he said on *Meet the Press* in August, "certainly we can spend a small percentage of that for the liberty and the freedom and the future of our own people in the United States." [97]

x

"Today in America," Robert Kennedy wrote in 1967, "we are two worlds." The world of the white middle class was reasonably pleasant. "But if we try to look through the eyes of the young slum-dweller — the Negro, and the Puerto Rican, and the Mexican-Amer-

ican — the world is a dark and hopeless place."[98] This was his own
startling capacity. "He could see things," said Cesar Chavez, "through
the eyes of the poor. . . . It was like he was ours."[99] A ghetto youth
told Robert Coles, "Kennedy . . . *is* on our side. We know it. He
doesn't have to say a word."[100]

Coles speculated why this should be so. It was partly, he thought,
that Kennedy imparted to the powerless a conviction of their own
cultural dignity, of "strengths that would enable them, given the
chance, to do something and be somebody." It was his absence of
glibness or condescension, an "activated urgent tension within him,
a seeking for expression and then finding it in the plight of other
tense people." It was their sense of him as a man who had "lived
with tragedy himself, felt suffering and could share that without
speaking it. It was in his language. It was in his eyes. It was in a
gesture. And they felt [he] could suffer with them and pick up their
suffering; and yet appreciate them as equals."[101]

And it was because of his experiencing nature. When he went into
Harlem or Watts, when he visited a sharecropper's cabin or an Indian
reservation, these were *his* children with bloated bellies, *his* parents
wasting away in dreary old age, *his* miserable hovel, *his* wretched
scraps for supper. He saw it all with personal intensity, as from the
inside. "I think Bobby knows precisely," a friend once said, "what it
feels like to be a very old woman."[102] Those he came among perceived this and gave him unreservedly their confidence and their
love. "Our first politician for the pariahs," Murray Kempton called
him;[103] "our great national outsider, our lonely reproach, the natural
standard held out to all rebels. That is the wound about him which
speaks to children he has never seen. He will always speak to children, and he will probably always be out of power."*

Lawrence Spivak, the *Meet the Press* impresario, asked Kennedy in
August 1967 whether he thought the American people had lost faith
in their leaders. "The people are terribly disturbed across this country," Kennedy replied, "as to what direction our country is moving

---

* Murray Kempton, "Bob Kennedy Voyages," *New York World-Telegram*, November
26, 1965. Kempton was more accurate than the Kennedy critics who charged that he
cultivated the poor for political reasons. The psephologist Richard Scammon estimated in 1967 that the poor represented less than 12 percent of the national vote
(Stewart Alsop, "Can Anyone Beat LBJ?" *Saturday Evening Post*, June 3, 1967). Many
of the poor were outside the political as well as the economic community. They had
no fixed place of abode or were too poor, too cynical, too apathetic, to get on the
registration rolls. Beyond that, as Stewart Alsop pointed out in 1967, Kennedy's
championship of the underclass was, "far too much for his own good," alienating "the
middle-class and middle-aged whites who make up a majority of the voters" (Stewart
Alsop, "Bobby Kennedy's Best Chance," *Saturday Evening Post*, June 3, 1967).

in . . . and whether they, as individuals, mean anything . . . whether their voices are ever going to mean anything or whether business has gotten so large, labor organizations so large that they care nothing for the individual. And even our universities and our educational system. So I think there is general dissatisfaction in our country, but not just with our political leaders."

Spivak asked how he felt himself. "I am dissatisfied with our society," Kennedy said. "I suppose I am dissatisfied with our country."[104]

36

Images

HE DID NOT KNOW the answers. But, more than other politicians of the day, he knew the questions. "Kennedy is on to something," wrote the New Leftist Andrew Kopkind. "He hovers over it like a pig in the *Perigord* sniffing a truffle. It is just below the surface; he can't quite see it; he doesn't know its size or shape or worth or even what it's called. He only knows it's there, and he is going to get it. Where does he look? Among the grape-pickers on strike in central California, in Cloth Market Square in Cracow, on the Ole Miss campus, in a Senate hearing room. And always with the same single-minded, almost frightening intensity. Perhaps the young know what it is; Kennedy spends an inordinate amount of time at schools and colleges talking with them. Maybe the poor know; he studies the condition of the urban ghettos. Is it in Latin America? He'll go and see. Is it in South Africa? Get him a visa."[1]

I

He was a divided man. One half was an incorrigible romantic. "When you have chosen your part," he underlined in his Emerson's *Essays*, "abide by it, and do not weakly try to reconcile yourself with the world. . . . Adhere to your own act, and congratulate yourself if you have done something strange and extravagant, and broken the monotony of a decorous age."[2] When the mood was on him, he permitted himself revolutionary fancies. "What do you think of Che Guevara?" he once asked Roger Baldwin, the old civil libertarian. "I think he is a bandit," Baldwin said. "What do you think?" Kennedy said, "I think he is a revolutionary hero."[3] He told an English journalist that, if he had not been born rich, he would have been a revolutionary.[4] This is what Alice Roosevelt Longworth saw when she said, "Bobby could have been a revolutionary priest."[5] No one understood practical politics better, said Richard Goodwin; "yet the imagining heart was always in the hills, leading some guerrilla army, without speeches or contaminating compromise, fighting to translate the utmost purity of intention into the power to change a nation or a world."[*]

If his heart was in the hills, his head was in the councils of state. In the predominant half of his nature he remained the realistic political leader. The ethic of responsibility prevailed over the ethic of ultimate ends. He wanted to be President, he believed in constitutional democracy, he abhorred violence, he could not have been a revolutionary. Still, something more than conventional politics was required. The process was plainly not working. It was not stopping the escalation of the war. It was not giving the poor a fair break or the minorities an equal opportunity. It was not dealing with rural squalor or urban decay. It appeared to be good only at keeping power arrangements as they were. Because it seemed useless for change, the poor, the minorities, the young were losing faith in it.

This was the predicament that increasingly consumed him. He was no longer, if he had ever been, a hater of people. "Of course," Murray Kempton wrote, "he knew how to hate; he hated on his father's behalf; he grew up to hate on his brother's," but the ordeals of life had "now left behind a man we recognize as having been unskilled at hating on his own."[6] "I don't think he was a hater at all," said Wes Barthelmes. "I think he had a great rage against injustice, and I think he had a rage against the impenetrability and the immovability of institutions."[7] This last was the awful question. He used to wonder, Goodwin recalled, "Could you really change the country even if you were President?"[8]

No one understood better how America was really run. "He knew," said his friend Allard Lowenstein, ". . . about worlds the rest of us didn't know about." He had mastered the textbook components of power in America — political organization, nominations, elections, legislatures, cabinets, courts, the unions, the press, television, academia. He had investigated the inner sanctums that radicals thought really controlled the system — industrialists, bankers, oil millionaires, multinational corporations, the Pentagon, the 'military-industrial complex,' the 'power elite.' And he had gone beyond both the text-

---

* Richard N. Goodwin, "A Day," *McCall's*, June 1970. The literature is full of such comments; for a further example, Warren Rogers and Stanley Tretick, "RFK," *Look*, July 9, 1968: "'What you really are is a revolutionary,' Stanley Tretick told him once. 'You should be in the hills with Castro and Che.' . . . 'I know it,' he said."

books and the radicals in learning about the underground streams through which so much of the actuality of American power darkly coursed: the FBI, the CIA, the racketeering unions and the mob, Hoffa, Giancana, Trafficante, the unseen forces in American life, their hidden penetration into and protection by the more visible realm — "an invisible empire," in Lowenstein's phrase, "allied to parts of an invisible government." The "enemy within" was wider and deeper than Kennedy himself had supposed a few years before. Still his book, dismissed at that time as overwrought, read later, Lowenstein observed, less as fanatic than prophetic.[9]

With all its faults, the old liberalism had at least arrayed itself against the immovable and impenetrable institutions. Kennedy had overcome most of his former scorn for liberals, even called himself a liberal on occasion and cheerfully addressed liberal groups he would have disdained a decade before. "I notice young Congressman Bill Green is here tonight," he said at an Americans for Democratic Action dinner in Philadelphia. Bill Green's father had been the cantankerous Democratic boss of Philadelphia in the days when the ADA, over his bitter opposition, helped elect Joseph Clark, Kennedy's Mississippi companion, as reform mayor. The boss's son was now a liberal congressman, supported by the ADA. "I think his father might have been shocked to see him here," Kennedy continued; then paused and added, "but *my* father might be shocked too."[10] In New York he charmingly explained his presence as the speaker at an ADA dinner by pointing at me and saying: "We fought and argued about all these issues; and he won; and here I am tonight."[11] It was a time of paradox. Adlai Stevenson and Robert Kennedy had not liked each other, but Dick Schaap concluded his 1967 book about Kennedy by suggesting that he might well end as "the Adlai Stevenson of the 1970s."[12] Eleanor Roosevelt had resisted him almost to the time of her death, but Kennedy had become, Franklin D. Roosevelt, Jr., said, "the torch-bearer of everything that my mother stood for and fought for."[13]

Yet the old liberalism had failed to beat the structures. And its distinctive institutions tended to leave out those too poor or demoralized to form organizations of their own. The programs of the New Deal, the Fair Deal, even of the New Frontier, Kennedy said in 1967, "put into effect with the finest of intentions, have been either inadequate or retrogressive."[14] At the Philadelphia ADA dinner Kennedy remarked that most of his audience, when it thought of organized labor, thought of the long struggle to establish labor's rights. "But youth looks with other eyes. They think of labor as grown sleek and bureaucratic with power . . . a force not for change but for

the status quo." The university had become a "corporate bureaucracy."[15] In different ways, welfare, public housing, farm
price supports, one creation after another of the old liberalism, had
congealed into props of the existing order. All this further reduced
the capacity of government to change things.

In September 1967 Daniel Patrick Moynihan urged the ADA to
join with conservatives in protecting "the social fabric of the nation"
against radical students and militant blacks. Kennedy took the opposite view. "I think," he said, "the ADA should just fold up and go
out of business. They're so out of touch with things. . . . Your generation should go out and start a new ADA that isn't dependent on
the unions for money and is engaged in direct action, instead of just
voting on resolutions."[16] (He later observed of Moynihan, "He knows
all the facts, and he's against all the solutions.")[17] Having neglected
in my White House years ever to ask John Kennedy to inscribe a
photograph, supposing always there was plenty of time, I did one
day ask Robert. The photograph duly arrived. He had inscribed it:
"With the highest regard of a fellow author, government employee,
~~liberal~~, Harvard graduate and a friend."

Conventional politics seemed impotent before the structures. The
great forces for change — the civil rights movement, the antiwar
movement, the nationalist movements of the Third World — represented direct action. If new institutions of power could be built
among the powerless, if the new movements could avoid violence
— this was why he so greatly valued Martin Luther King and Cesar
Chavez — change might come without tearing a fragile society
apart. He was too much a skeptic, or an Augustinean, to be altogether optimistic. "He always conveyed," Wes Barthelmes said,
". . . a bit of, I think, sadness as he talked about particular programs. . . . He sort of conveyed the futility of most means and the
uncertain glory of most ends. But I think that, if he had any com-
mandments, one . . . would be, 'It really is a secular sin not to
try.'"[18] His favorite song — one heard it so often blaring from some
unseen source in his New York apartment — was "The Impossible
Dream."

II

By November 1967 when Robert Kennedy had his forty-second birthday, he was the most original, enigmatic and provocative figure in
midcentury American politics. A man of intense emotion, he aroused
intense emotion in others. "Bobby had a psychic violence about
himself," said that perceptive actress Shirley MacLaine. "Let's be
violent with our minds and get this thing changed. Let's not be
violent with our triggers."[19] But intellectual violence, seeking the

root of the matter, could be frightening in frightened times. "He was a tortured guy," said Barthelmes, "and he was moved by the torture of others. That unsettled people."[20] Kennedy incarnated the idea of struggle and change. This moved many. It disturbed many. He gave hope to some groups in the country, generally the weak; threatened others, generally the strong. Some saw him as compassionate savior, some as ruthless opportunist, some as irresponsible demagogue plucking at the exposed nerves of the American polity — race, poverty, the war. Few were neutral, very few indifferent.

His movement beyond liberalism both fascinated and alarmed the young militants of the day. Robert Scheer, a radical journalist before he became *Playboy*'s expert on the mental lusts of Jimmy Carter, registered both reactions in an article for the New Left magazine *Ramparts* in 1967. Kennedy, he said, was "undoubtedly a very charming and alive man for a politician" but "dangerous" because he provided "the illusion of dissent without its substance. Hubert Humphrey is a bad joke to most young people, but Bobby is believable, and for that reason, much more serious. He could easily coopt prevailing dissent without delivering to it. . . . The Kennedy people have raised cooptation to an art form." After reading the *Ramparts* piece, which the author thought very tough, Kennedy merely told him he had some interesting points. "That was his whole fucking style, you know. . . . Any other politician would say, 'I want Scheer off the plane. What's a *Ramparts* guy doing on the plane?' . . . We used to have arguments all the fucking time and friendly jostling, and I never felt tense." "Also," said Scheer, "there was a certain kind of madness to him. . . . [And] I think there were certain things Kennedy believed. I think he gave a shit about Indians, for instance. . . . I thought he gave a shit about what was happening to black people"; still, at bottom, "a very orthodox political figure," in the end just another liberal.[21]

Not all liberals agreed. "Outside of Washington," Fred Dutton wrote him, "the usual old canards about you — 'too zealous,' 'ruthless,' 'narrowly ambitious for just himself rather than broader purposes and impulses' — remain far more entrenched than I had thought. . . . I frankly had thought this problem was behind you and am amazed at the extent it perseveres."[22] So the kindly veteran Gerald W. Johnson of the *Baltimore Sun*, now seventy-seven, described Kennedy in the *New Republic* as "a strong and dangerous man, driven by a maniacal energy . . . as ruthless as Torquemada."[23] The stereotypes — his father, his church, McCarthy, Hoffa, never-get-mad-get-even, the relentless prosecutor, wiretapper, grudge bearer — lingered especially among the "purist liberals," as James Wechsler called them, who seemed almost to fear power and prefer defeat to victory. This "fierce anti–Robert Kennedy obsession" led Wechsler to write a series for the *Progressive* entitled "Robert F. Kennedy: A Case of Mistaken Identity." "Some published critiques of him," Wechsler said, "should bear the warning usually associated with works of fiction: they bear little resemblance to the living character."[24] Liberal schizophrenia about Kennedy was brilliantly caught in Jules Feiffer's cartoons of the "Good Bobby" and the "Bad Bobby."[25]

The leadership of organized labor liked him no more than the purist liberals did, if for opposite reasons. There were important exceptions, like Reuther and Dubinsky; but the rest, as Don Ellinger of the Machinists, another exception, put it, "just didn't like him and that was all there was to it." George Meany called him that "jitterbug." "The big hacks would come in," said Wes Barthelmes, "and he'd question them about racial discrimination in unions. They wouldn't want to talk about anything but minimum wage."[26]

The business community suspected him as much as ever, despite valiant attempts by Thomas J. Watson, Jr., Douglas Dillon, Benno Schmidt and other friends to say that he wasn't all that sinister a fellow. Not only did he dislike self-congratulatory business banquets, said John Nolan, "but he was terrible at them. . . . He felt that what they were doing wasn't very important, and that impression was conveyed."[27] One year William Orrick invited him to attend the annual Bohemian Grove Encampment — "the greatest men's party that has been invented," as Herbert Hoover described it when he invited Joseph Kennedy in 1948.[28] Here was the American tycoon at play, with campfires, rituals, drink, dirty stories, drink, practical jokes, drink. Lyndon Johnson would have had a fine time. Kennedy detested every minute of it. "He was very difficult," Orrick said later. ". . . He just did not enjoy that company." And the company, Orrick added, was mostly "antagonistic toward him."[29] Dillon summed it up: "The general businessman's stereotype view of Bob Kennedy was even more inaccurate than Senator Kennedy's view of the business community."[30]

The far right hated him most of all: consider Frank A. Capell's *Robert F. Kennedy, Emerging American Dictator*; on the cover, a cut of Kennedy, a shadowy Castro looming behind him. "There can be no doubt of Bobby's pro-communist bias," Capell wrote. ". . . His dictatorial and ruthless methods combined with the power to implement them bode ill for the future. . . . Americans should BEWARE."[31] At a CIA seminar for Army officers in the spring of 1968, one of them, back from Vietnam, said, "You don't realize what it's like being sold out by these antiwar bastards, Bobby Kennedy, and the rest of the thimbleheads." A CIA instructor reproduced on the blackboard a

bumper sticker he had seen on the way to work: "First Ethel Now Us."[32] Westbrook Pegler, once a splendid if intemperate writer and a chum of Joseph P. Kennedy's, recorded for his old friend's son the hope that "some white patriot of the Southern tier will spatter his spoonful of brains in public premises before the snow flies."[33] Good old Clyde Tolson of the FBI's Edgar and Clyde, discussing Kennedy in the hearing of William Sullivan in 1968, said, "I hope that someone shoots and kills the son of a bitch."*

"Why do people hate me so?" Kennedy cried out to Dorothy Schiff in 1965[34] and to others from time to time later. The English journalist Margaret Laing told him she had met two people in the previous week who felt so strongly about him they wanted to hit him. "His astonishment on hearing about this was so complete that he thought he must have misunderstood." He said, "Hit me? You mean punch me? . . . No — you're kidding." He laughed in a way that was "almost a cough."[35] A reporter asked him to explain why he was thought ruthless. He paused; then said slowly, "I think that is what happens to you when you try to do things."[36] He bantered bitterly, almost obsessively, about his supposed ruthlessness. "Although he joked about the word," Theodore H. White wrote later, "it cramped his thoughts and public behavior."[37] He took from Emerson a consoling thought for his commonplace book: "God offers to every mind its choice between truth and repose. Take which you please — you can never have both."[38]

No one quite had the answer. "I do not understand the animus against Bobby," said William Benton. "What is the conceivable explanation? Where does it come from?"[39] "Could it be," asked Kenneth Galbraith later, "that he was the least known public figure of our time?"[40] "So many people have him absolutely wrong," observed Joseph Alsop, who lamented but adored him. "They think he is cold, calculating, ruthless. Actually, he is hot-blooded, romantic, compassionate."[41] Averell Harriman said, "It was impossible for him not to tell the truth as he saw it. I think that is why some people thought he was ruthless. At times the truth is ruthless."[42]

III

In the meantime, private life went on. There were more children — Matthew Maxwell Taylor Kennedy on January 11, 1965; Douglas Harriman Kennedy on March 24, 1967; homage to Taylor, Dillon

---

* William C. Sullivan, in interview by author, July 26, 1976. When I verified this quotation with Mr. Sullivan a few weeks before his own death, he added that Tolson said this a month or two before Kennedy was shot and killed; afterward, "I have wondered what thoughts went through Mr. Tolson's mind, if any" (Sullivan to author, October 1, 1977).

and Harriman, three impeccable members of the American establishment. Why these men, in the same years that Kennedy was in black ghettos and grapefields and Indian reservations? Did this prove political hypocrisy? social snobbery? a hope that disparate people could be united to meet common problems? or simply a man who moved from world to world?

The world of Hickory Hill was never more enjoyable. His older children were entering their teens. "I think that they'll develop their own lives," Kennedy told an interviewer.

> I talked to one of my sons about it one time and he said that he wanted, he was 12 years old, he wanted to make a contribution. And he said, "But I don't want to get involved in political life," and I said, "Well, what do you want to do?" . . . He loves animals and he said, "I want to make a contribution like Darwin and Audubon did." And so I think people work out their own lives . . . just as long as they understand that in the last analysis what is important [is] that they give something to others, and not just turn in on themselves.[43]

An overachiever himself since infancy, Kennedy was now more than ever the collector of overachievers. One never knew whom to expect at Hickory Hill — novelists, entertainers, columnists, decathlon champions, astronauts, football stars, diplomats, politicians, mountain climbers, international beauties, appearing in every age, sex, size, color. I remember the party in November 1966 to celebrate Harriman's seventy-fifth birthday. Ethel directed the guests to come in costumes commemorating an aspect of Harriman's life. Some appeared as railroad engineers in honor of the Union Pacific. Art Buchwald, remembering Harriman and the Yale crew, was a coxswain roaring jokes through a megaphone. George Plimpton came, irrelevantly, as an Arab. Kennedy himself found somewhere the distinctive style of long black overcoat and creased felt hat that Harriman had worn as wartime ambassador to Moscow. Hubert Humphrey was there (this was only a few months after the fox-in-the-chicken-coop crack, but Kennedy was not a feudist) and gave, I noted, the best toast of the evening; "he was witty, charming, relaxed and *brief*. One felt that he was quite at peace with himself and very cheerful and comfortable in the Kennedy environment."[44]

Then, in a sudden gesture, the dining room curtains were drawn back. Outside the bay windows in life-size wax replica were Roosevelt, Churchill and Stalin as at Yalta. "It was very cold," said Plimpton, "and [the Big Three] looked forlorn out there, like waifs, and awfully cold too, because of that spooky wax color. Someone had furled a scarf around Stalin's neck and Roosevelt had a beanie-type hat to make them look a bit more cheerful."[45] Odd, one reflected: fifteen years earlier Robert Kennedy, in his prize essay at Charlottesville on the iniquity of Yalta, had named Harriman as a leading villain.

There were the annual pet shows, with prizes for the longest nose, the shortest tail and other oddities, and Buchwald as chief judge and master of ceremonies. Since the Kennedys had more pets than anyone else, they won more prizes. Brumus, still monstrously in the picture, distinguished himself one year by lifting his leg and discharging on a nice old woman sitting placidly on the lawn. "Bobby went white," Buchwald recalled, "and he ran into the house. So when it came to a profile of courage in regard to Brumus, he was a coward." [46]

Correspondence of 1967: Edward Bennett Williams — once again, after the Hoffa years, a friend — to Robert Kennedy:

> I have been retained by Mr. Art Buchwald to represent him in the matter of the vicious and unprovoked attack made on him Wednesday, July 25, by the large, savage, man-eating, coat-tearing black animal owned by you and responding to the name, Broomass (phonetic).
>
> Mr. Buchwald has been ordered to take a complete rest by his physician until such time as he recovers from the traumatic neurosis from which he is suffering as a result of the attack. He will be in isolation at Vineyard Haven, Martha's Vineyard, Massachusetts for an indefinite period at a cost of $2,000 a month.
>
> He is concerned about the effect of exposing this ugly episode on your political future. . . . Since Broomass is black, the case is fraught with civil rights' undercurrents. [47]

So life ran along. His father was no better, but Kennedy ministered to him whenever he could, in winters at Palm Beach, in summers at Hyannis Port. There were admonitory letters from his mother: "In regard to our conversation about 'If I was . . .' as opposed to 'If I were . . .' I should like to offer the following quotations from a book of rules on grammar. . . ." [48] Or, from Rose Kennedy's secretary: "Your mother wanted you to know that the Balfour Resolution established the Jewish home in Jerusalem, but did not make it a Jewish state. She supposed that you know the difference, but some don't. If you don't, it should be explained to you." [49]

Though I left Washington in 1965, I seemed to see more of the Kennedys than ever before. In the autumn of 1966 I moved from Princeton to New York. Thereafter Angie Novello would often phone and say, "The Senator is coming up tonight and wonders whether you could meet him at such-and-such a place a little after ten." One would find a group, small or large, and talk to the early morning, the conversation shifting easily from the light to the serious, propelled by Robert's endless curiosity, irreverent wit and acerbic honesty. I went often, too, to Hyannis Port for long, lazy, sun-drenched summer days with the Kennedys or the Smiths. For all one's forebodings about the republic these were good and joyous times.

IV

He had always been a taker of risks from that day, so many years before, when he had thrown himself off the yawl into Nantucket Sound in his determination to learn to swim, and John Kennedy had said he had shown either a lot of guts or no sense at all, depending on how you looked at it. Now physical danger almost seemed a compulsion. After Dallas the Canadian government named the tallest unclimbed peak in North America, 14,000 feet high, in honor of his brother. The National Geographic Society proposed that the remaining Kennedy brothers join in the first assault on Mount Kennedy. The plane crash eliminated Edward. Robert reluctantly — he hated heights — decided in March 1965 that he would climb Mount Kennedy himself. His family could not shake his resolution. His mother's last words to him were, "Don't slip, dear."[50] But when his father "found a picture in the paper of Bobby in his mountain gear, he would angrily grind it into a ball," his nurse recalled, "and throw it across the room."[51]

Kennedy was rash but not reckless. Two veterans of the successful Mount Everest expedition of 1963, James Whittaker and Barry Prather, came with him. Whittaker asked Kennedy what he had been doing to get in shape for the climb. Kennedy said, "Running up and down the stairs and hollering 'help.'" Whittaker said to himself, "Oh, boy!" They flew north to the Yukon Territory, Kennedy reading a book by Churchill. They saw the peak, "a magnificent mountain," Kennedy recorded, "lonely, stark, forbidding." The professionals gave him a crash course in climbing. They helicoptered to a base camp at 8700 feet; then roped themselves together, Kennedy in the middle, and set out across glaciers and crevasses to the summit.

After two days, they saw the final cone standing vertically against the sky. Kennedy could not see how it was humanly possible to reach the top. Whittaker dug his ax into the steep ice ridge and began to pull himself, almost, it seemed to Kennedy, straight up. After sixty feet, he shouted back to Kennedy, "You're on belay. Now you climb!" Kennedy thought to himself, "What am I doing here?" But "I really only had one choice and went on." After some minutes Kennedy managed to join Whittaker, Prather following. "What do you think of it?" asked Whittaker, enjoying the hundred-fifty-mile view. "I don't want to look at anything," said Kennedy. "I just want to stay right here." When they reached the crest, Kennedy went ahead for the last two hundred feet. He deposited Kennedy memorabilia, crossed himself, stood a moment in inscrutable silence and then, with immense relief and exhilaration, made his way down.[52]

Professional climbers thought that, roped between Whittaker and Prather, Kennedy could hardly fail, and deprecated the achievement.[53] That may have been true enough physically. Psychologically it was a feat for a man with a terror of heights. "I didn't really enjoy any part of it," Kennedy said later. "Henceforth I'm going to stay on the first floor of my house."[54] He did not, of course. In winters he skied madly down precarious slopes in Vermont or Idaho. In summers he shot rapids in western rivers. On a trip down the Colorado, he took to diving into the rushing water. "I've done some foolish things," said George Plimpton, who was along, "but I wouldn't do anything as foolish as that. . . . Jim Whittaker was the first person up Mt. Everest; a terribly brave person, but I don't think he particularly wanted to jump into the water. . . . Yet here was the Senator [in] the wildest of the white water."[55]

In summers too he assembled friends and sailed along the New England coast. In September 1965, as the Kennedy schooner was buffeted by a thirty-knot wind off Long Island, a Coast Guard cutter hailed them to say that Kathleen Kennedy had received head injuries at a horse show. Kennedy, deciding that the quickest way to reach the hospital would be via the Coast Guard, plunged into ten-foot waves and, while everyone held his breath, swam fifty yards to the cutter. This at least made some sense.[56] But he was as impetuous in plunging into the chilling waters of Maine in heavy seas to save his brother's old sea jacket or his own dog Freckles.[57]

We all worried and speculated about this almost compulsive courtship of danger. No doubt the less athletic, like myself, exaggerated the hazards. John Glenn, who careened down the Salmon River with him in kayaks, said later, "I don't recall him taking what I would consider foolish risks. I think he was a pretty good judge of his own capacity on what he could do."[58] Still, as his skiing companion (and a professional) Thomas Corcoran, nephew of the old New Dealer, put it, "He obviously enjoyed approaching the brink of the impossible."[59]

"Men are often drawn to those aspects of nature which reflect their spirit," wrote Richard Goodwin, "and he loved rapids and the pounding surf, was drawn to motion and to turmoil."[60] But it was not only an affinity of temperament; it was some sort of moral necessity. James Dickey, explaining why he had written *Deliverance*, his novel of ordinary men under extreme pressure, recalled the poet John

Berryman saying "that it bothered him more than anything else that a man could live in this culture all his life without knowing whether he's a coward or not."[61] Kennedy thought it necessary to know. He marked the sentence in Emerson's *Essays*, "It was a high counsel that I once heard given to a young person, 'Always do what you are afraid to do.'"[62]

Courage, Robert Kennedy wrote, was the virtue John Kennedy "most admired,"[63] the virtue that, as Robert often quoted from Churchill, was "rightly esteemed the first of human qualities because it is the quality which guarantees all others."[64] John Kennedy once read aloud from the citation for Douglas MacArthur's Distinguished Service Cross: "On a field where courage was the rule, his courage was the dominant feature." Robert Kennedy said, "I would love to have that said about me."[65] This was why he liked Glenn and Whittaker, El Cordobés and José Torres, not to mention Martin Luther King and Cesar Chavez. As Pierre Salinger said, they were all "willing to lay their lives on the line for something they believed in, whether it was flying in space, or climbing a mountain, or facing a bull."[66]

Was there more to it than this? Sometimes it almost seemed a dance with death itself. "He believed in his powers," wrote Goodwin, "but was haunted by the omens of failure; wanted much from life but tempted death, even dared it."[67] Death remained a savage presence. In September 1966 Dean Markham, his cherished friend from Harvard days, and Ethel's brother George Skakel died in a plane crash in the mountains of western Idaho. I saw Robert and Ethel soon afterward at the Smiths. The evening was spent in Irish-wake style with uproarious reminiscence of their dead friends. Jean Smith said, "We're getting pretty used to this by now." Kennedy later told Goodwin, "Hackett and I have so much experience at this thing that we're offering a regular service for funerals. . . . We pick out a cheap casket to save the widow money. You know they always cheat you on the casket. We pick passages from the Bible and do all we can to ensure an interesting and inexpensive funeral."[68] He told Sorensen more somberly, "You had better pretend you don't know me. Everyone connected with me seems jinxed."[69]

Death was never far from his mind. "Not to be born is past all prizing best." My son Stephen's notes on a dinner in New York in the autumn of 1967:

RFK on God: "I don't know why God put us on earth. If I had my choice, I would never have lived. I had no control over it. But why should God put on earth some people who will go to the devil? I think Graham Greene's description of the defrocked priest in *The Power and the Glory* is marvelous." He asked everybody at the table whether they believed in God. I answered that I affirmed what my grandfather had said — "I do not believe in God as a corporeal being. Godliness is simply being kind and loving to people here on earth." Somebody asked RFK whether he believed in God.

"Yes," he said hesitantly, "I think. But one question which really shakes me, really shakes me — if God exists, why do poor people exist? Why does a Hitler arise? I can't give an answer for that. Only faith. . . .

"Yes, I do believe in an afterlife. Religion is a salve for confusion and misdirection. It gets people over the hump easier than what your grandfather said." [70]

He may not have wished to be on earth, but God had put him there, and he still believed in God, or at least believed in belief. I do not think he was courting death. It was rather that he had an almost insolent fatalism about life. No one understood better the terrible fortuity of existence. Men were not made for safe havens.

v

The paradoxes of Kennedy were a natural for the literary imagination. But, unlike Europe, where culture and the state share the same capitals, politics has generally been remote from literature in the United States. Except for an occasional Henry Adams, writers, journalists apart, have not often been stable-companions to statesmen. [71] Franklin Roosevelt acquired grand literary friends like Archibald MacLeish and Robert Emmet Sherwood, but this was after he became President. Most of John Kennedy's literary friends were postinaugural. Robert Kennedy as senator, however, sought out, interested, beguiled and sometimes repelled the writers of his day. (The other senator of the time with comparable tastes was Eugene McCarthy of Minnesota.)

George Plimpton was an old friend, an overachiever of impressive versatility and the only writer Kennedy knew who kept up with him in the melodrama of physical risk. He had been affectionately through labor racketeering and Watts with Budd Schulberg. Art Buchwald was an intimate of the house. Irwin Shaw was always great fun when he made his annual descent from Switzerland. Galbraith was a close counselor, John Bartlow Martin a loyal friend through the years. He had a good Irish time with newspapermen like Jimmy Breslin and Pete Hamill, who made much of the fact that they, unlike Kennedy, had really known what it meant to be poor. Once Kennedy and Breslin had an argument (about Lindsay), which Breslin concluded by saying, "Fuck you, Senator!" Kennedy looked at him and said, "You know, Jimmy, I used to think being poor built up character, and then I met you."[72] He liked James Wechsler, the liberal, and Jack Newfield, the radical, and Murray Kempton, the anarchist, because they were bright and entertaining and passionate.

In the summers Kennedy, his brother and assorted friends often sailed from Hyannis Port to see William Styron or Art Buchwald on Martha's Vineyard. Philip Roth, who met Kennedy on the Vineyard, thought him "witty and charming and very engaging."[73] Styron did not like Kennedy at first, but liked him more as the summers passed. "Unless you include the fact that he could put people off horribly," Styron said later, ". . . you're not going to get an honest picture."[74] Richard Yates, the novelist, who worked at Justice in 1963, did not like Kennedy. Nothing, even Kennedy's affinity for the wounded of the world, could appease Gore Vidal. Nor could James Baldwin, though in later years he acknowledged complexity in Kennedy,[75] ever get over that stormy session in 1963.

Truman Capote, a United Nations Plaza neighbor, was fascinated by Kennedy's power to make people jump — "meaning, if you move through a room and everything is galvanized on you and you're terribly aware of it, and people . . . move out of their way to make room for you . . . in effect, they 'jump.' I think Bobby had that power, but he rarely used it." They occasionally had drinks together. "I always felt that he was very uneasy with me and that his friendship with me was really based on my friendship with other people that he himself was fond of. . . . There was something exotic about me that he couldn't entirely accept; I mean, he was trying to accept it like a father doesn't want to accept long hair on a kid, or sideburns, but yet, he's sort of stuck with him." Walking his dog early one morning, he saw Kennedy talking sternly to two small boys. "Truman, come over here," Kennedy called. "You won't believe this. . . . I came out here to take my dog for a walk . . . and here come these two little kids smoking cigarettes." One said, "Honestly, honestly, Mr. Kennedy, I swear we'll never do it again." "It was," said Capote, "as if he was some sort of avenging angel who had fallen out of heaven upon them." Kennedy made them promise to give up cigarettes, and they ran like maniacs down the street. Then one swung around, raced back and said, "Can I have your autograph, Mr. Kennedy?"[76]

Life asked Saul Bellow to do a piece on Kennedy — an incident described in *Humboldt's Gift*. "I liked him," Bellow (as Charlie Citrine) said, ". . . perhaps against my better judgment. . . . His eyes were as blue as the void. . . . His desire was to be continually briefed . . . receiving what I said with a kind of inner glitter that did not tell me what he thought or whether he could use such facts."[77] One of the more cryptic items in the Kennedy literature was Donald Barthelme's semisatiric, doom-laden fantasy "Robert Kennedy Saved from Drowning" — a sequence of ambiguous dreamlike snapshots: "K. at His Desk"; "K. Reading the Newspaper"; "Sleeping on the Stones of Unknown Towns (Rimbaud)"; "Gallery-going" —

> K. looks at the immense, rather theoretical paintings. "Well, at least we know he has a ruler." The group dissolves in laughter. People repeat the remark to one another, laughing. The artist, who has been standing behind a dealer, regards K. with hatred.

"K. Penetrated with Sadness"; "A Friend Comments: K's Aloneness."

> K. in the water. His flat black hat, his black cape, his sword are on the shore. He retains his mask. His hands beat the surface of the water which tears and rips about him. The white foam, the green depths. I throw a line, the coils leaping out over the surface of the water. He has missed it. No, it appears that he has it. His right hand (sword arm) grasps the line that I have thrown him. . . . I pull him out of the water. He stands now on the bank, gasping.
> "Thank you."*

The mask worn to the end, the enigma remaining.

Kennedy, Norman Mailer thought, was the sheriff who could have been an outlaw. Mailer found himself "excited by precisely [Kennedy's] admixture of idealism plus willingness to traffic with demons, ogres, and overloads of corruption." Oddly Mailer, who might have relished Kennedy most, even perhaps understood him best, met him only once. His mouth "had no hint of the cruelty or calculation of a politician who weighs counties, cities, and states, but was rather a mouth ready to nip at anything which attracted its contempt or endangered its ideas." The blue of his eyes "was a milky blue like a marble so that his eyes, while prominent, did not show the separate steps and slopes of light some bright eyes show, but rather were gentle, indeed beautiful." Somehow "he had grown modest as he

---

* Donald Barthelme, "Robert Kennedy Saved from Drowning," in *Unspeakable Practices, Unnatural Acts* (New York, 1968). The story is striking because of its precision and insight. When I asked Mr. Barthelme about the provenance, he replied: "I never met Robert Kennedy nor did I talk to people who had. The story was begun while I was living in Denmark in 1965. . . . The only 'true' thing in it was Kennedy's remark about the painter. I happened to be in the gallery when he came in with a group; I think the artist was Kenneth Noland. Kennedy made the remark quoted about the ruler — not the newest joke in the world. The story was published in New American Review well before the assassination. I cannot account for the concluding impulse of the I-character to 'save' him other than by reference to John Kennedy's death; still, a second assassination was unthinkable at that time. In sum, any precision in the piece was the result of watching television and reading the New York Times" (Barthelme to author, July 16, 1977).

grew older, and his wit had grown with him." Mailer was struck by the "subtle sadness [that] had come to live in his tone of confidence. . . . He had come into that world where people live with the recognition of tragedy, and so are often afraid of happiness."[78]

Kennedy first met James Stevenson, the *New Yorker*'s gifted writer and cartoonist, at a Saint Patrick's Day reception Stevenson was covering for "The Talk of the Town." Stevenson was as shy as Kennedy. "I shook his hand, and I looked at him, and he looked at me, and neither of us said anything. We looked for a while, and then he went on. . . . But I immediately thought, 'Oh, that's an interesting person,' and then he was gone into the crowd." Kennedy thought Stevenson was interesting too and began to like seeing him. Carter Burden, then working in Kennedy's New York office, used to arrange their meetings. Stevenson, Burden recalled, "never would say anything. The senator would never say anything. But he was very comfortable in his presence." Stevenson himself said later, "You could feel a kind of direct communication without the chit-chat or whatever other people do." And: "He was much more alive than many people, let alone public figures, because he was constantly absorbing, reacting, responding. . . . And, of course, he was terribly funny." And: "You had the feeling you were looking at someone who could go in any direction, and that things could be accomplished that you really didn't believe before that could be accomplished. He carried around a kind of hope with him."[79] Their mysterious communion produced for "The Talk of the Town" the most exact and delicate prose ever written about Robert Kennedy. Thus Kennedy in a car bound for Bedford-Stuyvesant:

A taxi-driver recognizes Kennedy and yells, "Give it to 'em, Bobby!" Kennedy waves, then stares ahead again. He is deeply preoccupied now, at his most private. . . . He abandons, piece by piece, the outside world — he puts away the magazines, the cigar is forgotten, the offer of gum is unheard, and he is utterly alone. His silence is not passive; it is intense. His face, close up, is structurally hard; there is no waste, nothing left over and not put to use; everything has been enlisted in the cause, whatever it may be. His features look dug out, jammed together, scraped away. There is an impression of almost too much going on in too many directions in too little space; the nose hooks outward, the teeth protrude, the lower lip sticks forward, the hair hangs down, the ears go up and out, the chin juts, the eyelids push down, slanting toward the cheekbones, almost covering the eyes (a surprising blue). His expression is tough, but the toughness seems largely directed toward himself, inward — a contempt for self-indulgence, for weakness. The sadness in his face, by the same token, is not sentimental sadness, which would imply self-pity, but rather, at some level, a resident, melancholy bleakness.[80]

VI

Kennedy read history and biography and an occasional novel but most of all he loved poetry. Tennyson's *Ulysses* was a favorite —

> The lights begin to twinkle from the rocks:
> The long day wanes: the slow moon climbs: the deep
> Moans round with many voices. Come, my friends,
> 'Tis not too late to seek a newer world

— and provided title and epigraph for a book he published at the end of 1967. He told my son Stephen, "Richard Burton is a man's man. We recited our favorite poetry together."[81] This conjunction of ideas would have puzzled Lyndon Johnson. Kennedy was delighted once when he was able to finish a couplet that had slipped Burton's memory.[82] He invited Yevgeny Yevtushenko to his forty-second birthday party. Yevtushenko, like the rest, was mesmerized by the eyes, "two blue clots of will and anxiety. . . . They inhabit his face like two beings uninvolved in the general gaiety. Within these eyes, an exhausting hidden work transpires. . . . Like pale blue razor blades, they pierced through anyone in conversation." At the end of the evening, the two men stood alone in the hall, holding "antique crystal goblets in which tiny green sparks of champagne were dancing." Yevtushenko proposed a toast — to Robert's completion of his brother's work. He reminded Kennedy that, according to the Russian custom, the goblets must be emptied in a single gulp and smashed against the floor. Kennedy, a little embarrassed, said these were Ethel's heirlooms — "wives will be wives" — and produced instead some glasses from the kitchen. "It somewhat surprised me," Yevtushenko said, "that one could think about insignificant glasses when such a toast was being proposed, but, of course, wives will be wives." They downed the champagne and threw the glasses hard on the floor. The glasses did not break. By Russian tradition this meant the toast would not be fulfilled. "I have always been superstitious and a terrible foreboding passed through me. I looked at Robert Kennedy. He had turned pale. Probably he, too, was superstitious." I doubt it; he was more probably tired or bored. Later Yevtushenko wrote an awful poem about the Kennedy brothers.[83]

Yevtushenko was the poet whom the Soviet authorities liked sending abroad. Andrei Voznesensky was not in favor. The American Academy of Poets invited him to the United States in the spring of 1967, then received word that the authorities might not permit him to come. It was thought that Kennedy's intercession could help. I spoke to him, and he sent off a cable at once. After Voznesensky arrived, we had breakfast one morning in the New York apartment. "Voznesensky was serious, rueful and candid," I noted. ". . . Quite different from the somewhat flamboyant and exhibitionistic Yevtushenko." He and Kennedy talked about the frustrations of the young, about the Americans most read in the Soviet Union ("Salinger, Updike and Cheever," Voznesensky said), about Pasternak, Saint-Exupéry and John Kennedy. They liked each other very much.[84]

Allen Ginsberg, the bard of the counterculture, later chatted with Kennedy about the Soviet poets. Both Americans agreed they preferred Voznesensky. Kennedy thought him the more sensitive. Ginsberg, as Peter Edelman, who was present, recalled, explained that, "when he appeared on a platform with both of them, Yevtushenko had embraced him; but Voznesensky was clearly the more open person because they had kissed deeply and soulfully. Kennedy received that with equanimity."[85] Voznesensky's later poem about Kennedy — "June '68"[86] — was decidedly better than Yevtushenko's.

Ginsberg, a sweet and obsessed man, had come to explain to Kennedy his intricate theories about drugs, the Narcotics Bureau and the Mafia. Kennedy, who already knew about the syndicates, wanted to hear Ginsberg on the relations between the flower people, for whom Ginsberg was patron saint, and the young political militants. Ginsberg thought Kennedy "quizzical. . . . I liked him. He was friendly enough, he was available, he sounded serious. . . . He asked what were the serious possibilities of alliance between 'hippy' and 'black' groups. I was unsure. I asked if he'd ever tried grass. He said 'No,' whatever that meant." Toward the end, Ginsberg asked, "What do *you* think's going to happen to the country?" Kennedy said, "It'll get worse." Ginsberg departed, then remembered something "I'd forgotten to do which was most important." At that moment Kennedy passed through the outer office. Ginsberg asked whether he had two minutes to hear the Hare Krishna mantra. Kennedy said O.K. Ginsberg pulled out a harmonium and chanted: "Hare Krishna Hare Krishna Krishna Krishna Hare Hare Hare Rama. . . ." Kennedy said, "Now what's supposed to happen?" Ginsberg said it was "a magic spell for the preservation of the planet." Kennedy said, "You ought to sing it to the guy up the street," gesturing toward the White House. "He needs it more than I do."[87]

Among poets Kennedy had the closest and most complicated relations with Robert Lowell. "We never became good friends," Lowell said later. Still they interested each other. "I have always been fascinated by poets like Wyatt and Raleigh, who were also statesmen and showed a double inspiration," Lowell wrote Kennedy in early 1967,

> — the biggest of these must be Dante, who ruled Florence for a moment, and never could have written about Farinata and Manfredi without this experience. Large parts of the Commedia are almost a Ghibelline epic. Then there are those wonderful statesmen, like Lincoln and Edmund Burke, who were also great writers.
>
> Well, I do think you are putting into practice that kind of courage and ability that your Brother so subtly praised in his *Profiles*, and know how to be brave without becoming simple-minded. What more could one ask for in my slothful, wondering profession? [88]

Lowell had given Jacqueline Kennedy a Plutarch, which Robert, the poet later discovered, borrowed and read. Lowell was not surprised. "Bobby was very conscious of the nobility and danger of pride and fate." Remembering that Thomas Johnston had begun at the Yale Drama School, Lowell once asked him where in Shakespeare he would cast Kennedy. Johnston said at once, "Henry the Fifth." Lowell said that was trite. Kennedy objected to Lowell's objection. Henry V, Lowell reminded him, had not had that fortunate a life. He had died young, leaving in France a son who was murdered and an expeditionary force that was destroyed. Kennedy took down a volume of the Histories and read aloud the deathbed speech of Henry IV, where the old King, his crown taken prematurely by young Harry, said,

> For this the foolish over-careful fathers
> Have broke their sleep with thoughts, their brains with care
> Their bones with industry;
> For this they have engrossed and piled up
> The canker'd heaps of strange-achieved gold;
> For this they have been thoughtful to invest
> Their sons with arts and martial exercises. . . .
> For what in me was purchased,
> Falls upon thee in a more fairer sort.

Then Kennedy said, "Henry the Fourth, that's my father." At first Lowell thought this a non sequitur. As he reflected, he decided it was "very profound. . . . He meant he had a very difficult career coming from this difficult but very elevated forebear who made it possible. And he really *was* cast in the role of Henry the Fifth, not altogether a desirable one . . . perhaps a doomed one." [89]

"Doom was woven in your nerves," Lowell wrote in the best poem about Robert Kennedy,

... like a prince, you daily left your tower
to walk through dirt in your best cloth. Untouched,
alone in my Plutarchan bubble, I miss
you, you out of Plutarch, made by hand —
forever approaching our maturity. . . .
One was refreshed when you wisecracked through the guests,
usually somewhat woodenly, hoarsely dry,
pure Celt on the eastern seaboard. Who was worse stranded?
Is night only your torchlight wards gone black,
white wake on wave, pyre set for the fire that fell?[90]

37

The Dilemma

AUTUMN 1967: the war intensifying, the ghettos exploding, the campuses stirring, public frustration spreading, Lyndon Johnson massive and immovable in the White House. There was a turnabout again in the polls. "Senator Robert Kennedy of New York," Gallup reported at the end of September, "has made steady gains in political appeal over the last several months, while President Johnson's appeal has been fading." In July it had been 45–39 for Johnson. Now it was 51–39 for Kennedy.[1] In October Louis Harris gave it to Kennedy 52–32.[2]

I

The war: 112 Americans killed in action in 1964, 1130 in 1965, 4179 in 1966, 7482 in 1967; American forces in Vietnam growing from 23,000 at the end of 1964 to 525,000 at the end of 1967;[3] American bombers dropping more tons of explosives in Vietnam than on all fronts in the Second World War; negotiations forgotten; peace as distant as ever; Robert Lowell and Norman Mailer leading the armies of the night against the Pentagon; Robert Kennedy defending the protesters. While he thought it a "bad mistake" to shout down government officials, he could see ample reason for street demonstrations against the war. "When we talk about the violence and the people walking out and the lawlessness, there is no [other] way for people to express their point of view, and I think that is most unfortunate."[4]

Actually, while peaceniks were burning the Secretary of Defense in effigy, McNamara was in the Pentagon context a peacenik himself, resisting the importunities of the Joint Chiefs — and of the President too — to escalate the war and now the strongest advocate of negotiations within the cabinet. Henry Kissinger, who was working with McNamara on negotiations, gave me a graphic vignette of the summer of 1967, Johnson harrying McNamara in the cabinet room: "How can I hit them in the nuts? Tell me how I can hit them in the nuts."[5] In August, without clearing his testimony with the White House, McNamara told the Senate Military Affairs Committee that it was militarily pointless to keep on bombing the North. "The President was angry," recalled Major General Robert Ginsburgh, Walt Rostow's military assistant. "He decided to back the Joint Chiefs and ease McNamara out."[6]

Johnson blamed the usual suspects. "The Kennedys began pushing him harder and harder," he told Doris Kearns. "Every day Bobby would call up McNamara, telling how the war was terrible and immoral and that he had to leave. Two months before he left he felt he was a murderer and didn't know how to extricate himself. I never felt like a murderer, that's the difference. . . . After a while, the pressure got so great that Bob couldn't sleep at night. I was afraid he might have a nervous breakdown."[7]

I met Kennedy in New York in the late evening of November 29. The afternoon papers had McNamara leaving Defense to become head of the World Bank.

> RFK said that Bob had *not* had any intimation that Johnson had sent his name up for the Bank. He had had a general conversation with Johnson about it last spring . . . but that was all. Then suddenly he heard, not from the President but from a leak in London, that he was on his way out. I expressed incredulity at Bob's apparent acquiescence in this. Wouldn't any self-respecting man, I asked Bobby, have his resignation on the President's desk half an hour after he heard the London bulletin? Isn't that what you would do? . . . Why does he fall in with LBJ's plan to silence him and cover everything up? Bobby listened silently and a little gloomily. He said that he thought that was what would finally happen — that Bob would not take the World Bank job and would instead quietly resign from government. Obviously this is what Bobby had been urging him to do. When we broke up around one, we got copies of the morning *Times*. The headline: MCNAMARA TAKES WORLD BANK POST. Bobby was evidently surprised and sad.[8]

We all supposed that McNamara's dismissal removed the last hope at the summit of government for restraint in the war — a supposition reinforced when Johnson soon announced a leading hawk, Clark Clifford, as the new Secretary of Defense. Only McNamara, Kissinger told me in December, had kept negotiations alive. Kissinger added that he himself had come away from Washington "with a conviction that LBJ's resistance to negotiation verges on a sort of madness."⁹ Nor was madness a White House monopoly that autumn. Calling for a hand vote at a Catholic girls' college in November, Kennedy was appalled when a majority wanted more, not less, bombing. "Do you understand what that means?" he cried. ". . . It means you are voting to send people, Americans and Vietnamese, to die. . . . Don't you understand that what we are doing to the Vietnamese is not very different than what Hitler did to the Jews?"¹⁰

Later that month Tom Wicker asked him on *Face the Nation* whether, in light of the administration claim that the "great threat from Asian communism" made victory essential for the security of the United States, it did not follow that "perhaps we ought to do as much as needs to be done?" The United States, Kennedy replied, had originally gone into South Vietnam in order to permit the South Vietnamese to decide their own future. Plainly the South Vietnamese did not like the future held out by the Saigon regime. So we had moved on to the national security argument.

> Now we're saying we're going to fight there so that we don't have to fight in Thailand, so that we don't have to fight on the west coast of the United States, so that they won't move across the Rockies. . . . Maybe [the people of South Vietnam] don't want it, but we want it, so we're going in there and we're killing South Vietnamese, we're killing children, we're killing women, we're killing innocent people . . . because [the Communists are] 12,000 miles away and they might get to be 11,000 miles away.

He grew passionate, even eloquent.

> Do we have a right here in the United States to say that we're going to kill tens of thousands, make millions of people, as we have . . . refugees, kill women and children? . . . I very seriously question whether we have that right. . . . Those of us who stay here in the United States, we must feel it when we use napalm, when a village is destroyed and civilians are killed. This is also our responsibility. . . . The picture last week of a paratrooper holding a rifle to a woman's head, it must trouble us more than it does. . . .
> We love our country for what it can be and for the justice it stands for and what we're going to mean to the next generation. It is not just the land, it is not just the mountains, it is what this country stands for. And that is what I think is being seriously undermined in Vietnam.

There was, he said, "an unhappiness and an uneasiness within the United States at the moment, and there has to be an outlet for it."¹¹

II

One who perceived the need for an outlet most acutely was Allard Lowenstein, the perennially youthful activist, now thirty-eight years old, one of Eleanor Roosevelt's last protégés, a veteran of racial struggles in Mississippi and South Africa, an early opponent of the war. In the summer of 1967 Lowenstein had conceived the quixotic enterprise of organizing a movement to dump Johnson. It was, as Jack Newfield observed, rather like Castro and a handful of guerrillas in the Sierra Maestre planning to dump Batista.[12] And it could not succeed without a presidential candidate determined to end the war.

When Lowenstein began his quest for a candidate, liberals had an alternative strategy, advocated by Joseph Rauh. It was almost impossible, Rauh reasoned, to deny renomination to incumbent Presidents; no antiwar candidates were available anyway; so why not organize a fight for a peace plank at the 1968 convention? At first Robert Kennedy thought that this tactic would only make it easy for Johnson to sidetrack the doves by promising to let them help write the platform. In mid-September he seemed slightly more favorable. "If we go ahead on this," he told me, "at least it will make the year more interesting than it is likely to be otherwise." He added, as news from all over, that Jesse Unruh, the powerful organization Democrat in California, had turned against Johnson and that Mayor Daley in Chicago was "very deeply opposed" to the war, the son of his closest friend having recently been killed in Vietnam.[13]

A fortnight later I dined at Hickory Hill. Around ten-thirty that evening Lowenstein, Newfield and James Loeb came by. Lowenstein set forth in fervent language his certitude that Johnson could be beaten in the primaries. I brought up the peace plank. Lowenstein said that Johnson was more unpopular than the war. Kennedy said, "When was the last time millions of people rallied behind a plank?" Johnson, Lowenstein said, might even pull out if defeated in the early primaries. "I think Al may be right," Kennedy said. "I think Johnson might quit the night before the convention opens. I think he is a coward." Lowenstein and Newfield told Kennedy he must run. Older, cautious and wrong, Loeb and I argued that Kennedy was too precious a commodity to be expended in a doomed effort. "I would have a problem if I ran first against Johnson," Kennedy himself finally said. "People would say that I was splitting the party out of ambition and envy. No one would believe that I was doing it because of how I felt about Vietnam and poor people. I think Al is doing the right thing, but I think that someone else will have to be the first to run. It can't be me because of my relationship to Johnson."[14]

Lowenstein pursued his quest. He dashed around the country, sought out like-minded Democrats, set up local committees, prosecuted the search for a challenger. At Kennedy's suggestion he talked to George McGovern. McGovern said he had been urging Kennedy since 1965 to run against Johnson. As for himself, he was up for reelection in South Dakota and feared that his materialization as a peace candidate against the President might defeat him for the Senate in his still hawkish state.[15] Why not talk, McGovern told Lowenstein, to senators who did not have reelection problems — Lee Metcalf of Montana, Eugene McCarthy of Minnesota?

Later McGovern checked back with his nominees. Metcalf had dismissed the idea out of hand: "it's ridiculous." McGovern then said to McCarthy, "I sent some people up to talk to you, Gene, about running against Johnson." McCarthy said, "Yeah, I talked to them. I think I may do it." McGovern was "astounded. . . . But it was clear then that he was going to go. There must have been other people that had been talking to him before that I didn't know about."[16] This surmise was correct. The previous March, Thomas K. Finletter had given a small old-Stevensonian dinner for McCarthy in New York. McCarthy had held the dinner table with his eloquent denunciation of the war. "The only way to get Johnson to change," he had said, "would be for someone to run against him." He just might do it himself, he added, if no one else would. The Finletter guests offered to set up a small headquarters for him. McCarthy seemed interested, then backed off two days later, saying, "Well, maybe the time isn't right; play it by ear; live off the land; let's see what happens."[17]

Now, evidently, the time was right. My journal, October 18, 1967: "George McGovern called yesterday and said, among other things, that he thought Eugene McCarthy had about decided to go — i.e., that he would enter the New Hampshire, Wisconsin, California and Massachusetts primaries against Johnson." I was astounded too. McCarthy, who had had a brilliant career in the House in the fifties, had been a disappointing senator, especially after the frustration of his vice presidential hopes in 1964. He was immensely intelligent and attractive but had come to seem indolent, frivolous, cynical; also unduly responsive to the legislative requests of the Minneapolis–St. Paul banking community. His concern about the war, though now intense, had come distinctly later than Kennedy's and far later than McGovern's.

My journal: "Bobby called a few moments later." He had led the fight in the Senate that day for the abolition of the Subversive Activities Control Board.[18] (Among those voting to keep that inane agency in existence was McCarthy.) In the course of conversation, I passed along McGovern's arresting bulletin. McGovern:

> Within minutes Kennedy got hold of me. And he said, "Is that true?" And I said, "Yes, it is." And he was just terribly distressed about it because what became clear then is that he desperately wanted to keep that option open. . . . He said, "He's going to get a lot of support. I

can tell you right now, he'll run very strong in these primaries. He'll run strong in New Hampshire. . . . I'm worried about you and other people making early commitments to him because . . . it would make it hard for all of us later on if we wanted to make some other move." . . . I can't recall in the conversations I had with Bob anything that so much disturbed him as McCarthy's announcement. . . . I don't think Bob in his wildest imagination ever dreamed that McCarthy would announce. . . . I think he thought, "My God, I should have done this."[19]

By the time I saw Kennedy the next evening in New York, his agitation, if McGovern had it right, had subsided. "He seemed to feel," I noted, "that it might have some use, though he is well aware that McCarthy is not particularly friendly to him. The danger, of course, is that Gene will make himself the hero and leader of the anti-war movement and cast RFK as Johnny-come-lately. The hope is that he may increase the fluidity of the situation and draw the Johnson fire, making it possible for RFK to emerge as a less divisive candidate. We shall see."[20]

III

It was true that McCarthy and Kennedy were not particularly friendly. On the other hand, they were not at that point particularly unfriendly. The legend that the Kennedys had never forgiven McCarthy for his nomination of Adlai Stevenson in 1960 was a legend. Actually John Kennedy thought McCarthy's the best speech of the convention.[21] The two men, said O'Donnell, "were never soul-mates, but the President was a very practical fellow. He saw what Gene McCarthy was up to" in 1960 — a place on the national ticket — "and [felt] Gene had as much right as he had to do it."[22] Nothing was more natural than for McCarthy to strike an alliance with Johnson who, if he beat Kennedy for the nomination, would need another northern liberal Irish Catholic as his running mate.

In 1961 President Kennedy spoke at a fund-raising dinner in Minneapolis. The next morning local Democratic dignitaries proposed to accompany him to mass. "No," Kennedy said. "I only want Gene and Abigail with me because I know damn well they always go to church." Abigail McCarthy, an intelligent and strong-willed wife, later recalled, "He was interested in where F. Scott Fitzgerald had lived and he was frank about his own casual Catholicism. He teased me about the Missal I was carrying and said, 'That seems to be the thing now — Teddy carries one around that he can hardly lift.'" (When in 1964 someone showed Abigail McCarthy a photograph of their arrival with Kennedy at the church, she surprised herself by bursting into tears.)[23] Kennedy later sent McCarthy on missions abroad — to the Vatican to find out how strongly the Church really opposed the opening to the left in Italy; to Chile for an international conference of Christian Democrats. "My relations with President Kennedy during his term," McCarthy said in later years, "were all positive and friendly. . . . There was no great feud between me and Jack Kennedy."[24]

As for Robert Kennedy, he did think it odd that McCarthy should have nominated Stevenson so resoundingly when he was really — so we all supposed — for Johnson. Perhaps for that reason, but probably more because he favored Humphrey, he would have declined, if asked, to nominate McCarthy for the Vice Presidency in 1964.[25] But the two men did not know each other well, then or later. "I do not recall," said McCarthy in 1969, "that I had more than a half-dozen conversations with him."[26] Robert's scorn for McCarthy was less political or personal than senatorial. He thought McCarthy was undependable and uninvolved, too often absent when needed for a debate or a vote, too cozy on the Finance Committee with special interests. Friends of McCarthy's did not disagree.[27]

As Galbraith, who knew all three men well, put it, the Kennedys "carried their likes and dislikes very casually; they didn't allow them to interfere with life too much. . . . I think that Bobby felt very little antagonism toward Gene. . . . Personal dislike . . . was much more strongly felt by McCarthy."[28] McCarthy was "a petulant fellow anyway, as we all know," said O'Donnell, who liked McCarthy and whom McCarthy asked in December 1967 to manage his campaign, "and . . . dreamed up all these things that the President didn't like him."[29] In fact, McCarthy, like other brilliant, lonely men, looked down on practically everybody in his own profession. He admired poets but had little use for politicians.

Moreover, McCarthy and John Kennedy, men of the same age, faith, party, heritage, had been inevitable rivals from the day in 1949 when they first met in the House of Representatives. But Kennedy was rich, eastern, secular, glamorous, while Gene McCarthy was a small-town boy from the upper midwest who attended, and later taught at, a Benedictine college and spent the year in between in a monastery. McCarthy had made it on his own and, as Galbraith said, simply thought he was intellectually "better qualified than the Kennedys. He had worked harder, studied harder, was a better economist and knew more about philosophy, poetry and theology — the elements of an educated man — than did either Bobby or the President."[30] And, if he finally gave John Kennedy grudging respect for intelligence, will and imperturbability,[31] he found Robert's exercises in intellectual self-improvement comic. He found almost everything comic. He had an elegant, original, penetrating wit. Alas, he never turned it on himself.

Yet, for all this, McCarthy probably had no greater hostility for Robert Kennedy in 1967 than he had for the rest of the Senate. In March he told James Wechsler that he would be glad to support Kennedy as the strongest candidate against Johnson.[32] In the autumn, on a couple of occasions, he urged Kennedy to run. Had Kennedy told McCarthy that he would go, I have little doubt that McCarthy at this point would have supported him. Even at a later point, after McCarthy was well into New Hampshire, O'Donnell remembers Abigail McCarthy's saying, "If Bobby'd only run, we'd get out tomorrow morning."[33]

IV

His friends disagreed as to whether Bobby ought to run.

His staff thought he should. But Kennedy, while he respected them on issues, discounted them on politics. He believed them inexperienced, emotional and, like all staff people, institutionally committed to make their boss President. His older friends from the New Frontier were more skeptical. On October 8, Pierre Salinger convened a group in New York. Edward Kennedy and Stephen Smith were there. Robert was not. The general feeling, put most forcefully by Theodore Sorensen, was that Kennedy should wait for 1972. Kenneth O'Donnell, however, said that Johnson was a bully and a coward who would run away from a fight.[34] Though O'Donnell soon came privately to believe that Kennedy *should* run in 1968, he felt that only Kennedy could make the decision and that meetings about it were a waste of time.[35] This one, in any case, O'Donnell described as "just the most inconclusive bunch of crap thrown around."[36]

Richard Goodwin, who had not been at the New York meeting, soon weighed in with a long letter to Kennedy. He began, curiously, by saying that he would advise against his trying in 1968. Then the letter turned into a powerful argument for his doing so. "I have, in fact, little doubt," Goodwin wrote, "that you can beat Johnson almost everywhere; if you really run against him and for your vision of America." He disposed of the arguments against: that Kennedy would be regarded as a spoiler, hurt his own prospects, etc. "You may well be hurt more by supporting LBJ, since you will have to say a lot of things you don't believe." Nor would it be better to wait till 1972. If Johnson were reelected, he would devote all his attention to making sure Kennedy did not succeed him. When there was a tide in the affairs of men, it should be taken at the flood. "Your prospects," he concluded, "rest on your own qualities: the less true you are to them, and the more you play the [political] game, the harder it will be. People can forgive mistakes, ambition, etc., but they never get over distrust. That's the history of the Presidency in this century."[37]

That was one side. On November 3, in another long letter, Fred Dutton, an equally valued adviser, took the opposite side. *"You are not,"* Dutton began, *"as strong with a majority of the country as your upswing in the polls and particularly your topping of President Johnson might suggest."* The polls, said Dutton, proved Johnson's weakness, not Kennedy's strength. Dutton did think it *"increasingly possible — even close to probable, I believe — that LBJ will not run next year. . . .* In domestic politics I have never thought he is a fighter at all in a personal crunch — his toughness is of the locker room or cloakroom kind. But when he gets in the ring, even when he is trying to look tough, he wants to talk things out and whine a little, not slug it out." Still, an attempt to oust an incumbent President "will emphasize — 'act out' — with the public the negative qualities which they have thought they disliked about you in the past . . . ruthlessness, self-preoccupied ambition, etc." Of course, you might "decide LBJ is so dangerous and Vietnam so bad that he must be defeated and a Democratic election blocked regardless of who the GOP nominee may be. Your plunging in might be an act of conscience to some people. But it would likely also be political suicide for you."

Dutton thought Kennedy should

> even public[ly] fraternize with LBJ a little. . . . You make a mistake to indulge your view of him, however accurate. . . . At times, you are almost compulsive in your decisions, dislikes, etc. You still are not as disciplined a politician as President Kennedy. . . . Appeal to the middle class much more! . . . You and some of the more idea-oriented ones around you — as Arthur Schlesinger, Dick Goodwin, Adam, Peter, etc. — seem to want to mix it up in issue fights now instead of preempting the early future. . . . Above all, keep cool for now. Timing separates the great public men from merely the good ones.[38]

I sent Kennedy almost as long a letter the same day — how distracted he must have been by these weighty documents! I had now concluded that Lowenstein had been right and I wrong in September. If you don't run, I wrote, McCarthy will; and will thereby "become the hero of countless Democrats across the country disturbed about the war. . . . If you were to enter at some later point, there might well be serious resentment on the ground that you were a Johnny-come-lately trying to cash in after brave Eugene McCarthy had done the real fighting. In other words, McCarthy might tie up enough in the way both of emotion and even of delegates to make another anti-LBJ candidacy impossible." As for the theory that Kennedy's entry would split the Democrats and elect Nixon, this prospect, I argued, contained its own cure. "I think that you could beat LBJ in the primaries," and, if this led the Republicans to nominate Nixon, "so much the better. He is the one Republican candidate who would reunite even a divided and embittered Democratic party."[39]

I pressed the case at Hickory Hill over the first weekend in November. Kennedy's feeling remained, I noted afterward, that "it would be a great mistake for him to challenge Johnson at this point — that it would be considered evidence of his ruthlessness, his ambition and of a personal vendetta." He acknowledged the risk that McCarthy might be successful enough to prevent the emergence of another antiwar candidate, "but feels he has no alternative but to wait." Perhaps "as McCarthy beats LBJ in primaries," state leaders might ask him to run in the interests of party unity. He was refurbishing his national contacts; "Joe Dolan has been working almost full time on this for the last ten days." However, he did not think that local organizations mattered much except for Daley's in Illinois; "it is possible to win anywhere by running against the organization."[40] One felt that the conviction his motives would be misunderstood was the essential obstacle. "If his name had been something else," O'Donnell said later, "I think Bobby Kennedy would have announced for President in 1967 and taken Johnson on without any question whatsoever."[41]

McCarthy, though he had about decided, still had not declared. "All I hope for," he told Joseph Kraft, "is that Bob doesn't throw stones on the track when I'm running out there alone. But I understand that, if it goes, he's going to want to come in." Kraft relayed this to Kennedy, who said of course he would throw no stones, adding, "Tell him not to run a one-issue campaign."[42] Then, toward the end of November, McCarthy called on Kennedy to say he was about to announce. The meeting was brief and not entirely comfortable. "I didn't ask him what he was going to do," McCarthy said later. "I just said, 'I'm not worried as to whether I'm a stalking horse for you,' meaning that if Bobby were to enter later on I would not say I'd been tricked. I left it open to him. He didn't give me any encouragement or discouragement. He just accepted what I'd said."*

Kennedy told me that during their seven minutes together McCarthy had neither disclosed his own campaign plans nor shown any

---

* As reported by William H. Honan in *New York Times Magazine*, December 10, 1968; Richard T. Stout, *People* (New York, 1970), 76–77. Kennedy entered, and McCarthy in later years did suggest he had been tricked (see Eugene McCarthy, "Kennedy's Betrayal," in *The Sixties*, ed. L. K. Obst [New York, 1977], 258–260).

curiosity about Kennedy's thoughts on organization or issues or New Hampshire, the first primary. "He is a very strange fellow," Kennedy said. "After all, I don't want to blow my own trumpet, but I have had a little experience running primaries. But he didn't ask a single question." Kennedy would have advised McCarthy to say that he did not expect to win in New Hampshire and that he was entering in order to give Democrats a chance to discuss the issues. "He should walk the streets in every town, without entourage or fanfare. He should not talk about Vietnam very much, since he has all the peace votes anyway. He should run against the organization, against the Democratic establishment, against the big shots." Kennedy doubted that Johnson would get more than 20,000 votes and thought McCarthy had a "good chance" of actually taking New Hampshire.[43]

It was one of those hopeless encounters where two men meet, not much liking one another, one too proud to seek advice, the other too proud to volunteer it. The meeting confirmed Kennedy's belief that McCarthy would be no more serious as a candidate than he had latterly been as a senator. As for McCarthy, when he made his formal announcement on November 30, a reporter asked whether Kennedy might not take over his movement. McCarthy said candidly, "He might. It would certainly be nothing illegal or contrary to American politics if he or someone else were to take advantage of whatever I might do. . . . There's no commitment from him to stand aside all the way."[44]

Kennedy had just scribbled a postscript in a letter to Anthony Lewis in London: "Washington is dreadful — but what to do?"[45]

v

December 10, 1967. RFK finally summoned a council of war today on 1968. It took place at a Sunday luncheon at Bill vanden Heuvel's. Present, in addition to Bobby and Bill, were Ted Kennedy, Ted Sorensen, Dick Goodwin, Fred Dutton, Pierre Salinger and myself.

The serious discussion began with my putting the case for his running in 1968. I said . . . that, if McCarthy did moderately well in the primaries, he might expose Johnson without establishing himself; that state leaders would understand that their own tickets would go down if LBJ were at the head of the ticket; and that Bobby should then emerge as a candidate, rescue the party and end the war in Vietnam. The dissension in the Democratic party, I said, would tempt the Republicans to nominate Nixon, who would be the easiest candidate to beat.

Dick Goodwin then argued a slightly different case for running at once — entering the primaries, assuming a McCarthy withdrawal, and carrying the fight through to the convention.

RFK then said that the talk up to this point had revolved around himself and the party. He did think another factor should be weighed — i.e., the country. . . . He was not sure whether the country or the world could survive five more years of Johnson. Whether or not he was sure of winning, was there not a case for trying in 1968?

Ted Kennedy and, later, Ted Sorensen put the case for waiting. Ted Kennedy's view was that LBJ is sure of reelection and that Bobby is sure of nomination in 1972. Dick suggested that LBJ, if reelected, would use all his wiles and powers to prevent RFK's nomination. (Bobby interjected, "He would die and make Hubert President rather than let me get it.") Ted felt that he would try this, but his capacity to do damage would be limited. To do anything now, Ted said, would be to jeopardize, if not destroy, Bobby's future. It would be an unpopular step within the party; but no one should underestimate the amount of sentiment (presumably Kennedy sentiment) in the country, and this would still be there in 1972.

Ted Sorensen thought it would be futile to go into the primaries and impossible to buck up the McCarthy effort without implicating the Kennedys in his failure. RFK added that, if Gene had any success, he wouldn't get out, no matter what he might be saying at this point. . . . Pierre, who said afterward that he agreed philosophically with me and politically with Ted S., questioned the inevitability of LBJ's reelection. He said that no one should underestimate the total alienation from the President, especially among the young. . . .

Bobby said that he was by no means sure he would be any stronger in 1972 than he is now. He would have been around five years longer, taken more positions, estranged more interests, gained more enemies. Perhaps he would never be stronger than he is today. Suppose he entered the six primaries and won them; would he then get the nomination? Ted Sorensen said no; LBJ would still get it. RFK rather disagreed. A series of primary victories, he said, would have an impact on other delegations and change the picture. There was then some discussion about filing dates, etc.

It seemed evident that Bobby is sorely tempted. He would in a way like to get into the fight, and he also is deeply fearful of what another Johnson term might do to the world. But he said that practically he did not suppose he would have much chance of getting the nomination, though he thought, if he did, he could beat Nixon. He added that people in whom he had confidence were strongly opposed to his doing anything, nor, so far as he could see, were many people in the party for it. So he supposed he would do nothing, and nothing would happen. He said this regretfully and fatalistically. But he said no final decisions had to be made for a few more weeks, and in the meantime everyone should keep on brooding.[46]

VI

They were brooding at the White House too. December 4, 1967, John P. Roche, political scientist, former ADA chairman, now a presidential assistant, to Lyndon Johnson: "Bobby Kennedy is sponsoring a 'War of Liberation' against you and your Administration. To date, however, he has kept it as 'Phase I' — random guerrilla attacks. And I have been convinced that he himself has not made up his mind on whether to move on to 'Phase II' — organization of Main Force Units. I still don't think he knows what he is doing. . . . At the risk of sounding ironic, I would suggest that Bobby is no *more* decisive than you are when torn by conflicting sentiments." [47]

December 7, Johnson to David Lilienthal, who had been drafting economic development plans for Vietnam: "All you hear in the papers is about how bloodthirsty we are, nothing but killing and blood — not a word about the kind of thing *you* are doing. . . . Bobby Kennedy was on three TV shows last week." [48] December 8, Joseph Califano to Johnson: "Bill Moyers called to tell me . . . that Shriver believed Bobby Kennedy was getting ready to run against you in 1968." [49] And so on, week after week, the President's men trying to divine the intentions of the prince across the water.

The President himself was sure Kennedy would run. His advisers doubted it. January 16, James Rowe to Johnson: "I know you do not agree, but I am convinced that Bobby Kennedy has made a political judgment that he cannot take the nomination away from you." [50] January 26, Roche to Johnson:

> He risks total *obliteration* if he goes. . . . It is almost certain disaster in 1968 and 1972 vs. a live *possibility* in 1972. . . . If he goes in 1968, he will split the Democratic Party hopelessly, virtually guaranteeing a GOP victory. And he will never be able to put the pieces together for 1972 — a significant number of us would dedicate ourselves wholeheartedly to his political destruction. . . .
>
> Does he realize this? Of course he does. He is an arrogant little *schmuck* (as we say in Brooklyn), but nobody should underestimate his intelligence. [51]

VII

For Kennedy the decision posed moral as well as political questions. He deeply believed that Johnson's war, his growing neglect of poverty and racial justice and his violent personality were disasters for the republic. Was there not a case, as he had already suggested at the December war council, for running no matter whether he won or lost? Had he not said so often that an individual could make a difference? Had he not, like his brother before him, reserved the hottest place in hell for those who remained neutral in the face of injustice? Did he not regard courage as the transcendent virtue? "All of his own convictions, all of his own statements, all of his own feelings," said Thomas Johnston, "came back to really haunt him."[52] His friend Newfield wrote bluntly in the *Village Voice*:

> If Kennedy does not run in 1968, the best side of his character will die. He will kill it every time he butchers his conscience and makes a speech for Johnson next autumn. It will die every time a kid asks him, if he is so much against the Vietnam war, how come he is putting party above principle? It will die every time a stranger quotes his own words back to him on the value of courage.[53]

Newfield was prepared for expostulation on his next meeting with Kennedy. Instead Kennedy said simply, "My wife cut out your piece. She shows it to everybody." Newfield asked what he himself thought of the article. "I understand it," Kennedy replied. ". . . On some days I even agree with it. I just have to decide now whether my running can accomplish anything. . . . I don't want to drive Johnson into doing something really crazy. I don't want it to hurt the doves in the Senate who are up for reelection. I don't want it to be interpreted in the press as just part of a personal vendetta. . . . I just don't know what to do."[54]

The moral pressure was rising. A friend was quoted in *Newsweek*: "He cannot go to South Africa and stay out of New Hampshire."[55] In early January, Murray Kempton, a long-time admirer, wrote in a rather ominous way that he preferred McCarthy. "An obvious reason is that McCarthy has the guts to go. A less obvious but more significant reason is that I was not at all surprised that he would, and I'm not the least bit surprised that Kennedy wouldn't." Jules Feiffer:

> GOOD BOBBY. We're going in there and we're killing South Vietnamese, we're killing children, we're killing women. . . . we're killing innocent people because we don't want the war fought on American soil. . . .
> BAD BOBBY. I will back the Democratic candidate in 1968. I expect that will be President Johnson.
> GOOD BOBBY. I think we're going to have a difficult time explaining this to ourselves.[56]

Kennedy spoke at a college on Long Island. In the question period a student observed that the young people had come because they thought Kennedy a man of courage and integrity. "You tell us that you are in total disagreement with what the administration is doing in Vietnam, and that you feel that a great deal more important commitment should be made . . . toward the resolution of the urban problem, and yet you tell us in the next breath that it's your intention to support the incumbent President. . . . Whatever happened to the courage and the integrity?" Said Benno Schmidt, to whom Kennedy was telling the story, "That's a tough one." "I'll say it is," said Kennedy.[57]

At least they were polite on Long Island. At Brooklyn College students hoisted a placard: BOBBY KENNEDY: HAWK, DOVE OR CHICKEN? Kennedy complained to Dolan about it. Dolan thought it rather funny. "He wasn't in a mood to laugh about it. . . . He said, 'How would you like to make a speech and have somebody get up and say, "Chicken"?' . . . He was tormented by the fact that he ought to be running. And he wasn't going against his better judgment; he was going against his instincts because his judgment was not to run."[58] James Rowe warned Johnson: "He is under constant public and private attack from his own supporters and troops. The young are calling him a 'fink.' . . . The anti-Vietnam intellectuals are now turning their tongues loose on Kennedy — he is feeling the lash of their scorn."[59]

VIII

It was an argument between the two sides of his own nature — the romantic and the realist, the guerrilla in the hills and the councilor of state. The professionals — from James Rowe to Fred Dutton and Edward Kennedy — thought his entry would be an act of hara-kiri. The intellectuals who liked to pose as professionals — from John Roche to Theodore Sorensen — thought likewise. If Robert Kennedy yielded to the temptation, Joseph Alsop wrote, as Alsop's own cousin Theodore Roosevelt had yielded in 1912, "he will destroy himself. He will destroy his party. And he will bring into power, perhaps for many years to come, the extreme right wing of the Republican Party." It would be disastrous "if the most promising and most richly equipped man of his generation in American politics allows himself to be pressured into self-destruction . . . [by] pretended friends [Schlesinger and Goodwin] who are seeking to use him for their own peculiar purposes." Fortunately, Joe said, his "astute brother" Edward and his "wise adviser" Sorensen were providing counter-pressures.[*]

Only one professional — Unruh of California — was urging him on; and this, Kennedy recognized, was for Unruh's reasons as much as for Kennedy's. Jack English was even doubtful about New York. "With an incumbent President," he observed, "these old-line leaders

* Joseph Alsop, "Can Bob Kennedy Be Pressured into a Sacrificial Candidacy?" *Washington Post*, January 16, 1968. Alsop was an old and cherished friend of mine. The friendship was angrily interrupted by the Indochina War. In later years it has been happily restored.

would go with the Presidency because somebody wrote in a book or their mother told them, 'You know the rule says you don't dump an incumbent President.'... I was worried that ... he would be stymied, hurt, and embarrassed in his own state."[60]

Most disturbing of all was the lack of support from antiwar Democrats. At Kennedy's request Sorensen said to McGovern, "Look, George, this is a very serious thing you're urging Bob to do.... What evidence do you have that people would support him if he announced?" McGovern called Gaylord Nelson, Frank Church, Quentin Burdick of North Dakota, and several midwestern governors, including Harold Hughes of Iowa. He found, he told me, "universal reluctance to consider the possibility, and certainly no one was prepared to do anything about it." "The trouble," he added, "is that everyone seems only interested in taking care of himself. This is the atmosphere Johnson has created. Every one I talked to took that view: what would this do for me? No one was ready to stick his neck out. I was very much surprised. I expected a much better reaction."[61]

Kennedy felt a particular responsibility to fellow doves up for reelection. He said to Clark, "Joe, would it murder you if I got into this thing?" "Of course not," replied Clark, always the Philadelphia gentleman; "even if it did, I would urge you to do it." But Clark would not make a commitment to support him.[62] "There weren't members of the Senate urging him," said McGovern.[63] They well knew Johnson's capacity for retaliation. "I've talked to some of my colleagues in the Senate," Kennedy told Justin Feldman in New York, "and I can't find any support." He could be responsible, he calculated, "for the loss of six Senate seats." Actually, more than six doves were up in 1968: not only McGovern and Clark, but Morse, Gruening, Fulbright, Nelson, Church. Could Kennedy require them to choose between himself and an incumbent President of their own party? "I can't do that."[64] On January 29 McGovern, Rauh and I lunched in Washington. I mentioned that the Wisconsin people, Patrick Lucey and others, did not want Kennedy to enter their primary. McGovern said, "Well, if he did, it would give Gaylord Nelson heart failure." After a detailed rundown McGovern told us, "I cannot in all conscience recommend to Bobby that he run."[65]

Moreover, Johnson *was* President. Kennedy knew the resources of the office, and he feared the desperation of the incumbent.

In the course of his ruminations he expressed a desire to consult Walter Lippmann. I took him to the Lippmann apartment for a drink on a cold January afternoon. After some chat Kennedy said rather abruptly that he had to decide about the Presidency. He set forth his concern about the war, the cities, the poor, his sense that we were drifting toward disaster, his conviction that so many things might be done. Four more years of Johnson, he said, would be a "catastrophe" for the country; so would four years of Nixon. Yet what could he do? He had no support among the politicians. And he could not see in particular how he could overcome the advantage that Johnson had through his command of foreign policy. Nothing, Kennedy thought, would restrain Johnson from manipulating the war in whatever way would help him politically — he could escalate, deescalate, pause, bomb, stall, negotiate, as the domestic political situation required. "For example," Kennedy said to Lippmann, "suppose, in the middle of the California primary, when I am attacking him on the war, he should suddenly stop the bombing and go off to Geneva to hold talks with the North Vietnamese. What do I do then? Either I call his action phony, in which case I am lining up with Ho Chi Minh, or else I have to say that all Americans should support the President in his search for peace. In either case, I am likely to lose in California."[66] Lippmann, I noted, listened intently but "did not deliver a strong argument (as he had to me on New Year's Eve) for Bobby's running, perhaps because he was impressed by Bobby's presentation of the complexities of the situation, perhaps because he does not regard it as his job to offer advice to politicians." Finally Kennedy asked directly what he thought. Lippmann said, "Well, if you believe that Johnson's reelection would be a catastrophe for the country — and I entirely agree with you on this — then, if this comes about, the question you must live with is whether you did everything you could to avert this catastrophe." On the way to the airport afterward, Kennedy said he had washed 1972 out of his calculations. Life was wholly unpredictable. He could not make decisions now in terms of consequences four years from now.[67]

Two days later, he spent a political day in Westchester County. Returning to the city, he stopped by a cocktail party. Melina Mercouri, the Greek actress, swept down and asked him in her best histrionic style whether he wanted to go down in history as the senator who waited for a safer day. The remark struck home to the student of Edith Hamilton. Afterward Kennedy said, "My feeling now is that if one more politician, on the level of Unruh, asks me to run, I'll do it. . . . What bothers me is that I'll be at the mercy of events Johnson can manipulate to his advantage. . . . None of the doves in the Senate wants me to do it, and that plays a role. . . . The politicians who know something about it, they say it can't be put together. . . . Anyway, I'd rather run than not." He warned Jean Smith and Patricia Lawford at dinner, "This is going to cost you a lot of money."[68]

IX

But the politics of running seemed dismal. McCarthy was campaigning little and doing badly. Gallup's January sounding showed Democrats preferring Johnson over his challenger by 71–18. And, as Kennedy hovered on the brink of challenge, the polls were turning against him too. Johnson now led 52–40.[69] In late January, Louis Harris personally advised him against running.[70] "Bobby is coming to be regarded as an extremist," said Richard Scammon, who, as director of the Census Bureau, had counseled the Kennedy White House in 1963. ". . . A reputation for extremism in this country is a one-way ticket to oblivion."[71] Kennedy observed bleakly to Dorothy Schiff that he had "a lot of enemies" — business, labor, the newspapers, even most of the politicians. ("Who is for you?" she asked. He said, "The young, the minorities, the Negroes and the Puerto Ricans.")[72]

Some of us questioned the infallibility of both pols and polls. Ethel Kennedy listened disgustedly to Sorensen's enormously practical recital of the case against running and said: "Why, Ted! And after all those high-flown phrases you wrote for President Kennedy!"[73] "You can never tell what the reaction of a political leader will be until you announce," O'Donnell said, "because if you say, 'Should I run?' they'll all say no. But if you announce you are a candidate and then ask them, . . . they're put on a spot. . . . You're going to get a different answer and a decisive answer."[74] I made the same point more academically: "The tendency, with the polls and so on, is to view politics as a mechanical process. In fact, it is a chemical process. I think that his entry into the situation as a candidate would transform the situation and create new possibilities not presently foreseeable."[75] But the taciturn O'Donnell rarely offered advice to Kennedy unless directly asked; and my advice, while lavishly offered, lacked the professional imprimatur. Against the massed professional judgment only amateurs — Ethel, Jean Smith, Goodwin, Schlesinger and a rabble of enthusiasts around the country — were telling Kennedy it could be put together.

Then, on January 23, the North Koreans seized the American intelligence ship *Pueblo*. On January 25 Johnson called up 14,000 reserves. It was one of those rally-around-the-flag moments, and it renewed Kennedy's apprehension about the power foreign affairs conferred on the Presidency. "The ordeal continues," I noted when I saw him that day. "I have never seen RFK so torn about anything — I do not mean visibly torn, since he preserves his wryness and equanimity through it all, but never so obviously divided." That evening we dined with the Kennedys and the Benno Schmidts at the Caravelle. Robert went through the familiar arguments — ruthlessness, personal vendetta, no political support, lack of popularity in the country, etc. He said his brother was "the strongest opponent" of his running. "Ethel was urging him to run," Schmidt said later, ". . . and Arthur Schlesinger was articulating all the arguments in favor of Bob's running, as only Arthur can do. . . . telling him that if he got out and won two or three primaries the delegates would have to come to him."

Schmidt then articulated the arguments against Bob's running, as only Benno could do — did he have the delegates? could he get them? could he beat an incumbent President with power to control events? At one point Kennedy said, "I think if I run I will go a long way toward proving everything that everybody who doesn't like me has said about me . . . that I'm just a selfish, ambitious, little SOB that can't wait to get his hands on the White House." Ethel said, "Bob, you've got to get that idea out of your head; you're always talking as though people don't like you. People do like you, and you've got to realize that." Kennedy said, smiling, "I don't know, Ethel, sometimes in moments of depression, I get the idea that there are those around who don't like me." Benno Schmidt drove the Kennedys back to United Nations Plaza. Kennedy said, "I'm convinced that what you say is right."[76]

On January 28 he asked Goodwin and me to dine with him: "We have to settle this thing one way or another." But, when we arrived, we found him with one of his characteristic mélanges — John Glenn, Mike Nichols, Rod Steiger, George and Elizabeth Stevens, heaven knows who else. Obviously we were not about to settle anything. I imagine that he did not want to go over it all again with those he had prepared to disappoint. Two days later, he said at an off-the-record breakfast at the National Press Club, "I have told friends and supporters who are urging me to run that I would not oppose Lyndon Johnson under any conceivable circumstances." Mankiewicz, hoping to salvage something from the wreckage, persuaded him to substitute "foreseeable" for "conceivable" before releasing the statement.[77]

That appeared to be that. Bad Bobby had prevailed. Adam Walinsky gave notice. He had to do something about Johnson and the war, he told Kennedy, and, if Kennedy would not run, he was going to do it some other way.[78] Dolan said he could not vote for Johnson and requested sabbatical leave for the campaign.[79] Edelman decided to do the same thing.[80] Mankiewicz, regarding life as uncertain, put his trust in the unforeseeable. Lowenstein, thoroughly discouraged after two months with McCarthy, appeared at Kennedy's office the same day and announced, "I'm an unforeseeable circumstance." He begged Kennedy to reconsider. Kennedy, reciting his problems once again, said, "It can't be put together." Lowenstein said furiously that the honor and direction of the country were at stake. "I don't give a damn whether you think it can be put together or not. . . . We're going to do it without you, and that's too bad because you could have been President." He stalked out of the office. Kennedy followed, touched him on the shoulder: "I hope you understand I want to do it, and that I know what you're doing *should* be done; but I just can't do it." "I sniffled," Lowenstein recalled, "nodded yes, and walked out, watching him stand in the doorway in real pain."[81]

I was badly disappointed. Still, few questions seem to an historian purely moral. A political leader, as Burke pointed out long ago, differed from a moral philosopher; "the latter has only the general view of society; the former, the statesman, has a number of circumstances to combine with those general ideas, and to take into his consideration. Circumstances are infinite, are infinitely combined, are variable and transient. . . . A statesman, never losing sight of principles, is to be guided by circumstances." I felt that Kennedy had not lost sight of principles and that circumstances were changing every day. At bottom, I believed his instinct would triumph over his judgment.

But around the country antiwar activists said to hell with it and set to work for Eugene McCarthy. And in Vietnam, even while Robert Kennedy was breakfasting at the National Press Club, the Viet Cong and the North Vietnamese unleashed a surprise assault on South Vietnam, convulsing thirty provincial capitals and invading the American embassy in Saigon.

38

The Decision

T HE TET OFFENSIVE changed everything. No doubt the attackers suffered grievous losses and failed in major objectives. But they destroyed what remained of Lyndon Johnson's credit with millions of Americans. "We should expect our gains of 1967," General Westmoreland had said grandly four weeks before, "to be increased manyfold in 1968."[1] Now the enemy was swarming into the very courtyard of the American embassy. Tet showed that the administration had deceived either the American people or itself about the military prospect.* It inspired the Joint Chiefs of Staff, falling back on the only remedy it knew, to request 206,000 more troops. And it led to a basic reexamination of Vietnam by the new Secretary of Defense, Clark Clifford, who turned out to be not such a hawk after all.

I

"Half a million American soldiers," Robert Kennedy told a Chicago audience on February 8, "with 700,000 Vietnamese allies, with total command of the air, total command of the sea, backed by huge resources and the most modern weapons, are unable to secure even a single city from the attacks of an enemy whose total strength is about 250,000."[2]

Adam Walinsky and I had spent the previous afternoon with him going over the speech in his New York apartment. Here and there Richard Goodwin, who had done the original draft, and Walinsky

* A textbook prepared and used at West Point calls it an "intelligence failure ranking with Pearl Harbor. . . . The North Vietnamese gained complete surprise" (Lieutenant Colonel Dave R. Palmer, *Readings in Current Military History* [West Point, 1969], 103). For this citation and further analysis, see H. Y. Schandler, *The Unmaking of a President* (Princeton, 1977), 74–77.

had proposed alternate passages. "He just took the toughest one each time," as Walinsky said later[3] — and added some of his own. He put in nothing, as he had done so often in the past, to preserve his relations with the administration. He did not, as in the past, come out against unilateral withdrawal. He condemned the deception, condemned the bombing, condemned the insensate destruction, rejected the Saigon regime as incompetent and corrupt and called for "a settlement which will give the Vietcong a chance to participate in the political life of the country."[4]

It was his most passionate Vietnam speech so far. It excited passions. Kennedy told me a day or so later that Joseph Alsop, his old friend and mine, had left a message with Angie Novello: "In the last twelve hours I have talked to three friends of his, none of them particular friends of Lyndon Johnson, and each said to me that, after that speech, they were compelled to regard Bobby Kennedy as a traitor to the United States." For a moment, Alsop was closer to the national mood than Kennedy. The new Gallup poll, taken before Tet had sunk in, showed 61 percent hawks as against 23 percent doves; 70 percent wanted to continue the bombing.[5] Kennedy said to me, "It's just like Hitler — not a very good comparison — but I mean the way people who think themselves good and decent become accomplices. Do you suppose that ten years from now we will all look back and wonder how the American people ever went so far with something so terrible?"[6]

Early in March, Daniel Ellsberg, still at the Pentagon, tipped him off about the JCS request for more troops.[7] On March 7 Kennedy rose in the Senate to demand that "before any further major step is taken in connection with the war in Vietnam, the Senate be consulted." Escalation, he said, had been our invariable response to difficulty in Vietnam — and had invariably failed. We had always claimed that victory was just ahead. "It was not in 1961 or 1962, when I was one of those who predicted that there was a light at the end of the tunnel. There was not in 1963 or 1964 or 1965 or 1966 or 1967, and there is not now." How could escalation save a government so hopelessly corrupt that its own people would not fight for it? For the first time in public he struck directly, and bitingly, at Johnson. "When this [corruption] was brought to the attention of the President, he replied that there is stealing in Beaumont, Texas. If there is stealing in Beaumont, Texas, it is not bringing about the death of American boys." The claim that we had to fight in Vietnam to protect our own security reminded him, he said, of the justifications the Germans and Russians made when they overran Poland and the Baltic states at the start of the Second World War. In his most intense passage he returned to the question of moral responsibility. "Are we," he cried, "like the God of the Old Testament that we can decide in Washington, D.C., what cities, what towns, what hamlets in Vietnam are going to be destroyed?"[8]

Though he did not publicly advocate unilateral withdrawal, he had privately reached the conclusion that the United States had to get out. Harriman, arguing with William Walton, a long-time opponent of the war, said, "Your friend Bobby is not for cut-and-run as you are." Walton repeated this to Kennedy. Kennedy said, "Little does he know."[9] To Rowland Evans, still a hawk, Kennedy said, "We've got to get out of Vietnam. We've got to get out of that war. It's destroying this country."[10] To Thomas Watson of IBM: "I'd get out of there in any possible way. I think it's an absolute disaster. I think it's much worse to be there than any of the shame or difficulty that one would engender internationally by moving out. And so, with whatever kind of apologies and with whatever kind of grace I could conjure up, I'd get out of there in six months with all the troops the United States has."[11]

II

Tet changed everything, for Eugene McCarthy as well as for Lyndon Johnson. In January his campaign had been desultory and ineffectual. "I came here," François Mitterand told Kennedy, "thinking that Senator McCarthy was a stalking horse for you, but now I think he is a stalking horse for President Johnson."[12] In February, McCarthy suddenly found himself the leader of a crusade.

Found himself, and made himself: at last he was in New Hampshire and on the hustings, and he displayed a formidable instinct for the national mood. He was willful, impervious to advice, scornful of associates, but he was also strong, cool, slashing and gallant. People flocked to him, most of all the antiwar young. These had been Kennedy's people. In 1966 Murray Kempton had called Kennedy "our only politician about whom the young care."[13] Now they were leaving him. "I'm going to lose them," Kennedy said to Joseph Dolan, "and I'm going to lose them forever."[14]

The transformation of McCarthy into a serious candidate made Kennedy's position even more intolerable than before. He could not endorse Johnson. His choices now were to endorse McCarthy, which he could not bring himself to do, to remain neutral, which he could not do either, or to enter himself. Within a week of the National Press Club breakfast he was once again in palpable indecision. "How

Stevensonian can we all get!" I noted on February 7.[15] Kennedy, McCarthy told a reporter, is "like Enoch Arden. He won't go away from the window. He just comes back and taps on the window or scratches on the door. Every time I think he's gone to sea, he's back again the next morning."[16]

Pressures redoubled. Ethel and his sisters would not give up. Jacqueline Kennedy wrote out excerpts from William Graham Sumner's scathing anti-imperialist essay "The Conquest of the United States by Spain" and sent them to him.* Pete Hamill, finishing a novel in Ireland, sent an emotional letter: "In Watts I didn't see pictures of Malcolm X or Ron Karenga on the walls. I saw pictures of JFK. That is your obligation ... the obligation of staying true to whatever it was that put those pictures on those walls."[17]

His own conscience nagged at him. On February 10 he published in the *New York Times* a piece on "the malaise of the spirit" in America. The roots of despair, he said, fed at a common source: the failure of abundance to bring happiness. The gross national product rising over $800 billion a year counted "air pollution and cigarette advertising and ambulances to clear our highways of carnage ... special locks for our doors and jails for the people who break them ... Whitman's rifle and Speck's knife and television programs which glorify violence the better to sell toys to our children." But it did not count "the health of our youth, the quality of their education or the joy of their play ... the beauty of our poetry or the strength of our marriages, the intelligence of our public debate or the integrity of our public officials.... It measures everything, in short, except that which makes life worthwhile." He wrote of the "sense that as individuals we have far too little to say or do about these issues, which have swallowed the very substance of our lives." He concluded: "We seek to recapture our country.... And that is what the 1968 elections must really be about."[18]

Here was the platform. But where was the candidate who would recapture the country? He lingered in miserable indecision. Some advisers, like Fred Dutton, were changing their minds. Even Robert McNamara concluded that Johnson was no longer capable of objective judgment. "I think Bob has to run," he told Lawrence O'Brien at the end of February. "I don't see any other answer."[19] But the two Teds — Kennedy and Sorensen — remained in determined opposition. Robert asked Goodwin to talk to his brother. After a

---

* "There is not a civilized nation in the world which does not talk about its civilizing mission just as grandly as we do.... Each nation laughs at all the others.... They are all ridiculous by virtue of these pretensions, including ourselves" (Jacqueline Kennedy to RFK, n.d. [February 29, 1968], RFK Papers).

long evening in Boston in mid-February, Goodwin and Edward Kennedy agreed that Robert's chances of winning the nomination were about one in five. They agreed too, Goodwin reported to Robert, that "clearly all your instincts push you in this direction. If you don't do it, you won't feel good about it." Ted had said, "He usually follows his own instincts and he's done damn well." Goodwin asked what John Kennedy would have advised. "I'm not so sure about that," Ted Kennedy said, "but I know what Dad would have said. . . . Don't do it."

Then: "Jack would probably have cautioned him against it, but he might have done it himself."[20]

III

Lyndon Johnson's response to the urban riots of 1967 had been to appoint a Commission on Civil Disorders, with Governor Otto Kerner of Illinois as chairman. At the end of February 1968 the Kerner Commission submitted a powerful report. It portrayed a nation "moving toward two societies, one black, one white — separate but unequal" and proposed strong and specific action to reverse the "deepening racial division." Kennedy, skeptical when the commission was established, was impressed by its report. "The White House," as Daniel Patrick Moynihan wrote, "would not receive it."[21]

"This means," Kennedy said, "that he's not going to do anything about the war and he's not going to do anything about the cities either."[22] It was now as impossible to support Johnson at home as abroad. Kennedy was moving toward the decision. He asked Fred Dutton to see whether his brother was still opposed. "I think Bob is going to run," Edward Kennedy told Dutton resignedly, "and it's up to us to make some sense out of it." On March 5 the two Kennedys met with Dutton and Kenneth O'Donnell. "I left the meeting," said Dutton, "feeling that a decision had been made. . . . I think all four of us felt that Bob probably could not win, but that he now had to try."[23]

On March 9 he went to Iowa to speak at a dinner for Harold Hughes. He spent the late evening with Hughes and the Democratic governors of Missouri, Kansas, and North Dakota. The governors criticized Johnson and doubted whether he could carry their states. Thus obliquely encouraged — no governor promised any support — Kennedy went on to California. Cesar Chavez had been fasting in penance for violence provoked by his union's struggle for survival. Now the fast was ending, and his friends wanted Kennedy to be with him at the Mass of Thanksgiving.

John Seigenthaler accompanied him on the trip west. Seigenthaler told Kennedy he should not run. Kennedy said, "I recognize the logic of everything you say. . . . But I'd feel better if I were doing what I think ought to be done and saying what I know should be said."[24] Crowds pressed in on Kennedy as he made his way from the airport into Delano. People hugged and kissed him. "His hands were scratched," Chavez recalled, "where people were trying to touch him. . . . You could see the blood." The communion bread was passed. Kennedy shared his piece with Chavez. A television cameraman said, "Senator, this is perhaps the most ridiculous remark I've ever made in my life. Would you mind giving Cesar another piece of bread so we can get a picture." Kennedy said, "No. In fact, he should have a lot of bread now." He had rehearsed Spanish phrases for his remarks, and, said Chavez, "brought the house down. . . . Well, you can imagine Spanish with a Boston accent! . . . He looked down where I was sitting, and he said, 'Am I murdering the language?' I said, 'No. Go ahead.' It was a great day."[25]

Kennedy praised Chavez as a hero of our times; then to the crowd, "Let me say to you that violence is no answer." Chavez was too weak to speak. His speech was read for him:

When we are really honest with ourselves, we must admit that our lives are all that really belong to us. So it is how we use our lives that determines what kind of men we are. It is my deepest belief that only by giving our lives do we find life. I am convinced that the truest act of courage, the strongest act of manliness, is to sacrifice ourselves for others in a totally nonviolent struggle for justice. To be a man is to suffer for others. God help us be men.[26]

Beautiful words. On the plane back from Delano, Kennedy said, "Yes. I'm going to do it. If I can, I've got to try to figure out a way to get Gene McCarthy out of it; but, if I can't, I'm going to do it anyway."[27]

That Sunday evening Ethel tracked me down in Cambridge, where I was dining with Galbraith. Jubilation and relief sounded in her voice. Bobby had telephoned from the coast, she said. He would definitely run if McCarthy could be persuaded to withdraw after New Hampshire. He wanted any thoughts on how this could be brought about.[28] It was two days before the primary.

IV

He had already asked his brother to notify McCarthy of his intention. But Ted Kennedy had "decided on his own," Robert said later, "that he didn't want to tell him in the last days of the primary

campaign."[29] Perhaps Ted thought it an unfair burden on a preoccupied candidate. Perhaps he feared McCarthy might somehow turn
the news against his brother. In any event, Robert discovered that
McCarthy had not been informed. He then called Goodwin, who
had joined McCarthy's New Hampshire campaign. Goodwin, who
felt sufficiently under suspicion as a Kennedy man, was a reluctant
intermediary. When he finally transmitted the message on Monday,
McCarthy said, "Tell him to support me. I only want one term as
President. After that, he can take it over. . . . The presidency should
be a one-term office. Then the power could be in the institution. It
would not be so dependent upon the person."[30]

Persuading McCarthy to pull out would not be easy. Since Tet
there had been a stunning rise in the intensity of his support, especially in the universities and in the liberal community. My concern
was to minimize an anti-Kennedy explosion among those who, in his
season of vacillation, had turned to McCarthy and were now falling
in love with a new hero. I consulted Joseph Rauh, who had already
endorsed McCarthy but basically thought Kennedy would make a
stronger candidate and better President. If Kennedy came out for
McCarthy in New Hampshire, Rauh said, this would make it easier
for the pro-Kennedy people around McCarthy to urge his subsequent
withdrawal. On Monday I tried this out on Kennedy, now back in
Washington. "Bobby was unwilling to do this," I noted, "— partly,
I think, because it would put him in an odd position with Dick Daley
and other professionals, partly because he cannot bring himself to
say that he would really like Eugene McCarthy to be President of the
United States. So nothing happened."[31]

New Hampshire voted the next day. Kennedy, who was speaking
that night in the Bronx, asked me to meet him later in the evening
at "21." I went to a small dinner for Anthony Eden at Hamilton Fish
Armstrong's fine old house on West Tenth Street. Bill Moyers was
another guest. Afterward, Moyers drove me uptown. Johnson, he
said, was by now well sealed off from reality; the White House atmosphere was "impenetrable." The President explained away all
criticism as based on personal or political antagonism. Moyers used
the word "paranoid." His own personal debt to Johnson was so great,
he said, that it had taken him a long time to reach this conclusion
and even longer to say it, but he felt that "four more years of Johnson
would be ruinous for the country." He added that Johnson "flees
from confrontations. He is willing to take on people like Goldwater
and Nixon, to whom he feels superior. But he does not like confrontations when he does not feel superior." I asked Moyers about
Kennedy. Moyers said firmly that he thought it would be a great

mistake for him to go — that he could not win; that he would alienate the party; that, if he did nothing, he would be the inevitable successor.[32]

At "21," Kennedy looked subdued. McCarthy had won 42.2 percent of the New Hampshire vote and, because of oddities in the selection system, 20 of 24 delegates. "RFK," I noted, "felt that McCarthy's success had boxed him in. Obviously he could not now expect Gene to withdraw." Kennedy said, "Of course he feels that he gave me my chance to make the try, that I didn't and that he has earned the right to go ahead. I can't blame him. He has done a great job in opening the situation up." I suggested that he might endorse McCarthy on the theory that McCarthy could not conceivably get the nomination but could, with Bobby's support, show how vulnerable Johnson was. Then, when McCarthy played out his hand, he would be obliged to endorse Kennedy; and with the professionals now persuaded Johnson was a loser, the nomination would inexorably go to Kennedy. He was not impressed, citing again the hypocrisy of saying, or seeming to say, that he wanted McCarthy for President. We departed, perplexed and rather dejected. I gave him McCarthy's phone number in New Hampshire. Later they had a conversation — friendly, he told me, but not productive.[33]

The next day Kennedy scribbled a letter to Anthony Lewis in London:

> The country is in such difficulty and I believe headed for even more that it almost fills one with despair. But then when I realize all of that I wonder what I should be doing. But everyone who I respect with the exception of Dick Goodwin and Arthur Schlesinger have been against my running. My basic inclination and reaction was to try, and let the future take care of itself. However the prophecies of future doom if I took this course made to me by Bob McNamara and to a less extent Bill Moyers plus the politicians' almost unanimous feeling that my running would bring about the election of Richard Nixon and many other Republican right wingers because I would so divide and split the party and that I could not possibly win — all this made me hesitate — I suppose even more than that.
>
> But the last two days have seen Rusk before the Foreign Relations Committee, the New Hampshire primary — and in the last week it has been quite clear that Johnson is also going to do nothing about the riot panel report.
>
> So once again what should I do.
>
> By the time you receive this letter, both of us will know.
>
> If I am not off in the California primary Ethel and I will be coming to Ireland at the end of May. Why don't you join us. The Irish government is dedicating a memorial to President Kennedy on his birthday May 29th — you look a little Irish and it would be good for your black soul — and maybe mine also.[34]

V

He flew to Washington. Reporters at the airport asked him what he made of New Hampshire. He said, "I am actively reassessing the possibility of whether I will run against President Johnson." In Washington he called on McCarthy. The talk, he told me that evening, had been "friendly but did not go very far." He had congratulated McCarthy on his success and told him he was reappraising his own situation. McCarthy, Kennedy said, "had accepted that, said he had a perfect right to go ahead but that he did not intend to get out himself. He then repeated the bit about serving only one term as President — this apparently is a matter of constitutional principle for him — and saying that, 'while I'm not making an offer,' RFK might be the logical successor."[35] "There wasn't very much communication," Kennedy later observed to Jack Newfield. "I told him . . . I hoped we might work together in some coordinated way. I offered to support him in Wisconsin. But he didn't respond. He just said I could do what I wanted, and he would do what he wanted. . . . I would say he was cold to us."[36] Who really could blame him? "That Bobby; he's something, isn't he?" McCarthy said to Richard Stout of *Newsweek* the next morning.[37]

In the meantime Kennedy had called a council of war at Stephen Smith's New York apartment that afternoon, presumably to assist him in the task of reassessment we had been hearing so much about. A dozen people were there when I arrived. Ted Kennedy, looking a bit unhappy, crisply set forth various alternatives — from total inaction to total participation. Sorensen proposed that Kennedy continue speaking on the issues but remain neutral as between McCarthy and Johnson. Pierre Salinger said this would make Bobby look like a political opportunist. I suggested that Kennedy enter some primaries and support McCarthy in those he did not enter. Burke Marshall felt that Kennedy had to enter all the primaries he could. He later said to Sorensen and to me: "You, Ted, don't want Bob to run against Johnson; Arthur doesn't want him to run against McCarthy. I disagree with you both."[38]

It was seven o'clock. We turned on the evening news, knowing that Robert had taped an interview earlier that day with Walter Cronkite. Discussing his reasons for reassessment, he stopped only a hairline short of a declaration. "I don't know what we are meeting about," Ted Kennedy said. "He has made all the decisions already, and we're learning about them on television." At this point I had to go off somewhere and give a speech. By the time I got back, Robert and Ethel had arrived. The evening by now was thoroughly disorganized. Even though he had all but declared, Kennedy had, I thought, "a certain air of mingled gentleness and distraction, betraying, I imagine, the deep uncertainty he feels before he jumps." Dutton and O'Donnell were especially dissatisfied. "They talk about key people in the states," O'Donnell said. "What key people? Don't they realize that the day of the organization is over? Look what McCarthy did without an organization. . . . What Bobby has to decide is who is first. Personally I think Uncle Sam is first and RFK second. Sometimes I think they reverse the order. . . . Hell, maybe it would be better to wait for 1972, when a hundred thousand Americans will be killed in Vietnam; but I don't think so. The right thing to do is for him to run and not worry about the consequences."[39] Later they went off with Salinger to Toots Shor's and drank gloomily into the night.

Kennedy asked me to take Ethel back to United Nations Plaza. He said, "Are you happy about this?" I said I was. He looked searchingly at me and said, "But you have reservations, don't you?" I said, no, I had no reservations, but he must announce his decision soon. He said he planned to do this in the next thirty-six hours. He had told Mayor Daley, he added, that he had no choice but to challenge Johnson if there were no change in the war. Daley had spoken of getting the President to appoint a commission of eminent outsiders to review Vietnam policy. Kennedy was seeing Clark Clifford about this the next morning at eleven. Until he went through that exercise, he said, he could not announce.[40]

VI

The Vietnam commission was a curious idea. Daley was an unlikely progenitor, but he disliked the war and had mentioned the idea to Kennedy in Chicago on the day in February Kennedy had blasted the Vietnam policy. Sorensen had come to the same proposal independently in March. On the Monday before the New Hampshire primary he had discussed it with the President. During the meeting in the Smith apartment on Wednesday, Sorensen, to general amusement, received a message to call the White House. Johnson, he was told, was interested and wanted a list of possible names.[41]

What Kennedy really thought of it I do not know. I suppose he would have readily traded his candidacy for a guaranteed end to American involvement in Vietnam. But it was inconceivable — or so Dutton, O'Donnell and I thought — that Johnson would surrender control over foreign policy to an outside group. We feared Kennedy

was stumbling into a trap. Kennedy may have feared that too. But
he could not appear to reject an overture that purported both to
unify the party and move toward peace. "At the last minute Ted had
boxed us in," Dolan said later. ". . . [Kennedy] gritted his teeth about
it. He was clearly unhappy about it."[42] "I had a hard time convincing
Bob," Sorensen said the next day, "that even a commission would
have enough impact."[43]

Kennedy and Sorensen met Clifford the next morning in Washington. According to Clifford's notes, Kennedy began by saying that
he regarded the Vietnam policy as "a failure, and both because of
his conscience and pressure from others, he felt compelled to take
action." One way to correct the policy would be for him to become
a candidate; the other was to persuade Johnson to change the policy. Sorensen said that "if President Johnson would agree to make
a public statement that his policy in Vietnam had proved to be in
error, and that he was appointing a group of persons to . . . come up
with a recommended course of action, then Senator Robert Kennedy
would agree not to get into the race." Clifford said at once that
Johnson could not possibly announce that his Vietnam policy was a
failure. "Kennedy agreed with this and said that he felt the statement
need not go that far." An announcement of a basic reevaluation of
policy, by a commission consisting of persons he recommended,
would suffice.*

Clifford brought up the short-lived Democratic revolt against Truman in 1948 and said that Kennedy's chances of taking the nomination away from Johnson were "zero." With his own Vietnam reexamination doubtless in mind, he added cryptically that, if Kennedy
were counting on the war to get him the nomination, he might be
"grievously disappointed" as events developed. "There were a number of factors which remained under the President's control, such as
the decision when to start negotiations." And, if Kennedy somehow
got the nomination, it would so divide the party as to be valueless.
Kennedy replied that he had considered these points "and still felt
he would have to run unless President Johnson would agree to his
proposition."[44]

Returning to Capitol Hill, Kennedy decided to talk to George
McGovern. At the moment McGovern was host for one in a series
of periodic luncheons held by a group of liberal Democrats who
had served together in the House in the 1950s — Stewart Udall, Lee

---

* Kennedy's list, according to Clifford, was Roswell Gilpatric, Carl Kaysen, Edwin
Reischauer, Mike Mansfield, John Sherman Cooper, George Aiken, Generals Lauris
Norstad and Matthew Ridgway, Kingman Brewster and himself. With the possible
exceptions of Norstad and Brewster, all were on record against further escalation of
the war (Clark Clifford, memorandum of conversation with Senator Robert F. Kennedy and Theodore C. Sorensen, March 14, 1968, Johnson Papers).

Metcalf, Frank Thompson, Eugene McCarthy and himself. This time they had planned to celebrate McCarthy's success in New Hampshire. But McCarthy had not shown up; so, when Kennedy called, they invited him instead. McGovern, who had not seen Kennedy for a few weeks, was shocked. His face was "so drawn," his wrinkles were so deep; "he looked so much older."

Perhaps he had come hoping they would urge him to run. They urged him instead not to make any immediate decisions. Thompson, Metcalf and Udall said he should campaign for McCarthy in Wisconsin, the next primary. Kennedy said, "Gene McCarthy's not competent to be President"; if it were McGovern who had gone into New Hampshire, that would have been different. ("I never knew whether he would have stuck with that or not," McGovern said later. "It's very hard for people to keep commitments like that.")

Then, in "rather an impassioned way," according to Udall, Kennedy launched into an account of the situation as he saw it — the war, the racial divisions, the cities. "You could tell his mind was already made up. . . . I almost got the feeling that it was like a Greek tragedy in the sense that events themselves had been determined by fates setting the stage, and that there was really little choice left." McGovern thought him "almost oblivious to what we were saying. . . . I realized that he wasn't even communicating with us, that he was alone with his thoughts." Udall: "I could tell he really wasn't listening to what we said because he had heard the same thing from others, and the whole thing was very clear to him . . . with all the turmoil in the country and with what he felt was a need for definition of the issues and for the championing of the people who were unchampioned. . . . He was determined to follow his own convictions and to do what was true in terms of his own personality. I sensed that, if in the end he lost, he would feel he had done the right thing. . . . He was on fire." [45]

Over at the White House Clifford was discussing the Vietnam commission with Johnson, Humphrey and Abe Fortas, now on the Supreme Court. They all agreed in rejecting the proposition because, "no matter how the arrangement was handled, it would still appear to be a political deal"; because the President's other outside advisers — his own committee of Wise Men — would feel "completely ignored"; and because Johnson could not appoint Kennedy to such a commission without causing resentment on the Hill. Clifford so informed Kennedy, who asked whether it would make any difference if he himself were not a member of the group. Clifford thought, since the main objection was the appearance of a political deal, his removal from the list would make no difference. [46]

The White House promptly leaked its side of the story. Kennedy soon responded that the incident had made it "unmistakably clear to me that so long as Lyndon B. Johnson was President our Vietnam policy would consist of only more war, more troops, more killing and more senseless destruction of the country we were supposedly there to save. That night I decided to run for President."[47] In fact he had reached that decision a week or more earlier. Only a change in Vietnam policy could have canceled it.

To many, however, it looked as if he were deciding to enter only after McCarthy had shown the way in New Hampshire. For McCarthy himself it was Enoch Arden back with a vengeance. McCarthy workers sent furious telegrams and raged at him on television: ruthless Bobby all over again. On Friday morning, March 15, of the crowded week, Kennedy, musing with Jack Newfield and Haynes Johnson: "Not to run and pretend to be for McCarthy, while trying to screw him behind his back, that's what would really be ruthless. . . . I know I won't have much support. I understand I'm going in alone. . . . It is a much more natural thing for me to run than not to run. When you start acting unnaturally, you're in trouble. . . . I'm trusting my instincts now and I feel freer."[48]

VII

The same day, the evening, Hickory Hill. Vanden Heuvel and I flew down from New York on the seven o'clock shuttle. Arthur Goldberg, now ambassador to the United Nations, was on board. We talked about Kennedy. "He thought Bobby should 'tell the truth' — say he was the best candidate and go ahead."[49]

At Hickory Hill there was, inevitably, a party, scheduled earlier for other purposes — Buchwald, Hackett, Whittaker, the mountain climber, a customary mixture of Kennedy friends. Finally, around ten-thirty, Kennedy managed to pull a group of us — Sorensen, Dutton, George Stevens, Jr., Adam Walinsky and Jeff Greenfield of his senatorial staff, and Allard Lowenstein, the mystery guest — into a side room. ("I'm completely involved in the McCarthy campaign," Lowenstein had told Stevens earlier in the evening, "but," pointing at Kennedy, "that man ought to be President.")[50] In a few moments Kennedy left to take a call from Robert Lowell, who, though he had campaigned with McCarthy in New Hampshire, now wished Kennedy luck.

Both Sorensen and Walinsky had already written announcement statements. The drafts were read and haggled over. Kennedy objected to a listing of qualifications in one of the statements: "I don't like it. It sounds too much like one of those Nixon statements — I have visited 48 foreign countries and met 40 prime ministers, 28 kings, 15 foreign secretaries. . . ." Sorensen interrupted, "And two McCarthys." Walinsky and Greenfield regarded the Sorensen draft as pontifical and righteous. They complained particularly about his concluding sentence: "At stake is not simply the leadership of our party or even our country — it is our right to moral leadership on this planet." Alas, they failed to get it out. At one point I said to Sorensen, "It gives me more pleasure than I can say to see Walinsky and Greenfield look at you the way you and Goodwin looked at Galbraith and me in 1960." Fred Dutton later recalled the scene: "Arthur Schlesinger was very cool and detached; Ted Sorensen was waiting to have the final word; Adam Walinsky was his usual high-powered self."[51]

I wish I had been cool and detached. But the outraged reaction to the prospect of a Kennedy candidacy had made me more anxious than ever about conciliating the McCarthy people. Kennedy was concerned too. Indeed, as we talked, Edward Kennedy was on his way to meet McCarthy in Green Bay, Wisconsin, with a proposal for a joint Kennedy-McCarthy effort. Together Kennedy and McCarthy could obviously amass more anti-Johnson delegates than McCarthy could by himself. With the blessing of Blair Clark, McCarthy's campaign manager, Goodwin, who was still with McCarthy, had pursued the idea with Edward Kennedy. Clark then got McCarthy to agree to a meeting that night. However McCarthy, who seemed to his wife "preoccupied and depressed," told her nothing about any meeting with Ted Kennedy and went off to bed. When Abigail McCarthy learned of Kennedy's imminent arrival, she invaded her husband's room. McCarthy told her he did not want to see Ted Kennedy. He said, as she recalled it, "'I'm going to sleep,' . . . And, incredibly enough, he did."

Clark and Goodwin arrived, leaving Kennedy at a motel nearby. There followed what Abigail McCarthy called a "bitter interlude." She trusted neither Goodwin, whose attachment to the Kennedys was notorious, nor Clark, who had gone to Harvard with John Kennedy. Finally Clark said, "Abigail, either you wake Gene up or I will." McCarthy was awakened. About two in the morning Kennedy appeared. McCarthy made it clear he wished no help in Wisconsin. Kennedy started to open his briefcase and show the plan for a co-ordinated campaign. McCarthy would not let him take it out. He sardonically proposed that Robert Kennedy enter primaries in Louisiana, Florida and West Virginia. After forty-five minutes Kennedy left. McCarthy said to his wife, "That's the way they are. When it comes down to it, they never offer anything real."[52]

In Hickory Hill we had gone to bed around one-thirty. There was confusion as to where everyone would sleep. Kennedy said, "Well, you guys figure that out. I know where I'm sleeping." Then he looked up, with that wry smile, and said, "Well, that's ruthless. I guess I can't be ruthless any more."[53] The rest of us disposed ourselves in the bedrooms of absent children. About six I was awakened by someone hoarsely whispering, "Ted. Ted." It was Edward Kennedy looking for Sorensen. I asked how it had gone with McCarthy. He said, "Very unsatisfactory. His people had seemed to be for it, but he wouldn't go along. I think his wife was against it." He put it more succinctly to vanden Heuvel, whom he woke up next in his search: "Abigail said no." (In her memoir she denied any such role.)

An hour or so later I was awakened again — this time by Robert Kennedy, wandering in rather gloomily in pajamas. We talked about the failure of his brother's mission. He then said, "What do you think I should do?" I said that, if McCarthy would not collaborate, and if a Kennedy-McCarthy contest resulted in the election of Johnson delegates, it would take some people a long time to forgive him. He said morosely, "Well, I have to say something in three hours." I said, "Why not come out for McCarthy? Every McCarthy delegate will be a potential Kennedy delegate. He can't possibly win, so you will be the certain inheritor of his support."* He looked at me stonily and said, "I can't do that. It would be too humiliating. Kennedys don't act that way." He stayed a moment longer, then left the room.

I dressed and went downstairs to find Edward Kennedy, Sorensen and vanden Heuvel at the breakfast table. The morning sun was glinting through the French windows, and sounds of small children enlivened the house, but the atmosphere around the table was somber. Sorensen said, "Have you talked to him?" I said I had. Vanden Heuvel said, "Where is he now?" Sorensen said, "He is upstairs looking for someone else to wake up in the hope of finding someone who agrees with him." All of us thought it would be disastrous for him to go ahead. Ted Kennedy raised his arms in the air: "I just can't believe it. It is too incredible. I just can't believe that we are sitting around the table discussing anything as incredible as this." Vanden Heuvel and I proposed that, as an interim measure, he come out for McCarthy. My impression was that both Teds thought this

* In an interview with the *Boston Globe* at the end of the year, McCarthy was asked what he thought would have happened if Kennedy had not gone into the primaries. McCarthy: "I think he probably would have been nominated. . . . I would have beaten Johnson in 4 or 5 primaries and he would have looked weak. And the party wouldn't have gone for me. I don't think they would have. Bobby could have come in as the unifying force, who had not challenged the president" (*Boston Globe*, December 15, 1968).

might be preferable to his declaring for himself. Then Robert entered the room, still in his pajamas. He had heard the last part of our talk. He said, "Look, fellows, I can't do that. I can't come out for McCarthy. Let's not talk about that any more. I'm going ahead, and there is no point in talking about anything else." With that he left.

I proposed taking one more look at the situation. There must be some other course besides endorsing McCarthy or running himself. Teddy said, "No. He's made up his mind. If we discuss it any longer, it will shake his confidence and put him on the defensive. He has to be at his best at this god damned press conference. So we can't talk about it any more." He was right, of course. But all I could think of was a conversation seven years before in the same house when Robert Kennedy asked me to stop worrying his brother about the Bay of Pigs.

Soon he reappeared, now half-dressed. We reviewed the statement as redrafted by Sorensen in light of the Green Bay fiasco. The final text said firmly of McCarthy, "My candidacy would not be in opposition to his, but in harmony. . . . It is important now that he achieve the largest possible majorities in the Wisconsin, Pennsylvania and Massachusetts primaries. I strongly support his effort in those states and urge all my friends to give him their votes." Robert was now in excellent humor. At one point, inserting something into the statement, he said, "It doesn't make sense without that — not that anything we are doing today makes sense anyway." Later he said, "Let's put something in about healing the wounds of the country," then added, "by splitting the Democratic party into three pieces." He went upstairs to finish dressing. A barber appeared to deal with the famous hair. Ted Kennedy said, "Cut it as close as you can. Don't pay attention to anything he says. Cut off as much as you can." Little David Kennedy gravely asked me, "Is Daddy going to run for President?"

We drove into Washington. The announcement took place in the caucus room in the Old Senate Office Building, where John Kennedy at the same age had announced his candidacy eight years before. It went off well. Our spirits began to lift. He was trusting his instincts at last, and those who loved him all felt freer. All save one. A few days later, at a New York dinner party, Jacqueline Kennedy took me aside and said, "Do you know what I think will happen to Bobby?" I said no. She said, "The same thing that happened to Jack. . . . There is so much hatred in this country, and more people hate Bobby than hated Jack. . . . I've told Bobby this, but he isn't fatalistic, like me."[54]

39

The Journey Begins

IT WAS A STARTLING adventure. Except for the scene of announcement, it bore no resemblance to John Kennedy's carefully prepared quest eight years earlier. Robert Kennedy had no campaign staff, no national organization, no delegates, almost no promises of support. At the end, all the advisers around the breakfast table at Hickory Hill, even those who had favored his running, were against his going ahead at that particular moment and in that particular way. We all expected a bad reaction. For a moment it was even worse than we had anticipated.

I

"It is difficult for me," Dwight D. Eisenhower wrote an old friend, "to see a single qualification that the man has for the Presidency. I think he is shallow, vain and untrustworthy — on top of which he is indecisive."[1] An exultant Nixon thought the Democratic split gave the Republicans a "great, historic opportunity." He did not suppose Kennedy would beat the President — "Johnson could take the war away from him in a minute — and then what?" — but he twice mentioned Johnson's "poor health" to Richard Whalen, a Nixon speechwriter (and biographer of Joseph Kennedy), and said, "Bobby may kill him." If by any chance Kennedy got the nomination, "We can beat that little S.O.B."[2]

In the White House Lyndon Johnson received gratifying word from the Hill. As reported by his congressional liaison office:

SENATOR BYRD of West Virginia. Bobby-come-lately has made a mistake. I won't even listen to him. There are many who liked his brother — as Bobby will find out — but who don't like him.

SENATOR SCOOP JACKSON. I have just issued a statement expressing
100% support for President Johnson. . . . We may find ourselves in
a 1964 Goldwater situation if Bobby tears the party apart.
SENATOR GRUENING. Bobby has hurt himself by his actions, even with
the strong feeling against the war in Vietnam. . . .
EXTRANEOUS. . . . Bobby's candidacy further complicates matters for
those liberals running for re-election. . . . This is particularly the view
expressed by Wayne Morse and Abe Ribicoff.[3]

Even Gruening, even Morse, the two brave originals of the anti-
Vietnam fight, were unwilling in March 1968 to come out against
Johnson. Legislators who took that risk paid for it. "Senator Inouye
just called from Hawaii," the legislative staff advised Johnson, "to
warn us that [Congresswoman] Patsy Mink this afternoon will announce her support of Bob Kennedy. . . . Dan thinks it is time we
started to play hard ball with Patsy — that we cut her off the notification list of contracts, etc."[4]

Organization Democrats, like Mayor Daley, automatically rallied to
the President. Harry Truman, now eighty-three years old, sent word
that he was behind Johnson 100 percent and that his challengers
were "a damned bunch of smart alecks."[5] In New York Frank
O'Connor said that Kennedy "might well be endangering the future
of the country."[6] When Kennedy marched down Fifth Avenue in
the Saint Patrick's Day parade on the day of his announcement, there
were shouts from the Irish crowd of "Go back to Boston, ya bum!"
Harold Hughes of Iowa, so eloquently inveighing against Johnson
only a few nights before, fell into a deep silence. On March 24 the
New York Times, after a survey of state delegations, reported that
Johnson appeared to have more than 65 percent of the votes at the
convention.[7]

Kennedy's respectable Washington friends were dismayed. "General Maxwell Taylor pointed out at dinner one night recently," wrote
Maxine Cheshire, the society reporter of the *Washington Post,* "that
none of the three 'old fogies' who have sons of Senator Robert F.
Kennedy named in their honor are supporting his candidacy
today."[8] Harriman and Taylor were standing by Johnson, Dillon was
for Rockefeller.

The angriest reaction came naturally enough from the supporters
of McCarthy. Not all: some, like Galbraith, Rauh, Lowenstein, who
would have been for Kennedy, felt they had gone too far with
McCarthy to switch. But they understood Kennedy's problem and
hoped to influence their own candidate toward a coordinated anti-
Johnson campaign. Kennedy understood their problem too — after
all, he had caused it — and did not press. "In politics one sees a

man in many ways," Galbraith said later. "Those of us who, in its
lottery, were this year supporting Eugene McCarthy had occasion to
know [Kennedy's] generosity."[9] Late in March rumors spread that
Lowenstein, whom McCarthy had ignored, was dropping out of the
McCarthy campaign. Kennedy encountered Lowenstein on a bus
taking Democrats back to New York City from a party dinner in
Binghamton. They talked. Lowenstein said he was sticking with
McCarthy. After he returned to his seat a note was passed to him.
It read:

> For Al, who knew the lesson of Emerson and taught it to the rest of us:
> "They did not yet see, and thousands of young men as hopeful, now
> crowding to the barriers of their careers, did not yet see if a single man
> plant himself on his convictions and then abide, the huge world will
> come round to him."
>
> From his friend, Bob Kennedy.*

But these were the exceptions. Most McCarthy supporters took
their cue from their leader. Asked about Kennedy's entry, McCarthy
discoursed sarcastically, in the manner that rejoiced his followers, on
politicians "willing to stay up on the mountain and light signal fires
. . . and dance in the light of the moon, but none of them came
down."[10] McCarthy students denounced Kennedy as a ruthless opportunist. McCarthy professors were outraged. Two American historians, Lee Benson and James Shenton, took out an advertisement
in the *New York Times:*

> The movement that has made Senator McCarthy its symbol exemplifies
> rationality, courage, morality. The movement Senator Kennedy com-
> mands exemplifies irrationality, opportunism, amorality. . . . American
> intellectuals . . . must choose between morality and amorality, between
> McCarthy and Kennedy. . . .
>   PUBLICLY. UNEQUIVOCALLY. IMMEDIATELY.†

Old newspaper friends turned violently against him. James Wechsler attacked him mercilessly in the *New York Post*. "He didn't even let
Gene and the young people all around him have a few moments
to savor their victory," said Mary McGrory. "They were bitter and

---

* Jack Newfield, *Robert Kennedy* (New York, 1969), 237. Kennedy did not get the
quotation (from "The American Scholar") quite right. It is "for the career," not "of
their careers," "the" single man "planting" himself "indomitably on his instincts, and
there abide." But it was still pretty close for a tired man bumping along in a bus at
three in the morning after a political dinner in Binghamton.

† *New York Times*, March 20, 1968. For an analysis of the statement by another
American historian, see D. H. Fischer, *Historians' Fallacies: Toward a Logic of Historical
Thought* (New York, 1970). Fischer, whose interest was not in politics but in getting
historians to talk sense, concluded: "When words are used as they are by Benson and
Shenton, they become meaningless. Their statement is not merely false — it is solemn
and literal nonsense."

wounded by what Bobby did, and so was I."[11] Most stinging of all was Murray Kempton, characterizing Kennedy in a column entitled "Senator Kennedy, Farewell" as a coward who had come

> down from the hills to shoot the wounded. He has, in the naked display of his rage at Eugene McCarthy for having survived on the lonely road he dared not walk himself, done with a single great gesture something very few public men have ever been able to do: In one day, he managed to confirm the worst things his enemies have ever said about him. We can see him now working for Joe McCarthy, tapping the phones of tax dodgers, setting a spy on Adlai Stevenson at the UN, sending good loyal Arthur Schlesinger to fall upon William Manchester in the alleys of the American Historical Association.... I blame myself, not him, for all the years he fooled me.*

II

Even those of us who had anticipated an outburst were astonished by its virulence. To this historian it seemed a curious doctrine that priority in entering a political competition conferred the moral right to be the only liberal candidate. One recalled that sixteen years before a Democratic senator with stronger liberal credentials than McCarthy beat an incumbent Democratic President in New Hampshire and then went on to win every major primary. He came to the convention as the certified people's choice and led on the first two ballots. If anyone had ever "earned" the right to exclusive liberal support, one supposed, it had been Estes Kefauver in 1952. But at the convention the Democratic organization put across a candidate who had not entered a single primary. Despite Kefauver's priority, most liberals went to Adlai Stevenson without moral qualms. Now Stevensonian liberals were complaining bitterly because Kennedy had missed a single primary. It did not make much sense.

Kennedy made one last attempt to explain himself on *Meet the Press* the day after his announcement. "I don't think there is any question," he said, "that if I had gone into the race at an earlier time that it would have been felt by the press and by others that this was a personality struggle between President Johnson and myself.... What Senator McCarthy showed was that there was a deep division

---

* Murray Kempton, "Senator Kennedy, Farewell," *New York Post*, March 26, 1968. On June 11, 1968, Kempton wrote: "The language of dismissal becomes horrible once you recognize the shadow of death over every public man. For I had forgotten, from being bitter about a temporary course of his, how much I liked Senator Kennedy and how much he needed to know he was liked. Now that there is in life no road at whose turning we could meet again, the memory of having forgotten that will always make me sad and indefinitely make me ashamed" ("RFK — In Sorrow and Shame," *New York Post*, June 11, 1968).

and split within the Democratic Party and the country that had nothing to do with me." Five times he declared his intention "to cooperate in every way possible with Senator McCarthy."[12] Thereafter he dropped the subject and moved ahead. That night he flew to Kansas, where he had an old commitment to deliver the Alfred M. Landon Lecture at Kansas State University.

Conservative Kansas presumably backed the war and detested the rebel young. But, when he changed planes at Kansas City, Missouri, a cheering crowd broke through the barriers to get near him. Later that night, at Topeka, "they tore at Robert Kennedy," wrote Jimmy Breslin. "They tore the buttons from his shirt-cuffs. . . . They tore at his suit-buttons. They reached for his hair and his face. He went down the fence, hands out, his body swaying backwards so that they could not claw him in the face, and the people on the other side of the fence grabbed his hands and tried to pull him to them."[13]

The next morning, his hands trembling and his voice flat with nervousness, Kennedy gave the Landon Lecture before a crowd of fifteen thousand in the Kansas State fieldhouse in Manhattan. These were not eastern kids with long hair and beards, Breslin noted; they were "Kansas young with scrubbed faces and haircuts and ties."[14] Perhaps, Kennedy began, it had been a mistake to announce the "reassessment" of his position. "Yesterday there was a man from the Internal Revenue Service out reassessing my home." The stories about the Vietnam commission had misrepresented the differences between himself and Johnson. "The only difference was the makeup. I wanted Senators Mansfield, Fulbright and Morse, and the President, in his own inimitable way, he wanted General Westmoreland, John Wayne and Martha Raye."* Laughter; applause; then down to business.

"Every night," Kennedy said, "we watch horror on the evening news. Violence spreads inexorably across the nation, filling our streets and crippling our lives." The administration had no answer to the war — "none but the ever-expanding use of military force . . . in a conflict where military force has failed to solve anything." He jabbed the air with clenched fist, his voice intense and controlled. "Can we ordain to ourselves the awful majesty of God — to decide what cities and villages are to be destroyed, who will live and who will die, and who will join the refugees wandering in a desert of our own creation? . . . In these next eight months," he concluded, "we are going to decide what this country will stand for — and what

---

* Jules Witcover, *85 Days: The Last Campaign of Robert Kennedy* (New York, 1969), 101. Martha Raye had probably spent more time entertaining the troops in Vietnam than anyone except Bob Hope.

kind of men we are."[15] The Kansas State students, scrubbed faces and all, went wild. "The fieldhouse," wrote Jack Newfield, "sounded as though it was inside Niagara Falls; it was like a soundtrack gone haywire."[16] The reception was, if possible, even wilder in the afternoon at the University of Kansas. That evening on the plane east Kennedy said to Breslin, "You can hear the fabric ripping. If we don't get out of this war, I don't know what these young people are going to do. There's going to be no way to talk to them. It's very dangerous."[17]

Three days later he went to Alabama, where he spoke about racial justice (and where, when asked if he would accept second place on the ticket with Johnson, he replied, "I said I was for a coalition government in Saigon, not here").[18] Before March was over, he had visited sixteen states. Frenzy accompanied him everywhere. On television the tumultuous crowds frightened people for whom the pictures on the screen evoked the riots of the summer before. Reporters too were alarmed by the crush and hysteria. Some were also repelled by what they regarded as Kennedy's playing on his audiences by rhetorical excess. They had a point. He had got it into his head that the Saigon government was not drafting its eighteen-year-olds and continued saying this after Saigon had changed its policy. "When we are told to forego all dissent and division," he said at Nashville, in obvious reference to Johnson, "we must ask: who is truly dividing the country? It is not those who call for change, it is those who make present policy." That policy, Kennedy said, had driven young people to abandon their "public commitment of a few years ago for lives of disengagement and despair, turning on with drugs and turning off America." Kennedy, sniffed Richard Harwood of the *Washington Post*, "implied that the President is to blame for the alienation and drug addiction among American youth, for rebelliousness and draft resistance on American campuses, and for the 'anarchists' and rioters in American cities."[19]

Harwood was no doubt a little strong. So perhaps was Kennedy. I wrote him after his Nashville attack on Johnson, "It is a little early in the campaign for that. Let him get personal first."[20] One source of Kennedy's belligerence was what Mankiewicz called his "free-at-last syndrome."[21] The bonds that had so long repressed his concern about the country and his contempt for its leader were thrown off. He went on to California and said in Los Angeles, before one more screaming crowd, "The national leadership is calling upon the darker impulses of the American spirit — not, perhaps, deliberately, but through its action and the example it sets." The press agitation over this phrase — unfairly blamed on Walinsky; it had actually come

from material sent along by Goodwin — displayed the pomposity
that on occasion afflicted even the most reasonable of newspapermen. Of course Johnson and his Vietnam policy had called upon the
darker impulses of the American spirit. Who could have doubted
it? The very day Kennedy announced his candidacy Lieutenant
William Calley and Company C had yielded to precisely such impulses
in My Lai. But the statesmen of the traveling press were outraged.
Their word for Kennedy, Harwood wrote in an influential piece at
the end of the month, was "demagogue."[22]

III

Harwood suggested that the crowds were deliberately fomented as
part of "a strategy of revolution, of a popular uprising of such intensity and scale" that the convention would not dare turn Kennedy
down.[23] There was something to that. A candidate without organization or delegates had no choice but to demonstrate irresistible
popular appeal. "Our strategy," Walinsky said, "is to change the
rules of nominating a President. We're going to do it a new way. In
the streets."[24] "I have to win through the people," Kennedy told
Helen Dudar of the *New York Post*. "Otherwise I'm not going to
win."[25]

Jimmy Breslin described a day on the California trip. Kennedy
talked about Vietnam. "Our brave young men are dying in the
swamps of Southeast Asia," he said. "Which of them might have
written a poem? Which of them might have cured cancer? Which
of them might have played in a World Series or given us the gift of
laughter from the stage or helped build a bridge or a university?
Which of them would have taught a child to read? It is our responsibility to let these men live. . . . It is indecent if they die because of
the empty vanity of their country." His listeners "shrieked," wrote
Breslin. ". . . They lost control and began pushing forward. . . . It
took a half hour to get Kennedy out of the place. A half hour of
police pushing and the crowd pushing back and Kennedy trying to
smile while they pulled his hair and scratched his face. Women
screamed that their children were being crushed to death. Kennedy
had to pick up a small child who fell down between policemen."[26]

It went that way all day. At the end Breslin asked where Kennedy
thought he stood. "What about Daley?" "He's been very nice to me
personally," Kennedy said. "And he doesn't like the war. You see,
there are so many dead starting to come back, it bothers him." But
it's hard for him, Kennedy continued. "He has been a politician for
a long time. And party allegiance means so much to him. It's a

wrenching thing for him. We'll have to win the primaries to show the pols." "If you get Daley," he was asked, "where do you stand?" Kennedy said, "Daley means the ball game."[27]

And he had to have the crowds to win the primaries. But there was, I believe, more to it than that. He had gone ahead that Saturday morning at Hickory Hill surrounded by advisers exuding gloom. He had thereafter been savagely attacked, in many cases by people he respected and liked. Now the crowds reassured and sustained him. He told one associate, "I'm beginning to feel the mood of the country and the people and what they want."[28] He did not mind the tidal wave surging over him. Let people seize his cuff links, grab his hands, reach out to touch him. "They *loved* to touch his hair for some reason," said a California advance man. "He loved it. He seemed to sort of thrive on touching people back."[29] Bill Barry, his old friend from the FBI, took leave from his job as a bank vice president to travel with Kennedy. At the beginning, looking at his own and Kennedy's bleeding hands, he said, "I wish these people would be more courteous." Kennedy said, "They're here because they care for us and want to show us." "After that," Barry said later, "I never had any trouble adjusting to crowds. I found they wanted not just to touch a celebrity; they wanted to convey their feelings to him, and he accepted it for that."[30] Alan King said, "They're going to hurt you." Kennedy said mildly, "Well, so many people hate me that I've got to give the people that love me a chance to get at me."[31]

On March 23 the Gallup poll showed Democrats preferring Kennedy to Johnson by 44–41 (and Johnson to McCarthy by 59–29).[32] After a canvass of prospective delegations, *Newsweek* concluded at the end of the month that Lyndon Johnson "may be in real danger of being dumped by his own party."[33]

IV

Johnson had not been surprised by Kennedy's announcement. He wrote grimly in his memoirs, "I had been expecting it."[34] He had always known that, as in the classic Hollywood western, there would be the inevitable walkdown through the long silent street at high noon, and Robert Kennedy would be waiting for him.

For a moment he responded with bravado. Friends urged him on. Russell Wiggins of the *Washington Post* called his attention to passages in Richard Whalen's *Founding Father* displaying Joseph Kennedy as appeaser and defeatist: like father, like son. "One shouldn't carry the analogy too far," said Harry McPherson, transmitting Wiggins's message, "but as Joe once said, 'Bobby and I think alike.'"[35] J. Edgar Hoover kept the White House informed, reporting, for example, that Kennedy had tried to call Martin Luther King before announcing his candidacy.[36] On March 17, the day after Kennedy's announcement, Johnson flew to Minneapolis, pounded the podium and shouted about the war, "Make no mistake about it ... we are going to win." Two days later, addressing the National Foreign Policy Conference at the State Department, "We are the Number One Nation. And we are going to stay the Number One Nation."[37] But Tet had indeed changed everything. "I am shocked by the number of calls I have received today in protest against your Minneapolis speech," James Rowe wrote him. "Our people on the firing line in Wisconsin said it hurt us badly. . . . Everyone has turned into a dove."[38] It was true: Democratic leaders across the country; the press; even, on March 26, most of his own private Vietnam commission, the Wise Men, Acheson, Dillon, McGeorge Bundy, Cyrus Vance — all now abandoning the war.

On the surface an unaccustomed benignity suddenly overtook the President. He had been confiding to senior advisers since at least the previous October that he was "inclined" not to run in 1968.[39] He talked about this a good deal more now. In mid-March he asked McPherson to give three reasons why he should run. When McPherson said that no one else could get a program through Congress, Johnson said, "Wrong. Any one of 'em — Nixon, McCarthy, Kennedy — could get a program through next year better than I could. . . . Congress and I are like an old man and woman who've lived together for a hundred years. . . . We're tired of each other."[40] On March 27, he asked who Joseph Califano thought would get the nomination if he withdrew. Expecting an explosion, Califano said Kennedy. Johnson did not explode. "What's wrong with Bobby?" he said. "He's made some nasty speeches about me, but he's never had to sit here. Anyway, you seem to like his parties." Califano smiled nervously. The President continued: "Bobby would keep fighting for the Great Society programs. And when he sat in this chair he might have a different view on the war."[41]

Within, Johnson was in turmoil. He doubted the loyalty of his own government. There was, wrote his sycophantic press secretary George Christian (after Salinger, Reedy and Moyers, Johnson was taking no chances; he brought in a man who had worked for Price Daniel and John Connally in Texas), "a decidedly anti-Johnson tinge to what was called the Johnson Administration, and it was no secret to anyone."[42] It was a secret to Kennedy and McCarthy, but no matter; it was what the White House believed. On March 17 Lady Bird Johnson wrote in her diary, "I have a growing feeling of Prometheus Bound, just as though we were lying there on the rock, exposed to the vultures, and restrained from fighting back."[43]

A nightmare of paralysis had pursued Johnson since childhood. "I did not fear death so much," he later wrote, "as I feared disability. Whenever I walked through the Red Room and saw the portrait of Woodrow Wilson hanging there, I thought of him stretched out upstairs in the White House, powerless to move, with the machinery of the American government in disarray around him. And I remembered Grandmother Johnson, who had had a stroke and stayed in a wheelchair throughout my childhood, unable even to move her hands or to speak so that she could be understood."[44] Sometimes, he would dream that he was lying in his bed with his own head but with Wilson's shriveled body. Awakening in a cold sweat, he would take a flashlight, walk downstairs, touch Wilson's portrait; then, soothed, he could sleep again.[45]

Another recurrent dream went back to Grandfather Johnson's tales of cattle stampedes in the old southwest. He was alone on a vast plain, sitting in a tall, straight chair, cattle storming down on him. He tried to move but could not. "I felt that I was being chased on all sides by a giant stampede," he told Doris Kearns; ". . . I was being forced over the edge by rioting blacks, demonstrating students, marching welfare mothers, squawking professors, and hysterical reporters. And then the final straw. *The thing I feared from the first day of my Presidency was actually coming true. Robert Kennedy had openly announced his intention to reclaim the throne in the memory of his brother.* And the American people, swayed by the magic of the name, were dancing in the streets."[46]

Time was drawing short. He felt that the country would be divided and angry so long as he remained in the White House. He had, as his protégé Bobby Baker put it, a "deep fear of defeat." Lawrence O'Brien told him that McCarthy would beat him two to one in the Wisconsin primary on April 2.[47] If he waited till after Wisconsin, he would be a repudiated President, driven from public life by the voters of his own party. This was his last chance to make withdrawal appear his own decision rather than one forced upon him. At nine o'clock on the evening of March 31, Lyndon Johnson went on television ostensibly to address the nation on the war.

v

Robert Kennedy spent March 30 conducting a hearing of his Indian subcommittee in Flagstaff, Arizona. The next day he flew back to New York. He had asked Bill Barry to find him something to read.

Barry came up with *Alone*, Richard E. Byrd's memoir of a winter in Antarctica. Ethel said, "That's some book for him to read! . . . How's that going to relax him — *Alone?* "[48] When the plane landed shortly before ten o'clock at La Guardia, an ashen-faced John Burns rushed aboard. "The President is not going to run," he said. "You're kidding," said Kennedy. On the way to United Nations Plaza he was sunk in reverie. Finally he said to Richard Dougherty of the *Los Angeles Times*, "I wonder if he'd have done this if I hadn't come in."[49]

The mood when I arrived at the apartment a few minutes later was total bemusement. Kennedy looked terribly tired. Soon he was on the telephone, calling leaders around the country. A telegram was dispatched to Johnson proposing a meeting. Sorensen was at work on a statement for Kennedy's press conference the next morning; so, inevitably, was Walinsky. Around two in the morning, they were both read to Kennedy, who wearily asked me to "put them together." "Like old times," I noted. "I took the first page of Ted's, agreeing with Adam that stale rhetoric about 'we want neither the peace of the slave or the grave' should go (Adam's statement, however, was filled with equally stale rhetoric about there being more which unites us than divides us), and then added a page of my own."[50]

No one becomes more sentimental than a politician rejoicing in the withdrawal of a rival. Kennedy called Johnson's action "truly magnanimous."[51] McCarthy said Johnson deserved the "honor and respect of every citizen."[52] Johnson himself was less sentimental. When Abigail McCarthy impulsively called the White House after the broadcast, she was offended by the "note of suppressed triumph" in his voice — "the voice of a man who operated in the supreme confidence that he could outmaneuver anyone."[53] On April 2 McCarthy took 56 percent of the vote in Wisconsin; Johnson received hardly more than a third. His mood changed. He said angrily to reporters, in the style of Richard Nixon, "You fellows won't have me to pick on any more. You can find someone else to flog and insult."[54]

Told of Kennedy's request for a meeting, Johnson had said, "I won't bother answering that grand-standing little runt."[55] Nevertheless, the meeting on April 3 was friendly enough. Sorensen accompanied Kennedy; Walt Rostow and Charles Murphy were with Johnson. Johnson delivered a monologue about Vietnam, the Middle East, the budget. Kennedy inquired about Johnson's intentions in the campaign. Johnson said he planned to stay out of it but would let Kennedy know if he changed his mind. The President then began talking about John Kennedy, how well they had worked together, how he had tried to continue the Kennedy policies. But he hadn't succeeded: the young were disaffected despite all he had done for education; so were the blacks, despite all he had done for civil rights. Still, "as President Kennedy looked down at him every day from then until now, he would agree that he kept the faith." Kennedy finally said, "You are a brave and dedicated man." "I don't know," recalled Sorensen, "whether it was because he found it difficult to say or whether the emotion of the situation had overcome him, but it sort of stuck in his throat, and Johnson asked him to repeat it."[56]

When they left, Humphrey, who had been waiting in an anteroom, was ushered in. Johnson told his Vice President, according to the record of the meeting, that he had withdrawn because he "simply could not function on these great issues if he were subjected every day to attacks from Nixon, McCarthy, and Robert Kennedy." He did not ask Humphrey to run, though this assumption seemed to Humphrey to underlie the conversation.[57] Humphrey said he had considered the possibility of an immediate announcement but feared it might "demean" the President's statement. Johnson said that was up to him. He would rate Humphrey as A++ as a Vice President while he would rate himself as only B+. But, if Humphrey ran, "he must do a better job than he was able to get organized in Milwaukee" — a spiteful allusion to the Wisconsin result, for which the President evidently wished to hold Humphrey responsible. Humphrey said he had been in preliminary contact with Daley, Governor Richard Hughes of New Jersey and other party leaders. "They appear not yet to have made up their minds. He had the impression that they were not willing to be 'blitzed.'" Johnson said cruelly that "he thought it possible that, in the end, Daley and Hughes would go with Kennedy."[58]

Later in the day the President relaxed with old friends, Drew Pearson and David Karr. He was filled with self-pity. He kept talking about his "partnership" with John Kennedy. "Then my partner died, and I took over the partnership. I kept on the eleven cowhands [the cabinet]. Some of the tenderfeet left me. But I kept on." He repeated the line about John Kennedy's putative approval as he looked down from heaven. He was bitter about Robert Kennedy, blaming him for the Bay of Pigs. This, Johnson said, was where the credibility gap started, not over Vietnam. He went on and on. Karr found the performance "terrifying."[59]

VI

McGovern, Dutton, O'Donnell, Vance Hartke and others had predicted that Johnson lacked the guts to stay the course. One never quite believed them, but they were right, and now he was gone, a giant vanishing in a puff of smoke. Kennedy and McCarthy, instead of running against Johnson, were left to run against each other.

"It's narrowed down to Bobby and me," McCarthy told Jeremy Larner, his speechwriter, and Jonathan Schell, another gifted young journalist, the day after the Wisconsin primary. "So far he's run *with* the ghost of his brother. Now we're going to make him run *against* it. It's purely Greek: he either has to kill him or be killed by him. We'll make him run against Jack. . . . And I'm Jack." It was a Delphic thought. His audience was puzzled. McCarthy leaned back and laughed. Later Schell asked, "Did you understand that?" "Half," said Larner, who thought it fascinating stuff from a politician. "Well, I didn't," said Schell. Unwilling to work against Kennedy, Schell declined McCarthy's invitation to join his staff and returned to graduate studies at Harvard. Larner kept thinking about what McCarthy had said. "I still got only half, and there wasn't any more, then or later."[60] McCarthy may have meant that Robert Kennedy, having run against Lyndon Johnson's record, would be forced, with Johnson out, to run against aspects of John Kennedy's record. But McCarthy's "and I'm Jack" is, by this interpretation, total mystification. Perhaps McCarthy meant that he had more of John Kennedy's qualities — maturity, intellectuality, urbanity, control — than Robert had. Perhaps he meant nothing at all.

With Humphrey still waiting for a nondemeaning moment to announce, the visible drama lay in the series of contests between Kennedy and McCarthy, beginning in Indiana on May 7 and ending in New York on June 18. The Kennedy organization was taking shape. Edward Kennedy was his brother's closest adviser, always resourceful, energetic and protective. Stephen Smith, Robert told me, would be campaign manager, Sorensen, campaign director, O'Donnell, director of organization. I asked what in the world the difference was between campaign manager, campaign director and director of organization. "I don't know," he said, "but I've never put much stock in titles. There'll be enough for everyone to do."[61]

Salinger and Mankiewicz had the press. Goodwin was in charge of television. He had remained with McCarthy through Wisconsin. They parted on friendly terms. "I understood his motives and his loyalties," McCarthy said later. "And Dick was above board with me about them. Goodwin is like a professional ballplayer. You could trade him . . . and he wouldn't give away your signals to the other team."[62] Fred Dutton traveled with the candidate, offering frank counsel to Kennedy and placating Adam Walinsky and Jeff Greenfield, who, like all speechwriters, were in a state of periodic disgruntlement.* Then Lawrence O'Brien, after declining to run Hum-

---

* Greenfield also had a draft problem. He had warned Kennedy that, if his notification arrived, he would refuse to go, supposing that Kennedy would say that, in the

phrey's campaign, resigned as Postmaster General in April to join
Kennedy's, where he expected, and found, a leading role. David
Hackett headed the 'boiler room' operation as he had in 1960. Hackett brought in Fraser Barron from the poverty program for 'grassroots development' — the organization of the poor, of blacks, Mexican Americans, white Appalachians.

"There's danger, of course," Kennedy had told me, "in just using
people from 1960. Politics has changed a lot in the last eight
years."[63] Nevertheless, his seemed at the start the old politics of
motorcades, rally speeches and political organizations. McCarthy's
strength, as Goodwin said a day or two after he rejoined us, lay
in his understanding of the new politics of television and the
kids; Goodwin thought that the Kennedy people greatly underrated
McCarthy's seriousness and his political acuity.[64] Dutton and I shared
the fear that we were getting mired in the past. In early April I
circulated a memorandum to that effect called "The Old Politics and
the New." The post-1960 class felt this even more strongly. "The
classical political wisdom which is shaping this campaign," Thomas
Johnston soon wrote Kennedy, "is similar in all important essentials
to the advice which said you should not run this year." Your decision to run "was made by you, on your own, acting against this
advice. . . . You are at your strongest when you are most yourself. . . . The ultimate source of your political strength is your capacity to fire and shape the moral imagination of this country."[65]

Actually, the generational clash was overplayed by the press. Dutton later thought that the prevailing disorganization caused no serious troubles in the primaries.[66] I am sure he was right. Kennedy
himself arrived at a unique blend of the old and the new politics —
and both, in fact, were necessary in 1968. After California he intended to reorganize the campaign and place Stephen Smith in full
charge. For the time being there was too much else to do.

VII

With Johnson's withdrawal, Kennedy and McCarthy had lost their
most conspicuous issues: the unpopular President and, to some degree, the increasingly unpopular war, for Johnson had also on March

circumstances, he had better leave the staff. Kennedy said instead, "Well, Jeff, you
know if you go to jail, I'll see that you get treated right" — after all, he had once had
some influence over the prison system — "and besides, don't worry about it. A lot of
the greatest men in history have begun their careers by spending time in jails" (Peter
Edelman, in recorded interview by L. J. Hackman, July 15, 1969, 83, RFK Oral History
Program). A Kennedy aide said to David Halberstam of Greenfield, "That kid gets
his draft notice and we're the only campaign in town with a speech writer in Canada"
(Halberstam, "Travels with Bobby Kennedy," *Harper's*, July 1968).

31 abandoned major escalation and gestured toward negotiation. McCarthy affected to take it calmly and, for all I know, did. "Bobby has to shoot straight pool now," he told reporters, thereby deflecting attention to his rival. "When he was banking his shots off Lyndon it was a different game."[67] McCarthy's jabs often hit home. One felt a certain letdown in Kennedy, though of course he was tired after his transcontinental fortnight. (I told him he should begin to pace his campaign. He bridled a little and said, "I know I look tired, but I'm all right. I know the limits of my strength very well. There is no need to worry about that.")[68] He had enjoyed the quest. Now, in two weeks, the dragon was slain. There was, for a moment, a loss of steam and of theme.

Yet a theme remained — the theme that, along with the war, had absorbed him most in the Senate. For, more than anyone else in American politics, he had become the tribune of the underclass, the leader determined "to show," as he said, "that the individual *does* count in a society where he actually appears to count less and less,"[69] determined to overcome the alienations of American society, to bind the wounds of American life. As soon as he became a candidate he had reaffirmed this theme. "We are more divided now than perhaps we have been in a hundred years," he said on March 17. The great need was "to heal the deep divisions that exist between races, between age groups and on the war."[70] Now that he and McCarthy together had moderated the Vietnam policy and driven Johnson into retirement, he was free to move ahead where McCarthy could not easily follow — toward a coalition of the poor and powerless in the battle to bring the excluded groups into the national community. "I've got every establishment in America against me," he said on April 2.[71] "I want to work for all who are not represented," he told Charles Evers. "I want to be their President."[72]

A crucial component of any coalition would be the United Auto Workers. When Roy Reuther died in January 1968, Kennedy was the only one outside the family to sit with the Reuthers at the funeral. Victor Reuther (whose wife was on the Kennedy delegation in the District of Columbia), Leonard Woodcock, Douglas Fraser, Jack Conway, Paul Schrade and other UAW leaders worked for Kennedy in Indiana, Michigan and California. Walter Reuther delayed his decision because of an old friendship with Humphrey, but Victor was sure he would have supported Kennedy in the end.[73]

Another crucial figure in any coalition of the disestablished was Martin Luther King. Though Kennedy and King had kept their distance, events were bringing them closer together. In the spring of 1967 King had decided to oppose the war. In the summer the two men began collaboration, through intermediaries, on a new drive for economic and racial justice. Chatting with Marian Wright and Peter Edelman beside the pool at Hickory Hill, Kennedy had remarked, "The only way there's going to be change is if it's more uncomfortable for the Congress not to act than it is for them to act. . . . You've got to get a whole lot of poor people who just come to Washington and stay here until . . . Congress gets really embarrassed and they have to act." The next week Marian Wright presented the idea to King at a Southern Christian Leadership Conference retreat. This was the origin of the Poor People's Campaign of 1968. The Kennedy office was now working closely with Marian Wright and the organizers.*

When Johnson pulled out, King said to Walter Fauntroy, head of the SCLC's Washington office, "He's just doing like a Baptist preacher . . . you know, trying to get a vote of confidence. He'll pull back in later. But this country's through with him." King, Fauntroy recalled, was "*very* hopeful" that Kennedy would make it. He said, "We've got to get behind Bobby now that he's in."[74] Peter Edelman, citing Marian Wright: "King was prepared to endorse him."[75] Stanley Levison, the target of the wiretaps: "He said that while he hadn't publicly decided to take any stands yet, his mind was made up. He had decided that he would support Bobby Kennedy. . . . He felt that if he'd come this far, with the greater responsibility he could become one of the outstanding presidents. . . . No question: if he had lived, he would have supported Bobby Kennedy."[76]

On April 4 Kennedy began the Indiana campaign. He was scheduled in the evening to speak in the heart of the Indianapolis ghetto. Walter Sheridan and John Lewis had set up the meeting — John Lewis, the Freedom Rider, the SNCC chairman who had asked at the March on Washington which side the federal government was on but who had "started identifying" with Kennedy in later years as "the only political leader" addressing the "real issues of the United States" and who had offered his services as soon as Kennedy announced.[77] They had decided, Sheridan recalled, to put Kennedy "not only into the black community, but into the worst section of the black community." The Indianapolis mayor thought it dangerous;

---

* Peter Edelman, in recorded interview by L. J. Hackman, August 5, 1969, 331–333, RFK Oral History Program; Nick Kotz, *Let Them Eat Promises* (New York: Doubleday, Anchor reprint, 1971), 147, 161–165. Ten years later Andrew Young said, "I think now that Dr. King's assassination was directly related to the fear that officialdom had of his bringing large numbers of poor people to the nation's capital, demanding some response from them. . . . [At the time] I didn't see the Poor People's Campaign as the threat to Washington and the Establishment that I now see it was" (as interviewed in L. K. Obst, *The Sixties* [New York, 1977], 232, 236).

but, said Sheridan, "we had no real fears that there was going to be any problem."[78] In the afternoon Kennedy spoke at Muncie, where one of the last questions had come from a young black wondering whether Kennedy's apparent belief in the good faith of white people toward minorities was justified. Kennedy had said he thought it was. A few moments later, as they boarded the plane for Indianapolis, Pierre Salinger telephoned that Martin Luther King had been shot in Memphis. Perhaps they had better cancel the Indianapolis rally.[79]

Kennedy, on the plane, said to John J. Lindsay of *Newsweek,* "You know, it grieves me . . . that I just told that kid this and then walk out and find that some white man has just shot their spiritual leader." Soon they arrived in Indianapolis. Worse news: King was dead. Kennedy "seemed to shrink back," Lindsay thought, "as though struck physically." He put his hands to his face: "Oh, God. When is this violence going to stop?"[80] The chief of police warned the party not to go into the ghetto; he would not be responsible for anything that might happen.[81] Kennedy sent Ethel on to the hotel but was determined to keep his rendezvous. In the automobile he sat wrapped in thought. As his car entered the ghetto, the police escort left him.[82]

It was a cold, windy evening. People had been waiting in the street for an hour but were in a festive, political-rally mood. They had not heard about King. Kennedy climbed onto a flatbed truck in a parking lot under a stand of oak trees. The wind blew smoke and dust through the gleam of the spotlights.* "He was up there," said Charles Quinn, a television correspondent, "hunched in his black overcoat, his face gaunt and distressed and full of anguish."[83] He said, "I have bad news for you, for all of our fellow citizens, and people who love peace all over the world, and that is that Martin Luther King was shot and killed tonight." There was a terrible gasp from the crowd.

Robert Kennedy, speaking out of the somber silence of the ride from the airport, speaking out of aching memory, speaking out of the depth of heart and hope:

> Martin Luther King dedicated his life to love and to justice for his fellow human beings, and he died because of that effort.
> In this difficult day, in this difficult time for the United States, it is perhaps well to ask what kind of a nation we are and what direction we want to move in. For those of you who are black — considering the evidence there evidently is that there were white people who were

---

* From Lindsay to author, September 10, 1977. Mr. Lindsay added that he had recently by chance driven past the scene. "The winds still stirred the same trees but the hopes both Kennedy and King stirred in those days are largely gone from the national consciousness."

responsible — you can be filled with bitterness, with hatred, and a desire for revenge. We can move in that direction as a country, in great polarization — black people amongst black, white people amongst white, filled with hatred toward one another.

Or we can make an effort, as Martin Luther King did, to understand and to comprehend, and to replace that violence, that stain of bloodshed that has spread across our land, with an effort to understand with compassion and love.

For those of you who are black and are tempted to be filled with hatred and distrust at the injustice of such an act, against all white people, I can only say that I feel in my own heart the same kind of feeling. I had a member of my family killed, but he was killed by a white man. But we have to make an effort in the United States, we have to make an effort to understand, to go beyond these rather difficult times.

My favorite poet was Aeschylus. He wrote: "In our sleep, pain which cannot forget falls drop by drop upon the heart until, in our own despair, against our will, comes wisdom through the awful grace of God."

What we need in the United States is not division; what we need in the United States is not hatred; what we need in the United States is not violence or lawlessness, but love and wisdom, and compassion toward one another, and a feeling of justice towards those who still suffer within our country, whether they be white or they be black. . . .

We've had difficult times in the past. We will have difficult times in the future. It is not the end of violence; it is not the end of lawlessness; it is not the end of disorder.

But the vast majority of white people and the vast majority of black people in this country want to live together, want to improve the quality of our life, and want justice for all human beings who abide in our land.

Let us dedicate ourselves to what the Greeks wrote so many years ago: to tame the savageness of man and to make gentle the life of this world.

Let us dedicate ourselves to that, and say a prayer for our country and for our people.[84]

40

The Long Day Wanes

BACK IN THE HOTEL Kennedy called Coretta King. "I'll help in any way I can," he said. She said, "I'm planning to go to Memphis in the morning to bring back Martin's body." He said, "Let me fly you there. I'll get a plane down."[1] Southern Christian Leadership Conference officials told her this was a mistake; Robert Kennedy was running for President. Coretta King was not bothered. She remembered 1960, when Martin was in prison and John Kennedy was running for President. "Although they were political figures," she said later, ". . . they were human beings first, and their humanness reached out to the needs of other people."[2]

I

John Lewis had scheduled a meeting between Kennedy and a group of black militants after the Indianapolis rally. They waited for him now, filled, Lewis recalled, with "hostility and bitterness." When Kennedy finally arrived, one said angrily that "establishment people" were all the same: "Our leader is dead tonight, and when we need you we can't find you." Kennedy responded: "Yes, you lost a friend, I lost a brother, I know how you feel. . . . You talk about the Establishment. I have to laugh. Big business is trying to defeat me because they think I am a friend of the Negro." They talked on. Departing, the black leaders pledged their support.[3]

After the meeting, Kennedy seemed overwhelmed, despondent, fatalistic. Thinking of Dallas, perhaps also of Sophocles ("Death at last, the deliverer"), he said to Jeff Greenfield that King's death was not the worst thing that ever happened. Then he said, "You know that fellow Harvey Lee Oswald, whatever his name is, set something loose in this country." The first stories after Dallas, Greenfield remembered, had so miscalled Oswald. "That's the way he remembered [the name] because obviously he never took another look at it again." Early in the morning, restlessly roaming the hotel, he found Greenfield asleep on top of his bed and threw a blanket over him. Awakening, Greenfield said, "You aren't so ruthless after all." Kennedy said, "Don't tell anybody."[4]

That night fury raged in the ghettos of America. The next morning Kennedy kept an engagement to speak at the City Club in Cleveland. The Indianapolis remarks had been entirely his own. The Cleveland speech had contributions from Sorensen, from Walinsky, from Greenfield, all writing through the dreadful night.

Violence, Kennedy said in Cleveland, "goes on and on.... Why? What has violence ever accomplished? What has it ever created? No martyr's cause has ever been stilled by his assassin's bullet." Yet Americans seemed to be growing inured to violence. "We calmly accept newspaper reports of civilian slaughter in far off lands. We glorify killing on movie and television screens and call it entertainment. We make it easy for men of all shades of sanity to acquire whatever weapons and ammunition they desire.... We honor swagger and bluster and the wielders of force." And there was not only the violence of the shot in the night. Slower but just as deadly, he said, was "the violence of institutions.... This is the violence that afflicts the poor, that poisons relations between men because their skin has different colors. This is a slow destruction of a child by hunger ... the breaking of a man's spirit by denying him the chance to stand as a father and as a man among men." So much at least was clear: "Violence breeds violence, repression brings retaliation, and only a cleaning of our whole society can remove this sickness from our soul."[5]

There were riots in 110 cities; 39 people were killed, mostly black, more than 2500 injured; more than 75,000 National Guardsmen and federal troops in the streets. He flew back to Washington, a city of smoke and flame, under curfew, patrolled by troops. He walked through the black districts. "Burning wood and broken glass were all over the place," said Walter Fauntroy. ". . . The troops were on duty. A crowd gathered behind us, following Bobby Kennedy. The troops saw us coming at a distance, and they put on gas masks and got the guns at ready, waiting for this horde of blacks coming up the street. When they saw it was Bobby Kennedy, they took off their masks and let us through. They looked awfully relieved."[6]

On April 7 Martin Luther King was buried in Atlanta. Dignitaries

crowded the Ebenezer Baptist Church. Humphrey, Nixon, Rockefeller, McCarthy — all were there, all save the President himself.
Afterward there was a straggling march, five miles under the fierce
sun, from the church to Morehouse College. Kennedy hung his
jacket over his shoulder and walked with shirtsleeves rolled up. "It
struck me," noted John Maguire, the civil rights fighter, "that of *all*
the celebrities there, the only two people that were *constantly* cheered
wherever they walked ... were Sammy Davis, junior, and Robert
Kennedy."[7] Roy Jenkins, a friend from England, noted that the
Kennedy party got most of the offers of water and Coca-Cola from
the black crowd along the streets. Jenkins asked where Lyndon
Johnson was. Kennedy observed, without bravado, that lack of physical courage kept him away.[8]

Kennedy watched the crowds with disbelief: so few white faces
among them. Jimmy Breslin said, "You'd think even a few of them
would come out and just look, even for curiosity." "You'd think so,"
Kennedy said. "Then maybe this thing won't change anything at
all?" "Oh, I don't think this will mean anything," said Kennedy. He
turned to Charles Evers, walking beside him. "Do you think this will
change anything?" "Nothing," Charles Evers said. "Didn't mean
nothing when my brother was killed." "I know," Robert Kennedy
said.[9] I saw Jacqueline Kennedy after she returned from the funeral. "Of course people feel guilty for a moment," she said. "But
they hate feeling guilty. They can't stand it for very long. Then they
turn."[10]

II

Before leaving Atlanta, hoping to restore contact at a bitter time,
Kennedy held two meetings with black notables. One was with entertainers on the principle, verified in the reception accorded Sammy
Davis, Jr., that they exerted great influence on the black community. The meeting was a mess. Julian Bond, a young Georgia political
leader, observed it with disgust. "It became a matter of each of these
entertainers," he recalled later, "saying in what I thought was a *very*
egotistical way, how much they were doing for the movement." Bill
Cosby finally said, "This is a lot of shit! I'm going to leave" — and
left. Kennedy said little. When they broke up, he said to Bond,
"Julian, I bet you've been to a lot of meetings like this before, haven't
you?" Bond said, "Yes." Kennedy said, "I bet you don't want to go
to any more, do you?" Bond said, "No."[11]

The more serious meeting was with Martin Luther King's closest
associates — Andrew Young, Ralph Abernathy, Hosea Williams,
James Bevel. "There was a lot of undirected hostility in our group,"

Young said later. "People were just angry and bitter and grieving. . . . They decided to take it out on him." Bevel demanded to know whether Kennedy had a program for racial justice. Others joined the assault. "It was filled with profanity," said Young, "and when preachers get to cuss, they cuss good. It's kind of poetic. . . . He wasn't upset. He just handled himself very well. He refused to say he had a program. He said, 'Well, maybe we can get together and talk about that some time.' He said, 'I do have one or two ideas. But really I didn't come here to discuss politics. That would be in the worst taste.' He said, 'I just came to pay a tribute to a man that I had a lot of respect for.'"[12]

"It was very embarrassing," Young recalled, "because you got the impression . . . well, that in a way, he was more sensitive to the situation than some of us were." The atmosphere changed. Young himself had heretofore kept his distance from Kennedy. He felt it "dangerous to like people in power. . . . I had, up to that time, been refusing to admit that I'd even vote for him."[13] Now Kennedy's existence meant, as Abernathy, King's SCLC successor, said, "that white America does have someone in it who cares."[14] King's murder, recalled Williams, "left us hopeless, very desperate, dangerous men. I was so despondent and frustrated at Dr. King's death, I had to seriously ask myself . . . Can this country be saved? I guess the thing that kept us going was that maybe Bobby Kennedy would come up with some answers for the country. . . . I remember telling him he had a chance to be a prophet. But prophets get shot."[15]

III

On the plane back to Washington Kennedy sat with Nicholas Katzenbach. Their relationship had cooled during the wiretapping dispute in 1966. It had become even colder in the winter of 1967–68. Katzenbach, now under secretary of state and obliged to support the war, had questioned a Vietnam quotation Kennedy used in *To Seek a Newer World*. After vindicating his quotation, Kennedy had inquired sarcastically whether Katzenbach had noticed how "very kind and helpful" Ramsey Clark, now Attorney General, had recently been "in the renewed dispute with Hoover regarding the Department of Justice wiretapping. Such courage and integrity were appreciated." Katzenbach replied that "our intelligence people were guilty of inexcusably bad research" on the quotation (no surprise) and took strong exception to Kennedy's concluding lines, "a crack you would resent as deeply as I do."[16] None of this seemed important now. Kennedy touched Katzenbach's knee and said, "Forget about what happened and just erase the whole thing. Nothing more will be said."[17]

The primary contest resumed. Indiana was to vote on May 7. "Hoosiers," Kennedy was warned by John Bartlow Martin, a Hoosier himself, author of the best book on Indiana and now on the campaign staff, "are phlegmatic, skeptical, hard to move, with a 'show-me' attitude."[18] The state had once been a stronghold of the Ku Klux Klan. There was trouble between white and black workers in the cities. Eugene Pulliam's far-right *Indianapolis Star,* the most popular paper, assailed Kennedy brutally in daily cartoon and editorial. State officeholders and the party leaders backed the Democratic governor Roger Branigan, who headed the administration slate as a stand-in for the yet undeclared Hubert Humphrey. Except for the UAW, organized labor was for Branigan. The campuses were for McCarthy.

The assassination of Martin Luther King and the ensuing riots had sobered the nation. Kennedy spoke, as usual, about the miseries of the underclass and the duty to reclaim the miserable for the national community. He spoke too about the dangers of violence and the need for public order. This had begun as an adaptation to Indiana conservatism. At the end of March, Martin passed on a message from Professor Richard Wade, who had been sending over campaign workers from the University of Chicago. Wade, Martin told Kennedy, "reports that student petition circulators in Hammond encountered good response in both black ghetto and [white] backlash areas and attributes it to the feeling that, although you are pro–civil rights, you also come through as a strong executive capable of controlling disorders. Maybe as Schlesinger says ruthlessness has its uses after all."[19] After King's murder, Martin urged Kennedy to say "that violence and rioting cannot be tolerated, and emphasize it; then follow by saying neither can injustice be tolerated." Kennedy nodded, saying, "I can go pretty far in that direction. That doesn't bother me."[20] After all he had been for three and a half years the chief law enforcement officer of the land.

Kennedy had made the point about public order often before, and he now made it more often in Indiana. This briefly disturbed his younger staff, causing, Fred Dutton said later, "the most explicit debate of some substance that we had in the campaign."[21] The *New York Times* reported that, under the malign influence of Martin and Goodwin (who was innocent), Kennedy was beginning to sound like George Romney on law and order. The headline was "Kennedy: Meet the Conservative," and the effect was to suggest an opportunist tailoring his politics to his audience.[22] In fact, as Martin noted, Kennedy

> never once failed to show the other side of the coin — i.e., he never denounced violence and let it go at that but instead always followed it by denouncing injustice. It was a matter of emphasis, not substance.

Naturally in Negro areas he denounced injustice with great emphasis, and gave less emphasis to rioting; but he said both. At least, this was true every time I heard him speak, and I was watching for it carefully, because I knew the press would watch and I warned him to be sure and say both, and I would have picked him up quickly if he had missed a beat.[23]

Of course Kennedy wanted white as well as black votes. McCarthy also adjusted his campaign to Indiana, deleting references to "white racism" from his speeches.[24] But white voters were not led in either case to suppose that the candidates had weakened their stand on racial justice. Charles Quinn, the television correspondent, interviewed Poles and Lithuanians after Kennedy's Gary rally. "These people," Quinn concluded, "felt that Kennedy would really do what he thought was right for the black people but, at the same time, would not tolerate lawlessness and violence. The Kennedy toughness came through on that. . . . They were willing to gamble on this man, maybe, who would try to keep things within reasonable order; and at the same time, do some of the things that they knew really should be done."[25] Nor did blacks object. They were the main victims of riots, and in the end they voted overwhelmingly for Kennedy.

In retrospect the argument hardly appeared as earthshaking as it did at the time. Kennedy saw no incompatibility between racial justice and the rule of law; rather he thought that each reinforced the other and that only someone absolutely committed to equal rights could hope to stop violence in the ghetto. As for the blue-collar whites, they too felt government had forgotten them; they too were among the unrepresented. In later years, after intellectuals discovered their existence and gave them the awful name of 'ethnics,'* Kennedy's effort to keep the black and white working class at peace with each other in Indiana no longer seemed so sinister.

IV

"He always looked so alone," wrote John Bartlow Martin,

standing up by himself on the lid of the trunk of his convertible — so alone, so vulnerable, so fragile, you feared he might break. He was thin. He did not chop the air with his hands as his brother Jack had; instead he had a little gesture with his right hand, the fist closed, the thumb sticking up a little, and he would jab with it to make a point. When he got applause, he did not smile at the crowd, pleased; instead he looked down, down at the ground or at his speech, and waited till

---

* *Ethnic* means, simply, pertaining to a religious, racial, national or linguistic group. To confine this adjective, as current writers try to do, to people from eastern Europe is a gross solecism. A WASP is just as much an 'ethnic' as a Pole or a Slovak.

they had finished, then went on. He could take a bland generality and deliver it with such depth of feeling that it cut like a knife. Everything he said had an edge to it.[26]

David Halberstam heard him at a women's breakfast in Terre Haute. His formal remarks were "pedestrian." But, answering questions afterward, "he starts talking about the poor in America and gets carried away: 'The poor are hidden in our society. No one sees them any more. They are a small minority in a rich country. Yet I am stunned by a lack of awareness of the rest of us toward them.'"[27] Thomas Congdon of the *Saturday Evening Post* saw him in Vincennes at the luncheon of Civitan, a businessmen's club. The complacency of the Civitans provoked him. While they chewed away on their food, he began to talk of children starving, of "*American* children, starving in *America*." Then he said: "Do you know there are more rats than people in New York City?" The Civitans thought this hilarious and broke into guffaws. "Kennedy," wrote Congdon, "went grim and with terrible deliberateness said, '*Don't . . . Laugh. . . .*'" The room fell into confused silence.[28]

He spoke at the University of Indiana Medical School. "The national system of health care," he told an unmoved audience, "has failed to meet the most urgent medical needs of millions of Americans" — the rural and urban poor, the blacks, the Indians.[29] The applause was perfunctory. A black janitor called from the balcony, "We want Kennedy." On the floor students shouted back, "No we don't." Someone asked where the money for Kennedy's health program was coming from. Looking at the incipient M.D.'s about to enter lucrative careers, he snapped, "From you." Then:

> Let me say something about the tone of these questions. I look around this room and I don't see many black faces who will become doctors. You can talk about where the money will come from. . . . Part of civilized society is to let people go to medical school who come from ghettos. You don't see many people coming out of the ghettos or off the Indian reservations to medical school. You are the privileged ones. . . . It's our society, not just our government, that spends twice as much on pets as on the poverty program. It's the poor who carry the major burden in Vietnam. You sit here as white medical students, while black people carry the burden of the fighting in Vietnam.[30]

Afterward, shaking his head incredulously, he said to Halberstam, "They were so comfortable, so comfortable. Didn't you think they were comfortable?"[31]

Three days later in Valparaiso: "I have seen families with a dozen fatherless children trying to exist on less than $100 a month — and I have seen people in America so hungry that they search the local garbage dump for food."[32] Again the comfortable asked their questions. "You tell me something now," he told the students in the audience. "How many of you spend time over the summer, or on vacations, working in a black ghetto, or in Eastern Kentucky, or on Indian reservations?" He gave them Camus on tortured children.[33]

The campaign had quieter moments. He visited a day nursery a few steps away from James Whitcomb Riley's house in old Indianapolis. Most of the children were from broken homes. "Two little girls," wrote David Murray in the *Chicago Sun-Times,* "came up and put their heads against his waist and he put his hands on their heads. And suddenly it was hard to watch, because he had become in that moment the father they did not know. . . . You can build an image with a lot of sharpsters around you with their computers and their press releases. But lonely little children don't come up and put their heads on your lap unless you mean it."[34]

He returned always to his central theme. "He went yammering around Indiana," recalled Martin, "about the poor whites of Appalachia and the starving Indians who committed suicide on the reservations and the jobless Negroes in the distant great cities, and half the Hoosiers didn't have any idea what he was talking about; but he plodded ahead stubbornly, making them listen, maybe even making some of them care, by the sheer power of his own caring. Indiana people are not generous nor sympathetic; they are hard . . . but he must have touched something in them, pushed a button somewhere."[35]

On May 7 Kennedy received 42 percent of the Indiana vote, Branigan 31, McCarthy 27. Kennedy also beat Hubert Humphrey 62.5 percent to 37.5 the same night in the District of Columbia. McCarthy airily observed on television that it didn't really matter who came in first, second or third. "That's not what my father told me," the watching Kennedy said. Around midnight Kennedy went out for a late dinner. He ended talking till early in the morning with two young McCarthy volunteers he found in the airport coffee shop. They told him they planned to stick with McCarthy. After a long conversation, he said, a little sadly, "You're dedicated to what you believe, and I think that's terrific."[36] "He kind of neutralized me," one of them, Taylor Branch, said afterward. "I still worked for McCarthy, but I was drawn to Kennedy because of his flair and passion for the black people."[37]

V

Humphrey had announced his candidacy on April 27 — too late to qualify for Indiana but in time for the District of Columbia. He had spent the month after Johnson's withdrawal assembling his troops.

Walter Mondale and Fred Harris headed his campaign committee. They had been two of Kennedy's better friends in the Senate. Mondale owed his political career to Humphrey, and Kennedy understood his position. But the Harrises had been frequent and happy guests at Hickory Hill and Hyannis Port. Kennedy felt a sadness in losing them now. Of course, Harris had earlier declared for Johnson and was, like Mondale, a hawk on Vietnam. Yet Kennedy hoped that, after Johnson's withdrawal, Harris would rally to the causes he and Kennedy had espoused — the Indians, the Kerner Commission report, of which Harris had been a leading author, the powerless in general. Instead Harris stuck with the administration. (He grew radical later and wrote admiringly — though, in the circumstances, rather patronizingly — about Kennedy in his 1977 memoir *Potomac Fever*.)

How things had changed since Humphrey had bumped along in his bus in Wisconsin eight years before! Now he had representatives of every establishment in America for him — from George Meany to Henry Ford II, from the segregationist governor of Louisiana to the black mayor of Cleveland — and all the campaign funds he needed. It was perhaps small wonder that in his declaration he chirruped about "the politics of happiness . . . the politics of joy. And that's the way it's going to be, all the way, from here on in!" [38] Many found this insensitive three weeks after the murder of Martin Luther King, with the war still racking Vietnam and clouds of violence and hatred over America. Kennedy, who liked Humphrey, was genuinely appalled.

But labor, the party regulars and the south brought Humphrey great strength. He also had a long record of service to Jewish causes, and 60 percent of American Jews lived in the climactic primary states, California and New York. McCarthy too had strong Jewish support. The Jewish community saw him, said Adam Yarmolinsky, as "the professor who gave your bright son an A, and Bobby Kennedy was the tough kid on the block who beat up your son on his way to school." [39] When Kennedy came to New York the day after Indiana, a group of rabbis waited on him. "Why do I have so much trouble with the Jews?" he asked. "I don't understand it. Nobody has been more outspoken than I have. . . . Is it because of my father when he was in England? *That was thirty years ago.*" One of his visitors said that, after the Six Day War, American Jews needed "continual reassurance" about Israel. Kennedy wearily said he would make his position clear again on the west coast. [40]

He met with his delegate slate in the New York primary. Later his friend James Stevenson came by. "He has pushed himself to the limit," Stevenson wrote for the *New Yorker,* "but he does not mention his weariness. His face is gaunt, weathered; his eyes are sunken and red. He rubs his hand over his face again, as if to tear away the exhaustion. It is not something he has sympathy with, his hand is not consoling as it drags across his face — he is simply trying to get rid of an encumbrance."[41] Stevenson felt badly, he remarked later, about asking questions. "He wouldn't give you a slick answer; and that was so awful. I'd ask a question and then I'd feel like saying, 'Never mind. Don't answer that. Don't think about it,' because he'd pull himself together and he'd sort of wring his face and try to rub away all the weariness and exhaustion, and then he'd slowly come out with an answer that he's really thinking about because he wasn't inclined to just give a pre-mixed answer to anything."[42]

On the next day Kennedy addressed the UAW convention in Atlantic City, dined with his children at Hickory Hill and slept in Lincoln, Nebraska. In between, he flew to Hyannis Port to see his father. Rita Dallas noticed the deep lines in his face and thought he could hardly drag himself out of the plane. But he ran to his father, kissed him, then drove him to the compound, laughing, talking, elaborately emphatic in his gestures. "Mr. Kennedy never took his eyes off his son. He was completely absorbed by every move, every word." At luncheon Robert said, "Dad, I'm doing it just the way you would want me to — and I'm going to win." He stayed on till his father's nap. They said goodbye, the old man holding his son's hand with tight, lingering grip.[43]

Kennedy always snapped back quickly from fatigue; and he was beginning to move with the rhythms of the campaign. Early in Indiana reporters had thought him uncomfortable on the hustings. "Kennedy's manner," wrote David Broder of the *Washington Post,* "— the nervous, self-deprecating jokes; the trembling hands on the lectern; the staccato alternations of speech and silence; the sudden shifts of mood — all seem to betray an anticipation of hostility from the crowds."[44] They thought him even worse on television — either stiff in a studio or strident before crowds; too "hot," in Marshall McLuhan's terms, for so cool a medium.

In fact what conventional judgment saw as defects may well have been strengths. His very hesitations conveyed honesty, vulnerability. "Remember Bobby on television," wrote the Yippie Abbie Hoffman, "stuttering at certain questions, leaving room for the audience to jump in and help him agonize."[45] McLuhan himself thought Kennedy marvelously effective. "Now that Bob Kennedy has left that scene," he wrote later in the year, "it is easier to see how much bigger he was than the mere candidate role he undertook to perform. His many hidden dimensions appeared less on the rostrum than in his spontaneous excursions into the ghettos and in his easy rapport with the surging generosity of young hearts. He strove to do good by stealth and blushed to find it fame. It was this (reluctant hero) quality that gave integrity and power to his TV image."[46]

Before live audiences, the shortest way to disarm a crowd and demolish a stereotype, he had discovered, was to do what he did all the time anyway: joke about himself. There had been complaint when on his first visit he spoke, in the Massachusetts manner, of "Indian-er." Returning: "It's good to be here in Indiana. . . . Indian-uh! Some fellow from Massachusetts was in this state last week and pronounced it Indian-er. That was my younger brother. He looks like me. This fall we're going to elect a President who can pronounce Indiana."[47] His stump speech was taking form: jokes — "he has an even greater knack of wry humor than his brother had," Joseph Alsop thought;[48] references to local history and politics; then a problem defined, a rush of statistics, a rising note of urgency; his right fist smashing into his left palm, "I say that's not acceptable. . . . We can do better"; finally, a quotation from Shaw that John Kennedy had used in 1963 before the Irish Parliament and that Robert now rendered as "Some men see things as they are and say 'Why?' I dream of things that never were and say, 'Why not?'"[*]

"The style is neither elegant nor polished," Alsop wrote; "the statements are made staccato and there are frequent repetitions; yet what comes through most strongly is a sense of deep and true concern, a feeling that this man genuinely cares very greatly."[49] Of all the candidates he spoke most directly and concretely. His language cut far more deeply than the delphic abstractions of McCarthy or Humphrey's interminable anthology of liberal cliché. "Kennedy is, in fact," Stewart Alsop wrote, "a magnificent campaigner, capable of conveying conviction — the essence of the campaigner's art — with unique force. When he describes with passionate emphasis the awful weaknesses in our rich and complacent society, and then adds his laconic signature phrase — 'I don't find that satisfactory' — it is genuinely moving."[50]

As the weeks passed, he grew visibly more relaxed. His cocker spaniel Freckles was now always by his side. "If Freckles had a single admirer in the Kennedy campaign entourage," said John Douglas, a kindly man, "it was the candidate himself. . . . Most thought it to be a pest and an abomination."[51] They should have been glad he had

---

[*] The quotation was both inexact and out of context. It was spoken by the Serpent in *Back to Methuselah* in an effort to seduce Eve: "You see things; and you say 'Why?' But I dream things that never were; and I say 'Why not?'" JFK had it right in Dublin.

not brought Brumus. Once someone asked Richard Tuck, the campaign wit, why he was making such an effort over Freckles. "It may look like a dog to you," Tuck said, "but it's an ambassadorship to me." Kennedy thought this very funny. If Tuck erred thereafter, he would say, "You've just lost Madrid, Tuck."[52]

The campaign was acquiring a rollicking quality, typified most of all by the ride on the Wabash Cannonball, a sunny afternoon through north central Indiana on the old Wabash line. Trains had gone out of fashion in campaigns, but everyone — candidates, reporters, local politicians — missed whistle-stopping, and small Indiana towns, clustered around their depots, were ideal settings. A banjo group played the old song "The Wabash Cannonball" at each stop. Kennedy introduced Ethel, then gave his speech. Once he forgot to conclude with Shaw, and some reporters were left behind. They asked him never to do that again. Thereafter, "as George Bernard Shaw once said . . ." became the signal for the dash to the press bus. Between stops, reporters wrote their parody "The Ruthless Cannonball," singing it to the candidate himself before the end of the day:

> Now good clean Gene McCarthy came down the other track
> A thousand Radcliffe dropouts all massed for the attack,
> But Bobby's bought the right-of-way from here back to St. Paul,
> 'Cause money is no object on The Ruthless Cannonball. . . .
>
> So here's to Ruthless Robert, may his name forever stand,
> To be feared and genuflected at by pols across the land.
> Old Ho Chi Minh is cheering, and though it may appall,
> He's whizzing to the White House on The Ruthless Cannonball.

It went on too long — seven verses — and, after a time, some in the press wondered whether irreverence had gone too far. After a moment's silence, Kennedy said at last, "As George Bernard Shaw once said . . ." "Wild laughter," noted Thomas Congdon, "partly in relief that he had taken it lightly." "As George Bernard Shaw once said," Kennedy repeated, ". . . the same to you, buddy."[53]

VI

Complacent audiences nourished his eloquence, which was why his schedulers sent him to places like the Civitan Club and the University of Indiana Medical School. Country audiences nourished his humor. The Nebraska primary came a week after Indiana. The urban Jeff Greenfield watched the farm crowds. "They'd begin by folding their arms and looking up at him as if saying, 'OK, buddy . . . Let's hear the guff.' And then he'd begin by bantering, and . . . the myth of Robert Kennedy as a ruthless, machine guy totally began to break down."[54] "I like rural people," Kennedy told the urban Newfield. ". . . They gave me a chance. They listened to me."[55]

He was about as much a master of agricultural economics as his brother had been. During one speech a gust of wind blew a small piece of paper out of his hand. "Give me that back," he said. ". . . That's my farm program."[56] "You've got to elect me," he told farmers when he whistle-stopped across the state, "because I'm the best friend the farmer has. . . . I'm already doing more for the farmer than any of them, and if you don't believe me, just look down at my breakfast table. . . . We are consuming more milk and more bread and more eggs, doing more for farm consumption — than the family of any other candidate."[57]

He loved to parody the rituals of politics. He convulsed the press by saying in a conservative Indiana town, "Make like, not war. See how careful I am."[58] A reporter, examining the schedule in Nebraska, said it looked like an easy day ahead. "Yes," said Kennedy. "They've fired Marat/Sade as head of scheduling."[59] "You probably wonder," he told the crowd in the small town of Crete, Nebraska, "why I came to Crete. When I was trying to make up my mind whether to run for President, I discussed it with my wife and she said I should, because then I would be able to get to Nebraska. So I asked her why I should get to Nebraska, and she said, 'Because then you might have a chance to visit Crete!'" The crowd cheered. Kennedy: "All those who believe that, raise your hands," which all the kids did.[60] Arriving in one town in the midst, he was informed, of a Slovak festival, he asked how many Slovaks there were in the crowd. Apparently none. "Well, why are you having a Slovak festival?" "And they were laughing," recalled Greenfield. "They were trying to explain to him from the crowd what was going on. It was a totally honest, open, kind of sense of community, really; and that sense really developed all through Nebraska."[61]

Underneath the chaff the theme remained. On the day before the vote, a student at Creighton University in Omaha asked whether military service was not one way of getting young people out of the ghettos. "Here at a Catholic university," Kennedy said indignantly, "how can you say that we can deal with the problems of the poor by sending them to Vietnam? . . . Look around you. How many black faces do you see here, how many American Indians, how many Mexican-Americans? . . . If you look at any regiment or division of paratroopers in Vietnam, forty-five percent of them are black. How can you accept this? . . . You're the most exclusive minority in the world. Are you just going to sit on your duffs and do nothing?"[62]

He made his last Nebraska speech in the black section of Omaha. It had begun to pour, and the absurdity of it all overcame him. "Have any of you noticed that it's raining?" he asked. "Yes," the crowd — surprisingly large — shouted back. "If there are silly things to do," Kennedy said, "then this is the silliest. It's silly for you to be standing in the rain listening to a politician. . . . As George Bernard Shaw once said, 'Run for the buses.'"[63]

He liked Nebraska, said John Glenn, "the plain physical beauty of the countryside and the square fields and the plow patterns," and the greenness of the midwestern spring.[64] And he liked the farmers. "He really felt," said Peter Edelman, "that they, in a romantic kind of way, . . . were his kind of people." He had found "another kind of forgotten and alienated American, another person who thought that this system had just left him behind."[65] On May 14 he won the Nebraska primary with 51.5 percent of the vote. "The farmers," Jules Witcover wrote, "turned out for him in droves."[66] McCarthy, still precariously in the race, took 31 percent.

VII

Between them, Kennedy and McCarthy polled over 80 percent of the Nebraska vote — a smashing repudiation, Kennedy said, of the Johnson-Humphrey administration. "The people want to move in a different direction. We can't have the politics of happiness and joy when we have so many problems in our own country."[67] But Humphrey warily avoided the challenge. It remained a duel between McCarthy and Kennedy.

In fact, the two men agreed on most things. Their Senate records were comparable — Kennedy's better by ADA standards but McCarthy's entirely respectable.* They had differed occasionally — over federal gun control (which McCarthy opposed), over the Subversive Activities Control Board (which McCarthy supported). McCarthy was generally bolder in his attacks on the national security complex, continuing, as he had done since 1964, to accuse the CIA of having "taken on the character of an invisible government answering only to itself,"[68] and now promising, if elected, to fire J. Edgar Hoover. In foreign policy both denounced the idea of the United States as global policeman. "The worst thing we could do," Kennedy said in a "no more Vietnams" speech in Indiana, "would be

---

* Averaged over three years, Kennedy's ADA rating was 98 percent; McCarthy's, 78 percent (Robert Yoakum, "Kennedy & McCarthy: A Look at Some Votes," *New Republic*, May 11, 1968).

to take as our mission the suppression of disorder and internal upheaval everywhere it occurs."* McCarthy flavored his critique of
globalism with personal cracks at Rusk, McNamara, Rostow and other
appointees of John F. Kennedy.

Though the war receded somewhat as an issue after Johnson's
retreat from escalation, the killing went on, and both Kennedy and
McCarthy continued to assail the hawks. Whatever they may have
thought privately, neither came out for unilateral withdrawal. Humphrey, as George Christian, Johnson's last press secretary, wrote,
soon made "an effort to appear more dovish than he actually had
been in the Johnson administration war council."[69] He did not persuade those who recalled his emotional and bitter attack on critics of
the war.†

Where McCarthy and Kennedy diverged was not so much over
policies as over values. Both called for national reconciliation — who
did not? It was even Nixon's year for 'bringing us together' — but
they conceived the task in very different terms. The early primaries
encouraged Kennedy in his quest for an alliance of the disestablished
and unrepresented. "There has to be a new kind of coalition," he
told Newfield in Indiana, "to keep the Democratic party going, and
to keep the country together. . . . We have to write off the unions
and the South now, and replace them with Negroes, blue-collar
whites, and the kids." Poverty would be the tie that bound. "We
have to convince the Negroes and poor whites that they have common
interests. If we can reconcile those two hostile groups, and then add
the kids, you can really turn this country around."[70] In Nebraska he
added the farmers. Nor did he forget the Indians. In April he took
his Indian subcommittee to South Dakota — much to the irritation
of his staff who thought it a waste of precious time. He was in a
campaign, Fred Dutton reminded him, and he should knock off the
Injuns. Kennedy, who was saving his voice, scribbled a note: "Those
of you who think you're running my campaign don't love Indians the
way I do. You're a bunch of bastards."[71]

Richard Harwood of the *Washington Post* and other newspapermen
began to change their minds about the clamorous crowds. Maybe
there was something more to it than demagoguery. "We discovered
in 1968," Harwood said later, "this deep, almost mystical bond that

* He went on, in language reminiscent of his father's writings twenty years before,
to point to "the danger that in seeking universal peace, needlessly fearful of change
and disorder, we will in fact embroil ourselves and the world in a whole series of
Vietnams" (RFK, speech at the University of Indiana, Bloomington, April 24, 1968,
RFK Papers).

† Humphrey actually suggested that those who disagreed with Johnson and himself
on the war were racists — an accusation he reiterated as late as 1976 in *The Education
of a Public Man* (Garden City, N.Y., 1976), 486.

existed between Robert Kennedy and the Other America. It was a disquieting experience for reporters. . . . We were forced to recognize in Watts and Gary and Chimney Rock [Nebraska] that the real stake in the American political process involves not the fate of speechwriters and fund-raisers but the lives of millions of people seeking hope out of despair."[72]

He was rallying the unrepresented to recapture and reconstruct the nation, not to destroy it. Many called him a 'polarizer' and scoffed at the idea of Robert Kennedy, of all people, taking on a mission of reconciliation. McCarthy, according to his wife, "thought that Robert Kennedy could not be a unifying figure, that by his very nature he was divisive, that he aroused fears in the suburbanite and the middle class."[73] No doubt he did. So had Franklin Roosevelt; but by confronting problems Roosevelt had carried his nation farther toward reconciliation than a less 'divisive' President would have done.

"The conventional wisdom of political analysis," Richard Whalen in the Republican camp warned Nixon, "holds that Bobby Kennedy is hurt by his black following and seeming radicalism. Maybe so, but just suppose he were to say, 'I'm the *only* man who can deal with these people. I know where to draw the line. Choose me and I'll take charge.' . . . Bobby speaks plainly, you can hear *steel* in his sentences. Even people who say they hate him, if they are scared enough, will turn toward the sound of steel."* Kennedy had a unique ability, said Robert Coles, "to do the miraculous: attract the support of frightened, impoverished, desperate blacks, and their angry insistent spokesmen, and, as well, working-class white people."[74] "His greatest gift to the country," said Alexander Bickel, "would have been the respite these two groups would have granted him to seek solutions that cannot at anyone's hands come quickly."[75] Paul Cowan, a young radical who had started out for McCarthy, now saw Kennedy as "the last liberal politician who could communicate with white working class America."[76]

Thus his mission: to bridge the great schisms — between white and nonwhite, between affluent and poor, between age and youth, between the old and the new politics, between order and dissent, between the past and the future. It was an undertaking that, as Kennedy conceived it, required not only specific programs but active leadership. Kennedy had no doubt that Johnson had abused executive power in foreign affairs. But a general recession of presidential leadership, he believed, would increase the nation's impotence in the face of deep and angry national division. In the back of his mind

---

* Whalen added: "My attitude toward Kennedy had changed in the final months of his life. I didn't often agree with him, yet I had come to respect him. Once committed, he held nothing back" (R. J. Whalen, *Catch the Falling Flag* [Boston, 1972], 171–173).

was FDR during the depression. Only an activist Presidency and an affirmative national government, as he saw it, could pull together a divided people in a stormy time.

VIII

McCarthy had both a different constituency and a different theory of the Presidency. In a February article in *Look* called "Why I'm Battling LBJ," he had concentrated on the war and on constitutional questions. He had said not a word about racial justice or the fate of the dispossessed. His voting record on civil rights and poverty issues was good. But his speeches, Jack Newfield thought, revealed "a pattern of psychic distance from the poor. He is certainly not anti-Negro. . . . He is just not very interested in the other America."[77]

McCarthy was ill at ease with blacks, and they with him. The journalist Seymour Hersh resigned from his staff in Wisconsin because of the candidate's reluctance to go into the ghettos. Two weeks after King's murder, McCarthy met with black leaders in Indianapolis. One said his answers were too general; "people are asking direct questions and they need direct answers." Another said, "He didn't put himself into it"; especially compared to Kennedy — that "cat was able to relax."[78] Pride no doubt played a part. Abigail McCarthy spoke of "Gene's refusal to compete emotionally with Robert Kennedy."[79] McCarthy himself said later, "There wasn't political point in it for me. I couldn't get the Negro votes away from Bobby Kennedy."[80]

The poor were not his people. His people, as his wife described them, were "academia united with the mobile society of scientists, educators, technologists and the new post–World War II college class." He appealed to the churches and the suburbs, to civic-minded businessmen and enlightened Republicans, to "the best of Middle America," Abigail McCarthy thought, "in search of a new way of expressing itself as a consensus."[81] "Their common denominator," said Norman Mailer, with unjust but illuminating exaggeration, "seemed to be found in some blank area of the soul, a species of disinfected idealism which gave the impression when among them of living in a lobotomized ward of Upper Utopia." McCarthy, Mailer wrote, "did not look nor feel like a President. . . . No, he seemed more like the dean of the finest English department in the land."[82]

McCarthy himself summed it up in a moment of frivolous candor a few days before the vote in Oregon. The polls, he told an audience in Corvallis, showed Kennedy running best "among the less intelligent and less educated people in America. And I don't mean to fault them for voting for him, but I think that you ought to bear that in mind as you go to the polls here on Tuesday." [83] His speechwriters, Jeremy Larner and Paul Gorman, protested. "Was that unfair?" McCarthy asked. "I think it was," said Paul Gorman. "But it's true," said McCarthy. They asked what that meant. "Nothing!" he laughed. [84]

The constituency was real enough, but, unlike Kennedy's, it did not readily yield a campaign theme. Larner argued that McCarthy must "identify himself with a positive vision of America. McCarthy had such a vision, but he was content to go on expressing it in underdeveloped generalizations." His staff worked up a document proposing "public participation" as the theme and specifying how governing institutions could be brought under popular control. Their candidate professed himself delighted. But, said Larner, he then "went on just as before, saying practically nothing on the new politics. . . . There never was a McCarthy campaign in Indiana and Nebraska." [85]

If anything emerged as a theme, it was a critique of the Presidency itself. Johnson, McCarthy felt, had been "eroding and weakening" the other agencies of government. He had taken the war-making power away from Congress, had diminished the Supreme Court by appointing its Chief Justice to head the commission investigating John Kennedy's assassination, had politicized the Council of Economic Advisers, had enfeebled the Democratic National Committee and the national party. The trouble was his "personalization of the Presidency at the expense of our governmental and political institutions." [86] Actually this personalization had begun, McCarthy thought, with John Kennedy. Johnson had "done it defensively as things have got more and more out of control. Jack did it almost deliberately. He . . . conveyed the impression that all power radiated from the Presidency, which is not, and cannot be, the case in America." [87]

One remedy might be to limit the Presidency to a single term. Another was presidential self-restraint. Presidents must understand, McCarthy said in the campaign, "that this country does not so much need leadership, because the potential for leadership in a free country must exist in every man and every woman." [88] He called for the decentralization of the Presidency. "Has the integrity of Congress, of the cabinet and of the military," he asked, "been impinged upon by undue extensions of the executive power?" [89] The military seemed a curious inclusion.

McCarthy, one felt, was the first liberal candidate in the century to run *against* the Presidency; doing this, moreover, in times of turbulence that seemed to call for a strong Presidency to hold the country together. His views aroused doubts even among his own devoted workers. The political scientist Barry Stavis, for example, began to fear that, under a McCarthy administration, "the federal government would lose its power to protect exploited people.... McCarthy would not use the president's tools to combat a raise in the price of steel, for instance, or a strike by railroad workers' unions." "With such an attitude," Stavis wondered, "could he stop the military-industrial complex? . . . Could he stop the arms and space race? Could he get southern states to accept civil rights? Could he get northern states to accept open housing and integrated education?"[90]

IX

Because Kennedy and McCarthy agreed on so many other issues, those for whom neither the underclass nor the Presidency was a dominating consideration chose between them on personal grounds. Some felt a debt to McCarthy for venturing ahead while Kennedy had lingered behind. Others felt that Kennedy would be the stronger candidate against Humphrey and Nixon. Many made up their minds on impressions of temperament and character.

McCarthy attracted people by his detachment, his serenity, his very nonchalance. Property owners did not feel threatened by him. Where Kennedy, wrote David Halberstam, had "the look of a man who intended to rock the boat," McCarthy said radical things in a temperate way.[91] He was civilized, literate, quoted other people's verse and wrote his own. He disdained the ritual of politics. He acknowledged no obligations. He made no effort to conceal boredom. He was distant, private, imperious. An aura of mystery hung around him. Richard Harwood saw in him "vague intimations of an American de Gaulle." McCarthy, Harwood wrote, was the philosopher, Kennedy the evangelist. "McCarthy speaks in generalities and Kennedy speaks in specifics. [McCarthy] dwells on himself and his moment in history; Kennedy dwells on the tragedy of the poor. McCarthy . . . 'meanders' through his campaign; Kennedy drives on like a sprinter. McCarthy soothes; Kennedy arouses."[92]

Each in his way was fastidious. McCarthy was fastidious about wooing crowds. He thought Kennedy a rabble-rouser. He did not at first wish to go to Martin Luther King's funeral; it would be a "big vulgar public spectacle."[93] When Blair Clark told him he had to go, he was, his wife wrote, "adamantly opposed to the idea of walking" over to Morehouse College. At last, she said, "Gene began to feel the simple emotion of the situation and he . . . decided to get out of the car and walk the rest of the way."[94] Kennedy was fastidious about attacking individuals. J. Edgar Hoover would not have lasted thirty seconds in his administration, but he thought it cheap for McCarthy to seek easy cheers from academic audiences by promising to fire him.

I think it fair to say that, the closer people were to them, the more they liked Kennedy and the less they liked McCarthy. "Most politicians," wrote Halberstam, "seem attractive from a distance but under closer examination they fade.... Kennedy was different. Under closer inspection he was far more winning."[95] As for McCarthy, wrote Theodore H. White, "all through the year, one's admiration of the man grew — and one's affection lessened."[96]

Kennedy's staff, without exception so far as I know, adored him, spoke frankly to him, were received by him as friends. Richard Harwood could find "little camaraderie between McCarthy and those who enlisted in his campaign. They are functionaries rather than partners."[97] Books written by McCarthy aides are filled with uneasiness, with retrospective doubt, not seldom with hostility. "He treated us," wrote Barry Stavis, "not as *his* staff but as the organized segment of the American people demanding his presidency. Once an infuriated staff member threatened, 'I'm going to take that presidency and ram it down his throat.' ... Most of the staff did not like this relationship. They wanted to be part of *his* campaign, not part of a popular demand for him, and expected him to lead it and not observe it."[98] McCarthy's attitude, said Larner, was "that he was doing his supporters a favor by 'letting them use my name.'"[99]

Both Stavis and Larner commented on his resentment of criticism. "The biggest danger I foresaw with a McCarthy presidency," said Stavis, "was that if he made a mistake, it would be difficult to pressure him into correcting it.... His confidence and moralism might make him as hard to sway as Dean Rusk had been."[100] "Criticism was impermissible," wrote Larner, "no matter its source or its quality.... He could not tolerate disagreement or equality, could not, in fact, work directly and openly with others."[101] Both commented on his withdrawal behind a cozy circle of "snobs, sycophants, stooges, and clowns." Both were concerned by what Larner called "a deep-seated bitterness, never quite accounted for by immediate circumstance, a bitterness which made him down-rate individuals, even as he was calling for a national policy of generosity."[102]

A gnawing question, recalled Stavis, was "whether we really wanted Eugene McCarthy to be president."[103] Doubts troubled them at night, "but we acted, during the daytime," said Larner, "as if we were in the service of a wise and calm daddy, crudely attacked by a renegade brother who stirred up crowds with his long hair, squeaky voice, baby talk, and unfair money." They suffered and stayed. "There was finally in McCarthy's reserve, in all he left unsaid, a special air of mystery, a hint that he drew strength from a source beyond mere mortals like Kennedy and ourselves, a gift for grace that would tell him when and how to bring that strength to bear."[104]

One aide was deeply worried by McCarthy, deeply tempted by Kennedy. "But in the end," said Larner, "he stayed and suffered." The reason, the aide explained, was the one thing Kennedy could not offer. "I thought McCarthy had a secret. I thought one day the secret would explain it all."[105] Larner entitled his book *Nobody Knows*.

X

Distance magnified McCarthy's charms and Kennedy's infirmities. It was hard in later years, after McCarthy glided away as mysteriously as he had appeared, to recall the intensity of feeling on his behalf in 1968.

The rage among his supporters against Kennedy, far from subsiding, appeared to grow. Kennedy's mode of entry handed McCarthy, as Halberstam said, "the White Knight issue."[106] Then there was the underdog effect: Kennedy with his high-powered machine, personal fortune, Irish Mafia, puffed-up names from the past (Sorensen, Salinger, Schlesinger), against McCarthy with his Children's Crusade, youth and idealism, ministers and housewives, living on the land — the big battalions moving in on the guerrilla warriors.

Once Johnson had withdrawn, Kennedy replaced him as the hate object in sections of the intellectual left, at least in New York. Jack Newfield did not exaggerate when he wrote in May of "the deep hatred of Kennedy that is now so chic among liberal intellectuals."[107] I wrote about "the McCarthy hysteria" in my journal the same month: "I have never felt so much in my life the settled target of hostility. . . . I am hissed at practically every public appearance in this city. I have just been out to get the morning *Times,* and inevitably someone harangued and denounced me on Third Avenue — again a McCarthyite. I think these people are crazy."[108]

One memory sums up my impression of that embattled spring. George Plimpton, a strong if lonely Kennedy supporter, gave a party for McCarthy friends and invited me to present the case for our candidate. The atmosphere was icy. Questions. Someone asked about birth control and abortion. In fact, Kennedy approved birth control (for others), was sympathetic to the New York abortion law and had been helpful to Albert Blumenthal in getting it through the state legislature.* I mentioned this, then asked how McCarthy stood on these questions. No one knew, or cared. McCarthy could have come out for the auto-da-fé, one felt; this crowd would still prefer him to Kennedy. I finally asked what it was specifically they had against Kennedy. Nat Hentoff, a jazz critic and a goodhearted civil libertarian, launched into an emotional discourse — Joe McCarthy, Hoffa, Manchester, the old litany. "All I can say," he concluded, "is that he seems to me someone ruthless, vindictive, relentless, like, well — like Jean Valjean in *Les Misérables*." I said that Hentoff probably meant Javert. I was denied even this trivial victory. His loyal wife said, to general applause, "That is exactly the sort of pedantry we would expect from Kennedy people."

Robert Lowell speaking on Kennedy in Oregon: "He has just bought a hundred charisma suits. A charisma suit is made of cloth and cardboard; at the touch of a feather, at the touch of the weakest admirer, it pulls apart. But under the charisma suit is an anti-charisma Bobby-suit. It is made of cloth and steel wool. It doesn't tear at all and leaves metal threads in the rash admirer for months." Lowell added he would be "dishonorable if I didn't confess that I personally like and admire Senator Kennedy. . . . Still I wish to end up with invective; it's hard to forgive Kennedy his shy, calculating delay in declaring himself, or forgive the shaggy rudeness of his final entrance. . . . We cannot forgive Senator Kennedy for trying to bury us under a pile of gold."[109]

Yet, looking into the records, I conclude that Newfield and I were overreacting to the parochial environment of Manhattan. In fact, across the country, the arts and letters divided evenly enough between the two candidates. Elizabeth Hardwick Lowell later said of Kennedy, "I felt about him that he was one of the few people in public life who had truly changed and also that this possibility of change would continue. Many people were stopped by an image of Bobby Kennedy formed in the past. But with him I felt a wish to transcend that past. And actually at the end he had transcended his earlier self, whatever that was. Anyway he had gone beyond what we imagined him to have been and done so more than anyone I can think of."[110]

Among poets McCarthy had Lowell; Kennedy had John Berryman, John Ashbery, Donald Hall, Sandra Hochman. Among novelists McCarthy had William Styron, Mary McCarthy, Wilfrid Sheed. Kennedy had Norman Mailer, James Jones, Alison Lurie, Joyce Carol

* "A couple of times when I got into some trouble over the abortion law . . . he really went to bat for me" (Albert H. Blumenthal, in recorded interview by Roberta Greene, December 14, 1973, 47–48, RFK Oral History Program).

Oates, Irwin Shaw, Wright Morris, Truman Capote, Budd Schulberg. (Humphrey had Ralph Ellison and James T. Farrell.) Among playwrights McCarthy had Arthur Miller; Kennedy, William Inge and Alan Jay Lerner. McCarthy had Paul Newman and Joanne Woodward; Kennedy, Lauren Bacall, Henry Fonda, Shirley MacLaine, Warren Beatty. McCarthy had Ben Shahn; Kennedy had James Wyeth and Andy Warhol. McCarthy had Jules Feiffer; Kennedy, Charles Addams. McCarthy had J. K. Galbraith, Joseph Rauh, Allard Lowenstein, Barbara Tuchman, Erich Fromm; Kennedy had Archibald Cox, Paul Samuelson, Seymour Harris, Robert Coles, James Loeb.

It was interesting how many people who had tangled with Kennedy in the past rallied to him now. John Lewis of SNCC was only one example. J. Edward Day, who as Postmaster General had had his troubles with Kennedy; G. Robert Blakey, who had criticized his explanations of wiretapping; Joseph Mankiewicz, who had opposed him for the Senate — all endorsed him. To the dismay of those who still used the battle of the book as evidence of Kennedy's ruthlessness, William Manchester came out for him, calling him "not a brute . . . genuinely humane . . . the least understood man in the presidential arena."[111] ("When I read this spring that you were giving your support to Robert Kennedy," Jacqueline Kennedy wrote him later, "— I was absolutely startled — then so touched — and much more than that. . . . I want you to know that the last time I saw Bobby alive, we spoke of that. And it meant the same to him. . . . You gave him what he was pleading for for [from] others — a wiping off of the blackboard of the past — a faith in now — and a generosity of such magnitude and sacrifice.")[112] Alexander Bickel had written in the *New Republic* in 1961 that Kennedy was unfit to be Attorney General. "I campaigned for him in California," Bickel wrote in the *New Republic* in 1968, "and it meant a great deal to me, more than any prior political commitment or than any conceivable new one. I believed he had come to know better and more deeply than anyone how dangerously we are nearing a dead end, and I believed that he above all other public men would . . . stop war and heal suffering."[113]

Kennedy's greatest disappointment was the young. He used to say wistfully that McCarthy had the A students, and he envied McCarthy their devotion. Curiously he was stronger than McCarthy with the far-out young. In December 1967 Abbie Hoffman, in an effort to unite the hippies with the New Left, founded the Youth International Party, known popularly as 'Yippie!' with the intention of besieging the Democratic convention in Chicago. "Gene wasn't much," said Hoffman. "One could secretly cheer for him the way you cheer for the Mets. It's easy, knowing he can never win. But Bobby, there was the real threat. . . . *Come on,* Bobby said, *join the mystery battle against the television machine.* Participation mystique. Theater-in-the-streets. He played it to the hilt. . . . It was no contest. . . . Yippie! grew irrelevant. . . . By the end of May we had decided to disband Yippie! and cancel the Chicago festival."[114] They reinstated the festival later.

The other America was of course with him. Cesar Chavez was on his delegate slate in California. In Mississippi, Charles Evers and Oscar Carr, the enlightened planter, were co-chairmen of the Kennedy committee. "You may tell your husband," George Wiley, the fiery leader of the National Welfare Rights Organization, wrote Ethel Kennedy, "that I am personally very much in favor of his candidacy."[115] Michael Harrington, the author of *The Other America,* the heir of Norman Thomas, campaigned stoutly for him in Indiana and California. When he met the candidate, they talked about "why so many New York reformers hated Kennedy, a phenomenon that troubled and puzzled him. He told me with a sort of hurt disbelief how a peace activist in San Francisco had spit in his face and called him a fascist." Later they walked down a chilly, silent street toward Kennedy's hotel. An old black man ran up to Kennedy: "My wife's waiting back in the car and she wants to meet you." Kennedy pulled the lapels of his coat against his face and headed back in the cold wind. Harrington never saw him again. He wrote in his autobiography, "As I look back on the sixties, he was the man who actually could have changed the course of American history."[116]

Mid-April, George McGovern introducing Robert Kennedy to a crowd of five thousand who had waited an hour in the rain in Rapid City, South Dakota; a man, McGovern said, with "the absolute personal honesty of a Woodrow Wilson, the stirring passion for leadership of Andrew Jackson, and the profound acquaintance with personal tragedy of Lincoln." McGovern continued:

> You people know the affection and esteem I held for President Kennedy, but it is my carefully measured conviction that Senator Robert Kennedy, even more than our late beloved President, would now bring to the Presidency a deeper measure of experience and a more profound capacity to lead our troubled land into the light of a new day. . . . If he is elected President of the United States, he will, in my judgment, become one of the three or four greatest Presidents in our national history.*

* George McGovern, introduction of Robert Kennedy at Sioux Falls and Rapid City, South Dakota, April 16, 1968. Kennedy said of McGovern at Mitchell, South Dakota, on May 10: "Of all my colleagues in the United States Senate, the person who has the most feeling and does things in the most genuine way, without that affecting his life, is George McGovern. . . . That is truer of him than anyone else in the United States Senate." For full texts, see "Senator Robert Kennedy in South Dakota 1968 (A Mem-

XI

The Kennedy campaign, Charles Quinn said later, became a "huge, joyous adventure."[117] Even reporters found themselves caught up in the enterprise against all the rules of professionalism. "Quite frankly," said Tom Wicker, head of the *New York Times* Washington bureau, "Bobby Kennedy was an easy man to fall in love with," and he warned his own people against it.[118] Jules Witcover, who later wrote an excellent book about the campaign, spoke of Kennedy's "way of pulling individuals around him into his orbit, a strange disarming quality about him that somehow evoked sympathy."[119] By the time of Oregon, as Richard Harwood, initially the most hostile of all, said later, "We were getting partisan. We hadn't quite become cheerleaders but we were in danger of it."* One reporter asked to be taken off the campaign after California because he felt he could no longer be objective.[120]

Underneath the fun lay foreboding. A shadow had fallen across the happy day of the Wabash Cannonball when the train stopped at Logansport, and someone saw on top of a building, etched against the sky, a man with a gun. He turned out to be a policeman, but Thomas Congdon long remembered the "agonized" look on the face of Jerry Bruno, the advance man.[121] One evening a group of reporters sat around over drinks. Someone asked whether Kennedy had the stuff to go all the way. "Of course, he has the stuff to go all the way," replied John J. Lindsay of *Newsweek*, "but he's not going to go all the way. . . . Somebody is going to shoot him." There was "stunned silence" around the table. One by one, each journalist agreed. Lindsay said, "He's out there now waiting for him."[122]

Romain Gary, the French novelist, came to America that spring with his wife, the actress Jean Seberg. They lunched with Pierre Salinger after King's murder. "You know, of course," Gary said, "that your guy will be killed." Salinger froze, stared at Gary for a long moment, then said, "I live with that fear. We do what can be done, and that isn't much. He runs around like quicksilver." A month later, Gary met Kennedy himself. The novelist found Ken-

---

orial from Senator George McGovern)," (n.p., n.d. [1968]). See also George Mc-Govern, in recorded interview by L. J. Hackman, July 16, 1970, 57–62, RFK Oral History Program.

* Richard Harwood, in recorded interview by Jean Stein, September 6, 1968, 2, Stein Papers. Ben Bradlee, Harwood's editor, had originally given Harwood this assignment because he had been so "outspokenly skeptical of Bobby" (Benjamin C. Bradlee, *Conversations with Kennedy* [New York, 1975], 22).

nedy's boyishness and charm "much more apparent than the supposed ruthlessness" and singularly thought that, "when age and white hair come, he would look a bit like Cordell Hull." He said, "Somebody is going to try to kill you." Kennedy said that there were no guarantees against assassination. "You've just got to give yourself to the people and to trust them, and from then on . . . either [luck is] with you or it isn't. I am pretty sure there'll be an attempt on my life sooner or later. Not so much for political reasons. . . . Plain nuttiness, that's all."[123]

This was why Bill Barry was along. "It was not just a professional job with me," he told Jules Witcover. "It was something my life qualified me for. This would be my juggler's gift."[124] No one worked with less cooperation from his principal. Barry's main job was to get Kennedy through crowds. When the campaigning day was over, Kennedy refused protection. Barry tried surreptitious precautions, such as hiring off-duty policemen to stay in the hotel lobby. He dared not risk Kennedy's displeasure by putting them in corridors next to his room. When Kennedy learned of such extracurricular arrangements, he canceled them.

His attitude, Barry said later, "was that he was going to live his life and not be constantly fearful of what might happen. . . . He only accepted as much protection as he got because he liked me. . . . He wouldn't have had anybody if really left to his own choice." On April 11, in Lansing, Michigan, a police lieutenant notified Barry that a man with a rifle had gone into a building across from the hotel. Barry had Kennedy's car driven into the basement garage, so he could enter it without going out on the street. When they went to the garage, Kennedy was furious. He said, "Don't ever do this again. Don't ever change whatever we're doing until you talk to me, and I don't ever want to change it because I'm not afraid of anybody. If things happen, they're going to happen."[125] (The man with the rifle turned out to be an office worker bound for a hunting weekend.) Kennedy particularly objected to police escorts in ghettos.[126] As Lieutenant Jack Eberhardt of the Los Angeles Police Department put it, Kennedy "in no uncertain terms, told us he didn't care for our assistance. He felt that we were preventing him from getting a close rapport with his followers."[127]

There were several alarms — in Cleveland, in Salt Lake City, in California. Kennedy ignored them. Reporters nerved themselves to mention the danger. He told Charles Quinn, yes, he had thought about it, but he wasn't going to change his campaign because of it. Then his eyes got a faraway look, and he said, "You know, if I'm ever elected President, I'm never going to ride in one of those Goddamned cars"; he meant one of those bulletproof, bubble-top cars.[128] "If there is somebody out there who wants to get me," he told Warren Rogers, "well, doing anything in public life today is Russian roulette."[129]

In May, Bill Barry had his forty-first birthday. There was the usual Kennedy surprise party, poems, gags, joke presents, balloons, champagne. While someone read a poem, a balloon exploded with a loud bang. "It would have been a forgotten interlude," said Helen Dudar of the *New York Post*, "except for Kennedy's reaction; it was not a shellshock reflex and it almost proceeded in slow motion. The back of his hand came up toward his face, which was frowning, and he held his hand there, his head bent, for perhaps a count of ten. The party stood suspended in time for those seconds and then Kennedy came back from wherever he'd been and it resumed."[130]

Barry and Walter Sheridan continually discussed the security problem. "And we knew," said Sheridan, "that, really, there wasn't anything you could do about it because he was uncontrollable, and if you tried to protect him he'd get mad as hell."[131] Someone later asked Barry whether Kennedy was foolhardy. "I don't know whether that's a correct word," Barry said. ". . . He just didn't want to live in fear. So I think he was making a personal judgment of his own, based on his own life force."[132]

41

To Sail Beyond . . . the Western Stars,
Until I Die

IN MID-MAY Kennedy paid a flying visit to South Dakota, where Hubert Humphrey, a native son, and Eugene McCarthy, from Minnesota across the border, had also entered the primary. George McGovern, introducing him again, quoted a snatch from "The Impossible Dream." Kennedy wearily stumbled through his speech. Afterward he asked whether McGovern really thought the quest impossible. McGovern said, "No, I don't think it's impossible. I just . . . wanted the audience to understand that it's worth making the effort — whether you win or lose." Kennedy said, "Well, that's what I think." They met for breakfast in the morning. "He came in after taking his dog for a run nearby, and he looked rested and relaxed." McGovern said goodbye at the airport. "I remember that morning just being seized with a feeling of sadness. For some reason, he looked so small. Bob, at various times, appeared different sizes to me." An interesting observation; I had similar optical illusions about him. "Sometimes he seemed like a large man . . . I mean physically. At other times, he seemed very slight, small. It depended, I guess, on the angle of vision. But, as he walked away, he looked like such a frail and small person. I just had such a feeling of sadness as he got on the plane."[1]

I

South Dakota, like California, would vote on June 5; Oregon, a week earlier. Oregon: a pleasant, homogeneous, self-contained state filled with pleasant, homogeneous, self-contained people, overwhelmingly white, Protestant and middle class. Even the working class was middle class, with boats on the lakes and weekend cabins in the mountains. Oregonians were remote from problems but responsive to issues. McCarthy was the thoughtful, independent type they liked. And he had in Oregon, he said later, "a better organization than we had in any other state. It had been working for six months. . . . It was also the best financed of our efforts."* Most important of all, he had finally found the theme his campaign had lacked since Johnson's withdrawal. That theme was Robert Kennedy.

McCarthy had grown increasingly bitter toward Kennedy, conspicuously refusing to congratulate him, for example, after Indiana and again after Nebraska.[2] Even more than Kennedy's entry, his money and his presumed sense of entitlement, "the greatest disappointment," McCarthy said the following December, "was the kind of campaign the Kennedy people conducted against me," by which he meant primarily "that whole voting record thing they put out."[3] It is not clear why this should have outraged him so particularly. Voting records are sometimes hard to interpret, and McCarthy's own explanations for his 'bad' votes were complicated.[4] Moreover, as Arthur Herzog, who ran McCarthy's Oregon campaign, later wrote, "The charges were, in fact, partially true."[5] Nor were his own people all that careful. Three days before the California vote, the McCarthy for President Committee, taking a full page in the *San Francisco Chronicle,* charged wildly that Kennedy had been "directly involved" in the Bay of Pigs and "in the decisions that led us to intervene in the affairs of the Dominican Republic."[6]

Nevertheless McCarthy was deeply aggrieved over an indefensibly sloppy tabulation circulated by a free-lance New York group called Citizens for Kennedy. Salinger had promptly disclaimed this leaflet on behalf of Kennedy headquarters.[7] "Had it been our attack," said Stephen Smith, "we would have done a better job."[8] Kennedy himself, Frank Mankiewicz recalled, said he saw no point in "knocking McCarthy's voting record. . . . The people who were for Gene McCarthy were not for him because of his voting record. . . . If you could show them that McCarthy had voted wrong, let's say, on every issue in the past ten years that ADA felt was important, it wouldn't bother them."[9] Nevertheless, for all the disavowals, the inaccurate McCarthy voting record bobbed up in Indiana, in Nebraska and now in Oregon and California.

---

* Eugene McCarthy, *The Year of the People* (Garden City, N.Y., 1969), 145. In calling it the "best financed," McCarthy explained that he meant not that the Oregon campaign had the most money but that it had the money long enough in advance to plan the spending effectively.

McCarthy also resented Kennedy efforts to induce McCarthy students to desert through offers, as he understood it, "of more pay, educational advantages, and the like."[10] At the same time his own agents infiltrated the Kennedy campaign, though, Barry Stavis later acknowledged, "our very efficient intelligence network within the Kennedy circles yielded nothing of any particular value."[11] Thus competition debased both camps.

McCarthy was additionally angered by Kennedy's repeated public assertions of a desire to work together. This sounded generous, but, McCarthy correctly noted, "at no time in the campaign did he expressly say that he would support me . . . for the presidency."[12] The Kennedy people tried to work out joint slates in the District of Columbia, New York, Maryland and other states. McCarthy vetoed them all. He saw them as a way of channeling McCarthy delegates to Kennedy, not vice versa. His analysis was not unreasonable. I do not think that Kennedy could conceivably have brought himself to come out for McCarthy. Nor, after Indiana and Nebraska, was there any chance that McCarthy would come out for Kennedy. Despite Vietnam and the politics of joy, each, I would guess, preferred Hubert Humphrey to the other. Kennedy never indicated this publicly. McCarthy did. "My final judgment," McCarthy wrote later, "was expressed in a press interview on May 21 when I said that I could support Vice President Humphrey if he changed his position on Vietnam and *possibly* Senator Kennedy if there was a change in his campaign methods."[13]

The acceptability of a purified Humphrey alarmed the McCarthy staff. Jeremy Larner and Paul Gorman soon drafted a speech for McCarthy to give at the Cow Palace in San Francisco assailing Humphrey and Kennedy equally. On delivery, McCarthy amplified the Kennedy section. Then he read the sentences on Humphrey, Larner wrote later, "in a subdued, rapid tone. And when it came to the distinction between candidates for the war and against it, McCarthy said it in a garbled way that could not have made sense to his audience." Larner felt physically sick. He could no longer doubt "that McCarthy hated Bobby Kennedy, that on a personal level he preferred Humphrey."[14]

In Oregon McCarthy dwelt almost obsessively on Kennedy. He challenged him, repeatedly and scornfully, to debate. He criticized him for advocating federal gun control — a well-calculated criticism in a state that regarded gun ownership as one of the rights of man. With mordant and condescending wit, McCarthy ridiculed Kennedy's advisers, his astronaut, his dog. He liked to depict Kennedy himself as a spoiled child. "Bobby," he would say, "threatened to hold his breath unless the people of Oregon voted for him."[15] On the Saturday night before the vote he brought a screaming crowd at the Portland Coliseum to its feet again and again with personal sarcasm. It was his most rousing speech.[16] "The tone of mockery," Larner thought, "was sickening."[17]

II

Kennedy did not strike back. This restraint signified no affection for McCarthy, rather a desire not to antagonize McCarthy's supporters, especially the kids. He hoped to have them with him later. When he challenged anybody, it was Humphrey and his politics of joy. For the rest, he talked about the poor; he talked about jobs; he tried persiflage. None of it worked in Oregon.

He did not have a strong organization. Congresswoman Edith Green had agreed to run his campaign on condition that her bête noire David Hackett be forbidden to enter the state.[18] She had great popularity but no apparatus. On April 28, a month before the vote, the Kennedy headquarters consisted of two desks and three people.[19] The Teamsters, the strongest union in the state, had not forgotten the Rackets Committee investigations in Portland. The mayor of Portland, who had been indicted in 1959 as a result of these investigations, was, after acquittal, still mayor in 1968.*

Kennedy's personality was too intense for Oregon. His behavior offended unfathomable local sensitivities. Doing what he had done so often around the world, he yielded, for example, on May 24 to the temptation to take a dip in the Pacific. Apparently Oregonians never swam in the Pacific until August. They thought anyone who swam in May a showoff or a fool. Most damaging of all, as Kennedy muttered to Joseph Kraft, "there's nothing for me to get hold of."[20] Employment was high. Minorities — blacks, Indians, Chicanos — were 2 percent of the population. An observer watched him speak at an oscilloscope plant outside Portland. "The employees did not rise to his impassioned attack upon poverty, hunger, the ghettos and the decay of the cities. . . . Kennedy left the plant, shaking his head and declaring that these were the strangest workers he had ever met."[21] As Pierre Salinger said, "If you were going to carry on the central theme of Bob's campaign which had to do with poverty and blacks and [poor] people, this subject was absolutely falling on dead ears in Oregon. They couldn't care less. I mean the black ghetto in Portland was maybe five city blocks or something."[22] "Let's face it,"

---

* "No city in the U.S. regards [Robert Kennedy] with such suspicion, and in some cases with such downright enmity, as does Portland" (Paul O'Neil, "The No. 2 Man in Washington," *Life*, January 26, 1962).

Kennedy told a reporter, "I appeal best to people who have problems."[23]

His most serious mistake was his refusal to debate. He even dodged McCarthy personally when their schedules accidentally brought them together at the Portland Zoo. Jeremy Larner chased after Kennedy, saw a look of "exquisite hurt" on Kennedy's face, shouted "Coward." "For five minutes, I felt exhilarated," Larner wrote later.

> It was a lot of fun, like winning a game of touch football. I didn't think about it till I saw the film on TV back at the hotel. . . . When I heard my own voice pipe Coward, I got that flash of nausea that had come before in the Cow Palace. . . . I had become in my way pretty much like everyone else on either side: a gangster in a war of two mobs in the same family. The hurt look was no cry for pity: it was the registration that something had gone terribly wrong in our fight for territory. . . . Our family could only lose, and what did I know, what did I really know, that made me so eager to beat him?[24]

III

May 28: McCarthy, 44.7 percent; Kennedy, 38.8 percent. No Kennedy had ever before lost an election, except for the Harvard Board of Overseers. As the returns came in, Kennedy's two student organizers avoided him; they thought they had let him down. He put an arm around each and said, "I'm sorry I let you down."[25] Disdaining McCarthy's example, he sent a generous telegram of congratulation to the victor. To a reporter he said that Humphrey, not McCarthy, was the real winner. "I think what [McCarthy] wanted most was to knock me off. I guess he may hate me that much." He told the gloomy crowd at the hotel that he was going to reorganize his campaign; "I have decided to send Freckles home."[26]

The next morning he flew south to California. His press conference at the Los Angeles airport concentrated on Humphrey — on the Vice President's failure "to present his views to the voters of a single state," on the irrelevance of his "politics of joy" to "the conditions which are presently transforming our cities into armed camps." As for McCarthy, Kennedy said he would be happy to debate "with him and with the Vice President, or just the two of us."[27]

Of course winning would have been far better. Still, losing proved that Kennedy was human too. He seemed thereafter less ruthless, almost an underdog. "This defeat might be a help to your campaign — instead of a bitter blow," as Janet Auchincloss, Jacqueline Kennedy's mother, perceptively wrote him. "Somehow the first defeat — or setback — makes you a more sympathetic figure — and people will admire the courageous and graceful way you acknowledged it."[28]

As in Oregon, McCarthy was well organized and financed. Kennedy's California support, ranging from Jesse Unruh to Cesar Chavez, was multifarious and sometimes discordant. Unruh wanted to keep everything as much as possible in his own hands. Jack Conway and Walter Sheridan set up a parallel campaign for labor and the minorities. In the last weeks, when Stephen Smith had taken charge in Los Angeles and John Seigenthaler in San Francisco, a measure of coherence emerged (except that Seigenthaler brought along Paul Corbin, who organized Salinas under an assumed name, claiming to be a former lieutenant commander in the Navy. Corbin assured William Orrick that he could have been elected mayor of Salinas by the time he was through.)[29] Flying squads from the east — Alexander Bickel, Michael Harrington, Roger Hilsman, Michael Forrestal, Abba Schwartz, Edwin Reischauer, Roswell Gilpatric, George Plimpton, Pat Moynihan, Harry Golden, Adam Yarmolinsky, Marietta Tree, myself, others — invaded McCarthy strongholds in the colleges and the liberal community. Cesar Chavez registered Chicanos as never before. He also went to the campuses. When hecklers shouted, "Where was Kennedy when we were in New Hampshire?" Chavez would accurately reply, "He was walking with me in Delano!"[30]

I was on the train when Kennedy whistle-stopped through the Central Valley from Fresno to Sacramento. It was a brilliant, sunny, happy day at the end of May. The crowds were big and enthusiastic. At some stops they were mostly Mexican American. I was enchanted by the mixture of banter and intensity with which he beguiled and exhorted his hearers. At times he launched into wonderful travesties of stump speeches, pointing out how much more he could do for the local product, whatever it might be, because his family consumed more of it than any other family possibly could. Then he would turn with great seriousness to poverty, racial injustice, the war in Vietnam.[31]

Late that night he met with black militants in Oakland. "This may not be a pleasant experience," he warned John Glenn. "These people have got a lot of hostility and lots of reasons for it. When they get somebody like me, they're going to take it out on me. . . . But no matter how insulting a few of them may be, they're trying to communicate what's inside them."[32] After Kennedy talked, questions began. "It was a rough, gut-cutting meeting," said Seigenthaler, who was there, "in which a handful of people stood up and blistered white society and him as a symbol of white society."[33] Willie Brown, the black state assemblyman who was presiding, tried to compose things. They called him a "technicolor nigger." The Olympic decathlon champion, Rafer Johnson, who loved Kennedy, stood up to apologize for his people. They called him an Uncle Tom. Kennedy

asked him to sit down. Someone asked acidly what Kennedy really
thought of black people. He said he liked some, and some he did
not like; also he liked some white people and some he did not like.
"Look, man," said a local figure known as Black Jesus, "I don't want
to hear none of your shit. What the goddamned hell are you going
to do, boy. . . . You bastards haven't did nothing for us. We wants
to know, what are you going to do for us?"[34]

They drove back to San Francisco. Seigenthaler said he was sorry,
after such a long day, to have exposed Kennedy to this. Kennedy
said, "I'm glad I went. . . . I'll tell you why I'm glad I went. They
need to know somebody who'll listen."[35] Finally, almost to himself:
"after all the abuse the blacks have taken through the centuries,
whites are just going to have to let them get some of these feelings
out if we are all really ever going to settle down to a decent relationship."[36] It was a long journey from the meeting with James Baldwin
almost exactly five years before.

The next morning, they returned to West Oakland for a ghetto
rally. There were truculent shouts of "Free Huey!" from a squadron
of Black Panthers. To the surprise of the Kennedy party Black Jesus
was circulating cheerfully in the crowd. "I put a leaflet out that
Kennedy was coming in this area," he said later, "and that I wanted
him to be treated with the utmost respect." People told the Black
Panthers to shush as Kennedy began to speak. When he finished,
the crowd swarmed around the car, reaching out to touch him. The
car could not move, so, said Black Jesus, "I walked in front of the car
and raised my hand, and they parted so we could get through."[37]
The Reverend Hector Lopez, a community adviser: "Then a fascinating thing took place. . . . All of a sudden some Black Panthers got
out in front of the car and started shoving the people aside so the
car could carry on."[38] Willie Brown looked on with amazement.
"The same persons who were raising all the hell and asking all of the
very nasty questions and doing all of the loud screaming . . . were
the persons who were acting as his guards and . . . clearing the car
from the crowds."[39]

Black Americans, Hector Lopez reflected afterward, *believed* Kennedy. "What can you call Bobby? 'the last of the great liberals'? He
wouldn't have liked that. I know he wouldn't. I guess I'd have to
say he was 'the last of the great believables.'"[40]

IV

It was a tumultuous campaign. Kennedy was mobbed by admirers
in Watts, spat on by haters at San Francisco State. The McCarthy
campaign, aided by Humphrey money,[41] was hitting hard and clev-

erly at him. Television and radio spots questioned his courage and integrity. His early lead in the polls began to slip. He was holding his constituency — the blacks, the Chicanos, the poor — but suburban Democrats were sharply divided. The California Democracy had been strong for Adlai Stevenson. McCarthy appealed powerfully to the old Stevenson vote.

The debate, scheduled at last for June 1 in San Francisco, promised to be decisive. The attitude in the McCarthy camp, wrote Larner, was "lofty confidence." [42] Their candidate, they were sure, was cool, articulate, quick, the more mature, the more presidential. Kennedy knew he had to show both that he could handle McCarthy intellectually and, as ever, that he was not ruthless. His staff prepared the customary black, loose-leaf books on the issues. I do not know how carefully Kennedy went through them. But he did hold morning and afternoon briefing sessions that Saturday, meanwhile staring wistfully out the window at San Francisco sparkling in the sunlight below. It was a ravishing day. Around noon he went off to Fisherman's Wharf to campaign a little and savor the air.

We met again in the afternoon. I noted: "Bobby was in excellent form, funny, ironic and very much on the ball." [43] We wondered how to define the differences between the debaters. The Presidency seemed in 1968 too abstract as an issue. Race and the city? Kennedy had long believed that the only way to achieve integration was to give blacks the economic and psychological security that would enable them to become full members of a community. If a choice had to be made, this argued for the use of public money to improve the ghetto. A McCarthy position paper had advocated the dispersal strategy. Mankiewicz said, "Just to take people who are unemployed out of the ghetto and fill Orange County with them or Marin County isn't very helpful because they can't buy those houses anyway." [44]

McCarthy too had his briefing material. Thomas Finney, Clark Clifford's law partner, who had come west to run McCarthy's California campaign, tried to get him to focus on the debate. The managers were by now furious with their candidate's personal coterie. They called them the "astrologers," because they kept second-guessing the professionals. [45] Robert Lowell was the leading astrologer. "We tried to keep Robert Lowell away from McCarthy at very crucial times," said one McCarthy aide, "because we thought he always took the edge off." [46] We were all staying — McCarthyites and Kennedyites — at the Fairmont, San Francisco's traditional hotel for Democrats. I kept running into Lowell, a friend of twenty years, in the lobby. That Saturday morning Lowell, who thought that, if McCarthy lost, his people should support Kennedy and vice versa, had gone to see Kennedy. They had a fairly unsatisfactory talk. Kennedy, in Lowell's view, was making debater's points. Lowell said, "You mustn't talk to me this way." Kennedy said mildly that he guessed there was not much more to say. Lowell said, "I wish I could think up some joke that would cheer you up, but it won't do any good." Afterward he told McCarthy, "I felt like Rudolf Hess parachuting into Scotland."[47]

To the dismay of the professionals, the astrologers slipped through Finney's cordon and reached McCarthy before the debate. They made literary jokes and read poetry. "They castrated him at this point," said Thomas Morgan, McCarthy's press man. David Garth, his television man, rejoined, "No one is ever castrated if he doesn't want to be."[48] On the way to the studio Lowell and McCarthy composed a twentieth-century version of "Ode to St. Cecilia's Day."[49]

The debate was an anticlimax. At most three serious issues emerged. The rivals murkily disagreed as to whether the National Liberation Front should be brought into a coalition government before Vietnam negotiations began (as Kennedy thought McCarthy had been saying) or simply into the South Vietnamese political process (Kennedy's 1966 position). McCarthy went into his sequence about the government officials he planned to fire. Kennedy said, "I don't want to be playing games with people's reputations." The explosive moment came over the question of the ghettos. McCarthy argued for dispersal; Kennedy, for reconstruction. McCarthy called this "practical apartheid." Kennedy said he was all in favor of moving people out of the ghettos; but, at the present time, "to take these people out, put them in suburbs where they can't afford the housing, where their children can't keep up with the schools, and where they don't have the skills for the jobs, it is just going to be catastrophic." He added, "You say you are going to take ten thousand black people and move them into Orange County."[50] This sounded, and was, demagogic. It was not so demagogic as it sounded, however, because it had been Kennedy's consistent position. Later Kennedy told Mankiewicz that he started to say Marin County, but "I forgot at the last minute whether it was Marin or Merrin County [he did not want to demonstrate ignorance of the state by mispronunciation] so I didn't use that."[51] McCarthy replied weakly, saying that he had not understood Kennedy wished to concentrate so much on the ghetto. "One could only conclude," observed the *London Sunday Times* team, "that McCarthy had not been listening to what Kennedy had been saying for several years."[52] (On Tuesday McCarthy beat Kennedy in Orange County.)

Larner watched the debate with surprise. He thought McCarthy had the best of the opening exchange on Vietnam, but "Kennedy had had the guts to get up off the floor and fight it through, while McCarthy, dazed, was taking every punch."[53] At the end McCarthy made a slight comeback. Kennedy, for some reason, did not understand there were to be sum-ups and was caught off guard, and McCarthy was at his most eloquent. The press regarded the debate as a stand-off. But a stand-off, in the context of expectations, was a Kennedy victory. "It was clear," concluded Larner, "that Kennedy could take McCarthy head-on, with no fear of his magic powers. If McCarthy had something new and different going for him in American politics, it did not show in open competition with Kennedy."[54] McCarthy himself looked grim afterward. "He flubbed it!" cried Tom Finney. "He flubbed it!"[55] A *Los Angeles Times* telephone poll showed Kennedy the victor by 2½–1. The slide toward McCarthy stopped.

v

We had a delightful dinner afterward. Hodding Carter, the old Mississippi editor, now nearly blind, joined us. I remember the marked consideration Kennedy, with so much else on his mind, showed him. Then we went on to a fund-raising gala filled with Hollywood stars. Later I took Ethel, who was carrying her eleventh child, back to the Fairmont.* Robert went on to a party with the performers. The next day I flew east to attend a conference on Vietnam at the University of Chicago.

On Monday, June 3, Kennedy made a final dash around the state. He visited San Francisco's Chinatown, where Abba Schwartz had organized the Chinese Americans. As the car moved through cheering crowds, shots appeared to ring out. Kennedy continued to stand and wave while motioning a friend toward Ethel, who, pale and stricken, had slumped down in her seat. The shots were Chinese firecrackers. Impartially they visited the Japanese Cultural Center, where the Kennedys and Seigenthaler sang the Waseda song. Robert flew back to Los Angeles, spoke in Long Beach, went on to San Diego. By the end of the long day he was worn to the bone and near digestive collapse. He took Ethel and six children for the night to the Malibu beach house of John Frankenheimer, the film director.

He slept late. It was a sullen day. A cold fog hung over the ocean. Theodore White came out for luncheon; later Goodwin and Dutton. Kennedy played with his children on the beach; then

---

* Rory Elizabeth Katherine Kennedy was born on December 12, 1968.

plunged into the gray sea. For a moment his son David was caught in the undertow. Kennedy dived and brought him to the surface. The boy's forehead was bruised. When Goodwin arrived, he looked around for Kennedy. He found him by the pool, stretched out across two chairs, his lips slightly parted, motionless. Goodwin's stomach contracted with a spasm of fear. But he was sleeping, only sleeping. "God," Goodwin thought, "I suppose none of us will ever get over John Kennedy." [56]

White called CBS for its early projections. The first reports gave Kennedy 49 percent.* The candidate asked Dutton what could be done in the hours remaining to push it up to 50 percent. Then he yawned and went off for a nap. About six-thirty Frankenheimer drove him to the Hotel Ambassador. He sped furiously along the Santa Monica Freeway. "Take it easy, John," Kennedy said. "Life is too short." [57]

The polls closed earlier in South Dakota. George McGovern called in mid-evening. Kennedy had more votes, he said, than Humphrey and McCarthy, the local boys, combined. He had swept the farmers. He had swept the Indians. [58] In the Kennedy suite they talked about the next steps. Stephen Smith scheduled a series of meetings to work out the campaign reorganization. David Hackett had arrived from the boiler room with a breakdown of state delegations; Humphrey, 944 delegates; Kennedy, 524½; McCarthy, 204; 872 undecided. The objective by convention time was 1432½ for Kennedy, 1152½ for Humphrey. [59]

The key was McCarthy. Looking for a place to talk in the crowded suite, Kennedy finally took Goodwin into a bathroom. "I've got to get free of McCarthy," he said. "While we're fighting each other, Humphrey's running around the country picking up delegates. I don't want to stand on every street corner in New York for the next two weeks [contesting the New York primary]. I've got to spend that time going to the states, talking to delegates before it's too late. My only chance is to chase Hubert's ass all over the country. Maybe he'll fold. . . . Even if McCarthy won't get out, his people must know after tonight that I'm the only candidate against the war that can beat Humphrey." [60] Goodwin began calling the pro-Kennedy wing of the McCarthy movement — Kenneth Galbraith, Allard Lowenstein — asking whether they did not think the time had arrived to come over. Galbraith said, "I rather [think] this might be so. But I shouldn't do it without going and seeing Gene McCarthy first." [61] Lowenstein said he would come over if Galbraith did. [62]

---

* This turned out to be optimistic. The final result was 46.3 percent for Kennedy and 41.8 percent for McCarthy. The rest went to a Humphrey slate headed by the state attorney general.

Old friends milled around the suite — Newfield, Hamill and Breslin, Charles Evers and John Lewis. Kennedy talked to Budd Schulberg about the Watts Writers Workshop. "You've touched a nerve," he said. ". . . This workshop idea of yours is a kind of throwback to the Federal Theater and Writers Project of the New Deal. . . . I'd like to see it on a national scale with Federal help."[63] He spoke to Daley in Chicago, who wished him good luck and hinted that, if he won in California, Illinois would support him at the convention.[64] Jesse Unruh said it was approaching midnight: time to go down to the ballroom. Kennedy looked over suggestions for his victory statement. The draft said how great the McCarthy movement was. "I'd like to say something nice about him personally," Kennedy said.[65]

Failing to find Cesar Chavez, who had gone out in search of his wife, Kennedy asked Dolores Huerta instead to escort him downstairs. Now before the cheering crowd, he thanked those who had helped him: "ruthless" Steve Smith, Chavez, Unruh, Paul Schrade of the UAW, Rafer Johnson and Roosevelt Grier, the two devoted black athletes. "I want to express my gratitude to my dog Freckles, . . . I'm not doing this in any order of importance, but I also want to thank my wife Ethel. Her patience during this whole effort was fantastic."[66]

For a moment he was serious. "Here is [California] the most urban state of any of the states of our Union, South Dakota the most rural of any of the states of our Union. We were able to win them both. I think that we can end the divisions within the United States." He congratulated McCarthy and his followers and asked them to join with him "not for myself, but for the cause and the ideas which moved you to begin this great popular movement."

"What I think is quite clear," he said, "is that we can work together in the last analysis, and that what has been going on within the United States over a period of the last three years — the division, the violence, the disenchantment with our society; the divisions, whether it's between blacks and whites, between the poor and the more affluent, or between age groups or on the war in Vietnam — is that we can start to work together. We are a great country, an unselfish country, and a compassionate country. I intend to make that my basis for running." Jeremy Larner, listening over at the McCarthy hotel, thought it "his best speech of the campaign."[67] The crowd in the ballroom cheered and cheered. Then Kennedy left the room for a press conference, taking a short cut through the kitchen.

*   *   *

I heard the California returns at Saul Bellow's apartment in Chicago. Richard Wade and Frances FitzGerald were with us. When it was evident that Kennedy had won, I went back to my hotel and tried to call him in Los Angeles, but the line was always busy. I went to sleep. The phone rang. Wade said in a choked voice, "Turn on your television. He's been shot."

On Thursday evening I stood with many others in the soft summer night at La Guardia Airport waiting for the plane, as I had waited on a November night in Washington an eon before. Ethel Kennedy, incredibly composed, told me to come to the United Nations Plaza apartment. When I arrived, she was terrifyingly solicitous of stricken friends. Her first words to me were, "You were in Chicago for a meeting, weren't you? Wasn't it on Vietnam? How did it go?" Rose Kennedy was stoically there. Jean Smith and Eunice Shriver said good night to their mother. She said, "I'm so glad all you children are home again." [68]

At the hospital in Los Angeles, Jacqueline Kennedy had told Mankiewicz: "The Church is . . . at its best only at the time of death. The rest of the time it's often rather silly little men running around in their black suits. But the Catholic Church understands death. I'll tell you who else understands death are the black churches. I remember at the funeral of Martin Luther King. I was looking at those faces, and I realized that they know death. They see it all the time and they're ready for it . . . in the way in which a good Catholic is." Then she said, "We know death. . . . As a matter of fact, if it weren't for the children, we'd welcome it." [69]

Well before dawn Friday morning, lines began to form around St. Patrick's Cathedral. Through the night, friends of Robert Kennedy had stood honor guard by the casket. George McGovern noticed Richard Daley, his head bowed, the cords of his neck standing out, crying uncontrollably. [70] Tom Hayden, the revolutionary, a green cap from Havana sticking out of his pocket, wept silently by himself in a back pew. Jack Newfield remembered the quotation from Pascal that Camus had used in *Resistance, Rebellion and Death:* "A man does not show his greatness by being at one extremity, but rather by touching both at once." [71]

On Saturday morning came the mass at St. Patrick's, with Edward Kennedy's moving speech and the singing of "The Battle Hymn of the Republic." Then the train made its long journey south under the savage sun. I said to O'Donnell, "What marvelous crowds!" He said, "Yes, but what are they good for?" Coretta King looked at Ethel Kennedy and said — who had better earned the right to say it? — "I don't see how she has been able to go through this awful experience with such dignity." [72]

The train arrived in Washington. Night had fallen. Mourners

with twinkling candles followed the coffin into Arlington Cemetery. "There was," wrote a grieving Lady Bird Johnson, "a great white moon riding high in the sky." [73] But the cemetery itself was dark and shadowed. The pallbearers, not sure where to place the coffin, walked on uncertainly in the night. Averell Harriman finally said to Stephen Smith, "Steve, do you know where you're going?" Smith said, "Well, I'm not sure." Then Smith said, "I distinctly heard a voice coming out of the coffin saying, 'Damn it. If you fellows put me down, I'll show you the way.'"

NOTES
INDEX

Notes

See Acknowledgments for the location of manuscript collections and also for explanations of certain abbreviations. When two or more oral history interviews were held the same day, they are designated by a roman numeral following the date.

Prologue: 1968 *(page 1)*

1. Frank Church of Idaho, *Congressional Record,* July 30, 1968, S9713.
2. The UPI estimate; the Editors of United Press International and Cowles, *Assassination: Robert F. Kennedy — 1925-1968,* ed. Francine Klagsbrun and D. C. Whitney (New York, 1968), 201. See also *New York Times,* June 9, 1968.

1. The Family *(pages 3-20)*

1. James Joyce, *Ulysses* (New York, 1961), 34.
2. Rose Fitzgerald Kennedy's memoir, *Times to Remember* (Garden City, N.Y., 1974), is illuminating and indispensable; the verse from a Boston paper is on 10. For the Kennedy family, and Joseph P. Kennedy in particular, there are two serious and instructive, if hostile, biographies: R. J. Whalen, *The Founding Father: The Story of Joseph P. Kennedy* (New York, 1964); and D. E. Koskoff, *Joseph P. Kennedy: A Life and Times* (Englewood Cliffs, N.J., 1974).
3. Gail Cameron, *Rose* (New York: Dell reprint, 1972), 52.
4. As told by James A. Fayne, an old Kennedy associate, to R. J. Whalen, in Whalen, *Founding Father,* 44.
5. E. K. Lindley, "Will Kennedy Run for President?" *Liberty,* May 21, 1938; Whalen, *Founding Father,* 49.
6. Whalen, *Founding Father,* 104.
7. Joe McCarthy, *The Remarkable Kennedys* (New York, 1960), 53.
8. Michael Mooney to RFK, August 10, 1966, RFK Papers.
9. RFK to Mooney, August 18, 1966, RFK Papers.
10. Whalen, *Founding Father,* 74, 61.
11. Edward M. Kennedy, ed., *The Fruitful Bough: A Tribute to Joseph P. Kennedy* (privately printed, 1965), 112.
12. Ibid., 12.
13. McCarthy, *Remarkable Kennedys,* 66.
14. E. M. Kennedy, *Fruitful Bough,* 32.
15. Joseph P. Kennedy, *I'm for Roosevelt* (New York, 1936), 3.

16. Rose Kennedy, *Times to Remember,* 195.
17. McCarthy, *Remarkable Kennedys,* 58.
18. JPK to Felix Frankfurter, December 5, 1933, JPK Papers.
19. Raymond Moley, *The First New Deal* (New York, 1966), 381.
20. JPK to JPK, Jr., May 4, 1934, JPK Papers.
21. Arthur M. Schlesinger, Jr., *The Coming of the New Deal* (Boston, 1958), 467–468.
22. Raymond Moley, *After Seven Years* (New York, 1939), 288.
23. Harold L. Ickes, *The Secret Diary . . . : The First Thousand Days, 1933–1936* (New York, 1953), 173.
24. From a memorandum written by Rose Kennedy, evidently in 1940, JPK Papers.
25. Churchill to JPK, October 12, 1935, JPK Papers.
26. JPK to Churchill, October 19, 1935, JPK Papers.
27. Press conference no. 309, July 29, 1936, in Franklin D. Roosevelt, *Complete Presidential Press Conferences* (New York, 1972), vol. 8, 34–35.
28. Arthur Krock, *Memoirs: Sixty Years on the Firing Line* (New York, 1968), 332.
29. JPK, *I'm for Roosevelt,* 3, 7, 14, 93, 102–107.
30. JPK, address before the Democratic Businessmen's League of Massachusetts, October 24, 1936, Roosevelt Papers.
31. J. P. Kennedy, "The New Deal and Business," radio speech, October 21, 1936, JPK Papers. Kennedy wrote Roosevelt, October 24, 1936, that the speech had been partly written by a French journalist.
32. *New York Times,* January 28, 1957, quoted in William V. Shannon, *The American Irish* (New York, 1963), vii. The second sentence is ordinarily given as "My children were born here," but it makes much less sense that way.
33. JPK to Joseph I. Breen, March 16, 1937, JPK Papers.
34. Mary Bailey Gimbel, in interview by author, February 19, 1975.
35. Rose Kennedy, *Times to Remember,* 79.
36. Jean Stein and George Plimpton, eds., *American Journey* (New York, 1970), 35.
37. Rose Kennedy, *Times to Remember,* 148.
38. In the privately printed memorial volume *As We Remember Joe,* ed. John F. Kennedy (Cambridge, Mass., 1945), 4.
39. JPK, *As We Remember Joe,* 43–44.
40. JPK to Robert W. Bingham, November 11, 1935, JPK Papers.
41. Rose Kennedy, *Times to Remember,* 202.
42. Ibid., 121.
43. James MacG. Burns, *John Kennedy: A Political Profile* (New York, 1960), 28.
44. Myra McPherson, "'Losing Was Never Funny to Joseph P. Kennedy,'" *Washington Post,* November 19, 1969.
45. Rose Kennedy, *Times to Remember,* 143.
46. E. M. Kennedy, *Fruitful Bough,* 203.
47. Rose Kennedy, *Times to Remember,* 142, 144.
48. E. M. Kennedy, *Fruitful Bough,* 210–211.
49. Stein and Plimpton, *American Journey,* 35.
50. Burns, *John Kennedy,* 20.
51. Stein and Plimpton, *American Journey,* 35.
52. Charles Spalding, in recorded interview by L. J. Hackman, March 22, 1969, 70, RFK Oral History Program.
53. Felicia Warburg Roosevelt, *Doers and Dowagers* (Garden City, N.Y., 1975), 94.
54. Robert F. Kennedy, *The Pursuit of Justice* (New York: Harper & Row, Perennial Library reprint, 1964), 3.
55. Ralph Horton, Jr., in recorded interview by Joseph Dolan, June 1, 1964, 3–4, JFK Oral History Program.
56. JPK to JFK, October 10, 1934, December 5, 1934, April 29, 1935, JPK Papers.
57. Interview with Walter Cronkite, September 19, 1960, in Senate Commerce Committee, *Freedom of Communications,* pt. 3, *The Joint Appearances of Senator John F. Kennedy and Vice President Richard M. Nixon,* 87 Cong., 1 Sess. (1961), 54–55.
58. Stein and Plimpton, *American Journey,* 35.
59. Rose Kennedy, *Times to Remember,* 139.

60. K. LeMoyne Billings, in interview by author, July 8, 1975.
61. Charles Spalding, in interview by author, February 24, 1975.
62. Rose Kennedy, in recorded interview by Felicia Warburg, April 6, 1973, tape in possession of interviewer.
63. Rose Kennedy, *Times to Remember*, 162.
64. Spalding, in interview by author.
65. Rose Kennedy, *Times to Remember*, 163.
66. Jean Kennedy Smith, in interview by author, May 11, 1975.
67. Rose Kennedy to Edward Kennedy, April 18, 1945, JPK Papers.
68. Rose Kennedy to daughters-in-law, December 1, 1960, August 26, 1969, JPK Papers.
69. Rose Kennedy, *Times to Remember*, 148.
70. Charles Spalding, in recorded interview by John Stewart, March 14, 1968, 3–4, JFK Oral History Program.
71. Author's journal, March 7, 1962.
72. Franklin D. Roosevelt, Jr., in recorded interview by Jean Stein, December 9, 1969, Stein Papers.
73. "Ted Kennedy's Memories of JFK: A Conversation with Theodore Sorensen," *McCall's*, November 1973.
74. K. LeMoyne Billings, in recorded interview by Jean Stein, n.d., 13, Stein Papers.
75. E. M. Kennedy, *Fruitful Bough*, 214.
76. Stein and Plimpton, *American Journey*, 35.
77. William Manchester, *Portrait of a President* (Boston, 1962), 187.
78. JPK to JPK, Jr., May 4, 1934, JPK Papers.
79. In a memorandum, "Russia," n.d. but probably c. 1940, JPK Papers.
80. Rose Kennedy, *Times to Remember*, 172–173.
81. Ibid., 210.
82. Ibid., 173.
83. As told by Eunice Kennedy Shriver, in E. M. Kennedy, *Fruitful Bough*, 219.
84. Whalen, *Founding Father*, 171.
85. Waldrop to author, April 18, 1975.
86. As told by William vanden Heuvel in interview with author, May 1, 1975.
87. Joseph F. Dineen, *The Kennedy Family* (Boston, 1959), 110.
88. E. M. Kennedy, *Fruitful Bough*, 214.
89. Ibid., 33.
90. As told to Alexandra Emmet, c. 1961.

2. The Father *(pages 21–40)*

1. Rose Kennedy, *Times to Remember* (Garden City, N.Y., 1974), 102–103.
2. K. LeMoyne Billings, in interview by author, July 8, 1975.
3. Margaret Laing, *The Next Kennedy* (New York, 1968), 64–65.
4. RFK to JPK, April 29, 1935, JPK Papers.
5. Reproduced in the *Washington Post*, February 24, 1970.
6. Ann Geracimos, "Bobby Kennedy Was Here," *New York Herald Tribune*, October 11, 1964.
7. Camp Winona Report, n.d. [1937], JPK Papers.
8. Rose Kennedy, *Times to Remember*, 102.
9. JPK to JPK, Jr., March 2, 1937, JPK Papers.
10. Jack Newfield, *Robert Kennedy: A Memoir* (New York, 1969), 41–42.
11. R. E. Thompson and Hortense Myers, *Robert F. Kennedy: The Brother Within* (New York, 1962), 43.
12. William V. Shannon, *The Heir Apparent: Robert Kennedy and the Struggle for Power* (New York, 1967), 43.
13. D. E. Koskoff, *Joseph P. Kennedy: A Life and Times* (Englewood Cliffs, N.J., 1974), 90.
14. Manuscript of June 1937, JPK Papers.

15. R. J. Whalen, *The Founding Father: The Story of Joseph P. Kennedy* (New York, 1964), 197.
16. Robert I. Gannon, *The Cardinal Spellman Story* (New York: Pocket Books reprint, 1962), 142.
17. Kennedy is not mentioned in the biography of Ryan by F. L. Broderick, *The Right Reverend New Dealer: John A. Ryan* (New York, 1963). See also G. Q. Flynn, *American Catholics and the Roosevelt Presidency, 1932–1936* (Lexington, Ky., 1968).
18. James A. Farley, *Jim Farley's Story: The Roosevelt Years* (New York, 1948), 198.
19. Henry Morgenthau, Jr., diary, April 13, 1935, FDR Papers; Harold L. Ickes, *The Secret Diary . . . : The First Thousand Days, 1933–1936* (New York, 1953), 692.
20. Rose Kennedy, memoranda "The Roosevelts" and "A Description of the President," JPK Papers.
21. W. O. Douglas, *Go East, Young Man* (New York, 1974), 281.
22. Edward Kennedy, ed., *The Fruitful Bough: A Tribute to Joseph P. Kennedy* (privately printed, 1965), 73.
23. Farley, *Jim Farley's Story*, 114–115.
24. Arthur Krock, in recorded interview by Charles Bartlett, May 10, 1964, 3, JFK Oral History Program.
25. Rose Kennedy, memorandum, "London" [c. 1940], JPK Papers.
26. AP dispatch in *Portland Oregonian*, March 10, 1938.
27. Koskoff, *Joseph P. Kennedy*, 377.
28. George F. Kennan, *Memoirs, 1925–1950* (Boston, 1967), 91–92.
29. "From Baseball Captain to Ambassador," unsigned typescript, 14, JPK Papers.
30. Reverend J. Butterworth, "The Ambassador's Sons Help British Youth," n.d., JPK Papers; "Bobby Kennedy, 13, Proves Chip Off the Old (Diplomatic) Block," *New York Herald Tribune*, April 3, 1939.
31. *Washington Star*, November 18, 1969.
32. Nicholas Bethell, *The War Hitler Won* (Mount Kisco, N.Y.: Futura reprint, 1976), 280.
33. Edward Moore to Paul Murphy, January 24, 1939, JPK Papers.
34. Rose Kennedy, *Times to Remember*, 221.
35. Rose Kennedy to FDR, January 8, 1939, Roosevelt Papers.
36. JPK, "Experiences in London" (address at Oglethorpe University, May 24, 1941), 13, JPK Papers.
37. Koskoff, *Joseph P. Kennedy*, 158; see also J. M. Blum, *From the Morgenthau Diaries: Years of Crisis, 1928–1938* (Boston, 1959), 518.
38. Koskoff, *Joseph P. Kennedy*, 129.
39. Rose Kennedy, "Notes on Chamberlain" [c. 1940], JPK Papers.
40. Walter Lippmann to JPK, April 7, 1938, Lippmann Papers.
41. Blum, *Years of Crisis*, 518; Farley, *Jim Farley's Story*, 199.
42. J. P. Moffat, *The Moffat Papers*, ed. N. H. Hooker (Cambridge, 1956), 220–221; Koskoff, *Joseph P. Kennedy*, 158–159.
43. *New York World-Telegram*, October 25, 1938.
44. Franklin D. Roosevelt, *Public Papers . . . 1938* (New York, 1941), 564.
45. Koskoff, *Joseph P. Kennedy*, 149–150, 168.
46. JPK, "Experiences in London," 10–11.
47. JPK, "Summary of Strategic Situation," March 3, 1939, FDR Papers; cited in W. F. Kimball, *The Most Unsordid Act: Lend-Lease, 1939–1941* (Baltimore, 1969), 20–21.
48. Harold Ickes, *Secret Diary . . . . The Inside Struggle, 1936–1939* (New York, 1954), 707.
49. Joseph Alsop and Robert Kintner, *American White Paper: The Story of American Diplomacy and the Second World War* (New York, 1940), 68.
50. JPK to Rose Kennedy, n.d. [September 1939], JPK Papers.
51. Rose Kennedy to Ethel Kennedy, January 30, 1967, JPK Papers.
52. JPK to Rose Kennedy, September 24, 1939, JPK Papers.
53. RFK to his parents, n.d. [April 1942], Hyannis Port Papers.
54. RFK to Rose Kennedy, n.d., JPK Papers.

55. *Boston Record,* November 22, 1939.
56. RFK to his parents, n.d. [November 1940], Hyannis Port Papers.
57. See report card, February 21, 1941, JPK Papers.
58. RFK letter to his mother, n.d. [spring 1940], Hyannis Port Papers.
59. Reverend Gregory Borgstedt to JPK, March 8, 1940, JPK Papers.
60. Francis I. Brady, assistant headmaster, to Rose Kennedy, May 6, 1941, JPK Papers.
61. RFK to his parents, n.d. [October 1941], Hyannis Port Papers.
62. RFK to his parents, n.d. [January or February 1942], Hyannis Port Papers.
63. RFK to his parents, n.d. [November 1941], Hyannis Port Papers.
64. RFK to his parents, n.d. [January or February 1942], Hyannis Port Papers.
65. Interview with Edward M. Kennedy, February 23, 1973.
66. Whalen, *Founding Father,* 285.
67. W. L. Langer and S. E. Gleason, *The Challenge to Isolation, 1937–1940* (New York, 1952), 345.
68. Robert Murphy, *Diplomat among Warriors* (New York, 1964), 38.
69. *Spectator,* March 8, 1940; Alexander Kendrick, *Prime Time: The Life of Edward R. Murrow* (Boston, 1969), 192.
70. Raymond E. Lee, *London Journal . . . 1940–1941,* ed. James Leutze (Boston, 1971), 219, 241.
71. O. H. Bullitt, *For the President: Personal and Secret Correspondence Between Franklin D. Roosevelt and William C. Bullitt* (Boston, 1972), 437.
72. Koskoff, *Joseph P. Kennedy,* 239.
73. Bethell, *War Hitler Won,* 283, 285.
74. JPK to Rose Kennedy, March 14, 1940, JPK Papers.
75. Rose Kennedy to JPK, March 31 [1940], JPK Papers.
76. JPK to Rose Kennedy, April 26, 1940, JPK Papers.
77. Cordell Hull, *Memoirs* (New York, 1948), vol. 1, 766.
78. FDR to JPK, May 3, 1940, in Franklin D. Roosevelt, *His Personal Letters, 1928–1945,* ed. Elliott Roosevelt and Joseph P. Lash (New York, 1950), vol. 2, 1020.
79. JPK to Rose Kennedy, May 20, 1940, JPK Papers.
80. Koskoff, *Joseph P. Kennedy,* 263.
81. Rose Kennedy to JPK, letters in JPK Papers.
82. JPK to Rose Kennedy, September 10, 1940, JPK Papers.
83. Ibid.
84. Herbert Hoover, memorandum, May 15, 1945, Hoover Papers. See also Kennedy's discussion with James V. Forrestal in 1945 of Chamberlain's conviction "that America and the world Jews had forced England into the war." James V. Forrestal, *The Forrestal Diaries,* ed. Walter Millis (New York, 1951), 122.
85. JPK, draft autobiography, ch. 17, Landis Papers.
86. Koskoff, *Joseph P. Kennedy,* 166, 188.
87. [D. W. Brogan], "R.F.K.," *Times Literary Supplement,* August 1, 1968. The review was of course unsigned in that benighted period. I make this confident attribution on the basis of internal evidence.
88. FDR to JPK, August 28, 1940, in Roosevelt, *Personal Letters,* vol. 2, 1061; Koskoff, *Joseph P. Kennedy,* 268.
89. James F. Byrnes, *All in One Lifetime* (New York, 1958), 125.
90. Lee, *London Journal,* 115.
91. Krock, *Memoirs,* 335.
92. Rose Kennedy, "Visit to Washington" [c. 1941], JPK Papers.
93. Byrnes, *All in One Lifetime,* 126.
94. *Boston Globe,* November 10, 1940.
95. Charles A. Lindbergh, *Wartime Journals* (New York, 1970), 420.
96. JPK to Harvey Klemmer, August 4, 1941, JPK Papers.
97. RFK to JPK, January 18, 1941, JPK Papers.
98. RFK to Rose Kennedy, n.d. [January 1941], JPK Papers.
99. Koskoff, *Joseph P. Kennedy,* 309.
100. Kimball, *Most Unsordid Act,* 191; Koskoff, *Joseph P. Kennedy,* 309.

101. Harry Golden, "The Bobby Twins Revisited," *Esquire*, June 1965.
102. Transcript of interview with JPK by Charles Colebaugh and William Hillman of *Collier's*, November 26, 1940, JPK Papers.
103. JPK to Charles F. Adams, April 7, 1941, JPK Papers.
104. Frank Waldrop to author, April 18, 1975.
105. JPK to Adams.
106. Kennedy, "Experiences in London," passim.
107. Koskoff, *Joseph P. Kennedy*, 313–314.
108. Herbert Hoover to JPK, July 1, 1941, JPK to Hoover, July 11, 29, 1941, Hoover Papers.
109. JPK, Jr., to John T. Flynn, January 3, 1941, JPK Papers.
110. March 9, 1941; see flier in JPK Papers.
111. JPK to Landis, August 6, 1940, Landis Papers.
112. Burns, *John Kennedy*, 42.
113. John F. Kennedy, *Why England Slept* (New York: Doubleday, Dolphin reprint, 1962), 178, 184–185.
114. Joan Blair and Clay Blair, Jr., *In Search of J. F. K.* (New York, 1975), 113–144.

3. The War *(pages 41–61)*

1. RFK to his parents, n.d. [January or February 1942], Hyannis Port Papers.
2. RFK to Rose Kennedy, n.d. [spring 1942], Hyannis Port Papers.
3. Rose Kennedy to the children, January 5, 1942, JPK Papers.
4. RFK to his parents, n.d. [January 1942], Hyannis Port Papers.
5. Rose Kennedy to the children, February 2, 1942, JPK Papers.
6. Rose Kennedy to RFK, January 12, 1942, RFK Papers.
7. Mary Bailey Gimbel, in interview by author, February 19, 1975.
8. RFK to Rose Kennedy, December 7, 1942, Hyannis Port Papers.
9. RFK to JPK, n.d. [March or April 1943], Hyannis Port Papers.
10. Samuel Adams, in recorded interview by Jean Stein, May 1970, 2, Stein Papers.
11. RFK to his parents, n.d. [January 1953], Hyannis Port Papers.
12. *Dedham County Recorder*, September 10, 1943. For more on Clem Norton, see my introduction to *The Best and Last of Edwin O'Connor*, ed. Arthur M. Schlesinger, Jr. (Boston, 1970), 10.
13. RFK to JPK, December 13, 1942, Hyannis Port Papers.
14. RFK to Rose Kennedy, n.d. [September 1943], JPK Papers.
15. Patricia Kennedy Lawford, ed., *That Shining Hour* (n.p., 1969), 5.
16. RFK to JPK, n.d. [October 1943], Hyannis Port Papers.
17. Mary Bailey Gimbel, in interview by author; Jean Stein and George Plimpton, eds., *American Journey: The Times of Robert Kennedy* (New York, 1970).
18. Adams, in Stein interview, 2–4.
19. Lawford, *That Shining Hour*, 6.
20. RFK to Rose Kennedy, n.d. [early 1943], Hyannis Port Papers.
21. RFK to Rose Kennedy, n.d. [January 1944], Hyannis Port Papers.
22. JPK to Kathleen Kennedy, January 17, 1944, JPK Papers.
23. Adams, in Stein interview, 1; Mary Bailey Gimbel, in interview by Jean Stein, n.d., 2–3, Stein Papers.
24. John Knowles, *A Separate Peace* (New York, Bantam reprint, 1966), 6, 195.
25. David Hackett, in recorded interview by John Douglas, July 22, 1970, 4, RFK Oral History Program.
26. David Hackett, in interview by author, January 27, 1975.
27. Adams, in Stein interview, 1.
28. Cleveland Amory, "Curmudgeon-at-Large," *Saturday Review*, September 7, 1974.
29. Albert Norris to Rose Kennedy, January 3, 194[4], JPK Papers.
30. Norris to author, March 6, 1975.
31. Arthur B. Perry to JPK, January 15, 1944, JPK Papers.
32. JPK to FDR, December 7, 1941, March 4, 1942, JPK Papers. See also Franklin D. Roosevelt, *His Personal Letters, 1928–1945*, ed. Elliott Roosevelt and Joseph P. Lash (New York, 1950), vol. 2, 1289–1290.

33. JPK to Beaverbrook, August 12, 1942, Beaverbrook Papers.
34. David E. Koskoff, *Joseph P. Kennedy: A Life and Times* (Englewood Cliffs, N.J., 1974), 318, 577.
35. J. P. Lash, ed., *From the Diaries of Felix Frankfurter* (New York, 1974), 238.
36. Henry A. Wallace, *The Price of Vision: The Diary . . . 1942-1946*, ed. J. M. Blum (Boston, 1973), 144.
37. JPK to Kent, March 2, 1943, JPK Papers.
38. Kennedy to Beaverbrook, August 12, 1942, Beaverbrook Papers.
39. W. V. Shannon, *The American Irish* (New York, 1963), 357.
40. Constance Casey, in interview by author, May 14, 1975. The Caseys, who were political realists, understood the problem and were in later years good friends of John and Robert Kennedy's.
41. *Boston Herald*, August 16, 1942.
42. Edward M. Kennedy, ed., *The Fruitful Bough* (n.p., 1965), 242-243.
43. Arthur Krock, *Memoirs: Sixty Years on the Firing Line* (New York, 1968), 357.
44. JPK to Frank Kent, March 2, 1943, JPK Papers.
45. JFK to his parents, May 14, 1943, JPK Papers.
46. JFK to Henry ———, n.d. [August or September 1943], JPK Papers.
47. JFK to his parents, September 12, 1943, JPK Papers.
48. JFK to his parents, n.d. [November 1943], JPK Papers.
49. Susan Mary Alsop, *To Marietta from Paris, 1945-1960* (Garden City, N.Y., 1975), 90, 92.
50. Rose Kennedy, *Times to Remember* (New York, 1974), 285.
51. JPK to Rev. Maurice S. Sheehy, October 28, 1942, JPK Papers.
52. JFK to RFK, November 21, 1942, RFK Papers. General Lewis Hershey was, of course, director of the Selective Service System.
53. JFK to RFK, postmarked January 10, 1943, JPK Papers.
54. JPK to JFK, April 22, 1943, JPK Papers.
55. JFK to JPK, n.d. [autumn 1943], JPK Papers.
56. *Boston Record*, October 12, 1943.
57. JPK, Jr., to his parents, November 9, 1943, JPK Papers.
58. JFK to RFK, November 14, 1943, in R. J. Donovan, *PT 109* (New York, 1961), 236.
59. RFK to his parents, n.d. [March 1944], Hyannis Port Papers.
60. JPK to RFK, April 6, 1944, JPK Papers.
61. RFK to JPK, Jr., n.d. [spring 1944], JPK Papers.
62. RFK to parents, n.d. [spring 1944], RFK Papers.
63. The letter was addressed to "Mr. David Draftdodger Hackett," n.d. [April 1944], Hackett Papers.
64. To "Mr. David (the Brain) Hackett," postmarked April 3, 1944, Hackett Papers.
65. To "Pvt. David Low Hackett," postmarked September 5, 1944, Hackett Papers.
66. Rose Kennedy, *Times to Remember*, 30.
67. Kathleen Kennedy to JFK, July 29 [1943], in Rose Kennedy, *Times to Remember*, 292.
68. JPK, Jr., to JPK, April 10, 1944, JPK Papers.
69. JPK to Kathleen, April 1944, in E. M. Kennedy, *Fruitful Bough*, 208.
70. JPK to Kathleen, April 27, 1944, JPK Papers.
71. John F. Kennedy, ed., *As We Remember Joe* (Cambridge, Mass., 1945), 54.
72. Wallace, *Price of Vision*, 328.
73. Jean Kennedy Smith, in interview by author, September 7, 1975.
74. JPK to Beaverbrook, May 24, 1944, Beaverbrook Papers.
75. For more on Joseph P. Kennedy, Jr., see Hank Searls, *The Lost Prince: Young Joe, The Forgotten Kennedy* (New York, 1969); and Jack Olsen, *Aphrodite: Desperate Mission* (New York, 1970).
76. JFK, *As We Remember Joe*, 5.
77. Joe McCarthy, *The Remarkable Kennedys* (New York, 1960), 106.
78. JPK to Grace Tully, August 29, 1944, in Koskoff, *Joseph P. Kennedy*, 374-375.
79. *Boston Record*, May 9, 1957.

80. JPK to Beaverbrook, October 23, 1944, Beaverbrook Papers.
81. See, e.g., R. J. Whalen, *The Founding Father: The Story of Joseph P. Kennedy* (New York, 1964), 370; and Koskoff, *Joseph P. Kennedy*, 335.
82. Rose Kennedy, *Times to Remember*, 301.
83. RFK to Hackett, postmarked January 23, 1945, Hackett Papers.
84. Joan Blair and Clay Blair, Jr., *In Search of J. F. K.* (New York, 1976), 346.
85. RFK to his parents, n.d. [January 1945], Hyannis Port Papers.
86. Kathleen to family, February 27, 1945, JPK Papers.
87. RFK to Hackett, postmarked January 26, 1945, Hackett Papers.
88. RFK to Hackett, n.d. [November 1944], Hackett Papers.
89. RFK to Hackett, postmarked January 26, 1945, Hackett Papers.
90. RFK to Hackett, postmarked March 13, 1945, Hackett Papers.
91. RFK to Hackett, postmarked April 20, 1945, Hackett Papers.
92. RFK to Hackett, postmarked May 4, 1945, Hackett Papers.
93. JPK to RFK, January 8, 1945, RFK Papers.
94. RFK to his parents, n.d. [January 1945], Hyannis Port Papers.
95. JPK to RFK, January 29, 1945, JPK Papers.
96. RFK to his parents, n.d. [January 1945], Hyannis Port Papers.
97. JPK to RFK, March 28, 1945, March 31, 1945, May 21, 1945, JPK Papers.
98. JPK to Beaverbrook, October 23, 1944, Beaverbrook Papers.
99. JPK to Kathleen, May 1, 1945, JPK Papers.
100. "This story, which Bobby's father told for the first time last week . . ." (*Newsweek*, April 1, 1957).
101. Roy M. Mundorff, Commander, USNR, To Whom It May Concern, February 1, 1945, Hyannis Port Papers.
102. RFK to his parents, February 8, 1946, Hyannis Port Papers.
103. RFK to Billings, March 10 [1946], Hyannis Port Papers.
104. RFK to his parents, February 8, 1946, Hyannis Port Papers.
105. *The Forrestal Diaries*, ed. Walter Millis (New York, 1951), 134.
106. RFK to his parents, n.d. [March 1946], Hyannis Port Papers.
107. Charles Spalding, in recorded interview by L. J. Hackman, March 22, 1969, 4, RFK Oral History Program.
108. Joseph Dolan, in recorded interview by L. J. Hackman, April 11, 1970, RFK Oral History Program.

4. The Third Son *(pages 63–89)*

1. RFK to his mother, n.d. [May 1946], Hyannis Port Papers.
2. Paul Fay, *The Pleasure of His Company* (New York, 1966), 156–157.
3. Robert Kennedy, in recorded interview by John Stewart, July 20, 1967, 6–7, JFK Oral History Program.
4. RFK, in Stewart interview, 11.
5. Ralph G. Martin and Ed Plaut, *Front Runner, Dark Horse* (New York, 1960), 141.
6. Kenneth P. O'Donnell and David F. Powers, *"Johnny, We Hardly Knew Ye"* (Boston, 1972), 67–68.
7. K. LeMoyne Billings, in interview by author, July 8, 1975.
8. William Manchester, *Portrait of a President* (Boston, 1962), 21.
9. Otis N. Minot to JPK, April 9, 1947, JPK Papers.
10. Payson S. Wild to author, September 16, 1975.
11. R. E. Thompson and Hortense Myers, *Robert F. Kennedy: The Brother Within* (New York, 1962), 76.
12. Samuel Adams, in recorded interview by Jean Stein, May 1970, 7, Stein Papers.
13. George Plimpton, in recorded interview by Jean Stein, September 4, 1968, 1, Stein Papers.
14. John Knowles, in recorded interview by Roberta Greene, July 2, 1974, 5–7, RFK Oral History Program.
15. Anthony Lewis, in recorded interview by Jean Stein, November 7, 1968, 1, Stein

Papers; Lewis, in recorded interview by L. J. Hackman, July 23, 1970, RFK Oral History Program.

16. Eunice Kennedy Shriver, in recorded interview by Roberta Greene, April 29, 1971, 19-20, RFK Oral History Program; John Deedy, "Whatever Happened to Father Feeney?" *The Critic*, May-June 1973. The intercession of Cushing's successor, Cardinal Humberto Medeiros, brought about the removal of Feeney's excommunication in 1972 (*Time*, October 14, 1974).

17. Patricia Kennedy Lawford, ed., *That Shining Hour* (n.p., 1969), 17-18; George Sullivan in the *Boston Herald Traveler*, June 6, 1968; William V. Shannon, *The Heir Apparent* (New York, 1967), 44-45; Margaret Laing, *The Next Kennedy* (New York, 1968), 100.

18. William J. Brady, Jr., in recorded interview by Roberta Greene, November 5, 1974, 7, 23, RFK Oral History Program.

19. Kenneth O'Donnell, in recorded interview by Jean Stein, October 8, 1968, 7-9, Stein Papers; Jean Stein and George Plimpton, eds., *American Journey* (New York, 1970), 39.

20. Thompson and Myers, *Robert F. Kennedy*, 77.

21. Brady, in Greene interview, 30.

22. Stein and Plimpton, *American Journey*, 39.

23. Eunice Kennedy Shriver, in interview by Roberta Greene, April 29, 1971, 8-9, and January 13, 1972, 24, RFK Oral History Program.

24. O'Donnell, in Stein interview, 24.

25. George Sullivan, *Boston Herald Traveler*, June 6, 1968.

26. Thompson and Myers, *Robert F. Kennedy*, 76.

27. R. J. Whalen, *The Founding Father: The Story of Joseph P. Kennedy* (New York, 1964), 409.

28. Joseph P. Lash, ed., *From the Diaries of Felix Frankfurter* (New York, 1975), 311, 340.

29. *New York Times*, January 14, 1952.

30. W. O. Douglas, *Go East, Young Man* (New York, 1974), 200.

31. Joseph P. Kennedy, "The U.S. and the World," *Life*, March 18, 1946.

32. *Boston Sunday Advertiser*, May 25, 1947.

33. "Why Must There Be either War or the Marshall Plan?" draft article, evidently intended for *Life*, n.d. [spring 1948], JPK Papers.

34. Ibid.

35. Jefferson to Thomas Leiper, in Jefferson, *Writings*, ed. P. L. Ford (New York, 1895), vol. 9, 445-446.

36. "Why War or the Marshall Plan?"; another draft article, untitled and undated, in JPK Papers; see also J. P. Kennedy, "A Marshall Plan for the Americas," *PIC*, October 1948.

37. JPK to Beaverbrook, March 23, 1948, Beaverbrook Papers.

38. O'Donnell, in Stein interview, 2-3.

39. Ibid., 6-7. George Sullivan, the water boy, recalled Kennedy and O'Donnell "always discussing and arguing politics and their teammates eagerly joined in" (*Boston Herald Traveler*, June 6, 1968).

40. O'Donnell, in Stein interview, 4-5; Stein and Plimpton, *American Journey*, 38-39.

41. O'Donnell, in Stein interview, 4.

42. Kenneth O'Donnell, "Joseph Kennedy Felt the Generation Gap," *London Sunday Telegraph*, November 23, 1969. In *A Thousand Days: John F. Kennedy at the White House* (Boston, 1965), I describe an incident in which O'Donnell defended Franklin Roosevelt with such vigor that Mr. Kennedy, deeply angered, left the table (93). Robert Kennedy told me the story, but he evidently got the two occasions confused.

43. JPK to Beaverbrook, March 23, 1948, Beaverbrook Papers.

44. RFK's documentation is in the JPK Papers.

45. Felicia Warburg Roosevelt, *Doers and Dowagers* (Garden City, N.Y., 1975), 9.

46. RFK's handwritten diary is in the RFK Papers. A typed version in the JPK Papers

has errors of transcription. Subsequent RFK quotations about the trip, not otherwise identified, are from the diary.

47. Beaverbrook to JPK, March 12, 1948, Beaverbrook Papers.
48. RFK to his parents, n.d. [March 1948], Hyannis Port Papers.
49. G. E. Georgossy to Phil Reisman, March 31, 1948, Hyannis Port Papers.
50. Rose Kennedy to RFK, April 2, 1948, Hyannis Port Papers.
51. RFK to his parents, April 6, 1948, Hyannis Port Papers.
52. Alastair Forbes, "Upper Classmates," *Times Literary Supplement*, March 26, 1976; Forbes, "Camelot Confidential," *Times Literary Supplement*, June 13, 1975.
53. RFK to Patricia Kennedy, June 5, 1948, Lawford Papers.
54. RFK to his parents, June 30, 1948, Hyannis Port Papers.
55. RFK to his parents, from Grand Hotel, Stockholm, n.d. [July 1948], RFK Papers.
56. RFK to Paul Murphy, n.d. [July 1948], JPK Papers.
57. *Boston Advertiser*, January 16, 1949.
58. Thompson and Myers, *Robert F. Kennedy*, 46–47.
59. W. H. White to RFK, April 19, 1948, JPK Papers.
60. RFK to Paul Murphy, n.d. [April 1948], and cable, June 5, 1948, JPK Papers.
61. RFK to Patricia Kennedy, June 5, 1948, Lawford Papers.
62. E. G. Spies to Paul Murphy, June 7, 1948, JPK Papers.
63. Charles Spalding, in recorded interview by L. J. Hackman, March 22, 1969, 10, RFK Oral History Program.
64. The paper is in the Charles O. Gregory Papers, University of Virginia Law Library. I am much in debt to Frances Farmer, the librarian, for her generous assistance in gathering material about Robert Kennedy's Law School career.
65. RFK, "The Reserve Powers of the Constitution," JPK Papers.
66. Robert F. Kennedy, "A Critical Analysis of the Conference at Yalta, February 4–11, 1945," 9, 16, 19, 20, 34, 37.
67. JPK, "An American Policy . . . for Americans," JPK Papers.
68. *Congressional Record*, January 1, 1951, A8378–A8379; D. E. Koskoff, *Joseph P. Kennedy* (Englewood Cliffs, N.J., 1974), 353–356.
69. Whalen, *Founding Father*, 408.
70. RFK, in recorded interview by Anthony Lewis, December 4, 1964, I, 5, JFK Oral History Program; Bunche to RFK, March 16, 1951, Barrett Prettyman Papers.
71. RFK to Colgate W. Darden, March 7, 1951, University of Virginia Archives.
72. For the Bunche incident, I am indebted to Charles O. Gregory for a letter of September 13, 1975, and to Hardy Dillard for an interview on October 20, 1975. See also Laing, *Next Kennedy*, 120–121; Victor Lasky, *Robert F. Kennedy: The Myth and the Man* (New York: Pocket Books reprint, 1971), 69.
73. Frances Farmer to author, December 2, 1975.
74. Jean Kennedy Smith, in interview by author, September 15, 1975.
75. George Skakel to JPK, December 22, 1950, JPK Papers.
76. Lester David, *Ethel: The Story of Mrs. Robert F. Kennedy* (New York: Dell reprint, 1972), 53.
77. Jean Kennedy Smith, in interview by author, May 18, 1975.
78. Rose Kennedy, *Times to Remember* (Garden City, N.Y., 1974), 109.
79. RFK to Patricia Kennedy, n.d. [spring 1950], Lawford Papers.
80. Stein and Plimpton, *American Journey*, 39–40; see also Brady, in Greene interview, 34.
81. K. LeMoyne Billings, in interview by author, July 8, 1975.
82. K. LeM. Billings, in recorded interview by Jean Stein, n.d., 15, Stein Papers.
83. Eunice Kennedy Shriver, in Greene interview, 14, 33; Patricia Kennedy Lawford, in interview by author, July 2, 1975.

5. The Brothers: I *(pages 90–98)*

1. The exact dates were: Kathleen, July 4, 1951; Joe, III, September 24, 1952; Bobby, Jr., January 17, 1954; David, June 15, 1955, Courtney, September 9, 1956.

2. RFK, "U.S. Must Deal with New Asia," *Boston Post,* September 8, 1951.
3. James M. Landis to JPK, September 7, 1951, Landis Papers.
4. RFK to JPK, October 10, 1951, Hyannis Port Papers.
5. RFK's journal is in the RFK Papers. The account of this trip is drawn also from RFK, in recorded interview by John Stewart, July 20, 1967, RFK Papers; and from author's interview with Patricia Kennedy Lawford, July 2, 1975.
6. RFK to JPK, October 10, 1951, Hyannis Port Papers.
7. RFK to JPK, n.d. [October 1951], Hyannis Port Papers.
8. Theodore C. Sorensen, *Kennedy* (New York, 1965), 34.
9. JFK, "Report on His Trip to the Middle and Far East," speech over the Mutual Broadcasting Network, November 14, 1951, JPK Papers.
10. RFK, in Stewart interview, 21.
11. JPK, "Our Foreign Policy, Its Casualties and Prospects," address before the Economic Club of Chicago and broadcast over the Mutual Broadcasting Network, December 17, 1951, 2–4, 6–7, 9, 13, JPK Papers.
12. *New York Times,* November 25, 1952; R. J. Whalen, *The Founding Father* (New York, 1964), 436.
13. Ralph G. Martin and Ed Plaut, *Front Runner, Dark Horse* (New York, 1960), 195.
14. Kenneth P. O'Donnell and David F. Powers with Joe McCarthy, *"Johnny, We Hardly Knew Ye"* (Boston, 1972), 81.
15. Jean Stein and George Plimpton, eds., *American Journey: The Times of Robert Kennedy* (New York, 1970), 40–41; Kenneth O'Donnell, in recorded interview with Jean Stein, October 8, 1968, 12–13, Stein Papers.
16. Stein and Plimpton, *American Journey,* 40.
17. O'Donnell and Powers, *"Johnny,"* 83; Stein and Plimpton, *American Journey,* 41.
18. O'Donnell, in Stein interview, 16.
19. RFK, in Stewart interview, 30, 41–42.
20. Martin and Plaut, *Front Runner,* 176.
21. Stein and Plimpton, *American Journey,* 42.
22. O'Donnell and Powers, *"Johnny,"* 87.
23. Martin and Plaut, *Front Runner,* 165.
24. O'Donnell and Powers, *"Johnny,"* 88.
25. RFK, in Stewart interview, 29.
26. O'Donnell, in Stein interview, 15–16.
27. Lawrence F. O'Brien, *No Final Victories* (New York, 1974), 30.
28. Margaret Laing, *The Next Kennedy* (New York, 1968), 128–129.
29. K. LeMoyne Billings, in recorded interview by Jean Stein, n.d., 2, Stein Papers.
30. Burton Hersh, *The Education of Edward Kennedy: A Family Biography* (New York, 1972), 57.
31. O'Donnell, "Joseph Kennedy Felt the Generation Gap," *London Sunday Telegraph,* November 23, 1969.
32. *Boston Record,* May 9, 1957.
33. William O. Douglas, in recorded interview by John Stewart, November 9, 1967, 14, JFK Oral History Program.
34. Paul Dever to the author, c. 1953.
35. Billings, in Stein interview, 4.
36. John Seigenthaler, in an interview during the Democratic convention of 1960; Seigenthaler, in recorded interview by W. A. Geoghegan, July 1, 1964, 86, JFK Oral History Program.
37. Joe McCarthy, *The Remarkable Kennedys* (New York, 1960), 30.
38. JPK, in interview by Charles Spalding, February 24, 1975.
39. Arthur Krock, *Memoirs: Sixty Years on the Firing Line* (New York, 1968), 354.
40. Jacqueline Onassis, in interview by author, June 3, 1976.

6. The First Investigating Committee: Joe McCarthy *(pages 99–120)*

1. I have combined the stories as recalled by O'Donnell (Jean Stein and George Plimpton, eds., *American Journey: The Times of Robert Kennedy* [New York, 1970],

45) and O'Brien (Lawrence F. O'Brien, *No Final Victories* [Garden City, N.Y., 1974], 41–42).

2. Roy Cohn repeats this claim in *McCarthy* (New York, 1968); McCarthy's Marine commander, Glenn L. Todd, told Joan and Clay Blair, "I doubt that Joe met Kennedy in the South Pacific" (Joan Blair and Clay Blair, Jr., *The Search for J.F.K.* [New York, 1976], 298).

3. Jack Alexander, "The Senate's Remarkable Upstart," *Saturday Evening Post,* August 9, 1947.

4. I take this date from Cohn, *McCarthy*, 8.

5. Richard Rovere, *Senator Joe McCarthy* (London, 1960), 9.

6. JPK, in North American Newspaper Alliance interview with P. D. Garvan, May 20, 1951, Landis Papers.

7. Cohn, *McCarthy*, 47.

8. Theodore C. Sorensen, *The Kennedy Legacy* (New York, 1969), 41. In 1959 Kennedy told James MacGregor Burns that he had opposed Robert's joining the McCarthy staff (J. MacG. Burns, *John Kennedy: A Political Profile* [New York, 1960], 152). Kenneth O'Donnell wrote, "Jack had been strongly opposed to Bobby joining McCarthy's staff" (Kenneth P. O'Donnell and David F. Powers, *"Johnny, We Hardly Knew Ye"* [Boston, 1972], 111).

9. Cohn, *McCarthy*, 47, 66.

10. Patricia Kennedy Lawford, ed., *That Shining Hour* (n.p., 1969), 45–46.

11. John Kelso, in the *Boston Post*, April 12, 1953.

12. *Congressional Record,* May 14, 1953, 5078–5080. It was the *London Sunday Express* of April 19, 1953.

13. Senate Permanent Subcommittee on Investigations, *Control of Trade with the Soviet Bloc: Hearings*, pt. 1, 83 Cong., 1 Sess. (March 30–31, 1953), 29.

14. *Boston Post*, April 12, 1953.

15. F. D. Flanagan to RFK, April 2, 1953, RFK Papers.

16. LaVern Duffy, in interview by author, January 23, 1976.

17. Senate Permanent Subcommittee on Investigations, *Control of Trade with the Soviet Bloc: Hearings*, pt. 2, 83 Cong., 1 Sess. (May 4 and 20, 1953), 110, 125–126, 142.

18. *New York Journal-American*, May 6, 1953.

19. *Congressional Record,* May 19, 1953, 6448–6456.

20. Permanent Subcommittee on Investigations, *Control of Trade: Hearings*, pt. 2, 146.

21. Sherman Adams, *Firsthand Report* (New York: Popular Library reprint, 1962), 143.

22. *Boston Traveler*, May 26, 1953; *New York Times*, May 26, 1953; Cohn, *McCarthy*, 69–70.

23. Permanent Subcommittee on Investigations, *Control of Trade: Interim Report*, 83 Cong., 1 Sess. (1953), 13, 18.

24. *New York Times*, July 19, 1953.

25. *Washington Star*, August 26, 1954.

26. F. P. Carr to RFK, July 24, 1953, RFK Papers.

27. R. E. Thompson and Hortense Myers, *Robert F. Kennedy: The Brother Within* (New York: 1962), 112, 120–121; Robert F. Kennedy, *The Enemy Within* (New York: Popular Library reprint, 1960), 170, 291.

28. Stein and Plimpton, *American Journey*, 50.

29. Rovere, *Senator Joe McCarthy*, 51.

30. Thompson and Myers, *Robert F. Kennedy*, 121.

31. Kenneth O'Donnell, in recorded interview by Jean Stein, October 8, 1968, Stein Papers; Stein and Plimpton, *American Journey*, 49.

32. William McC. Blair, Jr., in recorded interview by Jean Stein, September 5, 1968, 2, Stein Papers.

33. Stein and Plimpton, *American Journey*, 50.

34. RFK to McCarthy, July 29, 1953, Symington to RFK, July 31, 1953, Jackson to RFK, July 31, 1953, RFK Papers.

35. Edward M. Kennedy, ed., *The Fruitful Bough* (n.p., 1965), 126.

36. RFK, memorandum, December 2, 1953, JPK Papers.

37. RFK, memorandum, November 16, 1953, JPK Papers.
38. K. LeMoyne Billings, in recorded interview by Jean Stein, n.d., 5, Stein Papers.
39. O'Brien, *No Final Victories*, 46.
40. Billings, in Stein interview; Billings, in interview by author, July 8, 1975.
41. *New York Times*, February 16, 1954.
42. RFK to the editor of the *Times*, February 17, 1954; copy to author, March 2, 1954, RFK Papers.
43. Sorensen, *Kennedy Legacy*, 36.
44. Herbert Hoover to RFK, February 19, 1954, Hoover Papers.
45. M. A. Jones to Louis B. Nichols, "Robert Francis Kennedy, Request to Meet the Director," July 20, 1955, RFK/FBI/FOIA release.
46. Charles E. Potter, *Days of Shame* (New York, 1965), 42.
47. Rovere, *Senator Joe McCarthy*, 37.
48. RFK to Hoover, n.d. [February 1954], Hoover Papers.
49. Potter, *Days of Shame*, 107.
50. Senate Permanent Subcommittee on Investigations, *Army Signal Corps — Subversion and Espionage: Hearings*, pt. 10, 83 Cong. 2 Sess. (March 11, 1954), 452–453.
51. Ibid., 458, 462.
52. Jones to Nichols, "Kennedy, Request to Meet Director."
53. RFK, in recorded interview by Anthony Lewis, December 4, 1964, IV, 23, RFK Papers.
54. John Kelso, in the *Boston Post*, April 4, 1954.
55. Cohn, *McCarthy*, 71. Jackson's interrogation is in Senate Special Subcommittee on Investigations, *Special Senate Investigation on Charges and Countercharges Involving: Secretary of the Army Robert T. Stevens, John G. Adams, H. Struve Hensel and Senator Joe McCarthy, Roy M. Cohn, and Francis P. Carr. Hearing*, pt. 63, 83 Cong. 2 Sess. (June 11, 1954), 2613–2617.
56. Michael Straight, *Trial by Television* (Boston, 1954), 192.
57. *New York Times*, June 12, 1954; *New York Daily News*, June 12, 1954.
58. LaVern Duffy, in interview by author, January 23, 1976.
59. Senate Special Subcommittee on Investigations, *Charges and Countercharges: Report*, 83 Cong., 2 Sess. (August 30, 1954), 88, 92, 94.
60. RFK, with James M. Landis, draft report, 229, Landis Papers.
61. "Speech Prepared for Delivery on the Senate Floor July 31, 1954," 2, 3, 5–6, JFK Papers.
62. RFK to Stuart Symington, June 1, 1957, RFK Papers.
63. Robert Griffith, *The Politics of Fear: Joseph R. McCarthy and the Senate* (Lexington, Ky., 1970), 308; Stennis to RFK, December 16, 1954, RFK Papers.
64. Welch to RFK, January 10, 1955, RFK to Welch, January 14, 1955, Welch to RFK, January 17, 1955, RFK Papers.
65. Mrs. Richard Metz to RFK, February 10, 1955, RFK to Mrs. Metz, February 15, 1955, RFK Papers.
66. Sorensen to RFK, December 8, 1954, RFK Papers; *Louisville Courier-Journal*, January 22, 1955.
67. *Louisville Courier-Journal*, January 22, 1955.
68. RFK to John L. McClellan, December 1, 1954, RFK Papers.
69. Senate Permanent Subcommittee on Investigations, *Army Personnel Actions Relating to Irving Peress: Report*, 84 Cong., 1 Sess. (July 14, 1955), 1, 35, 36, 42. Some writers — notably Victor Lasky and Ralph de Toledano — have also held Kennedy and the reconstituted committee responsible for a report of April 25, 1955, entitled *Army Signal Corps — Subversion and Espionage*, supporting the extravagant McCarthy theories about Fort Monmouth. Had they examined the report carefully, they would have noted the statement that it "was prepared for the period of the 83rd Con., 2d sess., under the chairmanship of Senator Joe McCarthy" and the disclaimer by Senators McClellan, Jackson and Symington that they could not "accept either the credit or responsibility for this report." The report was written when Kennedy was at the Hoover commission.
70. RFK to McClellan, November 23, 1954, RFK Papers.

71. RFK to Robert M. Harriss, January 31, 1955, RFK Papers.
72. See Moseley to RFK, January 20, 1955, RFK Papers. Moseley suggested that the destruction by the Customs Service of Soviet materials paid for by American research libraries "is causing great difficulties in our attempts to study Soviet developments closely."
73. RFK to Browder, March 16, 1954, RFK Papers.
74. Sylvia Berkowitz to Joseph Starobin, February 7, 1976; Starobin to the author, October 1, 1975, and February 20, 1976.
75. L. V. Boardman to A. Rosen, February 9, 1955, RFK/FBI/FOIA release.
76. James Juliana to RFK, January 28, 1955, RFK Papers. For RFK's attempt to get the investigation going, see his letter to Charles Tracy, January 18, 1955, and Tracy's reply, January 24, in RFK Papers.
77. Jones to Nichols, "Kennedy, Request to Meet Director."
78. *New York Times*, July 22–24, 1955; *Time*, August 1 and 8, 1955; Robert J. Donovan, *Eisenhower: The Inside Story* (New York, 1956), 332–334; Charles Bartlett, "Vignettes: Khrushchev, Kennedy and Kerr," *Washington Star*, September 13, 1959; Stein and Plimpton, *American Journey*, 54; Arthur Krock, *Memoirs* (New York, 1968), 308.
79. Lord Harlech, in recorded interview by Jean Stein, April 30, 1970, 14, Stein Papers; David Hackett, in interview by author, January 27, 1975.
80. Charles Spalding, in recorded interview by L. J. Hackman, March 22, 1969, 13, RFK Oral History Program.
81. Eunice Kennedy Shriver, in recorded interview by Roberta Greene, April 29, 1971, 41–42, 44, RFK Oral History Program.
82. Krock, *Memoirs*, 344.
83. *New York Times*, April 8, 1954.
84. Walter Lippmann, "Kennedy Destroys False Hopes," *Boston Globe*, April 12, 1954.
85. John F. Kennedy, "Foreign Policy Is the People's Business," *New York Times Magazine*, August 8, 1954.
86. Jacqueline Onassis, in interview by author, June 2, 1976.
87. Charles Spalding, in interview by author, February 24, 1975; Jacqueline Onassis, in interview by author.

7. Interlude: William O. Douglas and Adlai Stevenson *(pages 121–136)*

1. Rose Fitzgerald Kennedy, in interview by author, October 23, 1975.
2. W. O. Douglas, in recorded interview by Roberta Greene, November 13, 1969, 1, 3, RFK Oral History Program.
3. Mercedes [Douglas] Eichholz, in interview by author, April 26, 1975.
4. As described by RFK in early drafts of lectures about the trip, RFK Papers.
5. RFK to Mrs. Henry Kelly, March 18, 1955, RFK Papers.
6. Frederick W. Flott, in interview by author, February 26, 1975.
7. The sources for the trip, unless otherwise noted, are: RFK's diary, letters and speech drafts in the RFK papers; W. O. Douglas, *Russian Journey* (Garden City, N.Y., 1956); and the Flott and Eichholz interviews with the author.
8. Robert F. Kennedy, "The Soviet Brand of Colonialism," *New York Times Magazine*, April 8, 1956.
9. Douglas, in Greene interview, 7.
10. W. O. Douglas, in recorded interview by Jean Stein, n.d., 1, Stein Papers.
11. Douglas, in Greene interview, 9.
12. As remembered by Mercedes [Douglas] Eichholz, in interview by author.
13. Douglas, *Russian Journey*, 7, 238, 240.
14. RFK to David Lawrence, October 5, 1955, RFK Papers. When *U.S. News* interviewers asked him whether his views and Douglas's were the same, Kennedy answered diplomatically, "I have the greatest admiration for Justice Douglas — I would not presume to speak for him" (*U.S. News & World Report*, October 21, 1955).
15. *New York Journal-American*, September 15, 1955.

16. *New York Herald Tribune*, September 24, 1955.
17. RFK to P. W. Goetz, November 21, 1955, RFK Papers.
18. Angie Novello, in recorded interview by Jean Stein, September 17, 1968, 5–6, Stein Papers.
19. K. LeM. Billings, in recorded interview by Jean Stein, n.d., 9, Stein Papers.
20. "Lecture by the Hon. Robert F. Kennedy at Georgetown University," October 10, 1955, RFK Papers.
21. RFK to T. C. Streibert, January 23, 1956, RFK Papers.
22. Robert F. Kennedy, "The Soviet Brand of Colonialism," *New York Times Magazine*, April 8, 1956.
23. Robert F. Kennedy, "Colonialism within the Soviet Union," *Proceedings of the Fifty-Sixth Annual Meeting of the Virginia State Bar Association* (Richmond, 1956), 211; letter to *New York Times*, January 2, 1956.
24. RFK to Clive S. Gray, May 22, 1956, RFK Papers.
25. RFK, "Soviet Brand of Colonialism."
26. *New York Times*, March 19, 1956.
27. In RFK Papers.
28. Douglas, in Stein interview, 1.
29. Lord Harlech, in Stein interview, April 30, 1970, 14.
30. Interview with Robert Kennedy, "A Look Behind the Russian Smiles," *U.S. News & World Report*, October 21, 1955.
31. Mercedes [Douglas] Eichholz, in interview by author.
32. See letters to Senator McClellan from Acting Secretary of State Herbert Hoover, Jr., February 20, 1956, and Secretary of Commerce Sinclair Weeks, March 5, 1956; see also memorandum of Secretary of Defense Charles E. Wilson, March 5, 1956, in Senate Permanent Subcommittee on Investigations, *East-West Trade: Hearings*, pt. 1, app., 84 Cong., 2 Sess. (July 18, 1956), 273–284.
33. RFK, letter to editor, *American Metal Market*, April 5, 1956.
34. Permanent Subcommittee on Investigations, *East-West Trade: Hearings* (March 6, 1956), 251.
35. Ibid.: McClellan (February 15, 1956), 1; Ervin and Jackson (March 6, 1956), 203, 209.
36. Permanent Subcommittee on Investigations, *East-West Trade: Report*, 28.
37. Morse to RFK, August 2, 1956, RFK Papers.
38. RFK, in interview by John Bartlow Martin, December 7, 1966, Martin Papers.
39. RFK, in recorded interview by John Stewart, August 15, 1967, 58–60, JFK Oral History Program.
40. JPK to JFK, May 25, 1956, Sorensen Papers.
41. RFK, in Stewart interview, 51.
42. Kenneth P. O'Donnell and David F. Powers, *"Johnny, We Hardly Knew Ye"* (Boston, 1972), 118–119.
43. Author's journal, July 26, 1956.
44. Ibid., July 29, 1956.
45. O'Donnell and Powers, *"Johnny,"* 122.
46. Theodore C. Sorensen, *Kennedy* (New York, 1965), 88.
47. Kenneth O'Donnell, in interview by Jean Stein, October 8, 1968, 32–33, Stein Papers.
48. RFK to C. J. Bloch, September 5, 1956, RFK Papers.
49. O'Donnell and Powers, *"Johnny,"* 120.
50. Sorensen, *Kennedy*, 87–91; Arthur M. Schlesinger, Jr., *A Thousand Days* (Boston, 1965), 8–9; Ralph G. Martin and Ed Plaut, Front Runner, Dark Horse (New York, 1960), 90, 103.
51. O'Donnell, in Stein interview, 33.
52. O'Donnell and Powers, *"Johnny,"* 124.
53. Martin and Plaut, *Front Runner*, 107.
54. Ibid.
55. R. E. Thompson and Hortense Myers, *Robert F. Kennedy: The Brother Within* (New York, 1962), 184–185.

56. Jacqueline Onassis, in interview by author, June 3, 1976.
57. William McC. Blair, Jr., in recorded interview by Jean Stein, September 5, 1968, 3, 6, 8, Stein Papers; Hal Clancy, "In Politics or Investigating, Bob Kennedy Thinks Fast," *Boston Traveler*, January 11, 1961.
58. Newton N. Minow, in interview by author, August 30, 1974.
59. RFK, memorandum on Stevenson, January 25, 1957, 5–6, RFK Papers.
60. O'Donnell and Powers, *"Johnny,"* 126.
61. Harrison Salisbury, in recorded interview by Jean Stein, December 5, 1969, 2, Stein Papers.
62. RFK, memorandum on Stevenson, January 25, 1957, 1–2, RFK Papers.
63. RFK, memorandum, January 25, 1957, 5–6, RFK Papers.
64. Ibid., 4–5.
65. Ibid., 5–7.
66. Robert Kennedy, in interview with John Bartlow Martin, December 7, 1966, Martin Papers.
67. Stevenson to RFK, November 17, 1956, RFK Papers.
68. Salisbury, in Stein interview, 3.

8. Second Investigating Committee: Jimmy Hoffa *(pages 137–169)*

1. For good discussions, see John Hutchinson, *The Imperfect Union* (New York, 1970), ch. 14; and Daniel Bell, "The Scandals in Union Welfare Funds," *Fortune*, April 1954.
2. Clark Mollenhoff, *Tentacles of Power: The Story of Jimmy Hoffa* (Cleveland, 1965), 124.
3. Sam Romer, *The International Brotherhood of Teamsters* (New York, 1962), 35.
4. Murray Kempton, "The Salesman," *New York Post*, June 24, 1953.
5. *New York Times*, October 14, 1952; Mollenhoff, *Tentacles of Power*, 23.
6. Robert F. Kennedy, *The Enemy Within* (New York: Popular Library reprint, 1960), 16–17.
7. Romer, *Teamsters*, 6–8.
8. Edward Levinson, *Labor on the March* (New York, 1938), 13.
9. James R. Hoffa, *The Trials of Jimmy Hoffa: An Autobiography*, as told to Donald I. Rogers (Chicago, 1970), 105, 107–108.
10. B. J. Widick, *Labor Today* (Boston, 1964), 151.
11. John Bartlow Martin, *Jimmy Hoffa's Hot* (New York: Fawcett World, Crest, 1959), 31. This work, based in considerable part on interviews with Hoffa and Kennedy, is the paperback reprint of Martin's indispensable series "The Struggle to Get Hoffa," *Saturday Evening Post*, June 27–August 8, 1959. The paperback title was imposed by the publisher over the author's vigorous protest. For reasons of convenience I will cite the book — as Martin, *Hoffa* — rather than the magazine pieces. Mr. Martin has also made available to me the original manuscript of the *Post* series; this was considerably cut for publication. It will be cited as Martin, "Hoffa."
12. Ralph James and Estelle James, *Hoffa and the Teamsters* (Princeton, N.J., 1965), 114–116.
13. Martin, *Hoffa*, 24.
14. Paul Jacobs, *The State of the Unions* (New York, 1963), 48.
15. As conceded by Hoffa, *Trials*, 136; James R. Hoffa, *Hoffa: The Real Story*, as told to Oscar Fraley (New York, 1975), 65. Nevertheless Lewis was the labor leader Hoffa most admired; see Martin, "Hoffa," 147–148.
16. Martin, "Hoffa," 145.
17. Jacobs, *State of the Unions*, 56.
18. Carmine Bellino, in interview by author, February 3, 1976.
19. RFK, *Enemy Within*, 20.
20. Edwin Guthman, *We Band of Brothers* (New York, 1971), 1, 4–7.
21. Ibid., 9–10.

22. RFK, *Enemy Within*, 15; Nathan W. Shefferman, *The Man in the Middle* (Garden City, N.Y., 1961), 1–3.
23. Jean Kennedy Smith, in interview by author, September 15, 1975.
24. Mercedes [Douglas] Eichholz, in interview by author, April 26, 1975.
25. J. C. Goulden, *Meany* (New York, 1972), 233–235.
26. Ibid., 235.
27. John L. McClellan, *Crime without Punishment* (New York, 1962), 14.
28. RFK, memorandum, January 9, 1957, 1, RFK Papers.
29. RFK, memorandum, January 7, 1957, 1, RFK Papers.
30. RFK, *Enemy Within*, 33; Mollenhoff, *Tentacles of Power*, 146.
31. Ralph G. Martin and Ed Plaut, *Front Runner, Dark Horse* (New York, 1960), 191.
32. Martin, "Hoffa," 9.
33. Angie Novello, in recorded interview by Jean Stein, September 27, 1968, 2, Stein Papers.
34. Kenneth P. O'Donnell and David Powers, *"Johnny, We Hardly Knew Ye"* (Boston, 1972), 132–133.
35. John Bartlow Martin to author, October 19, 1975.
36. Pierre Salinger, *With Kennedy* (Garden City, N.Y., 1966), 16–18; Salinger, *Je suis un Américain: Conversations avec Philippe Labro* (Paris, 1975), 126–135; Salinger, in recorded interview by Theodore H. White, July 19, 1965, 2–9, JFK Oral History Program.
37. RFK, *Enemy Within*, 96.
38. Pierre Salinger, in recorded interview by L. J. Hackman, May 26, 1969, 22, RFK Oral History Program.
39. Walter Sheridan, in recorded interview by Jean Stein, July 23, 1968, 1–2, Stein Papers; Sheridan, *The Fall and Rise of Jimmy Hoffa* (New York, 1972), 33.
40. H. W. Flannery, "The *Other* Kennedy," *Ave Maria*, August 31, 1957; Martin, "Hoffa," 11.
41. Martin to author, October 19, 1975.
42. RFK, notes for close-up in *Life* [1957], 5, JPK Papers. There were 34 assistant counsels and investigators deployed on field investigations, reinforced by an average of 35 to 45 accountants and investigators from the General Accounting Office. Senate Select Committee on Improper Activities in the Labor or Management Field, *Final Report*, 86 Cong., 2 Sess. (March 31, 1960), 869.
43. Sheridan, in Stein interview, 1, 3–4.
44. LaVern Duffy, in interview by author, January 23, 1976.
45. McClellan, *Crime without Punishment*, 69–76.
46. RFK, in interview by Kenneth Brodney (Newhouse Newspaper Feature Syndicate), July 2, 1957, 11, 19, 22, JPK Papers.
47. Jean Stein and George Plimpton, eds., *American Journey* (New York, 1970), 55.
48. Ibid., 54.
49. Guthman, *We Band of Brothers*, 57.
50. Joseph A. Loftus, review of Robert F. Kennedy's *The Enemy Within*, *New York Times Book Review*, February 28, 1960.
51. John Seigenthaler, in recorded interview by W. A. Geoghegan, [July 1964], 1–12, 15, JFK Oral History Program; Seigenthaler, in recorded interview by Jean Stein, August 27, 1968, 1–5, Stein Papers.
52. RFK, *Enemy Within*, 29–30; Martin, "Hoffa," 43.
53. RFK, *Enemy Within*, 38.
54. Lahey to RFK, postmarked March 28, 1957, JPK Papers.
55. Goulden, *Meany*, 240.
56. *Cleveland Press*, March 6, 1957.
57. George M. Belknap to RFK, March 8, 1957, JPK Papers.
58. Mrs. W. D. Goode, Jr., to JFK, March 8, 1957, JPK Papers.
59. JPK to RFK, March 14, 1957, enclosing letter from Sargent Shriver, n.d., JPK Papers; *Chattanooga Times*, March 26, 1957.
60. Paul F. Healy, "Investigator in a Hurry," *Sign*, August 1957.

61. *New York Herald Tribune*, March 17, 1957.
62. *Toledo Blade*, March 3, 1957.
63. *Kansas City Star*, March 24, 1957.
64. *Louisville Courier-Journal*, March 24, 1957.
65. RFK, notes for *Life* close-up, 6.
66. K. LeM. Billings, in recorded interview by Jean Stein, n.d., 8, Stein Papers.
67. RFK, handwritten note, October 11 [1957], RFK Papers.
68. Mollenhoff to RFK, December 23, 1957, RFK Papers.
69. Paul O'Neil, "The No. 2 Man in Washington," *Life*, January 26, 1962.
70. RFK, notes for *Life* close-up, 3, 4.
71. *New York Herald Tribune*, August 13, 1957; S. P. Friedman, *The Magnificent Kennedy Women* (New York: Monarch reprint, 1964), 74; Lester David, *Ethel* (New York: Dell reprint, 1972), 87–88.
72. RFK, memorandum, September 8, 1958, RFK Papers. Cheyfitz evidently denied the story; see RFK, *Enemy Within*, 64, though this reference does not explain in full lurid detail what Cheyfitz was denying.
73. *New York Herald Tribune*, March 17, 1959.
74. RFK, handwritten note, April 27, 1957, RFK Papers.
75. RFK, handwritten note, December 15, 1957, RFK Papers.
76. Sheridan, *Fall and Rise*, 32.
77. RFK, *Enemy Within*, 44.
78. Ibid.
79. Ibid.
80. Martin, "Hoffa," 53.
81. Cheasty to RFK, May 3, 1957, RFK Papers.
82. Martin, *Hoffa*, 7.
83. RFK, *Enemy Within*, 48; Martin, *Hoffa*, 8. In reconstructing this dinner, I have relied on Kennedy's own accounts in contemporaneous notes and in his book; on Martin (*Hoffa*, 8–9; "Hoffa," 4–7), who interviewed both Kennedy and Hoffa shortly afterward; and on Mollenhoff (*Tentacles of Power*, 148–149), who interviewed Kennedy. The account in the first of Hoffa's autobiographies (*Trials*, 150) is exceedingly brief. His second autobiography is hopelessly unreliable. Hoffa, or his collaborator, describes an earlier meeting when, allegedly, Kennedy, Salinger and Bellino invaded Hoffa's Detroit office in the summer of 1956 and were humiliated by him. This is wholly imagined, though it may have been inspired by the visit of Salinger and Bellino to Detroit in the summer of 1957. As for the dinner, the second autobiography says that Kennedy ordered Cheyfitz from the room, asked Hoffa a series of personal questions, was worsted by him in Indian hand wrestling and did not even stay for dinner (*Hoffa: The Real Story*, 95–99). Hoffa claimed none of these things in his earlier autobiography. Most are refuted by other evidence.
84. February 21, 1957, RFK Papers.
85. Martin, *Hoffa*, 20–21; RFK, *Enemy Within*, 60–62; Mollenhoff, *Tentacles of Power*, 154–155. Hoffa's pushup crack was recorded by Kennedy in a handwritten note, RFK Papers.
86. Edward Bennett Williams, in interview by author, January 23, 1976.
87. Mollenhoff, *Tentacles of Power*, 202.
88. RFK, *Enemy Within*, 62–64.
89. Edward Bennett Williams, *One Man's Freedom* (New York, 1962), 221; Mollenhoff, *Tentacles of Power*, 198–207.
90. Novello, in Stein interview, 3; Mollenhoff, *Tentacles of Power*, 213–214.
91. Novello, in Stein interview, 3.
92. Senate Select Committee on Improper Activities in the Labor or Management Field (hereafter cited as Senate Select Committee), *Investigation of Improper Activities in the Labor or Management Field: Hearings*, 85 Cong., 1 Sess. (July 31, 1957), 3595–3596.
93. Martin, *Hoffa*, 35–36; Salinger, *With Kennedy*, 22–23; Bellino, in interview by

author, February 3, 1976; Pierre Salinger, in interview by author, November 13, 1975.

94. Martin, *Hoffa*, 49.
95. Senate Select Committee, *Hearings* (August 21, 1957), 5107–5108.
96. Ibid., 5109.
97. See the stories by Joseph Loftus, *New York Times*, August 21–24, 1957.
98. Senate Select Committee, *Hearings* (August 23, 1957), 52.
99. Martin, *Hoffa*, 60.
100. RFK, handwritten note, September 2 [1957], RFK Papers.
101. RFK, handwritten note, October 11 [1957], RFK Papers.
102. Martin, *Hoffa*, 61.
103. RFK, *Enemy Within*, 67–68.
104. Martin, *Hoffa*, 69.
105. RFK, memorandum, April 5, 1958, 4, RFK Papers.
106. Martin, *Hoffa*, 87.
107. RFK, *Enemy Within*, 77.
108. RFK, handwritten note, September 13 [1958], RFK Papers.
109. Martin, "Hoffa," 554.
110. Ibid., 731–732; RFK, *Enemy Within*, 157–158; Senate Select Committee, *Hearings*, 85 Cong., 2 Sess. (September 18, 1958), 15230–15231.
111. Martin, "Hoffa," 733.
112. Martin, *Hoffa*, 88.
113. RFK, handwritten note, December 19 [1958], RFK Papers.
114. Mollenhoff, *Tentacles of Power*, 325.
115. *International Teamster*, February 1959.
116. Angie Novello, in interview by author, December 11, 1975.
117. Hal Clancy, "Tough? Sensitive? Dedicated? — Depends on Whom You Ask," *Boston Traveler*, January 9, 1961.
118. Martin, "Hoffa," 137.
119. Martin, *Hoffa*, 11.
120. Martin to author, October 19, 1975.
121. Martin, *Hoffa*, 12.
122. Salinger, *With Kennedy*, 19.
123. Edward Bennett Williams, in recorded interview by Jean Stein, January 9, 1970, 9–10, Stein Papers.
124. Martin, "Hoffa," 142.
125. Hoffa, *Real Story*, 107, 115.
126. RFK, *Enemy Within*, 78.
127. Victor Lasky, *Robert F. Kennedy: The Myth and the Man* (New York: Pocket Book reprint, 1971), 119.
128. Novello, Bellino and Duffy, in interviews by author.
129. Hoffa, *Real Story*, 105.
130. Ibid., 117.
131. Stein and Plimpton, *American Journey*, 56–57.
132. RFK, handwritten note, January 15 [1959], RFK Papers.
133. Senate Select Committee, *Final Report*, 86 Cong., 2 Sess. (March 28, 1960), 570–731; RFK, *Enemy Within*, 158–159.
134. Ibid., 159.
135. Jacobs, *State of the Unions*, 50; see also Daniel Bell, "The Myth of Crime Waves," in *The End of Ideology* (New York: Collier reprint, 1962), 167–168.
136. Martin, "Hoffa," 749–750.
137. Senate Select Committee, *Hearings*, 85 Cong., 2 Sess. (August 20, 1958), 14061–14062.
138. RFK, *Enemy Within*, 238–240.
139. Senate Select Committee, *Final Report*, 514–569. For a summary of the Glimco file, see Hutchinson, *Imperfect Union*, 249–251.
140. R. S. Anson, *"They've Killed the President"* (New York, 1973), 227–228.

141. Sheridan, *Fall and Rise*, 17–18; RFK, *Enemy Within*, 87–89.
142. Senate Select Committee, *Hearings*, 86 Cong., 1 Sess. (February 20, 1959), 17042.
143. Donald R. Cressey, *Theft of the Nation* (New York: Harper & Row, Colophon reprint, 1969), 112, 189.
144. Senate Select Committee, *Hearings*, 86 Cong., 1 Sess. (June 9, 1959), 18672–18681.
145. The congressman was Roland Victor Libonati; Sheridan, *Fall and Rise*, 361–364; Cressey, *Theft of the Nation*, 271–272.
146. J. Edgar Hoover spoke of "Giancana's close relationship with Frank Sinatra" in a memorandum of May 10, 1962. Senate Select Committee to Study Governmental Operations with respect to Intelligence Activities, *Interim Report: Alleged Assassination Plots Involving Foreign Leaders*, 94 Cong., 1 Sess. (November 20, 1975), 133. See also Peter Lawford with Steve Dunleavy, "Peter Lawford Tells," *Star* (New York), February 17, 1976.
147. Cressey, *Theft of the Nation*, 220.
148. Sheridan, *Fall and Rise*, 28, 115, 282, 284; "Provenzano Comeback Reported," *New York Times*, December 6, 1975.
149. "Hoffa's Teamsters," *Life*, May 25, 1959.
150. James and James, *Hoffa and the Teamsters*, 66.
151. Senate Select Committee, *Hearings*, 85 Cong., 2 Sess. (September 16, 1958), 15092.
152. RFK, *Enemy Within*, 78.
153. Martin, "Hoffa," 749.
154. I take these terms from the fascinating study by Dwight C. Smith, Jr., *The Mafia Mystique* (New York, 1975), 89, 117, 322.
155. Walter Lippmann, "The Underworld, Our Secret Servant," *Forum*, January 1931.
156. Smith, *Mafia Mystique*, 121–122.
157. Burton Turkus and Sid Feder, *Murder, Inc.* (New York, 1951), 63.
158. RFK, *Enemy Within*, 228.
159. Senate Select Committee, *Final Report*, 488.
160. Mafiology is as murky a field as Kremlinology. The true believers include Estes Kefauver, *Crime in America* (Garden City, N.Y., 1951); President's Commission on Law Enforcement and Administration of Justice, *The Challenge of Crime in a Free Society* (Washington, 1967), especially 192–196; Cressey, *Theft of the Nation*; Peter Maas, *The Valachi Papers* (New York, 1968); Nicholas Gage, *The Mafia Is Not an Equal Opportunity Employer* (New York, 1971); and J. Edgar Hoover (later pronouncements). The skeptics include J. Edgar Hoover (early pronouncements); Turkus and Feder, *Murder, Inc.;* W. H. Moore, *The Kefauver Committee and the Politics of Crime, 1950–1952* (Columbia, Mo., 1974); Daniel Bell, "The Myth of the Cosa Nostra," *New Leader*, December 23, 1963; Murray Kempton, "Crime Does Not Pay," *New York Review of Books*, September 11, 1969; Eric Hobsbawm, "The American Mafia," *Listener*, November 20, 1969; F. A. J. Ianni, *A Family Business* (New York, 1972); J. L. Albini, *The American Mafia: Genesis of a Legend* (New York, 1972); Smith, *Mafia Mystique*. For what it is worth, I find the skeptics more persuasive. It is interesting to note that the gangster films of the early thirties — *Scarface* and *Little Caesar*, for example — do not depict their Italian protagonists as inhabiting the sentimental mythic realm later celebrated by Mario Puzo and Francis Ford Coppola in *The Godfather*. It may be that in the end publicity created an American Mafia. What hood could resist the fearsome prestige the myth makers were wishing on him?
161. RFK, *Enemy Within*, 229.
162. RFK, in recorded interview by Anthony Lewis, December 4, 1964, IV, 22, JFK Oral History Program.
163. Ronald W. May, "Organized Crime and Disorganized Cops," *Nation*, June 27, 1959.
164. Angie Novello, in interview by author, December 11, 1975.
165. RFK, memorandum, May 29, 1958, RFK Papers.
166. May, "Organized Crime and Disorganized Cops."

167. RFK, *Enemy Within*, 253.
168. Quoted in *Newsweek*, August 18, 1975. He made a similar remark to John Bartlow Martin, in "Hoffa," 749.

9. The Second Investigating Committee: Walter Reuther *(pages 170–191)*

1. Senate Select Committee on Improper Activities in the Labor or Management Field (hereafter cited as Senate Select Committee), *Investigation of Improper Activities in the Labor or Management Field: Hearings*, 85 Cong., 2 Sess. (August 20, 1957), 4963–4964.
2. Clark Mollenhoff, *Tentacles of Power* (Cleveland, 1965), 257–260.
3. *Newsweek*, July 22, 1957.
4. *Meet the Press*, April 28, 1957.
5. RFK, in interview by Kenneth Brodney, Newhouse Newspaper Feature Syndicate, July 2, 1957, 18, JPK Papers.
6. Charles Bartlett, "Vignettes: Khrushchev, Kennedy and Kerr," *Washington Star*, September 13, 1959.
7. LaVern Duffy, in interview by author, January 23, 1976.
8. Robert F. Kennedy, *The Enemy Within*, (New York: Popular Library reprint, 1960), 285–286.
9. Ibid., 287.
10. Pierre Salinger, in interview by author, November 13, 1975.
11. Arthur Watkins, *Enough Rope* (Englewood Cliffs, N.J., 1969), 192.
12. Richard Rovere, "The Last Days of Joe McCarthy," *Esquire*, August 1958.
13. RFK, memorandum, March 5, 1957, RFK Papers.
14. Ibid.
15. Robert E. Thompson and Hortense Myers, *Robert F. Kennedy: The Brother Within* (New York, 1962), 121.
16. Edwin Guthman, *We Band of Brothers* (New York, 1971), 24–25.
17. K. LeM. Billings, in recorded interview with Jean Stein, n.d., 7, Stein Papers.
18. RFK, handwritten note, May 4 [1957], RFK Papers.
19. Ibid.
20. *New York Times*, May 4, 1957.
21. RFK, *Enemy Within*, 254–255.
22. *Newsweek*, July 22, 1957.
23. Mundt, memorandum, July 15, 1957, RFK Papers.
24. RFK, *Enemy Within*, 257.
25. Ibid., 256; *Newsweek*, July 29, 1957; also *Newsweek* piece with RFK annotations, RFK Papers.
26. RFK, handwritten memorandum, October 11 [1957], RFK Papers.
27. RFK, *Enemy Within*, 259.
28. Frank Cormier and W. J. Eaton, *Reuther* (Englewood Cliffs, N.J., 1970), 344.
29. Senate Select Committee, *Hearings*, 85 Cong., 2 Sess. (March 29, 1958), 10165.
30. "Republicans Size Up Reuther's Union," *U.S. News & World Report*, February 29, 1960.
31. Jack Conway, in recorded interview by L. J. Hackman, April 10, 1972, 15–16, RFK Oral History Program.
32. RFK, *Enemy Within*, 260; Conway, in Hackman interview, 19.
33. RFK, *Enemy Within*, 261–262.
34. See W. H. Uphoff, *Kohler on Strike: Thirty Years of Conflict* (Boston, 1966), 106–107. For an opposing view, see Sylvester Petro, *Power Unlimited: The Corruption of Union Leadership* (New York, 1959), esp. ch. 4–5.
35. Senate Select Committee, *Hearings*, 85 Cong., 2 Sess. (March 26, 1958), 9962.
36. RFK, *Enemy Within*, 262–266.
37. Ibid., 254.
38. RFK, memorandum, January 8, 1958, 1–2, RFK Papers.
39. RFK, memorandum, February 22, 1958, 5, RFK Papers.

40. *U.S. News & World Report,* April 4, 1958.
41. Conway, in Hackman interview, 31–33.
42. The original letter appeared in the YPSL organ, *The Challenge,* in July 1934; Victor Reuther, *The Brothers Reuther* (Boston, 1976), ch. 17. See also RFK, memorandum, February 22, 1958, 1–4, RFK Papers; Cormier and Eaton, *Reuther,* 140.
43. RFK, memorandum, February 22, 1958, 5, memorandum, March 1, 1958, 1–2; RFK, *Enemy Within,* 275–276; Senate Select Committee, *Hearings,* 85 Cong., 2 Sess. (February 26, 1958), 8330.
44. Conway, in Hackman interview, 15–16.
45. Senate Select Committee, *Hearings,* 85 Cong., 2 Sess. (February 28, 1958), 8520.
46. RFK, memorandum, March 8, 1958, 2, RFK Papers.
47. Senate Select Committee, *Hearings,* 85 Cong., 2 Sess. (March 4, 1958), 8666.
48. Ibid. (March 19, 1958), 9525.
49. RFK, *Enemy Within,* 278.
50. RFK, memorandum, April 5, 1958, 6, RFK Papers.
51. Rauh to author, October 16, 1975.
52. RFK, *Enemy Within,* 279.
53. RFK, memorandum, March 1, 1958, 3, RFK Papers.
54. Rauh to author, October 16, 1975.
55. RFK, *Enemy Within,* 280.
56. RFK, memorandum, February 22, 1958, 1, RFK Papers.
57. RFK, *Enemy Within,* 282.
58. RFK, memorandum, April 5, 1958, 3.
59. Conway, in Hackman interview, 21.
60. John T. Dunlop, "The Public Interest in International Affairs of Unions," address before the American Bar Association, July 12, 1957 (mimeographed), 4–5.
61. RFK, *Enemy Within,* ch. 11.
62. Walter Sheridan, *The Fall and Rise of Jimmy Hoffa* (New York, 1972), 33. Shefferman gives a benign picture of his activities in *The Man in the Middle* (Garden City, N.Y., 1961).
63. Sheridan, *Fall and Rise,* 33, 50; Senate Select Committee, *Interim Report,* 85 Cong., 2 Sess. (March 24, 1958), 297–300; Daniel Bell, "Nathan Shefferman, Union Buster," *Fortune,* February 1958; Sheridan, in recorded interview by Jean Stein, July 23, 1968, 4, Stein Papers; Sheridan, in recorded interview by Roberta Greene, March 23, 1970, 34, RFK Oral History Program; Pierre Salinger, in recorded interview by L. J. Hackman, April 18, 1970, 73, RFK Oral History Program; Salinger, *Je suis un Américain* (Paris, 1957), 158.
64. RFK, *Enemy Within,* 207–209.
65. *New York Times,* November 2, 1957.
66. Warner Bloomberg, Jr., et al., "The State of the Unions," *New Republic,* June 22, 1959.
67. Goldberg to RFK, February 17, 1960, RFK Papers.
68. RFK, memorandum, March 20, 1959, RFK Papers.
69. Robert F. Kennedy, "An Urgent Reform Plan," *Life,* June 1, 1959.
70. RFK, handwritten note, September 28 [1958], RFK Papers.
71. Donald R. Larrabee, "Crooks Exposed, Not Prosecuted," *New Bedford Standard-Times,* October 4, 1959.
72. William G. Hundley, in recorded interview by J. A. Oesterle, December 9, 1970, 7–8, RFK Oral History Program.
73. RFK to Charles J. Lewin, August 1, 1959, RFK Papers.
74. RFK, memorandum, January 8, 1958, 3, RFK Papers.
75. RFK, memorandum, February 22, 1958, 4, RFK Papers.
76. RFK to Bowles, August 12, 1959, Bowles Papers.
77. *Newsweek,* August 10, 1959.
78. RFK to McClellan, September 10, 1959, RFK Papers.
79. Senate Select Committee, *Final Report,* 86 Cong., 2 Sess. (March 31, 1960), 868–869.

80. Walter Sheridan, in recorded interview by Roberta Greene, April 7, 1970, 7, 30–31, RFK Oral History Program.
81. RFK, *Enemy Within*, 284.
82. Felix Frankfurter, "Hands Off the Investigations," *New Republic*, May 21, 1924.
83. 273 U.S. 135.
84. Philip B. Kurland, "The Watergate Inquiry, 1973," in *Congress Investigates, 1972–1974*, ed. Arthur M. Schlesinger, Jr., and Roger Bruns (New York, 1975), 474.
85. Hugo Black, "Inside a Senate Investigation," *Harper's*, February 1936.
86. RFK, *Enemy Within*, 295–296.
87. *Watkins* v. *U.S.*, 354 U.S. 178.
88. RFK, *Enemy Within*, 161.
89. Pierre Salinger, in recorded interview by Theodore H. White, July 19, 1965, 10–11, JFK Oral History Program.
90. RFK, *Enemy Within*, 292.
91. John Bartlow Martin, "The Struggle to Get Hoffa," uncut manuscript, 3–12; Carmine Bellino, in interview by author, February 3, 1976.
92. Walter Sheridan, in recorded interview by Roberta Greene, March 23, 1970, 56–57, RFK Oral History Program.
93. RFK, *Enemy Within*, 175.
94. Martin to author, October 19, 1975.
95. Porter to RFK, August 23, 1957, RFK Papers.
96. Dan Wakefield, "Bob," *Esquire*, April 1962.
97. Victor Lasky, *Robert F. Kennedy: The Myth and the Man* (New York: Pocket Book reprint, 1971), 127.
98. James R. Hoffa, *The Trials of Jimmy Hoffa: An Autobiography*, as told to Donald I. Rogers (Chicago, 1970), 167.
99. Alexander M. Bickel, "Robert F. Kennedy: The Case Against Him for Attorney General," *New Republic*, January 9, 1961. See also Bickel, in recorded interview by Jean Stein, May 14, 1970, 2–3, Stein Papers.
100. Bickel, "Robert F. Kennedy."
101. Edward Bennett Williams, in interview by author, January 23, 1976.
102. RFK, handwritten note, November 2 [1957], RFK Papers.
103. Senate Select Committee, *Final Report*, 86 Cong., 2 Sess. (March 31, 1960), 868.
104. RFK, *Enemy Within*, 297–300.
105. Bickel, "Robert F. Kennedy."
106. John Seigenthaler, in recorded interview by W. A. Geoghegan [July 1964], 41–49, JFK Oral History Program.
107. Beaverbrook to RFK, May 11, 1961, Beaverbrook Papers.
108. RFK, speech at Notre Dame University, February 22, 1958, 1, RFK Papers.
109. RFK, *Enemy Within*, 307.
110. Murray Kempton, in recorded interview by Jean Stein, October 27, 1969, 3, Stein Papers.
111. Kempton, in interview by author, December 10, 1975.
112. RFK, *Enemy Within*, 261.
113. Walter Reuther, in recorded interview by Jean Stein, October 14, 1968, 4–5, Stein Papers.
114. Victor Reuther, in interview by author, August 10, 1971.
115. Peter Maas, "Robert Kennedy Speaks Out," *Look*, March 28, 1961.
116. *San Francisco Monitor*, July 19, 1957.
117. Kempton, in Stein interview, 4.
118. Murray Kempton, "The Uncommitted," *Progressive*, September 1960.

10. 1960 *(pages 192–221)*

1. Theodore C. Sorensen, *Kennedy* (New York, 1965), 35, 117.
2. RFK, memorandum to Theodore H. White on the manuscript of *The Making of a President, 1960*, March 23, 1961, RFK Papers.

3. David Hackett, in recorded interview by John Douglas, July 22, 1970, 13–14, RFK Oral History Program.
4. Pierre Salinger, in recorded interview by Theodore H. White, July 19, 1965, 28, JFK Oral History Program; Salinger, *With Kennedy* (Garden City, N.Y., 1966), 30.
5. Paul B. Fay, Jr., *The Pleasure of His Company* (New York, 1966), 6–7.
6. Fred Dutton, in recorded interview by L. J. Hackman, November 18, 1969, 15–16, RFK Oral History Program.
7. Harris Wofford, in recorded interview by Jean Stein, October 3, 1968, 7, Stein Papers.
8. Barrett Prettyman, Jr., in recorded interview by L. J. Hackman, June 5, 1969, 57, RFK Oral History Program.
9. Kenneth P. O'Donnell and David Powers, *"Johnny, We Hardly Knew Ye"* (Boston, 1972), 151.
10. Author's journal, March 25–27, 1960.
11. *Peoria Journal-Star*, February 14, 1960.
12. Joseph Alsop, in recorded interview by Roberta Greene, June 10, 1971, 55, RFK Oral History Program.
13. O'Donnell and Powers, *"Johnny,"* 157.
14. Hubert H. Humphrey, *The Education of a Public Man* (Garden City, N.Y., 1976), 208.
15. Ibid., 208.
16. Patrick Lucey, in recorded interview by L. D. Epstein, August 1, 1964, 37–38, JFK Oral History Program.
17. Jerry Bruno and Jeff Greenfield, *The Advance Man* (New York: Bantam reprint, 1972), 28, 35.
18. Jean Stein and George Plimpton, eds., *American Journey* (New York, 1970), 70.
19. Ibid., 69.
20. Joseph Dolan, in recorded interview by L. J. Hackman, April 10, 1970, 56–57, RFK Oral History Program; Stein and Plimpton, *American Journey,* 69.
21. Humphrey, *Education of a Public Man,* 208.
22. *National Review,* December 16, 1961.
23. Franklin Wallick, "I Remember Hubert," *United Auto Workers Washington Report,* January 16, 1978.
24. Lawrence F. O'Brien, *No Final Victories* (Garden City, N.Y., 1964), 65.
25. Walter Cronkite, in recorded interview by Jean Stein, December 2, 1969, 7, Stein Papers.
26. Minutes of "Meeting re West Virginia Primary — April 8, 1960," RFK Papers.
27. O'Donnell and Powers, *"Johnny,"* 161.
28. Minutes of meeting, April 8, 1960.
29. O'Donnell and Powers, *"Johnny,"* 166–167.
30. Salinger, in White interview, 59.
31. William Walton, in recorded interview by Jean Stein, September 22, 1968, 4, Stein Papers.
32. Franklin D. Roosevelt, Jr., in interview by author, February 12, 1975.
33. Charles Spalding, in recorded interview by L. J. Hackman, March 22, 1969, 31, RFK Oral History Program.
34. D. E. Koskoff, *Joseph P. Kennedy* (Englewood Cliffs, N.J., 1974), 399, 596; James MacGregor Burns, *John Kennedy* (New York, 1960), 196.
35. Fay, *Pleasure of His Company,* 9.
36. *Newsweek,* September 12, 1960.
37. Jeffrey Potter, *Men, Money and Magic: The Story of Dorothy Schiff* (New York, 1976), 261.
38. Rose Kennedy, in interview by author, May 25, 1976.
39. Franklin D. Roosevelt, Jr., in recorded interview by Jean Stein, December 9, 1969, 3–5, Stein Papers.
40. Ibid., 5.
41. O'Brien, *No Final Victories,* 68.

42. Humphrey, *Education of a Public Man*, 216-217.
43. Walter Lippmann to JFK, January 22, 1960, Lippmann Papers.
44. Sorensen, *Kennedy*, 141.
45. O'Brien, *No Final Victories*, 72.
46. Ibid., 73.
47. Victor Lasky, *J. F. K.: The Man and the Myth* (New York, 1963), 344.
48. Humphrey, *Education of a Public Man*, 475.
49. Theodore White, in interview by author, July 5, 1976.
50. O'Donnell and Powers, *"Johnny,"* 171.
51. Author's journal, May 14, 1960.
52. John Bartlow Martin, *Adlai Stevenson and the World* (New York, 1977), 499.
53. Author's journal, May 22, 1960.
54. Ibid., May 29, 1960.
55. Ibid., June 16, 1960.
56. Ibid.
57. RFK to author, June 17, 1960, Schlesinger Papers.
58. Theodore C. Sorensen, "Election of 1960," in *History of American Presidential Election*, ed. Arthur M. Schlesinger, Jr. (New York, 1971), vol. 4, 3460.
59. RFK, in recorded interview by author, February 27, 1965, 32-33, JFK Oral History Program.
60. Peter Lisagor, in recorded interview by R. J. Grele, April 22, 1966, 25, JFK Oral History Program.
61. *New York Times*, July 14, 1960.
62. John Seigenthaler, in recorded interview by W. A. Geoghegan [July 1964], 92-93, JFK Oral History Program.
63. Joseph Tydings, in recorded interview by Roberta Greene, May 3, 1971, 6, RFK Oral History Program.
64. Author's journal, July 14, 1960.
65. The Graham memorandum was subsequently printed by Theodore H. White as an appendix to *The Making of the President, 1964* (New York, 1965).
66. These and subsequent RFK quotations are from RFK, in recorded interview by author, February 27, 1965, JFK Oral History Program.
67. Jack Conway, in recorded interview by L. J. Hackman, April 10, 1972, 66-67, RFK Oral History Program.
68. Author's journal, June 12, 1960.
69. Clark Clifford, in recorded interview by L. J. Hackman, December 16, 1974, 23, JFK Oral History Program; Nancy Dickerson, *Among Those Present* (New York: Ballantine reprint, 1977), 47.
70. John Seigenthaler, in recorded interview by R. J. Grele, February 21, 1966, 30-31, JFK Oral History Program.
71. Jack Conway, in interview by L. J. Hackman, April 11, 1972, 2, RFK Oral History Program.
72. Edwin Guthman, *We Band of Brothers* (New York, 1971), 75.
73. O'Donnell and Powers, *"Johnny,"* 189.
74. Evelyn Lincoln, *Kennedy and Johnson* (New York, 1968), 92-93.
75. James H. Rowe, Jr., in interview by author, August 16, 1964.
76. O'Donnell and Powers, *"Johnny,"* 191-193.
77. Conway, in Hackman interview, 2-3.
78. RFK, in recorded interview by John Bartlow Martin, April 13, 1964, II, 5, JFK Oral History Program.
79. Ibid., 9.
80. Bobby Baker, with Larry King, *Wheeling and Dealing: Confessions of a Capitol Hill Operator* (New York, 1978), 113. Rowe, in interview by author.
81. Charles Spalding, in recorded interview by Jean Stein, January 22, 1970, 2, Stein Papers.
82. Charles Bartlett, in recorded interview by Jean Stein, January 9, 1970, 11, Stein Papers; John Seigenthaler, in Geoghegan interview [July 1964], 73.
83. Bartlett, in Stein interview, 14.

84. Theodore H. White, *The Making of the President, 1960* (New York, 1961), 383.
85. John Seigenthaler, in recorded interview by L. J. Hackman, July 1, 1970, 88, RFK Oral History Program.
86. Author's journal, July 15, 1960.
87. O'Donnell and Powers, "*Johnny,*" 199.
88. *Time*, October 10, 1960.
89. Richard Wade, in recorded interview by Roberta Greene, December 13, 1973, 9–10, 20.
90. Ralph Horton, Jr., in recorded interview by Joseph Dolan, June 1, 1964, 29, JFK Oral History Program.
91. Tydings, in Greene interview, 7–10.
92. Joan Braden, in recorded interview by D. J. O'Brien, October 11, 1969, 82, RFK Oral History Program.
93. Ralph de Toledano, *RFK: The Man Who Would Be President* (New York: New American Library, Signet reprint, 1967), 169.
94. Norman Mailer, *The Presidential Papers* (New York, 1963), 36.
95. Hugh Sidey, "Brother on the Spot," in *The Kennedy Circle,* ed. Lester Tanzer (Washington, 1961), 209.
96. As reported by Kennedy immediately afterward to Seigenthaler; Seigenthaler, in Grele interview, 168–169.
97. Merle Miller, ed., *Plain Speaking* (New York: Berkley reprint, 1974), 438.
98. Stewart Alsop, "Kennedy's Magic Formula," *Saturday Evening Post*, August 13, 1960.
99. Sidey, "Brother on the Spot," 207.
100. RFK, memorandum, February 14, 1959, RFK Papers.
101. Theodore Sorensen to RFK, December 14, 1959, RFK Papers.
102. RFK, "Georgia and Virginia," memorandum for the files, November 16, 1959, RFK Papers.
103. Author's journal, May 14, 1960.
104. RFK to JFK, June 24, 1960, JFK Papers.
105. Harris Wofford, in interview by Jean Stein, October 3, 1968, 4, Stein Papers.
106. Theodore H. White, 1960 convention notes, White Papers.
107. Author's journal, July 11, 1960; Stein and Plimpton, *American Journey*, 90.
108. Carl M. Brauer, *John F. Kennedy and the Second Reconstruction* (New York, 1977), 53.
109. John A. Williams, *The King God Didn't Save* (New York: Pocket Book reprint, 1971), 27.
110. This account is based on: Seigenthaler to the author, November 9, 1976; Wofford, in Stein interview, 9–20; Wofford, in recorded interview by Berl Bernhard, November 29, 1965, 24–28, JFK Oral History Program; Wofford's Christmas letter (mimeographed), January 7, 1964; and Seigenthaler, in Grele interview, 108–111.
111. J. K. Galbraith, *Ambassador's Journal* (Boston, 1969), 6.
112. Martin Luther King, Jr., "It's a Difficult Thing to Teach a President," *Look*, November 17, 1964; David L. Lewis, *King: A Critical Biography* (New York: Penguin, Pelican reprint, 1971), 129.
113. Brauer, *Kennedy and the Second Reconstruction*, 48.
114. Wofford, Christmas letter.
115. Murray Kempton, "His Brother's Keeper," *New York Post*, November 10, 1960.
116. *Atlanta Constitution*, September 10, 1960.
117. George McGovern, in recorded interview by Jean Stein, September 25, 1969, 1–2, Stein Papers; McGovern, in recorded interview by L. J. Hackman, July 16, 1970, 3–4, RFK Oral History Program.
118. *Time*, October 10, 1960.
119. Arthur Edson, "Bobby — Washington's No. 2 Man," AP feature for release April 14, 1963.
120. *Time*, October 10, 1960; Sidey, "Brother on the Spot," 209.
121. A thought I passed on to John Kennedy in a letter of August 30, 1960, Schlesinger Papers.

122. Wofford, Christmas letter; as amended by Theodore H. White, in interview by author, February 8, 1978.
123. Pierre Salinger, *With Kennedy* (Garden City, 1966), 51.
124. Jack Conway, in recorded interview by L. J. Hackman, April 11, 1972, 21–22, RFK Oral History Program.
125. Kempton, "His Brother's Keeper."
126. Churchill to JFK, November 11, 1960, JFK Papers.

11. To the Department of Justice *(pages 222–244)*

1. RFK, in recorded interview by John Bartlow Martin, February 29, 1964, 1, JFK Oral History Program.
2. John F. Kennedy, *A Compendium of Speeches, Statements, and Remarks Delivered During His Service in the Congress of the United States* (Washington, 1964), 1108.
3. Robert A. Lovett, in recorded interview by Dorothy Fosdick, July 20, 1964, 9, JFK Oral History Program.
4. RFK, memorandum, February 9, 1961, 10–11, RFK Papers.
5. Chester Bowles, *Promises to Keep* (New York, 1971), 299–300.
6. See W. A. Harriman and Elie Abel, *Special Envoy to Churchill and Stalin, 1941–1946* (New York, 1975), 78–105.
7. Author's journal, December 1, 1960.
8. Arthur M. Schlesinger, Jr., *A Thousand Days* (Boston, 1965), 149.
9. RFK, memorandum, 1–2.
10. Roswell Gilpatric, in recorded interview by D. J. O'Brien, May 5, 1970, 6, JFK Oral History Program.
11. RFK, in Martin interview, 12–13.
12. RFK, memorandum, 9.
13. Ibid., 5–6.
14. Ibid., 2–5.
15. Ibid., 13.
16. Dorothy Goldberg, *A Private View of a Public Life* (New York, 1975), 5.
17. RFK, in recorded interview by author, February 27, 1965, 53, JFK Oral History Program.
18. RFK, memorandum, 6–7, 2.
19. Author's journal, December 1, 1960.
20. Schlesinger, *Thousand Days*, 142–143, 162.
21. RFK, in interview by author, 50.
22. Author's journal, June 15, 1960.
23. RFK, memorandum, 7.
24. Hugh Sidey, "Brother on the Spot," in *The Kennedy Circle*, ed. Lester Tanzer (Washington, 1961), 186.
25. RFK, memorandum, 7; RFK, in interview by author, 49.
26. William O. Douglas, in recorded interview by Roberta Greene, November 13, 1969, 13–14, RFK Oral History Program; Douglas, in recorded interview by John Stewart, November 9, 1967, 13, JFK Oral History Program.
27. RFK, Martin interview, 21.
28. Drew Pearson to RFK, December 5, 1960, RFK Papers, 47–48.
29. RFK, in interview by author, 47–49.
30. Paul B. Fay, Jr., *The Pleasure of His Company* (New York, 1966), 11.
31. Clark Clifford, in recorded interview by L. J. Hackman, December 16, 1974, 65, JFK Oral History Program.
32. *New York Times*, November 23, 1960.
33. Author's journal, December 1, 1960.
34. RFK, in interview by author, 49.
35. William C. Sullivan, in interview by author, July 26, 1976.
36. Victor Navasky, *Kennedy Justice* (New York, 1971), 4.
37. W. O. Douglas, in introduction to Robert E. Thompson and Hortense Myers,

Robert F. Kennedy: The Brother Within (New York, 1962). In 1969 Douglas recalled it somewhat differently: "In the end, I urged him to do it, but I was trying to get him to think through what he wanted to do with his life and that maybe this was a turning point" (Douglas, in Greene interview, 13).

38. Sidey, "Brother on the Spot," 186.
39. John Seigenthaler, in recorded interview by R. J. Grele, February 22, 1966, 182, JFK Oral History Program.
40. RFK, in interview by author, 49.
41. Seigenthaler, in Grele interview, 183-201.
42. Peter Maas, "Robert Kennedy Speaks Out," *Look*, March 28, 1961.
43. RFK, memorandum, 7.
44. RFK, in recorded interview by Anthony Lewis, December 4, 1964, I, 14, JFK Oral History Program.
45. RFK, memorandum, 8.
46. Ibid., 9.
47. Ralph de Toledano, *RFK: The Man Who Would Be President* (New York: New American Library, Signet reprint, 1968), 178.
48. Alexander M. Bickel, "Robert F. Kennedy: The Case Against Him for Attorney General," *New Republic*, January 9, 1961.
49. Booth Mooney, *LBJ: An Irreverent Chronicle* (New York, 1976), 50.
50. For a sketch, see Charlotte Hays, "South of the Line of Succession," *Washington Post*, December 30, 1973.
51. RFK, in Lewis interview, I, 20-21.
52. Bobby Baker, with Larry King, *Wheeling and Dealing: Confessions of a Capitol Hill Operator* (New York, 1978), 120-121.
53. RFK, memorandum, 13.
54. Ibid.
55. Senate Judiciary Committee, *Robert F. Kennedy: Attorney-General-Designate, Hearing*, 87 Cong., 1 Sess. (January 13, 1961), 4-5, 30.
56. Ibid., 15.
57. Ibid., 22.
58. Ibid., 34.
59. *New York Times*, January 14, 1961.
60. Maas, "Robert Kennedy Speaks Out."
61. William Manchester, *Portrait of a President* (Boston, 1962), 60.
62. *San Francisco News*, July 12, 1957.
63. RFK, in Martin interview, 35-36.
64. Maas, "Robert Kennedy Speaks Out."
65. Alexander M. Bickel, in recorded interview by Jean Stein, May 14, 1970, 4, Stein Papers.
66. *Berlingske Tidende*, February 3, 1963; translation in RFK Papers.
67. Roy Wilkins, in recorded interview by Jean Stein, May 12, 1970, Stein Papers.
68. Arthur Edson, "Bobby — Washington's No. 2 Man," AP feature for release April 14, 1963.
69. Ovid Demaris, *The Director* (New York, 1975), 137.
70. Edward V. Bander to author, March 17, 1977.
71. Ramsey Clark, in recorded interview by L. J. Hackman, June 29, 1970, 35, RFK Oral History Program.
72. Robert Morgenthau, in recorded interview by Jean Stein, November 21, 1969, 1, Stein Papers.
73. Louis Oberdorfer, in recorded interview by Roberta Greene, February 5, 1970, 18, RFK Oral History Program.
74. John Seigenthaler, in recorded interview by R. J. Grele, February 22, 1966, 219, JFK Oral History Program.
75. Nicholas Katzenbach, in recorded interview by Anthony Lewis, November 16, 1964, 42, JFK Oral History Program.
76. Norbert Schlei, in recorded interview by John Stewart, February 20-21, 1968, 21, JFK Oral History Program.

77. Archibald Cox to author, January 3, 1977.
78. Victor Navasky, *Kennedy Justice* (New York, 1971), 182. This brilliant if sometimes overschematic book is indispensable for an understanding of Robert Kennedy as Attorney General. I have drawn heavily on it.
79. William Orrick, in recorded interview by L. J. Hackman, April 13, 1970, 63, RFK Oral History Program.
80. Louis Oberdorfer, in interview by author, December 23, 1976.
81. John Douglas, in recorded interview by L. J. Hackman, June 16, 1969, 17, 27, RFK Oral History Program.
82. Nicholas Katzenbach, in recorded interview by Jean Stein, August 7, 1968, 17, Stein Papers.
83. Clark, in Hackman interview, 34.
84. Edwin Guthman, *We Band of Brothers* (New York, 1971), 88.
85. Angie Novello to RFK, November 4, 1963, RFK Papers.
86. Paul O'Neill, "The No. 2 Man in Washington," *Life*, January 26, 1962.
87. Douglas, in Hackman interview, 16.
88. Morgenthau, in Stein interview, 1.
89. Patrick Anderson, "Robert's Character," *Esquire*, April 1965.
90. Joseph Dolan, in recorded interview by L. J. Hackman, April 11, 1970, 220, RFK Oral History Program.
91. Navasky, *Kennedy Justice*, 354.
92. RFK to Evelyn Eright, October 11, 1962, RFK Papers.
93. RFK, memorandum to assistant attorneys general, October 23, 1961, RFK Papers.
94. Navasky, *Kennedy Justice*, 355.
95. Ibid.
96. Jean Stein and George Plimpton, eds., *American Journey* (New York, 1970), 78.
97. Guthman, *We Band of Brothers*, 206.
98. Navasky, *Kennedy Justice*, 48.
99. Orrick, in Hackman interview, 27.
100. Clark, in Hackman interview, 9.
101. Guthman, *We Band of Brothers*, 86.

12. The Pursuit of Justice: J. Edgar Hoover *(pages 245–260)*

1. Victor Navasky, *Kennedy Justice* (New York, 1971), 8.
2. A. T. Mason, *Harlan Fiske Stone: Pillar of the Law* (New York, 1956), 152.
3. Max Lowenthal, *The Federal Bureau of Investigation* (New York, 1950), 298.
4. Mason, *Harlan Fiske Stone*, 153.
5. Fred J. Cook, "The FBI," *Nation*, October 18, 1958.
6. Senate Select Committee to Study Governmental Operations with respect to Intelligence Activities (hereafter cited as Church committee), *Hearings*, vol. 6, *Federal Bureau of Investigation*, 94 Cong., 2 Sess. (1976), 35, 409–415.
7. Church committee, *Final Report*, bk. II, *Intelligence Activities and the Rights of Americans*, 94 Cong., 2 Sess. (1976), 44.
8. George E. Allen, "J. Edgar Hoover Off-Duty," *Congressional Record*, October 13, 1972, S18165.
9. See interview with Allen in Ovid Demaris, "The Private Life of J. Edgar Hoover," *Esquire*, September 1975.
10. William C. Sullivan, "Personal Observations and Recommendations on Privacy," in *Privacy in a Free Society*, Final Report, Annual Chief Justice Warren Conference on Advocacy in the United States, June 1974, 94.
11. Demaris, "Private Life of Hoover."
12. W. W. Turner, *Hoover's FBI* (New York: Dell reprint, 1971), 29–30, 65.
13. Joseph L. Schott, *No Left Turns* (New York, 1975), 5–6, 130, 137, passim; William C. Sullivan, in interview by author, July 26, 1976; Ovid Demaris, *The Director* (New York, 1975), 81–82, 220.
14. W. C. Sullivan to J. E. Hoover, August 28, 1971, Sullivan Papers.

15. Schott, *No Left Turns,* 42, 249–250.
16. Turner, *Hoover's FBI,* xi.
17. Ibid., 4.
18. Roy Cohn, "Could He Walk on Water?" *Esquire,* November 1972.
19. W. C. Sullivan to Hoover, October 6, 1971, Sullivan Papers.
20. Demaris, *Director,* 77.
21. W. C. Sullivan, in interview by Jack Nelson of the *Los Angeles Times,* May 15, 1973; Frank J. Donner, "Hoover's Legacy," *Nation,* June 1, 1974.
22. Walter Pincus, "The Bureau's Budget," in *Investigating the FBI,* ed. Pat Watters and Stephen Gillers (New York, 1973), 70.
23. Dean Rusk, in testimony before the Senate Foreign Relations Committee, July 23, 1974; quoted in Church committee, *Final Report,* bk. III, *Supplementary Detailed Staff Reports on Intelligence Activities and the Rights of Americans,* 94 Cong., 2 Sess. (1976), 469–470.
24. Clarence M. Kelley on CBS program *Face the Nation,* August 8, 1976.
25. W. C. Sullivan to C. D. DeLoach, July 19, 1966, in Church committee, *Hearings,* vol. 6, 357.
26. W. C. Sullivan, in interview by author, July 26, 1976. The situation changed, Sullivan added (in a letter to me of October 1, 1977), when Ramsey Clark became Attorney General.
27. Church committee, *Final Report,* bk. II, 284.
28. *Olmstead* v. *United States,* 277 U.S. 438 (1928).
29. *Nardone* v. *United States,* 302 U.S. 379 (1937); *Nardone* v. *United States,* 308 U.S. 338 (1939).
30. FDR to Thomas H. Eliot, February 21, 1941, Roosevelt Papers.
31. Joseph Lash, *Roosevelt and Churchill, 1939–1941* (New York, 1976), 119.
32. FDR to Robert H. Jackson, May 21, 1940, Roosevelt Papers.
33. FDR to Eliot.
34. R. H. Jackson to Congressman Hatton Sumners, March 19, 1941, in Church committee, *Final Report,* bk. III, 280.
35. Francis Biddle, *In Brief Authority* (New York, 1962), 187.
36. Demaris, *Director,* 127.
37. Elsey to Truman, February 2, 1950, in Church committee, *Final Report,* bk. III, 283.
38. Brownell to Hoover, May 22, 1954, in Church committee, *Final Report,* bk. III, 296–297.
39. Sullivan, in interview by author.
40. RFK, in recorded interview by Anthony Lewis, December 4, 1964, IV, 22, JFK Oral History Program; RFK, in recorded interview by John Bartlow Martin, February 29, 1964, 51, JFK Oral History Program.
41. RFK, in Lewis interview, IV, 33–35.
42. Nicholas Katzenbach, in recorded interview by L. J. Hackman, October 8, 1969, 40, RFK Oral History Program.
43. RFK, in Lewis interview, IV, 35–36.
44. Author's journal, October 17, 1961.
45. Ibid., March 13, 1962.
46. William Manchester, *The Death of a President* (New York, 1967), 119.
47. Benjamin C. Bradlee, *Conversations with Kennedy* (New York, 1975), 228.
48. Demaris, *Director,* 288.
49. Biddle, *In Brief Authority,* 259.
50. Clark Clifford, in recorded interview by L. J. Hackman, February 4, 1975, 44, JFK Oral History Program.
51. Demaris, *Director,* 189–190.
52. RFK, in Lewis interview, V, 13; Demaris, *Director,* 185–186.
53. RFK, in Lewis interview, IV, 35.
54. Hoover, note on H. L. Edwards to John P. Mohr, "Attorney General's Effort to Get Into Bureau Gymnasium," February 1, 1961, RFK/FBI/FOIA release.
55. Schott, *No Left Turns,* 192–193.

56. Demaris, *Director,* 147.
57. Walter Sheridan, in recorded interview by Jean Stein, July 23, 1968, 7, Stein Papers.
58. Sullivan, in interview by author.
59. Edwin Guthman, *We Band of Brothers* (New York, 1971), 261.
60. Sullivan, in interview by author; Turner, *Hoover's FBI,* 89.
61. Ralph de Toledano, *J. Edgar Hoover: The Man in His Time* (New York: Manor Books reprint, 1974), 307.
62. Navasky, *Kennedy Justice,* 35.
63. William Barry, in recorded interview by Roberta Greene, October 22, 1969, 69, RFK Oral History Program.
64. Toledano, *J. Edgar Hoover,* 291.
65. Sullivan, in interview by author.
66. S. J. Ungar, *FBI* (Boston, 1976), 276.
67. John Seigenthaler, in recorded interview by R. J. Grele, February 23, 1966, 352, JFK Oral History Program.
68. Hoover to RFK, February 22, 1963, RFK Papers.
69. Turner, *Hoover's FBI,* 43.
70. William Barry, in recorded interview by Jean Stein, April 1970, 11, Stein Papers.
71. RFK, in Lewis interview, V, 13.
72. RFK, in recorded interview by John Bartlow Martin, April 13, 1964, I, 34, 36, JFK Oral History Program.

13. The Pursuit of Justice: The Mob *(pages 261–285)*

1. It is on this point, I take it, that I disagree most sharply with Victor Navasky's analysis of the contest; but I want to express again my debt to the insight and information in his valuable *Kennedy Justice* (New York, 1971).
2. Hoover to RFK, January 10, 1961, in Senate Select Committee to Study Governmental Operations with respect to Intelligence Activities (hereafter cited as Church committee), *Hearings,* vol. 6, *Federal Bureau of Investigation,* 94 Cong., 2 Sess. (1976), 822.
3. Church committee, *Hearings,* vol. 6, 58–59; W. C. Sullivan to J. E. Hoover, October 6, 1971, Sullivan Papers.
4. As told by Henry Brandon, in author's journal, December 4, 1961. When Brandon subsequently quoted the Attorney General in the *London Sunday Times* as saying that half the membership of the CPUSA consisted of FBI agents, an outraged patriot sent the clipping to Hoover. "It was good of you to make this item available to me," Hoover quickly replied, enclosing five statements expressing his own views of the Communist threat. See exchange in RFK/FBI/FOIA release, March 5, 8, 1962.
5. Transcript of interview in RFK Papers [1961].
6. I owe this interesting intelligence to the Freedom of Information Act, under which I received copies of: Hoover's notation on a memorandum to L. B. Nicholas, July 21, 1950; his instruction to the FBI Boston office, October 28, 1954; and his notation on a memorandum of June 15, 1962.
7. RFK, in speech given at the University of Georgia Law School; *New York Times,* May 7, 1961.
8. RFK, statement before the Senate Government Operations Committee, September 25, 1963 (mimeographed), 3, 23, RFK Papers.
9. D. P. Moynihan, "The Private Government of Crime," *Reporter,* July 6, 1961.
10. William W. Turner could find only one; Turner, *Hoover's FBI* (New York: Dell reprint, 1971), 152.
11. William C. Sullivan, in interview by author, July 26, 1976; Navasky, *Kennedy Justice,* 44.
12. Peter Maas, *The Valachi Papers* (New York: Bantam reprint, 1969), 28.
13. Ovid Demaris, *The Director* (New York, 1975), 141.

14. Ibid., 141–142; Pat Watters and Stephen Gillers, eds., *Investigating the FBI* (New York, 1973), 207.
15. Gerald Goettel, "Why the Crime Syndicates Can't Be Touched," *Harper's*, November 1960.
16. Turner, *Hoover's FBI*, 157. Emphasis added.
17. "Interview with J. Edgar Hoover," *U.S. News & World Report*, December 21, 1964.
18. Navasky, *Kennedy Justice*, 33–34.
19. Ralph de Toledano, *J. Edgar Hoover* (New York: Manor Books reprint, 1974), 263.
20. Ibid., 313.
21. S. J. Ungar, *FBI* (Boston, 1976), 391.
22. Kennedy's figure in 1963 ("Robert Kennedy Speaks His Mind," *U.S. News & World Report*, January 28, 1963). Navasky lists twenty-seven (*Kennedy Justice*, 50). Probably Kennedy left out the CIA.
23. Toledano, *J. Edgar Hoover*, 306–307.
24. RFK, in recorded interview by Anthony Lewis, December 4, 1964, IV, 30, 32, JFK Oral History Program.
25. Peter Maas, *The Valachi Papers* (New York: Bantam reprint, 1969), 54–55.
26. Navasky, *Kennedy Justice*, 32.
27. RFK, in Lewis interview, IV, 27–28.
28. William H. Hundley, in recorded interview by J. A. Oesterle, February 22, 1971, 129, 144–145, RFK Oral History Program.
29. *FBI Law Enforcement Bulletin*, September 1963.
30. RFK, in recorded interview by John Bartlow Martin, April 13, 1964, I, 35, JFK Oral History Program.
31. Navasky, *Kennedy Justice*, 32.
32. RFK, in Martin interview, I, 31.
33. Peter Maas, "Robert Kennedy Speaks Out," *Look*, March 28, 1961.
34. There are several accounts of this meeting: William Orrick, in recorded interview by L. J. Hackman, April 14, 1970, 269–271, RFK Oral History Program; Joseph Dolan, in recorded interview by L. J. Hackman, July 18, 1970, 268, RFK Oral History Program; Ramsey Clark, in recorded interview by L. J. Hackman, July 20, 1970, 117–119, RFK Oral History Program.
35. RFK, in Martin interview, I, 37; Theodore C. Sorensen, *The Kennedy Legacy* (New York, 1969), 242.
36. Nicholas Katzenbach, in recorded interview by L. J. Hackman, October 8, 1969, 54–55, RFK Oral History Program.
37. See Robert F. Kennedy, "Attorney General's Opinion on Wiretaps," *New York Times Magazine*, June 3, 1962.
38. Alexander M. Bickel, in recorded interview by Jean Stein, May 14, 1970, 5, Stein Papers.
39. *On Lee* v. *United States*, 343 U.S. 747, 760–761. For a convenient summary of the criticism, see the American Civil Liberties Union report *The Wiretapping Problem Today* (March 1962).
40. In 1968 the Omnibus Crime Control and Safe Streets Act attempted to establish standards for wiretapping.
41. Church committee, *Final Report*, bk. III, *Supplementary Detailed Staff Reports on Intelligence Activities and the Rights of Americans*, 94 Cong., 2 Sess. (1976), 301.
42. Katzenbach, in Hackman interview, 58.
43. RFK, in Martin interview, I, 38.
44. Toledano, *J. Edgar Hoover*, 316.
45. Church committee, *Hearings*, vol. 6, 166.
46. Evans to RFK, February 17, 1966, in Toledano, *J. Edgar Hoover*, 318–319.
47. Church committee, *Hearings*, vol. 6, 200.
48. Church committee, *Final Report*, bk. III, 329–330.
49. Toledano, *J. Edgar Hoover*, 317.
50. Demaris, *Director*, 146.

51. Sullivan, in interview by author.
52. William Hundley, in interview by author, August 4, 1976.
53. Edwin Guthman, *We Band of Brothers* (New York, 1971), 263.
54. Navasky, *Kennedy Justice*, 88–89.
55. Frank Mankiewicz, in recorded interview by L. J. Hackman, September 9, 1969, 21–22, RFK Oral History Program.
56. Navasky, *Kennedy Justice*, 91.
57. Hundley, in interview by author.
58. Katzenbach, in Hackman interview, 41.
59. Victor Lasky, *Robert F. Kennedy: The Myth and the Man* (New York: Pocket Book reprint, 1971), 428.
60. G. Robert Blakey to RFK, December 14, 1966, RFK Papers.
61. Church committee, *Final Report*, bk. III 368.
62. Ibid., 368–369.
63. Toledano, *J. Edgar Hoover*, 319.
64. Joseph W. Schott, *No Left Turns* (New York, 1975), 194.
65. Guthman, *We Band of Brothers*, 263–264.
66. Hundley, in interview by author; Hundley, in Oesterle interview, 45–46.
67. Katzenbach, in Hackman interview, 53.
68. Navasky, *Kennedy Justice*, 32.
69. Burke Marshall, "Can the FBI Rebuild Itself?" *Washington Post*, July 1, 1973.
70. John Douglas to RFK, December 27, 1966, RFK Papers.
71. Demaris, *Director*, 229.
72. Dolan, in Hackman interview, 287, 290, 291.
73. Demaris, *Director*, 145–146; Hundley, in Oesterle interview, 37–38.
74. Katzenbach, in Hackman interview, 41.
75. RFK, in Lewis interview, IV, 26.
76. J. Edgar Hoover, "The FBI's War on Organized Crime," *U.S. News & World Report*, April 18, 1966.
77. "Interview with J. Edgar Hoover," *U.S. News & World Report*, December 21, 1964.
78. And 677 in 1964. For statistics, see *Congressional Record*, March 11, 1969, S2642.
79. This letter, and another from Hunter on the same subject, were printed by Drew Pearson in the *Washington Post*, January 4, 1961; see Walter Sheridan, *The Fall and Rise of Jimmy Hoffa* (New York, 1972), 158–159, 165–166.
80. *New York Daily News*, December 21, 1960.
81. Navasky, *Kennedy Justice*, 413.
82. Sheridan, *Fall and Rise of Hoffa*, 193.
83. Ibid., 281.
84. Ibid., 217.
85. Ibid., 276–278.
86. Ibid., 280.
87. *Detroit Sunday News*, August 1, 1976.
88. Sheridan, *Fall and Rise of Hoffa*, 528.
89. *New York Times*, October 10, 1975.
90. Ralph James and Estelle James, *Hoffa and the Teamsters* (Princeton, 1965), 62.
91. Sheridan, *Fall and Rise of Hoffa*, 283, 459.
92. Ibid., 293, 460.
93. Ibid., 459.
94. Robert H. Jackson, "The Federal Prosecutor," *Journal of the American Judicature Society*, June 1940.
95. Navasky, *Kennedy Justice*, 417.
96. Ibid., 435–436.
97. Mortimer M. Caplin, "Special Racketeer Investigations," February 24, 1961, Johnson Papers; Navasky, *Kennedy Justice*, 49, 56, 60.
98. Maas, "Robert Kennedy Speaks Out."
99. Navasky, *Kennedy Justice*, 56–61; Louis Oberdorfer, in interview by author, December 29, 1976.

100. In congressional testimony in 1967, quoted by Herman Schwartz, "Six Years of Tapping and Bugging," *Civil Liberties Review*, Summer 1974.
101. Navasky, *Kennedy Justice*, 410.
102. Herbert Hoover, *Memoirs ... The Cabinet and the Presidency, 1920–1933* (New York, 1952), 276–277.
103. Elmer L. Irey with William J. Slocum, *The Tax Dodgers* (New York, 1948), 26, 35–36.

14. The Pursuit of Justice: Civil Rights *(pages 286–316)*

1. Peter Maas, "Robert Kennedy Speaks Out," *Look*, March 22, 1961.
2. Edwin Guthman, *We Band of Brothers* (New York, 1971), 181.
3. Harris Wofford, in recorded interview by Berl Bernhard, November 29, 1965, 10–11, JFK Oral History Program.
4. Senate Commerce Committee, *The Speeches of Senator John F. Kennedy: Presidential Campaign of 1960*, 87 Cong., 1 Sess. (1961), 432, 576.
5. Harris Wofford, "Memorandum to President-Elect Kennedy on Civil Rights — 1961," December 30, 1960, RFK Papers; Wofford to RFK, "On the Civil Rights Division," January 12, 1961, RFK Papers.
6. Carl M. Brauer, *John F. Kennedy and the Second Reconstruction* (New York, 1977), 66–67.
7. Martin Luther King, Jr., "It's a Difficult Thing to Teach a President," *Look*, November 17, 1964.
8. RFK, in recorded interview by Anthony Lewis, December 4, 1964, II, 4, 9, JFK Oral History Program. Burke Marshall also participated in the civil rights interviews on that date.
9. RFK, in recorded interview by John Bartlow Martin, February 29, 1964, 54, JFK Oral History Program.
10. John Seigenthaler, in recorded interview by R. J. Grele, February 22, 1966, 210, JFK Oral History Program.
11. Guthman, *We Band of Brothers*, 95–96.
12. RFK, memorandum, February 9, 1961, RFK Papers.
13. Burke Marshall, in recorded interview by Jean Stein, October 6, 1968, 2–3, Stein Papers.
14. RFK and Burke Marshall, in Lewis interview, II, 5.
15. Nicholas Katzenbach, in recorded interview by Anthony Lewis, November 16, 1964, 34–35, JFK Oral History Program.
16. John Seigenthaler, in recorded interview by R. J. Grele, February 23, 1966, 360–361, JFK Oral History Program.
17. Guthman, *We Band of Brothers*, 103–104.
18. RFK to the Board of Governors, Metropolitan Club, April 11 and September 19, 1961, RFK Papers; RFK to author, December 21, 1961, RFK Papers.
19. RFK, in Lewis interview, V, 22–23.
20. Alexander M. Bickel, *Politics and the Warren Court* (New York, 1972), 58–59.
21. Wofford to RFK, January 12, 1961, RFK Papers.
22. John T. Elliff, "Aspects of Federal Civil Rights Enforcement: The Justice Department and the FBI, 1939–1964," *Perspectives*, vol. 5 (1971), 621.
23. Victor Reuther, *The Brothers Reuther* (Boston, 1976), 281.
24. Elliff, "Civil Rights Enforcement," 643–647.
25. *To Secure These Rights: Report of the President's Committee on Civil Rights* (Washington, 1947), 123.
26. S. J. Ungar, *FBI* (Boston, 1976), 408.
27. William Sullivan, in interview by author, July 26, 1976; Senate Select Committee to Study Governmental Operations with respect to Intelligence Activities (hereafter cited as Church committee), *Hearings*, vol. 6, *Federal Bureau of Investigation*, 94 Cong., 1 Sess. (1975), 33.
28. Hoover, in interview by Dean Fischer, *Time*, December 14, 1970; see also interview by Ken Clawson, *Washington Post*, November 17, 1970.

29. James Wechsler, "The FBI's Failure in the South," *Progressive*, December 1963.
30. Elliff, "Civil Rights Enforcement," 650–651.
31. John Doar and Dorothy Landsberg, "The Performance of the FBI in Investigating Violations of Federal Laws Protecting the Right to Vote — 1960–1967" (1971), in Church committee, *Hearings*, vol. 6, esp. 895–896, 903–904, 928; Arlie Schardt, "Civil Rights: Too Much, Too Late," in *Investigating the FBI*, ed. Pat Watters and Stephen Gillers (Garden City, N.Y., 1973), 184; Alan Lichtman, "The Federal Assault Against Voting Discrimination in the Deep South, 1957–1967," *Journal of Negro History*, October 1969, 352.
32. RFK, in Lewis interview, V, 22–23.
33. RFK to Black, n.d. [March 1961], Black Papers. The paper was the *Shades Valley Sun*, March 2, 1961. Black replied that the editorial had probably been written by his friend Charles Feidelson (Black to RFK, March 28, 1961, Black Papers).
34. The text of the Law Day speech was printed in the *New York Times*, May 7, 1961.
35. Brauer, *Kennedy and the Second Reconstruction*, 98.
36. Nicholas Katzenbach, in recorded interview by Jean Stein, September 27, 1968, 2, Stein Papers.
37. Howell Raines, *My Soul Is Rested* (New York, 1977), 337.
38. Brauer, *Kennedy and the Second Reconstruction*, 152–153, 343.
39. Peter Maas, "Robert Kennedy Speaks Out," *Look*, March 28, 1961.
40. Seigenthaler, in Grele interview, February 23, 1966, 334.
41. *New York Times*, May 7, 1961.
42. *Boynton v. Virginia*, 364 U.S. 454 (1960).
43. Guthman, *We Band of Brothers*, 167; Rowe's testimony before Church committee, *Hearings*, vol. 6, 117.
44. Gary Thomas Rowe, Jr., *My Undercover Years with the Ku Klux Klan* (New York, 1976), 39–44; Church committee, *Hearings*, vol. 6, 127.
45. RFK, in Lewis interview, II, 7.
46. Maas, "Robert Kennedy Speaks Out."
47. Harris Wofford, in recorded interview by Jean Stein, October 3, 1968, 19, Stein Papers.
48. Seigenthaler, in Grele interview, February 22, 1966, 305–309.
49. RFK, in Lewis interview, II, 23.
50. Seigenthaler, in Grele interview, February 22, 1966, 314–324.
51. Guthman, *We Band of Brothers*, 171.
52. Ungar, *FBI*, 410; Victor Navasky, *Kennedy Justice* (New York, 1971), 22.
53. Jean Stein and George Plimpton, eds., *American Journey* (New York, 1970), 103.
54. Seigenthaler, in Grele interview, February 22, 1966, 325–333.
55. Raines, *My Soul Is Rested*, 309.
56. Brauer, *Kennedy and the Second Reconstruction*, 102.
57. Raines, *My Soul Is Rested*, 308–309.
58. Edwin Guthman in Pierre Salinger et al., eds., *An Honorable Profession* (Garden City, N.Y., 1968), 20–21.
59. RFK in Lewis interview, II, 20, 21; Guthman, *We Band of Brothers*, 178.
60. John Lewis, in recorded interview by Jean Stein, November 26, 1968, 3, Stein Papers.
61. William Orrick, in recorded interview by L. J. Hackman, April 14, 1970, 249–253, RFK Oral History Program.
62. RFK, in Lewis interview, III, 12–13.
63. August Meier and Elliott Rudwick, *CORE: A Study in the Civil Rights Movement, 1942–1968* (New York, 1973), 139.
64. John Maguire, in recorded interview by Jean Stein, October 17, 1969, 9, Stein Papers.
65. Raines, *My Soul Is Rested*, 123, 277.
66. RFK, in Lewis interview, III, 11.
67. Ibid.
68. Guthman, *We Band of Brothers*, 154–155.
69. Stein and Plimpton, *American Journey*, 96; Guthman, *We Band of Brothers*, 175.

70. Anthony Lewis and the *New York Times, Portrait of a Decade* (New York, 1964), 118.
71. Brauer, *Kennedy and the Second Reconstruction*, 119.
72. RFK, in Lewis interview, I, 21.
73. Elliff, "Civil Rights Enforcement," 651–652.
74. Doar and Landsberg, "Performance of the FBI," 905–906.
75. RFK, "Civil Liberties: Against the Rule of Force," address to Ansonia Independent Democratic Club, October 8, 1964, RFK Papers.
76. Burke Marshall, *Federalism and Civil Rights* (New York, 1964), 9.
77. Ibid., 11–12.
78. Raines, *My Soul Is Rested*, 228.
79. Vincent Harding, in recorded interview by Jean Stein, December 9, 1969, 4, Stein Papers.
80. RFK, in Lewis interview, III, 25.
81. Raines, *My Soul Is Rested*, 228.
82. Howard Zinn, *SNCC: The New Abolitionists* (Boston: Beacon reprint, 1965), 58–59.
83. James Forman, *The Making of Black Revolutionaries* (New York, 1972), 264.
84. Meier and Rudwick, *CORE*, 175.
85. Martin Luther King, Jr., "Dear Friend" fund-raising letter, November 11, 1961, Schlesinger Papers.
86. RFK, in Lewis interview, III, 27.
87. Forman, *Making of Black Revolutionaries*, 265–266.
88. Miriam Feingold, "Chronicling the 'Movement,'" *Reviews in American History*, March 1974, 155.
89. Samuel Lubell, "It's Bobby Who Roils Dixie," *Detroit Free Press*, July 13, 1961.
90. Marshall, *Federalism and Civil Rights*, 46–48.
91. RFK, in foreword to Marshall, *Federalism and Civil Rights*, viii.
92. Lichtman, "Federal Assault," 359.
93. Brauer, *Kennedy and the Second Reconstruction*, 158.
94. Burke Marshall to R. H. Barrett, January 3, 1964, Marshall Papers.
95. Thurgood Marshall, in recorded interview by Berl Bernhard, April 7, 1964, 22–24, JFK Oral History Program.
96. RFK, in Marshall, *Federalism and Civil Rights*, ix.
97. Brauer, *Kennedy and the Second Reconstruction*, 167.
98. Burke Marshall, "Equitable Remedies as Instruments of Social Change," speech at New York University, November 14, 1964, Marshall Papers.
99. R. H. Jackson, *The Supreme Court in the American System of Government* (Cambridge, Mass., 1955), 70–72.
100. Bickel, *Politics and the Warren Court*, 112–113.
101. Marshall, *Federalism and Civil Rights*, 81.
102. Marshall, "Equitable Remedies."
103. Pat Watters and Reese Cleghorn, *Climbing Jacob's Ladder: The Arrival of Negroes in Southern Politics* (New York, 1967), 231.
104. Walter Lord, *The Past That Would Not Die* (New York, 1965), 247.
105. RFK, in Lewis interview, III, 15–17.
106. Marshall, *Federalism and Civil Rights*, 31.
107. Wilkins to JFK, June 22, 1961, JFK Papers. The telegram is annotated: "Discussed with him."
108. RFK, in Lewis interview, III, 38; IV, 7.
109. Robert Sherrill, *Gothic Politics in the Deep South* (New York: Ballantine reprint, 1969), 212.
110. RFK, in recorded interview by John Bartlow Martin, May 14, 1964, I, 10, JFK Oral History Program.
111. RFK phone messages, June 1962, RFK Papers.
112. RFK, in Martin interview, I, 10.
113. See "Judicial Performance in the Fifth Circuit," *Yale Law Journal*, November 1963, n. 87. The Clarke County situation is well described by one of the gov-

ernment attorneys in Gerald M. Stern, "Judge William Harold Cox and the Right to Vote in Clarke County, Mississippi," in *Southern Justice*, ed. Leon Friedman (New York, 1965).

114. W. H. Cox to J. C. McLaurin, February 14, 1963, RFK Papers.
115. Cox to John Doar, October 16, 1963, RFK Papers.
116. RFK to Cox, November 18, 1963, RFK Papers.
117. *New York Times*, March 9, 1964.
118. Leon Jaworski to Burke Marshall, March 23, 1964, Marshall Papers.
119. "Judicial Performance," n. 71.
120. Marshall, *Federalism and Civil Rights*, 6.
121. Navasky, *Kennedy Justice*, 112.
122. Ibid., 251.
123. RFK to Mrs. Franklin D. Roosevelt, May 22, 1962, RFK Papers.
124. Navasky, *Kennedy Justice*, 270.
125. Mary J. Curzan, "A Case Study in the Selection of Federal Judges in the Fifth Circuit, 1953-1963." This Yale Ph.D. dissertation draws on confidential interviews and is available only in summary. It is discussed in Brauer, *Kennedy and the Second Reconstruction*, 123, 340; and in Navasky, *Kennedy Justice*, 269-270. Brauer points out that Kennedy's integrationist judges were largely appointed in the upper and western south, and that the somewhat rigid Curzan criteria had the curious result of classifying Judge Frank Johnson as a segregationist.
126. RFK, in Lewis interview, VI, 35.
127. Martin Luther King, Jr., "Equality Now," *Nation*, February 4, 1961.
128. Archibald C. Cox, address before the Associated Harvard Alumni, *Harvard Today*, June 1975.
129. John F. Kennedy, *Public Papers . . . 1961* (Washington, 1962), 256.
130. William S. White, *The Professional* (Boston, 1964), 228.
131. Evelyn Lincoln, *Kennedy and Johnson* (New York, 1968), 182.
132. John Seigenthaler, in recorded interview by L. J. Hackman, June 5, 1970, 26, RFK Oral History Program.
133. Brauer, *Kennedy and the Second Reconstruction*, 82.
134. RFK, in Lewis interview, IV, 23, 27.
135. "Interview of Attorney General by Bob Spivack . . . May 18, 1962," 11, RFK Papers.
136. Berl Bernhard, in interview by author, August 23, 1965.
137. Theodore M. Hesburgh, in recorded interview by Joseph E. O'Connor, March 27, 1966, 21, 23, JFK Oral History Program.
138. Ibid., 8-9.
139. Harris Wofford, in recorded interview by L. J. Hackman, February 3, 1969, 147, RFK Oral History Program.
140. RFK, in Lewis interview, VI, 38-39.
141. Hesburgh, in O'Connor interview, 20.
142. Martin Luther King, Jr., "Fumbling on the New Frontier," *Nation*, March 3, 1962.
143. Robert C. Weaver, in recorded interview by Daniel Patrick Moynihan, October 1, 1964, 210, JFK Oral History Program.
144. Wilkins to Dungan, November 19, 1962, JFK Papers.
145. Loren Miller, "Farewell to Liberals: A Negro View," *Nation*, October 20, 1962.
146. Bayard Rustin, "The Meaning of the March on Washington," *Liberation*, October 1963.
147. Meier and Rudwick, *CORE*, 180.
148. Forman, *Making of Black Revolutionaries*, 265, 546.
149. King, "It's Difficult to Teach a President."
150. Arthur Edson, Associated Press story on Robert Kennedy, April 14, 1963, RFK Papers; Clarence Mitchell, in recorded interview by John Stewart, February 1967, 46, JFK Oral History Program.
151. Marshall, in Bernhard interview, April 7, 1964, 16.
152. Stein and Plimpton, *American Journey*, 111.

153. Harris Wofford, in interview by L. J. Hackman, May 22, 1968, 63, RFK Oral History Program.

15. The Pursuit of Justice: Ross Barnett and George Wallace
(pages 317-342)

1. Loren Miller, "Farewell to Liberals: A Negro View," *Nation*, October 20, 1962.
2. Alexis de Tocqueville, *The Old Regime and the French Revolution*, ch. 4.
3. On *Meet the Press*, May 26, 1963; James Meredith, *Three Years in Mississippi* (Bloomington, Ind., 1966), 294.
4. John Bowers, "James Meredith at Columbia Law," *New York Herald Tribune*, April 3, 1966.
5. Walter Lord, *The Past That Would Not Die* (New York, 1965), 139.
6. Author's journal, September 17, 1962.
7. Transcript of conversation between RFK and Governor Ross Barnett, 12:20 P.M., September 25, 1962, Burke Marshall Papers.
8. RFK, in recorded interview by Anthony Lewis, December 4, 1964, VII, 6, JFK Oral History Program.
9. Edwin Guthman, *We Band of Brothers* (New York, 1971), 189; Lord, *Past That Would Not Die*, ch. 7; C. M. Brauer, *John F. Kennedy and the Second Reconstruction* (New York, 1977), ch. 7.
10. Guthman, *We Band of Brothers*, 189.
11. RFK, in Lewis interview, VII, 9-10.
12. Author's journal, September 28, 1962.
13. Ibid., September 29, 1962.
14. Transcript of conversation between JFK and Barnett, 2:30 P.M., September 29, 1962, Burke Marshall Papers.
15. Author's journal, September 29, 1962.
16. Norbert Schlei, in recorded interview by John Stewart, February 20-21, 1968.
17. Transcript of conversation between RFK and Barnett, 12:45 P.M., September 30, 1962, Burke Marshall Papers.
18. RFK, in Lewis interview, VII, 11.
19. Budd Schulberg, "R.F.K. — Harbinger of Hope," *Playboy*, January 1969.
20. Nicholas Katzenbach, in recorded interview by Anthony Lewis, November 29, 1964, 15-16, JFK Oral History Program.
21. Guthman, *We Band of Brothers*, 201; RFK, in Lewis interview, VIII, 5.
22. Meredith, *Three Years in Mississippi*, 194.
23. John F. Kennedy, *Public Papers . . . 1962* (Washington, 1963), 728.
24. Guthman, *We Band of Brothers*, 204-205.
25. Katzenbach, in Lewis interview, 30.
26. RFK, in Lewis interview, VII, 14, 17.
27. Lord, *Past That Would Not Die*, 3-4.
28. RFK, in Lewis interview, VII, 12.
29. Theodore C. Sorensen, *Kennedy* (New York, 1965), 487.
30. RFK, in Lewis interview, VII, 23.
31. Burke Marshall, *Federalism and Civil Rights* (New York, 1965), 68.
32. RFK, in Lewis interview, VII, 22, 25.
33. Lord, *Past That Would Not Die*, 231.
34. James Meredith to RFK, September 5, 1963, RFK Papers.
35. Guthman, *We Band of Brothers*, 205.
36. Ibid., 181.
37. John F. Kennedy, *Profiles in Courage* (New York, 1956), ch. 6.
38. RFK, in Lewis interview, VII, 10.
39. John F. Kennedy, *Public Papers . . . 1961* (Washington, 1962), 19.
40. James W. Silver, *Mississippi: The Closed Society* (New York, 1964), 144.
41. RFK to H. M. Ray, April 16, 1963, RFK Papers.
42. Guthman, *We Band of Brothers*, 188.

43. Ralph Dungan relaying the message from Rusk to JFK, October 3, 1962, JFK Papers.
44. Howard P. Jones to RFK, October 25, 1962, RFK Papers.
45. Arthur M. Schlesinger, Jr., *A Thousand Days* (Boston, 1965), 948.
46. William Goldsmith, *The Growth of Presidential Power* (New York, 1974), iii, 1665.
47. Martin Luther King, Jr., "It's a Difficult Thing to Teach a President," *Look*, November 17, 1964.
48. Martin Luther King, Jr., "Bold Design for a New South," *Nation*, March 30, 1963.
49. John F. Kennedy, *Public Papers . . . 1963* (Washington, 1964), 222.
50. King, "Bold Design for New South."
51. Burke Marshall to Leslie Dunbar, February 26, 1963, Marshall Papers.
52. RFK, in recorded interview by Anthony Lewis, December 22, 1964, 4, JFK Oral History Program.
53. Jean Stein and George Plimpton, eds. *American Journey* (New York, 1970), 114.
54. *Atlanta Inquirer*, April 14, 1962; D. W. Matthews and J. R. Prothro, *Negroes and the New Southern Politics* (New York, 1966), 240.
55. Martin Luther King, Jr., *Why We Can't Wait* (New York: New American Library, Signet reprint, 1964), 80–81.
56. Coretta King, *My Life with Martin Luther King, Jr.* (New York, 1970), 239–241.
57. King, *Why We Can't Wait*, 97.
58. Stein and Plimpton, *American Journey*, 118.
59. Alan F. Westin and Barry Mahoney, *The Trial of Martin Luther King* (New York, 1974), 149.
60. King, *Why We Can't Wait*, 103.
61. Stein and Plimpton, *American Journey*, 119.
62. David L. Lewis, *King: A Critical Biography* (New York: Penguin Books, Pelican reprint, 1971), 199.
63. Andrew Young, in recorded interview by Jean Stein, September 9, 1968, 4, Stein Papers.
64. Lewis, *King*, 202.
65. Matthews and Prothro, *Negroes*, 240.
66. King, *Why We Can't Wait*, 103.
67. Young, in Stein interview, 3; Stein and Plimpton, *American Journey*, 117–118.
68. Michael Dorman, *We Shall Overcome* (New York: Dell reprint, 1965), 197.
69. The *New Yorker* piece was incorporated in James Baldwin, *The Fire Next Time* (New York, 1963); the quotation is from p. 63.
70. RFK, in Lewis interview, December 22, 1964, 64.
71. James Baldwin, in recorded interview by Jean Stein, February 7, 1970, 1, Stein Papers.
72. Burke Marshall, in recorded interview by Jean Stein, October 6, 1968, 31.
73. August Meier and Elliott Rudwick, *CORE: A Study in the Civil Rights Movement, 1942–1968* (New York, 1975), 116, 143, 298, 408.
74. RFK, in recorded interview by Anthony Lewis, December 6, 1964, I, 7, JFK Oral History Program.
75. Baldwin, in Stein interview, 2–3.
76. Ibid., 4.
77. Kenneth B. Clark, in recorded interview by Jean Stein, January 30, 1970, 3, Stein Papers; James Wechsler, "RFK and Baldwin," *New York Post*, May 28, 1963.
78. Lena Horne and Richard Schickel, *Lena* (New York: New American Library, Signet reprint, 1966), 210.
79. Clark, in Stein interview, 4.
80. RFK, in Lewis interview, December 22, 1964, I, 68.
81. Kenneth B. Clark, in interview by author, September 9, 1976.
82. Horne and Schickel, *Lena*, 209.
83. *Meet the Press*, May 26, 1963; Meredith, *Three Years in Mississippi*, 299.
84. Wechsler, "RFK and Baldwin."
85. RFK, in Lewis interview, December 22, 1964, I, 67.

86. Author's journal, May 27, 1963; RFK, in recorded interview by John Bartlow Martin, April 30, 1964, III, 35, JFK Oral History Program.
87. Wechsler, "RFK and Baldwin."
88. Author's journal, May 27, 1963.
89. James Baldwin, Malcolm X, Martin Luther King talk with Kenneth B. Clark, in *The Negro Protest* (Boston, 1963), 3.
90. RFK, in Martin interview, III, 35–36.
91. Clark, in interview by author.
92. Committee on Equal Employment Opportunity, Minutes of the Seventh Meeting, May 29, 1963, Lyndon Johnson Papers.
93. Committee on Equal Employment Opportunity, Transcript of Eighth Meeting, July 18, 1963, 33–38; Jack Conway, in recorded interview by L. J. Hackman, April 11, 1972, 83, RFK Oral History Program; William V. Shannon, *The Heir Apparent* (New York, 1967), 49.
94. RFK, in Lewis interview, December 4, 1964, IV, 26.
95. RFK, memorandum, July 16, 1963, RFK Papers.
96. RFK, in Lewis interview, December 4, 1964, VI, 7.
97. Marshall Frady, *Wallace* (New York, 1968), 133, 142.
98. "Playboy Interview: Martin Luther King," *Playboy*, January 1965.
99. Brooks Hays to JFK, November 30, 1962, reporting on conversation with Sparkman, RFK Papers.
100. JFK to RFK, November 30, 1962, RFK Papers.
101. RFK, in Lewis interview, December 6, 1964, I, 42.
102. The observer was Benjamin Muse; Carl M. Brauer, *John F. Kennedy and the Second Reconstruction* (New York, 1977), 141.
103. RFK, in Lewis interview, December 6, 1964, I, 19–20.
104. The quotations are from "Transcript of Conversation between Attorney General Robert F. Kennedy and Governor Wallace, Montgomery, Alabama, April 25, 1963," RFK Papers. See also Frady, *Wallace*, 150–169.
105. JFK, *Public Papers . . . 1963*, 408.
106. Pierre Salinger, "Memorandum of Conversation between President Kennedy and Governor George Wallace . . . May 18, 1963," JFK Papers.
107. Burke Marshall, "Memorandum to the Members of the Cabinet re: University of Alabama," May 21, 1963, Marshall Papers.
108. RFK, in Lewis interview, December 1964, I, 30–31.
109. Ibid., 31–33.
110. Ibid., 36.
111. RFK to Frances Battle, June 11, 1963, RFK Papers.
112. Guthman, *We Band of Brothers*, 214–216; Sorensen, *Kennedy*, 492. There is also valuable material in Robert Drew's cinema verité film *Crisis — Behind a Presidential Commitment* (ABC News — Drew Associates, 1963).
113. Howell Raines, *My Soul Is Rested* (New York, 1977), 331, 341.
114. Frady, *Wallace*, 169–170; Guthman, *We Band of Brothers*, 214–217; *New York Times*, June 12, 1963; Dorman, *We Shall Overcome*, 335–365.
115. RFK, in Lewis interview, December 6, 1964, I, 43.
116. Robert Sherrill, *Gothic Politics in the Old South* (New York, Ballantine reprint, 1969), 331–332.
117. Robert F. Kennedy, Jr., *Judge Frank M. Johnson: A Study in Integrity* (New York, 1978), ch. 15.
118. RFK to David Kennedy and to Michael Kennedy, June 11, 1963, RFK Papers.

16. The Pursuit of Justice: Martin Luther King *(pages 343–367)*

1. RFK, in recorded interview by Anthony Lewis, December 4, 1964, VIII, 25; and December 6, 1964, II, 3, 5, JFK Oral History Program.
2. RFK, in Lewis interview, December 6, 1964, II, 7–9; RFK, in recorded interview by John Bartlow Martin, April 30, 1964, III, 39–40, JFK Oral History Program.

3. John F. Kennedy, *Public Papers . . . 1963* (Washington, 1964), 469.
4. James Meredith, *Three Years in Mississippi* (Bloomington, Ind., 1966), 310; Jack Mendelsohn, *The Martyrs* (New York, 1966), 78–80; article on Doar in *Milwaukee Sentinel*, August 12, 1963.
5. Charles Evers, in recorded interview by Jean Stein, n.d. [1968], 4, Stein Papers.
6. Charles Evers, "For Robert F. Kennedy," *New York Post*, October 23, 1964.
7. John Lewis, in recorded interview by Jean Stein, September 9, 1968, 1–3, Stein Papers.
8. William J. vanden Heuvel, in recorded interview by Jean Stein, November 10, 1968, 1–5, Stein Papers; Anthony Lewis and the *New York Times, Portrait of a Decade* (New York, 1964), 298–300.
9. Jack Newfield, *Robert Kennedy* (New York, 1969), 22–23.
10. RFK, in Lewis interview, December 22, 1964, 80–81.
11. Kennedy, *Public Papers . . . 1963*, 469.
12. Luther Hodges, in recorded interview by D. B. Jacobs, May 18, 1964, 92, JFK Oral History Program.
13. Victor Navasky, *Kennedy Justice* (New York, 1971), 99.
14. RFK, in Lewis interview, December 22, 1964, 4; December 4, 1964, VI, 19, VIII, 14.
15. For the background of the bill, see, from the administration's viewpoint, Norbert Schlei, in recorded interview by John Stewart, February 20–21, 1968, 52 ff., JFK Oral History Program; and Nicholas Katzenbach, in recorded interview by Anthony Lewis, November 16, 1964, 42 ff., JFK Oral History Program. From the viewpoint of the civil rights leadership, see Joseph Rauh, unpublished manuscript on the fight for the 1964 civil rights bill, Rauh Papers.
16. Marshall, in Lewis interview with RFK, December 4, 1964, 27.
17. Norbert Schlei to RFK, "Comments of the Vice President on the Civil Rights Legislative Proposals," June 4, 1963, RFK Papers; Schlei, in Stewart interview, 47–49.
18. Bobby Baker to Mike Mansfield, June 27, 1963, JFK Papers.
19. RFK, in Lewis interview, December 22, 1964, 5.
20. RFK, in Lewis interview, December 4, 1964, VIII, 24; and December 6, I, 3.
21. RFK, in Lewis interview, December 6, 1964, I, 52–53.
22. Burke Marshall, in recorded interview by Jean Stein, October 5, 1969, 4, Stein Papers.
23. RFK, in Lewis interview, December 6, 1964, I, 9–10.
24. Marshall, in Stein interview, October 6, 1968, 27–28.
25. Jean Stein and George Plimpton, eds. *American Journey* (New York, 1970), 123–124.
26. Rauh, unpublished manuscript, 4.
27. Rauh, "Memorandum Concerning Administration's Civil Rights Bill as Background for Meeting of Pro-Civil-Rights Groups on July 2, 1963," RFK Papers.
28. Jervis Anderson, *A. Philip Randolph: A Political Portrait* (New York, 1973), 323–325.
29. Author's journal, June 22, 1963.
30. I describe the meeting at length in *A Thousand Days* (Boston, 1965), 969–971.
31. Author's journal, June 22, 1963.
32. Joseph Rauh, in recorded interview by C. T. Morrissey, December 23, 1965, 103–104, JFK Oral History Program.
33. Martin Luther King, Jr., "It's a Difficult Thing to Teach a President," *Look*, November 17, 1964.
34. Walter Reuther, in recorded interview by Jean Stein, October 24, 1968, 12–13, Stein Papers; John Douglas, in recorded interview by L. J. Hackman, May 5, 1970, 2–3; Jack Conway, in recorded interview by L. J. Hackman, December 29, 1972, 8–9; Alan Raywid, in recorded interview by Roberta W. Greene, August 15, 1974, 34–35, RFK Oral History Program.
35. For details, see Douglas, in Hackman interview, and Raywid, in Greene interview.
36. Author's journal, August 15, 1963.

37. The original text may be found in Joanne Grant, *Black Protest: History, Documents and Analyses* (New York, 1968), 375-377.
38. Reuther, in Stein interview, 14.
39. James Forman, *The Making of Black Revolutionaries* (New York, 1972), 332, 335.
40. Ibid., 335-336.
41. Malcolm X, *Autobiography* (New York: Grove reprint, 1966), 280-281.
42. John A. Williams, *This Is My Country Too* (New York: New American Library, Signet reprint, 1966), 149.
43. Bayard Rustin, "The Washington March — a 10-Year Perspective," *Crisis*, September 1973.
44. Senate Select Committee to Study Governmental Operations with respect to Intelligence Activities (hereafter cited as Church committee), *Final Report*, bk. III, *Supplementary Detailed Staff Reports on Intelligence Activities and the Rights of Americans*, 94 Cong., 2 Sess. (1976), 105-109.
45. William Sullivan, in interview by author, July 26, 1976.
46. Quoted in Department of Justice, "Report of . . . Task Force to Review the FBI Martin Luther King, Jr., Security Assassination Investigation," January 11, 1977 (mimeographed), 165-166, 172.
47. Ibid., 113, 123.
48. Church committee, *Final Report*, bk. III, 82, 87-91.
49. Hoover to RFK, April 20, 1962, RFK Papers.
50. Don Oberdorfer, "King Adviser Says FBI 'Used' Him," *Washington Post*, December 15, 1975.
51. Senate Judiciary Committee, Internal Security Subcommittee, Levison Hearing, April 30, 1962, transcript in RFK Papers.
52. Stanley Levison, in interview by author, August 3, 1976.
53. Coretta Scott King, *My Life with Martin Luther King, Jr.* (New York, 1969), 345.
54. Church committee, *Final Report*, bk. III, 95-96; King to O'Dell, July 3, 1963, in Navasky, *Criminal Justice*, 143-144.
55. Church committee, *Final Report*, bk. III, 88.
56. Burke Marshall, in interview by author, August 1, 1976.
57. Levison, in interview by author.
58. Harris Wofford, in recorded interview by L. J. Hackman, February 3, 1969, 143, RFK Oral History Program.
59. John Seigenthaler, in recorded interview by R. J. Grele, February 22, 1966, 336-338, JFK Oral History Program.
60. RFK, in Lewis interview, December 4, 1964, VI, 2.
61. Church committee, *Hearings*, vol. 6, *Federal Bureau of Investigation*, 94 Cong., 1 Sess. (1975), 208.
62. Levison, in interview by author.
63. Department of Justice, "Report of Task Force to Review King," 125.
64. Church committee, *Hearings*, vol. 6, 170.
65. Sullivan, in interview by author.
66. Oberdorfer, *Washington Post*, December 15, 1975. See also David L. Lewis, *King: A Critical Biography* (New York: Penguin Books, Pelican reprint, 1957), 357: "Attorney Stanley Levison . . . advised Martin that his Vietnam position would bankrupt the organization."
67. Marshall, in interview by author; Carl Rowan, "King's 'Communist Adviser,'" *New York Post*, December 19, 1975.
68. Edward Jay Epstein, *Legend: The Secret World of Lee Harvey Oswald* (New York, 1978), 20, 263-264.
69. For accounts of the Kennedy meeting as reported by King to Young, see Leon Howell, "An Interview with Andrew Young," *Christianity and Crisis*, February 16, 1976; Howell Raines, *My Soul Is Rested* (New York, 1977), 430-431; Ovid Demaris, *The Director* (New York, 1975), 210; Church committee, *Final Report*, bk. III, 97.
70. Levison, in interview by author.
71. Howell, "Interview with Young"; Levison, in interview by author.
72. JFK, *Public Papers . . . 1963*, 574.

73. RFK to Senator Mike Monroney, July 23, 1963, RFK Papers. Identical letters went to a number of senators.
74. Sullivan, in interview by author.
75. Victor Navasky, "The Government and Martin Luther King," *Atlantic Monthly*, November 1970.
76. Church committee, *Final Report*, bk. III, 101–102.
77. David Wise, *The American Police State* (New York, 1976), 301.
78. *New York Times*, June 20 and 21, 1969.
79. So Evans told the Bureau; see Church committee, *Final Report*, bk. III, 103.
80. Rowan, "King's 'Communist Adviser.'"
81. Wise, *American Police State*, 301.
82. Howell, "Interview with Young."
83. Martin Luther King, Jr., *Why We Can't Wait* (New York: New American Library, Signet reprint, 1964), 147.
84. Levison, in interview by author.
85. Ibid.
86. Church committee, *Final Report*, bk. III, 132; Department of Justice, "Report of Task Force to Review King," 120, 176.
87. Nicholas Katzenbach, in recorded interview by L. J. Hackman, October 8, 1969, 61, RFK Oral History Program.
88. Sullivan, in interview by author.
89. Church committee, *Final Report*, bk. III, 143, 146.
90. RFK, in Lewis interview, December 4, 1964, VI, 7–8.
91. Church committee, *Final Report*, bk. III, 133.
92. Ibid., 108, 136–137.
93. Ibid., 137.
94. Ibid., 83.
95. *Time*, December 14, 1970.
96. *U.S. News & World Report*, December 7, 1964.
97. Church committee, *Final Report*, bk. III, 158–160.
98. Howell Raines, *My Soul Is Rested*, 428.
99. Howell, "Interview with Young."
100. Church committee, *Hearings*, vol. 6, 210.
101. Hugh Sidey, "L.B.J., Hoover and Domestic Spying," *Time*, February 10, 1975.
102. Church committee, *Final Report*, bk. III, 154.
103. Katzenbach, in Hackman interview, 69–70.
104. Kenneth Clark, in interview by author, September 9, 1976.
105. Church committee, *Final Report*, bk. III, 163–168; Howell, "Interview with Young."
106. Sullivan, in interview by author.
107. *Time*, December 14, 1970.
108. RFK, statement before the Senate Commerce Committee, July 1, 1963 (Department of Justice release), 14, RFK Papers.
109. Katzenbach, in Lewis interview, 63.
110. RFK, in Lewis interview, December 22, 1964, 15, 17.
111. RFK, statement before Senate Judiciary Committee, July 18, 1963 (Department of Justice release), 23–24, RFK Papers.
112. For a good account, see Dick Dabney, *A Good Man: The Life of Sam J. Ervin* (Boston, 1976), 210–217; also, Senate Judiciary Committee, *Hearings: Civil Rights — The President's Program*, 88 Cong. 2 Sess. (July–September 1963), passim.
113. Navasky, *Kennedy Justice*, 96–97.

17. The Politics of Justice *(pages 368–391)*

1. Peter Maas, "Robert Kennedy Speaks Out," *Look*, March 28, 1961.
2. Ramsey Clark, in interview by author, March 15, 1977.
3. Ovid Demaris, *The Director* (New York, 1975), 284; Courtney Evans to A. H. Belmont, "Re: Paul Corbin," February 1, 1962, RFK/FBI/FOIA release.

4. Hubert Humphrey to John Bailey, May 22, 1961, RFK Papers.

5. Author's journal, June 2, 1963.

6. Ibid., May 2, 1962.

7. RFK to John Bailey, July 19, 1961, RFK Papers.

8. Donald M. Wilson, in recorded interview by James Greenfield, September 2, 1964, 31, JFK Oral History Program.

9. Author's journal, September 18, 1962. See also Joseph Dolan, in recorded interview by L. J. Hackman, July 18, 1970, 299, RFK Oral History Program; John English, in recorded interview by Roberta Greene, November 3, 1969, 18–19, RFK Oral History Program.

10. Frank O'Connor, in recorded interview by Roberta Greene, June 19, 1970, 8–9, 16–18, RFK Oral History Program. Edwin Weisl, Lyndon Johnson's man in New York, reported to the Johnson White House in 1964 after a talk with Buckley, "In the last state election Attorney General Kennedy induced Buckley, against his better judgment, to support Henry [sic] Morgenthau, Jr., for Governor" (E. L. Weisl to Walter Jenkins, January 28, 1964, Johnson Papers).

11. RFK, in recorded interview by John Bartlow Martin, April 13, 1964, II, 61-62, JFK Oral History Program.

12. Milton Gwirtzman, in recorded interview by Roberta Greene, December 23, 1971, 6–9, RFK Oral History Program.

13. Author's journal, September 18, 1962.

14. RFK, in recorded interview by Anthony Lewis, December 22, 1964, 10–11, JFK Oral History Program.

15. Benjamin C. Bradlee, *Conversations with Kennedy* (New York, 1975), 71.

16. George McGovern, in recorded interview by L. J. Hackman, July 16, 1970, 13–17, RFK Oral History Program.

17. Joseph Dolan, in recorded interview by L. J. Hackman, July 18, 1970, 227, RFK Oral History Program.

18. Ibid., 326.

19. Robert Morgenthau, in letter to *New York*, April 19, 1971.

20. J. C. Goulden, *The Benchwarmers: The Private World of the Powerful Federal Judges* (New York: Ballantine reprint, 1976), 68.

21. Ibid., 70; H. W. Chase, *Federal Judges: The Appointing Process* (Minneapolis, Minn., 1972), 119.

22. RFK, in Lewis interview, December 4, 1964, IV, 1-2.

23. See the summary in a confidential report of the House Judiciary Committee, "Investigation of Judicial Behavior in the Tenth Circuit United States Court of Appeals," April 30, 1968, in Goulden, *Benchwarmers*, 255-256, 416.

24. Bobby Baker, with Larry King, *Wheeling and Dealing: Confessions of a Capitol Hill Operator* (New York, 1978), 84.

25. RFK, in Lewis interview, December 4, 1964, IV, 2; RFK, in Martin interview, April 13, 1964, I, 40.

26. RFK, in Lewis interview, December 4, 1964, IV, 2-5; Victor Navasky, *Kennedy Justice* (New York, 1971), 252-253.

27. Goulden, *Benchwarmers*, 68.

28. Ramsey Clark, in recorded interview by L. J. Hackman, July 7, 1970, 66-68, RFK Oral History Program; Navasky, *Kennedy Justice*, 262-263.

29. Clark, in Hackman interview, 71-74, 79; Navasky, *Kennedy Justice*, 263.

30. Author's journal, July 28, 1961.

31. Ibid., August 1, 1961.

32. Anthony Lewis, in recorded interview by L. J. Hackman, July 23, 1970, 28, RFK Oral History Program.

33. Nicholas Katzenbach, in recorded interview by Anthony Lewis, November 16, 1964, 78, JFK Oral History Program.

34. RFK, in Lewis interview, December 4, 1964, IV, 12-14; RFK, in Martin interview, April 13, 1964, I, 43; Archibald Cox, in interview by author, October 29, 1976.

35. Clark Clifford, in recorded interview by L. J. Hackman, February 4, 1975, 13–

14, JFK Oral History Program; RFK, in Lewis interview, December 4, 1964, IV, 14–15; Katzenbach, in Lewis interview, 66.
36. Author's journal, March 31, 1962.
37. RFK, in Lewis interview, December 4, 1964, IV, 14.
38. For accounts, see Katzenbach, in Lewis interview, 56–71; and Joseph Dolan, in recorded interview by Charles Morrissey, December 4, 1964, 98–102, JFK Oral History Program.
39. Kenneth O'Donnell and David Powers with Joe McCarthy, *"Johnny, We Hardly Knew Ye"* (Boston, 1972), 280.
40. Katzenbach, in Lewis interview, 71.
41. Author's journal, March 31, 1962.
42. Ibid., April 1, 1962.
43. RFK, memorandum, September 11, 1962, RFK Papers.
44. Dorothy Goldberg, *A Private View of a Public Life* (New York, 1975), 128.
45. Felix Frankfurter, in recorded interview by Charles McLaughlin, June 10, 1964, 52, JFK Oral History Program.
46. Author's journal, May 8, 1963.
47. Ibid., August 31, 1962.
48. John F. Kennedy, *Public Papers . . . 1961* (Washington, 1962), 19.
49. Raoul Berger, *Executive Privilege: A Constitutional Myth* (Cambridge, Mass., 1974), 239–240.
50. Clark Mollenhoff, *Washington Cover-Up* (Garden City, N.Y. 1962), 179–185.
51. *Meet the Press*, September 24, 1961, vol. 5, no. 37, 7.
52. Byron White to JFK, "Executive Privilege," February 6, 1962, RFK Papers. Emphasis added.
53. Clark Mollenhoff, *Despoilers of Democracy* (Garden City, N.Y., 1965), 21.
54. JFK to Porter Hardy, May 1, 1962, JFK Papers.
55. Berger, *Executive Privilege*, 240.
56. Ibid., 252.
57. Ronald Goldfarb, "Politics at the Justice Department," in *Conspiracy*, ed. John C. Raines (New York, 1974), 119–120.
58. RFK, in Martin interview, April 30, 1964, II, 17.
59. John Seigenthaler, in recorded interview by R. J. Grele, February 22, 1966, 268–269, JFK Oral History Program.
60. William Hundley, in recorded interview by J. A. Oesterle, February 17, 1971, 30–32, RFK Oral History Program.
61. Navasky, *Kennedy Justice*, 368–369.
62. Seigenthaler, in Grele interview, 270.
63. Hundley, in Oesterle interview, 31–32.
64. Ibid., 30.
65. Ibid.; Hundley, in interview by author, August 4, 1976.
66. Walter Sheridan, in recorded interview by Roberta Greene, May 1, 1970, 55–56, RFK Oral History Program.
67. Seigenthaler, in Grele interview, 266–268; Navasky, *Kennedy Justice*, 372–378.
68. John Douglas, in recorded interview by L. J. Hackman, June 11, 1970, 30, RFK Oral History Program.
69. RFK, memorandum, August 4, 1964, 3, RFK Papers; RFK, in Martin interview, April 30, 1964, III, 9–10.
70. Eric F. Goldman, *The Tragedy of Lyndon Johnson* (New York, 1969), 83.
71. RFK, memorandum, February 22, 1961, 2–3, RFK Papers; Hundley, Oesterle interview, December 9, 1970, 20.
72. RFK, memorandum, 3.
73. John Kennedy, in whatever mood of truth or teasing, told such a story to Joseph Alsop. No one then in the Department of Justice with whom I discussed the story had heard of it.
74. RFK, in Martin interview, February 29, 1964, 49–50.
75. RFK, in Martin interview, April 30, 1964, II, 14–15.

76. Ibid., II, 19; Brock Brower, *Other Loyalties* (New York, 1968), 98.
77. Jean Stein and George Plimpton, eds. *American Journey* (New York, 1970), 79–80.
78. RFK, in Martin interview, April 30, 1964, II, 18.
79. Igor Cassini, *I'd Do It All Over Again* (New York, 1977), 228–229.
80. Bradlee, *Conversations with Kennedy*, 170.
81. Stein and Plimpton, *American Journey*, 79, 81.
82. Peter Maas, in recorded interview by Jean Stein, September 15, 1969, 11, Stein Papers.
83. Justin Feldman, in recorded interview by Roberta Greene, October 23, 1969, 48, RFK Oral History Program.
84. Navasky, *Kennedy Justice*, 378.
85. Feldman, in Greene interview, 58–59; Navasky, *Kennedy Justice*, 388. The judge was Sylvester J. Ryan.
86. Feldman, in Greene interview, 61–66. For different language but the same points, see Navasky, *Kennedy Justice*, 389.
87. RFK to David Lawrence, September 3, 1963, RFK Papers.
88. RFK, in Martin interview, April 30, 1964, II, 21.

18. Justice and Poverty *(pages 392–416)*

1. Robert F. Kennedy, *The Pursuit of Justice* (New York, Perennial Press reprint, 1964), 9.
2. RFK, address at Marquette University, June 7, 1964, RFK Papers.
3. Kennedy, *Pursuit of Justice*, 9, 84. Emphasis added.
4. Peter Maas, "Robert Kennedy Speaks Out," *Look*, March 28, 1961.
5. Kennedy, *Pursuit of Justice*, 82.
6. Ramsey Clark, *Crime in America* (New York, Pocket Books reprint, 1971), 282–283.
7. *Gideon* v. *Wainwright*, 372 U.S. 335 (1963); Kennedy, *Pursuit of Justice*, 82.
8. Hoover, notation on UPI excerpt from UPI ticker, November 6, 1961; Hoover, notation on Courtney Evans to A. H. Belmont, "Travels of the Attorney General, June 25–30, 1962," July 3, 1962; both in RFK/FBI/FIOA release.
9. Victor Navasky, *Kennedy Justice* (New York, 1971), 10.
10. RFK, in recorded interview by John Bartlow Martin, April 30, 1964, II, 22–23, JFK Oral History Program.
11. James V. Bennett, *I Chose Prison* (New York, 1970), 183–184.
12. Patricia Kennedy Lawford, ed., *That Shining Hour* (n.p., 1969) 56.
13. So Robert Kennedy told me on December 21, 1962; noted in my journal of that date.
14. RFK, in Martin interview, April 30, 1964, I, 38–39.
15. Anthony Lewis, "A Tribute to Robert Francis Kennedy," delivered at unveiling of RFK portrait, Justice Department, July 22, 1975.
16. Navasky, *Kennedy Justice*, 38. For background on the case, see also James Wechsler, "Persecution," *New York Post*, November 28, 1962; and author to RFK, December 18, 1962, RFK Papers.
17. John Douglas, in recorded interview by L. J. Hackman, June 11, 1970, 31–32, RFK Oral History Program.
18. Abba Schwartz to author, memorandum, n.d. [1976], with supporting material from 1963. There is no point in identifying the scholar.
19. Francis Biddle to RFK, January 8, 1962; RFK to Biddle, January 22, 1962, RFK Papers.
20. Archibald Cox, in interview by author, October 29, 1976.
21. Burke Marshall, in recorded interview by Jean Stein, October 6, 1968, 24, Stein Papers.
22. RFK, in Martin interview, April 30, 1964, III, 31.
23. *Baltimore Sun*, July 23, 1928; reprinted in H. L. Mencken, *A Carnival of Buncombe* (New York, 1956), 160.
24. J. R. Moskin, "The Revolt against Rural Rule," *Look*, January 15, 1963.

25. John F. Kennedy, "The Shame of the States," *New York Times Magazine*, May 18, 1958.
26. Anthony Lewis, "Legislative Apportionment and the Federal Courts," *Harvard Law Review*, April 1958.
27. For useful discussions, see R. G. Dixon, Jr., *Democratic Representation: Reapportionment in Law and Politics* (New York, 1968), esp. ch. 6 and 177–178; R. C. Cortner, *The Apportionment Cases* (New York: Norton reprint, 1972), esp. ch. 5; and Royce Hanson, *The Political Thicket* (Englewood Cliffs, N.J., 1966), esp. ch. 3.
28. *New York Times*, March 28, 1962.
29. John F. Kennedy, *Public Papers . . . 1962* (Washington, 1963), 274.
30. *New York Times*, March 28, 1962.
31. Robert B. McKay, *Reapportionment: The Law and Politics of Equal Representation* (New York, 1970), 84.
32. Archibald Cox to RFK, December 20, 1962, RFK Papers.
33. These are in the RFK Papers.
34. James Clayton in the *Washington Post*; see Navasky, *Kennedy Justice*, 277.
35. RFK, "Gray v. Sanders. Notes for Oral Argument," RFK Papers.
36. Navasky, *Kennedy Justice*, 278.
37. Joseph Dolan, in recorded interview by L. J. Hackman, July 19, 1970, 402–403, RFK Oral History Program.
38. *Gray v. Sanders*, 372 U.S. 368, 381 (1963).
39. Navasky, *Kennedy Justice*, 303.
40. Cox to RFK, "Reapportionment Cases in Supreme Court," August 21, 1963, RFK Papers.
41. Cox to RFK, August 21, 1963, RFK Papers. See also Cox to RFK, "Reapportionment Cases," September 5, 1963, RFK Papers.
42. Navasky, *Kennedy Justice*, 316.
43. Cox, in interview by author, October 29, 1976.
44. RFK, in recorded interviea by Anthony Lewis, December 22, 1964, 54–55, JFK Oral History Program.
45. Archibald C. Cox, *The Role of the Supreme Court in American Government* (New York, 1976), 29.
46. See Ward E. Y. Elliott, *The Rise of Guardian Democracy: The Supreme Court Role in Voting Rights Disputes, 1845–1969* (Cambridge, Mass., 1974), esp. 218–236, 266, 273; and David Brady and Douglas Edmonds, "One Man, One Vote — So What?" *Trans-Action*, March 1967.
47. Cox, in interview by author.
48. Peter Maas, "Robert Kennedy Speaks Out," *Look*, March 28, 1961.
49. Lee Loevinger, in recorded interview by R. J. Grele, May 13, 1966, 6, JFK Oral History Program.
50. John Seigenthaler, in recorded interview by R. J. Grele, February 22, 1966, 276, JFK Oral History Program.
51. Hobart Rowen, *The Free Enterprisers: Kennedy, Johnson and the Business Establishment* (New York, 1964), 81.
52. Ramsey Clark, in recorded interview by L. J. Hackman, June 29, 1970, 26, RFK Oral History Program.
53. Loevinger, in Grele interview, 9–10.
54. William Orrick, in recorded interview by L. J. Hackman, April 14, 1970, 282, RFK Oral History Program.
55. Edwin Guthman, *We Band of Brothers* (New York, 1971), 237–238.
56. W. W. Rostow, *The Diffusion of Power, 1957–1972* (New York, 1972), 122.
57. Roger Blough, *The Washington Embrace of Business* (New York, 1975), 93–94. For general background, see Rostow, *Diffusion of Power*, chs. 13–15; and W. J. Barber, "The Kennedy Years: Purposeful Pedagogy," in *Exhortation and Controls: The Search for a Wage-Price Policy, 1945–1971*, ed. C. D. Goodwin (Washington, 1975), 135–191.
58. Roger Blough, "My Side of the Steel Story," *Look*, January 29, 1963.

59. Dorothy Goldberg, *A Private View of a Public Life* (New York, 1975), 112.
60. David McDonald, in recorded interview by Charles T. Morrissey, February 15, 1966, 15, JFK Oral History Program.
61. RFK, "Some Notes on the Steel Increase," April 20, 1962, 1-2, RFK Papers.
62. Grant McConnell, *Steel and the Presidency — 1962* (New York, 1963), 89.
63. Guthman, *We Band of Brothers*, 233.
64. RFK, in recorded interview by John Bartlow Martin, April 13, 1964, I, 35, JFK Oral History Program.
65. Ovid Demaris, *The Director* (New York, 1975), 288.
66. Douglas Dillon, in recorded interview by L. J. Hackman, June 18, 1970, 20-21, RFK Oral History Program.
67. RFK, in Martin interview, April 13, 1964, III, 5, 7-8, JFK Oral History Program.
68. *New York Times*, April 23, 1962.
69. Clark Clifford, in recorded interview by L. J. Hackman, February 4, 1975, 33, JFK Oral History Program.
70. RFK, "Some Notes on the Steel Increase," 5-6.
71. Blough, *Washington Embrace of Business*, 100-101.
72. Loevinger, in Grele interview, 19.
73. RFK, in Martin interview, April 30, 1964, I, 44.
74. "Interview with Robert Kennedy," *U.S. News & World Report*, January 28, 1963.
75. Author's journal, June 19, 1962.
76. Cox, in interview by author.
77. Ibid.
78. *Christian Science Monitor*, April 16, 1962.
79. *Wall Street Journal*, April 19, 1962.
80. Charles Reich, "Another Such Victory," *New Republic*, April 30, 1962.
81. Kennedy, *Public Papers . . . 1962*, 895.
82. RFK, in Martin interview, April 13, 1964, III, 5, 8.
83. William Goldsmith, *The Growth of Presidential Power* (New York, 1974), vol. 3, 1687.
84. Hugh Sidey, *A Very Personal Presidency: Lyndon Johnson in the White House* (New York, 1968), 71.
85. RFK, address at G.I. Forum, Chicago, August 23, 1963, 4-5, RFK Papers.
86. Joseph Dolan, in recorded interview by L. J. Hackman, July 18, 1970, 249, RFK Oral History Program.
87. Ramsey Clark, in recorded interview by L. J. Hackman, June 29, 1970, 37-38, RFK Oral History Program.
88. Clark, in Hackman interview, 48-50.
89. RFK, address before National Congress of American Indians, Bismarck, North Dakota, September 13, 1963, RFK Papers; RFK, commencement address, Trinity College, Hartford, Conn., June 2, 1963, RFK Papers.
90. Clark, in Hackman interview, 49.
91. Jack Newfield, *Robert Kennedy* (New York, 1969), 18.
92. Navasky, *Kennedy Justice*, 17.
93. Department of Justice release, "Text of Attorney General Robert F. Kennedy's Statement," April 6, 1961.
94. Richard Blumenthal, "The Bureaucracy: Antipoverty and the Community Action Program," in *American Political Institutions and Public Policy*, ed. A. P. Sindler (Boston, 1969), 134. Blumenthal's essay, derived from a Harvard honors thesis, draws extensively on unpublished interviews and is an essential source. Another invaluable work, Daniel Knapp and Kenneth Polk, *Scouting the War on Poverty: Social Reform Politics in the Kennedy Administration* (Lexington, Mass., 1971), is based on the files of the President's Committee on Juvenile Delinquency. Peter Marris and Martin Rein, *Dilemmas of Social Reform: Poverty and Community Action in the United States*, 2d ed. (Chicago, 1973), is a probing analysis and evaluation. See also L. A. Ferman, ed., "Evaluating the War on Poverty," *Annals of the American Academy of Political and Social Science*, September 1969, especially the papers by R. H. Davidson, L. B. Rubin, Sanford Kravitz and F. K. Kolodner. I

am also indebted for helpful comment to communications from Richard Boone (January 8, 1977), Lloyd Ohlin (January 14, 1977), William B. Cannon and Leonard J. Duhl (August 19, 1977).

95. Recorded transcript of conference sponsored by the Kennedy Library and Brandeis University, "The Federal Government and Urban Poverty," pt. 3, 30.
96. Richard A. Cloward and Lloyd Ohlin, *Delinquency and Opportunity: A Theory of Delinquent Gangs* (New York, 1960), 108–113, 124–127, 211.
97. See Daniel Patrick Moynihan, *Maximum Feasible Misunderstanding: Community Action in the War on Poverty* (New York, 1969), 170.
98. See Marris and Rein, *Dilemmas of Social Reform*, 132.
99. Blumenthal, "Bureaucracy," 133.
100. David Hackett, in interview by author, January 27, 1975.
101. David Hackett to RFK, November 8, 1963, RFK Papers.
102. Kenneth B. Clark, "Community Action and the Social Program of the 1960s," in *Toward New Human Rights: The Social Policies of the Kennedy and Johnson Administrations*, ed. David C. Warner (Austin, Tex., 1977), 99; see also Kenneth B. Clark, *Dark Ghetto: Dilemmas of Social Power* (New York, 1965).
103. Lloyd Ohlin to author, January 14, 1977.
104. Knapp and Polk, *Scouting the War on Poverty*, 75–77, 127.
105. See the account in Guthman, *We Band of Brothers*, 226–227.
106. Wesley Barthelmes, in recorded interview by Roberta Greene, June 5, 1969, 155–158, RFK Oral History Program.
107. Eunice Kennedy Shriver, in recorded interview by John Stewart, May 7, 1968, 26–27, JFK Oral History Program.
108. Richard Boone, in recorded interview by Jean Stein, September 20, 1968, 2–3, Stein Papers.
109. RFK, testimony before the Labor Subcommittee of the House Committee on Education and Labor; see Kennedy, *Pursuit of Justice*, 26–33.
110. RFK to LBJ, "Racial Violence in Urban Centers," August 5, 1964, RFK Papers.
111. E. Barrett Prettyman, Jr., in recorded interview by L. J. Hackman, June 5, 1969, 17–20, RFK Oral History Program.
112. Patrick Anderson, "Robert's Character," *Esquire*, April 1965.
113. Kenneth B. Clark and Jeanette Hopkins, *A Relevant War against Poverty: A Study of Community Action Programs and Observable Social Change* (New York, 1969), 4.
114. Arthur M. Schlesinger, Jr., *A Thousand Days* (Boston, 1965), 1011.
115. Boone, in Stein interview, 4.
116. Leonard Duhl, in recorded interview by Jean Stein, August 19, 1968, 3–4, Stein Papers.

19. The Kennedys and the Cold War *(pages 417–442)*

1. Kenneth P. O'Donnell and David F. Powers, *"Johnny, We Hardly Knew Ye"* (Boston, 1972), 278.
2. John F. Kennedy, *A Compendium of Speeches, Statements, and Remarks Delivered During His Service in the Congress of the United States*, Senate Document 79, 88 Cong., 2 Sess. (Washington, 1964), 710.
3. John F. Kennedy, *The Strategy of Peace* (New York, 1960), 81, 132–133.
4. The article appeared in *Foreign Affairs*, April 1960. Chester Bowles, in recorded interview by R. R. R. Brooks, February 2, 1965, 7, JFK Oral History Program.
5. Charles Spalding, in recorded interview by John Stewart, March 14, 1968, 5, JFK Oral History Program.
6. Joseph P. Kennedy, "A Foreign Policy for America" [1956], 2, 6, 9, JPK Papers. The article was never published, perhaps because Joseph Kennedy feared to complicate his son's political career.
7. Kennedy, *Strategy of Peace*, 7, 10–12.
8. Ibid., ch. 1.
9. RFK, in recorded interview by author, February 27, 1965, 57, JFK Oral History Program.

10. JFK to G. F. Kennan, February 13, 1958; see George F. Kennan, in recorded interview by Louis Fischer, March 23, 1965, 5–6, JFK Oral History Program.
11. JFK to Kennan, January 21, 1959; see Kennan, in Fischer interview, 11–12.
12. Author's journal, May 29, 1960.
13. Ibid., May 22, 1960.
14. John F. Kennedy, review of *Deterrent or Defense* by B. H. Liddell Hart, *Saturday Review*, September 3, 1960.
15. Kennedy, *Strategy of Peace*, 7.
16. "Deterrence and Survival in the Nuclear Age," reprinted by Joint Committee on Defense Production, 94 Cong., 2 Sess. (1976); see esp. 26–27.
17. Edgar M. Bottome, *The Missile Gap* (Rutherford, N.J., 1971), 93.
18. Henry A. Kissinger, *The Necessity for Choice: Prospects of American Foreign Policy* (New York, 1960), 1.
19. Kennan to JFK, August 17, 1960, JFK to Kennan, October 30, 1960, Kennan, in Fischer interview, 21–26.
20. Adlai Stevenson, report, November 7, 1960, pt. 2, 1, Schlesinger Papers.
21. Stevenson to Theodore Sorensen, December 30, 1960, Sorensen Papers.
22. N. S. Khrushchev, *Khrushchev Remembers: The Last Testament,* ed. Strobe Talbott (Boston, 1974), 489–490.
23. RFK to Dean Rusk, December 18, 1960, RFK Papers.
24. Kennan, in Fischer interview, 29–32.
25. N. S. Khrushchev, "For New Victories of the World Communist Movement," at the meeting of party organizations of the Higher Party School, the Academy of Social Sciences and the Institute of Marxism-Leninism, Moscow, January 6, 1961.
26. State Department, "Analysis of Moscow Conference 'Statement' and Khrushchev Speech of January 6," January 25, 1961, RFK Papers.
27. L. C. Gardner, Arthur M. Schlesinger, Jr., and Hans Morgenthau, *The Origins of the Cold War* (Lexington, Ky., 1970), 98.
28. RFK, in recorded interview by John Bartlow Martin, April 13, 1964, II, 22, JFK Oral History Program.
29. John F. Kennedy, *Public Papers . . . 1961* (Washington, 1962), 1–3.
30. *New York Times,* January 21, 1961.
31. Eleanor Roosevelt to JFK, January 24 [1961], JFK Papers.
32. JFK, in a speech at the University of Washington, Seattle, November 16, 1961; see JFK, *Public Papers . . . 1961,* 726.
33. See especially his speech in Berkeley, California, March 23, 1962; John F. Kennedy, *Public Papers . . . 1962* (Washington, 1963), 263–266.
34. John F. Kennedy, *Public Papers . . . 1963* (Washington, 1964), 462.
35. For an able account of these issues, see R. A. Aliano, *American Defense Policy from Eisenhower to Kennedy: The Politics of Changing Military Requirements, 1957–1961* (Athens, Ohio, 1975).
36. Bottome, *Missile Gap,* 179–184.
37. A. C. Enthoven and K. W. Smith, *How Much Is Enough? Shaping the Defense Program, 1961–1969* (New York: Harper Colophon reprint, 1972), 195.
38. Roswell Gilpatric, in recorded interview by D. J. O'Brien, June 30, 1970, 71, JFK Oral History Program.
39. Robert S. McNamara, *The Essence of Security: Reflections in Office* (New York, 1968), 58.
40. JFK, *Compendium of Speeches,* 929.
41. Maxwell Taylor, *The Uncertain Trumpet* (New York, 1960), 24.
42. See Bernard Brodie's brilliant *War and Politics* (New York, 1973), 126.
43. Author's journal, April 5, 1961.
44. Dean Acheson, in recorded interview by Lucius Battle, April 27, 1964, 12, 20, JFK Oral History Program.
45. Author's journal, May 10, 1961.
46. RFK, in recorded interview by author, February 27, 1965, 28, JFK Oral History Program.
47. John F. Kennedy, *The Strategy of Peace* (New York, 1960), 212.

48. JFK, *Public Papers . . . 1961*, 533–540.
49. RFK, dictated August 1, 1961, RFK Papers.
50. RFK, in Martin interview, April 13, 1964, I, 12.
51. L. C. McHugh, "Ethics at the Shelter Doorway," *America*, September 30, 1961.
52. Author's journal, November 24, 1961.
53. A. D. Sakharov, *Sakharov Speaks*, ed. Harrison Salisbury (New York, 1974), 33.
54. RFK, dictated September 1, 1961, RFK Papers.
55. Ibid.
56. Arthur M. Schlesinger, Jr., *A Thousand Days* (Boston, 1965), 398.
57. O'Donnell and Powers, *"Johnny, We Hardly Knew Ye,"* 299.
58. Author's journal, September 5, 1961.
59. Ibid., August 30, 1961.
60. RFK, dictated August 1, 1961, RFK Papers.
61. Author's journal, September 5, 1961.
62. Paul Blanshard, *Personal and Controversial: An Autobiography* (Boston, 1973), 243.
63. Barry Goldwater, *Why Not Victory?* (New York: Macfadden reprint, 1963), 16, 97, 118, 120, 122.
64. James Wechsler, "JFK (Contd.)," *New York Post*, September 22, 1961.
65. R. M. Slusser, *The Berlin Crisis of 1961* (Baltimore, Md., 1973). It is pathetic that Soviet historians are denied any opportunity to write honest contemporary history. In the absence of Soviet evidence, western historians have no alternative to speculation.
66. Ibid., x.
67. Theodore C. Sorensen, *Kennedy* (New York, 1965), 552.
68. JFK, *Public Papers . . . 1961*, 726–727.
69. JFK, *Compendium of Speeches*, 927–928.
70. Peter Lisagor, in recorded interview by R. J. Grele, May 12, 1966, 56, JFK Oral History Program.
71. RFK, in Martin interview, February 29, 1964, 28–31; also a subsequent interview on Adlai Stevenson, December 7, 1966, Martin Papers; also interview by author, February 27, 1965, 5, 13.
72. Averell Harriman, in recorded interview by author, June 6, 1965, 133–134, JFK Oral History Program.
73. David Halberstam, in recorded interview by Jean Stein, May 2, 1970, 5, Stein Papers.
74. John Kenneth Galbraith, in recorded interview by Jean Stein, July 30, 1968, 5, Stein Papers.
75. W. W. Attwood, *The Reds and the Blacks* (New York, 1967), 325.
76. Roswell Gilpatric, in recorded interview by D. J. O'Brien, August 12, 1970, 102, JFK Oral History Program.
77. Patrick Anderson, "Robert's Character," *Esquire*, April 1965.
78. RFK to Dean Rusk, May 16, 1963, RFK Papers. The editorial appeared on May 4, 1963.
79. Abba Schwartz, *The Open Society* (New York, 1968), 25.
80. Ibid., 4, 34.
81. Ibid., 40–44.
82. Angie Novello to Harold Reis, April 8, 1964, RFK Papers.
83. Schwartz, *Open Society*, 47–50.
84. William O. Douglas, in recorded interview by Roberta Greene, November 13, 1969, 18, RFK Oral History Program; Douglas, in recorded interview by Jean Stein, n.d., 1, Stein Papers.
85. *Kent v. Dulles*, 357 U.S. 118 (1958).
86. Schwartz, *Open Society*, ch. 5.
87. Ibid.; *Aptheker v. Secretary of State*, 348 U.S. 500 (1964).
88. Schwartz, *Open Society*, 28.
89. Galbraith to JFK, August 15, 1961, JFK Papers.
90. Chester Bowles to Adlai Stevenson, July 23, 1961, Bowles Papers.
91. RFK to JFK, August 15, 1961, RFK Papers.

92. RFK, in interview by author, February 27, 1965, 11–13.
93. Harriman, in interview by author, June 6, 1965, 80–82.
94. William Orrick, in recorded interview by L. J. Hackman, April 13, 1970, 154–155, RFK Oral History Program.
95. John Seigenthaler, in recorded interview by L. J. Hackman, June 5, 1970, 59, RFK Oral History Program.
96. Orrick, in Hackman interview, 162.
97. Donald Wilson, in recorded interview by W. W. Moss, March 13, 1972, 85, JFK Oral History Program.
98. RFK to JFK, May 30, 1962, RFK Papers.
99. See John F. Campbell's characterization in his generally intelligent book *The Foreign Affairs Fudge Factory* (New York, 1971), 54–59. For an account of the argument, see Schlesinger, *Thousand Days*, 875–881.
100. J. P. Davies, Jr., *Foreign and Other Affairs* (New York, 1964), 198.
101. Philip Bonsal, "Open Letter to an Author," *Foreign Service Journal*, February 1967.
102. Philip L. Graham to JFK, December 15, 1962, JFK Papers.
103. Orrick, in Hackman interview, 158–159.
104. RFK, in Martin interview, April 13, 1964, I, 19–20.
105. Eric Hobsbawm, "Why America Lost the Vietnam War," *Listener*, May 18, 1972.
106. R. J. Walton, *Cold War and Counter-Revolution: The Foreign Policy of John F. Kennedy* (New York, 1972), 233. See also Louise FitzSimons, *The Kennedy Doctrine* (New York, 1972).
107. Khrushchev. *Last Testament*, 491, 495.

20. The CIA and Counterinsurgency *(pages 444–467)*

1. RFK, memorandum, dictated June 1, 1961, RFK Papers.
2. Ibid.
3. Edwin Guthman, *We Band of Brothers* (New York, 1971), 113.
4. RFK, handwritten notes after Cuba Study Group meetings of May 1 and 11, 1961, RFK Papers.
5. Cuba Study Group, May 19, 1961, RFK Papers.
6. RFK, in recorded interview by John Bartlow Martin, March 1, 1964, II, 9, JFK Oral History Program.
7. Maxwell Taylor, in recorded interview by L. J. Hackman, October 22, 1969, 2, RFK Oral History Program.
8. "Conclusions of the Cuba Study Group," June 13, 1961, Schlesinger Papers; Maxwell Taylor, *Swords and Plowshares* (New York, 1972), ch. 14.
9. Taylor, *Swords and Plowshares*, 186.
10. RFK, in Martin interview, March 1, 1964, II, 33.
11. Taylor, in Hackman interview, 26.
12. Maxwell Taylor, in *That Shining Hour,* ed. Patricia Kennedy Lawford (n.p., 1969), 81–82.
13. Taylor, in Hackman interview, 16.
14. Taylor, *Swords and Plowshares*, 205.
15. Roswell Gilpatric, in recorded interview by D. J. O'Brien, August 12, 1970, 117, JFK Oral History Program.
16. Jerry Bruno, *The Advance Man* (New York: Bantam reprint, 1972), 20.
17. RFK, in recorded interview by Anthony Lewis, December 4, 1964, VII, 25, JFK Oral History Program.
18. *New York Times*, February 1, 1962.
19. Gilpatric, in O'Brien interview, June 30, 1970, 69, and August 12, 1970, 112, 116.
20. Paul B. Fay, Jr., *The Pleasure of His Company* (New York, 1966), 190.
21. John F. Kennedy, *Public Papers . . . 1961* (Washington, 1962), 735.
22. Joseph Rauh, in interview by author, October 26, 1976.

23. The full text is reprinted in Victor Reuther, *The Brothers Reuther* (Boston, 1976), 491–500.
24. See, for example, William E. Mallett, *The Reuther Memorandum: Its Applications and Implications* (n.p., Liberty Lobby, 1963).
25. Charles U. Daly, in recorded interview by C. T. Morrissey, April 5, 1966, 11–13, JFK Oral History Program.
26. For a critique, see Fred W. Friendly, *The Good Guys, the Bad Guys and the First Amendment* (New York, 1976), esp. ch. 3.
27. Author's journal, April 18, 1961.
28. RFK, in Martin interview, March 1, 1964, II, 13. See Haynes Johnson, *The Bay of Pigs* (New York, 1964), 75.
29. RFK, in handwritten notes after Cuba Study Group meetings of May 24 and June 11, 1961, RFK Papers.
30. Johnson, *Bay of Pigs*, 75–77, 222.
31. E. Howard Hunt, *Give Us This Day* (New York: Popular Library reprint, 1973), 188–189.
32. Fay, *Pleasure of His Company*, 188.
33. Tom Wicker, in interview by author, July 8, 1975. The *New York Times* was doing a series on the CIA.
34. Senate Select Committee to Study Governmental Operations with respect to Intelligence Activities (hereafter cited as Church committee), *Interim Report: Alleged Assassination Plots Involving Foreign Leaders*, 94 Cong., 1 Sess. (1975), 92.
35. Hunt, *Give Us This Day*, 38.
36. Cuba Study Group, May 19, 1961, RFK Papers.
37. Howard Hunt, "The Azalea Trail Guide to CIA," *National Review*, April 29, 1977.
38. Cuba Study Group, May 11, 1961, RFK Papers.
39. David Bruce and Robert Lovett, "Covert Operations," report to President's Board of Consultants on Foreign Intelligence Activities [1956], RFK Papers.
40. President's Board of Consultants on Foreign Intelligence Activities (hereafter cited as PBCFIA), report to President Eisenhower, December 20, 1956, RFK Papers.
41. William Manchester, *Portrait of a President* (Boston, 1962), 35.
42. PBCFIA, report to the Special Assistant for National Security, February 12, 1957, RFK Papers.
43. Joseph B. Smith, *Portrait of a Cold Warrior* (New York, 1976), 229–230, 240.
44. Howard P. Jones, *Indonesia: The Possible Dream* (New York, 1971), 143, 145.
45. In addition to J. B. Smith and H. P. Jones, see ch. 8 of David Wise and Thomas B. Ross, *The Invisible Government* (New York: Bantam reprint, 1965); and Ray S. Cline, *Secrets, Spies and Scholars* (Washington, 1976), 181–183.
46. White House meeting of PBCFIA with Eisenhower, December 16, 1958, RFK Papers.
47. Church committee, *Final Report*, bk. I, *Foreign and Military Intelligence*, 94 Cong., 2 Sess. (1976), 52.
48. Cuba Study Group, May 11, 1961, RFK Papers.
49. PBCFIA, report to Eisenhower, January 5, 1961, RFK Papers.
50. Cuba Study Group, May 11, 1961, RFK Papers.
51. Robert Lovett, in recorded interview by Dorothy Fosdick, August 17, 1964, 5, JFK Oral History Program.
52. Interview with Clark Clifford, in *Christian Science Monitor*, July 24, 1975; Clifford, in recorded interview by L. J. Hackman, December 17, 1974, 10, JFK Oral History Program.
53. Deborah Shapley, "Foreign Intelligence Advisory Board: A Lesson in Citizen Oversight?" *Science*, March 12, 1976.
54. Clifford, in Hackman interview, February 4, 1975, 43.
55. RFK, in Martin interview, March 1, 1964, II, 27–29.
56. Cuba Study Group, memorandum no. 4, "A Mechanism for the Planning and Coordination of Cold War Strategy," June 13, 1961, Schlesinger Papers.

57. This point is well made by Charles A. Cannon in an unpublished paper, "John F. Kennedy's 'Proving Ground of Democracy in Asia,'" read at the American Historical Association convention, December 29, 1975 (see esp. p. 6).

58. W. W. Rostow, "Guerrilla Warfare in Underdeveloped Areas," in *The Guerrilla — and How to Fight Him*, ed. T. N. Greene (New York, 1962), 55–58.

59. Special Group (Counterinsurgency), "Memorandum for the Record . . . the Special Group (CI) from January 18, 1962, to November 21, 1963," May 13, 1964, RFK Papers.

60. JFK, *Public Papers . . . 1961*, 232.

61. E. G. Lansdale, *In the Midst of Wars: An American's Mission to Southeast Asia* (New York, 1972), 84, 99.

62. E. G. Lansdale, in interview by author, December 30, 1976.

63. Lansdale, *In the Midst of Wars*, ix, 372–373, 376.

64. E. G. Lansdale, "Vietnam: Do We Understand Revolution," *Foreign Affairs*, October 1964.

65. Michael Forrestal, in interview by author, January 27, 1977.

66. Special Group (CI), "U.S. Overseas Internal Defense Policy," September 1962, RFK Papers.

67. Robert F. Kennedy, *To Seek a Newer World* (Garden City, N.Y., 1967), 116–117. Emphasis added.

68. Taylor, in Hackman interview, October 22, 1969, 21; Joint Chiefs of Staff (JCS) to Secretary of Defense, "Joint Counterinsurgency Concept and Doctrinal Guidance," April 5, 1962, RFK Papers.

69. Department of the Army, *U.S. Army Counterinsurgency Forces*, Field Manual no. 31-22, November 12, 1963; esp. appendix 2, "Examples of Civil Action," 96–97.

70. John Bartlow Martin, *U.S. Policy in the Caribbean* (Boulder, Colo., 1978), 364.

71. Herbert L. Matthews, *Return to Cuba* (Stanford University, n.d. [1964]), 15.

72. For a good account, see Maurice Halperin, *The Rise and Decline of Fidel Castro* (Berkeley, 1972), chs. 28–29. Since I had some harsh words about the author in *A Thousand Days*, may I say here that *The Rise and Decline of Fidel Castro* is a detached, incisive and valuable work.

73. *New York Times*, October 17, 1963.

74. John F. Kennedy, *Public Papers . . . 1963* (Washington, 1964), 184.

75. Romulo Betancourt, "The Venezuelan Miracle," *Reporter*, August 13, 1964.

76. Betancourt, "The Venezuelan Miracle"; Halperin, *Rise and Decline of Castro*, 318, 337–344.

77. Gilpatric, in O'Brien interview, August 12, 1970, 40.

78. C. V. Clifton, "Hail to the Chief!" *Army*, January 1964.

79. JCS to Special Group (CI), June 25, 1963, RFK Papers.

80. Robert F. Kennedy, *The Enemy Within* (New York: Popular Library reprint, 1960), 306.

81. Averell Harriman, in recorded interview by L. J. Hackman, March 13, 1970, 10, RFK Oral History Program.

82. W. V. Shannon, *The Heir Apparent* (New York, 1967), 50.

83. George Ball, in recorded interview by Jean Stein, n.d. [1968], 9, Stein Papers.

84. Ably discussed in D. S. Blaufarb, *The Counterinsurgency Era: U.S. Doctrine and Performance, 1950 to the Present* (New York, 1977), esp. 86–87, 289–292, 303–304.

85. Charles Maechling, Jr., "Our Internal Defense Policy: A Reappraisal," *Foreign Service Journal*, January 1969.

86. R. W. Komer to McGeorge Bundy, April 10, 1962, JFK Papers.

87. Quoted by Eqbal Ahmad, "The Theory and Fallacies of Counterinsurgency," *Nation*, August 2, 1971.

88. Peter De Vries, *I Hear America Swinging* (Boston, 1976).

89. Michael Forrestal, in recorded interview by Jean Stein, n.d. [1968], 5, Stein Papers.

90. RFK, in Martin interview, April 13, 1964, II, 50.

91. Taylor, in Hackman interview, 25.

92. Ralph Dungan to RFK, January 17, 1963, RFK Papers.

21. The Cuban Connection: I *(pages 468–498)*

1. Maxwell Taylor, *Swords and Plowshares* (New York, 1972), 181.
2. Kenneth P. O'Donnell and David F. Powers, *"Johnny, We Hardly Knew Ye"* (Boston, 1972), 274–275.
3. RFK, in recorded interview by John Bartlow Martin, April 30, 1964, III, 17, JFK Oral History Program.
4. The account of the Tractors for Freedom Committee is drawn from Milton Eisenhower, *The Wine Is Bitter* (Garden City, N.Y., 1963), chs. 15–16; Haynes Johnson, *The Bay of Pigs* (New York, 1964), bk. 4, ch. 3; and Victor Reuther, *The Brothers Reuther* (Boston, 1976), 440–445.
5. Johnson, *Bay of Pigs*, 273–275.
6. Author's journal, April 5, 1962.
7. Alvaro Sanchez, Jr., "Policy Memorandum," May 2, 1962, RFK Papers.
8. Johnson, *Bay of Pigs*, 303.
9. A transaction well described in Donovan's book *Strangers on a Bridge: The Case of Colonel Abel* (New York, 1964).
10. James B. Donovan, *Challenges: Reflections of a Lawyer-at-Large* (New York, 1967), 91.
11. RFK, memorandum dictated September 11, 1962, RFK Papers.
12. RFK to Kenneth O'Donnell for the President, April 19, 1961, RFK Papers.
13. RFK, memorandum dictated June 1, 1961, RFK Papers.
14. Chester Bowles, in recorded interview by R. R. R. Brooks, February 2, 1965, 33, JFK Oral History Program.
15. Senate Select Committee to Study Governmental Operations with respect to Intelligence Activities (hereafter cited as Church committee), *Interim Report: Alleged Assassination Plots Involving Foreign Leaders*, 94 Cong., 1 Sess. (1975), 142.
16. RFK, memorandum, June 1, 1961.
17. Author's journal, May 4–5, 1961.
18. RFK, memorandum, June 1, 1961.
19. Richard M. Nixon, "Cuba, Castro and John F. Kennedy," *Reader's Digest*, November 1964.
20. Mike Mansfield to JFK, May 1, 1961, JFK Papers.
21. Maxwell Taylor, in recorded interview by L. J. Hackman, October 22, 1969, 10, RFK Oral History Program.
22. Office of Legal Counsel, Department of Justice, "Constitutional and Legal Basis for So-Called Covert Activities of the Central Intelligence Agency," January 17, 1962, quoted in Church committee, *Final Report*, bk. I, *Foreign and Military Intelligence*, 94 Cong., 2 Sess. (1975), 36–37, 497. Emphasis added.
23. RFK to John McCone, January 19, 1962, RFK Papers.
24. Church committee, *Assassination Plots*, 148.
25. Schlesinger to Richard Goodwin, July 8, 1961, Schlesinger Papers.
26. Goodwin to JFK, September 6, 1961, Schlesinger Papers.
27. The White House meeting was described by the assistant to the head of the CIA unit working on Cuban operations. Bissell himself could not recall such a meeting, though he agreed in essence with the spirit of the instruction. Church committee, *Assassination Plots*, 141.
28. Goodwin to JFK, November 1, 1961, in Church committee, *Assassination Plots*, 139.
29. RFK, handwritten notes, November 7, 1961, RFK Papers.
30. Church committee, *Assassination Plots*, 140–141.
31. JFK to the Secretary of State, et al., November 30, 1961, in Church committee, *Assassination Plots*, 139.
32. Taylor, in Hackman interview, 11.
33. Church committee, *Assassination Plots*, 141.
34. Ibid., 142–143.
35. Ibid., 146–147.
36. According to one CIA source, it cost $100 million; J. B. Smith, *Portrait of a Cold*

Warrior (New York, 1976), 367. For a useful, though incomplete and occasionally inaccurate, account of the Miami operation, see Taylor Branch and George Crile III, "The Kennedy Vendetta," *Harper's*, August 1975.

37. Church committee, *Assassination Plots*, 146, 159.
38. Ramón Barquin, review of *Give Us This Day* by E. Howard Hunt, *Society*, July/August 1975.
39. E. Howard Hunt, *Give Us This Day* (New York: Popular Library reprint, 1973), 219-220.
40. David C. Martin, "The CIA's 'Loaded Gun,'" *Washington Post*, October 10, 1976; W. A. Corson, *Armies of Ignorance* (New York, 1977), 287-288.
41. Edward G. Lansdale, in interview by author, December 30, 1976.
42. Church committee, *Assassination Plots*, 144-146.
43. Branch and Crile, "Kennedy Vendetta."
44. Church committee, *Assassination Plots*, 146.
45. Ibid., 145-146.
46. Branch and Crile, "Kennedy Vendetta."
47. RFK, in recorded interview by John Bartlow Martin, April 30, 1964, III, 23, JFK Oral History Program.
48. Lansdale, in interview by author.
49. Hunt, *Give Us This Day*, 219.
50. Taylor, in Hackman interview, 12; Lansdale, in interview by author.
51. Martin, "CIA's 'Loaded Gun.'"
52. RFK to Maxwell Taylor, message dictated over telephone, August 9, 1962, RFK Papers.
53. Church committee, *Assassination Plots*, 147.
54. Edward Lansdale to RFK, October 15, 1962, RFK Papers.
55. Church committee, *Final Report*, bk. IV, *Supplementary Detailed Staff Reports on Foreign and Military Intelligence*, 94 Cong., 2 Sess. (1976), 128-131.
56. Church committee, *Assassination Plots*, 72-73.
57. Paul Meskil, "CIA Sent Bedmate to Kill Castro in '60," *New York Daily News*, June 13, 1976. Fiorini supported the Lorenz story. Later they fell out.
58. Nicholas Gage, *The Mafia Is Not an Equal Opportunity Employer* (New York, 1971), 64-65; Paul Meskil, "How U.S. Made Unholy Alliance with the Mafia," *New York Daily News*, April 23, 1975; R. S. Anson, "The CIA and the Mafia," *New Times*, May 30, 1975; and Anson, "Jack, Judy, Sam & Johnny," *New Times*, January 23, 1976.
59. Nicholas M. Horrock, "Maheu Says He Recruited Man for C.I.A. in Poison Plot," *New York Times*, July 31, 1975; Church committee, *Assassination Plots*, 307.
60. Church committee, *Assassination Plots*, 91, 97.
61. Ibid., 76-79, 126-131.
62. For Ruby's Cuba visits see *New York Daily News*, July 4, 1976, and W. S. Malone, "The Secret Life of Jack Ruby," *New Times*, January 23, 1978.
63. For this and other evidence pointing to the thesis that Trafficante was a double agent, see George Crile III, "The Mafia, the CIA and Castro," *Washington Post*, May 16, 1976.
64. The friend was Joseph Shimon. Church committee, *Assassination Plots*, 79, 82.
65. Ibid., 79-82.
66. Ibid., 83, 181-188.
67. Ibid., 83-85.
68. Ibid., 85-86.
69. Ibid., 15-16, 52-62.
70. Ibid., 65.
71. Ibid., 109-113.
72. Nixon, "Cuba, Castro and John F. Kennedy."
73. Philip Bonsal, *Cuba, Castro and the United States* (Pittsburgh, 1971), 93-94, 135, 174.
74. Hunt, *Give Us This Day*, 40.
75. Ehrlichman's notes were published in appendix 3 of the House Judiciary Com-

mittee's impeachment hearings; see Aaron Latham, "A Few Words About the 37th President," *New York*, July 28, 1975.

76. Corson, *Armies of Ignorance*, 345.
77. Church committee, *Assassination Flots*, 117–121.
78. *Boston Globe*, March 13, 1975.
79. Church committee, *Assassination Plots*, 148–150, 313.
80. Ibid., 136–137.
81. JFK to Gamal Abdel Nasser, March 2, 1961, in Mohammad Heikal, *The Cairo Documents* (Garden City, N.Y., 1973), 194–195.
82. RFK, notes dictated June 1, 1961, RFK Papers.
83. Church committee, *Assassination Plots*, 195–211; Richard Goodwin, "The Record of JFK and Political Assassinations," *Boston Globe*, November 24, 1975.
84. Arthur M. Schlesinger, Jr., *A Thousand Days* (Boston, 1965), 769.
85. Goodwin, "Record of JFK"; Church committee, *Assassination Plots*, 212–213; George Lardner, Jr., "JFK Rejected Trujillo Slaying, Aide Says," *Washington Post*, July 19, 1975.
86. RFK, notes dictated June 1, 1961.
87. Church committee, *Assassination Plots*, 138–139; Tad Szulc, "Cuba on Our Mind," *Esquire*, February 1974.
88. George Lardner, Jr., "Fear of Retaliation Curbed Anti-Castro Plots," *Washington Post*, July 21, 1975.
89. John F. Kennedy, *Public Papers . . . 1961* (Washington, 1962), 725.
90. George Smathers, in recorded interview by Donald M. Wilson, March 31, 1964, 6–7, 10–11, JFK Oral History Program; Jack Anderson, "Questions in the Closet," *New York Post*, January 19, 1971.
91. Church committee, *Assassination Plots*, 126–131.
92. Ibid., 131–134.
93. Richard Goodwin, in interview by author, June 11, 1977; also Tad Szulc, "The Politics of Assassination," *New York*, June 23, 1975.
94. Jack Anderson, "Washington Merry-Go-Round," for release March 3, 1967.
95. See Judith Exner, *My Story*, as told to Ovid Demaris (New York, 1977).
96. Church committee, *Assassination Plots*, 129.
97. Ibid., 129–130.
98. The first two sentences are from William Hundley, in interview by author, August 4, 1976; the third, from a story by Nicholas Gage, *New York Times*, April 13, 1976.
99. Giancana's intermediary was Charles English. See Nicholas Gage, quoting an FBI report, "Ex-Aides Say Justice Department Rejected a Sinatra Inquiry," *New York Times*, April 14, 1976.
100. Nicholas Gage, "Two Mafiosi Linked to CIA Treated Leniently by U.S.," *New York Times*, April 13, 1976.
101. "Peter Lawford Tells," *Star* (New York), February 17, 1976.
102. The report is summarized in Gage, *Mafia Is Not an Equal Opportunity Employer*, ch. 5.
103. "Peter Lawford Tells."
104. Author's journal, October 28, 1963.
105. See Gage piece, *New York Times*, April 14, 1976.
106. Hundley, in interview by author.
107. *New York Times*, April 16, 1976.
108. Church committee, *Assassination Plots*, 133.
109. Ibid., 133, 151, 153–155.
110. Ibid., 164–167.
111. See, e.g., Branch and Crile, "The Kennedy Vendetta."
112. Church committee, *Assassination Plots*, 120, 157, 159.

22. Robert Kennedy and the Missile Crisis *(pages 499–532)*

1. Pierre Salinger, *With Kennedy* (Garden City, N.Y., 1966), 198.
2. Benjamin C. Bradlee, *Conversations with Kennedy* (New York, 1975), 194.

3. James W. Symington, *The Stately Game* (New York, 1971), 144.
4. RFK, handwritten notes, November 7, 1961, RFK Papers.
5. RFK, memorandum dictated May 29, 1962, RFK Papers.
6. RFK, memorandum dictated July 21, 1962, RFK Papers.
7. RFK, in recorded interview by John Bartlow Martin, March 1, 1964, I, 40, 47, JFK Oral History Program.
8. RFK, in Martin interview, April 13, 1964, I, 9.
9. RFK, in Martin interview, March 1, 1964, I, 45.
10. RFK, memorandum dictated May 29, 1962, RFK Papers.
11. RFK to JFK, June 19, 1962, RFK Papers.
12. RFK to McGeorge Bundy, July 11, 1962, RFK Papers.
13. RFK, in Martin interview, March 1, 1964, I, 46.
14. Salinger, *With Kennedy*, 200.
15. Article by N. Karev in *Za Rubezhom*, June 9, 1962; reported in aerogram from American embassy, Moscow, June 19, 1962, RFK Papers.
16. Salinger, *With Kennedy*, 209.
17. RFK, in Martin interview, April 30, 1964, I, 10.
18. Llewellyn Thompson, in recorded interview by Elizabeth Donahue, March 25, 1964, 4, JFK Oral History Program.
19. RFK, memorandum dictated July 21, 1962, RFK Papers; RFK, Martin interview, April 30, 1964, I, 9.
20. Jean Daniel in *L'Express*, December 14, 1963; Daniel, "Unofficial Envoy," *New Republic*, December 14, 1963; Daniel, *Le Temps qui reste* (Paris, 1973), 159–160; N. S. Khrushchev, *Khrushchev Remembers: The Last Testament* (Boston, 1974), 511; Claude Julien in *Le Monde*, March 22, 1963.
21. N. S. Khrushchev, *Khrushchev Remembers* (Boston, 1970), 493–494.
22. Graham T. Allison, *Essence of Decision: Explaining the Cuban Missile Crisis* (Boston, 1971), 49.
23. Herbert L. Matthews, *Fidel Castro* (New York, 1969), 225.
24. Daniel, "Unofficial Envoy."
25. Khrushchev, *Khrushchev Remembers*, 494.
26. RFK, memorandum dictated September 11, 1962, RFK Papers.
27. Minutes of White House meeting, September 4, 1962, RFK Papers.
28. Louise FitzSimons, *The Kennedy Doctrine* (New York, 1972), 138; Clare Boothe Luce, "Cuba and the Unfaced Truth," *Life*, October 5, 1962; Hugh Thomas, *Cuba or the Pursuit of Freedom* (London, 1971), 1399–1400; Richard Rovere, "Washington Letter," *New Yorker*, October 10, 1962.
29. Robert F. Kennedy, *Thirteen Days: A Memoir of the Cuban Missile Crisis* (New York, 1971 ed., with afterword by Richard E. Neustadt and Graham T. Allison), 2.
30. I owe this story to R. Harris Smith, who turned it up in the course of his research on a biography of Allen W. Dulles (R. Harris Smith to author, December 22, 1973).
31. John F. Kennedy, *Public Papers . . . 1962* (Washington, 1963), 807, 897–898.
32. White House meeting, October 16, 1962, RFK Papers.
33. In *Thirteen Days* (9) he writes that he passed the note to the President. In a contemporaneous memorandum dictated October 31, 1962, in the RFK Papers, he said that he passed it to Sorensen. Perhaps Sorensen, who put Kennedy's manuscript in final shape for publication, made the change in an excess of modesty.
34. Maxwell Taylor, *Swords and Plowshares* (New York, 1972), 268.
35. RFK, handwritten notes, October 31, 1962, RFK Papers.
36. RFK, in interview by author, February 27, 1965, 6, JFK Oral History Program.
37. RFK, memorandum dictated November 30, 1962, RFK Papers.
38. Joseph Dolan, in recorded interview by L. J. Hackman, April 11, 1970, 1, RFK Oral History Program.
39. RFK, *Thirteen Days*, 15.
40. Ibid., 17.

41. Roswell Gilpatric, in recorded interview by D. J. O'Brien, May 27, 1970, 55, JFK Oral History Program.
42. "No Yearning to be Loved — Dean Acheson talks to Kenneth Harris," *Listener,* April 8, 1971.
43. Dean Acheson, "Dean Acheson's Version of Robert Kennedy's Version of the Cuban Missile Affair," *Esquire,* February 1969.
44. Leonard C. Meeker, memorandum of October 19, 1962, meeting of the Executive Committee of the National Security Council, 5, Schlesinger Papers. Meeker, an able and incorruptible public servant, was later ambassador to Rumania. We had worked together in the OSS during the Second World War.
45. RFK, memorandum dictated October 31, 1962, RFK Papers.
46. Meeker, memorandum, 5–6.
47. RFK, *Thirteen Days,* 16.
48. Meeker, memorandum, 7.
49. RFK, memorandum, October 31, 1962.
50. Jean Stein and George Plimpton, eds. *American Journey* (New York, 1970), 136–137.
51. U. Alexis Johnson, in recorded interview by William Brubeck, n.d. [1964], 45, JFK Oral History Program.
52. Minutes of National Security Council (NSC) meeting, 2:30–5:10 P.M., October 20, 1962, RFK Papers.
53. Theodore C. Sorensen, *Kennedy* (New York, 1965), 694.
54. Minutes of NSC.
55. Robert McNamara, in interview for David Wolper's television special on Robert Kennedy, October 23, 1969.
56. RFK, memoranda dictated October 31 and November 15, 1962, RFK Papers.
57. Taylor, *Swords and Plowshares,* 271.
58. Gilpatric, in O'Brien interview, August 12, 1970, 116.
59. Kenneth O'Donnell and David F. Powers, *"Johnny, We Hardly Knew Ye"* (Boston, 1972), 318.
60. R. J. Walton, *Cold War and Counterrevolution* (New York, 1972), 103, 116.
61. In addition to Walton, see, e.g., FitzSimons, *Kennedy Doctrine,* ch. 5; Ronald Steel, "Endgame," in *Imperialists and Other Heroes* (New York: Random House, Vintage reprint, 1973), 115–136; I. F. Stone, "What If Khrushchev Hadn't Backed Down?" in *In a Time of Torment* (New York, 1967), 18–27; Barton J. Bernstein, "The Cuban Missile Crisis," in *Reflections on the Cold War,* ed. L. H. Miller and R. W. Pruessen (Philadelphia, 1974), 108–142; Barton J. Bernstein, "'Courage and Commitment': The Missiles of October," *Foreign Service Journal,* December 1975.
62. McGeorge Bundy, on *Issues and Answers,* October 14, 1962.
63. Bernstein, "Cuban Missile Crisis," 137.
64. RFK, *Thirteen Days,* 45.
65. Walton, *Cold War and Counterrevolution,* 122.
66. Bohlen to JFK, October 18, 1962, in Charles E. Bohlen, *Witness to History, 1929–1969* (New York, 1973), 491.
67. Theodore Sorensen, memorandum, October 18, 1962, RFK Papers.
68. Sorensen, memorandum, October 20, 1962, RFK Papers.
69. RFK to JFK, October 24, 1962, RFK Papers.
70. Edwin Guthman, *We Band of Brothers* (New York, 1971), 126.
71. RFK, handwritten notes on White House meeting of October 24, 1962, RFK Papers. This passage, somewhat rewritten and polished, appeared in RFK, *Thirteen Days,* 48–49.
72. RFK, *Thirteen Days,* 48–49.
73. Theodore Sorensen, draft, October 18, 1962, RFK Papers.
74. Meeker, memorandum, 10.
75. Minutes of NSC.
76. RFK, in interview by John Bartlow Martin, December 7, 1966, Martin Papers.

77. Minutes of NSC.
78. O'Donnell and Powers, *"Johnny,"* 322.
79. Minutes of NSC.
80. O'Donnell and Powers, *"Johnny,"* 323.
81. RFK, memorandum, November 15, 1962.
82. Author's journal, October 22, 1962.
83. RFK, *Thirteen Days,* 28.
84. Harold Macmillan, *At the End of the Day, 1961-1963* (New York, 1973), 210-211.
85. Stone, *In a Time of Torment,* 21-22.
86. O'Donnell and Powers, *"Johnny,"* 310.
87. Ronald Steel, *Imperialists and Other Heroes,* 135; Bernstein, "'Courage and Commitment.'"
88. Sorensen, *Kennedy,* 711.
89. Macmillan, *End of the Day,* 199.
90. RFK, notes on White House meeting, October 24, 1962.
91. Michel Tatu, *Power in the Kremlin: From Khrushchev to Kosygin* (New York, 1969), 263.
92. Khrushchev to JFK, October 25, 1962. The final text is reprinted in *The Dynamics of World Power: A Documentary History of United States Foreign Policy, 1945-1973,* ed. Arthur M. Schlesinger, Jr., vol. 3, *Eastern Europe and the Soviet Union,* ed. Walter La Feber (New York, 1973), 699-703.
93. Macmillan, *End of the Day,* 211.
94. Khrushchev, *Last Testament,* 512.
95. RFK, *Thirteen Days,* 74.
96. Macmillan, *End of the Day,* 187, 212-213, 217; Macmillan, in interview by Robert McKenzie, "The Cuba Crisis," *Listener,* October 11, 1973.
97. Henry Kissinger, "Reflections on Cuba," *Reporter,* November 22, 1962.
98. Elie Abel, *The Missile Crisis* (New York: Bantam reprint, 1966), 168.
99. RFK, memorandum, November 15, 1962, RFK Papers.
100. Abel, *Missile Crisis,* 174.
101. Executive Committee of National Security Council, minutes, October 23, 1962, RFK Papers.
102. RFK, *Thirteen Days,* 75-76.
103. Allison, *Essence of Decision,* 225, 227.
104. RFK, *Thirteen Days,* 79.
105. Ibid., 79-81.
106. RFK to Dean Rusk, October 30, 1962, RFK Papers; and Anatoly Gromyko, "Diplomatic Efforts of the USSR to Liquidate the Crisis," *Voprosy Istorii,* August 1971 (Gromyko cites Soviet Foreign Ministry Archives). This was the second of two articles; the first, "Concoction of the Caribbean Crisis by the U.S. Government," ran in July 1971. The Gromyko series, a characteristic piece of Soviet 'history,' is noteworthy for the fact that the name N. S. Khrushchev nowhere appears.
107. So Dobrynin said to Averell Harriman in Hobe Sound, Florida, in the winter of 1969, as described by Harriman to me.
108. Khrushchev, *Last Testament,* 498.
109. Douglas Dillon, in recorded interview by L. J. Hackman, June 18, 1970, 13, RFK Oral History Program.
110. Abram Chayes, *The Cuban Missile Crisis: International Crises and the Role of Law* (New York, 1974), 98.
111. Robert McNamara to JFK, April 25 [1963], JFK Papers; Jupiter missiles were also removed from Italy.
112. Curtis E. LeMay, *America Is in Danger* (New York, 1968), 200.
113. Thomas, *Cuba or the Pursuit of Freedom,* 1414.
114. Lee Lockwood, *Castro's Cuba, Cuba's Fidel* (New York, 1967), 200.
115. A point I adapt from K. S. Karol, *Guerrillas in Power: The Course of the Cuban Revolution* (New York, 1970).
116. Author's journal, October 29, 1962.

117. Bradlee, *Conversations with Kennedy*, 1122.
118. Author's journal, October 30, 1962.
119. RFK, in recorded interview by John Bartlow Martin, May 14, 1964, II, 1, RFK Oral History Program.
120. RFK, memorandum dictated November 30, 1962, RFK Papers.
121. James B. Donovan, *Challenges* (New York, 1967), 100.
122. Abel, *Missile Crisis*, 190–191.
123. Sorensen, *Kennedy*, 720.
124. RFK, memorandum dictated November 15, 1962, RFK Papers.
125. RFK, memorandum dictated November 14, 1962, RFK Papers.
126. RFK, memorandum, November 30, 1962.
127. In RFK Papers.
128. RFK, memorandum, November 30, 1962.
129. Ibid.
130. RFK to Bolshakov, March 7, 1963; photocopy in RFK Papers. Alsop's column was in the *Washington Post*, November 5, 1962.
131. Fidel Castro, in interview with Claude Julien, *Le Monde*, March 22, 1963.
132. Bernstein, "'Courage and Commitment.'"
133. Robert McNamara to JFK, April 17, 1963, "Could the Defense Budget Be Cut to $43 Billion Without Weakening the Security of the United States?" JFK Papers.
134. Sorensen, *Kennedy*, 705.
135. Abel, *Missile Crisis*, 180.
136. RFK, *Thirteen Days*, 104–105.
137. Ibid., 106.
138. Tatu, *Power in the Kremlin*, 422.
139. Lord Harlech, in recorded interview by Jean Stein, April 30, 1970, 11, Stein Papers.
140. See Graham T. Allison, "Cuban Missiles and Kennedy Macho," *Washington Monthly*, October 1972.
141. Abel, *Missile Crisis*, 162.
142. Richard Nixon, "Cuba, Castro, and John F. Kennedy," *Reader's Digest*, November 1964.
143. Playboy Interview, *Playboy*, March 1977.
144. Khrushchev, *Khrushchev Remembers*, 500; *Last Testament*, 513–514.
145. Matthews, *Fidel Castro*, 225.
146. George S. McGovern, *Cuban Realities: May 1975*, report to the Committee on Foreign Relations, 94 Cong., 1 Sess. (1975), 14.
147. Robert McNamara, statement for release April 14 [1968], RFK Papers.
148. Adlai Stevenson to RFK, December 22, 1962, RFK Papers.
149. Khrushchev, *Khrushchev Remembers*, 500.
150. "Harold Macmillan and Lord Harlech Discuss with Robert MacNeil the Cuban Missile Crisis of 1962," *Listener*, January 30, 1969.
151. O'Donnell and Powers, *"Johnny,"* 283.
152. Stein and Plimpton, *American Journey*, 139.

23. The Cuban Connection: II *(pages 533–558)*

1. RFK, in recorded interview by John Bartlow Martin, April 30, 1964, III, 22–23, JFK Oral History Program; Senate Select Committee to Study Governmental Operations with respect to Intelligence Activities (hereafter cited as Church committee), *Interim Report: Alleged Assassination Plots Involving Foreign Leaders*, 94 Cong., 1 Sess. (1975), 148.
2. E. G. Lansdale to Special Group (Augmented), November 14, 1962, RFK Papers; Lansdale, in interview by author, December 30, 1976.
3. Maxwell Taylor, in recorded interview by L. J. Hackman, October 22, 1969, 9, RFK Oral History Program.
4. Lansdale, in interview by author.

5. John Nolan, in recorded interview by Frank DeRosa, April 25, 1967, 6, JFK Oral History Program.
6. Edwin Guthman, *We Band of Brothers* (New York, 1971), 131–132; Haynes Johnson, *The Bay of Pigs* (New York, 1964), 321.
7. Joseph Dolan, in recorded interview by Frank DeRosa, July 8, 1964, 2–3, JFK Oral History Program.
8. Victor Navasky, *Kennedy Justice* (New York, 1971), 335.
9. Lloyd N. Cutler, in recorded interview by Frank DeRosa, June 22, 1964, 23, JFK Oral History Program.
10. Robert F. Shea, in recorded interview by Frank DeRosa, July 1, 1963, 24–25, JFK Oral History Program.
11. Barrett Prettyman, "The Cuban Prisoner Exchange," n.d. [1963], 3, Prettyman Papers.
12. Ibid., 21.
13. Nolan, in DeRosa interview, 7–9.
14. Prettyman, "Cuban Prisoner Exchange," 34.
15. Kenneth O'Donnell and David Powers, *"Johnny, We Hardly Knew Ye"* (Boston, 1972), 275–276.
16. Richard Cardinal Cushing, in recorded interview by Jean Stein, n.d., 3, Stein Papers.
17. Dolan, in DeRosa interview, 6.
18. O'Donnell and Powers, *"Johnny,"* 276–277.
19. John F. Kennedy, *Public Papers . . . 1962* (Washington, 1963), 911–912.
20. *Newsweek,* January 7, 1963.
21. Church committee, *Assassination Plots,* 173.
22. Sterling Cottrell to McGeorge Bundy, January 22, 1963, RFK Papers.
23. Robert Hurwich, in recorded interview by John Plank, April 24–May 4, 1964, 151, JFK Oral History Program.
24. RFK to JFK, March 14, 1963, RFK Papers.
25. Bundy to Acting Director of Central Intelligence, March 14, 1963; RFK to JFK, March 26, 1963, RFK Papers.
26. Sir Herbert Marchant, "Cuba on the Brink — II," *London Sunday Telegraph,* October 22, 1967.
27. Roger Hilsman, *To Move a Nation* (Garden City, N.Y., 1967), 226; *New York Times,* April 21, 1963; George McCully, "Keating and the Debate Concerning the Soviet Buildup in Cuba," n.d. [September 1964], RFK Papers.
28. Benjamin C. Bradlee, *Conversations with Kennedy* (New York, 1975), 132–133.
29. Marquis Childs, *Witness to Power* (New York, 1975), 180.
30. RFK, handwritten notes, March 29, 1963, RFK Papers.
31. Church committee, *Final Report,* bk. V, *The Investigation of the Assassination of President John F. Kennedy: Performance of the Intelligence Agencies,* 94 Cong., 2 Sess. (1976), 11–13.
32. Church committee, *Assassination Plots,* 173.
33. McGeorge Bundy to RFK, May 16, 1963, RFK Papers.
34. Nolan, in DeRosa interview, 11–12.
35. Ibid., 22–26.
36. Lee Lockwood, *Castro's Cuba, Cuba's Fidel* (New York, 1967), 80.
37. M. C. Miskovsky to McCone, April 13, 1963, RFK Papers.
38. John Nolan to RFK, n.d. [April 1963], RFK Papers.
39. Nolan, in DeRosa interview, 10–13.
40. James B. Donovan, *Challenges* (New York, 1967), 92.
41. Ibid., 85.
42. Church committee, *Assassination Plots,* 172.
43. Ibid., 172–173, 337.
44. Ibid., 173.
45. McGeorge Bundy, in interview by author, July 18, 1977.
46. Church committee, *Assassination Plots,* 173, 337.

47. General Krulak appears under thin disguise as General Tartak in B. E. Ayer's account, *The War That Never Was* (Indianapolis, 1976), 15, 52–53. His CIA experience deeply upset Ayers; it broke up his marriage, created an impassioned identification with the Cubans and led to his resignation from both the Agency and the Army when interest passed in 1964 from Cuba to Vietnam. His book is at times highly emotional, though ordinarily prosaic when describing covert operations.
48. Dick Russell, "Little Havana's Reign of Terror," *New Times*, October 29, 1976.
49. Church committee, *Assassination of President Kennedy*, 12–13.
50. RFK, in Martin interview, III, 19, 24.
51. J. H. Crimmins to RFK, June 24, 1963, RFK Papers.
52. McGeorge Bundy to RFK, n.d. [soon after July 12, 1963], enclosing memorandum from CIA / Miami on Hendrix, RFK Papers.
53. John McCone to RFK, July 20, 1963, enclosing report provided by J. M. Bosch of a conversation with Somoza.
54. Memorandum of conversation between J. H. Crimmins, Felipe Rivero and Paulino Sierra, August 17, 1963, RFK Papers.
55. U.S. ambassador, Nicaragua, to Thomas Mann, April 3, 1964, Mann to ambassador, April 7, 1964, RFK Papers.
56. Church committee, *Assassination Plots*, 171.
57. Hugh Thomas, *Cuba* (London, 1971), 1286–1287; George Crile III, "The Riddle of AM LASH," *Washington Post*, May 2, 1976.
58. Church committee, *Assassination Plots*, 87; Church committee, *Assassination of President Kennedy*, 13–14.
59. Church committee, *Assassination of President Kennedy*, 14; Johnson, *Bay of Pigs*, 354.
60. Church committee, *Assassination of President Kennedy*, 17, 74–75; Church committee, *Assassination Plots*, 174.
61. Church committee, *Assassination Plots*, 88–89; Church committee, *Assassination of President Kennedy*, 17–19.
62. Church committee, *Assassination Plots*, 89–90; Church committee, *Assassination of President Kennedy*, 78–79. Tad Szulc believes that Howard Hunt may have been the CIA officer working with Cubela and Artime in this period (Szulc, *Compulsive Spy: The Strange Career of E. Howard Hunt* [New York, 1974], 90–97).
63. Thomas, *Cuba*, 1070.
64. Statement from office of Senator George McGovern, July 30, 1975; Church committee, *Assassination Plots*, 71.
65. The FBI deleted the words "in place" before sending copies of Morgan's deposition to the Attorney General, the Secret Service and the White House. For some reason no copy went to the CIA. Church committee, *Assassination of President Kennedy*, 83–85. For further detail, see Ronald Kessler and Laurence Stern, "Slain Mobster Claimed Cuban Link to JFK Death," *Washington Post*, August 22, 1976.
66. Jack Anderson, "Did Castro Arrange for Mob to Kill Kennedy?" *Washington Post*, March 24, 1977.
67. Nicholas Gage, "Mafia Said to Have Slain Rosselli Because of His Senate Testimony," *New York Times*, February 25, 1977.
68. George Crile III, "The Mafia, the CIA and Castro," *Washington Post*, May 16, 1976.
69. RFK, in Martin interview, III, 18.
70. Author's journal, December 13, 1963; notes taken by Richard Goodwin on meeting that day, Goodwin Papers.
71. Abba Schwartz, *The Open Society* (New York, 1968), ch. 5.
72. Ball to Rusk, December 13, 1963, RFK Papers.
73. RFK, in Martin interview, III, 18.
74. William Attwood, *The Reds and the Blacks* (New York, 1967), 142.
75. Church committee, *Assassination of President Kennedy*, 20.

76. William Attwood, "Memorandum on Cuba," September 18, 1963, RFK Papers.
77. Attwood, *Reds and Blacks*, 142.
78. Church committee, *Assassination Plots*, 173–174.
79. Attwood, *Reds and Blacks*, 143.
80. RFK, in Martin interview, III, 18–19.
81. Church committee, *Assassination Plots*, 174.
82. Jean Daniel, *Le Temps qui reste* (Paris, 1973), 149–152; Jean Daniel, "Unofficial Envoy," *New Republic*, December 14, 1963.
83. State Department, American Republics Division, "The Future of Cuba," November 7, 1963, RFK Papers.
84. McGeorge Bundy, memorandum re Attwood-Vallejo talks, November 12, 1963.
85. Attwood, *Reds and Blacks*, 144.
86. John F. Kennedy, *Public Papers . . . 1963* (Washington, 1964), 875–876.
87. Attwood, *Reds and Blacks*, 144; Church committee, *Assassination Plots, 174*.
88. Daniel, *Temps*, 158–162; Daniel, "Unofficial Envoy"; Daniel, "When Castro Heard the News," *New Republic*, December 7, 1963.
89. Daniel, *Temps*, 163–164; "When Castro Heard the News."
90. Author's journal, December 4, 1963.
91. Ibid., December 23, 1963.
92. Attwood, *Reds and Blacks*, 146.
93. Frank Mankiewicz and Kirby Jones, *With Fidel* (Chicago, 1975), 164, 166.
94. Maurice Halperin, *The Rise and Decline of Fidel Castro* (Berkeley, 1972), 344.
95. Hurwich, in Plank interview, 104–106.
96. Richard Helms to RFK, June 26, 1963, RFK Papers.
97. William Manchester, *The Death of a President* (New York, 1967), 46.
98. After John Kennedy's death, Bosch printed up his letter as a brochure with pictures (Russell, "Little Havana"). See also "Miami, Haven for Terror," *Nation*, March 19, 1977.
99. George Crile III, "Our Heritage — the Exile Cuban Terrorists," *Washington Post*, November 7, 1976.

24. Missions to the Third World *(pages 560–583)*

1. David Halberstam, *The Unfinished Odyssey of Robert Kennedy* (New York, 1968), 145–146.
2. Felix Houphouët-Boigny and Robert Kennedy, memorandum of conversation, August 7, 1961, RFK Papers.
3. Clark Mollenhoff, in recorded interview by Jean Stein, July 25, 1968, 17–18, Stein Papers.
4. RFK to JFK, September 5, 1961, RFK Papers.
5. Walt W. Rostow, *The Diffusion of Power* (New York, 1972), 202–203. I cheerfully yield to Rostow's memory of the presidential comment in preference to the version — "the hot breath of his disapproval" — in *A Thousand Days* (Boston, 1965), 573. Robert Kennedy himself had a variant memory: "He made some remark about [my] cross little face looking over his shoulder while he was doing this." RFK, recorded interview by John Bartlow Martin, April 13, 1964, II, 2, JFK Oral History Program.
6. RFK to JFK, December 7, 1961, RFK Papers.
7. RFK, in Martin interview, April 13, 1964, II, 2–3.
8. Author to RFK, July 1, 1963; Robert McNamara to Dean Rusk, July 11, 1963, RFK Papers.
9. Wayne Fredericks, in interview by author, January 10, 1977.
10. RFK to McGeorge Bundy, November 20, 1963, RFK Papers.
11. RFK, in Martin interview, April 13, 1964, II, 51–52. Ball recalled the memorandum as saying: "God watches every sparrow that may fall, but I don't see why we have to compete in that league." George Ball, in recorded interview by Jean Stein, n.d. [1968], 12, Stein Papers.

12. John Kenneth Galbraith, *Ambassador's Journal* (Boston, 1969), 303.
13. Anthony Lewis, "A Tribute to Robert Francis Kennedy," address at unveiling of RFK portrait, Department of Justice, March 14, 1975 (Washington, 1975).
14. Robert F. Kennedy, "Our Generation, Our World, Our Future," address at Nihon University, Tokyo, February 6, 1962, RFK Papers.
15. Robert F. Kennedy, *Just Friends and Brave Enemies* (New York, 1962), 53.
16. Edwin Reischauer, in recorded interview by D. J. O'Brien, April 25, 1969, 21, JFK Oral History Program.
17. RFK, *Just Friends*, 59.
18. John Seigenthaler, in recorded interview by R. J. Grele, February 23, 1966, 412, JFK Oral History Program.
19. American embassy, Tokyo, "Visit of Attorney General to Japan, February 4–10, 1962," February 28, 1962, RFK Papers.
20. RFK, *Just Friends*, 61.
21. Ibid., 66.
22. Seigenthaler, in Grele interview, 388.
23. Events at Waseda as observed by the Attorney General's interpreter, in "Visit of Attorney General to Japan," RFK Papers.
24. Reischauer to Rusk, February 8, 1962, RFK Papers.
25. Reischauer, in O'Brien interview, 22.
26. Ibid., 21.
27. "Visit of Attorney General."
28. Seigenthaler, in Grele interview, 389.
29. "Visit of Attorney General."
30. Kennedy's Kansai Trip," *Nihon Kezai*, February 12, 1962.
31. R. A. Mlynarchik to David Osborn, February 13, 1962, RFK Papers.
32. "Visit of Attorney General."
33. Reischauer, in O'Brien interview, 13–14.
34. Roger Hilsman, *To Move a Nation* (Garden City, N.Y., 1967), 363.
35. RFK, *Just Friends*, 113–114.
36. Ibid., 114–117; Howard Jones to Rusk, February 18, 1962, containing Kennedy's Mexican War heresy as transcribed in the tape recording of the meeting, RFK Papers.
37. Jones to Rusk, February 14 and February 22, 1962, RFK Papers.
38. N. S. Khrushchev, *Khrushchev Remembers: The Last Testament* (Boston, 1974), 312–322; RFK, *Just Friends*, 100.
39. RFK, in Martin interview, April 13, 1964, II, 33.
40. Howard P. Jones, in recorded interview by D. J. O'Brien, June 23, 1969–April 9, 1970, 41, JFK Oral History Program.
41. Khrushchev, *Last Testament*, 323.
42. Howard P. Jones, *Indonesia: The Possible Dream* (New York, 1971), 197, 199, 203.
43. W. Averell Harriman, in recorded interview by author, June 6, 1965, 16–17, JFK Oral History Program.
44. McGeorge Bundy to RFK, February 5, 1962, RFK Papers.
45. Howard Jones to Rusk, February 14, 1962, RFK Papers.
46. RFK to JFK and Rusk, February 14, 1962, RFK Papers.
47. RFK to Bundy, July 24, 1962, memorandum on conversation with Subandrio, July 20, 1962, RFK Papers.
48. Jones to RFK, May 3, 1962, RFK Papers.
49. Jones to Rusk, August 16, 1962, JFK Papers.
50. RFK to Mrs. William C. Battle, June 11, 1963, RFK Papers.
51. Khrushchev, *Last Testament*, 327.
52. Cindy Adams, *Sukarno: An Autobiography as Told to Cindy Adams* (Indianapolis, 1965), 271.
53. RFK, in Martin interview, April 13, 1964, II, 30.
54. Seigenthaler, in Grele interview, 399–408; RFK, in Martin interview, April 13, 1964, II, 35–37.

55. Jones, in O'Brien interview, 41.
56. Adams, *Sukarno*, 271.
57. Allen Pope to RFK, July 11, 1962, RFK Papers.
58. Galbraith, *Ambassador's Journal*, 305.
59. Author's journal, February 21, 1962.
60. Willy Brandt, *Begegnungen und Einsichten* (1976), as reported in the *New York Times*, August 18, 1976.
61. RFK, *Just Friends*, 153–161.
62. Author's journal, February 23, 1962.
63. RFK to JFK, "From Extensive Notes Made during My Discussion with Chancellor Adenauer," March 16, 1962, RFK Papers; RFK, *Just Friends*, 166–168.
64. RFK, in Martin interview, April 13, 1964, II, 41.
65. Anthony Lewis, in recorded interview by L. J. Hackman, July 23, 1970, 18, RFK Oral History Project.
66. For an extended discussion, see Arthur M. Schlesinger, Jr., "The Alliance for Progress: A Retrospective," in *Latin America: The Search for a New International Role*, ed. R. G. Hellman and H. J. Rosenbaum (New York, 1975), 57–92. The quotations are from Lawrence Harrison, "Waking from the Pan American Dream," *Foreign Policy*, Winter 1971–72, 1969; and Suzanne Bodenheimer, "Dependency and Imperialism: The Roots of Latin American Underdevelopment," in *Readings in U.S. Imperialism*, ed. E. T. Fann and D. C. Hodges (Boston, 1971), 177.
67. Eduardo Frei, "Urgencies in Latin America: The Alliance That Lost Its Way," *Foreign Affairs*, April 1967, 437–438, 442.
68. Richard Goodwin, "Our Stake in a Big Awakening," *Life*, April 14, 1967.
69. John F. Kennedy, *Public Papers . . . 1962* (Washington, 1963), 223.
70. William D. Rogers, *The Twilight Struggle: The Alliance for Progress and the Politics of Development in Latin America* (New York, 1967), 218–219.
71. Jerome Levinson and Juan De Onis, *The Alliance That Lost Its Way: A Critical Report on the Alliance for Progress* (Chicago, 1970), 71.
72. A. F. Lowenthal, "United States Policy Toward Latin America: 'Liberal,' 'Radical,' and 'Bureaucratic' Perspectives," *Latin American Research Review*, Fall 1973, 17.
73. Jean Daniel, "Unofficial Envoy," *New Republic*, December 14, 1963.
74. Frank Mankiewicz and Kirby Jones, *With Fidel* (Chicago, 1975), 200–202.
75. Herbert L. Matthews, *Return to Cuba* (Stanford University, n.d. [1964]), 15.
76. Daniel, "Unofficial Envoy."
77. Luis Muñoz-Marin, address before the AFL-CIO National Conference on Community Service, Chicago, May 3, 1962, 6 (mimeographed).
78. Ibid., 8.
79. JFK, *Public Papers . . . 1962*, 231, 495, 883.
80. John F. Kennedy, *Public Papers . . . 1961* (Washington, 1962), 172.
81. John F. Kennedy, *Public Papers . . . 1963* (Washington, 1964), 873–875.
82. Roberto Campos, in recorded interview by John E. Reilly, May 29–30, 1964, JFK Oral History Program.
83. Lincoln Gordon, in interview by author, October 17, 1974.
84. Gordon, in interview by author; Campos, in Reilly interview, 26.
85. Campos, in Reilly interview, 24–25.
86. Ibid., 33.
87. Gordon, in interview by author.
88. Ibid.
89. Ethel Kennedy to RFK, December 17, 1962, RFK Papers.
90. U.S. embassy, Brazil, memorandum of conversation of talk among Robert Kennedy, João Goulart and Lincoln Gordon, 11:15 A.M.–2:30 P.M., December 17, 1962, RFK Papers.
91. Campos, in Reilly interview, 45.
92. Gordon, in interview by author.
93. RFK, in Martin interview, April 30, 1964, I, 37.
94. RFK, in Martin interview, April 13, 1964, I, 51.

95. Robert F. Kennedy, *The Pursuit of Justice* (New York: Harper & Row, Perennial Library reprint, 1964), 116-117.

25. The Brothers: II *(pages 584-602)*

1. Richard N. Goodwin, "The Art of Assuming Power," *New York Times Magazine,* December 26, 1976.
2. John F. Kennedy, *Public Papers . . . 1962* (Washington, 1963), 889.
3. John F. Kennedy, foreword to Theodore C. Sorensen, *Decision-Making in the White House* (New York, 1963), xii.
4. Jean Stein and George Plimpton, eds. *American Journey* (New York, 1970), 161, 163-164.
5. Art Buchwald, in recorded interview by Roberta Greene, March 12, 1969, 11, RFK Oral History Program.
6. Edwin Guthman, *We Band of Brothers* (New York, 1971), 241-243; Pierre Salinger, *With Kennedy* (New York, 1966), 239-247.
7. David Brinkley to RFK, February 11, 1963, RFK Papers.
8. *Time,* December 29, 1961.
9. *Boston Record,* May 9, 1957.
10. John Seigenthaler, in recorded interview by W. A. Geoghegan, n.d. [July 1964], 94, JFK Oral History Program.
11. R. J. Whalen, *The Founding Father* (New York, 1964), 452.
12. Hugh Sidey, *John F. Kennedy, President* (New York, 1963), 21.
13. Stephen Smith, in interview by author, March 5, 1976.
14. Charles Spalding, in recorded interview by John Stewart, March 14, 1968, 102, JFK Oral History Program.
15. Rita Dallas with Jeanira Ratcliffe, *The Kennedy Case* (New York: Popular Library reprint, 1973), 71-72, 146.
16. Dallas, *Kennedy Case,* 36, 83.
17. Ibid., 119, 122.
18. Benjamin C. Bradlee, *Conversations with Kennedy* (New York, 1975), 167-170.
19. Dallas, *Kennedy Case,* 109, 141, 167.
20. RFK, in recorded interview by John Bartlow Martin, April 13, 1964, II, 20, JFK Oral History Program.
21. Peter Edelman, in recorded interview by L. J. Hackman, December 12, 1969, 58-59, RFK Oral History Program.
22. Author's journal, April 10, 1963.
23. RFK to Randolph Churchill, April 24, 1963, RFK Papers.
24. Author's journal, September 11, 1963.
25. Adlai Stevenson to Mary Lasker, May 21, 1962, in a forthcoming volume of Stevenson's letters, edited by Walter Johnson, to whose courtesy I am indebted for this letter.
26. Author's journal, August 6, 1962.
27. Buchwald, in Greene interview, 10.
28. John Glenn, in recorded interview by Roberta Greene, June 26, 1969, 5-6, RFK Oral History Program; Glenn, in recorded interview by Jean Stein, July 1, 1968, 1-2, 15-16, Stein Papers.
29. Rose Kennedy to RFK, October 4, 1961, RFK Papers.
30. RFK to author, July 25, 1963, Schlesinger Papers.
31. Stein and Plimpton, *American Journey,* 166-167.
32. Ramsey Clark, in interview by author, March 15, 1977.
33. Peter Maas, in recorded interview by Jean Stein, September 15, 1969, 4, Stein Papers.
34. Anthony Lewis, "A Tribute to Robert Francis Kennedy," address at the unveiling of RFK portrait, Department of Justice, March 14, 1975 (Washington, 1975).
35. Kenneth O'Donnell and David F. Powers, *"Johnny, We Hardly Knew Ye"* (Boston, 1972), 94.

36. Michael Forrestal, in recorded interview by Jean Stein, n.d. [1968], 13, Stein Papers.
37. RFK to author, June 4, 1962, Schlesinger Papers.
38. O'Donnell and Powers, *"Johnny,"* 282.
39. Gore Vidal, "The Best Man, 1968," *Esquire,* March 1963.
40. Interview with Gore Vidal, *Penthouse,* April 1975.
41. Anaïs Nin, *Diary,* vol. 4, *1944–1947,* ed. Gunther Stuhlman (New York, 1971), 105, 142.
42. *Penthouse* interview.
43. John English, in recorded interview by Roberta Greene, November 3, 1969, 4, 6, RFK Oral History Program.
44. Author's journal, November 12, 1961.
45. Budd Schulberg, "RFK — Harbinger of Hope," *Playboy,* January 1969.
46. Margaret Laing, *The Next Kennedy* (New York, 1968), 17.
47. Theodore C. Sorensen, *The Kennedy Legacy* (New York, 1969), 78.
48. RFK to JFK, April 3, 1963, RFK Papers.
49. Norman Cousins, *The Improbable Triumvirate* (New York, 1972), 114.
50. Author's journal, October 11, 1963. See O'Donnell and Powers, *"Johnny,"* 381.
51. Y. I. Nosenko, "Under the Eye of the KGB — a Former Police Chief Looks Back," *Listener,* May 22, 1975.
52. RFK, in Martin interview, May 14, 1964, I, 18.
53. RFK, in Martin interview, April 30, 1964, I, 3.
54. RFK, in Martin interview, December 7, 1966, Martin Papers.
55. George F. Kennan, in recorded interview by Louis Fischer, March 23, 1965, 95–96, JFK Oral History Program.
56. Mansfield, UPI interview by Mike Feinsilber, *Buffalo Evening News,* September 16, 1976.
57. O'Donnell and Powers, *"Johnny,"* 278.
58. RFK to JFK, March 14, 1963, RFK Papers.
59. RFK to JFK, November 12, 1963, RFK Papers. RFK had been reading Herbert Agar's *The Price of Union.*
60. RFK, memorandum dictated February 13, 1961, RFK Papers.
61. Stein and Plimpton, *American Journey,* 127.
62. Chester Bowles, in recorded interview by R. R. R. Brooks, February 2, 1965, 17, JFK Oral History Program.
63. Bradlee, *Conversations with Kennedy,* 142–143.
64. Robert McNamara, in interview for David Wolper's television special on Robert Kennedy, October 23, 1969.
65. Goodwin, "Art of Assuming Power."
66. Author's journal, December 16, 1963.
67. Pearl S. Buck, *The Kennedy Women: A Personal Appraisal* (New York, 1970), 83.
68. Charles Spalding, in recorded interview by L. J. Hackman, March 22, 1969, 41, RFK Oral History Program.
69. Roy Jenkins, *Nine Men of Power* (London, 1974), 215–216.
70. Tom Wicker, in recorded interview by Jean Stein, November 25, 1969, 7, Stein Papers.
71. Stein and Plimpton, *American Journey,* 129; Spalding, in Stewart interview, 7–8.
72. RFK, in Martin interview, April 13, 1964, I, 57.
73. Quoted by Ted Lewis, *New York Daily News,* June 6, 1968.
74. O'Donnell, *"Johnny,"* 278.
75. Stein and Plimpton, *American Journey,* 128.
76. Benjamin C. Bradlee, in recorded interview by Jean Stein, October 18, 1968, 3, Stein Papers.
77. John F. Kennedy, *Public Papers . . . 1962* (Washington, 1963), 259. This characteristic Irish American mood was classically expressed by George M. Cohan in his sardonic threnody "Life's a Very Funny Proposition After All."
78. Richard Neustadt, in recorded interview by Jean Stein, October 9, 1968, 3, Stein Papers.

79. *Newsweek,* March 18, 1963.
80. Roy Wilkins, in recorded interview by Berl Bernhard, August 13, 1964, 14, JFK Oral History Program.
81. As told me by Lord Harlech. Lord Longford eventually wrote a perceptive biography of Kennedy, but found little enough to say about the influence of Catholicism; see Lord Longford, *Kennedy* (London, 1976), 203-204.
82. Tom Wicker, *Kennedy without Tears* (New York, 1964), 61.

26. Corridors of Grief *(pages 603-620)*

1. RFK, in recorded interview by John Bartlow Martin, May 14, 1964, I, 28, JFK Oral History Program; RFK, in recorded interview by Anthony Lewis, December 4, 1964, I, 15-16, JFK Oral History Program.
2. RFK, in Martin interview, May 14, 1964, I, 27.
3. RFK, in Martin interview, April 30, 1964, I, 21; RFK, in Lewis interview, December 4, 1964, I, 17. He also discussed his Romney concern with Paul Fay but warned him not to talk about it (Fay, *The Pleasure of His Company* [New York, 1966], 259).
4. RFK, in Martin interview, May 14, 1964, I, 24.
5. Ibid.
6. Stephen Smith to the President et al., November 13, 1963, RFK Papers; Kenneth O'Donnell and David F. Powers, *"Johnny, We Hardly Knew Ye"* (Boston, 1972), 386-387.
7. Theodore H. White, *The Making of the President, 1964* (New York, 1965), 28.
8. Leonard Baker, *The Johnson Eclipse* (New York, 1966), 105-106.
9. Benjamin C. Bradlee, *Conversations with Kennedy* (New York, 1975), 217-218.
10. O'Donnell and Powers, *"Johnny,"* 5.
11. RFK, in Martin interviews, April 30, 1964, I, 5, May 14, 1964, I, 21.
12. Evelyn Lincoln, *Kennedy and Johnson* (New York, 1968), 205.
13. Sam Houston Johnson, *My Brother Lyndon* (New York, 1970), 117.
14. RFK, in Martin interview, May 14, 1964, I, 25.
15. Ibid., April 13, 1964, I, 59-60.
16. G. H. Gallup, ed., *The Gallup Poll: Public Opinion, 1935-1971* (New York 1972), vol. 3, 1800, 1845, 1850.
17. RFK, in Martin interview, May 14, 1964, I, 25.
18. Marquis Childs, *Witness to Power* (New York, 1975), 164-165.
19. Patrick Anderson, "Robert's Character," *Esquire,* April 1965.
20. John Douglas, in recorded interview by L. J. Hackman, June 16, 1969, 20, RFK Oral History Program.
21. Ramsey Clark, in recorded interview by L. J. Hackman, July 20, 1970, 8-11, RFK Oral History Program.
22. RFK, in recorded interview by William Manchester, May 16, 1964, 4-5, 8.
23. This and the following account, unless otherwise specified, is drawn from Robert Morgenthau, in recorded interview by Jean Stein, April 20, 1970, 1-2, Stein Papers; RFK, in Manchester interview, 12-33, 40-45; James Wechsler, "RFK's Ordeal" (an interview with Morgenthau), *New York Post,* December 3, 1963; Bill Davidson, "A Profile in Family Courage," *Saurday Evening Post,* December 14, 1963; William Manchester, *The Death of a President* (New York, 1967), 146, 195-196.
24. Edwin Guthman, *We Band of Brothers* (New York, 1971), 244.
25. Manchester, *Death of a President,* 378.
26. Author's journal, November 23, 1963.
27. Charles Spalding, in recorded interview by Jean Stein, January 22, 1970, 18-19, Stein Papers.
28. RFK, in Manchester interview, 46-47, 54-56.
29. Ibid., 63-65.
30. Ibid., 61.
31. Author's journal, November 25, 1963.

32. RFK Papers.

33. Author's journal, November 23, 1963.

34. Bradlee, *Conversations with Kennedy*, 243.

35. K. LeMoyne Billings, in recorded interview by Jean Stein, n.d. [1968], 18, Stein Papers.

36. John Seigenthaler, in recorded interview by L. J. Hackman, June 5, 1970, 31, RFK Oral History Program.

37. Pierre Salinger, in recorded interview by L. J. Hackman, May 26, 1969, 2, RFK Oral History Program.

38. Seigenthaler, in Hackman interview, 15.

39. Helen Keyes, in recorded interview by Jean Stein, October 9, 1968, 22, Stein Papers.

40. Mary Bailey Gimbel, in recorded interview by Jean Stein, n.d. [1968], 23, Stein Papers.

41. John Bartlow Martin, *Overtaken by Events* (Garden City, N.Y., 1966), 632.

42. Jean Stein and George Plimpton, eds., *American Journey* (New York, 1970), 146–147; Guthman, *We Band of Brothers*, 247.

43. Anderson, "Robert's Character."

44. Charles Spalding, in recorded interview by Jean Stein, January 22, 1970, 19, Stein Papers.

45. Nicholas Katzenbach, in recorded interview by L. J. Hackman, October 8, 1969, 75, RFK Oral History Program.

46. Edward Jay Epstein, *Legend: The Secret World of Lee Harvey Oswald* (New York, 1968), 15–17.

47. Senate Select Committee to Study Governmental Operations with respect to Intelligence Activities (Church committee), *Final Report*, bk. V, *The Investigation of the Assassination of President John F. Kennedy: Performance of the Intelligence Agencies*, 94 Cong., 2 Sess. (1976), 23, 33.

48. Vivian Cadden, "The Murder of President Kennedy," *McCall's*, March 1977.

49. Author's journal, December 9, 1963.

50. Walter Sheridan, in recorded interview by Roberta Greene, May 1, 1970, 3–4, RFK Oral History Program; Sheridan, *The Fall and Rise of Jimmy Hoffa* (New York, 1972), 300, 356, 408. The letter writer, Frank Chavez, Secretary-Treasurer of Local 901, was later murdered by one of his own bodyguards.

51. Author's journal, October 30, 1966.

52. Frank Mankiewicz, in recorded interview by L. J. Hackman, October 2, 1969, 69, RFK Oral History Program.

53. Anthony Lewis, "What Not to Do," *New York Times*, September 25, 1975.

54. The ballad was about Owen Roe O'Neill. Robert Kennedy quoted it at a St. Patrick's Day banquet in Scranton, Pennsylvania, in March 1964 (and elsewhere).

55. Handwritten notes [1964], RFK Papers.

56. Jacqueline Onassis, in recorded interview by author, June 2, 1976, 9.

57. I am indebted to Mary Bailey Gimbel for lending me one of Robert Kennedy's copies, dog-eared and heavily marked, of *The Greek Way*. He used the Norton Library paperback edition (1964) and always kept a copy in his dispatch case. The quotations are from 52, 53, 109, 116, 147, 158, 166.

58. Author's journal, August 14, 1966.

59. RFK to Angie Novello, n.d. [June 1967], RFK Papers.

60. Jeff Greenfield, in recorded interview by Roberta Greene, December 10, 1969, 99–100, RFK Oral History Program.

61. W. H. Auden, "The Christian Tragic Hero," *New York Times Book Review*, December 16, 1945.

62. "Favorite Quotations of John F. Kennedy and Robert F. Kennedy," under "Miscellaneous," RFK Papers. This was a loose-leaf volume, kept by Angie Novello under various headings and used by RFK for sustenance when he had to give a speech or write an article. He used the last three sentences in the third Camus quote as the epigraph for his own book *To Seek a Newer World* (New York, 1967), curiously omitting, however, the word *believer* in the last sentence.

63. Rita Dallas, *The Kennedy Case* (New York: Popular Library reprint, 1973), 259, 264.

27. Stranger in a Strange Land *(pages 621–645)*

1. Benjamin C. Bradlee, *Conversations with Kennedy* (New York, 1975), 194.
2. Doris Kearns, *Lyndon Johnson and the American Dream* (New York, 1976), 164.
3. Repeated by Johnson in a backgrounder on July 25, 1964, in Jack Valenti, *A Very Human President* (New York, 1975), 306.
4. Ralph Dungan, relaying message from Rusk to JFK, October 3, 1962, JFK Papers.
5. Charles Spalding, in recorded interview by Jean Stein, January 22, 1970, 8–9, Stein Papers.
6. Bobby Baker, with Larry King, *Wheeling and Dealing: Confessions of a Capitol Hill Operator* (New York, 1978), 244.
7. American Bar Association, Special Committee on Election Reform, "Symposium on the Vice-Presidency," *Fordham Law Review*, February 1977, 750. For some reason the ABA, wrong again, insists on the usage Vice-President. The title has no hyphen in the Constitution.
8. Evelyn Lincoln, *Kennedy and Johnson* (New York, 1968), 151, 153.
9. Kenneth P. O'Donnell and David F. Powers, *"Johnny, We Hardly Knew Ye"* (Boston, 1972), 8.
10. Lincoln, *Kennedy and Johnson*, 161, 186.
11. Bill Moyers, in television interview with David Susskind, October 13, 1974; Daniel Patrick Moynihan, interview, *Playboy*, March 1977.
12. Kearns, *Johnson*, 164.
13. O'Donnell and Powers, *"Johnny,"* 6; Baker, *Wheeling and Dealing*, 126.
14. Lyndon B. Johnson, *The Vantage Point* (New York, 1971), 539.
15. Spalding, in Stein interview, 13–14.
16. John Seigenthaler, in recorded interview by W. A. Geoghegan, n.d. [July 1964], 60–61, JFK Oral History Program; Seigenthaler, in recorded interview by L. J. Hackman, June 5, 1970, 60–61; Seigenthaler to RFK, January 28, 1963, RFK Papers.
17. RFK, in recorded interview by author, February 27, 1965, 37, 42–43, JFK Oral History Program.
18. Ramsey Clark, in recorded interview by L. J. Hackman, July 7, 1970, 93, RFK Oral History Program.
19. Nicholas Katzenbach, in recorded interview by L. J. Hackman, October 8, 1969, 30, RFK Oral History Program.
20. RFK, in interview by author, 37.
21. Pierre Salinger, in recorded interview by L. J. Hackman, May 26, 1969, 13, RFK Oral History Program.
22. O'Donnell and Powers, *"Johnny,"* 6.
23. Michael Janeway, "LBJ and the Kennedys," *Atlantic Monthly*, February 1972.
24. Helen Thomas, *Deadline: White House* (New York, 1975), 121.
25. Kearns, *Johnson*, esp. ch. 1 and author's postscript; also 17, 201.
26. Hubert Humphrey, *The Education of a Public Man* (New York, 1976), 307.
27. Memoranda by James Wechsler, November 27, 1963, and Joseph P. Lash, November 27, 1963, Wechsler Papers; memorandum by Dorothy Schiff, in Jeffrey Potter, *Men, Money and Magic: The Story of Dorothy Schiff* (New York, 1976), 280–281.
28. John Adams, of course; emphasis added; J. D. Feerick, *The Twenty-Fifth Amendment* (New York, 1976), 31.
29. Kearns, *Johnson*, 170.
30. Jim Bishop, *The Day Kennedy Was Shot* (New York: Bantam reprint, 1969), 350. Bishop and Robert Kennedy, it should be added, were old foes in consequence of pro-Hoffa articles Bishop had writtten in 1959.
31. RFK, in recorded interview by William Manchester, May 16, 1964, 26, RFK Papers.

32. RFK, in recorded interview by John Bartlow Martin, February 29, 1964, 30, JFK Oral History Program; RFK, in interview by author, 42.
33. RFK, in Martin interview, May 14, 1964, I, 30, JFK Oral History Program.
34. RFK, in Manchester interview, 50–51; Evelyn Lincoln's account to me, author's journal, March 25, 1964.
35. RFK, in Manchester interview, 51–53.
36. William Manchester, *The Death of a President* (New York, 1967), 477–478.
37. RFK, in Manchester interview, 70–71.
38. Author's journal, November 27, 1963.
39. RFK, in Manchester interview, 71–72.
40. LBJ to RFK, January 1, 1964, RFK to LBJ, January 3, 1964, RFK Papers.
41. Joseph L. Schott, *No Left Turns* (New York, 1975), 204–205.
42. Joseph Dolan, in recorded interview by L. J. Hackman, July 18, 1970, 296, RFK Oral History Program.
43. Author's journal, December 5, 1963.
44. RFK, in recorded interview by Anthony Lewis, December 4, 1964, IV, 38–40, JFK Oral History Program.
45. RFK, in Martin interview, April 13, 1964, I, 33.
46. RFK, in Lewis interview, December 4, 1964, V, 9; Kennedy made similar statements in a memorandum dictated August 6, 1964, RFK Papers.
47. RFK, in Lewis interview, December 4, 1964, V, 19–20.
48. Ken W. Clawson, "Praises Mitchell as 'Very Human': FBI's Hoover Scores Ramsey Clark, RFK," *Washington Post*, November 17, 1970.
49. Ralph de Toledano, *J. Edgar Hoover* (New York: Manor reprint, 1974), 301–302.
50. Tom Wicker, *JFK and LBJ* (Baltimore: Penguin reprint, 1970), 196.
51. Author's journal, December 13, 1963.
52. Ibid.; Richard N. Goodwin, "The Structure Itself Must Change," *Rolling Stone*, June 6, 1974.
53. Author's journal, December 14, 1963.
54. Author to RFK, December 15, 1963, RFK Papers.
55. Author's journal, December 23, 1963.
56. Walter Sheridan, in recorded interview by Roberta Greene, May 1, 1970, 4, RFK Oral History Program.
57. Howard P. Jones, *Indonesia: The Impossible Dream* (New York, 1971), 289.
58. Roger Hilsman, *To Move a Nation* (Garden City, N.Y., 1967), 393.
59. Jones, *Indonesia*, 295–297.
60. Ibid., 300.
61. McGeorge Bundy, in recorded interview by William W. Moss, January 12, 1972, 23, RFK Oral History Program.
62. Michael Forrestal to RFK, January 9, 1964, RFK Papers.
63. "Background Guidance on Attorney General's Visit to the Far East," January 10, 1964, RFK Papers.
64. Forrestal, in interview by author, January 27, 1977.
65. D. L. Osborn to State Department, February 14, 1964, RFK Papers; *New York Times*, January 19, 1964; Edwin Guthman, *We Band of Brothers* (New York, 1971), 248–249.
66. Jones, *Indonesia*, 301.
67. Forrestal to William Bundy, May 8, 1964, RFK Papers.
68. Cindy Adams, *My Friend the Dictator* (Indianapolis, 1967), 78.
69. Forrestal, in recorded interview by Jean Stein, [n.d., 1968], 12, Stein Papers; Forrestal, in interview by author.
70. Tunku Abdul Rahman to RFK, January 30, 1964, RFK Papers.
71. Guthman, *We Band of Brothers*, 251; Theodore H. White, *The Making of the President, 1964* (New York, 1965), 261.
72. Murray Kempton, "Pure Irish: Robert F. Kennedy," *New Republic*, February 15, 1964.
73. RFK, in Martin interviews, March 1, 1964, II, 21, May 14, 1964, II, 17.

74. "Spectrum of Courses of Action with Respect to Cuba," February 21, 1964, RFK Papers.
75. RFK, in Martin interview, May 14, 1964, I, 34; Johnson, *Vantage Point,* 184–187.
76. RFK, in Martin interview, May 14, 1964, I, 34–37; author's journal, March 11, 1964.
77. Adam Walinsky, in recorded interview by Thomas Johnston, November 25, 1969, 2–3, RFK Oral History Program; Jean Stein and George Plimpton, eds., *American Journey* (New York, 1970), 277–278. Muste was in fact seventy-nine years old.
78. Katzenbach, in Hackman interview, 15–16.
79. Kempton, "Pure Irish."
80. Murray Kempton, in recorded interview by Jean Stein, March 28, 1970, 3, Stein Papers.
81. Angie Novello, in interview by author, December 11, 1975.
82. Kenneth O'Donnell, in recorded interview by L. J. Hackman, May 6, 1969, 37, RFK Oral History Program.
83. William Hundley, in interview by author, August 4, 1976.
84. "The Federal Government and Urban Poverty," conference sponsored by the Kennedy Library and Brandeis University, June 16–17, 1973, transcript, vol. 3, 48. This is an indispensable source for the evolution of the war against poverty. I am also grateful to several participants for their kindness in letting me see unpublished papers: to William B. Cannon for the relevant portion of his manuscript "The Dangerous Abuse of the Lower Class"; to Richard W. Boone for "Reflections on Citizen Participation and the Economic Opportunity Act," a paper delivered on May 7, 1970; to Leonard J. Duhl for "Some Origins of the Poverty Program," a paper of April 6, 1967. These accounts are necessary to supplement and correct the arresting version provided by D. P. Moynihan in *Maximum Feasible Misunderstanding* (New York, 1969).
85. Cannon, "Dangerous Abuse," ch. 6, 16.
86. Johnson, *Vantage Point,* 74–75.
87. Richard Blumenthal, "The Bureaucracy: Antipoverty and the Community Action Program," in *American Political Institutions and Public Policy,* ed. A. P. Sindler, (Boston, 1969), 145–146.
88. Cannon, "Dangerous Abuse," ch. 6, 35.
89. Moynihan, *Maximum Feasible Misunderstanding,* 82; Adam Yarmolinsky, "The Beginnings of OEO," in *On Fighting Poverty,* ed. J. L. Sundquist (New York, 1969), 36.
90. Kenneth B. Clark and Jeanette Hopkins, *A Relevant War against Poverty: A Study of Community Action Programs and Observable Social Change* (New York, 1969), 4.
91. "The Federal Government and Urban Poverty," recorded transcript of conference sponsored by the Kennedy Library and Brandeis University, vol. 5, 8.
92. Richard Boone, in recorded interview by Jean Stein, September 20, 1968, 6, Stein Papers; Boone, "Reflections," 7–8; Cannon, "Dangerous Abuse," VI–46.
93. J. C. Donovan, *The Politics of Poverty* (New York, 1967), 35. Emphasis added.
94. *Atlanta Constitution,* May 27, May 28, 1964; *Robert F. Kennedy: Apostle of Change,* ed. Douglas Ross (New York: Pocket Book reprint, 1968), 74.
95. John Doar and Dorothy Landsberg, "The Performance of the FBI in Investigating Violations of Federal Laws Protecting the Right to Vote — 1960–1967" (1971) in Senate Select Committee to Study Governmental Operations with respect to Intelligence Activities (hereafter cited as Church committee), *Hearings,* vol. 6, *Federal Bureau of Investigation,* 94 Cong., 1 Sess. (1975), 929.
96. Haywood Burns, "The Federal Government and Civil Rights," in *Southern Justice,* ed. Leon Friedman (New York, 1965), 235; Pat Watters and Reese Cleghorn, *Climbing Jacob's Ladder* (New York, 1967), 139.
97. *Newsweek,* November 30, 1964.
98. Martin Luther King, Jr., "Hammer of Civil Rights," *Nation,* March 9, 1964.
99. Sally Belfrage, *Freedom Summer* (New York, 1965), 15.
100. James Weschsler, "The FBI's Failure in the South," *Progressive,* December 1963.

101. RFK, in Lewis interview, December 4, 1964, V, 21.
102. RFK to LBJ, June 5, 1964, Johnson Papers.
103. Church committee, *Final Report*, bk. III, *Supplementary Detailed Staff Reports on Intelligence Activities and the Rights of Americans*, 94 Cong., 2 Sess. (1976), 7–8.
104. Ibid., 65–66.
105. Hoover to McGeorge Bundy, July 25, 1961, in Church committee, *Final Report*, bk. II, *Intelligence Activities and the Rights of Americans*, 94 Cong., 2 Sess. (1976), 282.
106. Richard D. Cotter, "Notes toward a Definition of National Security," *Washington Monthly*, December 1975.
107. Sullivan, in interview by author, July 26, 1976; Ovid Demaris, *The Director* (New York, 1975), 326.
108. Gary Thomas Rowe, Jr., *My Undercover Years with the Ku Klux Klan* (New York, 1976), 53–54.
109. "Interview with J. Edgar Hoover," *U.S. News & World Report*, December 21, 1964.
110. Hoover to Katzenbach, September 2, 1965, Church committee, *Hearings*, vol. 6, 513–514; see also memorandum of December 19, 1967, 518–527, and *New York Times*, November 12, 1977. Deletions in the Church committee version of the September memorandum are supplied in Jack Nelson, "Will the Real KKK Please Stand Up?" *New York Post*, August 16, 1975, and in J. J. Berman and M. H. Halperin, eds., *The Abuses of the Intelligence Agencies* (Washington, 1975), 20. See also Church committee, *Final Report*, bk. III, 251–252; Sullivan, in interview by author; Harry Overstreet and Bonaro Overstreet, *The FBI in Our Open Society* (New York, 1969), 303.
111. For a listing, see Doar and Landsberg, "Performance of the FBI," in Church committee, *Hearings*, vol. 6, 938–940.
112. Demaris, *Director*, 327; Sullivan to author, October 1, 1977.
113. For the case against informers, see Frank Donner, "Political Informers," in *Investigating the FBI*, ed. Pat Watters and Stephen Gillers (Garden City, N.Y., 1973), 338–365, and Vern Countryman on 367–368; also Robert McAfee Brown, in *Conspiracy*, ed. John C. Raines (New York, 1975), intro.
114. Bernard Fensterwald thus denigrated Sam Baron. For Fensterwald's exchange with Kennedy, see Senate Subcommittee on Administrative Practice and Procedures, *Hearings: On Invasions of Privacy (Governmental Agencies)*, 89 Cong., 1 Sess., (March 3, 1965), 274.
115. *Dennis* v. *United States*, 183 F. 2nd 201, 224 (1950).
116. *Hoffa* v. *United States*, 385 U.S. 293 (1966).
117. Church committee, *Final Report*, bk. III, 243–244.
118. Ibid., 240.
119. Rowe, *Undercover Years*, 175.
120. Burke Marshall, "The Issues on Trial," in Raines, *Conspiracy*, 157–158.
121. Kearns, *Johnson*, 191.
122. RFK, in Lewis interview, December 22, 1964, 24, 26–27.
123. Kearns, *Johnson*, 183.
124. Victor Reuther, *The Brothers Reuther* (Boston, 1976), 430.
125. Walter Fauntroy, in recorded interview by Jean Stein, November 11, 1969, 13–14, Stein Papers.

28. The Vice Presidency *(pages 646–665)*

1. I am indebted to the Johnson Library for searching its records to the above effect.
2. Liz Carpenter, *Ruffles and Flourishes* (New York: Pocket Book reprint, 1971), 265–266.
3. Pierre Salinger, in recorded interview by L. J. Hackman, May 26, 1969, 16–17.
4. Doris Kearns, *Lyndon Johnson and the American Dream* (New York, 1976), 200.

5. Eric Goldman, *The Tragedy of Lyndon Johnson* (New York, 1969), 78–79.
6. Kenneth O'Donnell, in recorded interview by L. J. Hackman, April 3, 1969, 62–63, RFK Oral History Program.
7. Murray Kempton, "Pure Irish: Robert F. Kennedy," *New Republic*, February 15, 1964.
8. Ibid.
9. Walter Heller to LBJ, January 9, 1964, RFK Papers.
10. RFK, in recorded interview by John Bartlow Martin, May 14, 1964, I, 33, JFK Oral History Program.
11. Richard Goodwin, notes, March 1964, Goodwin Papers.
12. RFK, in Martin interview, I, 40.
13. McGeorge Bundy, in recorded interview by William W. Moss, January 12, 1972, 25, RFK Oral History Program.
14. Goldman, *Tragedy of Johnson*, 19.
15. RFK, in Martin interview, II, 4.
16. Ibid., I, 33.
17. Author's journal, March 19, 1964; author to RFK, March 20, 1964, RFK Papers.
18. Author's journal, March 25, 1964.
19. Tom Wicker, *JFK and LBJ* (Baltimore: Pelican reprint, 1970), 230.
20. Lyndon B. Johnson, *The Vantage Point* (New York, 1971), 92–94.
21. Kenneth P. O'Donnell and David F. Powers, *"Johnny, We Hardly Knew Ye"* (Boston, 1972), 391.
22. Sam Houston Johnson, *My Brother Lyndon* (New York, 1969), 159.
23. Joseph Dolan to RFK, December 3, 1963, RFK Papers. The Michigan committeewoman was Mildred Jeffrey.
24. Author's journal, December 5, 1963.
25. Kennedy sent the letter over to his brother Edward with a scribble: "Will you talk to him?" Peter Crotty to RFK, February 13, 1964, RFK to EMK, February 21, 1964, RFK Papers.
26. William Dunfey, in recorded interview by L. J. Hackman, December 15, 1971, 68–72, RFK Oral History Program.
27. Edwin Guthman, *We Band of Brothers* (New York, 1971), 254.
28. John Seigenthaler, in recorded interview by L. J. Hackman, June 5, 1970, 36, 45–46, RFK Oral History Program.
29. Joseph Dolan, in recorded interview by L. J. Hackman, April 10, 1970, 57, RFK Oral History Program.
30. Dunfey, in Hackman interview, 66–67.
31. O'Donnell, in Hackman interview, May 6, 1969, 71; Seigenthaler, in Hackman interview, 36; RFK, in Martin interview, I, 31–32.
32. Guthman, *We Band of Brothers*, 254.
33. Ibid., 254–256.
34. Ibid., 256–257, 270.
35. Author's journal, January 22, 1964.
36. George H. Gallup, ed., *The Gallup Poll: Public Opinion, 1935–1971* (New York, 1972), vol. 3, 1874–1875.
37. See Arthur M. Schlesinger, Jr., "On the Presidential Succession," *Political Science Quarterly*, Fall 1974.
38. Quoted in Ovid Demaris, *The Director* (New York, 1975), 170.
39. Charles Spalding, in recorded interview by L. J. Hackman, March 22, 1969, 68, RFK Oral History Program.
40. RFK, in Martin interview, II, 8–17.
41. Burton Hersh, *The Education of Edward Kennedy* (New York, 1972), 200–202.
42. Walter Sheridan, in recorded interview by Roberta Greene, May 1, 1970, 12, RFK Oral History Program.
43. Hersh, *Education*, 202.
44. William V. Shannon, "Said Robert Kennedy, 'Maybe We're All Doomed Anyway,'" *New York Times Magazine*, June 16, 1968.

45. Susan Wilson, "A Guide to Travelling with the Robert F. Kennedys," ms. in the author's possession, 4–5.
46. RFK, press conference, Berlin, June 26, 1964, transcript in RFK Papers.
47. John Moors Cabot to assistant secretary of state for European affairs, June 24, 1964, RFK Papers.
48. Guthman, *We Band of Brothers,* 275.
49. John Moors Cabot, in recorded interview by W. W. Moss, January 27, 1971, 22, JFK Oral History Program; William V. Shannon, *The Heir Apparent* (New York, 1967), 12.
50. Cabot, in Moss interview, 22.
51. Joseph Kraft, in recorded interview by Roberta Greene, March 7, 1970, 18, RFK Oral History Program; Shannon, *Heir Apparent,* 12.
52. Cabot, in Moss interview, 22.
53. Ibid., 23; RFK notes on conversation with Jozef Winiewicz, June 29, 1964, RFK Papers.
54. RFK, notes on conversation with Stefan Cardinal Wyszynski, June 30, 1964, RFK Papers.
55. Translation of article by Daniel Passent in RFK Papers.
56. Cabot, in Moss interview, 21.
57. Cabot to Harriman, July 2, 1964, RFK Papers.
58. Lee Stull to Harriman, July 6, 1964, RFK Papers.
59. Guthman, *We Band of Brothers,* 278.
60. Kearns, *Johnson,* 199–200.
61. Johnson, *My Brother Lyndon,* 161.
62. O'Donnell, in Hackman interview, April 3, 1969, 32–33; O'Donnell and Powers, *"Johnny,"* 393–394.
63. Benjamin Bradlee, in recorded interview by Jean Stein, October 18, 1968, 1–2, Stein Papers; Margaret Laing, *The Next Kennedy* (New York, 1968), 7.
64. *Newsweek,* July 6, 1964; Benjamin C. Bradlee, *Conversations with Kennedy* (New York, 1975), 24.
65. Author's journal, July 21, 1964.
66. Hubert Humphrey, *The Education of a Public Man* (Garden City, N.Y., 1976), 297–298.
67. Author's journal, June 9, July 23, 1964.
68. Jack Tarver to LBJ, July 26, 1964, Johnson Papers.
69. Johnson, *My Brother Lyndon,* 165.
70. Author's journal, July 23, 1964.
71. O'Donnell and Powers, *"Johnny,"* 396–397.
72. Guthman, *We Band of Brothers,* 280.
73. O'Donnell, in Hackman interview, May 6, 1969, 18.
74. Johnson, *Vantage Point,* 100, 576–577.
75. RFK, memorandum dictated August 4–6, 1964, RFK Papers.
76. Jack Valenti, *A Very Human President* (New York, 1975), 148.
77. Guthman, *We Band of Brothers,* 282.
78. O'Donnell and Powers, *"Johnny,"* 397; see Lawrence O'Brien, *No Final Victories* (Garden City, N.Y., 1974), 175.
79. RFK, memorandum.
80. Bundy, in Moss interview, 27.
81. RFK to Bundy, August 26, 1964, RFK Papers.
82. RFK, memorandum.
83. Richard Harwood and Haynes Johnson, *Lyndon* (New York, 1973), 73–74; David Wise, *The Politics of Lying* (New York, 1973), 294; Johnson, *My Brother Lyndon,* 167; Theodore H. White, *The Making of the President, 1964* (New York, 1965), 263–265; Guthman, *We Band of Brothers,* 281; RFK, memorandum.
84. RFK, memorandum.
85. Author's journal, August 2, 1964.
86. Clark Clifford, in recorded interview by L. J. Hackman, February 4, 1975, 39–41, JFK Oral History Program.

87. For details see: Senate Select Committee to Study Governmental Operations with respect to Intelligence Activities (hereafter cited as Church committee), *Final Report*, bk. III, *Supplementary Detailed Staff Reports on Intelligence Activities and the Rights of Americans*, 94 Cong., 2 Sess. (1976), 346–347; Church committee, *Hearings*, vol. 6, *Federal Bureau of Investigation*, 94 Cong., 1 Sess. (1975), 495–496; David Wise, *The American Police State* (New York, 1976), 287–288; Demaris, *Director*, 286.

88. Testimony of Leo T. Clark, *Washington Post*, January 26, 1975.

89. Sanford J. Ungar, *FBI* (Boston, 1975), 289.

90. William Barry, in recorded interview by Roberta Greene, October 22, 1969, 75, RFK Oral History Program.

91. Church committee, *Hearings*, vol. 6, 495–496, 509–510.

92. Seigenthaler, in Hackman interview, July 1, 1970, 95.

93. Humphrey, *Education*, 301–302.

94. O'Donnell and Powers, *"Johnny,"* 400.

95. Seigenthaler, in Hackman interview, July 1, 1970, 103–110.

96. Ibid., 113.

29. To the Senate *(pages 666–688)*

1. RFK to Peter Lisagor, August 17, 1964, RFK Papers.

2. John F. Kraft to Stephen Smith, August 20, 1964, Smith Papers.

3. Author's journal, August 2, 1964.

4. Author to RFK, August 5, 1964, RFK Papers.

5. Jack English, in recorded interview by Roberta Greene, November 25, 1969, 13, RFK Oral History Program.

6. Liz Carpenter to Johnson, memorandum reporting a telephone message from Harriman, August 9, 1964, Johnson Papers.

7. Stevenson to Marietta Tree, August 10, 1964, in John Bartlow Martin, *Adlai Stevenson and the World* (Garden City, N.Y., 1977), 812.

8. Stevenson to Mary Lasker, August 14, 1964, in Martin, *Stevenson*, 813.

9. Jeffrey Potter, *Men, Money and Magic* (New York, 1976), 290.

10. Gerald Gardner, *Robert Kennedy in New York* (New York, 1965), 82.

11. Author to Stephen Smith, September 15, 1964, Smith Papers.

12. Francis Biddle to the editor, *New York Times*, August 24, 1964, copy in RFK Papers.

13. See advertisements in *New York Times*, November 1, 2, 1964.

14. *New York Times*, October 27, 1964.

15. Justin Feldman, in recorded interview by Roberta Greene, November 26, 1969, 97, RFK Oral History Program.

16. Thomas Johnston, in recorded interview by L. J. Hackman, October 27, 1969, 198–199, RFK Oral History Program.

17. Guthman, *We Band of Brothers* (New York, 1971), 295; Gardner, *Kennedy*, 74–75.

18. Gardner, *Kennedy*, 44, 63.

19. Guthman, *We Band of Brothers*, 294.

20. Feldman, in Greene interview, 101.

21. Paul Corbin, in recorded interview by Jean Stein, n.d. [1968], 4, Stein Papers.

22. Douglas to RFK, October 8, 1964, Smith Papers.

23. John Douglas, in recorded interview by L. J. Hackman, June 16, 1969, 24–26, RFK Oral History Program.

24. Associated Press dispatch from Rochester, October 6, 1964.

25. Kraft to Smith, October 6, 8, 1964, RFK Papers.

26. *New York World-Telegram*, October 8, 1964.

27. Jean Stein and George Plimpton, eds., *American Journey* (New York, 1970), 180; Guthman, *We Band of Brothers*, 307–308.

28. Charles Evers, "For Robert Kennedy," *New York Post*, October 23, 1964.

29. *New York Times*, October 22, 1964.

30. Douglas Robinson, "Kennedy in Trouble with Italian-Americans," *New York Times*, October 13, 1964.
31. William Haddad to Stephen Smith, September 26, 1964, Smith Papers.
32. Neil Sheehan, "Keating Reported Gaining among Democratic and Liberal Jews," *New York Times*, October 3, 1964.
33. David Halberstam, "Keating Sees Cartel 'Deal,'" *New York Times*, September 21, 1964; Homer Bigart, "Kennedy Assails Keating Tactics," *New York Times*, September 23, 1964.
34. *Congressional Record*, March 4, 1963, 3333-3334. See also Guthman, *We Band of Brothers*, 101-102, 300-303; Victor Navasky, *Kennedy Justice* (New York, 1971), 349-352.
35. William Orrick, in recorded interview by L. J. Hackman, April 13, 1970, 128, RFK Oral History Program.
36. *New York Times*, September 23, 1964.
37. [James Stevenson], "Campaigning," in "The Talk of the Town," *New Yorker*, October 24, 1964.
38. Guthman, *We Band of Brothers*, 301.
39. *Newsweek*, November 16, 1964.
40. Mailer wrote his endorsement "A Vote for Bobby K." for the *Village Voice*. It is reprinted in Norman Mailer, *The Idol and the Octopus* (New York: Dell reprint, 1968), 242-245. The references to John Kennedy are from the same book, 112, 173.
41. Rita Dallas, *The Kennedy Case* (New York: Popular Library reprint, 1973), 288-289.
42. Gardner, *Kennedy*, 5, 9.
43. For accounts of RFK and television in the 1964 campaign, see Terry Smith, "Bobby's Image," *Esquire*, April 1965; George Lois, *George, Be Careful* (New York, 1972), 107-112; Fred Papert, in recorded interview by Roberta Greene, March 21, 1973, esp. 10-21, RFK Oral History Program; William vanden Heuvel and Milton Gwirtzman, *On His Own: RFK, 1964-1968* (Garden City, N.Y., 1970), 44; transcript of Columbia University appearance, RFK Papers.
44. *New York Times*, October 16, 1964.
45. Richard Wade, in recorded interview by Roberta Greene, December 13, 1973, 41-42, RFK Oral History Program.
46. Author's journal, October 30, 1966.
47. *New York Daily News*, October 19, 1964.
48. Kraft to Smith, October 28, 29, 30, 1964, Smith Papers.
49. *New York Times*, October 19, 1964.
50. Gardner, *Kennedy*, 93.
51. *New York Times*, October 28, 30, 1964; vanden Heuvel and Gwirtzman, *On His Own*, 49-52. For the FCPC executive director's version, see Bruce Felknor, *Dirty Politics* (New York, 1966), 175-196.
52. Harry Golden, "The Bobby Twins Revisited," *Esquire*, June 1965.
53. Murray Kempton, "Another Empty Chair," *New York World-Telegram*, October 28, 1964; Frank Borsky, "I Ran the Keating Obstacle Course," *New York Daily News*, October 28, 1964; Homer Bigart, "Keating vs. Kennedy: A Near Debate," *New York Times*, October 28, 1964; vanden Heuvel and Gwirtzman, *On His Own*, 52-54; Guthman, *We Band of Brothers*, 308-311.
54. Nancy Dickerson, *Among Those Present* (New York: Ballantine reprint, 1977), 149.
55. Stein and Plimpton, *American Journey*, 182.
56. Wes Barthelmes, in recorded interview by Roberta Greene, May 20, 1969, 6, RFK Oral History Program.
57. Ibid., 3.
58. Frank Mankiewicz, in recorded interview by L. J. Hackman, August 12, 1969, 42, RFK Oral History Program.
59. Mankiewicz, in Hackman interview, July 10, 1969, 49; Joseph Dolan, in recorded interview by L. J. Hackman, April 11, 1970, 115, RFK Oral History Program; Barthelmes, in Greene interview, May 20, 1969, 11; Peter Edelman, in recorded

interview by L. J. Hackman, July 15, 1969, 117, RFK Oral History Program.
60. Dolan, in Hackman interview, April 10, 1970, 11, 17.
61. Adam Walinsky, in recorded interview by Thomas Johnston, November 29, 1969, 34, RFK Oral History Program.
62. Walinsky, in Johnston interview, November 30, 1969, II, 2.
63. Barthelmes, in Greene interview, May 20, 1969, 7; June 5, 1969, 170.
64. Walinsky, in Johnston interview, November 29, 1969, 32; November 30, 1969 (2), 16, 25.
65. Martin Arnold, *New York Times,* June 7, 1968.
66. Walinsky, in Johnston interview, November 30, 1969, 54.
67. Remarks to Western New York Publishers Association, Painted Post, N.Y., October 9, 1965, RFK Papers.
68. Author's journal, January 19, 1965.
69. Nick Thimmesch and William Johnson, *Robert Kennedy at 40* (New York, 1965), 244; Walinsky, in Johnston interview, November 30, 1969, 53-57.
70. Barthelmes, in Greene interview, 14.
71. Edelman, in Hackman interview, January 3, 1970, 28.
72. Jacob Javits, in recorded interview by William vanden Heuvel, June 19, 1970, 11, RFK Oral History Program.
73. Javits, in recorded interview by Roberta Greene, June 7, 1973, 7, RFK Oral History Program.
74. *New York Times,* January 4, 1965; William V. Shannon, *The Heir Apparent* (New York, 1967), 75.
75. vanden Heuvel and Gwirtzman, *On His Own,* 64.
76. Stein and Plimpton, *American Journey,* 182-183.
77. vanden Heuvel and Gwirtzman, *On His Own,* 64.
78. Dolan, in Hackman interview, July 19, 1970, 414.
79. N.d., RFK Papers.
80. Fred Harris, *Potomac Fever* (New York, 1977), 119, 124, 126.
81. George McGovern, in recorded interview by L. J. Hackman, July 16, 1970, 19-20, 34.
82. Confidential source.
83. Fred Harris, in recorded interview by Roberta Greene, July 29, 1970, 5, RFK Oral History Program.
84. Jack Newfield, *Robert Kennedy* (New York, 1969), 58.
85. RFK to Byrd, August 8, 1965, RFK Papers.
86. RFK to Murphy, August 22, 1966, RFK Papers.
87. Shannon, *Heir Apparent,* 78-79.
88. Thus Meg Greenfield in the pro-Johnson *Reporter,* December 15, 1966, and elsewhere.
89. Harris, in Greene interview, 2.
90. Joseph Tydings, in recorded interview by Jean Stein, December 1969, 4-5, Stein Papers.
91. David Burke, in recorded interview by Jean Stein, October 17, 1968, 3-4, Stein Papers.
92. Javits, in vanden Heuvel interview, 14-15.
93. Richard Reeves, "Kennedy: 2 Years after His Election," *New York Times,* November 14, 1966.
94. RFK to LBJ, September 2, 1964, RFK Papers.
95. Burton Hersh, *The Education of Edward Kennedy* (New York, 1972), 233.
96. Milton Gwirtzman, in recorded interview by Roberta Greene, February 10, 1972, 28-29, RFK Oral History Program.
97. William C. Sullivan, in interview by author, July 26, 1976.
98. Harold W. Chase, *Federal Judges: The Appointing Process* (Minneapolis, 1972), 174.
99. Joseph Tydings, in recorded interview by Roberta Greene, September 29, 1971, 38-39, RFK Oral History Program.
100. Gwirtzman, in Greene interview, 43-44.
101. All letters in Johnson Papers.

102. Dolan, in Hackman interview, 33-34.
103. Clark to Watson, May 31, 1966, Johnson Papers. Actually Kennedy had considered Mansfield for judicial appointment when he was Attorney General and endorsed him now; Dolan, in Hackman interview, April 10, 1970, 37.
104. Javits, in vanden Heuvel interview, 12, 14.
105. Reeves, "Kennedy."
106. Roger Wilkins, in recorded interview by Jean Stein, n.d. [1968], 10-11, Stein Papers.
107. Harris, *Potomac Fever,* 105.
108. Robert S. Allen and Paul Scott in the *Birmingham News,* June 10, 1966.
109. Walter Mondale, in recorded interview by Roberta Greene, May 17, 1973, 19, RFK Oral History Program.
110. Harris, in Greene interview, 3.
111. Kenneth O'Donnell, in recorded interview by L. J. Hackman, April 3, 1969, 3, RFK Oral History Program.
112. McGovern, in Hackman interview, 32.
113. Barthelmes, in Greene interview, 27.
114. Mankiewicz, in Hackman interview, August 12, 1969, 5.
115. Donald Riegle with Trevor Armbrister, *O Congress* (New York: Popular Library reprint, 1972), 144.
116. Author's journal, January 19, 1965.
117. RFK to Lewis, n.d. [summer 1965], RFK Papers.
118. Margaret Laing, *The Next Kennedy* (New York, 1968), 31.

30. The Foreign Policy Breach: Latin America *(pages 689-700)*

1. William D. Rogers, *The Twilight Struggle: The Alliance for Progress and the Politics of Development in Latin America* (New York, 1967), 226.
2. Jerome Levinson and Juan De Onis, *The Alliance That Lost Its Way* (Chicago, 1970), 72-73.
3. Teodoro Moscoso, lecture 3 at Mills College, February 1965 (mimeographed), 12.
4. Quoted by Richard Goodwin, "Our Stake in a Big Awakening," *Life,* April 14, 1967.
5. Juscelino Kubitschek, "L'Alliance pour le progrès," *Historia,* hors série 33 (1973), 151.
6. Quoted by Evelyn Lincoln, *Kennedy and Johnson* (New York, 1968), 188.
7. Eric Goldman, *The Tragedy of Lyndon Johnson* (New York, 1969), 382.
8. Author's journal, April 30-May 2, 1965.
9. *Congressional Record,* May 6, 1965, 9761-9762.
10. Author's journal, July 16, 1965.
11. *Congressional Record,* June 23, 1965, 14566-14568.
12. Richard Goodwin, in interview by author, January 22, 1974. Since he had already cut another significant part of the draft because it had been leaked to Reston of the *Times,* the UN ended up with thin fare.
13. Valenti to LBJ, September 14, 1965, Johnson Papers.
14. Adam Walinsky, in recorded interview by L. J. Hackman, May 22, 1972, 21-25, RFK Oral History Program; Frank Mankiewicz, in recorded interview by Jean Stein, September 21, 1968, 4-8, Stein Papers; Mankiewicz, in recorded interview by L. J. Hackman, June 26, 1969, 5-8; Frank Mankiewicz and Tom Braden, "U.S. on Wrong Side in Dispute with Peru," column released March 13, 1969.
15. John Nolan, in recorded interview by Jean Stein, August 8, 1968, 12-14, Stein Papers.
16. William vanden Heuvel, "Notes on RFK South American Trip (2)," 8-12, vanden Heuvel Papers; *New York Times,* November 14, 1965.
17. Vanden Heuvel, "Notes," 12-17; William vanden Heuvel and Milton Gwirtzman, *On His Own* (Garden City, N.Y., 1970), 166-167.
18. Walinsky, in Hackman interview, 34.

19. Vanden Heuvel and Gwirtzman, *On His Own* 166; Peter Collier and David Horowitz, *The Rockefellers: An American Dynasty* (New York, 1976), 417.
20. Andrew J. Glass, "The Compulsive Candidate," *Saturday Evening Post*, April 23, 1966.
21. Transcript, "Senator Robert F. Kennedy . . . Conversation with Students in Hotel in Concepción," vanden Heuvel Papers; Glass, "Compulsive Candidate."
22. Martin Arnold, in recorded interview by Jean Stein, December 5, 1969, 9, Stein Papers. See also his account in *New York Times*, November 17, 1965.
23. John Seigenthaler, in recorded interview by Jean Stein, May 15, 1970, 27.
24. Vanden Heuvel and Gwirtzman, *On His Own*, 167–168; Glass, "Compulsive Candidate."
25. Arnold, in Stein interview, 16.
26. Margaret Laing, *The Next Kennedy* (New York, 1968), 291–292.
27. *New York Times*, November 19, 1965.
28. Robert Hopkins to author, December 1, 1965.
29. Arnold, in Stein interview, 21; William vanden Heuvel, in recorded interview by Jean Stein, February 28, 1970, 9, Stein Papers; Mildred Sage, in interview by author, June 9, 1977.
30. Angie Novello, in recorded interview by Jean Stein, September 27, 1968, 8, Stein Papers.
31. *New York Times*, November 23, 1965.
32. Glass, "Compulsive Candidate."
33. Ibid.; vanden Heuvel and Gwirtzman, *On His Own*, 175.
34. Sage, in interview by author.
35. Patricia Kennedy Lawford, ed., *That Shining Hour* (n.p., 1969), 67.
36. Sage, in interview by author.
37. Glass, "Compulsive Candidate."
38. "Impressions of RFK South American Tour," Johnson Papers. The anonymous author was in Lima when Kennedy arrived.
39. Glass, "Compulsive Candidate."
40. Jean Stein and George Plimpton, eds., *American Journey* (New York, 1970), 153.
41. Lincoln Gordon, in interview by author, October 14, 1974.
42. Paulo de Castro, "Robert Kennedy," *Correio da Manha* (Rio de Janeiro), November 23, 1965.
43. Walinsky, in Hackman interview, 31, 47–48.
44. *Meet the Press*, December 5, 1965.
45. Theodore C. Sorensen, *The Kennedy Legacy* (New York, 1969), 174.
46. Moyers to RFK, May 30, 1966, RFK Papers.
47. *Congressional Record*, May 9, 1966, 9609–9620, May 10, 1966, 9705–9716.

31. Vietnam Legacy *(pages 701–723)*

1. RFK, in recorded interview by John Bartlow Martin, April 13, 1964, I, 22, JFK Oral History Program.
2. Jann Wenner, in interview with Daniel Ellsberg, *Rolling Stone*, December 6, 1973.
3. JFK before the American Friends of Vietnam, Washington, June 1, 1956, *Vital Speeches*, August 1, 1956.
4. U.S. Department of Defense, *The Pentagon Papers*, Senator Gravel Edition Boston, 1971), vol. 1, 626.
5. Clark Clifford, "Memorandum on Conference between President Eisenhower and President-elect Kennedy and Their Chief Advisers on January 19, 1961," 3. The emphasis is in the original.
6. Macmillan to Eisenhower, April 9, 1961, copy transmitted by the British ambassador to JFK, April 9, 1961, JFK Papers.
7. RFK, memorandum dictated June 1, 1961, 3, RFK Papers.
8. Author's journal, May 14, 1962.
9. RFK, memorandum, 3.

10. *New York Times,* April 9, 1961.
11. *Pentagon Papers,* vol. 2, 8, 49.
12. Roswell Gilpatric, in recorded interview by D. J. O'Brien, May 5, 1970, 19, JFK Oral History Program.
13. RFK, memorandum dictated August 1, 1961, 1, RFK Papers.
14. Alexis Johnson, in recorded interview by William Brubeck, n.d. [1964], 33–34, JFK Oral History Program.
15. Maxwell Taylor, in recorded interview by L. J. Hackman, November 13, 1969, 47.
16. Arthur Krock, *In the Nation: 1932–1966* (New York, 1966), 324–325, 447.
17. Maxwell Taylor, *Swords and Plowshares* (Washington, 1972), 219, 225–226.
18. Taylor, in Hackman interview, 43.
19. *Pentagon Papers,* vol. 2, 92–93, 98, 108.
20. Taylor, in Hackman interview, 47.
21. Author's journal, November 13, 1961.
22. J. K. Galbraith, in interview by author, June 23, 1977.
23. William Bundy was the deputy assistant secretary of defense for international security affairs. I am indebted to Mr. Bundy for letting me read his valuable unpublished manuscript on United States policy in East Asia during the Kennedy-Johnson years. The quotation is taken from ch. 4.
24. George Ball, in interview by author, June 15, 1977.
25. *New York Times,* December 12, 1961.
26. Ibid., December 20, 1961.
27. Ibid., October 27, 1961.
28. Directorate for Information Operations, Department of Defense, March 22, 1972. See chart in R. H. Fifield, *Americans in Southeast Asia: The Roots of Commitment* (New York, 1973), 276.
29. Roger Hilsman, *To Move a Nation* (Garden City, N.Y., 1967), 426.
30. Roger Hilsman and Michael Forrestal, "A Report on South Vietnam," n.d. [January 1963], RFK Papers.
31. *Newsweek,* May 25, 1970.
32. *Pentagon Papers,* vol. 2, 123.
33. Robert Shaplen, "The Cult of Diem," *New York Times Magazine,* May 14, 1972.
34. Edward G. Lansdale, in interview by author, December 30, 1976.
35. Hilsman, *To Move a Nation,* 419, 439.
36. Michael Forrestal to Robert W. Komer, June 2, 1971, copy in Schlesinger Papers. This instinct was clearly expressed in General Lemnitzer's memorandum on "Counterinsurgency Operations in South Vietnam" to General Taylor, October 12, 1961 (*Pentagon Papers,* vol. 2, 650–651).
37. Roger Hilsman, in recorded interview by D. J. O'Brien, August 14, 1970, 20, RFK Oral History Program.
38. Charles Maechling, Jr., review of *Anatomy of Error* by Henry Brandon, *Foreign Service Journal,* March 1970.
39. Roswell Gilpatric, in interview by D. J. O'Brien, August 12, 1970, 1, RFK Oral History Program.
40. John F. Kennedy, *Public Papers . . . 1962* (Washington, 1963), 137, 228.
41. *Pentagon Papers,* vol. 2, 670–671.
42. Ball to McGeorge Bundy, May 1, 1962, Bundy to JFK, May 1, 1962, JFK Papers.
43. *Pentagon Papers,* vol. 2, 175–181.
44. Gilpatric, in O'Brien interview, August 12, 1970, 1; Ellsberg, in *Rolling Stone* interview.
45. *Viet Nam and Southeast Asia: Report of Senator Mike Mansfield, Senator J. Caleb Boggs, Senator Claiborne Pell, Senator Benjamin Smith to the Committee on Foreign Relations,* 88 Cong., 1 Sess. (February 24, 1963), 8.
46. Kenneth O'Donnell and David F. Powers, *"Johnny, We Hardly Knew Ye"* (Boston, 1972).
47. Henry Brandon, *Anatomy of Error* (London, 1970), 30.

48. John F. Kennedy, *Public Papers . . . 1963* (Washington, 1964), 244, 569, 652.
49. Louis Harris, *The Anguish of Change* (New York, 1973), 54.
50. Chalmers M. Roberts, *First Rough Draft* (New York, 1973), 195–196. Roberts was the very well informed diplomatic correspondent of the *Washington Post*.
51. O'Donnell and Powers, *"Johnny,"* 16.
52. *Washington Post*, August 3, 1970; Jack Anderson, "The Roots of Our Vietnam Involvement," *Washington Post*, May 4, 1975.
53. JFK, *Public Papers . . . 1963*, 652.
54. RFK, in Martin interview, April 30, 1964, II, 34.
55. *New York Times*, February 19, 1962.
56. Forrestal to RFK, November 7, 1962, RFK Papers.
57. RFK, in Martin interview, March 1, 1964, II, 13, 16–17.
58. RFK, in recorded interview by author, February 27, 1965, 8, JFK Oral History Program.
59. National Security Council meeting, August 28, 1963, RFK Papers.
60. Gilpatric, in O'Brien interview, May 5, 1970, 31.
61. RFK, in Martin interview, April 30, 1964, II, 34.
62. Charles Bartlett, in recorded interview by Jean Stein, January 9, 1970, 20, Stein Papers.
63. Michael Forrestal, in recorded interview by Jean Stein, n.d. [1968], 6, Stein Papers.
64. Hilsman, *To Move a Nation*, 501.
65. *Pentagon Papers*, vol. 2, 738.
66. Lodge to Rusk, September 11, 1963, RFK Papers.
67. RFK, in interview by author, 25–26; RFK, in Martin interviews, April 13, 1964, I, 62, April 30, 1964, II, 38.
68. See, e.g., the cable of September 17, 1963, *Pentagon Papers*, vol. 2, 743–745.
69. RFK, in Martin interview, April 30, 1964, II, 39.
70. Ibid., III, 3.
71. Brandon, *Anatomy of Error*, 30.
72. Bundy manuscript, ch. 9.
73. *Pentagon Papers*, vol. 2, 756. Emphasis added.
74. Taylor, *Swords and Plowshares*, 296–297.
75. Michael Forrestal, in interview by author, July 13, 1977. See also Averell Harriman, in recorded interview by author, June 6, 1965, 83–85, JFK Oral History Program.
76. O'Donnell and Powers, *"Johnny,"* 17.
77. JFK, *Public Papers . . . 1963*, 846.
78. *Pentagon Papers*, vol. 2, 170. See also the valuable analysis by Peter Dale Scott, "Vietnamization and the Drama of the Pentagon Papers" *Pentagon Papers* (Boston, 1972), vol. 5, esp. 224.
79. O'Donnell and Powers, *"Johnny,"* 18.
80. JFK, *Public Papers . . . 1963*, 421.
81. Hilsman, in O'Brien interview, 21; letter to *New York Times*, August 8, 1970.
82. Ellsberg, *Rolling Stone* interview; see also William vanden Heuvel and Milton Gwirtzman, *On His Own* (Garden City, N.Y., 1970), 243.
83. Shaplen, "Cult of Diem."
84. Bui Kien Thanh, "Mandarins of Vietnam," *International History Magazine*, January 1974.
85. Michael Forrestal, in interview by author.
86. Mieczyslaw Maneli, *War of the Vanquished* (New York, 1971), 146.
87. Maneli, *War*, 121–122; Maneli, "Vietnam '63 and Now," *New York Times*, January 27, 1975; Maneli, in interview by author, June 24, 1977.
88. Maneli, "Vietnam '63."
89. Maneli, *War*, 127, 134–135.
90. Ibid., 141–142.
91. Ibid., 141.

92. Maneli, in interview by author.
93. Bundy manuscript, ch. 9.
94. Arthur M. Schlesinger, Jr., *A Thousand Days* (Boston, 1965), 874.
95. JFK, *Public Papers . . . 1963*, 652.
96. The visitor was George T. Altman. See Altman's letter in the *Nation*, June 21, 1975.
97. *New York Times*, November 3, 1963.
98. Schlesinger, *Thousand Days*, 871.
99. National Security Council meeting, 4:20 P.M., October 29, 1963, RFK Papers.
100. *Pentagon Papers*, vol. 2, 789, 792.
101. Henry Cabot Lodge, *The Storm Has Many Eyes* (New York, 1973), 210.
102. Taylor, *Swords and Plowshares*, 301.
103. Schlesinger, *Thousand Days*, 997-998.
104. Lodge to JFK, November 6, 1963, RFK Papers.
105. Brandon, *Anatomy of Error*, 30.
106. Forrestal on an NBC show, in Roberts, *First Rough Draft*, 221.
107. *Boston Globe*, June 24, 1973.
108. Pierre Salinger, *Je suis un Américain* (Paris, 1975), 239.
109. Fifield, *Americans in Southeast Asia*, 276.
110. James M. Gavin, "We Can Get Out of Vietnam," *Saturday Evening Post*, February 24, 1968.
111. Mieczyslaw Maneli, "Encounters with John F. Kennedy and Discussions on 'Democratic Socialism,' Polish Policy, and Vietnam" (unpublished paper), 24.

32. The Breach Widens: Vietnam *(pages 725-742)*

1. LBJ to JFK, May 23, 1961, JFK Papers.
2. U.S. Department of Defense, *The Pentagon Papers*, Senator Gravel edition (Boston, 1971), vol. 2, 743.
3. Hubert Humphrey, *The Education of a Public Man* (Garden City, N.Y., 1976), 265.
4. Tom Wicker, *JFK and LBJ* (Baltimore: Penguin reprint, 1970), 205.
5. Bill Moyers, "Flashbacks," *Newsweek*, February 10, 1975.
6. Lyndon B. Johnson, *The Vantage Point* (New York, 1971), 45. Emphasis added.
7. John F. Kennedy, *Public Papers . . . 1963* (Washington, 1964), 569.
8. *Pentagon Papers*, vol. 3, 18.
9. *Pentagon Papers*, vol. 2, 171, 191.
10. Rusk to Lodge, December 6, 1963, Johnson Papers.
11. *New York Times*, January 2, 1964.
12. *Pentagon Papers*, vol. 2, 197-198.
13. Mieczyslaw Maneli, in interview by author, June 22, 1977.
14. Franz Schurman, P. D. Scott and Reginald Zelnick, *The Politics of Escalation in Vietnam* (Boston, 1966), 26.
15. McGeorge Bundy, ms. on U.S. policy in East Asia during the Kennedy-Johnson years, ch. 12.
16. RFK, in recorded interview by John Bartlow Martin, April 30, 1964, II, 30-31, RFK Oral History Program.
17. Sherman Kent for the Board of National Estimates, "Would the Loss of South Vietnam and Laos Precipitate a 'Domino Effect' in the Far East?" June 9, 1964; Mansfield to Johnson, June 9, 1964; RFK Papers.
18. RFK to LBJ, n.d. [June 11, 1964], Johnson Papers.
19. Jack Valenti, *A Very Human President* (New York, 1975), 141.
20. Transcript, Columbia University appearance, RFK Papers.
21. F. M. Kail, *What Washington Said: Administration Rhetoric and the Vietnam War* (New York, 1973), 104-105, 182.
22. Chester Cooper, "Fateful Day in Vietnam," *Washington Post*, February 11, 1975; see also Cooper, *The Lost Crusade* (New York, 1970), 256-260.
23. Charles Maechling, Jr., review of *Anatomy of Error* by Henry Brandon, *Foreign Service Journal*, March 1970.

24. Johnson, *Vantage Point*, 136.
25. Author's journal, May 6, 1965.
26. *Congressional Record*, May 6, 1965, 9760–9761.
27. Adam Walinsky, in recorded interview by Jean Stein, February 7, 1970, 9, Stein Papers.
28. RFK, advance text of commencement address at International Police Academy, July 9, 1965.
29. Marvin Watson to LBJ, July 8, 1965, Johnson Papers.
30. Fred Harris, in recorded interview by Roberta Greene, July 29, 1970, 3, RFK Oral History Program.
31. Bill Moyers, in recorded interview by Jean Stein, May 23, 1970, 2, Stein Papers.
32. Cherif Guellal, in recorded interview by Jean Stein, October 18, 1968, 1–7; Guellal, in interview by author, January 1, 1978.
33. Joseph Kraft, in recorded interview by Jean Stein, May 3, 1970, 6, Stein Papers; Kraft, in recorded interview by Roberta Greene, March 7, 1970, 37, RFK Oral History Program.
34. Guellal, in Stein interview, 12.
35. *Time*, October 22, 1965.
36. *New York Times*, October 29, 1965.
37. Transcript of RFK press conference, Los Angeles, November 5, 1965, RFK Papers.
38. *New York Daily News*, November 10, 1965.
39. *Chicago Tribune*, November 12, 1965.
40. Victor Lasky, *R.F.K.: The Myth and the Man* (New York: Pocket Books reprint, 1971), 302.
41. Adam Walinsky, in recorded interview by Thomas Johnston, November 30, 1969, 4, RFK Oral History Program.
42. Jack Newfield, *Robert Kennedy* (New York, 1969), 71; A. C. Brackman, *The Communist Collapse in Indonesia* (New York, 1969), 122.
43. *Congressional Record*, January 29, 1966. The editorial was from the *Washington Daily News*, January 28, 1966.
44. Author's journal, March 13, 1966.
45. David Kraslow and S. H. Loory, *The Secret Search for Peace in Vietnam* (New York, 1968), 132–133.
46. Stephen Schlesinger, "RFK–Hickory Hill," December 19, 1965.
47. See the still furious article by General Wallace M. Greene, "The Bombing 'Pause': Formula for Failure," *Air Force Magazine*, April 1976.
48. Author's journal, January 6, 1966.
49. RFK to LBJ, n.d. [January 1966], Johnson Papers.
50. LBJ to RFK, January 27, 1966, Johnson Papers.
51. McGovern to RFK, January 26, 1966, RFK Papers.
52. *Congressional Record*, January 31, 1966, 1602–1603.
53. Robert F. Kennedy, statement on Vietnam, February 19, 1966, RFK Papers.
54. *New York Times*, February 22, 1966.
55. McGeorge Bundy to David Ginsburg, March 13, 1967, Johnson Papers.
56. Draft in RFK Papers.
57. Bundy to RFK, February 21, RFK to Bundy, February 24, 1966, RFK Papers.
58. Unsigned memorandum to LBJ reporting conversation with Governor Connally, February 21, 1966, Johnson Papers.
59. *Chicago Tribune*, February 21, 1966.
60. Murray Kempton, "The Message Delivered," *New York World-Telegram*, February 23, 1966.
61. Moyers to LBJ, February 22, 1966, Johnson Papers; Patrick Anderson, *The President's Men* (New York, 1968), 346–347; William vanden Heuvel and Milton Gwirtzman, *On His Own* (Garden City, N.Y., 1970), 220–223.
62. David E. Lilienthal, *Journals*, vol. 6, *Creativity and Conflict, 1964–1967* (New York, 1967), 206.
63. Author's journal, February 27, 1966.

64. RFK, on *Face the Nation*, February 27, 1966.
65. RFK to Humphrey, January 27, 1966, RFK Papers.
66. RFK to Humphrey (handwritten), February 24, 1966, RFK Papers.
67. RFK, in Martin interview, May 14, 1964, II, 24.
68. Humphrey, on *Issues and Answers*, February 27, 1966.
69. Author's journal, February 27, 1966.
70. Vanden Heuvel and Gwirtzman, *On His Own*, 222.
71. *Congressional Record*, April 27, 1966, 9041.
72. See RFK-Humor file in Mankiewicz Papers.
73. Author's journal, July 20, 1966.
74. Ibid., July 24, 1966.
75. Ibid., August 7, 1966.
76. Spock to RFK, September 21, 1966, with RFK handwritten notes, RFK Papers.
77. *I. F. Stone's Weekly*, October 24, 1966.
78. Doris Kearns, *Lyndon Johnson and the American Dream* (New York, 1976), 251-252.
79. George Reedy, *The Twilight of the Presidency* (New York, 1970), 11.
80. Chester Bowles, *Promises to Keep* (New York, 1971), 535.
81. Kearns, *Johnson*, 316-317. I have taken the liberty of restoring a sentence that appeared in the manuscript but not in the book.
82. Richard Goodwin, in interview by author, June 11, 1977.
83. Author's journal, December 14, 1966.
84. Frank Mankiewicz, in recorded interview by L. J. Hackman, August 12, 1969, 3-4, RFK Oral History Program.
85. Kearns, *Johnson*, 253, 259.

33. The Breach Widens: South Africa, New York *(pages 743-758)*

1. RFK, handwritten first draft of South Africa article, 5, RFK Papers.
2. Wayne Fredericks, in interview by author, January 10, 1977.
3. *Manchester Union-Leader*, November 11, 1965.
4. Fredericks, in interview by author.
5. *New York Times*, May 28, 1966.
6. RFK draft, 6.
7. South Africa Department of Information, May 25, 1966. For this and other quotations I am indebted to a manuscript by Lawrence Ralston, "The Senator and the Republic: Robert Kennedy and the Many Ways of Looking at South Africa," vanden Heuvel Papers, and for clippings in the notebook kept by Thomas Johnston, Johnston Papers.
8. Thomas Johnston, in recorded interview by Jean Stein, October 19, 1969, 2, Stein Papers.
9. Thomas Johnston, in interview by author, June 21, 1977.
10. Ralston ms.
11. Robert F. Kennedy, "Suppose God Is Black," *Look*, August 23, 1966.
12. *Cape Times*, June 7, 1966.
13. Adam Walinsky, in recorded interview by L. J. Hackman, May 22, 1972, 88-89, RFK Oral History Program.
14. The full texts of Kennedy's South African speeches are to be found in a pamphlet published after his trip by the *Rand Daily Mail* entitled "Robert Kennedy in South Africa." The Day of Affirmation speech is on 7-12.
15. *Cape Times*, June 7, 1966.
16. Frank Taylor, "In South Africa, a Kennedy Turns Jeers to Cheers," *National Observer* (New York), June 13, 1966.
17. Ralston ms.
18. *Cape Times*, June 8, 1966.
19. RFK, "Suppose God Is Black."
20. Ralston ms.

21. Johnston, in interview by author.
22. Walinsky, in Hackman interview, 98–99.
23. Johnston, in Stein interview, 7.
24. RFK, handwritten notes, RFK Papers; RFK, "Suppose God Is Black."
25. William vanden Heuvel and Milton Gwirtzman, *On His Own* (Garden City, N.Y., 1970), 160.
26. "Robert Kennedy in South Africa," 29–31.
27. Ralston ms.
28. Quoted by Frank Taylor, *National Observer*, June 13, 1966.
29. Johnston, in interview by author.
30. Anthony Delius, "Daylight from Outside World — Why Kennedy Visit Made Such a Stir," *Cape Times*, June 10, 1966.
31. *Rand Daily Mail*, June 9, 1966.
32. American Embassy Country Team to State Department, July 26, 1966, RFK Papers.
33. Alan Paton, "Waiting for Robert: The Kennedy Visit," *Contact*, July 1966.
34. [James Stevenson], "Kennedy in Africa," in "The Talk of the Town," *New Yorker*, July 9, 1966.
35. *Cape Argus*, June 10, 1966.
36. *New York Times*, June 13, 1966.
37. *Cape Argus*, June 9, 1966.
38. RFK notes, RFK Papers.
39. Walinsky, in Hackman interview, 110.
40. RFK notes; RFK, "Suppose God Is Black."
41. Bowles to RFK, July 26, 1966, RFK to Galbraith, August 1, 1966, RFK Papers.
42. Walinsky, in Hackman interview, 117–118.
43. Joseph Kraft, "New York Safari," *Washington Post*, June 8, 1966.
44. [Stevenson], "Kennedy in Africa."
45. RFK, in recorded interview by John Stewart, August 15, 1967, 59–60, JFK Oral History Program.
46. John Burns, in recorded interview by Roberta Greene, November 25, 1969–February 25, 1970, 19, 66, RFK Oral History Program.
47. Fred Dutton to RFK, April 6, 1966, RFK Papers.
48. Richard Goodwin, a 1967 jotting; in interview by author, June 11, 1977.
49. Jack Newfield, *Robert Kennedy* (New York, 1969), 142.
50. To Albert H. Blumenthal, Ronnie Eldridge, Jack Newfield, Arthur M. Schlesinger, Jr., and others.
51. Marvin Watson to LBJ, June 6, 1966, Johnson Papers.
52. Vanden Heuvel and Gwirtzman, *On His Own*, 134.
53. William V. Shannon, *The Heir Apparent* (New York, 1967), 168–171; Newfield, *Kennedy*, 151–152; vanden Heuvel and Gwirtzman, *On His Own*, 138–139.
54. Newfield, *Kennedy*, 154.
55. Vanden Heuvel and Gwirtzman, *On His Own*, 140.
56. Samuel Silverman, in recorded interview by Roberta Greene, September 3, 1969, 2–4, RFK Oral History Program; Silverman, in recorded interview by Jean Stein, July 18, 1968, 2–3, Stein Papers.
57. Ronnie Eldridge, in recorded interview by Roberta Greene, April 21–July 13, 1970, 92, RFK Oral History Program.
58. Milton Gwirtzman, in recorded interview by Roberta Greene, February 10, 1972, 68, RFK Oral History Program.
59. Alfred Connable and Edward Silberfarb, *Tigers of Tammany* (New York, 1967), 358.
60. Penn Kimball, *Bobby Kennedy and the New Politics* (Englewood Cliffs, N.J., 1968), 102.
61. Silverman, in Stein interview, 9, in Greene interview, 10–11; vanden Heuvel and Gwirtzman, *On His Own*, 143; Kimball, *Bobby Kennedy*, 87; author's journal, June 23, 1966.

62. Author's journal, June 23, 1966.
63. Silverman, in Stein interview, 14; in Greene interview, 8.
64. Newfield, *Kennedy*, 155.
65. Justin Feldman, in recorded interview by Roberta Greene, February 4, 1970, 190, RFK Oral History Program.
66. Frank O'Connor, in recorded interview by Roberta Greene, June 19, 1970, 34, RFK Oral History Program.
67. Eugene Nickerson, in recorded interview by Roberta Greene, November 30, 1971, 12-13, RFK Oral History Program.
68. Jack English, in recorded interview by Roberta Greene, November 25, 1969, 44, RFK Oral History Program.
69. John Burns, in Greene interview, 267-268.
70. Author's journal, July 31, 1966.
71. English, in Greene interview, 47.
72. O'Connor, in Greene interview, 71, 74, 75.
73. Ibid., 69-71.
74. Vanden Heuvel and Gwirtzman, *On His Own*, 201.
75. Wilson to Robert Kintner, August 26, 1966, Johnson Papers.
76. In a piece of September 1966, reprinted in Jack Newfield, *Bread and Roses Too* (New York, 1971), 167, 169.
77. Newfield, *Kennedy*, 23.
78. Ibid., 24.
79. Ralph Blumenfeld, "Hugh Carey," *New York Post*, January 23, 1975.
80. Newfield, *Kennedy*, 24-25.
81. Ibid., 24-27.

34. Time of Troubles *(pages 759-777)*

1. For Gallup polls in August 1966 and early January 1967, see G. H. Gallup, ed., *The Gallup Poll: Public Opinion 1935-1971* (New York, 1972), vol. 3, 2023, 2046. The quotation is from the August 21 release. Harris polled to the same effect in November 1966.
2. Author's journal, October 30, 1966.
3. Victor Navasky, *Kennedy Justice* (New York, 1971), 357-358.
4. RFK to Katzenbach, July 13, 1966, RFK Papers.
5. Richard Goodwin, in interview by author, June 11, 1977.
6. *Washington Post*, December 18, 1966.
7. Henry Hall Wilson to LBJ, December 13, 1966, Johnson Papers.
8. "RFK-Humor" file, Mankiewicz Papers.
9. William Manchester, *Controversy* (Boston, 1976), 12.
10. The full text may be found in John Corry, *The Manchester Affair* (New York, 1967), 29-31.
11. Author to RFK, William Manchester and Evan Thomas, April 24, 1966, Schlesinger Papers.
12. Edwin Guthman, *We Band of Brothers* (New York, 1971), 313; Salinger to RFK, September 15, 1966, RFK Papers.
13. Manchester, *Controversy*, 20.
14. Ibid., 19.
15. Ibid., 45.
16. Evan Thomas, in recorded interview by Jean Stein, November 6, 1969, 2, Stein Papers.
17. Corry, *Manchester*, 97.
18. Manchester, *Controversy*, 4.
19. Corry, *Manchester*, 163.
20. Frank Mankiewicz, in recorded interview by L. J. Hackman, October 2, 1969, 25, RFK Oral History Program.
21. Corry, *Manchester*, 206-207.

22. For further reflections on this theme, see Arthur M. Schlesinger, Jr., "On the Writing of Contemporary History," *Atlantic Monthly*, March 1967.

23. RFK to Katharine Graham, n.d. [1966–67], RFK Papers.

24. Corry, *Manchester*, 199.

25. Mankiewicz, in Hackman interview, October 2, 1969, 46.

26. Murray Kempton, in recorded interview by Jean Stein, October 27, 1969, 13, Stein Papers.

27. Louis Oberdorfer, in recorded interview by Roberta Greene, February 12, 1970, 45–46, RFK Oral History Program; Lester David, *Ethel* (New York: Dell reprint, 1972), 99.

28. *Washington Post*, October 24, 25, 1966.

29. Jack Newfield, *Robert Kennedy* (New York, 1969), 128.

30. In vanden Heuvel papers.

31. William vanden Heuvel and Milton Gwirtzman, *On His Own* (Garden City, N.Y., 1970), 228–230.

32. Peter Osnos, "Kennedy Acclaimed on Oxford Visit," *Washington Post*, January 29, 1967.

33. Memorandum of conversation, RFK and François Mitterand, January 30, 1967, RFK Papers.

34. Memorandum of conversation, RFK and Couve de Murville, January 30, 1967, RFK Papers.

35. Memorandum of conversation, RFK and André Malraux, January 30, 1967, RFK Papers.

36. William vanden Heuvel, in recorded interview by Jean Stein, February 28, 1970, 2–3, Stein Papers.

37. Memorandum of conversation, RFK and Charles de Gaulle, January 31, 1967, RFK Papers.

38. Memorandum of conversation, RFK and Etienne Manac'h, January 31, 1967, RFK Papers.

39. David Kraslow and S. H. Loory, *The Secret Search for Peace in Vietnam* (New York: Random House, Vintage reprint, 1968), 177–178.

40. Memorandum of conversation, RFK and Kurt Georg Kiesinger, February 2, 1967, RFK Papers.

41. Memorandum of conversation, RFK and Willy Brandt, February 2, 1967, RFK Papers.

42. Memorandum of conversation, RFK and Amintore Fanfani, February 3, 1967, RFK Papers.

43. Memorandum of conversation, RFK and Giuseppe Saragat, February 3, 1967, RFK Papers.

44. Memorandum of conversation, RFK and Pope Paul VI, February 4, 1967, RFK Papers.

45. Kraslow and Loory, *Secret Search*, 201–203; vanden Heuvel and Gwirtzman, *On His Own*, 287; David Wise, *The Politics of Lying* (New York, 1973), 80, 369.

46. Chester L. Cooper, *The Lost Crusade* (New York, 1970), 503.

47. Rostow to LBJ, January 28, 1967, Johnson Papers.

48. Wise, *Politics of Lying*, 79–80, based on interviews with the relevant State Department people.

49. Mankiewicz, in Hackman interview, August 12, 1969, 62.

50. Ibid., 71.

51. Ibid., 6, 70–75; Mankiewicz, in recorded interview by Jean Stein, April 13, 1970, 1–6, Stein Papers; Peter Edelman, in recorded interview by L. J. Hackman, July 15, 1969, 65–70; Newfield, *Kennedy*, 131–132; Kraslow and Loory, *Secret Search*, 202–204.

52. *Time*, March 17, 1967. For Rostow's categorical denial of the SOB story and its withdrawal by Hugh Sidey of *Time*, see Denis O'Brien, *Murderers and Other Friendly People* (New York, 1973), 258, 262.

53. Guthman, *We Band of Brothers*, 315.

54. Mankiewicz, in Hackman interview, August 12, 1969, 82.
55. *New York Times,* October 16, 1967.
56. Ibid.
57. Newfield, *Kennedy,* 135–136; Jean Stein and George Plimpton, eds., *American Journey* (New York, 1970), 211–212.
58. W. C. Westmoreland, *A Soldier Reports* (Garden City, N.Y., 1976), 224.
59. Victor Lasky, *Robert F. Kennedy: The Myth and the Man* (New York: Pocket Books reprint, 1971), 469.
60. Dick Schaap, *R.F.K.* (New York: New American Library, Signet reprint, 1968), 22.
61. Newfield, *Kennedy,* 133.
62. Author, in recorded interview by Jean Stein, July 11, 1968, 7–8, Stein Papers.
63. John Burns, in recorded interview by Roberta Greene, November 25, 1969–February 25, 1970, 22, RFK Oral History Program.
64. Schaap, *R.F.K.,* 17–24. Schaap spent March 2 with the Kennedys.
65. Vanden Heuvel and Gwirtzman, *On His Own,* 254.
66. Schaap, *R.F.K.,* 27.
67. Bowles to Humphrey, May 25, 1967, Bowles Papers.
68. John Galloway, ed., *The Kennedys and Vietnam* (New York: *Facts on File,* 1971), 89.
69. *Congressional Record,* March 2, 1967, S2995–S3000.
70. Schaap, *R.F.K.,* 35.
71. Galloway, *Kennedys,* 97.
72. William Dunfey, in recorded interview by L. J. Hackman, December 15, 1971, 85, RFK Oral History Program; Lasky, *Robert F. Kennedy,* 471–472.
73. Fred Harris, in recorded interview by Roberta Greene, July 29, 1970, 7–8, RFK Oral History Program; Harris, *Potomac Fever* (New York, 1977), 147–148.
74. On *Meet the Press,* March 17, 1968.
75. Author's journal, April 18, 1967.
76. *Congressional Record,* April 25, 1967, 10617–10618.
77. Author's journal, April 26, 1967.
78. W. W. Rostow, *The Diffusion of Power, 1955–1972* (New York, 1972), 480, citing Albert H. Cantril, *The American People, Viet-Nam and the Presidency* (Washington, D.C.: American Political Science Association, 1970), 5.
79. *New York Post,* January 28, 1967.
80. *Washington Post,* November 14, 1966.
81. *Newsweek,* March 13, 1967.
82. RFK, remarks at Democratic State Committee Dinner, New York, June 3, 1967, RFK Papers.
83. Peter Maas, in recorded interview by Jean Stein, September 15, 1969, 14, Stein Papers.
84. Author's journal, April 18, 1967.
85. Ibid., May 20, 1967.

35. Tribune of the Underclass *(pages 778–800)*

1. "Favorite Quotations of John F. Kennedy and Robert F. Kennedy" (commonplace book), under "Miscellaneous," RFK Papers. The quotation, unattributed in the commonplace book, is from Keats's *Fall of Hyperion,* canto 1, l. 147. I am indebted to Emily Morison Beck, the editor of *Bartlett's Familiar Quotations,* for this identification.
2. Jack Valenti, *A Very Human President* (New York, 1975), 150.
3. Interview, Washington, D.C., June 5, 1963, in *Robert F. Kennedy: Apostle for Change,* ed. Douglas Ross (New York: Pocket Books reprint, 1968), 56. This is the best compilation of Kennedy's speeches and statements.
4. RFK to LBJ, "Racial Violence in Urban Centers," August 5, 1964, RFK Papers.
5. RFK, in recorded interview by Anthony Lewis, December 4, 1964, VII, 1–2, JFK Oral History Program.

6. John Seigenthaler, in recorded interview by L. J. Hackman, June 5, 1970, 50, RFK Oral History Program.
7. Author's journal, March 23, 1966.
8. Ralph Blumenfeld, "Bobby and Ike Clash over Riots," *New York Post*, August 18, 1965.
9. Richard Rovere, in interview by author, December 4, 1976.
10. Speech before National Council of Christians and Jews, April 28, 1965, in Ross, *Robert F. Kennedy*, 78.
11. RFK, address to Independent Order of Odd Fellows, Spring Valley, New York, August 18, 1965, RFK Papers.
12. Blumenfeld, "Bobby and Ike Clash."
13. Address to Third Annual WGHO Human Relations Award Dinner, Ellenville, New York, April 19, 1966, RFK Papers.
14. Memorandum of conversation, RFK and Pope Paul VI, February 4, 1967, RFK Papers.
15. Ralph Blumenfeld, "RFK Says Rights Leaders Must Share Riot Blame," *New York Post*, August 19, 1965.
16. Nick Kotz and Mary Lynn Kotz, *A Passion for Equality: George A. Wiley and the Movement* (New York, 1977), 253.
17. Martin Luther King, Jr., "Let Justice Roll Down," *Nation*, March 15, 1965.
18. *New York Post*, August 18, 1965.
19. Pat Watters, *Down to Now* (New York, 1971), 135.
20. Louis Lomax, "When 'Nonviolence' Meets 'Black Power,'" in *Martin Luther King, Jr.: A Profile*, ed. C. Eric Lincoln (New York, 1970), 170.
21. Andrew Young to William vanden Heuvel, May 14, 1969, vanden Heuvel Papers.
22. Pete Hamill, in recorded interview by Jean Stein, November 16, 1968, 17–18, Stein Papers.
23. John C. Donovan, *The Politics of Poverty* (New York, 1967), 74–78; Ross, *Robert F. Kennedy*, 146.
24. For RFK's attitude toward welfare, see "Dialogue: Robert Kennedy and Oscar Lewis," *Redbook*, September 1967; Ross, *Robert F. Kennedy*, ch. 6 and esp. his statement of May 19, 1968, "Solutions to the Problems of Welfare," on 551–558.
25. Kotz and Kotz, *Passion for Equality*, 249.
26. Adam Walinsky, in recorded interview by L. J. Hackman, May 22, 1972, 60.
27. Ross, *Robert F. Kennedy*, 555–556.
28. Peter Edelman, in recorded interview by L. J. Hackman, July 29, 1969, 227–230, RFK Oral History Program.
29. *Congressional Record*, October 2, 1974, S17987.
30. RFK, "Federal Role in Urban Affairs," statement before the Subcommittee on Executive Reorganization, August 15, 1966; reprinted in *Congressional Record*, January 23, 1967, esp. S667.
31. RFK, address to Day Care Council, RFK Papers.
32. RFK, address to Federation of Jewish Philanthropies, New York, January 20, 1966, RFK Papers.
33. RFK, "Federal Role in Urban Affairs," S667.
34. Robert F. Kennedy, "Crisis in Our Cities," *Critic*, October–November 1967.
35. Robert F. Kennedy, *To Seek a Newer World* (Garden City, N.Y., 1967), 21, 37.
36. RFK, "A Program for the Urban Crisis," in Ross, *Robert F. Kennedy*, 579.
37. Campaign statement, Binghamton, New York, September 8, 1964; text in Ross, *Robert F. Kennedy*, 61–62.
38. RFK, address to Borough President's Conference of Community Leaders, January 21, 1966, RFK Papers.
39. RFK, "Federal Role in Urban Affairs," S669–S670.
40. I take this analysis, and much that follows, from the illuminating recorded interview of Thomas Johnston by L. J. Hackman, January 21, 1970, esp. 247–250, RFK Oral History Program; supplemented by my interview with Johnston, June 21, 1977.

41. Michael Harrington, "The South Bronx Shall Rise Again," *New York,* April 3, 1978.
42. Johnston, in Hackman interview, 251–252; Bedford-Stuyvesant Restoration Corporation and Bedford-Stuyvesant D & S Corporation, *Annual Report,* 1968.
43. Johnston, in Hackman interview, 254.
44. Jack Newfield, *Robert Kennedy* (New York, 1969), 94.
45. Johnston, in interview by author; in Hackman interview, 254–255.
46. Johnston, in Hackman interview, 377–378, 389–390.
47. Johnston, in interview by author.
48. David E. Lilienthal, *Journals,* vol. 6, *Creativity and Conflict, 1964–1967* (New York, 1976), 302.
49. Benno Schmidt, in recorded interview by Roberta Greene, July 17, 1969, 1, 10–11, RFK Oral History Program.
50. Lilienthal, *Journals,* vol. 6, 444.
51. John Doar, in recorded interview by Jean Stein, July 31, 1968, 6, 13, Stein Papers.
52. Harrington, "The South Bronx Shall Rise Again."
53. Ross, *Robert F. Kennedy,* 581.
54. Newfield, *Kennedy,* 105.
55. Kenneth Clark, in recorded interview by Jean Stein, January 30, 1970, 9, Stein Papers.
56. In interview with Clive Barnes, *New York Times,* August 30, 1968.
57. Ronald B. Taylor, *Chavez and the Farm Workers* (Boston, 1975), 11–12.
58. Walter Reuther, in recorded interview by Jean Stein, October 24, 1968, 6–7, Stein Papers; Victor Reuther, *The Brothers Reuther* (Boston, 1976), 368–369.
59. Peter Edelman, in recorded interview by Jean Stein, March 6, 1969, 6, Stein Papers.
60. Jack Conway, in recorded interview by Jean Stein, August 21, 1968, 5, Stein Papers.
61. William vanden Heuvel and Milton Gwirtzman, *On His Own* (Garden City, N.Y., 1970), 102.
62. Taylor, *Chavez,* 160–167; Jacques Levy, *Cesar Chavez: Autobiography of La Causa* (New York, 1975), 204–205; vanden Heuvel and Gwirtzman, *On His Own,* 103.
63. Cesar Chavez, in recorded interview by Jean Stein, August 24, 1968, 2, Stein Papers; Chavez, in recorded interview by D. J. O'Brien, January 28, 1970, 7, RFK Oral History Program; Peter Matthiessen, *Sal Si Puedes: Cesar Chavez and the New American Revolution* (New York: Dell, Laurel reprint, 1973), 174–175.
64. Levy, *Cesar Chavez,* 449.
65. Chavez, in O'Brien interview, 1–2.
66. Dolores Huerta, in recorded interview by Jean Stein, September 8, 1968, 1–2, Stein Papers.
67. Edelman, in Hackman interview, July 15, 1969, 127; Edelman, in Stein interview, 8.
68. Jack Newfield in Patricia Kennedy Lawford, ed., *That Shining Hour* (n.p., 1969), 137–138; Newfield, *Kennedy,* 82–83; Jerry Bruno and Jeff Greenfield, *The Advance Man* (New York: Bantam reprint, 1972), 105–106.
69. Fred Harris, in recorded interview by Roberta Greene, July 29, 1970, 7, RFK Oral History Program.
70. Quoted by Fred Harris, *Congressional Record,* July 30, 1968, S9711.
71. Edelman, in Stein interview, 12; vanden Heuvel and Gwirtzman, *On His Own,* 108.
72. Steve Bell on ABC, June 9, 1968, reprinted in *An Honorable Profession,* ed. Pierre Salinger (Garden City, N.Y., 1968), 135.
73. Gertrude Claflin to Ethel Kennedy, June 6, 1968, in Salinger, *Honorable Profession,* 71.
74. Vine Deloria, Jr., *Custer Died for Your Sins: An Indian Manifesto* (New York: Avon reprint, 1969), 192, 272.

75. Nick Kotz, *Let Them Eat Promises* (New York: Doubleday, Anchor reprint, 1971), 3-4.
76. Carr, in O'Brien interview, 13.
77. Charles Evers, in recorded interview by Jean Stein, n.d. [1968], 17, Stein Papers.
78. Ibid.
79. Ibid., 18; Kotz, *Promises*, 2.
80. *New Orleans Times-Picayune*, June 9, 1968.
81. Roger Wilkins, in recorded interview by Jean Stein, n.d. [1968], 16, Stein Papers.
82. Edelman (who was present), in Stein interview, 9.
83. Gilbert A. Steiner, *The State of Welfare* (Washington, D.C., 1971), 222.
84. Califano to LBJ, April 17, 1967, Johnson Papers.
85. Steiner, *State of Welfare*, 223.
86. Daniel Patrick Moynihan, *The Politics of a Guaranteed Income* (New York, 1973), 118.
87. Joseph Califano, in recorded interview by Jean Stein, September 21, 1968, 5, Stein Papers.
88. Kotz, *Promises*, 8-9.
89. Robert Coles, in recorded interview by Jean Stein, August 2, 1968, 5-6, Stein Papers.
90. Kotz, *Promises*, 70-74, 77, 141, 170.
91. Ibid., 176-177, 179.
92. Daniel Patrick Moynihan, "The Democrats, Kennedy and the Murder of Dr. King," *Commentary*, May 1968.
93. José Torres, in recorded interview by Jean Stein, August 6, 1968, 2-4.
94. Lyndon B. Johnson, *The Vantage Point* (New York, 1971), 167-172.
95. Harry McPherson, *A Political Education* (Boston, 1972), 359-361.
96. Frank Mankiewicz, in recorded interview by L. J. Hackman, August 12, 1969, 87; Frank Mankiewicz and Tom Braden, *Washington Post*, February 24, 1970; Johnston, in interview by author.
97. *Meet the Press*, August 6, 1967.
98. RFK, "Crisis in Our Cities."
99. Chavez, in Stein interview, 2; in O'Brien interview, 16.
100. Quoted by Anthony Lewis, "A Tribute to Robert Francis Kennedy," Washington, July 22, 1975.
101. Robert Coles, in Stein interview, 2, 3, 9, 11, 16; Jean Stein and George Plimpton, eds., *American Journey* (New York, 1970), 278-279.
102. Newfield, *Kennedy*, 46.
103. Murray Kempton, "The Monument," *New York Post*, December 2, 1966.
104. *Meet the Press*, August 6, 1967.

36. Images *(pages 801-821)*

1. Andrew Kopkind in the *New Republic*, quoted by Penn Kimball, *Bobby Kennedy and the New Politics* (New York, 1968), 70.
2. Ralph Waldo Emerson, "Heroism," in *Essays*.
3. Roger Baldwin, in interview with author, May 15, 1975.
4. Margaret Laing, *The Next Kennedy* (New York, 1968), 32.
5. Jean Stein and George Plimpton, eds., *American Journey* (New York, 1970), 193.
6. Murray Kempton, "Pure Irish: Robert F. Kennedy," *New Republic*, February 15, 1964.
7. Wes Barthelmes, in recorded interview by Roberta Greene, May 20, 1969, 72, RFK Oral History Program.
8. Richard N. Goodwin, "A Day," *McCall's*, June 1970.
9. Allard Lowenstein, memorandum on Robert Kennedy, n.d. [summer 1967], kindly made available to me by Mr. Lowenstein.
10. "RFK-Humor" file, Mankiewicz Papers.
11. Julius Duscha, "Kennedy Man of Hour at ADA Dinner in N.Y.," *Washington Post*, January 28, 1966.

12. Dick Schaap, *R.F.K.* (New York: New American Library, Signet reprint, 1968), 121.
13. Franklin D. Roosevelt, Jr., in recorded interview by Jean Stein, December 9, 1969, 10, Stein Papers.
14. RFK, on the *Today* show, NBC, January 11, 1967, transcript in RFK Papers.
15. RFK, address to Americans for Democratic Action, February 27, 1967, RFK Papers.
16. Jack Newfield, *Robert Kennedy* (New York, 1969), 63.
17. Stein and Plimpton, *American Journey*, 185.
18. Barthelmes, in Greene interview, June 5, 1969, 171.
19. Shirley MacLaine, in recorded interview by Jean Stein, August 26, 1968, 2, Stein Papers.
20. Quoted by Milton Viorst, "The Skeptics," *Esquire*, November 1968.
21. Robert Scheer, "A Political Portrait of Robert Kennedy," *Ramparts*, February 1967; Scheer, in recorded interview by Jean Stein, September 6, 1969, 10–12, Stein Papers; Stein and Plimpton, *American Journey*, 196.
22. Fred Dutton to RFK, December 8, 1966, RFK Papers.
23. Gerald W. Johnson, "Whose Waterloo?" *New Republic*, February 10, 1968.
24. James Wechsler, "Robert F. Kennedy: A Case of Mistaken Identity," *Progressive*, June–July 1965.
25. *Village Voice*, February 2, 1967, and on occasion thereafter.
26. The Ellinger and Barthelmes quotations are from Viorst, "The Skeptics"; the Meany quotation is from Joseph C. Goulden, *Meany* (New York, 1972), 361.
27. John Nolan, in recorded interview by Roberta Greene, August 12, 1970, 57–58, RFK Oral History Program.
28. Herbert Hoover to JFK, April 10, 1948, Hoover Papers. For an overserious account, see G. W. Domhoff, *The Bohemian Grove and Other Retreats: A Study in Ruling-Class Cohesiveness* (New York, 1974).
29. William Orrick, in recorded interview by L. J. Hackman, April 14, 1970, 264–265, RFK Oral History Program.
30. Douglas Dillon, in recorded interview by L. J. Hackman, June 18, 1970, 39, RFK Oral History Program.
31. Frank A. Capell, *Robert F. Kennedy, Emerging American Dictator* (Zarephath, N.J., 1968), 4, 18.
32. Joseph B. Smith, *Portrait of a Cold Warrior* (New York, 1976), 399–400.
33. Finis Farr, *Fair Enough: The Life of Westbrook Pegler* (New Rochelle, N.Y., 1975), 221.
34. Jeffrey Potter, *Men, Money and Magic* (New York, 1976), 299.
35. Margaret Laing, *The Next Kennedy* (New York, 1968), 33.
36. Bruce Biossat, in *An Honorable Profession*, ed. Pierre Salinger (Garden City, N.Y., 1968), 144.
37. Theodore H. White, "The Wearing Last Weeks and a Precious Last Day," *Life*, June 21, 1968.
38. Emerson, "Intellect."
39. Benton to author, May 9, 1968, Schlesinger Papers.
40. Remarks in *In Memory of Robert Francis Kennedy: A Service . . . Memorial Church, Harvard University* (Cambridge, Mass., 1968), 6.
41. Joseph Alsop to James Stevenson, *New Yorker*, June 15, 1967.
42. Salinger, *Honorable Profession*, 121.
43. Excerpt from interview used in "The Journey of Robert F. Kennedy," David Wolper television production, aired February 17, 1970.
44. Author's journal, November 15, 1966.
45. Stein and Plimpton, *American Journey*, 164–165.
46. Art Buchwald, in recorded interview by Roberta Greene, March 12, 1969, 3, RFK Oral History Program.
47. Edward Bennett Williams to RFK, August 7, 1967, RFK Papers.
48. Rose Kennedy to RFK, November 16, 1967, RFK Papers.

49. Cynthia Stone to RFK, November 10, 1964, RFK Papers.
50. Robert F. Kennedy, "Our Climb Up Mt. Kennedy," *Life*, April 9, 1965; Robert F. Kennedy, "A Peak Worthy of the President," *National Geographic*, July 1965.
51. Rita Dallas, *The Kennedy Case* (New York: Popular Library reprint, 1973), 276.
52. James W. Whittaker, "The First Ascent," *National Geographic*, July 1965; RFK articles in *National Geographic* and *Life*; Whittaker in Patricia Kennedy Lawford, ed., *That Shining Hour* (n.p., 1969), 217; "Where Are They Now?" *Newsweek*, March 30, 1970.
53. See the reaction of the Explorers Club of New York as reported by Laing, *Next Kennedy*, 272.
54. Unidentified news ticker, RFK Papers.
55. George Plimpton, in recorded interview by Jean Stein, September 4, 1968, 6, Stein Papers.
56. See the account by Barrett Prettyman, who was along, in Lawford, *Shining Hour*, 204–207; and story by Ward Just, *Washington Post*, September 2, 1965.
57. Charles Spalding, in recorded interview by L. J. Hackman, March 22, 1969, 66, RFK Oral History Program; Lawford, *Shining Hour*, 216.
58. John Glenn, in recorded interview by Roberta Greene, June 26, 1969, 15, RFK Oral History Program.
59. Lawford, *Shining Hour*, 191.
60. Goodwin, "A Day."
61. James Dickey interview, *Paris Review*, Spring 1976.
62. Emerson, "Heroism."
63. Robert F. Kennedy, foreword to memorial edition of John F. Kennedy, *Profiles in Courage* (New York: Harper & Row, Perennial Library reprint, 1964), ix.
64. "Favorite Quotations of John F. Kennedy and Robert F. Kennedy," under "Miscellaneous," RFK Papers.
65. Kenneth P. O'Donnell and David F. Powers, *"Johnny, We Hardly Knew Ye"* (Boston, 1972), 14.
66. Stein and Plimpton, *American Journey*, 167.
67. Goodwin, "A Day."
68. Ibid.
69. Theodore C. Sorensen, *The Kennedy Legacy* (New York, 1969), 37.
70. Stephen Schlesinger, notes, autumn 1967.
71. The phrase is from *The Education of Henry Adams*, ch. 21.
72. As recalled by Jack Newfield in recorded interview, with Ronnie Eldridge, by Jean Stein, n.d. [1968], 17, Stein Papers.
73. Philip Roth, in recorded interview, with William Styron, by Jean Stein, November 13, 1969, 6, Stein Papers.
74. William Styron, in recorded interview, with Philip Roth, by Jean Stein, November 13, 1969, 6, Stein Papers.
75. James Baldwin, in interview by author, October 29, 1976.
76. Stein and Plimpton, *American Journey*, 168–169, 199.
77. Saul Bellow, *Humboldt's Gift* (New York, 1975), 113.
78. Norman Mailer, *Miami and the Siege of Chicago* (New York: New American Library, Signet reprint, 1968), 93, 200–201.
79. James Stevenson, in recorded interview by Jean Stein, November 13, 1969, 2–4, Stein Papers; Carter Burden, in recorded interview by Roberta Greene, February 13, 1974, 26–27, RFK Oral History Program.
80. [James Stevenson], "The Talk of the Town," *New Yorker*, June 15, 1968.
81. Stephen Schlesinger, notes on dinner, autumn 1967.
82. Thomas Johnston, in recorded interview by L. J. Hackman, February 9, 1970, 7, RFK Oral History Program.
83. Yevgeny Yevtushenko, "Under the Skin of the Statue of Liberty," *New York Times Magazine*, February 1970.
84. Author's journal, May 5, 1967.
85. Stein and Plimpton, *American Journey*, 188.

86. It is reprinted in Salinger, *Honorable Profession*, 122.
87. Allen Ginsberg, in recorded interview by Jean Stein, January 22, 1970, 6, 11–13; as amended by Mr. Ginsberg in a letter to me, August 1977. Peter Edelman, in recorded interview by Jean Stein, February 13, 1970, 2–6, Stein Papers; Stein and Plimpton, *American Journey*, 186–188.
88. Robert Lowell to RFK, February 25, 1967, RFK Papers.
89. Stein and Plimpton, *American Journey*, 36, 192–193.
90. Robert Lowell, "Robert Kennedy 1925–1968," *Notebook*, rev. ed. (New York, 1970), 197–198.

37. The Dilemma *(pages 822–841)*

1. *Washington Post*, October 1, 1967.
2. Louis Harris, *The Anguish of Change* (New York, 1973), 203.
3. The Defense Department statistics can be conveniently found in R. H. Fifield, *Americans in Southeast Asia* (New York, 1973), 274–276, and in H. Y. Schandler, *The Unmaking of a President: Lyndon Johnson and Vietnam* (Princeton, 1967), 32.
4. *Face the Nation*, November 26, 1967.
5. Author's journal, February 19, 1968.
6. Schandler, *Unmaking of a President*, 56, 61.
7. Doris Kearns, *Lyndon Johnson and the American Dream* (New York, 1976), 320–321.
8. Author's journal, November 29, 1967.
9. Ibid., December 7, 1967.
10. John Galloway, ed., *The Kennedys and Vietnam* (New York: *Facts on File*, 1971), 101–102.
11. *Face the Nation*, November 26, 1967.
12. Jack Newfield, *Robert Kennedy* (New York, 1969), 177. The story of Lowenstein's guerrilla movement, and indeed the whole political history of 1968, has been well told. For general treatments, see Theodore H. White, *The Making of the President, 1968* (New York, 1969); Lewis Chester, Godfrey Hodgson and Bruce Page, *An American Melodrama: The Presidential Campaign of 1968* (New York, 1969); David English and the staff of the *London Daily Express*, *Divided They Stand* (Englewood Cliffs, N.J., 1969); Norman Mailer, *Miami and the Siege of Chicago* (New York: Signet reprint, 1968).
    For the Kennedy viewpoint, see David Halberstam, *The Unfinished Odyssey of Robert Kennedy* (New York, 1968); Jules Witcover, *85 Days: The Last Campaign of Robert Kennedy* (New York, 1969); and the relevant parts of Newfield, *Kennedy*, and William vanden Heuvel and Milton Gwirtzman, *On His Own* (Garden City, N.Y., 1970).
    For the McCarthy viewpoint, see Eugene J. McCarthy, *The Year of the People* (New York, 1969); Abigail McCarthy, *Private Faces, Public Places* (Philadelphia: Curtis reprint, 1972); Jeremy Larner, *Nobody Knows: Reflections on the McCarthy Campaign of 1968* (New York, 1970); Richard T. Stout, *People* (New York, 1970); Ben Stavis, *We Were the Campaign* (Boston, 1969); Arthur Herzog, *McCarthy for President* (New York, 1969).
13. Author's journal, September 15, 1967.
14. Newfield's account in *Kennedy* (185–186) generally corresponds to my own notes and recollections. He forgets, however, that he and Lowenstein arrived with Loeb and did not find him there on arrival (in fact, they all drove out in Loeb's car); and he is incorrect in suggesting that Loeb was pro-Johnson. See also author's journal, September 23, 1967; Loeb to author, October 21, 1974; Loeb, in recorded interview by L. J. Hackman, May 25, 1972, 5–6, RFK Oral History Program.
15. George McGovern, in recorded interview by L. J. Hackman, July 16, 1970, 38, RFK Oral History Program.
16. Ibid., 47–48; see also George McGovern, *Grassroots* (New York, 1978), 111.
17. Stout, *People*, 59–60.

18. *Congressional Record*, October 17, 1967, S14894–S14897.
19. McGovern, in Hackman interview, 44–46.
20. Author's journal, October 19, 1967.
21. Kenneth O'Donnell and David Powers, *"Johnny, We Hardly Knew Ye"* (Boston, 1972), 186.
22. Kenneth O'Donnell, in recorded interview by L. J. Hackman, April 3, 1969, 26, RFK Oral History Program.
23. Abigail McCarthy, *Private Faces, Public Places*, 242.
24. Eugene McCarthy, *Year of the People*, 129; McCarthy, in interview with *Boston Globe*, December 22, 1968.
25. Theodore C. Sorensen, *The Kennedy Legacy* (New York, 1969), 129.
26. McCarthy, *Year of the People*, 131.
27. See, for example, the discussion of McCarthy in his second Senate term in Stout, *People*, 105–108, and in Herzog, *McCarthy*, 59–61.
28. J. K. Galbraith, in recorded interview by Jean Stein, September 19, 1969, 4, Stein Papers.
29. O'Donnell, in Hackman interview, 26.
30. Galbraith, in Stein interview, 4.
31. See Eugene McCarthy's eulogy of Kennedy in Herzog, *McCarthy*, 58.
32. McCarthy, *Year of the People*, 51.
33. O'Donnell, in Hackman interview, 28.
34. Joseph Dolan, in recorded interview by L. J. Hackman, April 10, 1970, 76, RFK Oral History Program.
35. Pierre Salinger, in recorded interview by L. J. Hackman, May 26, 1969, 51, RFK Oral History Program.
36. O'Donnell, in Hackman interview, 7.
37. Goodwin to RFK, n.d. [autumn 1967], RFK Papers.
38. Dutton to RFK, November 3, 1967, RFK Papers.
39. Author to RFK, November 3, 1967, RFK Papers.
40. Author's journal, November 5, 1967.
41. O'Donnell, in Hackman interview, 6.
42. Joseph Kraft, in recorded interview by Jean Stein, September 7, 1968, 24–25, Stein Papers; Kraft, in recorded interview by Roberta Greene, March 7; 1970, 65, RFK Oral History Program.
43. Author's journal, November 29, 1967.
44. Witcover, *85 Days*, 35.
45. RFK to Lewis, November 29, 1967, RFK Papers.
46. Author's journal, December 10, 1967.
47. Roche to LBJ, December 4, 1967, Johnson Papers.
48. David E. Lilienthal, *Journals*, vol. 6, *Creativity and Conflict* (New York, 1976), 529.
49. Califano to LBJ, December 8, 1967, Johnson Papers.
50. Rowe to LBJ, January 16, 1968, Johnson Papers.
51. Roche to LBJ, January 26, 1968, Johnson Papers.
52. Thomas Johnston, in recorded interview by L. J. Hackman, May 6, 1969, 57, RFK Oral History Program.
53. *Village Voice*, December 28, 1967.
54. Newfield, *Kennedy*, 196.
55. *Newsweek*, January 29, 1968.
56. Halberstam, *Unfinished Odyssey*, 58–59.
57. Benno Schmidt, in recorded interview by Roberta Greene, July 17, 1969, 31, RFK Oral History Program.
58. Dolan, in Hackman interview, 106–107.
59. Rowe, memorandum.
60. Jack English, in recorded interview by Roberta Greene, December 19, 1969, 19, RFK Oral History Program.
61. Author's journal, January 19, 1968.

62. *Congressional Record*, July 30, 1968, 39717.
63. McGovern, in Stein interview, 4.
64. Justin Feldman, in recorded interview by Roberta Greene, February 4, 1970, 147, RFK Oral History Program.
65. Author's journal, January 29, 1968.
66. Ibid., January 17, 1968.
67. Ibid.
68. Newfield, *Kennedy*, 200–203.
69. G. H. Gallup, ed., *The Gallup Poll, 1935–1971* (New York, 1972), vol. 3, 2104.
70. Harris, *Anguish of Change*, 207.
71. Quoted by A. J. Reichley, "He's Running Himself out of the Race," *Fortune*, March 1968.
72. Jeffrey Potter, *Men, Money and Magic* (New York, 1976), 308–309.
73. Frank Burns, in recorded interview by L. J. Hackman, April 17, 1970, 16, RFK Oral History Program.
74. English, in Greene interview, 23.
75. Author's journal, January 25, 1968.
76. Schmidt, in Greene interview, 33–36; author's journal, January 25, 1968.
77. Vanden Heuvel and Gwirtzman, *On His Own*, 293–294.
78. Adam Walinsky, in recorded interview by Jean Stein, February 7, 1970, 4, Stein Papers.
79. Dolan, in Hackman interview, April 11, 1970, 127.
80. Peter Edelman, in recorded interview by L. J. Hackman, July 15, 1969, 18, RFK Oral History Program.
81. Jean Stein and George Plimpton, eds., *American Journey* (New York, 1970), 223–224; Allard Lowenstein, in interview by author, February 26, 1973; Newfield, *Kennedy*, 204.

38. The Decision *(pages 842–857)*

1. *New York Times*, January 2, 1968.
2. RFK, address at Book and Author Luncheon, Chicago, February 8, 1968, RFK Papers.
3. Adam Walinsky, in recorded interview by Jean Stein, February 7, 1970, 5, Stein Papers.
4. RFK, Book and Author address.
5. G. H. Gallup, ed., *The Gallup Poll: Public Opinion, 1935–1971* (New York, 1972), vol. 3, 2106.
6. Author's journal, February 12, 1968.
7. As he told Jack Anderson, "Daniel Ellsberg," *Washington Post*, September 28, 1975.
8. *Congressional Record*, March 7, 1968, 5647–5648.
9. William Walton, in recorded interview by Roberta Greene, May 14, 1970, 45–46, RFK Oral History Program.
10. Rowland Evans, in recorded interview by Roberta Greene, July 30, 1970, 61, RFK Oral History Program.
11. Thomas J. Watson, Jr., in recorded interview by Roberta Greene, January 6, 1970, 21, RFK Oral History Program.
12. Quoted by Murray Kempton, "The Emperor's Kid Brother," *Esquire*, July 1968.
13. Murray Kempton, "The Monument," *New York Post*, December 2, 1966.
14. Joseph Dolan, in recorded interview by L. J. Hackman, April 11, 1970, 122, RFK Oral History Program.
15. Author's journal, February 7, 1968.
16. Arthur Herzog, *McCarthy for President* (New York, 1969), 104.
17. Jack Newfield, *Robert Kennedy* (New York, 1969), 208.
18. Robert F. Kennedy, "'Things Fall Apart; the Center Cannot Hold . . .'" *New York Times*, February 10, 1968.

19. Lawrence F. O'Brien, *No Final Victories* (Garden City, N.Y., 1974), 217.
20. Goodwin to RFK, "Conversation with TK on Feb. 13," n.d., RFK Papers.
21. D. P. Moynihan, *The Politics of a Guaranteed Income* (New York, 1973), 100. Moynihan's account is poetically correct, though Johnson in fact had the report analyzed and dismissed its recommendations on budgetary grounds; see the irritable account in Lyndon B. Johnson, *The Vantage Point* (New York, 1971), 172–173.
22. Jules Witcover, *85 Days: The Last Campaign of Robert Kennedy* (New York, 1969), 53.
23. Newfield, *Kennedy*, 211–212.
24. John Seigenthaler, in recorded interview by Jean Stein, May 15, 1970, 51, Stein Papers.
25. Jean Stein and George Plimpton, eds., *American Journey* (New York, 1970), 281–283.
26. Peter Matthiessen, *Sal Si Puedes* (New York: Dell, Laurel reprint, 1973), 176.
27. Peter Edelman, in recorded interview by Jean Stein, March 6, 1969, 20, Stein Papers.
28. Author's journal, March 10, 1968.
29. Witcover, *85 Days*, 57.
30. William vanden Heuvel and Milton Gwirtzman, *On His Own* (Garden City, N.Y., 1970), 302–303.
31. Author's journal, March 11, 1968.
32. Ibid., March 12, 1968.
33. Ibid.
34. RFK to Anthony Lewis, March 13, 1968, RFK Papers.
35. Author's journal, March 13, 1968.
36. Newfield, *Kennedy*, 218–219.
37. Richard T. Stout, *People* (New York, 1969), 185.
38. Author's journal, March 13, 1968.
39. Ibid.
40. Ibid.
41. Theodore C. Sorensen, *The Kennedy Legacy* (New York, 1969), 137.
42. Dolan, in Hackman interview, 164.
43. "Transcript of telephone conversation between DeVier Pierson at the White House and Ted Sorensen at Senator Robert F. Kennedy's office," March 14, 1968, Johnson Papers.
44. Clark Clifford, memorandum of conversation with Senator Robert F. Kennedy and Theodore C. Sorensen, March 14, 1968, Johnson Papers.
45. George McGovern, in recorded interview by L. J. Hackman, July 16, 1970, 52–56, RFK Oral History Program; McGovern, in recorded interview by Jean Stein, September 25, 1968, 5-6, Stein Papers; Stewart Udall, in recorded interview by Jean Stein, July 26, 1968, 9–11, Stein Papers (as edited by Mr. Udall in a letter to me, August 26, 1977); George McGovern, *Grassroots* (New York, 1978), 112.
46. Clifford, memorandum.
47. Statement by RFK, March 17, 1968, RFK Papers.
48. Newfield, *Kennedy*, 224–225.
49. Everything in this section, unless otherwise specified, is from author's journal, March 16, 1968.
50. George Stevens, Jr., in recorded interview by Roberta Greene, April 10, 1969, 69–70, RFK Oral History Program.
51. Fred Dutton, in recorded interview by Jean Stein, July 26, 1968, 13, Stein Papers.
52. Abigail McCarthy, *Private Faces, Public Places* (Philadelphia: Curtis reprint, 1972), 368–374; Herzog, *McCarthy*, 108–109; Blair Clark, in recorded interview by Jean Stein, November 26, 1969, 4-7, Stein Papers.
53. Jeff Greenfield, in recorded interview by Jean Stein, July 30, 1968, 22, Stein Papers.
54. Author's journal, April 2, 1968.

39. The Journey Begins *(pages 858–875)*

1. Eisenhower to Robert Cutler, March 26, 1968, *New York Post*, March 21, 1975.
2. Richard J. Whalen, *Catch the Falling Flag* (Boston, 1972), 96–97.
3. Mike Manatos to LBJ, March 16, 1968, Johnson Papers.
4. Ibid., March 19, 1968.
5. Sam Houston Johnson, *My Brother Lyndon* (New York, 1969), 242.
6. Jack Newfield, *Robert Kennedy* (New York, 1969), 226.
7. *New York Times*, March 24, 1968.
8. *Washington Post*, June 2, 1968.
9. J. K. Galbraith, "Robert F. Kennedy," *ADA World*, July 1968.
10. Richard T. Stout, *People* (New York, 1969), 186.
11. Milton Viorst, "The Skeptics," *Esquire*, November 1968.
12. *Meet the Press*, March 17, 1968.
13. Jimmy Breslin, "With Kennedy in Kansas," *New York Post*, March 18, 1968.
14. Jimmy Breslin, "Last Year in Manhattan (Kansas)," *New York*, June 9, 1969.
15. RFK, Alfred M. Landon Lecture, Manhattan, Kansas, March 18, 1969, RFK Papers.
16. Newfield, *Kennedy*, 234.
17. Breslin, "Last Year."
18. Newfield, *Kennedy*, 236.
19. Jules Witcover, *85 Days: The Last Campaign of Robert Kennedy* (New York, 1969), 109–110.
20. Author to RFK, March 27, 1968.
21. Witcover, *85 Days*, 119.
22. Richard Harwood, "Crowd Madness and Kennedy Strategy," *Washington Post*, March 28, 1968.
23. Ibid.
24. Newfield, *Kennedy*, 230.
25. Helen Dudar, "The Perilous Campaign," *New York Post*, June 5, 1968.
26. *New York Times*, March 25, 1968; Jimmy Breslin, "'Daley Means the Ball Game, Bobby Says," *Chicago Sun Times*, March 26, 1968.
27. Breslin, "'Daley Means the Ball Game.'"
28. Dun Gifford, in recorded interview by Jean Stein, August 7, 1968, 39–40, Stein Papers.
29. Jack Gallivan, in recorded interview by Jean Stein, n.d. [1968], 9–10, Stein Papers.
30. Witcover, *85 Days*, 114.
31. Alan King, in recorded interview by Jean Stein, May 1970, 5, Stein Papers.
32. G. H. Gallup, ed., *The Gallup Poll: Public Opinion, 1935–1971* (New York, 1972), vol. 3, 2112.
33. *Newsweek*, April 1, 1968.
34. Lyndon B. Johnson, *The Vantage Point* (New York, 1971), 538.
35. McPherson to LBJ, March 22, 1968, Johnson Papers.
36. "Report of the Department of Justice Task Force to Review the FBI Martin Luther King, Jr., Security and Assassination Investigations," January 11, 1977, 131.
37. Townsend Hoopes, *The Limits of Intervention*, rev. ed. (New York, 1973), 206.
38. James Rowe to LBJ, March 19, 1968, Johnson Papers.
39. Walt W. Rostow, *The Diffusion of Power* (New York, 1972), 521.
40. Harry McPherson, *A Political Education* (Boston, 1972), 428.
41. Joseph Califano, *A Presidential Nation* (New York, 1975), 211; Califano, in interview by author, July 25, 1971.
42. George Christian, *The President Steps Down* (New York, 1970), 259.
43. Lady Bird Johnson, *A White House Diary* (New York: Dell reprint, 1971), 706.
44. Johnson, *Vantage Point*, 425.
45. Doris Kearns, *Lyndon Johnson and the American Dream* (New York, 1976), 342.

46. Ibid., 28–29, 32, 343. Emphasis added.
47. Bobby Baker, with Larry King, *Wheeling and Dealing: Confessions of a Capitol Hill Operator* (New York, 1978), 103, also 32–33; Lawrence O'Brien, *No Final Victories* (Garden City, N.Y., 1974), 229.
48. William Barry, in recorded interview by Jean Stein, April 1970, 24, Stein Papers.
49. Witcover, *85 Days*, 126–127.
50. Author's journal, March 31, 1968.
51. Witcover, *85 Days*, 132.
52. Whalen, *Falling Flag*, 145.
53. Abigail McCarthy, *Private Faces, Public Places* (Curtis reprint, 1972), 257.
54. Johnson, *My Brother Lyndon*, 251–252.
55. Ibid.
56. The memorandum of the conversation in Johnson, *Vantage Point*, 539–542, differs in immaterial respects from "Notes on Meeting of the President with Senator Robert Kennedy, April 3, 1968," in Johnson Papers. Sorensen's recollections are in Theodore C. Sorensen, *The Kennedy Legacy* (New York, 1969), 146–147, and, with more detail, in Sorensen, in recorded interview by L. J. Hackman, March 21, 1969, 57–58.
57. Hubert Humphrey, *The Education of a Public Man* (Garden City, N.Y., 1976), 361.
58. W. W. Rostow, memorandum of conversation with the President, the Vice President, Charles Murphy, April 3, 1968, Johnson Papers.
59. Author's journal, April 4, 1968.
60. Jeremy Larner, *Nobody Knows: Reflections on the McCarthy Campaign of 1968* (New York, 1970), 63–64. Emphasis added.
61. Author's journal, March 19, 1968.
62. Jon Bradshaw, "Richard Goodwin: The Good, the Bad, and the Ugly," *New York*, August 18, 1975.
63. Author's journal, March 19, 1968.
64. Ibid., April 3, 1968.
65. Johnston to RFK, April 20, 1968, Schlesinger Papers.
66. Fred Dutton, in recorded interview by L. J. Hackman, November 18, 1969, 49, RFK Oral History Program.
67. Witcover, *85 Days*, 134.
68. Author's journal, April 3, 1968.
69. Robert S. Bird, "Robert F. Kennedy: At Home with the Heir Apparent," *Saturday Evening Post*, August 26, 1967.
70. *Meet the Press*, March 17, 1968.
71. Lewis Chester, et al., *An American Melodrama* (New York, 1969), 145.
72. Pierre Salinger, ed., *An Honorable Profession* (Garden City, N.Y., 1968), 125.
73. Victor Reuther, in interview by author, August 10, 1971; see also Frank Cormier and W. J. Eaton, *Reuther* (Englewood Cliffs, N.J., 1970), 391.
74. Walter Fauntroy, in recorded interview by Jean Stein, November 11, 1969, 25, Stein Papers.
75. Peter Edelman, in recorded interview by L. J. Hackman, July 29, 1969, 263, RFK Oral History Program.
76. Stanley Levison, in recorded interview by Jean Stein, November 21, 1962, 2, Stein Papers.
77. John Lewis, in recorded interview by Jean Stein, September 9, 1968, 5, Stein Papers.
78. Walter Sheridan, in recorded interview by Roberta Greene, August 5, 1969, 6, RFK Oral History Program.
79. John J. Lindsay, in recorded interview by Jean Stein, September 6, 1968, 10, Stein Papers; Walter Sheridan to author, October 17, 1977.
80. Lindsay, in Stein interview, 10–11; Lindsay to author, September 10, 1977.
81. William Barry, in recorded interview by Roberta Greene, March 20, 1969, 39, RFK Oral History Program.
82. Frank Mankiewicz, in recorded interview by Jean Stein, September 21, 1968, 26,

Stein Papers; Adam Walinsky, in recorded interview by Jean Stein, September 20, 1968, 25, Stein Papers.

83. Jean Stein and George Plimpton, eds., *American Journey* (New York, 1970), 256.
84. Remarks by Senator Robert F. Kennedy on the death of the Reverend Martin Luther King, rally in Indianapolis, Indiana, April 4, 1968.

40. The Long Day Wanes *(pages 876–902)*

1. Coretta Scott King, *My Life with Martin Luther King, Jr.* (London, 1970), 333–334.
2. Coretta Scott King, in recorded interview by Jean Stein, November 18, 1969, 1–2, Stein Papers.
3. John Lewis, in recorded interview by Jean Stein, September 9, 1968, 7, Stein Papers; William vanden Heuvel and Milton Gwirtzman, *On His Own* (Garden City, N.Y., 1970), 338.
4. Jeff Greenfield, in recorded interview by Roberta Greene, December 10, 1969, 96–99, RFK Oral History Program.
5. RFK, speech at Cleveland City Club, April 5, 1968, RFK Papers.
6. Jean Stein and George Plimpton, eds., *American Journey* (New York, 1970), 261.
7. Ibid., 258.
8. Roy Jenkins, *Nine Men of Power* (London, 1974), 208.
9. Jimmy Breslin, "Back to Earth," *New York Post*, December 27, 1968.
10. Author's journal, April 10, 1968.
11. Stein and Plimpton, *American Journey*, 260–261.
12. Ibid., 259–260.
13. Andrew Young, in recorded interview by Jean Stein, September 9, 1968, 12–13, Stein Papers; Joseph Lelyveld, "Our New Voice at the U.N.," *New York Times Magazine*, February 6, 1977.
14. Ralph Abernathy, in "R.F.K.," Metromedia Broadcast, June 6, 1969.
15. Stein and Plimpton, *American Journey*, 261.
16. RFK to Katzenbach, February 12, 1968, Katzenbach to RFK, February 16, 1968, RFK Papers.
17. Nicholas Katzenbach, in recorded interview by Jean Stein, September 27, 1968, 20, Stein Papers.
18. John Bartlow Martin to RFK, n.d. [April 1968], RFK Papers.
19. Martin to RFK, March 29, 1968, RFK Papers.
20. Martin, "RFK Notes [on the 1968 campaign]," June 28, 1968, 11–12, Martin Papers.
21. Fred Dutton, in recorded interview by L. J. Hackman, November 18, 1969, 52, RFK Oral History Program.
22. *New York Times*, April 28, 1968.
23. Martin, "RFK Notes," 26.
24. Barry Stavis, *We Were the Campaign* (Boston, 1969), 67–68.
25. Stein and Plimpton, *American Journey*, 248.
26. Martin, "RFK Notes," 36–37.
27. David Halberstam, "Travels with Bobby Kennedy," *Harper's*, July 1968.
28. Thomas B. Congdon, Jr., "Robert F. Kennedy, 1925–1968," *Saturday Evening Post*, June 29, 1968.
29. RFK, remarks at Indiana University Medical School, Indianapolis, April 26, 1968, RFK Papers.
30. Jules Witcover, *85 Days: The Last Campaign of Robert Kennedy* (New York, 1969), 165.
31. David Halberstam, *The Unfinished Odyssey of Robert Kennedy* (New York, 1968), 121.
32. RFK, remarks at Valparaiso University, April 29, 1968, RFK Papers.
33. Jack Newfield, *Robert Kennedy* (New York, 1969), 256–257.
34. Quoted in vanden Heuvel and Gwirtzman, *On His Own*, 345–347.
35. Martin, "RFK Notes," 36.
36. Newfield, *Kennedy*, 265.

37. Richard Stout, *People* (New York, 1970), 237.
38. Lewis Chester, et al., *An American Melodrama* (New York, 1969), 146.
39. Stein and Plimpton, *American Journey*, 196–197.
40. Vanden Heuvel and Gwirtzman, *On His Own*, 359–361.
41. [James Stevenson], "Notes and Comment," in "The Talk of the Town," *New Yorker*, June 15, 1968.
42. James Stevenson, in recorded interview by Jean Stein, November 13, 1969, 9, Stein Papers.
43. Rita Dallas, *The Kennedy Case* (New York: Popular Library reprint, 1973), 301–304; Theodore C. Sorensen, *The Kennedy Legacy* (New York, 1969), 25–26.
44. *Washington Post*, April 30, 1968.
45. "Free" [Abbie Hoffman], *Revolution for the Hell of It* (New York, 1968), 104.
46. Marshall McLuhan, "All of the Candidates Are Asleep," *Saturday Evening Post*, August 10, 1968.
47. Hal Higon, "Kennedy in the Midwest," *Chicago Tribune Magazine*, August 25, 1968.
48. Joseph Alsop, "Robert Kennedy on the Stump," *Washington Post*, May 16, 1968.
49. Ibid.
50. Stewart Alsop, "Bobby's Red Guards," *Saturday Evening Post*, May 4, 1968.
51. John Douglas, "Robert Kennedy and the Qualities of Personal Leadership," speech at Loyola University, Chicago, October 31, 1968, 9.
52. Halberstam, *Unfinished Odyssey*, 185.
53. Congdon, "Kennedy"; Witcover, *85 Days*, 160–163.
54. Jeff Greenfield, in recorded interview by Jean Stein, July 30, 1968, 18, Stein Papers; Greenfield, in recorded interview by Roberta Greene, January 5, 1970, 123–124, RFK Oral History Program.
55. Newfield, *Kennedy*, 261.
56. Halberstam, *Unfinished Odyssey*, 172.
57. Peter Edelman, in recorded interview by L. J. Hackman, August 5, 1969, 343, RFK Oral History Program; Theodore H. White, *The Making of the President, 1968* (New York, 1969), 171–172.
58. Newfield, *Kennedy*, 259.
59. Witcover, *85 Days*, 192.
60. Ibid., 191.
61. Greenfield, in Greene interview, 125.
62. Witcover, *85 Days*, 193–194.
63. Helen Dudar, "The Perilous Campaign," *New York Post*, June 5, 1968; Witcover, *85 Days*, 194.
64. John Glenn, in recorded interview by Roberta Greene, June 30, 1969, 40, RFK Oral History Program.
65. Edelman, in Hackman interview, 346.
66. Witcover, *85 Days*, 197.
67. Ibid., 198.
68. Eugene McCarthy, "Speaking Out," *Saturday Evening Post*, January 4, 1964.
69. In a eulogy of Humphrey, "Tragedy of Valor and Love," from the *Austin Daily Texan*, reprinted in *Congressional Record*, January 31, 1978, E274.
70. Newfield, *Kennedy*, 253.
71. Stein and Plimpton, *American Journey*, 286–287.
72. Richard Harwood, remarks at RFK Journalism Award's luncheon, Washington, D.C., May 14, 1976.
73. Abigail McCarthy, *Private Faces, Public Places* (Philadelphia: Curtis reprint, 1972), 373.
74. Robert Coles, "Ordinary Hopes, Ordinary Fears," in *Conspiracy: The Implications of the Harrisburg Trial for the Democratic Tradition*, ed. J. C. Raines (New York, 1974), 99–100.
75. Alexander M. Bickel, "Robert Kennedy as History," *New Republic*, July 5, 1969.
76. Paul Cowan, "Wallace in Yankeeland," *Village Voice*, July 18, 1968.
77. Jack Newfield, "A Look at Kennedy," mimeographed, n.d. [spring 1968].

78. Homer Bigart, "Negroes Are Cool to McCarthy as He Opens Indiana Campaign," *New York Times*, April 19, 1968.
79. A. McCarthy, *Private Faces*, 381.
80. Eugene McCarthy, in interview with *Boston Globe* Washington staff, *Boston Globe*, December 24, 1968.
81. Abigail McCarthy, "The McCarthy Campaign," *Atlantic Monthly*, August 1970.
82. Norman Mailer, *Miami and the Siege of Chicago* (New York; New American Library, Signet reprint, 1968), 92–93, 99.
83. *Salem* (Oreg.) *Statesman*, May 22, 1968.
84. Jeremy Larner, *Nobody Knows* (New York, 1970), 93.
85. Ibid., 76–77; Larner, "Nobody Knows . . . Part II," *Harper's*, May 1969.
86. Eugene J. McCarthy, "Why I'm Battling LBJ," *Look*, February 6, 1968.
87. *London Sunday Times*, August 18, 1968.
88. Eugene J. McCarthy, *The Year of the People* (Garden City, N.Y., 1969), 295.
89. McCarthy, speech in Cleveland, June 18, 1968, in *Washington Post*, June 19, 1968, *New York Daily News*, June 19, 1968.
90. Stavis, *We Were the Campaign*, 52, 134.
91. David Halberstam, "Travels with Bobby Kennedy," *Harper's*, July 1968.
92. Richard Harwood, "McCarthy and Kennedy: Philosopher vs. Evangelist," *Washington Post*, May 26, 1968.
93. Stein and Plimpton, *American Journey*, 257.
94. A. McCarthy, *Private Faces*, 397.
95. Halberstam, "Travels with Kennedy."
96. White, *Making of the President, 1968*, 79.
97. *Washington Post*, May 26, 1968.
98. Stavis, *We Were the Campaign*, 32.
99. Larner, *Nobody Knows*, 36.
100. Stavis, *We Were the Campaign*, 135.
101. Larner, *Nobody Knows*, 33.
102. Ibid., 33, 79–81, 144.
103. Stavis, *We Were the Campaign*, 51.
104. Larner, *Nobody Knows*, 76.
105. Ibid.
106. Halberstam, "Travels with Kennedy."
107. Jack Newfield, "The Arrogance of Class," *Village Voice*, May 2, 1968.
108. Author's journal, May 27, 1968.
109. Stein and Plimpton, *American Journey*, 269–270.
110. Elizabeth Hardwick, in recorded interview by Jean Stein, September 18, 1968, 4, Stein Papers, edited by Miss Hardwick in a letter of September 4, 1977.
111. William Manchester, "RFK: Not at All a Brute," *Baltimore Sun*, April 26, 1968.
112. Jacqueline Kennedy to William Manchester, June 17, 1968. I am indebted to Mr. Manchester for this letter.
113. Alexander M. Bickel, "The Kennedy Cause," *New Republic*, July 20, 1968.
114. "Free," *Revolution*, 104.
115. Nick Kotz and Mary Lynn Kotz, *A Passion for Equality: George A. Wiley and the Movement* (New York, 1977), 255.
116. Michael Harrington, *Fragments of the Century* (New York, 1973), 237–238, 243.
117. Charles Quinn, in recorded interview by Jean Stein, October 19, 1968, 15, Stein Papers.
118. Stein and Plimpton, *American Journey*, 319.
119. Witcover, *85 Days*, 224.
120. L. K. Obst, *The Sixties* (New York, 1977), 254.
121. Congdon, "Kennedy."
122. Stein and Plimpton, *American Journey*, 293; Newfield, *Kennedy*, 286.
123. Romain Gary, *White Dog* (New York, 1970), 192–196; Gary, in interview by Henry Raymont, *New York Times*, August 21, 1968; Pierre Salinger, *Je suis un Américain* (Paris, 1975), 247, 309–310; Stein and Plimpton, *American Journey*, 293–294.
124. Witcover, *85 Days*, 114.

125. William Barry, in recorded interview by Roberta Greene, March 19, 1969, 5–6, 9, RFK Oral History Program.
126. Barry, in Greene interview, March 20, 1969, 47.
127. Jack Eberhardt, in recorded interview by Jean Stein, August 23, 1968, 1, Stein Papers.
128. Stein and Plimpton, *American Journey*, 294–295; Quinn, in Stein interview, 19.
129. Warren Rogers and Stanley Trettick, "RFK," *Look*, July 9, 1968.
130. Dudar, "Perilous Campaign."
131. Walter Sheridan, in recorded interview by Roberta Greene, August 5, 1969, 90, RFK Oral History Program.
132. Barry, in Greene interview, 23.

41. To Sail Beyond . . . the Western Stars, Until I Die *(pages 903–916)*

1. Jean Stein and George Plimpton, eds., *American Journey* (New York, 1970), 273–274.
2. Jeremy Larner, *Nobody Knows* (New York, 1970), 91.
3. Eugene McCarthy, interview with *Boston Globe* Washington staff, *Boston Globe*, December 24, 1968.
4. See Eugene McCarthy, *The Year of the People* (Garden City, N.Y., 1969), 118–124.
5. Arthur Herzog, *McCarthy for President* (New York, 1969), 158.
6. *San Francisco Chronicle*, May 31, 1968.
7. *New York Times*, May 5, 1968.
8. Herzog, *McCarthy*, 158.
9. Frank Mankiewicz, in recorded interview by L. J. Hackman, September 30, 1969, 60–61, RFK Oral History Program.
10. McCarthy, *Year of the People*, 143.
11. Barry Stavis, *We Were the Campaign* (Boston, 1969), 62.
12. McCarthy, *Year of the People*, 143.
13. Ibid., 124. Emphasis added.
14. Larner, *Nobody Knows*, 94–97.
15. Lewis Chester, et al., *American Melodrama* (New York, 1969), 304.
16. Richard Stout, *People* (New York, 1970), 260; Jules Witcover, *85 Days: The Last Campaign of Robert Kennedy* (New York, 1969), 216–217.
17. Larner, *Nobody Knows*, 97.
18. Edith Green, in recorded interview by Roberta Greene, February 27, 1974, 2, RFK Oral History Program.
19. William vanden Heuvel and Milton Gwirtzman, *On His Own* (Garden City, N.Y., 1970), 367.
20. Joseph Kraft, in recorded interview by Roberta Greene, March 7, 1970, 72, RFK Oral History Program.
21. Ralph Friedman, "The Disenchanted Suburbia," *Nation*, May 8, 1972.
22. Pierre Salinger, in recorded interview by L. J. Hackman, April 18, 1970, 55, RFK Oral History Program.
23. Witcover, *85 Days*, 206.
24. Larner, *Nobody Knows*, 101–102.
25. M. S. Devorkin, "Kennedy Campaigning in Primary States" (B.S. thesis, Massachusetts Institute of Technology, June 1969), 114.
26. Witcover, *85 Days*, 221–222.
27. RFK, press conference statement, Los Angeles, May 29, 1968, RFK Papers.
28. Janet Auchincloss to RFK, n.d. [May 31, 1968], RFK Papers.
29. William Orrick, in recorded interview by L. J. Hackman, April 14, 1970, 258, RFK Oral History Program.
30. Walter Sheridan, in interview by Roberta Greene, August 13, 1969, 62, RFK Oral History Program; vanden Heuvel and Gwirtzman, *On His Own*, 104.
31. Author's journal, May 30, 1968.
32. Fred Dutton, in recorded interview by Jean Stein, July 26, 1968, 59.
33. Stein and Plimpton, *American Journey*, 305.

34. Curtis Lee Baker (Black Jesus), in recorded interview by Jean Stein, December 4, 1969, 5–6, Stein Papers.
35. John Seigenthaler, in recorded interview by Jean Stein, August 27, 1968, 38, Stein Papers.
36. Dutton, in Stein interview, 62.
37. Baker, in Stein interview, 7–8.
38. Hector Lopez, in recorded interview by Jean Stein, October 6, 1969, 6, Stein Papers.
39. Willie Brown, in recorded interview by Jean Stein, August 17, 1968, 6, Stein Papers.
40. Lopez, in Stein interview, 15.
41. Herzog, *McCarthy*, 177.
42. Larner, *Nobody Knows*, 111.
43. Author's journal, June 1, 1968.
44. Frank Mankiewicz, in recorded interview by L. J. Hackman, December 16, 1969, 53.
45. Herzog, *McCarthy*, 179.
46. Andreas Teuber; Stein and Plimpton, *American Journey*, 311.
47. Ibid., 309–310.
48. Stout, *People*, 271–272.
49. Stein and Plimpton, *American Journey*, 312.
50. *New York Times*, June 2, 1968.
51. Mankiewicz, in Hackman interview, December 16, 1969, 54.
52. Chester, *American Melodrama*, 345.
53. Larner, *Nobody Knows*, 117.
54. Ibid.
55. Herzog, *McCarthy*, 187.
56. Theodore H. White, "The Wearing Last Weeks and a Precious Last Day," *Life*, June 21, 1968; Richard N. Goodwin, "A Day," *McCall's*, June 1970.
57. R. B. Kaiser, *"R.F.K. Must Die"* (New York, 1970), 15.
58. George McGovern, in recorded interview by Jean Stein, September 25, 1968, 12, Stein Papers.
59. Hackett's estimate, misdated June 4, 1968, is in vanden Heuvel and Gwirtzman, *On His Own*, 390–392.
60. Richard N. Goodwin, "A Sentimental Tribute"; I quote here the original version rather than the one published as "A Day" in *McCall's*.
61. J. K. Galbraith, in recorded interview by Jean Stein, July 30, 1968, 14, Stein Papers.
62. Justin Feldman, in recorded interview by Roberta Greene, February 4, 1970, 160–162, RFK Oral History Program.
63. Budd Schulberg, "R.F.K. — Harbinger of Hope," *Playboy*, January 1969.
64. Salinger, *Je suis un Américain* (Paris, 1975), 312.
65. Goodwin, "A Day."
66. Witcover, *85 Days*, 262; Jack Newfield, *Robert Kennedy* (New York, 1969), 298.
67. Larner, *Nobody Knows*, 121.
68. Author's journal, June 6, 1968.
69. Mankiewicz, in Hackman interview, October 2, 1969, 56.
70. Gloria Steinem, "Link between the New Politics and the Old," *Saturday Review*, August 2, 1969.
71. Newfield, *Kennedy*, 303–304.
72. Author's journal, June 8, 1968.
73. Lady Bird Johnson, *A White House Diary* (New York: Dell reprint, 1971), 756.

Index